# 2nd International Conference on Mathematical Modeling in Physical Sciences



# (IC-MSQUARE 2013)

**Journal of Physics: Conference Series Volume 490**

## Prague, Czech Republic
## 1-5 September 2013

**Editors:**

**Elias C. Vagenas** **Dimitrios S. Vlachos**

**ISBN: 978-1-63266-715-1**
**ISSN: 1742-6588**

**Printed from e-media with permission by:**

Curran Associates, Inc.
57 Morehouse Lane
Red Hook, NY 12571

**Some format issues inherent in the e-media version may also appear in this print version.**



Printed by Curran Associates, Inc. (2014)

For permission requests, please contact the Institute of Physics
at the address below.

Institute of Physics
Dirac House, Temple Back
Bristol BS1 6BE UK

Phone: 44 1 17 929 7481
Fax: 44 1 17 920 0979

techtracking@iop.org

**Additional copies of this publication are available from:**

Curran Associates, Inc.
57 Morehouse Lane
Red Hook, NY 12571 USA
Phone: 845-758-0400
Fax: 845-758-2634
Email: curran@proceedings.com
Web: www.proceedings.com

# TABLE OF CONTENTS

Journal of Physics: Conference Series **490** (2014) 011001 doi:10.1088/1742-6596/490/1/011001

# Preface

The second International Conference on Mathematical Modeling in Physical Sciences (IC-MSQUARE) took place at Prague, Czech Republic, from Sunday 1 September to Thursday 5 September 2013.

The Conference was attended by more than 280 participants and hosted about 400 oral, poster, and virtual presentations while counted more than 600 pre-registered authors. The second IC-MSQUARE consisted of different and diverging workshops and thus covered various research fields where Mathematical Modeling is used, such as Theoretical/Mathematical Physics, Neutrino Physics, Non-Integrable Systems, Dynamical Systems, Computational Nanoscience, Biological Physics, Computational Biomechanics, Complex Networks, Stochastic Modeling, Fractional Statistics, DNA Dynamics, Macroeconomics.

The scientific program was rather heavy since after the Keynote and Invited Talks in the morning, three parallel sessions were running every day. However, according to all attendees, the program was excellent with high level of talks and the scientific environment was fruitful, thus all attendees had a creative time.

We would like to thank the Keynote Speaker and the Invited Speakers for their significant contribution to IC-MSQUARE. We also would like to thank the Members of the International Advisory and Scientific Committees as well as the Members of the Organizing Committee.

The Conference Chairmen

Theodosios Christodoulakis
University of Athens

Elias Vagenas
RCAAM of the Academy of Athens

Dimitrios Vlachos
University of Peloponnese

**Keynote Speaker**

- **Athanasios S. Fokas, University of Cambridge, UK**

**Invited Speakers**

- **Mairi Sakellariadou**, King's College, University of London, UK
- **Eugenio Bianchi**, Perimeter Institute, Canada
- **George M. Spyrou**, BRFAA, Greece
- **José Fernando Mendes**, Universidade de Aveiro, Portugal
- **Marika Taylor**, University of Southampton, UK

**Editors:**

- Dr Elias C. Vagenas, RCAAM of the Academy of Athens, Greece
- Prof Dimitrios S. Vlachos, University of Peloponnese, Greece

**Co-Editors:**

- Dr Sergio L. Cacciatori, Università dell'Insubria, Italy
- Prof Jose D. Edelstein, University of Santiago de Compostela, Spain
- Dr George M. Spyrou, BRF of the Academy of Athens, Greece
- Prof Jorge Pullin, Louisiana State University, United States
- Prof Douglas A. Singleton, California State University, Fresno, United States
- Prof Bernd Bruegmann, University of Jena, Germany

- Prof Saurya Das, University of Lethbridge, Canada
- Prof Kalman Varga, Vanderbilt University, United States
- Prof Young Min Song, Pusan National University, Korea, South
- Dr Julian Sienkiewicz, Warsaw University of Technology, Poland
- Dr Clement B. Ampadu, United States
- Prof Bojin Zheng, South-Central University for Nationalities, China, China
- Dr Babur M. Mirza, Quaid-i-Azam University, Pakistan
- Prof Jonas Fransson, Uppsala university, Sweden
- Dr Marco Zoli, University of Camerino, Italy
- Dr Odysseas T. Kosmas, University of Erlangen-Nuremberg, Germany
- Prof Jose M. Machado, São Paulo State University, Brazil
- Dr Colin Wilmott, Nottingham Trent University, United Kingdom
- Prof Liviu Chibotaru, University of Leuven, Belgium
- Dr Kaushik Ghosh, ST. Xavier's College (Autonomous), India
- Dr Federico Colecchia, Brunel University, United Kingdom
- Prof Emilio Elizalde, ICE-CSIC and IEEC, Spain
- Dr Lorenzo Iorio, Ministero dell'Istruzione, dell'Università e della Ricerca, Italy

# Peer review statement

All papers published in this volume of *Journal of Physics: Conference Series* have been peer reviewed through processes administered by the proceedings Editors. Reviews were conducted by expert referees to the professional and scientific standards expected of a proceedings journal published by IOP Publishing.

# Newton's method for solving a quadratic matrix equation with special coefficient matrices

**Sang-Hyup Seo, Jong Hyun Seo and Hyun-Min Kim***

Department of Mathematics, Pusan National University, Busan, 609-735, Republic of Korea

E-mail: hyunmin@pusan.ac.kr

**Abstract.** We consider the iterative method for solving a quadratic matrix equation with special coefficient matrices which arises in the quasi-birth-death problem. In this paper, we show that the elementwise minimal positive solvents to quadratic matrix equations can be obtained using Newton's method. We also prove that the convergence rate of the Newton iteration is quadratic if the Fréchet derivative at the elementwise minimal positive solvent is nonsingular. However, if the Fréchet derivative is singular, the convergence rate is at least linear. Numerical experiments of the convergence rate are given.(This is summarized a paper which is to appear in Honam Mathematical Journal.)

## 1. Introduction

We consider a quadratic matrix equation defined by

$$Q(X) = AX^2 + BX + C, \tag{1}$$

where $A, B$ and $C$ are real $n \times n$ matrices. In this paper, we study the quadratic matrix equation (1) for $A$ and $C$ are nonnegative matrices and $-B$ is a nonsingular $M$-matrix.

**Definition 1.** A positive solvent $S_1$ of the matrix equation $Q(X) = 0$ is an *elementwise minimal positive solvent* and a positive solvent $S_2$ of $Q(X) = 0$ is an *elementwise maximal positive solvent* if, for any positive solvent $S$ of $Q(X)$,

$$S_1 \leq S \leq S_2. \tag{2}$$

Similarly, if nonnegative solvents $S_1$ and $S_2$ satisfy (2) for any nonnegative solvent $S$, $S_1$ is called an *elementwise minimal nonnegative solvent* and $S_2$ is called an *elementwise maximal nonnegative solvent* [2, 3, 5].

## 2. Newton's Method

The purpose of this paper is to obtain the elementwise minimal nonnegative solvent of (1). In fact, nonlinear matrix equations like (1) often occur in some stochastic problems such as quasi-birth-and-death (QBD) processes [7, 8].

For the equation (1), Newton iteration can be simplified

$$\begin{cases} AH_iX_i + (AX_i + B)H_i = -Q(X_i), \\ X_{i+1} = X_i + H_i, \end{cases} \qquad i = 1, 2, \cdots. \tag{3}$$

The Fréchet derivative of the quadratic matrix equation (1) at $X$ in the direction $H$ is given by

$$Q'_X(H) = AHX + (AX + B)H. \tag{4}$$

The general approach for solving (3) is to solve the $n^2 \times n^2$ linear system derived by vec function and Kronecker product [6, 10] such as

$$\mathcal{D}_{X_i}\text{vec}(H) = \text{vec}(-Q(X_i))$$

where

$$\mathcal{D}_X = \left[(X^T \otimes A + I \otimes AX) + I \otimes B)\right]. \tag{5}$$

Now, we will give an assumption to (1).

**Assumption 2.** *For the quadratic matrix* (1)

I) *The coefficient matrices $A$ and $C$ are nonnegative and irreducible.*

II) $-B = rI - T_0$ is a nonsingular irreducible $M$-matrix [9] where $T_0 \geq 0$.

Applying Assumption 2, we obtain the following results.

**Corollary 3.** *Suppose the quadratic matrix equation satisfies Assumption 2. If there is a positive matrix $Y$ such that $Q(Y) \leq 0$, then for the Newton iteration (3) with $X_0 = 0$, the sequence $\{X_i\}$ is well defined, $X_0 < X_1 < X_2 < \cdots$, and converges to the elementwise minimal positive solvent $S$. Furthermore*

$$-\mathcal{D}_{X_i} = -\left[(X_i^T \otimes A + I_n \otimes AX_i) + I_n \otimes B)\right]$$

*is a nonsingular irreducible $M$-matrix at each iterate $X_i$ except $X_0$, and $-\mathcal{D}_S$ is an irreducible $M$-matrix.*

If $Q'_S$ is non-invertible, then $Q'_S$ has a null space $\mathcal{N} = \text{Ker}(Q'_S)$ and closed range $\mathcal{M} = \text{Im}(Q'_S)$. Suppose that the direct sum $\mathcal{N} \oplus \mathcal{M} = \mathbb{R}^{n\times n}$. Then we can define $\mathcal{P}_\mathcal{N}$ to be the projection onto $\mathcal{N}$ parallel to $\mathcal{M}$ and $\mathcal{P}_\mathcal{M} = I - \mathcal{P}_\mathcal{N}$.

**Theorem 4.** *If $\mathcal{D}_S$ is a singular $M$-matrix and the convergence rate of the Newton sequence $\{X_i\}$ in Corollary 3 is not quadratic, then $\|{Q'_{X_i}}^{-1}\| \leq c\|X_i - S\|^{-1}$ for all $i \geq 1$ and some constant $c > 0$. Moreover,*

$$\lim_{i\to\infty} \frac{\|\tilde{X}_{i+1}\|}{\|\tilde{X}_i\|} = \frac{1}{2}, \qquad \lim_{i\to\infty} \frac{\|\mathcal{P}_\mathcal{M}(\tilde{X}_i)\|}{\|\mathcal{P}_\mathcal{N}(\tilde{X}_i)\|^2} = 0.$$

Theorem 4 shows that $\{X_i\}$ converges linearly.

## 3. Numerical Experiment

**Example 5.** Consider the matrix equation (1) for a QBD process. We construct $n \times n$ matrices

$$A = W, \quad B = W - I_n, \quad \text{and} \quad C = W + \sqrt{\delta}I_n \tag{6}$$

where

$$W = \frac{1 - \sqrt{\delta}}{3(n-1)}(\mathbf{1}_{n\times n} - I_n)$$

for $0 < \delta < 1$. Then, $(3W + \delta I_n)\mathbf{1}_n = \mathbf{1}_n$. Note that as $\delta$ approaches zero, the problem becomes more unstable [1, 4, 7]. The matrices $A$, $B$ and $C$ satisfy the Assumption 2. So, the problem has the elementwise minimal positive solvent $S$ if it exists. The result is obtained with matrices $A$, $B$ and $C$ in (6) of size $n = 8$ and $n = 16$ with from $\delta = 10^{-1}$ to $\delta = 10^{-16}$.

The Figures 1, 2 and Table 1 show that the Newton sequence of the problem converges to a solvent linearly as $\delta$ approaches to zero whatever is $n$ because $-\mathcal{D}_S$ becomes nearly singular.

| $\delta$ | $n = 8$ | $n = 16$ | $\delta$ | $n = 8$ | $n = 16$ |
|---|---|---|---|---|---|
| $10^{-1}$ | 3.1623e−001 | 3.1623e−001 | $10^{-9}$ | 3.1623e−005 | 3.1623e−005 |
| $10^{-2}$ | 1.0000e−001 | 1.0000e−001 | $10^{-10}$ | 1.0000e−005 | 1.0000e−005 |
| $10^{-3}$ | 3.1623e−002 | 3.1623e−002 | $10^{-11}$ | 3.1626e−006 | 3.1626e−006 |
| $10^{-4}$ | 1.0000e−002 | 1.0000e−002 | $10^{-12}$ | 9.9996e−007 | 1.0037e−006 |
| $10^{-5}$ | 3.1623e−003 | 3.1623e−003 | $10^{-13}$ | 3.1628e−007 | 3.2818e−007 |
| $10^{-6}$ | 1.0000e−003 | 1.0000e−003 | $10^{-14}$ | 1.0125e−007 | 1.1738e−007 |
| $10^{-7}$ | 3.1623e−004 | 3.1623e−004 | $10^{-15}$ | 4.8129e−008 | 8.3301e−008 |
| $10^{-8}$ | 1.0000e−004 | 1.0000e−004 | $10^{-16}$ | 4.0916e−008 | 8.0053e−008 |

**Table 1.** The smallest eigenvalues of $-\mathcal{D}_S$

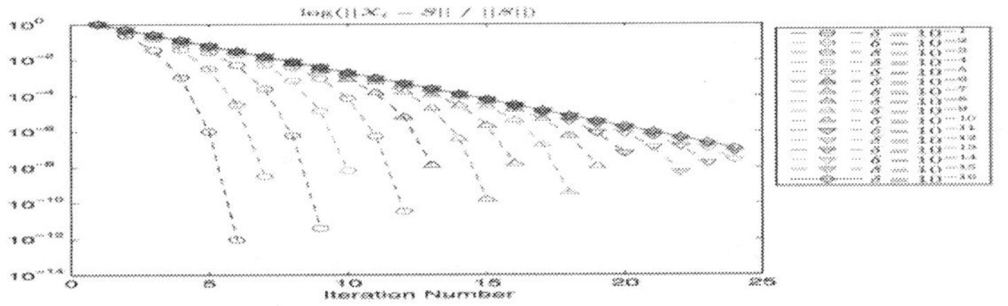


**Figure 1.** The convergence rate in Example 5 where $n = 8$

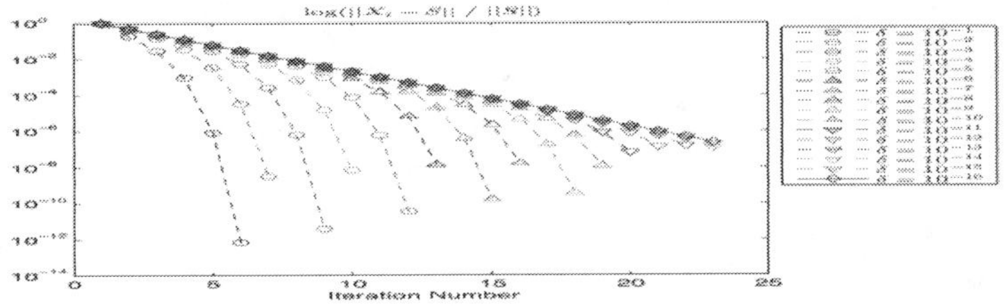


**Figure 2.** The convergence rate in Example 5 where $n = 16$

## References

[1] Dario A. Bini, Guy Latouche, and Beatrice Meini. *Numerical methods for structured Markov chains*. Oxford University Press Oxford, 2005.

[2] Chun-Hua Guo and Nicholas J. Higham. *Iterative solution of a nonsymmetric algebraic Riccati equation*. SIAM

Journal on Matrix Analysis and Applications, 29:396-412, 2007.

[3] Chun-Hua Guo and Peter Lancaster. *Analysis and modification of Newton's method for algebraic Riccati equations.* Mathematics of Computation, 67(223):1089-1105, 1998.

[4] C. He, B. Meini, N. H. Rhee, and K. Sohraby. *A quadratically convergent Bernoulli-like algorithm for solving matrix polynomial equations in Markov chains.* Electronic Transactions on Numerical Analysis, 17:151-167, 2004.

[5] Qi-Ming He and Marcel F.Neuts. *On the convergence and limits of certain matrix sequences arising in quasi-birth-and-death Markov chains.* Journal of Applied Probability, 38(2):519-541, 2001.

[6] W. Kratz and E. Stickel. *Numerical solution of matrix polynomial equations by Newton's method.* IMA Journal of Numerical Analysis, 7:355-369, 1987.

[7] G. Latouche. *Newton's iteration for nonlinear equations in Markov chains.* IMA Journal of Numerical Analysis, 14:583-598, 1994.

[8] G. Latouche and V. Ramaswami. *Introduction to Matrix Analytic Methods in Stochastic Modeling.* ASA-SIAM, 1999.

[9] George Poole and Thomas Boullion. *A survey on $M$-matrices.* SIAM review, 16(4):419-427, 1974.

[10] Jong Hyeon Seo and Hyun-Min Kim. *Solving matrix polynomials by Newton's method with exact line searches.* Journal of KSIAM, 12(2):55-68, 2008.

# Numerical Simulation of the Gravity Waves Generated by a Moving Obstacle

**Luděk Beneš**

Dept. Of Technical Mathematics, Faculty of Mechanical Engineering, Czech Technical University in Prague, Karlovo nám 13, 121 35, Prague 2, tel. +420 224357543

E-mail: `benes@marian.fsik.cvut.cz`

**Abstract.** The article deals with the numerical simulation of the flow pattern around a moving body in a stratified fluid. The flow is assumed to be unsteady, incompressible and stratified. The mathematical model is based on the Boussinesq approximation of the Navier–Stokes equations for viscous incompressible flow wit non–constant density. The resulting set of PDE's is then solved by the AUSM MUSCLE scheme in finite volume approximation. For the time integration the three stage BDF method of the second order is used. Three different obstacle models were tested.

## 1. Introduction

Stratification plays an important role in many industrial and environmental processes. Thoroughly study of the effect of stratification leads to a better modeling and finally to a better understanding of the problems conected with these processes. Since this is a very complicated issue, study of the simplified cases has a long tradition. One of the most popular cases is the study of a moving body in a stratified medium. The flow structure around a uniformly moving obstacle is the subject of intensive numerical and experimental studies. This investigation helps us to better understand the nature of the stratified flow and re-structuring with changing of the flow parameters. The processes occurring from the impulsive started body up to formation of internal waves are computed. The experimental and numerical studies of the generation of the gravity waves were proposed by e.g. (1),(2).

Since 2008 we have been dealing with the numerical study of the stratified flow. The flow past the body and over the hill for different Riachardson numbers using WENO, AUSM MUSCL and compact differences schemes was simulated see (4), (5), (3).

## 2. Mathematical model

The fluid is assumed to be incompressible, with non–constant density. As the mathematical model, the Boussinesq approximation of the Navier–Stokes equations was used. Density and pressure are divided into two parts: a background field (with subscript $_0$) plus a perturbation. The system of equations obtained is partly linearized around the average state $\rho_*$. The resulting set of equations can be written in the form

$$\frac{\partial \varrho}{\partial t} + \frac{\partial(\varrho u_j)}{\partial x_j} = -u_2 \frac{\partial \varrho_0}{\partial x_2}, \quad (1)$$

$$\frac{\partial u_i}{\partial t} + \frac{\partial(u_j u_i)}{\partial x_j} + \frac{1}{\varrho_*}\frac{\partial p}{\partial x_i} = \nu \frac{\partial^2 u_i}{\partial x_j \partial x_j} - \delta_{i,2} g, \quad (2)$$

$$\frac{\partial u_j}{\partial x_j} = 0, \quad (3)$$

where $W = [\varrho, u_1, u_2, p]^T$ is the vector of unknowns, $\varrho(x_1, x_2, t)$ denotes the perturbation of the density and $u_1$, $u_2$ are the velocity components, $p$ stands for the pressure perturbation and $g$ for the gravity acceleration. The $x_1$–axis is orientated in the direction of the motion and $x_2$ axis is perpendicular and pointing in the direction of the background density gradient.

## 3. Numerical scheme

The AUSM MUSCL scheme in the finite volume formulation has been used. The scheme is based on the artificial compressibility method in dual time. The spatial semidiscretization of the inviscid fluxes is achieved by the finite volume AUSM scheme. Quantities $(q)_{L/R}$ on the cell faces are computed using the MUSCL reconstruction with the Hemker-Koren limiter. The scheme is stabilized by the pressure diffusion. The viscous fluxes are discretized using central approach on a dual mesh (diamond type scheme).

The spatial discretization results in a system of ODE's solved by the second-order BDF formula. Arising set of nonlinear equations is then solved by the artificial compressibility method in the dual time $\tau$ by the explicit 3-stage second-order Runge-Kutta method. More about this scheme is written e.g in (5).

## 4. Computational setup

The presented computational setup corresponds to the experimental setup studied by Chaschechkin and Mitkin in (6).The gravity waves are generated by the moving of the thin horizontal strip $0.025 \times 0.002$ $m$. The computational domain is $0.5 \times 0.25m$. The obstacle is placed $0.3m$ from the left side (ranges in $< 0.3; 0.325 > m$) and in the middle height. At the time $t = 0$ the obstacle starts moving to the right (in the positive $x_1$ direction) with constant velocity $U^{ob} = 0.0017\ m/s$. The flow field is initially at rest with the exponential profile of stratification $\varrho_0 = \varrho_{00} \exp -\frac{x_2}{\Lambda}$, $\varrho_{00} = 1008.9\ kg/m^3$, $\Lambda = 47.735\ m$, the kinematic viscosity is $\nu = 10^{-6} m^2/s$.

Three different numerical approaches of the model are studied and compared to each other.

- The classical body fitted mesh surrounding the obstacle is used in the first case. Very simple Cartesian grids consisting of four parts can be used because of the simplicity of the problem.
- The second case is based on the volume penalization technique. The momentum equations are modified by adding of the term proportional to the difference between the fluid and obstacle velocities.
- The third model is simple variant of the immersed boundary method. The velocities in the cells lying in the obstacle are set to zero, while pressure and density are computed for the whole domain.

The Cartesian mesh of $500 \times 500$ cells was used. The resolution of the mesh is $1mm$ in the $x_1$ direction and $0.5mm$ in the $x_2$ direction. To verify independence of the solution on the mesh, the mesh two times refined in each direction was used.

## 5. Numerical results

Fig.1 shows the flow pattern in the form of the isolines of density perturbance $\varrho$ (left) and $u_2$–velocity component at the time $t = 75s$. The flow pattern is typical for transient internal waves past the body in stably stratified flow. The gravity waves are formed, the upstream disturbances are pronounced. Behind the obstacle strip with step–like density profile is formed.

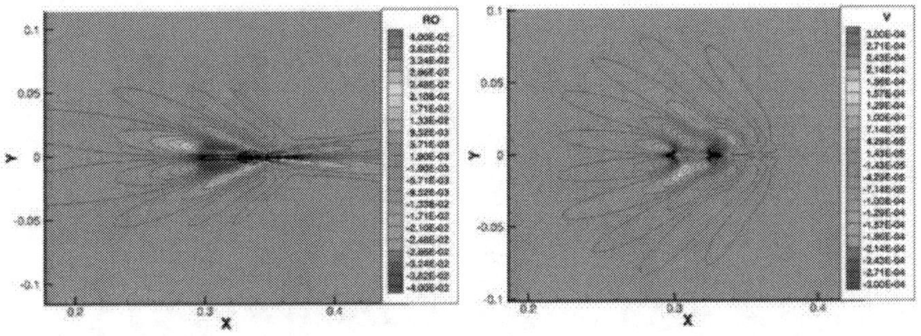


**Figure 1.** Flow pattern at time $t = 75s$. Left isolines of the density perturbance $\varrho$, right isolines of $u_2$ velocity component. Body fitted mesh.

| variable | multidomain | porous | immersed boundary |
|---|---|---|---|
| $\varrho$ | $4.34 \times 10^{-2}$ | $4.02 \times 10^{-2}$ | $4.04 \times 10^{-2}$ |
| difference | 0% | 7.3% | 6.9% |
| $u_1$ | $2.13 \times 10^{-3}$ | $2.11 \times 10^{-3}$ | $2.11 \times 10^{-3}$ |
| difference | 0% | 0.9% | .9% |
| $u_2$ | $3.23 \times 10^{-4}$ | $2.95 \times 10^{-4}$ | $2.97 \times 10^{-4}$ |
| difference | 0% | 8.7% | 8.0% |

**Table 1.** Maxima of the computed quantities and relative differences to the multidomain case.

In Fig.1, 2, the comparison of the different obstacle approaches in the form of the isolines of $u_2$–velocity component is given. The comparison shows a very good qualitative agreement. Small differences are in the wake behind the obstacle, which is stronger in the multidomain case. Immersed boundary case and permeable obstacle are practically the same.

Fig. 3 displays the perpendicular distribution of the density perturbance $\varrho$ in different distances. The wave length is the same in all models and is in a good agreement with theoretical prediction given by the Brunt-Väisälä frequency. Maxima and minima of all computed quantities are the highest in the multidomain approach mainly in front of and behind the obstacle. Results in the porous and immersed boundary cases are very similar. The boundary layer on the obstacle is well resolved in all approaches.

The maxima of the quantities in whole computational domain are summarized in the Tab.1. These maxima are compared to the multidomain approach and are approximately 8% lower in the immersed boundary and porous case.

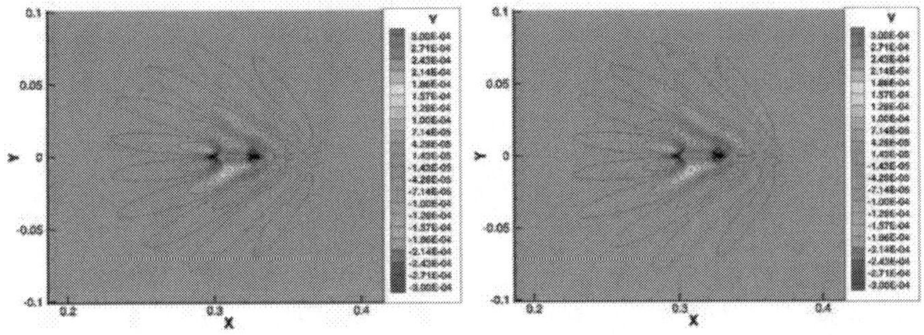


**Figure 2.** Isolines of $u_2$ velocity component, permeable obstacle model left, immersed boundary right.

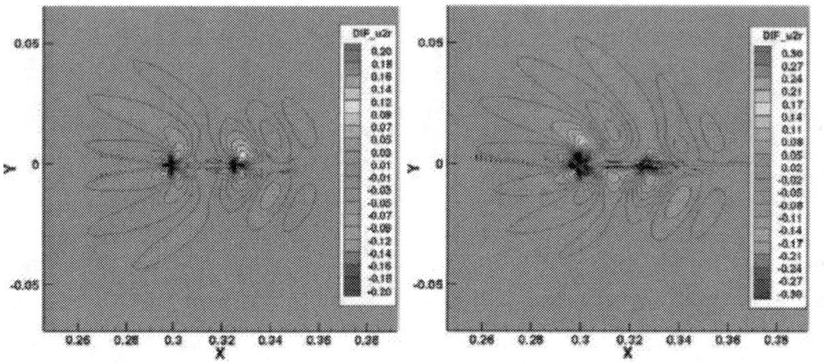

**Figure 3.** Comparison of the flow pattern at time $t = 75s$. Relative differences of the $u_2$ velocity component. Left immersed boundary - multidomain, right porous - multidomain.

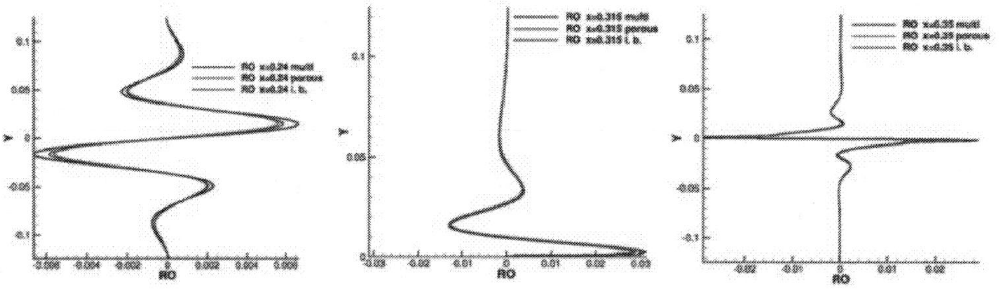

**Figure 4.** Vertical profiles of the density perturbation $\varrho$ in the distances [$x = 0.24m$,$x = 0.315m$, $x = 0.35m$.

## 6. Conclusion

The flow patterns produced by a moving vertical strip in a stratified fluid were computed. Three different obstacle models have been developed and compared. These results are in good mutual agreement. Quantitative and qualitative agreement of all methods was demonstrated. Differences in the maxima of the computed quantities that emerged between models require deeper investigation. Also sensitivity of the models on the quality of the mesh mainly close to the corners is interesting question. **Acknowledgments:** This work was supported by TACR Project TA01020428 and grant CTU SGS13/174/OHK2/3T/12.

## References


[1] Hanazaki H., Kashimoto K., Okamura T.: *Jets generated by a sphere moving vertically in a stratified fluid.* J. of Fluid Mech. **638**, 173–197, 2009

[2] Sutherland B., Linden P.: *Internal wave excitation from stratified flow over a thin barrier.* J. Fluid. Mech. **377**, 223-252, 1998.

[3] Bodnár T., Beneš L., Fraunié Ph., Kozel K.: *Application of compact finite–difference schemes to simulations of stably stratified fluid flows.* Applied Mathematics and Computation, Vol.219(7), 2012, p.3336–3353.

[4] Beneš L., Fürst J.: *Numerical simulation of the Stratified Flow Past a Body.* Numerical Mathematics and Advanced Applications. ENUMATH 2009. Berlin: Springer, 2010, p. 155-162.

[5] Beneš L., Fürst J., Fraunié Ph.: *Comparison of two numerical methods for the stratified flow.* J. Computers & Fluids, Vol:46(1) 2011 p. 148-154, ISSN 0045-7930.

[6] Chaschechkin Y.D., Mitkin V.V.: *Experimental study of a fine structure of 2D wakes and mixing past an obstacle in a continuously stratified fluid.* Dyn. Atmos. Oceans 34, 165–187, 2001

# Statistical analysis of laser-interferometric detector Dylkin-1 data and data on seismic activity

**R.S. Kirillov**[1,3,4], **V.V. Bochkarev**[1], **A.F. Skochilov**[2]

[1]Physics Department, Kazan Federal University, Russian Federation
[2] Scientific Center of Gravitational-Wave Research "Dulkyn", Academy of Sciences of the Republic of Tatarstan, Russian Federation

E-mail: kiroser@mail.ru

**Abstract.** This work presents statistical analysis of data collected from laser interferometric detector "Dylkin-1" and nearby seismic stations. The final goal of Dylkin project consists in creating detector of theoretically predicted gravitational waves produced by binary relativistic astrophysical objects. Currently, works are underway to improve sensitivity of detector by 2-3 orders. The goals of this research were to test isolation of detector from noise caused by seismic waves and to find out whether it is sensitive to variations in the gradient of gravitational potential (acceleration of free fall) caused by free Earth oscillations. Noise isolation has been tested by comparing energy of signals during significant seismic events. Sensitivity to variations in acceleration of free fall has been tested by means of cross-spectral analysis.

Current version of the "Dulkyn-1" detector[1] is a two-resonator laser system generating radiation with a wavelength of 3.39um. Frequency of the reference resonator bound to generation frequency of a He-Ne laser which is stabilized relative to the nonlinear hyper-narrow absorption resonance of a methane molecule (CH4) (relative instability of $10^{-14}$). The signal resonator is working in a free generation mode.

This configuration of the detector has been used to conduct an experiment to test the local position invariance principle, which is component of the Einstein's equivalence principle. One of its effects is gravitational shift of spectral lines, which is equivalent to dependence of a clock speed on the local gravitational potential $\varphi$.

Frequency difference of clocks with different physical nature has been measured in mentioned experiment. The result of experiment was approving of universality of the gravitational red shift with accuracy of 0.9%.

For experiments on the Earth's surface changes in gravitational potential caused by orbital movement of the Earth and influence of the Moon's gravitational field. Therefore, gravitational

[3] work was supported by grant RFBR № 12-05-31412
[4] Author to whom any correspondence should be addressed.

potential can be written in the following way: $\varphi(t,\boldsymbol{r})=\varphi_0+\Delta\varphi(t)+\boldsymbol{g}(t)\cdot\boldsymbol{r}$ , where $\boldsymbol{g}(t)$ – potential gradient. Changes in frequency of point clock (reference resonator, which is bound to methane molecule resonance frequency) are caused by $\Delta\varphi(t)$. Meanwhile, potential gradient $\boldsymbol{g}(t)$ also has a significant influence on frequency of signal resonator. If the principle of equivalence is right, changes in frequencies of point and extensive clocks, caused by value of $\Delta\varphi(t)$ will be equal. It means that difference in speed of clocks will be determined only by value of $\boldsymbol{g}(t)$.

It is well known that earthquakes induce the Earth's free oscillations[2], which in turn change gravitational potential and its derivatives on the Earth surface. This implies that correlation between a seismic data and readings of the detector could be observed on the free oscillations frequencies.

Magnitude squared coherence value has been calculated to estimate influence of the Earth's free oscillations on detector readings. This value is being obtained using two datasets and reflects their correlation on various frequencies. It can be written as follows:

$$C_{XY}(f)=\frac{\left|P_{XY}(f)\right|^2}{P_{XX}(f)\cdot P_{YY}(f)}, \qquad (1)$$

where X and Y – some datasets, P – estimate of power spectral density. $P_{XY}(f)$ – X and Y cross-spectrum. Welch's average modified periodogramm method has been used to calculate estimate of power spectral density. Signal has been divided into segments with length of about 5 days with 50% overlap. Each segment has been filtered with Hamming window.

Figures 1 and 2 contain magnitude squared coherence functions. Data sources are "Dulkyn-1" detector readings on one hand, and data collected from seismographs, on the other. Peaks connected to the Earth's spinning (12h and 24h) can be seen on both plots of figure 1. Besides of that, there are few long period components of some different nature that can be seen on plot calculated using the detector data.

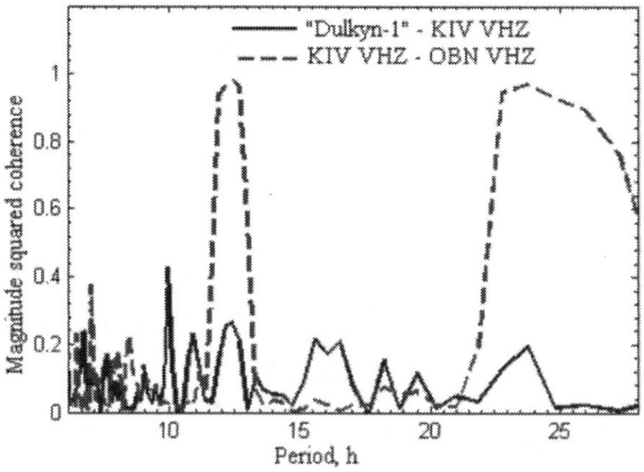


Figure 1. Long period components in detector and seismograph signals coherence function.

There are marks on figure 2 that correspond to some of the Earth's free oscillation modes. Values of the free oscillation mode periods can be found in article[3]. Therefore, in the "Dulkyn-1" data there are several long period components (including 12h and 24h components, induced by spinning of the Earth), and some modes of the Earth's free oscillations.

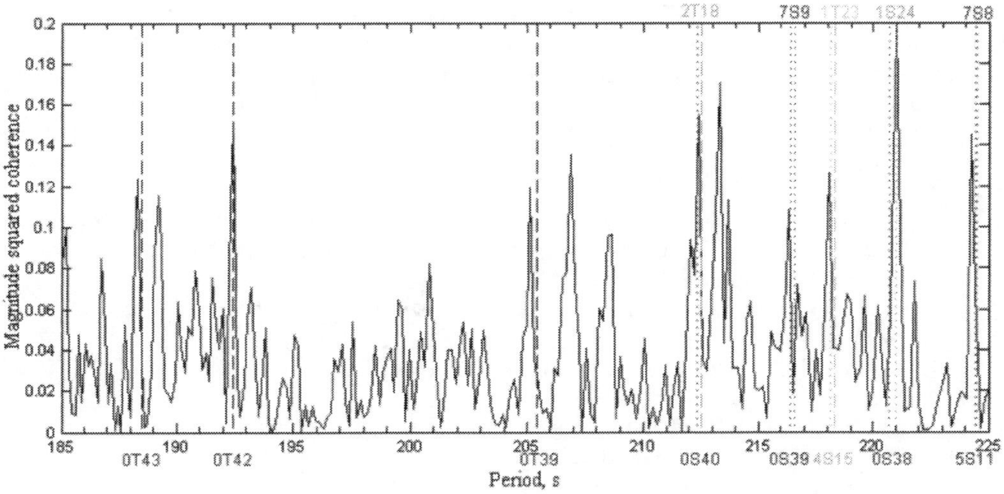


Figure 2. Modes of the Earth's free oscillations in detector and seismograph signals coherence function.

Seismic datasets used in this research has been obtained from http://www.iris.edu using BREQ-FAST (Batch REQuest-FAST)[4]. Name of dataset is formed from short name of station and name of seismograph channel. Datasets used for analysis collected from seismic stations of cities Kislovodsk (KIV) and Obninsk (OBN). Seismograph channel (VHZ) used in this work is vertical (Z) low frequency (0.1Hz – V) seismograph channel with high gain (H). Names of stations and channels can be obtained using IRIS DMC Metadata Aggregator service: http://www.iris.edu/mda. Abbreviation of channel name can be explained according to the SEED standard[5].

Another mechanism of interaction between "Dylkin-1" detector and seismic events is elastic waves induced by earthquakes, propagating into detector chamber. One of the goals of this research was to check vibration protection of the detector. For this purpose energies of the detector and seismograph signals has been calculated and compared during significant seismic events. On figure 3 there are energies of mentioned signals during the most powerful earthquake with magnitude of 5.8.

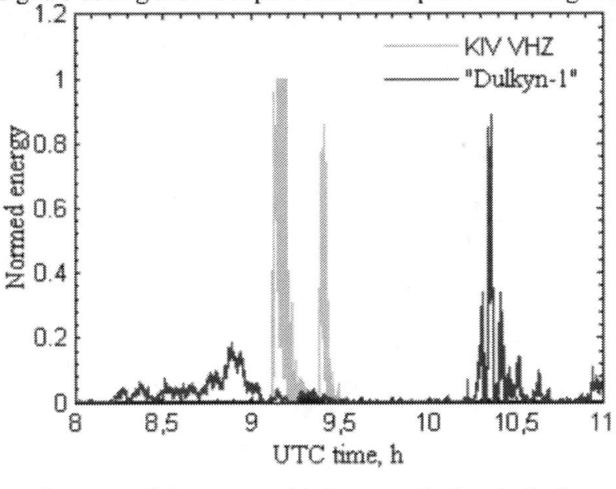


Figure 3. Normed energy of detector and seismograph signals during an earthquake.

From distance between the detector and seismograph and minimum propagation velocity for seismic wave one can calculate maximum time interval between arrivals on corresponding signals. For station used in this work this interval is about 10 minutes. Therefore, from this fact and figure 3 should be that this earthquake is not seen in detector data. The same result has been obtained for all significant seismic events. It means that for given detector sensitivity vibration protection system is working fine.

**References**

[1] A.R. Agachev, I.Yu. Belov, V.V. Bochkarev, R.A. Daishev, S.V. Mavrin, Z.G. Murzakhanov, A.F. Skochilov, Yu.P. Chugunov, O.P. Shindyaev, Journal of Experimental and Theoretical Physics, 2010, Volume 110, Issue 1, pp 1-6.

[2] F.D. Stacey and P.M. Davis, Physics of the Earth, Cambridge University Press: New York, 2008, pp. 255-261.

[3] A.M. Dziewonski, D.L. Anderson, Preliminary reference Earth model, Phys. Earth Planet Inter., Volume 25, Issue 4, 1981, pp. 294-356.

[4] BREQ_FAST Request Manual, http://www.iris.edu/manuals/breq_fast.htm

[5] SEED Reference Manual, FDSN, 2012, http://www.fdsn.org/seed_manual/SEEDManual_V2.4.pdf

# LSSVM based initialization approach for parameter estimation of dynamical systems

**Siamak Mehrkanoon, Rien Quirynen, Moritz Diehl and Johan A.K. Suykens**

KU Leuven, ESAT-SCD, Kasteelpark Arenberg 10, B-3001 Leuven (Heverlee), Belgium

E-mail:
{Siamak.Mehrkanoon,Rien.Quirynen,Moritz.Diehl,Johan.Suykens}@esat.kuleuven.be

**Abstract.** In this study the estimation of parameters in dynamical systems governed by parameter-affine ordinary differential equations is explored. The described method by Mehrkanoon et al.~ in [1] is utilized as an initialization of the nonlinear optimization problem for parameter estimation. In contrast to existing convex initialization approaches [2] that use a first order Euler discretization, we do not require any integration method to simulate the dynamical system. Furthermore, a denoising scheme using LSSVM is proposed to first filter the measured data then proceed with the filtered signals for parameter estimation problem. Experimental results demonstrate the efficiency of the proposed method, compared to alternative approaches on different examples from the literature.

## 1. Introduction

Mathematical models are widely used in various fields of application to describe physical systems. These models involve some unknown parameters that require to be estimated and the performance of the model largely depends on the way it is parameterized. In this study the estimation of parameters in dynamical systems governed by ordinary differential equations (ODEs) is explored. In general, during the process of parameter estimation one tries to make the differences between simulation results and the observational data as small as possible using an optimization algorithm. However, depending on the nature of the model, the optimization problem can be non-convex, potentially leading to multiple local minima. In the case of parameter-affine models, a convex formulation approach for initialization of the parameter estimation problem is described in [2]. In this approach in order to keep the approximate problem convex, one relies on a simple Euler discretization of the system. Recently Mehrkanoon et al.~ [1] proposed a different approach based on Least Squares Support Vector Machines (LSSVM) for estimating the unknown parameters in ODEs. As opposed to the approach described in [2], the method introduced in [1] does not need to use any integration method to simulate the dynamical system. Therefore the drawbacks of using the Euler method concerning its stability region are removed. The aim of this paper is to first employ the method described in [1] to obtain an initial guess for the parameters and then to solve the original nonconvex problem. The latter is done using a multiple shooting discretization and constrained Gauss-Newton to solve the nonlinear programming problem (NLP). Moreover a scheme based on LSSVM is proposed to pre-process the data and use the filtered data in the convex approximate problem.

## 2. Problem statement

Consider a dynamical system of the form: $\frac{dX}{dt} = F(t, X, \theta)$, $X(0) = X_0$. In what follows, $X = [x_1, ..., x_m]^T$ is the state vector of the system, $\theta = [\theta_1, ..., \theta_p]^T$ are the unknown parameters of the system and $X_0$ are initial values. In order to estimate the unknown parameters $\theta$, the state variable $X(t)$ is observed at $N$ time instants $\{t_1, ..., t_N\}$, i.e.~ $Y(t_i) = X(t_i) + E_i$, $i = 1, ..., N$, where $\{E_i\}_{i=1}^N$ are independent measurement errors with zero mean. The objective is to determine appropriate parameter values so that errors between the model prediction and the measured data are minimized.

## 3. Brief overview of the existing initialization approaches

Parameter estimation is typically formulated as following non-convex optimization problem:

$$\begin{aligned} \underset{X(t_0),...,X(t_N),\theta}{\text{minimize}} \quad & \frac{1}{2}\sum_{i=0}^{N} \|Y(t_i) - X(t_i)\|_2^2 \\ \text{subject to} \quad & X(t_{k+1}) = X(t_k) + \int_{t_k}^{t_{k+1}} F(\tau, X(\tau), \theta)\, d\tau, \quad k = 0, ..., N-1, \end{aligned} \tag{1}$$

Due to the nonconvexity coming from the nonlinear model, a Newton type method can only find locally optimal solutions. Depending on the initialization, one can obtain a different local solution. In case of parameter-affine models attempts have been made to provide a good initial guess through a convex optimization approach. A Least Squares Prediction Error Method (PEM) proposed in [3] is formulated as a convex problem to provide such an initial guess for (1), with parameter-affine function $F$, as follows:

$$\begin{aligned} \underset{X(t_0),...,X(t_N),\theta}{\text{minimize}} \quad & \frac{1}{2}\sum_{i=0}^{N} \|Y(t_i) - X(t_i)\|_2^2 \\ \text{subject to} \quad & X(t_{k+1}) = Y(t_k) + T_s F(t_k, Y(t_k), \theta), \quad k = 0, ..., N-1, \end{aligned} \tag{2}$$

where $T_s$ is the sampling time. However the solution obtained by the PEM approach can be biased if the process and measurement noise are not modeled appropriately. Therefore the method does not perform well in the presence of noisy data and one needs to filter the residual errors. On the other hand, the authors in [2] proposed a so-called Least Squares Convex Approach (CA). In contrast to the PEM approach, there is no need for filtering the residual error in CA formulation [2]:

$$\begin{aligned} \underset{X(t_0),...,X(t_N),\theta}{\text{minimize}} \quad & \frac{1}{2}\sum_{i=0}^{N} \|Y(t_i) - X(t_i)\|_2^2 \\ \text{subject to} \quad & X(t_{k+1}) = X(t_k) + T_s F(t_k, Y(t_k), \theta), \quad k = 0, ..., N-1. \end{aligned} \tag{3}$$

However in this approach in order to keep the optimization problem (3) convex, one still has to rely on a simple Euler discretization of the system. Recently, Mehrkanoon et al.~ [1] proposed an approach based on Least Squares Support Vector Machines (LSSVM) [4] for parameter estimation of ODEs. Closed-form approximate models for the state and its derivative are first derived from the observed data by means of LSSVM. The time-derivative information is then substituted into the given dynamical system, reducing the parameter estimation problem into an algebraic optimization problem. The problem is formulated as the following convex optimization problem:

$$\underset{\theta}{\text{minimize}} \quad \frac{1}{2}\sum_{i} \left\| \frac{d}{dt}\hat{X}(t_i) - F(t_i, \hat{X}(t_i), \theta) \right\|_2^2 \tag{4}$$

where $\hat{X}(t_i)$ and $\frac{d}{dt}\hat{X}(t_i)$ are obtained using the LSSVM model (see [1, 5] for more details). As opposed to the previous approaches, one does not need to use any integration method to simulate the dynamical system. Therefore the drawbacks of using the Euler method, concerning its stability region, are removed. We refer further to this approach as LSSVM.

## 4. Pre-processing using LSSVM

Data pre-processing plays a very important role in many applications. We will make use of the LSSVM regression ability to reduce the effect of noise and as a result having a smoother signal to proceed with. Given observational data $Y(t) = [y_1(t), \ldots, y_m(t)]^T$, and assuming a model of the form $\hat{y}_k(t) = w_k^T \varphi(t) + b_k$ for $y_k(t)$, $k = 1, \ldots, m$, the LSSVM regression is formulated as the following optimization problem [4]:

$$\begin{array}{ll} \underset{w_k, b_k, e_k(t_1), \ldots, e_k(t_N)}{\text{minimize}} & \frac{1}{2} w_k^T w_k + \frac{\gamma_k}{2} \sum_{i=1}^{N} e_k(t_i)^2 \\ \text{subject to} & y_k(t_i) = w_k^T \varphi(t_i) + b_k + e_k(t_i), \quad i = 1, \ldots, N, \end{array} \tag{5}$$

where $\gamma_k \in \mathbb{R}^+, b_k \in \mathbb{R}$, $w_k \in \mathbb{R}^h$, $\varphi(\cdot) : \mathbb{R} \rightarrow \mathbb{R}^h$ is the feature map and $h$ is the dimension of the feature space. The dual solution is then given by

$$\left[\begin{array}{c|c} \Omega + \gamma^{-1} I_N & 1_N \\ \hline 1_N^T & 0 \end{array}\right] \left[\begin{array}{c} \alpha^k \\ \hline b_k \end{array}\right] = \left[\begin{array}{c} y^k \\ \hline 0 \end{array}\right] \tag{6}$$

where $\Omega_{ij} = K(t_i, t_j) = \varphi(t_i)^T \varphi(t_j)$ is the $(i, j)$-th entry of the positive definite kernel matrix. $1_N = [1, \ldots, 1]^T \in \mathbb{R}^N$, $\alpha^k = [\alpha_1^k, \ldots, \alpha_N^k]^T$, $y^k = [y_k(t_1), \ldots, y_k(t_N)]^T$ and $I_N$ is the identity matrix. The model in dual form becomes: $\hat{y}_k(t) = w_k^T \varphi(t) + b_k = \sum_{i=1}^{N} \alpha_i^k K(t_i, t) + b_k$ where $K$ is the kernel function. The signal $\hat{y}_k(t)$ can be considered as a denoised version of the $y_k(t)$ measurements. The idea now would be to replace the measurements $Y(t)$ in PEM and CA formulations (2) and (3) by $\widehat{Y}(t)$ where $\widehat{Y}(t) = [\hat{y}_1(t), \ldots, \hat{y}_m(t)]^T$. These schemes will be referred to as PEM+LSSVM and CA+LSSVM respectively.

## 5. Numerical Results

Five initialization approaches i.e.~ PEM, PEM+LSSVM, LSSVM, CA and CA+LSSVM are considered for providing a good starting point for solving the original nonlinear least squares parameter estimation problem (1). Then the constrained Gauss-Newton method (CGN) is applied for solving the multiple shooting formulation in (1).

**Problem 5.1:** Consider the Lorenz equation

$$\frac{dx_1}{dt} = \theta_1(x_2 - x_1), \quad \frac{dx_2}{dt} = x_1(\theta_2 - x_3) - x_2, \quad \frac{dx_3}{dt} = x_1 x_2 - \theta_3 x_3 \tag{7}$$

where $\theta = (\theta_1, \theta_2, \theta_3)$ are the unknown parameters of the system. The initial values at $t = 0$ are $(-9.42, -9.34, 28.3)$ and the correct parameter values to be estimated are $\theta = (10, 28, 8/3)$.

**Problem 5.2** Consider Barne's problem [1]:

$$\frac{dx_1}{dt} = \theta_1 x_1 - \theta_2 x_1 x_2, \quad \frac{dx_2}{dt} = \theta_2 x_1 x_2 - \theta_3 x_2, \tag{8}$$

with initial state values $(1.00, 0.3)$ and $\theta = (\theta_1, \theta_2, \theta_3) = (0.86, 2.07, 1.81)$ as the true unknown parameters of the system.

Numerical results illustrating the performance of different initialization methods in terms of Mean Squared Error (MSE) can be found in Figure 1(a) and (b), respectively using 300 measurements at 0.04s (Lorenz) and 200 measurements at 1s sampling time (Barne). In Figure 1,

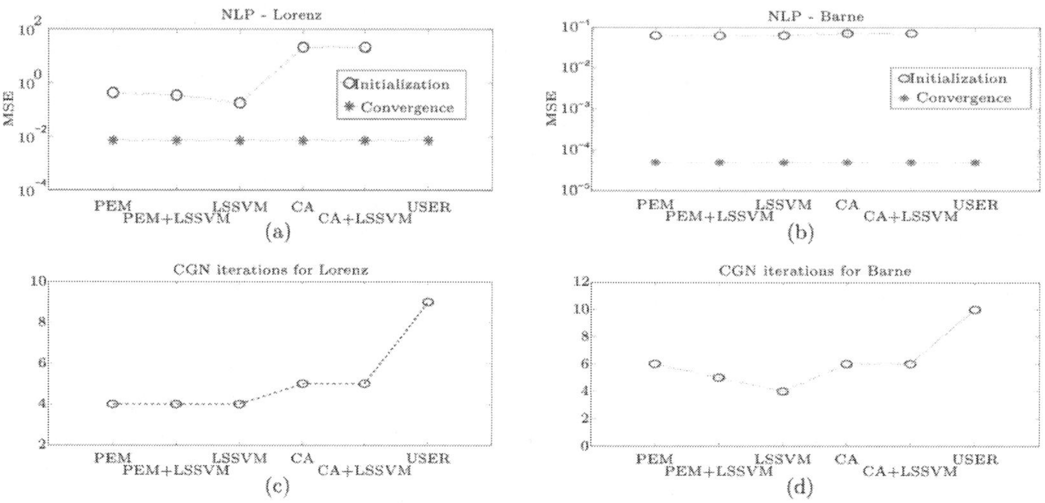


**Figure 1.** (a) and (b) The obtained Mean Squared Error for Problem 5.1 and 5.2 respectively. (c) and (d) Number of required Newton iterations needed to satisfy the given tolerance.

the case where the user is providing some starting point, based upon available prior knowledge, is referred to as USER initialization approach. The implementations and simulations were carried out in MATLAB and for the discretization of (1), the ACADO integrators were used as presented in [6]. Figure 1 shows that LSSVM based initialization is comparable to that of conventional convex initialization approaches and for Barne system it requires the least number of iterations to converge.

## 6. Conclusion

In this paper we presented an alternative method based on LSSVM for the initialization of nonlinear least squares parameter estimation problems. As opposed to conventional approaches the proposed method does not need to simulate the given dynamical system in order to provide a good approximate solution.

## Acknowledgments

This work was supported by: • Research Council KUL: GOA/10/09 MaNet, PFV/10/002 (OPTEC), several PhD/postdoc & fellow grants • Flemish Government: ∘ IOF: IOF/KP/SCORES4CHEM; ∘ FWO: PhD/postdoc grants, projects: G.0320.08 (convex MPC), G.0558.08 (Robust MHE), G.0557.08 (Glycemia2), G.0588.09 (Brain-machine), G.0377.09 (Mechatronics MPC); G.0377.12 (Structured systems) research community (WOG: MLDM); ∘ IWT: PhD Grants, projects: Eureka-Flite+, SBO LeCoPro, SBO Climaqs, SBO POM, O&O-Dsquare • Belgian Federal Science Policy Office: IUAP P7/ (DYSCO, Dynamical systems, control and optimization, 2012-2017) • IBBT • EU: ERNSI, FP7-EMBOCON (ICT-248940), FP7-SADCO (MC ITN-264735), ERC ST HIGHWIND (259 166), ERC AdG A-DATADRIVE-B • COST: Action ICO806: IntelliCIS • Contract Research: AMINAL • Other: ACCM. Johan Suykens is a professor at the KU Leuven, Belgium.

## References

[1] Mehrkanoon S, Falck T, Suykens J A K 2012 Parameter Estimation for Time Varying Dynamical Systems using Least Squares Support Vector Machines, *In Proc. of the 16th IFAC Symposium on System Identification (SYSID 2012), Brussels, Belgium*, pp. 1300-1305.

[2] Bonilla J, Diehl M, De Moor B and Van Impe J 2008 A Nonlinear Least Squares Estimation Procedure without Initial Parameter Guesses, *Proceedings of the 47th IEEE Conference on Decision and Control* Cancun, Mexico, pp. 5519-5524.

[3] Ljung L 1999 *System Identification, Theory for the Users*, Prentice Hall, NJ.

[4] Suykens J A K, Van Gestel T, De Brabanter J, De Moor B and Vandewalle J 2002 *Least Squares Support Vector Machines* (World Scientific, Singapore.)

[5] Mehrkanoon S and Suykens J A K 2012 LS-SVM approximate solution to linear time varying descriptor systems *Automatica*, **48**, 2502-2511.

[6] Quirynen R, Gros S, Diehl M 2013 Fast auto generated ACADO integrators and application to MHE with multi-rate measurements, *Proceedings of the European Control Conference*, accepted.

# An example of quaternion parameterization for dynamical simulations

**Valeria Artale, Cristina L.R. Milazzo, Angela Ricciardello**

Faculty of Engineering and Architecture, University of Enna "Kore",
Cittadella Universitaria, 94100 Enna, Italy

E-mail: valeria.artale@unikore.it, cristina.milazzo@unikore.it,
angela.ricciardello@unikore.it

**Abstract.** The dynamical simulation of rigid bodies can be gathered from the classical Newton-Euler differential equations, which commonly make use of the Euler angles parametrization. In this work, the initial value problem associated with motion is presented in terms of quaternion formulation instead of the Euler one. The reason why the quaternion parametrization is proposed lies on the possibility of avoiding singularities that can occur by considering Euler angles. Moreover, the strength of quaternions is represented by the linearity of their formulation, the easiness of their algebraic structure and, overall, on their stability and efficiency. Our proposed application is the mathematical modelling of a small Unmanned Aerial Vehicle dynamics. In particular a multirotor with six blades has been taken into account, its mathematical model is deduced and a comparison between the results obtained by implementing our formulation and the classical one is produced.

## 1. Introduction

The main issue related to rigid body dynamical simulation is represented by its mathematical modelling. An efficient mathematical model should be able to describe the movement of the body in space as well as possible. In literature, Euler angles are often applied to define its orientation and therefore are considered as state variables. Although the formulation based on Euler angles is easy to be understood and implemented, its matrix representation has an innate singularity in the parametrization, thus a loss of degrees of freedom can occur [1]. For this reason alternative approaches should be evaluated. This work focuses on quaternions, highlighting the effectiveness and the advantages of this choice on applying them in aerospace engineering field, [2]. The relevant feature of quaternions consists of avoiding the singularities and of being well-suited to represent the attitude of a rigid body over time. Quaternions are commonly used in several application fields: from computer game development to robotics, from 3D virtual words to control theory, [3],[4]. In this paper the mathematical model of a small Unmanned Aerial Vehicle is introduced. UAVs are aircraft without a human pilot on board, they are employed for both military and civil applications mainly for rescue, inspection and surveillance, [4]- [7]. Due to the universal trend of the evolution of this industry, an accurate description of the dynamics of these vehicles is required and is becoming a primary concern for industrial applications. In this work, UAV dynamics is presented in terms of quaternion formulation instead of the Euler one, both our formulation and the classical one have been implemented and, thus, the gained results have been compared and discussed.

## 2. Mathematical Model

A free-moving rigid body in space is a mechanical system having six degrees of freedom and its position has two components: linear and angular, respectively. The linear position is achieved through the Cartesian coordinates of the origin of the mobile frame with respect to the inertial one. On the other hand, in many applications, the angular position of a rigid body is described by means of three specific rotation angles, the so called *Euler angles*. In other words, the position of the mobile frame, whose origin lies on the centre of mass of the rigid body, is described by means of a set of rotations starting from an inertial reference frame. There are a lot of possible sequences of successive rotations but the most commonly used, especially in aerospace engineering and computer graphics, consists of a rotation about the z-axis, then about the y-axis and finally about the x-axis of the inertial frame, also known as *roll-pitch-yaw* convention. The composition of the previous transformations allows to obtain the well-known rotation matrix that connects the components of the same vector in both of the frames of reference, the inertial one and the mobile one fixed on rigid body. So, the transformation related to the position and the angular velocity from the body to the inertial frame are obtained respectively by

$$\mathbf{R} = \begin{bmatrix} C\theta C\psi & C\psi S\theta S\phi - C\phi S\psi & C\phi C\psi S\theta + S\phi S\psi \\ C\theta S\psi & C\phi C\psi + S\theta S\phi S\psi & C\phi S\theta S\psi - C\psi S\phi \\ -S\theta & C\theta S\phi & C\theta C\phi \end{bmatrix} \qquad \mathbf{W}^{-1} = \begin{bmatrix} 1 & S\phi T\theta & C\phi T\theta \\ 0 & C\phi & -S\phi \\ 0 & Sc\theta S\phi & C\phi Sc\theta \end{bmatrix} \quad (1)$$

where $\phi, \theta, \psi$ are the roll, pitch and yaw angles and, for the sake of clarity, the letters $C$, $S$, $T$, $Sc$ are used to express the cosine, sine, tangent and secant functions, respectively. Although Euler angles are easy to be understood, implemented and visualized, there is a particular circumstance in every rotation sequences in which a degree of freedom is lost. This happens when two axes are aligned with each other, that is when $\theta_{gl} = \pi/2 + k\pi, k \in \mathbb{Z}$. This situation is commonly known as *gimbal lock* and it can be more appreciated in the transformation matrix $\mathbf{W}^{-1}$ defined if and only if $\theta \neq \theta_{gl}$. To overcome this problem it is possible to consider a different parametrization for the rigid body orientation in space. For this reason, quaternion formulation is introduced. According to Euler's theorem, a rotation in space can be described by a rotation about an axis over a fixed angle. So, if $\boldsymbol{u} = (u_1, u_2, u_3)$ is the unit vector of the axis and $\alpha$ is the rotation angle, a quaternion is defined as $\boldsymbol{q} = (q_0, q_1, q_2, q_3)$, where $q_0 = \cos(\alpha/2)$, $q_1 = \sin(\alpha/2)\, u_1$, $q_2 = \sin(\alpha/2)\, u_2$ and $q_3 = \sin(\alpha/2)\, u_3$, also they satisfy the condition $q_0^2 + q_1^2 + +q_2^2 + q_3^2 = 1$. The quaternion approach is fully equivalent to the Euler one, but it is more easy and efficient from a computational point of view and, above all, it does not exhibit the gimbal lock phenomenon. Our aim consists in describing the motion a hexagon shaped aircraft or hexacopter, with three pairs of counter-rotating propellers arranged on vertices the hexagon, [8]. Supposed the drone as a rigid body, its dynamics is deduced from the classical Newton - Euler equations but in terms of quaternions as shown in [9]. Taking into account all the internal and external influences, the translational and rotational components of the motion read

$$m\,\ddot{\boldsymbol{\xi}} = \mathbf{F}_g + \mathbf{Q}\,\mathbf{T}_B, \qquad \ddot{q} = d\,(\mathbf{S}\,\boldsymbol{\nu})/d\,t. \quad (2)$$

in which $m$ is the mass of the drone, $\boldsymbol{\xi} = (x, y, z)$ represents its position vector with respect to the inertial frame, $\mathbf{F}_g$ is the gravitational force, $\mathbf{T}_B$ is the total thrust, $\mathbf{Q}$ is the orthogonal transformation matrix from the body frame to the inertial one, $\mathbf{S}$ is the velocity transformation matrix and $\boldsymbol{\nu} = (p, q, r)$ is the angular velocity solution of $\mathbf{I}\,\dot{\boldsymbol{\nu}} + \boldsymbol{\nu} \times (\mathbf{I}\,\boldsymbol{\nu}) + \boldsymbol{\Gamma} = \boldsymbol{\tau}_B$ where $\mathbf{I}$ is diagonal inertial matrix, $\boldsymbol{\Gamma}$ represents the gyroscopic effects and $\boldsymbol{\tau}_B = (\tau_\phi, \tau_\theta, \tau_\psi)$ the roll, pitch and yaw moment torque vector. Given an initial condition, the mathematical model is sufficient to describe the evolution of the aircraft, but it does not give information on the final position of the drone. It is therefore essential to associate with the mathematical model a control technique enabling to maneuver the drone, to manage the flight and to decide in advance which position will occupy the aircraft. Attitude and altitude control consists in conveniently modifying angular velocity of the six rotors. For this purpose an original control technique based on quaternions has been implemented [8].

## 3. Simulation Study

In order to validate the model and to emphasize the efficiency of the quaternion parametrization, a lot of experimental testing has been performed. Quaternions are computationally more efficient, because their simplicity, linearity and lack of singularities and they require fewer mathematical operations moreover. All these features are reflected in a gain of time, that is significant in real flight simulations. Figure 1 shows the execution time versus the number of points in which the temporal interval is divided: dashed line is related to the Euler formulation, while the continuous line to the quaternions formulation. It is evident the gain of time with increasing the number of points, despite the system (2) counts two more equations with respect to the system using the Euler parametrization.

Moreover, while Euler formulation suffers from the gimbal lock, quaternion parametrization

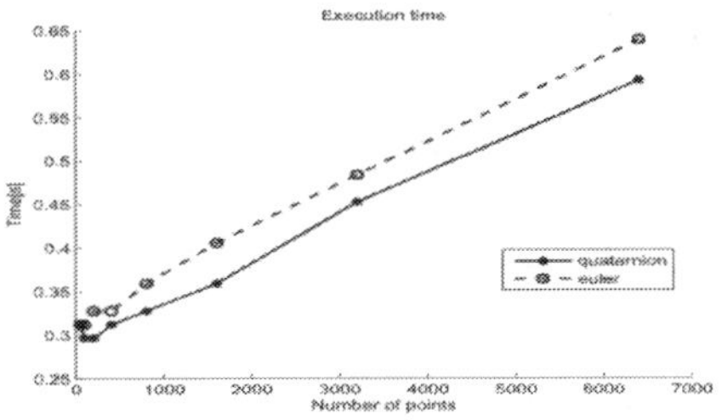


**Figure 1.** Comparison of execution time between Euler (dashed) and quaternions (continuous) formulation.

does not encounter it: as shown in Figure 2, when pitch angle is close to $\theta_{gb}$, roll and yaw angles exhibit a jump; whereas in Figure 3 quaternion component are continuous and smooth in time.

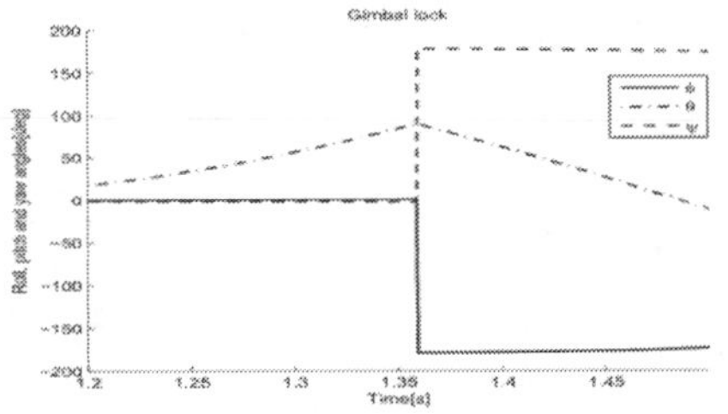


**Figure 2.** Evolution of Euler angles.

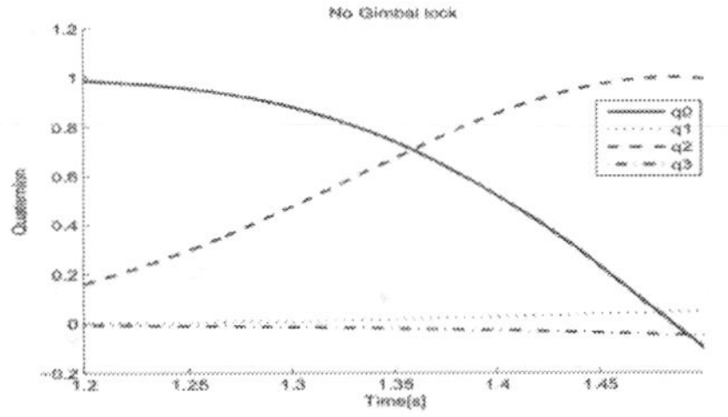


**Figure 3.** Evolution of quaternion.

Simulation results of the attitude and altitude controller in terms of quaternion are now presented. Analogous results can be obtained by means of Euler angles approach. Let us observe that environmental disturbances, such as wind, and aerodynamical effects are not incorporated in simulation model. The initial condition for equations (2) has been taken as follows: $\boldsymbol{\xi}_0 = (0, 0, 0)$ and $\boldsymbol{q}_0 = (1, 0, 0, 0)$. Target position and orientation are: $\boldsymbol{\xi}_d = (0, 1, 1)$ and the quaternion $\boldsymbol{q}_d$ corresponding to $(\phi, \theta, \psi)_d = (\pi/4, \pi/3, \pi/6)$. In Figure 4 the components of position vector are shown. Desired position is reached after less than 20 seconds and after that the hexacopter maintains hovering condition. Figure 6 shows the evolution of the quaternion, describing the orientation of the body, with respect to time, while Figure 7 shows the evolution of the quaternion error describing the deviation between the actual quaternion and the desired one.

## 4. Conclusion

In this work an initial value problem is presented in terms of quaternion formulation. A comparison between Euler and quaternion approach has been driven, highlighting the efficiency

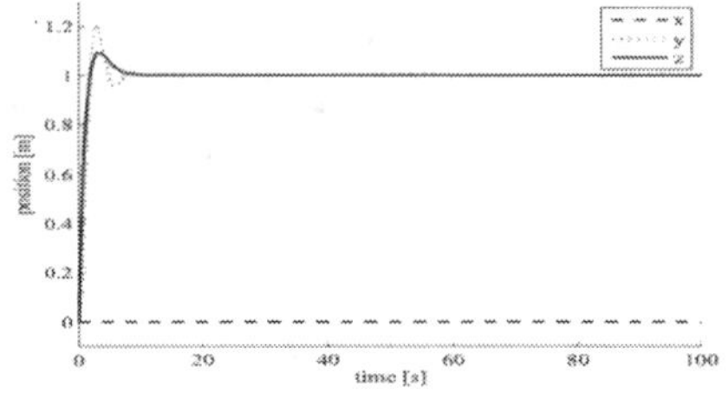


**Figure 4.** Position $x$, $y$ and $z$

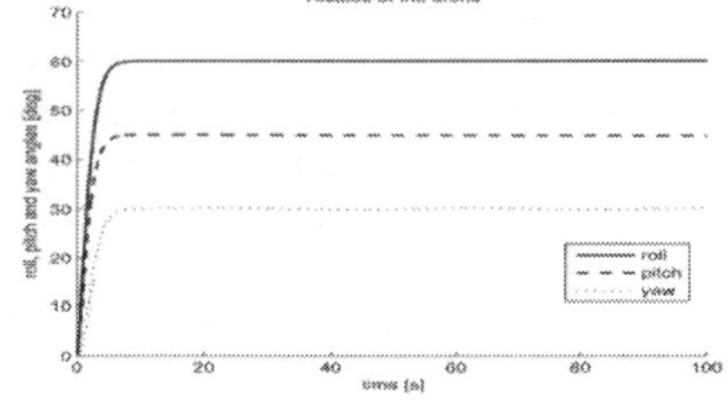


**Figure 5.** Attitude of the hexacopter.

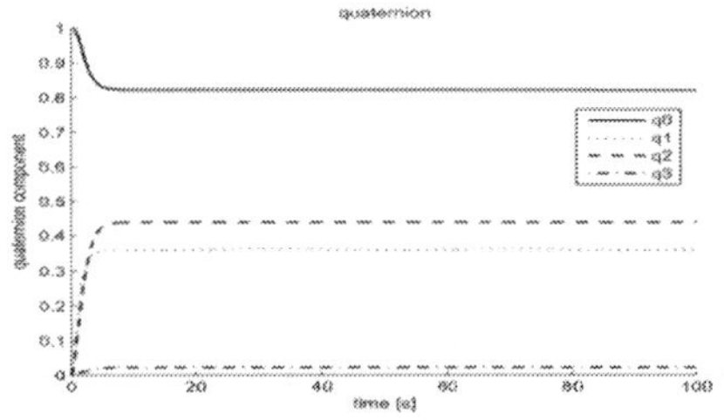


**Figure 6.** Quaternion versus time.

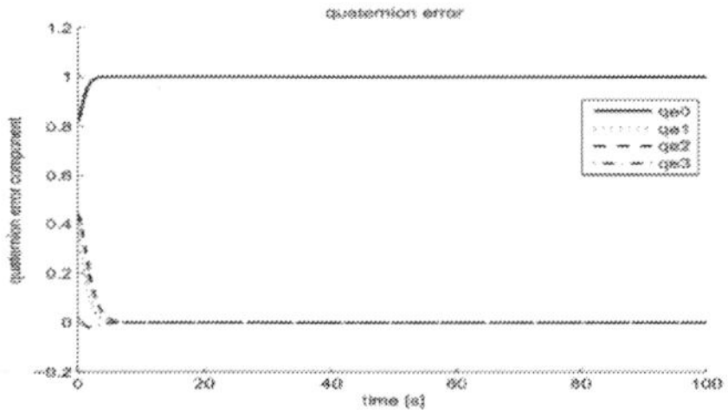


**Figure 7.** Quaternion error versus time.

of the second method from a computational point of view. Our application consists in describing and controlling a small UAV. The small computational time and small error in integration provide good answers in real flight simulations for an easier management and maneuverability.

## Acknowledgments
Project supported by the PO. FESR 2007/2013 subprogram 4.1.1.1 Prog. "Mezzo Aereo a controllo remoto per il Rilevamento del TErritorio - MARTE" Grant No.10772131.

## References
[1] K. Großekatthöfer, Z. Yoon, *Introduction into quaternions for spacecraft attitude representation*, TU Berlin, 2012.

[2] R. F. Stengel, *Flight Dynaimcs*, Princeton University Press, 2004.

[3] J.B Kuipers, *Quaternions and Rotation Sequences: A Primer With Applications to Orbits, Aerospace, and Virtual Reality*, Princeton University Press, 1999.

[4] P. Mc Kerrow (2004) Modelling the Draganflyer four-rotor helicopter, *In Proceedings of the 2004 International Conference on Robotics & Automation, N. Orleans, LA: IEEE* 3596–3600.

[5] B. S. Morse, C. H. Engh, M. A. Goodrich (2010) UAV video coverage quality maps and prioritized indexing for wilderness search and rescue, *HRI'10 Proceedings of the 5th ACM/IEE International Conference on Human-Robot interaction.*

[6] A. Rango, A. Laliberte, J.E. Herrick, C. Winters, K. Havstad, C. Steele, D. Browning, "Unmanned aerial vehicle-based remote sensing for rangeland assessment, monitoring and management", *J. Appl. Remote Sens.*, vol. 3, no. 1, 2009.

[7] S. Waharte, N. Trigoni, "Supporting search and rescue operations with UAVs", *EST'10 Proceedings of the 2010 International Conference on Emerging Security Technologies*, 2010.

[8] A. Alaimo, V. Artale, C. Milazzo, A. Ricciardello, L. Trefiletti, "Mathematical Modeling and Control of a Hexacopter", *Unmanned Aircraft Systems (ICUAS), 2013 International Conference on, pp. 1043 – 1050, ISBN:978-1-4799-0815-8, D.O.I. 10.1109/ICUAS.2013.6564793 (2013).*

[9] V. Artale, C.L.R. Milazzo, A. Ricciardello "Mathematical Modeling of Hexacopter", Applied Mathematical Sciences, Vol. 7, no. 97, 4805 - 4811 (2013).

# How reliable is the ReaxFF Potential for Describing the Structure of Alkanethiols on Gold? A Molecular Dynamics Study

**V Vasumathi and Maria Natalia D S Cordeiro**

REQUIMTE, Department of Chemistry and Biocemistry, University of Porto, Rua do Campo Alegre 687, 4168-007, Porto, Portugal

E-mail: vasuphy@gmail.com and ncordeir@fc.up.pt

**Abstract.** The structures of self-assembled monolayers (SAMs) of short (methyl) and long (hexyl) chain alkyl thiols on the clean gold (111) surface were modelled using for the Au-S interactions either the reactive ReaxFF potential or the well known non-reactive Morse potential, while for the Au-Au interactions either the ReaxFF potential or an embedded-atom method (EAM). Analysis of the MD trajectories of possible SAM structures suggests that disordering of interfacial Au atoms is definitely driven by the gold-sulphur interactions. Our MD results reveal a novel structure where two methanethiol molecules are bound to a gold adatom that has been lifted from the surface at 300 K, and the same kind of RS-Au-SR motif was also observed for hexanethiol at 600 K but not at 300 K. What is more, the above motif is only observed for the reactive ReaxFF potential. Moreover, these results are in clear agreement with recent experiments and more costly first principles-based MD simulations. These findings strongly support the use of reactive potentials such as ReaxFF for gathering an accurate description of Au-S interactions in inexpensive classical MD simulations.

## 1. Introduction

Self-assembled monolayers (SAMs) of alkanethiols on metal surfaces provide many unique features that make them attractive for various important applications [1-3], including electrochemical sensors, biosensors, patterning of surfaces, friction and lubrication control, micro-fabrication, protective coatings and molecular electronics. A detailed knowledge of the structure of the molecule metal interface is needed to understand the growth mechanism [4] and the electrostatic properties [5] of this kind of materials. However, the structural features of thiols absorbed on the gold(111) crystal surface are still under debate. For instance, theoretical quantum chemical studies [6,7] reported that the preferred binding site for the sulphur of methylthiolate on the flat Au(111) surface is the bridge site, whereas experimental investigations by X-ray photoelectron diffraction [8] and normal incidence X-ray standing wave [9] suggest a preference for the ontop site. Apart from these ontop and bridge sites, there is another type of binding often known as the RS-Au-SR motif (two S atoms are joined by an intermediate Au adatom), where a gold atom is lifted from the surface (so called adatom) and shared by two thiols. The role of the underlying gold substrate in the chemisorption of thiols has been recently recognized in both surface and nanoparticles [4,10,11,12]. X-ray standing wave experiments have shown the occurrence of adatoms [4,10] which form the RS-Au-SR motif on the metal surface.

Besides the importance of the adatom in film growth and in lifting the Au surface herringbone rebuilding has been also investigated by scanning tunnelling microscopy [4]. At both low [4] and high [11] coverage of thiols, experiments concluded that the RS-Au-SR motif does occur. This has been sustained and predicted by density functional theory (DFT)-based molecular dynamics (MD) simulations [11, 13-15] in what regards methanethiol and hexanethiol on gold. Recently, similar RS-Au-SR motifs have been identified at the surface of p-carboxybenzenethiol protected gold nanoparticles [16]. However, there is no report on the existence of RS-Au-RS motif in SAMs adsorbed on gold surfaces using classical molecular dynamics (MD) simulations. This is certainly due to the lack of a reliable potential for the Au-S interactions that might lead to this novel motif. Very recently, Jarvi et al. developed interatomic reactive potentials (ReaxFF) [17] for Au-SRH systems and predicted the RS-Au-SR motif by DFT calculations. It is not still clear whether the potential models proposed so far for classical MD simulations such as the ReaxFF can account for the available experimental results or not. Therefore, in the present study, we have carried out classical molecular dynamics simulation to probe the existence of the RS-Au-SR motif pro short and long chain alkanethiols adsorbed onto the Au (111) surface, by comparing the results using two different interaction potential for Au-S namely a reactive potential (ReaxFF) [17] and a non-reactive potential (Morse) [18].

## 2. Simulation Details

In this work, we study the adsorption onto the Au (111) surface of SAMs of methylthiolate and hexylthiolate (denoted hereafter as MT and HT, respectively). The Au (111) surface was modelled using five atomic layers, the bottom two being kept fixed and the remaining three layers being mobile. The surface comprised 180 atoms with lateral dimensions of $15.18 \times 16.00 \AA^2$ . Based on experimental results [8], the alkanethiols are bounded to the surface through the S atom on-top binding-sites, and form a densely packed layer arranged into a $(\sqrt{3} \times \sqrt{3})R30^o$ lattice. We have used also here the experimental lattice constant (= 2.88 $\AA$) for the gold surface. Thus, the initial configuration of the SAMs were built up as single MT or HT components by placing a total of 12 SAM molecules bound to the gold surface on top sites with an intermolecular spacing of 4.99 $\AA$. Periodic boundary conditions were applied in the plane parallel to the surface and nonperiodic in the plane normal to the surface with a 50 $\AA$ vacuum along the z direction. All classical MD simulations were carried out using the LAMMPS program package [19]. Our aim is to compare the structural properties of SAMs when using two interaction potentials of Au-S namely: reactive potentials (ReaxFF) [17] (FF1) and non-reactive Morse potentials [18] (FF11). In FFI , the Au-Au, Au-S and between hydrocarbons interactions are described using reactive ReaxFF potentials. For FFII, the Au-Au interaction is described using the Embedded Atom Model potential (EAM) [20], the Au-S interaction is described using the Morse-potential and finally, the CHARMM27 [21] force field is used to describe the interactions between hydrocarbons. The LennardJones (LJ) parameters of the interacting atoms have been combined according to the Lorentz Berthelot rule. Moreover, harmonic bond stretching, angle bending and dihedral angle terms were taken from the CHARMM27 force field. We have used the following simulation protocol for the above two kinds of force fields. Firstly, each resulting complex structure was subjected to 5000 steps of steepest descent minimization of the potential energy, followed by 5000 steps of conjugate gradient minimization. Then, the minimized structures were gradually heated from 0 to 300 K with 50 K increments in short MD runs of 50 ps, using a time step integration of 0.1 fs for FFI and 1.0 fs for FFII. The final conformations from the above runs were used as input structures for NVT ensemble simulations of 3-5 ns with a time step of 0.1 fs for FFI and 1.0 fs for FFII. The temperature was held fixed at 300 K (or at 600 K) using a heat bath [22, 23] with a coupling time constant of 0.1 fs/1.0 fs. In the MD simulations with FFII, long-range electrostatics were handled using the Particle-Particle Particle-Mesh (PPPM) method [24] with a real-space cut-off distance set to 12 $\AA$ and an error tolerance to 105, while the short-range LJ interactions were smoothly shifted to zero by the CHARMM force method between 10 Å and 12 Å.

## 3. Results and Discussions

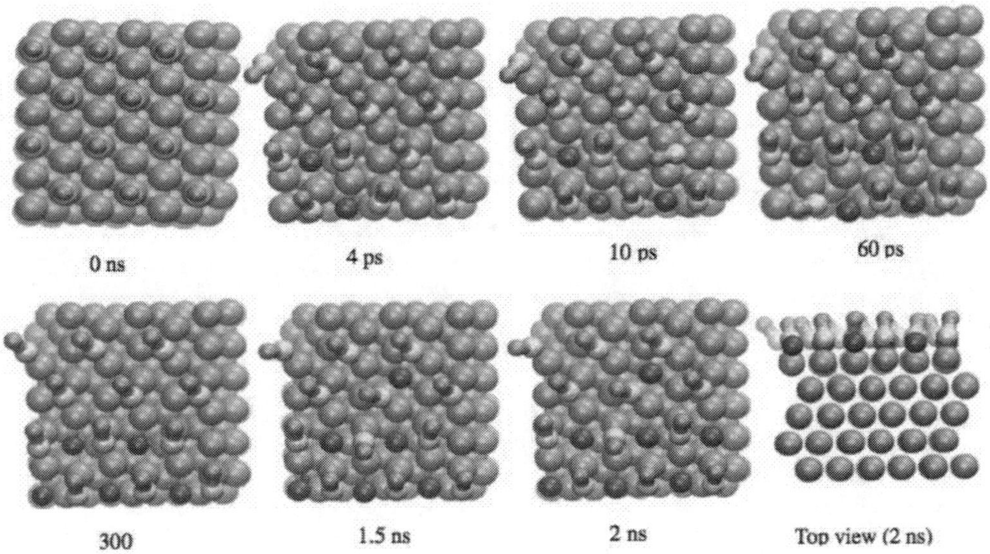


**Figure 1.** Snapshots of the equilibrated structures of the MT on gold surface (different colours refer to different atoms in the system, namely: blue - carbon, purple - Au adatom, yellow - sulphur and golden-gold atom).

In this section, we present the results of our simulations of MT/HT on Au(111) surface using the two different interaction potentials for Au-S. We are interested in comparing the results of disordering of interfacial Au atoms driven by the reactive and non-reactive gold-sulphur interactions, in particular the existence of the RS-Au-SR motif. Our simulation results showed that such motif appears when using FFI but not for FFII even at high temperatures. We first consider the results of MT on Au (111) surface at 300K with the potential of FFI. Figure 1 shows the instantaneous snapshots representing the structure observed during the simulation of MT on the Au (111) surface. From these figures, one can observe the RS-Au-SR motif for MT on Au(111), where a single Au atom is pulled out from the surface by 2.0A leaving a surface vacancy and shared by two MTs. Interestingly, at the beginning (say 4ps) of simulation we observe 2 MT-Au-MT, but later on five more adatoms were created at various instances and forming MT-Au-MT, as can be observed from the root mean square displacement (RMSD) plot of Au adatoms shown in figure 2 and also the instantaneous snapshots shown in figure 1. Further, fluctuations/stepwise rise in figure 2 clearly dictates the Au adatom creations, i.e. the Au adatoms are created as time goes on and they do not return to the surface instead they stay there stabilised by the interactions between the two thiols. Hence, the MT-Au-MT motif configurations remained in equilibrium for the duration of the simulation. The structure brought down from the last MD configuration consists of 5 bridges and 7 MT-Au-MT (see figure 1).

We have calculated the Au-S bond length for the bridge sites (2.5 Å) and for the MT-Au-MT motif (2.3 Å and 2.6 Å) configurations, being these results in good agreement with the literature [15]. Coming to the results of HT on the Au (111) surface, the HT-Au-HT motif was not arising at 300 K instead only when raising at higher temperatures say 600 K (see figure 3). As we

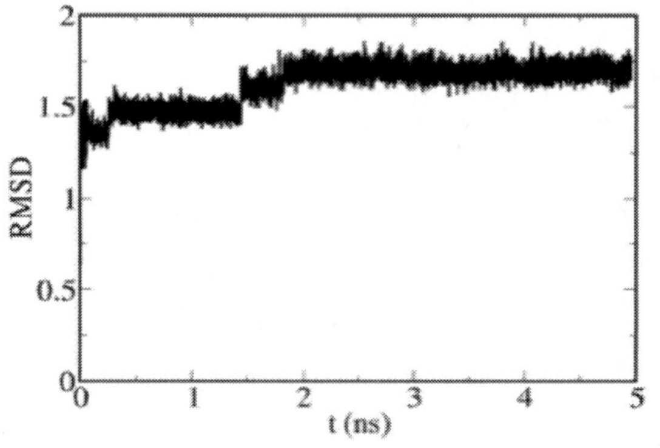


**Figure 2.** RMSD of gold adatoms.

observed in MT, here also the adatoms are created as time goes on and those are equilibrated after 1.5 ns. Further, from these simulations, we observe that the HT chain hampered the dynamics, and as a consequence, the amount of inter-conversion between different structures was strongly reduced when compared with MT (data not shown here). Due to the frequent structural inter-conversion for MT, the tilt angle (say 32) is higher than the tilt angle (say 22) for HT, but these tilt angles are also consistent with experimental [15] and theoretical values [25]. Hence, our simulation results suggest that ReaxFF potential can reproduce the novel motif RS-Au-SR as well as other structural properties of alkanethiols on the gold surface.

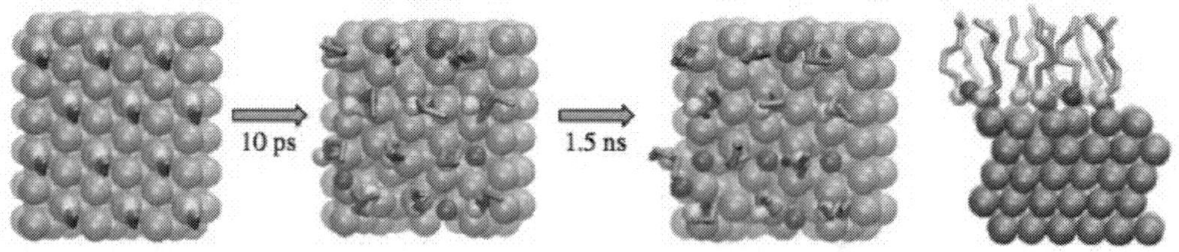


**Figure 3.** Snapshots of the equilibrated structures of the HT on the gold surface (different colours refer to different atoms in the system, namely: blue - carbon, purple - Au adatom, yellow - sulphur and golden - gold atom).

## 4. Conclusions

Using classical MD simulation, we have studied the structural properties of short and long chain alkanethiols on the gold(111) surface using two different interaction potential for Au-S, namely the reactive ReaxFF potential and the non-reactive Morse potential. In particular, we have investigated the existence of novel motifs of the type RS-Au-SR in MT/HT on the surface using the two different Au-S potentials. Our simulation results show that, the novel motif was observed in MT on Au (111) surface at room temperature and it was not observed in HT on Au(111) at room temperature though it was observed at higher temperature. The calculated structural parameters say tilt angle, and Au-S bond lengths for bridge sites and for the RS-Au-SR motifs are in well agreement with the literature. Furthermore, the novel motif of RS-Au-SR was observed only when applying the reactive potential but not for the non-reactive potential. These findings strongly support the use of the reactive potentials such as ReaxFF for gathering an accurate description of Au-S interactions in inexpensive classical MD simulations.

**Acknowledgments**
This work is supported by project PTDC/CTMNAN/112241/2009, and grant No. PestC/EQB/LA0006/2011, financed by FEDER through COMPETE - Programa Operacional Factores de Competitividade and by FCT - Fundacao para a Ciencia e a Tecnologia.

**References**

1 J. J. Gooding, F. Mearns, W. R. Yang and J. Q. Liu 2003 Electroanalysis 15 81.

2 M. C. Daniel, and D. Astruc 2004 Chem. Rev. 104 293.

3 J. T. Koepsel, and W. L. Murphy 2012 Chembiochem 13 1717.

4 P. Maksymovych, D. C. Sorescu and J. T. Yates, Jr 2006 Phys. Rev. Lett. 97 146103.

5 V. De Renzi, R. Rousseau, D. Marchetto, R. Biagi, S. Scandolo and U. Del Pennino 2005 Phys. Rev. Lett. 95 046804.

6 M .C. Vargas, P. Giannozzi, A. Selloni, and G. Scoles 2001 J. Phys. Chem.105 9509.

7 T. Hayashi, Y. Morikawa and H. Nozoye 2001 J. Chem. Phys. 114 7615.

8 H. Kondoh, M. Iwasaki, T. Shimada, K. Amemiya, T. Yokoyama and T. Ohta 2003 Phys. Rev. Lett. 90 066102.

9 M. G. Roper, M.P. Skegg, C.J. Fisher, J.J. Lee, V.R. Dhanak, D.P. Woodruff, Robert G. Jones 2004 Chem. Phys. Lett. 389 87.

10 Miao Yu, N. Bovet, Christopher J. Satterley, S. Bengio, Kevin R. J. Lovelock, P. K. Milligan, Robert G. Jones, D. P. Woodruff, and V. Dhanak 2006 Phys. Rev. Lett. 97 166102.

11 R. Mazzarello, A. Cossaro, A. Verdini, R. Rousseau, L. Casalis, M. F. Danisman, L. Floreano, S. Scandolo, A. Morgante, and G. Scoles 2007 Phys. Rev. Lett. 98 016102.

12 H. Hakkinen, M. Walter , and H. Gronbeck 2006 J. Phys. Chem. B 110 9927.

13 A. Nagoya and Y. Morikawa 2007 J. Phys. Cond. Matter 19 365245.

14 H. Gronbeck and H. Hakkinen, 2007 J. Phys. Chem. B 111 3325.

15 A. Cossaro, R. Mazzarello, R. Rousseau, L. Casalis, A. Verdini, A. Kohlmeyer, L. Floreano, S. Scandolo, A. Morgante, M. L. Klein, G. Scoles 2008 Science 321 943.

16 P. D. Jadzinsky, G. Calero, C. J. Ackerson, D. A. Bushnell and R. D. Kornberg 2007 Science 318 430.

17 T.T. Jarvi., A.C.T . van Duin., K. Nordlund. and W. A. III .Goddard 2011 J. Phys. Chem. A, 115 10315.

18 B.J .Henz, T. Hawa and M. R. Zachariah 2008 Langmuir 24 773.
[

19 S. Plimpton 1995 J. Comput. Phys. 117 1.

20 S. M. Foiles, M.I. Baskes, M.S. Daw, 1986 Phys. Rev. B 33 7983.

21 S. E. Feller and A. D MacKerell 2000 J. Phys. Chem. B 104 7510.

22 S. Nose 1984 Mol. Phys. 52 255.

23 W. G. Hoover 1985 Physical Review A 31 1695.

24 S. J. Plimpton, R. Pollock, and M. Stevens 1997 Proceedings of the Eighth SIAM conference on Parallel Processing for Scientific Computing, Minneapolis, MN.

25 B. Szefczyk, B. R. Franco, J. A. N. F Gomes, and M. N. D. S. Cordeiro 2010 J. Mol. Struct.-Theochem 946 83.

# A surrogate model enables a Bayesian approach to the inverse problem of scatterometry

**S. Heidenreich, H. Gross, M.-A. Henn, C. Elster and M. Bär**
Physikalisch-Technische Bundesanstalt, Abbestr. 2-12, 10587 Berlin, Germany
E-mail: sebastian.heidenreich@ptb.de

**Abstract.** Scatterometry is an indirect optical method for the determination of photomask geometry parameters from scattered light intensities by solving an inverse problem. The Bayesian approach is a powerful method to solve the inverse problem. In the Bayesian framework estimates of parameters and associated uncertainties are obtained from posterior distributions. The determination the probability distribution is typically based on Markov chain Monte Carlo (MCMC) methods. However, in scatterometry the evaluation of MCMC steps require solutions of partial differential equations that are computationally expensive and application of MCMC methods is thus impractical. In this article we introduce a surrogate model for scatterometry based on polynomial chaos that can be treated by Bayesian inference. We compare the results of the surrogate model with rigorous finite element simulations and demonstrate its convergence. The accuracy reaches a value of lower than one percent for a sufficient fine mesh and the speed up amounts more than two order of magnitudes. Furthermore, we apply the surrogate model to MCMC calculations and we reconstruct geometry parameters of a photomask.

## 1. Introduction

Scatterometry provides a fast indirect optical method for the determination of photomask grating profile parameters that are obtained from scattered light intensities using inverse methods [1, 2]. In recent investigations profile parameters of rough absorber lines are reconstructed [5] from extreme ultraviolet light (EUV) measurements by employing a maximum likelihood method for parameter estimations [3, 4, 6]. This method provides accurate profile parameters with uncertainties of several nanometers. To further decrease uncertainties different measurements can be combined consistently by Bayesian interference.

In this paper, focus on an EUV photomask that consists of periodic absorber lines and a multilayer structure (see Fig. (1a) for details).

## 2. Inverse problem in scatterometry

In general the propagation of electromagnetic waves is described by Maxwell's equations. For a grating geometry that consists of straight absorber lines, Maxwell's equations reduce to the two dimensional Helmholtz equation: $\Delta u(x,y) + k^2(x,y)\, u(x,y) = 0$, where $u$ is the transversal field component and $k(x,y)$ is the wave number. The resulting boundary value problem is solved by a finite element method (FEM) which is computationally expensive. We have used the software package DIPOG [1] for our calculations.

[1] developed at WIAS; http://www.wias-berlin.de/software /DIPOG

The mathematical model is defined by a map of geometry parameters onto light intensities. In particular, the set of parameters $p_1....p_N$ fixes a specific grating geometry and the solution of the boundary value problem with respect to the chosen geometry defines the map $f_1(\mathbf{p})...f_M(\mathbf{p})$: $\mathbf{p} \mapsto f_j(\mathbf{p})$. Here, the index $j$ labels the diffraction orders of light diffracted intensities $f_j$. Experimental data are modeled by the forward map $f_j$ and additional noise, i.e. $y_j = f_j + \epsilon_j$, where $\epsilon_j \sim \mathcal{N}(0, \sigma_j)$ describes Gaussian noise with zero mean. The variance $\sigma_j^2 = (a \cdot f_j(\mathbf{p}))^2 + b^2$, in which $b$ models the background noise strength and $a$ models noise of power fluctuations of the incident beam during the measurement [4]. The inverse problem in scatterometry is defined as the reconstruction of geometry parameter values $p_i$ from $y_j$.

## 3. Bayesian Approach

The Bayesian approach provides a statistical method to solve the inverse problem based on Bayes theorem,

$$\pi(\theta; \mathbf{y}) = \frac{\mathcal{L}(\theta; \mathbf{y})\pi_0(\theta)}{\int \mathcal{L}(\theta; \mathbf{y})\pi_0(\theta)\mathbf{d}\theta}. \tag{1}$$

Here, the distribution $\pi_0$ includes prior knowledge about the parameter and $\mathcal{L}$ determines the likelihood function. For scatterometry we choose prior knowledge $\pi_0 = U[\theta^{min}, \theta^{max}]$, where $U$ is a uniform distribution and the vector $\theta$ consists of $\{\mathbf{p}, a, b\}$. As in our previous studies we choose the likelihood function [3]

$$\mathcal{L}(\mathbf{p}, a, b; \mathbf{y}) = \prod_{j=1}^{N} \frac{1}{\sqrt{2\pi\sigma_j^2(\mathbf{p}, a, b)}} \exp\left[-\frac{(y_j - f_j(\mathbf{p}))^2}{2\sigma_j^2(\mathbf{p}, a, b)}\right], \sigma_j^2 = (a \cdot f_j(\mathbf{p}))^2 + b^2. \tag{2}$$

The expectation value of parameter $\theta_i$ is given by the average with respect to the posterior measure, i.e., $\bar{\theta}_i = E[\theta_i]_\pi$ and the variance by $var(\theta_i) = E[\theta_i^2]_\pi - E[\theta_i]_\pi^2$. In the Bayesian framework parameters are frequently estimated by MCMC sampling techniques which typically have slow convergence. For every sampling step the likelihood function and in turn the forward model has to be evaluated. Clearly, such an approach is not efficient for the inverse problem in scatterometry. In this respect we will apply a polynomial chaos method to construct a surrogate model that dramatically speeds up forward calculations.

## 4. Polynomial chaos

Polynomial chaos (PC) was introduced in 1938 by Wiener [7]. He used polynomial approximations for the Gaussian stochastic process. Later, Cameron and Martin [8] proved that every stochastic process with a finite second moment can be approximated by an infinite convergent sum of orthogonal polynomials. The approach became popular for uncertainty quantification after Xiu and Karniadakis [9] have generalized polynomial chaos by the introduction of the Wiener-Askey scheme for flow problems. The Wiener-Askey scheme relates stochastic processes to orthogonal polynomials which ensures good convergence results in the expansion. In this perspective we will focus on the non-intrusive variant of generalized polynomial chaos method in which proprietary software can be used as a black box for rigorous evaluations of the forward model necessary to determine expansion coefficients. In particular, the forward map for uncertain input geometry parameter $p_i$ is approximated by a polynomial series

$$f_j(p_{1...n}) = f_j(\mu_{1...n}) + \sum_{i=1}^{n}\sum_{k=1}^{R} f_{jk}^i(\mu_{1...n})\Psi_k(\omega_i) + ... \tag{3}$$

where $\mu_i$ is the mean value of the related uniform prior distribution and $\omega_i$ a random number. Specifically, a random parameter drawn from a prior distribution $U[\theta_i^{min}, \theta_i^{max}]$ is given by $\theta_i = \mu_i + \omega_i$. Therefore, the range of random variables $\omega_i$ is set by its prior distribution, i.e., $\omega_i \in U[\theta_i^{min} - \mu_i, \theta_i^{max} - \mu_i]$. For simplifications we introduce the notation $U(\omega_i) = U[\theta_i^{min} - \mu_i, \theta_i^{max} - \mu_i]$.

According to the Wiener-Askey scheme [9] we choose Legendre polynomials as basis functions $\psi_k$. The choice of basis functions specifies the kind of chaos, here, to Legendre chaos. The expansion coefficients are determined by projections onto the chosen Legendre functions $\Psi_k$, i.e.,

$$f_{jk}^{\alpha}(\mu_{1...n}) = \frac{\int f_j(p_{1...n})\Psi_k(\omega_i)U(\omega_i)d\omega_i}{\int \Psi_k^2(\omega_i)d\omega_i} \approx \frac{1}{\int \Psi_k^2(\omega_i)U(\omega_i)d\omega_i} \sum_k \lambda_k^i f_j(\omega_i^k)U(\omega_i^k), ..., \quad (4)$$

The integral in the denominator is solved analytically and the integral in the nominator is approximated by quadrature rules. Quadrature rules provide a mesh (index $k$) and corresponding weights $\lambda_k^i$.

## 5. Results and conclusion

For the application of the surrogate model we first show the accuracy and convergence of the approximation scheme. In our study, we used prior distributions for the parameters: $U[133, 147]nm$ for the bottom CD, $U[74.4, 85.6]nm$ for the height, $U[80, 90]^\circ$ for the side-wall angle, $U[0.01, 0.1]$ for $a$ and $U[0.001, 0.1]\%$ for $b$ (noise parameters). We calculated the Legendre chaos expansion of the forward map up to the 10th order, drew a geometry from the prior distribution, calculated the corresponding light diffracted intensities with the rigorous FEM solver and compared these results with results from the PC surrogate model. Fig. (1b) shows a good agreement of the approximation scheme with FEM based calculations for one chosen geometry. To cover a wider parameter range we considered 100 random geometries and calculated the $L^2$ norm of the difference between FEM results and PC surrogate approximations. As a measure of the overall error we took the maximum value of the deviations for chosen geometries. Fig. (1c) shows the convergence of PC surrogate approximations with increasing number of grid points. It is found to be much better than results obtained with a surrogate model based on simple nearest neighborhood interpolations (NN). For further investigations we fixed the resolution to 608 supporting points, where the relative deviation of efficiencies is lower than 1%.

Finally, we have reconstructed geometry parameters from simulated noisy data sets $y_i$. We included the PC surrogate model in a MCMC scheme and estimated geometry parameters with $10^4$ sampling steps. Fig. (1d) shows eleven reconstructions for the side-wall angle (SWA). All reconstructions are consistent with the reference value and the confidence interval of 95 % is small; about $0.7^\circ$.

In this paper we introduced a polynomial chaos based surrogate model to demonstrate its suitability to speed up Bayesian inferences. We have shown the accuracy and convergence of the surrogate model. Moreover, the PC surrogate model was used for reconstructions of the geometry parameters from numerical generated noisy data sets. The reconstructions are sufficiently precise which makes the scheme applicable for solving the inverse problem in scatterometry based on MCMC methods. Typically computational times of the proposed method compared to rigorous FEM based calculations are depicted in table (1). The computational cost of pre-calculations is about $20h$ due to the determination of expansion coefficients. Polynomial chaos surrogate methods provide an effective method to speed up Bayesian inferences using MCMC calculations.

[1] H. Huang, and F Terry Jr., Thin Solid Films **455**, 828–836 (2004).
[2] C. Raymond, M. Murane, S. Prins, S. Sohail, H. Navi, J. Mc Neil, and J. Horsch, J. Vac. Sci. Technol. **B15**, 361–368 (1997).
[3] M.-A. Henn, H. Gross, F. Scholze, M. Wurm, C. Elster, and M. Bär, Optics Express,**20**, 12771 - 12786 (2012).
[4] M-A.Henn, S. Heidenreich, H. Gross, A. Rathsfeld, F. Scholze, and M. Bär, Optics Letters, **37** (24), 5229 (2012).
[5] H. Gross, M-A. Henn, S. Heidenreich, A. Rathsfeld, and M. Bär, Appl. Optics **51**, 7384 (2012).
[6] S. Heidenreich, M.-A. Henn, H. Gross, B. Bodermann, M. Bär, SPIE 87890T (2013).
[7] Wiener, N, American Journal of Mathematics, 60(4), 897-935, 1938.
[8] Cameron R H, Martin W T, Annals of Mathematics, 48(2), 385-392, 1947.
[9] Xiu D, Karniadakis G E, Journal of Computational Physics, 187, 137-167, 2002.

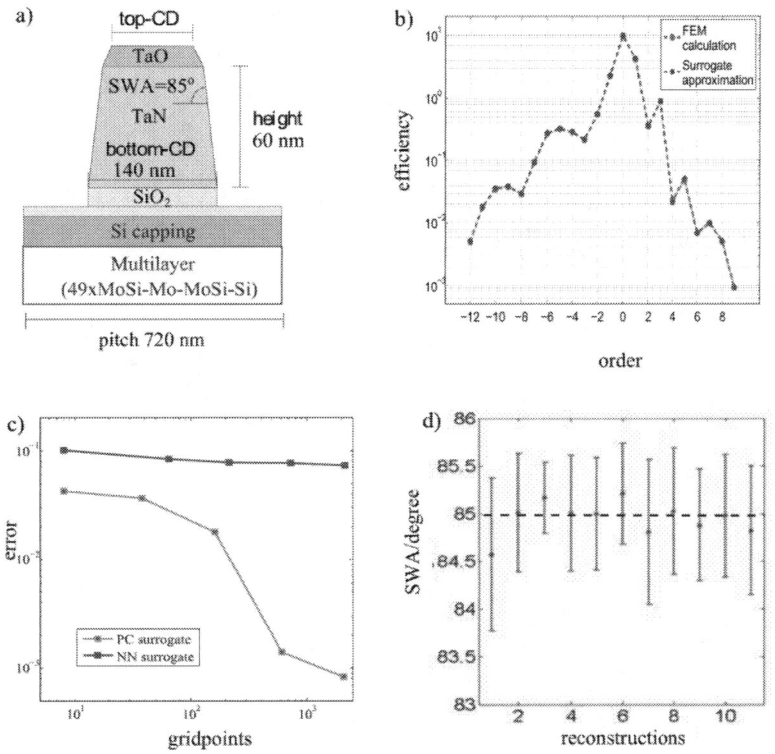


**Figure 1.** a) Cross section of one period of an EUV photomask which we used in our study. b) Polynomial chaos based surrogate model compared with rigorous finite difference (FEM) calculations. The relative error is lower than 3 percent. c) Convergence rate of polynomial chaos based surrogate model (PC) compared to nearest neighborhood method (NN). The error is given as the maximum deviation in the $L^2$ norm of the surrogate model from FEM calculations for 100 sample geometries. d) Results for eleven reconstructions of the side wall-angle from noisy data using the MCMC sampling based on PC surrogate method. The error bars show the 95% confidence interval and the dashed line shows the reference value.

| Method | Function evaluations | FEM | PC |
|---|---|---|---|
| Sensitivity Analysis | $3 \times 10^4$ | $41d$ | $20h + 0.1s$ |
| Bayesian MCMC | $\sim 10^5$ | $139d$ | $20h + 47min$ |
| Maximum Likelihood | 30 | $1h$ | $20h + 0.84s$ |
| Least squares | 15 | $0.6h$ | $20h + 0.42s$ |

**Table 1.** Speed up for different methods. The PC-surrogate needs always $20h$ for the determination of expansion coefficients.

# Zipf exponent of trajectory distribution in the hidden Markov model

**V.V. Bochkarev and E.Yu. Lerner**

Kazan Federal University, Kremlevskaya str.18, Kazan 420018, Russia

E-mail: vbockarev@mail.ru, eduard.lerner@gmail.com

**Abstract.** This paper is the first step of generalization of the previously obtained full classification of the asymptotic behavior of the probability for Markov chain trajectories for the case of hidden Markov models. The main goal is to study the power (Zipf) and nonpower asymptotics of the frequency list of trajectories of hidden Markov frequencys and to obtain explicit formulae for the exponent of the power asymptotics. We consider several simple classes of hidden Markov models. We prove that the asymptotics for a hidden Markov model and for the corresponding Markov chain can be essentially different.

In mathematical linguistics, one of the most widely known frequency laws is the Zipf law [1]. According to this law, there is a power dependence of the probability of the $r$th word in an ordered frequency list on $r$. The exponent in the classical Zipf law equals $-1$. In recent years, opportunities for this research became much wider, owing to the appearance of new large data repositories, in particular, Google Books Ngram corpus [6]. The analysis of large data corpora shows that the power dependence with an exponent close to $-1$ can approximately describe only frequencies of most commonly used words [4]. At the same time, frequencies of the main and peripheral vocabularies can be approximately described by a power law whose exponent is modulo greater than 1 (its typical values lie within the interval 1.7–2). Moreover, in languages with hieroglyphic scripts the asymptotic decrease of frequencies is faster than power law [5]. The development of a model explaining these phenomena is an actual problem.

There were many attempts to explain the frequency distribution of word usage with the help of probabilistic frequencies of text generation. One has considered both the "monkey model" with independent letters, and Markov models.In [2] one describes the dependence of the asymptotic behavior of the frequency distribution on the structure of the transition probability matrix and obtains an explicit formula for the exponent. Note that unlike the monkey model [3], where there is a great abundance (because all words consisting of the same letters have equal probabilities), in Markov models, owing to dependencies between states, there is no such abundance. We can observe even a less entropy (and, consequently, a less abundance) in a hidden Markov model (HMM), where some hidden symbols merge to form the same visible ones. In future, the use of HMM will allow one to construct more realistic text generation models, taking into account the syllabic and morphological word structure. Moreover, within HMM one can also take into account the presence of misprints and mistakes in word corpora (according to [6], up to 30 percent of unique word forms in Google Books Ngram appear as a result of word recognition errors). Another linguistic application of HMM is the statistics of various sentences. In the latter case, graphs of Markov chains (MC) represent Chomsky syntactic structures, while HMM do their realizations as sentences. Our goal is to study the power and nonpower asymptotics of the frequency list of trajectories of HMM and to obtain explicit formulae for the power asymptotics exponent. Note that the value of this exponent appears to be greater than one, which corresponds to true values, if we estimate the transition probability matrix in accordance with the word frequencies in Google Books Ngram. In particular, these models with lighter distribution tails (in comparison with the classical Zipf model with the exponent of $-1$) better predict frequencies of peripheral vocabularies and the small quantity of hapax legomena.

Let us give exact definitions. Consider a Markov chain (MC) whose state set is $E = \{E_0, \ldots, E_n\}$; here the state $E_0$ is absorbing, while the rest ones are nonrecurrent, i.e., having started with any state, we finally get at $E_0$. Consider a hidden Markov model (HMM), where each non-absorbing state is associated with a random variable that takes on values in some alphabet $X = \{x_1, \ldots, x_m\}$; thus we get words representing a sequence of letters of the alphabet $X$. A word ends, when we reach the state $E_0$. If the initial distribution $a = (a_1, \ldots, a_n)$ is known, then each word has a certain probability, and the sum of these probabilities equals one. We are interested in the distribution of probabilities over words, namely, we are interested in decrease rates of word probabilities in their complete list sorted in the non-increasing order of probabilities.

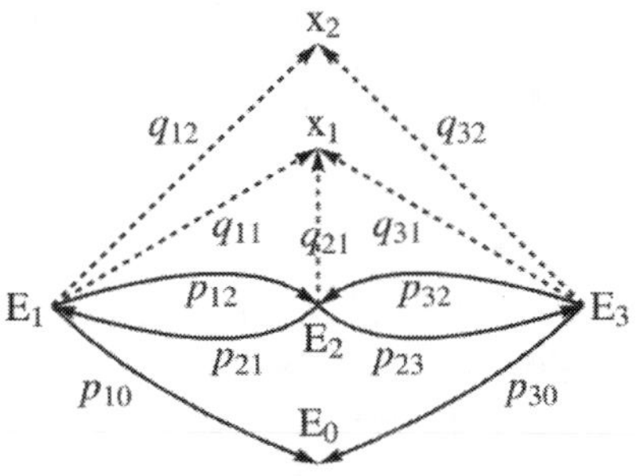


**Figure 1.** An example of the graph of an HMM with three states with a two-letter alphabet, $q_{21} = 1$.

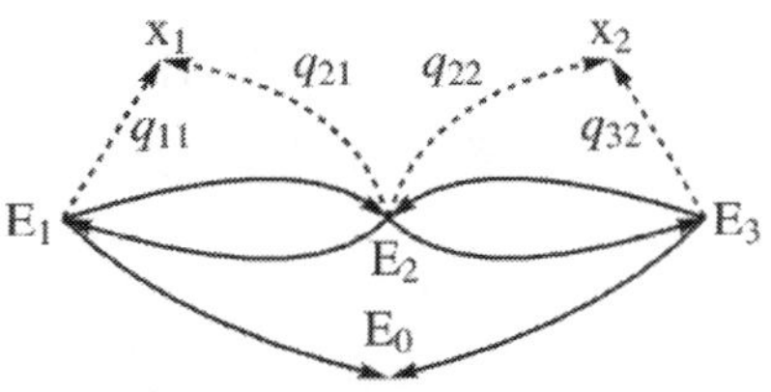


**Figure 2.** Another example of the graph of an HMM (modifying the example shown in Fig. 1), for which the hidden word is uniquely defined from the exposed one (see Theorem 3); here $q_{11} = q_{32} = 1$.

For example, let us consider a HMM, whose graph is given in Fig. 1. Here vertices correspond to states of the MC, solid arcs do to nonzero elements of the matrix of transition probabilities, and weights shown next to arcs are values of the corresponding elements of the matrix. Dashed arcs outgoing from non-absorbing vertices correspond to various variants of values of random variables associated with vertices, and weights shown next to them are probabilities of taking on the corresponding values. Evidently, for any non-absorbing vertex the sum of weights of all outgoing solid arcs, as well as that of all dashed ones, equals one. In particular, in Fig. 1, $q_{21} = 1$. In this example, assuming that at the initial time moment we certainly are situated at state $E_2$, possible words represent any sequences consisting of an even number of letters of the alphabet $\{x_1, x_2\}$, where odd positions are necessarily occupied by the symbol $x_1$. In particular, the probability of the word $(x_1, x_1)$ equals $p_{21}q_{11}p_{10} + p_{23}q_{31}p_{30}$, that of the word $(x_1, x_2)$ does $p_{21}q_{12}p_{10} + p_{23}q_{32}p_{30}$, and so on.

Introduce some denotations. We are interested in the decrease order of the function $p(\cdot)$, where $p(r)$ is the probability to obtain the $r$th word in the sorted list. Let $G_0$ be the graph of an MC (in Fig. 1 this graph consists of solid arcs), and let $G$ be the subgraph of $G_0$ consisting of all its vertices and arcs, except the vertex $E_0$ and arcs entering it.

Evidently, if the graph $G$ contains at least one cycle, then (and only in this case) the number of words is infinite. We are interested just in this case, namely, in the order of the asymptotics of the function $p(r)$.

Let $P_G$ be a substochastic matrix of transition probabilities of the MC corresponding to the graph $G$ (i.e., the transition matrix where the absorbing state is not taken into account) and let $P_G(\beta)$ be obtained from $P_G$ by raising each its element to the power $\beta$. Earlier in [2] we considered ordinary MC (rather than HMM), i.e., the case when $X = \{E_1, \ldots, E_n\}$ and each random variable is a deterministic identical map. We classified all possible variants of essentially different asymptotics of the function $p(r)$ and consider some techniques for finding parameters of these asymptotics for MC. In particular, we have proved the following assertion (we use the standard $O$-symbols, including the symbol $\Theta$ for denoting the asymptotic order and the symbol $\Omega$ for the lower estimate of the order):

**Theorem 1** *Assume that the initial distribution $a$ is such that the probability to reach (at least once) each state $E_i$, $i = 1, \ldots, n$, is greater than zero. Then the following alternatives are possible:*

1. *If the graph $G$ contains a vertex which go through to two different simple cycles, then $p(r) = \Omega(r^{-1/\beta})$, where $\beta$ is a real value, with which the greatest modulo eigenvalue of the matrix $P_G(\beta)$ equals one. Note that such $\beta$ exists, it is unique and belongs to the interval $(0, 1)$. Moreover, $p(r) = o(r^{-1/\gamma})$ for any $\gamma > \beta$. Note that the exact power order (i.e., the equality $p(r) = \Theta(r^{-1/\beta})$) takes place if and only if any simple path in the graph $G$ belongs to no more than one strongly connected component $H$ of this graph, for which the matrix $P_H(\beta)$ has unit eigenvalue.*
2. *If the graph $G$ contains cycles, and each vertex of the graph $G$ belongs to no more than one simple cycle, then $p(r) = \Omega(\alpha^r)$ and $p(r) = o(r^{-\lambda})$, where $\lambda$ is any positive number and $\alpha \in (0, 1)$ is a constant depending on the matrix $P$ (i.e., $p(r)$ is decreasing faster than any power function, but slower than some exponential one). Moreover, the exact exponential asymptotics ($p(r) = \Theta(\exp(-\nu r))$) for some $\nu > 0$ (see [2] for the calculation of $\nu$) takes place if and only if any path in the graph $G$ goes through vertices of no more than one cycle.*

Thus, for example, for a Markov chain with the graph $G$ given in Fig. 1 (we pay no attention to dashed arcs) there is only one strongly connected component $p(r) = \Theta(r^{-1/\beta})$, where $\beta$ is the root of the equation $p_{12}^{\beta}p_{21}^{\beta} + p_{23}^{\beta}p_{32}^{\beta} = 1$. The power asymptotics is typical; it defines the Zipf law [1] for frequencies of the occurrence of various words. Theorem 1 implies that if the matrix of transition probabilities is very sparse (more exactly, conditions of Theorem 1.2 are fulfilled), then $p(r)$ is decreasing faster than the power function. This property is valid for hieroglyphic scripts, in particular, for the Chinese language [5].

For a vertex $v$ we understand a syllable $w(v)$ as a part of a word in an HMM which can be obtained by tracing some simple cycle in the graph $G$ beginning at $v$. For example, for the HMM illustrated in Fig. 1 there are only two syllables in the form $w(E_2)$, namely, $(x_1, x_1)$ and $(x_1, x_2)$. We say that syllables $w_1(v)$ and $w_2(v)$ are essentially distinct, if no (multiple) concatenation of one syllable gives another one. For example, all letterwise distinct syllables of the same length are essentially distinct, while syllables composed of letters of the Latin alphabet $(o, l, e)$ and $(o, l, e, o, l, e, o, l, e)$ are not so.

**Theorem 2** *Assume that the initial distribution $a$ is such that the probability to reach (at least once) each state $E_i$, $i = 1, \ldots, n$, is greater than zero. For an HMM we have the bound $p(r) = \Omega(r^{-\beta'})$ for certain $\beta'$ if and only if there exists a vertex $v$, for which one can find at least two essentially distinct syllables $w_1(v)$ and $w_2(v)$.*

Evidently, conditions of Theorem 2 are fulfilled for the HMM illustrated in Fig. 1.

Note that the fulfillment of conditions of Theorem 1.1 does not guarantee the fulfillment of conditions of Theorem 2. Similarly, the fulfillment of conditions of Theorem 2 does not guarantee the fulfillment of conditions of Theorem 1.1, i.e., the power order of the asymptotics of "exposed" and "hidden" words, generally speaking, is inherited neither from MC to HMM, nor vice versa.

Let us now consider the question on the exact order of the power asymptotics. Denote by the symbol $Q$ a stochastic $n \times m$-matrix, whose elements $q_{ij}$ ($i \in \{1, \ldots, n\}$, $j \in \{1, \ldots, m\}$) are probabilities that the random variable corresponding to the state $E_i$ takes on the value $x_j$. Denote by $\widehat{P}(\beta)$ the $n \times n$-matrix, whose elements $\widehat{p}_{ij}$ ($i, j \in \{1, \ldots, n\}$) obey the formula $\widehat{p}_{ij} = p_{ij}^{\beta} \sum_{k=1}^{m} q_{jk}^{\beta}$.

**Theorem 3** *Let conditions of Theorem 2 be fulfilled. For simplicity, we assume that the graph $G$ consists of one strongly connected component. In addition, we assume that (with the considered initial distribution) for each exposed word $w$ (i.e., a sequence of letters of the alphabet $X$ or, in other words, values of random variables associated with states $E_i$), the realizing it hidden word (i.e., the set of states $E_i$ visited by us) is defined uniquely. Then $p(r) = \Theta(r^{-1/\beta})$, where $\beta$ is a real value, with which the greatest modulo eigenvalue of the matrix $\widehat{P}(\beta)$ equals one (it exists and belongs to the interval $(0, 1)$).*

Note that the HMM shown in Fig. 1 does not satisfy conditions of Theorem 3; for example, one can obtain the word $(x_1, x_1)$ in two ways, namely, with the help of either the hidden word $(E_2, E_1)$ or that $(E_2, E_3)$. We can satisfy conditions of Theorem 3 by slightly changing directions of dashed arcs as is shown in Fig. 2. In this case any exposed word, whose $2l$th position is occupied by $x_1$, corresponds to a hidden word, whose $2l$th position is occupied by $E_1$; any exposed word, whose $2l$th position is occupied by $x_2$, corresponds to a hidden word, whose $2l$th position is occupied by $E_3$. In addition, odd positions in a hidden word are occupied by $E_2$. In this case by Theorem 3 we have $p(r) = \Theta(r^{-1/\beta})$, where $\beta$ is defined from the equation

$$(p_{12}^{\beta} p_{21}^{\beta} + p_{23}^{\beta} p_{32}^{\beta})(q_{21}^{\beta} + q_{22}^{\beta}) = 1.$$

The condition for recovering a hidden word from an exposed one seems to be not very strong, however, it is just the case (due to the abundance of the language information) for the statistics of recognized (with errors) wordforms in the Google Books repository.

**Corollary 1** *Assume that an HMM satisfies conditions of Theorem 2 (including the possibility to reach each state) and for hidden words its MC has the exponential asymptotics. Then $p(r) = \Theta(r^{-1/\beta})$, where $\beta$ is a real value, with which the greatest modulo eigenvalue of the matrix $\widehat{P}(\beta)$ equals one.*

Let us now consider approaches to the calculation of $\beta$ in the case, when the recovery of a hidden word from an exposed one may appear to be impossible. Let $w(v)$ be some syllable in an HMM; assume that one can obtain this syllable in several ways when tracing a cycle (cycles) in the graph $G$, beginning at a vertex $v$. We understand the weight of such a tracing $c$ as the product of weights of all edges that enter in the cycle $c$ multiplied by the product of probabilities of the corresponding values from the alphabet $X$ in the corresponding state $E_i$. We understand the weight of a syllable $Pr(w(v))$ as the sum of weights calculated over all such tracings.

For example, for the HMM illustrated in Fig. 1 there are two syllables of type $w(E_2)$, namely, $(x_1, x_1)$ and $(x_1, x_2)$; we have $Pr(x_1, x_1) = p_{21}q_{11}p_{12} + p_{23}q_{31}p_{32}$, while $Pr(x_1, x_2) = p_{21}q_{12}p_{12} + p_{23}q_{32}p_{32}$. For the HMM illustrated in Fig. 2 there are four syllables, namely, all possible two-letter combinations.

**Theorem 4** *Let conditions of Theorem 2 be fulfilled. Moreover, we assume that all cycles in the graph $G$ contain some vertex $v$ and all letterwise distinct syllables $w(v)$ are essentially distinct. Then $p(r) = \Omega(r^{-1/\beta})$, where $\beta$ is determined as the root of the equation $\sum Pr(w(v))^{\beta} = 1$, and the sum is calculated over all possible syllables $w(v)$.*

Theorems 2, 3, and 4 cover particular cases of rather sparse matrices $P$ or $Q$. Note that even in this case the asymptotics of the frequency list of hidden and visible states can be essentially different. Thus, in Theorem 3 it is proved that the exponent of the power asymptotics of visible states can be determined with the help of the matrix $P'(\beta)$, while the exponent of the power asymptotics of hidden states (see Theorem 1) can be determined with the help of the matrix $P(\beta)$. Since, evidently, $\widehat{p}_{ij} \geq p_{ij}^{\beta}$, for hidden states this exponent is (modulo) not less than that for visible ones (we assume that hidden states also have power asymptotics). An essentially different asymptotic behavior is also possible (e.g., Corollary 1). In a general case, frequencies of hidden words do not necessarily decrease faster than those of visible ones. Thus, for example, under conditions of Theorem 4, if several hidden syllables correspond to the same visible ones, then exponents of the power asymptotics satisfy the converse inequality.

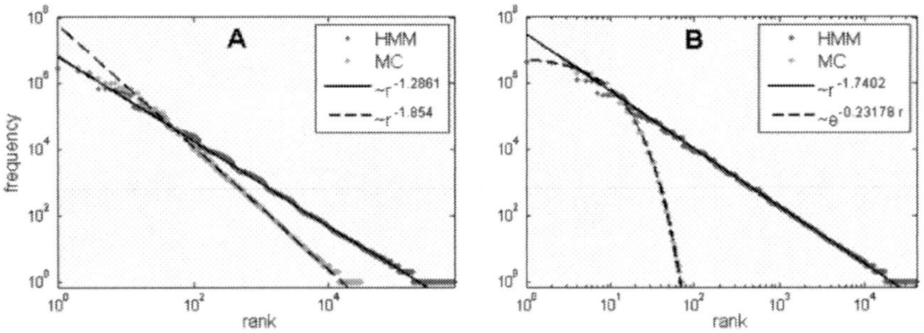


**Figure 3.** Examples of the frequency graph of an HMM and the corresponding MC. The total number of generated (nonunique) words equals 25 million.

In Fig. 3 we represent results of simulation experiments that demonstrate the essential distinction of asymptotics of frequencies of visible and hidden words; everywhere the number of hidden and visible states equals three. In case A, $P_G = \begin{pmatrix} 0.0007 & 0.007 & 0.6823 \\ 0.3493 & 0.0014 & 0.3493 \\ 0.007 & 0.6923 & 0.0007 \end{pmatrix}$, $Q = \begin{pmatrix} 0.5 & 0.3 & 0.2 \\ 0.3 & 0.5 & 0.2 \\ 0.2 & 0.3 & 0.5 \end{pmatrix}$, in case B, $P_G = \begin{pmatrix} 0 & 0.5 & 0 \\ 0 & 0 & 0.5 \\ 0.5 & 0 & 0 \end{pmatrix}$, $Q = \begin{pmatrix} 0.75 & 0.15 & 0.1 \\ 0.15 & 0.75 & 0.1 \\ 0.1 & 0.15 & 0.75 \end{pmatrix}$. In case A, both the frequencies of hidden words and visible ones have power asymptotics, but their exponents are essentially different (they are calculated in accordance with Theorem 1 and results of the simulation experiment, correspondingly). In case B, frequencies of hidden words decrease exponentially, while those of visible ones have power asymptotics.

Thus, we have established explicit formulae for exponents of the power asymptotics for some variants of HMM. We show that asymptotics of frequency lists of an HMM and the corresponding MC can be essentially different.

## References

[1] Baayen R H 2001 *Word Frequency Distributions* (Dordrecht: Kluwer Academic Publishers)
[2] Bochkarev V V and Lerner E Yu 2012 Zipf and non-Zipf Laws for Homogeneous Markov Chain *Preprint* arXiv:1207.1872
[3] Bochkarev V V and Lerner E Yu 2012 *Russian Mathematics* **50** 25
[4] Gerlach M and Altmann E G 2013 *Physical Review* **X**(3), 021006
[5] Lu L, Zhang Zi-Ke and Zhou T 2012 Scaling Laws in Human Language *Preprint* arXiv:1202.2903.
[6] Michel JB, Shen JB, Aiden AP, Veres A, Gray MK, Pickett JP, Hoiberg D, Clancy D, Norvig P, Orwant J, Pinker S, Nowak MA and Aiden EL 2011 *Science* **331** 176

# Deviations in the Zipf and Heaps laws in natural languages

**Vladimir V. Bochkarev, Eduard Yu.Lerner, Anna V. Shevlyakova**

Kazan Federal University, Kremlevskaya str.18, Kazan 420018, Russia r

E-mail: vladimir.bochkarev@kpfu.ru

**Abstract.** This paper is devoted to verifying of the empirical Zipf and Hips laws in natural languages using Google Books Ngram corpus data. The connection between the Zipf and Heaps law which predicts the power dependence of the vocabulary size on the text size is discussed. In fact, the Heaps exponent in this dependence varies with the increasing of the text corpus. To explain it, the obtained results are compared with the probability model of text generation. Quasi-periodic variations with characteristic time periods of 60-100 years were also found.

## 1. Introduction

Dependence of the vocabulary size (total number of different words) on the increasing text size is indistinct nowadays though great text corpora are available. Understanding of the processes governing the appearance and usage of new words and their dependence on the text size is essential. This problem is widely discussed in different linguistic and non-linguistic contexts. The empirical laws which focus on the distribution of word frequencies are known as Heaps and Zipf laws. Here we study the relation between the two laws analysing the evolution of written language. Statistical analysis is used to test the validity of these laws using an extremely large text corpus Google Books Ngram which provides a great set of written texts over a period of more than four centuries.

## 2. Connection between the Zipf and Heaps laws

According to the Zipf law, word usage frequency is defined as a power function $p_r \sim r^{-\beta}$, where $r$ – is a word rank, in other words it is a word number in a list of words ranked according to their frequency [1]. The Zipf law is closely connected with the Heaps law which says that the size of lexicon $N$ (the number of particular words in a text) in a text or set of texts of size $L$ is determined by an expression $N \sim L^{-k}$. Different probability models of text generation (if it`s assumed that the Zipf law is fulfilled) result in a simple relation between the exponents β и $k$:.

$$k = \beta^{-1} \tag{1}$$

Initially, the Zipf law was tested on a relatively small corpus of texts where the values of exponent β were close to 1. However, the Heaps law was initially derived while analysing news items. A $k$ exponent was estimated as close to 0.5.

A creation of a large text corpus Google Books Ngram which contains data on more than 9 languages over a great time interval offered an opportunity to study static dependencies of words usage [2,3,4]. The Zipf law is tested in detail on the texts written in the main European languages in [4]. It was shown that the Zipf law in its classical form isn't fulfilled but two power sections can be seen in the frequency distribution plot. The exponent of the 1st section (for frequently used words) is close to 1 and the exponent of the second section (for non-frequent words) is much higher and varies from 1.7 to 2.5 for different languages. The last values correlate better with the power values of the Heaps law and expression (1). The typical word frequency distribution is shown in fig. 1.A.

Let`s consider the following simple model of text generation. Let us have a (finite or infinite) set of possible words, which usage probabilities at a subsequent stage is an a-priori known and equal to $p_i$. How many words will be used on average at least once in a text of $L$ length? To answer this question let us use the method of indicators. Consider the random variable which takes on value 1 if the $i^{th}$ word from our lexicon is used at least once and value 0 if it isn`t used at all. The probability that the $i^{th}$ word will not be used at least once is $(1-p_i)^L$, consequently our random variable takes value 1 with the probability $1-(1-p_i)^L$ and its mathematical expectation is equal to this number. The mathematical expectation of such random variables is the average number of N words. This yields the following:

$$N(L)=\sum_i \{1-(1-p_i)^L\} \quad (2)$$

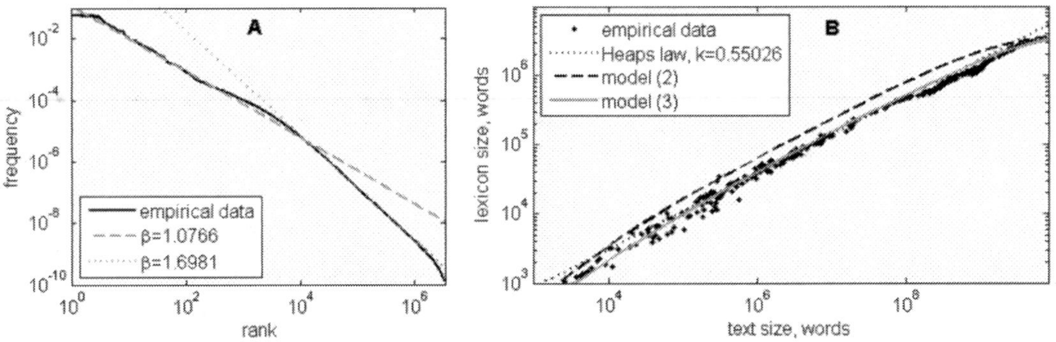


**Figure 1. A.** Word frequency distribution, the common base of English words, year 2000 The solid line represents the empirical data, the dashed and dotted lines – the power approximation; **B.** The dependence of the lexicon size on the text size of the common base. The dots represent the empirical data, the dotted line – the power law approximation; The dashed and solid lines – the calculation according to the models (2) and (3)

## 3. The comparison of the modelled and empirical data

Fig 1.B shows the dependence of the lexicon size on the text size according to the Google Books Ngram corpus data for the common base of the English language. The number of different word forms used in the given year (the letter case wasn't differentiated) and the total number of words contained in the text base were calculated. The word forms consisting only of the letters of the Latin alphabet were considered. The approximation of the empirical data using power law dependence is also shown in the figure. The power exponent was calculated using the least square criterion and was 0.5503. As it can be seen, the empirical data is poorly described by the exponential function. It should be noted that the correlation with the power law for the other languages presented in the Google Books Ngramis worse than for the English language.

The modelling of the expected lexicon size was performed using formula (2). It`s necessary to know word-usage frequencies to calculate the expected lexicon size using formula (2). The calculation of frequencies of rare words is relatively problematic, mostly of words used in earlier time periods because the number of texts in Google Books Ngram where they can be found is rather small. Empirical frequencies of words used in the year 2000 were analysed to estimate probabilities $p_k$ because the biggest number of texts is dated back to this year. Thus, we can estimate the frequencies of the greatest number ($3.97\square10^6$) of unique words. The counting results according to formula (2) are represented in the figure by the dashed line. It can be seen that the modelled dependence is close to the empirical one but it lies slightly above it. To enable better match of the modelled and experimental data, the model was improved.

Traditionally, the words are divided into content and functional words. The latter indicate a grammatical relationship and their frequency directly depend on the sentence structure. However, the percentage of functional words can be relatively high. For example, the percentage of functional words in the English language changed from about 0.48 to 0.57 in 1800-2000 (the list of English words was used during the calculation which can be found in [5]). It can be assumed that all the function words will be used in a set of texts covering the wide range of topics and the number of content words will grow up with the increase of the text size. This results in the following modified model:

$$N(L) = N_{serv} + \sum_{i \in I} \{1 - (1 - p_i)^{\zeta L}\}, \tag{3}$$

where $N_{serv}$ is the number of functional words, $I$ – the set of content word numbers in a common list, $\zeta$ – the part of content words in a text. In the modelling process, the parameter $\zeta$ is defined for each year according to the Google Books Ngram data, after that the expected lexicon size is calculated using formula (3).

It can be seen that the model (3) yields the best approximation of the empirical data. The exponent obtained during the fitting procedure of the modelled curve using the power dependence (on the section from $10^3$ to $10^{10}$) is 0.5674. It is close to the value mentioned above obtained during the fitting of the empirical data. At the same time, a serious divergence of the modelled and empirical curves can be seen. For example, variations in the large values domain can hardly be explained using a simple model.

Consequently, the observed divergence can be due to imperfect model or result from the dynamic processes in a language. The analysis of the data was performed for different time intervals to verify which of the possibilities is realised. The data selected using the 50 year sliding time window was fitted by power dependence. The dependence variations of the Heaps exponent with the time obtained for the English, Russian, German, and French languages are shown in fig. 2A. Two features of the obtained curves should be mentioned. Firstly, the descending trend is typical of all the languages. Secondly, quasi-periodic variations with characteristic time periods of 60-100 years can be seen in the plots.

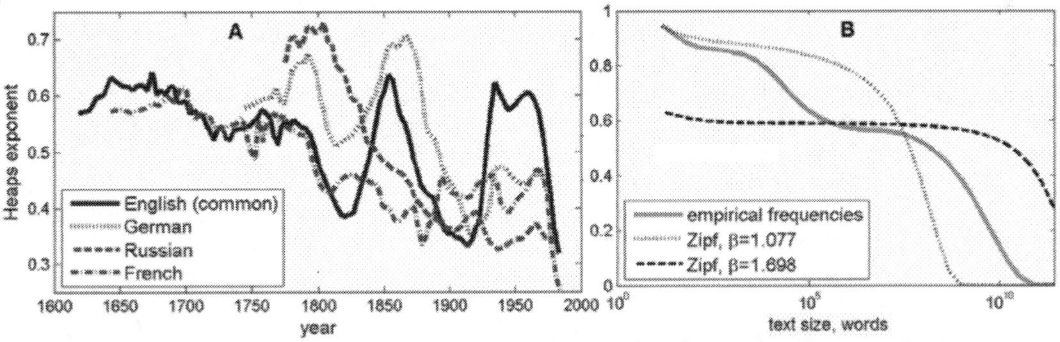


**Figure 2. A.** Change of the Heaps exponent with the time for the four European languages; **B.** Modelled values of the Heaps exponent at different corpus size. The solid line represents the empirical frequency distribution (English, 2000), the dashed, dot-and-dash and dotted lines represent the Zipf distribution with different exponents

On average the number of texts in the Google Books Ngram base is much higher for later time periods. So the question is: can the observed variations of the Heaps exponent with the time can be due to the increase of the average size of the corpus or are they do to some dynamic processes in the language? The modelling results of the expected Heaps exponent at different text size are shown in fig. 2B. The modelling was performed using empirical word-usage frequencies and the modelled frequencies correlated with the Zipf law. The dependence $N(L)$ is approximated using the power law

$N \sim L^{-k(L)}$ for each small section, where $k(L)$ is assumed to be a slowly varying function. The required power exponent $k(L)$ can be calculated by the following formula

$$k(L) = -\frac{\partial \ln N}{\partial \ln L} \tag{4}$$

The Heaps exponents were modelled by (2,4) using empirical word-usage frequencies of the common English base for the year 2000 (the solid line). A descending trend and the Heaps exponent tending to zero at $L$□∞ can be observed in the plot. It`s not surprising because our model has the finite potential lexicon and the dependence $N(L)$ should tend to saturation. Two steps are seen in the plot. The dashed and dotted lines in the fig. 2B represent the modelling results assumed that the word-usage frequencies correspond with the Zipf law $p_k = Ak^{-\beta}$ (with the exponent β equal to 1.077 and 1.698 correspondingly) with the finite number of 3.97□$10^6$ words. The curve tends asymptotically to zero as in the previous case. At that, almost a horizontal section of the local Heaps exponent can be seen in the plot close to the value $k(L) = 1/\beta$ as it was expected from the above mentioned thoughts about the connection of the exponents of the Zipf and Heaps laws. Thus, it can be expected that two flat sections of the first curve correspond with the two almost power sections in the word frequency plot (fig. 1A).

Comparison of fig. 2A and fig. 2B shows that the Heaps exponent decreases with time due to text size increase. But both model (2) and model (3) can (in contrast to the given in fig.3) give only monotonically decreasing dependencies for the Heaps exponent. Thus, the quasi-periodic variations of the Heaps exponent observed in fig. 2.A are most likely due to dynamic processes in the language.

## 4. Conclusion

Thus, the Heaps law is fulfilled restrictedly for small texts and texts related to a short historical period. At that, the Heaps exponent varies with characteristic interval of 60—100 years which reflects the dynamic processes in a language.

This work was supported by the Russian Foundation for Basic Research (grant №12-06-00404a).

## References

[1] Baayen, R.H.: Word Frequency Distributions. Springer, Berlin (2002)

[2] J.-B. Michel, Y.K. Shen, A.P. Aiden, et. al., Quantitative Analysis of Culture Using Millions of Digitized Books, Science 331, 176 (2011).

[3] A.M. Petersen, J.N. Tenenbaum, S. Havlin, H.E. Stanley, M. Perc, Languages cool as they expand: Allometric scaling and the decreasing need for new words, Sci. Rep. 2 (2012) 943.

[4] M. Gerlach and E.G. Altmann, Stochastic Model for the Vocabulary Growth in Natural Languages, Physical Review X, 3, 021006 (2013)

[5] J.M. Hughes, N.J. Foti, D.C. Krakauer, D.N. Rockmore, Quantitative patterns of stylistic influence in the evolution of literature, PNAS, 2012, vol. 109, no. 207682-7686.

# Skew Divergence-Based Fuzzy Segmentation of Rock Samples

**Bruno M. Carvalho†, Edgar Garduño‡, Iraçú O. Santos†**

†DIMAp - UFRN, Campus Universitário, S/N, Lagoa Nova, Natal, RN, 59.072-970 - Brazil
‡DCC, IIMAS, UNAM, Cd. Universitaria, C.P. 04510, Mexico City, Mexico

E-mail: bruno_m_carvalho@yahoo.com, edgargar@ieee.org, irassu@ppgsc.ufrn.br

**Abstract.** Digital image segmentation is a process in which one assigns distinct labels to different objects in a digital image. The MOFS (Multi Object Fuzzy Segmentation) algorithm has been successfully applied to the segmentation of images from several modalities. However, the traditional MOFS algorithm fails when applied to images whose composing objects are characterized by textures whose patterns cannot be successfully described by simple statistics computed over a very restricted area. Here, we present an extension of the MOFS algorithm that achieves the segmentation of textures by employing adaptive affinity functions that use the Skew Divergence as a measure of distance between two distributions. These affinity functions are called adaptive because their associated area (neighborhood) changes according to the characteristics of the texture being processed. We performed experiments on mosaic images composed by combining rock sample images which show the effectiveness of the adaptive skew divergence based fuzzy affinity functions.

## 1. Introduction

Digital image segmentation is the process of assigning distinct labels to different objects in a digital image. The task of segmenting out an object from its background in an image becomes particularly hard for a computer when, instead of the intensity values, what distinguishes the object from the background is some textural property.

Textures are generally described as the feel or shape of a surface or substance, and in the areas of imaging they are characterized by statistical distributions of the observed intensities. Textures can be classified as structural or non-structural [1], where structural textures have a repeated pattern associated with a texture periodicity [2], and non-structural or statistical textures do not possess this characteristic and may be described by statistical distributions.

In this article, we are concerned with the segmentation of images formed by a composition of samples of natural non-structural textures, by using a semi-automatic, interactive region-growing segmentation method, called *fuzzy segmentation* [3, 4]. The fuzzy segmentation method is used with adaptive fuzzy affinity functions, whose scales are related to the statistical distribution of the textural patterns they are associated with, and the Skew divergence to compute how close two pixels are in texture space.

## 2. Related Work

One example of an unsupervised texture segmentation technique is the one proposed by Unser and Eden [5], that extracts local texture properties using local linear transforms, estimates local statistics at the output of a filter bank and generates a multi-resolution sequence using an iterative Gaussian smoothing algorithm. The texture features are then reduced to a single component that is thresholded to produce the segmentation.

There are also several model-based approaches [6, 7] that deal with the texture segmentation problem by modeling the intensity field of textures as a Markov or a Gauss-Markov random field to represent the local spatial dependencies between pixel intensities. However, these techniques are usually very computationally intensive because they require a large number of iterations to converge. Lehmann [8] proposed an alternative approach that models two-dimensional textured images as the concatenation of two one-dimensional hidden Markov auto-regressive (HMM-AR) models, one for the lines and one for the columns.

## 3. Fuzzy Segmentation

Because of the general nature of the approach used, we refer to elements of the set $V$ to be segmented as *spels*, which is short for *spatial elements* [9]. The spels can be pixels of an image (as in [4]), but they can also be dots in the plane (as in [10]), or any variety of other things, such as feature vectors (as in [11]). Thus, the theory and algorithm mentioned here can also be applied to data clustering [12] in general.

The goal is to partition $V$ into a specified number of objects, but in a fuzzy way, by assigning a spel to a particular object with a grade of membership (a number between 0 and 1) to that object. In order to do that, we assign, to every ordered pair $(c, d)$ of spels, a real number not less than 0 and not greater than 1, which is referred to as the *fuzzy connectedness* of $c$ to $d$.

In the original approach [4], fuzzy connectedness is defined in the following general manner. We call a sequence of spels a *chain*, and its *links* are the ordered pairs of consecutive spels in the sequence. The strength of a link is also a fuzzy concept. We say that the *$\psi$-strength* of a link is the value of a *fuzzy spel affinity* function $\psi : V^2 \rightarrow [0, 1]$. A chain is formed by one or more links and the *$\psi$-strength of a chain* is the $\psi$-strength of its weakest link; the $\psi$-strength of a chain with only one spel in it is 1 by definition.

The fuzzy connectedness from $c$ to $d$ is computed using a *fuzzy connectedness* function $\mu_\psi : V^2 \rightarrow [0, 1]$ defined by the $\psi$-strength of the strongest chain from $c$ to $d$. We can then define the *$\psi$-connectedness map* $f$ of a set $V$ for a seed spel $o$ as the picture formed by the fuzzy connectedness values of $o$ to $c$ ($f(c) = \mu_\psi(o, c)$), for all $c \in V$.

An *$M$-semisegmentation* of $V$ is a function $\sigma$ that maps each $c \in V$ into an $(M + 1)$-dimensional vector $\sigma^c = (\sigma_0^c, \sigma_1^c, \cdots, \sigma_M^c)$, such that $\sigma_0$ is non-negative but not greater than 1, for each $m$ $(1 \leq m \leq M)$, the value of $\sigma_m^c$ is either 0 or $\sigma_0^c$, and, for at least one $m$ $(1 \leq m \leq M)$, $\sigma_m^c = \sigma_0^c$. Finally, an *$M$-segmentation* of $V$ is an $M$-semisegmentation such that $\sigma_0^c > 0$ for every spel $c \in V$. There is an unique *$M$-segmentation* asociated with a set $V$ and sets of seed spels $V_m$, for $1 \leq m \leq M$, that is n fact the segmentation of that set.

For more information related to the fuzzy segmentation algorithm, the reader should refer to [4], where the FAST-MOFS algorithm [4] is introduced. The FAST-MOFS is a greedy algorithm that computes the grades of membership of all spels to the objects rounding them to three decimal places. This allows us to use an array as a priority queue, instead of a heap, thus lowering the computational complexity of the algorithm [4].

### *3.1. Fuzzy Segmentation of Textures*

The way we specified $\psi_m$ and $V_m$ $(1 \leq m \leq M)$ for segmenting an image is the following. We click on some spels in the image to identify them as belonging to the $m$th object, and $V_m$ is formed by these points and their eight neighbors. We then define two Gaussian density functions with the means and standard deviations computed based on the average values and the absolute differences of pairs of spels in $V_m$. In [4] the fuzzy affinity function between two spels was then defined as the average value of these two Gaussians that receive as parameters the average value of the two spels $c$ and $d$ and the absolute difference between them.

The first term of the fuzzy affinity function described above encodes a local degree of homogeneity while the second term captures some inhomogeneity information, that can be construed as a very basic textural information.

In this work, in order to include more information about textural properties in the segmentation process, we use adaptive fuzzy affinity functions whose neighborhood areas are selected according to the distribution of the intensities surrounding the selected seed spels. This selection defines a scale which characterizes the correspondent texture. As mentioned before, the original fuzzy affinity functions used the intensity values of two spels $c$ and $d$ to compute the strength of the link connecting $c$ to $d$. Here, we collect, opposed to what was done in [4], information of the intensities in the neighborhood area (scale) defined for the texture in question. For more details about that, the reader should refer to [13].

There are several techniques for detecting the window size that characterizes the texture periodicity pattern [2] and even orientation [14] of structural textures. However, such area is not well-defined in the case of non-structural textures, that can exhibit localized distributions of intensities in smaller scales and very different distributions observed at larger scales.

The algorithm used for choosing the appropriate scale works by computing statistics for a neighborhood of size $3 \times 3$ (i.e., with diameter equal to 1) surrounding the seed spels, and increasing this neighborhood until the differences of the statistics collected on the neighborhoods of two consecutive sizes are smaller than predefined thresholds. The neighborhood sizes used for each object can also be part of the user input to the segmentation method.

To give the algorithm a notion of distance between the distributions surrounding two pixels, i.e., between the histograms of the pixels belonging to the neighborhoods of these two pixels, we use a measure based on the Kullback-Leibler Divergence [16], namely, the Skew Divergence [15], that is used to compute how far two probability distributions are, and are respectively given by:

$$\mathcal{KL}(p \parallel q) = \sum_x p(x) log \frac{p(x)}{q(x)}, \qquad (1)$$

$$\mathcal{S}(p \parallel q) = \mathcal{KL}(q \parallel \alpha p + (1-\alpha)q). \qquad (2)$$

The Skew Divergence (2) circumvents an indetermination problem that happens with the Kullback-Leibler Divergence (1) when there is an index $x$ such that $p(x) > 0$ and $q(x) = 0$, and it does that by mixing the two distributions according to the parameter $\alpha$. In the experiments shown in this paper, this parameter was set to 0.77. Note, from Equations (1) and (2), that both divergences are asymmetric. The general definition of the fuzzy affinity functions employed in MOFS does not limit the usage of asymmetric functions, but one can use the average value of the divergences in both directions as a symmetric measure, if wanted.

## 4. Results and Discussion

We tested our method on mosaics that were composed by putting together parts of three image samples of different rocks. Each rock sample has a statistical pattern that is recognizable by a human, but poses a challenge for segmentation algorithms. The samples and the mosaics were downloaded from the *The Prague Texture Segmentation Data Generator and Benchmark* [17]. Three of the produced mosaics (with $512 \times 512$ pixels each) can be seen in Figure 1, alongside their respective segmentations. In these segmentations, the color of one spel codifies two informations, as the brightness indicates the grade of membership of that spel to the object indicated by the spel's hue.

The three segmentations were performed on a personal computer equipped with 2 Intel Core$^{TM}$ i7 3.07GHz, with 6GB of RAM running Linux, and took between 1.9$s$ and 2.25$s$, i.e.,

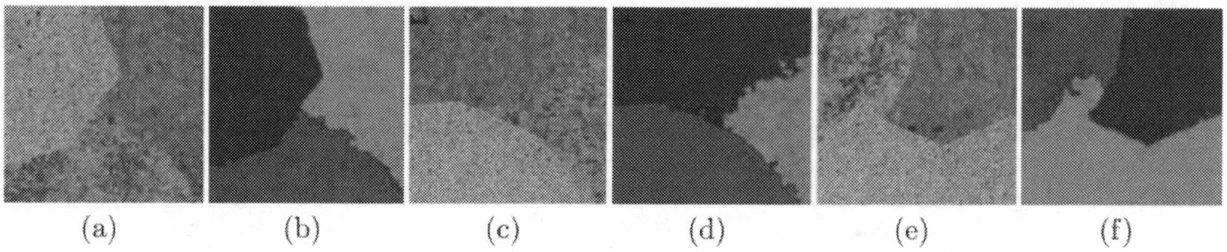

Figure 1. Examples of mosaics produced by combining two Brodatz Textures (a), (c) and (e), and their correspondent connectedness maps (b), (d) and (f).

between $7.25\mu s$ and $8.58\mu s$ per spel. The accuracy reported for the three mosaics shown are 97.58%, 96.43% and 96.01%. It can be seen from the results shown in Figure 1 that the results provide good segmentations for all three textures present in the three mosaics.

## 5. Conclusion

In this article, we showed how the fuzzy segmentation method can be used to segment images with textural properties by employing adaptive affinity functions and the Skew Divergence. The adaptive fuzzy affinity functions allows the method to use different neighborhood sizes for textures with different characteristics, while the Skew Divergence gives a robust measure of distance between the distributions of the areas surrounding two spels.

As of future work, we are currently working on automatizing the segmentation process for this specific application, thus allowing the method to work without the need of an user indicating the seed spels, as well as on extending this method to color images.

## Acknowledgments

We would like to thank the support of the grants FAPERN PRONEM No 610006/2010, and CNPq Universal No 486951/2012-0.

## References


[1] Haralick R M 1979 *Proceedings of The IEEE* **67** 786–804
[2] Jan S R and Hsueh Y C 1998 *Pattern Recogn. Lett.* **19**(5-6) 439–446 ISSN 0167-8655
[3] Herman G T and Carvalho B M 2001 *IEEE Trans. Pattern Anal. Mach. Intel.* **23** 460–474 ISSN 0162-8828
[4] Carvalho B M, Herman G T and Kong T Y 2005 *Disc. Appl. Math.* **151** 55–77 ISSN 0166-218X
[5] Unser M and Eden M 1989 *IEEE Trans. Pattern Anal. Mach. Intel.* **11** 717–728
[6] Manjunath B S and Chellappa R 1991 *IEEE Trans. Pat. Anal. Mach. Intell.* **13**(5) 478–482 ISSN 0162-8828
[7] Robertson R, Villebrun E and Hoeher P 1995 *IEEE Int. Conf. on Communications* vol 2 pp 1009–1013
[8] Lehmann F 2011 *IEEE Trans. Pattern Anal. Mach. Intell.* **33**(1) 16–29 ISSN 0162-8828
[9] Herman G 1998 *Geometry of Digital Spaces* (Springer)
[10] Ahuja N 1982 *IEEE Trans. Pat. Anal. Mach. Intell.* **3** 336–343
[11] Duda R and Hart P 1973 *Pattern Classification and Scene Analysis* (New York: John Wiley & Sons)
[12] Jain A, Murty M and Flynn P 1999 *ACM Computer Surveys* **31** 264–323
[13] Carvalho B M, Souza T S and Garduño E 2012 *Proceedings of SAC '12* (New York, NY, USA: ACM) pp 51–53 ISBN 978-1-4503-0857-1
[14] Lin H C, Wang L L and Yang S N 1997 *Pattern Recogn. Lett.* **18**(5) 433–443 ISSN 0167-8655
[15] Lee L 1999 *Proceedings of the 37th annual meeting on Computational Linguistics* ACL '99 (Stroudsburg, PA, USA: Association for Computational Linguistics) pp 25–32 ISBN 1-55860-609-3
[16] Kullback S and Leibler R A 1951 *Annals of Mathematical Statistics* **22** 49–86
[17] Haindl M and Mikes S 2008 *ICPR* (IEEE) pp 1–4 ISBN 978-1-4244-2175-6

# Stability analysis for virus spreading in complex networks with quarantine and non-homogeneous transition rates

**L. A. Alarcon-Ramos[1], A. Schaum[1], C. Rodríguez Lucatero[2] and R. Bernal Jaquez[1]**

[1]Departamento de Matematicas Aplicadas y Sistemas
[2]Departamento de Tecnologias de la informacion
Universidad Autonoma Metropolitana-Cuajimalpa

E-mail: {lalarcon, aschaum, crodriguez, rbernal}@correo.cua.uam.mx

**Abstract.** Virus propagations in complex networks have been studied in the framework of discrete time Markov process dynamical systems. These studies have been carried out under the assumption of homogeneous transition rates, yielding conditions for virus extinction in terms of the transition probabilities and the largest eigenvalue of the connectivity matrix. Nevertheless the assumption of homogeneous rates is rather restrictive. In the present study we consider non-homogeneous transition rates, assigned according to a uniform distribution, with susceptible, infected and quarantine states, thus generalizing the previous studies. A remarkable result of this analysis is that the extinction depends on the weakest element in the network. Simulation results are presented for large free-scale networks, that corroborate our theoretical findings.

## 1. Introduction

The virus or disease propagation models are important because it is possible to extend their use in other areas, for example, the spreading of rumors or data in computers networks [3, 5, 6], and the analysis of these models allow us to apply certain policies or mechanisms, in order to control the spread of a disease.

A common method to combat the spread of a disease is the isolation of individuals who are infected, by placing them in a state of Q-Quarantine, which is added to the well-know models SI or SIS [1, 2]. For example SIQS, SIQR and SEIQR models are presented in [6, 8, 7], where every individual in the population is classified according to the disease status with homogeneous transition parameters, and where system's thresholds, equilibrium points and their stability are discussed. However, in this paper we consider the model SIQS, where individual states are modeled by a discrete-time Markov chain, but with non-homogeneous transition probabilities. As long as complex networks theory allows exploring the dynamics of propagation in structures generated by relationship established between individuals, we will follow this approach studying any kind of network but taking, in the simulations, the particular case of a network in which nodes' degree follows a Power-Law distribution according to the algorithm proposed in [4]. Besides our theoretical work, simulations are performed in order to corroborate our findings.

The paper is organized as follows: In Section 2, the mathematical model SIQS process is introduced. In Section 3 global extinction conditions are derived. In Section 4, numerical simulations are presented that corroborate our theoretical results. Finally, conclusions are presented in Section 5.

## 2. SIQ Model

Consider the SIQS model for virus spread in complex network of $N$ individuals or nodes, as described in [8], but adapted for each individual $i$ in the network ($i = 1, 2, \ldots, N$), as shown in the state diagram in Fig. 1. The transitions between states depend of non-homogeneous parameters, which are described below for each individual $i$ of the network where: $\mu_i$ is the recovery probability, $\tau_i$ is the probability that $i$ enters quarantine, and $\zeta_i$ is the probability that $i$ is not being infected by interacting with its neighbors.

The interactions consist that in each time step, the node $i$ will perform $\lambda_i$ average attempts to contact with probability $\gamma_i$ with each of its neighbors. The nodes $i$ and $j$ may be in quarantine at time $t$ with probability $q_i(t)$ and $q_j(t)$ respectively, so that the probability of contact on each attempt would be given by $\eta_{ij}(t) = \gamma_i(1 - q_i(t))(1 - q_j(t))$. Therefore, the probability of $i$ performing at least one contact is given, according to a binomial distribution, by

$$r_{ij}(t) = 1 - [1 - \eta_{ij}(t)]^{\lambda_i}. \tag{1}$$

The probability that the node $i$ is being infected due to contact with $j$ is given by $1 - \beta_i r_{ij}(t) p_j(t)$, where $p_j(t)$ is the probability that the node $j$ is infected at time $t$. Considering that the interactions between $i$ and its neighbors are independent, the probability that $i$ is not infected by its neighbors is

$$\zeta_{ij}(t) = \prod_{i=1}^{N} \left[1 - a_{ij}\beta_i r_{ij}(t) p_j(t)\right], \tag{2}$$

where $a_{ij}$ are the entries of the adjacency matrix $A \in \mathbb{R}^{N\times N}$ that represents the existing connections between the $N$ nodes of the network. According to the transitions diagram (see Fig. 1), the following mathematical model is obtained:

$$\begin{aligned} p_i(t+1) &= (1-\tau_i)p_i(t) + (1-\zeta_i(t))(1-q_i(t)-p_i(t)), \\ q_i(t+1) &= \tau_i p_i(t) + (1-\mu_i)q_i(t), \quad i = 1, 2, \ldots N, \\ 1 &= s_i(t+1) + p_i(t+1) + q_i(t+1), \end{aligned} \tag{3}$$

where $s_i(t)$, $p_i(t)$ and $q_i(t)$ are the probability of being susceptible, infected and in quarantine at time $t$, respectively.

To determine the fixed points of the system of equations (3) for each node $i$, substitute $p_i(t+1) = p_i(t) = p_i$, $q_i(t+1) = q_i(t) = q_i$, $s_i(t+1) = s_i(t) = s_i$, and obtein after some algebra

$$[p_i, q_i, s_i,]^T = \left(\frac{1}{1 + (1/\tau_i + 1/\mu_i)\omega_i}\right)\left[\frac{\omega_i}{\tau_i}, \frac{\omega_i}{\mu_i}, 1\right]^T, \quad \omega_i = 1 - \zeta_i. \tag{4}$$

From equation (4), we identified two solutions: the parameter independent extinction state $[0, 0, 1]^T$ given by $\omega_i = 0$, and the parameter dependent endemic state $0 < \omega_i \leq 1$.

## 3. System dynamics

We are interested in determining the conditions for which the extinction state $[p_i, q_i, s_i]^T = [0, 0, 1]^T$ is a global attractor for any node $i$. First, we consider the following bound

$$1 - \zeta_{ij}(t) \leq \sum_{j=1}^{N} a_{ij}\beta_i r_{ij}(t) p_j(t). \tag{5}$$

proved in [9, 10] and then determine a linear dynamics that bounds the nonlinear dynamics given by (3). Doing so leads take following result

**Theorem 1.** The extinction state of the SIQS model (3), is a global attractor for any node $i$ of network, if the following condition holds

$$\sigma(S) \in \{\lambda \in \mathbb{C} : |\lambda| < 1\}, \quad \mu_i > 0, \tag{6}$$

where $S = I_N - T + BA$, $T = diag(\tau_i)$, $B = diag(\beta_i)$, and $\sigma$ denotes the spectrum of a matrix.

**Proof:** Substitute (5) and (1) into the first equation of (3)

$$\begin{aligned} p_i(t+1) &\leq (1-\tau_i)p_i(t) + \sum_{j=1}^{N} a_{ij}\beta_i p_j(t)\Big(1 - [1-\eta_{ij}(t)]^{\lambda_i}\Big)\Big(1 - p_i(t) - q_i(t)\Big)\Big) \\ &= (1-\tau_i)p_i(t) + \sum_{j=1}^{N} a_{ij}\beta_i p_j(t)\Big(1 - p_i(t) - q_i(t) - [1-\eta_{ij}(t)]^{\lambda_i}(1 - p_i(t) - q_i(t))\Big) \\ &\leq (1-\tau_i)p_i(t) + \beta_i \sum_{j=1}^{N} a_{ij}p_j(t). \end{aligned}$$

Accordingly the solutions of (3) are bounded by the solutions of the linear dynamics

$$\begin{bmatrix} p(t+1) \\ q(t+1) \end{bmatrix} \leq \begin{bmatrix} S & 0 \\ T & M \end{bmatrix} \begin{bmatrix} p(t) \\ q(t) \end{bmatrix} = \mathbf{A} \begin{bmatrix} p(t) \\ q(t) \end{bmatrix}, \tag{7}$$

where $M = diag(1-\mu_i)$, $p(t) = [p_1(t), \ldots, p_N(t)]^T$ and $q(t) = [q_1(t), \ldots, q_N(t)]^T$. The linear dynamics given by the matrix $\mathbf{A}$ are asymptotically stable if and only if $\sigma(\mathbf{A}) \in \{\lambda \in \mathbb{C} : |\lambda| < 1\}$, but due to the triangular form of the matrix $\mathbf{A}$, it is sufficient that

$$\sigma(S) \in \{\lambda \in \mathbb{C} : |\lambda| < 1\}, \qquad \textbf{QED}.$$

For our purposes, we suppose that $\tau$ and $\beta$ are distributed uniformly in the complex network, such that $\tau_i \in [\tau_{min}, \tau_{max}]$ and $\beta_i \in [\beta_{min}, \beta_{max}]$, $i = 1, 2, \ldots, N$. Then, for any vector $p = [p1, p2, \ldots, p_N]^T$, the *ith* entry of the product $Sp$ is given by

$$[Sp]_i = (1-\tau_i)p_i + \beta_i \sum_{j=1}^{N} a_{ij}p_j \leq (1-\tau_{min})p_i + \beta_i \sum_{j=1}^{N} a_{ij}p_j \leq (1-\tau_{min})p_i + \beta_{max} \sum_{j=1}^{N} a_{ij}p_j. \tag{8}$$

Note that we have bounded (7) more strictly. Taking the second term and according to the Theorem 1, it is sufficient to consider $\sigma(S)$ for $T = \tau_{min}I_N$, and therefore $\tau_{min} > \rho(BA)$, where $\rho(\cdot)$ is the spectral radius. Finally, doing the same for the third term in the inequality, it is sufficient to consider $\sigma(S)$ for $T = \tau_{min}I_N$ and $B = \beta_{max}I_N$, and therefore $\tau_{min} > \beta_{max}\rho(A)$.

The inequality (8) says that $\tau_{min}$ and $\beta_{max}$ play an important role for the conditions of (3), bounded by linear dynamics (7). These conditions are summarized as follows: *i)* $\rho(\mathbf{A}) > 1$, *ii)* $\tau_{min} > \rho(BA)$ and *iii)* $\tau_{min} > \beta_{max}\rho(A)$. The first one is a necessary condition, while the others are sufficient conditions to assure the extinction state of the system (3) be a global atractor.

## 4. Experiments

We performed a set of simulations of (3) on a complex network generated according to [4], with $m_o = 5$ and $m = 2$. As it been above, it seems that the recovery probability $\mu_i$ and the average number of attempts to contact a neighbor don't play an important role in the dynamics of (3), so we considered $\mu_i = 0.5$, $\lambda_i = 10000$ (simulating *reactive process*), and $\beta_i \in (0.055, 0.065)$ distributed uniformly. For our matrix $A$, $\rho(A) = 8.3439$, therefore $\tau_{min} > 0.542$. However $\rho(BA) = 0.508$, so that $\tau_{min} > 0.508$, as it is shown in figures 2 and 3 for several initial conditions. In all figures, we used $\rho_p(t)$ and $\rho_q(t)$ that represent the average probability of finding an individual infected or in quarantine respectively. To be more precise, consider $\rho_p(t) = \frac{1}{N}\sum_{i=1}^{N} p_i(t)$.

Although Fig. 2 and 3 show that the system given by (3) reaches the extinction state, if we take a value of $\tau_{min} < 0.508$, for example $\tau_{min} = 0.480$, the system also reaches the extinction state. To illustrate this case in Fig. 4, the response for several values of $\tau_{min}$ are shown. Note that $\tau_{min}$ is a bifurcation parameter with threshold $\tau_{min_c} \approx 0.43$, as it can see from Fig. 4.

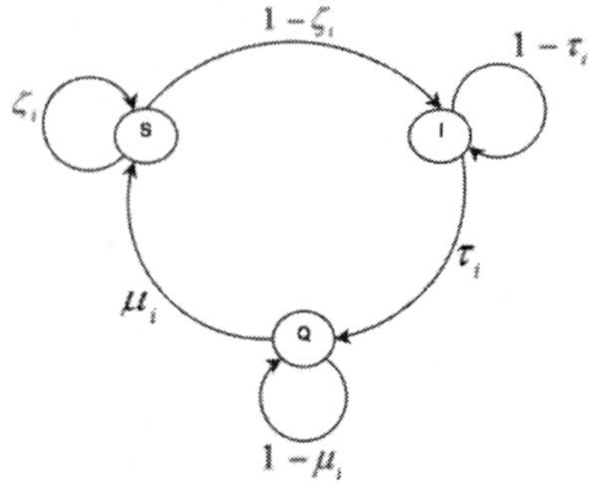


**Figure 1.** State transition diagram associated to the SIQS model.

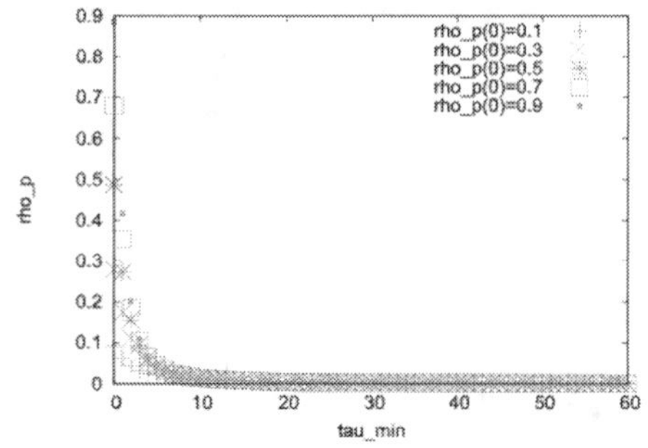


**Figure 2.** For all curves, $\rho_q(0) = 0.1$, $\tau_{min} > 0.542$ and $\tau_{max} < 0.570$.

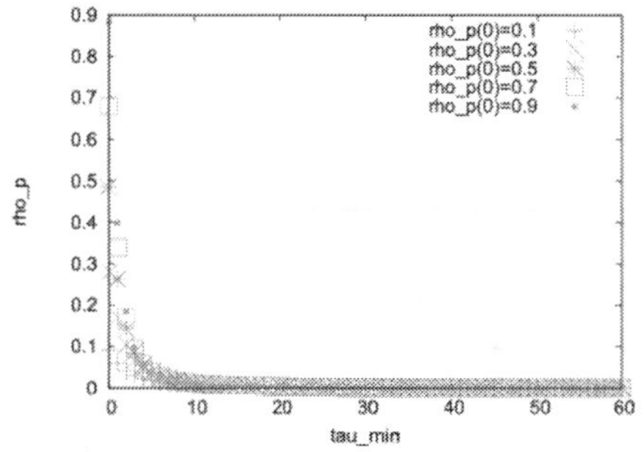


**Figure 3.** For all curves, $\rho_q(0) = 0.1$, $\tau_{min} > 0.508$ and $\tau_{max} < 0.570$.

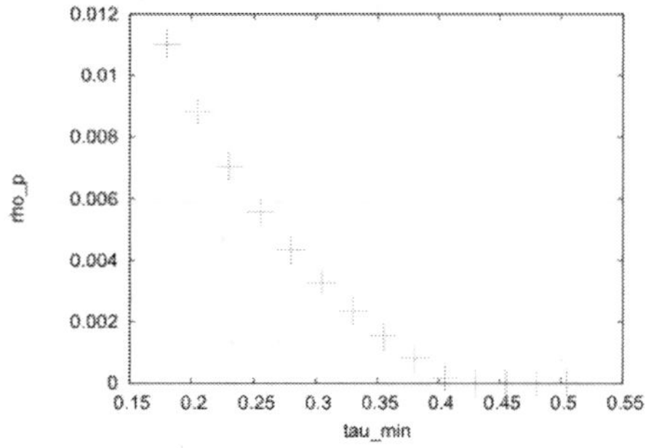


**Figure 4.** $\rho_p$ *vs* $\tau_{min}$ for large values of $t$.

## 5. Conclusions

We analyzed the stability of the extinction state of virus spreading in a complex network with Power-Law connectivity, quarantine mechanism, and non-homogeneous state transition probabilities. Sufficient conditions for the stability of the extinction state are obtained, in terms of the weakest individuals given by their probability putting in quarantine ($\tau_{min}$) and the probability of infection ($\beta_{max}$). However, the necessary condition provides a greater margin, being $\tau$ the only bifurcation parameter in our analysis, from which the system reaches the extinction state or endemic state.

## References

[1] Allen L. 2008 *Mathematical Epidemiology* (Lecture Notes in Mathematics/Mathematical Biosciences Subseries) ed F Brauer, P van den Driessche and J Wu pp 81-130

[2] Gomez S, Arenas A, Borge-Holthoefer J, Meloni S and Moreno Y 2010 *EPL* **89** 38009p1-6

[3] Nekovee M, Moreno Y, Bianconi G and Marsili M 2007 *Physica A* **374** 457-470

[4] Barabási A L and Albert R 1999 Emergence of scaling in random networks *Science* **286** 509-512

[5] Moreno Y, Nekovee M and Pacheco A 2004 *Phys. Rev. E* **69** 0661301-7

[6] Chakrabarti D, Leskovec D, Faloutsos C, Madden S, Guestrin C and Faloutsos M *IEEE INFOCOM 2007* 1316-24

[7] Kumar Mishra B and Kumar Singh A 2012 *Mathematical Problems in Engineering* **2012** 1-13

[8] Hethcote H, Zhien M 2002 and Shengbing L 2002 *Mathematical Biosciences* **180** 141-160

[9] Wang Y, Chakrabarti D, Wang C and Faloutsos C 2003 *Reliable Distributed Systems 2003, Proceedings. 22nd International Symposium on* 25-34

[10] Chakrabarti D, Wang Y, Wang C, Leskovec J and Faloutsos C 2008 *ACM Trans. Inf. Syst. Secur.* **10** pp 1-26

# Fuzzy spaces topology change and BH thermodynamics

**C A S Silva**[1] **and R R Landim**[2]

[1] Instituto Federal de Educação Ciência e Tecnologia da Paraíba (IFPB), Campus Campina Grande - Rua Tranquilino Coelho Lemos, 671, Jardim Dinamérica I

[2] Departamento de Física, Universidade Federal do Ceará - Caixa Postal 6030, CEP 60455-760, Fortaleza, Ceará, Brazil

E-mail: `calex@fisica.ufc.br`

**Abstract.** What is the ultimate fate of something that falls into a black hole? From this question arises one of the most intricate problems of modern theoretical physics: the black hole information loss paradox. Bekenstein and Hawking have been shown that the entropy in a black hole is proportional to the surface area of its event horizon, which should be quantized in a multiple of the Planck area. This led G.'t Hooft and L. Susskind to propose the holographic principle which states that all the information inside the black hole can be stored on its event horizon. From this results, one may think if the solution to the information paradox could lies in the quantum properties of the black hole horizon. One way to quantize the event horizon is to see it as a fuzzy sphere, which posses a closed relation with Hopf algebras. This relation makes possible a topology change process where a fuzzy sphere splits in two others. In this work it will be shown that, if one quantize the black hole event horizon as a fuzzy sphere taking into account its quantum symmetry properties, a topology change process to black holes can be defined without break unitarity or locality, and we can obtain a possible solution to the information paradox. Moreover, we show that this model can explain the origin of the black hole entropy, and why black holes obey a generalized second law of thermodynamics.

## 1. Introduction

Black hole (BH) thermodynamics has been discussed in the context of topology change, as conceived for some classes of quantum spaces, called fuzzy spheres [1]. It has been argued that a model based on the topology change of fuzzy manifolds can be used to shed some light on the origin of the BH entropy, including why BH evaporation process obeys the Generalized Second Law of Thermodynamics(GSL). Moreover, it has been suggested that topology change for fuzzy spaces can relieve the information problem.

Fuzzy spheres consist in one of the most simplest example of noncommutative spaces, and appear as vacuum solutions in Euclidean gravity [3]. It is obtained when we quantize the usual sphere $S^2$ replacing the commutative algebra of functions on this manifold by the noncommutative algebra of matrices. The use of fuzzy spheres in the context of BH physics is mostly motivated by the Bekenstein's limit [4]. Since fuzzy spheres, are obtained from quantization of a compact space, they are described by finite dimensional matrices, in a way that the number of independent states defined on the fuzzy sphere is limited, and the entropy associated with these states is finite, in agreement with the Bekenstein's limit [4]. Another important feature of fuzzy spheres is its close relationship with Hopf algebras, which allow us to define a linear operation (the coproduct of a Hopf algebra) on a fuzzy sphere $S_F^2$ and compose two fuzzy spheres preserving algebraic properties intact. This operation produces a topology change process where a fuzzy sphere splits into two others [5], and can be used as a good mathematical model to BH topology change [1, 2].

In this paper, we will revise the main results of the references [1, 2], where the BH thermodynamics is put in connection with topology change of fuzzy manifolds. It will be shown that a model based on the topology change of fuzzy manifolds can be used to shed some light on BH evaporation.

## 2. Fuzzy spaces topology change and BH thermodynamics

To describe the fuzzy sphere topology change, we have that under the quantization procedure, functions defined on $S^2$ are replaced by matrices on $S_F^2$ [3]. In this way, let a matrix $\hat{M}$ describing a wave function on $S_F^2$, the Hopf coproduct $\Delta : S_F^2(j) \rightarrow S_F^2(K) \otimes S_F^2(L)$ acts on $\hat{M}$ as

$$\Delta_{(K,L)}(\hat{M}) = \sum_{\mu_1,\mu_2,m_1,m_2} C_{K,L,J;\mu_1,\mu_2} C_{K,L,J;m_1,m_2} \times M_{\mu_1+\mu_2,m_1+m_2} e^{\mu_1 m_1}(K) \otimes e^{\mu_2 m_2}(L) \, , \quad (1)$$

where $C$'s are the Clebsh-Gordan coefficients and $e^{\mu_i m_j}$'s are basis for a matrix space defined on the fuzzy sphere [5].

The meaning of the eq. (1) is that a wavefunction $\hat{M} \in S_F^2(J)$ splits into a superposition of wavefunctions on $S_F^2(K) \otimes S_F^2(L)$. In this way, the information in $\hat{M}$ is divided between two regions of the spacetime, i.e, the two fuzzy spheres with spins $K$ and $L$ respectively. In this way, the basic assumptions of this article are introduced as follows: (i) If one use the fuzzy sphere Hilbert space as the ones of the BH, the maximum of information about the BH that an outside observer can obtain would be encoded in wave functions defined on the fuzzy sphere Hilbert space. (ii) We shall find out, through the Hopf coproduct $\Delta$, a topology change process for the BH. In this process the information about the BH initial state, will be divided into two spacetime regions. One of them is a fuzzy sphere with spin $K$, which we shall consider as the original world and name it "the main world". The other one is a fuzzy sphere with spin $L$ which we shall name "the baby world". (iii) Since the baby world arises in the BH interior, an observer in the main world can not access the degrees of freedom there. In this way, from his standpoint, the BH will appear to evolve from a pure to a mixed state described by a density matrix $\hat{\rho}$. It enable us to define an entropy, measured by the observer in the main world, associated to the BH horizon. The coproduct $\Delta$ is an unitary operation, in a way that information is globally preserved in the fuzzy BH topology change.

In order to analyze how the topology change process drives the BH evaporation, as seen by an observer in our world, we will investigate how the fuzzy topology change drives the BH area transitions. We will admit that the selection rules for the BH area transitions are the ones for the topology change. These rules are obtained from the eq. (1), when one traces over the degrees of freedom in the baby universe. The probability amplitude for a BH n-steps transition can be obtained from eq. (1). In ref. [1], it has been shown that the splitting process (1) obeys cluster decomposition (locality). Then, we have that different steps $J \rightarrow J - 1/2$ (the lowest step in the BH evaporation) are independent events. In this way, the probability amplitude of n steps occur in the black evaporation process is given by the product of the probability amplitudes of each one of this steps occurring by itself. It is easy to show that the probability amplitude for n steps in the BH evaporation is given by [2]

$$a_{Jn} = \frac{2J+1}{2J-n+1} \, . \quad (2)$$

In order to analyze the BH area transitions, we will introduce a canonical ensemble in which our system (the BH) can occupy different area microstates. The idea of using these types of ensembles goes back to Krasnov [6]. In this framework, the probability amplitude for the BH evaporate is given by $a_{Jn} = e^{-\beta\delta A_{Jn}}$, where $\beta$ is a temperature-like parameter dual to the BH area. If one identifies this probability amplitude with the expression given by the equation (2), we will have that the value of the BH area in the J-state will be written as

$$A_J = \beta^{-1} l_p^2 \ln(2J+1)\ . \tag{3}$$

The logarithmic dependence of the BH area spectrum on $J$, in the expression (3), tell us that the decrease in the horizon area is continuous at large values of $J$, and discrete to small values of $J$, when the BH approaches the Planck scale.

In this way, if one models a BH horizon by a fuzzy sphere and consider its quantum symmetry properties, a topological change process which can be used to solve the black hole information loss paradox is obtained. In this process a BH event horizon, modeled by a fuzzy sphere with spin J , splits into two others. The fuzzy sphere splitting can be used to describe a black hole evaporation process in which information about the black hole initial state is divided between two topologically disconnected regions: the main and the baby world. Nor unitarity or locality is broken in the evolution of the whole system. On the other hand, an observer in the main world sees the topology change process occurs in a non-unitary way, due to the impossibility of access the degrees of freedom in the baby world. It is possible, but not necessary, that information returns via quantum gravity tunneling at the final stages of BH evaporation. In this point, the baby universe ceases to exist, and the BH evolution as seen by an observer outside the BH is unitary. In fact, as we shall see in the next section, due to the discreteness of BH area spectrum (3), we have that the BH emission spectrum must have some deviations from the strictly thermality when the BH approaches the Planck scale. This point has been also discussed in the reference [7], where the discreteness of the BH area spectrum in the Planck regime has been put into connection with the obedience to a Generalized Uncertainty Principle(GUP) by the radiation emitted by the BH.

## 3. Entropy emitted during the evaporation process

The entropy of a system measures one's lack of information about its actual internal configuration. Suppose that everything we know about the internal configuration of the system is that it may be found in any of a number of states, with probability $p_n$ for the nth state. Then the entropy associated with the system is given by Shannon's well-known relation $S = -\sum p_n ln p_n$.

The probability for a BH to emit a specific quantum should be given by the expression (2), in which we must yet include a gray-body factor $\Gamma$ (representing a scattering of the quantum off the spacetime curvature surround the BH). Thus, the probability $p_n$ to the BH goes n steps down in the area ladder is proportional to $\Gamma(n)\ e^{-\frac{\delta A_{Jn}}{4}}$. Moreover, the discrete area spectrum (3) implies a discrete line emission from a quantum BH. For a Schwarzschild black hole, the radiation emitted will be at frequencies given by $\omega_{Jn} = \frac{1}{2\sqrt{\pi}}(\sqrt{ln(2J+1)} - \sqrt{ln(2J+1-n)})$.

To gain some insight into the physical problem, we shall consider a simple toy model suggested by Hod [8]. To begin with, it is well known that, for massless field, $\Gamma(M\omega)$ approaches 0 in the low-frequency limit, and has a high-frequency limit of 1. A rough approximation of this effect can be archived by introducing a low frequency cutoff at some $\omega = \omega_c$. That is, $\Gamma(\bar{\omega}) = 0$ for $\bar{\omega} < \bar{\omega}_c$, and $\Gamma(\bar{\omega}) = 1$ otherwise, where $\bar{\omega} = M\omega$.

The ratio $R = |\dot{S}_{rad}/\dot{S}_{BH}|$ of entropy emission rate from the quantum BH to the rate of BH entropy decrease is given by:

$$R = \left| \frac{\sum_{i=1}^{N_s} \sum_{n=1}^{2J} C\Gamma(n) e^{-\frac{\delta A_{Jn}}{4}} ln\left[C\Gamma(n)e^{-\frac{\delta A_{Jn}}{4}}\right]}{\sum_{i=1}^{N_s} \sum_{n=1}^{2J} C\Gamma(n) e^{-\frac{\delta A_{Jn}}{4}} \ \frac{\delta A_{Jn}}{4}\Big)} \right| , \tag{4}$$

where C is a normalization factor, defined by the normalization condition:

$$\sum_{i=1}^{N_s}\sum_{n=1}^{2J} C\Gamma(n)e^{-\frac{\delta A_{Jn}}{4}} = 1. \tag{5}$$

For the effective number of particle species emitted ($N_s$), we will take into account the various massless modes emitted. We will consider

$$N_s = \begin{cases} 2J+1 & \text{for} \quad 2J+1 < 112\,, \\ 112 & \text{for} \quad 2J+1 \geq 112\,. \end{cases}$$

In this way $N_s$ is upper limited by the number of modes of massless particles in nature which make the dominant contribution to the black-hole spectrum (the 1/2, 3/2, 5/2 neutrino modes, the 1 and 2 photon modes, and the 2 and 3 graviton modes [8]), and by the size of the fuzzy sphere Hilbert space.

At this point it would be interesting to discuss the possible ways a BH could end. Specifically, if the BH will evaporate completely or give rise to a remnant. As we have said, due to the form of the spectrum (3), some deviations from the thermal behavior must occur in the final stages of BH evaporation. In the reference [7], these deviations have been related with a GUP.

Some authors have argued that that the total BH evaporation is prevented by the GUP in exactly the same way that the Uncertainty Principle prevents the hydrogen atom from total collapse. In this way, a BH remnant must appear in the final of BH evaporation [9]. However, the remnant hypothesis suffers from some troubles. The main problem is, since the initial BH has been arbitrarily massive, the remnant must be capable of carrying an arbitrarily large amount of information. In this way, there must be an infinity of distinct, long-lived, species of remnant, with mass comparable with Planck mass, for each BH which could possibly be formed. As has been discussed in [10], it seems hard to reconcile the remnant infinite degeneracy with causality and unitarity.

On the other hand, as have been emphasized by [11], these problems are alleviated in topology change case, as the baby universes do not manifest themselves directly in the parent universe, and a finite number of long lived (but not eternal) remnants can arise from small BHs in the final stages of BH evaporation. In this context, the information stored in the remnant could return to the external universe in an unitary process. In this way, in our work, we shall consider that BH will evaporate completely.

In the fig. 3, we have plotted down R taking $\bar{\omega}_c \simeq 0.2$ (the location of the peak in the total power spectrum [8]). With this frequency cutoff, the minimal non-null value to the quantum number $J$, in order to have $\Gamma \neq 0$, is $J = 6.0$. In this point, the BH must evaporate completely. From the graphic for $R$, we have that the non-unitary evolution of the BH geometry in the main world, due to the topology change process, imposes obedience to a "second law of thermodynamics" on the BH evolution process, since R is ever larger than (or equal) to unity. The value of $R$ approaches the value of 1.3 at the large $J$ limit in agreement with known semiclassical results [12]

It is important to notice that the entropy emitted from the BH decreases as the area spacing increases. The striking consequence of this is the possibility that, since the BH radiation becomes less and less entropic as the evaporation process takes place, some information about the black hole initial state could leak out from its interior and be accessible to an observer in our universe, where we can perform measures. The possibility of information leakage from a BH with a discrete area spectrum is already pointed out by Hod [8].

Concerning this point, information could leak out from the BH during the final stages of evaporation through quantum tunneling. In fact, in a famous article, Parikh and Wilczek [13], by analyzing Hawking radiation as tunneling, have been shown that the radiation spectrum

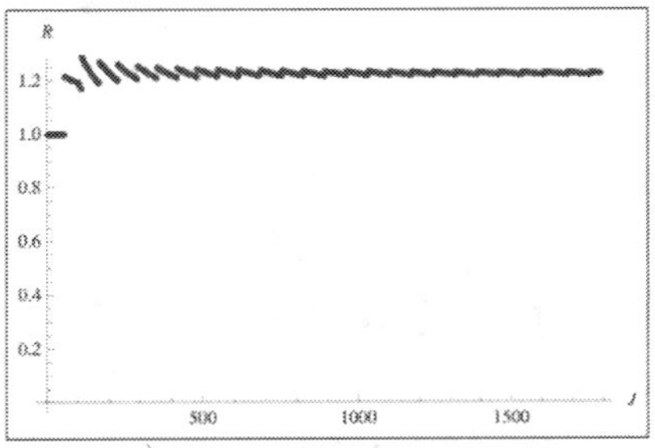


cannot be strictly thermal because of the BH back reaction. This result has been confirmed recently in the reference [14], where the authors have found the existence of correlations among Hawking radiations which are described as hidden messengers in BH evaporation, in a way that information can be recovered during the emission process. Other interesting approach is due to Corda, who interpreted BH quasi-normal models in terms of quantum levels and the Hawking quanta in terms of "jumps" like the levels [15], and recently found a time dependent Schrodinger equation for the system composed by Hawking radiation and BH quasi-normal models which permits pure states to evolve in pure states [16].

**Acknowledgments**

The authors thank to Coordenação de Aperfeiçoamento de Pessoal de Ensino Superior-CAPES(Brazil), Conselho Nacional de Desenvolvimento Científico e Tecnológico - CNPQ(Brazil), and Instituto Federal de Educação, Ciência e Tecnologia da Paraíba, Campus Campina Grande, for the financial support.

**References**

[1] C.A.S. Silva, Phys.Lett.B 677 (2009), 318.
[2] C.A.S. Silva, R.R. Landim, Europhys.Lett. 96 (2011) 10007.
[3] J. Madore, Class.Quant.Grav. 9 (1992) 69.
[4] J.D. Bekenstein, Lett. Nuovo Cim. 4 (1972) 737.
[5] A.P. Balachandran, S. Kürkçüoğlu, Int.J.Mod.Phys.A 19 (2004) 3395.
[6] K.V. Krasnov, Phys. Rev. D 55, (1997) 3505.
[7] C. A. S. Silva and R. R. Landim, Europhys. Lett. 100 (2012) 10002.
[8] S. Hod, Phys.Rev.D 61 (2000) 124016.
[9] R. J. Adler, P. Chen and D. I. Santiago, Gen. Rel. Grav. 33 (2001) 2101 [gr-qc/0106080].
[10] G. t Hooft, Nucl. Phys. B 256 (1985) 727.
[11] S. D. H. Hsu, Phys. Lett. B 644 (2007) 67 [hep-th/0608175].
[12] W. H. Zurek, Phys. Rev. Lett. 49, (1982), 1683.
[13] M. K. Parikh and F. Wilczek, Phys. Rev. Lett. 85, 5042 (2000)
[14] B. Zhang, Q.-Y. Cai, M. S. Zhan, and L. You, Int. Journ. Mod. Phys. D 22 1341014 (2013.
[15] C. Corda, Int. Journ. Mod. Phys. D 21, 1242023
[16] C. Corda, arXiv:1304.1899 [gr-qc].

# Modeling the diffusion magnetic resonance imaging signal inside neurons

[1]D. V. Nguyen, [1]J. R. Li, [2]D. S. Grebenkov, [3]D. Le Bihan

[1] INRIA Saclay-Equipe DEFI, CMAP, Ecole Polytechnique, 91128, Palaiseau, France

[2] LPMC, CNRS – Ecole Polytechnique, 91128, Palaiseau, France

[3] Neurospin - CEA Saclay, Gif-sur-Yvette, France

E-mail: vdnguyen@cmap.polytechnique.fr, jingrebecca.li@inria.fr

**Abstract.** The Bloch-Torrey partial differential equation (PDE) describes the complex transverse water proton magnetization due to diffusion-encoding magnetic field gradient pulses. The integral of the solution of this PDE yields the diffusion magnetic resonance imaging (dMRI) signal. In a complex medium such as cerebral tissue, it is difficult to explicitly link the dMRI signal to biological parameters such as the cellular geometry or the cellular volume fraction. Studying the dMRI signal arising from a single neuron can provide insight into how the geometrical structure of neurons influences the measured signal. We formulate the Bloch-Torrey PDE inside a single neuron, under no water exchange condition with the extracellular space, and show how to reduce the 3D simulation in the full neuron to a 3D simulation around the soma and 1D simulations in the neurites. We show that this latter approach is computationally much faster than full 3D simulation and still gives accurate results over a wide range of diffusion times.

## 1. Introduction

Diffusion magnetic resonance imaging (dMRI) is an imaging modality that gives a measure of the average displacement of water molecules in biological tissue during a diffusion time in the range of tens of milliseconds [1]. In a complex medium such as cerebral tissue, it is difficult to explicitly link the dMRI signal to biological parameters such as the cellular geometry or the cellular volume fraction. Studying the dMRI signal arising from a single neuron can provide insight into how the geometrical structure of neurons influences the measured signal.

Neurons are made of a neuronal body called the soma to which are attached long protrusions called neurites (axons and dendrites). The size of the soma is on the order of $10\mu$m, the diameter of the dendrite segments can range from a few $\mu$m to less than half a $\mu$m, and the total length of all the dendrite segments is on the order of several mm [2]. The Bloch-Torrey partial differential equation (PDE) describes the complex transverse water proton magnetization due to diffusion-encoding magnetic field gradient pulses. The integral of the solution of this PDE yields the diffusion magnetic resonance imaging (dMRI) signal [3, 4, 5]. In a first step, we focus on the dMRI signal arising from the interior of a single neuron, without considering the water exchange between neurons and the extracellular space.

## 2. Theory

Let $\Omega \subset \mathbb{R}^3$ be the spatial domain that defines the geometry of a neuron. We denote the time profile of the diffusion-encoding magnetic field gradient by $f(t)$ and its linear spatial dependence by $\mathcal{G}(\mathbf{r}) = \mathbf{g} \cdot \mathbf{r}$, where the vector $\mathbf{g}$ contains the amplitude and direction information of the magnetic field gradient. The water proton magnetization $M(\mathbf{r}, t)$ satisfies the Bloch-Torrey PDE:

$$\frac{\partial}{\partial t} M(\mathbf{r}, t) = -I\gamma\, f(t)\, \mathcal{G}(\mathbf{r})\, M(\mathbf{r}, t) + \nabla \cdot (D^0 \nabla M(\mathbf{r}, t)), \quad \mathbf{r} \in \Omega, \tag{1}$$

where $\gamma = 2.67513 \times 10^8\,\mathrm{rad\,s^{-1}T^{-1}}$ is the gyromagnetic ratio of the water proton, $I$ is the imaginary unit, and $D^0$ the intrinsic diffusion coefficient. The magnetization $M(\mathbf{r}, t)$ is a function of position $\mathbf{r}$ and time $t$, and depends on the diffusion gradient vector $\mathbf{g}$ and the time profile $f(t)$. We will solve the Bloch-Torrey PDE for the Pulsed Gradient Spin Echo (PGSE) sequence [6], consisting of two rectangular pulses of duration $\delta$, separated by a time interval $\Delta - \delta$ (Fig. 1).

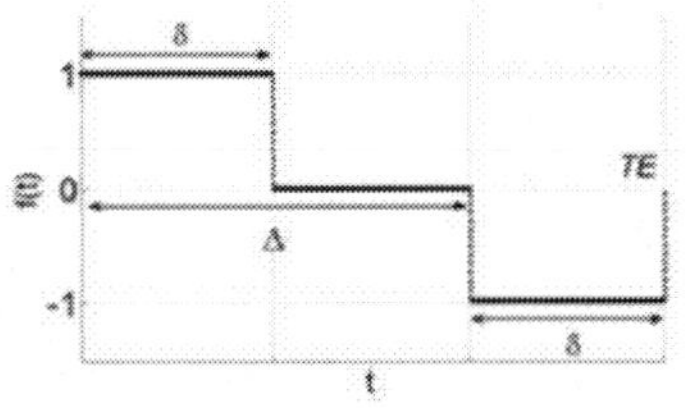


Figure 1: The temporal profile $f(t)$ of a standard PGSE sequence with two rectangular gradient pulses.

For the signal arising only from the neuron, without water exchange with the extracellular space, the boundary condition is:

$$\nabla M(\mathbf{r}, t) \cdot \mathbf{n}(\mathbf{r}) = 0, \quad \mathbf{r} \in \partial\Omega, \tag{2}$$

where $\mathbf{n}(\mathbf{r})$ is the normal to the boundary $\partial\Omega$ at $\mathbf{r}$. Assuming a uniform excitation of the magnetization in the imaging voxel, the initial condition is:

$$M(\mathbf{r}, 0) = 1. \tag{3}$$

The dMRI signal is measured at echo time $t = TE \geq \Delta + \delta$ and is given as the integral of $M(\mathbf{r}, TE)$, normalized by the mass:

$$S^{3D}(b) := \frac{1}{|\Omega|} \int_{\mathbf{r} \in \Omega} M(\mathbf{r}, TE)\, d\mathbf{r},$$

where the $b$-value, $b(\mathbf{g}) = \gamma^2 |\mathbf{g}|^2 \delta^2 (\Delta - \delta/3)$, is used when $\mathbf{g}$ varies only in amplitude (while staying in the same direction). In a homogeneous medium, the signal attenuation is simply $e^{-D^0 b}$.

The model described above will be called the full 3D model.

### 3. The 3D+1D model

We assume the neuron is made up of two sub-domains: the soma $\mathcal{S}$ and the neurites $\mathcal{N}$, such that $\Omega = \mathcal{S} \bigcup \mathcal{N}$. To simplify the presentation, we consider that the neurites attached to the soma are the union of straight cylinder segments: $\mathcal{N} = \cup_k \mathcal{B}_k$, where $\mathcal{B}_k = \mathcal{T}_k \times \mathcal{O}_k$, $\mathcal{T}_k$ is a 1D line segment of length $l_k$, parallel to the unit vector $\mathbf{w}_k$, and $\mathcal{O}_k$ is the disk of radius $r$ perpendicular to $\mathbf{w}_k$. Figure 2a shows an example of a neuron consisting of a spherical soma attached to two dendrite trees.

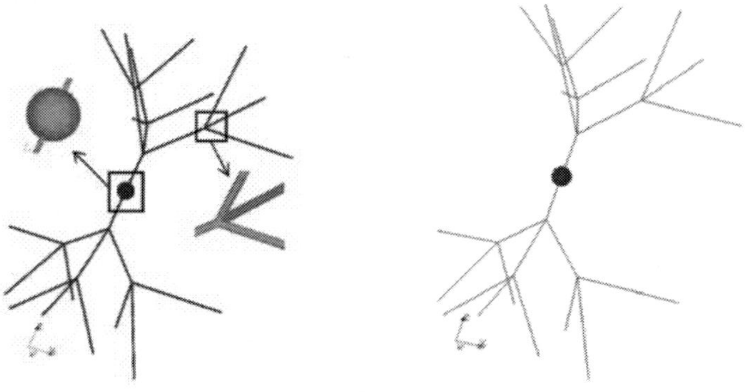

(a) (b)

Figure 2: Full 3D neuron (2a) and its 3D+1D model (2b).

We now formulate an approximation of the full 3D model (Eqs. (1, 2)) where we replace the 3D PDE in the neurites by a 1D PDE on the union of the *linked* line segments $\mathcal{N}^{1D} = \cup_k \mathcal{T}_k$. Let $\mathcal{T}_k = \mathbf{p}_k + s\mathbf{w}_k$, $s \in [0, l_k]$, be the natural parameterization of the line segment $\mathcal{T}_k$, $\mathbf{p}_k$ being one endpoint of the segment. We define the 1D magnetization on the segment $\mathcal{T}_k$ as the average value of the magnetization over $\mathcal{O}_k$:

$$M_k^{1D}(s,t) := \frac{1}{|\mathcal{O}_k|} \int_{(\mathbf{p}_k + s\mathbf{w}_k) \times \mathcal{O}_k} M(\mathbf{r}, t) d\mathbf{r}.$$

It is easy to show that $M_k^{1D}(s,t)$ satisfies the 1D Bloch-Torrey PDE:

$$\frac{\partial}{\partial t} M_k^{1D}(s,t) = -I\gamma\, f(t)\, \mathcal{G}(\mathbf{p}_k + s\mathbf{w}_k)\, M_k^{1D}(s,t) + D^0 \frac{\partial^2}{\partial s^2} M_k^{1D}(s,t). \tag{4}$$

Two conditions have to be imposed on any intersection point $\mathbf{r}$ of several line segments, $\mathbf{r} = \cap_{k\in\mathcal{K}} \mathcal{T}_k$. The first condition is the continuity of the magnetization:

$$M_k^{1D}(\mathbf{r},t) = M_{k'}^{1D}(\mathbf{r},t),\; k, k' \in \mathcal{K}. \tag{5}$$

The second condition is the Kirchhoff law ensuring the conservation of flux:

$$\sum_{k\in\mathcal{K}} D^0 \frac{\partial}{\partial s'} M_k^{1D}(\mathbf{r},t) = 0, \tag{6}$$

where $s' = s$ if $\mathbf{r} = \mathbf{p}_k$ is the starting point of the segment, $s' = -s$ otherwise.

Then we formulate the interface conditions linking the soma to the neurites. Suppose $\mathcal{I}$ is the intersection between the soma $\mathcal{S}$ and the neurite segment $\mathcal{B}_k$ (there may be several such segments that extend out from the soma). Without loss of generality, we assume the parameterization variable $s = 0$ at the intersection is the starting point of the line segment $\mathcal{T}_k$, then it can be shown that the following conditions hold:

$$\frac{1}{|\mathcal{I}|} \int_{\mathcal{I}} M_{\mathcal{S}}(\mathbf{r},t) d\mathbf{r} = M_k^{1D}(0,t), \tag{7}$$

and

$$\frac{1}{|\mathcal{I}|} \int_{\mathcal{I}} D^0 \nabla M_{\mathcal{S}}(\mathbf{r},t) \cdot \mathbf{w}_k = D^0 \frac{\partial}{\partial s} M_k^{1D}(0,t), \tag{8}$$

where $\mathbf{w}_k$, parallel to $\mathcal{T}_k$, is also the normal vector pointing outward from the soma at $\mathcal{I}$. From Eqs. (3, 7), we can deduce that the initial condition on the 1D neurites should be $M_k^{1D}(s,0) = 1$. We call the model described above the 3D+1D model.

In the 3D+1D model, the dMRI signal is obtained from the 3D solution in the soma, $M_{\mathcal{S}}$, and the 1D solution in the neurites, $M_k^{1D}$, $k = 1, 2, \ldots$

$$S^{3D+1D}(b) := \frac{1}{|\mathcal{S}| + \pi r^2 \sum_k l_k} \left( \int_{\mathcal{S}} M_{\mathcal{S}}(\mathbf{r}, TE) d\mathbf{r} + \pi r^2 \sum_k \int_0^{l_k} M_k^{1D}(s, TE)\, ds \right).$$

We emphasize that the solutions in the soma and in each of the line segments are *linked* to each other via the interface conditions of the PDE formulation.

## 4. Numerical results

The dMRI signal of the 3D+1D model was obtained and compared to that of the full 3D model. Both were computed using the code described in [7] on a Dell Precision M4700 laptop (Intel(R) Core(TM) i7-3740QM CPU @ 2.70GHz). We set the PGSE sequence to have a fixed pulse duration $\delta = 2.5$ms and varied $\Delta$. The dMRI signal at

8 $b$-values, $b = 0, 100, 500, 1000, 1500, 2000, 2500, 3000\text{s/mm}^2$, was computed. The same intrinsic diffusion coefficient $D^0 = 3 \cdot 10^{-3}\text{mm}^2/\text{s}$ was used for all the simulations. We constructed a sample neuron geometry (shown in Fig. 2a) consisting of a spherical soma and two dendrite trees. We set the soma radius to $R = 10\mu\text{m}$ and the neurite radius to $r = 1\mu\text{m}$ [2]. The length of the dendrite segments varies from $50\mu\text{m}$ to $112.5\mu\text{m}$ and the total length of the two dendrite trees is $2272\mu\text{m}$. This neuron comprises of 63% neurites and 37% soma by volume.

The 3D simulation in the full neuron took about 30 minutes on a finite elements mesh with 11314 vertices, where the mesh in the soma contains 1666 vertices. The simulation of the 3D+1D model (while keeping the same 3D mesh in the soma) took about 5 minutes.

We show the dMRI signals of the full 3D model and the 3D+1D model in Fig. 3a for $\Delta = 2.5\text{ms}$ and in Fig. 3b for $\Delta = 100\text{ms}$, in the gradient direction $\mathbf{g}/|\mathbf{g}| = (1,1,1)/\sqrt{3}$. We computed the relative difference in the $L^2$ norm between the signals to be 7% at the shorter diffusion time ($\Delta = 2.5\,\text{ms}$) and 1% at the longer diffusion time ($\Delta = 100\,\text{ms}$). As expected, the 1D approximation on the neurites works better when the transverse diffusion is negligible. For a given neurite radius, the longer the diffusion time, the better the approximation of the full 3D model by the 3D+1D model (compare Fig. 3a and Fig. 3b). For a given diffusion time, the approximation is better at lower gradient strengths: at $\Delta = 100\text{ms}$, the two signals look indistinguishable in Fig. 3b, where only $b \in [0, 3000]\,\text{s/mm}^2$ is shown. However, at very high b-values, $b \in [0, 20000]\,\text{s/mm}^2$, shown in Fig. 3c, some differences in the signals become visible. When the neurite radius is larger, the transverse diffusion is more significant than when the neurite radius is smaller. So when the neurite radius is increased to $r = 2\mu\text{m}$, the difference in the signals is larger than when $r = 1\mu\text{m}$ (see Fig. 3d). The difference between the dMRI signals of the full 3D model and the 3D+1D model is 28% at $\Delta = 2.5\text{ms}$ and 3% at $\Delta = 100\text{ms}$ (compare to 7% and 1% for $r = 1\mu\text{m}$). It is worth noting that the transverse diffusion in cylinders can be analytically included into the dMRI signal because many analytical results exist for the dMRI signal inside impermeable disks (see [5] and references therein).

## 5. Conclusion

We proposed a 3D+1D model to compute the dMRI signal arising from a single neuron by replacing the 3D simulation in the full neuron by a 3D simulation in the soma and 1D simulations in the neurites and linking them by the appropriate interface conditions. We showed that the 3D+1D model is computationally much faster to simulate than the full 3D model and still gives accurate results under a wide range of parameters. The 3D+1D model is accurate when the contribution to the dMRI signal from transverse diffusion in the neurites is negligible, i.e., at longer diffusion times, smaller neurite radii, and lower gradient strengths. In addition to computational advantages, the 3D+1D model, being more amenable to theoretical analysis due to the existence of (semi-)analytical results about diffusion in line segments, disks, and spheres, may be potentially exploited to produce semi-analytical expressions for the dMRI signal in neurons.

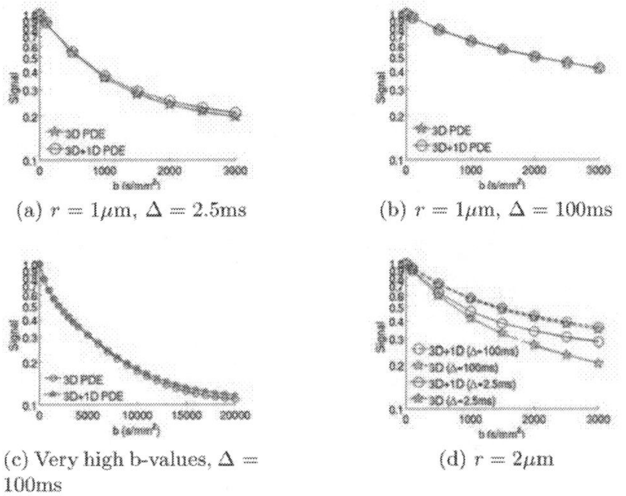


(a) $r = 1\mu$m, $\Delta = 2.5$ms (b) $r = 1\mu$m, $\Delta = 100$ms

(c) Very high b-values, $\Delta =$ 100ms (d) $r = 2\mu$m

Figure 3: The dMRI signals inside the neuron shown on Fig. 2a in the gradient direction $\mathbf{g}/|\mathbf{g}| = (1,1,1)/\sqrt{3}$ at two diffusion times: $\Delta = 2.5$ms (3a) and $\Delta = 100$ms (3b) for neurites with radius $r = 1\mu$m, at very high b-values (3c), and for neurites with radius $r = 2\mu$m (3d)


## Acknowledgments

This work is funded by the ANR (Agence Nationale de la Recherche), project 'SIMUDMRI'.


## References


[1] D. Le Bihan and H. Johansen-Berg, "Diffusion mri at 25: Exploring brain tissue structure and function," *NeuroImage*, vol. 61, pp. 324–341, June 2012.

[2] N. Spruston, "Pyramidal neurons: dendritic structure and synaptic integration," *Nat Rev Neurosci*, vol. 9, pp. 206–221, Mar 2008.

[3] H. Torrey, "Bloch equations with diffusion terms," *Physical Review Online Archive (Prola)*, vol. 104, no. 3, pp. 563–565, 1956.

[4] W. S. Price, A. V. Barzykin, K. Hayamizu, and M. Tachiya, "A model for diffusive transport through a spherical interface probed by pulsed-field gradient nmr," *Biophysical Journal*, vol. 74, pp. 2259–2271, May 1998.

[5] D. Grebenkov, "Nmr survey of reflected brownian motion," *Reviews of Modern Physics*, vol. 79, no. 3, pp. 1077–1137, 2007.

[6] E. O. Stejskal and J. E. Tanner, "Spin diffusion measurements: Spin echoes in the presence of a time-dependent field gradient," *The Journal of Chemical Physics*, vol. 42, no. 1, pp. 288–292, 1965.

[7] D. V. Nguyen, D. Grebenkov, and J.-R. Li, "An efficient finite-elements code to simulate diffusion mri signals in complex tissue geometries." submitted, see http://www.cmap.polytechnique.fr/ jingrebeccali/preprints.html, 2013.

# Quantization of the canonical tensor model and an exact wave function

**Naoki Sasakura**

Yukawa Institute for Theoretical Physics, Kyoto University, Kitashirakawa, Sakyo-ku, Kyoto 606-8502, Japan

E-mail: sasakura@yukawa.kyoto-u.ac.jp

**Abstract.** Tensor models in various forms are being studied as models of quantum gravity. Among them the canonical tensor model has a canonical pair of rank-three tensors as dynamical variables, and is a pure constraint system with first-class constraints. The Poisson algebra of the first-class constraints provides an algebraically consistent way of discretizing the Dirac algebra for general relativity. This paper successfully formulates the Wheeler-DeWitt quantization of the canonical tensor model. Formally one can obtain wave functions of the "universe" by solving the partial differential equations representing the constraints. For the simplest non-trivial case, the unique wave function is exactly and globally obtained. Although this case is far from being realistic, the wave function is physically interesting; locality is favored, and there exists a locus of configurations with features of the beginning of the universe.

## 1. Introduction

Tensor models [1] were first proposed to describe simplicial quantum gravity in dimensions higher than two, with the hope to extend the success of matrix models for the two dimensional case. However, because of some unfavored features, these original tensor models have been unsuccessful so far. On the other hand, the difficulties have been overcome by the advent of the colored tensor models [2], and various interesting analytical results have been derived [3]. Yet, since the dominant simplicial manifolds in colored tensor models are branched polymers [4], a serious open question remains on how wide-spread spaces like our universe can be realized in tensor models. A resolution may be obtained by considering time direction more seriously. In fact, the Causal Dynamical Triangulation (CDT), which is simplicial quantum gravity with a causal time direction, has numerically been shown to generate wide-spread simplicial spaces [5], while the above developments of colored tensor models are in Euclidean signature.

Being motivated by the above consideration, the author of the present paper has proposed a tensor model in the canonical formalism, dubbed canonical tensor model [6, 7, 8]. It has a canonical pair of rank-three tensors as its dynamical variables, and is defined as a pure first-class constraint system. The first-class constraint Poisson algebra guarantees the consistency of "many-fingered time" evolutions, and the fact that it has structure functions makes the dynamics non-trivial. In view of these features, the canonical tensor model is very similar to general relativity, and in fact it is possible to derive the Dirac constraint algebra [9] for general relativity from the constraint algebra of the canonical tensor model by taking a formal locality limit [6]. Since the Dirac constraint algebra plays a major role in geometrodynamics, it would be a highly interesting possibility that the canonical tensor model reproduces general relativity in a certain, yet unknown, infrared limit. For this to be seen, first of all, a space-like object must be dynamically generated from the canonical tensor model, and more specifically the aforesaid formal locality limit must be derived from dynamics. The purpose of the present paper is to take a first step in this direction by formulating the quantization of the canonical tensor model, and obtaining a wave function of the "universe" for the simplest case [10].

## 2. Quantization of the canonical tensor model

This paper deals with the Wheeler-DeWitt quantization of the minimal canonical tensor model [10], the classical form of which has been introduced in [8]. The dynamical variables of the model are a canonical pair of rank-three tensors, $M_{abc}, P_{abc}$ $(a, b, c = 1, 2, \ldots, N)$. In the quantization, they are lifted to Heisenberg operators as

$$[\hat{M}_{abc}, \hat{P}_{def}] = i \sum_{\sigma} \delta_{a\,\sigma(d)} \delta_{b\,\sigma(e)} \delta_{c\,\sigma(f)}, \quad [\hat{M}_{abc}, \hat{M}_{def}] = [\hat{P}_{abc}, \hat{P}_{def}] = 0, \tag{1}$$

where the summation over $\sigma$ is for all the permutations of $def$. The operators are assumed to satisfy the properties corresponding to the classical case,

$$\hat{M}_{abc} = \hat{M}^{\dagger}_{abc} = \hat{M}_{bca} = \hat{M}_{cab} = \hat{M}_{bac} = \hat{M}_{acb} = \hat{M}_{cba}, \tag{2}$$

$$\hat{P}_{abc} = \hat{P}^{\dagger}_{abc} = \hat{P}_{bca} = \hat{P}_{cab} = \hat{P}_{bac} = \hat{P}_{acb} = \hat{P}_{cba}, \tag{3}$$

where $^{\dagger}$ denotes the Hermitian conjugate of operators.

The model has three kinds of constraints, $\hat{\mathcal{J}}_{[ab]}$, $\hat{\mathcal{D}}$, $\hat{\mathcal{H}}_a$. The operator ordering in the constraints can be determined by imposing the Hermiticity and the first-class closure of the constraint algebra [10]. The results are

$$\hat{\mathcal{J}}_{[ab]} = \frac{1}{4} \left( \hat{M}_{bcd} \hat{P}_{acd} - \hat{M}_{acd} \hat{P}_{bcd} \right), \tag{4}$$

$$\hat{\mathcal{D}} = -\frac{1}{6} \left( \hat{M}_{abc} \hat{P}_{abc} - i \frac{N(N+1)(N+2)}{2} \right), \tag{5}$$

$$\hat{\mathcal{H}}_a = \frac{1}{2} \left( \hat{M}_{abc} \hat{M}_{bde} \hat{P}_{cde} - i \frac{(N+2)(N+3)}{2} \hat{M}_{abb} \right). \tag{6}$$

These constraint operators form a first-class constraint algebra given by

$$[\hat{H}(T_1), \hat{H}(T_2)] = i\hat{J}([\hat{T}_1, \hat{T}_2]), \tag{7}$$

$$[\hat{J}(V), \hat{H}(T)] = i\hat{H}(VT), \tag{8}$$

$$[\hat{J}(V_1), \hat{J}(V_2)] = i\hat{J}([V_1, V_2]), \tag{9}$$

$$[\hat{\mathcal{D}}, \hat{H}(T)] = i\hat{H}(T), \tag{10}$$

$$[\hat{\mathcal{D}}, \hat{J}(V)] = 0, \tag{11}$$

where

$$\hat{H}(T) = T_a \hat{\mathcal{H}}_a, \tag{12}$$

$$\hat{J}(V) = V_{[ab]} \hat{\mathcal{J}}_{[ab]}, \tag{13}$$

$$\hat{J}(\hat{V}) = \hat{V}_{[ab]} \hat{\mathcal{J}}_{[ab]}, \tag{14}$$

$$\hat{T}_{bc} = T_a \hat{M}_{abc}. \tag{15}$$

Here, operators are distinguished from classical variables by $\hat{}$.

The above first-class algebraic closure of the quantum constraints guarantees the existence (at least locally) of the Wheeler-DeWitt wave functions $\Psi$, which are solutions to the Wheeler-DeWitt equations,

$$\hat{\mathcal{H}}_a \Psi = \hat{\mathcal{J}}_{[ab]} \Psi = \hat{\mathcal{D}} \Psi = 0, \tag{16}$$

in appropriate representations of operators.

## 3. The exact wave function for $N = 2$

This section solves the Wheeler-DeWitt equations (16) explicitly for the simplest non-trivial case. Consider a representation of the operators in terms of $M$ as

$$\Psi = \Psi(M), \tag{17}$$

$$\hat{M}_{abc} = M_{abc}, \tag{18}$$

$$\hat{P}_{abc} = -i\Delta(abc)\frac{\partial}{\partial M_{abc}}, \tag{19}$$

where $\Delta(abc)$ is a multiplicity factor defined by

$$\Delta(abc) = \begin{cases} 6 \text{ for } a = b = c, \\ 2 \text{ for } a = b \neq c, b = c \neq a, c = a \neq b, \\ 1 \text{ for } a \neq b, b \neq c, c \neq a. \end{cases} \tag{20}$$

This factor is needed to consistently take account of the properties (1) and (2). Then the Wheeler-DeWitt equations (16) are a set of first-order partial differential equations for $\Psi(M)$.

The total number of the first-order partial differential equations in (16) is given by $N + N(N-1)/2 + 1 = (N^2 + N + 2)/2$, while the number of degrees of freedom of $M_{abc}$ is $N(N+1)(N+2)/6$. Thus, for $N = 2$, the Wheeler-DeWitt equations have a unique solution, granted that it is also globally consistent. It is elementary to obtain the solution, and the result is [10]

$$\Psi(M) = c_0 \frac{\sqrt{\epsilon_{ac}\epsilon_{bd}\epsilon_{eg}\epsilon_{fh}\epsilon_{e'g'}\epsilon_{f'h'} M_{aef} M_{bgh} M_{ce'f'} M_{dg'h'}}}{M_{acd}M_{bde}M_{bef}M_{afc} - M_{acd}M_{bde}M_{aef}M_{bfc}} \tag{21}$$

with a numerical constant $c_0$.

Since the wave function (21) has infinite peaks where the denominator vanishes, there exists a kind of preference for such configurations. One can prove that, by using the kinematical $SO(N)$ transformations generated by $\hat{\mathcal{J}}_{[ab]}$, such configurations can always be transformed to the diagonal form,

$$M_{abc} = m_a \delta_{ab} \delta_{ac}, \tag{22}$$

with real $m_a$.

The canonical tensor model can be interpreted as describing the Hamiltonian dynamics of fuzzy spaces [11]. In this interpretation, $M_{abc}$ corresponds to the structure constants defining a fuzzy space characterized by an algebra,

$$f_a \star f_b = M_{abc} f_c, \tag{23}$$

among functions $f_a$ ($a = 1, 2, \cdots, N$) on a fuzzy space. Physically speaking, functions can be regarded as fuzzy "points" (in an appropriate basis of functions), and in the case of the diagonal form (22), the "points" are mutually independent, because

$$f_a \star f_b = m_a \delta_{ab} f_a. \tag{24}$$

Thus, the divergent configurations of the wave function (21) correspond to fuzzy spaces in which locality is maximized.

The configurations where the wave function vanishes can also be characterized as follows. When the numerator of (21) vanishes, one can show that there exists a real vector $v_a$ such that $M_{abc}v_b v_c$ vanishes. In this sense, the configuration $M$ is degenerate at the locus, which would agree with the interpretation that the locus is the beginning of the "universe", since our universe should have started from a point-like state (or a very small state).

## 4. Summary and discussion

In this paper, the Wheeler-DeWitt quantization of the minimal canonical tensor model has successfully been formulated. The classical constraints have been lifted to quantum ones, and their commutation algebra has been shown to be first-class. These constraints form a consistent set of Wheeler-DeWitt equations, and the wave function of the "universe" can be obtained by solving them. Indeed, the unique wave function for $N = 2$ has explicitly been obtained and its physical interpretation has been discussed. Although the case $N = 2$ is far from being realistic, the wave function shows physically interesting features such as that locality is favored, and that there exists a locus of configurations which have characteristics of the beginning of the universe.

An obvious question for future study is whether the physically interesting properties of the wave function found for $N = 2$ can be generalized for $N \geq 3$ or not. In particular, the emergence of locality is essentially important for the constraint algebra of the canonical tensor model to be identified with the Dirac or the hypersurface deformation algebra of general relativity.

## References


[1] J. Ambjorn, B. Durhuus and T. Jonsson, "Three-Dimensional Simplicial Quantum Gravity And Generalized Matrix Models," Mod. Phys. Lett. A **6**, 1133 (1991); N. Sasakura, "Tensor Model For Gravity And Orientability Of Manifold," Mod. Phys. Lett. A **6**, 2613 (1991); N. Godfrey and M. Gross, "Simplicial Quantum Gravity In More Than Two-Dimensions," Phys. Rev. D **43**, 1749 (1991).

[2] R. Gurau, "Colored Group Field Theory," Commun. Math. Phys. **304**, 69 (2011) [arXiv:0907.2582 [hep-th]].

[3] R. Gurau and J. P. Ryan, "Colored Tensor Models - a review," SIGMA **8**, 020 (2012) [arXiv:1109.4812 [hep-th]].

[4] R. Gurau and J. P. Ryan, "Melons are branched polymers," arXiv:1302.4386 [math-ph].

[5] J. Ambjorn, A. Goerlich, J. Jurkiewicz and R. Loll, "Quantum Gravity via Causal Dynamical Triangulations," arXiv:1302.2173 [hep-th].

[6] N. Sasakura, "Canonical tensor models with local time," Int. J. Mod. Phys. A **27**, 1250020 (2012) [arXiv:1111.2790 [hep-th]].

[7] N. Sasakura, "Uniqueness of canonical tensor model with local time," Int. J. Mod. Phys. A **27**, 1250096 (2012) [arXiv:1203.0421 [hep-th]].

[8] N. Sasakura, "A canonical rank-three tensor model with a scaling constraint," Int. J. Mod. Phys. A **28**, 1350030 (2013) [arXiv:1302.1656 [hep-th]].

[9] R. L. Arnowitt, S. Deser and C. W. Misner, "The Dynamics of general relativity," gr-qc/0405109; P. A. M. Dirac, "The Theory of gravitation in Hamiltonian form," Proc. Roy. Soc. Lond. A **246**, 333 (1958); B. S. DeWitt, "Quantum Theory of Gravity. 1. The Canonical Theory," Phys. Rev. **160**, 1113 (1967); S. A. Hojman, K. Kuchar and C. Teitelboim, "Geometrodynamics Regained," Annals Phys. **96**, 88 (1976).

[10] N. Sasakura, "Quantum canonical tensor model and an exact wave function," Int. J. Mod. Phys. A **28**, 1350111 (2013) [arXiv:1305.6389 [hep-th]].

[11] N. Sasakura, "Tensor models and 3-ary algebras," J. Math. Phys. **52**, 103510 (2011) [arXiv:1104.1463 [hep-th]]; N. Sasakura, "An Invariant approach to dynamical fuzzy spaces with a three-index variable," Mod. Phys. Lett. A **21**, 1017 (2006) [hep-th/0506192].

# Towards a Stochastic Cellular Automata Model of Log Wood Combustion

**Klaus Lichtenegger**[1], **Wilhelm Schappacher**[2], **Babette Hebenstreit**[1,3], **Christoph Schmidl**[1], **Ernst Höftberger**[1]

[1]Bioenergy 2020+ GmbH, Gewerbepark Haag 3, A-3250 Wieselburg-Land,
[2]Institute of Mathematics and Scientific Computing, Karl-Franzens-Universität, A-8010 Graz
[3]Energy Engineering, Department of Engineering Sciences and Mathematics, Luleå University of Technology, 971 87 Luleå, Sweden

E-mail: `klaus.lichtenegger@bioenergy2020.eu`

**Abstract.** Describing the combustion of log wood and others solid fuels with complex geometry, considerable water content and often heterogenous struture is a nontrivial task. Stochastic Cellular Automata models offer a promising approach for modelling such processes. Combustion models of this type exhibit several similarities to the well-known forest fire models, but there are also significant differences between those two types of models. These differences call for a detailed analysis and the development of supplementary modeling approaches. In this article we define a qualitative two-dimensional model of burning log wood, discuss the most important differences to classical forest fire models and present some preliminary results.

**Keywords:** cellular automata, stochastic modeling, combustion

## 1. Introduction: The Role of Log Wood

Biomass is an important source of renewable energy; the efficient use of biomass will be essential to achieve the transition to a lifestyle based on renewable resources. In order to reach the European 20-20-20 goals [1], about half of the demand for renewable energy will have to be covered by biomass. More than half of this amount will presumably be used in domestic small-scale combustion systems.

While pellet technology – promising better control and lower emissions than traditional wood combustions – is on the rise, log wood is still an important fuel and will presumably remain to be so for decades to come. A main advantage of log wood is an economic one: Since less processing is required, log wood has a lower price than pellets. Also stoves tend to be cheaper. At the same time, less volume is needed for storing wood logs than for storing wood chips, an even cheaper renewable fuel. Also the handling of wood logs is easier than that of wood chips.

The combustion of wood logs, however, is a complex process which involves transport processes, chemical reactions and heat transfer processes within complex geometry. There has been put quite some effort in the development of models for combustion of woody particles, in particular shrinking-core and reacting-core models [2, 3, 4, 5] and CFD-related approaches based on partial differential equations [6, 7]. These models grasp some important aspects of wood combustion, but they cannot (yet) incorporate the often complex geometry of wood logs. Complementary modeling approches, as the one presented in this article, may help to improve this situation.

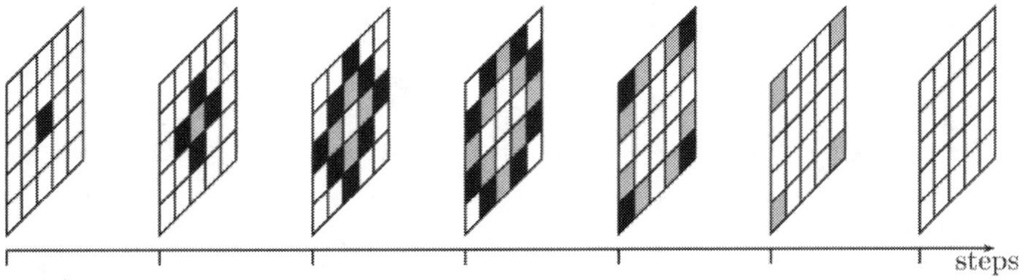


**Figure 1.** Sketch of the Cellular Automata approach on a [5 × 5]-grid for seven time steps.

## 2. Cellular Automata: Forest Fire Models and Beyond

Cellular Automata (CA) are recognized as an important modeling tool in many fields of science. In the CA approach, space is defined as an $N$-dimensional grid (with $N = 1$ or $N = 2$ being the most popular choices) of cells. Time is defined as a sequence of (equidistant) time steps.

The cell states are usually chosen from a limited set of states. Time evolution is described by simple but universal update rules, see fig. 1. These rules typically only make use of the current cell state and the states of cells in a small neighbourhood. This results in a limited speed of the transport of information, naturally implementing the principle of locality.

A particularly famous class of CA-based models are forest fire models, which can be formulated as deterministic [8, 9, 10] or stochastic models [11]. These models may exhibit a number of fascinating properties, including critical behaviour and self-organized criticality. With suitable re-interpretation, these models can also be applied to disease spreading and other propagation phenomena.

At first glance, such forest fire models may seem well-suited to describe, at least qualitatively, also the combustion of solid fuels. At second glance, however, there are some striking differences between those systems:

- Oxygen supply is rarely a limiting factor for forest fires. As a consequence, fire propagation is most efficient for a high density of "forest" cells. In contrast to that, oxygen supply is critical in the log wood case, in particular in the ignition phase. A large surface-to-volume ratio is beneficial for ignition; a "fragmented" piece of wood has a much higher chance of catching fire than a solid block. Without sufficient oxygen supply, one may have charring, but not burning of the char.
- Forest fires are typically modeled via nearest-neighbour interactions. This is reasonable since a cell usually describes a patch of vegetation of considerable size, and it is known that firebreaks can indeed be an efficient way to stop the propagation of forest fires. (Still, in situations with strong wind, flying sparks can lead to fire propagation which defies a description with nearest-neighbour interaction.)
  For wood combustion, radiation effect can be important, and ignition does not necessarily depend on direct contact. Thus the local nearest-neigbour-approach will have to be supplemented by an additional mechanism for fire propagation.

## 3. A CA Model for Log Wood Combustion

Wood combustion is a complex process which is composed of heat-up, drying, pyrolysis and char burnout. A detailed modeling of the physical and chemical processes tends to become quite involved [12]. In this article we aim at a simple effective model which still yields a reasonable qualitative description of the key processes.

| type of celll | index $k$ | color | $E^{\rm trans}_{{\rm min},k}$ | $E^{\rm released}_{k\to k+1}$ | $p^{\rm trans}_{k\to k+1}$ | $R^{\rm oxy}_{k\to k+1}$ | color mod. by energy |
|---|---|---|---|---|---|---|---|
| boundary | -2 | | - | - | - | - | |
| stone | -1 | | - | - | - | - | |
| air/empty | 0 | | - | - | - | - | |
| wet wood | 1 | | 1 | $-1$ | 1/2 | - | |
| drying wood | 2 | | 1 | $-1$ | 1/2 | - | |
| drying wood | 3 | | 1 | $-1$ | 1/2 | - | |
| dry wood | 4 | | 2 | $-1$ | 1/2 | - | |
| char | 5 | | 2 | 2 | 1 | - | |
| char | 6 | | 2 | 2 | 1 | - | |
| char | 7 | | 3 | 2 | 1 | $\surd$ | |
| burning char | 8 | | 0 | 4 | 1 | - | |
| ash | 9 | | 0 | 0 | 1/5 | $\surd$ | |

**Table 1.** Characteristics of different cell states $w_{i,j}$. A state is essentially defined by its transition properties: the minimum energy $E^{\rm trans}_{{\rm min},i}$ required for a transition, the net energy $E^{\rm released}_{k\to k+1}$ released in the transition, the transition probability $p^{\rm trans}_{k\to k+1}$ and a flag $R^{\rm oxy}_{k\to k+1}$ to indicate whether oxygen or free space (in this model a neighbour cell with $k = 0$) is required for the transition. Note that in this scheme, $10 \equiv 0$, i.e. ash is converted to air/empty space. In addition we show the color code for combined graphical representations of $\boldsymbol{W}$ and $\boldsymbol{E}$, as the one shown in fig. 2.

The present model is formulated in terms of two matrices: $\boldsymbol{W}$ encodes the status of the woody fuel, $w_{i,j} \in \{-2, -1, 0, \ldots, 8, 9\}$. $\boldsymbol{E}$ encode the density of discrete "energy packages" which describe in a combined way excess heat and temperature level, $e_{i,j} \in \{0, 1, 2, \ldots\}$. Thermal conversion of fuel is described by a transition $w_{i,j} = k \to k+1$ which changes the energy content $e_{i,j} = n \to n + E^{\rm released}_{k\to k+1}$. A specific transition $k \to k+1$ may require a minimum energy content $E^{\rm trans}_{{\rm min},k}$, may require an empty neighbour cell (if oxygen or empty space is needed for the underlying process) and may occur only with a certain probability $p^{\rm trans}_{k\to k+1}$. The list of cell states is given in tab. 1.

Most of the parameters are rough estimates at the present stage, but a fundamental requirement for the model to make sense is the condition $\sum_{k=1}^{9} E^{\rm released}_{k\to k+1} > 0$. This is necessary in order to describe an exothermic reaction which can sustain itself.

The energy packages are subject to a biased random walk, with the probability to move upwards (0.17) being slightly larger than the probability to move left or right (0.15 each) and noticebly larger than the probability to move downwards (0.13). Ignition is initiated by an external energy input at one corner of the grid during the first few timesteps.

The model also contains an extremely simplified description of gravity. If a fuel structure is not connected to stone ($w_{i,j} = -1$), the whole structure moves downwards by one spatial step per timestep. Any effects of torque or other stability issues are not yet included.

A snapshot of a sample simulation run is shown in fig. 2, the net energy release during such a run in fig. 3.

**Acknowledgements**
The model has been developed in the project FLOWS-II at the competence center Bioenergy 2020+. The competence center Bioenergy 2020+ is supported in the framework of *COMET – Competence Centers for Excellent Technologies* by BMVIT, BMWFJ and the provinces of Burgenland, Niederösterreich and Steiermark. The program is operated by the FFG.

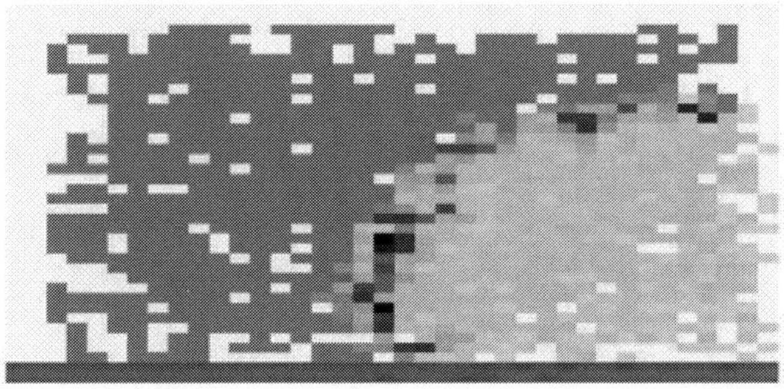

**Figure 2.** Snapshot of a typical simulation run on a $[40 \times 40]$-grid in a combined representation of $\boldsymbol{W}$ and $\boldsymbol{E}$. The fire has been ignited by external energy input in the lower left corner and now spreads, creating a zone of dried and charred wood. The color code is defined in tab. 1

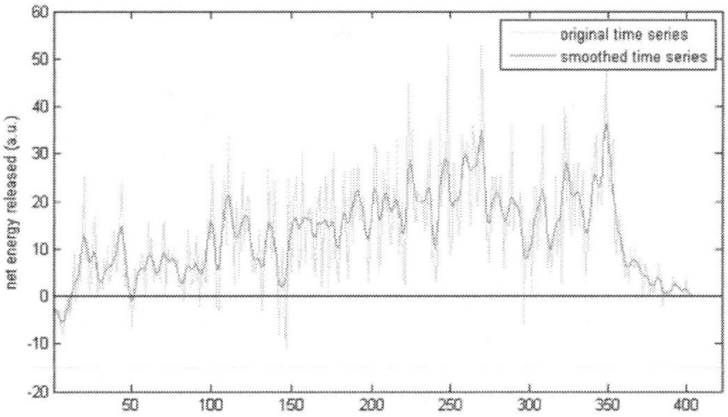


**Figure 3.** Net energy released in drying and conversion processes for a typical simulation run on a $[40 \times 40]$-grid. Since short-time fluctuations are large, the time series has been smoothed by weighted averaging over consecutive values. Note that the energy release is negative in the first phase (when external energy input is required for ignition) and that even later it can approach and even drop below zero. Such events may indicate critical phases in the combustion process where the fire is at risk of being extinguished.

## References


[1] Brussels European Council 2007 March 8/9 *Presidency Conclusions*, see `http://ec.europa.eu/clima/policies/package/index_en.htm` for details and supplementary documentation
[2] A. Quedraogo, J. C. Mulligan, J. G. Cleland, Comb. & Flame 114 (1998) pp. 1-12
[3] B. Peters, C. Bruch, Chemosphere 42 (2001) pp. 481-90
[4] C. Bruch, B. Peters, T. Nussbaumer, Fuel 82 (2003) 729-38
[5] F. He, F. Behrendt, Comb. & Flame 158 (2011) pp. 2500-11
[6] V. Novozhilov, B. Moghtaderi, D. F. Fletcher, J.H. Kent,o Fire Safety Journal 27 (1996) pp. 69-84
[7] R. Mehrabian, S. Zahirovic, R. Scharler, I. Obernberger et al., Fuel Proc. Tech. 95 (2012) pp. 96-108
[8] P. Bak, K. Chen, C. Tang, Phys. Lett. A 147 (1990) pp. 297-300.
[9] K. Chen, P. Bak, M. H. Jensen, Phys. Lett. A 149 (1990) pp. 207-210
[10] S. Clar, B. Drossel, F. Schwabl, Phys. Rev. E 50 (1994) 2, pp. 1009-18
[11] K. Lichtenegger, W. Schappacher, Int. J. o. Mod. Phys. C 20 (2009) 8, pp. 1247-1269
[12] Y. Haseli, J.A. van Oijen, L.P.H. de Goey, Biores. Tech. 102 (2011) pp. 9772-82

# Local Refinements in Classical Molecular Dynamics Simulations

**Konstantin Fackeldey and Marcus Weber**
Zuse Institute Berlin (ZIB), Takustr. 7, 14196 Berlin, Germany
E-mail: fackeldey@zib.de

**Abstract.** Quantum mechanics provide a detailed description of the physical and chemical behavior of molecules. However, with increasing size of the system the complexity rises exponentially, which is prohibitive for efficient dynamical simulation. In contrast, classical molecular dynamics procure a coarser description by using less degrees of freedom. Thus, it seems natural to seek for an adequate trade-off between accurateness and computational feasibility in the simulation of molecules. Here, we propose a novel method, which combines classical molecular simulations with quantum mechanics for molecular systems. For this we decompose the state space of the respective molecule into subsets, by employing a meshfree partition of unity. We show, that this partition allows us to localize an empirical force field and to run locally constrained classical trajectories. Within each subset, we compute the energy on the quantum level for a fixed number of spatial states (ab initio points). With these energy values from the ab initio points we have a local scattered data problem, which can be solved by the moving least squares method.

## 1. Introduction

In the last few decades the simulation of biomolecules by classical molecular dynamics as well as quantum mechanical methods have been developed with great success. However, the huge computational complexity is still a burden. With the objective of coping with this problem, multiscale methods have been developed, combining quantum mechanical simulations with classical molecular simulation in order to reach a convenient trade off between precision and complexity. These multiscale methods are referred to as QM/MM methods, which have their origin in the pioneering work by Warshel and Levitt [1]: The molecule is geometrically decomposed into at least two parts, where the usually smaller part is calculated by quantum methods and the other (larger) domain by classical molecular dynamics or molecular mechanics (cp Fig 1). The corresponding Hamiltonian of this decomposed method is called effective Hamiltonian ($H_{\text{eff}}$) and is the sum of a pure quantum Hamiltonian, a pure classical Hamiltonian and a QM/MM Hamiltonian, whereas the latter is accounting for the interaction between the two scales. Since the pioneering work in 1976, different applications and variants have been developed e.g.[2, 3, 4, 5]. However it is well known, that special care has to be taken for the QM/MM boundary, i.e. the junction of the QM- and the MM-region.

## 2. Local Quadratic Approximation Algorithm

In order to abandon this boundary problem, we propose a method which abstains from a spatial decomposition of the molecule by considering its conformation space. Instead of decomposing

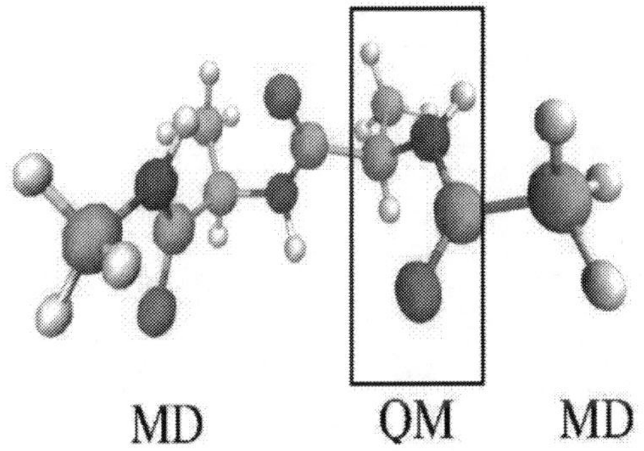


**Figure 1.** Spatial decomposition of a molecule into a QM and a MM region.

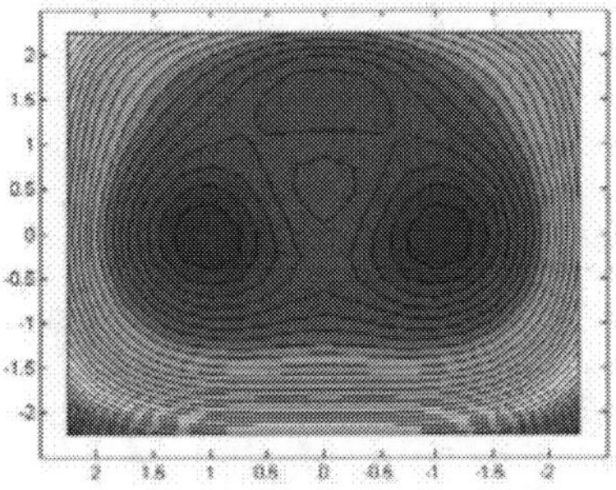

**Figure 2.** Sample potential used in the example.

the geometry of the molecule, we aim at a pointwise evaluation of ab initio energies and a local quadratic approximation, which is, thus, leading to a decomposition of the state space. More precisely, we seek for a decomposition of the potential energy surface $V(q)$ of a molecule with $d$ atoms being at position $q \in \mathbb{R}^{3d}$ into $n$ local subsets $V_i$, i.e.

$$V_i = V - \beta^{-1} \ln(\varphi_i) \quad \text{such that} \sum_{i=1}^{n} e^{-\beta V_i} = e^{-\beta V} \tag{1}$$

where $\beta = 1/(k_B T)$ is the inverse temperature and $k_B$ the Boltzmann constant and $\{\varphi_i\}$ is a partition of unity [6, 7], i.e. $\sum_{i=1}^{n} \varphi_i(q) = 1$ and $\varphi_i \geq 0 \quad \forall i$. In order to obtain such a partition a set of nodes $x_i$ has to be selected such that to each $x_i$ a radial basis function $W_i$ can be given, i.e.

$$W_i(x) = \|x_i - x\|^2, \; i = 1, ..., n, \tag{2}$$

for which the Euclidean norm of a vector $v = (v_1, ..., v_s)$ is given by $\|v\| = \sqrt{v_1^2 + ... + v_s^2}$. We remark that also other choices for the norm are possible.

Since these functions do not form a partition of unity, we follow Shepard's approach [8]:

$$\varphi_i := \frac{W_i}{\sum_{j=1}^{s} W_j}, \; i = 1, ..., n. \tag{3}$$

to gain a partition of unity.

Closely related to this decomposition scheme is the question of the approximation quality. For certain regions in the conformation space, a classical MM force field is quite a good approximation, whereas other regions can be found where the classical and the ab initio description differ. As a consequence, a decomposition in the above fashion for coupling QM and MM, should account for this aspect. We therefore suggest a method, which designs a decomposition with regard to the approximation quality. The following algorithm is not restricted to QM/MM coupling strategies, it can also be used as an initial decomposition in the context of meshfree methods in a pure classical context [9].

We assume, that the potential energy surface is twice differentiable. This seems to be a strong requirement, since it excludes poles. However, most sampling schemes, like the Metropolis Hasting Monte Carlo explicitly avoid these regions. More precisely, these regions are not or very rarely visited and can, thus, be excluded. Let $\varepsilon \in \mathbb{R}$ be a positive number (tolerance), $n$ the maximum number of sampling points and $x$ be an initial state. Then in each step of a "for-loop" from 1 to $n$, we compute:

**Algorithm 1** Local Quadratic Approximation of the ab-intio Points

**Require:** ab-initio points $(x_1, ..., x_n)$, energy values $(e_1, ..., e_n)$, gradients $(g_1, ..., g_n)$, Hessians $(h_1, ..., h_n)$, sampling point $x_{\text{old}}, f_{\text{old}}$
  $x_{\text{new}} := x_{\text{old}} + r$
  find closest ab-initio point $x_i$ to $x_{\text{new}}$
  **if** $\|x_i - x_{\text{new}}\| < \varepsilon$ **then**
    $f_{\text{new}} = e_i + g^T(x_i - x_{\text{new}}) + \frac{1}{2}(x_i - x_{\text{new}})^T \cdot h_i(x_i - x_{\text{new}})$
  **else**
    add $x_{\text{new}}$ to ab initio points and compute $f_{\text{new}} = e(x_{\text{new}}), g_{\text{new}} = \nabla e(x_{\text{new}}), h_{\text{new}} = H(e(x_{\text{new}}))$
    n=n+1
  **end if**
  **if** $\exp(-\beta(f_{\text{new}} - f_{\text{old}})) <$ rand **then**
    $x_{\text{old}} = x_{\text{new}}$
    $f_{\text{old}} = f_{\text{new}}$
  **end if**

Let us now give some comments on the algorithm. At the beginning the energy on the quantum level has to be computed for $n$ states (ab initio points). Starting from one of these points ($x_{\text{old}}$) we generate a new point $x_{\text{new}}$. Let $x_i$ be the closest ab-initio point to $x_{\text{new}}$ with a distance smaller than $\varepsilon$. Then we evaluate the Taylor expansion of $x_i$ at $x_{\text{new}}$ to obtain $f_{\text{new}}$ and apply the Metropolis criterion [10] on the approximated energy $f_{\text{new}}$. If this candidate fails we generate a new starting point outgoing from this failed candidate. If $\{x_j | \|x_j - x_{\text{new}}\| < \varepsilon, j = 1, ..., n\} = \emptyset$, candidate $x_i$ becomes a new ab initio point, that is the energy, gradient and the hessian have to be computed on the quantum level and the number of ab initio points increases ($n = n + 1$). We can interpret this algorithm as a discretization method for a partition of unity.

## 3. Example

We apply the method to a synthetic energy landscape [11] which is given by the potential energy function for $q = (q_1, q_2)$

$$\begin{aligned} e(q) = \; & 3\exp\left(-q_1^2 - (q_2 - \tfrac{1}{3})^2\right) - 3\exp\left(-q_1^2 - (q_2 - \tfrac{5}{3})^2\right) - 5\exp\left(-(q_1 - 1)^2 - q_2^2\right) \\ & -5\exp\left(-(q_1 + 1)^2 - q_2^2\right) + 0.2q_1^4 + 0.2\left(q_2 - \tfrac{1}{3}\right)^4. \end{aligned} \tag{4}$$

From Fig. 2 it can be seen that the two dimensional energy landscape has two basins of equal depth centered at $(-1, 0)$ and $(1, 0)$ and a smaller basin centered at $(0, 5/3)$. This 2-dimensional energy landscape should be considered as the "true" potential, which has be approximated locally by quadratic functions. In the end we checked the quality of our approximation method by testing the symmetry of the sampling: positive and negative $x$-values should be sampled with the same probability due to the symmetry of the potential energy function. See Table 1 for the results. The finer the discretization the better this symmetry is reproduced. We also checked the maximal deviation of the true function value from the approximated ones (again Table 1). A discretization parameter of $\varepsilon = 0.1$ leads to good results. The good results also depend on an adequate number of sampling points. About 40000 sampling points are needed to reproduce symmetry (blue/tiny points Fig. 3). Note, that the ab initio points in Fig 3 are given a priori and are thus not distributed according to the potential. With our method, we could reduce the number of 40000 function evaluations on the "ab initio level" to only about 580 (red/bold points in Figures 3).

| $\varepsilon$ | % | error |
|---|---|---|
| 0.1 | 49.82 | 0.1053 |
| 0.2 | 47.47 | 0.6381 |
| 0.3 | 45.77 | 1.935 |

**Table 1.** Different tolerances $\varepsilon$, measure for symmetry and error.

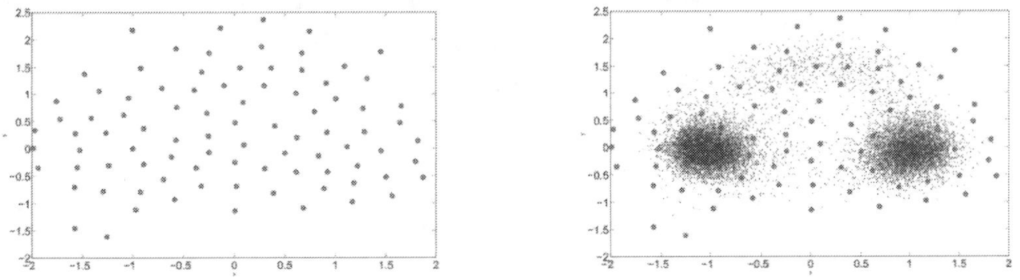


**Figure 3.** Schematic sketch: *Left:* Ab initio points $\{x_i\}$ being the center of the local approximation spaces $\{V_i\}$. *Right:* Blue: Sampling points according to the local approximation. The potential Fig. 2 could be reproduced.

## 4. Summary

We have proposed a method which aims at an efficient coupling between the quantum and the classical scale via the ab initio points and a local quadratic approximation. The approach claims that the potential energy surface is twice differentiable. For an artificial potential we could show, that our scheme reduces the number of ab initio points. In the future we plan to apply our scheme to real world applications like the conformational space of a peptide or receptor ligand binding system.

## References

[1] Warshel A and Levitt M 1976 *J. Mol. Biol.* **103** 227
[2] Gao J 1996 *Acc. Chem. Res.* **29** 298 – 305
[3] Lin H and Truhlar D G 2007 *Th. Chem. Acc.* **117** 185–199
[4] Kerdcharoen T, Liedl K R and Rode B M 1996 *Chemical Physics* **211** 313 – 323
[5] Exner T 2011 *International Journal of Quantum Chemistry* **111** 1002 – 1012
[6] Weber M 2006 *Meshless Methods in Conformation Dynamics* Doctoral thesis Department of Mathematics and Computer Science, Freie Universität Berlin
[7] Fackeldey K, Bujotzek A and Weber M 2012 A meshless discretization method for markov state models applied to explicit water peptide folding simulations *Meshfree Meth PDE VI* (*LNCSE* vol 89) (Springer) pp 141 – 154
[8] Shepard D 1968 A two-dimensional interpolation function for irregularly spaced data *Proceeding of the 1968 23rd ACM national conference* pp 517– 524
[9] Bujotzek A, Schuett O, Nielsen A, Fackeldey K and Weber M 2013 Efficient conformational analysis by partition-of-unity coupling to appear in J Chem Math
[10] Metropolis N, Rosenbluth A W, Rosenbluth M N, Teller A H and Teller E 1953 *J Chem Phys* **21** 1087–1092
[11] Schütte C and Huisinga W 2003 Biomolecular conformations can be identified as metastable sets of molecular dynamics *Computational Chemistry* (*Handbook Num. Anal.* vol 10) ed LeBris C (Elsevier) pp 699–744

# Boundary value problems and medical imaging

**Athanasios S Fokas**[1,2] **and George A Kastis**[2]
[1] Department of Applied Mathematics and Theoretical Physics, University of Cambridge, Cambridge, CB30WA, UK
[2] Research Center of Mathematics, Academy of Athens, Soranou Efessiou 4, Athens 11527, Greece

E-mail: T.Fokas@damtp.cam.ac.uk

**Abstract.** The application of appropriate transform pairs, such as the Fourier, the Laplace, the sine, the cosine and the Mellin transforms, provides the most well known method for constructing analytical solutions to a large class of physically significant boundary value problems. However, this method has several limitations. In particular, it requires the given PDE, domain and boundary conditions to be separable, and also may *not* be applicable if the given boundary value problem is non-self-adjoint. Furthermore, it expresses the solution as either an integral or an infinite series, neither of which are uniformly convergent on the boundary of the domain (for nonvanishing boundary conditions), which renders such expressions unsuitable for numerical computations. Here, we review a method recently introduced by the first author which *can* be applied to certain nonseparable and non-self-adjoint problems. Furthermore, this method expresses the solution as an integral in the complex plane which is uniformly convergent on the boundary of the domain. This method, which also suggests new numerical techniques, is illustrated for both evolution and elliptic PDEs. Athough this method was first applied to certain *nonlinear* PDEs called *integrable* and was originally formulated in terms of the so-called *Lax pairs*, it can actually be applied to *linear* PDEs without the need to analyse the associated Lax pair. The existence of Lax pairs is used here in order to motivate a related development, namely the emergence of a novel formalism for analysing certain inverse problems arising in medical imaging. Examples include PET and SPECT.

## 1. Boundary Value Problems

### *1.1. Evolution PDEs of the Half-Line*

In order to introduce the new method we consider the simplest possible initial-boundary value (IBV) problem for an evolution PDE, namely the heat equation on the half-line:

$$u_t = u_{xx}, \quad 0 < x < \infty, \quad 0 < t < T, \quad T > 0, \tag{1}$$

where

$$u(x,0) = u_0(x), \quad 0 < x < \infty, \quad u(0,t) = g_0(t), \quad 0 < t < T. \tag{2}$$

The functions $u_0(x)$ and $g_0(t)$ are given functions with appropriate smoothness and $u_0(x)$ decays as $x \to \infty$.

The above IBV problem can be solved by the sine-transform pair,

$$\hat{f}(\lambda) = \int_0^\infty sin(\lambda x) f(x) dx, \quad \lambda > 0; \quad f(x) = \frac{2}{\pi}\int_0^\infty sin(\lambda x) f(\lambda) d\lambda, x > 0. \tag{3}$$

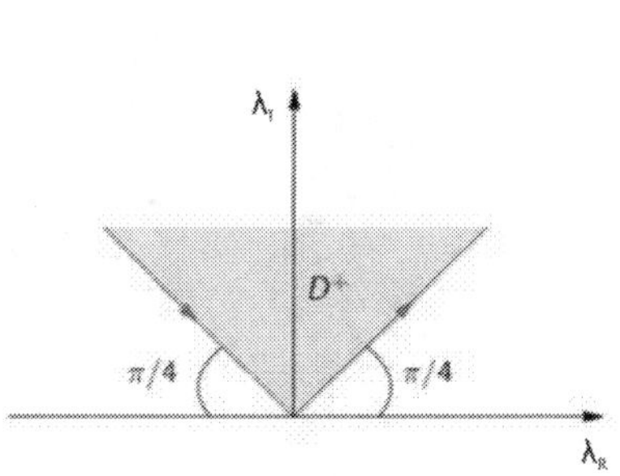


**Figure 1.** The domain $D^+$ for the heat equation.

Employing this transform we find

$$u(x,t) = \frac{2}{\pi}\int_0^\infty e^{-\lambda^2 t}\sin(\lambda x)\Big[\int_0^\infty \sin(\lambda\xi)u_0(\xi)d\xi - \lambda\int_0^t e^{\lambda^2 s}g_0(s)ds\Big]d\lambda. \tag{4}$$

The above representation suffers from the generic disadvantage that is associated with *every* representation obtained via a classical transform, namely it is *not* uniformly convergent at the boundary. Indeed, if the right-hand-side of (4) converged uniformly at $x = 0$, then we could take the limit $x \to 0$ inside the integral and we would obtain $u(0,t) = 0$ instead of $u(0,t) = g_0(t)$.

The unified method yields

$$u(x,t) = \frac{1}{2\pi}\int_{-\infty}^\infty e^{i\lambda x-\lambda^2 t}\hat{u}_0(\lambda)d\lambda - \frac{1}{2\pi}\int_{\partial D^+} e^{i\lambda x-\lambda^2 t}\left[\hat{u}_0(-\lambda) + 2i\lambda G_0(\lambda^2)\right]d\lambda, \tag{5}$$

where the functions $\hat{u}_0(\lambda)$ and $G_0(\lambda)$ are defined by

$$\hat{u}_0(\lambda) = \int_0^\infty e^{-i\lambda x}u_0(x)dx, \quad \mathsf{Im}\lambda \le 0, \quad G_0(\lambda) = \int_0^T e^{\lambda s}g_0(s)ds, \quad \lambda \in \mathbb{C}, \tag{6}$$

and the contour $\partial D^+$ is the boundary of the domain $D^+$ shown in Fig. 1.

It is straightforward to show [1, 2] that the right-hand-side of (5) is indeed uniformly convergent at $x = 0$. Furthermore, the only $(x,t)$ dependence of the right-hand-side of (5) is in the form $e^{i\lambda x-\lambda^2 t}$, thus it is immediately obvious that this representation satisfies the heat equation.

The experienced reader may worry about the dependence of $G_0(\lambda)$ on T, which contradicts causality (the solution of an evolution equation *cannot* depend on future data). However, using analyticity arguments, it can be shown [1, 2] that $G_0(\lambda)$ can be replaced by the function $G_0(\lambda,t)$, where

$$G_0(\lambda,t) = \int_0^t e^{\lambda s}g_0(s)ds, \quad \lambda \in \mathbb{C}. \tag{7}$$

The limited applicability of the standard transforms becomes evident by considering the Stokes equation on the half-line:

$$u_t + u_{xxx} + u_x = 0, \quad 0 < x < \infty, \quad 0 < t < T, \quad T > 0, \tag{8}$$

with the initial and boundary conditions defined in (2).

It can be rigorously established [3, 4] that there does *not* exist an appropriate x-transform for this problem, *i.e.* there does *not* exist the analogue of the sine transform for a linear evolution

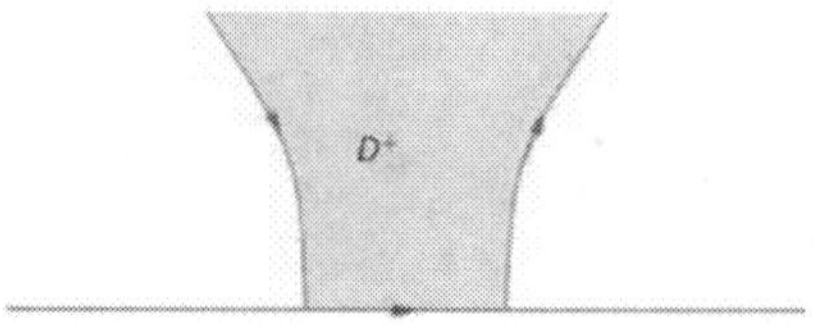


**Figure 2.** The domain $D^+$ for the Stokes equation.

PDF involving a third order derivative. One may attempt to solve the above IBV problem with the Laplace transform in t. But then one has to make the unnatural initial assumption of $T = \infty$, and furthermore one has to solve a cubic algebraic equation!

The unified method yields

$$u(x,t) = \frac{1}{2\pi}\int_{-\infty}^{\infty} e^{i\lambda x-(i\lambda-i\lambda^3)t}\hat{u}_0(\lambda) - \frac{1}{2\pi}\int_{\partial D^+} e^{i\lambda x-(i\lambda-i\lambda^3)t}\tilde{g}(\lambda)d\lambda, \tag{9}$$

where the function $\hat{u}_0(\lambda)$ is defined in (6), $\tilde{g}(\lambda)$ is defined by

$$\tilde{g}(\lambda) = \frac{1}{\nu_1-\nu_2}[(\nu_1-\lambda)\hat{u}_0(\nu_2)+(\lambda-\nu_2)\hat{u}_0(\nu_1)] + (3\lambda^2-1)G_0(\omega(\lambda)), \tag{10}$$

with $G_0(\lambda)$ defined in (6) and $\omega(\lambda)$ given by $\omega(\lambda) = i\lambda - i\lambda^3$; the contour $\partial D^+$ is the boundary of the domain defined by

$$D^+ = \{\text{Re }\omega(\lambda) < 0\} \cap \mathbb{C}^+ \tag{11}$$

and shown in Fig. 2. The complex numbers $\nu_1$ and $\nu_2$ are the two nontrivial transformations $\lambda \to \nu_1(\lambda)$, $\lambda \to \nu_2(\lambda)$ which leave $\omega(\lambda)$ invariant, *i.e.* they are the two nontrivial roots of the equation $\omega(\lambda) = \omega(\nu(\lambda))$:

$$\nu_j^2 + \lambda\nu_j + \lambda^2 - 1 = 0, \quad j = 1, 2. \tag{12}$$

The unified method involves the following three steps:

Step 1: *Rewrite the given PDE in a divergence form, or equivalently in terms of a closed differential form.*

For the heat equation we find

$$(e^{-i\lambda x+\lambda^2 t}u)_t - (e^{-i\lambda x+\lambda^2 t}(u_x + i\lambda u))_x = 0, \quad \lambda \in \mathbb{C}. \tag{13}$$

Equivalently, the following differential one-form is closed:

$$W(x,t,\lambda) = e^{-i\lambda x+\lambda^2 t}[udx + (u_x + i\lambda u)dt]. \tag{14}$$

If the PDE is valid in a given domain $\Omega$ then Green's theorem immediately implies the following *global relation* (GR):

$$\int_{\partial\Omega} W = 0. \tag{15}$$

For the heat equation on the half-line the GR becomes:

$$e^{\lambda^2 T}\hat{u}(\lambda,T) = \hat{u}_0(\lambda) - i\lambda G_0(\lambda^2) - G_1(\lambda^2), \quad \text{Im}\lambda \le 0, \tag{16}$$

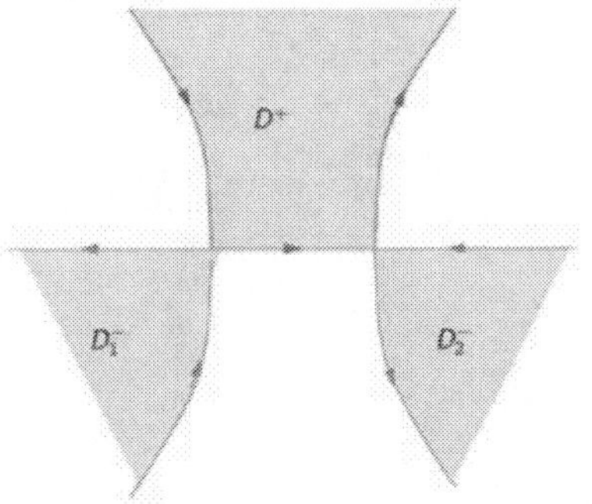


**Figure 3.** The analysis of the GR for the Stokes equation.

where $G_0(\lambda)$ is defined in (6) and the functions $G_1(\lambda)$ and $\hat{u}(\lambda, T)$ are defined by

$$G_1(\lambda) = \int_0^T e^{\lambda s} u_x(0,s)ds, \quad \lambda \in \mathbb{C}; \quad \hat{u}(\lambda, T) = \int_0^\infty e^{-i\lambda x} u(x,T)dx, \ \mathrm{Im}\lambda \leq 0. \tag{17}$$

Step 2: *Integral representation of the solution.*
The representation can be obtained either using the Fourier transform on the half-line and then deforming the relevant integral from the real line to the complex $\lambda$-plane, or using the spectral analysis of the associated Lax pairs [2]. For the heat equation we find

$$u(x,t) = \frac{1}{2\pi}\int_{-\infty}^{\infty} e^{i\lambda x - \lambda^2 t}\hat{u}_0(\lambda)d\lambda - \frac{1}{2\pi}\int_{\partial D^+} e^{i\lambda x - \lambda^2 t}[G_1(\lambda^2) + i\lambda G_0(\lambda^2)]d\lambda, \tag{18}$$

where $\partial D^+$ is shown in Fig. 1.
Step 3: *Elimination of the unknown boundary values.*
This can be achieved by using the GR and by employing all transformations in the complex $\lambda$-plane which leave the associated $\omega(\lambda)$ invariant. For the heat equation, $\omega(\lambda) = \lambda^2$, thus replacing in the GR $\lambda$ with $-\lambda$, we find

$$e^{\lambda^2 T}\hat{u}(-\lambda, T) = \hat{u}_0(-\lambda) + i\lambda G_0(\lambda^2) - G_1(\lambda^2), \quad \mathrm{Im}\lambda \geq 0. \tag{19}$$

Solving this equation for $G_1$ and using the fact that $\hat{u}(-\lambda, T)$ does not contribute to the solution $u(x,t)$ [1, 2], we find that (18) becomes (5).

For the Stokes equation, the associated GR is

$$G_2 + i\lambda G_1 = (\lambda^2 - 1)G_0 - \hat{u}_0(\lambda) - e^{(i\lambda - i\lambda^3)T}\hat{u}(\lambda, T), \quad \mathrm{Im}\lambda \leq 0. \tag{20}$$

Replacing in this equation $\lambda$ by $\nu_1$ and by $\nu_2$ we find two equations, both of which are valid in $D^+$, see Fig. 3. Then, neglecting the contribution of $\hat{u}(\nu_1, T)$ and $\hat{u}(\nu_2, T)$, we find

$$G_1 \sim -i(\nu_1 + \nu_2)G_0 + i\frac{\hat{u}_0(\nu_1) - \hat{u}_0(\nu_2)}{\nu_1 - \nu_2}, \quad G_2 \sim -(1 + \nu_1\nu_2)G_0 + \frac{\nu_2\hat{u}_0(\nu_1) - \nu_1\hat{u}_0(\nu_2)}{\nu_1 - \nu_2}. \tag{21}$$

*Numerical Implementation*
For the case that $\hat{u}_0(\lambda)$ and $G_0(\lambda)$ can be computed explicitly, it is straightforward to compute

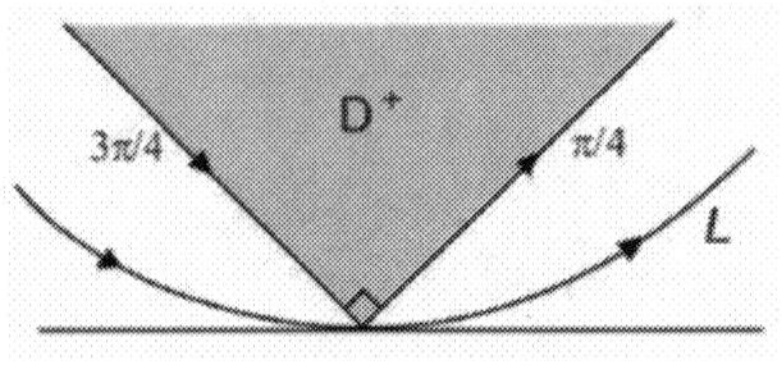


**Figure 4.** The contour $L$ for the heat equation.

numerically the solution. Consider, for example, the heat equation with the following initial and boundary condition:

$$u_0(x) = x\exp(-a^2x), \quad g_0(t) = \sin bt, \quad a, b > 0. \tag{22}$$

Then, (4) becomes

$$u(x,t) = \frac{1}{2\pi}\int_{\partial D^+} e^{i\lambda x - \lambda^2 t}\left[\frac{1}{(i\lambda+a)^2} - \frac{1}{(-i\lambda+a)^2} - \lambda\left(\frac{e^{(\lambda+ib)t}-1}{\lambda+ib} - \frac{e^{(\lambda-ib)t}-1}{\lambda-ib}\right)\right]d\lambda. \tag{23}$$

On the contour $\partial D^+$, $e^{i\lambda x}$ decays exponentially for large $\lambda$, whereas $e^{-i\lambda^2 t}$ oscillates. However, deforming $\partial D^+$ to the contour $L$, see Fig. 4, we achieve decay as $\lambda \to \infty$ in both $e^{i\lambda x}$ and $e^{-i\lambda^2 t}$. Thus, the deformed integral can be computed numerically most efficiently [5].

### *1.2. Evolution PDEs on the Interval*

The heat and the Stokes equations with $\{0 < x < 1, 0 < x < T\}$ are analysed in [6]. For both these equations, the unified method yields $u(x,t)$ in terms of *integrals* in the complex $\lambda$-plane (as opposed to infinite series). It should be emphasized that it is *impossible* to express the solution of a typical IBV problem for the Stokes equation in terms of an infinite series. Therefore, the usual statement that a *finite* domain corresponds to a *discrete* spectrum is *not* valid in general (unless the associated problem is self-adjoint).

### *1.3. Elliptic PDEs in the Interior of a Polygon*

For elliptic PDEs, including the Laplace, the modified Helmholtz and Helmholtz equations, formulated in the interior of a polygon, the unified method: (a) for certain simple domains, like the interior of an equilateral triangle, provides the analytical solution for a variety of boundary value problems, for which apparently the classical approaches fail. (b) For an arbitrary polygon, it provides a powerful approach for computing numerically the associated generalized *Dirichlet to Neumann map*, *i.e.* computing the unknown boundary value in terms of the given boundary data.

Consider for example the modified Helmholtz equation

$$\frac{\partial^2 u}{\partial x^2} + \frac{\partial^2 u}{\partial y^2} - \beta^2 u = 0, \quad \beta > 0, \quad x,\ y \in \mathbb{R}. \tag{24}$$

It is straightforward to verify that the following differential form $W$ is closed:

$$W(x,y,\lambda) = e^{-i\frac{\beta}{2}\left(\lambda z - \frac{\bar{z}}{\lambda}\right)}\left\{\left[-u_y + \frac{\beta}{2}\left(\lambda + \frac{1}{\lambda}\right)u\right]dx + \left[u_x + \frac{\beta}{2}\left(i\lambda + \frac{1}{i\lambda}\right)u\right]dy\right\}, \quad \lambda \in \mathbb{C}\setminus\{0\}. \tag{25}$$

Hence, if $\partial\Omega$ denotes the boundary of $\Omega$ then the following GR is valid:

$$\int_{\partial\Omega} W(x,y,\lambda) = 0, \quad \lambda \in \mathbb{C} \setminus \{0\}. \tag{26}$$

In the particular case of a square with corners at (1,1), (1,-1),(-1,-1), (-1,1), (26) becomes

$$\sum_{j=1}^{4} \hat{u}_j(\lambda) = 0, \quad \lambda \in \mathbb{C} \setminus \{0\}, \tag{27}$$

where

$$\hat{u}_1(\lambda) = e^{\frac{\beta}{2}(i\lambda + \frac{1}{i\lambda})} \int_{+1}^{-1} e^{\frac{\beta}{2}(\lambda + \frac{1}{\lambda})y} \left[ u_x^{(1)} + \frac{\beta}{2}\left(i\lambda + \frac{1}{i\lambda}\right) u^{(1)} \right] dy, \tag{28}$$

and similarly for $\{u_j(\lambda)\}_2^4$. Equation (27) involves four unknown functions, however (27) is valid for *all* $\lambda \in \mathbb{C} \setminus \{0\}$. By expanding the unknown functions in terms of Legandre polynomials and by choosing appropriate collocation points $\lambda = \lambda_j$, we can solve the GR in an efficient way with spectral accuracy [7, 8].

## 2. Medical Imaging

Equation (13) suggests the introduction of the "potential" $e^{-i\lambda x + \lambda^2 t}\mu$, satisfying

$$\left(e^{-i\lambda x+\lambda^2 t}\mu\right)_x = e^{-i\lambda x+\lambda^2 t}u, \quad \left(e^{-i\lambda x+\lambda^2 t}\mu\right)_t = e^{-i\lambda x+\lambda^2 t}(u_x + i\lambda u) \tag{29}$$

These equations provide a *Lax pair* formulation for the heat equation.

The initial value problem of the heat equation can be solved by the Fourier transform. The above Lax pair suggests that this transform can be constructed via the spectral analysis of the first of equations (29). Indeed, it was shown in [9] that the spectral analysis of the equation

$$(\partial_x - i\lambda)\,\mu(x,\lambda) = u(x), \tag{30}$$

yields the 1-D Fourier transform; similarly the spectral analysis of the equation

$$(\partial_{x_1} + i\partial_{x_1} - \lambda)\,\mu(x_1,x_2,\lambda) = u(x_1,x_2), \quad (x_1,x_2) \in \mathbb{R}^2, \tag{31}$$

yields the 2-D Fourier transform. This has lead to the emergence of a new method for deriving transforms, or equivalently for inverting certain integrals. The power of this new method was illustrated by R Novikov [10], who was able to invert the so-called *attenuated Radon transform*. This transform plays the same crucial role in SPECT that the Radon transform plays in CT and in PET [11, 12, 13, 14].

## References


[1] Fokas A S and Spence E A Synthesis as opposed to separation of variables 2012 *SIAM Rev.* **54** 291

[2] Deconinck B, Trogdon T and Vasan V The method of Fokas for solving linear partial differential equations *SIAM Rev.* (to appear)

[3] Smith D A Spectral theory of ordinary and partial linear differential operators on finite intervals 2011 (PhD Thesis: University of Reading)

[4] Papanicolaou G 2011 An example where separation of variable fails *J. Math. Anal. Appl.* **373** 739

[5] Flyer N and Fokas A S A hybrid analytical-numerical method for solving evolution partial differential equations. I. The half-line 2008 *Proc. R. Soc.* **464** 1823

[6] Fokas A S and Pelloni B A transform method for evolution PDEs on the interval 2005 *IMA J. Appl. Math.* **75** 564

[7] Fornberg B and Flyer N A numerical implementation of Fokas boundary integral approach: Laplace's equation on a polygonal domain 2011 *Proc. R. Soc. A* **467** 2983

[8] Smitheman S A and Fokas A S A numerical technique for polygonal domains: I interior (submitted).

[9] Fokas A S and Gelfand I M Integrability of linear and nonlinear evolution equations, and the associated nonlinear Fourier transforms 1994 *Lett. Math. Phys.* **32** 189

[10] Novikov R G An inversion formula for the attenuated X-ray transformation 2002 *Ark. Mat.* **40** 145

[11] Fokas A S, Iserles A and Marinakis V Reconstruction algorithm for single photon emission computed tomography and its numerical implementation 2006 *J. R. Soc. Interface* **3** 45

[12] Kastis G A, Gaitanis A, Fernandez Y, Kontaxakis G and Fokas A S Evaluation of a spline reconstruction technique: comparison with FBP, MLEM and OSEM 2010 *IEEE Nuclear Science Symposium Conf. Record (Knoxville)* p 3282

[13] Kastis G A, Gaitanis A, Skouras T, Fokas A S Evaluation of a spline reconstruction technique for SPECT: comparison with FBP and OSEM 2011 *IEEE Nuclear Science Symposium Conf. Record (Valencia)* p 3950

[14] Kastis G A, Kyriakopoulou D, Gaitanis A, Fernandez Y , Hutton B F and Fokas A S Evaluation of the Spline Reconstruction Technique for PET 2013 *Med. Phys.* (submitted)

# How many is different? Answer from ideal Bose gas

**Jeong-Hyuck Park**
Department of Applied Mathematics and Theoretical Physics, Cambridge, CB3 0WA, England
Department of Physics, Sogang University, Seoul 121-742, Korea[1]

E-mail: park@sogang.ac.kr

**Abstract.** How many $H_2O$ molecules are needed to form water? While the precise answer is not known, it is clear that the answer should be a finite number rather than infinity. We revisit with care the ideal Bose gas confined in a cubic box which is discussed in most statistical physics textbooks. We show that the isobar of the ideal gas zigzags on the temperature-volume plane featuring a *boiling-like* discrete phase transition, provided the number of particles is equal to or greater than a particular value: 7616. This demonstrates for the first time how a finite system can feature a mathematical singularity and realize the notion of 'Emergence', without resorting to the thermodynamic limit.

## 1. Introduction

Emergence is a generic notion that quantitative increase leads to qualitative change. It is often said that the whole is greater than the sum of its parts. It is a key idea in condensed matter physics as well as in statistical physics, such that a classic paper by Anderson goes with the title, *More is different* [1]. Then the question we address in this note is *How many is different?* To answer the question, we consult the quantum statistical physics where the key quantity is a partition function,

$$Z = \mathrm{Tr}\left[e^{-H/(k_{\mathrm{B}}T)}\right] . \qquad (1)$$

The energy eigenvalues of the Hamiltonian are generically quantized with respect to the size of the volume. Consequently the partition function depends on both the temperature, $T$, and the volume, $V$. Once we know the exact expression of the partition function we may compute all the physical quantities which are typically given by the derivatives of the logarithm of the partition function.

For a finite system, the partition function and the derivatives are all analytic and hence all finite, never diverge. For example, $C_V$, the specific heat at constant volume is finite: no singularity arises. The conventional way to realize a singularity is then to assume the thermodynamic limit where the volume and the number of particles go to infinity, $V \to \infty$, $N \to \infty$, while the density, $N/V$, is kept fixed. In this limit, the partition function may become non-analytic and feature a singular behaviour [2]. It might seem that only infinity system could feature a singularity and realize a discrete phase transition, to coin a phrase, "More is the same: infinitely more is different" [3].

However, strictly speaking, infinite limits are hardly realistic and exist only in theory [4, 5, 6]. Experiments or our daily experiences, like the boiling water in the kitchen, do appear not just

[1] Sabbatical leave of absence.

to mimic but actually to manifest discrete phase transitions. As a matter of historical fact in 1937, at the Van der Waals memorial meeting the audience could not agree on the question, whether partition function for a finite system could explain a sharp phase transition or not. So the chairman of the session, Kramers, put it to a vote![2]

Nevertheless, despite the old controversy, in a modern rather axiomatic approach adopted mainly by mathematicians, phase transitions are defined simply in the thermodynamic limit only, with the "mathematically-natural" conventional answer *infinity* to the question, *How many is different?* The present note highlights an alternative finite answer, like $N = 7616$ reported in the works of the author with Sang-Woo Kim and Imtak Jeon [7, 8, 9].[3] The emphasis is put on physical implications rather than technical derivations.

## 2. Discrete phase transition of a finite system under constant pressure

A crucial ingredient in our approach to realize a discrete phase transition from a finite system is to *keep not the volume but the pressure constant.* The temperature derivative at fixed pressure acting on an arbitrary function of $T$ and $V$ reads

$$\left.\frac{\partial}{\partial T}\right|_P = \left.\frac{\partial}{\partial T}\right|_V - \left[\frac{\partial_T P(T,V)}{\partial_V P(T,V)}\right]\left.\frac{\partial}{\partial V}\right|_T . \tag{2}$$

When this differential operator acts on *e.g.* the logarithm of the analytic partition function of a finite system, the temperature and the volume derivatives on the right hand side of the equality cannot generate any mathematical singularity. However, when $\partial_V P(T,V)$ vanishes, namely at the 'spinodal' points, it can be singular. In fact, the usual argument against the mathematical singularity from a finite system assumes the volume to be fixed. Alternatively, if we fix the pressure constant, a singularity may occur. That is to say, for a finite system, $C_V$ is finite never diverges, but $C_P$ may become singular,

$$C_V \ll \infty \qquad \text{versus} \qquad C_P = \infty \qquad \text{for} \qquad N \ll \infty . \tag{3}$$

Physically, if we fully fill a rigid box with water and heat it up, the temperature will increase but the water hardly evaporates: no discontinuous phase transition to occur in this case. Nevertheless, opening the lid will set the pressure constant or at 1 atm, and the water will surely start to boil at 100 degree Celsius: no need to take the thermodynamic limit!

Our main point is that a singularity or a discrete phase transition may occur even for a finite system, provided we impose the constraint of keeping the pressure constant, rather than fixing the volume or the density. The next question is then the existence of the spinodal points where $\partial_V P(T,V) = 0$. Our claim is that the spinodal curve originates essentially due to the *identical nature of particles* without need of any extra interactions. As we describe below, a careful study of an ideal Bose gas indeed reveals the spinodal points and hence a *boiling-like* discrete phase transition for finite $N$.

## 3. Ideal Bose gas can boil under constant pressure if $N \geq 7616$

We are interested in a canonical partition function, $Z_N(T,V)$, of the ideal Bose gas confined in a cubic box with a definite number of particles, $N$. It is a standard quantum mechanical system discussed in most of statistical physics textbooks, *e.g.* [10]. We aim to address the precise dependence on $N$. The commonly adopted approximation of the partition function is

$$Z_N \sim \frac{(Z_{N=1})^N}{N!} , \tag{4}$$

[2] The author is not aware of the result of this 'democratic' decision.

[3] A short video presentation is also available on-line:
`http://www.youtube.com/watch?v=rNmOvgQsGnU&list=FLfVQVNKya2JpncnEthOOMjA&feature=mh_lolz`

such that, *e.g.* the division by $N!$ solves the Gibbs paradox. This approximation gives for a generic derivative,

$$\partial \ln Z_N \sim N\partial \ln Z_{N=1}\,, \tag{5}$$

and hence the number of particles appears as an overall parameter corresponding to an extensive quantity. However, the above approximation would be only correct if the $N$ particles occupied always $N$ distinct quantum states. But if some of the particles are in a common state, the devision by $N!$ is too much and the approximation requires improvement. In fact, the correction can be organized as a determinant of a certain $N \times N$ matrix, $\Omega_N$, and the *exact* formula of the canonical partition function can be expressed as

$$Z_N = \mathrm{Det}\,(\Omega_N) \times \frac{(Z_{N=1})^N}{N!}\,. \tag{6}$$

For the details we refer the readers to [7, 8]. The point is that with the exact expression, $N$ is no longer a strictly-extensive parameter: physical quantities depend *non-linearly* on the number of particles. Especially a thermodynamic instability and hence zigzagging of the isobar on $(T, V)$-plane *emerge* provided the number of particles is equal to or greater than 7616 [7].

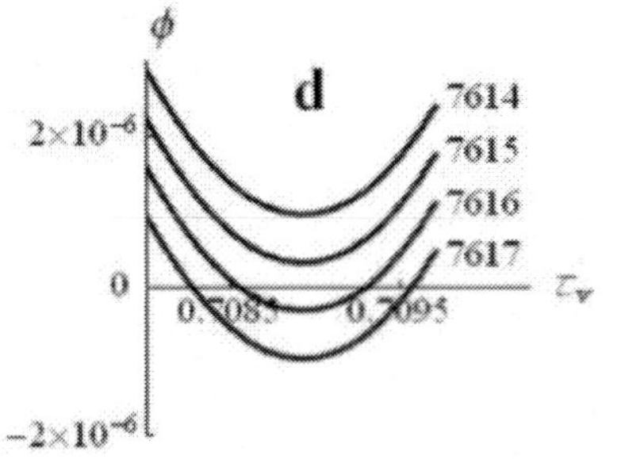


**Figure 1.** A supercomputer-powered numerical analysis on the exact partition function reveals that a thermodynamically unstable region *emerges* on $(T, V)$-plane when $N \geq 7616$.

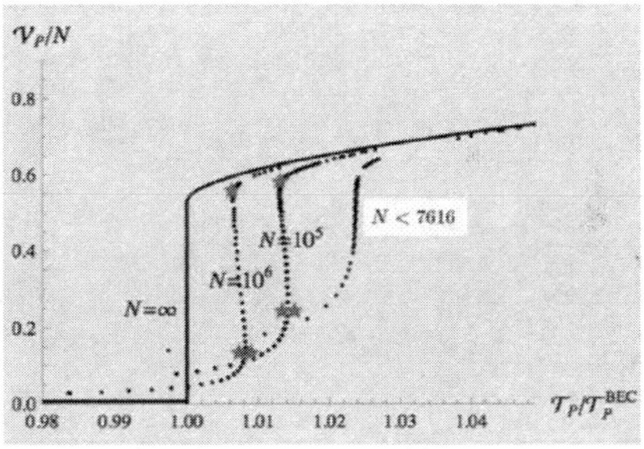


**Figure 2.** The isobar of the ideal Bose gas zigzags on $(T, V)$-plane when $N \geq 7616$, which features a *boiling-like* discrete phase transition.

In the figures above, utilizing the scale symmetry of the system, we set dimensionless quantities:

$$\phi = -\left(\tfrac{V^2}{Nk_{\mathrm{B}}T}\right)\partial_V P\,, \quad \mathcal{T}_V = k_{\mathrm{B}}T\left(\tfrac{V}{N}\right)^{\frac{2}{3}}\left(\tfrac{2m}{\pi^2\hbar^2}\right), \quad \mathcal{V}_P = V\left(P\tfrac{2m}{\pi^2\hbar^2}\right)^{\frac{3}{5}}, \quad \mathcal{T}_P = k_{\mathrm{B}}TP^{-\frac{2}{5}}\left(\tfrac{2m}{\pi^2\hbar^2}\right)^{\frac{3}{5}}. \tag{7}$$

Further, for $N \gg 7616$, the supercooling $(*)$ and the superheating $(**)$ points are analytically [9],

$$\begin{aligned}
&\mathcal{T}_P^*/\mathcal{T}_P^{\rm BEC} \simeq 1 + \tfrac{\pi^3}{60}\left[\left(\mathcal{T}_P^{\rm BEC}\right)^5/\mathcal{T}_V^{\rm BEC}\right]^{\frac{1}{2}} N^{-\frac{1}{3}}\,, && \mathcal{T}_P^{**}/\mathcal{T}_P^{\rm BEC} \simeq 1 + \tfrac{1}{150}\left(\tfrac{\pi^{15}}{15}\right)^{\frac{1}{4}}\left(\mathcal{T}_P^{\rm BEC}\right)^{\frac{5}{2}} N^{-\frac{1}{4}}\,,\\
&\mathcal{V}_P^* \simeq \left(\mathcal{T}_V^{\rm BEC}/\mathcal{T}_P^{\rm BEC}\right)^{\frac{3}{2}}\left(N + \tfrac{\pi}{4}\mathcal{T}_V^{\rm BEC} N^{\frac{2}{3}}\ln N\right), && \mathcal{V}_P^{**} \simeq 8\left(\tfrac{15}{\pi^3}\right)^{\frac{3}{4}}\left(\mathcal{T}_P^{\rm BEC}\right)^{-\frac{3}{2}} N^{\frac{3}{4}}\,,\\
&\mathcal{T}_V^* \simeq \mathcal{T}_V^{\rm BEC}\left(1 + \tfrac{\pi}{6}\mathcal{T}_V^{\rm BEC} N^{-\frac{1}{3}}\ln N\right), && \mathcal{T}_V^{**} \simeq 4\left(\tfrac{15}{\pi^3}\right)^{\frac{1}{2}} N^{-\frac{1}{6}}\,,
\end{aligned} \tag{8}$$

where $\mathcal{T}_P^{\rm BEC}$ and $\mathcal{T}_V^{\rm BEC}$ denote the "BEC" constants,

$$\mathcal{T}_P^{\rm BEC} = \left(\tfrac{64}{\pi^3}\right)^{\frac{1}{5}}\left[\zeta(\tfrac{5}{2})\right]^{-\frac{2}{5}} \simeq 1.02781\,, \qquad \mathcal{T}_V^{\rm BEC} = \tfrac{4}{\pi}\left[\zeta(\tfrac{3}{2})\right]^{-\frac{2}{3}} \simeq 0.671253\,. \tag{9}$$

## 4. Discussion

The above boiling-like discrete phase transition is an *emergent phenomenon* of the finitely many bosonic identical particles, which we *ab initio* derived from the first principles in statistical physics. The singularity is due to the spinodal curve that also sharply defines the phase diagram itself [8], *c.f.* [11].

Our results seem to suggest that, *a generic liquid-gas phase transition occurs essentially due to the identical nature of particles* rather than additional interactions.

The critical number 7616 can be viewed as a quantum mechanically determined characteristic of 'cube', the geometric shape of the box. Boxes of different shapes will have their own critical numbers. For example, for 'sphere' it is 10458.

It will be experimentally challenging to find a corresponding critical number for each molecule in Nature. We believe there must be a finite definite answer to the question, *How many* $H_2O$ *molecules are needed to form water that features boiling phenomenon with zigzagged isobar?*


## Acknowledgments

The author wishes to thank Hugh Osborn at DAMTP, University of Cambridge for hospitality and encouragement during the sabbatical visit. This work was supported by the Sogang University Research Grant of 201310030.01.


## References


[1] P. W. Anderson, *Science* 177, 393 (1972).
[2] C. N. Yang and T. D. Lee, *Phys. Rev.* 87, 404 (1952); 87, 410 (1952).
[3] L. P. Kadanoff, *J. Stat. Phys.* 137, 777 (2009).
[4] B. U. Felderhof, *Nature* 225, 17 (1970).
[5] D. H. E. Gross, cond-mat/0201235.
[6] J. Butterfield, *Foundations of Physics* 41, 6 (2011).
[7] J.-H. Park and S.-W. Kim, *Phys. Rev.* A 81, 063636 (2010).
[8] J.-H. Park and S.-W. Kim, *New J. Phys.* 13, 033003 (2011).
[9] I. Jeon, S.-W. Kim and J.-H. Park, *Phys. Rev.* A 84, 023636 (2011).
[10] K. Huang, *Statistical Mechanics* (Wiley, New York, 1963).
[11] L. Casetti, M. Pettini and E. G. D. Cohen, Phys. Rept. **337** (2000) 238.

# NOνA neutrino experiment status

**Filip Jediný**[1]
Deparment of Dosimetry and Application of Ionizing Radiation
Faculty of Nuclear Sciences and Physical Engineering
Czech Technical University in Prague
Břehová 7, Prague, Czech Republic

jedinfil@fjfi.cvut.cz

**Abstract**. The primary goal of the NOνA neutrino oscillation experiment is to study the probabilities of transformation of muonic-neutrinos into electron-neutrinos. The experiment is currently under construction and will use a 700 kW accelerator-based NuMI beam (Neutrinos at the Main Injector) and two detectors. The Near Detector (329 t at Fermi National Accelerator Laboratory, Illinois) and the Far Detector (14 kt, Ash River, Minnesota) are aligned to 14 mrad off-axis and separated by 810 km. They are made of active liquid scintillator and readout by avalanche photo-diodes. Recent results from world-wide neutrino experiments indicate that NOνA is in the position to determine the neutrino mass hierarchy as it is also searching for the first hints of CP violation in neutrino sector. The design, the goals and the current status of the NOνA experiment are presented here with the current estimates of its sensitivity to the mass hierarchy measurement.

## 1. Introduction

Neutrino physics, with many interesting topics and questions reaching beyond Standard Model, is being more and more appealing for particle physicists. Despite the huge progress made in recent years, there are still many open questions, regarding i.e. the mass ordering, the mixing angles and the *CP* violation in neutrino sector.

The NuMI Off-Axis $\nu_e$ Appearance experiment (NOνA) was designed to answer some of these questions [1]. Being currently built in North America, it's about to be one of the world's leading neutrino experiments and the flagship of the Intensity Frontier – particle physics program of Fermi National Laboratory in Batavia, IL (USA).

NOνA is a long-baseline, two-detector neutrino oscillation experiment looking mostly for appearance of $\nu_e$ in well-defined NuMI beam of $\nu_\mu$. Both its detectors are being currently under construction and the beam was recently upgraded to be capable of delivering 700 kW of beam power, with optional change between neutrino and antineutrino beam.

Recent results from both long-baseline and reactor-based neutrino experiments confirm non-zero value of $\theta_{13}$ [2-5] and non-maximal $\sin^2 2\theta_{13}$ [6-8], which shows that NOνA is in a good position to measure precisely for the first time $\Delta m^2_{23}$ (the mass ordering of three mass eigenstates); to get the first data on the CP violation delta phase; to accurately measure $\theta_{13}$ mixing angle through probability of appearance of $\nu_e$ in $\nu_\mu$ beam and to determine the octant of $\theta_{23}$.

[1] On behalf of the NOνA collaboration

## 2. NOvA experiment

To carry out a measurement of accelerator neutrino oscillation, a so-called second generation experiment has to have an intense beam of neutrinos and a massive detector several hundreds of kilometers away. NOvA detects the NuMI beam (same as MINOS [8]) with Far Detector (FD) 810 km from the target and Near Detector (ND) 1 km from the target.

### 2.1. NuMI neutrino beam

The source of neutrinos to detect in NOvA detectors is the NuMI beam – Neutrinos at the Main Injector. The neutrinos are created by extracting 120 GeV protons from the Main Injector facility at Fermilab, and colliding them with a graphite target, which results in production of secondary mesons ($\pi^+$ particles can decay into $\mu^+$ and $\nu_\mu$). To meet the NOvA physics goals, the NuMI beam was upgraded from 300 kW to 700 kW of nominal beam power. This is achieved by reducing the 10 $\mu$s-pulse cycle time of the Main Injector from 2.2 s to 1.3 s via slip-stacking in the recycler ring; increasing the intensity per cycle with 12 Booster batches instead of 11 by installing new RF stations and a new injection kicker magnet; and upgrading the target and horns to accommodate the increased proton intensity.

As a result, there will be a 10 $\mu$s beam spill every 1.33 s in an intensity of $4.9 \times 10^{13}$ protons per pulse (corresponding to $6.0 \times 10^{20}$ protons on target per year of running). From pion decay kinematics, we know the neutrino energy will depend on the characteristic decay angle between the neutrino and the parent pion in the laboratory frame. For 14 mrad off-axis, most pion decays result in neutrinos with E = 2 GeV, some with energy smearing around that value. Therefore, the NOvA detectors placed 14 mrad off the NuMI beam axis will measure a narrow band beam peaked very near the $\nu_\mu$ □ $\nu_e$ oscillation maximum at 810 km (E ~1.6 GeV), as shown in Fig. 1. The narrow beam energy spectrum strongly reduces the background from feed-down of higher energy neutral-current (NC) neutrino events, the dominant background for $\nu_\mu$ □ $\nu_e$ oscillation searches with the detectors placed on-axis.

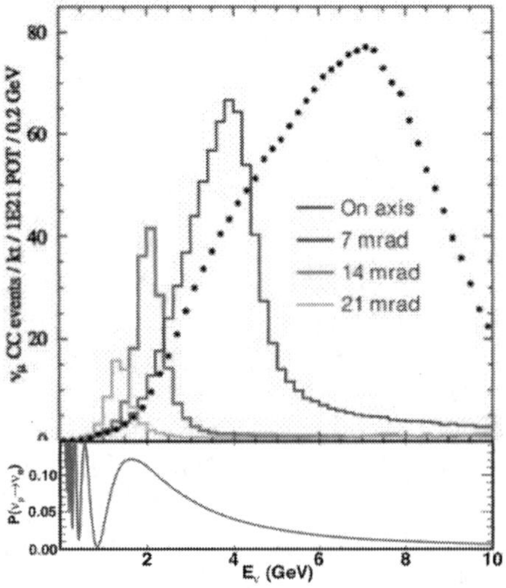


**Figure 1.** The energy spectra dependence on angle together with the ν oscillation probability profile.

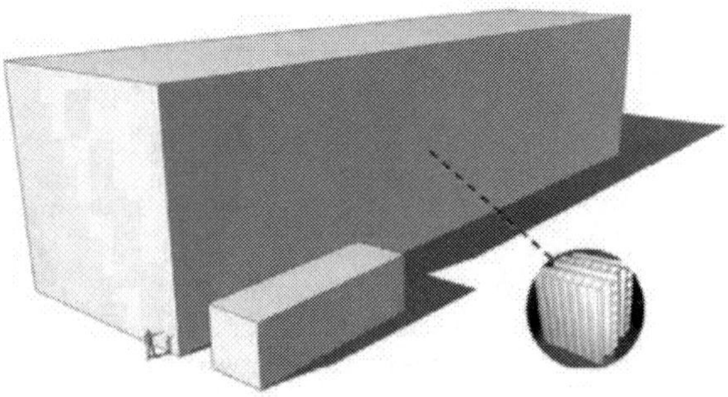

**Figure 2.** NOvA Far and Near detectors with a cut-away of alternating extruded PVC planes.

### 2.2. NOvA detectors

The NOvA detectors are finely segmented, 64 % active tracking calorimeters. The segmentation and the overall mechanical structure of the detectors are provided by a lattice of extruded PVC cells with cross sectional size 6 cm × 4 cm. Each cell extends the full width or height of the detector (15.6 m in the FD, 4.1 m in the ND) and is filled with mineral oil mixed with liquid scintillator (pseudocumene). Secondary particles from neutrino interactions excite the scintillator and the light produced by the scintillator is collected and transported to the end of the cell by a wavelength-shifting fiber that terminates on a pixel of a 32-channel avalanche photodiode (APD). Figure 2 shows a sketch of the FD and ND along with a cut-away view of the PVC lattice.

Each of the 928 layers of the FD has 384 cells, for ~344 000 total channels of readout. The ND has 206 layers each with 96 cells plus a muon range stack at the downstream end made by interleaving steel plates with standard detector layers, totalling 18 000 channels.

There is also a 0.2 kton prototype Near Detector On Surface (NDOS) which served for DAQ, cooling, electronics and firmware tuning and as a proof of concept and still runs and provides valuable feedback for design enhancements and operational experience

### 2.3. NOvA physics goals

NOvA will begin its oscillation data run in ν mode - that is with the NuMI horn configured to focus positive secondary hadrons. While FD construction and NuMI commissioning will still be underway during the first year of data taking, NOvA can nonetheless reach a 5σ C.L. observation of $\theta_{13}$-driven $\nu_\mu$ □ $\nu_e$ oscillations after one year, assuming the normal mass hierarchy. The baseline NOvA exposure is $3.6\times10^{21}$ POT, which can be accumulated in six years at design intensities, with a 14-kton FD. For the sensitivities shown below, this exposure is assumed divided evenly between ν and $\bar{\nu}$ running, a split that works well for many parameter scenarios but is nonetheless adjustable. The analysis techniques used here are those described in the NOvA Technical Design Report [1]. Updated analyses for use in the first NOvA results are under active development.

Outside of these primary goals, NOvA will also look for evidence of new physics through comparisons of $\nu_\mu$ □ $\nu_\mu$ and $\bar{\nu}_\mu$ □ $\bar{\nu}_\mu$, provide constraints on sterile neutrino models by measuring the total flux of active neutrinos at its downstream detector, monitor for supernova neutrino activity, perform neutrino-nucleus cross section measurements with a narrow-band beam, and pursue a variety of non-neutrino topics including searches for magnetic monopoles and hidden sector particles.

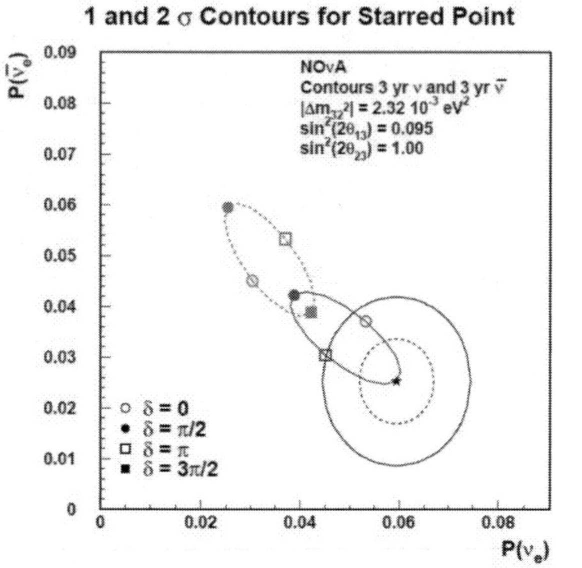


**Figure 2.** An example of possible result after 3+3 years of neutrino/antineutrino runs, excluding the inverted mass hierarchy at $2\sigma$.

## 3. Status of the NOνA experiment

The commissioning of the NOνA experiment is successfully approaching its completion. The FD laboratory currently holds 21 of 28 completed blocks, from which more than a half was filled with scintillator. Over 2 kt of detector (4.17 blocks) are equipped with APDs, read out and fully operational. The Near Detector cavern now accommodates the complete muon catcher and two regular blocks, which will continue being installed and should be completed in summer 2014.

The NuMI beam returned on September 4, 2013, with temporary capacity of 500 kW of beam power. After the Booster RF system upgrades, it will deliver the full 700 kW power in summer 2014. All reconstruction and analysis tools are in place and ready to be fed with data to produce early results on first runs from 2014.

## 4. Summary

NOνA is on track to make many important contributions to neutrino physics such as the measurement of the mixing angle $\theta_{13}$, determination of the neutrino mass hierarchy, search for neutrino CP violation, determination of $\theta_{23}$ octant and more precise measurements of $\Delta m^2_{23}$ and $\sin^2(2\theta_{23})$. Recent results from worldwide neutrino experiments are very encouraging for the NOνA goals as they provide evidence for a non-zero $\theta_{13}$ and non-maximal $\sin^2(2\theta_{23})$.

The Far detector block construction is being finished and more than 2kt of its mass have been instrumented with APDs and are capable of particle detection. Construction of the full 14 kt is expected to be completed by early 2014. The upgraded NuMI is now back on and will deliver 500 kW of beam power before the recycler ring will be upgraded to handle 700 kW beam. The Near detector muon catcher is in place, the first blocks are being assembled and the detector should be ready before mid 2014. The NOνA NDOS prototype has been taking and analyzing data since October 2010 and continues to operate and provide critical feedback for design enhancements and operational experience.

## References

[1] D. S. Ayres et al. "NOνA", FERMILAB-DESIGN-2007-01 (2007)
[2] F. An et al. "DAYA-BAY", Phys.Rev.Lett. 108, 171803 (2012), 1203.1669.
[3] J. Ahn et al. "RENO", Phys.Rev.Lett. 108, 191802 (2012), 1204.0626.

[4] Y. Abe et al. "Double Chooz", Phys.Rev. D86, 052008 (2012), 1207.6632.
[5] D. Dwyer "Daya Bay", talk given at the Neutrino 2012 Conference, June 3-9, 2012, Kyoto, Japan, http://neu2012.kek.jp/.
[6] K. Abe et al., Phys.Rev.Lett. "T2K" 107, 041801 (2011), 1106.2822.
[7] R. Nichol "MINOS" (2012), talk given at the Neutrino 2012 Conference, June 3-9, 2012, Kyoto, Japan, http://neu2012.kek.jp/
[8] P. Adamson et al., "MINOS" Phys. Rev. Lett. 107, 181802 (2011), hep-ex/1108.0015v1.

# Performing edge detection by Difference of Gaussians using q-Gaussian kernels

**L Assirati[1], N R Silva[2,1], L Berton[2], A A Lopes[2], O M Bruno[1,2]**

[1] Scientific Computing Group, São Carlos Institute of Physics, University of São Paulo (USP), cx 369 13560-970 São Carlos, São Paulo, Brazil - www.scg.ifsc.usp.br
[2] Institute of Mathematics and Computer Science, University of São Paulo (USP), Avenida Trabalhador são-carlense, 400 13566-590 São Carlos, São Paulo, Brazil

E-mail: assirati@usp.br, nubiars@icmc.usp.br, lberton@icmc.usp.br, alneu@icmc.usp.br, bruno@ifsc.usp.br

**Abstract.** In image processing, edge detection is a valuable tool to perform the extraction of features from an image. This detection reduces the amount of information to be processed, since the redundant information (considered less relevant) can be disconsidered. The technique of edge detection consists of determining the points of a digital image whose intensity changes sharply. This changes are, for example, due to the discontinuities of the orientation on a surface. A well known method of edge detection is the Difference of Gaussians (DoG). The method consists of subtracting two Gaussians, where a kernel has a standard deviation smaller than the previous one. The convolution between the subtraction of kernels and the input image results in the edge detection of this image. This paper introduces a method of extracting edges using DoG with kernels based on the q-Gaussian probability distribution, derived from the q-statistic proposed by Constantino Tsallis. To demonstrate the method's potential, we compare the introduced method with the tradicional DoG using Gaussians kernels. The results showed that the proposed method can extract edges with more accurate details.

## 1. Introduction

Image processing is designated as any type of signal processing where the input is an image and the output can be another image or a set of features extracted from the input image. Once the computer vision involves the identification and subsequent classification of certain objects in a given image, edges detection is an essential tool in image analysis. When performing edge detection on an image, there is a reduction of the amount of information to be processed because the redundant information (considered less relevant) can be disconsidered.

The segmentation by edge detection is based on two important concepts: similarity and discontinuity. Thus, the algorithms look for points (or curves and contours) of the digital image where the intensity changes abruptly. This sudden change in intensity may occur for various reasons, as example, the orientation discontinuities in a surface and changes in brightness and illumination in a scene. Applications for the edge detection method are found in various fields of science: medicine [1], engineering and satellite images [2], robotics and machine vision [3].

There are several methods for edge detection, like: Canny, Sobel, Prewitt, and based on Gaussian masks (kernels), as Laplacian of Gaussian (LoG) and Difference of Gaussian (DoG) [4]. The DoG method generally uses classical Gaussians in its approach. But this work suggests the use of q-Gaussian for the composition of the mask that will be applied to the image to extract its edges. The q-Gaussian probability distribution comes from the q-algebra introduced by Tsallis. The composition of the DoG filter using q-Gaussian kernels is motivated by results obtained in other fields of science [5]. The potential of the introduced method is demonstrated by comparing with the traditional method for DoG. One can notice that the presented method is able to perform edge extraction with greater detail.

## 2. Laplacian of Gaussian vs Difference of Gaussians

Consider the one-dimensional Gaussian distribution: $f(x,\mu,\sigma) = \frac{1}{\sqrt{2\pi\sigma^2}} \exp\left(-\frac{(x-\mu)^2)}{2\sigma^2}\right)$, with $-\infty < x < \infty$, and $\sigma > 0$, where $\mu$ is the mean, $\sigma$ is the standard deviation and $\sigma^2$ is the variance.

If we take the second derivative of the one-dimensional Gaussian function considering $\mu = 0$, we obtain the Ricker wavelet: $\psi(x) = \frac{1}{\sqrt{2\pi\sigma^2}} \frac{1}{\sigma^4}(x^2 - \sigma^2) \exp \frac{-x^2}{2\sigma^2}$.

In the two-dimensional the Gaussian distribution becomes:

$$f(x,y,\sigma) = \frac{1}{2\pi\sigma^2} \exp\left(-\frac{(x^2+y^2)}{2\sigma^2}\right), -\infty < x,y < \infty, \sigma > 0. \tag{1}$$

The Laplacian of Gaussian LoG is a multidimensional generalization of the Ricker wavelet. To obtain it we need to take the two-dimensional Laplacian of the Gaussian distribution:

$$LoG(x,y) = -\frac{1}{\pi\sigma^4}\left[1 - \frac{x^2+y^2}{2\sigma^2}\right] \exp -\frac{x^2+y^2}{2\sigma^2} \tag{2}$$

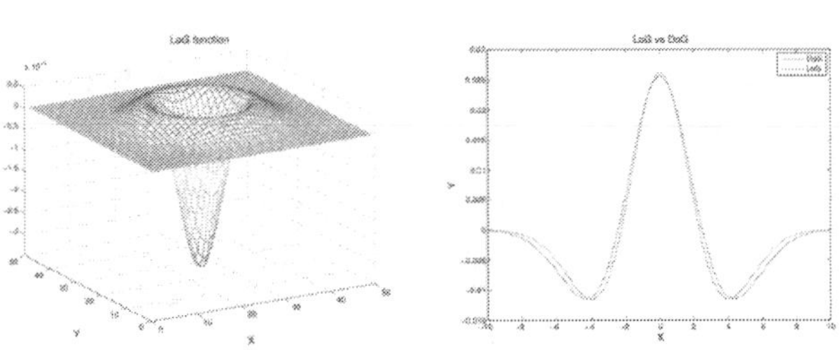

(a) (b)

**Figure 1.** (a) LoG Function, $\sigma = 2.5$, (b) Difference of Gaussians vs Laplacian of Gaussians 1D.

However, in practice the Laplacian of Gaussian (LoG), Figure 1(a), is approximated by the Difference of Gaussians function (DoG) since this reduces the computational costs for two or more dimensions. The DoG is obtained by performing the subtraction of two Gaussian kernels where a kernel must have a standard deviation slightly lower than the previous one. Figure 1(b) compares the LoG function with $\sigma = 2.5$ with the DoG function using kernels with $\sigma_1 = 2.5$ and $\sigma_2 = 2.15$. The convolution of the DoG filter with the input image generates the edge detection for this image.

## 3. q-Gaussian

In 1988, Tsallis proposed the non additive statistical mechanics, entitled "Q-statistic" [6]. This theory suggests that different systems require different tools of analysis, appropriated to the particularities of this system. The informational tool entropy, applied to the information theory by Shannon [7] is defined as: $S(x) = -\sum_{x=0}^{W} p(x)\log p(x)$, where $p(x)$ is the occurrence probability, and $W$ is the total number of probabilities.

The generalization proposed by Tsallis gives the definition of the q-entropy: $S_q(x) = \frac{1}{q-1}\left(1 - \sum_{x=0}^{W} p^q(x)\right)$, where $p(x)$ is the occurrence probability, $W$ is the total number of probabilities and $q$ is an adjustable parameter, freely variable. The correct choice of certain $q$ parameters can evidence important characteristics of the system. When $q \to 1$, one retrieves the standard entropy.

The q-Gaussian probability distribution comes from the maximization of the Tsallis entropy under appropriate constraints [8]. Again, when $q \to 1$, one retrieves the Gaussian distribution. The q-Gaussian is defined as:

$$G_q(x) = \frac{1}{C_q\sqrt{2\sigma^2}} \exp_q \frac{-x^2}{2\sigma^2}, \tag{3}$$

with $\exp_q(x) = [1 + (1-q)x]^{\frac{1}{1-q}}$ and

$$C_q = \begin{cases} \frac{2\sqrt{\pi}\Gamma(\frac{1}{1-q})}{(3-q)\sqrt{1-q}\Gamma(\frac{3-q}{2(1-q)})} & \text{if } -\infty < q < 1 \\ \sqrt{\pi} & \text{if } q = 1 \\ \frac{\sqrt{\pi}\Gamma(\frac{3-q}{2(q-1)})}{(3-q)\sqrt{q-1}\Gamma(\frac{1}{q-1})} & \text{if } -\infty < q < 1. \end{cases}$$

The same way as the classical Gaussian has a Two-Dimensional version, we can derive the multidimensional generalization to q-Gaussian. The bi-dimensional q-Gaussian is defined by $G_q(x,y) = \frac{\exp_q -(x^2+y^2)/(2\sigma^2)}{2C_q^2\sigma^2}$. It is important to note that curves with the same parameter $\sigma$ can have its shape changed adjusting the $q$ parameter, adapting it to the peculiarities of the problem in which it is employed. Figures 2(a), 2(b) and 2(c) show some representatives of the family of 2D q-Gaussian.

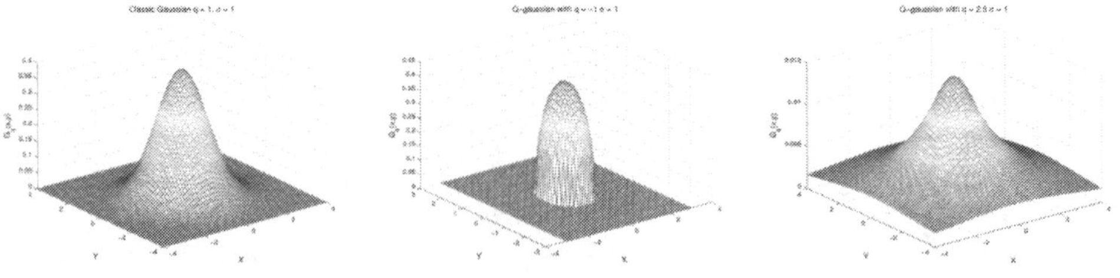

(a) $q = 1$, $\sigma = 1$ (b) $q = -1$, $\sigma = 1$ (c) $q = 2.5$, $\sigma = 1$

**Figure 2.** Q-Gaussian Function: Same $\sigma$, but different shapes.

## 4. Method and Results

This study introduces the use of DoG method using q-Gaussian kernels as an alternative to traditional use of Gaussian kernels in edge detection. Following the metric proposed by the DoG filter, standard deviations $\sigma_1$ and $\sigma_2$ are setted, with $\sigma_2$ smaller than $\sigma_1$. After the filter having the appropriate size, we should set it with the input image in gray scale. After the convolution we identify the edges by using the "zero cross" detector.

The results for edge detection using the method DoG with q-Gaussian kernels show up rich in detail when compared to the method DoG with Classic Gaussian kernels because the q-Gaussian probability distribution have the adjustable parameter $q$. This parameter allow us to define the degree of detail that we seek in our detection. Figures 3(a), 3(b) and 3(c) show results obtained from q-Gaussian using $\sigma_1 = 0.2$ and $\sigma_2 = 0.1$.

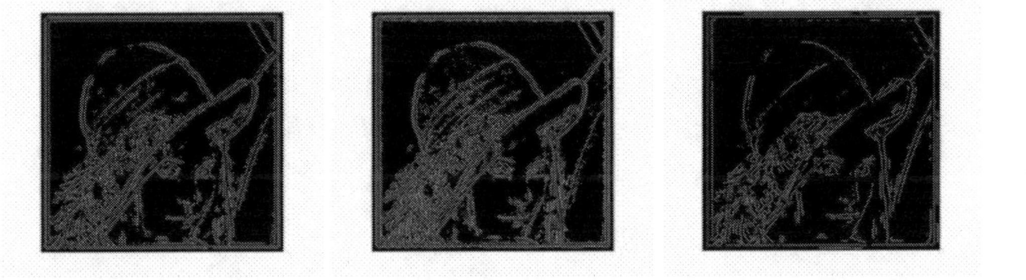

(a) $q = 1$ (b) $q = -1$ (c) $q = 2.5$

**Figure 3.** (a)DoG with Classic Gaussian kernels, (b) and (c) DoG with q-Gaussian kernels.

## 5. Conclusions

The results presented in this work show that using q-Gaussian kernels proves to be an excellent alternative to the classical Gaussian kernels. Compared to the DoG filter with kernels using the normal distribution of probabilities, we note that we gain in details of edge detection. That is because in addition to the variable parameter $\sigma$, responsible for more or less blurring (Gaussian blur), we also have the entropic index $q$, variable and responsible for the shape of q-Gaussian, being able to get more details that the traditional approach when both have the same blur. The extensiveness or not extensiveness of the entropy depends on the system characteristics. Thus, it can be extended for certain values of $q$. In this point, we can apply this concept to our work. By using the q-Gaussian method, the entropic index $q$ allow us adjust the function used in the filter to get the details and results that are more relevant.

## Acknowledgments

Lucas Assirati acknowledges the Confederation of Associations in the Private Employment Sector (CAPES). Núbia R. Silva, Lilian Berton and Odemir M. Bruno are grateful for São Paulo Research Foundation, grant Nos.: 2011/21467-9, 2011/21880-3 and 2011/23112-3. Bruno also acknowledges the National Council for Scientific and Technological Development (CNPq), grant Nos. 308449/2010-0 and 473893/2010-0.

## References

[1] Gudmundsson M, El-Kwae E and Kabuka M 1998 *Medical Imaging, IEEE Transactions on* **17** 469–474
[2] Augusto G, Goltz M and Demísio J 1984 1044–1045
[3] Jain R, Kasturi R and Schunck B 1995 *Machine vision*
[4] Gonzalez R and Woods R 2011 *Digital Image Processing* (Pearson Education) ISBN 9780133002324
[5] Soares I J A 2013 *Desenvolvimento de uma base de funções paramétricas para interpolação de imagens médicas.* Ph.D. thesis University of São Paulo - Faculdade de Filosofia, Ciências e Letras de Ribeirão Preto
[6] Tsallis C 1988 *Journal of Statistical Physics* **52** 479–487
[7] Shannon C E 1948 *The Bell System Technical Journal* **27** 379–423–623–656
[8] Tsallis C 2011 *Entropy* **13** 1765–1804

# Modelling study of magnetic and concentration phase transition in ultrathin antiferromagnetic films

**Afremov Leonid, Petrov Aleksandr**

Departament of Theoretical and Experimental Physics, The School of Natural Sciences, Far Eastem Federal University, 8, Sukhanova St., 690950, Vladivostok, Russian Federation

e-mail: afremov.ll @ dvfu.ru, aleksandr.al.petrov@gmail.com

**Abstract.** Using the method of the "average spin" a modelling study of magnetic and concentration phase transition in ultrathin antiferromagnetic of different crystalline structure has been carried out. It has been shown, that relative change of Neel temperature is subject to the power law with negative index which doesn't depend on the film's crystal kind. The calculation of the dependence of phase transition critical concentration in diluted magnetic material on the film thickness has been made out. The legitimacy of the use of the method developed for modelling of magnetic and concentration phase transition in different nanostructures is certified by accordance between the results of calculations and the experimental data.

## 1. Introduction

Increasing of magnetic phase transition temperature in antiferromagnetic ultrathin films with growth of its thickness was marked in many researches [see, for example, 1 - 4]. And also, as was shown in the work [2], experimentally observable dependence of Neel temperature $T_N$ on the thickness of the film $N$ can be described with the help of the following correlation:

$$\varepsilon(D) = \frac{T_N(D\to\infty)-T_N(D)}{T_N(D\to\infty)} = \left(\frac{\xi_0}{D}\right)^{\lambda} \tag{1}$$

where $T_N(N \to \infty)$ is the temperature of "lump sample", $\xi_0$ is the correlation length at $T = 0\ K$, the argument $\lambda$ is related with critical value of spin magnetic moments correlation $\nu$ as $\lambda = 1/\nu$. It should be marked out also that the critical temperature of magnetic phase transition in diluted magnetic materials depends very much on the concentration of "magnetic atoms" [3, 5].

A modeling study of effect of finiteness of the film thickness and the concentration of "magnetic atoms" on the temperature of magnetic phase transition $T_{Neel}$ in diluted antiferromagnetic material has been carried out by us in this article. For solution of the problem let us use the model of "average spin" developed by us [5, 6], the basic provision of which are formulated below.

## 2. The model

- Ultrathin film consist of $N$ monolayers of diluted magnetic material with the concentration of "magnetic atoms" $p$;

- The direct exchange interaction occurs only between the nearest neighbors; moreover, the interaction fields $h$ between the spin magnetic moments of the atoms are distributed in a random manner;
- Spin magnetic moments are oriented along an axis $Oz$ (approximation of the Ising model).

According to the approximation of the above model, the equations defining the temperature of magnetic phase transition $T_{Neel}$ in antiferromagnetic ultrathin film can be put in the following form:

$$1 = \sum_{k_1=0}^{z_{11}} \sum_{l_1=0}^{k_1} \sum_{k_2=0}^{z_{12}} \sum_{l_2=0}^{k_2} C_{z_{11}}^{k_1} C_{k_1}^{l_1} C_{z_{12}}^{k_2} C_{k_2}^{l_2} p^{k_1+k_2}(1 - p)^{z_{11}+z_{12}-k_1-k_2} \frac{(2l_1 - k_1) + (2l_2 - k_2)x_2}{2^{k_1+k_2}} \tanh\left(\frac{(2l_1 - k_1) + (2l_2 - k_2)i_{1,2}}{t_{Neel}}\right)$$

$$x_n = \sum_{k_1=0}^{z_{nn}} \sum_{l_1=0}^{k_1} \sum_{k_2=0}^{z_{n,n-1}} \sum_{l_2=0}^{k_2} \sum_{k_3=0}^{z_{n,n+1}} \sum_{l_3=0}^{k_3} C_{z_{nn}}^{k_1} C_{k_1}^{l_1} C_{z_{n,n-1}}^{k_2} C_{k_2}^{l_2} C_{z_{n,n+1}}^{k_3} C_{k_3}^{l_3} p^{k_1+k_2+k_3}(1 - p)^{z_{nn}+z_{n,n-1}+z_{n,n+1}-k_1-k_2-k_3} \frac{(2l_1 - k_1)x_n + (2l_2 - k_2)x_{n-1} + (2l_3 - k_3)x_{n+1}}{2^{k_1+k_2+k_3}} \tanh\left(\frac{(2l_1 - k_1)i_{n,n} + (2l_2 - k_2)i_{n-1,n} + (2l_3 - k_3)i_{n,n+1}}{t_{Neel}}\right) \quad (2)$$

$$x_N = \sum_{k_1=0}^{z_{NN}} \sum_{l_1=0}^{k_1} \sum_{k_2=0}^{z_{N,N-1}} \sum_{l_2=0}^{k_2} C_{z_{z_{NN}}}^{k_1} C_{k_1}^{l_1} C_{z_{z_{N,N-1}}}^{k_2} C_{k_2}^{l_2} p^{k_1+k_2}(1 - p)^{z_{N,N}+z_{N,N-1}-k_1-k_2} * \frac{(2l_1 - k_1)x_N + (2l_2 - k_2)x_{N-1}}{2^{k_1+k_2}} \tanh\left(\frac{(2l_1 - k_1)i_{N,N} + (2l_2 - k_2)i_{N,N-1}}{t_{Neel}}\right)$$

where $1 < n < N$, $x_n = (\partial\mu_n/\partial\mu_1)|_{(\mu_1\to 0,\dots\mu_n\to 0,\dots,\mu_N\to 0)}$, $i_{n-1,n} = m_n J_{n-1,n}/(m_1 J_{1,1})$, $i_{n,n} = m_n J_{n,n}/(m_1 J_{1,1})$, $i_{n,n+1} = m_n J_{n,n+1}/(m_1 J_{1,1})$, $t_{Neel} = k_B T_{Neel}/(m_1 J_{1,1})$ ; $m_n$ – the magnetic moment, $z_{n,n}$ – the number of the nearest neighbours and $J_{n,n}$ is the exchange constant in the layer with the number $n$, $J_{n-1,n}$, $J_{n,n+1}$, $z_{n-1,n}$, $z_{n,n+1}$ are the constants of exchange interaction and the number of the nearest neighbours between the atoms of neighboring layers.

Using the correlation (2) we can investigate the effect of the "magnetic atoms" concentration p, the film thickness $N$ and its crystalline structure on the Neel temperature.

It's remarkable that the simultaneous equations (2) can be resolved using the Newton method. The equations (2), notwithstanding their complicated form, have been put in an algorithmically comfortable form, allow to carry out modeling studies of magnetic properties of the materials basing on the knowledge of the film's crystalline structure and the numerical value of the exchange integral $J_{m,k}$ only. The rule for selecting of nearest neighbours of each atom $z_{m,k}$ is an important part of the method. When changing of this rule we can easily correct the modeling method for magnetic systems with different crystallographic structure.

## 3. Modelling study of magnetic phase transition in undiluted ($p = 1$) antiferromagnetic material

The calculation results for dependence of critical temperature of phase transition on the thickness of ultrathin undiluted ($p = 1$) antiferromagnetic film, the crystalline structure and the orientation of crystallographic axes are shown at the figure 1. According to expectations, with increasing of monolayers number the Neel temperature rather quickly reaches the values adequate to a "lump sample". For example for a simple cubic lattice the extreme relative Neel temperature $t_N(N \to \infty) =$

5.05, for a body-centered lattice it is $t_N(N \to \infty) = 7.01$ и for a face-centered is $t_N(N \to \infty) = 11.04$.

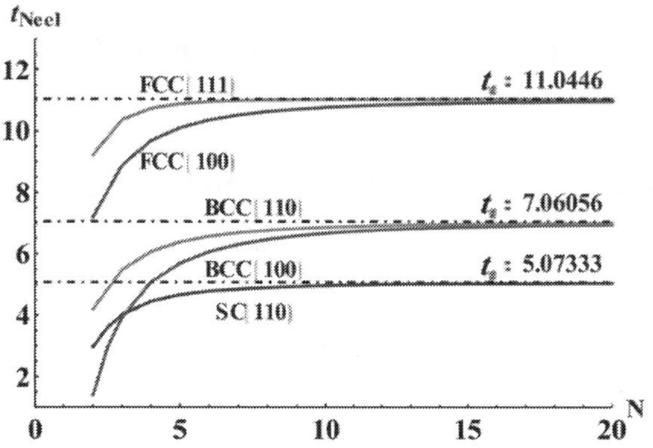


**Figure 1.** The dependence of relative Neel temperature $t_N = k_B T_N/(m_1 J_{1,1})$ on the films thickness $N$ (in monolayers) for different crystalline lattice and crystallographic axes directions.

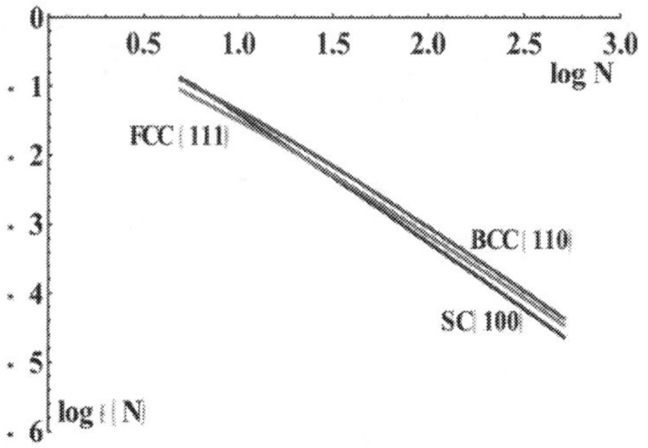


**Figure 2.** The dependence of relative change of the Neel temperature $\varepsilon(N)$ on the film's thickness in a logarithmic scale.

The results represented at the Figure 1, can be put in a logarithmic scale (see the Figure 2), where $\varepsilon(N) = (T_N(N \to \infty) - T_N(N))/T_N(N \to \infty)$ is a relative change of the Neel temperature.

Evidently, the relative change of the Neel temperature $\varepsilon(N)$ can be well described with the help of the correlation (1) with the argument $\lambda$ values presented in the table below. It results from the table that the values of $\lambda$ calculated by us, are rather close to the experimental data obtained in respect of the films $CoO/SiO_2$ in which $CoO$ has a crystalline structure kind of $NaCl$ with a FCC lattice.

**Table 1.** The theoretical and experimental values of $\lambda$.

| Crystalline lattice | Argument $\lambda$ |
|---|---|
| Calculated values | |
| Simple cubic (SU (100)) | 1.802 |
| Body-centered (BCC (100)) | 1.599 |
| Body-centered (BCC (110)) | 1.609 |
| Face-centered (FCC (100)) | 1.581 |
| Face-centered (FCC (111)) | 1.585 |
| Experimental values obtained in respect of the films $CoO/SiO_2$ | |
| $\lambda_{exp} = 1.55 \pm 0.05$[1], $\lambda_{exp} = 1.60 \pm 0.1$[2] | |

## 4. A modelling study of magnetic phase transition in diluted antiferromagnetic material

The investigations results for concentration dependence of the magnetic phase transition temperature for films of different thickness and crystalline lattice are represented at the Figure 3. The picture illustrates quite a natural result: while decreasing the concentration of "magnetic atoms" the interaction between them falls down, and as a consequence the Neel temperature decreases. Moreover, under some critical concentration the Neel temperature falls down to zero and the system transits from an ordered state to an unordered (paramagnetic) state.

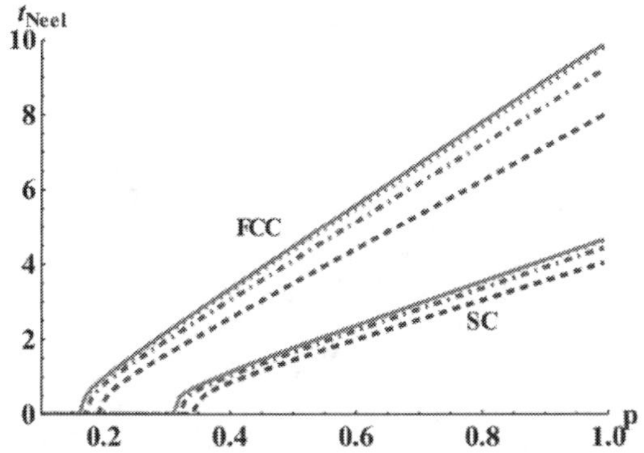


**Figure 3.** The dependence of the reduced Neel temperature on the concentration of “magnetic atoms” and on the films thickness $N$ (in monolayers) for different crystalline lattice (the lower line: $N = 2$, the middle line: $N = 3$, the upper lines $N = 5$ and $N = 8$ coincide).

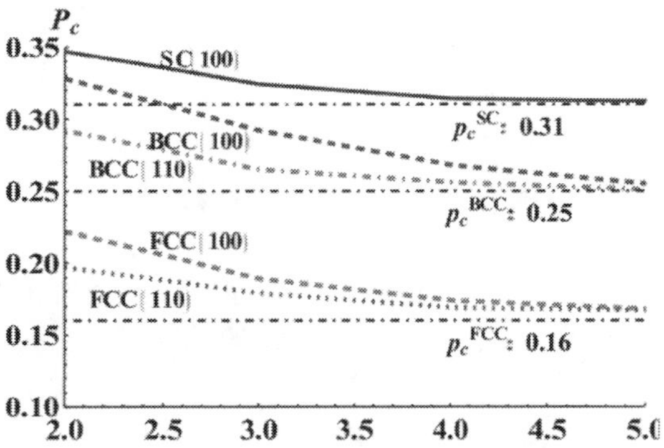


**Figure 4.** The dependence of the critical concentration on the films thickness $N$ (in monolayers) for different crystalline lattice.

From the figure 4 we can see that the critical concentration depends on the film thickness and its crystalline lattice, that is related with changing of the number of the nearest neighbors. Therefore, while increasing the number of monolayers the curve $p_c = p_c(N)$ asymptotically approach to the percolation thresholds of the relative lattices: $p_c(N \to \infty) = 0.31$ for a simple cubic lattice, $p_c^{BSS}(N \to \infty) = 0.25$ for a body-centered lattice and $p_c^{FSS}(N \to \infty) = 0.167$ for a face-centered lattice [1, 7].

## 5. Conclusion

A modeling study of magnetic phase transition in diluted magnetic materials has been carried out in the framework of a rather simple model developed by us earlier [6]. The efficiency of the model is defined both by the small number of varying parameters ($J_{m,k}$, $z_{m,k}$ и p) and by the simple procedure of choosing of the nearest neigbours $z_{m,k}$ and the values of exchange integrals $J_{m,k}$.

## 6. Acknowledgements

The work is supported by grant of Ministry of Education and Science. Reference number is Government Order-2013/287, code 2.8649.2013.

## References


[1] Ambrose T and Chien C L 1996 *Physical Review Letters* **76** (10) 1743-1746

[2] Ambrose T and Chien C L 1996 *Journal of Applied Physics* **79** 5920-5922

[3] Faz C A, Bland J A C, Lauhoff G 2008 *Reports on Progress in Physics* **71** 1-78

[4] Gradmann U 2001 *Encyclopedia of Materials: Science and Technology* 5832-5837

[5] Belokon V, Nefedev K J 2001 *Journal of Experimental Theoretical Physics* **93** 136-142

[6] Afremov L L, Il'yushin I G 2013 *Advanced Materials Research* **683** 69-72

[7] Sadeh B, Doi M, Shimizu T, Matsui M 2000 *Journal of the Magnetics Society of Japan* **24** (4/2) 511–514

# Regular graph construction for semi-supervised learning

**Didier A. Vega-Oliveros, Lilian Berton, Andre Mantini Eberle, Alneu de Andrade Lopes, Liang Zhao**

Instituto de Ciências Matemáticas e de Computação, Universidade de São Paulo - Campus de São Carlos, CP 668 - São Carlos, SP - Brazil

E-mail: davo,lberton,andre,alneu,zhao@icmc.usp.br

**Abstract.** Semi-supervised learning (SSL) stands out for using a small amount of labeled points for data clustering and classification. In this scenario graph-based methods allow the analysis of local and global characteristics of the available data by identifying classes or groups regardless data distribution and representing submanifold in Euclidean space. Most of methods used in literature for SSL classification do not worry about graph construction. However, regular graphs can obtain better classification accuracy compared to traditional methods such as $k$-nearest neighbor ($k$NN), since $k$NN benefits the generation of hubs and it is not appropriate for high-dimensionality data. Nevertheless, methods commonly used for generating regular graphs have high computational cost. We tackle this problem introducing an alternative method for generation of regular graphs with better runtime performance compared to methods usually find in the area. Our technique is based on the preferential selection of vertices according some topological measures, like closeness, generating at the end of the process a regular graph. Experiments using the global and local consistency method for label propagation show that our method provides better or equal classification rate in comparison with $k$NN.

## 1. Introduction

Semi-supervised learning (SSL) uses large amount of unlabeled data and available labeled data to build classifiers applyied to real problems. As SSL requires less human effort and gives higher accuracy, it is of great interest [10], [2]. Among the current SSL methods, graph based approaches have emerged and highlighted, specially, when no parametric information is available about the data distribution.

Several graph-based methods were developed and much of them are similar to each other. Zhu (2005) [10] arguments that it is more important to construct a good graph than to choose among the methods. However, graph construction is not a well studied area. Only recently, the issue of graph construction has received attention [8], [4], [5].

The most common method used for graph construction is neighbor graphs, for example $k$-Nearest Neighbors ($k$NN) graph, where each item is connected to its $k$ nearest neighbors under some distance measure. As $k$NN method greedily connects the $k$ nearest neighbors to each vertex and may return graphs where some vertices have more than $k$ neighbors, Jebara et al. (2009) [4] proposed the $b$-matching, which ensures the graph is regular (every vertex with $b$ neighbors) and by experimental results suggest that a regular graph can achieve better classification results compared to $k$NN. Huang and Jebara (2007) [3] developed an implementation based on belief propagation, but the guaranteed running time of the implementation is $O(bn^3)$. In some cases, like in the work of Ozaki et al. (2011) [6], building a $b$-matching graph is inviable in terms of computational cost.

In the supervised context, nearest neighbor classification does not work properly in high-dimensional space. Radovanovic et al. (2010) [7] argue that this happen because a hub is an example close to many other examples in the (high-dimensional) example space. They state that such hubs inherently emerge in high-dimensional data as a side effect of the "curse of dimensionality". Ozaki et al., (2011) [6] extend this argument and made an observation that a hub in the data space also makes a hub in the $k$NN graph, since $k$NN graph construction greedily connects a pair of vertices, if the corresponding vertex is among the $k$ closest neighbors of the other example in the original space.

To test this hipotesis, that regular graph can be better for SSL, we introduce a new method for generation of graphs with no hubs. Our method has quadratic time complexity, as $k$NN algorithm. To evaluate if this technique is better than other that generates hubs, like $k$NN, we compare the classification accuracy between them using Local and Global Consistency (LGC) algorithm [9] for the label propagation task. The classification results from UCI [1] and Chapelle [2] data sets show the presented method achieves results better or equal than $k$NN method.

The remainder of this paper is organized as follows. Section 2 defines basic concepts. Section 3 provides the details of the graph construction method introduced and the experimental validation results for the algorithm on benchmark datasets. Concluding remarks are then provided in Section 4.

## 2. Definitions

Given a set of $n$ examples, $X = \{x_1, \ldots, x_n\}$, semi-supervised classification methods utilize $l$ labeled examples $\{(x1, y1), \ldots, (xl, yl)\}$ and the remaining $u = n - l$ unlabeled examples $\{x_{l+1}, \ldots, x_{l+u}\}$ to infer the missing labels $\{y_{l+1}, \ldots, y_{l+u}\}$ corresponding to the unlabeled examples. For using a graph-based algorithm it is necessary the estimation of a weighted undirected sparse graph $G$ derived from the input data $X$. In this paper we are interested in how to construct a regular graph from $X$.

A graph $G = (V, E)$ is formed by a set $V$ of vertices (nodes) and a set $E$ of edges (links) that connect pairs of vertices. The cardinality of $V$ is usually denoted by $n$, the cardinality of $E$ by $m$. If two vertices are joined by an edge, they are adjacent and we call them neighbors. Often it is useful to associate numerical values (weights) to the edges or vertices of a graph $G$. Edge weights can be represented as a function $w : E \rightarrow \Re$ that assigns to each edge $e \in E$ a weight $w(e)$. In the context of this work, edge weights describe similarity between the adjacent vertices.

A graph $G$ can be described by the adjacency matrix $P$, a $N \times N$ square matrix whose entry $p_{ij}$ $(i, j = 1, \ldots, N)$ is equal to 1 when the link $p_{ij}$ exists, and 0 otherwise. The degree $d_i$ of a node $i$ is the number of edges incident with it, and is defined in terms of the adjacency matrix $P$ as $d_i = \sum_{j \in N} p_{ij}$. If a node has a degree much bigger than the others nodes, it is called *hub*. The averaged degree for a network is defined as $\langle d \rangle = \frac{1}{N} \sum_{n \in N} d_n$. Graph-based methods are in general transductive, that means it only works on the labeled and unlabeled training data, and not handle unseen data.

**Algorithm 1** Sequencial $k$NN method

**input:** data base $X$, $k_{max}$
**symbols:** $P$ - adjacency matrix; $R$ - set of vertices order by a relevance criterium;
$G$ - set of vertices degree initialized as $G : [0, \ldots, 0]_{1 \times N}$

```
compute kNN ← getK-NearestNeighbor(X)
compute R ← getOrderByRelevance(X)
v_i, v_j, k = 1
repeat
   v_i ← getNextNode(R)
   v_j ← getNearestPossibleNeighbor(v_i, k, k_max, kNN)
   if v_j ≠ NULL
      connect(P, v_i, v_j, G)
   else
      k ← k + 1
until ∃ v_i | G(v_i) < k_max
output: P
```

## 3. Regular graph construction

We introduce a new method for regular graph construction called *Sequencial kNN* (S-$k$NN). The method consists in create connections incrementaly, from $k = 1$ to a maximum $k_{max}$ value. In the process a vertex is chosen by a relevance criterium and establish a connection with the disponible nearest neighbor. The relevance criterium order the vertices by a Complex Network measure. Here we use the measure *closeness*: $C_i = \frac{1}{\sum_{x_j \in V} \|x_i - x_j\|}$, where $\| \, . \, \|$ is some distance kernel, and we use Euclidian distance.

Algorithm 1 describes the steps for the graph construction. After computing the $k$ nearest neighbors vector for the vertices and order it by a relevance criterium, we take vertices from the ordered vector and try to connect each vertex to the nearest neighbor into the $k_{max}$ neighbors, that have a degree smaller than $k$. If it is not possible, we increment $k$ and then, we repeat the process. The algorithm ends when all the vertices have degree bigger or equal than $k_{max}$. If it does not happen, the vertex that have degree smaller than $k_{max}$ will connect to the $k$ nearest neighbor with smallest degree. The complexity is the same as $k$NN algorithm.

The experiments were carried out on ten data sets. The first seven are from UCI Machine Learning Repository [1] and the last three are from Chapelle et al. (2006) [2]. For USPS, $\text{DIGIT}_1$ and $\text{COil}_2$ we apply Principal Component Analysis (PCA) to all data sets reducing the dimensions to 50. The matrix was symmetrized as follows $P_{ij} = max(P_{ij}, P_{ji})$. To generate the weighted graph $W$ we use the binary weighting approach, where $W = P$. The labeled points was randomly selected from all the points. The parameter $k_{max}$ was varied from 1 to 20. Averaged classification accuracy of 30 runs is used as the evaluation measure and the results are shown in Table 1.

**Table 1.** Averaged accuracy with different labeled points.

| Data set (# labeled points) | # Instances | # Atributes | # Classes | % Accuracy $k$NN | % Accuracy S-$k$NN |
|---|---|---|---|---|---|
| Zoo (7) | 101 | 16 | 7 | 72.489 ± 11.614 | **76.361 ± 9.59** |
| Zoo (21) | | | | **74.875 ± 5.239** | 73.4 ± 4.75 |
| Iris (3) | 150 | 4 | 3 | 89.918 ± 4.78 | **91.931 ± 4.581** |
| Iris (9) | | | | 89.177 ± 6.732 | **90.226 ± 8.048** |
| Glass (6) | 214 | 9 | 6 | 37.115 ± 8.125 | **37.894 ± 7.895** |
| Glass (18) | | | | 46.448 ± 6.072 | **47.969 ± 7.434** |
| Breast Cancer(2) | 286 | 9 | 2 | 96.279 ± 1.192 | **96.755 ± 0.698** |
| Breast Cancer (6) | | | | 91.535 ± 8.336 | **94.322 ± 4.51** |
| Ecoli (8) | 336 | 7 | 8 | **67.903 ± 8.663** | 66.283 ± 8.817 |
| Ecoli (24) | | | | 76.367 ± 4.835 | **76.809 ± 3.706** |
| Blood transfusion(2) | 748 | 4 | 2 | 52.788 ± 14.899 | **52.906 ± 17.879** |
| Blood transfusion (6) | | | | **66.417 ± 8.902** | 65.425 ± 11.096 |
| Yeast(10) | 1484 | 8 | 10 | 29.398 ± 5.760 | **30.474 ± 8.434** |
| Yeast (30) | | | | 41.490 ± 3.762 | **41.815 ± 4.310** |
| USPS (10) | 1500 | 241 | 2 | **83.543 ± 4.596** | 82.930 ± 3.356 |
| USPS (100) | | | | **89.609 ± 3.169** | 89.516 ± 1.947 |
| $\text{DIGIT}_1$ (10) | 1500 | 241 | 2 | **89.619 ± 7.488** | 87.225 ± 7.554 |
| $\text{DIGIT}_1$ (100) | | | | 97.033 ± 1.121 | **97.307 ± 0.683** |
| $\text{COIL}_2$(10) | 1500 | 241 | 2 | **65.731 ± 5.19** | **65.738 ± 6.178** |
| $\text{COIL}_2$(100) | | | | 96.938 ± 1.95 | **97.297 ± 1.243** |

Figure 1 shows the degree distribution for S-$k$NN and $k$NN graphs built using the Breast-Cancer data set with $k = 7$. We notice that S-$k$NN method has almost all vertices with a degree equal 7, less than 250 vertices have degree equal $8, 9, 10$. It generated 2597 edges with the averaged degree equal 7.6. The $k$NN method has vertices with vary different degree where less than 200 vertices have degree equal 7, the remainig have degree from 8 to 36. It generated 3548 edges with the averaged degree equal 10.4. Figure 2 shows a graph built using the Glass data set with $k = 5$. Bigger the points, bigger the vertice degree. The $k$NN graph has more bigger points compared to S-$k$NN.

## 4. Conclusion

From the experiments results we notice that the introduced method achieves better or equals results than $k$NN algorithm. This indicates that regular graphs also have good classification accuracy in graph-based

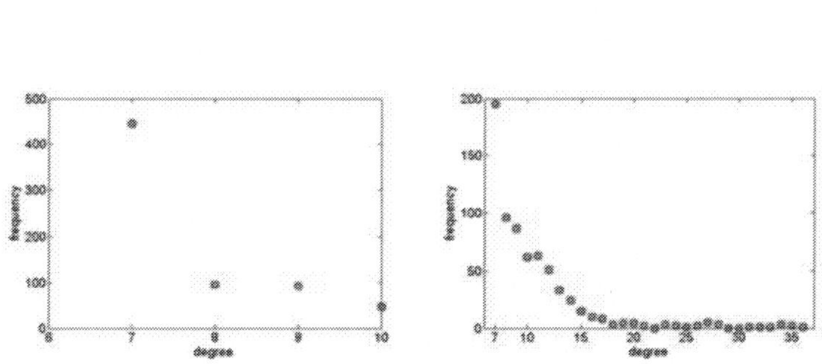

**Figure 1.** Degree distribution for S-$k$NN and $k$NN graphs built using the Breast-Cancer data set, $k = 7$.

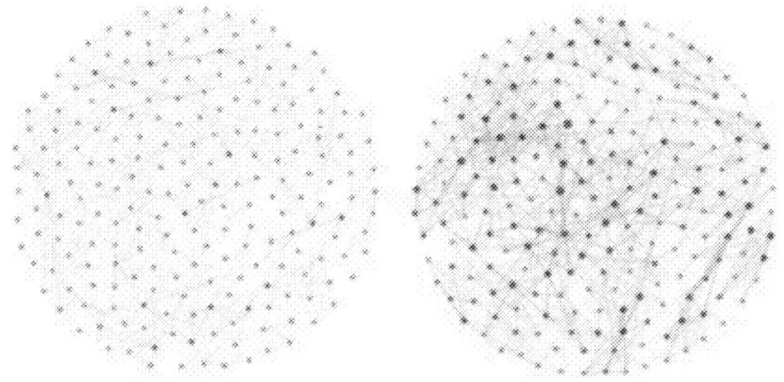

**Figure 2.** S-$k$NN and $k$NN graphs built using the Glass data set, $k = 5$.

SSL. For future work we will do statistic tests to detect if there are differences among algorithms. We also will compare the results to $b$-matching method and test other measures for the relevance criterium.

## 5. Acknowledgments

Grant 2011/21880-3, Sao Paulo Research Foundation (FAPESP) and National Council for Scientific and Technological (CNPq). The opinions, assumptions, and conclusions or recommendations expressed in this material are those of the author(s) and do not necessarily reflect the views of FAPESP and CNPQ.

## References

[1] Bache, K.; Lichman, M. (2013). *UCI Machine Learning Repository* [http://archive.ics.uci.edu/ml]. Irvine, CA: University of California, School of Information and Computer Science.

[2] Chapelle, O.; Schlkopf, B.; Zien, A. editors (2006) *Semi-Supervised Learning*. MIT Press, Cambridge, MA.

[3] Huang, B.; Jebara, T. (2007) *Loopy belief propagation for bipartite maximum weight b-matching*. Int. Workshop on Artificial Intelligence and Statistics.

[4] Jebara, T.; Wang, J.; Chang, S.F. (2009) *Graph construction and b-matching for semi-supervised learning*. In Proceedings of the 26th Annual International Conference on Machine Learning, p. 441-448.

[5] Maier, M.; Luxburg, U. (2009) *Influence of graph construction on graph-based clustering measures*. The Neural Information Processing Systems, v. 22, p. 1025-1032.

[6] Ozaki, K.; Shimbo, M.; Komachi, M.; Matsumoto, Y. (2011) *Using the Mutual k-Nearest Neighbor Graphs for Semi-supervised Classification of Natural Language Data*. In Proceedings of the 15th Conference on Computational Natural Language Learning, p. 154-162.

[7] Radovanovic, M.; Nanopoulos, A; Ivanovic, M. (2010) *Hub in space: popular nearest neighbors in high-dimensional data*. Journal of Machine Learning Research, v.11.

[8] Wang, F.; Zhang, C. (2008) *Label propagation through linear neighborhoods*. IEEE Transactions on Knowledge and Data Engininеering, v. 20, p. 55-67.

[9] Zhou, D.; Bousquet, O.; Lal, T. N.; Weston, J.; Schlkopf, B. (2004) *Learning with local and global consistency*. In Advances in Neural Information Processing Systems, v. 16, p. 321-328: MIT Press.

[10] Zhu, X. (2005) *Semi-supervised learning literature survey*. Technical report 1530 - Computer Sciences, University of Wisconsin-Madison.

# An Intuitive Dashboard for Bayesian Network Inference

**Vikas Reddy, Anna Charisse Farr, Paul Wu, Kerrie Mengersen, Prasad K.D.V. Yarlagadda**

2 George St, Queensland University of Technology, QLD 4000, Australia

E-mail: v1.reddy@qut.edu.au, a.farr@qut.edu.au

**Abstract.** Current Bayesian network software packages provide good graphical interface for users who design and develop Bayesian networks for various applications. However, the intended end-users of these networks may not necessarily find such an interface appealing and at times it could be overwhelming, particularly when the number of nodes in the network is large. To circumvent this problem, this paper presents an intuitive dashboard, which provides an additional layer of abstraction, enabling the end-users to easily perform inferences over the Bayesian networks. Unlike most software packages, which display the nodes and arcs of the network, the developed tool organises the nodes based on the cause-and-effect relationship, making the user-interaction more intuitive and friendly. In addition to performing various types of inferences, the users can conveniently use the tool to verify the behaviour of the developed Bayesian network. The tool has been developed using QT and SMILE libraries in C++.

## 1. Introduction

In recent years, Bayesian Networks/Directed graphical models [1] are being applied in diverse fields, such as computational biology [2], ecology [3], defence [4], and robotics [5]. Bayesian networks (BNs) provide an intuitive way of representing and visualising the structure of a probabilistic model. They aid in solving numerous learning and inference problems in complex models conveniently.

A typical BN $\mathcal{G} = \{V, E\}$ constitutes $V$ vertices that are connected by $E$ directed edges. Let $\mathbf{x}$ be a set of discrete random variables, denoted by $\mathbf{x} = \{x_1, x_2, \ldots, x_K\}$. In a graphical model, each vertex represents a random variable or a set of random variables, while the directed links express causal probabilistic relationships between the variables. Specifically, depending on the direction of the edge, we denote certain nodes as parent or child with respect to other nodes. For example, in Figure. 1, node $x_3$ is the parent of $x_4$ and child of $x_1$. The joint distribution of directed graphs is given by:

$$p(\mathbf{x}) = \prod_{n=1}^{N} p(x_n|\mathbf{x}_{\pi_n}) \quad (1)$$

The joint distribution is the product of the conditional distribution of each node $n$ conditioned only on its parents $\mathbf{x}_{\pi_n}$. In the example shown in Figure. 1, the joint distribution factorises as $p(x_1, x_2, x_3, x_4) = p(x_1)p(x_2|x_1)p(x_3|x_1)p(x_4|x_3)$. There are numerous BN software packages such as Hugin[1], Genie[2], BayesiaLab[3], Netica[4] which could be used for developing, inferencing and learning of the network.

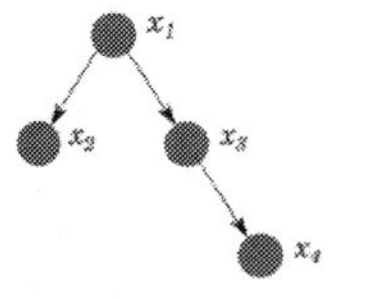


Figure 1: A toy example of a Bayesian Network.

Although, such BN packages provide a good graphical interface for users who design and develop Bayesian networks for various applications, the intended end-users of these networks, who could be from a non-technical background (*e.g.* airport operators, physicians, chemists), may not necessarily find such an interface appealing and comfortable to use. At times, comprehending such networks could be overwhelming, particularly when the number of nodes in the network is large (more than 10). Furthermore, performing inferences on such a network could be even difficult as it may not be straightforward to perceive how changes to a particular node value will affect the rest of the nodes it is connected to (*e.g.* as a parent or child node).

To illustrate this phenomenon, we consider a fairly complex real-world BN model recently proposed for wayfinding at airports [6]. The Wayfinding Bayesian Network Model (WBNM) explored the effects that human and environmental factors have on effective wayfinding in airports. It investigated the main influences on these factors as well as the most important elements of communication, the built environment, and what effect if any, do gender, airport familiarity and anxiety have on effective wayfinding in airports. A screen-shot of the model rendered using Genie software package is shown in Figure 2. Using the package, the end-user can infer the posterior probabilities of the nodes of the network (shown for few nodes at the bottom-right in Figure 2). It is evident that such a visualisation could be cumbersome while doing inferences. For example, if the end-user changed the evidence of one of the nodes and wanted to see its impact on the other nodes of the network, the user has to carefully track the node connections to see the influences. Such a process could be painstaking and time-consuming for the end-user.

To circumvent the above-mentioned limitations, this paper presents an intuitive dashboard, which provides an additional layer of abstraction, enabling the end-users to easily perform inferences over the BNs. The developed tool is called INSIGHTS, as it aims to pro-actively aid the user to clearly understand the model and the underlying causal relationships between its various factors (nodes). The tool organises the nodes based on the cause-and-effect relationship, making the user-interaction more intuitive and friendly. The rest of the paper is organised as follows: the details of the developed tool INSIGHTS v1.0, is provided in Section 2, while Section 3 discusses the results and capabilities of the developed tool followed by conclusion in Section 4.

## 2. INSIGHTS

The proposed tool INSIGHTS has the ability to run in real-time and automatically compute and display the marginal probabilities of the states of the nodes in a more intuitive manner than that shown in Figure 2. Screen-shots of the dashboard are shown in Figure 3. Such an intuitive dashboard enables the end-users to easily perform various types of inferences on the model. The tool internally uses SMILE[5] and QT libraries [7] to perform the Bayesian inferences and GUI visualisations, respectively. The SMILE library has a set of APIs that not only read and edit BNs but also perform inferences on them. Most of the inference algorithms are based on the approximate stochastic sampling techniques.

INSIGHTS requires the end-user to load two input files - BN model and ancillary text files. The BN model file is developed by using one of the existing Bayesian software packages such as Genie. The current version of the tool is able to parse standard BN file formats supported by Hugin,

[1] http://www.hugin.com/
[2] http://genie.sis.pitt.edu
[3] http://www.bayesia.com/
[4] http://www.norsys.com
[5] http://genie.sis.pitt.edu/index.php/downloads

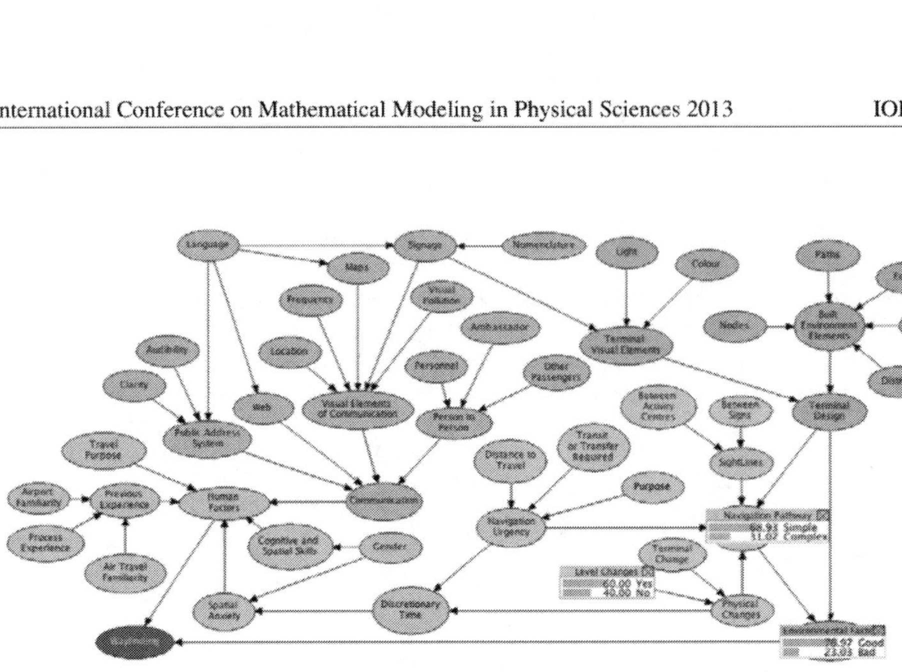


Figure 2: An example of current visualisation of the wayfinding model [6] using a popular BN software package Genie. End-users can inspect the posterior probabilities of the various nodes of the network, which get overlaid on top of the nodes as depicted at the bottom-right corner. Such a visualisation is hard to comprehend and perform inferences on the network.

Genie and Netica. The whole BN network is parsed and stored internally as a tree structure. Based on end-users response (typically button clicks), relevant sections of the BN are displayed on the dashboard. The other input, ancillary text file, contains the list of key nodes one wishes to monitor and will be displayed on the dashboard at the start-up as shown in Figure 3(a).

## 3. Results

The wayfinding model [6] was initially developed in Genie. All the necessary parameters such as the node states and relevant probability tables (conditional and marginal) were defined. As mentioned earlier, INSIGHTS requires two input files for its execution. The BN file representing the wayfinding model and an ancillary text file, which in this context, contained the name of one node 'Wayfinding' as it is the key node that the end-user is interested in (see Figure 2).

On execution of INSIGHTS, the user sees a dashboard displaying the key nodes (depicted as push-buttons) along with the marginal probabilities of their pre-defined states (represented by progress bars) as shown in Figure 3(a). Here the node 'Wayfinding' has two states 'Effective' and 'Ineffective' with values of 81% and 19%, respectively. By clicking the 'Wayfinding' push-button located to the left of the progress bars leads to a new screen similar to Figure 3(b), which displays its two parents, namely 'Environmental' Factors and 'Human Factors, along with the marginal probabilities of their states.

The users can further drill down to see all the other nodes that influence the value of wayfinding. Furthermore, they can set new evidence for any of the nodes and inspect their impact on the other nodes that are linked to them (either directly or indirectly) by clicking on the 'New Evidence' button (see the pop-up windown in Figure 3(b)). The 'Reset' button is used to clear the evidences. Such a feature, could be conveniently used by researchers to validate the behaviour of the developed model under various values of evidences, particularly at the extremities, and simulate 'what-if' scenarios.

The navigation display panel on the left side of the screen shows how the user has traversed the network to reach to the current display screen. Similarly, the blue text line at the bottom of the screen reveals the buttons clicked to reach to the current display screen. Both features get updated dynamically every time user clicks the buttons. The '<<' button at the bottom is used to roll back a step. As a result, with mere clicks the user could traverse the whole network

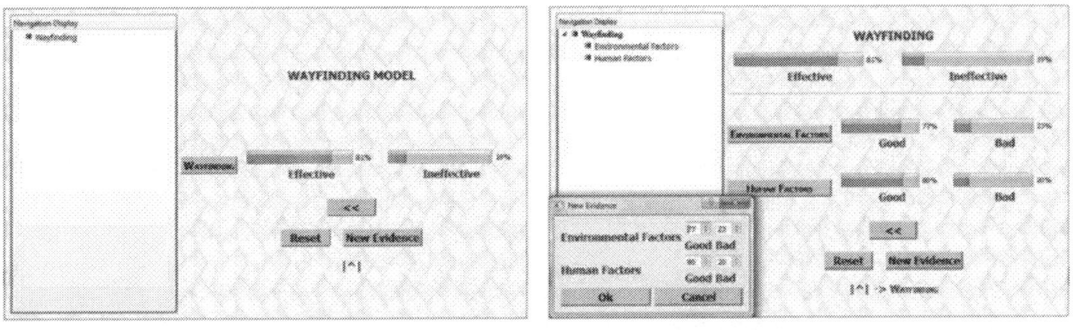


(a) (b)

Figure 3: Screen-shots of INSIGHTS at various points during its operation. **(a)** Screen shot at start-up, showing the posterior probability of the resultant node measuring the wayfinding effectiveness of a building [6]. **(b)** obtained after clicking the 'Wayfinding' push-button. It displays the two factors affecting the wayfinding - 'Environment Factors' and 'Human Factors', along with their state-probabilities. As noted, the end-user can also add new evidence and inspect how it could affect the performance of the overall network.

in an intuitive manner, without looking at the actual BN model (consisting of nodes and arcs). Clustering of nodes in this manner (parent-child relationship) will provide better understanding of the BN model and to study the influences between its nodes.

## 4. Conclusion

Current BN packages typically overlay the results on top of the graph making it difficult for end-users to see patterns, trends, or exceptions in the model results. The proposed tool INSIGHTS provides an easy and intuitive interface for end-users to perform inferences on the BNs. These visualisations allow end-users to see the results, interdependencies and influences between factors in the BN model in a more lucid manner.

In summary, the developed tool INSIGHTS has 3 key benefits. It can: (i) organise and render the nodes of the network based on a cause-and-effect relationship; (ii) be used as a diagnostic tool to find the underlying causes of a certain phenomenon/behaviour conveniently; and (iii) perform various 'what-if' scenarios by adding new evidence quite easily. As part of future work, we would like to extend this tool to automatically learn marginal/conditional probability tables from the relevant data for each node variable of the network.

*Acknowledgments*

The Airports of the Future research project is supported under the Australian Research Councils Linkage Projects funding scheme (LP0990135). The authors also acknowledge the contribution of the industry stakeholders involved in this project. More details on the project can be found at: www.airportsofthefuture.qut.edu.au.

## References

[1] Pearl J 1988 *Probabilistic Reasoning in Intelligent Systems: Networks of Plausible Inference* (Morgan Kaufmann)
[2] Friedman N, Linial M, Nachman I and Pe'er D 2000 *Journal of computational biology* **7** 601–620
[3] Kragt M E 2009 *A beginners guide to Bayesian network modelling for integrated catchment management* (Landscape Logic)
[4] Starr C and Shi P 2004 *An introduction to bayesian belief networks and their applications to land operations* (DSTO Systems Sciences Laboratory)
[5] Lazkano E, Sierra B, Astigarraga A and Martínez-Otzeta J M 2007 *Robotics and Autonomous Systems* **55** 253–265
[6] Farr A C, Kleinschmidt T, Yarlagadda P K and Mengersen K 2012 *Transport Reviews (In Press)* Version of record first published: 17 Aug 2012
[7] Blanchette J and Summerfield M 2008 *C++ GUI programming with Qt 4* (Prentice Hall)

# Bäcklund transformation and solitonic solutions for a parametric coupled Korteweg-de Vries system

**Luis Cortés Vega**[1], **Alvaro Restuccia**[2], **Adrián Sotomayor**[1]

[1] Departamento de Matemáticas, Universidad de Antofagasta, Antofagasta, Chile
[2] Departamento de Física, Universidad de Antofagasta, Antofagasta, Chile

E-mail: lcortes@uantof.cl,arestu@usb.ve,asotomayor@uantof.cl

**Abstract.** We analize a parametric coupled KdV system and we find a Bäcklund transformation. For a positive value of the parameter the system reduces to two KdV decoupled equations. For negative value of the parameter the system has non trivial coupling and presents multisolitonic solutions generated by the Bäcklund transformation. We compare the results with the already known in the literature.

## 1. Introduction

Historically, Bäcklund transformations arose as a useful tool to be used in differential geometry and differential equations [1, 2], for example, in the study of surfaces in classical differential geometry [3], and can be understood in a natural way as a generalization of contact transformations [4, 5].

More recently, Bäcklund transformations were used to find new solutions from known ones in the context of non linear partial differential equations [6, 7, 8]. It seems also that there exists a direct relation between such transformations and the integrability properties of the associate problems, in particular those obtained from the inverse scattering method for solving them.

Coupled Korteweg-de Vries (KdV) systems have generated great interest by virtue of its outstandings properties and wide range of applications [9, 10, 11, 12, 13].

In this work we find a Bäcklund transformation for a parametric coupled KdV system with relevant physical applications, following the approach developed by Whalquist and Estabrook [7]. We then generate multisolitonic solutions and we compare our results with those reported in the literature.

## 2. The Bäcklund transformation for a coupled Korteweg-de Vries system

We consider a particular coupled Korteveg-de Vries system, describing a two-layer liquid model studied in references [11, 14, 15, 16]. The system is formulated in terms of two independent functions $u(x,t)$ and $v(x,t)$ and given by the following partial differential equations:

$$u_t + uu_x + u_{xxx} - vv_x = 0 \tag{1}$$

$$v_t + u_x v + v_x u + v_{xxx} = 0. \tag{2}$$

We notice that by redefining $u$ and $v$ as $-6u$ and $6v$ respectively we obtain the system considered in [11]. Both systems (1),(2) and the one in [11] are completely equivalent.

We will follow the approach of Whalquist and Estabrook [7] in order to obtain a Bäcklund transformation for the system given by (1) and (2).

Moreover, we will consider a generalization of the system given by (1) and (2) in terms of an arbitrary parameter $\lambda$, more explicitly we will analize the system

$$u_t + uu_x + u_{xxx} + \lambda vv_x = 0 \tag{3}$$

$$v_t + u_x v + v_x u + v_{xxx} = 0. \tag{4}$$

By a redefinition of $v$ we may reduce the value of $\lambda$ to be $+1$ or $-1$. Both systems are not equivalent. The case $\lambda = -1$ corresponds the the system given by (1) and (2).

We notice that the system given by (3) and (4) is invariant under a Galileo transformation of the coordinates $x, t$. In fact, if

$$\begin{aligned} x &\to x + ct \\ t &\to t \end{aligned} \tag{5}$$

and

$$\begin{aligned} u &\to u - c \\ v &\to v \end{aligned} \tag{6}$$

then (3) and (4) remain invariant for any value of $\lambda$.

The system (1),(2) may be derived from a lagrangian

$$\mathcal{L} = -\frac{1}{2}w_x w_t - \frac{1}{6}(w_x)^3 - \frac{1}{2}w_{xxx}w_x + \frac{1}{2}w_x(y_x)^2 + \frac{1}{2}y_x y_t + \frac{1}{2}y_{xxx}y_x \tag{7}$$

expressed in terms of the fields $w(x,t)$ and $y(x,t)$ related to the original fields $u(x,t), v(x,t)$ by the following relations:

$$u = w_x \quad , \quad v = y_x.$$

The hamiltonian and associated Poisson structure may be derived from the above lagrangian.

The field equations obtained from the lagrangian (7) by taking variations on $w$ and $y$ are the following, once an integration on $x$ is performed,

$$\begin{aligned} &w_t + \tfrac{1}{2}(w_x)^2 + w_{xxx} + \tfrac{1}{2}\lambda(y_x)^2 + C(t) = 0 \\ &y_t + y_{xxx} + w_x y_x + \widetilde{C}(t) = 0 \end{aligned} \tag{8}$$

for $\lambda = -1$, $C(t)$ and $\widetilde{C}(t)$ are integration constants which depend only on the $t$ variable. Hence they may be eliminated by a redefinition of $w$ and $y$:

$$\begin{aligned} w(x,t) &\to w(x,t) + \textstyle\int_0^t C(\tau)d\tau \\ y(x,t) &\to y(x,t) + \textstyle\int_0^t \widetilde{C}(\tau)d\tau. \end{aligned} \tag{9}$$

We end up with the following system in terms of $w(x,t), y(x,t)$, equivalent to the original system (1),(2) for $\lambda = -1$,

$$w_t + \frac{1}{2}(w_x)^2 + w_{xxx} + \lambda\frac{1}{2}(y_x)^2 = 0 \tag{10}$$

$$y_t + y_{xxx} + w_x y_x = 0. \tag{11}$$

We propose the following Bäcklund transformation, whose integrability conditions yield (10),(11) for any value of $\lambda$:

$$w_x + w'_x = 2\eta - \frac{1}{12}(w - w')^2 - \frac{\lambda}{12}(y - y')^2, \tag{12}$$

$$\begin{aligned} w_t + w'_t = \tfrac{1}{6}(w - w')(w_{xx} - w'_{xx}) + \tfrac{\lambda}{6}(y - y')(y_{xx} - y'_{xx}) - \tfrac{1}{3}{w_x}^2 \\ -\tfrac{1}{3}{w'_x}^2 - \tfrac{1}{3}w_x w'_x - \tfrac{\lambda}{3}{y_x}^2 - \tfrac{\lambda}{3}{y'_x}^2 - \tfrac{\lambda}{3}y_x y'_x, \end{aligned} \tag{13}$$

$$y_x + y'_x = 2\mu - \frac{1}{6}(w - w')(y - y'), \tag{14}$$

$$\begin{aligned} y_t + y'_t = \tfrac{1}{6}(w - w')(y - y')_{xx} + \tfrac{1}{6}(w - w')_{xx}(y - y') \\ - \left(\tfrac{2}{3}w_x y_x + \tfrac{2}{3}w'_x y'_x + \tfrac{1}{3}w_x y'_x + \tfrac{1}{3}w'_x y_x\right). \end{aligned} \tag{15}$$

We define

$$Q_1(w, y) \equiv w_t + \frac{1}{2}{w_x}^2 + w_{xxx} + \frac{1}{2}\lambda {y_x}^2 \tag{16}$$

$$Q_2(w, y) \equiv y_t + w_x y_x + y_{xxx}. \tag{17}$$

It then follows that

$$(12)_{xx} + (13) = 0$$

implies

$$Q_1(w, y) + Q_1(w', y') = 0, \tag{18}$$

$$-(12)_t + (13)_x = 0$$

implies

$$\left[Q_1(w, y) - Q_1(w', y'))\right](w - w') + \lambda\left[Q_2(w, y) - Q_2(w', y')\right](y - y') = 0, \tag{19}$$

$$(14)_{xx} + (15)_x = 0$$

implies

$$Q_2(w, y) + Q_2(w', y') = 0, \tag{20}$$

and

$$-(14)_t + (15)_x = 0$$

implies

$$\left[Q_1(w, y) - Q_1(w', y'))\right](y - y') + \left[Q_2(w, y) - Q_2(w', y')\right](w - w') = 0. \tag{21}$$

Consequently, if

$$(w - w')^2 - \lambda(y - y')^2 \neq 0, \tag{22}$$

then a solution of (12),(13),(14),(15) implies

$$\begin{aligned} Q_1(w, y) = 0, Q_2(w, y) = 0 \\ Q_1(w', y') = 0, Q_2(w', y') = 0 \end{aligned}$$

that is, a solution of the system (10),(11).

Moreover, if $w', y'$ is a solution of the system (10),(11), then a solution $w, y$ of (12),(13),(14),(15) is a solution of (10),(11). That is, given a (may be a trivial) solution of (10),(11) by resolving (12),(13),(14),(15) we obtain a new solution of the original system (10),(11).

We notice that condition (22) when $\lambda$ is negative, in particular $\lambda = -1$, implies $(w, y) \neq (w', y')$ and vice versa. That is, if $(w, y) \neq (w', y')$ are solutions of (12),(13),(14),(15) then $(w, y)$ and $(w', y')$ are different solutions of (10),(11).

When $\lambda$ is positive, by redefining $y$ we may reduce to consider $\lambda = 1$. In that case equations (10),(11) reduce to two decoupled equations in terms of $(w + y)$ and $(w - y)$. The equations are exactly the KdV equation for each variable.

Using the above Bäcklund transformation and starting from trivial solution of the coupled system one can generate solitonic solutions. These solutions agree with the already reported multisolitonic solutions in [11].

## 3. Conclusions

We obtained, following the approach given in [7], a Bäcklund transformation for a parametric coupled system, including as a special case the system studied by [11]. We can then generate multisolitonic solutions, they agree with the already reported in the literature.

**Acknowledgments**

L. C. , A. R. and A. S. are partially supported by Project Fondecyt 1121103, Chile.

## References

[1] E. Goursat , "Le problème de Bäcklund", Memorial Sci. Math. , Fasc. 6 (Gauthier-Villars, Paris, (1925).
[2] A. R. Forsyth , "Theory of Differential Equations", Dover, Volume 6, (1959).
[3] L. Eisenhart , "A treatise on the Differential Geometry of Curves and Surfaces ", Dover, (1960).
[4] R. Hermann , "Geometric Theory of Non-Linear Differential Equations, Bäcklund Transformations and Solitons", Parts A and B , Math. Sci. Press, (1977).
[5] F. A. Pirani, D. C. Robinson and W. F. Shadwick, "Local Jet Bundle Formulation of Bäcklund Transformations, with Applications to Non Linear Evolution Equations, D. Reidel Publishing Company, (1979).
[6] G. L. Lamb Jr., J. Math. Phys. 15, 2157 (1974).
[7] H. D. Whalquist and F. D. Estabrook, Phys. Rev. Lett. 31, 1386-90 (1973).
[8] P. G. Drazin, "Solitons", London Mathematical Society Lecture Note Series 85, Cambrigde University Press, (1983).
[9] R. Hirota and J. Satsuma, Phys. Lett. A 85, 407 (1981).
[10] Deng-Shan Wang, Appl. Math. and Comp. 216, 1349-1354 (2010).
[11] Yang Jian-Rong and Mao Jie-Jian, Commun. Theor. Phys. 49, 22-26 (2008).
[12] A. Restuccia and A. Sotomayor, "The Hamiltonian structure of a coupled system derived from a supersymmetric breaking of Super KdV equations", arXiv:1306.0528 (2013).
[13] A. Restuccia and A. Sotomayor, "Stability of solitonic solutions of Super KdV equations under Susy breaking conditions", arXiv:1207.5870 (2012).
[14] J. A. Gear and R. Grimshaw, Stud. Appl. Math. 70, 235 (1984); J. A. Gear, Stud. Appl. Math. 72, 95 (1985).
[15] S. Y. Lou, B. Tong, H. C. Hu and X. Y. Tang, J. Phys. A: Math. Gen. 39, 113 (2006).
[16] V. A. Brazhnyi and V. V. Konotop, Phys. Rev. E 72, 026616 (2005).

# Bosonization and operatorial extensions of supersymmetric Korteweg-de Vries equations

**Adrián Sotomayor**[1], **Mercedes Fernández**[1], **Alvaro Restuccia**[2]

[1] Departamento de Matemáticas, Universidad de Antofagasta, Antofagasta, Chile
[2] Departamento de Física, Universidad de Antofagasta, Antofagasta, Chile

E-mail: asotomayor@uantof.cl,mfernandez@uantof.cl,arestu@usb.ve

**Abstract.** We discuss the bosonization of supersymmetric KdV as a first step on the construction of operatorial extensions of that system. We obtain the hamiltonian structure of those operatorial extensions using extended Miura tranformations. We also discuss their integrability properties in terms of extended Gardner tranformations. We give explicit examples of some operatorial extensions.

## 1. Introduction

Some years ago Mathieu [1] obtained an interesting supersymmetric extension of Korteweg-de Vries (KdV) equation with associated infinite local and non-local conserved quantities [1, 2, 3, 4]. That property was deduced following and idea of Gardner [5].

Such extension ($N = 1$ SKdV), equivalent to the Super KdV equations obtained in [6] by reduction from the super Kadomtsev-Petviaschvili hierarchy, consists on a system of coupled partial differential equations for a commuting and an anticommuting field, taking values at the even and odd part respectively of a given Grassmann algebra. Basic ingredients in that extension are a superfield, a covariant derivative and a preferred generator in the algebra. Later on, in a similar way by increasing the number of preferred elements of the algebra, supersymmetric extensions of KdV were obtained [7, 8, 9, 10]. All such supersymmetric extensions have at least one associated Hamiltonian.

Recently, expanding the corresponding fields in terms of the generators of the algebra, via a bosonization approach [11], were obtained new interesting solutions for such supersymmetric extensions [12, 13].

In [11] an operatorial extension of KdV, which contains several systems as particular cases including $N = 1$ SKdV, preserving the property of having infinite local conserved quantities was given. In the following, we present a new operatorial extension of KdV with infinite local conserved quantities. We obtain its Hamiltonian structure, following the Dirac procedure for constraints systems [14, 15, 16, 17].

The hamiltonian analysis may be performed directly in terms of the odd variables but also one may give a bosonized hamiltonian formulation in terms of even variables. Both formulations are completely equivalent.

## 2. General algebraic framework

We consider an algebra of even and odd elements. The even ones belong to a commutative algebra $\mathcal{P}$ with unit while the odd ones satisfy

$$\begin{aligned} &\mathcal{Q}\mathcal{P} \subset \mathcal{Q} \\ &[\mathcal{Q}, \mathcal{P}] = 0 \\ &[\mathcal{Q}, \mathcal{Q}] \subset \mathcal{P}. \end{aligned} \tag{1}$$

The odd part of the algebra $\mathcal{Q}$ is characterized by the condition:

$$[q, \mathcal{Q}] \neq 0, \tag{2}$$

for any $q \in \mathcal{Q}$. That is, given $q$ there always exists $\hat{q} \in \mathcal{Q}$ such that $[q, \hat{q}] \neq 0$.

In fact, otherwise $q$ would be an even element.

Examples of this algebraic structure are:

(i) Operatorial Phase Space.
Given $q$ and its conjugate momenta $\frac{\partial}{\partial q}$ then: $[q,p] = 1$, $[q,q] = 0$, $[p,p] = 0$. We identify the even part $\mathcal{P}$ with elements proportional to the unit, while $q$ and $p$ belong to the odd part of the algebra $\mathcal{Q}$.

(ii) Grassmann algebras.
In this case the odd elements satisfy $\mathcal{Q}\mathcal{Q} \subset \mathcal{P}$ and in addition for any $q_1, q_2$: $q_1q_2 + q_2q_1 = 0$. In particular, the odd elements are nilpotent.

(iii) Extensions of the Grassmann algebras.
The algebra satisfies: $\mathcal{Q}\mathcal{P} \subset \mathcal{Q}, \mathcal{Q}\mathcal{Q} \subset \mathcal{P}, [\mathcal{Q}, \mathcal{P}] = 0$.
This is a particular case of (1) but different from (i) and includes (ii). It is different from (i) because in the phase space $qq$ is not proportional to the unit and hence it does not satisfy $\mathcal{Q}\mathcal{Q} \subset \mathcal{P}$. It is more general than (ii) because we do not require $q_1q_2 + q_2q_1 = 0$. Particular violations of this condition occur if we consider, for example, the matrix realization

$$q_1 = \begin{pmatrix} 0 & 0 & 0 & 0 \\ 1 & 0 & 0 & 0 \\ 0 & 0 & 0 & 0 \\ 0 & 0 & 1 & 0 \end{pmatrix} \quad , \quad q_2 = \begin{pmatrix} 0 & 0 & 0 & 0 \\ 0 & 0 & 0 & 0 \\ 1 & 0 & 0 & 0 \\ 0 & a & 1 & 0 \end{pmatrix}, \quad a \in \mathbb{R}.$$

We get

$$q_1q_1 = 0 \ , \ q_2q_2 = 0 \ , \ q_1q_2 = \begin{pmatrix} 0 & 0 & 0 & 0 \\ 0 & 0 & 0 & 0 \\ 0 & 0 & 0 & 0 \\ 1 & 0 & 0 & 0 \end{pmatrix} \ , \ q_2q_1 = aq_1q_2.$$

If $a = -1$, we obtain an example of a Grassmann algebra. If $a \neq -1$ we have a deformation of the Grassmann algebra.

## 3. The Lagrangian of an extension of Super Miura Korteweg-de Vries (Super MKdV) equations

We introduce a Lagrangian formulated in terms of fields $w$ and $\eta$ valued on $\mathcal{P}$ and $\mathcal{Q}$ respectively, which satisfy (1):

$$\mathcal{L}(w,\eta) = \frac{1}{2}\dot{w}w' + \frac{1}{2}[\dot{\eta},\eta] - \frac{1}{2}(w'')^2 - \frac{1}{2}(w')^4 - \frac{1}{2}[\eta,\eta']^2 - \frac{1}{2}[\eta'',\eta'] - \frac{3}{2}(w')^2[\eta',\eta]. \tag{3}$$

The fields equations are obtained by taking variations with respect to $w$ and $\eta$. By taking variations with respect to $w$ we obtain

$$\dot{v} + v''' - 6v^2v' - 3\big(v\,[\eta', \eta]\big)' = 0 \tag{4}$$

where $v \equiv w'$.

By taking variations with respect to $\eta$ we obtain

$$\langle [\delta\eta, F] \rangle_{x,t} = 0 \tag{5}$$

where

$$F = \dot{\eta} + \eta''' - 3v^2\eta' - 3vv'\eta + \eta'\,[\eta, \eta'] + \frac{1}{2}\eta\,[\eta, \eta''] \tag{6}$$

and $<>_{x,t}$ denotes integration on $x$ and $t$.

(5) implies $[Q, F] = 0$ for any odd element and from (2) we get

$$F = 0 \tag{7}$$

as the field equation associated to general variations of the odd field $\eta$. If $\mathcal{P}, \mathcal{Q}$ generates a Grassmann algebra then the terms $\eta'\,[\eta, \eta']$ and $\frac{1}{2}\eta\,[\eta, \eta'']$ in (6) are zero and the field equations are exactly the $N = 1$ Super MKdV equations obtained from the $N = 1$ Super KdV equations by performing a Miura transformation.

We now proceed to construct the Hamiltonian structure of the system (4) and (7).

## 4. Hamiltonian structure

The conjugate momenta to $w$ and $\eta$ will be denoted by $p$ and $\mu$ respectively. We obtain directly from their definitions

$$\begin{aligned} \phi &\equiv p - \tfrac{1}{2}v = 0 \\ \psi &\equiv \mu - \tfrac{1}{2}\eta = 0 \end{aligned} \tag{8}$$

These are primary constraints of the phase space formulation. It turns out, following the Dirac approach to determine the constrained structure of the phase space, that they are the only constraints of the theory.

They satisfy the following Poisson bracket relations:

$$\begin{aligned} \{\phi(x), \phi(y)\}_{PB} &= -\partial_x\delta(x, y) \\ \{\phi(x), \psi(y)\}_{PB} &= 0 \\ \{\psi(x), \psi(y)\}_{PB} &= -\delta(x, y), \end{aligned}$$

consequently (8) are second class constraints.

The Poisson structure of the constrained phase space is then given by the Dirac brackets defined by

$$\{F, G\}_{DB} = \{F, G\}_{PB} - \Big\langle \big\langle \{F, \varphi_i(\tilde{x})\}_{PB} \left\{\varphi_i(\tilde{x}), \varphi_j(\tilde{\tilde{x}})\right\}_{PB} \left\{\varphi_j(\tilde{\tilde{x}}), G\right\}_{PB} \big\rangle_{\tilde{x}} \Big\rangle_{\tilde{\tilde{x}}}$$

where $i = 1, 2$ and $\varphi_1 \equiv \phi, \varphi_2 \equiv \psi$ (summation on $i$ and $j$ is understood).

After performing some calculations we obtain the Dirac bracket algebra for the fields $w$ and $\eta$. It turns out that one may directly evaluate the algebra of the fields $v \equiv w'$ and $\eta$. We obtain

$$\begin{aligned} \{v(x), v(y)\}_{DB} &= \partial_x\delta(x, y) \\ \{v(x), \eta(y)\}_{DB} &= 0 \\ \{\eta(x), \eta(y)\}_{DB} &= \delta(x, y). \end{aligned}$$

The Hamiltonian associated to the Lagrangian (3) is obtained via a Legendre transformation. We get

$$\mathcal{H} \equiv p\dot{w} + \frac{1}{2}[\dot{\eta},\eta] - \mathcal{L} = \frac{1}{2}(v')^2 + \frac{1}{2}v^4 + \frac{1}{2}[\eta,\eta']^2 + \frac{1}{2}[\eta'',\eta'] + \frac{3}{2}v^2[\eta',\eta].$$

The canonical field equations

$$\begin{aligned} \dot{v} &= \{v, H\}_{DB} \\ \dot{\eta} &= \{\eta, H\}_{DB} \end{aligned} \tag{9}$$

where $H = \langle\mathcal{H}\rangle_x$, exactly agree with the lagrangian field equations (4),(7), as it should be.

We now consider the most general Miura transformation. A dimensional analysis shows that it has the form

$$\begin{aligned} u &= v' + v^2 - [\eta,\eta'] \\ \xi &= \eta' + v\eta. \end{aligned} \tag{10}$$

It then follows the field equations for $u$ and $\xi$. We obtain

$$\begin{aligned} \dot{u} &= \{u, H\}_{DB} = -u''' + 6uu' - 3[\xi,\xi''] \\ \dot{\xi} &= \{\xi, H\}_{DB} = -\xi''' + 3(\xi u)' + \theta' + v\theta \end{aligned} \tag{11}$$

where

$$\theta = [\eta,\eta']\eta' + v[\eta,\eta'']\eta.$$

Under the restricted algebra satisfying:

$$\mathcal{QP} \subset \mathcal{Q}, \mathcal{QQ} \subset \mathcal{P}, [\mathcal{Q},\mathcal{P}] = 0 \tag{12}$$

we have

$$\theta = 0,$$

even when the odd elements are not necessarily nilpotent. The resulting equations (11) under the restricted algebra (12) were analyzed in [11]. It was proven, via a Gardner transformation, that they have an infinite sequence of local conserved quantities. In fact, in that case we have

$$\begin{aligned} u_t &= -u''' + 6uu' + 3[\xi'',\xi] \\ \xi_t &= -\xi''' + 3(u\xi)', \end{aligned} \tag{13}$$

$$\begin{aligned} z_t &= (-z'' + 3z^2 + 3[\sigma',\sigma])' + \epsilon^2(2z^3 + 3z[\sigma',\sigma])' \\ \sigma_t &= (-\sigma'' + 3z\sigma)' + \epsilon^2 3(z^2\sigma' + zz'\sigma + \sigma'[\sigma',\sigma]), \end{aligned} \tag{14}$$

$$\begin{aligned} u &= z + \epsilon z' + \epsilon^2(z^2 + [\sigma',\sigma]) \\ \xi &= \sigma + \epsilon\sigma' + \epsilon^2 z\sigma. \end{aligned} \tag{15}$$

(14) and (15) are the Gardner equations and associated Gardner transformations respectively.

After simplifying by crossing out derivatives in $x$ and using the inverse Gardner transformation, the first four nontrivial conserved quantities for the operator-extended KdV system (13) are:

$$\begin{aligned} H_0 &= \int u dx \\ H_2 &= \int (u^2 + [\xi',\xi]) dx \\ H_4 &= \int \left(2u^3 + (u')^2 + 4u[\xi',\xi] + [\xi'',\xi']\right) dx \\ H_6 &= \int \Big(5u^4 + 10u(u')^2 + (u'')^2 + 15u^2[\xi',\xi] - 2u[\xi'',\xi'] \\ &- 8u[\xi''',\xi] + 3[\xi',\xi]^2 + [\xi''',\xi'']\Big) dx. \end{aligned} \tag{16}$$

## 5. Conclusions

We have then introduced an extension of the Super KdV equations, obtained its Hamiltonian and its Poisson structure on the constrained phase space, and shown it has an infinite sequence of conserved quantities. We notice that the case of the operatorial phase space analyzed in [11] does not satisfy (12) hence the Hamiltonian structure discussed here does not apply to that case.

### Acknowledgments

A. S., M. F. and A. R. are partially supported by Project Fondecyt 1121103, Chile.

## References

[1] P. Mathieu, J. Math. Phys. 29, 2499 (1988).
[2] P. Dargis, P. Mathieu, Phys. Lett. A 176, 67 (1993).
[3] S. Andrea, A. Restuccia, A. Sotomayor, J. Math. Phys. 46, 103517 (2005).
[4] S. Andrea, A. Restuccia and A. Sotomayor, Phys. Lett. A 376, 245251 (2012).
[5] R. M. Miura, C. S. Gardner, and M. D. Kruskal, J. Math. Phys. 9, 1204 (1968).
[6] Yu. I. Manin, A. O. Radul, Commun. Math. Phys. 98, 65 (1985).
[7] P. Labelle and P. Mathieu, J. Math. Phys. 32, 923 (1991).
[8] S. Bellucci, E. Ivanov and S. Krivonos, J. Math. Phys. 34, 3087 (1993).
[9] F. Delduc and E. Ivanov, Phys. Lett. B 309, 312-319 (1993).
[10] Z. Popowicz, Phys. Lett. B 459, 150-158 (1999).
[11] S. Andrea, A. Restuccia and A. Sotomayor, J. Math. Phys. 42, 2625 (2001).
[12] X. N. Gao and S. Y. Lou, Phys. Lett. B 707, 209 (2012).
[13] X. N. Gao, S. Y. Lou and X. Y. Tang, JHEP 05, 029 (2013).
[14] P. A. M. Dirac, "Lectures on Quantum Mechanics", Belfer Graduate School Monograph Series No.2, Yeshiva University, New York, (1964).
[15] Y. Nutku, J. Math. Phys. 25 (6) , June (1984).
[16] G. W. Kentwell, J. Math. Phys. 29, 46 (1988).
[17] A. Restuccia and A. Sotomayor, "The Hamiltonian structure of a coupled system derived from a supersymmetric breaking of Super KdV equations", arXiv:1306.0528 (2013).

# Modeling complex network epidemics using nodes' roles

**D.S. Vlachos and K. J. Parousis-Orthodoxou**

Department of Computer Science and Technology,
Faculty of Sciences and Technology,
University of Peloponnese,
GR-221 00 Tripolis, Greece

E-mail: dvlachos@uop.gr

**Abstract.** In this work we investigate the usefulness of node roles of complex systems in dynamical processes. More specifically, we use the characteristic equations of an SIS epidemic spreading in order to simulate an actual epidemic spreading. The equations that are used, encapsulate the degree correlations in comparison to the role distribution and the results are compared with an actual epidemic release simulation.

## 1. Introduction

The study of complex systems is acquiring more and more interest in the past decade. A field that has been researched thoroughly, are the scale free networks. The key to the degree correlations on a complex network hase been the pioneering effort of Barabasi and Albert [1], [2], with their work on the "preferential attachment" rule. As a direct metric of the degree, the correlations that emerge from it, have vastly impacted the way of examining complex systems in general.

More recently, the scientific community has vigorously tried to analyze the repeated occurrences of local subgraphs in complex networks, named motifs [3]. These patterns are formed from clusters of nodes that have identical connectivity patterns. the motif correlations usually emerge by using as reference the Erdos-Renyi random network [4], [5]. This approach has been followed in many fields, but mainly in the field of biology [6].

The first attempts to characterize the nodes of a complex network with respect to their roles have gathered a lot of attention [7], [8] and [9]. This paper shall be used in order to demonstrate the attempt to analyze networks in the roles of the nodes that reside in them, and the merits of this analysis. This will be achieved by using the *role distribution* described in the work [10], in order to extract correlations between these roles and use them to model a dynamical process. The choice of dynamical process for this test, is an SIS epidemic spreading model, which shall be simulated in two different networks.

## 2. The SIS epidemic spreading model

The SIS epidemic spreading model, [11], [12] and [13], describes the spreading of an epidemic in a network by factoring in that in any time step there are two states in which a node can be. It can be either *Susceptible* or *Infected.* The transposition of the nodes between these two states is controlled by the characteristic general equations of the model:

$$I_k^{t+1} = I_k^t + \lambda \Theta_k^t S_k^t - \mu I_k^t$$

$$S_k^{t+1} = S_k^t - \lambda \Theta_k^t S_k^t + \mu I_k^t$$

where:

$I_k^t$ are the infected nodes at time step $t$ with degree $k$.
$S_k^t$ are the susceptible nodes at time step $t$ with degree $k$.
$\lambda$ is the infection rate of the epidemic.
$\mu$ is the recovery rate of the epidemic.
$\Theta_k^t$ is the quantity that changes for the two distinct cases of degree analysis.

These equations contain a quantity, $\Theta_k^t$, which differs in the cases of no correlations between the nodes exist, or the network has been analyzed with respect to the degrees of the nodes and their interconnectivity.

### 2.1. Uncorrelated case

In this case, the network has not been analyzed and the only information that we have is the degree of each node. The quantity of $\Theta_k^t$ in this case gives the probability of the disease to infect a node with degree $k$ and it can be factored in the general equations as:

$$\Theta_k^t = 1 - \left(1 - \sum_{k'} \frac{k' P(k')}{< k >} I_{k'}^t\right)^k$$

### 2.2. Degree correlated case

In this scenario, it is assumed that a further analysis has be done to the network, and the correlations of the nodes are imprinted in the **joint degree distribution**, $P(k, k')$, of the network, which shows the probability of choosing arbitrarily an edge with endpoints $k$ and $k'$ respectively (equivalent to $P(\rightarrow k')$). In this case, the quantity of $\Theta_k^t$ becomes:

$$\Theta_k^t = 1 - \left(1 - \sum_{k'} \frac{< k > P(k, k')}{k P(k)} I_{k'}^t\right)^k$$

### 2.3. Role analysis

Instead of looking at the network in terms of the degree of its nodes, we focus on the role of each node and the type of connection it might have with the other nodes of similar or different role. As explained in the work [10], after performing a role analysis in the network, we get a set of vectors which contain the number of connections of a role to all the others, denotes as $CV(r, r')$. Also we get the number of nodes that each role occupies as $N_r$. Summing all values of each vector, we get the number of connections of the role to all the others, ergo its degree $k_r$.

*2.3.1. General equations*

The same reasoning that was applied in the degree cases above, will produce similar results for the general equations of the epidemic spreading, but now from the point of view of roles. So:

$$I_r^{t+1} = I_r^t + \lambda\Theta_r^t S_r^t - \mu I_r^t$$
$$S_r^{t+1} = S_r^t - \lambda\Theta_r^t S_r^t + \mu I_r^t$$

$I_r^t$ are the infected nodes at time step $t$ which have a role $r$.

$S_r^t$ are the susceptible nodes at time step $t$ which have a role $r$.

$\Theta_r^t$ is the probability of the disease to infect a node with role $r$ at time step $t$.

*2.3.2. Role correlations*

The **conditioned role probability**, $P(r'|r)$ describes the probability of an edge that starts from a node with role $r$ to end to a node with role $r'$. This probability can be analyzed as the ratio of the probability to pick arbitrarily an edge from a node of role $r$ to a node of role $r'$, $P(r \to r')$, over the probability of an edge to start from a node with role $r$, $P(r \to)$. So:

$$P(r'|r) = \frac{P(r \to r')}{P(r \to)}$$

but $P(r \to r') = \frac{N_r E_{r,r'}}{E}$ and $P(r \to) = \frac{N_r k_r}{E}$ which means that the above equation becomes:

$$P(r'|r) = \frac{CV(r,r')}{k_r}$$

It is interesting to note that the role correlations are shown to be independent of the number of nodes that occupy each role.

Since the above quantity is calculated with the connectivity vectors, $\Theta_r^t$ becomes:

$$\Theta_r^t = 1 - \left(1 - \sum_{r'} \frac{CV(r,r')}{k_r} I_{r'}^t\right)^{k_r}$$

## 3. Networks to experiment

*3.1. Model created network*

The first network that is examined, is generated by the model described in [14], and shows a modified Barabasi-Albert model that generates small-world networks with scale-free properties. The network is generated with preferential attachment and with the appearance of local links on new nodes. The specific network, is generated with 2000 nodes.

*3.2. Real network*

One of the most interesting datasets that is distributed in the field of complex systems, is the network that describes the topology of the Western States Power Grid of the United States, compiled by Duncan Watts and Steven Strogatz [15]. That network has 4941 nodes and it is undirected.

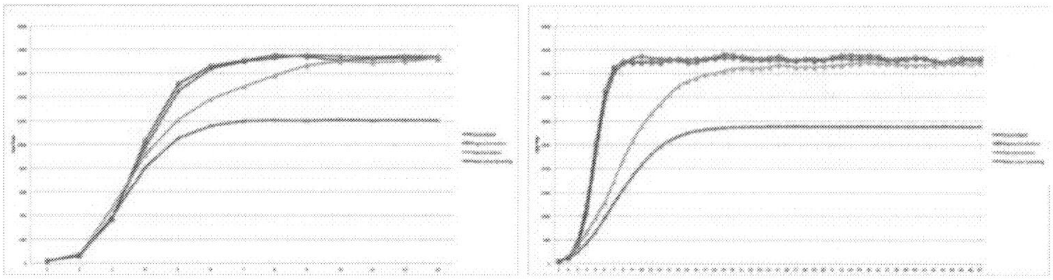

(a) The model created network. (b) The real network.

Figure 1: SIS epidemic spreading simulations for the two networks.

## 4. Results and interpretation

The process allows us to demonstrate the behavior of an SIS epidemic spreading in the above networks, by factoring in certain information. In the graphs that we demonstrate, the curves represent the following:

- The case where the number of infected nodes is calculated by factoring in only the degree of the nodes of each network.
- The case where the number of infected nodes is calculated by factoring in the probability distribution of a connection between two degree values.
- The case where the number of infected nodes is calculated by factoring in the connectivity vectors of the roles that are assumed by the nodes of the networks.
- The case where the number of infected nodes is measured by releasing an infection within the networks and monitoring the actual spreading.

The last case presents the realistic scenario of an epidemic spreading in the network. Figures 1b and 1a show the measurements of the above cases in comparison to each other.

It is seen from Figure 1 that the actual spreading in both networks is stabilized in a lower value of infected nodes than all the other simulation cases. It is also seen that in both networks, the uncorrelated case and the case with degree correlations have very similar curves and progress almost identically, whereas in the case of role correlations, the curve differentiates and progresses significantly closer to the actual spreading, even though the value that it is stabilized is closer to the two first cases. This proves that when factoring role correlations in a dynamical procedure, the results give a more accurate approach of the inner dynamics of certain networks that cannot be explored differently.

## Acknowledgment

The project is implemented under the Operational Program "Education and Lifelong Learning" and funded by the European Union (European Social Fund) and by Greek national resources.

## References

[1] A.-L. Barabasi, R.Albert, Science 286 (1999) 509.
[2] A.-L. Barabasi, R.Albert, H. Jeong, Physica A 272 (1999) 173.
[3] R. Milo, S. Shen-Orr, S. Itzkovitz, N. Kashtan, D. Chklovskii, U. Alon, Science 298 (2002) 824-827.
[4] R. Milo, S. Itzkovitz, N. Kashtan, R. Levitt, S. Shen-Orr, I Ayzenshtat, et al., Science 303 (2004) 1538-1542.
[5] Erds P., Rnyi, A. (1959) Publ. Math. 6, 290-297
[6] E. Alm, A.P. Arkin, Biological networks, Current Opinion in Structural Biology 13 (2) (2003) 193202.
[7] R. Guimera, L.A.N. Amaral, Nature 433 (2005) 895 - 900

[8] R. Guimera, S. Mossa, A. Turtschi, L.A.N. Amaral, Proceedings of National Academy of Science USA 102 (2005) 7794 - 7799.
[9] R. Guimera, M. Sales-Pardo, L.A.N. Amaral, Nature Physics, 3 (2007) 63-69
[10] K J Parousis-Orthodoxou, M Stamos and D S Vlachos 2013 J. Phys.: Conf. Ser. 410 012080 doi:10.1088/1742-6596/410/1/012080
[11] Anderson, R. M. ed. (1982) Population Dynamics of Infectious Diseases: Theory and Applications, Chapman and Hall, London-New York.
[12] Brauer, F. and Castillo-Chvez, C. (2001). Mathematical Models in Population Biology and Epidemiology. NY: Springer.
[13] Hethcote, H. W. (2000). "The mathematics of infectious diseases." Society for Industrial and Applied Mathematics, 42, 599 653.
[14] Xu Xin-Ping, Liu Feng and Li Wei, Chinese Physics Letters, Volume 23, No 3, 2005.
[15] D. J. Watts and S. H. Strogatz, "Collective dynamics of 'small-world' networks", Nature 393, 440-442 (1998).

# Evolutionary Algorithm for Optimal Vaccination Scheme

**K.J. Parousis-Orthodoxou and D.S. Vlachos**

Department of Computer Science and Technology,
Faculty of Sciences and Technology,
University of Peloponnese,
GR-221 00 Tripolis, Greece

E-mail: kp0100@uop.gr

**Abstract.** The following work uses the dynamic capabilities of an evolutionary algorithm in order to obtain an optimal immunization strategy in a user specified network. The produced algorithm uses a basic genetic algorithm with crossover and mutation techniques, in order to locate certain nodes in the inputted network. These nodes will be immunized in an SIR epidemic spreading process, and the performance of each immunization scheme, will be evaluated by the level of containment that provides for the spreading of the disease.

## 1. Introduction

The field of complex systems, has taken a lot of attention the recent years ([1], [2] and [3]), because of the number of applications that can be found in everyday life. Some examples of complex systems that are have been analyzed over the literature, are the Internet (in router level), the World Wide Web [4], the protein interaction network [5], the power grid of a country, the colonization of ants and several other examples. Many of these networks have shown very interesting properties, with the most important being the small-world effect [6] and the scale-free property [7].

The dynamical operations that can be applied in various complex systems, are the cornerstone of the research in the field. Many studies have focused on the approaches of the epidemic spreading ([8], [9] and [10]) of various types (SIS, SIR etc) and in different types of networks (scale-free, random etc). Also, many of these studies have been concentrated in locating optimal vaccination schemes to control the spreading of a disease or information in distinct network topologies ([11], [12], [13] and [14]). These two different cases might be focused on maximizing the diffusion (information) or minimizing it (disease) within the network.

Another dynamical operation, which has been proven to be an extremely useful tool in obtaining optimal solutions to various problems, are the evolutionary algorithms. These type of process, mimics to a certain point, the process of evolution that has been met in most ecosystems. Using certain concepts such as mutation, chromosome or crossover, these algorithms can be modified in such a way as to locate global extrema and minimize the time of the operation ([15] and [16]).

There have been some attempts to merge the two fields, either for optimizing the effect of the application of a genetic algorithm in a problem, or by using a genetic algorithm in order to explore a certain property of a network. An example of the first type is the attempt to distribute the population of a genetic algorithm onto the nodes of a complex network, and allow the crossover and mutation processes only between the population members that reside on these nodes [17]. The second type of merging of the two pioneering topics involves using a certain type of genetic algorithm as a tool that balances the exploration-exploitation trade off, in order to explore a specific property of a network [18], and the study that is presented here falls into that category.

## 2. Algorithm Description

In this section, we will analyze the parts of the algorithm that operate simultaneously. Our algorithm is divided in two categories, the genetic algorithm and the epidemic spreading process. Aiming for the optimal vaccination scheme, the genetic algorithm creates random vaccination strategies in a binary form. The fitness criteria of the genetic algorithm are given by runs of SIR epidemic spreading model, which is described below.

### *2.1. The SIR Model*

The specific model of epidemic spreading operates with basic principle that there are three pools which contain all the nodes of the network. At the beginning of the spreading all the nodes (except from the vaccinated ones) are in the Susceptible pool. As the spreading progresses, the nodes are transferred with probability $p_i = 0.76$ to the Infected pool. Then with probability $p_c = 0.002$ are transferred to the Recovered pool. The key component of the SIR model is the fact that, as soon as a node has recovered from the infection, it can no longer carry it again. This is why the three states of the model are Susceptible Infected Recovered. The vaccinated nodes are considered to be in the Recovered state as soon as the infection starts to spread. For the purposes of fairness to the vaccination schemes that are produced, the patient zero of the spreading, is chosen in the beginning of the process and is the same for all fitness evaluations that are performed during the run of the genetic algorithm.

### *2.2. The Fitness Function*

The fitness function produces outcomes that are based on the cost of each vaccination scheme, which factors the number of vaccinated nodes of the scheme and the number of infected nodes during the spreading process. The cost of each vaccine is 5 whereas the cost of treating each infected node is 1. Therefore, the dominating scheme in each generation should use an optimal number of vaccines, that would allow for a minimal percentage of infected population. Note that since the cure/vaccine ratio is 1/5, it is never profitable to vaccinate more than 1/5 of the population. Therefore, it is in each case, the maximum allowed vaccination size.

### *2.3. The Genetic Algorithm*

The process of the genetic algorithm initializes with locating certain basic attributes of the network that is inputted (number of nodes, degrees etc). These values will define some of the parameters of the process, such as the population (number of nodes). Then, the algorithm fills the population with vaccination schemes in binary form (0 for the non-vaccinated nodes and 1 for the vaccinated ones) and evaluates their fitness with the use of the function described above. The fitness is then stored uniformly in a list that will be used as reference for obtaining the fittest. Note that one of the random parameters for the vaccination scheme is the number of nodes that will be vaccinated.

When the initial population is set, the algorithm begins the evolutionary process. We choose two nodes with the use of the *roulette function*. That is, by choosing randomly, but with greater possibilities for the fittest of the population. This ensures that the choice will be fair, since the fittest have an advantage, but also not excluding the share of the population that did not perform so well and might yield more promising results if coupled with someone else.

After the two 'parents' are chosen, they crossover with probability $p_{cross} = 0.7$ and produce two offsprings with half their chromosomes (single point crossover in the middle of the chromosome). If crossover is not performed, the mutation process will be explored with probability $p_{mut} = 0.002$ for each of the initially chosen members of the population. If the mutation is a go, one of the bits of the chromosome are reversed and the result is forwarded to the next generation. In each case, if one of the operations is performed (crossover,mutation) the resulting chromosomes take the place of the initially chosen, whereas, if no operation is performed, the original chromosomes are transfered into the population intact. This function continues until the entire population is rebuilt. The number of times that the function is executed is half the population size (since it produces two members at each time step).

When the population for each time step is completed, the algorithm creates the uniform fitness for it, and restarts the process for the next time step. This carries on for user defined generations. After the final generation, the algorithm locates and returns the optimal generation scheme for the specific input parameters and its cost.

## 3. Experimental Procedure and Results

In this section, we will present the networks that were used for the experimental procedure of the algorithm, the results they yielded as well as an analysis of the vaccination schemes that were ultimately chosen. Each network has been tested 50 times and the results allow us to locate the best fitted schemes.

The experiments that are carried out, are performed in three distinct types of networks, which mean to cover a broad spectrum of the field. The first network that is presented, is produced by the preferential attachment model of Barabasi and Albert [7]. This infamous model in complex networks, produces undirected networks with the PA rule, which gives scale-free networks. The number of nodes for this graph, as well as the rest, is 500, whereas the number of edges differ from graph to graph and in this case they are 1491. The mean number of edges is 3. The BA graph has given some interested results regarding the optimal vaccination schemes in each of the 50 runs. Figure 1 shows the number of vaccinated nodes in each of the 50 runs that were performed in the BA network. The largest number of vaccinated nodes is 89, which means that only for vaccination, the cost of this scheme is 445. It is assumed that in that specific scheme, the epidemic was contained. In 14 of the cases, the no of vaccinated nodes was between 42 and 89. In all the rest (36), it was below 5.

The second network is based in the creation of triangles [19]. The newly introduced node, chooses at random among the preexisting edges of the network and it then connects to both ends of this edge. The control parameter q decides if the old edge between the two preexisting end-nodes, stays or not. This model operates in such a way as to create network with embedded small-world property. In the network examined in this paper, the control parameter q has the value of 0.5, which means that approximately half of the edges in question, are erased. This network is also undirected and the number of edges are 735.

For this network, the results are more condensed.As it shown in Figure 2, the no of vaccinated nodes is wrapped around 30 (varies between 7 and 51). This means that the algorithm manages to locate a fairly good vaccination scheme almost every time.

The third and final network that is used in this work is an Erdos-Renyi random network [20]. For the network at hand, the probabillity to establish a link is $p_L = 0.02$ which returns a network of 2530 edges.

For the ER network, it is clear in Figure 3 that the vaccination schemes keep the number of vaccinated nodes low. Half of these schemes returned as best option, 0 vaccinated nodes. Also, the maximum value of vaccinated nodes is 3, which is found only in four occasions. Because

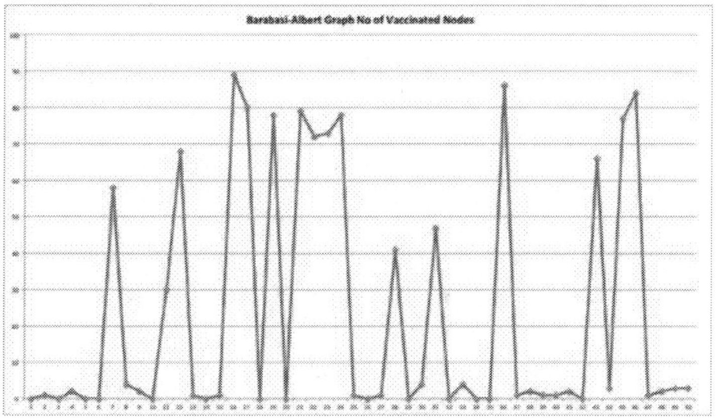

**Figure 1.** The number of vaccinated nodes for the 50 runs of the algorithm, in the BA network.

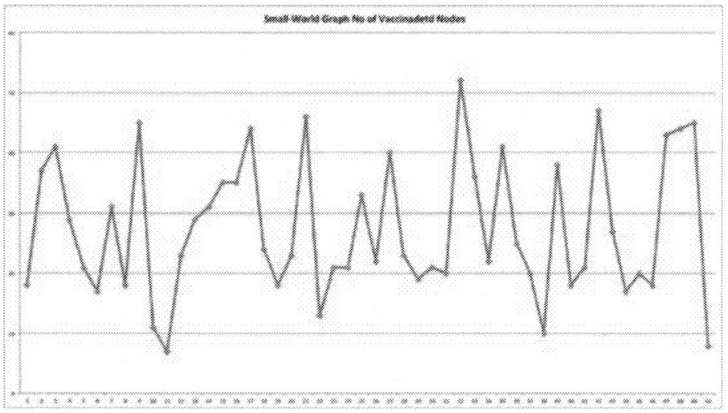

**Figure 2.** The number of vaccinated nodes for the 50 runs of the algorithm, in the small-world network.

of the significantly greater number of edges than the other networks, the epidemic spreading manages to find a way through the network and this just increases the cost for any vaccination scheme.

## 4. Conclusions and Future Work

The work that has been analyzed in this paper, is an effort to combine an evolutionary process with a dynamical process in complex networks, in order to obtain an optimal scheme for various types of networks. The results where promising for the structured networks, whereas the behavior of the algorithm in the different types of networks shows that the process converges to desirable values. For the future, the algorithm must be tested more extensively and the nodes that were chosen in each network, must be analyzed further with respect to their role in it.


## Acknowledgement

The project is implemented under the Operational Program "Education and Lifelong Learning" and funded by the European Union (European Social Fund) and by Greek national resources.

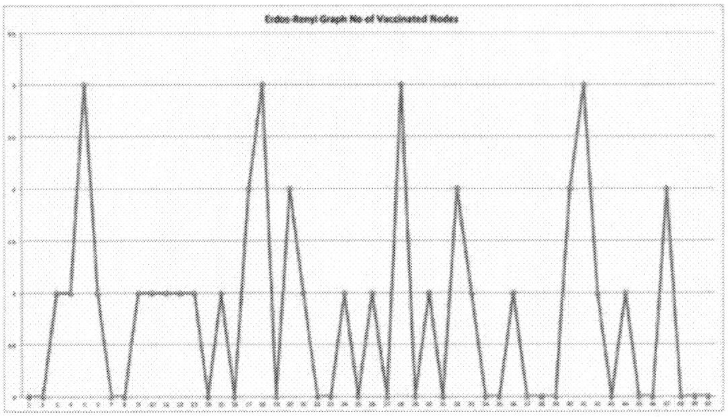

**Figure 3.** The number of vaccinated nodes for the 50 runs of the algorithm, in the ER random network.


[1] S. H. Strogatz, "Exploring complex networks", Nature, vol. 410, pp. 268-276, March 2001.
[2] R. Albert and A. - L. Barabasi, "Statistical Mechanics of Complex Networks", Review of Modern Physics, vol. 74 pp. 47-91, January 2002.
[3] X. F. Wang, "Complex networks: topology, dynamics and synchronization", Int. J. Bifurcation & Chaos, vol.12 no. 5 pp. 885-916, May 2002.
[4] R. Albert H. Jeong, and A.-L. Barabasi, Nature 401, 130 (1999).
[5] Jing-Dong J. Han, Nicolas Bertin, Tong Hao, et al, "Evidence for dynamically organized modularity in the yeast proteinprotein interaction network" Nature 430, 88-93 (1 July 2004).
[6] L. A. N. Amaral, A. Scala, M. Barthlmy, and H. E. Stanley, "Classes of small-world networks", PNAS vol. 97 no. 21, 1114911152.
[7] Albert-Laszlo Barabasi, Reka Albert, "Emergence of Scaling in Random Networks", Science 15 October 1999, Vol. 286 no. 5439 pp. 509-512 .
[8] Romualdo Pastor-Satorras and Alessandro Vespignani, "Epidemic Spreading in Scale-Free Networks", Phys. Rev. Lett. 86, 32003203 (2001).
[9] Marian Boguna and Romualdo Pastor-Satorras, "Epidemic spreading in correlated complex networks", Phys. Rev. E 66, 047104 (2002).
[10] Y. Moreno, R. Pastor-Satorras, A. Vespignani, "Epidemic outbreaks in complex heterogeneous networks", The European Physical Journal B - Condensed Matter and Complex Systems, April 2002, Volume 26, Issue 4, pp 521-529
[11] Romualdo Pastor-Satorras and Alessandro Vespignani, "Immunization of complex networks", Phys. Rev. E 65, 036104 (2002).
[12] N. Madar, T. Kalisky, R. Cohen, D. ben-Avraham, S. Havlin, "Immunization and epidemic dynamics in complex networks", The European Physical Journal B - Condensed Matter and Complex Systems, March 2004, Volume 38, Issue 2, pp 269-276.
[13] P. Holme, "Efficient local strategies for vaccination and network attack", EPL (Europhysics Letters) Volume 68 Number 6 P. Holme 2004 Europhys. Lett. 68 908.
[14] J. Gomez-Gardenes, P. Echenique, Y. Moreno, "Immunization of real complex communication networks", The European Physical Journal B - Condensed Matter and Complex Systems, January 2006, Volume 49, Issue 2, pp 259-264.
[15] Melanie Mitchell, "An Introduction to Genetic Algorithms", 06 February 1998.
[16] Davis, L., "Genetic algorithms and simulated annealing", Research notes in artificial intelligence (ISSN 0268-7526), 1987.
[17] D S Vlachos and K J Parousis-Orthodoxou, "Evolutionary Processes in Complex Networks and Small Worlds", Journal of Physics: Conference Series Volume 410 conference 1
[18] Anca Gog, D. Dumitrescu, Bat Hirsbrunner, "Community Detection in Complex Networks Using Collaborative Evolutionary Algorithms", Advances in Artificial Life, 2007
[19] Zhongzhi Zhang, Shuigeng Zhou, Zhen Shen, From regular to growing small-world networks, Physica A 385 (2007) 765-772.
[20] Erdos, P.; Renyi, A. (1959). "On Random Graphs. I". Publicationes Mathematicae 6: 290297.

# Small-angle scattering from three-phase systems: Investigation of the crossover between mass fractal regimes

**Eugen M Anitas[1,2], Alexander Yu Cherny[1], Vladimir A Osipov[1] and Alexander I Kuklin[1,3]**

[1]Joint Institute for Nuclear Research, Dubna 141980, Moscow region, Russia
[2]Horia Hulubei National Institute of Physics and Nuclear Engineering, RO-077125 Bucharest, Romania
[3]Laboratory of Advanced Research on Membrane Proteins, MIPT, Moscow, Russia

E-mail: anitas<at>theor.jinr.ru

**Abstract.** In this paper, we construct a three-phase model (that is, a system consisting of three homogeneous regions with various scattering length densities), which illustrate the behavior of small-angle scattering (SAS) scattering curves. Here two phases are a deterministic fractal embedded in another deterministic mass fractal, and they altogether are further embedded in a third phase, which can be a solution or solid matrix. We calculate SAS intensities, derive expressions for the crossover position (that is, the point where the power-law scattering exponent changes) as a function of control parameters, including size, concentration, and volumes of each phase. The corresponding SAS intensities from these models describe a succession of power-law regimes in momentum space where both regimes correspond to mass fractals. The models can be applied to SAS data where the absolute value of the scattering exponent of the first power-law regime is higher than that of the subsequent second power-law regime, that is, the scattering curve of "convex" kind near the crossover position.

## 1. Introduction

Small-angle scattering (SAS) [1, 2] experimental data (X-ray or neutron) often show a succession of power-law regimes, on a double logarithmic scale, whose scattering exponents are taking arbitrarily values from $-4$ to $-1$. Although this type of behavior is usually associated with a multilevel structure, the explanation of the crossover position is a more intricate problem and a method to extract the structural properties about each phase is lacking. Experimentally it has been shown that the crossover position in a multi-phase system can depend on the variation of the scattering length density (SLD) of the component phases [3] and a theoretical framework for describing these type of results has been developed in Ref. [4] where this dependence of the crossover position was confirmed for the particular transition between the Porod and mass fractal [5] regimes.

Recently, a practical formula which can be used to fit experimental SAS data from multi-phase systems has been derived [6]. It can be applied to SAS data which show a succession of power-law regimes: $I(q) \propto q^{\tau}$, where $\tau = D$ for mass fractals ($0 < D < 3$), and $\tau = 6 - D$ for surface fractals ($2 < D < 3$) [7]. However, for fractal structures embedded in other fractal structures, the underlying physical structure of each phase and their spatial correlations are usually difficult to be clearly understood. In this paper we extend the results obtained in [4] to the transition between two mass fractal regimes, we calculate the corresponding scattering intensity and estimate the crossover position.

**Figure 1.** (Color online) The initiator and the first two iterations of the fractals composing region 2 (upper panel: Cantor fractal) and respectively, region 1 (lower panel: Menger-like fractal). The three-phase system consists of region 1 absorbing region 2.

## 2. Models

The three-phase model consists of a uniform matrix with scattering length density (SLD) $\rho_0$ (region 3) which contains mass fractal regions with SLD $\rho_1$ (region 1) which absorb other mass fractals with SLD $\rho_2$ (region 2). A similar model in which a mass fractal is embedded in a ball has been developed in [4]. In Ref. [6] we relate this model to the *type I* structure "fractal inside fractal".

The construction process of the fractal used here is very similar to that of the generalized Cantor and Vicseck fractals [8,9]. For the both fractals we start with a cube of edge $l_0$. We choose a Cartesian system of coordinates with the origin in the cube center and the axes being parallel to the cube edges. The iteration rule is to replace the cube of edge $l_0$ with twenty one smaller cubes of edge length $l_1 = l_0/3$. The positions of the cubes with edge $l_1$ correspond to a Menger sponge structure with an additional cube in the center. At the $m$th iteration the edge of the resulting cubes is $l_m = l_0/3^m$.

The normalized scattering amplitude (form factor) of an object is $F(\boldsymbol{q}) = 1/V \int_V e^{-i\boldsymbol{q}\cdot\boldsymbol{r}} \mathrm{d}\boldsymbol{r}$, where $V$ is the volume of the object and $F(0) = 1$. Then for regions 1 and 2, the normalized form factors of the fractals at the $m$th iteration can be written as [8–10] $F_{1,2}^{(m)}(\boldsymbol{q}) = F_0(\boldsymbol{q}l_0/3^m) \prod_{i=0}^{m} G_{1,2i}(\boldsymbol{q})$ where $F_0(\boldsymbol{q})$ is the form factor of a cube of unit edge length, and $G_{1,2m}$ are the generative functions of the fractals with the property that $G_{1,20}(\boldsymbol{q}) \equiv 1$. They are determined by the positions of the centers of the cubes for each iteration, and for the region 1 (Fig. 1, lower panel) it is given by

$$G_{1m}(\boldsymbol{q}) = (1 + 4\Gamma_m(\boldsymbol{q}) + 8\Lambda_m(\boldsymbol{q}))/21, \tag{1}$$

where $\Gamma_m(\boldsymbol{q}) = \cos(l_m q_x)\cos(l_m q_y) + \cos(l_m q_x)\cos(l_m q_z) + \cos(l_m q_y)\cos(l_m q_z)$ and $\Lambda_m(\boldsymbol{q}) = \cos(l_m q_x)\cos(l_m q_y)\cos(l_m q_z)$. For the region 2 we consider as generator the same cube with edge $l_0$ and we choose the generative function corresponding to a Cantor set (Fig. 1, upper panel) and we have [8]

$$G_{2m}(\boldsymbol{q}) = \cos(l_m q_x)\cos(l_m q_y)\cos(l_m q_z). \tag{2}$$

Therefore, we shall observe the transition in the SAS intensity from the region 1, which is a Menger sponge-like fractal, to the region 2 (Cantor set): $q^{-D_1} \to q^{-D_2}$, where $D_1 = 2.77$ and $D_2 = 1.89$ are the corresponding fractal dimensions. In order to assure a complete inclusion of the region 2 into region 1, the iteration number of region 2 shall be higher than the iteration number of region 1.

## 3. Scattering intensity

The scattered intensity for the multiphase system can be written as [4, 6]

$$I(q) = I(0)\left\langle \left|\alpha F_1^{(m)}(\boldsymbol{q}) + (1-\alpha)F_2^{(p)}(\boldsymbol{q})\right|^2 \right\rangle, \tag{3}$$

where $I(0) = nV_{\rm r}^2(\rho_1 - \rho_0)^2\alpha^{-2}$ and $\alpha$ is the contrast parameter given by

$$\alpha = \left(1 + \frac{\rho_2 - \rho_1}{\rho_1 - \rho_0}\frac{V_{\rm f}}{V_{\rm r}}\right)^{-1}. \tag{4}$$

A more accurate description of SAS from a physical system should involve size polydispersity of the scatterers. Here we consider an ensemble of the three-phase fractal with different sizes taken at random and distributed according to a log-normal distribution function with relative variance $\sigma_{\rm r}$ [9]. Thus, the average in Eq. (3) is taken both over angles and sizes.

## 4. Results and Discussions

The numerical results for the scattering intensity using Eq. (3), but neglecting the correlations (which can play a significant role only at low $q$ [6]), are shown in Fig. 2 for different values of the contrast parameter $\alpha$. One can clearly see the appearance of the crossover and we can estimate its position by a simple approximation in the limit of strongly developed polydispersity [6]

$$\langle |F_{1,2}^{(m)}(\boldsymbol{q})|^2\rangle \propto \begin{cases} 1, & q \lesssim 2\pi/l_0, \\ (ql_0/2\pi)^{-D_{1,2}}, & 2\pi/l_0 \lesssim q \lesssim 2\pi/l, \\ \dfrac{1}{u}(ql/2\pi)^{-4}, & q \gtrsim 2\pi/l, \end{cases} \tag{5}$$

where $u$ is of the order of $(l_0/l)^{D_{1,2}}$.

For a given $q$, the amplitude of region 1 $F_1^{(m)}(\boldsymbol{q})$ and the amplitude of region 2 $F_2^{(m)}(\boldsymbol{q})$ in the r.h.s. of Eq. (3) can be of *different orders of magnitude*, provided the *corresponding exponents are not equal*. Then only one of the amplitudes dominates. Therefore in a good approximation, the crossover position can be found by equating the corresponding intensities $\alpha^2\langle|F_1^{(m)}(\boldsymbol{q}_{\rm c})|^2\rangle \simeq (1-\alpha)^2\langle|F_2^{(p)}(\boldsymbol{q}_{\rm c})|^2\rangle$, where one can use the estimations from Eq. (5). This gives us the following expression for the crossover position

$$q_{\rm c} \simeq \frac{2\pi}{l_0}\left(\frac{(1-\alpha)^2}{\alpha^2}\right)^{1/(D_2-D_1)}. \tag{6}$$

One can clearly observe the Guinier region (which corresponds to both region 1 and region 2), fractal regions of the region 1 and region 2, and a Porod region (Fig. 2), where we choose $\rho_2/\rho_1 = 0.9$ for all the plots, $\sigma_{\rm r} = 0.4$ for the relative dispersion, and fractal iteration number $m_1 = 6$, and respectively $m_2 = 7$. Then, the values of $\alpha$ depend on the ratio $\rho_0/\rho_1$. For arbitrarily values of $\alpha$ one obtains a direct and *smooth* transition between fractal regions. The crossover positions are estimated from Eq. (6) and their variation is depicted in Fig. 2b.

Choosing the SLDs so that the contrast $\rho_2 - \rho_1$ between the region 1 and region 2 is suppressed or setting the volume of region 2 small enough, $\alpha$ gets closer to one and as a consequence, a part of the curve corresponding to region 2 is "absorbed" by the curve corresponding to region 1. Therefore, at sufficiently small values of $|1-\alpha|$, the fractal region of the region 2 may not be observed at all (when $\alpha \gtrsim 0.9$ at the chosen values of parameters).

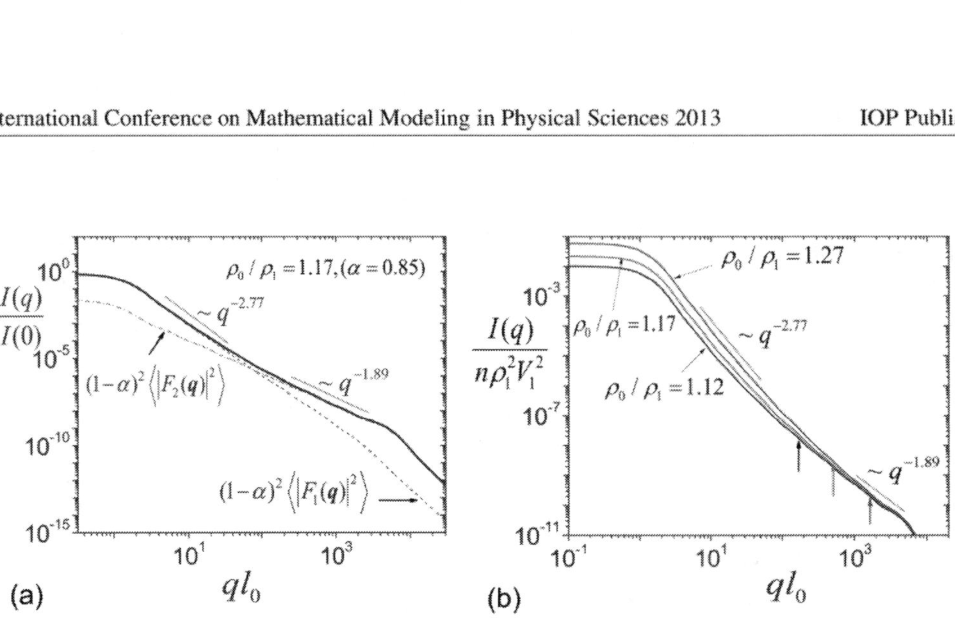


**Figure 2.** (Color online) The scattering intensities from the three-phase model, where region 1 (Menger-like mass fractal) absorbs region 2 (Cantor mass fractal). The parameters are described in the text. (a) The scattering intensity (black: solid line) and the contributions of the region 2 (blue: dashed) and region 1 (red: dashed-dot) fractals. (b) Contrast variation of the scattering intensity. The vertical arrows represent the crossover position as estimated from Eq. (6). Below the crossover point, the contribution of region 2 is negligible, while above it, the contribution of region 1 is negligible.

## 5. Conclusions

We develop a theoretical model which describes SAS from multi-phase fractal systems, where one mass fractal (one phase) is embedded into another mass fractal (a second phase) and altogether the combined system is further put into a solution or a solid matrix (a third phase). We derive an analytical expression for the SAS intensity from a mixture of such multi-phase systems. The scattering intensity is characterized by the presence of two successive mass fractal regimes with various scattering exponents and by the appearance of a crossover between these power-law regimes. We show that the crossover position is controlled by the effective contrast parameter which in turn, depends on the relative values of the SLD of each phase and their volumes, and estimate its position.

The developed model can be extended to describe scattering from various complex structures embedded into each other, such as mass fractal into a surface fractal, surface into a surface or surface into a mass fractal.


## Acknowledgments

We acknowledge financial support from JINR - IFIN - HH projects and from JINR grant No. 13-302-02.


## References


[1] Feigin L A and Svergun D I 1987 *Structure Analysis by Small-Angle X-Ray and Neutron Scattering* (New York, London: Plenum press)
[2] Glatter O and Kratky O 1982 *Small-angle X-ray Scattering* (London: Academic Press)
[3] Lebedev D V, Filatov M V, Kuklin A I, Islamov A K, Kentzinger E, Pantina R, Toperverg B P and Isaev-Ivanov V V 2005 *FEBS Letters* **579** 1465–1468
[4] Cherny A Y, Anitas E M, Kuklin A I and Osipov V A 2012 *J. Phys: Conf. Ser.* **393** 012031–1–012031–6
[5] Mandelbrot B 1983 *The Fractal Geometry of Nature* (USA: W.H. Freeman)
[6] Cherny A Y, Anitas E M, Kuklin A I and Osipov V A 2013 *arXiv:1307.2351 [cond-mat.stat-mech]*
[7] Martin J E and Hurd A J 1987 *J. Appl. Cryst.* **20** 61–78
[8] Cherny A Y, Anitas E M, Kuklin A I, Balasoiu M and Osipov V A 2010 *J. Appl. Cryst.* **43** 790–797
[9] Cherny A Y, Anitas E M, Kuklin A I and Osipov V A 2011 *Phys. Rev. E* **84** 036203–1–036203–11
[10] Cherny A Y, Anitas E M, Kuklin A I, Balasoiu M and Osipov V A 2010 *J. Surf. Inves.* **4** 903–907

# Modeling the Galactic Magnetic Field and its Application in verifying a Pulsar Origin of Very High Energy Cosmic Rays

**Pantea Davoudifar, Keihanak Rowshan Tabari**

Research Institute for Astronomy and Astrophysics of Maragha(RIAAM)-Maragha, Iran, P.O. Box:55134-441

E-mail: dfpantea@riaam.ac.ir

**Abstract.** Deflection of Cosmic Ray charged particles under the influence of magnetic fields (Galactic and Extragalactic) causes a nearly isotropic distribution of their observed fluxes especially in lower energy ranges. Anyhow, as very high energy cosmic rays experience less deflections in their paths, they may point out the direction of their sources within a few degrees. We used a Galactic magnetic field model to study the possible Galactic sources of these cosmic rays. The propagation of cosmic rays in this Galactic magnetic field is simulated to estimate average deflection angles into their straight-line paths from their sources. Pulsars with suitable characteristics are selected and deflection regions around them are defined. Compared with the observational data (i.e. detected directions of observed CRs), the possibility of a Galactic origin of ultra high energy cosmic rays is examined. We defined deflection angles in terms of energies for sources in a distance $d$ into center and anti-center directions. The probability of observing cosmic rays of different energies from the direction of a source in a distance $d$ is studied and the possibility of a pulsar origin of very high energy cosmic rays due to some recent models, is discussed.

## 1. Introduction

The observation of a *knee* (1958) in Cosmic Ray(CR)'s flux, together with supper *GZK* (*Greisen-Zatsepin-Kuzmin*) particles gave rise to models in which acceleration mechanisms and the power of astronomical accelerators (i.e. supernovas, pulsars, Galactic shocks, etc [1, 2, 3]) were discussed. As supernovae models at the prior researches were not able to produce the observed energies [4, 5, 6], the models developed to situations in which particles were originated from a very young pulsar and accelerated up to Very High Energies (VHE) in the pulsar wind [7, 8, 9, 10, 11]. Pulsar origin of CR particles has studied in many researches [6, 7, 10, 11, 12, 13, 14, 15, 16, 17, 18, 19, 20] in which a variety of models were considered to explain both the mechanisms of particle acceleration in the Interstellar Medium (ISM) and the origin of the observed CR's flux [6, 7, 8, 9, 19, 20, 21].

Continuing the theory first started by *Faraday* and observing the Faraday Rotation Measures (FRM) of polarized electromagnetic waves from different sources, estimations of weak Magnetic Fields (MF) of the Galaxy [22, 23, 24] was made possible. These MFs were estimated to have values up to $6\mu G$ around the Sun and up to $10\mu G$ in the inner Galaxy [23, 25, 26, 27, 28, 29, 30].

Particles in the Galaxy have gyroradiuses as: $r_g = \frac{E_{cr}}{ZeB}$. Simple calculations has showed that the gyroradius of VHE CRs (i.e. $\sim 10^{18} - 10^{19}$ and higher) in the Galactic MF is much less than the typical distance to a young neutron star ( $8kpc$) [21] which has the capability of accelerating CR protons up to around $10^{21}eV$ [5, 12]. On the other hand, due to GZK cutoff [31, 32], it is usually accepted that the possible EG sources of VHE CRs should lie within about 100 *Mpc*. In a previous work [33, 34] it was showed that assumed EG particles can deflect up to 10 degrees into the direction of their sources. Anyhow, Sigl et al.[35] has showed that there is an inconsistency between the arrival directions of Fly's Eye and Yakutsk events and the possible EG sources within this distance. If there is a pulsar origin for VHE CRs, the galactic directions of these CRs should point out the direction of their Galactic sources within a few degrees.

## 2. The candidate pulsars

Possible large MFs of a young pulsars ($10^{12} \sim 10^{13}$ G [11] of young pulsars, $10^{14} \sim 10^{15}$ G of magnetars [10]) with low barycentric periods gives less cooling times as [10, 19]:

$$T = \frac{P^2{}_{ms}}{2B_{12}} \times 10^{15} \tag{1}$$

where $B_{12}$ is the strength of MF in $10^{12}$ G. Using data from ATNF Pulsar Catalogue [36, 37]; 255 objects out of 2210 objects with a specified distance were selected for which the barycentric period of the pulsar, $p_0$, was about $10ms$ and less. A second category was selected with 69 pulsars with surface magnetic flux density in the range $10^{13} - 10^{16}G$. Nevertheless for these Galactic pulsars the cooling times are not in the range specified by Wibig and Wolfendale [10]. Within first category, 234 objects are from inner Galaxy and 21 from outer Galaxy with mean distances: $d$=4.6$kpc$ (inner Galaxy) and $d$=2.5$kpc$ (outer Galaxy).

## 3. The specifications of this simulation

We assumed the arrival directions of particles in GMF to be the Galactic directions of selected pulsars. The particles considered to be protons as above the *ankle*, the composition of observed CRs seems to tend to lower mass particles. Anyhow it is a sophistical assumption, because it is possible that the observed high energy CRs are light nuclei produced in the ISM through interactions of even heavier primary CRs, here obviously we have omitted this case in our first step.

To compare the results, a collection of high energy observed events in the energy ranges $10^{18} - 10^{19}$, $10^{19} - 10^{20}$ and $10^{20} - 10^{21}eV$ from data catalogues of Haverah Park and Volcano Ranch [38], SUGAR [39], Auger [40], Fly's Eye [41], Yakutsk [42, 43, 44] and AGASA [45, 46] have selected. For the propose of this simulation, protons of energies $10^{18}$, $10^{19}$, $10^{20}$ and $10^{21}$ been ejected into GMF with the initial Galactic direction of the object (i.e. pulsar $(l,b)$). The light path of a particle, $d(b)$, from the pulsar catalogue then is simply its distance [36, 37].

For a proton the gyroradius $R$ is equal to: $R(pc) = \frac{1.08E(PeV)}{B(\mu G)}$ considering particles traveling $\mathcal{D}$ parsecs by deflection in the varying MF of the Galaxy in the form of [47]: $B = B_0 e^{\frac{-\mathcal{L}}{3000}}$, where the regular magnetic field component is along the $z$ axis, $\mathcal{L}$ is the disk height in $kpc$ and regular MF decays exponentially with a decay length of $3000kpc$.

To calculate the actual deflected paths of the particles, we considered a Random Walk Method (RWM) with a varying Length Factor (LF) [48] and calculated the actual deflected path, $\mathcal{D}$, for each event. To reach the average values of $\mathcal{D}$, 100 simulations been made for each energy.

The GMF been considered to have a *Kolmogrov* type with a turbulent energy spectrum of :

$$S(k) \sim k^{-\delta} \begin{cases} for\ one\ dimension\ \delta = \frac{5}{3} \\ k = \frac{2\pi}{\lambda}, wave\ number, (or\ \frac{2\pi}{L}, L\ is\ Eddy\ Size) \end{cases}$$

with varying scales between $L_{min}$=20 $pc$ and $L_{max}$=200 $pc$[49]. In a previous work [33, 34] it was shown that for *Auger* events the value of $k$ is about 0.006 $Mpc^2yr^{-1}$, consistence with a *Kolmogrov* type spectrum.

The time delay of observed particles is just due to their deflected paths in varying GMF:

$$\tau = \alpha(\mathcal{D} - d) years \quad (2)$$

where $\mathcal{D}$ and $d$ are in parsecs and $\alpha$ is ~3.26 in an assumption in which all particles are traveling at the speed of light.

### 3.1. Deflection Angles

For each selected object, the time delay distributions for different energies ($10^{18}$, $10^{19}$, $10^{20}$ and $10^{21}$ $eV$) is calculated. In Galactic condition, the spread in deflection angles is always comparable to the average deflection angles and the average time delay is given as follow [50, 51, 52, 53]:

$$\tau(E, d) \approx \frac{\theta^2(E, d)d}{4} \quad (3)$$

These results are sensitive to LFs and the strength of MF. For LFs of orders of 10 $pc$ and $B_0$=10 and 5 $\mu G$, with varying *Kolmogrov* scales between $L_{min}$=1 $pc$ and $L_{max}$=100 $pc$, our results are showed for two typical pulsar located at ~ center direction: J1751-2857 ($l$=0.65, $b$=-1.12), d=1.44 $kpc$, $p_0$ =3.9 $ms$ and J0621+25 ($l$=187.31, $b$=4.93), d=4.25 $kpc$, $p_0$ =2.7 $ms$ in ~ anti-center direction [36, 37]. The result of this simulation for time delays and deflection angles are listed in table 1.

**Table 1.** The result of this simulation for J1751-2857 and J0621+25 for different energies in EeV

| Object Name | | $\mathcal{D}(kpc)$ | | | | $\tau(yr)$ | | |
|---|---|---|---|---|---|---|---|---|
| | 1 | 10 | $10^2$ | $10^3(EeV)$ | 1 | 10 | $10^2$ | $10^3(EeV)$ |
| $J0621+25$ | 16.6 | 4.26 | 4.25 | 4.25 | $4.01 \times 10^4$ | 35.7 | 0.37 | $3.50 \times 10^{-3}$ |
| $J1751-2857$ | 1.73 | 1.44 | 1.44 | 1.44 | $9.30 \times 10^2$ | 7.80 | 0.08 | $7.02 \times 10^{-4}$ |
| | | $\theta(degree)$ | | | | $d(kpc)$ | | |
| | 1 | 10 | $10^2$ | $10^3(EeV)$ | | | | |
| $J0621+25$ | 6.2 | 0.18 | 0.02 | less than 0.01 | 4.25 | | | |
| $J1751-2857$ | 1.6 | 0.15 | 0.02 | less than 0.01 | 1.44 | | | |

For these two objects (as an example) the diagrams of deflection regions were also drawn (Figure 1).

## 4. The result

We have completed the results of this simulation for 80 objects with the barycentric periods, $p_0$, about 10$ms$ and less with $0 < l, b < 90$. This results are shown in figure 2. For a constant energy of protons, deviation angle has a power law relation with $d$ (the distance of object) as $\theta(d) = a + b \times d^c$ in which $c$ for the objects of figure 2 calculated to be $0.7 \pm 0.1$. As a result, it is showed that for a variety of distances the deflection angles are related to both energy and distance as expected (Figure 2), but the structure of Galactic MF is also of special importance. In fact a more precise evaluation of Galactic structure and considering turbulent MFs in some cases cause larger values of deflection angles.

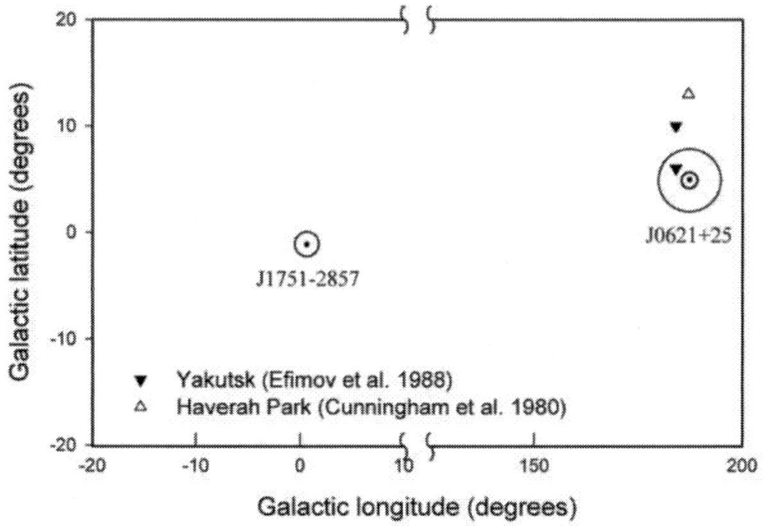


**Figure 1.** Deflection angles around two sample pulsars. Inner circles represent deflection of protons of $10^{20}eV$ and outer circles represent deflection of protons of $10^{19}eV$.

With the selected parameters of this simulation it is showed that for protons with energies $\sim 10^{20}eV$ and above, the average deflection angles are less than 1 degrees. This result means that for these particles having a Galactic pulsar origin, their arrival directions should exactly point to a pulsar or pulsars in the Galaxy. For lower energies, in the range $10^{19} - 10^{20}$, the results are shown in figures 3 with the assumption that all of them were protons.

Our preliminary results shows that for a nearly homogeneous distribution of Yakutsk [42, 43, 44], Volcano Ranch, Haverah Park [38], Auger [40] and AGASA [45, 46] there is a consistency between Galactic directions of 8 objects (out of 80) located at higher distances. This result did not verify for energies above $10^{20}eV$. For fast pulsars with distances lower than $\sim 4kpc$ no correlation been observed.

If higher mass particles being considered, higher energies and higher values of deflection angles will resulted, so for the PSNR model [10] another possibility is that the particles are not protons and have higher mass numbers. For a fortiori conclusion, a whole sky survey should be completed.

## Acknowledgments

This work has been supported financially by Research Institute for Astronomy and Astrophysics of Maragha (RIAAM) under research project No. 1/2352.

## References

[1] Meli A., Becker J. K., and Quenby J. J., 2008, *Astron. Astrophys.*, **492**, 323-336
[2] , Moskalenko I. V., Stawarz L., Porter T. A., and Cheung Chi C., 2009, *Astrophys. J.*, **693**, 1261-1274
[3] , Berezhko E. G., 2009, *Astrophys. J.* ,**698**, L138-L141
[4] Lagage P. O., Cesarsky C. J., 1983, *Astron. Astrophys.*, **125**, 249-257
[5] Hillas A. M., 1984, *Ann. Rev. Astron. Astrophys.*, **22**, 425-444
[6] Berezhko E. G., 2008, *J. Phys. Conf. Series*, **120** ,062003
[7] Berezhko E. G. 1991, *Proc. Int. Cosmic Ray Conf.*, **2**, 436-439

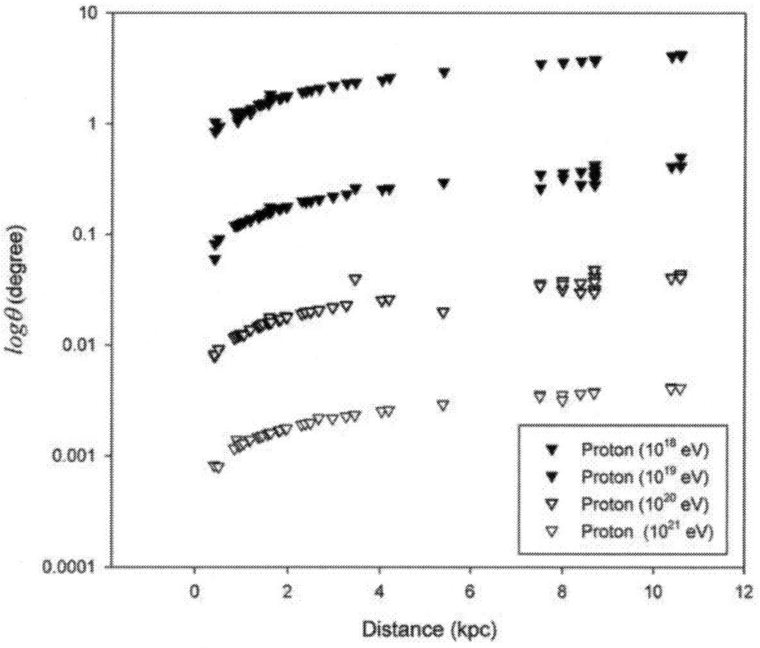


**Figure 2.** Average values of deflection angles versus distance for objects with $0 < l, b < 90$ (Simulated for protons).


[8] Berezhko E. G., 1993, *Proc. Int. Cosmic Ray Conf.*, **2**, 348-351
[9] Berezhko E. G., 1994, *Astron. Lett.*, **20** (1), 75-79
[10] Wibig T. and Wolfendale A. W., 2011, *Open Astron. J*,**4**, 211-217
[11] Fang Ke. Kotera K. and Olinto A. V., 2012, *Astrophys. J.*, **750**, 118
[12] Ostriker J. P., and Gunn J. E., 1969, *Astrophys. J.*, **157**, 1395-1417
[13] Karakula S. Osborne J. L., and Wdowczyk J., 1974, it J. Phys. G: Math., Nucl., Gen., **7** (3), 437-443
[14] Bell A.R., 1992, *Mon. Not. R. Astron. Soc.*, **257**, 493
[15] Usov V. V., 1992, *Nature*, **357**, 472-474
[16] Chi X., Checng K. S., and Young E. C. M., 1996, *Astrophys. J.*,**459**, L83-L86
[17] Blasi P., Epstein R. I., and Olinto A. V., 2000, *Astrophys. J.* **533**, L33-L36
[18] Giller M. and Lipski M., 2002, *J. Phys. G: Nucl. Part. Phys.*, **28**, 1275-1286
[19] Bednarek W., and Bartosik M., 2004, *Astron. Astrophys.*, **423**, 405-413
[20] Bhadra A., 2005, *Proc. Int. Cosmic Ray Conf.*, **3**, 117-120
[21] Blasi P., Epstein R. I., and Olinto A. V., 2000, *Astrophys. J.* **533**, L123-L126
[22] Jacques P., Vallee and Kronberg P. P., 1975, *Astron. Astrophys.*, **43**, 233-242
[23] Vallee J. P., Simard-Normandin M., and Bignell R. C., 1988, *Astrophys. J.*, **331**, 321-324
[24] Stil J. M., Taylor A. R., and Sunstrum C., 2011, *Astrophys. J.*, **726**, 4
[25] Vallee J. P., 1991, *Astrophys. J.*, **366**, 450-454
[26] Indrani C., Deshpande A. A., 1999, *New Astron.*, **4**, 33-40
[27] Beck R., Brandenberg A., Moss D., Shukurov A., and Sokoloff D., 1996, *Annu. Rev. Astron. Astrophys.* **34**, 155-206
[28] Beck R., 2001, *Space Sci. Rev.*, **99**, 243-260
[29] Beck R., 2009, *Astrophys. Space Sci. Trans.*, **5** ,43-47
[30] Mao S. A., et al., 2010, *Astrophys. J.* ,**714**, 1170-1186
[31] Zatsepin G. T., and Kuzmin V. A., 1966, *J. Exp. Theor. Phys. Lett.*, **4**, 78-80
[32] Greisen K., 1966, *Phys. Rev. Lett*, **16**(17), 748-750
[33] Davoudifar P., Fatemi S. J., Clay R., and Whelan B., 2011, *J. Sciences*, **22**(1), 75-84

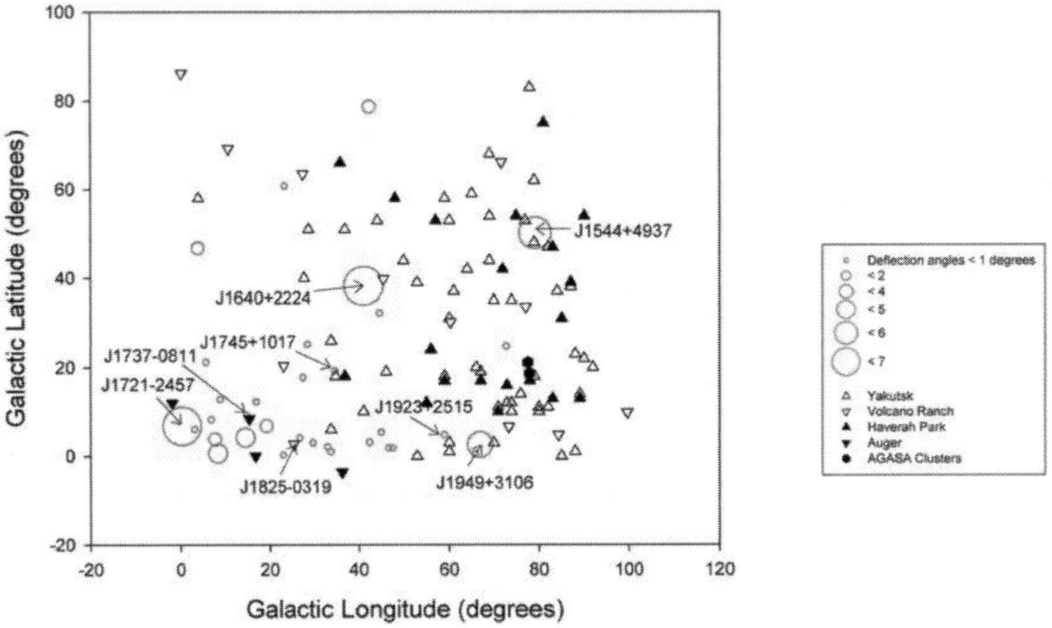


**Figure 3.** Galactic arrival direction of VHE CRs [38, 39, 40, 41, 42, 43, 44, 45, 46] and Galactic pulsar positions [36, 37] for $0 < l < 90$ and $0 < b < 90$ degrees. The deflection regions also are considered (present simulation).

[34] Davoudifar P., 2011, *Proc. Int. Cosmic Ray Conf.*, **2**, 230-233
[35] Sigl G., Schramm D. N., and Bhattacharjee P., 1994, *Astropart. Phys.*, **2** (3), 401-414
[36] Manchester R. N., Hobbs G. B., Teoh A., and Hobbs M., 2005, *Astron. J.*, **129**, 1993-2006
[37] http://www.atnf.csiro.au/research/pulsar/psrcat
[38] Cunningham G., et al., 1980, *Catalogue of Highest Energy Cosmic Rays-Giant Extensive Air Showers*, **1**
[39] Winn M. M., Ulrichs J., Peak L. S., McCusker C. B. A., and Horton L., 1986, *Catalogue of Highest Energy Cosmic Rays-Giant Extensive Air Showers*, **2**
[40] Abreu P., et al., 2010, Astropart. Phys., **34**, (5), 314-326
[41] Bird D. J., et al., 1995, *Astrophys. J.*, **441**, 144
[42] Efimov N. N., Egorov T. A., Krasilnikov D. D., Pravdin m. I., and Sleptsov I. Ye., 1988, *Catalogue of Highest Energy Cosmic Rays-Giant Extensive Air Showers*, **3**
[43] Glushkov A. V., Efimov N. N., Mikhailov A. M., 1991, *Proc. Int. Cosmic Ray Conf.*, **2**, 113-116
[44] Pravdin M. I., et al., 2005, *Proc. Int. Cosmic Ray Conf.*, **7**, 243-246
[45] Medina Tanco G. A., 1997,*Proc. Int. Cosmic Ray Conf.*, **4**, 481-484
[46] Takeda M., et al., 1999, *Astrophys. J.*, **522**, 225-237
[47] Stanev T., Seckel D., and Engel R., 2003, *Phys. Rev. D*, **68**, 102004
[48] Strong A. W., Moskalenko I. V., and Reimer O., 2000, *Astrophys. J.*, **537**, 763-784
[49] Giacinti G., Kachelriess M., Semikoz D. V., and Sigl G., 2011, *Proc. Int. Cosmic Ray Conf.*, **2**, 43-46
[50] Waxman E., and Coppi P., 1996, *Astrophys. J.*, **464**, L75-L78
[51] Waxman E., Miralda-Escude J., 1996, *Astrophys. J.*, **472**, L89-L92
[52] Miralda-Escude J., and Waxman E., 1996, *Astrophys. J.*, **462**, L59-L62
[53] Bhattacharjee P., 2000, *Phys. Rep.*, **327**, 109-247
[54] Giller M. and Lipski M., 2001, *Proc. Int. Cosmic Ray Conf.*, **6**, 2092-2095

# A microscopic modeling of the instant coffee effect

**M Isoda and Y Nishimori**

Department of Physics, Faculty of Education, Kagawa University, Saiwai-cho 1-1, Takamatsu 760-8522, Japan

E-mail: misoda@ed.kagawa-u.ac.jp

**Abstract.** So-called the instant coffee effect is well known in the field of the physics education. The effect is explained that the sound yielded by touching the cup with a spoon is shifted to low-pitched by adulterating bubble owing to putting a spoon of instant coffee into hot water. The phenomenon has been interpreted with the averaged density and compressibility of the fluid in the macroscopic relation for the sound velocity, $v = \sqrt{(\kappa\rho)^{-1}}$. We introduce the linear coupled oscillator model with finite oscillators including the impurity air-mass oscillator. The model may well reproduce the increase in the shift of the eigen frequency accompanying with the amount of bubble.

## 1. Introduction

The *instant coffee effect* is well recognized in the field of physics education [1]. The effect is described as follows: As a spoon of instant coffee is put into the hot water in a cup or a glass, the sound yielded by touching the cup with a spoon gradually changes to higher in pitch.

The effect was, as far as we know, initially reported by Farrell et al. [2]. More than ten years later, Crawford has performed the minute experiment and given a quantitative discussion based on the macroscopic oscillation theory [3]. Following the theory, this sound effect due to bubble sprung out from the granular instant coffee comes from the fundamental mode, which stands with one fourth of the wavelength on the water column in the vessel, where the bottom of water works as a fixed edge and the top of water as a free edge. Two studies have also been performed as the some kind of advanced studies at high school in Japan. There the time evolution of frequency spectrum was observed [4, 5], and it was clarified that the sound velocity decreases with increasing the quantity of instant coffee [5]. Recently, the time evolution of the fundamental frequency mode was clearly observed through a clever experiment by using a tall glass and the experiment shows that the mode is the fundamental eigen frequency standing on the water column in the tall glass [6].

On the other hand, such effect due to bubbles in sound propagation is expected to have been investigated in the field of fluid technology. As far as we know, van Wijngaarden [7] derived the sound velocity in bubbly water as a function of the volume ratio of bubble based on the phenomenological theory and had shown the concave behavior of it [8].

The consideration of the effect by a microscopic modeling will be interested for getting the insight for the properties of each frequency mode. For the purpose, we introduce a linear coupled oscillator model as a simplified microscopic model. As a result, it is obtained that the eigen frequency is softened by adulterating bubbles.

## 2. A linear coupled oscillator model

We introduce a microscopic model to describe the sound wave propagating along the principal axis of the water column in a glass. There the bottom of the column is considered as a fixed end for oscillation and the top of it is free end.

The model is given as a linear coupled oscillator with $N$ oscillators, which are connected via spring each other. The oscillators are assumed to be constituted with $N-1$ water-masses with the mass $M$ and a single air-mass with the mass $m$. In this setting, the number of oscillators $N$ determines the volume ratio of bubbles in the water as $N^{-1}$, which is called as the void-ratio. The air-mass is positioned at $j=[N/2]$ with the Gauss symbol [ ], where $j$ is numbered from the bottom. The spring constant connecting two water-mass to its both edges or connecting on the water-mass and the fixed end at the bottom is denoted by $K$ and it connecting one water-mass and one air-mass by $k$. The spring constant at the top is distinguished as $K'$, because the top end is assumed to be free. The model oscillator is depicted in Fig.1.

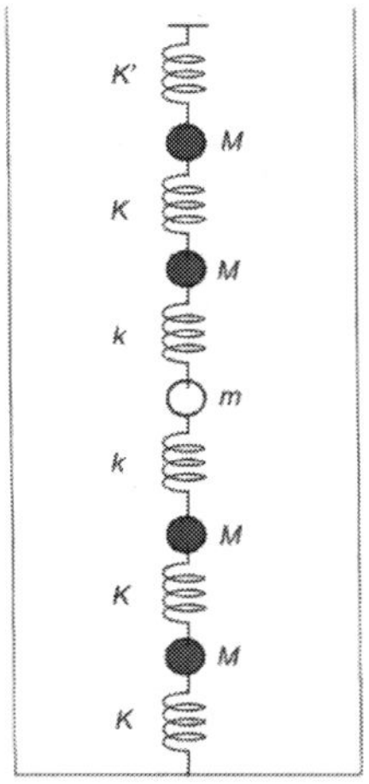


**Figure 1.** A linear coupled oscillator model of water including bubbles for describing the impurity effect of sound propagation. The symbol ● denotes the water-mass with the mass $M$ and the symbol ○ the impurity air-mass with the mass $m$. The spring constants, $k, K$, and $K'$ are given in text.

The equation of motion on the displacement $x_j$ of $j$-th oscillator is expressed as

$$M_j \ddot{x}_j = -K_j(x_j - x_{j-1}) + K_{j+1}(x_{j+1} - x_j), \tag{1}$$

where $M_j$ and $K_j$ denote the mass of $j$-th oscillator and the spring constant of $j$-th spring, as a general expression, both counted from the fixed end corresponding to the bottom of the water column. Under the assumed condition in the present model, $M_j$ and $K_j$ are represented as

$$M_j = m\delta_{j,[\frac{N}{2}]} + M(1-\delta_{j,[\frac{N}{2}]}), \tag{2}$$

$$K_j = k(\delta_{j,[\frac{N}{2}]} + \delta_{j,[\frac{N}{2}]+1}) + K'\delta_{j,N+1} + K(1-\delta_{j,[\frac{N}{2}]})(1-\delta_{j,[\frac{N}{2}]+1})(1-\delta_{j,N+1}). \tag{3}$$

The dimensionless parameters like $\hat{m} \equiv m/M$ and $\hat{k} \equiv k/K$ are introduced.

The differential equation (1) is actually solved for $N = 3 \sim 7$. By assuming the solution of type $x_j = c_j \exp^{i\omega t}$ with the time variable $t$. The obtained $N$ algebraic equations are solved numerically, under the condition of $c_0 = 0$ and $c_{N+1} = 0$. And for the free end, $K' = 0$ is assumed hereafter. The eigen frequency $\omega$ is normalized as $\hat{\omega} = \omega/\omega_0$ by $\omega_0 = \sqrt{K/M}$, which

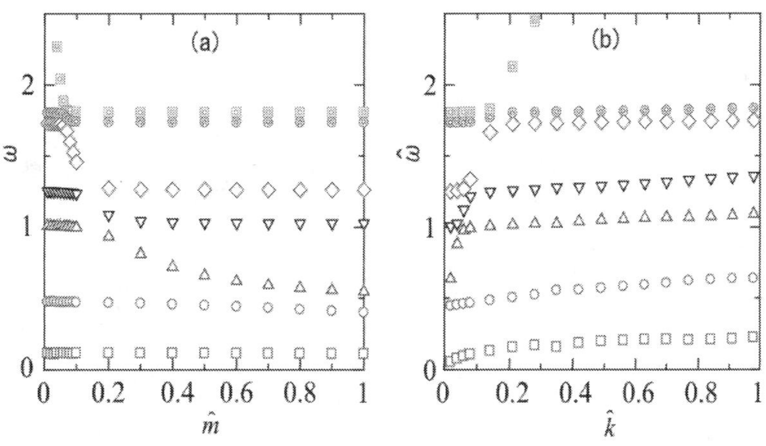


**Figure 2.** Seven eigen frequencies for $N = 7$ system as a function of $\hat{m}$ at $\hat{k} = 0.1$ (a) and of $\hat{k}$ at $\hat{m} = 0.1$ (b).

is the angular frequency in the harmonic oscillator with a single water-mass. The ratio of mass $\hat{m}$ is chosen as the value smaller than unity, because the density of air is 800 times smaller than that of water. On the other parameter $\hat{k}$, it is not directly estimated. Here we consider the quantity of the spring constant $k$ divided by the suitable length like the natural length of the spring $\ell$,

$$k/\ell = \Delta F/\ell\Delta x = (\Delta F/\ell^2)/(\ell^2\Delta x/\ell^3) \sim \Delta p/(\Delta V/V) \sim \kappa^{-1}, \tag{4}$$

where $V \equiv \ell^3$, $\Delta F$ is the force applied to the spring and $\Delta x$ the displacement. Therefore, $\hat{k}$ is proportional to the ratio of the compressibilities of water $\kappa_{\text{water}}$ and air $\kappa_{\text{air}}$, as $\hat{k} = k/K \propto \kappa_{\text{water}}/\kappa_{\text{air}}$. The compressibility of air $\kappa_{\text{air}}$ is about 15,000 times larger than that of water $\kappa_{\text{water}}$, then the $\hat{k}$ may be expected to be sufficiently smaller than unity. The calculated frequencies $\hat{\omega}$ for $N = 7$ are shown in Fig.2(a) and (b), as a function of $\hat{m}$ at fixed $\hat{k} = 0.1$ and of $\hat{k}$ at fixed $\hat{m} = 0.1$, respectively.

We here come back to our original purpose to get the void-ratio dependence of eigen frequencies. In the present model, the length of coupled oscillator is different for the different number of oscillators $N$, which corresponds to $N\ell$ for the case of free top end of $K' = 0$. To unify the system length, that is the wavelength of standing wave, the obtained eigen frequencies for $N$-oscillator system $\omega_N$ are rescaled as $\hat{\Omega}_N \equiv \frac{N}{N_{\text{max}}}\hat{\omega}_N$. To compare with the experiment, the deviation of frequency from the purely water system without the impurity air-mass has to be estimated. Then the eigen frequencies of $N$-oscillator system without air-mass are calculated and the rescale procedure has been done as same as the air-mass included system. The rescaled frequency for purely water system is denoted as $\hat{\Omega}_N^{\text{w}}$. With this quantity, the deviation of frequency from the purely water system is obtained as $\Delta\hat{\Omega}_N = \hat{\Omega}_N - \hat{\Omega}_N^{\text{w}}$. In Fig.3, the quantity for the lowest frequency mode, designated by the superscript (1), is shown as a function of $N^{-1}$, for the case of $\hat{k} = 0.1$ and $\hat{m} = 0.1$.

The decreasing tendency of the deviation of frequency is, as shown in Fig.3, reproduced in the present finite size calculation under the present model. This decrease is qualitatively consistent with the experimental result [5] and the phenomenological theory [7].

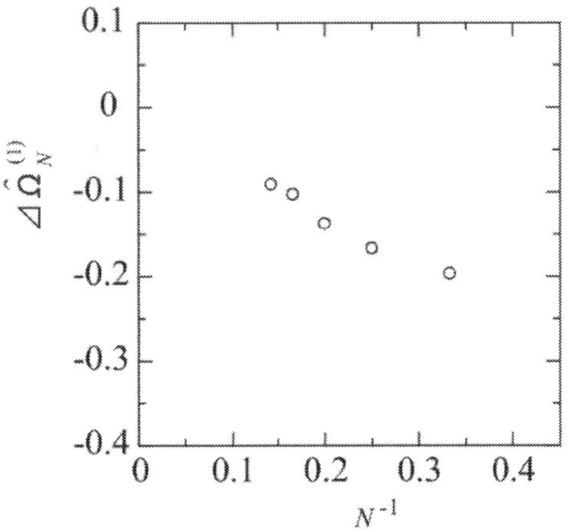


**Figure 3.** Sift of rescaled lowest eigen frequency by replacing a water-mass by an air-mass, $\Delta\hat{\Omega}_N^{(1)}$, as a function of $N^{-1}$, corresponding to the void-ratio. The parameters are chosen as $\hat{k} = 0.1$ and $\hat{m} = 0.1$.

## 3. Discussions and conclusions

The linear coupled oscillator model including the air-mass as an impurity has been proposed to give the microscopic description of the *instant coffee effect.* The systems with finite number of oscillators have been calculated to get the eigen frequencies, numerically. The decrease of the fundamental frequency by adulterating bubbles increases with the volume-ratio of bubble, in accord with the experiment. Although the present calculated results are given only for the case of the central position of the impurity air-mass, it has been confirmed that the result is not modified remarkably.

The present coupled oscillator model may be discussed by connecting with the macroscopic relation for the velocity $v = (\sqrt{\kappa\rho})^{-1}$ by applying the effect of the impurity oscillator into the compressibility $\kappa$ and the density $\rho$. The procedure and the discussion will be given in the separate issue.

## Acknowledgments

We gratefully acknowledge Jumpei Okamoto for sending his prepublication data and for discussions and Isao Sawada for discussions.

## References

[1] Walker J 1975 *The Flying Circus of Physics With Answers* (New York: Wiley & Sons).
[2] Farrell W E, McKenzie D P and Parker R L 1969 *Proc. Camb. Phil. Soc.* **65** 365.
[3] Crawford F S 1982 *Am. J. Phys.* **50** 398.
[4] Tsukamoto K 2007 *Butsuri Kyoiku* **55** 291 in Japanese.
[5] Okamoto J unpublished.
[6] Planinšič G and Etkina E 2012 *The Physics Teacher* **50** 460.
[7] van Wijngaarden L 1972 *Annu. Rev. Fluid Mech.* **4** 369.
[8] Kyotoh H, Maeda N and Sakurai T 2004 *Nagare* **23** 27 in Japanease.

# R-Matrix Analysis of Structures in Economic Indices: from Nuclear Reactions to High-Frequency Trading

**Frank W K Firk**

The Henry Koerner Center for Emeritus Faculty,
Yale University,
149 Elm St. New Haven, CT 06520–8368, USA

E-mail: fwkfirk@aol.com

**Abstract.** It is shown that the R-matrix theory of nuclear reactions is a viable mathematical theory for the description of the fine, intermediate and gross structure observed in the time-dependence of economic indices in general, and the daily Dow Jones Industrial Average in particular. A Lorentzian approximation to R-matrix theory is used to analyze the complex structures observed in the Dow Jones Industrial Average on a typical trading day. Resonant structures in excited nuclei are characterized by the values of their fundamental strength function, (average total width of the states)/(average spacing between adjacent states). Here, values of the ratios (average lifetime of individual states of a given component of the daily Dow Jones Industrial Average)/(average interval between the adjacent states) are determined. The ratios for the observed fine and intermediate structure of the index are found to be essentially constant throughout the trading day. These quantitative findings are characteristic of the highly statistical nature of many-body, strongly interacting systems, typified by daily trading. It is therefore proposed that the values of these ratios, determined in the first hour-or-so of trading, be used to provide valuable information concerning the likely performance of the fine and intermediate components of the index for the remainder of the trading day.

## 1. Introduction

A basic parameter in systems that exhibit energy- or time-dependent resonant-like structures, both quantum and classical, is the ratio (average width of the individual states)/(average spacing of the states), denoted by $\langle\Gamma\rangle/\langle D\rangle$ (in energy units) or the ratio (average lifetime of the individual states)/(average time interval between states), here denoted by $\langle\Delta t\rangle/\langle T\rangle$ (in time-units). If $\langle\Gamma\rangle/\langle D\rangle < 1$, the parameters that characterize the individual states can be determined. In quantum mechanical systems, resonant structures are analyzed using the R-matrix theory of Wigner and Eisenbud [1]; it is an exact theory with parameters that are understood at a fundamental level. In the present analysis, the (classical) structures observed in the daily Dow Jones Industrial Average (DJIA) are analyzed using Lorentzian forms as good approximations to the complex R-matrix forms.

As discussed previously by the author [2], a striking similarity is observed between the forms of the fine, intermediate, and gross structure of photonuclear states in the Giant

Electric Dipole Resonance region of medium-mass nuclei, and the fine, intermediate, and gross structure observed in the fluctuations (states) of daily economic indices

In nuclei, the gross, intermediate and fine structures are attributed to the evolution of nuclear reactions from the underlying *single-particle* states to *few-particle-few-hole* states to *many-particle-many-hole* compound-nuclear states, respectively. In the fluctuations of an economic index, the three structural forms are associated with *long*, *medium*, and *short-range* time-correlations in the complex trading options that take place among the innumerable elements that contribute to the index. The fine structure is a characteristic feature of High-Frequency Trading (HFT).

Traditional trading strategies lack a dynamical model of the *mathematical structure* associated with a typical economic index; the *lifetime* of each state is of critical importance.

## 2. A Time-Dependent Lorentzian: the Breit-Wigner resonance form

In a seminal paper, Thomas [3] developed the exact R-matrix theory of resonant quantum states using arguments that were of practical importance in the analysis of observed spectra. His method has formed the basis of a great deal of the subsequent work in the field of nuclear spectroscopy.

The Thomas R-function is

$$R_{c'c''} = \Sigma_\lambda \gamma_{\lambda c'}\gamma_{\lambda c''} /(E - E_\lambda - i\Gamma_\lambda^e/2), \qquad (1)$$

where the $\gamma$'s are reduced width amplitudes for the levels $\lambda$ and the channels c', c'' and $\Gamma_\lambda^e$ is the width of the "eliminated" channels (in general, the unknown reaction channels). At sufficiently low neutron energies, equation (1) leads to the following concise form for the neutron total cross section in the region of a single level $\lambda$,

$$\sigma_{tot} = \sigma'/(1 + x^2) + 2ka\sigma' x/(1 + x^2) + 4\pi a^2 \text{ (the Breit-Wigner equation)}, \qquad (2)$$

where $x = (2/\Gamma)(E - E_\lambda)$, E is the incident neutron energy, $E_\lambda$ is the resonance energy, $\Gamma$ is the total width of the resonance ($\Gamma = \Gamma_n + \Gamma_\gamma (= \Gamma_\lambda^e)$), the sum of the elastic scattering width and the radiation width, $\sigma'$ (the peak cross section when on resonance), k is the neutron wave number and a is the effective nuclear radius.

The structures of typical daily economic indices, reported in 10-second intervals, do not have the richness of form observed in high-resolution studies of low-energy neutron-nucleus interactions. An approximate form of the R-matrix theory is therefore justified in the analysis of an economic index such as the daily DJIA.

The neutron total cross section given in equation (2) has a leading term $\sigma'/(1 + x^2)$; it is a *Lorentzian function* in the energy domain. To model the structure in the daily DJIA the following time-dependent form is used:

$$\text{lzn}(X) = M_0/(1 + X^2) \qquad (3)$$

where $X = (2/\Delta t)(t - t_0)$, $t$ is the time variable, $t_0$ is the central time of the symmetric function, $M_0$ is the maximum value of the function, and $\Delta t$ is the full width at half maximum of the function.

## 3. An Analysis of the DJIA data on April 3, 2012

We are concerned with an analysis of the DJIA reported on April 3, 2012; the analysis requires an understanding of the basic components of the index; they are:

1. Gross structure with a characteristic time-scale of many hours,
2. Intermediate structure that appears in two distinct patterns: I, with an average adjacent interval of about 60 minutes and II, with an average adjacent interval of about 10 minutes,
3. Fine structure with a characteristic time-scale of 1- to 2-minutes. It is this component that is associated with HFT.

In a limited interval, the two intermediate and the fine structure components are illustrated in figure 1:

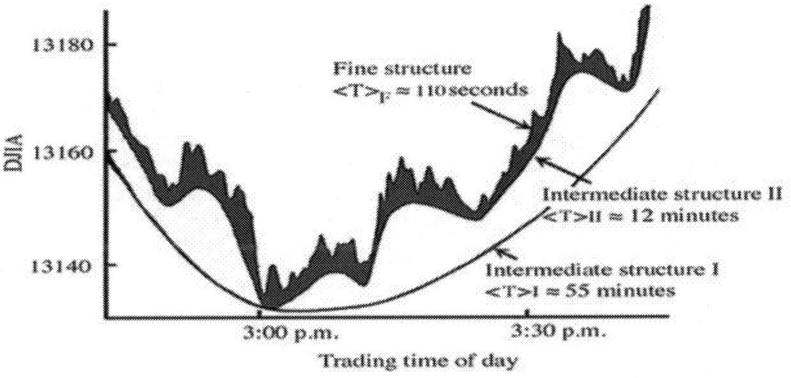


Figure 1. Intermediate components and the fine structure component required to analyze the DJIA. Data are shown in 30-second intervals. The average intervals of the states are $<T>_I \approx$ 55 min, $<T>_{II} \approx$ 12 min, and $<T>_F \approx$ 110 sec.

3.1. Intermediate structure I

An iterative, least-squares method is used to analyze the overlapping Lorentzian states. The procedure is continued until a satisfactory fit is obtained. The value of $<\Delta t>/<T>_I \approx 0.6$. The small number of intermediate I states during the trading day limits the accuracy of this value.

3.2. Intermediate structure II

The value of $<T>_{II}$ for the intermediate II structure is obtained by averaging over the entire period of the trading day; it is $<T>_{II} = 12$ minutes. An iterative fit to the data gives a value of $<\Delta t>/<T>_{II} \approx 0.6$.

3.3. Fine structure

The fit to the fine structure in the time interval 2:45 to 3:45 p.m. is shown in figure 2.

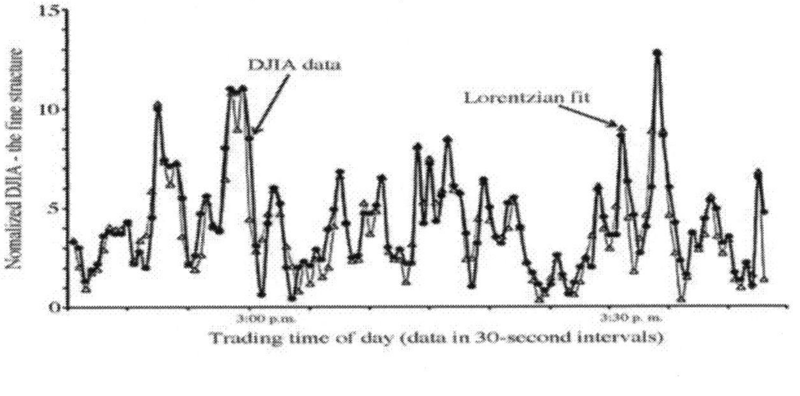


**Figure 2**. The thin line illustrates the fit to the fine structure data from 2:45 to 3:45 p.m.

The value of the strength function for these states is $\langle \Delta t \rangle / \langle T \rangle_F = 0.52$.

## 4. Conclusions

An analysis of the DJIA on April 3, 2012 has been carried out using multi-state Lorentzian functions to model the index. The HFT states are excellent examples of discrete states with spacing and lifetime distributions that are analogous to the statistical distributions that characterize low energy neutron resonance reactions (Porter [4]). The parameters of the defining Lorentzians are used to obtain the fundamental ratios $\langle \Delta t \rangle / \langle T \rangle$ of the basic components of the index; they are found to be remarkably constant throughout a given trading day. These characteristic constants are associated with the statistical nature of daily trading. The HFT strength function is largely unaffected by the uncontrollable, time-dependent market forces that can have such dramatic effects on the value of an economic index when studied over periods of many hours or days. The value of the fine structure strength function depends upon the average load, bandwidth, processing capacity, and delays in buy and sell orders (the *latency*) associated with HFT on a given day.

It is proposed that the values of $\langle \Delta t \rangle / \langle T \rangle$ for the fine and intermediate structure of the states of the DJIA, on a given day, be obtained by analyzing the data in the first one- to two-hours of trading, and that these values be used as *predictors* of their values for the remaining trading time of the day. This procedure assumes that there is sufficient statistical inertia in the fine and intermediate structure associated with trades on the given day; the present analysis provides support for this assumption.

## References


[1] Wigner E P and Eisenbud L 1947 *Phys. Rev.* **72** 29

[2] Firk F W K 2012 *arXiv.org*: 1203.6021[q–fin.GN]

[3] Thomas R G 1955 *Phys. Rev.* **97** 224

[4] Porter C E 1965 *Statistical Theories of Spectra: Fluctuations* (New York: Academic Press)

# New nonlocal symmetries and conservation laws of the (1+1)-dimensional Sine-Gordon equation

**Zufeng Liang**[1], **Xiaoyan Tang**[2] **and Senyue Lou**[3,4]

[1] Department of Physics, Hangzhou Normal University, Hangzhou 310036, China

[2] Department of Physics and Astronomy, Shanghai Jiao Tong University, Shanghai 200240, China

[3] Shanghai Key Laboratory of Trustworthy Computing, East China Normal University, Shanghai 200062, China

[4] Faculty of Science, Ningbo University, Ningbo, 315211, China

E-mail: `xytang@sjtu.edu.cn`

**Abstract.** A system of linearized equations of the (1+1)-dimensional Sine-Gordon (SG) equation and its Bäcklund transformation are given to determine not only nonlocal symmetries but also nonlocal conservation laws of the SG equation. Through the parameter expansion procedure, a sequence of infinitely many nonlocal symmetries and a sequence of infinitely many nonlocal conservation laws are obtained.

## 1. Introduction

Symmetries and conservation laws are focus topics in the studies of nonlinear differential equations, and they have a very close relation, for instance, one can expect one or more conservation laws from one known symmetry [1]. Usually, the Lie group approach involving the prolongation structures, or equivalently the symmetry approach, is a powerful method to find classical local Lie point symmetries [2, 3]. To find nonlocal symmetries and even higher order local symmetries, one has to use different ways. For instance, nonlocal potential symmetries can be obtained through the local symmetries of the potential form of the given system, residual symmetries actually related to Bäcklund transformations (BTs) are obtained via the truncated Painlevé expansion [4]. Other types of well studied nonlocal symmetries include the so-called eigenfunction symmetries [5, 6, 7, 8], and other nonlocal symmetries related to BTs [9].

Conservation laws have a variety of applications, such as to describe conserved physical quantities like mass, energy and momentum. They are also used in the stability analysis and in the design of numerical methods, and so on. Many effective methods have been developed to construct conservation laws, such as the celebrated Noether's theorem [10], the conservation law multiplier approach [11] and the characteristic method [12].

Our recent paper [13] has discovered infinitely many nonlocal symmetries and infinitely many local and nonlocal conservation laws of the Sine-Gordon (SG) equation via a parameter expansion procedure. Here, following the same scheme, we present more infinitely many nonlocal symmetries and nonlocal conservation laws of the (1+1)-dimensional SG equation

$$u_{xt} = \sin u, \tag{1}$$

which has an auto-Bäcklund transformation

$$u_x - v_x = 2\lambda \sin\left(\frac{u}{2} + \frac{v}{2}\right), \tag{2a}$$

$$u_t + v_t = \frac{2}{\lambda} \sin\left(\frac{u}{2} - \frac{v}{2}\right), \tag{2b}$$

with $v$ satisfying the same SG equation

$$v_{xt} = \sin v. \tag{3}$$

In section 2, using the parameter expansion procedure, a new sequence of infinitely many nonlocal symmetries is generated from one special nonlocal symmetry, originally obtained in Ref. [13]. In a similar way, a series of infinitely many nonlocal conservation laws is presented in section 3. The last section is devoted to summary and discussions.

## 2. Infinitely many nonlocal symmetries related to BT

A system of four linearized equations is obtained as

$$\sigma^u_{xt} - \sigma^u \cos u = 0, \tag{4a}$$

$$\sigma^v_{xt} - \sigma^v \cos v = 0, \tag{4b}$$

$$\sigma^u_x - \sigma^v_x - 2\delta \sin\left(\frac{u}{2} + \frac{v}{2}\right) - \lambda \cos\left(\frac{u}{2} + \frac{v}{2}\right)(\sigma^u + \sigma^v) = 0, \tag{4c}$$

$$\sigma^u_t + \sigma^v_t + \frac{2\delta}{\lambda^2} \sin\left(\frac{u}{2} - \frac{v}{2}\right) - \frac{1}{\lambda} \cos\left(\frac{u}{2} - \frac{v}{2}\right)(\sigma^u - \sigma^v) = 0, \tag{4d}$$

by requiring Eqs. (1)-(3) are form invariant under the infinitesimal transformation $u \to u + \epsilon\sigma^u$, $v \to v + \epsilon\sigma^v$, and $\lambda \to \lambda + \epsilon\delta$, where $\epsilon$ is an infinitesimal parameter. It is noted that the symmetry $\sigma^u$ determined by Eq. (4) is nonlocal because two functions $u$ and $v$ are related through BT. In Ref. [13], three special special solutions of Eq. (4) have been obtained and one sequence of infinitely many nonlocal symmetries is generated from the first special nonlocal symmetry via the parameter expansion method. Here, in a similar way, a more complicated series of nonlocal symmetries of the SG equation is obtained from the nonlocal symmetry

$$\sigma^u = qe^{\lambda p} \tag{5}$$

with $p$ and $q$ determined by

$$p_x = \cos\left(\frac{u}{2} + \frac{v}{2}\right),\ p_t = \frac{1}{\lambda^2}\cos\left(\frac{u}{2} - \frac{v}{2}\right),\ q_x = \frac{\sin\left(\frac{u}{2} + \frac{v}{2}\right)}{e^{\lambda p}},\ q_t = \frac{\sin\left(\frac{u}{2} - \frac{v}{2}\right)}{\lambda^2 e^{\lambda p}}. \tag{6}$$

In order to construct infinitely many nonlocal symmetries from the special nonlocal symmetry (5) with (6), we have to first expand the functions $p$ and $q$ as $p = \sum_{i=0}^{\infty} p_i\delta^i$ and $q = \sum_{i=0}^{\infty} q_i\delta^i$ with an arbitrary expansion constant $\delta$, then insert them into (5) with $\lambda$ replaced by $\lambda + \delta$, and finally expand the nonlocal symmetry as

$$\sigma^u = \sum_{j=0}^{\infty} q_j\delta^j \times \exp\left((\lambda + \delta)\sum_{i=0}^{\infty} p_i\delta^i\right) \equiv \sum_{i=0}^{\infty} \sigma^u_i \delta^i, \tag{7}$$

where $\sigma^u_i$ is computed through $\sigma^u_i = \frac{d^n\sigma^u}{d\delta^n}|_{\delta=0}$. It is remarkable that all the expansion coefficients $\sigma^u_i$ constitutes a sequence of infinitely many nonlocal symmetries of the SG equation.

Though it is difficult to write down a general explicit expression for $\sigma_i^u$, it is easy to obtain their explicit forms one after one from the general formula (7). For the lack of space, we just present the first three explicit nonlocal symmetries $\sigma_0^u$, $\sigma_1^u$ and $\sigma_2^u$,

$$\sigma_0^u = q_0 e^{\lambda p_0}, \tag{8}$$

$$\sigma_1^u = (q_1 + p_0 q_0 + \lambda p_1 q_0) e^{\lambda p_0}, \tag{9}$$

$$\sigma_2^u = \frac{1}{2}\left(2q_2 + 2q_1(\lambda p_1 + p_0) + q_0(\lambda p_1 + p_0)^2 + 2q_0(\lambda p_2 + p_1)\right) e^{\lambda p_0}. \tag{10}$$

The system of equations determining $p_i$, $q_i$ and $v_i$ $(i = 0, 1, 2)$ are obtained by substituting the $p$'s and $q$'s expansions and the $v$'s expansion $v = \sum_{i=0}^{\infty} v_i \delta^i$ into (6) and (2) with $\lambda$ replaced by $\lambda + \delta$, and then equating zero all the coefficients of different orders of $\delta$, which reads

$$p_{0x} = \cos\left(\frac{u}{2} + \frac{v_0}{2}\right),\quad p_{0t} = \frac{1}{\lambda^2}\cos\left(\frac{u}{2} - \frac{v_0}{2}\right), q_{0x} = \frac{\sin\left(\frac{u}{2} + \frac{v_0}{2}\right)}{e^{\lambda p_0}},\quad q_{0t} = \frac{\sin\left(\frac{u}{2} - \frac{v_0}{2}\right)}{\lambda^2 e^{\lambda p_0}},$$

$$p_{1x} = -\frac{v_1}{2}\sin\left(\frac{u}{2} + \frac{v_0}{2}\right),\quad p_{1t} = -\frac{2}{\lambda^3}\cos\left(\frac{u}{2} - \frac{v_0}{2}\right) + \frac{v_1}{2\lambda^2}\sin\left(\frac{u}{2} - \frac{v_0}{2}\right),$$

$$q_{1x} = \frac{v_1}{2e^{\lambda p_0}}\cos\left(\frac{u}{2} + \frac{v_0}{2}\right) - \frac{\lambda p_1 + p_0}{e^{\lambda p_0}}\sin\left(\frac{u}{2} + \frac{v_0}{2}\right),$$

$$q_{1t} = \frac{v_1}{2\lambda^2 e^{\lambda p_0}}\cos\left(\frac{u}{2} - \frac{v_0}{2}\right) + \frac{2 + \lambda^2 p_1 + \lambda p_0}{\lambda^3 e^{\lambda p_0}}\sin\left(\frac{u}{2} - \frac{v_0}{2}\right),$$

$$p_{2x} = -\frac{v_1^2}{8}\cos\left(\frac{u}{2} + \frac{v_0}{2}\right) - \frac{v_2}{2}\sin\left(\frac{u}{2} + \frac{v_0}{2}\right),$$

$$p_{2t} = \left(\frac{3}{\lambda^4} - \frac{v_1^2}{8\lambda^2}\right)\cos\left(\frac{u}{2} - \frac{v_0}{2}\right) - \left(\frac{v_1}{\lambda^3} - \frac{v_2}{2\lambda^2}\right)\sin\left(\frac{u}{2} - \frac{v_0}{2}\right),$$

$$q_{2x} = \frac{v_2 - \lambda p_1 v_1 - p_0 v_1}{2e^{\lambda p_0}}\cos\left(\frac{u}{2} + \frac{v_0}{2}\right) - \frac{8\lambda p_2 + 8p_1 - 8\lambda p_1 p_0 - 4\lambda^2 p_1^2 - 4p_0^2 + v_1^2}{8e^{\lambda p_0}}\sin\left(\frac{u}{2} + \frac{v_0}{2}\right),$$

$$\begin{aligned} q_{2t} = {} & \frac{\lambda(v_2 - p_0 v_1) - 2v_1 - \lambda^2 p_1 v_1}{2\lambda^3 e^{\lambda p_0}}\cos\left(\frac{u}{2} - \frac{v_0}{2}\right) \\ & - \frac{4\lambda^4 p_1^2 + 8\lambda^3(p_1 p_0 - p_2) + \lambda^2(8p_1 + 4p_0^2 - v_1^2) + 16\lambda p_0 + 24}{8\lambda^4 e^{\lambda p_0}}\sin\left(\frac{u}{2} - \frac{v_0}{2}\right), \end{aligned}$$

with $v_i$ $(i = 0, 1, 2)$ given by

$$v_{0x} = u_x - 2\lambda\sin\left(\frac{u}{2} + \frac{v_0}{2}\right),\quad v_{0t} = -u_t + \frac{2}{\lambda}\sin\left(\frac{u}{2} - \frac{v_0}{2}\right),$$

$$v_{1x} = -\lambda v_1\cos\left(\frac{u}{2} + \frac{v_0}{2}\right) - 2\sin\left(\frac{u}{2} + \frac{v_0}{2}\right),\quad v_{1t} = -\frac{v_1}{\lambda}\cos\left(\frac{u}{2} - \frac{v_0}{2}\right) - \frac{2}{\lambda^2}\sin\left(\frac{u}{2} - \frac{v_0}{2}\right),$$

$$v_{2x} = -(v_1 + \lambda v_2)\cos\left(\frac{u}{2} + \frac{v_0}{2}\right) + \frac{\lambda v_1^2}{4}\sin\left(\frac{u}{2} + \frac{v_0}{2}\right),$$

$$v_{2t} = m\left(\frac{v_1}{\lambda^2} - \frac{v_2}{\lambda}\right)\cos\left(\frac{u}{2} - \frac{v_0}{2}\right) + m\left(\frac{2}{\lambda^3} - \frac{v_1^2}{4\lambda}\right)\sin\left(\frac{u}{2} - \frac{v_0}{2}\right).$$

It is not difficult to verify that Eqs. (8)-(10) are all nonlocal symmetries of the SG equation by the direct substitution of them with above equations into the symmetry equation (4a).

## 3. Infinitely many nonlocal conservation laws

The linearized equations (4) can also provide nonlocal conservation laws of the SG equation. Three special nonlocal conservation laws and one sequence of infinitely many nonlocal conservation laws from one of them have been obtained in Ref. [13]. In this section, another series of infinite number of nonlocal conservation laws is produced from

$$\left[\lambda^2 e^{-\lambda p}\sin\left(\frac{u}{2}+\frac{v}{2}\right)\right]_t+\left[e^{-\lambda p}\sin\left(\frac{u}{2}-\frac{v}{2}\right)\right]_x=0. \tag{11}$$

Following the parameter expansion scheme, substituting the expansions of $v$ and $p$ into Eq. (11) with $\lambda$ replaced by $\lambda+\delta$, and then setting zero the coefficients of different orders of $\delta$, a series of infinite number of nonlocal conservation laws is readily generated. The first one has the form of (11) with $p$ and $v$ replaced by $p_0$ and $v_0$, respectively, and the next two read

$$\begin{aligned}&\left[\lambda^2 v_1 e^{-\lambda p_0}\cos\left(\frac{u}{2}+\frac{v_0}{2}\right)-2\lambda e^{-\lambda p_0}(\lambda^2 p_1+\lambda p_0-2)\sin\left(\frac{u}{2}+\frac{v_0}{2}\right)\right]_t\\&\quad+\left[v_1 e^{-\lambda p_0}\cos\left(\frac{u}{2}-\frac{v_0}{2}\right)-2e^{-\lambda p_0}(\lambda p_1+p_0)\sin\left(\frac{u}{2}-\frac{v_0}{2}\right)\right]_x=0,\\&\left[4\lambda e^{-\lambda p_0}(2v_1-\lambda(v_1p_0-v_2)-\lambda^2 p_1 v_1)\cos\left(\frac{u}{2}+\frac{v_0}{2}\right)+e^{-\lambda p_0}(4\lambda^4 p_1^2-8\lambda^3(p_2-p_0p_1)\right.\\&\quad\left.-\lambda^2(v_1^2+24p_1-4p_0^2)-2\lambda p_0+1)\sin\left(\frac{u}{2}+\frac{v_0}{2}\right)\right]_t+\left[4e^{-\lambda p_0}(\lambda p_1 v_1+p_0 v_1-v_2)\cos\left(\frac{u}{2}-\frac{v_0}{2}\right)\right.\\&\quad\left.-e^{-\lambda p_0}(-4\lambda^2 p_1^2+8\lambda(p_2-p_1p_0)+8p_1-4p_0^2+v_1^2)\sin\left(\frac{u}{2}-\frac{v_0}{2}\right)\right]_x=0,\end{aligned}$$

where the equations determine $p_i$ and $v_i$ $(i=0,1,2)$ are given in the previous section.

## 4. Summary and Discussion

Using the parameter expansion method, a sequence of infinite number of nonlocal symmetries and a sequence of infinitely many nonlocal conservation laws related to BT are obtained from one particular parameter dependent nonlocal symmetry and one nonlocal conservation law, respectively, for the SG equation. This scheme can be applied to any other nonlinear partial differential equation possessing BT. In addition, one can further not only investigate the invariant solutions by using new nonlocal symmetries through localization procedure, but also search for new integrable systems from new nonlocal conservation laws.

## Acknowledgement

The work is the financially supported by the National Natural Science Foundation of China (No. 11275123 and No. 11175092), Shanghai Knowledge Service Platform for Trustworthy Internet of Things (No. ZF1213), Talent Fund and K. C. Wong Magna Fund in Ningbo University.

## References

[1] Jia M, Gao Y and Lou S Y 2010 *Phys. Lett. A* **374** 1704
[2] Olver P J 1993 *Applications of Lie Groups to Differential Equations* (New York: Springer)
[3] Tang X Y and Lou S Y 2002 *Chin. Phys. Lett.* **19** 1
[4] Gao X N, Lou S Y and Tang X Y 2013 *J. High Energy Phys.* **05** 029
[5] Lou S Y 1994 *J. Math. Phys.* **35** 2390
[6] Oevel W and Schief W 1994 *Rev. Math. Phys.* **6** 1301
[7] Lou S Y and Hu X B 1997 *J. Phys. A* **30** L95
[8] Hu X R, Lou S Y and Chen Y 2012 *Phys. Rev. E* **85** 056607
[9] Lou S Y, Hu X R and Chen Y 2012 *J. Phys. A: Math. Theor.* **45** 155209
[10] Noether E 1971 *Trans. Theor. Stat. Phys.* **1** 186
[11] Naz R 2012 *J. Appl. Math.* **2012** 871253
[12] Steudel H Z 1962 *Naturforsch A* **17** 129
[13] Tang X Y, Liang Z F and Lou S Y 2013 Infinitely many nonlocal symmetries and conservation laws for the (1+1)-dimensional Sine-Gordon equation *arXiv* 1308.3018

# Optimization strategies in the modelling of SG-SMB applied to separation of phenylalanine and tryptophan

**Leôncio Diógenes Tavares Câmara**

Instituto Politécnico da Universidade do Estado do Rio de Janeiro (IPRJ-UERJ), Dep. de Engenharia Mecânica e Energia, DEMEC, Nova Friburgo-RJ, Brazil.

diogenescamara@gmail.com, dcamara@iprj.uerj.br

**Abstract.** The solvent-gradient simulated moving bed process (SG-SMB) is the new tendency in the performance improvement if compared to the traditional isocratic solvent conditions. In such SG-SMB process the modulation of the solvent strength leads to significant increase in the purities and productivity followed by reduction in the solvent consumption. A stepwise modelling approach was utilized in the representation of the interconnected chromatographic columns of the system combined with a lumped mass transfer model between the solid and liquid phase. The influence of the solvent modifier was considered applying the Abel model which takes into account the effect of modifier volume fraction over the partition coefficient. Correlation models of the mass transfer parameters were obtained through the retention times of the solutes according to the volume fraction of modifier. The modelling and simulations were carried out and compared to the experimental SG-SMB separation unit of the amino acids Phenylalanine and Tryptophan. The simulation results showed the great potential of the proposed modelling approach in the representation of such complex systems. The simulations showed great agreement fitting the experimental data of the amino acids concentrations both at the extract as well as at the raffinate. A new optimization strategy was proposed in the determination of the best operating conditions which uses the phi-plot concept.

## 1. Introduction

The simulated moving bed (SMB) process has been implemented successfully in the pharmaceutical industry to perform difficult separation of molecules. The simulated moving bed (SMB) is a continuous process of interconnected high-performance liquid chromatographic (HPLC) columns which can be used to separate enantiomeric mixtures with high purity and low solvent consumption. The process consists in the simulated countercurrent movement of solid adsorbent phase by switching the positions of the two inlet (Feed and solvent) and the two outlet (Extract and raffinate) streams in the clockwise direction. The more and less adsorbable molecules of enantiomers are collected at the extract and raffinate stream, respectively.

The solvent-gradient simulated moving bed process (SG-SMB) is the new tendency in the performance improvement if compared to the traditional isocratic solvent conditions [1-2]. In such SG-SMB process the modulation of the solvent strength leads to significant increase in the purities and productivity followed by reduction in the solvent. The solvent strength in the SG-SMB process is altered through the volume fraction of the modifier ($\phi$) as can be seen in Fig. 1. Nam *et al.* [1] carried out SG-SMB experiments in the performance evaluation of phenylalanine and tryptophan separation which was compared to isocratic solvent conditions. Its operating conditions were determined from the SG-SMB optimization tool based on genetic algorithm.

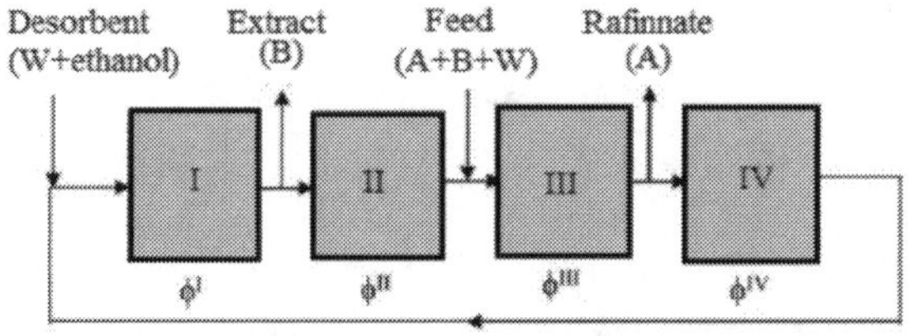


Figure 1 – Schematic representation of SG-SMB

The determination of the operation conditions of SG-SMB is a difficult task [2] and only numerical methods has been utilized as carried out by Nam et al. [1]. Therefore in this work it was proposed a new methodology to achieve the process operating conditions without heavy numerical calculations. A novel optimization strategy was proposed which considers the Henry's constants as a function of the modifier volume fraction of the desorbent near the feed entrance. The influence of the solvent modifier was considered applying the Abel model which takes into account the effect of modifier volume fraction over the partition coefficient.

## 1. Modelling and optimization approaches

The chromatographic columns (Fig. 2A) of the SG-SMB process are modeled as a discrete representation of N mixed cells in series combined with lumped mass transfer solid-liquid model,

$$\frac{dC_i^p}{dt} + \frac{dq_i^p}{dt} = (C_{i0}^p - C_i^p)\sigma_i^p \tag{1}$$

in which $C_i^p$, $C_{i0}^p$, $q_i^p$ and t represents the concentration of compound i in the liquid phase, the concentration of compound i in the liquid phase at the entrance of the mixed cell, the concentration of compound i in the solid adsorbent phase and the time, respectively, all at column p. In Eq. 1 $\sigma_i^p = F/\epsilon V$ with F, $\epsilon$ and V representing the volumetric flow rate, the total porosity and the volume of the mixed cell, respectively. It was utilized the Abel's model for the determination of the partition coefficient (K)

$$K = \frac{p_1}{(1+p_2\phi)^{p_3}} \tag{2}$$

where $p_1$, $p_2$ and $p_3$ are the model parameters of either tryptophan or phenylalanine. The Mazzotti et al. [3] approach is the classical procedure determining the operating conditions of isocratic SMB process under linear mass transfer equilibrium. The classical Mazzotti's approach [3] was extended analogously to solvent gradient condition (SG-SMB) assuming the Henry's constant a function of the modifier volume fraction.

$$m_{II} > H_{(-)}(\phi_{II}) \quad \text{and} \quad H_{(+)}(\phi_{III}) > m_{III} \tag{3}$$

which can be organized as follow, where $m_{II}$, $m_{III}$, $H_{(-)}$ and $H_{(+)}$ are flow rates ratio at section II and II and the Henry's constants of the less and more adsorbed compounds, respectively.

$$m_{II} - H_{(-)}(\phi_{II}) > 0 \quad \text{and} \quad H_{(+)}(\phi_{III}) - m_{III} > 0 \tag{4}$$

The terms $\phi_{II}$ and $\phi_{III}$ are, respectively, the modifier volume fraction at section II and III. At the feed stream position the above Eq.(4) can be combined to give

$$H_{(+)}(\hat{\phi}) + H_{(-)}(\hat{\phi}) = m_{II} + m_{III} \tag{5}$$

The average modifier volume fraction,

$$\hat{\phi} = \frac{\phi_{II}+\phi_{III}}{2} \tag{6}$$

is determined from volume balance of modifier at the feed node.

The optimization approach follows the next steps: i) a feed flow rate and the volume fraction of modifier at desorbent ($\phi_0$) are specified; ii) as there is the volume fraction of the adsorbent ($\phi_0 = \phi_{II}$) it can be obtained the volume fraction at section III ($\phi_{III}$) and also the average volume fraction of modifier ($\hat{\phi}$); iii) from the equation find the flow rates of desorbent, extract and raffinate with switch time to reach the equality of Eq. (5); iv) the phi-plot, which is the plot of Eqs. (4) versus phi ($\phi$), is evaluated to ensure the inequalities are true; The higher the inequalities of Eqs. (4) the better is the chance to achieve complete separation; v) the SG-SMB simulator is run to provide the modifier volume fraction profile along the columns and also the new phi-plot. The operating conditions can be altered to determine better results for the inequalities of Eqs. (5). In general only the steps i to iv are necessary to achieve the operating conditions for complete separation (purities at the extract and raffinate higher than 99.9%).

## 3. Results and discussions

The novel modelling and optimization procedure was correlated to the experimental data of Nam et al. [1] which utilized the SG-SMB process to separate the amino-acids phenylalanine and tryptophan. In a first moment it was evaluated the optimization strategy routine in the determination of the operating conditions of the SG-SMB process. From Fig. 2A to 2C can be seen the improvement in the extract and raffinate purities as the inequalities of Eq. (5) become higher than zero.

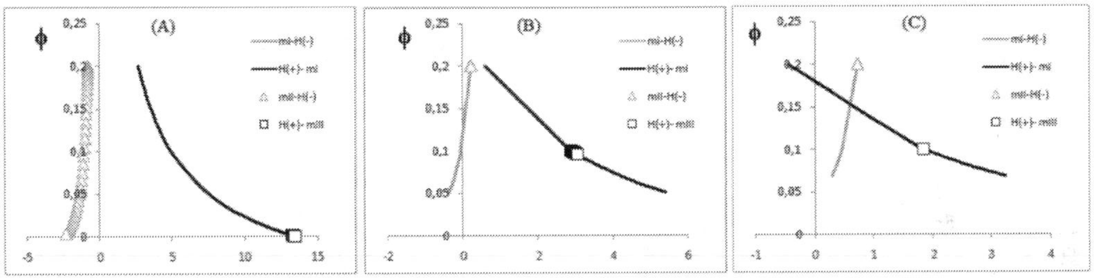


Figure 2 – Phi-plots related to pour (A), intermediary (B) and good separation (C); flow rates of feed, desorbent, extract and raffinate equal to 5,8,3 and 3 mL/min, respectively; porosity equal to 0.55; $\phi_0 = 0.2$

The purity values can be found in Table 1 so increasing the switch time leads to better separation performance of SG-SMB process.

Table 1 – Separation performance of SG-SMB

| Switch time (min) | 10 (Fig. 2A) | 20(Fig. 2B) | 25(Fig. 2C) |
|---|---|---|---|
| Eq.(5) diff. | 4.47 | 1.35 | -0.22 |
| Raffinate purity (%) | 0.00 | 98.95 | 99.92 |
| Extract purity (%) | 31.43 | 97.94 | 99.98 |
| Productivity (g/h.L) | 0.1036 | 0.346 | 0.364 |

The high difference in terms of the left and right hand side of Eq. (5) indicates that the chosen operating conditions of Fig. 2A is not ideal leading to a poor separation of the solutes. In the other hand the low difference value of Eq. (5) leads to a high separation performance (Fig. 2C). The operating conditions of Nam et al. [1] were utilized which confirms the experimental operating conditions utilized to achieve a great separation performance (See Fig. 3). The Fig. 3 shows the good agreement between the present modelling and the experiments of Nam et al. [1] which validates the applied modelling approach. The present optimization routine was easily able to achieve the experimental conditions obtained by Nam et al. [1].

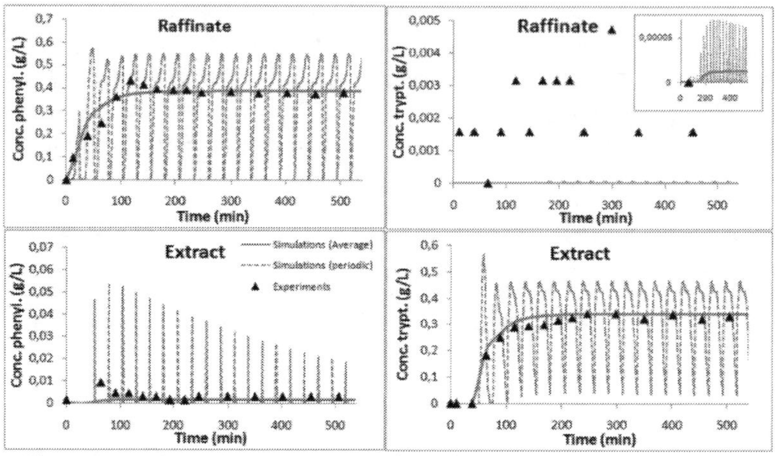

Figure 3 – Amino acids concentration along time at raffinate and extract streams

The next Fig. 4 shows the simulation results for concentrations of phenylalanine (less adsorbed), tryptophan (more adsorbed) and the volume fraction of ethanol (modifier) along the SG-SMB columns. As can be noted the modifier volume fraction reduces drastically at the feed as there is no injection of ethanol at this point. The Fig. 4 profile indicates the great separation performance as there are only the more adsorbed and the less adsorbed substances leaving the extract (E) and the raffinate (R), respectively.

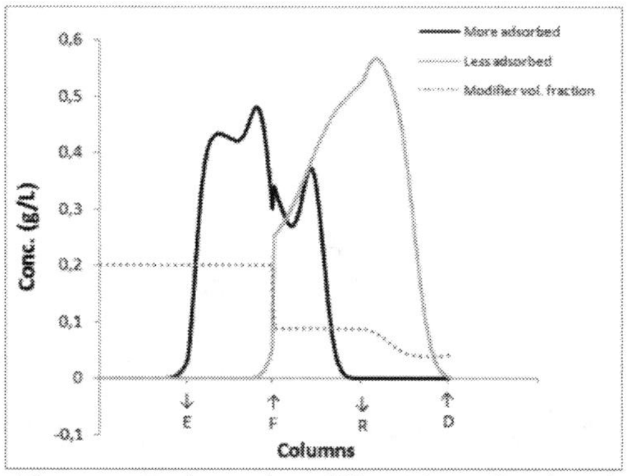

Figure 4 – Amino acids concentration and modifier volume fraction along the columns

## 3. Conclusions

The novel optimization approach showed potential determining the operating conditions of SG-SMB process without the utilization of numerical search algorithms. The new approach utilizes the concept of phi-plots to determine the ideal process conditions to achieve great separation performance.


## Acknowledgements

The author thanks to UERJ, Faperj, CNPq and Capes for the financial support.


## References


[1] Nam, H-G., Jo, S-H., Park, C., Mun, S. *Process Biochemistry*, 47 (2012) 401-409.

[2] Antos, D. Seidel-Morgenstern, A. *Chemical Engineering Process*, 56 (2001) 6667-6682.

[3] Mazzotti, M., Storti, G., Morbidelli, M., *Journal of Chromatography A*, 769 (1997) 3-24.

# Fermion-Antifermion Asymmetry

**Gunn Quznetsov**

**Abstract.** It is calculated that in single-particle states events with a fermion have positive energy and occurrences with an antifermion have negative energy. In double-particle states events with pair of antifermions have negative energy and events with pair of fermions and with fermion-antifermion pair have positive energy. An event with positive energy transfers this energy photons which carries it on recorders observers. Observers know that this event occurs, not before it happens. But event with negative energy should absorb this energy from observers. Consequently, observers know that this event happens before it happens. Since time is irreversible then only the events with positive energy can occur.

Chelyabinsk State University, Chelyabinsk, Russia

E-mail: gunn@mail.ru, quznets@yahoo.com

## 1. Introduction

Here I consider behavior of fermions and anti-fermions (for instance: [1], [2]) within the Quantum Theory Event-Probability Interpretation [3], [4, pp.61–109]. I define operators of the creating and elimintion by usual way (for instance: [5], [7], [9]). Thus it appears that events of the single fermion creating, and event of the of the couple of fermions creating have positive energy. Events of the couple fermion-anti-fermion creatng can have positive energy. But events of the single anti-fermion creating and an event of the couple of anti-fermions creating have negative energy.

But a occurence of events with negative energy contradicts the principle of irreversibility of time (which is proved in detail in [4, pp.5–62]). Therefore, anti-fermions can be created together with fermions, only.

Let $t$, $x_1$, $x_2$,$x_3$ be real numbers, and let $\mathbf{x} := \langle x_1, x_2, x_3 \rangle$.

Let $\mathcal{A}$ be some dot event. Let $\rho(t, \mathbf{x})$ be a probability density of $\mathcal{A}$. and let[1] $\rho(t, \mathbf{x}) = 0$ if $t > \pi \mathrm{c}/\mathrm{h}$ and/or $|\mathbf{x}| > \pi \mathrm{c}/\mathrm{h}$. In that case here exists $4 \times 1$-complex matrix function $\varphi(t, \mathbf{x})$ such that

$$\varphi^{\dagger}(t, \mathbf{x})\varphi(t, \mathbf{x}) = \rho(t, \mathbf{x}) \tag{1}$$

and $\varphi(t, \mathbf{x})$ obeys some generalization of the Dirac equation [6]. The Dirac equation for free fermion does have the following form:

$$\left(\frac{1}{\mathrm{c}}\frac{\partial}{\partial t} - \sum_{s=1}^{3} \beta^{[s]}\frac{\partial}{\partial x_s} - \mathrm{i}\frac{\mathrm{h}}{\mathrm{c}} n\gamma^{[0]}\right)\varphi(t, \mathbf{x}) = 0;$$

here $n$ is a natural number and

$$\beta^{[k]} := \begin{bmatrix} \sigma_k & 0_2 \\ 0_2 & -\sigma_k \end{bmatrix} \gamma^{[0]} := \begin{bmatrix} 0_2 & 1_2 \\ 1_2 & 0_2 \end{bmatrix}, 0_2 := \begin{bmatrix} 0 & 0 \\ 0 & 0 \end{bmatrix}, 1_2 := \begin{bmatrix} 1 & 0 \\ 0 & 1 \end{bmatrix};$$

[1] $\mathrm{c} := 299792458, \mathrm{h} := 6.6260755^{-34}$

here $k \in \{1,2,3\}$, $\sigma_k$ – Pauli's matricies. In this case operator $\widehat{H}_0$ is *the free Dirac Hamiltonian* if

$$\widehat{H}_0 := \mathrm{c}\left(\sum_{s=1}^{3}\beta^{[s]}\mathrm{i}\frac{\partial}{\partial x_s}+\frac{}{\mathrm{c}}n\gamma^{[0]}\right).$$

Let $\mathbf{k}$ be a vector $\langle k_1, k_2, k_3\rangle$ where $k_s$ are natural numbers and let $\omega(\mathbf{k}) := \sqrt{k_1^2+k_2^2+k_3^2+n^2}$ .

Let functions $e_1(\mathbf{k})(2\mathrm{c}/\mathrm{h})^{3/2}\exp(-\mathrm{i}(\mathrm{h}/\mathrm{c})\mathbf{kx})$ and $e_2(\mathbf{k})(2\mathrm{c}/\mathrm{h})^{3/2}\exp(-\mathrm{i}(\mathrm{h}/\mathrm{c})\mathbf{kx})$ be orthogonal eigenvectors of $\widehat{H}_0$ with eigenvalues $(+\mathrm{h}\omega(\mathbf{k}))$, and functions $e_3(\mathbf{k})(2\mathrm{c}/\mathrm{h})^{3/2}\exp(-\mathrm{i}(\mathrm{h}/\mathrm{c})\mathbf{kx})$ and $e_4(\mathbf{k})(2\mathrm{c}/\mathrm{h})^{3/2}\exp(-\mathrm{i}(\mathrm{h}/\mathrm{c})\mathbf{kx})$ be orthogonal eigenvectors of $\widehat{H}_0$ with eigenvalues $(-\mathrm{h}\omega(\mathbf{k}))$.

## 2. Single-Particle States

Let M be some unitary space. Let $\widetilde{0}$ be the zero element of M, $\widehat{0}$ be the zero operator on M. $\widehat{1}$ be the identy operator on M.

Let linear operators $b_{s,\mathbf{k}}$ ($s \in \{1,2,3,4\}$) act on all elements of this space. And let these operators fulfill the following conditions:

$$\left\{b_{s,\mathbf{k}}^{\dagger}, b_{s',\mathbf{k}'}\right\} := b_{s,\mathbf{k}}^{\dagger}b_{s',\mathbf{k}'} + b_{s',\mathbf{k}'}b_{s,\mathbf{k}}^{\dagger} = (\mathrm{h}/(2\pi))^3\,\delta_{\mathbf{k},\mathbf{k}'}\delta_{s,s'}\widehat{1},$$

$$\left\{b_{s,\mathbf{k}}, b_{s',\mathbf{k}'}\right\} = b_{s,\mathbf{k}}b_{s',\mathbf{k}'} + b_{s',\mathbf{k}'}b_{s,\mathbf{k}} = \left\{b_{s,\mathbf{k}}^{\dagger}, b_{s',\mathbf{k}'}^{\dagger}\right\} = \widehat{0}.$$

Hence,

$$b_{s,\mathbf{k}}b_{s,\mathbf{k}} = b_{s,\mathbf{k}}^{\dagger}b_{s,\mathbf{k}}^{\dagger} = \widehat{0}.$$

There exists element $\widetilde{F}_0$ of M such that $\widetilde{F}_0^{\dagger}\widetilde{F}_0 = 1$ and for any $b_{s,\mathbf{k}}$: $b_{s,\mathbf{k}}\widetilde{F}_0 = \widetilde{0}$. Hence, $\widetilde{F}_0^{\dagger}b_{s,\mathbf{k}}^{\dagger} = \widetilde{0}$.

Let

$$\psi_s(\mathbf{x}) := \sum_{\mathbf{k}}\sum_{r=1}^{4} b_{r,\mathbf{k}}e_{r,s}(\mathbf{k})\exp\left(-\mathrm{i}\frac{\mathrm{h}}{\mathrm{c}}\mathbf{kx}\right).$$

These operators obey the following conditions:

$$\begin{aligned}\left\{\psi_s^{\dagger}(\mathbf{x}), \psi_{s'}(\mathbf{x}')\right\} :=&\quad \psi_s^{\dagger}(\mathbf{x})\psi_{s'}(\mathbf{x}') + \psi_{s'}(\mathbf{x}')\psi_s^{\dagger}(\mathbf{x}) = \delta(\mathbf{x}-\mathbf{x}')\,\delta_{s,s'}\widehat{1}.\\ \psi_s(\mathbf{x})\widetilde{F}_0 &= \widetilde{0},\ \left\{\psi_s(\mathbf{x}),\ \psi_{s'}(\mathbf{x}')\right\} = \left\{\psi_s^{\dagger}(\mathbf{x}), \psi_{s'}^{\dagger}(\mathbf{x}')\right\} = \widehat{0},\\ \psi_s(\mathbf{x})\psi_{s'}(\mathbf{x}') &= \psi_s^{\dagger}(\mathbf{x})\psi_{s'}^{\dagger}(\mathbf{x}') = \widehat{0}.\end{aligned}$$

Let

$$\Psi(t,\mathbf{x}) := \sum_{s=1}^{4}\varphi_s(t,\mathbf{x})\psi_s^{\dagger}(\mathbf{x})\widetilde{F}_0.$$

These function obey the following condition:

$$\Psi^{\dagger}(t,\mathbf{x}')\Psi(t,\mathbf{x}) = \varphi^{\dagger}(t,\mathbf{x}')\varphi(t,\mathbf{x})\delta(\mathbf{x}-\mathbf{x}').$$

Hence,

$$\int d\mathbf{x}'\cdot\Psi^{\dagger}(t,\mathbf{x}')\Psi(t,\mathbf{x}) = \rho(t,\mathbf{x}).$$

Let a Fourier series of $\varphi_s(t,\mathbf{x})$ has the following form:

$$\varphi_s(t,\mathbf{x}) = \sum_{\mathbf{p}}\sum_{r=1}^{4} c_r(t,\mathbf{p})e_{r,s}(\mathbf{p})\exp\left(-\mathrm{i}\frac{\mathrm{h}}{\mathrm{c}}\mathbf{px}\right).$$

In that case:

$$\underline{\Psi}(t,\mathbf{p}) := \left(\frac{2\pi\mathrm{c}}{\mathrm{h}}\right)^3\sum_{r=1}^{4} c_r(t,\mathbf{p})b_{r,\mathbf{p}}^{\dagger}\widetilde{F}_0.$$

If

$$\mathcal{H}_0(\mathbf{x}) := \psi^{\dagger}(\mathbf{x})\widehat{H}_0\psi(\mathbf{x}) \tag{2}$$

then $\mathcal{H}_0(\mathbf{x})$ is called a Hamiltonian $\widehat{H}_0$ density.

Because

$$\widehat{H}_0\varphi(t,\mathbf{x}) = \mathrm{i}\frac{\partial}{\partial t}\varphi(t,\mathbf{x})$$

then

$$\int d\mathbf{x}'\cdot\mathcal{H}_0(\mathbf{x}')\,\Psi(t,\mathbf{x}) = \mathrm{i}\frac{\partial}{\partial t}\Psi(t,\mathbf{x}). \tag{3}$$

Therefore, if

$$\widehat{\mathrm{H}} := \int d\mathbf{x}'\cdot\mathcal{H}_0(\mathbf{x}')$$

then $\widehat{\mathrm{H}}$ acts similar to the Hamiltonian on space . And if

$$E_\Psi\left(\widetilde{F}_0\right) := \sum_{\mathbf{p}} \underline{\Psi}^\dagger(t,\mathbf{p})\,\widehat{\mathrm{H}}\underline{\Psi}(t,\mathbf{p})$$

then $E_\Psi\left(\widetilde{F}_0\right)$ is an energy of $\Psi$ on vacuum $\widetilde{F}_0$.

Operator $\widehat{\mathrm{H}}$ obeys the following condition:

$$\widehat{\mathrm{H}} = \left(\frac{2\pi\mathrm{c}}{\mathrm{h}}\right)^3\sum_{\mathbf{k}}\mathrm{h}\omega(\mathbf{k})\left(\sum_{r=1}^{2}b^\dagger_{r,\mathbf{k}}b_{r,\mathbf{k}} - \sum_{r=3}^{4}b^\dagger_{r,\mathbf{k}}b_{r,\mathbf{k}}\right).$$

This operator is not positive defined and in this case

$$E_\Psi\left(\widetilde{F}_0\right) = \left(\frac{2\pi\mathrm{c}}{\mathrm{h}}\right)^3\sum_{\mathbf{p}}\mathrm{h}\omega(\mathbf{p})\left(\sum_{r=1}^{2}|c_r(t,\mathbf{p})|^2 - \sum_{r=3}^{4}|c_r(t,\mathbf{p})|^2\right).$$

This problem is usually solved in the following way (for instance: [7, p.54]): Let:

$$v_1(\mathbf{k}) := \gamma^{[0]}e_3(\mathbf{k}), v_2(\mathbf{k}) := \gamma^{[0]}e_4(\mathbf{k}), d_{1,\mathbf{k}} := -b^\dagger_{3,-\mathbf{k}}, d_{2,\mathbf{k}} := -b^\dagger_{4,-\mathbf{k}}.$$

In that case:

$$\psi_s(\mathbf{x}) := \sum_{\mathbf{k}}\sum_{r=1}^{2}\left(b_{r,\mathbf{k}}e_{r,s}(\mathbf{k})\exp\left(-\mathrm{i}\frac{\mathrm{h}}{\mathrm{c}}\mathbf{kx}\right) + d^\dagger_{r,\mathbf{k}}v_{r,s}(\mathbf{k})\exp\left(\mathrm{i}\frac{\mathrm{h}}{\mathrm{c}}\mathbf{kx}\right)\right)$$

$$\widehat{\mathrm{H}} = \left(\frac{2\pi\mathrm{c}}{\mathrm{h}}\right)^3\sum_{\mathbf{k}}\mathrm{h}\omega(\mathbf{k})\sum_{r=1}^{2}\left(b^\dagger_{r,\mathbf{k}}b_{r,\mathbf{k}} + d^\dagger_{r,\mathbf{k}}d_{r,\mathbf{k}}\right) - 2\sum_{\mathbf{k}}\mathrm{h}\omega(\mathbf{k})\,\widehat{1}.$$

The first term on the right side of this equality is positive defined. This term is taken as the desired Hamiltonian. The second term of this equality is infinity constant. And this infinity is deleted (?!) [7, p.58]

But in this case $d_{r,\mathbf{k}}\widetilde{F}_0 \neq \widetilde{0}$. In order to satisfy such condition, the vacuum element $\widetilde{F}_0$ must be replaced by the following:

$$\widetilde{F}_0 \to \widetilde{\Phi}_0 := \prod_{\mathbf{k}}\prod_{r=3}^{4}\left(\frac{2\pi\mathrm{c}}{\mathrm{h}}\right)^3 b^\dagger_{r,\mathbf{k}}\widetilde{F}_0.$$

But in this case:

$$\psi_s(\mathbf{x})\,\widetilde{\Phi}_0 \neq \widetilde{0}.$$

And condition (3) isn't carried out.

In order to satisfy such condition, operators $\psi_s(\mathbf{x})$ must be replaced by the following:

$$\psi_s(\mathbf{x}) \to \phi_s(\mathbf{x}) := \sum_{\mathbf{k}} \sum_{r=1}^{2} (b_{r,\mathbf{k}} e_{r,s}(\mathbf{k}) \exp\left(-\mathrm{i}\frac{\mathrm{h}}{\mathrm{c}}\mathbf{kx}\right) + d_{r,\mathbf{k}} v_{r,s}(\mathbf{k}) \exp\left(\mathrm{i}\frac{\mathrm{h}}{\mathrm{c}}\mathbf{kx}\right)).$$

Hence,

$$\widehat{\mathrm{H}} = \left(\frac{2\pi\mathrm{c}}{\mathrm{h}}\right)^3 \sum_{\mathbf{k}} \mathrm{h}\omega(\mathbf{k}) \sum_{r=1}^{2} \left(b_{r,\mathbf{k}}^{\dagger} b_{r,\mathbf{k}} - d_{r,\mathbf{k}}^{\dagger} d_{r,\mathbf{k}}\right).$$

And again we get negative energy.

Let's consider the meaning of such energy: An event with positive energy transfers this energy photons which carries it on recorders observers. Observers know that this event occurs, not before it happens. But event with negative energy should absorb this energy from observers. Consequently, observers know that this event happens before it happens. This contradicts Theorem 3.4.2 [8]. Therefore, events with negative energy do not occur.

Hence, over vacuum $\widetilde{\Phi}_0$ single fermions can exist, but there is no single antifermions.

## 3. Two-Particle States

A two-particle state is defined the following field operator (for instance: [9]):

$$\psi_{s_1,s_2}(\mathbf{x},\mathbf{y}) := \begin{vmatrix} \phi_{s_1}(\mathbf{x}) & \phi_{s_2}(\mathbf{x}) \\ \phi_{s_1}(\mathbf{y}) & \phi_{s_2}(\mathbf{y}) \end{vmatrix}.$$

In that case:

$$\widehat{\mathrm{H}} = 2\mathrm{h}\left(\frac{2\pi\mathrm{c}}{\mathrm{h}}\right)^6 \left(\widehat{\mathrm{H}}_a + \widehat{\mathrm{H}}_b\right)$$

where

$$\widehat{\mathrm{H}}_a := \sum_{\mathbf{k}} \sum_{\mathbf{p}} (\omega(\mathbf{k}) - \omega(\mathbf{p})) \sum_{r=1}^{2} \sum_{j=1}^{2} \times$$
$$\times \{v_j^{\dagger}(-\mathbf{k}) v_j(-\mathbf{p}) e_r^{\dagger}(\mathbf{p}) e_r(\mathbf{k}) \times \left(b_{r,\mathbf{p}}^{\dagger} d_{j,-\mathbf{k}}^{\dagger} d_{j,-\mathbf{p}} b_{r,\mathbf{k}}\right) + \left(d_{r,-\mathbf{p}}^{\dagger} b_{j,\mathbf{k}}^{\dagger} b_{j,\mathbf{k}} d_{r,-\mathbf{p}}\right) +$$
$$+ v_j^{\dagger}(-\mathbf{p}) v_j(-\mathbf{k}) e_r^{\dagger}(\mathbf{k}) e_r(\mathbf{p}) \left(-b_{r,\mathbf{k}}^{\dagger} d_{j,-\mathbf{p}}^{\dagger} d_{j,-\mathbf{k}} b_{r,\mathbf{p}}\right) + \left(-b_{r,\mathbf{p}}^{\dagger} d_{j,-\mathbf{k}}^{\dagger} d_{j,-\mathbf{k}} b_{r,\mathbf{p}}\right)\}$$

and

$$\widehat{\mathrm{H}}_b := \sum_{\mathbf{k}} \sum_{\mathbf{p}} (\omega(\mathbf{k}) + \omega(\mathbf{p})) \sum_{r=1}^{2} \sum_{j=1}^{2} \times$$
$$\times \{v_j^{\dagger}(-\mathbf{p}) v_j(-\mathbf{k}) v_r^{\dagger}(-\mathbf{k}) v_r(-\mathbf{p}) \left(-d_{r,-\mathbf{k}}^{\dagger} d_{j,-\mathbf{p}}^{\dagger} d_{j,-\mathbf{k}} d_{r,-\mathbf{p}}\right) + \left(-d_{r,-\mathbf{p}}^{\dagger} d_{j,-\mathbf{k}}^{\dagger} d_{j,-\mathbf{k}} d_{r,-\mathbf{p}}\right) +$$
$$+ e_r^{\dagger}(\mathbf{k}) e_r(\mathbf{p}) e_j^{\dagger}(\mathbf{p}) e_j(\mathbf{k}) \left(b_{r,\mathbf{k}}^{\dagger} b_{j,\mathbf{p}}^{\dagger} b_{j,\mathbf{k}} b_{r,\mathbf{p}}\right) + \left(b_{r,\mathbf{p}}^{\dagger} b_{j,\mathbf{k}}^{\dagger} b_{j,\mathbf{k}} b_{r,\mathbf{p}}\right)\}.$$

(here $v_j^{\dagger}(-\mathbf{k}) v_j(-\mathbf{p}) e_r^{\dagger}(\mathbf{p}) e_r(\mathbf{k})$, $v_j^{\dagger}(-\mathbf{p}) v_j(-\mathbf{k}) e_r^{\dagger}(\mathbf{k}) e_r(\mathbf{p})$, $v_j^{\dagger}(-\mathbf{p}) v_j(-\mathbf{k}) v_r^{\dagger}(-\mathbf{k}) v_r(-\mathbf{p})$ and $e_r^{\dagger}(\mathbf{k}) e_r(\mathbf{p}) e_j^{\dagger}(\mathbf{p}) e_j(\mathbf{k})$ are real and positive)

If velosities are small then the following formula is fair.

$$\widehat{\mathrm{H}} = 4\mathrm{h}\left(\frac{2\pi\mathrm{c}}{\mathrm{h}}\right)^6 \left(\widehat{\mathrm{H}}_a + \widehat{\mathrm{H}}_b\right)$$

where

$$\widehat{\mathrm{H}}_a := \sum_{\mathbf{k}} \sum_{\mathbf{p}} (\omega(\mathbf{k}) - \omega(\mathbf{p})) \times \sum_{r=1}^{2} \sum_{j=1}^{2} \left( d_{j,-\mathbf{p}}^{\dagger} b_{r,\mathbf{k}}^{\dagger} b_{r,\mathbf{k}} d_{j,-\mathbf{p}} - b_{j,\mathbf{p}}^{\dagger} d_{r,-\mathbf{k}}^{\dagger} d_{r,-\mathbf{k}} b_{j,\mathbf{p}} \right)$$

and

$$\widehat{\mathrm{H}}_b := \sum_{\mathbf{k}} \sum_{\mathbf{p}} (\omega(\mathbf{k}) + \omega(\mathbf{p})) \sum_{j=1}^{2} \sum_{r=1}^{2} \left( b_{j,\mathbf{p}}^{\dagger} b_{r,\mathbf{k}}^{\dagger} b_{r,\mathbf{k}} b_{j,\mathbf{p}} - d_{j,-\mathbf{p}}^{\dagger} d_{r,-\mathbf{k}}^{\dagger} d_{r,-\mathbf{k}} d_{j,-\mathbf{p}} \right).$$

Therefore, in any case events with pairs of fermions and events with fermion-antifermion pairs can occur, but events with pairs of antiftrmions can not happen.

## 4. Conclusion

Therefore, an antifermion can exists only with a fermion.

## 5. References


[1] Sodickson L, Bowman W, Stephenson J 1961 *Physical Review* **124** (6): 18511861.
[2] Dirac P 1930 *Proceedings of the Royal Society* A **126** (801): 360365.
[3] Quznetsov G 2011 *AIP Conf. Proc.* 1327, 460
[4] Quznetsov G 2013 *Logical foundation of fundamental theoretical physics* (Lambert Academic Publ.)
[5] Weinberg, S 1995 *The Quantum Theory of Fields*, Volume 1: *Foundations.* (Cambridge University Press)
[6] Quznetsov G 2009 *Progress in Physics* **2** 96
[7] Peskin E and Schroeder V 1995 *An Introduction to Quantum Field Theory* (Perseus Books Publishing, L.L.C.)
[8] Quznetsov G 2006 *Logical Foundation of Theoretical Physics* (Nova Sci. Publ.) 34
[9] Ziman J 1969 *Elements of Advanced Quantum Theory* (Cambridge) 32.

# Semi-analytical model of hydrogen ionization by strong laser pulse at low field frequencies

**Alexander G. Galstyan[1], Yuri V. Popov[2], Ochbadrakh Chuluunbaatar[3], Bernard Piraux[4]**

[1]Faculty of Physics, Lomonosov Moscow State University, Russia
[2]Skobeltsyn Institute of Nuclear Physics, Moscow State University, Moscow, Russia
[3]Joint Institute for Nuclear Research, Dubna, Russia
[4]Institute of Condensed Matter and Nanosciences, Universite catholique de Louvain,Louvain-la-Neuve, Belgium

E-mail: galstyan@physics.msu.ru

**Abstract.** We consider the interaction of hydrogen atom with a very intense low frequency laser pulse. The Henneberger-Kramers representation of the time-dependent Schrödinger equation is the most appropriate one for this purpose. It is shown that in the case of very low frequencies, the quantum dispersion of the electron wave packet plays a dominant role in the dynamics of the atom.

## Introduction

Numerous theoretical and experimental research works on the interaction of quantum systems with an electromagnetic field have been performed lately. This is mainly due to recent advances in laser technologies allowing one to obtain very high intensities. However, in modern theories, the case of intense low frequency fields is still not very well described.

A quantum system exposed to an electromagnetic field can change its state by absorbing or emitting photons. Their energy corresponds in atomic units ($\hbar = e = m_e = 1$) to the frequency $\omega_0$ of the external field. In the case of low frequencies the period of laser pulse with $N$ cycles $T = (2\pi N)/\omega_0$ becomes really huge.

Usually one solves numerically the time-dependent Schrödinger equation (TDSE) with initial conditions by means of a finite difference grid method or a spectral method. This leads to a system of ordinary first order differential equations in time. However, such numerical approach is no longer tractable if the frequency of the laser pulse is very small and its intensity very high. Indeed, the amplitude of the quiver motion becomes so big that the size of the spatial grid or of the basis becomes prohibitively large making the numerical calculations impossible to perform even on supercomputers. In this case, we need mathematical and physically reliable models to understand deeply the interaction mechanisms. In this contribution, we consider such a model.

**Theory**

The TDSE that describes the interaction of a pulsed electric field of period $T$ with an hydrogenlike atom writes as follows:

$$\left[i\frac{\partial}{\partial t} - \frac{1}{2}\left(-i\vec{\nabla} + \frac{1}{c}\,\vec{e}A(t)\right)^2 + \frac{Z}{r}\right]\tilde{\Psi}(\vec{r},t) = 0, \quad \tilde{\Psi}(\vec{r},0) = \tilde{\varphi}_0(r) = \sqrt{\frac{Z^3}{\pi}}e^{-Zr}. \tag{1}$$

$Z$ designates the nuclear charge. The vector-potential associated to the electric field is given by $1/c\ A(t) = -\omega_0^{-1}\sqrt{I/I_0}\sin^2(\pi\frac{t}{T})\sin(\omega_0 t)$, $0 \le t \le T$, and $\vec{e}$ is the field polarization vector. The electron wave packet satisfies the normalization condition:

$$\int d\vec{r}|\tilde{\Psi}(\vec{r},t)|^2 = 1,$$

that guaranties the conservation of the total probability. In our numerical calculations, the peak intensity $I = 1\cdot 10^{14}\ W/cm^2$ ($I_0 = 3.5\cdot 10^{16}\ W/cm^2$ is the atomic unit of intensity), the photon frequency $\omega_0 = 0.00228$ in atomic units, and $N = 5$. It is easy to see that for any fixed time $t$, the vector-potential vanishes like $\omega_0^2$, if $\omega_0 \to 0$.

The electromagnetic field is a gauge-invariant. Any gauge transformation of electromagnetic potentials can be compensated by a corresponding phase transformation of the wave function $\tilde{\Psi}(\vec{r},t)$. In such a way, we obtain different gauge-equivalent forms of Eq. (1). For example, if we use the Henneberger-Kramers (HK) representation [1]

$$\tilde{\Psi}(\vec{r},t) = \exp\left[b(t)(\vec{e}\vec{\nabla}_r) - \frac{i}{2}\int_0^t dt'(1/cA(t'))^2\right]\tilde{\Phi}_{HK}(\vec{r},t), \quad b(t) = -\frac{1}{c}\int_0^t A(t')dt',$$

we obtain from (1)

$$\left[i\frac{\partial}{\partial t} + \frac{1}{2}\triangle_r + \frac{Z}{|\vec{r} - \vec{e}b(t)|}\right]\tilde{\Phi}_{HK}(\vec{r},t) = 0; \quad \tilde{\Phi}_{HK}(\vec{r},0) = \tilde{\varphi}_0(r). \tag{2}$$

Note, that at any time $t$, the wave packet is normalized to unity

$$\int d\vec{r}\ |\tilde{\Phi}_{HK}(\vec{r},t)|^2 = 1.$$

Here, we analyze the amplitude $\mathcal{A}_0$ which determlines the probability for the atom to stay in its ground state. For this purpose, we define

$$< \varphi_0|\tilde{\Phi}_{HK}(t) >= e^{i(Z^2/2)t}\mathcal{A}_0(t)$$

and obtain from (2)

$$\frac{\partial}{\partial t}\mathcal{A}_0(t) = -iZe^{-i(Z^2/2)t} < \varphi_0\left|\frac{1}{r} - \frac{1}{|\vec{r} - \vec{e}b(t)|}\right|\tilde{\Phi}_{HK}(t) >, \quad \mathcal{A}_0(0) = 1. \tag{3}$$

In this particular case the integration volume is always limited by $r \sim Z^{-1}$.

Near $t \le t_0 \sim 0$ ($\sqrt{I/I_0}\ \omega_0/(2N)^2\ t_0^3 \ll 1$), the amplitude is still vanishingly small. But the value $b(t)$ being proportional to $\omega_0^{-2}$ grows quickly in a course of time, and the ionization begins. A further quick growth of $b(t)$ leads to the disappearance of the term $1/|\vec{r} - \vec{e}b(t)|$ in (3). The electron leaves the vicinity of the nucleus, follows the field oscillation and simultaneously undergoes the free evolution (dispersion). At the root-point $t_1$, the $b(t_1)$ is zero again (see Fig. 1 left), and the amplitude stays constant during very short time around $t \sim t_1$. Afterward, the term $1/|\vec{r} - \vec{e}b(t)|$ disappears again.

For $t \geq t_0$ the free solution of Eq. (2) can be written in the form

$$\tilde{\Phi}_{HK}(\vec{r}, t) = \int d^3r' G_0(\vec{r} - \vec{r}\,', t - t_0)\tilde{\Phi}_{HK}(\vec{r}\,', t_0) \tag{4}$$

with

$$G_0(\vec{r}, t) = \int \frac{d^3k}{(2\pi)^3} \exp\left[i\vec{k}\cdot\vec{r} - i\frac{k^2}{2}t\right]\theta(t).$$

Inserting (4) in (3) we obtain

$$\mathcal{A}_0(t) \approx \mathcal{A}_0(t_0) - iZe^{-i(Z^2/2)t_0}\int_0^{t-t_0} d\xi e^{-i(Z^2/2)\xi}\int d^3r' \int \frac{d^3r}{r}\varphi_0(r)G_0(\vec{r} - \vec{r}\,', \xi)\tilde{\Phi}_{HK}(\vec{r}\,', t_0) =$$

$$\mathcal{A}_0(t_0) - \left(\frac{Z}{\pi}\right)^{5/2} e^{-i(Z^2/2)t_0}\int d^3k \frac{1 - \exp[-i(k^2/2 + Z^2/2)(t - t_0)]}{(Z^2 + k^2)^2}\Phi_{HK}(\vec{k}, t_0). \tag{5}$$

where $\Phi_{HK}(\vec{k}, t_0)$ is a Fourier transform of $\tilde{\Phi}_{HK}(\vec{r}\,', t_0)$. We write approximately

$$\Phi_{HK}(\vec{k}, t_0) \approx \frac{8\sqrt{\pi Z^5}}{(k^2 + Z^2)^2}e^{i(Z^2/2)t_0}$$

In this case for $t > t_0$ we have

$$\mathcal{A}_0(t) \approx \frac{8Z^5}{\pi^2}e^{-i(Z^2/2)(t-t_0)}\int d^3k \frac{\exp[-i(k^2/2)(t - t_0)]}{(Z^2 + k^2)^4}, \quad \mathcal{A}_0(t_0) \approx 1 \tag{6}$$

Eq. (6) is valid for the time period $t_0 \leq t < t_1$. In a very narrow time-period near $t \sim t_1$ we can fix $\mathcal{A}_0(t) \approx \mathcal{A}_0(t_1)$ from (6). For $t > t_1$ (second half-period of the first optical cycle) we can repeat the scheme (4) - (6), but instead of (5) we obtain

$$\mathcal{A}_0(t > t_1) = \mathcal{A}_0(t_1) - \left(\frac{Z}{\pi}\right)^{5/2} e^{-i(Z^2/2)t_1}\int d^3k \frac{1 - \exp[-i(k^2/2 + Z^2/2)(t - t_1)]}{(Z^2 + k^2)^2}\Phi_{HK}(\vec{k}, t_1), \tag{7}$$

etc. The amplitude $\mathcal{A}_0(t_1)$ is very small and continues to decrease at $t > t_1$. At the end of pulse after 5 optical cycles we obtain $|\mathcal{A}_0(T)|^2 = 5 \times 10^{-11}$.

## Conclusion

What is the physical picture implied? Two distinct physical processes determine the motion of the bound electron in the oscillating electric field. First, the quiver motion of this electron. In the absence of Coulomb potential, the electron oscillates in the field with an amplitude given by the ratio of the peak electric field and the square of the frequency in atomic units. In the present case, this amplitude is very large, of the order of 30 000 atomic units. The second process is the dispersion of the electron wave packet. For low frequencies, the dispersion time is extremely long. After each half cycle, the electrons that go back to the nucleus interact only weakly and during a rather short time with the nucleus so that the probability to stay in the ground state is hardly affected. So, at very small frequencies, the only role of the field is to push out the electron from its stationary state at the very beginning of the process. Afterward, the electron wave packet freely disperses. By contrast, for higher frequencies and shorter pulse durations, the role of the field becomes more and more dominant, while dispersion is much less important.

## References

[1] H. A. Kramers, Les particules elementaires, Report to the Eighth Solvay Conference (Editions Stoops, Brussels, 1950); W. C. Henneberger, Phys. Rev. Lett. **21**, 838 (1968).

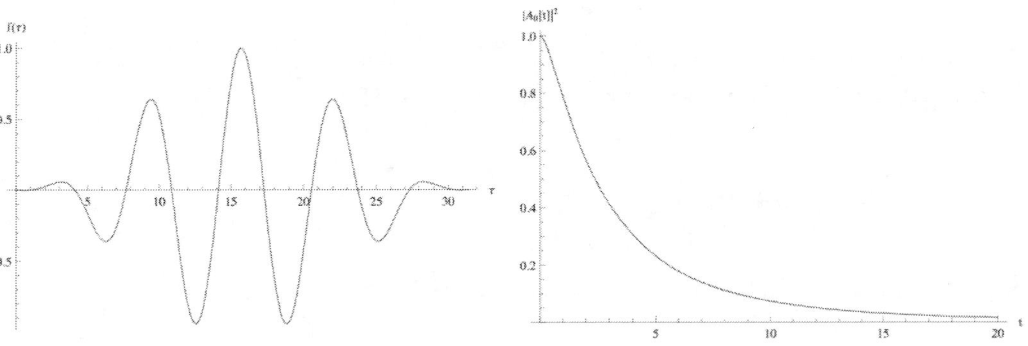


**Figure 1.** On the rhs: x-axis, $\tau = \omega_0 t$, y-axis, the reduced $b(t) = \omega_0^{-2}\sqrt{I/I_0}\, f(\tau)$. $|\mathcal{A}_0(t)|^2$ from (6) on the lhs. In this time scale $t_0 \approx 0$.

# Molecular Interaction study of Multi-Functional Group of Alcohols through Dielectrics Using TDR

**B D Achole[1#] and V P Pawar[2]**
[1]Department of Physics, Shri Havagiswami Mahavidyalaya, Udgir, Maharashtra, India-413517.
[2]Department of Physics and Electronics, Maharashtra Udayagiri Mahavidyalaya, Udgir, Maharashtra , India-413517

E-mail: b_achole1234@rediffmail.com

**Abstract.** The complex permittivity spectra for tert-butyl alcohol(TB) with n-propanol(nP) were determined over the frequency range of 10 MHz to 20 GHz using the time domain reflectometry (TDR) in the temperature range of 25°C to 55°C for 11 different concentrations of the system. The static dielectric constant ($\varepsilon_0$) and relaxation time ($\tau$) have been determined these spectra using the Debye model. Excess properties and Kirkwood correlation factor of the mixtures have been determined. The excess permittivity is found to be positive in the tart-butyl alcohol rich region and negative in the n-Propanol rich region. However, the excess inverse relaxation time has different trend. The static dielectric constants for the mixtures have also been fitted with the modified Bruggeman model by assuming an additional parameter in the model.

## 1. Introduction

The measurement and characterization of the response of a material to an applied electric field is called dielectric analysis. Dielectric analysis measures changes in the properties of a material as it is subjected to a periodic electric field. The dielectric relaxation of binary polar liquids provides information regarding their interactions. In this work, we report the dielectric study of tert-butyl alcohol (TB) with n-propanol (nP) mixture; both are hydroxyl group molecules.

## 2. Experimental

TB and nP (AR grade, Merck Pvt. Ltd., West Mumbai, India) were used without further purification. The solutions were prepared at 11 different volume percentage of nP in TB from 0 % to 100 % in steps of 10 %. Using these volume percents the mole fraction is calculated as

$x_1 = (v_1\rho_1/m_1) / [(v_1\rho_1/m_1) + (v_2\rho_2/m_2)]$

where $m_i$, $v_i$, and $\rho_i$ represent the molecular weight, volume percent, and density of the $i^{th}$ (i=1, 2) liquids, respectively. The density and molecular weight of the liquids are as follows:

TB: Density=0.8093 $gmcm^{-3}$ · Mol. Wt. = 74.12

nP : Density= 0.7997 $gmcm^{-3}$, Mol. Wt. = 60.11

The complex permittivity spectra were studied using the time domain reflectometry [1]-[6] method. The Hewlett Packard HP 54750 sampling oscilloscope with HP 54754A TDR plug in module has been

[1] Department of Physics, Shri. Havagiswami Mahavidyalaya, Udgir-413 517, Maharashtra, India.
(Mobile: +91-9422652892, E-mail: b_achole1234@rediffmail.com)

used. The temperature controller system with water bath and a thermostat has been used to maintain the constant temperature within the accuracy limit of ±1°C.

## 3. Data Analysis

The time dependent data were processed to obtain complex reflection coefficient spectra $\rho^*(\omega)$ over the frequency range from 10 MHz to 20 GHz using Fourier transformation [7], [8] as

$$\rho^*(\omega) = (c/j\omega d)\,[p(\omega)/q(\omega)] \quad (1)$$

where p (ω) and q (ω) are Fourier transforms of $[R_1(t) - R_x(t)]$ and $[R_1(t) + R_x(t)]$ respectively, c is the velocity of light, ω is angular frequency, d is the effective pin length and $j = \sqrt{-1}$.

The experimental values of $\varepsilon^*(\omega)$ are fitted with the Debye equation [9]

$$\varepsilon^*(\omega) = \varepsilon_\infty + \frac{\varepsilon_0 - \varepsilon_\infty}{1 + j\omega\tau} \quad (2)$$

with $\varepsilon_0$, $\varepsilon_\infty$ and τ as fitting parameters. A nonlinear least-squares fit method [10] was used to determine the values of dielectric parameters.

## 4. Results and Discussion

The obtained values of $\varepsilon_0$, $\varepsilon_\infty$ and τ are listed in Table 1. The excess permittivity $\varepsilon^E$ is defined as

$$\varepsilon^E = (\varepsilon_0 - \varepsilon_\infty)_m - [(\varepsilon_0 - \varepsilon_\infty)_1\, x_1 + (\varepsilon_0 - \varepsilon_\infty)_2\, x_2\,] \quad (3)$$

where x- mole fraction and suffices m, 1, 2 represents mixture, liquid 1(TB) and liquid 2(nP) respectively.

The excess permittivity may provide qualitative information about multimers formation in the mixture as follows: $\varepsilon^E = 0$ , the liquid 1 and 2 do not interact ; $\varepsilon^E < 0$ , the liquid 1 and 2 interaction in such a way that the total effective dipoles get reduced and $\varepsilon^E > 0$, indicates the liquid 1 and 2 interaction such that the total effective dipole moment increases.

Similarly, the excess inverse relaxation time is defined as

$$(1/\tau)^E = (1/\tau)_m - [(1/\tau)_1\, x_1 + (1/\tau)_2\, x_2] \quad (4)$$

where $(1/\tau)^E$ is excess inverse relaxation time which represents the average broadening of dielectric spectra. The inverse relaxation time analogy is taken from spectral line broadening in the resonant spectroscopy [12]. The information regarding the dynamic of liquid 1 and 2 interactions from this excess property is as: $(1/\tau)^E = 0$; there is no change in the dynamics of liquid 1 and 2 interaction, $(1/\tau)^E < 0$ ; the liquid 1 and 2 interaction produces a field such that the effective dipoles rotate slowly and $(1/\tau)^E > 0$ ; the liquid 1 and 2 interaction produces a field such that the effective dipoles rotate quickly .

The experimental values of both the excess parameters were fitted to the Redlich-Kister equation [13], [14] as

$$A^E = (x_1 x_2) \sum_n B_n (x_1 - x_2)^n \quad (5)$$

where A is either $\varepsilon^E$ or $(1/\tau)^E$. By using these $B_n$ values, $A^E$ values were calculated. The 'g' values for pure liquid may be obtained by the expression

$$\frac{4\Pi N\mu^2\rho}{9kTM} g = \frac{(\varepsilon_0 - \varepsilon_\infty)(2\varepsilon_0 + \varepsilon_\infty)}{\varepsilon_0(\varepsilon_\infty + 2)^2} \quad (6)$$

where μ is dipole moment in gas phase, ρ is density, M is molecular weight, k is Boltzmann's constant and N is Avogadro's number. The dipole moments for TB and nP in gas phase are taken as 1.69 D and 1.68 D respectively [15]. Temperature dependent $g^{eff}$ and $g_f$ for the system is also shown in Fig. 1, (a) and (b).

**Table 1.** Temperature Dependent Dielectric Relaxation Parameters for TB + NP mixture*

| $x_2$ | T= 25 °C | T= 35 °C | T= 45 °C | T= 55°C |
|---|---|---|---|---|
| | | $\varepsilon_0$ | | |
| 0.0000 | 11.89(0) | 11.32(0) | 10.87(0) | 09.34(0) |
| 0.1197 | 15.06(4) | 13.87(9) | 12.62(3) | 10.99(1) |
| 0.2342 | 16.54(2) | 14.50(7) | 13.27(5) | 11.40(2) |
| 0.3439 | 17.69(3) | 15.39(2) | 13.97(2) | 11.97(4) |
| 0.4492 | 18.24(7) | 16.03(2) | 14.70(8) | 12.55(2) |
| 0.5502 | 18.89(3) | 17.36(3) | 15.52(4) | 13.17(4) |
| 0.6473 | 19.54(9) | 18.03(4) | 16.26(5) | 13.84(1) |
| 0.7406 | 20.02(2) | 18.60(9) | 16.92(9) | 14.97(3) |
| 0.8303 | 20.73(7) | 19.29(3) | 17.60(8) | 15.53(4) |
| 0.9167 | 21.28(3) | 19.89(2) | 18.47(4) | 16.15(1) |
| 1.0000 | 21.94(0) | 20.18(0) | 19.02(0) | 16.83(0) |
| | | $\varepsilon_\infty$ | | |
| 0.0000 | 3.26(1) | 3.09(0) | 2.80(0) | 2.61(1) |
| 0.1197 | 2.89(4) | 2.74(1) | 2.70(3) | 2.50(0) |
| 0.2342 | 2.65(3) | 2.59(3) | 2.55(4) | 2.39(3) |
| 0.3439 | 2.38(8) | 2.34(8) | 2.29(8) | 2.14(4) |
| 0.4492 | 2.08(2) | 2.01(4) | 2.01(3) | 1.98(7) |
| 0.5502 | 1.86(7) | 1.81(2) | 1.86(9) | 1.79(2) |
| 0.6473 | 1.64(9) | 1.59(1) | 1.60(7) | 1.52(8) |
| 0.7406 | 1.54(4) | 1.38(9) | 1.32(2) | 1.28(3) |
| 0.8303 | 1.33(3) | 1.18(3) | 1.08(4) | 1.04(1) |
| 0.9167 | 1.27(8) | 0.97(4) | 0.89(3) | 0.81(4) |
| 1.0000 | 0.81(0) | 0.78(0) | 0.69(0) | 0.63(0) |
| | | $\tau$ (ps) | | |
| 0.0000 | 0.0000 | 0.0000 | 0.0000 | 0.0000 |
| 0.1197 | 0.1197 | 0.1197 | 0.1197 | 0.1197 |
| 0.2342 | 0.2342 | 0.2342 | 0.2342 | 0.2342 |
| 0.3439 | 0.3439 | 0.3439 | 0.3439 | 0.3439 |
| 0.4492 | 0.4492 | 0.4492 | 0.4492 | 0.4492 |
| 0.5502 | 0.5502 | 0.5502 | 0.5502 | 0.5502 |
| 0.6473 | 0.6473 | 0.6473 | 0.6473 | 0.6473 |
| 0.7406 | 0.7406 | 0.7406 | 0.7406 | 0.7406 |
| 0.8303 | 0.8303 | 0.8303 | 0.8303 | 0.8303 |
| 0.9167 | 0.9167 | 0.9167 | 0.9167 | 0.9167 |
| 1.0000 | 1.0000 | 1.0000 | 1.0000 | 1.0000 |

*$x_2$ is the mole fraction of nP in TB. Number in bracket represent error in the corresponding value.

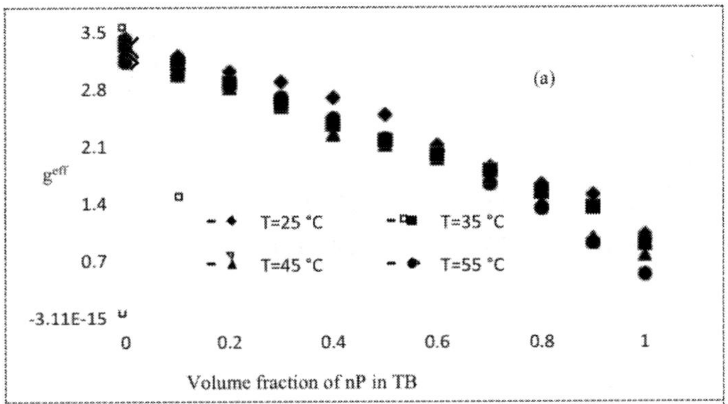


**Figure 1(a).** Variation of $g^{eff}$ versus Volume fraction of nP in TB at various temperatures.

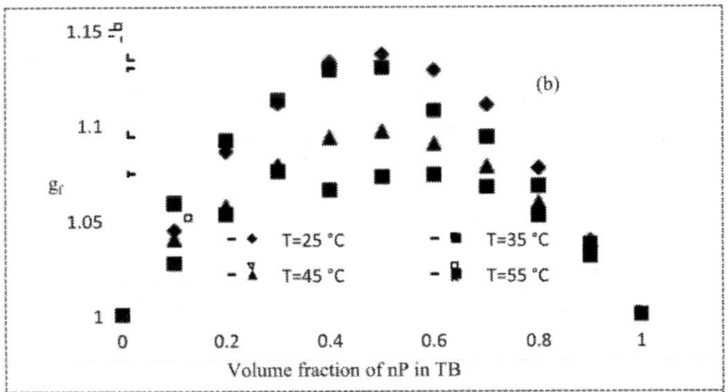


**Figure 1. (b)** Variation of $g_f$ versus Volume fraction of nP in TB at various temperatures.

## 5. Conclusion

The study has provided the permittivity parameters for the mixture of binary polar liquids. The molecules of these compounds have several hydrogen bonding sites and enter into intra or inter molecular hydrogen bonding giving rise to different conformations. This provides a unified approach to understand the interaction between two constituents in the liquid mixtures, using different models.

## 6. References


[1] Patil S. P., and Mehrotra S. C., *J. Chem. Eng. Data*, 44, 875-878 **(1999).**

[2] R. H. Cole, J. G. Berbarian, S. Mashimo, G.Chryssikos, A.Burns and E. Tombari, J. Appl. Phys., 66 (1989) 793.

[3] V. P. Pawar and S. C. Mehrotra, *J. Sol. Chem*,. 31 (2002) 559-576.

[4] S. M. Puranik, A. C. Kumbharkhane and S. C. Mehrotra, *J. Chem. Soc. Faraday Trans.*, 87 (1991) 1569.

[5] V. P. Pawar and S. C. Mehrotra, *J. Mol. Liq.*, 95 (2002) 63-74.

[6] S. M. Puranik, A. C. Kumbharkhane and S. C. Mehrotra, *J. Microwave Power and EM. Energy*, 26 (1991) 196.

[7] C. E. Shannon, Proc. IRE, 37 (1949) 10.

[8] H. A. Samulan, Proc. IRE, 39 (1951) 175.

[9] P. Debye, Polar molecules, Chemical Catalog, New York (1929).

[10] P. R. Bevington, Data reduction and error analysis for the physical sciences, Mc-Graw Hill, New York, (1969).

[11] S. C. Mehrotra and J. E. Boggs, *J. Chem. Phys.*, 66 (1977) 5306.

[12] M.I. Aralaguppi, T.M. Aminabhavi, R.H. Balungi, S.S. Joshi, *J. Phys. Chem*, 95 (1991) 5299.

[13] A. C. Kumbharkhane, S. M. Puranik and S. C. Mehrotra, *J. Sol. Chem.*, 22 (1993) 219.

[14] H. Frolhich, Theory of dielectrics, Oxford University Press, London, (1949).

[15] R. C. Weast, Handbook of Chemistry and Physics, $64^{th}$ ed., CRC Press: Boca Ratan, Florida, (1983-84).

# Multilevel models of polycrystals using crystal plasticity: investigation of hardening laws influence on the macro effects of cyclic loading

**P S Volegov, P V Trusov, A I Shveykin**

Perm National Research Polytechnic University, Mathematical Modeling of Systems and Processes, 29, Komsomolsky ave., Perm, Russian Federation

E-mail: crocinc@mail.ru

**Abstract**. The problem of constructing a physically based hardening laws of mono- and polycrystalline samples in multi-level theories using crystal plasticity is considered, these hardening laws should allow describing the process of the defect structure evolution of the material due to the intensive inelastic deformations. It is also should be applicable to the description of simple and cyclic loading. An approach to the construction of a general and a particular form of hardening law is proposed, which takes into account the interaction of full and split dislocations with each other, forming and destruction of dislocation barriers, annihilation of dislocations during reverse loading. Using the obtained hardening law, the known experimental effects of simple and cyclic loading are described.

## 1. Introduction

Changes in the physical and mechanical properties of the specimen during deformation in complex cyclic path is a consequence of a substantial restructuring of the micro- and mesostructure of the material, mainly - a consequence of a significant evolution of the dislocation structure of the material [1]. Describe such processes without studying and establishing appropriate mathematical models that explicitly take into account the physical root causes of the material's microstructure evolution at large deformations. Directly into the structure of crystal plasticity relations description of the microstructure evolution is introduced through specific relationships that determine the change of the critical shear stress on the slip systems on a set of parameters defined on the basis of physical analysis (shears, temperature, stacking fault energy etc.), which are commonly called hardening law [2-3]. The above explains the considerable attention in crystal plasticity theories, which is paid to the modification of hardening law, in particular - in connection with the new experimental data.

The aim is to study the effects produced by polycrystalline representative macro volume of material under complex and cyclic loading (and the transition from one to another type of loading) as a consequence of changes occurring at the level of the dislocation structure in the process of loading, and attempt to modify the laws hardening so way that they can physically transparently describe these changes and effects.

## 2. Two-level constitutive model for inelastic deformations of polycrystals

As a top (macro-) level, we consider representative volume of the material, and the lower level means the level of the individual crystallites. Next, to simplify the upper level (macro-representative)

will be called the macro level, and the lower (separate single crystals with ideal crystal lattice) will be called meso level.

The constitutive model of the macro-level is the following set of equations (hereinafter macro-parameters are indicated in capital letters, the similar meso parameters – in lower case):

$$\begin{cases} \boldsymbol{\Sigma}^{R} \equiv \dot{\boldsymbol{\Sigma}} + \boldsymbol{\Omega}^{T}\cdot\boldsymbol{\Sigma} + \boldsymbol{\Sigma}\cdot\boldsymbol{\Omega} = \boldsymbol{\Pi} : \mathbf{D}^{e} = \boldsymbol{\Pi} : (\mathbf{D} - \mathbf{D}^{in}), \\ \boldsymbol{\Omega} = \boldsymbol{\Omega}\left(\boldsymbol{\omega}_{(i)}, \boldsymbol{\pi}_{(i)}, \boldsymbol{\sigma}_{(i)}\right), i = 1,\ldots,N, \\ \boldsymbol{\Pi} = \boldsymbol{\Pi}(\boldsymbol{\pi}_{(i)}, \mathbf{o}_{(i)}), i = 1,\ldots,N, \\ \mathbf{D}^{in} = \mathbf{D}^{in}(\mathbf{d}^{in}_{(i)}, \boldsymbol{\pi}_{(i)}, \boldsymbol{\omega}_{(i)}), i = 1,\ldots,N, \end{cases} \tag{1}$$

here $\boldsymbol{\Sigma}$ – Cauchy stress tensor, $\boldsymbol{\Pi}$ – elastic moduli tensor, $\mathbf{D}$, $\mathbf{D}^{e}$, $\mathbf{D}^{in}$ – strain rate tensor, its elastic and inelastic parts, index «R» means independent of reference system choice derivative [1], $\boldsymbol{\Omega}$ – spin tensor; $\boldsymbol{\pi}_{(i)}, \boldsymbol{\sigma}_{(i)}, \mathbf{d}^{in}_{(i)}, \boldsymbol{\omega}_{(i)}, \mathbf{o}_{(i)}$ - elastic constant tensor, stress tensor, elastic and inelastic parts of strain rate tensor, spin and the orientation of $i$-crystallite, $N$ – number of crystallites forming a representative macro-level.

At the meso level (the level of the crystallite) in the two-level model using the following system of relations (crystallite number is omitted):

$$\begin{cases} \boldsymbol{\sigma}^{r} \equiv \dot{\boldsymbol{\sigma}} - \boldsymbol{\omega}\cdot\boldsymbol{\sigma} + \boldsymbol{\sigma}\cdot\boldsymbol{\omega} = \boldsymbol{\pi} : \mathbf{d}^{e} = \boldsymbol{\pi} : (\mathbf{d} - \mathbf{d}^{in}), \\ \mathbf{d}^{in} = \sum_{i=1}^{K} \dot{\gamma}^{(i)} \mathbf{m}^{(i)}_{(S)}, \\ \dot{\gamma}^{(i)} = \dot{\gamma}_0 \left| \dfrac{\tau^{(i)}}{\tau_c^{(i)}} \right|^{1/n} H(\tau^{(i)} - \tau_c^{(i)}),\ i = 1,\ldots,K, \\ \dot{\tau}_c^{(i)} = f(\gamma^{(j)}, \dot{\gamma}^{(j)}),\ i, j = 1,\ldots,K, \end{cases} \tag{2}$$

here $\boldsymbol{\sigma}$ – Cauchy stress tensor, $\mathbf{d}, \mathbf{d}^{e}, \mathbf{d}^{in}$ – strain rate tensor, its elastic and inelastic parts, $\gamma^{(i)}$, $\tau_c^{(i)}$ – accumulated shear and the critical shear stress on the $i$-th slip system, $\mathbf{m}^{(i)}_{(S)}$ – symmetric part of the orientation tensor of the $i$-th slip system, $\mathbf{m}^{(i)}_{(S)} = 1/2(\mathbf{b}^{(i)}\mathbf{n}^{(i)} + \mathbf{n}^{(i)}\mathbf{b}^{(i)})$, $\mathbf{b}^{(i)}, \mathbf{n}^{(i)}$ – unit vectors in the direction of the Burgers vector and the normal to the slip plane; $\dot{\gamma}_0, n$ – material constants: the characteristic shear rate and rate sensitivity of the material, $\tau^{(i)}$ – acting slip system shear stress, $\tau^{(i)} = \mathbf{b}^{(i)}\mathbf{n}^{(i)} : \boldsymbol{\sigma}$, $H(\cdot)$ – Heaviside function, $K$ – the number of slip systems for this type of crystal lattice, $\mathbf{o}$ – tensor of the current orientation of the crystallographic coordinate system to the fixed laboratory system.

For scale transition we used generalized Voigt hypothesis, according to which the velocity gradient of movement for each crystallite is equal to the macro-velocity gradient $\hat{\nabla}\mathbf{v} = \hat{\mathbf{W}}$.

## 3. Hardening model

Hardening is divided into "non-oriented" and "oriented". The first describes the hardening regardless of the direction of deformation (intersection of dislocations, plaits, braids, dislocation barriers), and a hardening increases the critical shear stress at once on many slip systems (or even all at once). The second is related to the accumulation of elastic energy to "pursed dislocations" (at different barrier) and this energy may be (fully or partially) released at the change the direction of deformation. The second type, in general, can be described by the kinematic hardening, or due to simultaneous changes in the critical shear stress on the opposite slip systems.

We received both general and particular form of hardening laws of mono- and polycrystalline, allows to describe the formation and destruction of dislocation barriers, the annihilation of dislocations (and so describes Bauschinger effect), and additional hardening, resulting from the interaction of intragranular and grain boundary dislocations [3].

As the basic law is considered a power hardening law in type of:

$$\dot{\tau}^{(k)}_{c\,b} = f^{(k)}\left(\gamma^{(i)}, \dot{\gamma}^{(i)}\right) = \psi E \left( \sum_{i=1}^{24} a_i^{(k)} \left( \frac{\gamma^{(i)}}{\sum_{j=1}^{24} \gamma^{(j)}} \right)^{\psi-1} \dot{\gamma}^{(i)} \right), k = \overline{1,24}, \ \psi > 1, \gamma^{(i)} \geq 0, \tau^{(k)}_{c\,b}(0) \quad \tau^{(k)}_{c\,b0} \qquad (3)$$

which takes into account the interaction of forest dislocations and modified to reflect the complexity of the previous loading.

Assuming additivity of the critical shear stress rates on the slip system due to different mechanisms of hardening, the power law (3) is supplemented by terms that take into account the basic mechanisms of obstacles during plastic deformation, left out the first (power) term:

$$\dot{\tau}^{(k)}_{c} = f^{(k)}\left(\gamma^{(i)}, \dot{\gamma}^{(i)}\right) + f^{(k)}_{bar}\left(\gamma^{(i)}, \dot{\gamma}^{(i)}; \alpha_1^{(i)}, \alpha_2^{(i)}, \ldots, \alpha_n^{(i)}\right) + f^{(k)}_{annih}\left(\gamma^{(i)}, \dot{\gamma}^{(i)}; \beta_1^{(i)}, \beta_2^{(i)}, \ldots, \beta_m^{(i)}\right), \ i,k = \overline{1,24}, \qquad (4)$$

here $\alpha_1^{(i)}, \alpha_2^{(i)}, \ldots, \alpha_n^{(i)}; \beta_1^{(i)}, \beta_2^{(i)}, \ldots, \beta_m^{(i)}$ – sets of internal variables describing appropriate mechanisms (in general; here the second term describes additional hardening due to reactions to the split dislocations, and third allows to consider a decrease of the critical shear stress for reverse slip through dislocation annihilation.

An additional hardening function $f^{(i)}_{bar}$ is taken in the form of:

$$f^{(i)}_{bar}\left(\gamma_{SFE}, \dot{\gamma}^{(i)}, \gamma^{(j)}\right) = \sum_{k=1}^{6} \xi_{ik} \tau_c^{(i)} \mathrm{H}\left(1 - \frac{\gamma_{SFE}}{\gamma^{*}_{SFE}}\right) \left( \int_0^t f^{(i)}_{bar} d\tau + f_0^{(i)} \right)^{-1} \dot{\gamma}^{(i)} \left( \sum_{j \neq i}^{N^*} \gamma^{(j)} + \gamma_0^b \right) \mathrm{H}\left( \int_0^t f^{(i)}_{bar} d\tau \quad \tau^{(i)}_{c\,fr}, \right) \qquad (5)$$

where $\gamma_{SFE}$ – stacking fault energy (SFE) of the material, $\gamma^{*}_{SFE}$ – critical SFE, beyond which this mechanism relies insignificant for this material, $N^*$ – the number of slip systems, coupled to given, $\tau_c^{(i)}$ – current (full) critical stress, $\tau^{(i)}_{c\,fr}$ – critical stress for barrier destruction, $\gamma_0^b$ – small constant, $\xi_{ik}$ – material constants, taking into account the strength of each of the six types of barriers.

Oriented hardening, which is realized by "pursed" by obstacles dislocation annihilation, due to the changing the deformation direction is also considered; detail the physics of the annihilation process and factors affecting the decrease of the critical shear stress on the slip systems as a result of the annihilation of dislocations is considered [3]. To account for the released elastic energy in relation for $f^{(i)}_{annih}$ puts an additional factor that takes into account the complexity of the loading on all of the slip systems (here is an example for the for fcc lattice):

$$f^{(i)}_{annih}\left(\beta_1, \beta_2, \ldots, \beta_m\right) = \frac{d\tau^{(i)}_{annih}}{dt} = -\xi_2 \tau^{(i)}_{annih} \frac{\gamma^{(i)}}{\sum_j \gamma^{(j)}} \dot{\gamma}^{(i)} \left(\gamma^{(i+12)} + \gamma_0^a\right), \ \tau^{(i)}_{annih}\Big|_{t=0} = \tau^{(i)}_{c0} \qquad (6)$$

here $\gamma_0^a$ – small constant, $\xi_2$ – material constant.

## 4. Numerical results

In Fig. 1 is shown a diagram of the cyclic uniaxial loading polycrystalline aggregate using modified relations (3) – (6), the physical and mechanical parameters of the model correspond to the technically pure copper. Nonlinear effects associated with the formation and destruction of dislocation barriers, there do not appear in the smallness of deformations; clearly visible out on the stationary trajectory of deformation.

Fig. 2 shows stress-strain diagram for polycrystalline aggregate when considering the term (6), which describes the decreasing of the critical stress on the slip system, due to the annihilation of dislocations during pursed reverse loading. The calculations were performed for two cycles in tension-compression. Clearly visible reduction of the yield strength when the sign change of deformation:

from 32 MPa initially to 28 MPa after the first change of deformation direction, and from 34 MPa to 30 MPa in the second cycle.

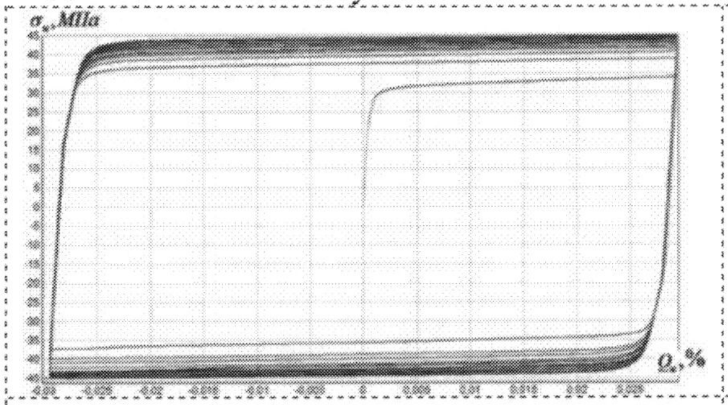

**Figure 1.** The stress - strain diagram during cyclic deformation of polycrystalline aggregate, 20 cycles total.

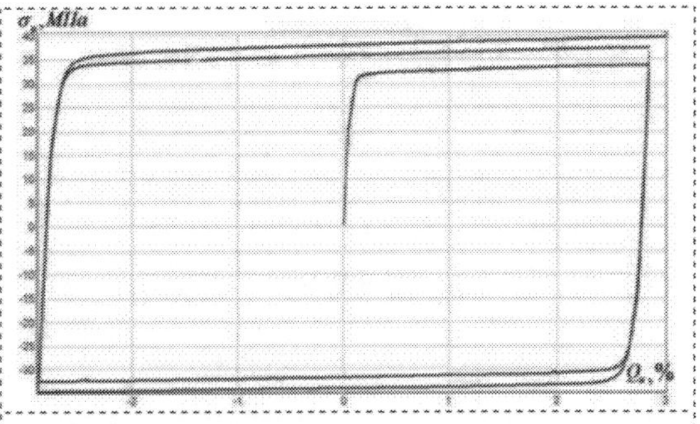

**Figure 2.** The stress - strain diagram for cyclic deformation of polycrystalline aggregate with terms (8) and (9).

Fig. 3 shows the dependence of various stress-strain diagram for polycrystalline aggregate if using in the hardening law additional term in the form (5), with the values specified in the caption. Clearly visible nonlinearities appearing in the diagram due to the effect of "blocking" slip systems by sessile dislocation when accumulated to a certain critical value, and accordingly releasing these systems from the deformation process. In the moments of one system closing and before the activate another systems the share of elastic deformation in full deformation rises sharply, resulting in a steep increase in stress on the diagram, with a further deformation there is gradual diagram alignment by activate new or additional slip systems.

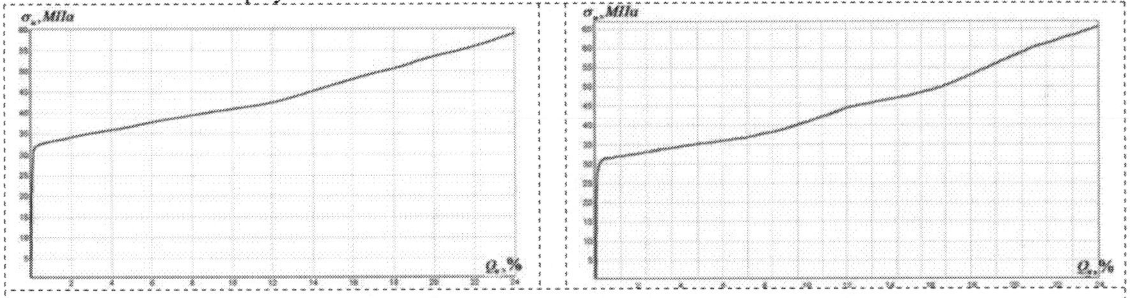

**Figure 2.** The stress – strain diagram under uniaxial compression of polycrystalline aggregate, a) $f_0^{(i)}=1.05$, $\gamma_0^b=5\cdot10^{-5}$, b) $f_0^{(i)}=1.02$, $\gamma_0^b=2{,}5\cdot10^{-5}$.

## 5. Acknowledgments

This work was supported by RFBR (grants №12-08-01052-a, №12-08-33082-mol_a_ved, №12-01-31094-mol_a, №13-01-96006-r_ural_a), the President Grants №MK-3989.2012.1, №MK-390.2013.1, Federal Target Program "Scientific and scientific-pedagogical personnel of innovative Russia in 2009 - 2013 years" (activity 1.2.2, Agreement 14.B37.21.0382).

## References

[1] Trusov P V, Shveykin A I, Nechaeva E S and Volegov P S 2012 *Physical Mesomechanics* **15** 155-175

[2] Trusov P V and Volegov P S 2011 *Vestnik PNIPU. Mehanika* **3** 146-197

[3] Trusov P V, Volegov P S and Yanz A Yu *Scientific and technical statements of St.Petersburg State Polytechnical University. Physics and mathematics* 2010 **2(98)** 110-119

# The Complex Angle in Normed Spaces

**VOLKER THÜREY**
**Rheinstr. 91**
**28199 Bremen, Germany** [1]

**Abstract.** We consider a generalized angle in complex normed vector spaces. Its definition corresponds to the definition of the well known Euclidean angle in real inner product spaces. Not surprisingly it yields complex values as 'angles'. This 'angle' has some simple properties, which are known from the usual angle in real inner product spaces. But to do ordinary Euclidean geometry real angles are necessary. We show that even in a complex normed space there are many pure real valued 'angles'. The situation improves yet in inner product spaces. There we can use the known theory of orthogonal systems to find many pairs of vectors with real angles, and to do geometry which is based on the Greeks 2000 years ago.

*Keywords and phrases*: complex normed space, generalized angle, complex Hilbert space.

There are some attempts to generalize the well known Euclidean angle of real inner product spaces to real normed spaces, see, for instance, the references [2], [6] [7], [9], [10] [11], [12]. Also angles in complex inner product spaces were considered, see [3], [4], [8]. In this paper for the first time a complex-valued 'angle' is defined for all complex normed spaces, and we investigate its properties. We believe that this concept of a complex-valued angle is superior to the enforcement of pure real angles, since real angles may suppress some true properties of a complex normed vector space.

It is well known that the inner product $< . | . >$ in an inner product space $(X, < . | . >)$ over the complex field $\mathbb{C}$ can be expressed by its norm, i.e. for arbitrary $\vec{x}, \vec{y} \in X$ we can write $< \vec{x} \mid \vec{y} > = (1/4) \cdot \left[ \|\vec{x}+\vec{y}\|^2 - \|\vec{x}-\vec{y}\|^2 + \mathbf{i} \cdot \left( \|\vec{x}+\mathbf{i}\cdot\vec{y}\|^2 - \|\vec{x}-\mathbf{i}\cdot\vec{y}\|^2 \right) \right]$, where the symbol '**i**' means the imaginary unit. We use this fact, and we normalize all vectors $\vec{x}, \vec{y} \neq \vec{0}$ to generate a continuous product in all complex normed vector spaces $(X, \|\cdot\|)$, which is just the inner product in the special case of a complex inner product space.

**Definition 1.** *Let $\vec{x}, \vec{y}$ be two arbitrary elements of $X$. In the case of $\vec{x} = \vec{0}$ or $\vec{y} = \vec{0}$ we set $< \vec{x} \mid \vec{y} > := 0$, and if $\vec{x}, \vec{y} \neq \vec{0}$ (i.e. $\|\vec{x}\| \cdot \|\vec{y}\| > 0$) we define the complex number*

$$< \vec{x} \mid \vec{y} > :=$$

$$\|\vec{x}\| \cdot \|\vec{y}\| \cdot \frac{1}{4} \cdot \left[ \left\| \frac{\vec{x}}{\|\vec{x}\|} + \frac{\vec{y}}{\|\vec{y}\|} \right\|^2 - \left\| \frac{\vec{x}}{\|\vec{x}\|} - \frac{\vec{y}}{\|\vec{y}\|} \right\|^2 + \mathbf{i} \cdot \left( \left\| \frac{\vec{x}}{\|\vec{x}\|} + \mathbf{i} \cdot \frac{\vec{y}}{\|\vec{y}\|} \right\|^2 - \left\| \frac{\vec{x}}{\|\vec{x}\|} - \mathbf{i} \cdot \frac{\vec{y}}{\|\vec{y}\|} \right\|^2 \right) \right] .$$

□

[1] T: 49 (0)421/591777, E-mail: volker@thuerey.de

It is easy to show that the product fulfils the conjugate symmetry ($<\vec{x}|\vec{y}> = \overline{<\vec{y}|\vec{x}>}$), where $\overline{<\vec{y}|\vec{x}>}$ means the complex conjugate, the positive definiteness ($<\vec{x}|\vec{x}> \geq 0$ and $<\vec{x}|\vec{x}>= 0$ only for $\vec{x}=\vec{0}$), and the homogeneity for real numbers, ($<r\cdot\vec{x}\,|\,\vec{y}> = r\cdot <\vec{x}\,|\,\vec{y}>$), and the homogeneity for pure imaginary numbers, ($<r\cdot\mathbf{i}\cdot\vec{x}\,|\,\vec{y}> = r\cdot\mathbf{i}\cdot <\vec{x}\,|\,\vec{y}>$), for $\vec{x},\vec{y}\in X$, $r\in\mathbb{R}$. We express the norm of $\vec{x}$ by $\|\vec{x}\| = \sqrt{<\vec{x}\,|\,\vec{x}>}$. Further we have the identity $<e^{\mathbf{i}\cdot\varphi}\cdot\vec{x}\,|\,\vec{x}> = e^{\mathbf{i}\cdot\varphi}\cdot <\vec{x}\,|\,\vec{x}>$, but for distinct $\vec{x},\vec{y}\in X$ generally it holds the inequality $<e^{\mathbf{i}\cdot\varphi}\cdot\vec{x}\,|\,\vec{y}> \neq e^{\mathbf{i}\cdot\varphi}\cdot <\vec{x}\,|\,\vec{y}>$.

If we switch for a moment to real inner product spaces $(X, <.\,|\,.>_{real})$ we recall that we have for all $\vec{x},\vec{y}\neq\vec{0}$ the usual Euclidean angle $\angle_{Euclid}(\vec{x},\vec{y}) = \arccos\left(\frac{<\vec{x}\,|\,\vec{y}>_{real}}{\|\vec{x}\|\cdot\|\vec{y}\|}\right)$.

For all pairs $\vec{x},\vec{y}\neq\vec{0}$ from a complex normed vector space $(X,\|\cdot\|)$ we define an 'angle'. It coincides with the definition of the Euclidean angle in real inner product spaces.

**Definition 2.** *Let $\vec{x},\vec{y}\in X\backslash\{\vec{0}\}$, where $(X,\|\cdot\|)$ is a complex normed space. We define*

$$\angle(\vec{x},\vec{y}) \;:=\; \arccos\left(\frac{<\vec{x}\,|\,\vec{y}>}{\|\vec{x}\|\cdot\|\vec{y}\|}\right).$$

*This number $\angle(\vec{x},\vec{y})\in\mathbb{C}$ is called the angle of the pair $(\vec{x},\vec{y})$.* □

We state that the angle $\angle(\vec{x},\vec{y})$ is defined for all $\vec{x},\vec{y}\neq\vec{0}$. Since we deal with complex vector spaces it is not surprising that we get complex numbers as 'angles'. For the definition we need the extension of the cosine and arccosine functions on complex numbers. See as a reference, e.g., the books [1] or [5]. We use two subsets $\mathcal{A}$ and $\mathcal{B}$ of the complex plane $\mathbb{C}$, where $\mathcal{A} := \{a+\mathbf{i}\cdot b\in\mathbb{C}\mid 0<a<\pi,\ b\in\mathbb{R}\}\cup\{0,\pi\}$, $\mathcal{B} := \mathbb{C}\backslash\{r\in\mathbb{R}\mid r<-1 \text{ or } r>+1\}$. We have two known homeomorphisms $\cos:\mathcal{A}\xrightarrow{\cong}\mathcal{B}$ and $\arccos:\mathcal{B}\xrightarrow{\cong}\mathcal{A}$.

The cosines of the 'angles' are out of the 'complex square' $\mathcal{SQ} := \{r+\mathbf{i}\cdot s\in\mathbb{C}\mid -1\leq r,s\leq +1\}\subset\mathcal{B}$. The values of the 'angles' are from its image $\arccos(\mathcal{SQ})$, which forms a symmetric hexagon (with concave sides) in $\mathcal{A}$. Its center is $\pi/2$, two corners are $0$ and $\pi$. We get a third if we use one of four possibilities of $\pm1\pm\mathbf{i}$, for instance

$$\arccos(1+\mathbf{i}) = (\pi/2)-(1/2)\cdot\left[\arccos\left(\sqrt{5}-2\right)+\mathbf{i}\cdot\log\left(\sqrt{5}+2+2\cdot\sqrt{\sqrt{5}+2}\right)\right] \approx 0.90-\mathbf{i}\cdot 1.06\,.$$

This 'angle' has eight comfortable properties (An 1) - (An 8) which are known from the Euclidean angle $\angle_{Euclid}$ in real inner product spaces.

**Theorem 1.** *In a complex normed space $(X,\|\cdot\|)$ the angle $\angle$ has the following properties:*

- (An *1*) *$\angle$ is a continuous map from $\left(X\backslash\{\vec{0}\}\right)^2$ onto a subset of* $\arccos(\mathcal{SQ})\subset\mathcal{A}$.

  *For elements $\vec{x},\vec{y}\neq\vec{0}$ it holds that*
- (An *2*) $\angle(\vec{x},\vec{x})=0$,
- (An *3*) $\angle(-\vec{x},\vec{x})=\pi$,
- (An *4*) $\angle(\vec{x},\vec{y})=\overline{\angle(\vec{y},\vec{x})}$,
- (An *5*) *for real numbers $r,s>0$ we have $\angle(r\cdot\vec{x},s\cdot\vec{y})=\angle(\vec{x},\vec{y})$,*
- (An *6*) $\angle(-\vec{x},-\vec{y})=\angle(\vec{x},\vec{y})$,
- (An *7*) $\angle(\vec{x},\vec{y})+\angle(-\vec{x},\vec{y})=\pi$.
- (An *8*) *For any two linear independent vectors $\vec{x},\vec{y}$ of $(X,\|\cdot\|)$ there is a continuous injective map $\Theta:\mathbb{R}\longrightarrow\mathcal{A}$, $t\mapsto\angle(\vec{x},\vec{y}+t\cdot\vec{x})$. The limits are*

  $\lim_{t\to-\infty}\Theta(t)=\pi$ *and* $\lim_{t\to\infty}\Theta(t)=0$.

In the following table we list some angles and their cosines. We choose two elements $\vec{x}, \vec{y} \neq \vec{0}$ of a complex normed space $(X, \|\cdot\|)$, and six suitable real numbers $a, b, r, s, v, w$, with $-\frac{\pi}{2} \leq a, v \leq \frac{\pi}{2}$ and $-1 \leq r, s \leq 1$, such that $\angle(\vec{x}, \vec{y}) = \frac{\pi}{2} + a + \mathbf{i} \cdot b \in \mathcal{A}$, $\cos(\angle(\vec{x}, \vec{y})) = r + \mathbf{i} \cdot s \in \mathcal{B}$, and $\angle(\mathbf{i} \cdot \vec{x}, \vec{y}) = \frac{\pi}{2} + v + \mathbf{i} \cdot w \in \mathcal{A}$.
Note that the cosines of all angles in the table (third column) have the same modulus $\sqrt{r^2 + s^2}$.

| pair of vectors | their angle $\angle$ | the cosine of $\angle$ | the angle for $\vec{x} = \vec{y}$ | its cosine for $\vec{x} = \vec{y}$ |
|---|---|---|---|---|
| $(\vec{x}, \vec{y})$ | $\frac{\pi}{2} + a + \mathbf{i} \cdot b$ | $r + \mathbf{i} \cdot s$ | $0$ | $1$ |
| $(-\vec{x}, \vec{y})$ | $\frac{\pi}{2} - a - \mathbf{i} \cdot b$ | $-r - \mathbf{i} \cdot s$ | $\pi$ | $-1$ |
| $(\vec{y}, \vec{x})$ | $\frac{\pi}{2} + a - \mathbf{i} \cdot b$ | $r - \mathbf{i} \cdot s$ | $0$ | $1$ |
| $(-\vec{y}, \vec{x})$ | $\frac{\pi}{2} - a + \mathbf{i} \cdot b$ | $-r + \mathbf{i} \cdot s$ | $\pi$ | $-1$ |
| $(\mathbf{i} \cdot \vec{x}, \vec{y})$ | $\frac{\pi}{2} + v + \mathbf{i} \cdot w$ | $-s + \mathbf{i} \cdot r$ | $\frac{\pi}{2} - \mathbf{i} \cdot \log\left[\sqrt{2}+1\right]$ | $\mathbf{i}$ |
| $(\vec{y}, \mathbf{i} \cdot \vec{x})$ | $\frac{\pi}{2} + v - \mathbf{i} \cdot w$ | $-s - \mathbf{i} \cdot r$ | $\frac{\pi}{2} + \mathbf{i} \cdot \log\left[\sqrt{2}+1\right]$ | $-\mathbf{i}$ |
| $(\vec{x}, \mathbf{i} \cdot \vec{y})$ | $\frac{\pi}{2} - v - \mathbf{i} \cdot w$ | $s - \mathbf{i} \cdot r$ | $\frac{\pi}{2} + \mathbf{i} \cdot \log\left[\sqrt{2}+1\right]$ | $-\mathbf{i}$ |
| $(\mathbf{i} \cdot \vec{y}, \vec{x})$ | $\frac{\pi}{2} - v + \mathbf{i} \cdot w$ | $s + \mathbf{i} \cdot r$ | $\frac{\pi}{2} - \mathbf{i} \cdot \log\left[\sqrt{2}+1\right]$ | $\mathbf{i}$ |

With given two vectors $\vec{x}, \vec{y} \in X$ we consider as before the angle $\angle(\vec{x}, \vec{y}) = \frac{\pi}{2} + a + \mathbf{i} \cdot b$ with suitable real numbers $a, b$. Now we express the complex number $\angle(\mathbf{i} \cdot \vec{x}, \vec{y})$ in dependence of $a$ and $b$. For the presentation the real valued cosine and hyperbolic cosine are used, and their inverses arccosine and area hyperbolic cosine. The result is as follows:

**Theorem 2.** *In a complex normed space $(X, \|\cdot\|)$ we take two elements $\vec{x}, \vec{y} \neq \vec{0}$. We assume the angle $\angle(\vec{x}, \vec{y}) = \frac{\pi}{2} + a + \mathbf{i} \cdot b \in \mathcal{A}$, i.e. $-\frac{\pi}{2} \leq a \leq \frac{\pi}{2}$. (If $a = -\frac{\pi}{2}$ or $a = \frac{\pi}{2}$ since $\angle(\vec{x}, \vec{y}) \in \mathcal{A}$ it follows $b = 0$.) We use abbreviations $\mathsf{H}_-$ and $\mathsf{H}_+$. We get*

$$\angle(\mathbf{i} \cdot \vec{x}, \vec{y}) = \frac{\pi}{2} + \frac{1}{2} \cdot \left[-\operatorname{sgn}(b) \cdot \arccos(\mathsf{H}_-) + \mathbf{i} \cdot \operatorname{sgn}(a) \cdot \operatorname{arcosh}(\mathsf{H}_+)\right], \quad \textit{where}$$

$$\mathsf{H}_\pm := \sqrt{\left[\cos^2\left(\frac{\pi}{2} + a\right) + \cosh^2(b) - 2\right]^2 + 4 \cdot \cos^2\left(\frac{\pi}{2} + a\right) \cdot \cosh^2(b)} \pm \left[\cos^2\left(\frac{\pi}{2} + a\right) + \cosh^2(b) - 1\right].$$

It is worthwhile to look at special cases. We consider a real angle $\angle(\vec{x}, \vec{y}) = \frac{\pi}{2} + a$ (i.e. $b = 0$), and an angle on the vertical axis $x = \frac{\pi}{2}$, this means $\angle(\vec{x}, \vec{y}) = \frac{\pi}{2} + \mathbf{i} \cdot b$ (i.e. $a = 0$).

**Corollary 1.** *For a pure real angle $\angle(\vec{x}, \vec{y}) = \frac{\pi}{2} + a$ with $-\frac{\pi}{2} \leq a \leq \frac{\pi}{2}$, i.e. $b = 0$, we get a complex angle $\angle(\mathbf{i} \cdot \vec{x}, \vec{y})$ with a real part $\pi/2$,*

$$\begin{aligned} \angle(\mathbf{i} \cdot \vec{x}, \vec{y}) &= \frac{\pi}{2} + \mathbf{i} \cdot \frac{1}{2} \cdot \operatorname{sgn}(a) \cdot \operatorname{arcosh}\left[2 \cdot \cos^2\left(\frac{\pi}{2} + a\right) + 1\right] \\ &= \frac{\pi}{2} + \mathbf{i} \cdot \operatorname{sgn}(a) \cdot \log\left[\sqrt{\cos^2\left(\frac{\pi}{2} + a\right) + 1} + \left|\cos\left(\frac{\pi}{2} + a\right)\right|\right]. \end{aligned}$$

**Corollary 2.** *For an angle $\angle(\vec{x}, \vec{y}) = \frac{\pi}{2} + \mathbf{i} \cdot b$ (i.e. $a = 0$) we get a pure real angle $\angle(\mathbf{i} \cdot \vec{x}, \vec{y})$,*

$$\angle(\mathbf{i} \cdot \vec{x}, \vec{y}) = \frac{\pi}{2} - \frac{1}{2} \cdot \operatorname{sgn}(b) \cdot \arccos[3 - 2 \cdot \cosh^2(b)] = \frac{\pi}{2} - \arcsin[\sinh(b)] = \arccos[\sinh(b)].$$

Now we look for real valued angles in complex vector spaces. We can prove that in a complex normed vector space $(X, \|\cdot\|)$ for each pair $\vec{x}, \vec{y} \neq \vec{0}$ of non-zero vectors there is a real number $\varphi \in [0, 2\pi]$ such that the pair $(e^{i\cdot\varphi}\cdot\vec{x}, \vec{y})$ has a pure real angle, i.e. $\angle(e^{i\cdot\varphi}\cdot\vec{x}, \vec{y}) \in \mathbb{R}$.

This fact ensures the existance of many real valued angles even in complex normed spaces. The situation improves further in the special case of complex vector spaces provided with an inner product $<.|.>$. We discuss the case of complex inner product spaces $(X, <.|.>)$ now.

To do ordinary Euclidean geometry we need real valued angles. The idea is simple. We take an orthogonal basis $\mathsf{T}$ of $(X, <.|.>)$, and we generate $\mathcal{L}(\mathbb{R})(\mathsf{T})$, the set of all finite real linear combinations of elements of $\mathsf{T}$. Let

$$\mathcal{L}(\mathbb{R})(\mathsf{T}) := \left\{ \sum_{i=1}^{n} r_i\cdot\vec{x}_i \mid n \in \mathbb{N},\ r_1, r_2, \ldots, r_n \in \mathbb{R},\ \vec{x}_1, \vec{x}_2, \ldots, \vec{x}_n \in \mathsf{T} \right\}.$$

If $X$ is a Hilbert space, i.e. it is complete, we even can use Cauchy sequences from elements of $\mathcal{L}(\mathbb{R})(\mathsf{T})$. It means that we take the closure $\overline{\mathcal{L}(\mathbb{R})(\mathsf{T})}$ of $\mathcal{L}(\mathbb{R})(\mathsf{T})$ in $X$. This creates a real linear subspace $\overline{\mathcal{L}(\mathbb{R})(\mathsf{T})}$ of $X$, where all angles are real.

After that we consider complex inner product spaces $X$ of finite complex dimension $n \in \mathbb{N}$. Their real dimension is $2\cdot n$, and we state that the maximal dimension of a real subspace of $X$ with all-real angles is $n$. The real span $\mathcal{L}(\mathbb{R})(\mathsf{T})$ of an orthogonal basis $\mathsf{T} \subset X$ yields an example.

Finally we remark that real angles are useful to do classical Euclidean geometry. We are still considering complex inner product spaces $(X, <\cdot|\cdot>)$, and we mention three well-known theorems of usual geometry. Let $\vec{x}, \vec{y} \neq \vec{0}$ be elements of $X$. We obtain the three identities $\angle(\vec{x}, \vec{y}) = \angle(\vec{x}, \vec{x}+\vec{y}) + \angle(\vec{x}+\vec{y}, \vec{y})$, (Addition of angles), $\angle(\vec{x}, \vec{y}) + \angle(-\vec{x}, \vec{y}-\vec{x}) + \angle(-\vec{y}, \vec{x}-\vec{y}) = \pi$, (The sum of inner angles in a triangle is $\pi$), $\|\vec{x}-\vec{y}\|^2 = \|\vec{x}\|^2 + \|\vec{y}\|^2 - 2\cdot\|\vec{x}\|\cdot\|\vec{y}\|\cdot\cos(\angle(\vec{x}, \vec{y}))$, (The 'Law of Cosines' ), if and only if the angle $\angle(\vec{x}, \vec{y})$, and therefore all angles in the above terms are pure real.

An extended paper with all proofs and additional informations is in preparation.

**Acknowledgements:** We like to thank Prof. Dr. Eberhard Oeljeklaus and Jeremy Goodsell for careful reading of the paper and suggesting some improvements, as well Verena Thürey and Wilfried Henning for technical aid.

## References

[1] I.N. Bronstein, K.A. Semendjajew, *Taschenbuch der Mathemetik*, 19. Auflage, Harri Deutsch (1980)

[2] C.R. Diminnie, E.Z. Andalafte and R.W. Freese, *Angles in Normed Linear Spaces and a Characterization of Real Inner Product Spaces*, Math. Nachr. **129** (1986), 197-204

[3] A. Froda, *Sur l'angle complexe, orienté, de deux vecteurs d'un espace unitaire*, Atti Accad. Naz. Lincei, Cl. Sci. Fis. Mat. Natur. Rend. Lincei(8) Mat. Appl. **30** (1961), 845-853

[4] A Galántai, C.J. Hegedüs, *Jordan's Principal Angles in Complex Vector Spaces*, Numer. Linear Algebra Appl. **13** (2006), 589-598

[5] Alan Jeffrey, *Complex Analysis and Applications*, 2. Edition, Chapman and Hall/CRC(2006)

[6] P. M. Miličić, *Sur le* g-*angle dans un espace norme*, Matematik Vesnik **45** (1993), 43-48

[7] P. M. Miličić, *On the B-Angle and* g-*Angle in Normed Spaces*, J. Inequal. Pure Appl. Math. **8**, Iss. 4 (2007), 1-18

[8] Klaus Scharnhorst, *Angles in Complex Vector Spaces*, Acta Appl. Math. **69** (2001), 95-103

[9] Volker W. Thürey, *A Generalization of the Euclidean Angle*, J. Convex Anal. **20**, No 4 (2013), 1025-1042

[10] Volker W. Thürey, *Angles and a Classification of Normed Spaces*, Ann. Funct. Anal. **4**, Iss. 1 (2013), 114-137

[11] J.E. Valentine, C. Martin, *Angles in Metric and Normed Linear Spaces*, Colloq. Math. **34** (1976), 209-217

[12] J.E. Valentine, S.G. Wayment, *Wilson Angles in Linear Normed Spaces*, Pacific J. Math. **36**, No 1 (1971), 239-243

# Feynman integral treatement of the Hyperbolical potential with a centrifugal term approximation

**A. Diaf**[1], **M. Lassaut**[2]

[1]Laboratoire de l'Énergie et des Systèmes Intelligents, Université de Khemis Miliana, Route de Thénia, Khemis Miliana, Algérie.

[2]Institut de Physique Nucléaire IN2P3-CNRS and Université Paris-Sud, F-91406, Orsay CEDEX, France.

E-mail: `s_ahmed_diaf@yahoo.fr; lassaut@ipno.in2p3.fr`

**Abstract.** We obtain an analytical expression for the energy eigenvalues of the Hyperbolical potential using an approximation of the centrifugal term. In order to obtain the $l$-states solutions, we use the the Feynman path integral approach to quantum mechanics. We show that by performing nonlinear space-time transformations in the radial path integral, we can derive a transformation formula that relates the original path integral to the Green function of a new quantum solvable system. The explicit expression of bound state energy is obtained and the eigenfunctions are given in terms of hypergeometric functions. The present results are consistent with those obtained by others methods.

## 1. Introduction

The calculation of exact solutions of the radial Schrödinger equation (rSe) in some physical potential models has been an important research area. Unfortunately, for an arbitrary $l$-state ($l \neq 0$), the radial Schrödinger equation (rSe) does not admit an exact solution. In this case, the rSe is solved numerically [1], or by approximation methods [2]. Different methods have been used in this sense such as the standard method [3] the asymptotic iteration method [4] and the exact quantization rule method [5]. An example of exponential-type potentials that we want to treat is the Hyperbolical potential, it has the form:

$$V^{Hyp}(r) = D\left[1 - \sigma_0 \coth(\alpha r)\right]^2 \tag{1}$$

Here $D$, $\alpha$ and $\sigma_0$ are positive parameters.

The purpose of this paper is to study $l$-states solution of the hyperbolical potential within the Feynman path integrals formalism in order to improve the previous results [6]. The method we propose consists of using the approximation [7] $\frac{1}{r^2} \approx \frac{1}{a^2}\left(C_0 + \frac{1}{e^{\frac{r}{a}}-1} + \frac{1}{(e^{\frac{r}{a}}-1)^2}\right)$, where $C_0$ is a dimensionless parameter, for the centrifugal term. This approximation is based on the expansion of the centrifugal term in a series of exponentials depending on the intermolecular distance r and keeping terms up to second order. The organization of the paper is as follows. In section 2, we calculate the $l$-wave eigensolutions for the Hyperbolical potential using the Duru-Kleinert method of path integral formalism. In section 3, we present our numerical results for certain values of the quantum numbers $n$ and $l$. Conclusions are drawn in section 4.

## 2. The path integral for the Hyperbolical potential

The propagator related to the Hyperbolical potential, between two time-space points $(\vec{r}\,', t')$ and $\left(\vec{r}\,'', t''\right)$ is written [8]:

$$K(\vec{r}\,'', t''; \vec{r}\,', t') = \frac{1}{4\pi r'' r'} \sum_{l=0}^{\infty} (2l+1) K_l(r'', t''; r', t') P_l(\cos\theta). \tag{2}$$

$P_l(\cos\theta)$ is the Legendre polynomial with $\theta \equiv (\vec{r}\,'', \vec{r}\,')$ and

$$K_l(r'', t''; r', t') = \lim_{N\to\infty} \int \prod_{j=1}^{N} \exp\left[\frac{i}{\hbar} S_j\right] \prod_{j=1}^{N} \left[\frac{m}{2\pi i\hbar\varepsilon}\right]^{\frac{1}{2}} \prod_{j=1}^{N-1} dr_j. \tag{3}$$

with $S_j = \frac{m}{2\varepsilon}(\Delta r_j)^2 - \varepsilon V_{eff}(r_j)$ where

$$V_{eff}(r_j) = \frac{\hbar^2 l(l+1)}{2m r_j r_{j-1}} + D\left[1 - \sigma_0 \coth(\alpha r_j)\right]^2 \tag{4}$$

$\Delta r_j = r_j - r_{j-1}$, $\varepsilon = t_j - t_{j-1}$, $t' = t_0$ and $t'' = t_N$,

To solve the above equation for non-zero angular momentum states, we need to apply the approximate scheme to the centrifugal term given by Ikhdair in ref [7]

$$\frac{1}{r^2} \approx 4\alpha^2 \left[C_0 + \frac{e^{-2\alpha r}}{1 - e^{-2\alpha r}} + \left(\frac{e^{-2\alpha r}}{1 - e^{-2\alpha r}}\right)^2\right] \tag{5}$$

where the parameter $C_0 = 0.0823058167837972$ is a proper shift found by the expansion procedures. Therefore the equation (4) becomes

$$\begin{aligned} V_{eff}(r_j) &= \frac{2\alpha^2\hbar^2 l(l+1)}{m}\left[C_0 + \frac{e^{-2\alpha r_j}}{1 - e^{-2\alpha r_j}} + \left(\frac{e^{-2\alpha r_j}}{1 - e^{-2\alpha r_j}}\right)^2\right] \\ &\quad + D\left[1 - \sigma_0 \coth(\alpha r_j)\right]^2 \end{aligned} \tag{6}$$

we can also write :

$$V_{eff}(r_j) = -A\coth(\alpha r_j) + \frac{B}{\sinh^2(\alpha r_j)} + C, \tag{7}$$

where $A = 2D\sigma_0$, $B = \frac{\hbar^2}{8m} l(l+1)(2\alpha)^2 + D\sigma_0^2$ and $C = \frac{\hbar^2 l(l+1)C_0}{2m}(2\alpha)^2 + D\left(1 + \sigma_0^2\right)$

In order to reach the expression of the space-time transformed propagator, we introduce the following space coordinate change

$$r = \left(\frac{1}{\alpha}\right) \operatorname{arccoth}\left(2\coth^2 q - 1\right). \tag{8}$$

The propagator expression becomes

$$\begin{aligned} \widehat{K}_l(q'', q'; s'') &= \exp\left[\frac{i}{\hbar} s''(E - C + A)\left(\frac{1}{\alpha}\right)^2\right] \\ &\quad \times \int \mathcal{D}q(s) \exp\left[\frac{i}{\hbar}\int_0^{s''}\left(\frac{m}{2}\dot{q}^2 - \frac{\hbar^2}{2m}\left[\frac{\eta(\eta-1)}{\sinh^2 q} - \frac{v(v-1)}{\cosh^2 q}\right]\right) ds\right] \end{aligned} \tag{9}$$

$$= \exp\left[\frac{i}{\hbar} s''(E - C + A)\left(\frac{1}{\alpha}\right)^2\right] K_l^{MPT}(q'', q'; s'') \tag{10}$$

with $\eta = \frac{1}{2} \pm \sqrt{1 + 8mB(\frac{1}{\alpha})^2/\hbar^2}$ and $\upsilon = \frac{1}{2} \pm \sqrt{-2m(\frac{1}{\alpha})^2(E - C - A)/\hbar^2}$ and $K_l^{MPT}$ is the path integral of the modified Pöschl-Teller potential, which is a known solved problem [9].

The bound states and the wave functions are explicitly given by [10]:

$$\chi_{l,n}^{(k_1,k_2)}(q) = N_n^{(k_1,k_2)}(\sinh q)^{2k_2-1/2}(\cosh q)^{-2k_1+3/2} \quad (11)$$
$$\times {}_2F_1(-k_1 + k_2 + k, -k_1 + k_2 - k + 1; 2k_2; -\sinh^2 q) \quad (12)$$

and

$$E_n^{MPT} = -\frac{\hbar^2}{2m}\left[2(k_1 - k_2 - n) - 1\right]^2 \quad (13)$$

with $N_n^{(k_1,k_2)} = \frac{1}{\Gamma(2k_2)}\left(\frac{(2k-1)\Gamma(k_1+k_2-k)\Gamma(k_1+k_2+k-1)}{\Gamma(k_1-k_2+k)\Gamma(k_1-k_2-k+1)}\right)^{1/2}$, $k = k_1 - k_2 - n$.

The results (11) and (13) are used in the following section to determine the spectrum and wave functions of the Hyperbolical potential.

### 2.1. *Energy spectrum and wave functions*

Calculating $K_l(q'', q'; s'')$ allows us to obtain the Green function. Since this is known, the whole energy spectrum is obtained from its poles, and from the residues at the poles we obtain the corresponding wave functions.

Substituting (11) in (10), we get

$$\widehat{K_l}(q'', q'; s'') = \sum_{n=0}^{N_m} \exp\left\{\frac{i}{\hbar}s''\left[(E - C + A)(2a)^2 - E_n^{MPT}\right]\right\}\chi_{l,n}^{(k_1,k_2)}(q'')\chi_{l,n}^{*(k_1,k_2)}(q') \quad (14)$$

Then, by integrating this latter over the pseudo-time parameters $s''$, the Green function $G$ is written as:

$$G_l(r'', r'; E) = \sum_{n=0}^{N_m} \frac{\chi_{l,n}^{MR(k_1,k_2)}(r'')\chi_{l,n}^{MR*(k_1,k_2)}(r')}{E_{n,l}^{MR} - E} \quad (15)$$

As in [11, 12], satisfying the boundary conditions for $r \to 0$ and $r \to \infty$ gives:

$k_1 = \frac{1}{2}\left[(1 + \frac{1}{2}\left(s + 2n + 1)\right) + 2mA(\frac{1}{\alpha})^2/(\hbar^2(s + 2n + 1)))\right]$,

$k_2 = \frac{1}{2}\left(1 + \sqrt{1 + 8mB(\frac{1}{\alpha})^2/\hbar^2}\right) = \frac{1}{2}(1 + s)$.

Substituting these values of $k_1$, $k_2$ in (11) and (13), and by considering $u = \frac{1}{2}[1 + \tanh(2\alpha r)]$, we get

$$\chi_{l,n}^{Hyp(k_1,k_2)}(r) = \sqrt{\alpha}N_n^{(k_1,k_2)}(u - 1)^{(1/2)-k_1+n}u^{k_1-1-(1/2)s-n}{}_2F_1(-n, 2k_1 - n - 1; s + 1; \frac{1}{1-u}) \quad (16)$$

The energy spectrum is obtained from the poles,

$$E_{n,l}^{Hyp} = \alpha^2 E_n^{MPT} - A + C \quad (17)$$
$$= -\left[\frac{\hbar^2\alpha^2(s + 2n + 1)^2}{8m} + \frac{2mA^2}{\hbar^2\alpha^2(s + 2n + 1)^2}\right] + C \quad (18)$$

**Table 1.** Eigenvalues $-E_{n,l}$ of the Hyperbolical potential in atomic units ($\hbar = m = 1$) with $D = 10$ and $\sigma_0 = 0.1$.

| States | $\alpha$ | Present | Ikhdair et al[7] | Lucha et al[1] |
|---|---|---|---|---|
| 2p | 0.1 | 2.61885 | 2.61874 | 2.61935 |
| | 0.15 | 3.90570 | 3.90544 | 3.90645 |
| | 0.2 | 5.00378 | 5.00331 | 5.00457 |
| | 0.25 | 5.88668 | 5.88594 | 5.88725 |
| 3p | 0.1 | 4.73552 | 4.73540 | 4.73638 |
| | 0.15 | 6.04569 | 6.04543 | 6.04649 |
| | 0.2 | 6.91710 | 6.91663 | 6.91733 |
| | 0.25 | 7.48474 | 7.48400 | 7.48358 |
| 3d | 0.1 | 3.62734 | 3.62699 | 3.62769 |
| | 0.15 | 5.29484 | 5.29404 | 5.29510 |
| | 0.2 | 6.47635 | 6.47492 | 6.47598 |
| 4p | 0.1 | 6.00298 | 6.00287 | 6.00390 |
| | 0.15 | 7.11552 | 7.11526 | 7.11589 |
| 4d | 0.1 | 5.33164 | 5.33129 | 5.33216 |
| | 0.15 | 6.73663 | 6.73583 | 6.73642 |
| 4f | 0.1 | 4.69036 | 4.68965 | 4.69058 |
| | 0.15 | 7.43683 | 6.42992 | 6.43112 |

## 3. Results and discussions

With our method, the energy $E_{n,l}^{Hyp}$ of the Hyperbolical potential, for angular momentum ($l \neq 0$), has been calculated. Indeed, to show the accuracy of the approximation scheme, we calculate the energy eigenvalues for various $n$ and $l$ quantum numbers. Our main results for the Hyperbolical potential (Eq. (18)) are displayed in the Table 1 and compared with the numerical values of Lucha and Schöberl [1] and with the Numerov-Uvanov method [7]. It is found that our energy eigenvalues are in good agreement with those obtained numerically [1], and better than those obtained by using Numerov-Uvanov method [7].

## 4. Conclusion

In this paper, we have given analytic solutions for the Hyperbolical potential by using an approximation scheme for the centrifugal term potential. This approach enables us to find the $l$-dependent solutions and the corresponding energy eigenvalues for different screening parameters of the Hyperbolical potential.

Finally, the set of results we have ended with shows that the path integral approach constitutes a reliable alternative to the Schrödinger formalism for solving such problems.

## References


[1] W. Lucha and F.F. Schöberl, Int. J. Mod. Phys. C 10, (1999) 607.
[2] C. L. Pekeris, Phys. Rev. 45, 98 (1934); C. Berkdemir, Nucl. Phys A 770, (2006) 32.
[3] G.F. Wei, C.Y. Long, S.H. Dong, Phys. Lett. A 372, (2008) 2592.
[4] O. Bayrak and I. Boztosun, 2006 J. Phys. A: Math. Gen. 39 6955.
[5] W. C. Qiang and S. H. Dong, 2007 Phys. Lett. A 363 169.
[6] F. Taskin, G. Koak, Chinese Phys. B 19, (2010)090314.
[7] S. M. Ikhdair and R. Sever, Ann. Phys. (Berlin) 18, No. 4, 189 197 (2009).
[8] D. C. Khandekar, S.V. Lawande, K.V. Bhagwat, Path Integral Methods and their Applications, World Scientific, Singapore, 1986.
[9] M. Böhm and G. Junker, J . Math. Phys. A 117, (1986) 375.
[10] C. Groshe J. Phys. A. Math. Gen. 22, (1989) 5073, A. Frank, K. Wolf J. Math. Phys. 26(1985) 5.
[11] A. Diaf and A. Chouchaoui, Phys. Scr. 84 (2011) 015004.

# Time series modeling with pruned multi-layer perceptron and 2-stage damped least-squares method

**Cyril Voyant[1,2], Wani Tamas[2], Christophe Paoli[2], Aurélia Balu[2], Marc Muselli[2], Marie-Laure Nivet[2], Gilles Notton[2]**

1-CHD Castelluccio, radiophysics unit, 20000 Ajaccio (France)

2-University of Corsica, *UMR* CNRS *6134 SPE*, Campus Grimaldi, BP 52, 20250 Corte, (France)

Cooresponding author, Cyril Voyant : voyant@univ-corse.fr

**Abstract. A Multi-Layer Perceptron (MLP) defines a family of artificial neural networks often used in TS modeling and forecasting. Because of its "black box" aspect, many researchers refuse to use it. Moreover, the optimization (often based on the exhaustive approach where "all" configurations are tested) and learning phases of this artificial intelligence tool (often based on the Levenberg-Marquardt algorithm; LMA) are weaknesses of this approach (exhaustively and local minima). These two tasks must be repeated depending on the knowledge of each new problem studied, making the process, long, laborious and not systematically robust. In this paper a pruning process is proposed. This method allows, during the training phase, to carry out an inputs selecting method activating (or not) inter-nodes connections in order to verify if forecasting is improved. We propose to use iteratively the popular damped least-squares method to activate inputs and neurons. A first pass is applied to 10% of the learning sample to determine weights significantly different from 0 and delete other. Then a classical batch process based on LMA is used with the new MLP. The validation is done using 25 measured meteorological TS and cross-comparing the prediction results of the classical LMA and the 2-stage LMA.**

## 1. Background

The primary goal of time series (TS) analysis is forecasting, i.e. using the past to predict the future [1]. This formalism is used in many scientific fields like econometrics, seismology or meteorology. Lot of methods are dedicated to the prediction of discrete phenomena, one of the most popular is the artificial neural network (ANN) [1,2]. From a mathematical point of view, ANN is a function defined as the composition of other functions [3]. Members of the class of such functions are obtained by varying parameters, (as connections or weights). A Multi-Layer Perceptron (MLP) defines a family of functions often used in TS modeling [1]. In this model, neurons are grouped in layers and only forward connections exist. A typical MLP consists of an input, hidden and output layers, including neurons, weights and a transfer functions. Each neuron (noted $i$) transforms the weighted sum (weight $w_{ij}$, bias $b_i$) of inputs ($x_j$) into an output ($y_i = f(\sum_{j=1}^{n} x_j w_{ij} + b_i)$) using a transfer function ($f$). The goal of this method is to determine weights and bias for a given problem. A complex process is necessary to adapt connections using a suitable training algorithm (often based on the Levenberg-Marquardt algorithm; LMA [4,5]). The training step is dependent of the number of inputs, layers and hidden nodes. The better configuration defines the optimized MLP [1]. This step is the weaknesses of this approach because no consensus or scientific rules exist, often the use of the exhaustive approach (where "all" configurations are tested) is the only usable and must be repeated for new studied problem, making the process, long, laborious and not systematically robust. This "black box" aspect leads any researchers to refuse to use it. In this paper a pruning process allowing to automatically selecting inputs is proposed. This method allows, during the training phase, to carry out a selecting method activating (or not) inter-nodes connections. With this process, the optimization step becomes self-acting and the parsimony principle is kept. Less the MLP is complex more it is efficient [3].

## 2. Materials and methods

We propose to use iteratively the popular LMA also known as the damped least-squares method to activate the inputs and neurons ($m$ weights and bias) [6]. A first pass is applied to 10% of the learning sample. For each step, a system of $m$ non-linear equations with $m$ unknowns is solved (see equation 1 in case of 1 hidden layer MLP where $O$ and $I$ are the outputs and inputs, $W^1$, $B^1$ and $W^2$, $B^2$ the weights and bias matrices of the hidden and output layer).

$$O^i = W^2.(\tanh(W^1.I^i + B^1)) + B^2 \text{ with } 1<i<m, \tag{1}$$

After this first phase, each weights and bias ($\omega_{i\in[1,m]}$) are represented by probability distributions. A statistical test based on the bootstrap distribution is used to determine if the first moment of each $\omega_i$ is significantly different from 0 [7]. Before to initiate the second pass, the network is customized and connections related to each $\omega_i$ non-significantly different from 0 are canceled. Then a classical batch process based on LMA is used with the new MLP. We validated our method with 5 hourly meteorological TS (wind direction *WD*, wind speed *WS*, Global radiation *Glo*, Humidity *Hum* and temperature *Tem*), each one measured in 5 French sites (Ajaccio, Bastia, Corte, Marseille and Nice). Note that no pretreatment are operated and that different non-stationarities and periodicities are present. For all TS and locations, we used 3200 measures for the training and 400 for the cross comparison of the classical LMA and the 2-stage LMA during the year 2008.

### 2.1. First pass

The first stage begins with the generation of $N$ (10% of the total data used during training randomly chosen) systems of $m$ nonlinear equations with $m$ unknowns (MLP constructed with $m$ weights and bias). The method chosen for solve this problem is the LMA method and the ad-hoc objective function $F$ (mean square error between calculations and measures). It is an approximation of the Gauss-newton method, the result of the kth iterations ($\omega_k$) corresponding to the local minimum of the function $F$ is generated by the linear set of equations (1) [4–6,8]:

$$(J(\omega_k)^T.J(\omega_k) + \lambda_k I).d_k = -J(\omega_k)^T.F(\omega_k) \tag{2}$$

$J$ denotes the Jacobian matrix of $F$ and the scalar $\lambda_k$ controls both the magnitude and variation $d_k$ ($d_k = \Delta\omega = \omega_{k+1} - \omega_k$). After the $N$ solving, all the weights and bias are represented by a distribution which will be used during the second stage of the second stage of the methodology. Note that these distributions are not all normal (according to the Jarque-Bera test).

### 2.2. Second pass

Before to initiate the second pass, the network is customized and connections related to each $\omega_i$ non-significantly different from 0 are canceled. We use confidence intervals from the bootstrap distributions (4000 samples) of the weights and bias parameters [7]. The rules of selection (directly linked to the $\alpha$ value of significance model) is based on the product of the two endpoints $t_1$ and $t_2$ defined respectively by the $\alpha/2^{th}$ and $(1-\alpha/2)^{th}$ percentiles of the distribution. If $t_1 . t_2 < 0$ the weight (or bias) is considered as non-significantly different from 0, else it is considered different from 0. The connection of the MLP corresponding to the first case ($t_1 . t_2 < 0$) are cancelled, other are kept. The pruned MLP (noted pMLP) is then trained with the classical LMA.

## 3. Results

In the Figure 1a is represented the box plot of the nRMSE distribution concerning the five meteorological parameters and the five studied cities. For each case, seven runs are operated, so 175 manipulations are performed with the pruned methodology describe above (at left in the Figure 1 and noted pMLP) and the standard approach (at right in the Figure 1 and noted MLP). The chosen architecture is the same for all cases: 7 inputs representing the seven first lags of the meteorological parameter tested and 2 hidden nodes (only one hidden layer). In this figure, we see that the first, the second and the third quartiles are equivalent; thereby we understand that a lot of connections and weights are superfluous. The results of all the simulations are represented in the Figure 1b. Only the points positioned in the upper zone are related to "pMLP is better than MLP" cases are plotted. It appears the points are closer to the y=1 curve in the top area rather than in the bottom area, but this observation seems insignificant.

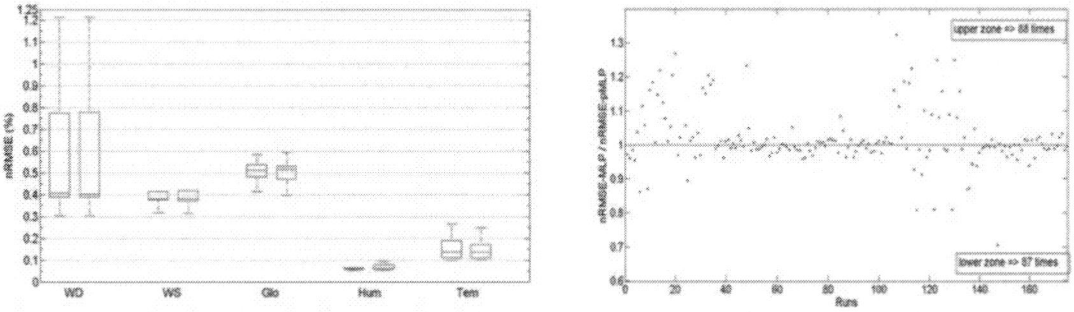


**Figure 1: a. nRMSE distribution comparison related to pMLP (at left) and MLP (at right).
b. ratio of the nRMSE generated by MLP and pMLP**

The table 1 exposes the results for all locations and parameters of the MLP and pMLP approach, the minimum of the nRMSE and nMAE [1] through the seven runs.

| Data | *City* | MLP | | pMLP | | |
|---|---|---|---|---|---|---|
| | | ***nRMSE*** | ***nMAE*** | **Ratio pruning** | ***nRMSE*** | ***nMAE*** |
| WD | *Aja* | 0.765 | 0.463 | 0.27 | **0.762** | **0.461** |
| | *Bas* | 0.393 | 0.257 | 0.27 | **0.388** | **0.252** |
| | *Cor* | **1.191** | 0.941 | 0.21 | 1.194 | **0.931** |
| | *Mar* | **0.387** | **0.256** | 0.20 | 0.388 | 0.257 |
| | *Nic* | **0.304** | **0.199** | 0.20 | 0.305 | 0.211 |
| WS | *Aja* | **0.399** | **0.302** | 0.11 | 0.410 | 0.308 |
| | *Bas* | 0.374 | **0.277** | 0.18 | **0.370** | 0.279 |
| | *Cor* | **1.148** | 0.910 | 0.22 | 1.183 | **0.902** |
| | *Mar* | **0.377** | 0.264 | 0.21 | **0.377** | **0.263** |
| | *Nic* | 0.318 | 0.206 | 0.23 | **0.314** | **0.202** |
| Glo | *Aja* | 0.525 | 0.413 | 0.17 | **0.472** | **0.372** |
| | *Bas* | **0.439** | **0.323** | 0.21 | 0.456 | 0.370 |
| | *Cor* | **0.298** | **0.214** | 0.21 | 0.323 | 0.250 |
| | *Mar* | 0.416 | 0.346 | 0.21 | **0.378** | **0.302** |
| | *Nic* | **0.455** | **0.380** | 0.20 | 0.465 | 0.386 |
| Hum | *Aja* | **0.064** | 0.049 | 0.22 | **0.064** | **0.048** |
| | *Bas* | **0.061** | **0.044** | 0.20 | **0.061** | 0.045 |

| | | | | | | |
|---|---|---|---|---|---|---|
| | *Cor* | **0.055** | **0.036** | 0.23 | 0.056 | 0.037 |
| | *Mar* | **0.053** | **0.036** | 0.23 | **0.053** | **0.036** |
| | *Nic* | **0.083** | **0.059** | 0.19 | **0.083** | **0.059** |
| Tem | *Aja* | **0.099** | **0.077** | 0.29 | 0.101 | **0.077** |
| | *Bas* | 0.113 | 0.080 | 0.25 | **0.111** | **0.079** |
| | *Cor* | 0.224 | 0.158 | 0.29 | **0.207** | **0.133** |
| | *Mar* | 0.111 | 0.079 | 0.28 | **0.109** | **0.073** |
| | *Nic* | 0.147 | **0.108** | 0.21 | **0.146** | 0.111 |

**Table 1: nRMSE and nMAE minima for all location and all parameters, in bold the better results between MLP and pMLP**

pMLP is very slightly better than MLP, in 60% of cases the nMRSE and the nMAE are the lowest. Note that the pruning concerns about 20% of weights and bias.

## 4. Conclusions

The 2-pass approach improves slightly the forecasting quality. In average, 20% of the connections are removed with this approach. According to the parsimony principle, these simplifications increase the generalization capacity and should allow building a robust predictor [2].

This first study done in a quasi-optimized case (7 inputs and 2 hidden neurons) precedes a more general one, where a standard MLP (more than 15 inputs and 15 hidden nodes) will be studied. Indeed, we have shown that the pruning method presented here is able to simplify the network while the performance is roughly equivalent. Applying the 2-pass approach in a 15x15 MLP should allow to optimize it without apply the exhaustive test where all the architectures are try out. Moreover, for users it is a totally transparent methodology, suitable for all TS and faster than classical optimization process. According to the conclusion of this study, it possible that results based on the 2-stage approach may be better than the classical approach based on the "1-stage" LM algorithm.

## References


[1] Voyant C, Muselli M, Paoli C, Nivet M-L. Numerical weather prediction (NWP) and hybrid ARMA/ANN model to predict global radiation. Energy 2012;39:341–55.

[2] Mellit A, Kalogirou SA, Hontoria L, Shaari S. Artificial intelligence techniques for sizing photovoltaic systems: A review. Renew Sustain Energy Rev 2009;13:406–19.

[3] Cybenko G. Approximation by superpositions of a sigmoidal function. Math Control Signals Syst 1989;2:303–14.

[4] Fan J, Pan J. A note on the Levenberg–Marquardt parameter. Appl Math Comput 2009;207:351–9.

[5] Yoo CK, Sung SW, Lee I-B. Generalized damped least squares algorithm. Comput Chem Eng 2003;27:423–31.

[6] Brusset H, Depeyre D, Petit J-P, Haffner F. On the convergence of standard and damped least squares methods. J Comput Phys 1976;22:534–42.

[7] Kreiss J-P, Paparoditis E. Bootstrap methods for dependent data: A review. J Korean Stat Soc 2011;40:357–78.

[8] Dias FM, Antunes A, Vieira J, Mota A. A sliding window solution for the on-line implementation of the Levenberg–Marquardt algorithm. Eng Appl Artif Intell 2006;19:1–7.

# Mathematical supply-chain modelling: Product analysis of cost and time

**D J Easters**

Department of Fashion Management, Marketing & Communication, School of Art & Design, Nottingham Trent University, Nottingham NG1 4BU, UK

E-mail: dj@fashioneasters.com

**Abstract**. Establishing a mathematical supply-chain model is a proposition that has received attention due to its inherent benefits of evolving global supply-chain efficiencies. This paper discusses the prevailing relationships found within apparel supply-chain environments, and contemplates the complex issues indicated for constituting a mathematical model. Principal results identified within the data suggest, that the multifarious nature of global supply-chain activities require a degree of simplification in order to fully dilate the necessary factors which affect, each sub-section of the chain. Subsequently, the research findings allowed the division of supply-chain components into sub-sections, which amassed a coherent method of product development activity. Concurrently, the supply-chain model was found to allow systematic mathematical formulae analysis, of cost and time, within the multiple contexts of each sub-section encountered. The paper indicates the supply-chain model structure, the mathematics, and considers how product analysis of cost and time can improve the comprehension of product lifecycle management.

## 1. Introduction & Background

Since 1990 [1] evolving global supply-chain efficiencies has become a key driver in the race for retail prominence [2]; this notion was highlighted by [3], who said *"modern competition has shifted from company vs company, to supply-chain vs supply-chain"*. Recent global retail players [2] have shown awardable attributes and have provided other companies with unfathomable judgments on whether to manage responsive or efficiency supply-chains, when adopting stages for product development processes [4]. However, found within other current literature, many academics believe that there is a lack of applied strategic perspective on global sourcing [5] and a lack of performance measuring techniques [6], contributing to organisations being unable to fully encompass the analysis of a complete product lifecycle [7], throughout its supply-chain conception [8].

On that account and that of the absent mathematical supply-chain model [6]; it seems appropriate to aim for a suitable solution to this required dimension, of the modern day global supply-chain strategy. It has therefore, been suggested that there is a current need for the complex supply-chain components to be reconfigured, to allow a more comprehendible and holistic overview of the activities involved [6]. Consequently, this paper ventures towards a holistic model of the apparel supply-chain, whilst giving a concurrent mathematical algorithm, to enable future performance measures to transmogrify.

## 2. Theoretical PACTUM Model©: Incorporated research data and analysis

The collected data from the various research samples was converged with the methodical approaches to model formulation outlined by [9], together with constituents of the Six Sigma DMAIC problem solving practices [10]. This convergence of data and hypotheses gave a clear and systematic approach to the steps needed for the development of such a complex synergy. Therefore, instead of arranging quotations, graphs and charts as research data, the results from the methodology used in this research is presented in a graphical model simulation. This system of graphical representation of data allowed for the variables to be consolidated and simplified. Research conducted by [9] suggested that statistical learning techniques are a valuable tool for detecting complex relationships and interaction variables, which was indeed helpful in the mathematical model conception.

These approaches to research data management and data representation enabled the mathematical assessment of *effort* encountered within the complex apparel supply-chain, which is identified in figures 4-11. Consequently, the *'Product Analysis of Cost & Time Using Mathematical Modelling' (PACTUM Model©)* method is supported by the literature, primary data collected and was found to allow systematic mathematical formulae analysis, of cost and time, within the multiple contexts of each supply-chain sub-section encountered. This has then led to a holistic mathematical formula for calculating overall and divisions of *effort* (using the indexed sub-sections found in section 2.2.6.) whilst analysing each sub-section and/or entirety of a products lifecycle, when considering cost and time as the initial calculating factors.

### *2.1. PACTUM Model© mathematical formula*

In order to understand the holistic mathematical model, the below equations describe the sum for calculating the *highest value* of total supply-chain effort:

*2.1.1. Formula 1 & 2.* The PACTUM Model© allows the calculation of *effort* (***E***), within each sub-section of the supply-chain to be measured; with the *sub-sections* of the supply-chain defined as (***i***); *costs* defined as (***c***); *time* is defined as (***t***) and the function of these sub-sections defined as (***F***):

$$E_i = F_i(c_i, t_i) \tag{1}$$

$$E_i = F_i(c_i, t_i) = \frac{1}{\sqrt{c_i^2 + t_i^2}} \tag{2}$$

*2.1.2. Formula 3.* In consideration of the holistic mathematical formula of the PACTUM Model©; formula 3 below describes how *effort* can be measured within a single products sub-section lifecycle. The values will ultimately range from *high effort* to *low effort*. Identified within indexed section 2.2.6., these values can be assigned by adding the effort values of each dependent variable:

$$\left(\sum_i E_i\right) \tag{3}$$

*2.1.3. Formula 4.* Significantly by observing the non-assumption that the effort value cannot be equalled too or larger than one, it can therefore be said that **1** is *the highest effort value* achievable. Most importantly, the lowest value assigned can only be determined by the nature of the organisations own differential supply-chain functions which must be defined by the model adopter, according to their own operational cost values, time frames and subsequent effectiveness:

$$E_i = \frac{1}{\sqrt{c_i^2 + t_i^2}} \tag{4}$$

### *2.2. PACTUM Model© graphical illustrations*

Represented in a simple yet intuitive manner, by figure 4, the entire formation of the PACTUM Model© shows the converged functions, variables and relationships found within global supply-chains and product development process. However, it could be argued that this simplified representation of such complex synergies is lacking more problematical ascendants of current global supply-chain functions. To address this point, it has been considered that by the very nature of such complexities, the illustrated PACTUM Model© allows model adopters a more fluid approach, to supply-chain analysis, when considering their own differential organisational structures and product development strategies; Hence, allowing the variable supply-chain functions to be customised rather than prescribed.

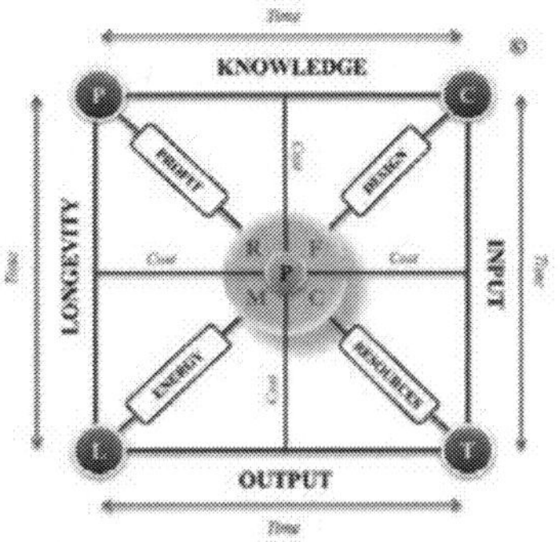


**Figure 4.** Holistic overview of PACTUM Model© The central point being (P) represents the actual product being developed.

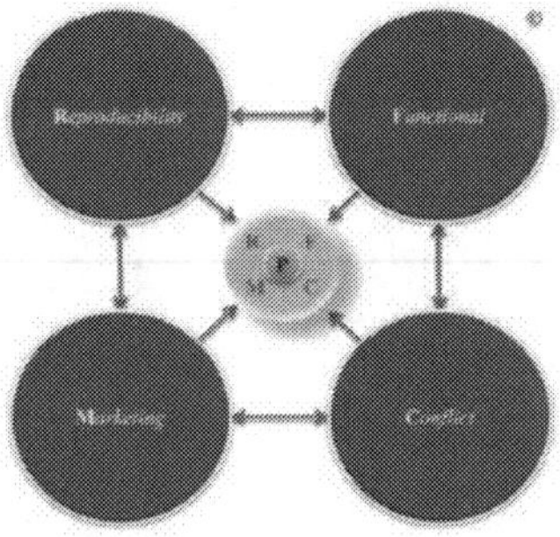

**Figure 5.** Illustration showing the four main independent variables of the supply-chain that affect the product during its entire lifecycle.

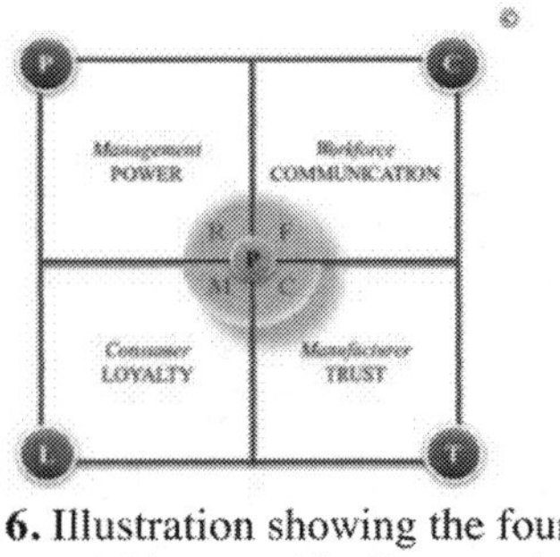


**Figure 6.** Illustration showing the four main dependent variables specifically associated with relationships and job functions of the supply-chain that are necessary to produce and sell a product during its lifecycle.

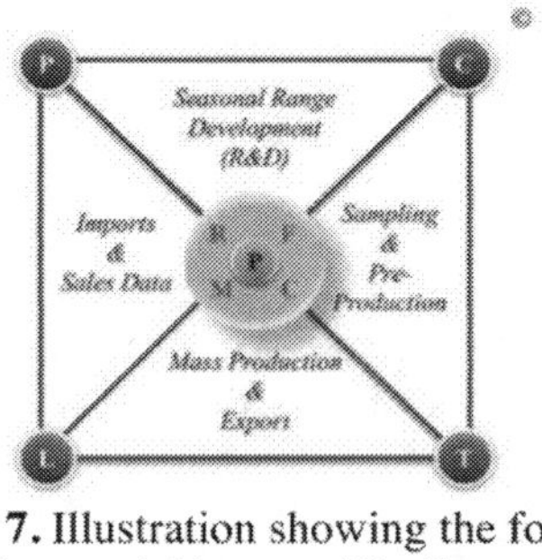


**Figure 7.** Illustration showing the four main dependent variables specifically associated with the processes found within the supply-chain that are necessary to produce and sell a product during its lifecycle.

*2.2.1. PACTUM Model©: Dependent and independent variables.* If we consider that figures 6 & 7 were laid on top of one another, it would be possible to appreciate that both the management and workforce have elements of human functional capabilities and engagement factor levels within the *seasonal range development* stages of the products lifecycle; simultaneously the workforce and manufacturer both have elements of human functional capabilities and engagement factor levels within the *sampling and pre-production* stages; the manufacturer and consumer both have elements of human functional capabilities and engagement factor levels within the *mass production and export* requirements; and the consumer and the management both have elements of human functional capabilities and engagement factor levels within the *import* necessity and subsequent *sales data*.

At this stage it is worth pointing out that the PACTUM Model© is able to uniquely divide the purpose of each supply-chains functional capabilities conducted; allowing an adopter of this model to define their own specific supply-chain practices (by using the suggested index in section 2.2.6. as guidance) and then tailor the functional elements/values of each sub-section, to their own organisational structure and product development strategies.

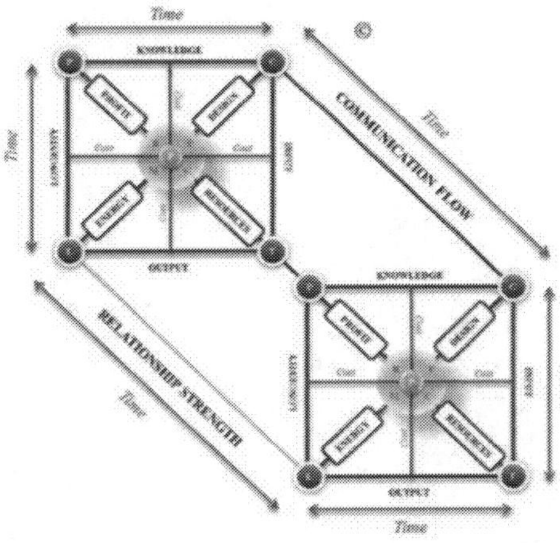


**Figure 8.** Illustration showing how the PACTUM Model© can be configured to represent a supply-chain of numerous organisational components and demonstrates the relationship of retailer and supplier engagement; with the top PACTUM Model© representing the calculations attributed to Primary Retail organisations and with the bottom PACTUM Model© representing calculations attributed to Secondary Supplier organisation.

*2.2.2. PACTUM Model©: Dependent effort functions.* Written around the edge (vertically and horizontally) of figure 4 are the *dependent effort functions: Knowledge, Input, Output* and *Longevity*. These dependent effort functions are required in order to fulfill the variables outlined in figures 5, 6 and 7.

*2.2.3. PACTUM Model©: Core dependent product functions.* Illustrated by figure 4 are the core product functions that all other previously mentioned variables and functions are dependent upon, with these being: *Design, Resources, Energy and Profit*. These core dependent product functions are placed within the sub-sections of the PACTUM Model© that represent their attributed independent and dependent variables and functionalities as formerly illustrated and discussed.

*2.2.4. PACTUM Model© sub-section index system.* The following sections of this paper shows how the PACTUM Model© uses graphical representation, of each sub-section, to define the calculations by using a right-angled triangle to represent the variable values. For example, within the calculus of the *effort value* for the *knowledge/design* (***d***) sub-section, this is defined as ($E_i$), costs as ($c_i$) and time as ($t_i$); and similarly within the *input/design* (***s***) sub-section and so on and so forth for all other sub-sections. Shown in the PACTUM Model© diagram (figure 4), the two right angles of the triangle denote either the sub-section(s) *functional costs* ($c_i$) or the sub-section(s) *functional time* ($t_i$) values, found within the supply-chain parameters. Consequently, the hypotenuse of the triangle (noted as not being of significant numerical / financial value) does however, allow for a visual and mathematical calculation of *effort* ($E_i$); which provides the hypotenuse a value which significantly correlate to that of the values assigned to the *costs* and *time* of the sub-section functionalities.

Accordingly, the man-made index (shown in section 2.2.6.) displays the variables relating to the main sub-sections of the supply-chain; these variables have been identified through the primary and secondary research. It should be noted that the variables within each of the eight indexed sub-sections, would inevitably be commensurate with organisational structure and specific supply-chain networks, yet there is sufficient flexibility to allow model adopters to enter their own differential financial and time frame calculations.

*2.2.5. PACTUM Model© sub-section index.* In relation to figure 9, the index below describes the arbitrary values and supply-chain functions which assess the *effort* grading system. The index also indicates the main titles given to the individual sub-sections which have been divided by using the supply-chains dependent variables:

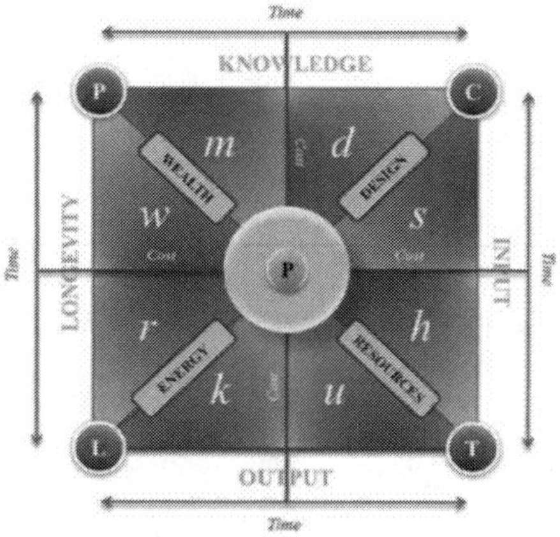


*2.2.6. Sub-section index.*

$\boldsymbol{m}$ = Primary management
$\boldsymbol{d}$ = Primary R&D
$\boldsymbol{s}$ = Primary pre-production
$\boldsymbol{h}$ = Secondary pre-production
$\boldsymbol{u}$ = Secondary mass production
$\boldsymbol{k}$ = Third party logistics
$\boldsymbol{r}$ = Primary logistics
$\boldsymbol{w}$ = Primary product sales data
$\boldsymbol{c}$ = Costs
$\boldsymbol{t}$ = Time
$\boldsymbol{E}$ = Effort

**Figure 9.** Illustration of how the PACTUM Model© can visually show the highest *effort* values, assigned to the *costs* and *time* of each sub-section of the supply-chain encountered.

*2.2.7. PACTUM Model© cost and time formulas.* Using the model representation in figure 9 the PACTUM Model© is able to assess the defined indexed sub-sections, either by individual products, specific product categories, product ranges and total product output of an organisation.

Shown in figure 9 and using the index (in section 2.2.6.), it should be noted that the axis of the triangle associated with *costs* ($\boldsymbol{c}$) is the actual monetary value (£) and changes according to the assessment required. For example: an individual product category may have a value associated with the total category budget. Similarly the axis of the triangle associated with *time* ($\boldsymbol{t}$) is the actual time frame value which changes according to the assessment required for the period. Therefore these values of *costs* and *time* are variable due to the nature of the predetermined product assessment and the nature of the organisation encountered.

Subsequently, the *effort total* ($\boldsymbol{E_T}$) calculus for the supply-chain is dependent on the differential variables attributed to each organisation assessed and is expressed in formula 5. It is worthy of pointing out however, that these sub-section indexed variables may require high levels of *cost* and *time* in order to perform the functions required by each sub-section depending on the type of product being developed and supply-chain networks used. Illustrated by figures 10 & 11, the graphical PACTUM Model© simulations show how different product lifecycle practices can highlight variable degrees of *high effort* and *low effort* values, within each specific sub-section indexed.

$$E_T = \sum_i E_i = \sum_n \frac{1}{\sqrt{c_i^2 + t_i^2}} \qquad (5)$$

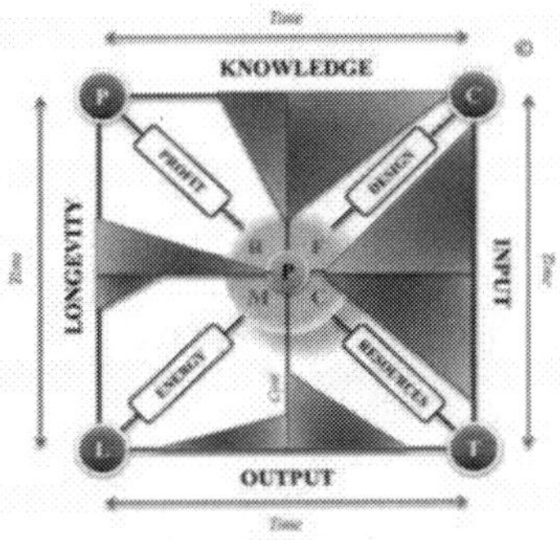


**Figure 10.** Illustration and an example of how the PACTUM Model© can visually show the effort values, of cost and time, aligned to each sub-sections of the supply-chain.

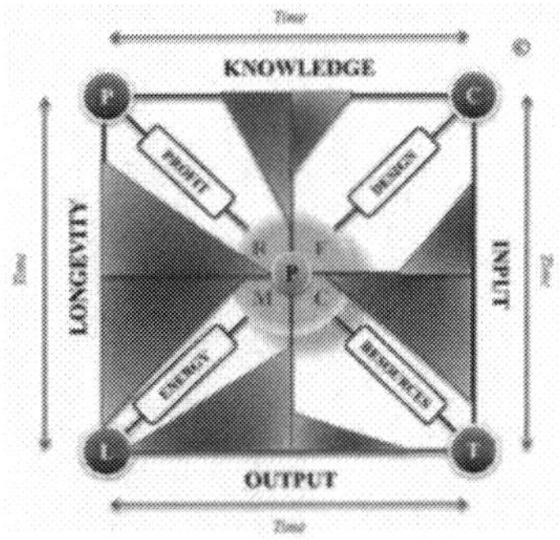


**Figure 11.** Illustration and an example of how the PACTUM Model© can visually show the effort values, of cost and time, aligned to each sub-sections of the supply-chain.

## 2. Conclusion & Recommendations

The methodical approaches to model formulation outlined by [9] and [10] were found to give a clear starting point and systematic approaches to the steps needed for the development of such a complex synergy. Thus, by highlighting the most significant *core dependent product functions, dependent effort functions, dependent variables* and *independent variables* found to be used within current supply-chain strategies; the PACTUM Model© represented by figure 4 allows potential model adopters to divide their differential organisational *functions, practices, employment roles and overheads,* into structured indexed sub-sections (considering their own specific cost and time values for each indexed sub-section). Significantly, this new method for calculating and comprehending product lifecycle functional capabilities within global fashion supply-chains; does allow a multitude of future research avenues which could provide accomplished academics and industry specialists incentive to develop collaborative links in order to produce a computer generated version of the PACTUM Model© (Intellectual Property rights apply).

Conclusively, the mathematical formulas will enable some established business leaders and academics to address the holistic nature of organisational variables, attributed to the *effort values* found within differential supply-chain environments and product development practices; to be calculated, analysed, evaluated, re-configured and critiqued in more depth than previously envisaged.

## References

[1] Tyler D, Heely J, Bhamra T 2006 *J. Fash. Mktg. & Mgt.* **10** 316
[2] Hayes S, Jones N 2006 *J. Fash. Mktg. & Mgt.* **10** 282
[3] Caridi M, Perego A, Tumino A 2013 *An Int. J.* **20** 25-6
[4] Lam J K C, Postle R 2006 *Int. J. Clthg. Sci. & Tech.* **18** 266
[5] Trent R, Monczka R M 2003 *Int. J. Phys. Dist. & Logs. Mgt.* **33** 609-27
[6] Lee Y, Kincade D H 2003 *J. Fash. Mktg. & Mgt.* **7** 31-2
[7] Easters D J 2011 *Int. J. Fash. Des. Tech & Edu.* **5/1** 135-43
[8] Mattila H, Kind R, Ojala N 2002 *J. Fash. Mktg. & Mangt.* **6** 340-51
[9] Wilson C, Wilson R 2012 *Tex. Inst. J.* **3** 26-8
[10] De Mast J, Lokkerbol J 2012 *Int. J. Production Economics* **139** 604-14

# Interaction between a drift and a fractional power of a Laplacian in semi-group theory

**Rémi Léandre**

Laboratoire de Mathématiques. Université de Franche-Comté. Route de Gray. 25030. Besançon. FRANCE

E-mail: Remi.Leandre@univ-fcomte.fr

**Abstract.** We give an interaction between a drift and a fractional power of a degenerated Laplacian such that the involved semi-group has a density by using the Malliavin Calculus for boundary processes translated by ourself in semi-group theory in [1].

## 1. Introduction and statement of the main theorem

We consider $m$ vector fields on $R^d$ with bounded derivatives at each order $X_1, .., X_m$ and the diffusion generator

$$L = \frac{\partial}{\partial s} + 1/2 \sum_{i>0} (X_i)^2 \tag{1}$$

on $R^{d+1}$. We could add a drift in (1), but it is done to simplify the proof. We consider a vector field on $R^d$ $D$ with bounded derivatives at each order. Bismut [2] considers the generator

$$A = D - 1/2\sqrt{-2L} \tag{2}$$

For the theory of fractional powers of Laplacian, we refer to the book of Yosida [3]. Let us recall quickly its definition. $L$ generates a semi-group $P_s$ acting on bounded continuous functions $f$ on $R^{d+1}$:

$$\frac{\partial}{\partial t} P_t f(s, x) = L P_t(s, x) \tag{3}$$

Then

$$\sqrt{-2L} = C \int_0^\infty s^{-3/2}(P_s - I) ds \tag{4}$$

$A$ generates a Markovian semi-group $\exp[tA]$ acting on continuous functions $f$ on $R^{1+d}$:

$$\frac{\partial}{\partial t} \exp[tA] f(s, x) = A \exp[tA] f(s, x) \tag{5}$$

There is a stochastic representation of this semi-group (See [2]). Let $(B_1, .., B_m, z_t)$ be a Brownian motion on $R^{m+1}$ starting from the origin. Let $L_t$ be the local time associated to

$z_t$ ([4]) and $A_t$ its right inverse process. We introduce the stochastic differential equation in Stratonovitch sense issued from $x$:

$$dx_t = \sum_1^m X_i(x_t)dB_i + D(x_t)dL_t \tag{6}$$

We consider the subordinated process $(x_{A_t}, A_t + s)$. Unlike $x_t$, this process is not continuous but is still a Markov process. We have the main relation

$$\exp[tA]f(s,x) = E[f(A_t + s, x_{A_t})] \tag{7}$$

This paper follows the probabilistic intuition which comes from this stochastic representation of the semi-group. But in [2] and [5], only stochastic differential equations appear which explain that we can expulse the probabilistic language of [2] and [5].

The natural question is to know if the semi-group has an heat-kernel:

$$\exp[tA]f(s,x) = \int_{R^{1+d}} q_t(s,x,s',x')f(s',x')ds'dx' \tag{8}$$

This problem was solved by Bismut by using the Malliavin Calculus and a stochastic representation of it ([2]) in the elliptic case. We applied Bismut's technic to state an Hoermander theorem for fractional powers of Laplacians in [5] We have translated Malliavin Calculus of Bismut for Boundary processes in semi-group theory in [1] and state a regularity result for the semi-group associated to $A$ in the elliptic case. We do now an Hoermander type hypothesis. We put

$$G_1 Y = Y \tag{9}$$

$$G_l Y = \cup_{i \geq 0} \cup_{Z \in G_{l-1}} ([Z, X_i]) \cup G_{l-1} Z \tag{10}$$

We put:

$$E_l = \cup_{j \leq l} \cup_{i > 0} (G_j X_i) \tag{11}$$

The following theorem was proved in [5] by using the Calculus of Boundary Process of Bismut. We prove it again by using the Malliavin Calculus of Bismut type in semi-group theory of [1]:

**Theorem 1** Let us suppose that the uniform Hoermander's hypothesis is checked:

$$\inf_{x \in R^d, \|f\|=1} \sum_{Y \in E_{l_0}} < Y, f >^2 + < [D, Y], f >^2 > C > 0 \tag{12}$$

Then the heat-kernel on $R^{d+1}$ $q_t(0,x,s,y)$ exists.

**Remark:**It should be possible to show that $(s,y) \to q_t(0,x,s,y)$ is smooth.

**Remark:**It is possible to replace (12) by the general hypothesis (3.5) of [5].

## 2. The main ingredient of the proof

Let $E_d = R^{1+d} \times G_d \times M_d$ where $G_d$ denotes the set of invertible matrices on $R^d$ and $M_d$ the set of symmetric matrices on $R^d$. $(s,x,U,V)$ is the generic element of $E_d$. $V$ is called the Malliavin matrix.

On $E_d$ we consider the vector fields:

$$\hat{D} = (0, D, DD(x)U, 0) \tag{13}$$

$$\hat{X}_i = (0, X_i, DX_i(x)U, 0) \tag{14}$$

$$\hat{Y} = (0,0,0,\sum_i^m < U^{-1}X_i, . >^2) \tag{15}$$

We consider the Malliavin generator $\hat{L}$ on $E_d$:

$$\hat{L} = \frac{\partial}{\partial s} + 1/2\sum_{i=1}^m (\hat{X}_i)^2 + \hat{Y} \tag{16}$$

and the square root associated $\sqrt{-\hat{L}}$. This semi-group on this bigger space is defined according the line of (5).

We consider

$$\hat{A} = \hat{D} - 1/2\sqrt{-2\hat{L}} \tag{17}$$

and the Malliavin semi-group $\exp[t\hat{A}]$.

An adaptation of one of main result of Léandre [1] is the following:

**Theorem 2** Let us suppose that the Malliavin condition is checked: for all $p \in N$, all $s > 0$

$$\exp[t\hat{A}][V^{-p}1_{[0,s_0]}](0,x,I,0) < \infty \tag{18}$$

then

$$\exp[tA]f(0,x) = \int_{R^{1+d}} f(s,y)q_t(s,y)dsdy \tag{19}$$

where $q_t(s,y) \geq 0$.

**Remark:**The proof follows the proof of Theorem 2.1 of [1]. Following the general strategy of the Malliavin Calculus, it is enough to show the theorem to get integration by parts formulas. If $f$ is with compact support

$$|\exp[tA](df)(s,x)| \leq C\|f\|_\infty \tag{20}$$

where $C$ depends only from the support of $f$ and $\|f\|_\infty$ denotes the supremum norm of $f$. For that, we integrate by parts under the underlying diffusion $P_s$ and the Brownian motions $B_i$ as in part 3 of [1]. This allows to remove the space derivatives of $f$. In order to remove the time derivatives in $df(s,x)$, we integrate by parts on the subordinators $A_t$ as it was done in part 4 of [1]. The main difference with [1] is that $\hat{D}$ appears when we take the variation of the subordinated semi-group. It is the only change in the abstract theorem of part 5 of [1]. When we get this abstract theorem, the drift $D$ will appear another time in the inversion of the Malliavin matrix $V$.

## 3. Inversion of the Malliavin matrix in semi-group theory

Let be

$$F_l(x,U,\xi) = \sum_{Y \in G_l} < U^{-1}Y(x), \xi >^2 \tag{21}$$

where $\xi$ is of modulus one. A simple adaptation of Lemma 3 of [6] shows:

**Lemma 3** Let us suppose that

$$\exp[t_0\hat{A}][1_{[0,s_0]}; F_l(I,.,\xi) > Ct_0^\alpha](0,x_0,U_0,0) > C > 0 \tag{22}$$

for all $x_0 \in R^d$, $\|U_0\| < t_0^{-\epsilon}$ for a small $\epsilon$ and some positive $\beta$. Then (22) remains true on an interval of length $t_0^\beta$ for another $\beta$.

Since $\hat{A}$ is Markovian, $\exp[t\hat{A}]$ is represented by a stochastic process $X_t$ following the same line of the representation of $\exp[tA]$ by a stochastic process. $X_t$ is a Markov jump process. It has a Levy measure [7]. The main remark is the following: if the Levy measure of a jump process is enough concentrated in small jumps, there are a lot of small jumps.We get:

**Definition 4** If $f$ is a function from $R^+$ into $R^+$, we consider the Levy measure associated with the Malliavin matrix where $\xi$ is of norm 1:

$$\mu_\xi(f) = C\int_0^{s_0} \frac{ds}{s^{3/2}} \hat{P}_s[f(V'(\xi) - V(\xi))](0, x, U, V) \tag{23}$$

We recall (See Lemma 3 of [6]):

**Lemma 5** Let us suppose that $F_1(x, U, \xi) \geq \rho$ for $|U| + |U^{-1}| < \rho^{-\epsilon}$ for some small $\epsilon$. Then $\mu_\xi[z > \rho^\alpha] \geq C\rho^\beta$ for some positive $\alpha$ and some negative $\beta$.

The theorem will follow as in the proof of Theorem 1 (38), (39), (40) in [6] if we show the next proposition:

**Proposition 6** Let us suppose that $|U| + |U^{-1}| < \rho^{-\epsilon}$ for some small $\epsilon$ . There exists $\alpha$ such that

$$\exp[\rho^\alpha \hat{A}][F_1(x, U, \xi) \geq \rho; 1_{[0,s_0]}](0, x, U, 0) > C > 0 \tag{24}$$

We can state an analog of Lemma 2 of [6]:

**Lemma 7** Let us suppose that

$$\exp[\rho^\alpha \hat{A}][F_l(x, U, \xi) \geq \rho^\beta; 1_{[0,s_0]}](0, x_0, U_0, 0) > C > 0 \tag{25}$$

where $|U_0| + |U_0^{-1}| < \rho^{-\epsilon}$ for some small $\epsilon$.. Then (25) remains true for $l-1$ for others $\alpha$ and $\beta$.

By using this lemma, it is enough to show the following propopsition in order to show Proposition 6:

**Proposition 8** If we take $l_0 + 2$, (25) is checked if $|U| + |U^{-1}| < \rho^{-\epsilon}$ for some small $\epsilon$.

**Proof of proposition 8** Let us suppose that

$$F_{l_0}(x_0, U_0, \xi) \leq \rho^\beta \tag{26}$$

and

$$F_{l_0+2} \leq \rho^{\beta_1} \tag{27}$$

By Hypothesis (12 ), we can find a $Y \in E_{l_0}$ such that

$$< [D, Y](x_0), \xi >> C > 0 \tag{28}$$

We choose $C > 0$ to simplify the exposition and we choose $\epsilon = 0$ in order to simplify the expostion of the proof.

We remark that if $u < t$ and if $\gamma < 1/2$ that

$$\exp[u\hat{A}][1_{[t^\gamma,\infty[}](0, x_0, U_0, 0) \leq Ct^r \tag{29}$$

for some $r > 0$. So it is enough to estimate

$$\exp[u\hat{A}][F_{l_0}(.,.,\xi) > \rho^\alpha; 1_{[0,t^\gamma]}](0, x_0, U_0, 0) \tag{30}$$

for some well choosed $\alpha$ We put

$$G(x, U, \xi) =< U^{-1}Y, \xi > g(\frac{F_{l_0+2}(x, U, \xi)}{\rho^{\beta_1}}) \tag{31}$$

where $g$ is a smooth function from $R^+$ into $[0,1]$ equals to 1 in a neighborhood of 0 and to 0 in a neighborhood of the infinity.

Let us suppose that

$$|x-x_0|+|U-U_0|<C\rho^{\beta/2} \tag{32}$$

Let us estimate for $s'\leq t^\gamma$

$$\sqrt{-\hat{L}}[G(.,.,\xi)1_{[0,t^\gamma]}](s',x,U,0) \tag{33}$$

For that we look at

$$f(s)=\hat{P}_s[G(.,.,\xi)1_{[0,t^\gamma]}](s',x,U,0) \tag{34}$$

where $\hat{P}_s$ is the semi-group generated by $\hat{L}$.

$$f'(s)=\hat{P}_s[\hat{L}[G(x',U',\xi)1_{[0,t^\gamma]}]](s',x,U,0) \tag{35}$$

We distinguish if

$$|x-x'|+|U-U'|<C.\rho^{\beta/2} \tag{36}$$

or not. If yes

$$|\hat{L}[G(x',U',\xi)1_{[0,t^\gamma]}]|\leq\rho^{\beta/2}1_{[0,t^\gamma]} \tag{37}$$

If not we remark that

$$\hat{P}_s[|x-x'|+|U-U'|\geq\rho^{\beta/2}](s',x,U,0)\leq Cs\rho^{-\beta} \tag{38}$$

In conclusion, we deduce that

$$|\sqrt{-\hat{L}}[G(.,.,\xi)1_{[0,t^\gamma]}](s',x,U,0)|\leq\rho^{\beta/2}t^{\gamma/2}+\rho^{-\beta}\rho^{-2\beta_1}t^{\frac{3\gamma}{2}} \tag{39}$$

Let us consider the case

$$|x-x_0|+|U-U_0|>C\rho^{\beta/2} \tag{40}$$

In such a case

$$|f'(s)|\leq C\rho^{-2\beta_1}1_{[0,t^\gamma]} \tag{41}$$

Therefore

$$|\sqrt{-\hat{L}}[G(.,.,\xi)1_{[0,t^\gamma]}](s',x,U,0)|\leq C\rho^{-\beta_1}t^{\gamma/2}1_{[0,t^\gamma]} \tag{42}$$

On the other hand

$$|\sqrt{-\hat{L}}[|U'-U_0|^2+|x'-x_0|^2;1_{[0,t^\gamma]}](s',x,U,0)|$$
$$\leq A1_{[0,t^\gamma]}\int_0^{t^\gamma}ds\frac{s}{s^{3/2}}\leq Ct^{\gamma/2}1_{[0,t^\gamma]} \tag{43}$$

This shows that

$$exp[u\hat{A}][|U-U_0|^2+|x-x_0|^2;1_{[0,t^\gamma]}](0,x_0,U_0,0)\leq Cut^{\gamma/2} \tag{44}$$

Therefore

$$exp[u\hat{A}][|U-U_0|+|x-x_0|>\rho^{\beta/2};1_{[0,t^\gamma]}](0,x_0,U_0,0)\leq ut^{\gamma/2}\rho^{-\beta} \tag{45}$$

By putting all together, we deduce that if $\gamma<1/2$

$$|\exp[u\hat{A}][\sqrt{-\hat{L}}[G(.,.,\xi)1_{[0,t^\gamma]}]](0,x_0,U_0,0)|$$
$$C(\rho^{\beta/2}t^{\gamma/2}+\rho^{-\beta}\rho^{-2\beta_1}t^{\frac{3\gamma}{2}}+ut^{\gamma/2}\rho^{-2\beta_1}\rho^{-\beta}) \tag{46}$$

Let us now estimate

$$\exp[u\hat{A}][\hat{D}G(x,U,\xi)1_{[0,t^\gamma]}](0,x_0,U_0,0) \tag{47}$$

We suppose first of all that

$$|x-x_0|+|U-U_0|<C\rho^{\beta/2} \tag{48}$$

In such a case we have a lower bound in $C>0$ of the expression. If the previous inequality is not checked we have an estimate by using the previous considerations in $Cut^{\gamma/2}\rho^{-\beta_1}\rho^{-\beta}$ By using the semi-group property for $\exp[u\hat{A}]$ we deduce that if $u<t^\gamma$

$$g'(u)=\frac{\partial}{\partial u}\exp[u\hat{A}][G(.,.,\xi)1_{[0,t^\gamma]}](0,x_0,U_0,0)\geq$$
$$C-Ct^{3\gamma/2}\rho^{-\beta_1}\rho^{-\beta}-C(\rho^{\beta/2}t^{\gamma/2}+\rho^{-\beta}\rho^{-2\beta_1}t^{\frac{3\gamma}{2}} \tag{49}$$

We distinguish if

$$|g(0)|<C\rho^{3\beta} \tag{50}$$

or not. If (50) is not checked, we can apply Lemma (7). If not we have $\beta^1<\beta$. We deduce that

$$g(t)\geq Ct-C\rho^{3\beta}-C\rho^{-\beta}\rho^{-2\beta_1}t^{1+3\gamma/2}-C\rho^{\beta/2}t^{1+\gamma/2} \tag{51}$$

We choose $t=C_1\rho^\beta$ for a big $C_1$.From (51), we deduce that

$$g(C_1\rho^\beta)\geq C\rho^\beta-C\rho^{-\beta}\rho^{-2\beta_1}\rho^{(\beta)(1+3\gamma/2)} \tag{52}$$

to choose $\gamma$ close from $1/2$ and $\beta_1$ very small in order to deduce that

$$g(C_1\rho^\beta)\geq C\rho^\beta \tag{53}$$

We deduce that

$$\exp[C_1\rho^\beta\hat{A}][<U^{-1}Y(x),\xi>>C\rho^\beta](0,x_0,U_0)>C>0 \tag{54}$$

Therefore the result holds.

◇.

**Remark:** The principle of the proof is very simple: we establish a criterium in order to show that the Levy measure associated to $V(\xi)$ is very concentrated in small jumps. If the Levy measure is very concentrated in small jumps, there are a lot of small jumps which obliges that $V(\xi)$ to be not very small. If this criterium is not satisfied, another criterium will obliged it to be satisfied. In [6], this criterium comes from the interaction between two Levy measures. Here it come from the interaction between a Levy measure and the drift $D$.

## References


[1] Léandre R 2011, *Inter. Jour. of Diff. Equ.*. article ID 575383.
[2] Bismut J.M. 1983 *Annales Scientifiques E.N.S.* **17** 507.
[3] Yosida K 1977 *Functional analysis* (Berlin: Springer).
[4] Ikeda N and Watanabe S 1981 *Stochastic differential equations and diffusions processes* (Amsterdam: North-Holland)
[5] Léandre R 1984 *Une extension du théoreme de Hoermander a divers processus de sauts* ( PHD Thesis, Besançon, France: Université de Franche-Comté).
[6] Léandre R 2012 *XII International Carpathian control conference (Podbanske)*, ( IEEE ISBN 978-1-4577-1866-3; IEEE: Xplore), p 421.
[7] Ishikawa Y 2013 *Stochastic Calculus of variations for jump processes* (Berlin: De Gruyter)

# A heterogeneous fleet vehicle routing model for solving the LPG distribution problem: A case study

**S Onut, M R Kamber, G Altay**
Yildiz.Technical University, Mechanical Faculty, Department of Industrial Engineering, 34349, Besiktas, Istanbul, Turkey

E-mail: onut@yildiz.edu.tr, mkamber@yildiz.edu.tr, altaygokan@hotmail.com

**Abstract:** Vehicle Routing Problem (VRP) is an important management problem in the field of distribution and logistics. In VRPs, routes from a distribution point to geographically distributed points are designed with minimum cost and considering customer demands. All points should be visited only once and by one vehicle in one route. Total demand in one route should not exceed the capacity of the vehicle that assigned to that route. VRPs are varied due to real life constraints related to vehicle types, number of depots, transportation conditions and time periods, etc. Heterogeneous fleet vehicle routing problem is a kind of VRP that vehicles have different capacity and costs. There are two types of vehicles in our problem. In this study, it is used the real world data and obtained from a company that operates in LPG sector in Turkey. An optimization model is established for planning daily routes and assigned vehicles. The model is solved by GAMS and optimal solution is found in a reasonable time.

**Keywords:** Vehicle routing; Heterogeneous fleet; Distribution; LPG sector

## 1. Introduction

Tremendous advance of production, transportation and communication means eased to access knowledge, technology and resulted in a rough competition among companies. In order to survive in this competition, companies should carry out their activities more efficiently, faster and less costly. They examine their processes and try to eliminate unnecessary costs. Distribution of goods is a kind of such activities. Companies make studies and investments in this field to get the processes under control.

Increased globalization and offshore sourcing has created a massive increase in the complexity of supply lines and transportation networks. This has led firms to give more importance to the transportation and distribution function and its associated long term design decisions. Essentially transportation is tried to consider as a whole process to benefit from the all modes of the transportation. Transportation decisions include transportation mode selection, shipment size, vehicle routing, and scheduling, all of which are directly related to the location of transshipment points, warehouses, customer and factories [1].

According to the time horizon for the problems considered in the transportation decisions literature, the studies can be gathered under the strategic, tactical and operational decision levels. In this study we concentrated on the operational decision level. Operational level consists of daily and real time decisions and requires transactional data. Operational (short-term) decisions are performed in a highly dynamic environment where the time factor plays an important role. It includes the routing and dispatching of vehicles and crews; the allocation of scarce resources.

In a vehicle routing problem (VRP) a fleet of m vehicles is available to visit the customers. Vehicle k has a capacity of $Q_k$ and the duration of its route cannot exceed $D_k$. The VRP consists in designing m vehicle routes of minimum total cost such that (i) each route starts and ends at the depot; (ii) each customer is visited by exactly one route; and (iii) the total demand of the customers visited by route k does not exceed $Q_k$ and its duration does not exceed $D_k$. The VRP is a hard combinatorial optimization problem. [2]

In the VRP, while most solution methods have assumed a single depot and a homogeneous fleet, real-world problems usually include multiple terminals and a finite set of vehicles with non-uniform capacity. Moreover, the best solution for the VRP that minimizes the total distance traveled by the vehicles to visit all the customers often includes undesirable high waiting times. A more suitable VRP cost function should be a combination of fixed vehicle utilization costs and variable operational costs with the latter ones including distance and travel times, waiting time and service time costs [3].

The heterogeneous fleet vehicle routing problem (HVRP) is a problem of simultaneously determining the composition and routing of a heterogeneous fleet of vehicles in order to service a pre-specified set of customers with known delivery demands from a central depot. The number of vehicles of each type is assumed to be unlimited. The HVRP consists of designing a set of vehicle routes, each starting and ending at the depot, and such that each customer is visited exactly once, the total demand of a route does not exceed the capacity of the vehicle assigned to it, and the total cost is minimized [4].

The HVRPs have been studied since mid 80s. Some papers published in this field are [5], [6], [7], [8].

The paper is organized as follows. Section 2 presents the heterogeneous fleet vehicle routing model for a LPG distribution company. The results of the proposed model are reported in section 3 and also Section 3 concludes the paper.

## 2. A heterogeneous fleet vehicle routing model for a LPG distribution company

Distribution of materials is a challenging problem faced by major industrial and construction companies. In the same manner, in our study it is used the real world data for this kind of distribution or routing problem and obtained from a company (we called as ABC Company) that operates in Liquefied Petroleum Gas (LPG) sector in Turkey. ABC Company is one of the biggest distributors of LPG sector in Turkey. The logistics department of the company can see the current gas levels of the stations by software, and calculates the daily demands. The distribution routes are determined daily, heuristically. The purpose of this study is finding optimal routes for the daily demands of 13 stations in city X. The optimization model is constructed considering customer demands, station locations, vehicle types and vehicle capacities. Optimization model is solved by GAMS for the real data for the demands in the 2nd of April 2012 obtained from the company.

Tanks are used for distributing LPG to the stations in the company. There are two types of tanks, namely T1 and T2. T1 and T2 have a capacity of 10.5 and 15 tones of LPG, respectively. There are 9 T2 tanks and one Type-1 tank in the fleet of the company.

The real data of daily routes and their costs for the previous days are taken from the company first. Then the variable costs of the roads between stations are determined. Also, the coordinates of the stations are taken from the company and are used as inputs, and the distances between stations are calculated by using the website [9]

The main assumptions of the model are as follows. It is assumed that there is no time restriction. Traffic disruptions are ignored. The same unit costs are determined for both directions between stations. A station may be visited by more than one vehicle. The transportation cost per kilometer for vehicle T1 is 70 cents while it is 100 cents for vehicles T2. The company owns the tanks, so there is no difference in fixed cost whether the vehicles are assigned to any route or not.

The depot is assigned to the first node, and each customer is also assigned to a node. The number of customers visited is limited to 3 because the company aims to distribute more gas to any station at one visit.

The notations used and the mathematical model of the problem is as follows:

Notations:

- $c_{ijk}$ : the variable cost of the arc between the node i and node j
- $m_k$ : the unit cost of transportation for the kth vehicle (cent/km)
- $u_{ij}$ : the distance between the node i and node j
- $x_{ijk}$ : 0-1 variable for the usage of the arc between the nodes i and j by kth vehicle
- $Q_k$ : the capacity of kth vehicle
- $d_j$ : demand of the node j
- $e_i$ : demand of the node i

Objectives:

$$min \sum_{i=1}^{14} \sum_{j=1}^{14} \sum_{k=1}^{10} x_{ijk} * c_{ijk} \quad (1)$$

Constraints:

$$\sum_{k=1}^{10} \sum_{j=1}^{14} x_{ijk} = 1, \quad \text{for } i = \{2,3,4,\ldots,14\} \quad (2)$$

$$\sum_{k=1}^{10} \sum_{i=1}^{14} x_{ijk} = 1, \quad \text{for } i = \{2,3,4,\ldots,14\} \quad (3)$$

$$\sum_{j=1}^{14} x_{1jk} = 1, \quad \text{for } \forall k \quad (4)$$

$$\sum_{i=1}^{14} x_{i1k} = 1, \quad \text{for } \forall k \quad (5)$$

$$\sum_{i=1}^{14}\sum_{j=1}^{14} x_{ijk} * d_j \leq Q_k, \qquad \text{for } \forall k \tag{6}$$

$$\sum_{i=1}^{14}\sum_{j=1}^{14} x_{ijk} * d_j = \sum_{i=1}^{14}\sum_{j=1}^{14} x_{ijk} * e_i, \qquad \text{for } \forall k \tag{7}$$

$$x_{ijk} + x_{jik} \leq 1 \; for \; i = \{2,3,4,\dots,14\}, j = \{2,3,4,\dots,14\} \; and \; \forall k \tag{8}$$

$$\sum_{i=2}^{14}\sum_{j=2}^{14} x_{ijk} \leq 2, \qquad \text{for } \forall k \tag{9}$$

$$x \in \{0,1\} \tag{10}$$

$$c_{ijk} = m_k * u_{ij}, \qquad \text{for } \forall i,j \; ve \; k \tag{11}$$

Equation (1) is the objective function. The objective is to minimize the total transportation cost. Equation (2) and equation (3) state that each customer should be visited once by any vehicle if the customer assigned to route of that vehicle. Equation (4) and equation (5) ensure that each vehicle assigned to any route should leave and arrive at depot once. Equation (6) is the capacity constraint for each vehicle. Equation (7) forbids the allocation of the same node by different vehicles. Equation (8) prevents using any assigned route in reverse direction. Equation (9) states that at most 3 customers should be visited by any vehicle. Equation (10) says that x is a binary variable. Equation (11) states that the total transportation cost between the node i and the node j for vehicle k is equal to multiplication of the unit transportation cost of the vehicle and the distance of between the nodes.

### 3. Results and conclusions

The model of the problem is solved optimally by using GAMS software in about 15 minutes. The total cost is calculated as 47791.4 cents. One type-1 vehicle and six type-2 vehicles are assigned to the routes. The routes for the vehicles are as follows:

Vehicle 1 (type 1): 1-11-9-1
Vehicle 2 (type 2): 1-14-7-1
Vehicle 3 (type 2): 1-6-13-1
Vehicle 4 (type 2): 1-12-4-2-1
Vehicle 5 (type 2): 1-3-8-1
Vehicle 7 (type 2): 1-10-1
Vehicle 9 (type 2): 1-5-1

The model decided not to use vehicles 6, 8, and 10. However, considering the fact that, 9 vehicles are identical, there are alternative solutions.

In this paper we presented the heterogeneous vehicle routing problem which is capable of handling a variety of real-life requirements. In our study, an optimization model for determining the daily routes for real customers is constructed. The optimal solution is found. Because the company has been setting the routes heuristically, the optimal solution found by the model is absolutely better than or equal to a heuristic solution. In the real world operational systems are always customer specific. For this reason our problem and proposed model has proven to be an interesting research topic with many possibilities for future research

**References**


[1] Chow W S, Madu C N, Kuei C-H, Lu M H, Lin C and Tseng H 2008 Supply chain management in the US and Taiwan: An empirical study *Omega* 36(5); 665-679.

[2] Cordeau J-F and Maischberger M 2012 A parallel iterated tabu search heuristic for vehicle routing problems *Computers & Operations Research* 39 2033-2050.

[3] Dondo R and Cerda J 2007 A cluster-based optimization approach for the multi-depot heterogeneous fleet vehicle routing problem with time Windows *European Journal of Operational Research* 176 1478-1507.

[4] Lima C M R R, Goldbarg M C and Goldbarg E F G 2004 A Memetic Algorithm for the Heterogeneous Fleet Vehicle Routing Problem *Electronic Notes in Discrete Mathematics* 18 171-176.

[5] Golden B Assad A Levy L Gheysens F 1984 The fleet size and mix vehicle routing *Computers and Operations Research* 11 49-66.

[6] Yaman H 2006 Formulations and valid inequalities for the heterogeneous vehicle routing problem. *Mathematical Programming* 106 365-390.

[7] Desroches M Verhoog T W 1991 A new heuristic for the fleet size and mix vehicle routing problem *Computers and Operations Research* 18 263-274

[8] Wassan N A Osman I H 2002 Tabu search variants for the mix fleet vehicle routing problem *Journal of the Operational Research Society* 53 768-782

[9] http://www.freemaptools.com/how-far -is-it between.htm.

# On equations with infinitely many derivatives: integral transforms and the Cauchy problem

**Humberto Prado and Enrique G Reyes**

Departamento de Matemática y Ciencia de la Computación, Universidad de Santiago de Chile, Casilla 307 Correo 2 Santiago, Chile

E-mail: humberto.prado@usach.cl; ereyes@fermat.usach.cl; e_g_reyes@yahoo.ca

**Abstract.** We analyze initial value problems for ordinary differential equations with infinitely many derivatives such as (linearized versions of) nonlocal field equations of motion appearing in particle physics, nonlocal cosmology and string theory. We show that the corresponding initial value problem on a half line is well-posed and that it requires only a finite number of initial conditions. We also investigate nonlinear pseudo-equations defined by functions of the Laplace operator, that is, nonlinear partial differential equations with infinitely many derivatives.

## 1. Introduction

In this report we outline some recent developments in the analytic understanding of differential equations with an infinite number of derivatives, also called *nonlocal equations.* These equations have appeared recently as field equations of motion in particle physics, in string theory, and in gravity and cosmology (see for instance [13, 7, 16, 1, 3, 4, 12]). An important example is

$$p^{a\,\partial_t^2}\phi = \phi^p\,, \qquad a > 0\,. \tag{1}$$

Equation (1) describes the dynamics of the open $p$-adic string for the scalar tachyon field (see [1, 12, 14, 15] and references therein) and it can be understood, at least formally, as an equation in an infinite number of derivatives if we expand the left hand side as a power series in $\partial_t^2$.

Our first aim is to investigate analytic properties of nonlocal *linear* ordinary differential equations of the form

$$f(\partial_t)\phi = J(t)\,, \qquad t > 0\,, \tag{2}$$

including the initial value problem. The importance of such a study has been recently stressed by Aref'eva and Volovich [1]: classical versions of Big Bang cosmological models contain singularities at the beginning of time, and therefore the time variable appearing in the field equations should vary only over a half line. Then, we wish to move on to *partial* differential equations in an infinite number of derivatives such as

$$\Delta e^{-c\Delta}\phi = U(x,\phi)\,, \quad c > 0\,, \tag{3}$$

on Euclidean space and on compact Riemannian manifolds $(M, g)$. "Wick rotated" versions of these equations appear naturally in cosmology and string theory, see for instance [4, 5, 8, 16].

Our approach is, as explained in [8, 9, 10], to emphasize the role played by integral transforms in the analysis of Equations (2) and (3). This point of view allows us to prove that if some a priori data connected with our functional calculus is specified, the initial value problem for Equation (2) is well-posed and it requires only a *finite number* of initial conditions. It also allows us to move on to equations such as (3) posed on on Euclidean space or compact Riemannian manifolds, as we show in [8, 10].

## 2. Lorentzian functional calculus and linear equations

We fix a number $\omega \in \mathbb{R}$ and we consider the Banach space $L^\infty_\omega(\mathbb{R}_+)$ of all complex-valued and exponentially bounded functions on $\mathbb{R}_+$, and the *Widder space*, $C^\infty_W(\omega,\infty)$, defined as the image of the Laplace transform

$$\phi \in L^\infty_\omega(\mathbb{R}_+) \mapsto \mathcal{L}(\phi)(s) = \int_0^\infty e^{-st}\phi(t)dt \in C^\infty_W(\omega,\infty) .$$

As proven in [2], functions in $C^\infty_W(\omega,\infty)$ extend analytically to the region $Re(s) > \omega$ and the Laplace transform $\mathcal{L}(\phi)$ converges for $Re(s) > \omega$. Moreover, $\mathcal{L}$ is an isometric isomorphism from $L^\infty_\omega(\mathbb{R}_+)$ onto $C^\infty_W(\omega,\infty)$.

*Definition 1. Let $f$ be an entire function and let $\mathcal{R}$ be the space of analytic functions on $\mathbb{C}$. We consider the subspace $D_f$ of $L^\infty_\omega(\mathbb{R}_+) \times \mathcal{R}$ consisting of all the pairs $(\phi, r)$ such that $\widehat{(\phi,r)} = f\,\mathcal{L}(\phi) - r$ belongs to the Widder space $C^\infty_W(\omega,\infty)$. The domain of $f(\partial_t)$ as a linear operator on $L^\infty_\omega(\mathbb{R}_+) \times \mathcal{R}$ is $D_f$. If $(\phi,r) \in D_f$ then $f(\partial_t)\,(\phi,r) = \mathcal{L}^{-1}(\ \widehat{(\phi,r)}\ ) = \mathcal{L}^{-1}(f\,\mathcal{L}(\phi) - r)$.*

Now, let us assume that a function $r \in \mathcal{R}$ has been fixed. We think of an equation such as $f(\partial_t)(\phi,r) = J$ as an equation for $\phi$ such that $(\phi,r) \in D_f$, and we simply write $f(\partial_t)\phi = J$.

*Definition 2. Let us fix a function $r \in \mathcal{R}$. We say that $\phi \in L^\infty_\omega(\mathbb{R}_+)$ is a solution to Equation $f(\partial_t)\phi = J$ if and only if*

*(i) $\widehat{\phi} = f\,\mathcal{L}(\phi) - r \in C^\infty_W(\omega,\infty)$ ; ( i.e., $(\phi,r) \in D_f$ );*
*(ii) $f(\partial_t)(\phi) = \mathcal{L}^{-1}(f\,\mathcal{L}(\phi) - r) = J$ .*

**Theorem 1.** *Let us fix an entire function $f$ and a function $J \in L^\infty_\omega(\mathbb{R}_+)$. We assume that the function $(\mathcal{L}(J)+r)/f$ is in Widder space. Then, the linear equation $f(\partial_t)\phi = J$ can be uniquely solved for $\phi \in L^\infty_\omega(\mathbb{R}_+)$. The solution is given by the explicit formula*

$$\phi = \mathcal{L}^{-1}\left(\frac{\mathcal{L}(J)+r}{f}\right) . \qquad (4)$$

## 3. The initial value problem

Formula (4) tells us that –expanding the analytic function $r$ appearing in (4) as a power series– $\phi$ depends in principle on an infinite number of arbitrary constants. However, this fact *does not* mean that the equation itself is superfluous, as (4) depends essentially on $f$ and $J$. We think of $r$ as a "generalized initial condition".

*Definition 3. A generalized initial condition for the equation*

$$f(\partial_t)\phi = J \qquad (5)$$

*is an analytic function $r_0$ such that $(\phi, r_0) \in D_f$ for some $\phi \in L^\infty_\omega(\mathbb{R}_+)$. A generalized initial value problem is an equation such as (5) together with a generalized initial condition $r_0$. A* *solution to a given generalized initial value problem* $\{(5), r_0\}$ *is a function* $\phi$ *satisfying the conditions of Definition 2 with* $r = r_0$.

Thus, given a generalized initial condition, we find a unique solution for (5) much in the same way as given *one* initial condition we find a unique solution to a first order linear ODE. *There is no reason to believe that for a given* $r$ *the solution* (4) *will be analytic*: we can only conclude that the solution is an integrable exponentially bounded function, and it follows that classical initial value problems do not even exist in full generality!

*Definition 4.* *A classical initial value problem is an equation* $f(\partial_t)\phi = J$ *together with a finite set of conditions* $\phi(0) = \phi_0, \phi'(0) = \phi_1, \cdots, \phi^{(k)}(0) = \phi_k$. *A solution to a classical initial value problem is a pair* $(\phi, r_0) \in D_f$ *satisfying the conditions of Definition 2 with* $r = r_0$ *such that* $\phi$ *is differentiable at zero and the* $k+1$ *conditions above for* $\phi(0)$ *hold.*

**Theorem 2.** *Let* $f$ *be entire and take* $J$ *in* $L^\infty_\omega(\mathbb{R}_+)$ *such that* $\mathcal{L}(J)/f \in C^\infty_W(\omega, \infty)$. *Fix also a number* $N \geq 0$, *a finite number of points* $\omega_i$ *to the left of* $Re(s) = \omega$, *and (if* $N > 0$*) a finite number of positive integers* $r_i$, $i = 1, ..., N$. *Set* $K = \sum_{i=1}^N r_i$. *Then, generically, given* $K$ *initial conditions,* $\phi_0, \dots, \phi_{K-1}$, *there exists an analytic function* $r_0$ *such that*

($\alpha$) $\dfrac{r_0}{f}$ *has a finite number of poles* $\omega_i$ *of order* $r_i$, $i = 1, \dots, N$ *to the left of* $Re(s) = \omega$ ;

($\beta$) $\dfrac{\mathcal{L}(J) + r_0}{f} \in C^\infty_W(\omega, \infty)$ ;

($\gamma$) $\left|\dfrac{r_0(s)}{f(s)}\right| \leq \dfrac{M}{|s|^p}$ *for some* $p \geq 1$ *and* $|s|$ *sufficiently large.*

*Moreover, the unique solution* $\phi$ *to* $f(\partial_t)\phi = J$ *with* $r = r_0$ *is of class* $C^K$ *and it satisfies* $\phi(0) = \phi_0, \dots, \phi^{(K-1)}(0) = \phi_{K-1}$.

The proof of Theorem 2 (see [9]) depends on the ability to differentiate certain function of $J$ and $f$, which we present in [9], at least $K$ times, and on the assumption that a system of $K$ linear equations has a solution. In this sense we are solving initial value problems "generically". The proof also shows that it is essential to give *a priori* a finite number of points $\omega_i$ in order to have classical initial value problems. If no points $\omega_i$ are present, the solution to $f(\partial_t)\phi = J$ is simply $\phi = \mathcal{L}^{-1}(\mathcal{L}(J)/f)$, a formula which fixes completely the values of all the derivatives of $\phi$ at zero.

## 4. On nonlocal equations of elliptic type

Now we consider equations of the form (3). More generally, in [10] we investigate equations of the form

$$f(\Delta)u = U(x, u(x)) \,, \tag{6}$$

in which $f$ is in principle an arbitrary *measurable* function defined on the whole Euclidean space $\mathbb{R}^n$ or on a compact Riemannian manifold $(M, g)$. As examples, we note that Barnaby, [3], studied Equation (6) in a Lorentzian context using his "method of steps" and delay differential equations, and that in the paper [6] Dragovich considers (6) in which $f$ is the Riemann $\zeta$-function.

At this level of generality neither pseudo-differential operators, [11], nor generalized functions are used: we attach a meaning to the nonlocal operator $f(\Delta)$ directly by using Fourier transform, and we solve Equation (6) on a family of Hilbert spaces $\mathcal{H}^\beta(f)$, $\beta > 0$, of differentiable functions on which the operator $f(\Delta) + I$ acts naturally. For example, [8], we define the action of the operator $L_c = \Delta e^{-c\Delta} - Id$ which appears in the analysis of (3) as

$$L_c u = -\mathcal{F}^{-1}\left(\mathcal{F}(u) + |\xi|^2 e^{c|\xi|^2}\mathcal{F}(u)\right) .$$

These Fourier-based definitions allow us to show the existence of *regular* solutions to Equation (3) and more generally to equations of the type (6). We quote from [10]:

**Theorem 3.** *Let us assume that $f$ is such that the function $s \mapsto f(-s^2)$ is non-negative, and that there exist real numbers $\beta, R, M > 0$ satisfying*

$$M(1+|\xi|^2)^{\frac{\beta}{2}} \le f\left(-|\xi|^2\right) \text{ for all } \xi \text{ with } |\xi| > R$$

*for $\beta > \frac{n}{2}\left(\frac{\alpha-1}{\alpha}\right)$. Suppose that $U$ is a spherically symmetric function with respect to $x$, and assume also that there exist a constant $\alpha > 1$, and functions $h \in L^2(\mathbb{R}^n)$, $g \in L^{\frac{2\alpha}{\alpha-1}}(\mathbb{R}^n)$ such that the following two inequalities hold:*

$$|U(x,y)+y| \le C(|h(x)|+|y|^\alpha), \quad \left|\frac{\partial}{\partial y}(U(x,y)+y)\right| \le C\left(|g(x)|+|y|^{\alpha-1}\right)$$

*for some constant $C > 0$. Then, there exist $0 < \epsilon < 1$ and $0 < \rho_\epsilon < 1$ such that whenever $\|h\|_{L^2(\mathbb{R}^n)} < \rho_\epsilon$, there is a smooth radial solution $u$ to Equation (6) satisfying $\|u\|_{L^{2\alpha}(\mathbb{R}^n)} \le \epsilon$.*

## Acknowledgments

H.P. has been partially supported by FONDECYT grant # 1130554 and E.G.R. has been partially supported by FONDECYT grant # 1111042. Both authors thank their friend P. Górka (Warsaw University of Technology) for his enthusiastic collaboration in some of the topics touched upon in this report.

## References

[1] Aref'eva I Ya and Volovich I V, Cosmological Daemon. Preprint, arXiv:1103.0273v2 [hep-th] 21 Jun 2011.
[2] Arendt W, Batty C J K, Hieber M and Neubrander F, "Vector-valued Laplace transforms and Cauchy problems". Birkhäuser, 2001.
[3] Barnaby N, A new formulation of the initial value problem for nonlocal theories. *Nuclear Physics B* 845 (2011), 1–29.
[4] Barnaby N and Kamran N, Dynamics with infinitely many derivatives: the initial value problem. *J. High Energy Physics* 2008 no. 02, Paper 008, 40 pp.
[5] Calcagni G, Montobbio M and Nardelli G, Localization of nonlocal theories. *Physics Letters B* 662 (2008), 285-289.
[6] B. Dragovich, Zeta-nonlocal scalar fields. *Theoret. Math. Phys.*, 157 (2008), 1671–1677.
[7] Eliezer D A and Woodard R P, The problem of nonlocality in string theory. *Nuclear Physics B* 325 (1989), 389–469.
[8] Górka P, Prado H and Reyes E G, Generalized euclidean bosonic equation. In: *Spectral Analysis of Quantum Hamiltonians: Spectral Days 2010.* Birkhaüser Series 'Operator Theory: Advances and Applications', Vol. 224, (2012) 147-169.
[9] Górka P, Prado H and Reyes E G, The initial value problem for ordinary differential equations with infinitely many derivatives. *Class. Quantum Grav.* 29 (2012) 065017 (15pp).
[10] Górka P, Prado H and Reyes E G, On a General Class of Nonlocal Equations. *Ann. Henri Poincaré* 14 (2013), 947–966.
[11] L. Hörmander, The analysis of linear partial differential operators III. Springer-Verlag, Berlin, 1985.
[12] Moeller N and Zwiebach B, Dynamics with infinitely many time derivatives and rolling tachyons. *J. High Energy Physics* 2002, no. 10, Paper 34, 38 pp.
[13] Moffat J W, Ultraviolet Complete Electroweak Model Without a Higgs Particle. *Euro. Phys. J. Plus* 126:53 (2011).
[14] Vladimirov V S, The equation of the $p$-adic open string for the scalar tachyon field. *Izvestiya: Mathematics* 69 (2005), 487–512.
[15] Vladimirov V S and Volovich Ya I, Nonlinear dynamics equation in p-adic string theory. *Teoret. Mat. Fiz.* 138 (2004), 355–368; English transl., *Theoret. Math. Phys.* 138 (2004), 297–309.
[16] Witten E, Noncommutative geometry and string field theory. *Nuclear Physics B* 268 (1986), 253–294.

# Thermal Creep Flow of Helium Gas at Cryogenic Temperatures

**G Babac**[1,2] **and J M Reese**[2]
[1]Institute of Energy, Istanbul Technical University, 34469, Maslak-Istanbul, TURKEY
[2]Department of Mechanical & Aerospace Engineering, University of Strathclyde, Glasgow G1 1XJ, UK

E-mail: babac@itu.edu.tr

**Abstract.** We present a molecular dynamics (MD) simulation of nanoscale gas flow due to thermal creep at cryogenic temperatures. Helium is considered because its low liquefying temperature allows a wide range of cryogenic analysis. The thermal creep flow along a nanochannel is generated by applying temperature differences along the channel. Pressure and density variations are measured at various rarefaction conditions, covering the slip flow to the free molecular regimes. Thermo molecular pressure difference (TMPD) values are also calculated. Our results are compared with those in the literature.

## 1. Introduction

As is well known, the transport properties of gases at the micro/nanoscale differ from those at the macroscale because of differences in the governing flow characteristics. Nanoscale gas flows are generally analyzed as free molecular flows, and for these kinds of flows between reservoirs at different pressures and temperatures the Knudsen law holds:

$$\frac{p_h}{\sqrt{T_h}} = \frac{p_c}{\sqrt{T_c}}, \tag{1}$$

where the $h$ and $c$ subscripts indicate quantities at the hot and cold reservoirs. According to the Knudsen law, in a system in the free molecular flow regime, a temperature gradient induces a pressure difference. This is called the "thermo-molecular pressure difference" (TMPD), which enables several effects including thermal creep and the operating principle of the Knudsen pump. The TMPD has been determined in the literature for different gases, system dimensions and temperature gradients [2-4]. For the TMPD at low temperatures, the Weber-Schmidt equation [5-6] represents experimental data over a wide pressure range, utilizing the results obtained by Knudsen.

In this paper, nanochannel gas flows are examined under cryogenic temperature gradients. The pressure and temperature evolutions are analyzed. The TMPD values we obtain are compared with the TMPD generated by the same temperature difference but in a conventional temperature range, and also with the results from the Weber-Schmidt equation. The flow system is simulated using the molecular dynamics (MD) routines in OpenFOAM [7-9], which were developed in the authors' research group.

## 2. Molecular Dynamics Simulation

In the current study, monatomic helium gas confined in a rectangular channel is simulated using MD. The channel is thermally in contact with a hot ($T_H$) and a cold ($T_L$) reservoir. A schematic of the simulated system is in figure 1, in which all the geometry and gas atoms are coloured depending on the surrounding temperature. The application of a Berendsen thermostat [10] in the reservoir zones ensures that their temperature remains constant around target values.

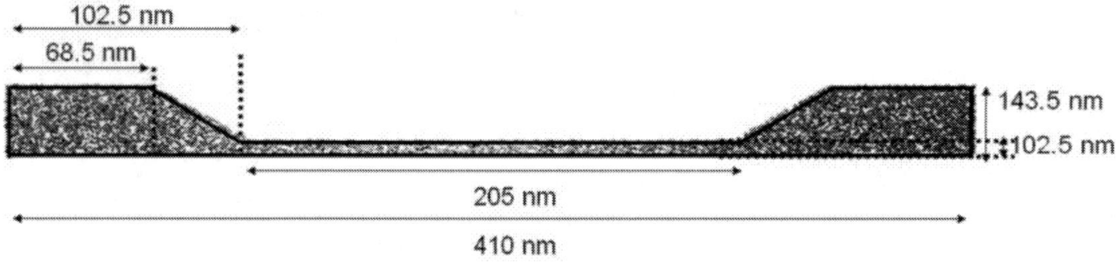


Figure 1: Schematic of the simulated flow system. The hot reservoir is at the left, and the cold reservoir at the right.

In the MD simulations, the helium atoms interact with each other through a Lennard-Jones (LJ) potential:

$$U_{LJ} = 4\varepsilon\left[\left(\frac{\sigma}{R_{ij}}\right)^{12} - \left(\frac{\sigma}{R_{ij}}\right)^{6}\right], \tag{2}$$

where $R_{ij}$ is the distance between particles $i$ and $j$, $\varepsilon$ is the depth of the potential well, and $\sigma$ is the finite distance at which the inter-particle potential is zero. For helium-helium interactions we set $\varepsilon_{He-He} = 1.9654\times10^{-22} J$ and $\sigma_{He-He} = 2.2023\times10^{-10} m$ from Ref [11]. The potential is truncated and shifted at a distance $R_{cut} = 6.6\times10^{-10} m$. Diffusive wall boundary conditions, with a linear temperature gradient, are applied along the system.

## 3. Results and Discussions

### *3.1. Thermo-molecular pressure difference*

Simulations are run for different temperature gradients and different Knudsen numbers. For the first case, the temperature of the cold reservoir is kept at 100 K and the hot reservoir is at 300 K, and the TMPD is calculated for Kn=0.5, 1, 5, and 10.

In the second case, the temperature is kept at 10 K and 100 K for the cold and hot reservoirs, respectively. (The liquefaction temperature of He is around 4.3 K and the critical temperature is around 5.2 K. The classical gas limit is also checked for the system and 10 K is chosen as the temperature of cold reservoir, keeping the system within the classical limit so that quantum degeneracy effects are negligible.)

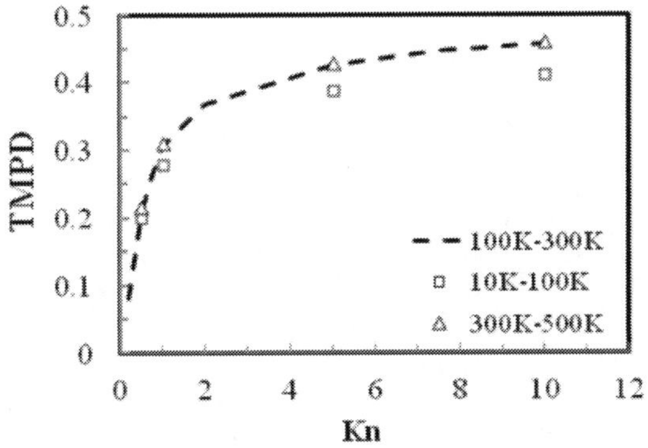


Figure 2: The normalized TMPD values under different temperature gradient

In figure 2, the dashed line represents the system behaviour in the cryogenic range, under a temperature difference of 200 K ($T_L$=100 K and $T_H$=300 K). TMPD values are calculated in this temperature range for the Kn=0.1, 0.2, 0.5, 0.75, 1, 2, 5, 7.5, and 10. The triangular symbols in figure 2 represent the TMPD values at conventional temperatures ($T_L$=300 K and $T_H$=500 K) but also with a 200 K temperature difference. In this case, the simulation were completed for Kn=0.5, 1, 5, and 10.

As is predicted by kinetic theory, normalized TMPD values increase with increasing Knudsen number and converge to 0.5 in free molecular flow. The TMPD values in the cryogenic (100-300 K) and conventional (300-500 K) temperature ranges merge with each other. This is consistent with the results in the literature [3]. We have therefore shown that TMPD values only depend on the temperature difference applied, not the absolute values of the temperature.

For the second case, a lower temperature difference is applied: 90 K ($T_L$=10 K and $T_H$=100 K). The rectangular symbols in figure 2 represent the TMPD values in this case, and it can be seen that a smaller pressure difference is obtained for this smaller temperature gradient. The applied temperature difference is the main determiner of TMPD, rather than the absolute temperature values.

### *3.2. Weber-Schmidt (WS) Equation*

For pressure and temperature evolution in free molecular conditions at cryogenic temperatures, the Weber-Schmidt equation [5-6] represents experimental data over a wide pressure range. The Weber-Schmidt equation is applicable to helium-3 as well as to helium-4, and it is given as follows for helium gases [6]:

$$\ln\left(\frac{p_h}{p_c}\right)=\frac{1}{2}\ln\left(\frac{T_h}{T_c}\right)+0.18131\ln\left(\frac{y_h+0.1878}{y_c+0.1878}\right)+0.41284\ln\left(\frac{y_h+1.8311}{y_c+1.8311}\right)-0.15823\ln\left(\frac{y_h+4.9930}{y_c+4.9930}\right) \quad (3)$$

In this equation, $y$ is defined as $y=r/\lambda$, where $r$ is the radius of the channel and $\lambda$ is the molecular mean free path. The $y_h$ and $y_c$ values are calculated for the hot and the cold regions. The other constants in equation (3) were obtained by fitting to the experimental data. In our case we have a rectangular domain, so $r$ is chosen as the height of the channel, $r$=102.5 nm, and the mean free path is calculated as $\lambda=k_B T/\sqrt{2}\pi\sigma^2 p$ for each case, where $k_b$ is the Boltzmann's constant. The calculated pressure ratio for different Knudsen numbers at cryogenic temperatures is then compared with the Weber-Schmidt results in figure 3 for Kn=0.5, 1 and 5.

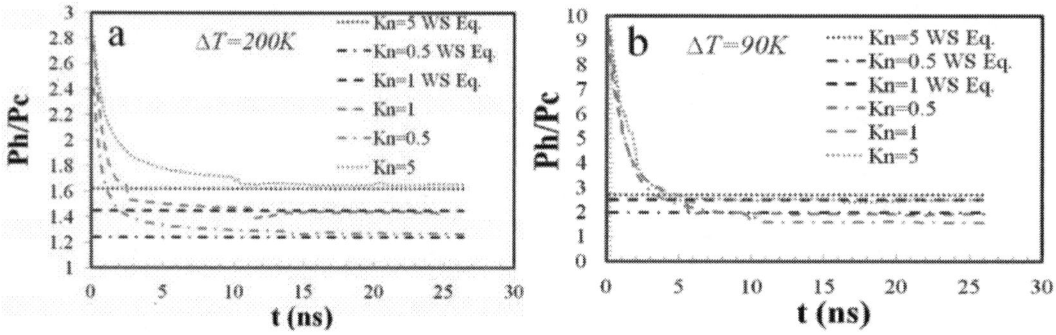


Figure 3: Pressure ratio calculated by MD (gray lines) and the Weber-Schmidt equation (black lines) versus time (ns). The results for Kn=5, 1 and 0.5 are represented using dotted, dashed and dot-dashed lines, respectively.

The Weber-Schmidt equation represents a system at steady-state; these results are represented as black lines in figure 3, with dotted, dashed and dot-dashed lines for Kn=5, 1 and 10, respectively. Our MD simulation results reach a steady-state after 10 to 15 ns. When the simulations reach steady-state, the MD values are quite closer to the values predicted by the Weber-Schmidt equation. The best comparison between the two is obtained for a $\Delta T$=200 K temperature difference, while the MD values for $\Delta T$=90 K are slightly underpredicted.

## 4. Conclusion

Monatomic helium gas in a nanochannel under different temperature gradients has been simulated using molecular dynamics. Thermo-molecular pressure differences in the cryogenic ranges have been compared with those in conventional temperature ranges. For the same temperature difference, the TMPD values were very close to each other, so operating the system in cryogenic temperature ranges does not affect the TMPD values. As a secondary case, the system has been analyzed under a smaller temperature difference, with the temperature of the cold reservoir reduced to 10 K. The smaller temperature difference introduces a smaller pressure difference, as is consistent with the literature. The calculated pressure values have been compared with those from the Weber-Schmidt equation, and better agreement was shown for the higher temperature difference case.

## References


[1] Karniadakis G, Beskok A. and Aluru N 2005 *Microflow and Nanoflows: Fundamentals and Simulation,* Springer Science Business Media, New York.

[2] Cardenas M R, Graur I, Perrier P and Meolans J G 2011 Phys Fluids, **23** 031702.

[3] Han Y L, Muntz E P, Alexeenko A, Young M 2007 Nanosc Microsc Therm **11** 151

[4] G. Babac, N. Dongari and J. M. Reese 2012 Thermal transpiration of nanoscale gas flow *Proc. 28th Int. Symp. on Rarefied Gas Dynamics* 9-13 July Zaragoza(Spain) **1501** 946.

[5] Weber S and Schmidt G 1936 Leiden Commlln. 246c.

[6] Watkins R A, Taylor W L and Haubach W J, 1967 J. Chem. Phys. **46** 1007

[7] Macpherson G B and Reese J M 2008 Mol Simulat **34(1)** 97.

[8] Macpherson G B, Borg M K and Reese J M 2007 Mol Simulat **33** 1199.

[9] Borg M K, Macpherson G B and Reese J M 2010 Mol Simulat **36(10)** 745.

[10] Hünenberger P H, 2005 Adv Polym Sci **173** 105.

[11] Bird R B, Stewart W E and Lightfoot E N 2006 *Transport Phenomena. Revised 2nd Edition,* John Wiley & Sons, West Sussex

# Apply a hydrological model to estimate local temperature trends

**Masao Igarashi**[1], **Tatsuya Shinozawa**[1]

[1] College of Bioresource Sciences, Nihon University, 1866 Kameino Fujisawa Kanagawa 252-0880, Japan

E-mail: igarashi.masao@nihon-u.ac.jp

**Abstract.** Continuous times series $\{f(x)\}$ such as a depth of water is written $f(x) = T(x)+P(x)+S(x)+C(x)$ in hydrological science where $T(x), P(x), S(x)$ and $C(x)$ are called the trend, periodic, stochastic and catastrophic components respectively. We simplify this model and apply it to the local temperature data such as given E. Halley (1693), the UK (1853-2010), Germany (1880-2010), Japan (1876-2010). We also apply the model to $CO_2$ data. The model coefficients are evaluated by a symbolic computation by using a standard personal computer. The accuracy of obtained nonlinear curve is evaluated by the arithmetic mean of relative errors between the data and estimations. E. Halley estimated the temperature of Gresham College from 11/1692 to 11/1693. The simplified model shows that the temperature at the time rather cold compared with the recent of London. The UK and Germany data sets show that the maximum and minimum temperatures increased slowly from the 1890s to 1940s, increased rapidly from the 1940s to 1980s and have been decreasing since the 1980s with the exception of a few local stations. The trend of Japan is similar to these results.

## 1. Introduction

Meteorological offices, including the UK Met Office (UKMO), Deutscher Wetterdienst (DWD) and Japan Meteorological Agency provide the mean of local monthly maximum and minimum temperature values. The mean monthly values of $CO_2$ (ppm) at Manualoa and Ayari are also provided by, such as United States' National Aeronautics and Space Administration (NASA). Many agencies show the trend curve of them. Here we provide a simple and robust method to estimate a year trend of these data.

We propose a following nonlinear model (1) to evaluate the trends of temperature and $CO_2$ data at the observed points by ignoring the stochastic and catastrophic components of the hydrological model,

$$f(x) = T(x) + P(x) = \underbrace{a_1 + a_2 x}_{\text{Trend component}} + \underbrace{a_3 \sin\left(a_4 + \frac{\pi x}{n}\right)}_{\text{Periodical component}} \qquad (1)$$

The unknown coefficients $a_i, \ = \ 1,2,3,4$ and $n$ are usually evaluated by the numerical computation based on the least square method and Newton iteration method. For simplicity, we apply a symbolic computation to obtain the coefficients under the fixed integer $n$. As the temperature periodically changes, so the integer $n$ is estimated 6 for the month variable $x$. Hence we fixed the integer $n$ between $1 \leqq n \leqq 12$. The best $n$ among them is determined by the arithmetic mean of the sum of the relative errors between the observed and estimated temperatures.

The temperatures and $CO_2$ data sets are obtained from the following paper and URLs, respectively:

(i) E. Halley 1694 *Phil. Trans.Royal Society of London* **18** p 185

(ii) `http://www.metoffice.gov.uk/climate/uk/stationdata/`

(iii) `http://www.dwd.de/bvbw/appmanager/bvbw/`

(iv) `http://cdiac.ornl.gov/trends/co2/sio-mlo.html/`

(v) `http://ds.data.jma.go.jp/ghg/kanshi/obs/co2/`

Reducing $f(x)$ to the upper $f_u(x)$ and lower $f_l(x)$ linear line respectively, we can estimate the annual temperature trend and also count the number of points included between $f_u(x)$ and $f_l(x)$.

## 2. Materials and Method

We describe our proposed method on the basis of data from the Oxford meteorological station.

### *2.1. Data*

The UK Met Office provides monthly mean maximum (MAX) and minimum temperatures (MIN) data from January 1853 to December 2011. We classify the data according to decadal periods (e.g. 1860-2010, 1870-2010, $\cdots\cdots$, 2000-2010) to allow simple comparison with results from other stations. The monthly mean maximum and minimum temperatures (°C) for the period 01/1860−12/2010 are expressed as follows:

MAX=(6.8, 5.2, 8.4, 10.6, 17, 17, 19.3, 17.9, 15.6, 13.4, 7.4, 4.4, 3.8,$\cdots$, 8.2, 2.7)

MIN=(0.8, -1.6, 1.9, 1.7, 7.7, 9.5, 10.3, 10.9, 7.0, 7.3, 2.1, 0.7*, -0.7, $\cdots$, 2.9, -2.2)

For example, the first record, 6.8 (MAX) and 0.8 (MIN), indicates the monthly mean maximum and minimum temperatures for January 1860. The last record, 2.7 (MAX) and -2.2 (MIN), refers to the monthly mean maximum and minimum temperatures for December 2010. Each data set consists of $151\times12 = 1812$ data points. Any missing data are interpolated using the neighbouring values. The interpolated MAX and MIN values are indicated by an asterisk. For example, 0.7* (MIN) represents a missing datum and is interpolated from the neighboring values as (2.1-0.7)/2=0.7. Meteorological stations for which numerous data were missing (e.g. Chiven) or have been closed (e.g. Ringway) are not considered.

### *2.2. Model*

We apply the following nonlinear curve to the data from each meteorological station:

$$y = f(x) = a_1 + a_2 x + a_3 \sin(a_4 + \pi x/n) \tag{2}$$

where $a_1$, $a_2$, $a_3$, $a_4$ are unknown parameters and the integer $n$ is fixed between $1 \leqq n \leqq 12$. For simplicity, we use Mathematica `FindFit` function to evaluate the coefficients. The model is applied to the data of $CO_2$.

### *2.3. Criterion*

To choose the best integer $n$ between $1 \leqq n \leqq 12$, we introduce the criterion $\mathrm{CR}_n$ [1]:

$$\mathrm{CR}_n = \frac{\sum_{k=1}^{m} \dfrac{|\widehat{y}_k - y_k|}{\max\left(|\widehat{y}_k|, |y_k|\right)}}{m} \tag{3}$$

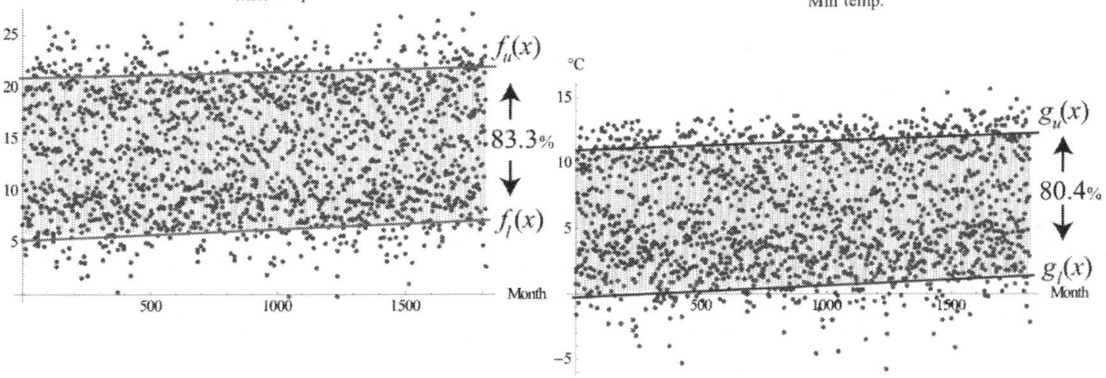


**Figure 1.** Data (dots), nonlinear curves (red and blue) and trend lines of Oxford

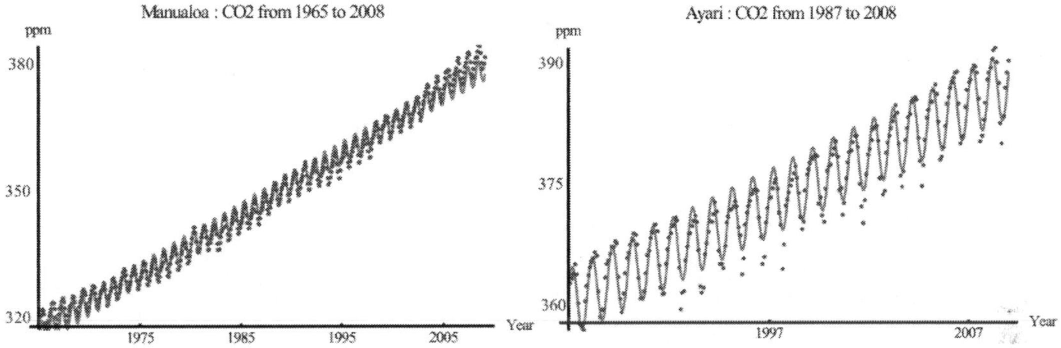


**Figure 2.** Data (dots) and nonlinear curves of Manualoa and Ayari

where $m$ is the number of data points, and $\{\widehat{y_k}\}$ and $\{y_k\}$ $(k = 1, 2, \cdots\cdots, m)$ are the observed and estimated temperatures ($CO_2$), respectively. $CR_n$ is the arithmetic mean of the sum of the relative errors between the observed and estimated temperatures. Hence, if $CR_n$ is small, $y_k = f(k)$ in (2) agrees well with the observed temperatures [2]. Conversely, if $CR_n$ is large, $f(k)$ does not agree well with the observed temperatures. Thus, an appropriate integer $n$ is selected to obtain the smallest value of $CR_n$.

## 3. Results

We show the typical nonlinear curve with related matters. $f(x)$ and $g(x)$ are the nonlinear curve of maximum and minimum temperatures. The reduced linear line $f_u(x)$ and $f_l(x)$ of $f(x)$ show the upper and lower trend of maximum temperature. And also the reduced linear line $g_u(x)$ and $g_l(x)$ show the upper and lower trend of minimum temperature. In the case of $CO_2$, these are denoted $h(x)$, $h_u(x)$ and $h_l(x)$.

Figure 1 shows the data (dots), the obtained curve and trend lines of Oxford maximum and minimum temperature. 83.3% of data points are contains between the line of $f_u(x)$ and $f_l(x)$. Figure 2 shows the results of $CO_2$ for Manualoa and Ayari.

**Oxford : 01/1860-12/2010**

$$f(x) = 13.33 + 0.0006069x - 7.6198\sin(0.9854 + \pi x/6), \quad \text{CR}_6 = 0.041$$
$$f_u(x) = 20.95 + 0.00728x, \quad f_l(x) = 5.710 + 0.00728x \text{ , } 83.3\% \text{ data are included}$$
$$g(x) = 5.53 + 0.0007190x - 5.5151\sin(0.8349 + \pi x/6), \quad \text{CR}_6 = 0.066$$
$$g_u(x) = 11.05 + 0.00862x, \quad g_l(x) = 0.019 + 0.0086 \text{ , } 80.4\% \text{ data are included}$$

**Aachen : 01/1900-12/2010**

$$f(x) = 13.11 + 0.0008218x - 8.9991\sin(1.01189 + \pi x/6), \quad \text{CR}_6 = 0.053$$
$$f_u(x) = 4.11 + 0.0008218x, \quad f_l(x) = 22.11 + 0.0008218x \text{ , } 83.4\% \text{ data are included}$$
$$g(x) = 5.45 + 0.001209x - 6.7895\sin(0.8779 + \pi x/6), \quad \text{CR}_6 = 0.067$$
$$g_u(x) = -1.33 + 0.001209x, \quad g_l(x) = 12.24 + 0.001209x \text{ , } 81.1\% \text{ data are included}$$

**Tokyo : 01/1880-12/2010**

$$f(x) = 18.19 + 0.001299x - 10.2015\sin(0.8175 + \pi x/6), \quad \text{CR}_6 = 0.032$$
$$f_u(x) = 28.39 + 0.001299x, \quad f_l(x) = 7.9943 + 0.001299x, \ 82.9\% \text{ included}$$
$$g(x) = 8.79 + 0.002823x + 11.3499\sin(3.9396 + \pi x/6), \quad \text{CR}_6 = 0.053$$
$$g_u(x) = 20.14 + 0.002823x, \quad g_l(x) = -2.5532 + 0.002823x \text{ , } 84.2\% \text{ included}$$

**Manualoa : 01/1965-12/2008**

$$h(x) = 315.951 + 0.1264x - 2.87594\sin(44.5759 - \pi x/6), \quad \text{CR}_6 = 0.0017$$
$$h_u(x) = 318.827 + 0.1264x, \quad h_l(x) = 313.075 + 0.1264x \text{ , } 68.3\% \text{ included}$$

**Ayari : 01/1987-12/2008**

$$h(x) = 349.489 + 0.1460x + 5.9508\sin(2.8080 - \pi x/6), \quad \text{CR}_6 = 0.0045$$
$$h_u(x) = 355.440 + 0.1460x, \quad h_l(x) = 343.538 + 0.1460x \text{ , } 77.6\% \text{ included}$$

## 4. Conclusion

We introduce a nonlinear model and apply it to the data of local temperature of the UK (33-station), Germany (52-station) and Japan (57-station). The nonlinear curves are practically evaluated for decadal year data set of all stations and $\text{CR}_6$ always gives the smallest value. It illustrates the robustness of the proposed method. The annual trends of maximum and minimum temperatures are estimated by $f_u(x)$ and $g_u(x)$. For example the annual trend of maximum temperature of Tokyo is estimated 1.5 °C up per 100-year, because of $0.00129 \times 12 \times 100 = 1.5$.

We also show the trend of $CO_2$ of Manualoa and Ayari. The value $\text{CR}_6$ is more improved if we introduce $a_5x^2$ term in (2).

## References

[1] M. IGARASHI and E. NUNOHIRO and J. G. PARK 2009 *I. J. Geology* **3** 97

[2] S. D. Conte and Carl de Boor 1972 Elementary Numerical Analysis (McGraw-Hill Book Company) p 10

# A secondary solution of the Schrödinger and Klein-Gordon equations in modeling atomic, molecular and electrodynamic systems

**Alexandru Popa**

National Institute for Laser, Plasma and Radiation Physics, Laser Department, P.O. Box MG-36, Bucharest, Romania 077125

**Abstract.** We study atomic and molecular (AM) stationary systems and electrodynamic (ED) systems, composed of an electron interacting with an electromagnetic field. We show that the Schrödinger and Klein-Gordon equations written for these systems have a secondary solution, which is the wave function associated to a classical system. For AM systems, the wave surfaces ($\Sigma$ surfaces) and their normals ($C$ curves) are solutions of the Hamilton-Jacobi equation, written for the same system. The $\Sigma$ surfaces have a periodical motion and the $C$ curves are closed. The integral relation of the Schrödinger equation on the $C$ curve has a solution identical to the wave function associated to the classical motion. This solution leads to the generalized Bohr quantization relation and to the generalized de Broglie relations, which are valid in the space of the electron coordinates. An identical wave function verifies the Klein-Gordon equation, in the case of the EM systems. The above properties lead to a central field method for calculation of the energetic values and symmetry properties for AM systems, whose accuracy is comparable to the accuracy of the Hartree-Fock method. They also lead to an accurate method for modeling EM systems, which is verified by experimental data from the literature. The above properties are deduced without using any approximation.

## 1. Introduction

Despite the fact that AM and ED systems are very different, our analysis shows that they share common features. In both cases a solution associated to the classical motion results, without any approximation, from the mathematical properties of both, the Schrödinger and Klein-Gordon equations. Another important common feature of AM and ED systems is that in both cases the solutions exhibit periodicity properties. These properties have a practical importance, because they lead to accurate models for the calculation of properties of AM and ED systems. We prove these properties with the aid of our previous results, published in the papers [1]-[6]. The calculation models were verified on numerous experimental data from literature. In Refs. [7]-[9] we presented models for the calculation of the energetic relations and of the symmetry properties of numerous atoms and diatomic molecules, whose accuracy is comparable to the accuracy of the Hartree-Fock method. In Refs. [6] and [10]-[14] we presented accurate calculations of the spectral and geometric distributions of the radiations generated at the interaction between very intense laser beams and electron plasmas or atomic gases.

## 2. Secondary solution of the Schrödinger equation

We shall analyze a discrete system composed of $N$ mobile points (electrons) and $N'$ fixed points (nuclei). Our analysis is made in the space $R^{3N}$ of the electron coordinates, which are denoted by $q_j$, where $j$ takes values between 1 and $3N$. We denote by $q = (q_1, q_2, ..., q_N)$ the coordinates of a point in the space $R^{3N}$. The equations are written in the International System.

We consider the following initial hypotheses. (h1) The system is closed and stationary (i.e. the total energy, denoted by $E$, is constant and the potential energy, denoted by $U$, does not depend explicitly on time). (h2) The total energy has real negative values (i.e. the system is in a bound state). (h3) The behavior of the system is completely described by the Schrödinger equation. (h4) The relativistic and magnetic effects are neglected. (h5) The nuclei are fixed on average positions and their motion is neglected.

We prove that the mathematical properties of this system lead exactly to a secondary solution for the wave function of the system, as follows [1]-[5].

The Schrödinger equation $-i\hbar\partial\Psi/\partial t - \left(\hbar^2/2m\right)\sum_j \left(\partial^2\Psi/\partial q_j^2\right) + U\Psi = 0$ is equivalent to the system composed by the wave equation $\sum_j \left(\partial^2\Psi/\partial q_j^2\right) - \left(1/v_w^2\right)\partial^2\Psi/\partial t^2 = 0$ and the equation

$$\Psi = \Psi_0 \exp(-iEt/\hbar) \tag{1}$$

which reflects the stationarity property of the system. In the above equations $\Psi$, $m$, $t$ and $i$ are, respectively, the wave function, the electron mass, the time and the imaginary constant, while $\hbar$ is the normalized Planck constant ($\hbar = h/2\pi$). The function $\Psi_0$ is the time independent wave function, which is the solution of the equation $-\left(\hbar^2/2m\right)\sum_j \left(\partial^2\Psi_0/\partial q_j^2\right) + (U - E)\Psi = 0$, and the quantity $v_w$ is given by the relation

$$v_w = \pm|E|/\sqrt{2m(E-U)} \tag{2}$$

The functions $\Psi$ and $\Psi_0$ are of the form $\Psi = \Psi(q, t, E, c)$ and $\Psi_0 = \Psi_0(q, E, c)$ [see [15], page 330], where $E$ and $c = (c_1, c_2, \ldots, c_{3N-1})$ are the eigenvalues of the constants of motion.

The wave surfaces of the system are the same as the characteristic surfaces of the wave equation [16], [17]. The characteristic surface of the wave equation, denoted by $\Sigma$, is given by the following equation: [17]-[18]:

$$\chi(q, t, E, c) = 0 \tag{3}$$

where $\chi$ is a single valued function, called *characteristic function*, which satisfies the *characteristic equation* $\sum_j (\partial\chi/\partial q_j)^2 - (1/v_w^2)(\partial\chi/\partial t)^2 = 0$

The equation (3) leads to the following equation of the $\Sigma$ surfaces, $f(q, E, c) = |E|t - p\pi/k$, where $k$ is a real constant, p is an integer and $f(q, E, c)$ is a single valued function (the complete integral), which verifies the time independent Hamilton-Jacobi equation $\sum_j (\partial f/\partial q_j)^2 + 2m(U - E) = 0$. Since the velocity of a point of the $C$ curve, which results from the time independent Hamilton-Jacobi equation, is $\overline{v} = (1/m)\nabla f(q, E, c)$ [19], it follows that the $C$ curve is normal to the $\Sigma$ surface.

In Refs. [3] and [4] we have deduced following properties of the $\Sigma$ surfaces. The motion of the $\Sigma$ surface is periodic and the $C$ curves are closed. An arbitrary point $P \in \Sigma$ moves on a closed $C$ curve with velocity $|v_w|$, in only one direction, and the amplitude of the wave function $\Psi$ in that point varies periodically. The point P moves synchronously with the $\Sigma$ surface. The periodic motion of the $\Sigma$ surface is illustrated by the following equation: $f(q, E, c) = |E|t - p|E|\tau_w$ for $p\tau_w \leq t < (p+1)\tau_w$, where $\tau_w$ is the period of the wave motion and p is a positive integer ($p = 0$ for the first period, $p = 1$ for the second period, and so on). The function $f$ is bounded, namely $0 \leq f(q, E, c) < f_M$, and the maximum value of $f$ is

$$f_M = |E| \cdot \tau_w \tag{4}$$

It follows that the point $P$ passes successively through all the values of the function $f$ when it runs on the curve $C$.

The reduced action function along the $C$ curve, denoted by $S_0$, is given by the equation $S_0(q, E, c) = f(q, E, c) + pf_M$. Note that $S_0$ increases continuously along the curve C. The variation of the function $S_0$ along the closed curve $C$, denoted by $\Delta_C S_0$, is given by:

$$\Delta_C S_0 = f_M \tag{5}$$

Two motions are associated with the $C$ curve, the motion of a point of the quantum wave and the motion of a classical point, which results from the Hamilton-Jacobi equation. The corresponding velocities are, respectively, $v_w$ and $v$. We have proven [3], [5] the following relation, which reflect the connection between the two motions:

$$vv_w = |E|/m \tag{6}$$

The relation between the periods of the two motions is $\tau_w = 2\tau$.

In Refs. [3] and [5] we have proven a general equation which connects the quantum and classical solutions. With the substitution

$$\Psi_0 = \exp(i\sigma/\hbar) \tag{7}$$

where $\sigma$ is a complex valued function of the electron coordinates, we have proven the following integral relation of the Schrödinger equation on the $C$ curve [see Eq. (37) from [5]]:

$$\int_C \sum_j \left(\frac{\partial \sigma}{\partial q_j}\right)^2 dt - i\hbar \int_C \left(\sum_j \frac{\partial^2 \sigma}{\partial q_j^2}\right) dt = \int_C \sum_j \left(\frac{\partial S_0}{\partial q_j}\right)^2 dt \tag{8}$$

An integral relation admits a specific supplementary solution (in addition to the solution coming from the initial equation that has been integrated, which in our case is $\Psi_0$), called generalized solution [18]. We proved [5] that the generalized solution of Eq. (8) is $\sigma_g = S_0$. By virtue of this equation, the time independent wave function corresponding to the generalized solution is:

$$\Psi_{0g} = A\exp(iS_0/\hbar) \tag{9}$$

The corresponding time dependent wave function results from Eqs. (1) and (9), as follows:

$$\Psi_g = A\exp(iS/\hbar) \quad \text{where} \quad S = S_0 - Et \tag{10}$$

Since $S$ and $S_0$ are, respectively, the action and reduced action corresponding to the $C$ curve, it follows that $\Psi_g$ is the wave function associated to the classical motion on the $C$ curve. Since $\Psi_{0g}$ is single valued, it must have the same value when we go along the curve $C$ and arrive again in the initial point. This property leads to the generalized Bohr quantization relation:

$$\Delta_C S_0 = nh \tag{11}$$

where $n$ is a positive integer. By virtue of Eqs. (4), (5) and (11), we have:

$$|E|\tau_w = nh \tag{12}$$

Since the wavelength and angular frequency of the Broglie wave are, respectively, $\lambda_w = \tau_w \cdot v_w$ and $\omega_w = 2\pi/\tau_w$, from Eqs. (6) and (12) we obtain the de Broglie relations:

$$\bar{p} = n\hbar\bar{k}_w \text{ and } |E| = n\hbar\omega_w \tag{13}$$

where $\overline{p} = m\overline{v}$ and $\overline{k}_w$ is the wave vector of the de Broglie wave, which is given by relation $\overline{k}_w = (2\pi/\lambda_w)(\overline{v}_w/v_w)$.

Is follows that there is a de Broglie wave associated to the motion on the $C$ curve, which results from the Hamilton-Jacobi equation. This wave is characterized by the wave vector $\overline{k}_w$ and the angular frequency $\omega_w$. We note that the motion on the $C$ curve is associated also with the wave function $\Psi_g$, which results from the integral relation of the Schrödinger equation on the $C$ curve [see Eq. (10)].

## 3. Secondary solution of the Klein-Gordon equation

We consider a system composed of an electron interacting with an elliptic polarized electromagnetic field in the general case, when the electron velocity and the phase of the field are arbitrary. In Ref. [6] we proved that the Klein-Gordon equation, written for this system, which is $\left[c^2\left(-i\hbar\nabla + e\overline{A}\right)^2 - (i\hbar\partial/\partial t)^2 + (mc^2)^2\right]\Psi = 0$, is verified exactly by the wave function associated with the classical motion of the electron in electromagnetic field, which is

$$\Psi = A\exp(iS/\hbar) \tag{14}$$

The existence of this solution, which is a classical type wave function associated to the classical trajectory of the electron in the electromagnetic field, justifies the accuracy of the classical models presented in Refs. [6] and [10]-[14].

## 4. Conclusions

A solution of the type $A\exp(iS/\hbar)$ which is associated to the classical motion results, without any approximation, from the mathematical properties of both, the Schrödinger and Klein-Gordon equations. The solution $A\exp(iS/\hbar)$ in the case of AM systems is associated with the geometric elements of the wave described by the Schrödinger equation, which result from the Hamilton-Jacobi equation, written for the same system. The same solution in the case of ED systems is associated to the classical electron trajectory which results from the relativistic Hamilton-Jacobi equation. There is a similitude between our result and the double solution theory of de Broglie, despite the fact that the two treatments are completely different. In the double solution theory [20], it is shown that Schrödinger's equation admits a second wave solution, which is a classical-type wave function, as in our analysis.

## References


[1] Popa A 1998 *J. Phys. Soc. Jpn.* **67** 2645
[2] Popa A 1999 *J. Phys. Soc. Jpn.* **68** 763
[3] Popa A 1999 *J. Phys. Soc. Jpn.* **68** 2923
[4] Popa A 2003 J. Phys. A: Math. Gen. **36** 7569
[5] Popa A 2005 J. Chem.Phys. **122** 244701
[6] Popa A 2011 Phys. Rev. A **84** 023824
[7] Popa A 2008 Eur. Phys. J. D **49** 279
[8] Popa A 2009 Eur. Phys. J. D **54** 575
[9] Popa A 2011 Mol. Phys. **109** 575
[10] Popa A 2008 J. Phys. B: At. Mol. Opt. Phys. **41** 015601
[11] Popa A 2009 J. Phys. B: At. Mol. Opt. Phys. **42** 025601
[12] Popa A 2012 Las. Part. Beams **30** 591
[13] Popa A 2013 IEEE J. Quant. Elect. **49** 522
[14] Popa A 2013 IEEE T. Plasma Phys., in print
[15] Messiah A 1965, *Quantum Mechanics*, Vol. 1, (Amsterdam: North-Holland Publishing Company)
[16] Courant R and Hilbert D 1962 *Methods of Mathematical Physics*, Vol. II, (New York: Wiley-Interscience)
[17] Smirnov V I 1964 *A Course of Higher Mathematics* Vol. IV, (New York: Pergamon Press)
[18] Zauderer E 1983 *Partial Differential Equations of Applied Mathematics* (New York: John Wiley and Sons)

[19] Landau L et Lifschitz M 1980 *Mechanics* (Oxford: Butterworth Heinemann)
[20] de Broglie, L 1956 *Une tentative d'interprtation causale et nonlinare de la Mcanique ondulatorie: la thorie de la double solution* (Paris: Gauthier-Villars)

# Kinetic Monte Carlo Simulation of the oscillatory catalytic CO oxidation using a modified Ziff-Gulari-Barshad model.

**Indrajit Sinha**[1] **and Ashim K Mukherjee**[2]

[1] Applied Chemistry, Indian Institute of Technology (B. H. U.), Varanasi, INDIA
[2] Applied Chemistry, Indian Institute of Technology (B. H. U.), Varanasi, INDIA

E-mail: ashim.apc@iitbhu.ac.in

**Abstract**. The oxidation of CO on Pt-group metal surfaces has attracted widespread attention since a long time due to its interesting oscillatory kinetics and spatiotemporal behavior. The use of STM in conjunction with other experimental data has confirmed the validity of the surface reconstruction (SR) model under low pressure and the more recent surface oxide (SO) model which is possible under sub-atmospheric pressure conditions [1]. In the SR model the surface is periodically reconstructed below a certain low critical CO-coverage and this reconstruction is lifted above a second, higher critical CO-coverage. Alternatively the SO model proposes periodic switching between a low-reactivity metallic surface and a high-reactivity oxide surface. Here we present an overview of our recent kinetic Monte Carlo (KMC) simulation studies on the oscillatory kinetics of surface catalyzed CO oxidation. Different modifications of the lattice gas Ziff-Gulari-Barshad (ZGB) model have been utilized or proposed for this purpose. First we present the effect of desorption on the ZGB reactive to poisoned irreversible phase transition in the SR model. Next we discuss our recent research on KMC simulation of the SO model. The ZGB framework is utilized to propose a new model incorporating not only the standard Langmuir-Hinshelwood (LH) mechanism, but also introducing the Mars-van Krevelen (MvK) mechanism for the surface oxide phase [5]. Phase diagrams, which are plots between long time averages of various oscillating quantities against the normalized CO pressure, show two or three transitions depending on the CO coverage critical threshold (CT) value beyond which all adsorbed oxygen atoms are converted to surface oxide.

## 1. Introduction

Reactions rate oscillations have been observed in the catalytic oxidation of CO on Pt-group metal surfaces. Investigations conducted with LEED, STM etc. have shown that at low pressures an adsorbate-induced phase transition, which is controlled by critical adsorbate coverages, may explain such rate oscillations [1-2]. On the other hand, at higher pressures (up to atmospheric) the "oxide model" based on the Mars–van Krevelen (MvK) mechanism has been used to explain rate oscillations [2-3].

The adsorbate-induced transition has been incorporated successfully into the ZGB simulation model by Albano to obtain periodic oscillatory behaviour, caused by transitions of the surface reconstructions coupled with reaction coverages [4-5]. On the other hand Noussiou and Provata in order to account for rate oscillations at atmospheric pressures modified the ZGB simulation model to include an additional mechanistic step of surface oxide formation along with the standard Langmuir-Hinshelwood (LH) mechanism [3].

The present authors have investigated the effect of CO desorption on the first order irreversible phase transition reported by Albano in his surface reconstruction model [6]. The authors have also carried out Monte Carlo simulation of the surface oxide model, referred to earlier, incorporating not only the standard Langmuir-Hinshelwood (LH) mechanism, but also introducing the Mars-van Krevelen (MvK) mechanism for the surface oxide phase [7].

## 2. **Model and Simulation Procedure**

The simulation in either of the models is based on the ZGB algorithm, incorporating the Langmuir-Hinshelwood mechanism for catalytic reaction and has certain common features [8]. The model system is a two-dimensional square lattice of side $L(=100)$. Periodic boundary conditions are employed throughout the simulation. Each simulation starts from an empty lattice. The system evolves in units of Monte Carlo steps. One Monte Carlo step (MCS) of our simulation involves $L^2$ trials. A trial begins by choosing either CO or $O_2$ for adsorption attempt on the surface. Reactants CO and $O_2$ are adsorbed on the surface of the catalyst with normalized probabilities $P_{CO}$ and $P_{O_2}$ (that is $P_{CO}+P_{O_2}=1$) proportional to the respective reactant partial pressures. If the trial selects a CO molecule from the gas phase then it may adsorb onto a vacant lattice site. On the other hand if an oxygen molecule is selected, it first dissociates into two atoms and then may get adsorbed onto two neighboring vacant sites. As per the LH mechanism, if adjacent sites are occupied by O and CO molecules, then they react to form $CO_2$ leaving behind two empty sites. The reaction scheme followed in the oxide model is as given below:

$CO\,(g) + s \rightarrow CO_{ads}$ (1) $\quad O_2(g) + 2s \rightarrow 2O_{ads}$ (2)

$O_{ads} \rightarrow ox$ under the condition $[\Theta_{O(ads)} > CT]$ (3) $\quad CO_{ads} + O_{ads} \rightarrow CO_2(g) + 2s$ (4)

$CO\,(g) + ox \rightarrow CO_2(g) + s$ (5) $\quad CO_{ads} + ox \rightarrow CO_2(g) + 2s$ (6)

Here the (ads) and (g) indices denote the adsorbed and gaseous species, respectively, while 's' and 'ox' denote the vacant lattice and oxide sites, respectively. In the SR model the reaction scheme consists of steps (1), (2) and (4) only.

The critical threshold value in the SO model is defined as the adsorbed oxygen coverage above which all adsorbed oxygen atoms convert into surface oxide sites (step 3). After the formation of the above oxide sites the production of $CO_2$ may take place by any of the three mechanisms (steps (4)-(6)), including the Mars-van Krevelen (MvK) mechanism (step 5). In the SR model once CO coverage $\Theta_{CO}$ exceeds $\geq 0.485$, all sites in the model catalyst surface transform to the non-reconstructed phase. The sticking coefficient of oxygen becomes 1, the normal ZGB algorithm is initiated and $\Theta_{CO}$ decreases because of reactions between neighboring adsorbed CO and O. Whenever the condition $\Theta_{CO} \leq 0.1$ is achieved, all sites in the system are assumed to transform to the reconstructed phase. Thereafter, as mentioned earlier, only CO adsorption occurs at a rate proportional to $P_{CO}$ until the condition $\Theta_{CO} \geq 0.485$ is achieved again. This completes one cycle of the adsorbate coverage induced transition.

## 3. **Results and Discussion**

### 3.1. *Surface reconstruction Model*

In order to understand the effect of desorption on this discontinuous transition, we now present results for systems with k > 0 [6]. It is found that for smaller sizes (L=60) with increase in k there is broadening in the bimodal distribution to such an extent, that for k = 0.04, a unimodal distribution sweeps across smoothly from the reactive side to the partially poisoned side as $P_{CO}$ is increased (Fig 1). Focusing only on the distribution at the finite size coexistence point $P^{c}_{CO}(L)$ for systems at different desorption rates, it is observed that for a given large enough system size (L=230), the transition retains the bimodal distribution character (Fig. 2) even at high desorption rates (k=0.04). This suggests that the phase transition remains first-order and only becomes weaker with increase in desorption rates. For a given desorption rate the strength of first-order phase transition increases with size.

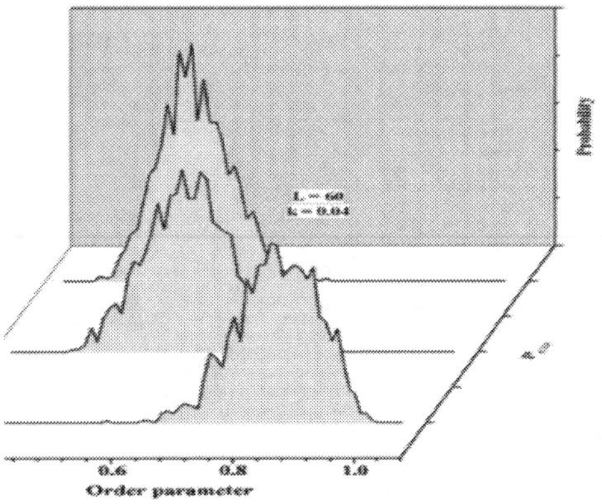


Fig 1 Order parameter probability distribution $\overline{\Theta_{CO}}$ for L=60 and desorption rate k = 0.04

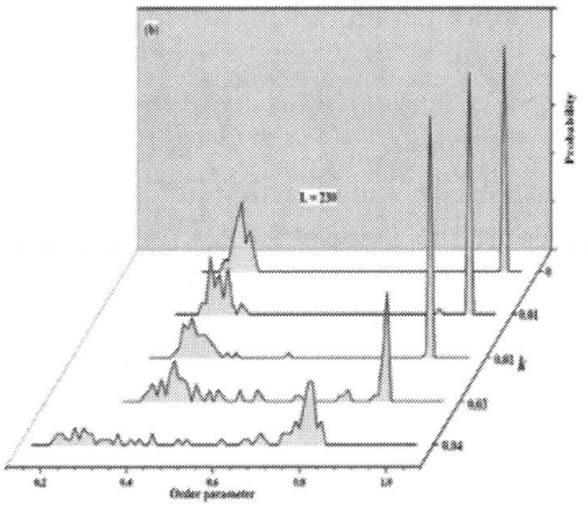

Fig 2 Order parameter probability distribution $\overline{\Theta_{CO}}$ for L= 230 at increasing desorption rates

3.2. *Oxide Model*

The phase diagram obtained in our "oxide" model, which includes the Mars-van Krevelen (MvK) mechanism, is shown in figure 3 for critical threshold (CT) = 0.4. It is observed from the curve that the reactive state undergoes a continuous transition from one reactive state to another at $T_1$. Above $T_1$ the nature of rate oscillations changes gradually with $P_{CO}$. The continuous transition ($T_1$) in terms of $\overline{\Theta_{O(ad)}}$ is indicative of the change in mechanism of CO oxidation. In this regime the mechanism of CO oxidation oscillates between the MvK + LH mechanism exhibited by the oxide phase and the LH only mechanism of adsorbed CO with adsorbed oxygen. It is found that for $CT \geq 0.5$ the phase diagram shows another transition $T_2$ (before$T_3$) where $\overline{\Theta_{ox}}$ decreases to zero (ig 4). If $\overline{\Theta_{ox}}$ is considered the order parameter then $T_2$ appears to be a continuous transition. With increase in CT, the maximum $P_{CO}$ at which the oxide phase transition ($T_2$) can happen, decreases, that is $T_2$ shifts leftwards with CT. The change in nature of $CO_2$ rate oscillations with increase in $P_{CO}$ at CT = 0.55 is depicted in figure 5. With increase in CT the transition$T_2$ shifts to the left, therefore the distance between $P_{CO}$= 0.45 and $T_2$ decreases. In other words, with increase in CT, at a given $P_{CO}$ value the probability of surface oxide formation decreases. We observe from the figure that the amplitude and period of the rate oscillations increase as $P_{CO}$ is increased gradually from 0.45 to 0.5 and the rate oscillations die out or become negligible at $P_{CO}$= 0.52, since this is above the transition $T_2$.

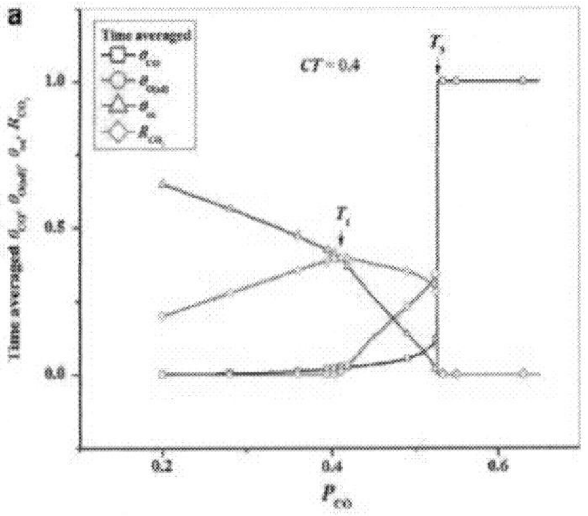


Fig 3 Phase diagram showing plots of $\overline{R_{CO_2}}, \overline{\Theta_{CO}}, \overline{\Theta_{O_2(ad)}}, \overline{\Theta_{Ox}}$ vs. $P_{CO}$ at $CT = 0.4$

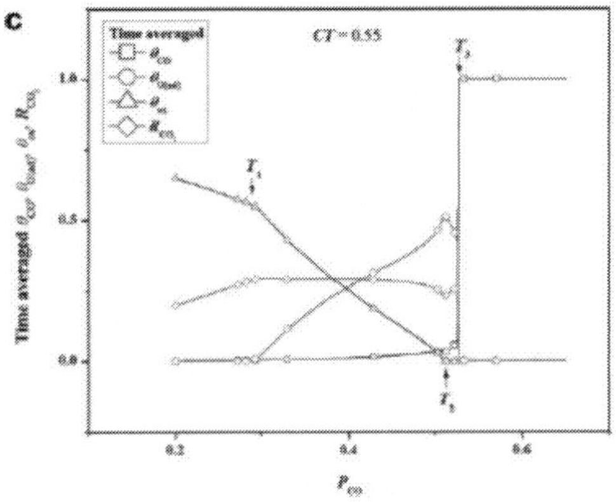


Fig 4 Phase diagram showing plots of $\overline{R_{CO_2}}, \overline{\Theta_{CO}}, \overline{\Theta_{O_2(ad)}}, \overline{\Theta_{Ox}}$ vs. $P_{CO}$ at $CT = 0.55$

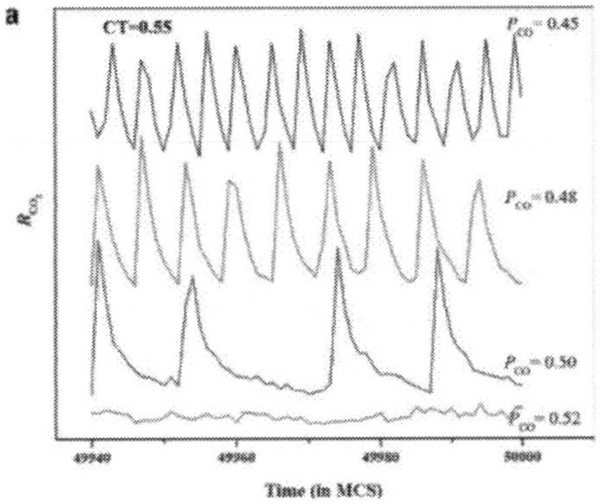


Fig 5. Change in $R_{CO_2}$ oscillatory behaviour with increase in $P_{CO}$ at $CT = 0.55$

## References


1. Imbihl R and Ertl G 1995 *Chem. Rev.* **95** 697.
2. Hendriksen B L M, Bobaru S C and Frenken J W M 2005 *Catal. Today* **105** 234.
3. Noussiou V K and Provata A 2008 *Chem. Phys.* **348** 11.
4. Albano E V 1997 *Langmuir* **13**, 4013.
5. Albano E V 1998 *J. Chem. Phys.* **109** 7498.
6. Sinha I and Mukherjee A K 2012 *J. Stat. Phys.* **146** 669.
7. Sinha I and Mukherjee A K 2012 *Chem. Phys. Lett.* **553** 30.
8. Ziff R M, Gulari E and Barshad Y 1986 *Phys. Rev. Lett.* **56** 2553.

# Arbitrary l-state solutions of the Feynman propagator for the Rosen-Morse Potential with a centrifugal term approximation

**Ahmed Diaf and Mohamed Hachama**

Production Editor, *Journal of Physics: Conference Series*, IOP Publishing, Dirac House, Temple Back, Bristol BS1 6BE, UK

E-mail: jacky.mucklow@iop.org

**Abstract.** We present analytic solutions of the Feynman propagator with the Rosen-Morse potential. To do that, an approximation of the centrifugal potential is used and nonlinear space-time transformations are applied. A relation between the original path integral and the Green function of a new quantum soluble system is derived. Explicit expressions of the bound state energy spectra and the eigenfunctions are obtained and compared to those of Schrodinger formalism.

## 1. Introduction

It is well known that only a limited number of potentials admit exact solutions of the Schrödinger equation [1, 2], especially in the three-dimensional space when the angular momentum $l \neq 0$. This is due to the presence of the centrifugal term $(1/r^2)$. Motivated by the success in obtaining approximately the bound state solutions of the Feynman propagator with the Manning-Rosen potential [5, 6], we attempt here to solve the arbitrary $l$-wave propagator with the Rosen-Morse potential. An improved approximation scheme for the centrifugal term is used as in [7, 8, 9] and non linear space-time transformations are applied.

The Rosen-Morse $(RM)$ potential is given by

$$V^{RM}(r) = -\frac{V_1}{\cosh^2(r/a)} + V_2 \tanh(r/a). \tag{1}$$

Here $V_1$ and $V_2$ are two strength parameters; the screening parameter $a$ has a dimension of length. This potential has been one of the most useful and convenient models to study the energy eigenvalues of diatomic molecules. It is known that for this potential, the Schrödinger equation can be solved exactly when $l = 0$ either directly [10] or by using the Feynman path integral formalism [11]. Unfortunately, for arbitrary $l$-state, the radial Schrödinger equation does not admit exact solutions. In this case, some authors have used the approximation scheme proposed by Lu [7] to study analytically arbitrary $l$-wave bound states of the Schrödinger equation and the relativistic wave equation for the Rosen-Morse potential [8, 9].

The purpose of this paper is to study $l$-states solutions of the Rosen-Morse potential within the Feynman path integrals formalism as done in the recent works [8, 9] in which the

Nikiforov-Uvarov method is used. The method we propose consists in using the centrifugal term approximation porposed by Lu [7], $\frac{1}{r^2} \approx \frac{1}{r_0{}^2}\left(C_0 + \frac{C_1}{e^{\frac{r}{a}}+1} + \frac{C_2}{(e^{\frac{r}{a}}+1)^2}\right)$, where the parametrs $(C_i)$ can be expressed in terms of the specific potential parameters, and $r_0 = -\tanh^{-1}(\frac{V_2}{V_1})$ corresponds to the minimum of the potential.

The rest of the paper is organized as follows. In section 2, we present the centrifugal term approximation and apply it to the Rosen-Morse propagator. Then, using the Duru-Kleinert method we determine the $l$-wave eigensolutions for the Rosen-Morse potential within the path integral formalism. Finally, in Section 3, we give concluding remarks.

## 2. Path integral for the the Rosen-Morse propagator

In spherical coordinates, the propagator related to the Rosen-Morse potential, between two time-space points $(\vec{r}\,', t')$ and $(\vec{r}\,'', t'')$, is written [12]:

$$K(\vec{r}\,'', t''; \vec{r}\,', t') = \frac{1}{4\pi r'' r'} \sum_{l=0}^{\infty} (2l+1) K_l(r'', t''; r', t') P_l(\cos\theta). \tag{2}$$

$P_l(\cos\theta)$ is the Legendre polynomial with $\theta \equiv (\vec{r}\,'', \vec{r}\,')$ and

$$K_l(r'', t''; r', t') = \lim_{N\to\infty} \int \prod_{j=1}^{N} \exp\left[\frac{i}{\hbar} S_j\right] \prod_{j=1}^{N} \left[\frac{m}{2\pi i\hbar\varepsilon}\right]^{\frac{1}{2}} \prod_{j=1}^{N-1} dr_j. \tag{3}$$

Here $S_j = \frac{m}{2\varepsilon}(\Delta r_j)^2 - \varepsilon V_{eff}(r_j)$, where

$$V_{eff}(r_j) = \frac{\hbar^2}{2m} \frac{l(l+1)}{r_j r_{j-1}} - \frac{V_1}{\cosh^2(r_j/a)} + V_2 \tanh(r_j/a) \tag{4}$$

and $\Delta r_j = r_j - r_{j-1}$, $\varepsilon = t_j - t_{j-1}$, $t' = t_0$ and $t'' = t_N$. Moreover, setting $T = t'' - t'$, we can write:

$$K_l(r'', r', T) := K_l(r'', t''; r', t'). \tag{5}$$

To calculate the propagator $K_l$ given by equation (3) for $l \neq 0$ states, we apply the following approximate scheme to the centrifugal term [7] :

$$\frac{l(l+1)}{r_j{}^2} \approx \frac{l(l+1)}{r_0{}^2}\left[C_0 + C_1 \frac{-e^{\frac{-2r_j}{a}}}{1+e^{-\frac{2r_j}{a}}} + C_2\left(\frac{-e^{\frac{-2r_j}{a}}}{1+e^{-\frac{2r_j}{a}}}\right)^2\right], \tag{6}$$

where

$$C_0 = 1 - \left(\frac{1+e^{-2(r_0/a)}}{2(r_0/a)}\right)^2 \left(\frac{8(r_0/a)}{1+e^{-2(r_0/a)}} - 3 - 2(r_0/a)\right),$$

$$C_1 = -2\left(1+e^{2(r_0/a)}\right) \times \left[3\left(\frac{1+e^{-2(r_0/a)}}{2(r_0/a)}\right) - (3+2(r_0/a))\left(\frac{1+e^{-2(r_0/a)}}{2(r_0/a)}\right)\right],$$

$$C_2 = \left(1+e^{2(r_0/a)}\right)^2 \left(\frac{1+e^{-2(r_0/a)}}{2(r_0/a)}\right)^2 \times \left(3 + 2(r_0/a) - \frac{4(r_0/a)}{1+e^{-2(r_0/a)}}\right).$$

By substituting the centrifugal term in (4) by its approximation (6), we get

$$V_{eff}(r_j) = A\tanh(r_j/a) - \frac{B}{\cosh^2(r_j/a)} + C, \tag{7}$$

with $A = (C_1 - C_2)\frac{l(l+1)}{4mr_0^2} + V_2$, $B = C_2\frac{l(l+1)}{8mr_0^2} + V_1$ and $C = (C_2 - C_1)\frac{l(l+1)}{4mr_0^2} + \frac{l(l+1)}{2mr_0^2}C_0$.

### 2.1. Space-time transformations

In the following, we study the potential (7) with the Duru-Kleinert method [3]. The path integral for the potential (7) is

$$K_l(r'',r';T) = \int \mathcal{D}r(t)\exp\left[\int_{t'}^{t''}\left(\frac{m}{2}\dot{r}^2 - V_{eff}(r)\right)dt\right]. \tag{8}$$

By achieving the point canonical transformation and the time transformation [4],

$$\begin{cases} r &= f(q) \\ dt &= [f'(q)]^2 d\mathfrak{s}, \end{cases} \tag{9}$$

with

$$f(q) = a\,.\,arctanh\left(2(\tanh q)^2 - 1\right). \tag{10}$$

The propagator expression becomes [6]:

$$\widehat{K}_l(q'',q';s'') = \exp\left[\frac{i}{\hbar}s''(E+C-A)a^2\right]\times$$
$$\int \mathcal{D}q(\mathfrak{s})\exp\left[\frac{i}{\hbar}\int_0^{s''}\frac{m}{2}\dot{q}^2 - \frac{\hbar^2}{2m}\left[\frac{\eta(\eta-1)}{\sinh^2 q} - \frac{\upsilon(\upsilon-1)}{\cosh^2 q}\right]d\mathfrak{s}\right]$$
$$= \exp\left[\frac{i}{\hbar}s''(E+C-A)a^2\right]K_l^{MPT}(q'',q';s'') \tag{11}$$

with $\eta = \frac{1}{2} \pm \sqrt{-2ma^2(E+C+A)/\hbar^2}$, $\upsilon = \frac{1}{2} \pm \sqrt{1+8mBa^2/\hbar^2}$, and $K_l^{MPT}$ is the path integral of the modified Pöschl-Teller potential, which is a known solved problem [13, 14, 15] for witch the bound states wave functions are explicitly given by [15]:

$$\chi_{l,n}^{(k_1,k_2)}(q) = N_n^{(k_1,k_2)}(\sinh q)^{2k_2-1/2}(\cosh q)^{-2k_1+3/2}\times {}_2F_1(-k_1+k_2+k,-k_1+k_2-k+1;2k_2;-\sinh^2 q), \tag{12}$$

where $k = k_1 - k_2 - n$, and

$$N_n^{(k_1,k_2)} = \frac{1}{\Gamma(2k_2)}\left(\frac{(2k-1)\Gamma(k_1+k_2-k)\Gamma(k_1+k_2+k-1)}{\Gamma(k_1-k_2+k)\Gamma(k_1-k_2-k+1)}\right)^{1/2}, \tag{13}$$

The eigenvalues are expressed by

$$E_n^{MPT} = -\frac{\hbar^2}{2m}\left[2(k_1-k_2-n)-1\right]^2. \tag{14}$$

### 2.2. Energy spectrum and wave functions

Calculating $\widehat{K}_l(q'',q';s'')$ allows us to obtain the Green function $G$. Once this is known, the whole energy spectrum is obtained from the poles. We obtain the corresponding wave functions from the residues at the poles.

Substituting (12) in (11), we get

$$\widehat{K}_l(q'',q';s'') = \sum_{n=0}^{N_m}\exp\left\{\frac{i}{\hbar}s''\left[(E+C-A)(a)^2 - E_n^{MPT}\right]\right\}\chi_{l,n}^{(k_1,k_2)}(q'')\chi_{l,n}^{*(k_1,k_2)}(q')$$
$$+\int_0^\infty dK\exp\left\{\frac{i}{\hbar}s''[(E+C-A)(a)^2 - \frac{\hbar^2K^2}{2m})]\right\}\chi_K^{(k_1,k_2)}(q'')\chi_K^{*(k_1,k_2)}(q'). \tag{15}$$

Then, by integrating this latter over the pseudo-time parameters $s''$, the Green function $G$ given in [6] becomes

$$G_l(r'',r';E)=\sum_{n=0}^{N_m}\frac{\chi_{l,n}^{RM(k_1,k_2)}(r'')\chi_{l,n}^{RM*(k_1,k_2)}(r')}{E_{n,l}^{RM}-E}+\int_0^{\infty}dK\frac{\chi_{l,K}^{RM}(r'')\chi_{l,K}^{RM*}(r')}{(\hbar^2K^2/2ma^2)-A-C-E}. \quad (16)$$

As in [4, 11], satisfying the boundary conditions for $r\to 0$ and $r\to\infty$ gives:

$$k_1=\frac{1}{2}\left(1+\sqrt{1+8mBa^2/\hbar^2}\right)\equiv\frac{1}{2}(1+s)\,,k_2=\frac{1}{2}\left(1+\frac{1}{2}(s-2n-1)-2mAa^2/(s-2n-1)\right).$$

By substituting these last values in (12) and (14), and considering $u=\frac{1}{2}\left[1+\tanh\frac{r}{a}\right]$, we get

$$\chi_{l,n}^{RM(k_1,k_2)}(r)=\frac{N_n^{(k_1,k_2)}}{\sqrt{a}}u^{k_2-(1/2)}(1-u)_2^{(s/2)-k_2-n}\times F_1(-n,s-n;2k_2;u), \quad (17)$$

where

$$E_{n,l}^{RM}=-\frac{\hbar^2(s-2n-1)^2}{8ma^2}-\frac{2mA^2a^2}{\hbar^2(s-2n-1)^2}+C. \quad (18)$$

The constants entering in this expression have been defined in Section 2.

If we set $\delta=-\frac{1}{2}(1+s)$, the energy $E_{n,l}^{RM}$ turns to

$$E_{n,l}^{RM}=\frac{l(l+1)C_0}{2mr_0^2}+V_2-\frac{\hbar^2}{2ma^2}\times\left[\frac{(n+1)^2+\delta(2n+1)+(\delta^2+\delta)+mAa^2/\hbar^2}{(n+\delta+1)}\right]^2. \quad (19)$$

This expression is perfectly identical to that given in [8, 9], obtained with the framework of the Schrödinger formalism.

## 3. Concluding remarks

In this paper, we have calculated the energy levels of the Feynman kernel with the Rosen-Morse potential. Use is made of an improved approximation scheme for the centrifugal term and a nonlinear space-time transformations in the radial path integral. Explicit expressions of the wave functions are obtained in terms of Jacobi polynomials and the energy eigenvalues are expressed in function of the potential parameters. We have shown that our results are identical with those obtained with the Schrödinger formalism. In future works, we plan to apply this technique to other exponential-type potentials.

## References


[1] L. D. Landau and E. M. Lifshitz (1977) Quantum Mechanics, Non-Relativistic Theory, Pergamon, Oxford.
[2] L. I. Schiff (1955) Quantum Mechanics 3rd edition, McGraw-Hill, New York.
[3] H. Kleinert, Path integrals in Quantum Mechanics, Statistics and Polymer Physics, World scientific, Singapore, 2009.
[4] C. Groshe, F. Steiner, Handbook of Feynman Path Integrals, Springer, Berlin, 1998.
[5] A. Diaf, A. Chouchaoui and R.L. Lombard, Ann. Phys. 317, (2005) 354.
[6] A Diaf and A Chouchaoui, Phys. Scr. 84, (2011) 015004.
[7] Jun Lu, Phys. Scr. 72, (2005) 349.
[8] F. Taskin, Int. J. Theor. Phys. 48, (2009) 26922697.
[9] S.M. Ikhdair, J. Math. Phys. 51, (2010) 023525.
[10] M. M. Nieto, Phys. Rev. A 17, (1978) 1273.
[11] H. Kleinert, I. Mustapic, J. Math. Phys. 33 (1992) 643-662.
[12] D.C. Khandekar, S.V. Lawande, K.V. Bhagwat, Path Integral Methods and their Applications, World Scientific, Singapore, 1986.
[13] Böhm M and Junker G J . Math. Phys. A 117, (1986) 375.
[14] Inomata, A., Kuratsuji, H., Gerry, C.C.: Path Integrals and Coherent States SU(2) and SU(1,1). World Scientific, Singapore, 1992.
[15] C. Groshe J. Phys. A. Math. Gen. 22, (1989) 5073.

# Interaction forces between nanoparticles in Lennard-Jones (L-J) solvents

**Indrajit Sinha[1] and Ashim K Mukherjee[1]**

[1]Applied Chemistry, Indian Institute of Technology (B. H. U.),Varanasi,INDIA

E-mail: isinha.apc@iitbhu.ac.in

**Abstract.** Molecular simulations, such as Monte Carlo (MC) and molecular dynamics (MD) have been recently used for understanding the forces between colloidal nanoparticles that determine the dispersion and stability of nanoparticle suspensions. Herein we review the current status of research in the area of nanoparticles immersed in L-J solvents. The first study by Shinto et al. used large smooth spheres to depict nanoparticles in L-J and soft sphere solvents. The nanoparticles were held fixed at a particular interparticle distance and only the solvents were allowed to equilibrate. Both Van-der-waals and solvation forces were computed at different but fixed interparticle separation. Later Qin and Fitchthorn improved on this model by considering the nanoparticles as collection of molecules, thus taking into the account the effect of surface roughness of nanoparticles. Although the inter particle distance was fixed, the rotation of such nanoparticles with respect to each other was also investigated. Recently, in keeping with the experimental situation, we modified this model by allowing the nanoparticles to move and rotate freely. Solvophilic, neutral and solvophobic interactions between the solvent atoms and those that make up the nanoparticles were modelled. While neutral and solvophobic nanoparticles coalesce even at intermediate distances, solvophilic nanoparticles are more stable in solution due to the formation of a solvent shield.

## 1. Introduction

Nanoparticle surfaces need to be passivated or stabilized to enable the formation of stable nanoparticle dispersions. Such stabilization may be electrostatic or steric or electrosteric or simply the solvent itself or a ligand acts as the stabilizer. Inter nanoparticle forces due to various sources therefore determine the state of dispersion and the stability of nanoparticle suspensions. Properties of such nanofluids depend on the inter-nanostructure forces, as well as the interactions between the nanostructures and the surrounding medium [1, 2]. The focus of the present communication is to review the research in the area of effect of solvent on inter nanoparticle interaction forces.

## 2. Forces between smooth nanoparticles in L-J Fluids

In the first study by Shinto et al [3] the solvation forces between two large smooth spheres in an L-J fluid was simulated using MD. Interactions between the solvent molecules were given by a dimensionless shifted 12-6 L-J potential expression. In contrast the interaction between solvent particles and nanoparticle were computed using the shifted 10-4-3 L-J potential. Solvophilic or solvophobic particle-solvent molecule interactions were considered. The L-J solvent dimensionless density was taken to be 0.59 while the dimensionless temperature was chosen to be 1.2. The nanoparticle was assumed to be 10 times the size of the monatomic solvent molecule. The solvation forces were computed at fixed inter-nanoparticle distances which ranged from $0\sigma$ to $5.0\sigma$ where $\sigma$ is the diameter of a solvent molecule. For the solvophobic particle-solvent interaction model, the solvation force profile exhibited a strong attractive trend (with decrease in interparticle distance) without display of any oscillatory behaviour (Fig. 1a). The presence of the solvophobic nanoparticles disrupts attractive solvent-solvent interactions and makes difficult the formation of any dense solvent molecule layer near the nanoparticle surface. On the other hand, the solvent density near the solvophilic surface was found to be higher than that near the solvophobic surface. The solvation forces oscillate with a periodicity of about solvent molecule diameter due to the packing effect. This periodicity in oscillatory forces was observed up to three solvent molecular diameters. Solvation forces could be of comparable or greater magnitude to the van der Waals attraction between the nanoparticles.

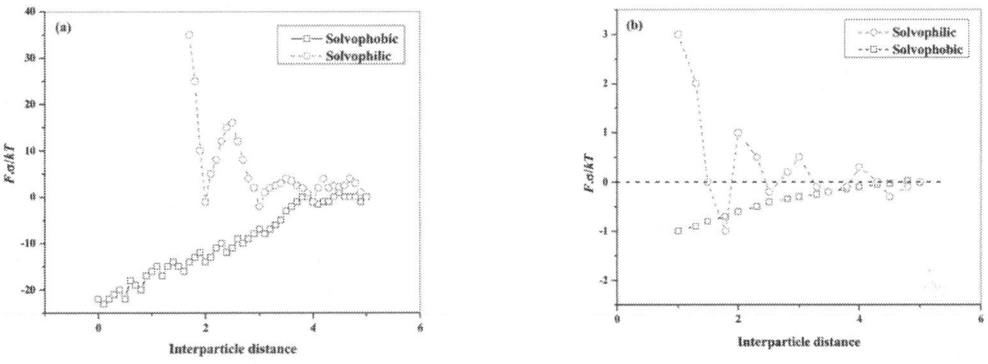


Figure1. A comparison of solvation forces for a) smooth [3] and b) rough [4] nanoparticles.

These solvation force predictions are similar to those to those given by earlier MC and MD simulations for fluids confined between flat and infinite surfaces. The solvent molecules tend to form layers around nanoparticles which prevent the nanoparticles from aggregating. However, when the separation distance between the large particles is smaller than the diameter of the small particles so that no small particles can fit in the gap between the large particles, a net attractive force, the so-called depletion force, between the large particles is expected. The oscillatory nature of the solvation forces is also due to this reason. Properly packed layered solvent molecule structure formation induces a repulsive barrier. Between formation of integral number of packed layers weak attractive depletion forces do occur resulting in observation of oscillatory forces.

## 3. Effect of surface roughness on solvation forces in L-J Fluids

A very important aspect of inter nanoparticle forces is the balance between solvation and van der Waals forces. For smooth nanoparticles, the above mentioned study by Shinto et al. demonstrated that solvation forces could be of comparable or greater magnitude to the van der Waals attraction between the nanoparticles. On the other hand surface roughness of the nanoparticles can also significantly affect the magnitude of solvation forces influencing the balance between solvation and van der Waals forces. It was with this objective that Qin and Fichthorn [4] simulated nanoparticles as rigidly fixed L-J clusters. To maintain such nanoparticles in solid state the energy parameter in the LJ expression for solid-solid interactions $\left(\varepsilon^{solid-solid}\right)$ is taken to be 10 times the solvent-solvent $\left(\varepsilon^{solvent-solvent}\right)$ interaction energy parameter. The diameters of solid and solvent molecules are taken to be the same and the typical NVT ensemble molecular dynamics simulations were carried out. Solvophilic or solvophobic nanoparticle-solvent molecule interactions were modelled. Since the solvophilic case is characterized by good attraction between the solvent and solid molecules constituting the nanoparticles, therefore this case is modeled by taking the solid-fluid molecule interaction energy parameter as $\varepsilon_{solvophilic}^{solid-solvent} = 5\varepsilon^{solvent-solvent}$. Weaker interactions between molecules making solvophobic nanoparticles and solvent molecules are modeled by taking $\varepsilon_{solvophobic}^{solid-solvent} = 0.2\varepsilon^{solvent-solvent}$. Figure 1b shows oscillatory solvation forces for solvophilic nanoparticles and monotonic, attractive forces for solvophobic nanoparticles. Similar to results obtained by Shinto et al, here also the solvation forces are found to be comparable or greater than the van der Waals forces. Surface roughness affects van der Waals forces the least and that only for very small interparticle distances. For solvophobic nanoparticles, as the interparticle separation is increased, surface roughness causes fluctuations in the monotonic attractive trend of forces. But in the solvophilic case, surface roughness influences the phase shift of the oscillatory solvation-force profile, as well as the amplitude. This means that the solvation force for a fixed separation can be either attractive or repulsive, depending on the relative orientation of the nanoparticles. Thus in a later work [5] the same authors demonstrated that oscillations diminish or disappear with inter particle distance if the nanoparticles are allowed to rotate although the separation between them is kept fixed.

## 4. Effect of movement of nanoparticles on solvation forces

In a recent communication [6] our group has modified the model given by Qin and Fichthorn. As in reference [4], here also to include surface roughness we consider the nanoparticles to be clusters of molecules. Solvophilic, neutral and solvophobic interactions between the solvent molecules and those that make up the nanoparticles are considered. Simulations were done using typical NVT ensemble Metropolis algorithm. In this investigation the nanoparticles are free to move as well rotate and thus to find the state of minimum free energy for the given state of the system. This is made possible by considering Monte Carlo (MC) moves for both the molecules making up the nanoparticles and the solvent. The dimensionless temperature is taken to be 1.0. Ensemble average solvation and van der Waal's forces are computed once equilibrium inter nanoparticle distances are achieved.

A simulation is initiated from a state of two nanoparticles at a specified distance. The solvent molecules are added randomly to achieve the L-J solvent dimensionless density 0.64. This study emphasizes finding the interparticle distance at which the two nanoparticles are stable against coalescence. For example, the solvophilic case the configuration starting from initial interparticle distance $\delta_i = 1.5\sigma$ stabilizes at $\delta_{eq} = 1.0\sigma$. A thin solvent shield is found to stabilize the nanoparticle suspension. In contrast to solvophilic case, for the neutral interaction case solvent–solvent molecular interactions and solvent–solid (or nanoparticle) molecule interactions are equal. Therefore, two nanoparticles are effectively solvated or stable against coalescence only at larger $\delta_i$ in comparison to solvophilic case. For the solvophobic case the solvent–solvent interactions are much stronger than solvent–solid interactions. Here the nanoparticles are stable against coalescence at still larger $\delta_i$ as compared to solvophilic or the neutral systems. Further, one also observes that solvophilic, neutral and solvophobic systems starting from the same $\delta_i$ distance may stabilize to different average $\delta_{eq}$ for separate simulation runs. Therefore, stable inter nanoparticle distance configurations seem to represent local minima. In contrast to regular oscillatory behavior found in references 3 and 4, here only weak but irregular oscillatory behaviour of the solvation forces for solvophilic nanoparticles are observed. In the present study force profiles for different nanoparticle orientations and distances are phase shifted relative to one another, so that the solvation forces at a fixed center-of-mass distance can be attractive for some particle configurations and repulsive for others. This suggests that if nanoparticles are allowed to freely rotate and move, they select configurations that minimize the free energy and reduce the solvation forces. For the stable solvophobic nanoparticles case, the solvent–solvent interaction forces stronger than solvent–solid interactions and this is manifested by the slightly positive bias of the solvation forces in the solvophobic system.

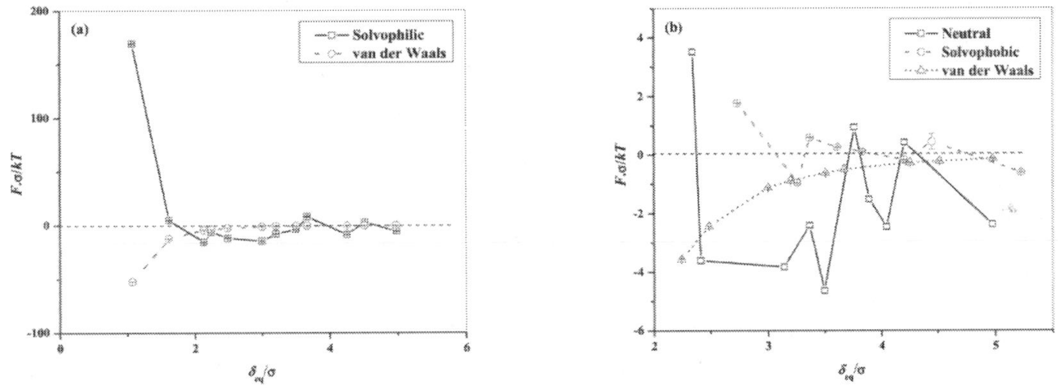


Figure 2. (a) Solvation and van der Waals forces with increasing interparticle distance $\delta_{eq}$ for solvophilic system

(b) Solvation forces with increasing interparticle distance ($\delta_{eq}$) for the neutral and solvophobic systems.

## References

[1] Wasan S D, Nikolova A and Moudgil B 2005 *Powder Technol.* **153** 135
[2] Pilkington G A and Briscoe W H 2012 *Adv. Colloid Interfac.* **179-182** 68
[3] Shinto H, Miyahara M and Higashitani K 1999 *J Colloid Interf. Sci.* **209** 79
[4] Qin Y and Fichthorn K A 2003 *J. Chem. Phys.* **119** 9745
[5] Qin Y and Fichthorn K A 2006 *Phys. Rev. E* **73** 020401R
[6] Verma S K, Mukherjee A K and Sinha I 2013 *Chem. Phys. Lett.* **572** 85

# Structure of genuine and conjugate Fermi holes and the first Hund rule in He-like systems

**Tokuei Sako**

Laboratory of Physics, College of Science and Technology, Nihon University, 7-24-1 Narashinodai, Funabashi, 274-8501 Chiba, Japan

E-mail: `sako@phys.ge.cst.nihon-u.ac.jp`

**Abstract.** The singly-excited states of He and He-like atomic ions in the manifold of the $n = 3$ principal shell have been studied replying on the full configuration interaction wave function focusing on the origin of the first Hund rule. The probability density distributions of the singlet-triplet pairs of states for the three orbital configurations of $(1s)(3s)$, $(1s)(3p)$, and $(1s)(3d)$ have been examined in detail in the internal space $(r_1, r_2, \phi_-)$. The structure of the genuine and *conjugate Fermi holes* for the $(1s)(3s)$ and $(1s)(3p)$ configurations has shown similar characteristics to those for the respective $(1s)(2s)$ and $(1s)(2p)$ configurations examined in an earlier study. A significantly smaller size of the genuine and conjugate Fermi holes for the $(1s)(3d)$ singlet-triplet pair of states rationalizes the much smaller singlet-triplet energy gap of this pair than those for the $(1s)(3s)$ and $(1s)(3p)$ pairs of states.

## 1. Introduction

Hund's rule [1, 2, 3] is one of the most fundamental rules in atomic physics, that predicts the ordering of energy levels possessing the same orbital configuration yet different spin and orbital angular momentum quantum numbers. The first Hund rule, concerning the spin multiplicity, is particularly universally valid not only for atomic systems but for molecules [4] and artificial atoms [5, 6]. Despite the past long standing debates [7, 8, 9, 10, 11] focusing mainly on the relative importance of the energy components, such as the electron-nuclear attraction energy vs. the electron-electron repulsion energy, the search for its origin persists primary due to a lack of precise knowledge of the electronic structure of atoms for different spin states.

We have recently shown that for the singly-excited $S$ and $P$ states arising from the $(1s)(2s)$ and $(1s)(2p)$ configurations in the He and He-like atomic ions the origin of Hund's rule can be rationalized on the basis of the structure of the so-called genuine and *conjugate Fermi holes* [12, 13]. In the present study we have extended our analysis for different angular momentum states of singly-excited manifold of He-like systems and thus presented a unified understanding for the origin of this historical rule.

## 2. Theoretical model and computational methodology

In our previous study the so-called two-dimensional Helium model has been used in which the spatial degrees of freedom of the two electrons are confined to a two-dimensional $xy$ plane with the nucleus place at the origin. The Hamiltonian of this two-dimensional He-like systems reads in atomic units as

$$\mathcal{H}_Z = -\frac{1}{2}\sum_{i=1}^{2}\nabla_i^2 - \sum_{i=1}^{2}\frac{Z_n}{|\vec{r_i}|} + \frac{1}{|\vec{r}_1 - \vec{r}_2|}, \tag{1}$$

where $\vec{r_i} = (x_i, y_i)$ $[= r_i(\cos\phi_i, \sin\phi_i)]$ for $i = 1, 2$, while $Z_n$ represents the nuclear charge. Our previous study showed that this *two-dimensional helium model* reproduces all the characteristic features of the energy spectrum of the real 3D helium atom [12]. By reducing the dimensionality and thus the number of the degrees of freedom the *internal* part of the wave functions can be easily visualised. This allows an unambiguous manifestation of the origin of the Hund rule. By introducing a scaling for the coordinates with the nuclear charge $Z_n$ such as $\vec{s_i}\equiv Z_n\vec{r_i}$ the Hamiltonian becomes

$$\mathcal{H}_Z/Z_n^2 = -\frac{1}{2}\sum_{i=1}^{2}\nabla_{s,i}^2 - \sum_{i=1}^{2}\frac{1}{|\vec{s_i}|} + \frac{1}{Z_n}\frac{1}{|\vec{s}_1 - \vec{s}_2|}. \tag{2}$$

This scaled Hamiltonian indicates that the one-electron part, namely, the kinetic energy plus the electron-nuclear attraction energy, is independent of $Z_n$ while not the electron-electron repulsion energy. The factor $\frac{1}{Z_n}$ in front of the last term in the right-hand side of Eq. (2) guarantees that the electron-electron interaction becomes smaller relative to the one-electron energy as $Z_n$ decreases. Two-dimensional Cartesian Gaussian-type functions of the form

$$\chi^{\vec{a},\zeta}(\vec{r}) = x^{a_x}y^{a_y}\exp[-\zeta(x^2 + y^2)] \tag{3}$$

have been used to expand the one-electron orbitals for the scaled Hamiltonian (2). A detailed description of the basis set may be found elsewhere [13].

The energy and wave function of the Hamiltonian (2) has been obtained by diagonalising the full configuration interaction (full CI) matrix. The *internal space* of the two-electron He-like systems may be defined by the three variables $(r_1, r_2, \phi_-)$ [12, 13] where $r_i\equiv|\vec{r_i}|$ for $i$=1,2 and $\phi_-\equiv(\phi_1 - \phi_2)/2$. The angle variable complementary to $\phi_-$ defined by $\phi_+\equiv(\phi_1 + \phi_2)/2$ is associated with the total angular momentum $L$. The probability density distribution in the internal space $\rho(r_1, r_2, \phi_-)$ is thus obtained from the full CI wave function $\Psi(r_1, \phi_1, r_2, \phi_2)$ by integrating the square norm of $\Psi$ with the $\phi_+$ variable. The detail of the procedure may be found in an earlier paper [12].

## 3. Fermi and conjugate Fermi holes

The genuine and conjugate Fermi holes are defined by the difference between the probability density distributions of the singlet state and of the corresponding triplet state comprising the Hund's pair of states in the limit of $Z_n = \infty$ [12, 13]. By considering this large limit of the nuclear charge they are uniquely defined by the pair of zeroth-order one-electron orbitals that are the eigenfunctions of the one-electron part of the Hamiltonian (2). It is noted that in this large $Z_n$ limit the singlet and triplet wave functions give exactly the same electron density distributions. Therefore, their difference in the probability density distributions in the internal space is caused solely by the difference in the symmetry properties of their spatial wave functions with respect to exchanging the spatial coordinates of electron 1 and 2, i.e., symmetric and antisymmetric for the singlet and triplet wave functions, respectively. The difference in the topological structure between the genuine and conjugate Fermi holes allows us an insightful visualization of the variation of the singlet and triplet states, and thus the origin of the lower energy of the triplet state than the counter part singlet state.

The genuine and Fermi holes for the $(1s)(3s)$ and $(1s)(3p)$ configurations have been plotted in Figs. 1 (a) and (b), respectively. The blue and red surfaces displayed in this figure represent, respectively, the regions where the probability density of the singlet state is larger than that of

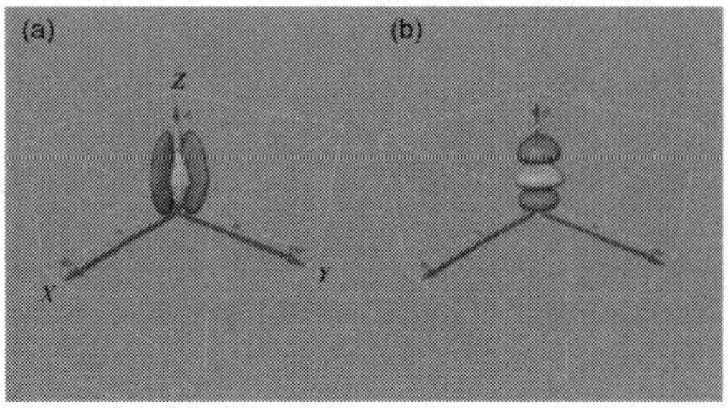


**Figure 1.** The genuine and conjugate Fermi holes in the internal space $(r_1, r_2, \phi_-)$ for the $(1s)(3s)$ and $(1s)(3p)$ singlet-triplet pairs of states [(a) and (b), respectively] of He-like systems. The $X$, $Y$, and $Z$ axes represent $r_1$, $r_2$, and $\phi_-$, respectively. The probability density at the displayed surface is $5.0\times10^{-4}$.

the triplet state, and vice versa. The density at the surfaces is $5.0\times10^{-4}$. Inside the blue surfaces the probability density of the triplet state is much smaller than that of the corresponding singlet state. Therefore, these regions represent the so-called standard Fermi holes. On the other hand, inside the red surfaces the probability density of the singlet state in turn is much smaller than the triplet state, that have been defined and called as the *conjugate Fermi holes* in earlier studies [12, 13]. The existence of these conjugate Fermi holes as well as the genuine ones can be rationalized as follows.

The spatial part of the wave functions of the singlet-triplet pair of states for the $(1s)(nl)$ configuration in the limit of $Z_n \to \infty$ can be described as

$$\Psi_{nl}^{\pm} = \frac{1}{\sqrt{2}}[\psi_{1s}(\vec{r}_1)\psi_{nl}(\vec{r}_2) \pm \psi_{nl}(\vec{r}_1)\psi_{1s}(\vec{r}_2)], \tag{4}$$

where the symmetric and antisymmetric wave functions, $\Psi_{nl}^{+}$ and $\Psi_{nl}^{-}$, represent, respectively, the singlet and triplet states. By choosing $\vec{r}_1 = \vec{r}_2$ for the triplet $\Psi_{nl}^{-}$ wave function, the first and second terms in the bracket of the right hand side of Eq. (4) cancel with each other. This results in the well-known Fermi holes as indicated by the blue regions in Fig. 1. In the case of the singlet $\Psi_{nl}^{+}$ wave function this cancellation doesn't occur for $\vec{r}_1 = \vec{r}_2$ because of its symmetric nature. However, another type of 'cancellation' can occur for the singlet wave function: As is well known the $\psi_{nl}$ orbital has at least one node [radial node(s) for $n > 2$ and/or angular node(s) for $l > 0$] where the amplitude becomes zero. Therefore, this orbital changes its sign when the argument crosses the node point. By arbitrary choosing two points, designated by $\vec{r}_{\rm A}$ and $\vec{r}_{\rm B}$, so that they cross this node point and putting them into the singlet wave function as $\Psi_{nl}^{+}(\vec{r}_{\rm A}, \vec{r}_{\rm B})$, the first and second terms in the bracket of the right hand side of Eq. (4) cancel with each other since $\psi_{nl}(\vec{r}_{\rm A})$ and $\psi_{nl}(\vec{r}_{\rm B})$ have different signs. In the case of the triplet $\Psi_{nl}^{+}$ wave function this cancellation doesn't occur since the minus sign in front of the second term compensates the change of the sign in the $\psi_{nl}$ orbital. Therefore, in the vicinity of $(\vec{r}_{\rm A}, \vec{r}_{\rm B})$ the triplet wave functions should have much larger probability density than the singlet wave functions. This guarantees the existence of the conjugate Fermi holes as indicated by the red surfaces in Fig. 1.

The genuine and conjugate Fermi holes displayed in Fig. 1 look quite similar to those for the $(1s)(2s)$ and $(1s)(2p)$ configurations examined in detail in earlier studies [12, 13]. The main difference between the two configurations, $(1s)(2s)$ and $(1s)(3s)$, and between $(1s)(2p)$ and $(1s)(3p)$ is that the outer orbitals, $(3s)$ and $(3p)$, in the higher principle shell of $n = 3$ have an additional radial node than the corresponding $(2s)$ and $(2p)$ orbitals in the lower shell. The reason for the topological similarity in their genuine and conjugate holes in spite of the existence of this additional radial node is that genuine and conjugate Fermi holes exist in the regions where the two orbitals, $\psi_{1s}$ and $\psi_{nl}$, constituting the wave function of Eq. (4) spatially overlap with each other. Therefore, since the node point of this additional node in the higher $n$ shell is far away from the small $r$ region where the maximum density of the tight $(1s)$ orbital resides, it does hardly contribute to the formation of genuine and conjugate Fermi holes. Thanks to the similarity in the structure of the genuine and conjugate Fermi holes between the $(1s)(2s)$ and $(1s)(3s)$ configurations and between $(1s)(2p)$ and $(1s)(3p)$, the mechanism of the origin of the first Hund rule operating in the $(1s)(2s)$ and $(1s)(2p)$ configurations described in detail in earlier papers [12, 13] can be applied to the present $(1s)(3s)$ and $(1s)(3p)$ cases.

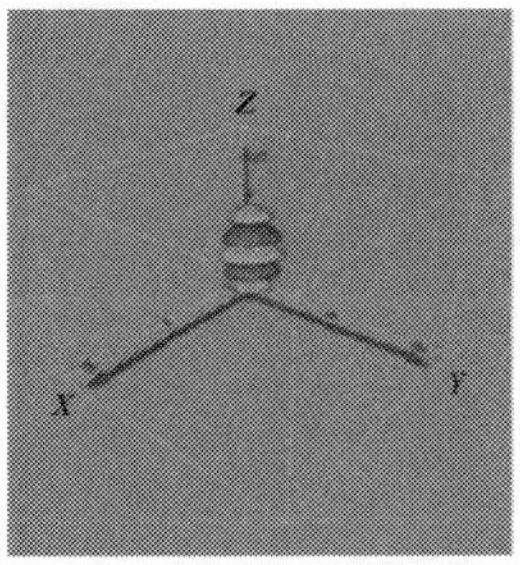


**Figure 2.** The genuine and conjugate Fermi holes in the internal space $(r_1, r_2, \phi_-)$ for the $(1s)(3d)$ singlet-triplet pair of states of He-like systems. The probability density at the displayed surface is $5.0\times10^{-5}$. See the caption to Fig. 1 for other remarks.

The genuine and conjugate Fermi holes for the $(1s)(3d)$ configuration have been displayed in Fig. 2. It is noted that the density at the displayed surface in this figure is $5.0\times10^{-5}$, 10 times smaller than that for Fig. 1. Therefore, the integrated probability density in the genuine and conjugate holes for this $(1s)(3d)$ configuration is much smaller than those for the $(1s)(3s)$ and $(1s)(3p)$ configurations. Since the density in the genuine and conjugate holes determines the difference between the singlet and triplet states, the significantly smaller density of the $(1s)(3d)$ configurations indicates a much smaller energy difference than the cases for the $(1s)(3s)$ and $(1s)(3p)$ configurations. This is consistent with the experimental observation that the singlet-triplet energy gap for this $(1s)(3d)$ pair of states is significantly smaller than those for the $(1s)(3s)$ and $(1s)(3p)$ pairs of states.

## 4. Acknowledgments

The present study has been supported in parts by Grants-in-Aid for Scientific Research (No 23550025) from the Ministry of Education, Culture, Sports, Science and Technology (MEXT) and by Nihon University Strategic Projects for Academic Research. The author would like to thank Prof. Geerd H. F. Diercksen and Prof. Josef Paldus for helpful discussion.

## References

[1] Hund F 1925 *Z. Phys.* **33** 345
[2] Hund F 1925 *Z. Phys.* **34** 296
[3] Hund F 1927 *Linienspektren und periodisches System der Elemente* (Springer, Berlin) pp. 98, 124
[4] Colpa J P and Brown R E 1973 *Mol. Phys.* **26** 1453
[5] Sajeev Y, Sindelka M, and Moiseyev N 2008 *J. Chem. Phys.* **128** 061101
[6] Sako T, Paldus J, and Diercksen G H F 2010 *Phys. Rev. A* **81** 022501
[7] Davidson E R 1964 *J. Chem. Phys.* **41** 656-658
[8] Davidson E R 1965 *J. Chem. Phys.* **42** 4199-4200
[9] Katriel J and Pauncz R 1977 *Adv. Quantum Chem.* **10** 143-185
[10] Boyd R J 1984 *Nature* **310** 480-481
[11] Oyamada T, Hongo K, Kawazoe Y, and Yasuhara H 2010 *J. Chem. Phys.* **133** 164113
[12] Sako T, Paldus J, Ichimura A, and Diercksen G H F 2011 *Phys. Rev. A* **83** 032511
[13] Sako T, Paldus J, Ichimura A, and Diercksen G H F 2012 *J. Phys. B* **45** 235001

# The evolution of wealth transmission in human populations: a stochastic model

**G. Augustins[1], L. Etienne[2], J-B. Ferdy[3], R. Ferrer[2], B. Godelle[2], E. Pitard[4], F. Rousset[2]**

[1] Laboratoire d'Ethnologie et de sociologie comparative, Université Paris X
[2] Institut des Sciences de l'Evolution de Montpellier, Université Montpellier 2
[3] Laboratoire Evolution et Diversité Biologique, Université Toulouse 3
[4] Laboratoire de Physique Charles Coulomb, Université Montpellier 2-CNRS

E-mail: estelle.pitard@univ-montp2.fr

**Abstract.** Reproductive success and survival are influenced by wealth in human populations. Wealth is transmitted to offsprings and strategies of transmission vary over time and among populations, the main variation being how equally wealth is transmitted to children. Here we propose a model where we simulate both the dynamics of wealth in a population and the evolution of a trait that determines how wealth is transmitted from parents to offspring, in a darwinian context.

## 1. Introduction

It has been shown that in Eurasian societies, there exists a link between family types and ideological values, such as individualism, liberalism or equality [1]. In this paper we are particularly interested in the way wealth is transmitted through generations in human populations. One is interested for example to the case of primogeniture (or unigeniture), i.e when the wealth is given only to one preferred child, and this choice would be preferred to a case of more equal treatment i.e when the wealth is distributed evenly between children [2]. These conditions are numerous, and can be for example cultural, historical or economical factors.

Here we are interested in the economical factors that influence this choice. Indeed, it is known that wealth has an influence on demography, and more precisely on fecundity. It also has an influence on the social status, which comprises the probability to marry, and hence, to have descendants. It is also known that wealth of the parents influence the financial status of their children: statistically, the richer the parents, the richer the children [3, 4]. These facts are documented by several case studies, which can be in some cases very complex to interpret.

In the case of rural, traditional populations, the above facts seem to be less complex, and tend to show that the probability to reproduce increases as wealth increases. In these populations, wealth is measured in units of land, as money was not necessarily an indicator for richness, nor was even needed.

Concerning the wealth transmission rule in human populations, historical studies show that they are very variable, and depend on the population considered [5]. For example, in rural regions in France before industrialization, the rules are diverse: egalitarian in Limousion but primogeniture in the Basque region. They are also variable in time. Take for example the population of settlers in New England in the 17th century: their rule was first egalitarian, having a large amount of land available. But after several generations, land becoming scarce, they changed the rule and chose primogeniture, as Alexis de Toqueville reports in his book "Democracy in America" (1835): "In Massachusetts, estates are very rarely divided; the eldest son generally takes the land, and the others go to seek their fortune in the wilderness. The law has abolished the right of primogeniture, but circumstances have concurred to re-establish it".

Our aim is to derive a simple model that allows to draw a link between the wealth distribution inside a population and how wealth transmission rule evolves. We will start by constructing an economical model based on specific rules in a traditional, agricultural, closed population, that allows to quantify economical inequalities. Then we choose a reproduction model, where reproduction is conditioned on wealth. The transmission (or devolution) model consists in introducing a parameter $\rho$ which is the part given to the first, preferred, offspring. The last ingredient is the mutation rate, which allows $\rho$ to vary randomly over the generations. This parameter is also subject to selection, through the value of the parent's wealth. The evolution of $\rho$ is considered and results in two distinct regimes. In the case of strong economical inequalities, equal transmission of wealth is preferred, whereas in the case of less inequalities, primogeniture is preferred.

## 2. An economical model of wealth based on land

Here we focus on wealth distribution of closed, rural societies. In these agricultural societies, wealth $w$ is the amount of land owned, on which crops can be grown. The model is described as follows.

At time $t$, each individual $i$ owns some land (= wealth) $w_i(t)$. The total land of the community is supposed to be conserved over time: if there are $N$ individuals, then $w_{tot} = \sum_{i=1}^{N} w_i(t)$ is a constant.

The production of crops -typically every year- results in a quantitative harvest $x_i(t)$, which is subject to some environmental noise of variance per unit of land $\sigma^2$, depending on climate variations, quality of soil and other potential factors. It is also assumed that the harvest can be measured in units of land (actually it may be thought that both land and harvest can be measured in the same unit type, such as the number of seeds needed, then harvested after the plant has grown). For simplicity, we also assume that the productivity is equal to 1, so that the harvest is finally equal on average -over the noise- to $w_i(t)$. This can be written as:

$$x_i(t) = w_i(t) + \sqrt{\sigma^2 w_i(t)}\eta_i(t).$$

Here $\eta_i(t)$ is a gaussian noise of zero mean and variance 1, uncorrelated in time.

It has been observed that such communities choose a tax system, in a way to decrease variations of wealth between individuals. In such communities with little interaction with some outside world, money was not the correct tool for transactions. A certain amount of wealth -for example, seeds- is taken to each individual proportionally to his harvest, with rate $\alpha$, and redistributed equally among all members of the community. The amount of harvest remaining after tax is: $x'_i(t) = (1-\alpha)x_i(t) + \frac{\alpha}{N}\sum_{i=1}^{N} x_i(t)$

For each land owner, there is also a maintenance cost for the land, corresponding for example to the cost of buying the correct amount of seeds needed for the next crop. We will again estimate this quantity in the same unit as the wealth and the harvest, and assume that it is simply equal to $w_i(t)$. Finally, what remains of the harvest after tax and maintenance cost is: $h_i(t) = x'_i(t) - w_i(t)$.

The quantity $h_i(t)$ can actually be positive or negative: by chance, some individuals produce more, while others are not able to maintain their land. One can check that the average over the noise of the quantity $\sum_{i=1}^{N} h_i(t)$ is equal to zero for all times. A way to erase debts and assure sustainability of the whole community is a redistribution of land among individuals. An individual with positive $h_i(t)$ can convert it into buying land, while an individual with negative $h_i(t)$ can sell some land to compensate for bad harvest. In all cases, one can write the wealth at time $t+1$ as: $w_i(t+1) = w_i(t) + h_i(t)$.

After such exchanges, the land is redistributed, but with time, some stationary probability distribution $P(w)$ is reached. Typically such an economical cycle (with a crop each year) lasts a generation, namely approximately 25 years.

One can use the preceding equations to write a continuous time evolution of wealth $w(t)$:

$$\frac{\partial w}{\partial t} = \alpha(w_0 - w(t)) + (1-\alpha)\sigma\sqrt{w(t)}\eta(t)$$

with $w_0 = w_{tot}/N$.

From this Langevin-type equation of evolution, one can derive, in the Stratanovich sense [6], as we have here an extrinsic noise- the corresponding Fokker-Planck equation for $P(w,t)$:

$$\frac{\partial P(w,t)}{\partial t} = -\frac{\partial}{\partial w}[(\alpha(w_0 - w) + \frac{(1-\alpha)^2\sigma^2}{4})P(w,t)] + \frac{1}{2}\frac{\partial^2}{\partial w^2}[(1-\alpha)^2\sigma^2 wP(w,t)]$$

The stationary solution comes directly as:

$$P_{st}(w) \propto \frac{1}{w^{1-\mu}} \exp[-(\mu - \frac{1}{2})\frac{w}{w_0}]$$

with $\mu - \frac{1}{2} = \frac{2\alpha w_0}{(1-\alpha)^2\sigma^2}$.

The resulting distribution functions have an exponential tail, except for a situation with no tax and no redistribution ($\alpha = 0$) where the distribution of wealth is a power law. At fixed tax rate $\alpha$ (Figures 1 and 2), the distribution gets broader as the environmental variance increases ($\sigma^2 \to \infty$ also leads to a power law). If the environmental variance $\sigma^2$ is fixed (results not shown), reducing the tax rate $\alpha$ makes the distribution broader as well, inequalities between individuals increase.

Note that other economical models are studied to the literature, and mainly lead to power-law type of wealth distribution, such as in [7]. In this last study, the economical context is different, individuals exchange wealth through trading, and the stochastic noise describes the growth or decrease of wealth due to investment in stock markets. This noise term is then of the form $w_i(t)\eta_i(t)$ -instead of $\sqrt{w_i(t)}\eta_i(t)$ in our case-, to insure invariance of the evolution equation under a change of monetary units.

## 3. Evolution model for the transmission of wealth

Given the economical model described above, we are now interested in the way individuals transmit wealth to their offsprings. We used a numerical simulation scheme, that follows over time the reproduction of individuals and the transmission of wealth to the children, for many generations. The scheme goes as follows:

*(i)* the economical cycle (described above) first runs over 25 years, namely a generation.

*(ii)* then reproduction occurs: an individual has a probability $p(w)$ to have children. $p(w)$ is an increasing function of $w$, typically a sigmoid going from 0 to 1 as $w$ varies from 0 to $\infty$. At this stage, wealth acts as a selection pressure for reproduction.

*(iii)* if there is reproduction, then the number of chidren $n$ is drawn from a Poisson distribution with mean $f$, which is the fecundity parameter.

*(iv)* during the devolution step, the father dies (non-overlapping generations) and transmits all his wealth to his children. We choose the following rule of transmission: the fraction of wealth given to child $j$ is:

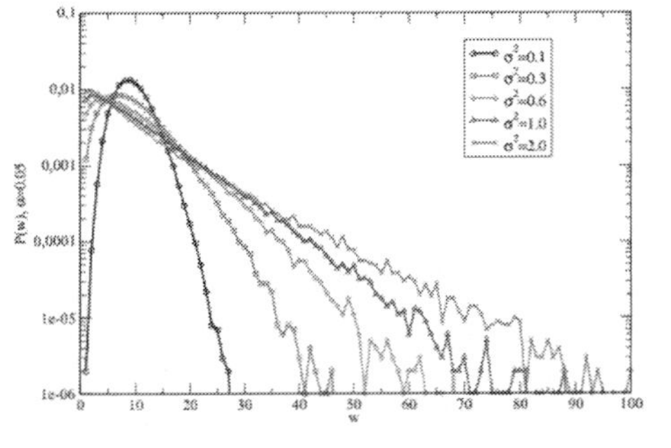


**Figure 1.** Simulation results for $P(w)$ with fixed tax rate, $\alpha = 0.05$. Inequalities increase when environmental variance $\sigma^2$ increases (from left to right).

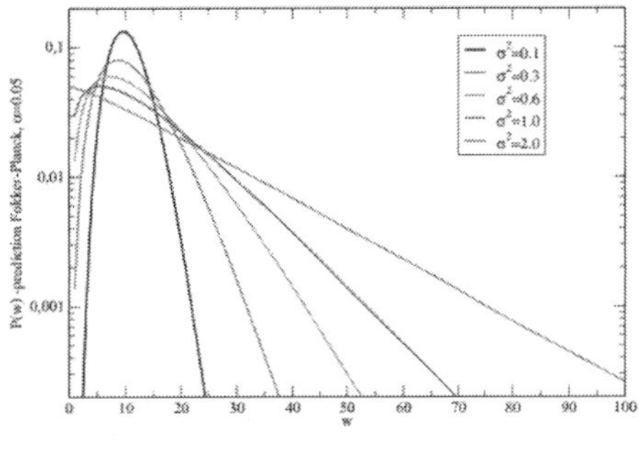


**Figure 2.** Theoretical results for $P(w)$ with the same parameters as in Figure 1.

$$h_j = \rho w \frac{(1-\rho)^{j-1}}{1-(1-\rho)^n}, j = 1, ..n.$$

Here the parameter $\rho$ is the fraction of father's land ($w$) given to the firstborn (preferred) child. If $\rho = 1$ the firstborn child gets all the father's land (primogeniture). If $\rho = 0$ all children receive the same fraction of the father's land: $h_j = w/n$, the transmission is egalitarian. In the case where there is no offspring, wealth is redistributed among the population so as to preserve wealth ratios between individuals. Then the economical cycle starts again for 25 years, and the whole process continues.

It has been argued [5] that the type of wealth transmission (modeled here by $\rho$) is a trait subject to evolution, in the darwinian sense. In the model, it is a heritable trait. However, because this trait is expected to adjust to the wealth distribution, we let it evolve during the simulation:

$$\rho_{child} = \rho_{father}(1 + s)$$

The parameter $s$ is a mutation rate, drawn from a gaussian distribution of zero mean and small variance (typically $10^{-3}$): on average children transmit their wealth just as their father did, but mutation creates random changes between father's and children's $\rho$. As the simulation goes on, starting from an initial value equal to $\rho = 1/2$ for all individuals, we measured the final, stationary value of $\rho$ averaged over all individuals.

The results (Figure 3) reveal the existence of two regimes of selection of $\rho$. When the environmental variance is weak, the fraction $\rho$ given to the firstborn child increases, which means that primogeniture emerges as the rule of transmission. When environmental variance is high, the fraction $\rho$ given to the firstborn child decreases and one tends to the regime of egalitarian transmission. These features can actually be well understood in the framework of our model, and rely much on the basic assumption that wealth is a condition for successful reproduction.

In the case of narrow $P(w)$ ($\sigma^2$ small or $\alpha$ large), many individuals are moderately rich or poor. If $\rho$ is too small, many offsprings will not reproduce because they are too poor. Because $\rho$ is determined by the survivors, it evolves to larger values and leads to favor primogeniture. In

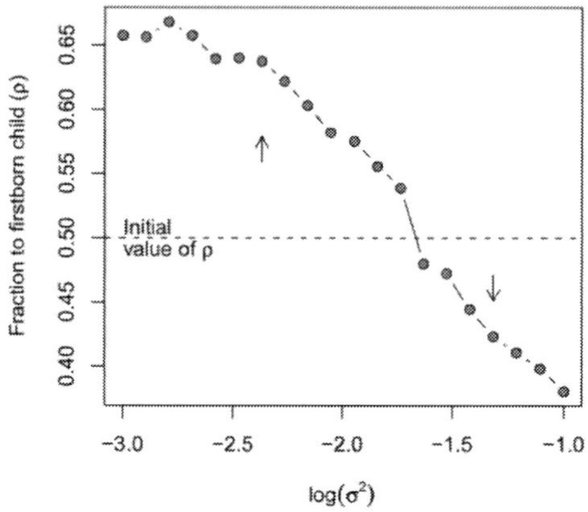


**Figure 3.** Evolution of the fraction given to first child $\rho$, after many generations, for different values of the environmental noise $\sigma^2$. Here, 50 independent simulations were run for 50000 generations and $\alpha = 0.04, f = 1.75, N = 10^3$.

the case of broad $P(w)$ ($\sigma^2$ large or $\alpha$ small), the population is composed of many poor people but there is also a significant fraction of very rich people. For the rich, if $\rho$ is too large, less offsprings will reproduce than if $\rho$ is small because even if this case, they are sufficiently rich. $\rho$ is better small for the rich, so on average it evolves to smaller values and more egalitarian transmission.

## 4. Conclusion

Our results show that economical systems that decrease inequalities promote primogeniture. On the contrary, economical systems that increase inequalities promote more egalitarian transmission rule, which is a priori counter-intuitive. This last result is new and we do not expect it to depend on the type of trading (it is a priori sufficient that the wealth distribution is broad and close to a power-law). However, if the value of $\rho$ is more egalitarian on average, it can hide strong disparities between the poor and the rich. This could lead to further work, in particular in order to understand how wealth is redistributed -or not- among families. Moreover, comparison with demographic data will be needed [1]. Finally, one can imagine a more general model, where the fecundity could also be an evolving parameter: this could help to understand the conditions for a demographic transition in time.

## References

[1] Todd E 2013 *L'origine des systèmes familiaux*, Gallimard. Todd E 1999 *La diversité du monde: famille et modernité*, Seuil.
[2] Borgerhoff Mulder M and al. 2009 *Science* **326** 682.
[3] Voland E 1990 *Behav Ecol Sociobiol* **26** 65.
[4] Hopcroft R 2006 *Evolution and Human Behavior* **27** 104.
[5] Hrdy S and Judge D 1993 *Human Nature* **4** 1.
[6] van Kampen NG 1981 *J Stat Phys* **24** 175.
[7] Bouchaud J-P and Mézard M 2000 *Physica* A **282** 536.

# Modelling and Simulation Based on Matlab/Simulink: A Press Mechanism

**R Halicioglu, L C Dulger, A T Bozdana**
Gaziantep University, Faculty of Engineering, Mechanical Engineering Department, Gaziantep 27310, TURKEY

E-mail: rhalicioglu@gantep.edu.tr, dulger@gantep.edu.tr, bozdana@gantep.edu.tr

**Abstract.** In this study, design and kinematic analysis of a crank-slider mechanism for a crank press is studied. The crank-slider mechanism is the commonly applied one as direct and indirect drive alternatives in practice. Since inexpensiveness, flexibility and controllability are getting more and more important in many industrial applications especially in automotive industry, a crank press with servo actuator (servo crank press) is taken as an application. Design and kinematic analysis of representative mechanism is presented with geometrical analysis for the inverse kinematic of the mechanism by using desired motion concept of slider. The mechanism is modelled in MATLAB/Simulink platform. The simulation results are presented herein.

## 1. Introduction

Forming process is a mechanical deformation without removing or adding any part is significant in industry. Sheet metal and bulk metal forming processes are often used in automotive sectors [1-2]. The process' device is hydraulic and mechanical presses. While hydraulic presses provide directly flexibility, conventional mechanical presses are more reliable and accurate, provide flexibility with their mechanisms such as linkages and screw [3-5]. Studies have condensed on servo presses with evolution of servo motor for several decades. The servo press is a mechanical press that has been actuated by servo motors. The most important advantage of servo presses possesses the positive properties of hydraulic and conventional mechanical press [4-5]. Miyoshi [6] represented the recent appearance of servo presses was an epoch-making event that would change and expand the possibility of press production techniques. Also he gives Komatsu's servo press types and comparison of the press types' motion. Osakada et al [4] have presented a detailed study on servo presses technology and metal forming. Kiper [5] has prepared a review study on servo presses in the industry. Their another study is about input design of servo hybrid presses with cardan gear [7]. Tokuz and Jones [8] have offered a hybrid actuator which is composed from one servo driven actuator and one conventional motor (has constant velocity). Their hybrid configuration is considered as first hybrid actuator system, and is fundamental of hybrid servo presses [8-9]. Sun et al [10] have discussed on AC servo presses' further develop trends and key problems. Although there are a lot of press mechanisms in literature and industry, commonly used mechanism in industry is *crank-slider* mechanism. *Servo crank presses* are preferred generally because of their suitable sizes, ease of manufacturing, long and adjustable strokes, cheapness, etc. Servo presses are been by mechanisms. Because of designing of a mechanism, kinematic and dynamic analyses are required. The analysis can be modeled numerical or by having a software. Serbest et al [11] have modeled a motion of a person's sitting and upping via

Matlab/SimMechanics tool box. Liu et al [12] have designed a dynamic model of slider-crank mechanism without external forces by Matlab/Simulink. Nagchaudhuri [13] has studied on slider-crank mechanism too. Fung et al [14] have presented kinematic and dynamic analyses of two different slider-crank mechanisms.

In the study, a motion scenario adaptable to servo crank press has been selected and kinematic analysis of the scenario on the mechanism has been applied. A Simulink model is presented with simulation results.

## 2. Free Motion Concept

The most important difference between conventional and servo presses is flexibility of the motion. While motion characteristic of conventional presses is provided by mechanism design, motion characteristic of servo presses is provided by control of the actuator. Infinite motion scenario can be created by control unit from a mechanism. The property gain flexibility of motion such as hydraulic presses. Flexible motion of servo presses is presented as free motion concept by Miyoshi in 2004 [6]. Comparison of the motion characteristic between the conventional press and the free motion press can be seen in figure 1. Process time interval in conventional presses is longer than servo presses. Press manufacturers have many motion selection parameters such as crank, knuckle soft, silent, link, pulsating, programmable, pendulum, coining, stop, iterative, etc [15-17]. Some of the servo press motion concepts are given at figure 2.

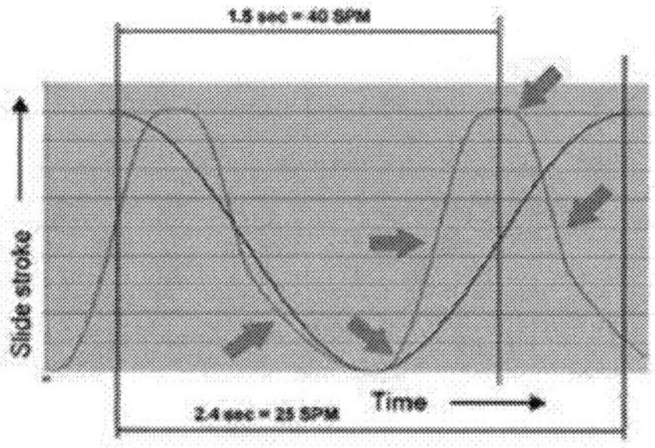


**Figure 1.** Comparison of a conventional eccentric-drive press (blue) and servo-mechanical press (red) [18].

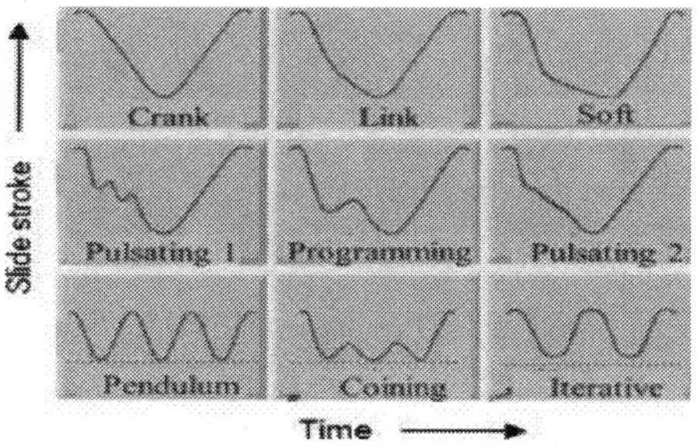


**Figure 2.** Some motion concepts of servo presses [4].

## 3. Case Study

### *3.1. Press Mechanism and Motion Scenario*

Servo crank press is selected for the study. Definition of the mechanism is that crank length, crank angle, rod length, rod angle, slider position are given with parameters $r$, $\theta, l$, $\beta$, $y$ respectively such as figure 3. The slider crank mechanism's rod-crank ratio is selected as six. The stroke of mechanism is 700mm, in-line mechanism's crank length must be 350mm, and the rod length is six time of it as 2100 mm. The soft motion scenario is created by MATLAB/Curve Fitting tool box [19] is given in figure 4.

### *3.2. Kinematic Analysis*

Vectorial expression of mechanism is given as equation (1).

$$\vec{OB} = \vec{OA} + \vec{AB} \tag{1}$$

$$y = r\cos\theta + l\cos\beta \tag{2}$$

$$0 = r\sin\theta - l\sin\beta \tag{3}$$

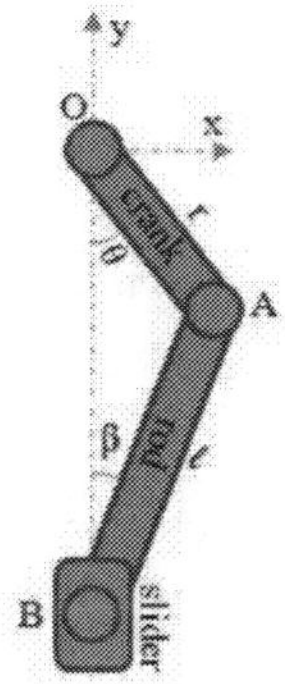


**Figure 3.** Used mechanism model of Servo crank press mechanism for analysis

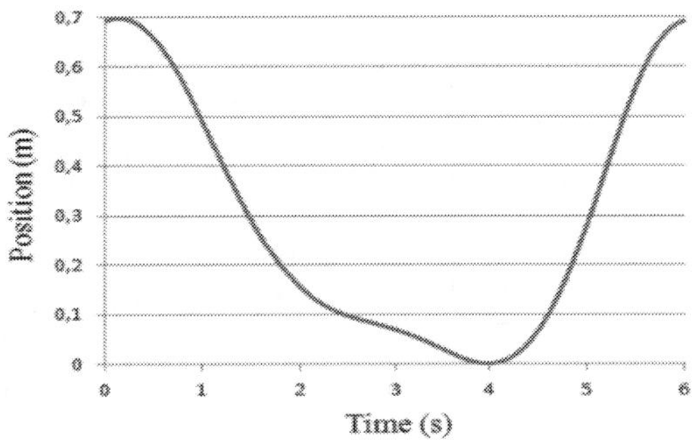


**Figure 4.** Soft motion scenario (slider stroke position-time graphic)

First and Second derivatives of equations (1) and (2) are taken and written in a matrix form to give velocities and accelerations such as equation (4) and (5);

$$\begin{bmatrix} l\sin\beta & 1 \\ l\cos\beta & 0 \end{bmatrix}\begin{bmatrix} \dot{\beta} \\ \dot{y} \end{bmatrix} = \begin{bmatrix} -r\dot{\theta}\sin\theta \\ r\dot{\theta}\cos\theta \end{bmatrix} \tag{4}$$

$$\begin{bmatrix} l\sin\beta & 1 \\ l\cos\beta & 0 \end{bmatrix}\begin{bmatrix} \ddot{\beta} \\ \ddot{y} \end{bmatrix} = \begin{bmatrix} -(r\ddot{\theta}\sin\theta + r\dot{\theta}^2\cos\theta + l\dot{\beta}^2\cos\beta) \\ r\ddot{\theta}\cos\theta - r\dot{\theta}^2\sin\theta_2 + l\dot{\beta}^2\sin\beta \end{bmatrix} \tag{5}$$

Inverse kinematic analysis is necessary to gain the crank position for desired slider motion profile. Geometrical analysis is performed for the inverse kinematic of the mechanism.

$$x^2 = r^2 - (y - l\cos\beta)^2 = l^2 - (y - r\cos\theta)^2 \quad , \quad x = r\sin\theta \tag{6}$$

$$\cos\theta = \left[\frac{r^2 + y^2 - l^2}{2yr}\right] \tag{7}$$

Stroke "s" of crank press mechanism is written as;

$$s = y - l + r \tag{8}$$

If y is substituted from equation (7) to equation (8);

$$\theta = \cos^{-1}\left[\frac{r^2 + (s + l - r)^2 - l^2}{2r(s + l - r)}\right] \tag{9}$$

Equation (9) gives crank angle. The first and second derivatives of equation (9) give the crank velocity and acceleration respectively.

## 4. Modelling of the Mechanism

Model of the mechanism is presented by MATLAB/Simulink tool box [19] using the equations (2-5) as figures 5 and 6. The equations can be solved directly by using the model. Figures 7 represents the servo motion of crank mechanism and figure 8 gives the slider's output motion is desired scenario.

## 5. Conclusions

The crank-slider mechanism is modelled by using Matlab/Simulink platform. It is possible to extend motion scenarios for slider with servo inputs. An example motion is explored with inverse kinematics. Simulation results are presented for its possible application. They give the desired motion concept.

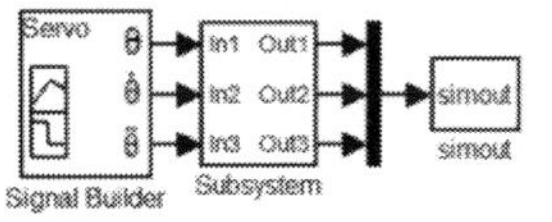


**Figure 5.** Simulink model

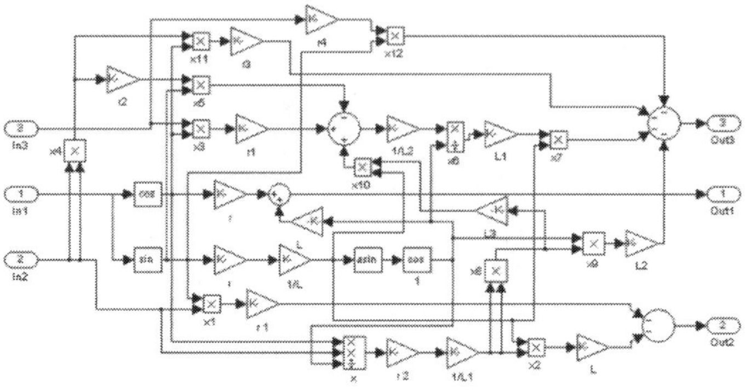

**Figure 6.** Open state of subsystem model

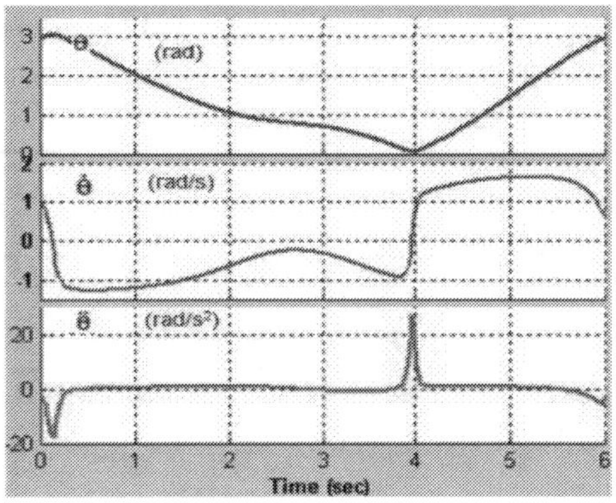


**Figure 7.** The motor servo inputs

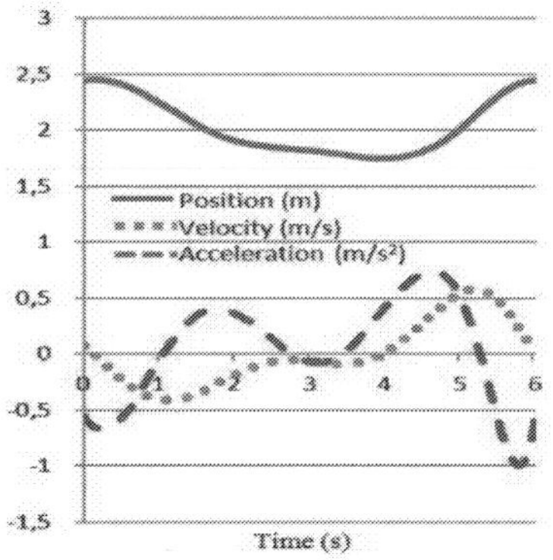


**Figure 8.** Slider motion (simout model output)

## 6. Acknowledgment
The authors would like to thank the Ministry of Science, Industry and Technology in Turkey for the financial support under SANTEZ project (Number: 01422.STZ.2012-I)

## 7. References
[1] Hosford W F and Caddell R M 2011 *Metal forming: mechanics and metallurgy* (New York ;Cambridge Uni. Press)
[2] Merklein M, Koch J, Schneider T, Opel S, and Vierzigmann U 2010 *Int. J. Material Forming* **3(1)** 347-350
[3] Lange K 1985 *Handbook of metal forming* (McGraw-hill)
[4] Osakada K, Mori K, Altan T and Groche P 2011 *CIRP Annals-Manufacturing Technology* **60** 651-672
[5] Kiper G 2012 *Servo mechanical presses-literature review* (Bursa;Coşkunöz Metal Form)
[6] Miyoshi K 2004 *In Proc. 3rd Int. Seminar on Precision Forging*, (Nagoya, Japan: JSTP) p 69-75
[7] Kiper G 2012 *Study on input design of servo mechanic press* (Bursa;Coşkunöz Metal Form)
[8] Tokuz L C and Jones J R 1992 *Hybrid Machine Modelling and Control Ph.D. Dissert.* (Liverpool: Polytechnic)
[9] Kütük M E and Dülger L C 2012 *Elec. J. Machine Technol.* **9(1)** 61-72
[10] Sun Y S, Zhou X H, Wang H G, Wei L M, Huang K S and He H S 2008 *Forging & Stamping Technol.* **33(4)** 1-8
[11] Serbest K, Çilli M and O Eldoğan 2012 *4th Nation. Conf. on BioMechanics* (September,Sakarya) p 205-212
[12] Liu M, Cao Y, Zhang Q and Zhou H 2010 *Int. Conf. on Comp. App. and Sys. Model.* (ICCASM-IEEE) **9** 557-563
[13] Nagchaudhuri A 2002 *Proceedings of IMECE 02* (November, New Orleans, Louisiana) p 17-22
[14] Fung R F, Chiang C L and Chen S J 2009 *Applied Mathematical Modelling* **33** 2411-2420
[15] Amada Co. Ltd. HP: http://www.amada.co.jp/english/
[16] Boerger D 2006 Better control, better flexibility *Fabricating & Metalworking* Apr. 16-17.
[17] Simpac Co. Ltd. http://www.simpac.co.kr/english/
[18] Osborn, A and Paul S 2008 *Metal Forming Magazine* **8** 18-23
[19] MathWorks Inc. 2011 *Matlab* (MathWorks)

# Dipole simulation method for two-dimensional potential problems in exterior regions and periodic regions

**Hidenori Ogata**

Department of Communication Engineering and Informatics,
Graduate School of Informatics and Engineering,
The University of Electro-Communications,
1-5-1 Chofugaoka, Chofu, Tokyo 182-8585 JAPAN.

E-mail: ogata@im.uec.ac.jp

**Abstract.** In this paper, we propose a dipole simulation method for two-dimensional potential problems. The proposed method is a modification of the charge simulation method (the method of fundamental solutions), which is a numerical solver for potential problems, and it gives an approximate solution as a superposition of the potentials due to electric dipoles positioned outside the problem region instead of point charges as in the charge simulation method. We find that the proposed method achieves high accuracy under some conditions from a numerical example. Besides, it is advantageous to the charge simulation method in approximating complex analytic functions. We also show the dipole simulation method for two-dimensional potential problems with one-dimensional periodicity.

## 1. Introduction

Consider the two dimensional potential problems, that is, Dirichlet problems of the Laplace equation

$$\begin{cases} \triangle u = 0 & \text{in } \Omega \\ u = f & \text{on } \partial\Omega\,, \quad \Omega \subset \mathbb{R}^2 \quad \text{(exterior region)} \\ u \text{ is bounded} & \text{in } \Omega \end{cases} \tag{1}$$

where $f$ is a given function on $\partial\Omega$. The charge simulation method (the method of fundamental solution method), which is well-known numerical solver for potential problems [3], gives an approximate solution of (1) in the form[1]

$$u(z) \approx u_N(z) = Q_0 - \frac{1}{2\pi}\sum_{j=1}^{N} Q_j \log|z - \zeta_j|, \tag{2}$$

where $\zeta_j \in \mathbb{C}$, $j = 1, 2, \ldots, N$, are given *charge points* and *charges* $Q_j \in \mathbb{R}$, $j = 0, 1, 2, \ldots, N$ are subject to the constraint $\sum_{j=1}^{N} Q_j = 0$ and determined so that $u_N$ satisfies the collocation

[1] Hereafter we identify $\mathbb{R}^2$ with $\mathbb{C}$.

condition that $u_N(z_i) = f(z_i)$, $i = 1, 2, \ldots, N$ for given *collocation points* $z_i \in \partial\Omega$, $i = 1, 2, \ldots, N$. In terms of physics, $u_N$ is a superposition of the potentials due to point charges outside the problem region. We remark that the approximate solution $u_N$ satisfies the Laplace equation $\Delta u_N = 0$ exactly in $\Omega$. The charge simulation method is used in various fields of science and engineering such as numerical conformal mappings [1], wave problems [4, 5] and so on.

In this paper, we propose the *dipole simulation method* for two-dimensional potential problems in exterior regions[2]. In the proposed method, the solution is approximated by a superposition of the dipole potentials.

## 2. Dipole simulation method

The approximate solution of the dipole simulation method for the potential problem (1) is given by

$$u(z) \approx u_N(z) = C_0 + \frac{1}{2\pi}\sum_{j=1}^{N} p_j \operatorname{Re}\left\{\frac{n_j}{z-\zeta_j}\right\} \tag{3}$$

where $\zeta_j\eta_j \in \mathbb{C} \setminus \overline{\Omega}$, $j = 1, 2, \ldots, N$, are given *dipole points*, $n_j \in \mathbb{C}$, $j = 1, 2, \ldots, N$, are given *dipole directions* satisfying $|n_j| = 1$ and $C_0, p_j \in \mathbb{R}$, $j = 1, 2, \ldots, N$ are *dipole magnitudes* which are subject to the constraint $\sum_{j=1}^{N} p_j = 0$ and determined by the collocation condition that $u_N(z_i) = f(z_i)$, $i = 1, 2, \ldots, N$ for given *collocation points* $z_i \in \partial\Omega, i = 1, 2, \ldots, N$. $u_N$ satisfies the Laplace equation $\Delta u_N = 0$ exactly in $\Omega$. In terms of physics, $u_N$ is a superposition of the potentials due to the dipoles outside the problem region.

The dipole simulation method has the following advantage to the charge simulation method in approximating complex analytic functions such as numerical conformal mapping. Let $f(z)$ be a complex function analytic in a region $\Omega(\subset \mathbb{C})$. Regarding that $\operatorname{Re} f(z)$ is harmonic in $\Omega$, the charge simulation method approximates $f(z)$ by

$$f(z) \approx Q_0 - \frac{1}{2\pi}\sum_{j=1}^{N} Q_j \log(z-\zeta_j), \quad Q_0, Q_j \in \mathbb{R}, \quad \left(\sum_{j=1}^{N} Q_j = 0\right), \quad \zeta_j \in \mathbb{C}\setminus\overline{\Omega}, \quad j = 1, \ldots, N.$$

In actual computations, the complex analytic function $\log(z - \zeta_j)$ is computed as the principal value, that is, the branch such that $-\pi < \arg(z - \zeta_j) \leq \pi$, and there appears a discontinuity of $2\pi\mathrm{i}$ on the half infinite line $\{\, \zeta_j + t \mid -\infty < t \leq 0 \,\}$. This discontinuity makes the computation difficult and we need some techniques to avoid the discontinuity. On the other hand, the dipole simulation method approximates $f(z)$ by the rational function

$$f(z) \approx C_0 + \frac{1}{2\pi}\sum_{j=1}^{N} \frac{p_j n_j}{z-\zeta_j}, \quad C_0, p_j \in \mathbb{R},\ n_j \in \mathbb{C}\ (\,|n_j| = 1\,),\ \zeta_j \in \mathbb{C}\setminus\overline{\Omega}, \quad j = 1, \ldots, N.$$

This approximate function is free of the discontinuity due to the principal values of the complex logarithmic functions and we compute it easily compared with the approximation by the charge simulation method.

*Numerical Example* We show a numerical example of the dipole simulation method. The computations were carried out on a HP Compaq 8000 Elite SFF PC personal computer with a 3.00GHz Intel Core2 Duo CPU using programs coded in C++ with double precision working. The example is the problem (1) of the boundary data $f = x/(x^2+y^2)$ in the exterior of an ellipse

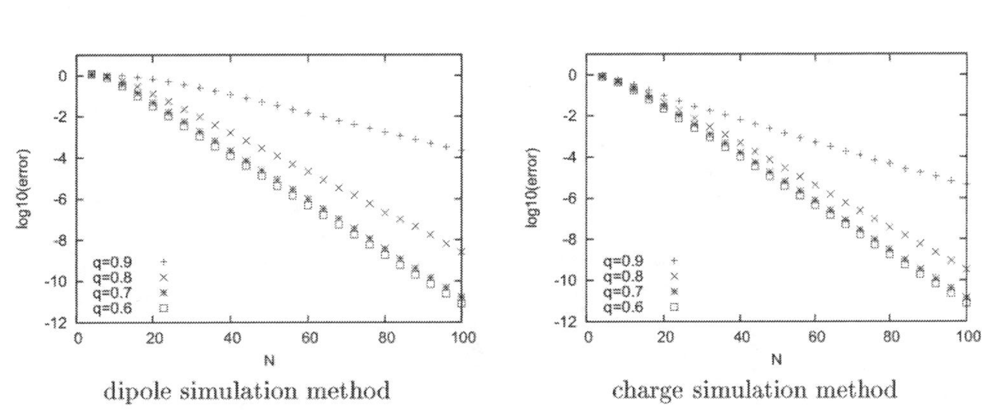


**Figure 1.** The error estimates of the dipole simulation method and the charge simulation method.

$\Omega = \{ (x,y) \in \mathbb{R}^2 \mid x^2/A^2 + y^2/B^2 > 1 \}$ ( $A = 1$, $B = 0.5$ ). We computed an approximate solution of the problem by the dipole simulation method and the charge simulation method with the dipole points (the charge points) $\zeta_j$, the dipole direction $n_j$ and the collocation points $z_j$ respectively given by

$$\zeta_j = J(q\rho\omega^{j-1}), \quad z_j = J(\rho\omega^{j-1}), \quad n_j = \frac{\omega^{j-1}J'(q\rho\omega^{j-1})}{|J'(q\rho\omega^{j-1})|}, \quad j = 1,2,\ldots,N, \tag{4}$$

where $\omega = \exp(2\pi\mathrm{i}/N)$, $q$ is an *assignment parameter* such that $\rho^{-1} < q < 1$, $\rho = \sqrt{(A+B)/(A-B)}$ and $J(\cdot)$ is the Joukowski transform $J(\zeta) = (c/2)(\zeta + 1/\zeta)$, $c = \sqrt{A^2 - B^2}$. We computed the error estimates $\epsilon_N \equiv \max_{z\in\partial\Omega} |u_N(z) - u(z)|$, where the maximums are estimated on 256 points distributed uniformly on the boundary $\partial\Omega$, and show the results in Figure 1. From the figures, we find that the magnitudes of the errors of the dipole simulation method is approximately the same as the magnitudes of the errors of the charge simulation method, and the errors of the dipole simulation method decay exponentially as $N$ increases.

## 3. Problems in periodic regions

We show a modification of the dipole simulation method for two-dimensional potential problems with one-dimensional spatial periodicity. We consider the problem of two-dimensional electric field with conductors in one-dimensional periodic array parallel to the real axis, and assume that the electric field is uniform at infinity in the direction perpendicular to the conductor array. We write the electric potential as $\phi = -E(\operatorname{Im} z) + u$, where $E$ is the magnitude of the uniform electric field at infinity. Then, $u$ satisfies the boundary value problem

$$\begin{cases} \triangle u = 0 & \text{in } \Omega \\ u = E(\operatorname{Im} z) & \text{on } \partial D_n \quad (\, n \in \mathbb{Z} \,) \\ u \sim \text{constant} & \text{as } \operatorname{Im} z \to \pm\infty, \end{cases} \tag{5}$$

[2] The dipole simulation method was originally proposed in [2] for the two-dimensional potential problems in disk regions and our work is an extension of his work for problems in general regions.

where $D_0$ is a simply-connected interior region, $D_n = D_0 + na(\, n \in \mathbb{Z},\ a > 0\,)$ and $\Omega = \mathbb{C} \setminus \{\bigcup_{n\in\mathbb{Z}} \overline{D_n}\}$. Considering that $u$ includes a periodic function, we approximate $u$ by

$$\begin{aligned} u(z) \approx u_N(z) &= C_0 + \frac{1}{2a}\sum_{j=1}^{N} p_j \operatorname{Re}\left\{n_j \cot\left[\frac{\pi}{a}(z-\zeta_j)\right]\right\} \\ &= C_0 + \frac{1}{2\pi}\sum_{j=1}^{N} p_j \operatorname{Re}\left\{\frac{n_j}{z-\zeta_j} + n_j \sum_{n\neq 0}\left(\frac{1}{z-\zeta_j-na} + \frac{1}{na}\right)\right\}, \end{aligned} \tag{6}$$

where $\zeta_1, \zeta_2, \ldots, \zeta_N \in D_0$ are the given dipole points, $n_1, n_2, \ldots, n_N \in \mathbb{C}$ are the given dipole direction such that $|n_j| = 1,\ j = 1, 2, \ldots, N$ and $C_0, p_1, p_2, \ldots, p_N \in \mathbb{R}$ are determined by the collocation condition and the constraint $\sum_{j=1}^{N} p_j = 0$. The figure (a) in Figure 2 shows the electric field with cylindrical conductors $D_n = \{\, z \in \mathbb{C} \mid |z - na| < \rho \,\},\ n \in \mathbb{Z}\ (\, a = \rho\,)$ in a periodic array computed by the dipole simulation method with the dipole points $\zeta_j$, the dipole directions $n_j$ and the collocation points $z_j$ respectively given by $\zeta_j = q\rho\omega^{j-1},\ n_j = \omega^{j-1},\ z_j = \rho\omega^{j-1},\ j = 1, 2, \ldots, N$, where $q$ is an assignment parameter such that $0 < q < 1$. We also computed the error estimates $\epsilon_N = \max_{z\in\partial D_0} |u_N(z) - u(z)|$, which are shown in the figure (b) of Figure 2. From the figure (b), the error $\epsilon_N$ decays exponentially as $N$ increases.

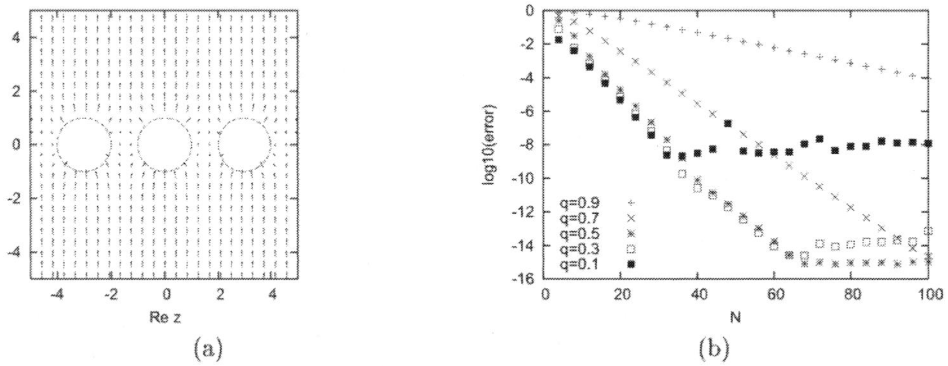


**Figure 2.** (a) The electric field with cylindrical conductors in a periodic array and (b) the error estimates of the dipole simulation method.

## Acknowledgments
This study is supported by JSPS KAKENHI Grant number 22540116.

## References
[1] Amano K 1994 *J. Comput. Appl. Math.* **53** 353
[2] Katsurada M 1989 *J. Fac. Sci. Univ. Tokyo Sect. IA, Math.* **36** 135
[3] Murashima S 1983 *Charge simulation Method and Its Applications* (Tokyo: Morikita) (in Japanese)
[4] Sánchez-Sesma F R 1981 *Arch. Mech.* **33** 167
[5] Sánchez-Sesma F R and Rosenblueth E 1979 *Earthq. Eng. Struct. Dyn.* **7** 441

# The Evolution of Quantum Measuring Devices

**Yehuda Roth**

Oranim Academic College, Q. Tivon 36006, Israel

E-mail: yudroth@gmail.com

**Abstract.** A quantum measuring device is introduced through a projective operator of any complete set of states that span the Hilbert space. Consequently, even a "bizarre" basis such as a basis of states composed of superpositions between location states, is legitimate despite its incomprehensible interpretation of a particle located in some places simultaneously.
The collapse scenario that lies in the essence of any quantum measuring device, suggests that measurement is actually an interpretation process that translate reality into the predefined concepts determined by the particular selection of the basis of states. The very fact that there are bases that contradict "common sense" suggests that our brain by serving as a measuring and interpreting "device", selects only unique measuring processes. We suggest a procedure of nonlinear recursive maps that dominant an evolution of states toward few selected bases of states.

## 1. Introduction

Let us start with the motivated example that in some aspects resemble the description of the Feynman quantum double slits experiment [1]. We consider a Hilbert space that can be spanned by the two location states entitled as $x_1$ and $x_2$. In representing (and potentially measure) a corresponding particle states, a conceivable basis might be the $|x_1\rangle$ and $|x_2\rangle$ states that simply describes the particle as being in one of the two locations. On the other hand, the states $|\pm\rangle = \frac{1}{\sqrt{2}}(|x_1\rangle \pm |x_2\rangle)$ which also generate a physically legitimate basis, possess the "unconceivable" interpretation of a particle that stays in both locations simultaneously. An attempt to solve this unimaginable description is by performing a direct measurement of the apparently real particle location. Although each measurement will indeed revile a location, this measurement corresponds with collapsing the $|\pm\rangle$ original states into one of the $|x_{1,2}\rangle$ states [2]-[7] and therefore it cannot serve as a true reviling measurement of the real nature of the $|\pm\rangle$ states. A more appropriate interpretation for the $|\pm\rangle$ states is the measurement of the $|x_1\rangle$-$|x_2\rangle$ relative phase (0 or $\pi$). Unfortunately, it seems that in the vocabulary of ordinary perceptions, the concept *relative phase between location states of* ***a single particle*** is absent. It seems to exist only in abstract theories such as Quantum-Mechanics. Moreover, even the so called appropriate interpretation, namely *the relative phase between the locations* is more likely a description of a relative state between the real well imaginable concepts namely the particle location, rather then a description of the $|\pm\rangle$ states true nature. We note that not all superposition between location states correspond with a strange interpretation. In extreme situations such as in superposition between infinite-coordinate-Hilbert space, an equal phase superposition between locations states, namely the state $|p\rangle = \sum_x e^{-kx}|x\rangle$ give rise to a completely new concept- that is, the *momentum* concept which is easily associated with the well conceivable velocity concept.

We can conclude this argument by observing that although nature allows us many alternatives in describing a physical word that is formally described with the Hilbert space, yet, only few options are conceivable to the human mind. In this paper we demonstrate only the nonlinear procedure that leads to the unification of specific measuring tolls over the other.

## 2. The Single Parameter Complementary Maps

Complementary maps are defined as coupled recursive maps that determine the process of the concepts generation.

Assume we have a map defined by the function $f_R(\alpha_n)$ with a *single* control parameter $R$ such that

$$\alpha_{n+1} = f_R(\alpha_n), \quad x \in [0,1] \tag{1}$$

where all values of $\alpha_n$ are within the interval $[0,1]$.

We introduce the term maps coherence with the definition of the **complementary map** as follows

$$\beta_{n+1} \equiv 1 - f_R(1-\beta_n), \tag{2}$$

where it is easy to see that $\forall n \; \beta \in [0,1]$. For example, the $\alpha$-$\beta$ maps for the logistic map are:

$$\alpha_{n+1} = R\alpha_n(1-\alpha_n) \qquad \beta_{n+1} = 1 - R\beta_n(1-\beta_n). \tag{3}$$

In general the two complementary maps $\alpha_n$ and $\beta_n$ can evolve independently provided that the initial variables $\alpha_0$ and $\beta_0$ are determined separately. Otherwise, when the initial conditions are initially coordinated such that $\alpha_0 + \beta_0 = 1$ the maps are considered to be *initially unitary correlated.*

## 3. Concepts Evolution in a 1/2-Spin Space

Various models and in particular the Spin-Glass-Model[8], associate the brain activity with interacting spins model. We adapt this approach.

We show a concept generation in a single spin system.

Assume the 1/2-spin-space. We observe the initial condition basis of states:

$$|0\rangle = |\nwarrow\rangle = A_0|\uparrow\rangle + B_0|\downarrow\rangle, \quad |\tilde{0}\rangle = |\searrow\rangle = -B_0|\uparrow\rangle + A_0|\downarrow\rangle \tag{4}$$

and under the maps $n$-iterations the initial coefficients $A_0$ $B_0$ evolve into the $n$-coefficients $A_n$ $B_n$ to generate the states

$$|n\rangle = A_n|\uparrow\rangle + B_n|\downarrow\rangle, \qquad |\tilde{n}\rangle = -B_n|\uparrow\rangle + A_n|\downarrow\rangle \tag{5}$$

where the coefficients that follow the recursive maps function $f$ are:

$$\alpha_n = |A_n|^2 \quad \beta_n = |B_n|^2 \tag{6}$$

with the recursive maps of eqs. 1 and 2. We demonstrate our formalism with the logistic formula:

$$\alpha_{n+1} = R\alpha_n(1-\alpha_n), \qquad \beta_{n+1} = 1 - R\beta_n(1-\beta_n) \tag{7}$$

The logistic maps posses fixed points such that $\forall\; n \; \alpha_n = 0$ and $\beta_n = 1$ meaning that if we start with the initial condition $|0\rangle = |\uparrow\rangle$ and $|\tilde{0}\rangle = |\downarrow\rangle$ then the states will be constantly fixed. Thus, a very limited observer who experience reality under logistic maps dominated brain, notifies the uniqueness of the $up - down$ states among the other rotated states and he associate them with the concept *vertical* states.

Recursive maps such as the logistic map have also the tendency of reaching constant values, provided that they are not in the chaotic regimes. In our example of the logistic maps, if $R < 3$ we obtain that *for* $n \to \infty$, $\alpha_n \to \frac{R-1}{R}$ and $\beta_n \to \frac{1}{R}$. Consequently we obtain states that except for the $\alpha_n = 0, 1$-values, always converge into the defined basis of states:

$$|\infty\rangle_R = \sqrt{\frac{R-1}{R}}\,|\uparrow\rangle + \sqrt{\frac{1}{R}}\,|\downarrow\rangle, \quad |\tilde{\infty}\rangle_R = -\sqrt{\frac{1}{R}}\,|\uparrow\rangle + \sqrt{\frac{R-1}{R}}\,|\downarrow\rangle \tag{8}$$

To simplify our discussion furthermore, we assume that $R = 2$ to obtain:

$$|\infty\rangle_{1/2} = \sqrt{\frac{1}{2}}\,|\uparrow\rangle + \sqrt{\frac{1}{2}}\,|\downarrow\rangle \equiv |\rightarrow\rangle, \quad |\tilde{\infty}\rangle_{1/2} = -\sqrt{\frac{1}{2}}\,|\uparrow\rangle + \sqrt{\frac{1}{2}}\,|\downarrow\rangle \equiv |\leftarrow\rangle \tag{9}$$

Going back to the limited observer who view and interpret the word in according to the elementary $R = 2$-logistic-map. Now in addition to the *vertical* basis, his brain is also capable of generating the *horizontal* concepts. These two concepts are the only way that our "limited" observer experiences the word. Other options collapse by his brain interpretation tools into one of these two options.
In general we can select complementary maps that converge into the constant values $A_\infty$ and $B_\infty$ (see eqs. 5). This yields always to the same universal states representation that is eventually associated by the observer with a concept.

### *3.1. Double Bifurcation Recursive Maps and Concepts Evolution*

Let us slightly extend our analysis to the double period bifurcation maps behavior [9], that is, maps that converge into the two sets:

$$\tilde{\alpha}_- = \lim_{n\to\infty} \alpha_n, \quad \tilde{\alpha}_+ = \lim_{n\to\infty} \alpha_{n+1} \quad \tilde{\beta}_- = \lim_{n\to\infty} \beta_n = 1 - \tilde{\alpha}_- \quad \tilde{\beta}_+ = \lim_{n\to\infty} \beta_{n+1} = 1 - \tilde{\alpha}_+ \tag{10}$$

Although now each map converges into two values, it is possible to redefine a double $g$-maps which retrieves a single fixed point

$$g_R = f\left(f_R\left(\alpha_n\right)\right) \equiv f_R \circ f_R \qquad \tilde{g}_R \equiv \left(1 - f_R\right) \circ \left(1 - f_R\right), \tag{11}$$

yielding

$$\alpha_{n+2} = g_R\left(\alpha_n\right) \qquad \beta_{n+2} = \tilde{g}_R\left(\beta_n\right) \tag{12}$$

where the composed $g$-maps corresponds with a time interval which is twice the original $f$-time. Under the $g$-maps we obtain two types of sets; The firsts converges into the $(\alpha_-, \beta_-)$ values and the others reach the output $(\alpha_+, \beta_+)$
It is easy to identify each set as follows: Suppose that we start with initial conditions that are already fixed at the final states $\tilde{\alpha}_-$ or $\tilde{\alpha}_+$. For $n = 0, 1, 2, 3...$ the maps split into two types:

$$\begin{array}{l} \textbf{Even series iteration}: \quad \alpha_{2n+2} = g_R\left(\alpha_{2n}\right) \quad \textit{The series } \alpha_0,\ \alpha_2,\ \alpha_4... \\ \beta_{2n+2} = 1 - g_R\left(\beta_{2n}\right) \textit{The series } \beta_0,\ \beta_2,\ \beta_4... \end{array} \tag{13}$$

$$\begin{array}{l} \textbf{Odd series iteration}: \quad \alpha_{2n+3} = g_R\left(\alpha_{2n}\right) \quad \textit{The series } \alpha^{(1)},\ \alpha^{(3)},\ \alpha^{(5)}... \\ \beta_{2n+3} = 1 - g_R\left(\beta_{2n}\right) \quad \textit{The series } \beta_1,\ \beta_3,\ \beta_5... \end{array} \tag{14}$$

where the even and odd series converge into the values $\left(\tilde{\alpha}_-, \tilde{\beta}_-\right)$ and $\left(\tilde{\alpha}_+, \tilde{\beta}_+\right)$, respectively.
Since the two maps operate separately, we associate the maps with independent states $|o\rangle$ and $|e\rangle$ types, respectively.
For the logistic map $\alpha_{n+1} = R\alpha_n\left(1 - \alpha_n\right)$ we obtain:

$$(\alpha_-, \beta_-) = \left(\frac{2R-4}{R}, \frac{4-R}{R}\right) \qquad (\alpha_+, \beta_+) = \left(\frac{2}{R}, \frac{R-2}{R}\right) \tag{15}$$

In general the initial states converge into multi kinds of representations, for example:

$$\begin{array}{llll} \textbf{Even basis}: & |\infty\rangle_e = \sqrt{\alpha_+}\,|\uparrow\rangle + \sqrt{\beta_+}\,|\downarrow\rangle & |\tilde{\infty}\rangle_e = -\sqrt{\beta_+}\,|\uparrow\rangle + \sqrt{\alpha_+}\,|\downarrow\rangle \\ \textbf{Odd basis}: & |\infty\rangle_o = \sqrt{\alpha_-}\,|\uparrow\rangle + \sqrt{\beta_-}\,|\downarrow\rangle & |\tilde{\infty}\rangle_o = -\sqrt{\beta_-}\,|\uparrow\rangle + \sqrt{\alpha_-}\,|\downarrow\rangle \end{array} \quad (16)$$

where for the logistic map we have four representations:

**1. Standard basis** : Refers to $\alpha_0 = 0$ and $\beta_0 = 1$ $\quad \forall n, \tilde{n}\ |n\rangle = |\uparrow\rangle\ \ |\tilde{n}\rangle = |\downarrow\rangle$

**Bases for** $\alpha \neq 0,\ \beta \neq 1$
**2. Single value basis**
For $1 \leq R \leq 3$ $\quad |\infty\rangle_s = \sqrt{\frac{R-1}{R}}\,|\uparrow\rangle + \sqrt{\frac{1}{R}}\,|\downarrow\rangle \quad |\tilde{\infty}\rangle_s = -\sqrt{\frac{1}{R}}\,|\uparrow\rangle + \sqrt{\frac{R-1}{R}}\,|\downarrow\rangle$
**Double bifurcation bases.**
For $3 \leq R \leq 1+\sqrt{6}$

$$\begin{array}{lll} \textbf{3. Even basis},\ n = 0,2,4,... & |\infty\rangle_e = \sqrt{\frac{2}{R}}\,|\uparrow\rangle + \sqrt{\frac{R-2}{R}}\,|\downarrow\rangle & |\tilde{\infty}\rangle_e = -\sqrt{\frac{R-2}{R}}\,|\uparrow\rangle + \sqrt{\frac{2}{R}}\,|\downarrow\rangle \\ \textbf{4. Odd basis},\ n = 1,3,5,... & |\infty\rangle_o = \sqrt{\frac{2R-4}{R}}\,|\uparrow\rangle + \sqrt{\frac{4-R}{R}}\,|\downarrow\rangle & |\tilde{\infty}\rangle_o = -\sqrt{\frac{4-R}{R}}\,|\uparrow\rangle + \sqrt{\frac{2R-4}{R}}\,|\downarrow\rangle \end{array} \quad (17)$$

Numerical calculations show[10] that even in the bifurcation regime $(3 \leq R \leq 1+\sqrt{6})$, if we start with the initial condition $\alpha_0 = \alpha_\infty = \frac{R-1}{R}$, although we are in the double-bifurcation-regime, the maps will remain in the same single value $\alpha_{n\to\infty} = \frac{R-1}{R}$ as in eq. 7. It follows that an observer who experience maps in the bifurcation regime is familiar with all the four bases as expressed in eq. 17 including the *Single Value Basis*.

## 4. summary

An important insight of quantum mechanics concerns the way we experience reality. Clearly, by selecting a unique measuring device, an observer enforces reality to exhibits a face that is consistent with his basic understanding that is scalded in his measuring tool preference. In other words, the observer edits reality according to his interpretation tools.

In this paper we explored the procedure that leads to that specific observer selection and we showed that nonlinear maps can converge into values that lead to the unique measuring tools definitions. In that sense our procedure provides a formalism of how concepts concerning reality can be formed in the observer mind.


[1] R. Feynman, R. Leighton and M. Sands, *The Feynman Lectures on Physics*, Vol. 3, 1966. Library of Congress Catalog Card No. 63-20717
[2] R. Penrose, , *The Road to Reality: A Complete Guide to the Laws of the Universe*, Ch. 2, (Vintage Books, 2004)
[3] A. Bassi, *J. Phys.: Conf. Series* **67**, 012013 (2007)
[4] A. Bassi and D.G.M. Salvetti:, *Journ. Phys. A: Math. Theor*, **40**, 9859 (2007).
[5] A. Bassi, *Physics Today*, **44**, 36 (1991)
[6] S. L. Adler et al *J. Phys. A: Math. Theor.*, **40** 15083-15098 (2007)
[7] G. C. Ghirardi, A. Rimini, T. Weber, *Phys. Rev. D*, **34** 470-491 (1986) 34, 470491 (1986)
[8] D.l L. Stein, C. M. Newman *Spin Glasses and Complexity (Primers in Complex Systems)*, Ch. 1, (Princeton university, 2013)
[9] J.D. Crawford, *Rev. Mod. Phys.* **63**, 991 (1991)
[10] It can be tested on every elementary data base program
[11] Y. Roth *EPL*, (2008),**82**, 10006
[12] Y. Roth Int. J. Theor. Phys,IJTP-D-12-00155, (2012)

# Flow simulation and topological rearrangements of ordered foams

**J S Espinoza Ortiz**[1] **, H Belich Jr**[2]

[1]Departamento de Física, Universidade Federal de Goiás, Catalão, GO 75704-020, Brasil.

[2]Depto de Física e Química, Universidade Federal do Espírito Santo, ES 29060-900, Brasil.

E-mail: `jsespino@physics.org`

**Abstract.** Flow through a narrow bent channel may induce topological rearrangements in a two-dimensional monodispersed dry liquid foam. We use the Cellular Potts Model to simulate a foam under a variable driving force in order to investigate the strain-rate response from these rearrangements. We observe a set of foams' behaviors ranging from *elastic*, *viscoelastic* to *fluid* regime. Bubble's topological rearrangements are localized and their cumulative rearrangements change linearly with time, thus non avalanches critical behavior is found. The strain-rate affects the rate of topological rearrangements, its dependence on the drag force is nonlinear, obeying a Herschel-Bulkley like relationship below the foam's flow point.

## 1. Introduction

Foams consist of two continuous liquid fluid phases walls enclosing gas cells where a surfactant stabilizes the liquid walls against rupture by reducing the liquid-gas surface energy. Dry foams contain less than 5% liquid per volume, the liquid forms channels of finite width (*Plateau borders*) rounding the sharp edges and corners of polyhedral cells. In two-dimensions ( *2D*) dry foams consist of polygonal cells, while bubbles in wet foams are nearly circular [1]. Bubbles in foams have an unusual set of physical properties [2]: They deform easily with bubbles adjusting their positions and shapes in order to reach a local minimum of total interfacial area [3]. Bubbles evolve slowly by diffusion of gas from smaller few-sided bubbles to larger many-sided bubbles, leading to coarsening. Generally a foam's configuration is very close to a true equilibrium configuration except when a local topological change involving bubble disappearance or irreversible bubbles neighbor switching sides ($T_1$) forces the foam system to evolve from a metastable configuration to a new one[4].

Foam rheological properties depend on the liquid's fraction, bulk viscosity, surface viscosity, surface tension and geometry. Foams behave like solids under small applied shear stress, its elastic modulus is very small because elasticity originates from the small surface tension of the soap films. For intermediate stress (beyond a yield stress), foams behave like viscoelastic fluids, while for large deformations the energy barrier to $T_1$s is relatively negligible and the foam flows like a liquid[2]. In order to understand the mechanical response of foams under shear it may be necessary to consider different aspects of foam physics, especially the role of $T_1$s in shear and flow, as well as the relation between the dynamics of individual bubbles and the bulk. The spatial and temporal distributions of $T_1$s under increasing applied stress will determine the mechanical (inelastic) properties of the foam and how it releases energy during flow. We use the

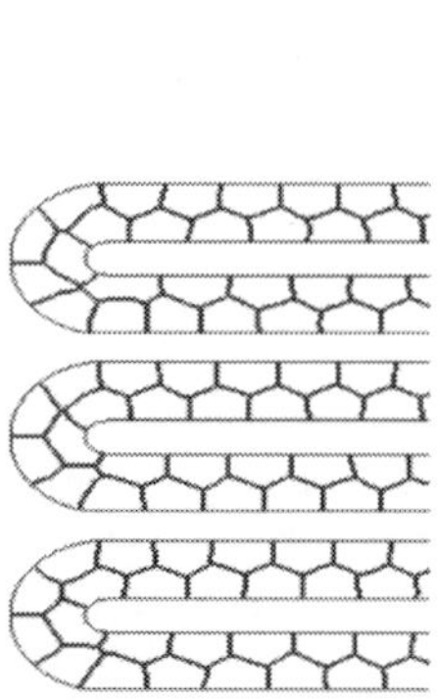

**Figure 1.** The bent channel with bubbles moving counterclockwise dragged by a force. The spot in the middle shows a typical $T_1$ event.

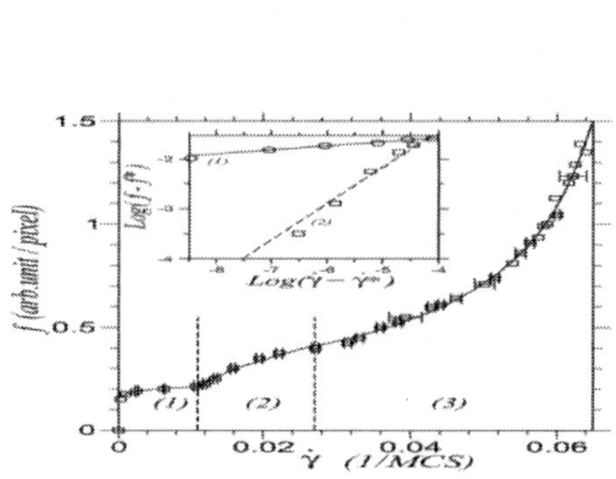


**Figure 2.** The dependence of the driving force on the strain-rate for the *elastic* ( *1*), *viscoelastic* ( *2*) and *fluid* ( *3*) regimes. The inset shows a log-log plot of the same order parameters showing a power law dependence for regimes ( *1*) and ( *2*).

Cellular Potts Model ( *CPM*) to investigate the strain-rate response of *2*D dry ordered foams subject to drag forces in a bent channel. We present the *CPM* algorithm and the numerical simulation in section 2. Next, in section 3 are presented our numerical results and discussions. Conclusions are finally presented in section 4.

## 2. The Cellular Potts model and the Numerical Simulation

The *CPM* represents bubbles on a lattice by a spin variable $\sigma_i$ within bubbles, while film boundaries are domain walls between regions of different spins. The basic foam Free Energy ( $\mathcal{E}$) consists of tree terms: The first one is the exchange energy between neighboring spins at bubbles boundaries, representing the surface tension. The next term describe the gas compressibility, an energy penalty for changes in the areas of bubbles, where $i$ represents a site with coordinates $(x_i, y_i)$, $\mathcal{J}_{i,j}$ is the energy per unit contact length between neighboring bubbles $\sigma_i$ and $\sigma_j$ , $\lambda$ is the inverse of the gas compressibility, $a_n$ is the area of the $n$th bubble, $A_n$ its corresponding area under no compression. The third term considers a driving force that biases the probability of a spin reassignment in some given direction, where $\overrightarrow{f}$ is the driving field and $\overrightarrow{\delta l}$ is a unit vector pointing along the direction of the channel [5],

$$\mathcal{E} = \sum_{i,j}^{neighbors\,(i,j)} \mathcal{J}_{i,j}\left(1 - \delta_{\sigma_i \sigma_j}\right) + \lambda \sum_{n} (a_n - A_n)^2 + \sum_{i} \overrightarrow{f}\,(x_i, y_i) \cdot \overrightarrow{\delta l}\ .$$

The foam evolves via Monte Carlo Metropolis dynamics at zero temperature. A site is chosen at random and a trial reassignment of its value to one of its neighboring sites. This step is is accepted if the energy change is negative. No change in the energy is accepted with probability 1/2. Time is measured in units of *Monte Carlo Steps* ( *MCS*), consisting of as many spin attempts as the number of lattice sites. We use fourth-nearest neighbors interactions on a square lattice to minimize anisotropy .

We consider a narrow bent channel to drive bubbles through the curve region. Here we restrict ourselves to dry ordered foam at zero temperature, although the *CPM* is readily amenable for general cases such as wet and disordered foams at non-zero temperature [4, 6, 7]. We work in a lattice of $378 \times 90$ pixels$^2$ with a racetrack channel of 23 pixels width with semicircular caps. The radius of the outer channel boundary is 45 pixels while the inner channel boundary radius is 22 pixels. A linear channel of length 288 pixels separates the curved regions. We arrange the initial bubbles consistently with the boundary conditions. The pattern is generated by partitioning the network into equal size domains, each containing $22 \times 20$ pixels$^2$. The number of bubbles

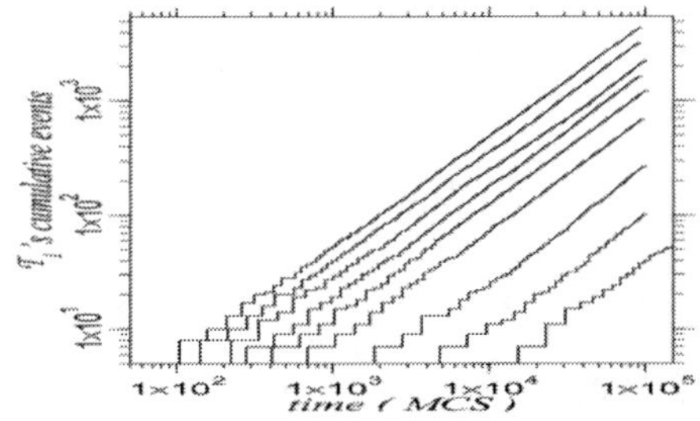


**Figure 3.** The cumulative $\mathcal{T}_1$s linear behavior versus time in *MCS*, for different $f$ values decreasing from the top to the bottom.

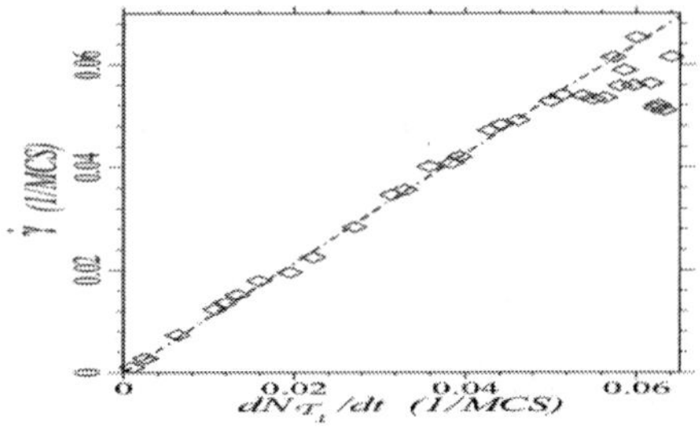


**Figure 4.** Comparison between the strain-rate and the rate of cumulative topological events, a linear dependence is described by the slope approximately equal to 1.0675.

determines each bubble's area, we fix all target areas to $A_n$ = total-area/number-of-bubbles and choose the area constraint $\lambda$ to guarantee that the bubble area dispersion is smaller than 0.075. We use the following parameter values: $n = 56$, the coupling parameter $\mathcal{J}_{ij} = 2.0$, the area constraint $\lambda = 2 \times 10^{-4}$ and the target area $A_n = 485$ pixel$^2$. The initial pattern is obtained by turning off the driving term and relaxing the foam for around 100 *MCS* to produce 31 bubbles in the outer layer and 25 in the inner one. Simulated foams are now driving by a constant applied force, under these conditions the relaxation time is no longer than 30 *MCS*, provided the bubble area dispersion is as specified above. We consider the driving force values (in *arb.unit*) into a range of $(0, 1.4)$ , and for an specific $f$ value we compute the $\mathcal{T}_1$s events by comparing bubbles neighbors side after each *MCS* and relaxation.

## 3. Results and Discussions

We found that $\mathcal{T}_1$s events take place at specific locations in the bent channel, with a systematic offset from the center of the bend, independent of the driving force value. Figure 1 shows a typical bubble rearrangement at $2.39\,rad$, in the upper left-side channel, for a driving force equal to $0.5(arb.unit/pixel)$, occurred precisely at $2025\,MCS$.

We discuss foams macroscopic physics, by computing bubbles' mean velocities by tracking them during many circuits around the channel, for a time interval of around $10^5\,MCS$. The velocity fluctuations normal to the direction of motion are negligibly small. The strain-rate $\dot{\gamma}$ is computed as in Couette's experiment[8] , it is proportional to the mean velocity, *i.e.*, $\dot{\gamma} \propto \triangle\omega = V_{\overrightarrow{\delta l}}/D$, where $\triangle\omega$ is the relative angular velocity of the bubbles in the two layers, $V_{\overrightarrow{\delta l}}$ is the tangential mean velocity and $D$ is the bubble diameter. Figure 2 shows a rheogram-plot of the driving force as a function of the strain-rate. Here, we can distinguish three different regimes:( *1*) $\dot{\gamma} \leq 0.0105$ ($f \leq 0.1946$), ( *2*) $0.0105 < \dot{\gamma} < 0.027$ and ( *3*) $\dot{\gamma} \geq 0.027 (f \geq 0.40)$. There are many constitutive models availables to describe rheological fluid behavior, Herschel-Bulkley like foams behavior has been reported in foams experiments [9, 10, 11]. The model employs three rheological parameters: the yield stress, the consistency index and the power-law index. We use a similar power law relationship to describe the bubbles regimes ( *1* – *2*) as shown in the inset of Figure 2, i.e; $f_i = f_i^\star + K_i(\dot{\gamma} - \dot{\gamma}_i^\star)^{\psi}$, the parameters are computed by extrapolation using the Lavenberg-Marquadt method [12]. We found for regime ( *1*) :$f_1^\star \approx 0.0125$, $\dot{\gamma}_1^\star \approx 0.00012$, $K = 0.286$ and $\psi_1 = 0.08$. It is called *elastic*, because it also shows a jamming behavior. The regime ( *2*) is called *viscoelastic* with parameters: $f_2^\star \approx 0.1946$ , $\dot{\gamma}_2^\star \approx 0.0105$, $K_2 = 4.351$ and $\psi_2 = 0.73$. The third regime is called a *fluid* regime and obeys a more complex rational function of the form $h(\dot{\gamma})/g(\dot{\gamma})$, where both functions are polynomials of fourth order.

Let us come back to the microscopic rearrangements. We found that $T_1$s events are roughly periodic with approximately 100 per round trip. There are no $T_1$s avalanche events so that its cumulative amount grow linearly with *MCS* time, irrespective of the value of the driving force applied, as shown in Figure 3. Following we compute the slope for these curves, which represents the rate of number events occurring at specific positions in the bended channel.They are related to the viscous gradient of $T_1$s spatial distribution as reported in [13]: it relates the advective charge induced by the imposed shear at strain-rate $\dot{\gamma}$ to the cumulative relaxation of stress associated with $T_1$s. We compare these two different time scale processes in Figure 4 by plotting the strain-rate as a function of the rate of topological events. A linear behavior predominates in regions $(1-2)$ and even in a considerable part of region $(3)$, with the slope approximately equal to 1.0675, in agreement with [13].

## 4. Conclusions

Our *CPM* results agree qualitatively with observed experimental foam behavior, since we consistently find *elastic*, *viscoelastic* and *fluid* type responses, depending on the applied dragging force. Perhaps our more important result is to show that the dragging force is naturally related to the strain-rate by a power law relationship with a given exponent. We used a similar Herschel-Bulkey expression in order to describe the *elastic* and *viscoelastic* foam regimes. For the later we found the exponent 0.73, closer to the 2/3 as computed from analytic approximations and it has indeed been measured [14, 15, 16]. But some experiments suggest this value is in the range (0.25 , 1.0) [9, 17]. The strain-rate response arises from topological rearrangements which occurred at specific locations at the channel bend regions and no avalanche phenomena are found. This lack of self-organized criticality has also been reported in both monolayer and bulk foams experiments on *2* and *3 D*, including some simulations [18, 19]. We found that the rate of localized rearrangements events is proportional to the (low) shear-rate, in agreement with kinematic expression reported in [13].

## Acknowledgments

JSEO is grateful to Dr J A Glazier and Dr Y Hayakawa for fruitful discussions. This work is supported by the Goiás Research Foundation - FAPEG.

## References

[1] Kraynik A M 1988 *Annu Rev Fluid Mech* **20** 325
[2] Weaire D and Hutzler S 2001 *The Physics of Foams* (Oxford University Press)
[3] Ono I K, Tewari S, Langer S A and Liu A J 2003 *Phys Rev E* **67** 061503
[4] Jiang Y, Swart P J, Saxena A, Asipauskas M and Glazier J A 1999 *Phys Rev E* **59** 5819
[5] Jiang Y and Glazier J A 1996 *Philos Mag Lett* **74** 119
[6] F. Graner F and J.A. Glazier J A 1992 *Phys Rev Lett* **69** 2013
[7] J.A. Glazier and F. Graner 1993 *Phys Rev E* **47** 2128
[8] Janiaud E and Graner F 2005 *J Fluid Mech* **532** 243
[9] Langlois V J, Hutzler S and Weaire D 2008 *Phys Rev E* **78** 021401
[10] Barry J D, Weaire D and Hutzler S 2010 *Rheol Acta* **49** 687
[11] Katgert G,Latka A, Möbius M E and Hecke vam M 2009 *Phys Rev E* **79** 066318
[12] Press W H,TeukolskyS A,Vetterling W T and Flannery B P 2007 *Numerical Recipes-The Art of Scientific Computing* (Cambridge University Press)
[13] Kabla A and Debrégeas G 2003 *Phys Rev Lett* **90** 258303
[14] Bretherton F P 1961 *J Fluid Mech* **10** 166
[15] Lauridsen J, Twardos M, and Dennin M 2002 *Phys Rev Lett* **89** 098303
[16] Cantat I, Kern N and Delannay R 2004 *Europhys Lett* **65** 726
[17] Dennin M 2005 *Coll Surf A* **263** 66
[18] Dennin M, and Knobler C M 1997 *Phys Rev Lett* **78** 2485
[19] Tewari S, Schiemann D, Durian D J, Knobler C M,Langer S A and Liu A J 1999 **60** 4385

# On inverse problem of identification the cavity in the elastic cylinder.

**Olga Bocharova**

Southern Scientific Center of Russian Academy of Sciences, Rostov-on-Don, Russia

E-mail: olga_rostov1983@mail.ru

**Abstract.** In the present work we consider the inverse problem of identification the spherical cavity of small relative size in the isotropic elastic cylinder. Based on the methods of perturbation theory, analytical formulas for determining the parameters of the cavity have been obtained using information concerning corrections to the natural frequencies. A series of numerical experiments has been carried out.

## 1. Introduction

Problems of identification defects in elastic bodies are very important and belong to a class of inverse problems of the theory of elasticity. Cavities are widespread type of defects which arise usually in the stages of fabrication and operation of constructional elements. Defects of this nature are stress concentrators and often greatly reduce the carrying capacity of structures. Investigation of vibrations of bodies with localized inhomogeneities allows to reveal the influence of inhomogeneities of this type on the dynamic stress concentration, to predict the destruction of structures in the early stages. Note that, in practice, the most important is information about the type of the defect (cavity or crack) and its characteristic dimensions, because this information is a major and decisive from the point of view of fracture mechanics. Defect type and its size can be determined according to the change of the resonance frequency, assuming that the defect is small in comparison with the characteristic geometric dimensions of the body.

## 2. Statement of the problem and solution methods

It is possible to obtain corrections for the first resonance frequency in the analytical form using perturbation theory [1]. In [2] shows the general scheme of obtaining corrections for the anisotropic elastic body, weakened by internal cavity is shown. In [3] the approach described in [2], has been used to determine corrections to the resonance frequencies in the problem of steady-state longitudinal oscillations with a frequency $\omega$ of the isotropic elastic cantilever fixed rod with length $l$, weakened by the small cavity. In paper [3] the model of an inhomogeneous rod has been used, such that the presence of a cavity characterized by the dependence of the cross sectional area on the longitudinal coordinate in the form of:

$F(x)=F_0(1-\varepsilon\eta(x))$, where $\varepsilon=\dfrac{r_0^2}{R^2}$ - the small parameter

$r_0$ - the cavity radius, $R$ - radius of the rod cross section,

$F_0$ - cross sectional area of the rod without defects

$\eta(x)$ - positive function with compact support, that simulates the presence of the cavity:

$$\eta(x) = \begin{cases} >0, & x \in [a,b] \\ =0, & x \notin [a,b] \end{cases} \quad [a,b] \subset [0,l]$$

Natural frequencies and natural modes have been presented in the form of:

$$u^2 = u_0 + \varepsilon u_1 + o(\varepsilon^2) \quad \omega^2 = \omega_0^2 + \varepsilon \omega_1 + o(\varepsilon^2)$$

For the rod, weakened by spherical cavity, centered at the point $c_0$ of radius $r_0$ the following formula to calculate the corrections to the natural frequencies has been obtained:

$$\omega_{1n} = -\frac{8}{3}\omega_{0n}^2 \frac{r_0}{l} \cos\left(2\omega_{0n}\frac{c_0}{l}\right) \tag{1}$$

where $\omega_{0n}$ - natural frequencies of longitudinal vibrations of the rod without cavity.

Based on (1) the formula for finding the center of the cavity by using the information on corrections to the first and second frequencies has been obtained:

$$c_0 = \pi^{-1} l \arccos\left(\pm\frac{1}{6}\sqrt{\frac{\omega_{12}}{\omega_{11}} + 27}\right) \tag{2}$$

where, the sign is selected «+», if $\omega_{11} < 0$, and sign «–», if $\omega_{11} > 0$.

After finding $c_0$, $r_0$ is determined from the relation (1).

### 3. The results of numerical experiments

The described approach has been used to finding the parameters of a spherical cavity in the rod (with axis length $l = 1m$, radius of the circle $R = 0.1m$), weakened by spherical cavity (with radius $r_0 = 0.2R = 0.02m$), located on the symmetry axis of the rod.

**Table 1.** Dimensionless wave number for the rod model

| | rod without cavity | $c_0 = 0.1$ | $c_0 = 0.2$ | $c_0 = 0.3$ | $c_0 = 0.8$ | $c_0 = 0.9$ |
|---|---|---|---|---|---|---|
| $\kappa_1$ | 1.57077 | 1.56913 | 1.56937 | 1.56974 | 1.57212 | 1.57236 |
| $\kappa_2$ | 4.71180 | 4.70874 | 4.71329 | 4.71656 | 4.71014 | 4.71472 |

Table 1 shows the values of the first and second dimensionless wave numbers $\kappa = \omega l\sqrt{\frac{\rho}{E}}$ for rod without cavity and rod , weakened by spherical cavity, centered at the point $c_0$.

**Table 2.** Recovery results of the center and radius of the cavity for the rod model

| | $c_0 = 0.1$ | $c_0 = 0.2$ | $c_0 = 0.3$ | $c_0 = 0.8$ | $c_0 = 0.9$ |
|---|---|---|---|---|---|
| $c_0^*$ | 0.0996 | 0.1976 | 0.0294 | 0.7977 | 0.8990 |
| $r_0^*$ | 0.0202 | 0.0202 | 0.0201 | 0.0200 | 0.0200 |

Table 2 shows the results of recovery center $c_0^*$ and the radius $r_0^*$ of the spherical cavity on the corrections to the wave numbers. It should be noted that the error recovery center of the cavity does not exceed $1.2\%$, the error recovery of the cavity radius does not exceed $0.9\%$.

Next, in the present study experiment has been carried out to find the spherical cavity parameters in the cylinder (with axis length $l = 1m$, radius of the circle $R = 0.1l = 0.1m$), weakened by spherical cavity (with radius $r_0 = 0.2R = 0.02m$), located on the symmetry axis of the rod.

The considered problem has been solved using the finite element package ANSYS. Program has been written in the macro language APDL for ANSYS to calculate the natural frequencies of the cylinder depending on the position of the center of the cavity. Volumetric finite element SOLID 95 has been used in modeling.

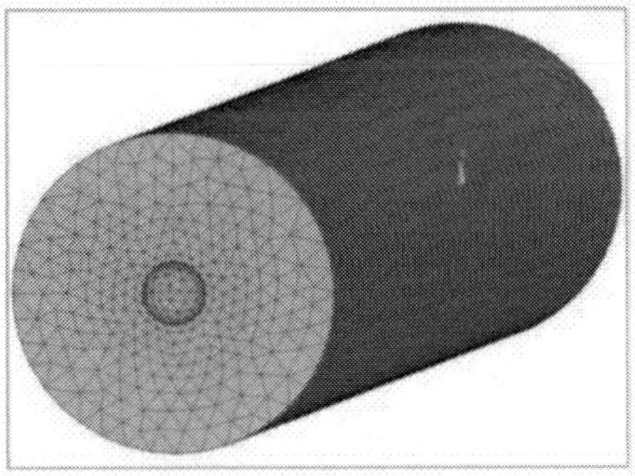

**Figure 1.** The cross section of the cylinder, weakened by spherical cavity

**Table 3.** Dimensionless wave number for the cylinder model

| | cylinder without cavity | $c_0 = 0.1$ | $c_0 = 0.2$ | $c_0 = 0.3$ | $c_0 = 0.8$ | $c_0 = 0.9$ |
|---|---|---|---|---|---|---|
| $\kappa_1$ | 1.57827 | 1.57469 | 1.57457 | 1.57593 | 1.57951 | 1.57988 |
| $\kappa_2$ | 4.71410 | 4.70571 | 4.71336 | 4.71866 | 4.70879 | 4.71595 |

The natural frequencies calculated by ANSYS have been brought to a dimensionless form.

Table 3 shows the values of the first and second dimensionless wave numbers for cylinder without cavity and cylinder, weakened by spherical cavity, centered at the point $c_0$.

Based on formulas (1)-(2) by using information on corrections to the wave numbers, the computational experiment of reconstruction parameters of a spherical cavity in the cylinder has been carried out. Table 4 shows the results of recovery center $c_0^*$ and the radius $r_0^*$ of the spherical cavity.

Note that in this case the error recovery parameters of the cavity increases as compared with the one-dimensional model of the rod (maximum error recovery center of the cavity is 24%, the maximum error recovery of the cavity radius is 26%).

**Table 4.** Recovery results of the center and radius of the cavity for the cylinder model ($R = 0.1l$)

| | $c_0 = 0.1$ | $c_0 = 0.2$ | $c_0 = 0.3$ | $c_0 = 0.8$ | $c_0 = 0.9$ |
|---|---|---|---|---|---|
| $c_0^*$ | 0.0756 | 0.161 | 0.2218 | 0.7159 | 0.8715 |
| $r_0^*$ | 0.0259 | 0.0272 | 0.0244 | 0.0211 | 0.0202 |

Note, the decreasing ratio of the cylinder radius to the axis length leads to increase accuracy of reconstruction parameters cavity.

Table 5 shows the results of recovery for cylinder (with axis length $l = 1m$, radius of the circle $R = 0.05l = 0.05m$), weakened by spherical cavity (with radius $r_0 = 0.2R = 0.01m$).

**Table 5.** Recovery results of the center and radius of the cavity for the cylinder model ($R = 0.05l$)

| | $c_0 = 0.1$ | $c_0 = 0.2$ | $c_0 = 0.3$ | $c_0 = 0.8$ | $c_0 = 0.9$ |
|---|---|---|---|---|---|
| $c_0^*$ | 0.09 | 0.164 | 0.22 | 0.723 | 0.878 |
| $r_0^*$ | 0.0128 | 0.0126 | 0.0124 | 0.01 | 0.0098 |

## 4. Conclusion

The proposed approach provides the most accuracy of reconstruction cavity parameters, in the case when the cavity center is located close to the free end of the cylinder. In the case when the cavity center is close to the center axis of the cylinder, error recovery parameters cavity increases considerably.

**Acknowledgments**

This work was supported by the President of Russian Federation (grant MK-6213.2012.1).

## References

[1] Naife A. 1984 *Introduction to perturbation methods (in Russian)* (Moscow: Mir) p. 535

[2] Vatulyan A O 2007 *Inverse problems in solid mechanics. (in Russian)* (Moscow: Fizmatlit) p. 223

[3] Bocharova O V, Vatulyan A O, Zharkov R S 2006 *Izv. Vyssh. Uchebn. Zaved. Sev. Kavkaz. Region, Estetsv. Nauki (in Russian)* **2** pp. 28-32

# Non-lineal equations for the Raman spectroscopy analysis

**M R Joya**[1] **C E Paris**[2] **J Barba-Ortega**[1]

[1] Departamento de Física, Universidad Nacional de Colombia, Bogotá, Colombia.

[2] Empresa Brasileira de Pesquisa Agropecuária, Embrapa Instrumentação, São Carlos, São Paulo, Brazil.

E-mail: `mrinconj@unal.edu.co`

**Abstract.** We study some equations used in the analysis of the materials of the Raman spectrum. Some equations are, phonon lifetime, related to the full width at half maximum (FWHM), the size of the nanoparticles, as explained with the equation of the intensity of Raman Spectrum; the Grüneisen parameter isothermal and isobaric explained by the frequency shift of the Raman scattered light; it is directly related to the structural properties of the material. The frequency shift may be dependent upon the application of temperature, hydrostatic pressure and the composition of the material itself. Furthermore, it is possible to obtain the behavior of anharmonic material using the Raman scattering frequency shift.

## 1. Introduction

Raman spectroscopy probing the molecular and crystal lattice vibrations and hence being bottom up approach of the matter appears as one of the most powerful techniques in the study of nanomaterials and nanostructures. [1, 2, 3, 4, 5, 6, 7, 8].

Previously we had assumed that the dipole moment of the molecule, $\mathbf{P}$, in an electric field is:

$$\mathbf{P} = \mu_0 + \alpha.\mathbf{E} \tag{1}$$

where $\mu_0$ is the permanent dipole moment, $\alpha$ is the polarizability, and $\mathbf{E}$ is the electric field of the light wave. However, when the intensity of the incident light is sufficiently large, the induced oscillation of the dipole moment becomes nonlinear:

$$\mathbf{P} = \mu_0 + \alpha.\mathbf{E} + \beta.\mathbf{E}.\mathbf{E} + \gamma.\mathbf{E}.\mathbf{E}.\mathbf{E} \tag{2}$$

The second-order nonlinear phenomena are described by the second term of the equation. One of the best known nonlinear optical phenomenon of the third-order is the stimulated Raman scattering.

The variation in both the Stokes and anti-Stokes component intensities with temperature for a particular experimental. These data are in very good agreement with quantum mechanical predictions for the intensities temperature variation given here as:

$$I_{st} \propto \frac{\hbar}{2\omega_s} (\omega_l - \omega_s)^4 \left[1 + \left(e^{\frac{\hbar\omega_s}{kT}} - 1\right)^{-1}\right] \tag{3}$$

$$I_{a-st} \propto \frac{\hbar}{2\omega_s}(\omega_l+\omega_s)^4\left(e^{\frac{\hbar\omega_s}{kT}}-1\right)^{-1} \tag{4}$$

Here $\omega_s$, is the phonon frequency, $\omega_l$ is the line center frequency of the laser excitation source, T is the sample temperature measured by optical, and $\hbar$ and $k$ are the Planck and Boltzmann constants.The proportionality constant in equations (3) and (4) include the variation in the polarization with normal coordinate separation and the collection efficiency of the optics, and are essentially equal for both the Stokes and anti-Stokes components. The close agreement between the measured and theoretical temperature dependence suggests that the sample is optically thin to the laser, Stokes, and anti-Stokes radiation [10].

The Stokes to anti-Stokes intensity ratio

$$\frac{I_{st}}{I_{a-st}} = \left(\frac{\omega_l-\omega_s}{\omega_l+\omega_s}\right)^4 e^{\frac{\hbar\omega_s}{kT}} \tag{5}$$

can be a direct measure of the sample temperature. Indeed, this has been used as an unambiguous measure of sample temperature in a range of materials by many investigators [10].

The lifetime of phonons is an important physical quantity, since it reveals detailed information on the anharmonic phononphonon interactions in a solid. Even in a perfect crystal at low temperature, phonons have a finite lifetime due to the decay, through anharmonic interactions, into combinations of phonons of different energy. Selection of the excitation wavelength depends on several factors.The Raman differential scattering cross-section varies inversely with the fourth power of the excitation wavelength by the proportionality formula:

$$\sigma \propto (\widetilde{\nu}_0 - \widetilde{\nu}_{vib})^4 \tag{6}$$

where $\widetilde{\nu}_0$ is the wavenumber of the incident radiation $(1/\lambda_0)$ and $\widetilde{\nu}_{vib}$ is the wavenumber of the vibrational mode $(\nu_{vib}/c)$.

The photon lifetimes as a function of temperature can be accurately measured by Raman spectroscopy. However, the temperature dependent Raman scattering measurement can be used to estimate not only the temperature variation of the phonon lifetimes, but also the anharmonicity in the vibrational potential accurately. The finite values of the lifetimes are predominantly due to phonon-phonon interaction, which is highly temperature-dependent, but also has a contribution due to the scattering by crystal defects, and boundary scattering (phonon propagation is limited by the sample dimension) which is almost temperature-independent [9].

The anharmonicity present in the vibrational potential for the two modes is estimated by means of numerically calculating the anharmonic constants. Furthermore, the temperature-independent broadening due to the inherent disordered nature of the crystal and boundary scattering is also estimated for the two modes.

In the quasiharmonic model, phonon modes behave harmonically with lifetimes, but their frequencies are altered by the effects of volume on the interatomic potential. Pure anharmonicity originates with phonon-phonon interactions, which increase with temperature. Pure anharmonicity contributes to shifts in phonon frequencies, but also causes phonon damping and lifetime broadening of phonon peaks. To separate the peak shifts caused by pure anharmonicity from the shifts caused by quasiharmonicity, we treat the mode frequency $\omega_j = \omega_j(V,T)$ as a function of volume and temperature.

$$\left(\frac{\partial \ln \omega_j}{\partial T}\right)_P = -\frac{\beta}{\kappa}\left(\frac{\partial \ln \omega_j}{\partial P}\right)_T + \left(\frac{\partial \ln \omega_j}{\partial T}\right)_V \tag{7}$$

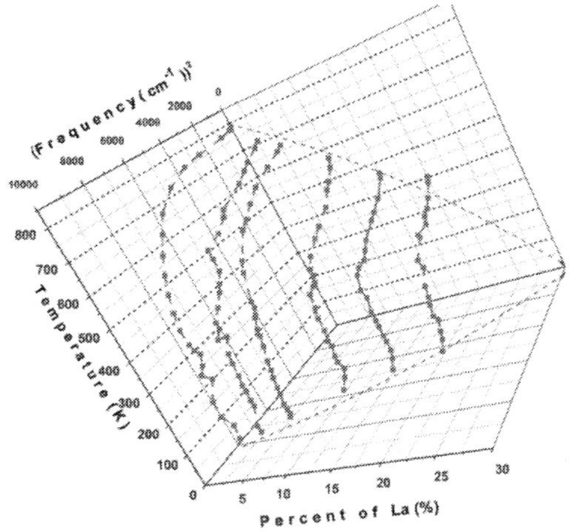


**Figure 1.** Low-frequency region of the Raman spectra for different La sample. Observe the mode softening evolution and the redshift of E1(TO) with increasing applied of temperature and concentration of La.

where $j$ is the phonon mode index, $\beta$ is the volume thermal expansivity, and $\kappa$ is the isothermal compressibility. This is a general method to separate phonon quasiharmonicity from phonon anharmonicity and is not unique for rutile structures,for example. The left hand side gives the directly measurable temperature-dependent isobaric frequency shift, including both quasiharmonic and pure anharmonic behavior. The first term on the right-hand side, the isothermal frequency shift as a function of pressure, is the quasiharmonic contribution, which is also measurable. By defining a mode Grüneisen parameter as the proportionality of the relative change of the mode frequency to the relative change of volume ,$\gamma_j = -\frac{\partial(\ln \omega_j)}{\partial(\ln V)}$ The second term on the right-hand side of Eq. (7) is the frequency shift from the pure anharmonicity, which depends on temperature and not volume. From the difference between the isobaric and the isothermal frequency shifts, the pure anharmonicity can be determined experimentally [11]. f

## 2. Experimental

Raman scattering measurements were performed using a Jobin-Yvon T64000 triple monochromator with a CCD detector. The exciting light from an argon laser was focused on the samples with an optical microscope using a long working distance 50X objective. For the low temperature measurements, in the 10 K to 300 K range, a helium closed cycle cryostat was used while for high temperature, in the 300 K to 850 K range, an optical micro furnace was used. The frequency values of each vibrational mode were obtained by fitting asymmetric Lorentzian line shapes.

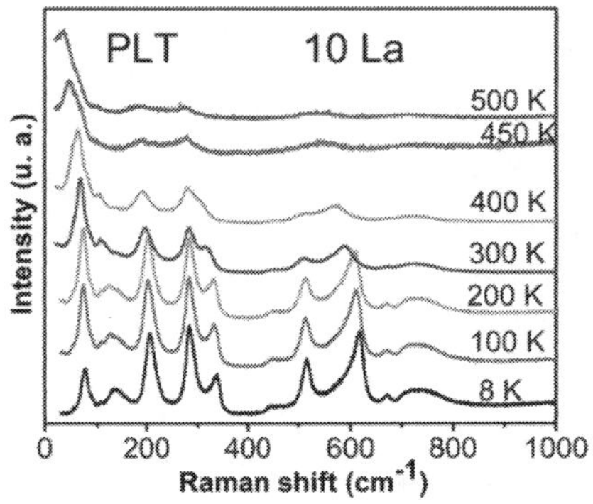


**Figure 2.** Raman frequencies of *PLT* (10% La) for different values of applied temperature.

## 3. Results and Discussion

The observed softening of the mode E1, shown in Fig. 1. These temperature dependent changes can be attributed to the anharmonicity in the vibrational potential, which leads to the decay of optical phonons into low-energy phonons. Furthermore, the inherent disorder in the crystal provides decay channels and leads to broadening of the phonon. This term is almost temperature independent. The FWHM of Raman lines, the inverse of the phonon lifetime, has been measured repeatedly in the past in particular for diamond-type crystals. The main attention was focussed on the temperature dependence [3], and on the dependence on isotopic composition [3, 4]. While the temperature dependence of the FWHM is governed mainly by the change in the phonon occupation numbers, its pressure dependence originates only from the pressure induced changes in the phonon dispersion relation and from the change of the anharmonic force constants. The pressure-induced change in the linewidths is a comparably small effect. Therefore, special precautions have to be taken to determine the correct FWHM of the phonon modes. The finite resolution of the spectrometer has to be taken into account. The experimental lineshape is usually a convolution of a Lorentzian-type phonon with the Gaussian spectrometer profile. The resulting lineshape is the so-called Voigt lineshape. To determine the spectrometer profile, the spectrum of a nearby the line was measured for each Raman spectrum [12].

The variation in the fullwidth at half maximum intensity (FWHM) with temperature is shown in Fig.2.The temperature-dependent Raman scattering indicates that the decrease in the phonon lifetime with increasing temperature is predominantly due to the decay of strongly interacting optical phonons into weakly interacting low-energy phonons. A contribution for the phonon lifetime also comes from the scattering by inherent crystal defects and from the boundary scattering, but this contribution is found to be temperature-independent.

## References

[1] K. Kneipp, H. Kneipp, I. Itzkan, R. R. Dasari, and M. S. Feld,Chem. Rev. **99**, 2957 (1999).
[2] J. V. Miller and E. G. Bartick, Appl. Spectroscopy, **55** 1729 (2001).
[3] M. R. Joya, P. S Pizani, Appl. Phys. Lett. textbf97 031903 (2010).

[4] M. R. Joya, J. Barba-Ortega, P.S. Pizani, J. Appl. Phys.**113**, 013512 (2013).
[5] M. R. Joya, J. Barba-Ortega, P.S. Pizani,Cerâmica. **58**,313 (2012).
[6] M. R. Joya, J. Barba, P.S. Pizani,DYNA,**79**,137 (2012).
[7] M. R. Joya, J. Barba-Ortega, DYNA, **78** 216 (2011).
[8] P. P. Kalantri, R. R. Somani and D. T. Makhija, Der Chem. Sinica, **1** 1 (2010).
[9] S. Anand, P. Verma, K.P. Jain, S. C. Abbi, Phys. **B** 226, 331 (1996).
[10] H. Herchen and M. A. Cappelli, Phys. Rev. **B** 43, 11740 (1991)
[11] T.Lan, C.w.Li, B.Fultz, Phys. Rev. **B** 86, 134302 (2012).
[12] A. Debernardi, C. Ulrich, M. Cardona, and K. Syassen, Phys. Stat. Sol. (b) **223**, 213 (2001).

# Elastic properties and lattice dynamics of ruthenium at high pressures

**A V Lugovskoy, M P Belov, Yu Kh Vekilov and O M Krasilnikov**

National University of Science and Technology "MISIS", Department of Theoretical Physics and Quantum Technologies, Leninskiy pr. 4, Moscow, Russia

E-mail: andrey.lugovskoy@gmail.com

**Abstract.** The elastic properties and structural stability in ruthenium under pressure are investigated. The analysis is performed in the framework of Landau theory and nonlinear elasticity. For this purpose the definition of effective elastic constants (EC) of n-th ($n \geq 2$) order characterizing elastic properties of loaded crystal and the relations between effective EC and corresponding EC of Bragger type for *hcp* crystals is given. The conditions of *hcp* lattice stability to the uniform shear strain under the pressure P are expressed in terms of the second order effective EC. The method of effective EC calculations for *hcp* crystals under hydrostatic pressure is presented. The equation of state and EC of second and third order and phonon dispersion relations in high-symmetry directions in the pressure range of 0 – 600 GPa are calculated in the framework of the density functional theory (DFT) and the density functional perturbation theory (DFPT) respectively. EC are in the good agreement with available experimental data and increase monotonically with pressure, no softening or stability condition violation are observed. Softening of phonon frequencies near the Brillion zone center is also not observed.

## 1. Introduction

Ruthenium is a transitional 4d metal with the hexagonal closed packed (*hcp*) structure at normal conditions. A sequence of the *bcc* metals (V, Mo, W) demonstrates a possibility of structural transitions at high pressure associated with the EC and phonon frequencies softening. Hexagonal closed packed structure is common for transitional metals, however the most calculations were performed for metals with cubic lattice. To our knowledge data on elastic constants and phonon dispersion of Ru at pressure are poor. Elastic constants of second order for the number of metals with *hcp* structure (including ruthenium) were calculated ab initio [1] at $P = 0$ ($T = 0$ K) and the single ($\widetilde{C}_{44}$) EC was determined from the Raman frequencies up to 60 GPa [2]. In work [3] the elastic constants of second and third order of Os and Ru were calculated at $P = 0$ ($T = 0$ K) using Lennard-Jones potential. The phonon dispersion in high-symmetry directions at ambient pressure were measured by the neutron scattering and calculated using the DFT in work [4]. The purpose of these research is to investigate the elastic properties and phonon stability of the *hcp* ruthenium at high pressure. The elastic constants of second and third order of *hcp* ruthenium in the wide pressure range (0 – 600 GPa) are calculated by the DFT from the energy – strain relation. The phonon dispersions in high-symmetry directions in the same pressure range are determined by the DFPT.

## 2. Elastic constants of loaded crystal and stability conditions

Thermodynamic definition of the elastic constants of $n$-th order ($n \geq 2$) for the unloaded crystal is presented in work [5]. For the isothermal constants we have:

$$C_{ijkl...} = \frac{1}{V_0}\left(\frac{\partial^n F}{\partial\eta_{ij}\partial\eta_{kl}\cdot\cdot}\right)_0 \tag{1}$$

Here $F$ is the Helmholtz free energy, $V_0$ is the volume in the initial state, $\eta_{ij}$ is the components of the Lagrangian finite strain tensor [6]. The isothermal elastic constants of the loaded crystal (effective elastic constants [7]) can be defined the same way as it is done in work [8]

$$\tilde{C}_{ijkl...} = \frac{1}{V_0}\left(\frac{\partial^n G}{\partial\eta_{ij}\partial\eta_{kl}\cdot\cdot}\right)_0, \tag{2}$$

where $G$ is the Gibbs free energy (thermodynamic potential at given temperature and external loading). The change of $G$ at the hydrostatic pressure $P$ caused by the deformation $\eta$, $\Delta G = G(P,T,\eta) - G(P,T,0)$ can be presented in form $\frac{\Delta G}{V_0} = \frac{\Delta F}{V_0} + P\frac{\Delta V}{V_0}$; here $\Delta F$ and $\Delta V$ are the change in free energy and volume at this strain respectively. The effective elastic constants (2) in contrast to (1) include both the change in free energy $F$ in presence of small strain near the initial loaded state and the work against external stress by the forces caused by the strain itself. If strain isn't applied, definitions (1) and (2) are identical. When the effective constants $\tilde{C}_{ijkl..}$ are used all the relations of the theory of elasticity can be applied including Born's stability conditions which are identical in both loaded and unloaded states.

From the requirements of a positively defined quadratic form of the thermodynamic potential we obtain the stability conditions of a crystal to a uniform strain. For crystals with hexagonal symmetry we have $\tilde{C}_{33}(\tilde{C}_{11}+\tilde{C}_{12})-2\tilde{C}_{13}^2>0$, $\tilde{C}_{11}-\tilde{C}_{12}>0$, $\tilde{C}_{44}>0$; here $\tilde{C}_{\alpha\beta}$ are the elastic constants of second order in the Voigt's notation. These conditions can be violated if temperature or pressure are changed. This violation can cause the structural transformation into spontaneously deformed state, stability of which is ensured by the nonlinear elasticity of the lattice (elastic constants of 3, 4 and further orders).

Let us express the effective elastic constants $\tilde{C}_{\alpha\beta}$ and $\tilde{C}_{\alpha\beta\gamma}$ in terms of the corresponding Bragger type elastic constants, calculated in loaded state as it is done in [8] for cubic crystals. The crystals of hexagonal symmetry ($622, 6/mmm, \bar{6}m2, 6mm$) have five elastic constants of second and ten elastic constants of third order [6]. The obtained results are presented in Table 1. Thus for the calculation of the effective elastic constants $\tilde{C}_{\alpha\beta.}$ it is required to find pressure and elastic constants $C_{\alpha\beta.}$ from the expansion of free energy $F$ in components $\eta_{ij}$ (see [8], formula (6)).

**Table 1.** Relations between $\tilde{C}_{\alpha\beta...}$ and $C_{\alpha\beta...}$.

| | | |
|---|---|---|
| $\tilde{C}_{11}=C_{11}-P$ | $\tilde{C}_{111}=C_{111}+3P$ | $\tilde{C}_{144}=C_{144}-P$ |
| $\tilde{C}_{12}=C_{12}+P$ | $\tilde{C}_{112}=C_{112}-P$ | $\tilde{C}_{155}=C_{155}+P$ |
| $\tilde{C}_{13}=C_{13}+P$ | $\tilde{C}_{113}=C_{113}-P$ | $\tilde{C}_{222}=C_{222}+3P$ |

| | | |
|---|---|---|
| $\tilde{C}_{44}=C_{44}-P$ | $\tilde{C}_{123}=C_{123}+P$ | $\tilde{C}_{333}=C_{333}+3P$ |
| $\tilde{C}_{66}=(\tilde{C}_{11}-\tilde{C}_{12})/2=$ $(C_{11}-C_{12}-2P)/2$ | $\tilde{C}_{133}=C_{133}-P$ | $\tilde{C}_{344}=C_{344}+P$ |

### 3. Calculation details and results discussion

To calculate all five second order and ten third order elastic constants ten various deformation schemes were applied for each of the initial configurations. The initial loaded state was defined by the volume $V_0$ . Deformed lattice vectors were expressed by relation $r_i=\alpha_{ij}R_j$ , where strain gradient $\alpha_{ij}$ is expressed in terms $\eta_{ij}$ (see [8], formula (7)), $r$ and $R$ are the lattice vectors components in stressed and initial states respectively. For the calculation of the full energy of *hcp* Ru ($T=0$ K) at various values of $V_0$ and the deformation $\eta_{ij}$ VASP [9] package was used. The pressure and the elastic constants $C_{\alpha\beta\gamma}$ were determined as the coefficients of the expansion of free energy $F$ in components $\eta_{ij}$ or their combinations (see [8], formula (6)) by the mean square method (30 points with step 0.003 in the range of $\eta=\pm0.045$).

The phonon dispersions at the studied pressure range were calculated by DFPT through QuantumESPRESSO code (QE) [10].

Exchange-correlation effects were treated by the generalized gradient approximation with PW91 parameterization [11]. To describe ion-electron interactions the projector augmented wave method was applied [12] in VASP and Vanderbilt ultrasoft pseudopotential in QE.

For the integration of Brillouin zone the $\Gamma$- centered Monkhorst-Pack grids [13] were used with 28x28x18 and 16x16x10 meshes in VASP and QE respectively to take into account the symmetry of the studied crystal. The plane wave cutoff energy equaled to 600 eV in VASP and 50 Ry in QE.

All the initial loaded states of crystal in the range $V_0/V(0)=1\div0.59$ were fully relaxed. In the VASP calculations the Methfessel-Paxton smearing [14] with broadening of 0.2 eV was used for relaxation. For total energy calculations the tetrahedron method with Blochl [15] corrections was applied. In all QE calculations the Marzari-Vanderbilt cold-smearing scheme [16] was used.

The obtained lattice parameters are in good agreement with both experimental and theoretical results of other authors.

The results of second order EC calculations are given in table 2. As it follows from the table all three stability conditions, mentioned in section 2 are satisfied throughout the whole pressure range studied.

**Table 2.** Equation of state and EC of second order of *hcp* Ru for various values of P ($T=0$K ). Pressure and $C_{\alpha\beta}$ are given in GPa.

| $V_0$, Å$^3$ | $P$ | $\tilde{B}$ | $\tilde{C}'$ | $\tilde{C}_{11}$ | $\tilde{C}_{12}$ | $\tilde{C}_{13}$ | $\tilde{C}_{33}$ | $\tilde{C}_{44}$ |
|---|---|---|---|---|---|---|---|---|
| 13.621 | 0.072 | 332.5 | 197.3 | 574.2 | 179.5 | 210.5 | 643.4 | 187.7 |
| 12.203 | 47.080 | 521.1 | 289.6 | 914.3 | 335.1 | 366.3 | 1008.0 | 267.1 |
| 10.880 | 127.300 | 576.3 | 419.6 | 1416.0 | 576.5 | 605.2 | 1545.0 | 376.3 |
| 9.651 | 259.300 | 1204.0 | 602.8 | 2149.0 | 943.3 | 969.2 | 2331.0 | 526.2 |
| 8.512 | 471.800 | 1794.0 | 852.7 | 3203.0 | 1498.0 | 1526.0 | 3471.0 | 724.5 |
| 7.977 | 621.400 | 2177.0 | 1005.0 | 3883.0 | 1873.0 | 1899.0 | 4216.0 | 841.5 |

The value of $C_{\alpha\beta}$, corresponding to $P=0$ are in good agreement with the experimental and theoretical data [1, 17].

We also calculated the phonon dispersions in the whole studied pressure range, however we present only date for P=0 and 600 GPa (see figure 1). Phonon dispersion at P=0 GPa is in good agreement

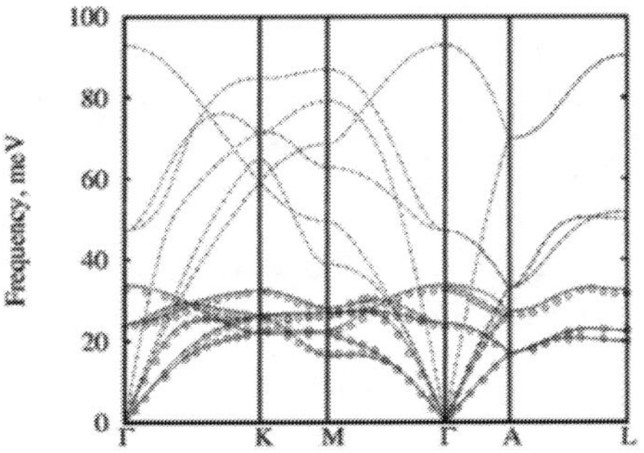


**Figure 1.** The phonon dispersion of the Ru for various directions in reciprocal space. Curves are theoretical results (T=0 K): solid - P = 0 GPa, dot-dashed– P = 600 GPa. The open circles are the experimental room temperature data [4].

with experimental data and reproduces peculiarities near the *M*-point, unique for Ru and described in [4]. These are essentially dispersionless acoustic branch of transverse vibrations for more than 1/3 of the $\Gamma$-*M* direction and the presence of almost degenerate branches at *M*. The drop of the longitudinal modes discussed in [4] is pronounced weaker in present research which can be explained by a poorer consideration of Fermi-surface geometry (smaller k-points mesh size) in our calculation. Phonon modes become harder with increasing pressure while the mentioned peculiarities become less and less pronounced and quickly disappear which finally results in spectrum similar to the one shown for P=600 GPa on figure 1.

It is seen that *hcp* structure of ruthenium remains mechanically and dynamically stable at pressure range of 0 – 600 GPa. Moreover all the elastic constants of second order in this range increase monotonically (crystal lattice becomes more durable) with pressure,

The third order EC $\widetilde{C}_{\alpha\beta\gamma}$ are the important parameters, characterizing nonlinear mechanical response of crystal at high pressures and finite strains. In work [3] elastic constants $\widetilde{C}_{\alpha\beta\gamma}$ were calculated at $P=0$ using Lennard-Jones potential. It is obvious, that such a calculations produce only approximate values. Different values of $\widetilde{C}_{\alpha\beta\gamma}$ calculated at various pressures in these research are given in table 3. All the obtained $\widetilde{C}_{\alpha\beta\gamma}$ are negative and their absolute values increase monotonically with pressure in the studied pressure range. The only exception is the value of $\widetilde{C}_{123}$.

**Table 3.** Third order elastic constants of hcp Ru at various pressures (*T=0 K*). All EC values are given in $10^{-2}$ GPa.

| $P$, GPa | $-\widetilde{C}_{111}$ | $-\widetilde{C}_{222}$ | $-\widetilde{C}_{333}$ | $-\widetilde{C}_{133}$ | $-\widetilde{C}_{113}$ | $-\widetilde{C}_{112}$ | $-\widetilde{C}_{123}$ | $-\widetilde{C}_{144}$ | $-\widetilde{C}_{155}$ | $-\widetilde{C}_{344}$ |
|---|---|---|---|---|---|---|---|---|---|---|
| 0.072 | 75.24 | 67.90 | 69.31 | 15.63 | 7.075 | 8.336 | 2.384 | 2.770 | 10.97 | 17.04 |
| 47.080 | 110.20 | 98.74 | 100.30 | 22.79 | 10.48 | 12.07 | 2.551 | 4.067 | 15.30 | 25.00 |
| 127.300 | 159.90 | 142.30 | 143.80 | 32.82 | 14.92 | 17.14 | 2.389 | 5.792 | 21.15 | 36.11 |
| 259.300 | 228.10 | 201.60 | 204.00 | 47.26 | 21.34 | 24.45 | 2.706 | 8.176 | 29.01 | 51.34 |
| 471.800 | 318.20 | 280.20 | 285.50 | 66.66 | 29.64 | 34.59 | 2.825 | 11.14 | 39.23 | 71.47 |
| 621.400 | 371.60 | 326.90 | 334.10 | 78.96 | 34.41 | 42.09 | 2.468 | 12.82 | 45.14 | 83.38 |

## Conclusion

The technique of n-th (n≥2) order elastic constants calculation based on nonlinear elasticity for the *hcp* lattice is briefly described. The elastic constants of second and third order and phonon dispersions for the *hcp* Ru were calculated in the wide pressure range. The analysis of the results indicates the mechanical and dynamical stability of *hcp* Ru throughout the whole pressure range studied. The results are in good agreement with the available experimental and theoretical data.

The study was supported by The Ministry of Education and Science of Russian Federation (project 14.A18.21.0893) and by Russian Foundation For Basic Research (grants 13-02-00338 and 13-02-00844).

## References

[1] Fast L, Wills J M, Johansson B and Eriksson O 1995 *Phys. Rev. B* **51** 17431–17438
[2] Olijnyk H, Jephcoatand A P and Refson K 2001 *Europhys. Lett.* **53**(4) 504
[3] Pandey D K, Singh D and Yadawa P K 2009 *Platinum Met. Rev.* **53** 91–97
[4] Heid R, Pintschovius L, Reichardt W and Bohnen K P 2000 *Phys. Rev.* B **61** 12059–12062
[5] Brugger K 1964 *Phys. Rev.***133** A1611–A1612
[6] Wallace D C 1970 *Solid State Physics* vol **25** pp 301–404
[7] Martin J W 1975 *J. Phys. C Solid State* **8** 2858
[8] Krasilnikov O M, Vekilov Y K and Mosyagin I Y 2012 *J. Exp. Theor. Phys.* **115** 237–241
[9] Kresse G and Furthmuller J 1996 *Phys. Rev. B* **54** 11169–11186
[10] Giannozzi P et al., http://www.quantum-espresso.org
[11] Blochl P E 1994 *Phys. Rev. B* **50** 17953–17979
[12] Kresse G and Joubert D 1999 *Phys. Rev. B* **59** 1758–1775
[13] Monkhorst H J and Pack J D 1976 *Phys. Rev. B* **13** 5188–5192
[14] Methfessel M and Paxton A T 1989 *Phys. Rev. B* **40**, 3616
[15] Blochl P E, Jepsen O and Andersen O K 1994 *Phys. Rev. B* **49** 16223–16233
[16] Marzari N, Vanderbilt D and De Vita A and Pyne D 1999 *Phys. Rev. Lett.* **82**, 3296.
[17] Simmons G and Wang H 1971 Single Crystal Elastic Constants and Calculated Aggregate Properties: A Handbook (MIT Press, Cambridge)

# Numerical solution of the problems for plates on partial internal supports of complicated configurations

**Dang Quang A**
Institute of Information Technology, VAST, 18 Hoang Quoc Viet, Hanoi, Vietnam
E-mail: dangqa@ioit.ac.vn

**Truong Ha Hai**
Thainguyen University of Information and Communication Technology, Thainguyen, Vietnam
E-mail: haininhtn@gmail.com

**Abstract.** Very recently in the work "Simple Iterative Method for Solving Problems for Plates with Partial Internal Supports, *Journal of Engineering Mathematics* , DOI: 10.1007/s10665-013-9652-7 (in press)", we proposed a numerical method for solving some problems of plates on one and two line partial internal supports (LPIS). In the essence they are problems with strongly mixed boundary conditions for biharmonic equation. Using this method we reduced the problems to a sequence of boundary value problems for the Poisson equation with weakly mixed boundary conditions, which are easily solved numerically. The advantages of the method over other ones were shown.

In this paper we apply the method to plates on internal supports of more complicated configurations. Namely, we consider the case of three LPIS and the case of the cross support. The convergence of the method is established theoretically and its efficiency is confirmed on numerical experiments.

## 1. Introduction

In this paper we consider problems of rectangular plates with three line partial internal supports (LPIS) and a cross internal support. The geometry of the problems is depicted in Fig. 1 (case of three LPIS) and Fig. 2 (case of a cross support). Suppose that the plates are subjected to a uniformly distributed load ($q$), their bottom and top edges are clamped, while the left and right edges are simply supported. Then due to the two-fold symmetry the problems are reduced to the solution of the biharmonic equation $\Delta^2 u = f$ for the deflection $u(x, y)$ in a quadrant of the plates, where $f = q/D$, $D$ is the flexural rigidity of the plates, under corresponding boundary conditions.

Essentially, they are strongly mixed boundary value problems for the biharmonic equation in the sense that there is one or more points of change of types of boundary conditions in edges of the rectangle. Therefore, they belong to the class of the problems with boundary singularities. A brief review of the methods concerning these problems is given in our recent work [1], where we proposed an iterative method, combining the reduction of the equation order [2], [3] and domain

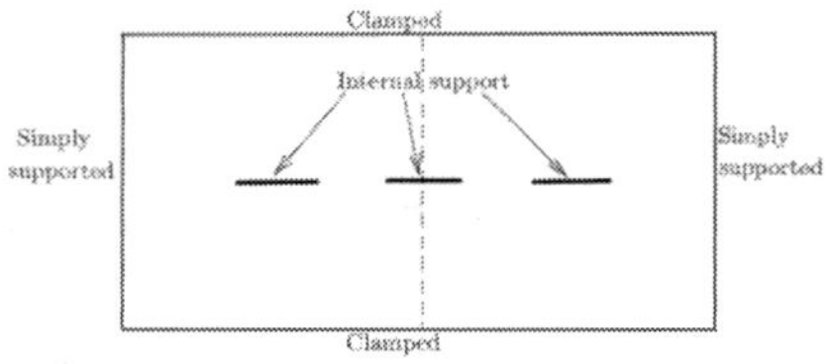


**Figure 1.** The plate with three internal line supports.

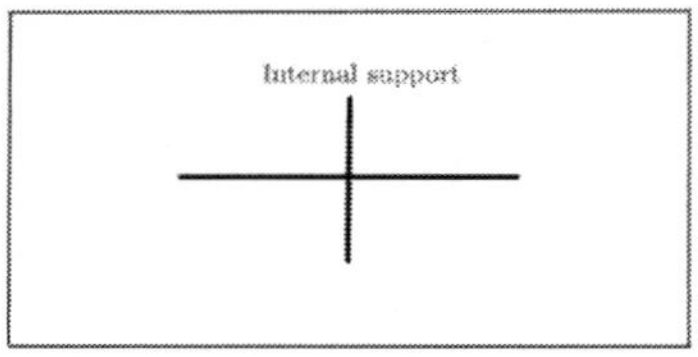


**Figure 2.** The plate with a cross internal support

decomposition [4] for the problems of plates on one and on two LPIS. It should be noticed that the motivation for our research was the paper of Sompornjaroensuk and Kiattikomol [6]. The present work is a further development of the method proposed in [1] to the problems of plates on three LPIS and on a cross support. The convergence of the method is established theoretically and verified on numerical examples.

## 2. Iterative method for solving the problem for plates with three LPIS

### *2.1. Description of the method*

We consider the problem with boundary conditions in general form

$$
\begin{aligned}
\Delta^2 u &= f \quad \text{in } \Omega, \\
u &= g_0 \quad \text{on } S_B \cup S_D \cup S_F \cup S_G, \ \frac{\partial u}{\partial \nu} = g_1 \quad \text{on} \quad \Gamma \setminus S_G, \\
\Delta u &= g_2 \quad \text{on } S_G, \ \frac{\partial}{\partial \nu} \Delta u = g_3 \quad \text{on } S_A \cup S_C \cup S_E,
\end{aligned}
\tag{1}
$$

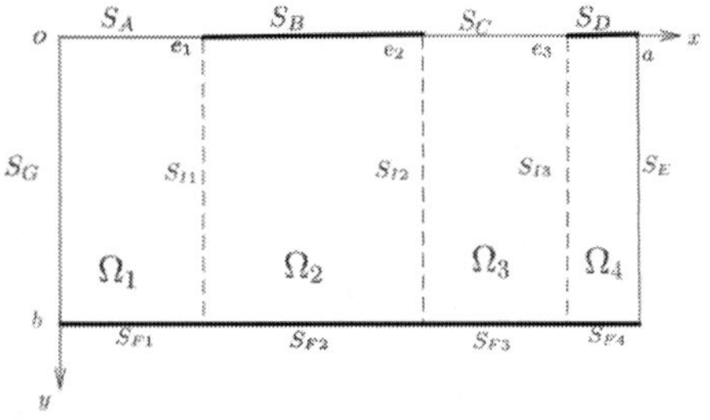


**Figure 3.** Domain decomposition for the problem with three LPIS considered in a quadrant of plate

where $\Omega$ is the rectangle $(0, a) \times (0, b)$, $S_A, S_B, S_C, S_D, S_E, S_F$ and $S_G$ are parts of the boundary $\Gamma = \partial\Omega$ as shown in Fig. 3, $\Delta$ is the Laplace operator, $f$ and $g_i$ $(i = \overline{0,3})$ are functions given in $\Omega$ and on parts of the boundary $\Gamma$, respectively.

In the case if all boundary functions $g_i = 0 \, (i = \overline{0,3})$ the problem models the bending of a quardrant of a rectangular plate on three LPIS.

For conciseness we put $\Gamma_1 = S_B \cup S_D \cup S_F$. Next we set $\Delta u = v$ in $\Omega$, $v\,|_{\Gamma_1} = \varphi$.

For solving the problem we divide the domain $\Omega$ into three subdomains $\Omega_1$, $\Omega_2$, $\Omega_3$ and $\Omega_4$ by the lines $x = e_1$, $x = e_2$ and $x = e_3$, and denote the interfaces of these subdomains by $S_{I_1}, S_{I_2}$ and $S_{I_3}$ as depicted in Fig. 3.

Consider the following combined iterative method with the idea of simultaneous iterative update of $\varphi$ on $\Gamma_1$ and $\xi_i = \partial v_2/\partial\nu_2$, $\eta_i = \partial u_2/\partial\nu_2$ on the interfaces $S_{I_i}$ $(i = 1, 2)$, $\xi_3 = \partial v_4/\partial\nu_4$, $\eta_3 = \partial u_4/\partial\nu_4$ on the interface $S_{I_3}$. Here and afterwards we denote $u_i = u|_{\Omega_i}$, $v_i = v|_{\Omega_i}$ and $\nu_i$ denotes the outward normal to the boundary of $\Omega_i$ $(i = 1, ..., 4)$.

Step 1. Given $\varphi^{(0)} = 0$ on $S_B \cup S_D \cup S_F$; $\xi_i^{(0)} = 0$, $\eta_i^{(0)} = 0$ on $S_{I_i}$, $(i = 1, 2, 3)$.

Step 2. Knowing $\varphi^{(k)}$, $\xi_i^{(k)}$, $\eta_i^{(k)}$, $(k = 0, 1, ...)$, $(i = 1, 2, 3)$ solve consecutively problems for $v_2^{(k)}$ and $u_2^{(k)}$ in $\Omega_2$, problems for $v_4^{(k)}$ and $u_4^{(k)}$ in $\Omega_4$, problems for $v_1^{(k)}$ and $u_1^{(k)}$ in $\Omega_1$ and problems for $v_3^{(k)}$ and $u_3^{(k)}$ in $\Omega_3$:

$$\begin{cases} \Delta v_2^{(k)} = f & \text{in } \Omega_2, \\ v_2^{(k)} = \varphi^{(k)} & \text{on } S_B \cup S_{F2}, \\ \dfrac{\partial v_2^{(k)}}{\partial \nu_2} = \xi_i^{(k)} & \text{on } S_{I_i}, \ (i = 1, 2), \end{cases} \qquad \begin{cases} \Delta u_2^{(k)} = v_2^{(k)} & \text{in } \Omega_2, \\ u_2^{(k)} = g_0 & \text{on } S_B \cup S_{F2} \\ \dfrac{\partial u_2^{(k)}}{\partial \nu_2} = \eta_i^{(k)} & \text{on } S_{I_i}, \ (i = 1, 2), \end{cases} \tag{2}$$

$$\begin{cases} \Delta v_4^{(k)} = f & \text{in } \Omega_4, \\ v_4^{(k)} = \varphi^{(k)} & \text{on } S_D \cup S_{F_4}, \\ \dfrac{\partial v_4^{(k)}}{\partial \nu_4} = \xi_3^{(k)} & \text{on } S_{I_3}, \\ \dfrac{\partial v_4^{(k)}}{\partial \nu_4} = g_1 & \text{on } S_E, \end{cases} \qquad \begin{cases} \Delta u_4^{(k)} = v_4^{(k)} & \text{in } \Omega_4, \\ u_4^{(k)} = g_0 & \text{on } S_D \cup S_{F_4}, \\ \dfrac{\partial u_4^{(k)}}{\partial \nu_4} = \eta_3^{(k)} & \text{on } S_{I_3}, \\ \dfrac{\partial u_4^{(k)}}{\partial \nu_4} = g_1 & \text{on } S_E, \end{cases} \tag{3}$$

$$\begin{cases} \Delta v_1^{(k)} = f & \text{in } \Omega_1, \\ \dfrac{\partial v_1^{(k)}}{\partial \nu_1} = g_3 & \text{on } S_A, \\ v_1^{(k)} = \varphi^{(k)} & \text{on } S_{F1}, \\ v_1^{(k)} = g_2 & \text{on } S_G, \\ v_1^{(k)} = v_2^{(k)} & \text{on } S_{I_1}, \end{cases} \qquad \begin{cases} \Delta u_1^{(k)} = v_1^{(k)} & \text{in } \Omega_1, \\ \dfrac{\partial u_1^{(k)}}{\partial \nu_1} = g_1 & \text{on } S_A, \\ u_1^{(k)} = g_0 & \text{on } S_{F1} \cup S_G, \\ u_1^{(k)} = u_2^{(k)} & \text{on } S_{I_1}, \end{cases} \tag{4}$$

$$\begin{cases} \Delta v_3^{(k)} = f & \text{in } \Omega_3, \\ \dfrac{\partial v_3^{(k)}}{\partial \nu_3} = g_3 & \text{on } S_C, \\ v_3^{(k)} = \varphi^{(k)} & \text{on } S_{F3}, \\ v_3^{(k)} = v_2^{(k)} & \text{on } S_{I_2}, \\ v_3^{(k)} = v_4^{(k)} & \text{on } S_{I_3}, \end{cases} \qquad \begin{cases} \Delta u_3^{(k)} = v_3^{(k)} & \text{in } \Omega_3, \\ \dfrac{\partial u_3^{(k)}}{\partial \nu_3} = g_1 & \text{on } S_C, \\ u_3^{(k)} = g_0 & \text{on } S_D \cup S_{F3}, \\ u_3^{(k)} = u_2^{(k)} & \text{on } S_{I_2}, \\ u_3^{(k)} = u_4^{(k)} & \text{on } S_{I_3}, \end{cases} \tag{5}$$

Step 3. Compute the new approximation

$$\begin{aligned}
&\xi_1^{(k+1)} = (1-\theta)\xi_1^{(k)} - \theta\frac{\partial v_1^{(k)}}{\partial \nu_1}, \quad \eta_1^{(k+1)} = (1-\theta)\eta_1^{(k)} - \theta\frac{\partial u_1^{(k)}}{\partial \nu_1} \text{ on } \ S_{I_1}\\
&\xi_2^{(k+1)} = (1-\theta)\xi_2^{(k)} - \theta\frac{\partial v_3^{(k)}}{\partial \nu_3}, \quad \eta_2^{(k+1)} = (1-\theta)\eta_2^{(k)} - \theta\frac{\partial u_3^{(k)}}{\partial \nu_3} \text{ on } \ S_{I_2}\\
&\xi_3^{(k+1)} = (1-\theta)\xi_3^{(k)} - \theta\frac{\partial v_3^{(k)}}{\partial \nu_3}, \quad \eta_3^{(k+1)} = (1-\theta)\eta_3^{(k)} - \theta\frac{\partial u_3^{(k)}}{\partial \nu_3} \text{ on } \ S_{I_3}\\
&\varphi^{(k+1)} = \varphi^{(k)} - \tau\Big(\frac{\partial u_i^{(k)}}{\partial \nu_i} - g_1\Big) \text{ on } \ \Gamma_1,
\end{aligned} \tag{6}$$

where $\theta$, $\tau$ are iterative parameters to be chosen for the fast convergence of the iterative method.

The convergence of the above iterative method is proved in the same way as for the cases of one and of two LPIS in [1].

### 2.2. Numerical example

For numerical realization of the above iterative method each subdomain is covered by an uniform grid with the same number of nodes in the $y-$direction. On these grids the mixed BVPs for the Poisson equation (2)-(5) are discretized by difference schemes of second order approximation. After that the obtained systems of difference equations are solved by the method of complete reduction [5, Chapt. 3]. For computing the normal derivatives in (6) we also use formulas of second order error. We perform iterative process (2)-(6) until $\| u^{(k+1)} - u^{(k)} \|_\infty \leq \varepsilon$, where $\varepsilon$ is a given accuracy taken of the same order as $O(\hat{h}^2)$, $\hat{h}$ being the step size of the grid.

The results of testing the convergence of the method on some exact solutions $u(x,y)$ with the corresponding boundary conditions and the right side function for the grids of $65 \times 65$ nodes on each subdomain show that for the values $\theta = 0.95$, $\tau = 0.55$ and some different combinations of $e_1, e_2, e_3$ after $40-60$ iterations the deviation of the approximate solution from the exact solution is less than $10^{-4}$. Therefore, in sequel we use these values of $\theta$ and $\tau$ .

Now, we apply the proposed iterative method to the problem of the rectangular plates depicted in Fig. 3, where $u(x,y)$ is the deflection function, $f = q/D$, $D$ being flexural rigidity of the plate which is expressed as $D = Eh^3/12(1-\nu^2)$, while $h$ is the plate thickness, and $\nu$ and $E$ are the Poisson's ratio and the Young's modulus of the plate, respectively.

The iterative process (2)-(6) for solving the problem for plates with three partial internal supports in domain $\Omega = [0, \pi/2] \times [0, \pi/2]$ on the grid $65 \times 65$ nodes for each subdomain with the given accuracy $10^{-4}$, $h = 0.5$, $q = 0.3$, the iteration parameters $\theta = 0.95$, $\tau = 0.55$ converges after 48 iterations. Fig. 4 presents the normalized deflection surfaces of a quadrant and of the whole plate for $e_1/\pi = 0.15, e_2/\pi = 0.20, e_3/\pi = 0.45$.

## 3. Problem for a plate on a cross internal support

### 3.1. Description of the iterative method

Now we consider the following problem

$$\begin{aligned}
&\Delta^2 u = f \quad \text{in } \Omega,\\
&u = g_0 \quad \text{on } S_B \cup S_C \cup S_E, \ \frac{\partial u}{\partial \nu} = g_1 \quad \text{on} \quad S_A \cup S_B \cup S_C \cup S_D \cup S_E,\\
&\Delta u = g_2 \quad \text{on } S_F, \ \frac{\partial}{\partial \nu}\Delta u = g_3 \quad \text{on } S_A \cup S_D,
\end{aligned} \tag{7}$$

where $\Omega$ is the rectangle $(0,a) \times (0,b)$, $S_A, S_B, S_C, S_D, S_E$ and $S_F$ are parts of the boundary $\Gamma = \partial\Omega$ as shown in Fig. 5.

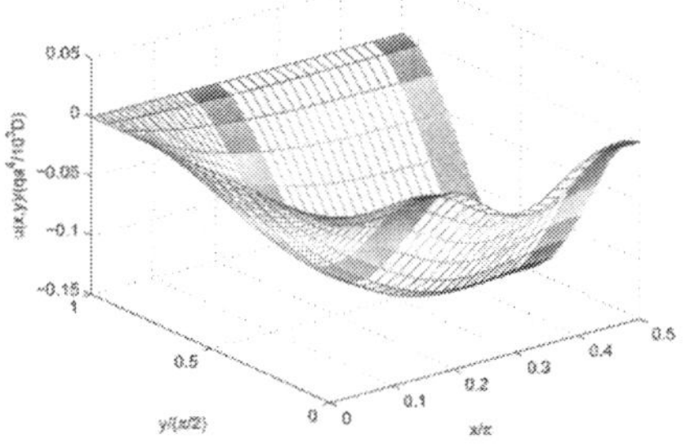

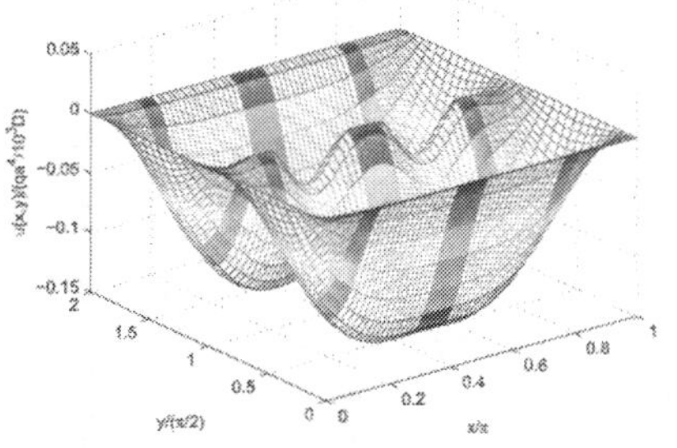

**Figure 4.** The surfaces of deflection of a quadrant (left) and of the whole plate (right)

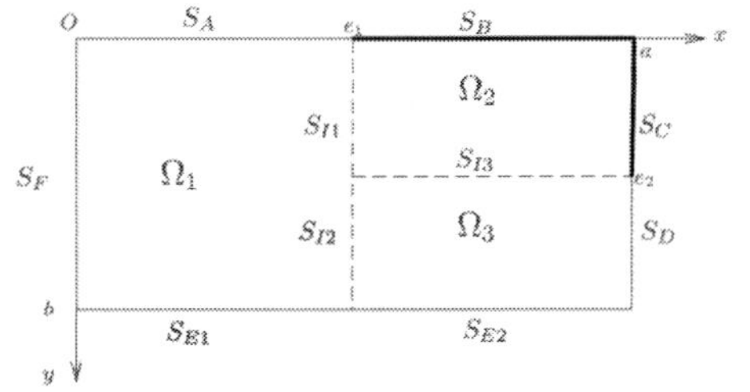

**Figure 5.** Domain decomposition for the problem with an internal cross support considered in a quadrant of plate

In the case if all boundary functions $g_i = 0\,(i = \overline{0,3})$ the problem models the bending of a quardrant of a rectangular plate.

For solving the problem we divide the domain $\Omega$ into three subdomains $\Omega_1$, $\Omega_2$ and $\Omega_3$ by the lines $x = e_1$ and $y = e_2$, and denote the interfaces of these subdomains by $S_{I_1}, S_{I_2}$ and $S_{I_3}$ as depicted in Fig. 5. As usual, we denote $u_i = u|_{\Omega_i}$, $v_i = v|_{\Omega_i}$ and $\nu_i$ denotes the outward normal to the boundary of $\Omega_i$ $(i = 1,2,3)$

Consider the following *combined iterative method* for the problem (7):

Step 1. Given $\varphi^{(0)} = 0$ on $S_B \cup S_C \cup S_E$; $\xi_i^{(0)} = 0$, $\eta_i^{(0)} = 0$ on $S_{I_i}$, $(i = 1,2,3)$.

Step 2. Knowing $\varphi^{(k)}$, $\xi_i^{(k)}$, $\eta_i^{(k)}$, $(k = 0,1,...)$, $(i = 1,2,3)$ solve consecutively problems for $v_1^{(k)}$ and $u_1^{(k)}$ in $\Omega_1$, problems for $v_2^{(k)}$ and $u_2^{(k)}$ in $\Omega_2$, and problems for $v_3^{(k)}$ and $u_3^{(k)}$ in $\Omega_3$:

$$\begin{cases} \Delta v_1^{(k)} &= f \quad \text{in } \Omega_1, \\ \dfrac{\partial v_1^{(k)}}{\partial \nu_1} &= g_3 \text{ on } S_A, \\ v_1^{(k)} &= \varphi^{(k)} \quad \text{on } S_{E1}, \\ v_1^{(k)} &= g_2 \quad \text{on } S_F, \\ \dfrac{\partial v_1^{(k)}}{\partial \nu_1} &= \xi_i^{(k)} \text{ on } S_{I_i},\ (i=1,2), \end{cases} \qquad \begin{cases} \Delta u_1^{(k)} &= v_1^{(k)} \quad \text{in } \Omega_2, \\ \dfrac{\partial u_1^{(k)}}{\partial \nu_1} &= g_1 \text{ on } S_A, \\ u_1^{(k)} &= g_0 \quad \text{on } S_{E1} \cup S_F \\ \dfrac{\partial u_1^{(k)}}{\partial \nu_1} &= \eta_i^{(k)} \text{ on } S_{I_i},\ (i=1,2), \end{cases} \tag{8}$$

$$\begin{cases} \Delta v_2^{(k)} &= f \quad \text{in } \Omega_2, \\ v_2^{(k)} &= \varphi^{(k)} \quad \text{on } S_B \cup S_C, \\ \dfrac{\partial v_2^{(k)}}{\partial \nu_2} &= \xi_3^{(k)} \quad \text{on } S_{I_3}, \\ v_2^{(k)} &= v_1^{(k)} \quad \text{on } S_{I_1}, \end{cases} \qquad \begin{cases} \Delta u_2^{(k)} &= v_2^{(k)} \quad \text{in } \Omega_2, \\ u_2^{(k)} &= g_0 \quad \text{on } S_B \cup S_C, \\ \dfrac{\partial u_2^{(k)}}{\partial \nu_2} &= \eta_3^{(k)} \quad \text{on } S_{I_3}, \\ u_2^{(k)} &= u_1^{(k)} \quad \text{on } S_{I_1}, \end{cases} \tag{9}$$

$$\begin{cases} \Delta v_3^{(k)} &= f \quad \text{in } \Omega_3, \\ v_3^{(k)} &= v_2^{(k)} \quad \text{on } S_{I_3}, \\ v_3^{(k)} &= \varphi^{(k)} \quad \text{on } S_{E2}, \\ v_3^{(k)} &= v_1^{(k)} \quad \text{on } S_{I_2}, \\ \dfrac{\partial v_3^{(k)}}{\partial \nu_3} &= g_3 \quad \text{on } S_D, \end{cases} \qquad \begin{cases} \Delta u_3^{(k)} &= v_3^{(k)} \quad \text{in } \Omega_3, \\ u_3^{(k)} &= u_2^{(k)} \quad \text{on } S_{I_3}, \\ u_3^{(k)} &= g_0 \quad \text{on } S_{E2}, \\ u_3^{(k)} &= u_1^{(k)} \quad \text{on } S_{I_2}, \\ \dfrac{\partial u_3^{(k)}}{\partial \nu_3} &= g_1 \quad \text{on } S_D, \end{cases} \tag{10}$$

Step 3. Compute the new approximation

$$\begin{aligned} \xi_1^{(k+1)} &= (1-\theta_1)\xi_1^{(k)} - \theta_1 \frac{\partial v_2^{(k)}}{\partial \nu_2}, \quad \eta_1^{(k+1)} = (1-\theta_2)\eta_1^{(k)} - \theta_2 \frac{\partial u_2^{(k)}}{\partial \nu_2} \text{ on } S_{I_1} \\ \xi_2^{(k+1)} &= (1-\theta_1)\xi_2^{(k)} - \theta_1 \frac{\partial v_3^{(k)}}{\partial \nu_3}, \quad \eta_2^{(k+1)} = (1-\theta_2)\eta_2^{(k)} - \theta_2 \frac{\partial u_3^{(k)}}{\partial \nu_3} \text{ on } S_{I_2} \\ \xi_3^{(k+1)} &= (1-\theta_1)\xi_3^{(k)} - \theta_1 \frac{\partial v_3^{(k)}}{\partial \nu_3}, \quad \eta_3^{(k+1)} = (1-\theta_2)\eta_3^{(k)} - \theta_2 \frac{\partial u_3^{(k)}}{\partial \nu_3} \text{ on } S_{I_3} \\ \varphi^{(k+1)} &= \varphi^{(k)} - \tau\Big(\frac{\partial u_i^{(k)}}{\partial \nu_i} - g_1\Big) \text{ on } S_B \cup S_C \cup S_E, \end{aligned} \tag{11}$$

where $\theta_1, \theta_2$ and $\tau$ are iterative parameters to be chosen for guaranteeing the convergence of the iterative process.

### 3.2. Numerical example

As for the problem of plate on three LPIS we verify the convergence of the discrete analogy of the iterative process (8)-(11) on some exact solutions for some sizes of the cross support. Performed experments show that the convergence rate depends on the sizes $(e_1, e_2)$ and the values of the iteration parameters $\theta_1, \theta_2$ and $\tau$. In application to the problem when $e_1/\pi = 0.25, e_2/(\pi/2) = 0.4$ with the iterative parameters $\theta_1 = 0.75, \theta_2 = 0.9$ and $\tau = 0.55$ for reaching the accuracy $10^{-4}$ the number of iterations needed is 46. The surfaces of deflection of a quadrant and of the whole plate in this case are depicted in Fig. 6.

## 4. Concluding remarks

In this paper we developed an iterative method for solving the problems for a plate having three line partial internal supports and for a plate having a cross internal support. The idea of the method is to reduce the problems to a sequence of problems for the Poisson equation with strongly mixed boundary conditions, which can be efficiently solved by a domain decomposition method. The convergence of the method is proved and tested on examples. The computations by the method are simple due to the use of our software for solving the Poisson equation with different weakly mixed boundary conditions. The method can be applied to the problems of plates on internal supports of more complex configurations such as a system of more than three LPIS, a system of two symmetric cross supports, a system of a cross and line internal supports or a system of parallel internal line supports.

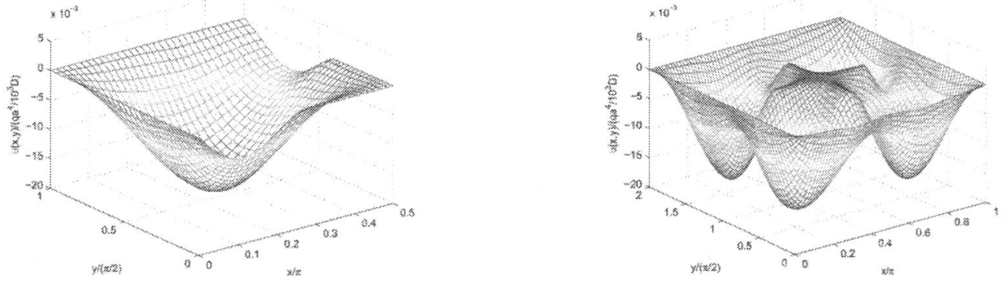

**Figure 6.** The surfaces of deflection of a quadrant (left) and of the whole plate (right) for $e_1/\pi = 0.25, e_2/\pi = 0.40$


## Acknowledgments

This work is supported by Vietnam National foundation for Science and Technology Development (NAFOSTED) under the grant 102.99-2011.24.


## References


[1] Dang Q. A., Truong H.H. (2013), Simple iterative method for solving problems for plates with partial internal supports, Journal of Engineering Mathematics, DOI: 10.1007/s10665-013-9652-7 (in press).

[2] Dang Q. A (2006) Iterative method for solving the Neumann boundary value problem for biharmonic type equation, Journal of Computational and Applied Mathematics, 96: 634 -643.

[3] Dang Q. A, Le T. S. (2009) Iterative method for solving a problem with mixed boundary conditions for biharmonic equation, Advances in Applied Mathematics and Mechanics, 1: 683-698.

[4] Dang Q. A, Vu V. Q. (2012) A domain decomposition method for strongly mixed boundary value problems for the Poisson equation, In book: Modeling, Simulation and Optimization of Complex Processes (Proc. 4th Inter. Conf. on HPSC, 2009, Hanoi, Vietnam), Springer, 65-76.

[5] Samarskii A. and Nikolaev E. (1989) Numerical methods for grid equations, v. 1: Direct Methods, Birkhäuser, Basel.

[6] Sompornjaroensuk Y. and Kiattikomol K. (2008) Exact analytical solutions for bending of rectangular plates with a partial internal line support, Journal of Engineering Mathematics, 62: 261-276.

# Generalization of the Wide-Sense Markov Concept to a Widely Linear Processing

**Juan Antonio Espinosa-Pulido**[1], **Jesús Navarro-Moreno** [2], **Rosa María Fernández-Alcalá** [2], **Juan Carlos Ruiz-Molina** [2], **Antonia Oya-Lechuga** [2] **and Nuria Ruiz-Fuentes** [2]

Department of Statistics and Operations Research, University of Jaén, Campus Las Lagunillas, 23071 Jaén, Spain

E-mail: [1]jaespu@gmail.com, [2]{jnavarro, rmfernan, jcruiz, aoya, nfuentes}@ujaen.es

**Abstract.** In this paper we show that the classical definition and the associated characterizations of wide-sense Markov (WSM) signals are not valid for improper complex signals. For that, we propose an extension of the concept of WSM to a widely linear (WL) setting and the study of new characterizations. Specifically, we introduce a new class of signals, called widely linear Markov (WLM) signals, and we analyze some of their properties based either on second-order properties or on state-space models from a WL processing standpoint. The study is performed in both the forwards and backwards directions of time. Thus, we provide two forwards and backwards Markovian representations for WLM signals. Finally, different estimation recursive algorithms are obtained for these models.

## 1. Introduction

Next, we present the main definitions, notation and auxiliary results. We also present an example which motivates the necessity of the new concept introduced.

The proofs of the results given along the paper can be found in [1].

Bold capital letters will be used to refer to matrices and bold lower-case letters will be used to refer to vectors. The row $j$ of any matrix $\boldsymbol{A}(\cdot)$ will be denoted by $\boldsymbol{A}_{[j]}(\cdot)$ and the $n$-vector of zeros by $\mathbf{0}_n$. Furthermore, the superscripts $*$, $T$ and $H$ represent the complex conjugate, transpose and complex transpose, respectively.

Let $\{x_t, t \in \mathbb{Z}\}$ be a zero-mean complex random signal with correlation function $r(t,s) = E[x_t x_s^*]$ and complementary correlation function $c(t,s) = E[x_t x_s]$. The linear minimum-mean square error (MMSE) estimator of $x_t$ based on the set of observations $\{x_{t_1}, x_{t_2} \ldots, x_{t_m}\}$ will be denoted by $\hat{x}(t|t_1, t_2, \ldots, t_m)$ and we will refer to it as the strictly linear (SL) estimator.

The Markov condition on a signal $\{x_t, t \in \mathbb{Z}\}$ establishes the following identity for the conditional probability:

$$P(x_t \leq x | x_{t_1}, x_{t_2}, \ldots, x_{t_m}) = P(x_t \leq x | x_{t_1})$$

for all $x$ and $t > t_1 > \ldots > t_m$. Doob [2] introduced a weaker concept based on the SL estimator which has received great attention in the literature. A signal $x_t$ is called WSM if $\hat{x}(t|\tau \leq s) = \hat{x}(t|s)$ for any $s < t$. Such signals have remarkable properties. For example, a signal $x_t$ is WSM if, and only if, the function $\bar{k}(t,s) = r(t,s)r^{-1}(s,s)$ has the triangular property, i.e.

$$\bar{k}(t,s) = \bar{k}(t,\tau)\bar{k}(\tau,s), \qquad t \geq \tau \geq s \tag{1}$$

Another characterization in terms of so-called Markovian state-space models can be found in [2]. It is shown that a signal $\{x_t, t \geq 0\}$ is WSM if, and only if, it has a state-space model of the form

$$x_{t+1} = \bar{k}(t+1,t)x_t + u_t \tag{2}$$

where $u_t$ is a white noise uncorrelated with $x_0$. Doob's definition was later generalized in [3] in the following sense: $x_t$ is a WSM signal of order $n \geq 1$ if $\hat{x}(t|\tau \leq s) = \hat{x}(t|s, s-1, \ldots, s-n+1)$ for any $s < t$. The authors also studied the second-order properties of such signals.

All these studies have a common characteristic: the information supplied by the complementary correlation function is ignored, i.e., the results are derived assuming implicitly that the signal is proper ($c(t,s) = 0$). Nowadays, the research activity in the field of the complex-valued signal is more and more focused on the better performing and less familiar WL processing. It has proved to be a more useful approach than SL processing when complex-valued random signals are in general improper (i.e., they are correlated with their complex conjugates). In this setting the SL MMSE estimator is replaced by the WL MMSE estimator, denoted by $\hat{x}^{WL}(t|t_1, t_2, \ldots, t_m)$, which uses the information of the augmented vector of observations $[x_{t_1}, x^*_{t_1}, x_{t_2}, x^*_{t_2}, \ldots, x_{t_m}, x^*_{t_m}]^T$. The immediate question that arises is whether the classical concept of WSM signals remains valid in the WL processing approach. The following example gives us the answer.

**Example 1** *Consider a signal $\{x_t, t \geq 0\}$ with correlation function $r(t,s) = \frac{1}{2}(e^{3|t-s|} + e^{|t-s|})$ and complementary correlation function $c(t,s) = \frac{1}{2}(e^{3|t-s|} - e^{|t-s|})$. It is easy to check that $r(t,s)$ does not satisfy the triangular property* (1) *and then, the signal cannot be modeled by a representation of the form* (2). *However, as we will show later, it is possible to find a state-space representation for such a signal given by* (5). *Thus, the classical WSM condition is clearly insufficient in the improper case to find a state-space representation for the signal involved.*

## 2. WLM Signals

From Example 1 we extract the following consequence: the classical definition of a WSM signal must be extended to deal with improper signals, this new concept must be characterized adequately to avoid the drawback shown in this example and new results about modeling are necessary to exploit the information available in both $x_t$ and $x^*_t$. Next, we introduce such a definition in a WL processing setting.

**Definition 1** *A complex-valued signal $\{x_t, t \in \mathbb{Z}\}$ is said to be WLM of order $n \geq 1$, briefly a WLM($n$) signal, if the following condition holds*

$$\hat{x}^{WL}(t|\tau \leq s) = \hat{x}^{WL}(t|s, s-1, \ldots, s-n+1)$$

*for any $s < t$.*

Notice that this concept extends both the classical notion of WSM introduced by [2] and the later generalization given in [3].

In the rest of the section we provide different characterizations of WLM($n$) signals. For that, we need to introduce some additional notation. Denote the augmented forwards vector of order $n \geq 1$ of $x_t$ as the $2n$-vector

$$\mathbf{x}_t = [x_t, x^*_t, x_{t-1}, x^*_{t-1}, \ldots, x_{t-n+1}, x^*_{t-n+1}]^T$$

and its correlation function by $\boldsymbol{R}(t,s) = E[\mathbf{x}_t\mathbf{x}_s^H]$. From now on, we assume that $\det\{\boldsymbol{R}_t\} \neq 0$ with $\boldsymbol{R}_t := \boldsymbol{R}(t,t)$. Moreover, we define the normalized correlation function as

$$\boldsymbol{K}(t,s) = \boldsymbol{R}(t,s)\boldsymbol{R}_s^{-1}$$

Similarly, we define the augmented backwards vector of order $n \geq 1$ of $x_t$ as the $2n$-vector

$$\mathbf{x}_t^b = [x_{t+n-1}, x_{t+n-1}^*, x_{t+n-2}, x_{t+n-2}^*, \ldots, x_t, x_t^*]^T$$

The following results establish the relation between the signals $x_t$ and their augmented forwards and backwards versions. We start first with the augmented forwards vector and we give a test similar to (1) for a signal being WLM($n$).

**Theorem 1** *The following statements are equivalent:*

*(i)* $\{x_t, t \in \mathbb{Z}\}$ *is a WLM(n) signal.*

*(ii) For* $s < t$, *the WL MMSE estimator of* $\mathbf{x}_t$ *on the basis of the set* $\{\mathbf{x}_\tau, \mathbf{x}_\tau^*, \tau \leq s\}$ *is of the form*

$$\hat{\mathbf{x}}^{WL}(t|\tau \leq s) = \boldsymbol{K}(t,s)\mathbf{x}_s$$

*(iii) For* $t \geq \tau \geq s$,

$$\boldsymbol{K}(t,s) = \boldsymbol{K}(t,\tau)\boldsymbol{K}(\tau,s) \tag{3}$$

Now, we suggest a characterization based on the augmented backwards vector. This result also shows the independence from the time direction of the Markov condition.

**Theorem 2** *The following statements are equivalent:*

*(i)* $\{x_t, t \in \mathbb{Z}\}$ *is a WLM(n) signal.*

*(ii)* $\hat{x}^{WL}(t|\tau \geq s) = \hat{x}^{WL}(t|s, s+1, \ldots, s+n-1)$ *for any* $s > t$.

*(iii) For* $s > t$, *the WL MMSE estimator of* $\mathbf{x}_t^b$ *on the basis of the set* $\{\mathbf{x}_\tau^b, \mathbf{x}_\tau^{b^*}, \tau \geq s\}$ *is of the form*

$$\hat{\mathbf{x}}^{b^{WL}}(t|\tau \geq s) = \boldsymbol{K}(t+n-1, s+n-1)\mathbf{x}_s^b$$

## 3. Modeling of WLM Signals

We aim to provide different ways of modeling for WLM($n$) signals. We present a new characterization in which the equivalence between a WLM signal of order $n$ and their forwards and backwards representations is given. Such representations show that a WLM($n$) signal depends only on the $n$ preceding or subsequent states and their conjugates.

**Theorem 3** *A signal* $\{x_t, 0 \leq t \leq m\}$ *is a WLM(n) if, and only if, it has the forwards and backwards representations*

$$x_{t+1} = \boldsymbol{k}_t^T\mathbf{x}_t + w_t, \qquad t \geq n-1 \tag{4}$$

$$x_t = \boldsymbol{k}^{b^T}_{t+1}\mathbf{x}_{t+1}^b + w_{t+1}^b, \qquad t \leq m-n+1$$

*where* $\boldsymbol{k}_t$, $\boldsymbol{k}_t^b$ *are* $2n$*-vectors, and* $w_t$, $w_t^b$ *are doubly white noises such that*

$$E[w_t\mathbf{x}_{n-1}] = \mathbf{0}_{2n}, \qquad t \geq n-1$$

$$E[w_t^b\mathbf{x}_{m-n+1}^b] = \mathbf{0}_{2n}, \qquad t \leq m-n+1$$

**Example 1 (continued)** *It is not difficult to check that* $x_t$ *is a WLM(1) signal by using property* (3). *Hence, applying the theorem above, it has the state-space representation*

$$x_{t+1} = \frac{1}{2}(e^3+e)x_t + \frac{1}{2}(e^3-e)x_t^* + w_t \tag{5}$$

*with* $w_t$ *a doubly white noise uncorrelated with* $x_0$ *and* $x_0^*$.

## 4. Estimation Problem of WLM Signals

Once the modeling problem has been solved for WLM($n$) signals, we address the MMSE estimation problem of such signals under a WL processing approach. The forwards and backwards representations given in Theorem 3 notably simplify the design of different recursive estimation algorithms. To this end, we use the Kalman recursions on the forwards representation to provide the solution for the prediction and filtering problems.

Suppose that we observe a WLM($n$) signal $\{x_t, 0 \leq t \leq m\}$ via the process

$$y_t = h_t x_t + v_t, \qquad 0 \leq t \leq m$$

with $v_t$ a doubly white noise such that $E[v_t v_t^*] = n_{1,t}$ and $E[v_t v_t] = n_{2,t}$ with $n_{1,t} > |n_{2,t}|$. Moreover, we assume that $v_t$ is uncorrelated with $x_s$ and $x_s^*$ for all $t, s$.

Consider the 2-vector $\mathbf{y}_t = [y_t, y_t^*]^T$, the $2 \times 2n$ matrix

$$\boldsymbol{H}_t = \begin{bmatrix} h_t & 0 & 0 & \cdots & 0 \\ 0 & h_t^* & 0 & \cdots & 0 \end{bmatrix}$$

the $2n \times 2n$ matrix $\boldsymbol{Q}_t = \boldsymbol{R}_{t+1} - \boldsymbol{K}_t \boldsymbol{R}_t \boldsymbol{K}_t^H$ and the $2 \times 2$ matrix

$$\boldsymbol{N}_t = \begin{bmatrix} n_{1,t} & n_{2,t} \\ n_{2,t}^* & n_{1,t} \end{bmatrix}$$

Denote the WL filtered estimator of $x_t$ by $\hat{x}_t^{WL}$ and the one-step-ahead predictor of $x_{t+1}$ by $\hat{x}_{t+1|t}^{WL}$, both obtained on the basis of the information provided by the set $\{y_0, y_0^*, \ldots, y_t, y_t^*\}$, and consider their associated errors $p_t = E[|x_t - \hat{x}_t^{WL}|^2]$ and $p_{t+1|t} = E[|x_{t+1} - \hat{x}_{t+1|t}^{WL}|^2]$. Also denote the estimate of $\mathbf{x}_{n-1}$ obtained from the information provided by $[y_{n-1}, y_{n-1}^*, \ldots, y_0, y_0^*]^T$ by $\hat{\mathbf{x}}_{n-1}$ and its associated error by $\boldsymbol{P}_{n-1}$. By combining the forwards representation (4) and the classical Kalman filter we present the following Algorithm.

**Algorithm 1** WL filter and prediction

**Require:** $\mathbf{y}_t$, $\boldsymbol{H}_t$, $\boldsymbol{N}_t$, $\boldsymbol{K}_t$, $\boldsymbol{Q}_t$, $\boldsymbol{g} = [1, 0, \ldots, 0]^T$, $\hat{\mathbf{x}}_{n-1}$, and $\boldsymbol{P}_{n-1}$
**Ensure:** $\hat{x}_{t+1|t}^{WL}$, $\hat{x}_{t+1}^{WL}$, $p_{t+1|t}$, and $p_{t+1}$

**for** $t \geq n-1$ **do**
  $\hat{\mathbf{x}}_{t+1|t} \leftarrow \boldsymbol{K}_t \hat{\mathbf{x}}_t$
  $\boldsymbol{P}_{t+1|t} \leftarrow \boldsymbol{K}_t \boldsymbol{P}_t \boldsymbol{K}_t^H + \boldsymbol{Q}_t$
  $\boldsymbol{F}_{t+1} \leftarrow \boldsymbol{P}_{t+1|t} \boldsymbol{H}_{t+1}^H [\boldsymbol{H}_{t+1} \boldsymbol{P}_{t+1|t} \boldsymbol{H}_{t+1}^H + \boldsymbol{N}_{t+1}]^{-1}$
  $\hat{\mathbf{x}}_{t+1} \leftarrow \hat{\mathbf{x}}_{t+1|t} + \boldsymbol{F}_{t+1}[\mathbf{y}_{t+1} - \boldsymbol{H}_{t+1} \hat{\mathbf{x}}_{t+1|t}]$
  $\boldsymbol{P}_{t+1} \leftarrow \boldsymbol{P}_{t+1|t} - \boldsymbol{F}_{t+1} \boldsymbol{H}_{t+1} \boldsymbol{P}_{t+1|t}$
  $\hat{x}_{t+1|t}^{WL} \leftarrow \boldsymbol{g}^T \hat{\mathbf{x}}_{t+1|t}$
  $\hat{x}_{t+1}^{WL} \leftarrow \boldsymbol{g}^T \hat{\mathbf{x}}_{t+1}$
  $p_{t+1|t} \leftarrow \boldsymbol{g}^T \boldsymbol{P}_{t+1|t} \boldsymbol{g}$
  $p_{t+1} \leftarrow \boldsymbol{g}^T \boldsymbol{P}_{t+1} \boldsymbol{g}$
**end for**

## 5. References

[1] Espinosa-Pulido J A, Navarro-Moreno J, Fernández-Alcalá, R M and Ruiz-Molina J C 2012 Widely Linear Markov Signals *EURASIP J. Adv. Signal Process.* **256** doi:10.1186/1687-6180-2012-256

[2] Doob J L 1953 *Stochastic Processes.* New York: John Willey

[3] Kasprzyk A, Szczotka W 2006 Covariance Structure of Wide-Sense Markov Processes of Order $k \geq 1$ *Applicat. Mathemat.* **33** (2) 129–143

# Glass phase in anisotropic surface model for membranes

**Hiroshi Koibuchi**
Ibaraki National College of Technology, Nakane 866, Hitachinaka, Ibaraki 312-8508, Japan
E-mail: koibuchi@mech.ibaraki-ct.ac.jp

**Andrey Shobukhov**
Faculty of Computational Mathematics and Cybernetics, Lomonosov Moscow State University, 119991, Moscow, Leninskiye Gory, MSU, 2-nd Educational Building, Russia

**Abstract.** A Finsler geometric surface model for membranes is studied by using the Monte Carlo simulation technique on connection-fixed triangle lattices with sphere topology. An in-plane three-dimensional unit vector $\sigma$ is assumed to be the in-plane tilt variable of the surface. The interaction with $\sigma$ is described by the XY-model Hamiltonian. Since this variable $\sigma$ is considered as a vector field on the surface, a Finsler metric is defined by using $\sigma$. We find that the model has three different phases. They change from the para-magnetic phase to the ferromagnetic and to the glass phases when the strength of the $XY$ interaction increases. Both the para-magnetic and the glass phases are characterized by random configuration of $\sigma$; the variable $\sigma$ randomly fluctuates in the para-magnetic phase while it is randomly frozen in the glass phase. We also find that the surface becomes spherical in both phases. On the contrary, in the ferro-magnetic phase the surface shape becomes tubular or discotic due to the anisotropic bending rigidity and surface tension coefficient, which are dynamically generated by ordered configurations of $\sigma$.

## 1. Introduction

Anisotropic shape can be seen in biological membranes such as liposomes and vesicles [1] or in liquid crystal elastomers (LCEs) membranes [2, 3]. In those membranes, the strength of the surface force such as the surface tension $\gamma$ and bending rigidity $\kappa$ is expected to be dependent on whether these molecules are aligned or not. Indeed, the surface anisotropy can be obtained by modifying the bending rigidity $\kappa$ to be dependent on the internal direction of surface in the Ginzburg-Landau or Helfrich-Polyakov (HP) model for membranes [4, 5]. However, the modification of $\kappa$ in those models by hand gives only a constant $\kappa$ over the surface although it should depend on the direction.

Another technique to make $\kappa$ anisotropic is to assume a Finsler metric on the surface [6, 7]. In this approach, $\kappa$ depends not only on the direction but also on the position on the surface. Since the Finsler geometric model is defined in a continuous form, its treatment such as the discretization is not always uniquely determined. In the anisotropic model of Ref. [6], we assumed that (i) the tilt variable $\sigma$ belongs to $\mathbf{S}^2$: unit sphere, and its tangential components are used as a vector field to define the Finsler metric, (ii) the Heisenberg spin model Hamiltonian $S_0$ is assumed for $\sigma$, (iii) the Finsler metric $g^F_{ab}$ is constructed by deforming the Euclidean metric

$\delta_{ab}$, (iv) $g^F_{ab}$ is assumed for both the bond potential $S_1$ and the bending energy $S_2$, while $\delta_{ab}$ is assumed for $S_0$.

In this presentation, we show the results obtained for the model, which is different from the one in [6]. It is defined as follows: (i) $\sigma \in \mathbf{S}^1$: $\sigma$ has the unit length, it belongs to the tangential plane and its values are assigned to the vertices, (ii) the XY model Hamiltonian $S_0$ is assumed for $\sigma$, (iii) $g^F_{ab}$ is constructed by modifying the induced metric $g_{ab}=\partial_a X^\mu \partial_b X^\mu$, (iv) $g^F_{ab}$ is assumed not only for $S_1$ and $S_2$ but also for $S_0$.

## 2. Model

The model is defined on the triangulated surface in $\mathbf{R}^3$, which is characterized by the three numbers $(N, N_B, N_T)$, which are the total numbers of vertices, bonds and triangles.

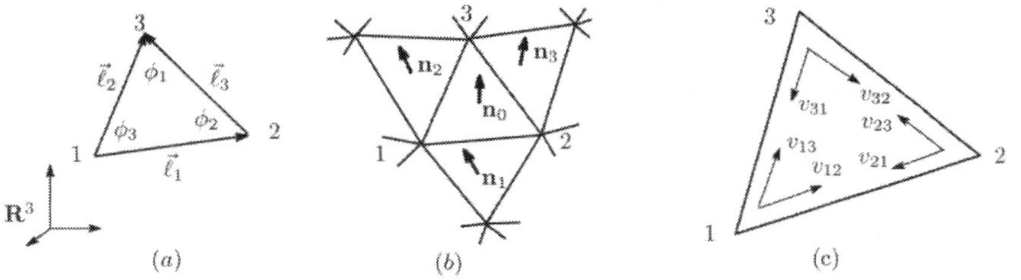


**Figure 1.** (a) Edge vectors $\vec{\ell}_i$ and internal angles $\phi_i$ of a triangle in $\mathbf{R}^3$, (b) a triangle 123 and the three neighboring triangles, and (c) the parameters $v_{ij}$ defined by Eq. (4).

The Hamiltonian is given by

$$S(X,\sigma) = \lambda S_0 + S_1 + \kappa S_2, \tag{1}$$

where $S_0$ is an energy for the directors (or tilts) $\sigma_i$. The $\sigma_i$ has values in $\mathbf{S}^1$, which is the unit circle on the tangential plane at the vertex $i$. The tangential plane is defined such that the unit normal vector $\tilde{\mathbf{n}}_i$ at the vertex $i$ is given by $\tilde{\mathbf{n}}_i = \mathbf{N}_i/|\mathbf{N}_i|$, $\mathbf{N}_i = \sum_j \mathbf{n}_j a_j$, where $\mathbf{n}_j$ and $a_j$ are the unit normal vector and the area of the triangle $j$ linked to the vertex $i$.

The internal energy $S_0$, the Gaussian bond potential $S_1$ and the bending energy $S_2$ are defined as follows:

$$\begin{aligned}
S_0 &= \frac{1}{6}\sum_{\Delta} S_{0\Delta}/a_\Delta, \quad S_1 = \frac{1}{6}\sum_{\Delta} S_{1\Delta}/a_\Delta, \quad S_2 = \frac{1}{6}\sum_{\Delta} S_{2\Delta}/a_\Delta, \\
S_{0\Delta} &= \left(v_{12}v_{13}^{-1}\ell_2^2 + v_{21}v_{23}^{-1}\ell_3^2\right)(1-\sigma_1\cdot\sigma_2) + \left(v_{23}v_{21}^{-1}\ell_1^2 + v_{32}v_{31}^{-1}\ell_2^2\right)(1-\sigma_2\cdot\sigma_3) \\
&\quad + \left(v_{13}v_{12}^{-1}\ell_1^2 + v_{31}v_{32}^{-1}\ell_3^2\right)(1-\sigma_3\cdot\sigma_1) \\
&\quad -\vec{\ell}_1\cdot\vec{\ell}_2\,(\sigma_2-\sigma_1)\cdot(\sigma_3-\sigma_1) - \vec{\ell}_2\cdot\vec{\ell}_3\,(\sigma_1-\sigma_3)\cdot(\sigma_2-\sigma_3) \\
&\quad -\vec{\ell}_3\cdot\vec{\ell}_1\,(\sigma_3-\sigma_2)\cdot(\sigma_1-\sigma_2), \\
S_{1\Delta} &= \gamma_1\ell_1^2\ell_2^2 + \gamma_2\ell_2^2\ell_3^2 + \gamma_3\ell_3^2\ell_1^2 - 2\left(\ell_1^2\ell_2^2\cos^2\phi_3 + \ell_2^2\ell_3^2\cos^2\phi_1 + \ell_3^2\ell_1^2\cos^2\phi_2\right), \qquad (2) \\
S_{2\Delta} &= \kappa_1\ell_1^2(1-\mathbf{n}_0\cdot\mathbf{n}_1) + \kappa_2\ell_2^2(1-\mathbf{n}_0\cdot\mathbf{n}_2) + \kappa_3\ell_3^2(1-\mathbf{n}_0\cdot\mathbf{n}_3) \\
&\quad -\vec{\ell}_1\cdot\vec{\ell}_2\,(\mathbf{n}_0-\mathbf{n}_1)\cdot(\mathbf{n}_0-\mathbf{n}_2) - \vec{\ell}_2\cdot\vec{\ell}_3\,(\mathbf{n}_0-\mathbf{n}_2)\cdot(\mathbf{n}_0-\mathbf{n}_3) \\
&\quad -\vec{\ell}_3\cdot\vec{\ell}_1\,(\mathbf{n}_0-\mathbf{n}_3)\cdot(\mathbf{n}_0-\mathbf{n}_1),
\end{aligned}$$

where $2\cos^2\phi_3$ in $S_{1\Delta}$ can also be written as $2\cos^2\phi_3 = (\ell_1^2+\ell_2^2-\ell_3^2)/\ell_1\ell_2$. The symbols $\ell_i (i=1,2,3)$ are the edge lengths of $\Delta$, and $\mathbf{n}_i (i=0,1,2,3)$ denote the unit normal vectors of triangles (Fig. 1(b)). The symbol $a_\Delta$ is the area of triangle $\Delta$ defined by $a_\Delta = (1/2)\ell_1\ell_2 \sin\phi_3 = (1/2)\ell_3\ell_1 \sin\phi_2 = (1/2)\ell_2\ell_3 \sin\phi_1$. The symbol $\sum_\Delta$ denotes the sum over all $\Delta$. The coefficients $\gamma_i$ and $\kappa_i$ in Eq. (2) are defined by

$$\begin{aligned} \gamma_1 &= v_{13}v_{12}^{-1} + v_{12}v_{13}^{-1}, \quad \gamma_2 = v_{32}v_{31}^{-1} + v_{31}v_{32}^{-1}, \quad \gamma_3 = v_{21}v_{23}^{-1} + v_{23}v_{21}^{-1}, \\ \kappa_1 &= v_{13}v_{12}^{-1} + v_{23}v_{21}^{-1}, \quad \kappa_2 = v_{12}v_{13}^{-1} + v_{32}v_{31}^{-1}, \quad \kappa_3 = v_{21}v_{23}^{-1} + v_{31}v_{32}^{-1}. \end{aligned} \tag{3}$$

The parameters $v_{ij}$ are given by

$$v_{ij} = 1 + [\sigma_{ij}], \quad \sigma_{ij} = N_v \, |\vec{\sigma}_i \cdot \mathbf{t}_{ij}|, \tag{4}$$

where $[x]$ represents $\mathrm{Max}\,[n \in \mathbf{Z} | n \le x]$, $N_v = 100$, and $\mathbf{t}_{ij}$ is the unit tangential vector from the vertex $i$ to the vertex $j$. The partition function $Z$ is given by $Z(\lambda, \kappa) = \sum_\sigma \int' \prod_{i=1}^N dX_i \exp[-S(X, \sigma)]$, where $\int' \prod_{i=1}^N dX_i$ is the multiple three-dimensional integrations, that are performed by fixing the center of mass of the surface to the origin of $\mathbf{R}^3$, and $\sum_\sigma$ denotes the sum over tilt variables.

The assumed Finsler metric $g_{ab}^F$ is given by $g_{ab}^F = \begin{pmatrix} \ell_1^2 v_{12}^{-2} & \vec{\ell}_1 \cdot \vec{\ell}_2 v_{12}^{-1} v_{13}^{-1} \\ \vec{\ell}_1 \cdot \vec{\ell}_2 v_{12}^{-1} v_{13}^{-1} & \ell_2^2 v_{13}^{-2} \end{pmatrix}$, which reduces to the discrete induced metric if $v_{ij} = 1$ for all $ij$. The discrete Hamiltonians $S_0$, $S_1$ and $S_2$ in Eq. (2) with the coefficients in Eq. (3) are obtained from the continuous Hamiltonians just like in the model of Ref. [6].

## 3. Monte Carlo results

**Figure 2.** Snapshots for (a) $\lambda = 0.2$ (sphere), (b) $\lambda = 1.5$ (tube), (c) $\lambda = 3$ (disk), (d) $\lambda = 17$ (glass). $N = 5762$, $b = 300$, and $N_v = 100$. The view point of the lower snapshot is rotated about $\pi/4$ around the vertical axis of the upper one. Red brushes denote the in-plane variable $\sigma$.

We show snapshots for $\lambda = 0.2$, $\lambda = 1.5$, $\lambda = 3$, and $\lambda = 17$ (Figs.2(a)-2(d)), where the first one belongs to the high temperature phase and the remaining three belong to the low temperature phase. The random configurations in (a) and (d) correspond to the high temperature and the glass configurations, respectively. The magnetization $M/N$ defined by $M/N = (1/N)\,|\sum_i \sigma_i| = (1/N)\,|(\sum_i \sigma_i^x, \sum_i \sigma_i^y, \sum_i \sigma_i^z)|$ reflects whether the spin variables *spatially* align or not (Fig. 3(a)). To the contrary, the following $[M]/N$ can reflect whether the spins *timely* align or not (Fig. 3(b)):

$$[M]/N = (1/N) \left| \sum_i M_i \right|, \quad M_i = \frac{1}{n_s} \left| \left( \sum_n \sigma_i^x(n), \sum_n \sigma_i^y(n), \sum_n \sigma_i^z(n) \right) \right| \tag{5}$$

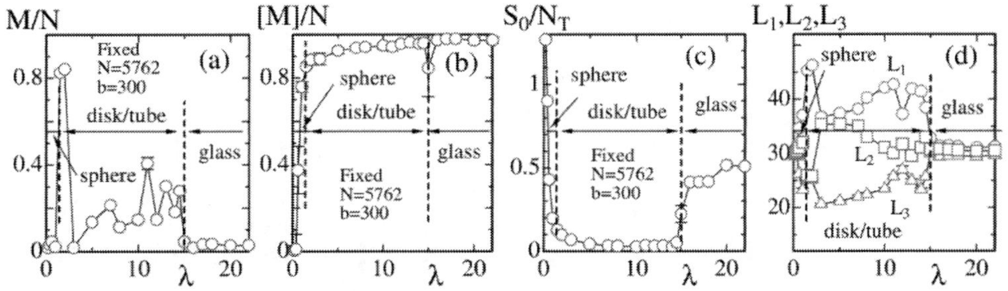


**Figure 3.** (a) $M/N$ vs. $\lambda$, (b) $[M]/N$ vs. $\lambda$, (c) $S_0/N_T$ vs. $\lambda$, and (d) $L_i(i=1,2,3)$ vs. $\lambda$, where $L_1$ is the maximal linear extension and $L_2$, $L_3$ are diameters of the surface section perpendicular to the line from which $L_1$ is obtained.

where $n_s = \sum_n 1$ and $(1/n_s)\sum_n$ is the time series average of samples $\{\sigma_i(n)\}$, which is obtained at the vertex $i$ with the Monte Carlo (MC) time $n$. The magnetization $M/N$ is numerically obtained by performing the lattice average firstly and the time average finally, while $[M]/N$ is obtained with the time average firstly and with the lattice average finally. We expect $M/N \to 0$ if $\sigma_i$ is spatially random, while $[M]/N \to 0$ if $\sigma_i$ is timely random. Thus, the glass phase is characterized by $M/N \to 0$ and $[M]/N \to 1$. The glass phase is expected for $\lambda \geq 15$ (Figs. 3(a), (b)). In the high temperature phase, we see $M/N \to 0$ and $[M]/N \to 0$, which implies that $\sigma$ is random both spatially and timely. In the low temperature phase for $\lambda \geq 1.5$, $[M]/N \to 1$ is expected because $\sigma$ is timely unchanged or aligned whenever the surface is in the tube or disk configuration, while $M/N \to 1$ ($M/N \to 0$) is expected in the tube (disk) configuration.

At the phase boundary between the glass and disk/tube phases, $S_0/N_T$ and the maximal (minimal) linear extension $L_1$ ($L_3$) appear to change discontinuously, where $L_1$ is the maximal surface length along the line going through the center of surface, while $L_2$ is the maximal diameter of the sectional ellipse which is perpendicular to the line corresponding to $L_1$, and $L_3$ is the ellipse diameter at the line vertical to both of the lines corresponding to $L_1$ and $L_2$. We see $L_1 \simeq L_2 \simeq L_3$ in the glass phase (Fig. 3(d)), which implies that the surface is spherical in this phase. The surface shape in the disk/tube phase seems to be dependent on the initial condition in contrast to the model in Ref. [6].

The glass phase of the model is expected to be not stable but quasi-stable. Indeed, the mean field analysis of the Finsler XY model defined by $S = \lambda S_0$ with $\ell_i = 1$ on the regular square lattice indicates that the free energy has a quasi-stable state corresponding to the glass phase. This will be reported elsewhere. The author (HK) acknowledges Hideo Sekino in Toyohashi University of Technology for comments.

## References

[1] H.-G. Döbereiner and U. Seifert, Europhys. Lett. **36**, 325 (1996).
[2] Xiangjun Xing, Ranjan Mukhopadhyay, T. C. Lubensky, and Leo Radzihovsky, Phys. Rev. E. **68** (2003) 021108(1-17).
[3] Xiangjun Xing and Leo Radzihovsky, Annals of Phys. **323** (2008) 105.
[4] L. Radzihovsky, in Statistical Mechanics of Membranes and Surfaces, Second Edition, edited by D. Nelson, T. Piran, and S. Weinberg, (World Scientific, 2004) p.275.
[5] M. Bowick and A. Travesset, Phys. Rep. **344** (2001) 255.
[6] H. Koibuchi "A Finsler geometric model for membranes on triangulated surfaces", ic-msquare2102, Journal of Physics Conference Series Vol. 410 (2013) 012056 (4 pages).
[7] H. Koibuchi and H. Sekino, "Monte Carlo studies of a Finsler geometric surface model", Physica A (in press), http://arxiv.org/abs/1208.1806.

# Pattern recognition issues on anisotropic smoothed particle hydrodynamics

**Eraldo Pereira Marinho**
Univ Estadual Paulista (UNESP/IGCE), Department of Computing, Rio Claro, SP, Brazil
E-mail: emarinho@rc.unesp.br

**Abstract.** This is a preliminary theoretical discussion on the computational requirements of the state of the art smoothed particle hydrodynamics (SPH) from the optics of pattern recognition and artificial intelligence. It is pointed out in the present paper that, when including anisotropy detection to improve resolution on shock layer, SPH is a very peculiar case of unsupervised machine learning. On the other hand, the free particle nature of SPH opens an opportunity for artificial intelligence to study particles as agents acting in a collaborative framework in which the timed outcomes of a fluid simulation forms a large knowledge base, which might be very attractive in computational astrophysics phenomenological problems like self-propagating star formation.

## 1. Introduction

Smoothed particle hydrodynamics (SPH) has been a successful computer simulation paradigm originated in computational astrophysics since 1977 [2, 5]. Nowadays SPH is used in other areas and has gained significant improvement in accuracy and stability, not only on simulating compressible shock as also on performing high resolution incompressible fluid, solids etc, e.g. [4].

One essential issue in SPH is anisotropy, which arises naturally on performing adaptive interpolation, e.g. [7], where the dimensionality reduction to detect critical surfaces, as shock layers might be included. Anisotropy is an important subject in pattern recognition for feature extraction methods [1].

There are encouraging works on the application of artificial intelligence to reproduce real systems, mainly in the context of complex systems as in economics and many other social and life sciences, e.g. [3]. The collective phenomena of star formation, unveiled in the intermittent pattern in the spiral galaxies arms, have been proposed by means of the stochastic self-propagating star formation model, in the sense that star formation is contagious, e.g. [8].

The present work is a brief discussion on some aspects of SPH circumstanced by the concepts of pattern recognition and artificial intelligence. Several details are omitted to fulfill the limited space with no significative lost of focus on reviewing the theory in the context of intelligent computing.

## 2. SPH database and space representation

The SPH database comprises $N$ instances, usually thousand hundreds or even millions particles, indexed by a descriptor table $\mathcal{P}_N$. Each particle is addressed by a unique label, or descriptor

$i$, and as much as possible the particle object is referred as just $i$ or $i$-particle. Any particle attribute, say $A$, is addressed by sub-indexing the same with the particle label, $A_i$.

Since a specific SPH problem is headed by the mathematical, physical and computational models, the adopted methods might be included in the database in the form of classes and module libraries, described in a commonly used data model language as for instance the XML, which may also improve the information interchange between different SPH simulation bases.

The usual 3D space description in SPH uses hierarchical spatial tessellation, as for instance by means of octrees, e.g. [6], which are the 3D version of quadtrees. Other tree-based spatial tessellation schemes are also adopted. For instance [7] proposes an approach to easily adapt earlier versions of octree-based SPH codes to covariance-based octree tessellation to improve anisotropic kernel computations, e.g. [11].

## 3. Starring pattern recognition and AI in SPH

The $k$-nearest neighbor is a mathematical relation $\mathcal{N}_k \subseteq \mathcal{P}_N \times \mathcal{P}_N$, which associates $i \in \mathcal{P}_N$ with a subset $\mathcal{N}_k(i) = \{i_1, \ldots, i_k\} \subseteq \mathcal{P}_N$, so that $j \in \mathcal{N}_k(i)$ if, and only if, the adopted distance $d(\vec{x}_i, \vec{x}_j)$ from $i$ to $j$ obeys the inequality $d(\vec{x}_i, \vec{x}_j) \leq \max\{d(\vec{x}_i, \vec{x}_l) | \, l \in \mathcal{N}_k(i)\}$. The KNN algorithm is the method by which the relation $\mathcal{N}_k$ is populated by ordered pairs $(i, j)$ in $\subseteq \mathcal{P}_N \times \mathcal{P}_N$, given the particle-descriptor table $\mathcal{P}_N$.

The $\mathcal{N}_k$ relation is asymmetric and reflexive. The later comes from the fact that each point is the nearest neighbor of itself – this is called improper neighbor. The former comes from the fact that if $a$ is the proper nearest neighbor (not a reflection) of $b$, not necessarily $b$ is the nearest neighbor of $a$. For example, $a$ could be closer to a third point $c$ than $b$, which is too faraway from any other point but $a$.

The KNN asymmetry reflects imperfections on writing simpler forms of the SPH conservation equations, which require particle commutation symmetry. To workaround the asymmetry issue is necessary to introduce the *symmetric closure* of the KNN relation, which is known as effective neighbors

$$\mathcal{E}_k = \mathcal{N}_k \cup \{(i, j) \in \mathcal{P}_N \times \mathcal{P}_N \mid (j, i) \in \mathcal{N}_k\}. \tag{1}$$

Of course, the KNN algorithm requires a predesign metric in the 3D space. If the metric is invariant under rotation, the KNN relation is isotropic. On the other hand, the metric relation is said anisotropic. For instance the Mahalanobis metric, as adopted in the KNN algorithm proposed by [7], is anisotropic and is used to reveal biased structures like the arms in spiral galaxy images.

The Mahalanobis distance $\xi_{ij}$ is defined in terms of the covariance tensor $\mathbf{\Sigma}$:

$$\xi_{ij} = (\vec{x}_i - \vec{x}_j)\mathbf{\Sigma}^{-1}(\vec{x}_i - \vec{x}_j)^{\mathrm{T}}, \tag{2}$$

where $(\vec{x}_i - \vec{x}_j)^{\mathrm{T}}$ is the transpose of the matrix representation of the relative position vector $(\vec{x}_i - \vec{x}_j)$.

Of course, equation (2) is not the only way of defining anisotropic distance in SPH. For instance, the positive-definite stress tensor $\mathbf{T}$ might be eventually used to define the non-normalized anisotropic distance $\xi_{ij}$:

$$\xi_{ij} = (\vec{x}_i - \vec{x}_j)\mathbf{T}^{-1}(\vec{x}_i - \vec{x}_j)^{\mathrm{T}}. \tag{3}$$

According to equations (2) or (3), the outermost boundary for the $k$-nearest neighbors of the $i$-particle is an ellipsoid centered in the query position $\vec{x}_i$, whose principal axes are set by the respective tensor eigenvectors [7].

A cognitive interpretation for the well-known SPH interpolation formula can be illustrated as follows: given an $i$-labeled particle, say $i$-particle, one may suppose this particle has to make an estimation, $\tilde{A}_i$, of a local fluid quantity, $A_i$, after hearing votes, e.g. [1], from its effective neighbors what impression they get regarding the same quantity.

A democratic decision is made if the $i$-particle weights the individual suggestions from its informants, giving more importance to the closest ones. The importance, or weight, comes from a compact-support smoothing kernel, which drops to zero outside the influence zone defined by the effective neighbors and grows up as gets closer to $i$, reaching its maximum for $i$ itself.

Each $i$-particle, $i = 1, \ldots, N$, has its own effective neighbors, $\mathcal{E}_k(i)$. The $i$-particle asks each $j$-particle in $\mathcal{E}_k(i)$ for suggestions, which answers accordingly to the predefined protocol, $A_j m_j/\rho_j$, whose reliability is expressed by a weight, or smoothing kernel $W_{ij}$.

The $i$-particle gets a conclusive perception $\tilde{A}_i$ from its locality by adding together all of the weighted votes, $W_{ij}A_j m_j/\rho_j$, received from its $k$-nearest neighbors:

$$\tilde{A}_i = \sum_{\forall j \in \mathcal{E}_k(i)} W_{ij} A_j \frac{m_j}{\rho_j}, \tag{4}$$

where $W_{ij} = W(\vec{x}_i - \vec{x}_j)$ is the smoothing kernel, whose analytical profile might be an issue regarding accuracy and stability on SPH simulations, but this particular subject will not be discussed here.

Similar election procedure applies on estimating the interpolated gradient, $\vec{\nabla}_i A_i$, yielding

$$\vec{\nabla}_i A_i = \sum_{\forall j \in \mathcal{E}_k(i)} \vec{\nabla}_i W_{ij} A_j \frac{m_j}{\rho_j}, \tag{5}$$

where $\vec{\nabla}_i W_{ij} = \vec{\nabla}_i W(\vec{x}_i - \vec{x}_j)$ is the smoothing-kernel gradient.

If the kernel is symmetric, one finds from the effective neighbors symmetry that $i \in \mathcal{E}_k(j) \Leftrightarrow j \in \mathcal{E}_k(i)$, and also finds $W_{ij} = W_{ji} \neq 0$, and $\vec{\nabla}_i W_{ij} = -\vec{\nabla}_j W_{ji}$ if and only if $(i, j) \in \mathcal{E}_k$.

Densities are required to perform SPH interpolations, as in equations (4) and (5), and they are estimated from equation (4) itself by means of a self-consistent replacement $A_j \to \rho_j$, yielding

$$\tilde{\rho}_i = \sum_{\forall j \in \mathcal{E}_k(i)} W_{ij} m_j = \rho_i. \tag{6}$$

The SPH fluid equations of motion are derived from the actual fluid equations, and they must be solved by means of some integration scheme regarding accuracy and stability. The timed outcomes from the integration scheme express discrete states of the particle description. Depending on the time-integration method, each particle knows a brief history of its previous states.

The way as the SPH equations are presented usually requires rearrangement to attend to subsidiary information concerning physics, chemistry etc. For instance, in most astrophysical problems, the SPH momentum conservation equation can be written as

$$\frac{\mathrm{d}\vec{v}_i}{\mathrm{d}t} = - \sum_{\forall j \in \mathcal{E}_k(i)} \vec{\nabla}_i W_{ij} \Pi_{ij} + \vec{F}_i, \tag{7}$$

where the $\Pi_{ij}$-factor carries the pressure coefficients, which might even include anisotropic pressure as the elastic stress tensor and the Maxwell stress tensor, e.g. [6]. The $\vec{F}_i$-vector term is a non-hydrodynamic acceleration as for instance the gravity field $\vec{F}_i = -\vec{g}_i$ on $i$-particle.

Time-integration scheme plays the role of particle actuators modifying their local environment, in response to the information received from their effective neighbors. Every particle contributes to a global knowledge, which might attend to a subsidiary simulation, as for example the qualitative results of self-propagating star formation, e.g. [8], the SPH-data history constitutes a knowledge-base [10] or even a more pretentious context as in live tissue simulations [9].

Each SPH particle recognizes its surroundings by means of its effective neighbors using pattern detection techniques to identify the neighborhood morphology and consequent critical surfaces. However, particles obey a set of transition rules, according to the physics model, to decide what action they have to do against their local environment.

From the theory of intelligent agents, SPH particles might be classified as *simple reflex agents* [10], acting as environment modifiers in function of what they percept in their surrounds through their effective neighbors. The particles act under the local physical conditions in response to the input they receive from their effective neighbors, ignoring the long term history of all their actions and percepts. Regarding the adopted time integration scheme embedded as actuators, only the knowledge of a recent past is required.

## 4. Conclusion

More than a numerical simulation technique, SPH is a very complex system that can be studied not only under applied mathematics techniques but also under the light of intelligent computing, where particles are individuals cooperatively working in behalf of a collective objective of mimicking the fluid behavior.

The SPH spirit resides in computationally reproducing the continuous fluid flow using free particles. A fluid particle moves like a marker, accordingly to the lagrangian equations of motion. Each particle is a data structure storing the specific fluid properties as density, pressure, position, velocity etc. Any particle knows its surroundings through its $k$-nearest neighbors (KNN), which play the role of sensors, or informants. The information mechanism is known as KNN-based kernel interpolation, which might be interpreted as a weighted voting, from the machine learning viewpoint.

## References


[1] R. O. Duda and P. E. Hart, *Pattern Classification and Scene Analysis*, New York: Wiley & Sons, 1973.

[2] R. A. Gingold and J. J. Monaghan, *Smoothed particle hydrodynamics - Theory and application to non-spherical stars*, Monthly Noticies of the Royal Astronomical Society **181** (1977), 375–389.

[3] Volker Grimm, Eloy Revilla, Uta Berger, Florian Jeltsch, Wolf M. Mooij, Steven F. Railsback, Hans-Hermann Thulke, Jacob Weiner, Thorsten Wiegand, and Donald L. DeAngelis, *Pattern-oriented modeling of agent-based complex systems: Lessons from ecology*, Science **Vol. 310** (2005), no. 5750, 987–991.

[4] X. Guo, Y. Wei, Z. Jin, and D. Guo, *Simulation research on diamond cutting of mold steel using SPH method*, Society of Photo-Optical Instrumentation Engineers (SPIE) Conference Series, Society of Photo-Optical Instrumentation Engineers (SPIE) Conference Series, vol. 8202, November 2011.

[5] L. B. Lucy, *A numerical approach to the testing of the fission hypothesis*, Astronomical Journal **82** (1977), 1013–1024.

[6] E. P. Marinho, C. M. Andreazza, and J. R. D. Lépine, *SPH simulations of clumps formation by dissipative collisions of molecular clouds*, Astronomy and Astrophysics **379** (2001), no. 3, 1123–1137.

[7] Eraldo P. Marinho and Carmen M. Andreazza, *Mecánica Computacional. Computational Geometry (A)*, vol. XXIX, Mecánica Computacional, no. 60, ch. Anisotropic K-nearest Neighbor Search Using Covariance Quadtree, pp. 6045–6064, Asociación Argentina de Mecánica Computacional, Buenos Aires, Argentina, November 2010, Open Journal Systems.

[8] T. Mineikis and V. Vansevičius, *Disk Galaxy Models Driven by Stochastic Self-Propagating Star Formation*, Baltic Astronomy **19** (2010), 111–120.

[9] Matthias Müller, Simon Schirm, and Matthias Teschner, *Interactive blood simulation for virtual surgery based on smoothed particle hydrodynamics*, Technol. Health Care **12** (2004), no. 1, 25–31.

[10] Stuart Russell and Peter Norvig, *Artificial intelligence: A modern approach, 3rd edition*, 3rd ed., Prentice Hall Copyright ©2010, Dec 1 2009, ISBN-10: 0-13-604259-7, ISBN-13: 978-0-13-604259-4, Format: Cloth.

[11] Thanh N. Tran, Ron Wehrens, and Lutgarde M.C. Buydens, *Knn-kernel density-based clustering for high-dimensional multivariate data*, Computational Statisticas & Data Analysis **51** (2006), 513–525.

# A unified theory of quantum Hall effect and high temperature superconductivity

**Shigeji Fujita**[1], **Akira Suzuki**[2]

[1] Department of Physics, University at Buffalo, State University of New York, Buffalo, New York 14260-1500, USA

[2] Department of Physics, Faculty of Science, Tokyo University of Science, Shinjyuku-ku, Tokyo 162-8601, Japan

**Abstract.** The quantum Hall effect (QHE) and high temperature superconductivity (HTSC) have remarkable common features. They occur only in two-dimensional (2D) solids. The critical temperature $T_c$ of some HTSC exceeds 160 K while the room temperature QHE is observed in graphene. The cause of both QHE and HTSC is the phonon exchange attraction. We develop a theoretical model for the QHE in terms of the composite bosons (fermions), each containing an electron and an odd (even) number of fluxons (magnetic flux quanta). The composite particles (boson, fermion) are bound by the phonon exchange attraction. If the Bose-Einstein condensation (BEC) of the composite (c) - bosons occurs, then the system exhibits zero resistivity and the associated Hall conductivity plateau. The Hall conductivity is calculated rigorously without averaging. The mystery of the fractional charge carried by the c-bosons is resolved in our model.

## 1. Introduction

In 1983 Laughlin introduced a Laughlin ground-state wave function for a system of $N$ electrons at the Landau level occupation ratio (also called filling factor) $\nu = 1/3$, and studied the elementary excitations (quasiparticles) over the ground-state [1]. He predicted that the quasiparticle has a *fractional charge*:

$$e^* = e/3\,, \quad e = \text{electron charge.} \tag{1}$$

This prediction appears to have been later confirmed in the magnetotransport experiments by Clark *et al.* and others [2]. To interpret the experimental data it is convenient to introduce composite (c) - particles (bosons, fermions). The c-boson (fermion), each containing an electron and an odd (even) number of *flux quanta (fluxons)*, were introduced by Zhang *et al.* [3] and others (Jain [4]) for the description of the fractional QHE (Fermi liquid). The fractional charge is also known in quantum chromodynamics. Quarks have charges 1/3 and 2/3 in unit of $e$ [5]. But quarks are fermions. Bosons can be constructed from two fermions while no fermions can be constructed from bosons. Hence fermions are considered as more elementary building blocks of material. Any c-fermion in solid state physics is thought to have the charge magnitude $e$. Hence the fractional charge carried by c-bosons is a little strange, which must be grounded in the full quantum statistical theory. Besides, an unlimited set of charges: $1/2$, $1/5$, $\cdots$ conceived for QHE is unnatural. The quarks' charges are limited to 1/3 and 2/3 by the SU3 symmetry.

The prevalent theories [6] based on the Laughlin wave function in the Schrödinger picture deal with the QHE at 0 K and immediately above. The system ground-state does not carry a current, however. It is desirable to treat the QHE below and above the critical temperature $T_c$ in a unified manner. Beside, the prevalent theories have limitations:

- The high-field limit is taken at the outset. The integer QHE at $\nu \equiv P/Q$ are observed for small integer $P$ only. The question why the QHE is not observed for high $P$ (weak field) cannot be answered. We better describe the phenomena for all fields.
- The Hall resistivity $\rho_H$ value $(Q/P)(h/e^2)$ is obtained in a single stroke. To obtain $\rho_H$ we need two separate measurements of the Hall field $E_H$ and the current density $j$. We must calculate $(E_H, j)$ and take the ratio $E_H/j$ to find $\rho_H$ ($\equiv E_H/j$).

The main purposes of the present work is to report that a unified theory can be developed for the QHE and HTSC, that the Hall resistivity $\rho_H$ can be calculated rigorously without averaging, and that the fractional QHE can be described within the frame-work of c-particles model without using Laughlin's theory and results about the fractional charges.

## 2. The Hamiltonian

Fujita and Okamura developed a quantum statistical theory of the QHE [7]. We follow this theory. See this reference for more details.

There is a remarkable similarity between the QHE *and* the High-Temperature Superconductivity (HTSC), both occurring in two-dimensional (2D) systems as pointed out by Laughlin [8]. We regard the *phonon exchange attraction* as the causes of both QHE and HTSC. Starting with a reasonable Hamiltonian, we calculate everything, using quantum statistical method.

The countability concept of the fluxons, known as the *flux quantization*:

$$B = \frac{N_\phi}{A}\frac{h}{e} \equiv n_\phi \frac{h}{e}, \tag{2}$$

where $A$ = sample area, $N_\phi$ = fluxon number (integer) and $h$ = Planck constant, is originally due to Onsager [9]. The magnetic (electric) field is an axial (polar) vector and the associated fluxon (photon) is a half-spin fermion (full-spin boson). The magnetic (electric) flux line cannot (can) terminate at a sink, which supports the fermionic (bosonic) nature of the associated fluxon (photon). No half-spin fermion can annihilate itself because of angular momentum conservation. The electron spin originates in the relativistic quantum equation (Dirac's theory of electron) [10]. The discrete (two) quantum numbers ($\sigma_z = \pm 1$) cannot change in the continuous limit, and hence the spin must be conserved. The countability and statistics of the fluxon are fundamental particle properties. We postulate that *the fluxon is a half-spin fermion with zero mass and zero charge.* The fluxons are similar to neutrinos. The fluxon (neutrino) occurs in electron (nucleon) dynamics. Hence fluxon and neutrino are regarded as distinct from each other.

We assume that the magnetic field $\boldsymbol{B}$ is applied perpendicular to the interface. The 2D Landau level energy,

$$\varepsilon = \hbar\omega_c(N_L + 1/2), \quad \omega_c \equiv eB/m^* \quad N_L = 0, 1, 2, \cdots, \tag{3}$$

with the states $(N_L, k_y)$, have a great degeneracy. The cyclotron frequency $\omega_c$ contains the electron effective mass $m^*$. The Center-of-Mass (CM) of *any* c-particle moves as a fermion (boson). The eigenvalues of the CM momentum are limited to 0 or 1 (unlimited) if it contains an odd (even) number of elementary fermions. This rule is known as the *Ehrenfest-Oppenheimer-Bethe's* (EOB's) *rule* [11]. Hence the CM motion of the composite containing an electron and $Q$ fluxons is bosonic (fermonic) if $Q$ is odd (even). The system of c-bosons condenses below some critical temperature $T_c$ and exhibits a superconducting state while the system of c-fermions shows a Fermi liquid behavior.

A longitudinal phonon, acoustic or optical, generates a density wave, which affects the electron (fluxon) motion through the charge displacement (current). The exchange of a phonon between electrons and fluxons can generate an *attractive* transition.

Bardeen, Cooper and Schrieffer (BCS) [12] assumed the existence of Cooper pairs [13] in a superconductor, and wrote down a Hamiltonian containing the "electron" and "hole" kinetic energies and the pairing interaction Hamiltonian with the phonon variables eliminated. We start with a BCS-like Hamiltonian $\mathcal{H}$ for the QHE [7]:

$$\begin{aligned}\mathcal{H} &= \sum_{\mathbf{k}}{}' \sum_{s} \varepsilon_{\mathbf{k}}^{(1)} n_{\mathbf{k}s}^{(1)} + \sum_{\mathbf{k}}{}' \sum_{s} \varepsilon_{\mathbf{k}}^{(2)} n_{\mathbf{k}s}^{(2)} + \sum_{\mathbf{k}}{}' \sum_{s} \varepsilon_{\mathbf{k}}^{(3)} n_{\mathbf{k}s}^{(3)} \\ &- \sum_{\mathbf{q}}{}' \sum_{\mathbf{k}}{}' \sum_{\mathbf{k}'}{}' \sum_{s} v_0 \left[ B_{\mathbf{k}'\mathbf{q}\,s}^{(1)\dagger} B_{\mathbf{k}\mathbf{q}\,s}^{(1)} + B_{\mathbf{k}'\mathbf{q}\,s}^{(1)\dagger} B_{\mathbf{k}\mathbf{q}\,s}^{(2)\dagger} + B_{\mathbf{k}'\mathbf{q}\,s}^{(2)} B_{\mathbf{k}\mathbf{q}\,s}^{(1)} + B_{\mathbf{k}'\mathbf{q}\,s}^{(2)} B_{\mathbf{k}\mathbf{q}\,s}^{(2)\dagger} \right], \qquad (4)\end{aligned}$$

where $n_{\mathbf{k}s}^{(j)} = c_{\mathbf{k}s}^{(j)\dagger} c_{\mathbf{k}s}^{(j)}$ is the number operator for the "electron" (1) ["hole" (2), fluxon (3)] at momentum $\mathbf{k}$ and spin $s$ with the energy $\varepsilon_{\mathbf{k},s}^{(j)}$, with annihilation (creation) operators $c$ $(c^\dagger)$ satisfying the Fermi anti-commutation rules:

$$\{c_{\mathbf{k}s}^{(i)}, c_{\mathbf{k}'s'}^{(j)\dagger}\} \equiv c_{\mathbf{k}s}^{(i)} c_{\mathbf{k}'s'}^{(j)\dagger} + c_{\mathbf{k}'s'}^{(j)\dagger} c_{\mathbf{k}s}^{(i)} = \delta_{\mathbf{k},\mathbf{k}'}\delta_{s,s'}\delta_{i,j}\,, \quad \{c_{\mathbf{k}s}^{(i)}, c_{\mathbf{k}'s'}^{(j)}\} = 0\,. \qquad (5)$$

The fluxon number operator $n_{\mathbf{k}s}^{(3)}$ is represented by $a_{\mathbf{k}s}^\dagger a_{\mathbf{k}s}$ with $a$ $(a^\dagger)$ satisfying the anti-commutation rules:

$$\{a_{\mathbf{k}s}, a_{\mathbf{k}'s'}^\dagger\} = \delta_{\mathbf{k},\mathbf{k}'}\delta_{s,s'}\,, \quad \{a_{\mathbf{k}s}, a_{\mathbf{k}'s'}\} = 0\,. \qquad (6)$$

The phonon exchange attraction can create electron-fluxon composites. We call the conduction-electron composite with an odd (even) number of fluxons c-boson (c-fermion). The electron (hole)-type c-particles carry negative (positive) charge. The pair operators $B$ in Eq. (4) are defined by

$$B_{\mathbf{k}\mathbf{q},s}^{(1)\dagger} \equiv c_{\mathbf{k}+\mathbf{q}/2,s}^{(1)\dagger} a_{-\mathbf{k}+\mathbf{q}/2,-s}^\dagger\,, \quad B_{\mathbf{k}\mathbf{q},s}^{(2)} \equiv a_{-\mathbf{k}+\mathbf{q}/2,-s} c_{\mathbf{k}+\mathbf{q}/2,s}^{(2)}\,. \qquad (7)$$

The prime on the summation in Eq. (4) means the restriction:

$$0 < \varepsilon_{\mathbf{k}s}^{(j)} < \hbar\omega_{\mathrm{D}} \quad \omega_{\mathrm{D}} = \text{the Debye frequency}. \qquad (8)$$

The pairing interaction terms in Eq. (4) conserve the charge. The term $-v_0 B_{\mathbf{k}'\mathbf{q}\,s}^{(1)\dagger} B_{\mathbf{k}\mathbf{q}\,s}^{(1)}$, where $v_0 \equiv |V_{\mathbf{q}} V_{\mathbf{q}}'| (\hbar\omega_0 A)^{-1}$, $A$ = sample area, is the pairing strength, generates a transition in electron-type c-fermion states. Similarly, the exchange of a phonon generates a transition between hole-type c-fermion states, represented by $-v_0 B_{\mathbf{k}'\mathbf{q}\,s}^{(2)\dagger} B_{\mathbf{k}\mathbf{q}\,s}^{(2)\dagger}$. The phonon exchange can also pair-create (pair-annihilate) electron (hole)-type c-boson pairs, and the effects of these processes are represented by $-v_0 B_{\mathbf{k}'\mathbf{q}\,s}^{(1)\dagger} B_{\mathbf{k}\mathbf{q}\,s}^{(2)\dagger} \left(-v_0 B_{\mathbf{k}\mathbf{q}\,s}^{(1)} B_{\mathbf{k}\mathbf{q}\,s}^{(2)}\right)$.

The Cooper pair is formed from two "electrons" (or "holes"). Likewise the c-bosons may be formed by the phonon-exchange attraction from c-fermions and fluxons. If the density of the c-bosons is high enough, then the c-bosons will be condensed and exhibit a superconductivity.

To treat superconductivity we modify the pair operators in Eq. (7) as

$$B_{\mathbf{k}\mathbf{q},s}^{(1)\dagger} \equiv c_{\mathbf{k}+\mathbf{q}/2,s}^{(1)\dagger} c_{-\mathbf{k}+\mathbf{q}/2,-s}^{(1)\dagger}\,, \quad B_{\mathbf{k}\mathbf{q},s}^{(2)} \equiv c_{-\mathbf{k}+\mathbf{q}/2,-s}^{(2)} c_{\mathbf{k}+\mathbf{q}/2,s}^{(2)}\,. \qquad (9)$$

Then, the pairing interaction terms in Eq. (4) are formally identical with those in the generalized BCS Hamiltonian [14]. If we assume that only zero momentum Cooper pairs ($q = 0$) are generated, then the Hamiltonian $\mathcal{H}$ in Eq. (4) is reduced to the original BCS Hamiltonian, ref. [12], Eq. (2.14).

We first consider integer QHE. We choose a conduction electron and a fluxon for the pair. The c-bosons, having the linear dispersion relation:

$$\varepsilon^{(j)} = w_0 + \frac{2}{\pi} v_{\mathrm{F}}^{(j)} p \,, \tag{10}$$

can move in all directions in the plane with the constant speed $(2/\pi)v_{\mathrm{F}}^{(j)}$. A brief derivation of Eq. (10) is given in Appendix. The supercurrent is generated by $\mp$ c-bosons monochromatically condensed, running along the sample length. The supercurrent density (magnitude) $j$, calculated by the rule: $j =$ (carrier charge $e^*$) $\times$ (carrier density $n_0$) $\times$ (drift velocity $v_{\mathrm{d}}$), is given by

$$j \equiv e^* n_0 v_{\mathrm{d}} = e^* n_0 \frac{2}{\pi} \left| v_{\mathrm{F}}^{(1)} - v_{\mathrm{F}}^{(2)} \right| \,. \tag{11}$$

The Hall field (magnitude) $E_{\mathrm{H}}$ equals $v_{\mathrm{d}}B$. The magnetic flux is quantized:

$$B = n_\phi \Phi_0 \,, \quad \Phi_0 \equiv e/h \,, \tag{12}$$

where $n_\phi \equiv N_\phi / A$ is the fluxon density. Hence the Hall resistivity $\rho_{\mathrm{H}}$ is given by

$$\rho_{\mathrm{H}} \equiv \frac{E_{\mathrm{H}}}{j} = \frac{v_{\mathrm{d}} B}{e^* n_0 v_{\mathrm{d}}} = \frac{1}{e^* n_0} n_\phi \left( \frac{h}{e} \right) = h/e \,. \tag{13}$$

We assume that *the c-fermion has a charge magnitude e*. For the integer QHE, $e^* = e$, $n_\phi = n_0$, thus we obtain $\rho_{\mathrm{H}} = h/e^2$, the correct plateau value observed for the principal QHE at $\nu = 1$.

The supercurrent generated by equal numbers of $\mp$ c-bosons condensed monochromatically is neutral. This is reflected in our calculations in Eq. (11). In the calculation of $\rho_{\mathrm{H}}$ in Eq. (13), we used the *unaveraged* drift velocity difference $(2/\pi)|v_{\mathrm{F}}^{(1)} - v_{\mathrm{F}}^{(2)}|$, which is significant. *Only* the unaveraged drift velocity $v_{\mathrm{d}}$ cancels out exactly from numerator/denominator, leading to an exceedingly accurate plateau value.

We now extend our theory to include elementary fermions (electron, fluxon) as members of the c-fermion set. We can then treat the QHE in a unified manner, using the same Hamiltonian $\mathcal{H}$.

## 3. Fractional Quantum Hall Effect

We assume that *any* c-fermion has the effective charge $e^*$ equal to the electron charge (magnitude) $e$:

$$e^* = e \quad \text{for } \textit{any} \text{ c-fermion.} \tag{14}$$

After studying the low-field QH states of c-fermions, we obtain

$$n_\phi^{(Q)} = n_{\mathrm{e}}/Q, \quad Q = 0, 2, 4, \cdots \,, \tag{15}$$

for the density of the c-fermions with $Q$ fluxons, where $n_{\mathrm{e}}$ is the electron density. All fermionic QH states (points) lie on the classical-Hall straight line passing the origin with a constant slope when $\sigma_{\mathrm{H}}$ is plotted as a function of $B^{-1}$. The density $n_\phi^{(Q)}$ is proportional to the magnetic field $B$. As the magnetic field is raised, the separation between the LL becomes greater, and the higher-$Q$ c-fermion is more difficult to form energetically. This condition is unlikely to depend on the statistics of the c-particles. Hence Eq. (15) should be valid for all integers, odd or even.

We take the case of $Q = 3$. The c-boson containing an electron and three (3) fluxons can be formed from a c-fermion with two (2) fluxons and a fluxon. If the c-bosons are Bose-condensed, then the supercurrent density $j$ is given by Eq. (11). Hence we obtain

$$\rho_{\mathrm{H}} \equiv \frac{E_{\mathrm{H}}}{j} = \frac{v_{\mathrm{d}} B}{e^* n_0 v_{\mathrm{d}}} = \frac{n_\phi^{(3)}}{e^* n_0} \left( \frac{h}{e} \right) = \frac{1}{3} \frac{h}{e^2}, \tag{16}$$

where we used $e^* = e$ from Eq. (14), and $n_\phi^{(3)}/n_0 = 1/3$, a bosonic extension of Eq. (15).

The principal fractional QHE occurs at $\nu = 1/3$, where the Hall resistivity value is $h/(3e^2)$ as shown in Eq. (16). A set of weaker QHE occur on the lower field side at

$$\nu = \frac{1}{3}, \frac{2}{3}, \cdots . \tag{17}$$

The QHE behavior at $\nu = P/Q$ for any odd $Q$ is similar. We illustrate it by taking integer QHE with $\nu = P$. The field magnitude becomes smaller with increasing $P$. The LL degeneracy is proportional to $B$, and hence $P$ LL's must be considered. First consider the case $P = 2$. Without the phonon-exchange attraction the electrons occupy the lowest two LL's with spin. See Fig. 1 (a). The electrons at each level form c-bosons. In the superconducting state

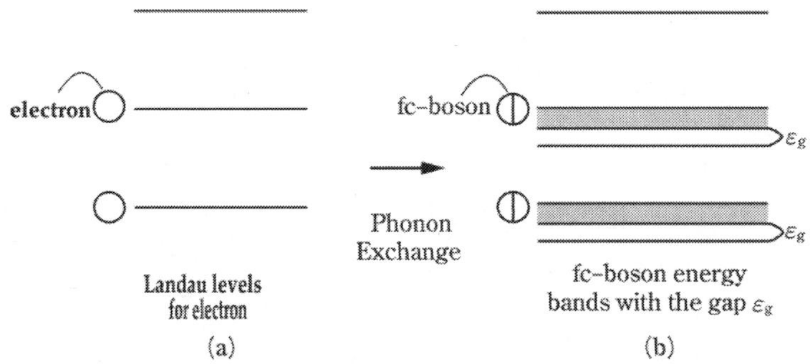


**Figure 1.** The electrons which fill up the lowest two LL's, shown in (a) form the QH state at $\nu = 2$ in (b) after the phonon-exchange attraction and the BEC of the c-bosons.

the supercondensate occupy the monochromatically condensed state, which is separated by the superconducting gap $\varepsilon_{\mathrm{g}}$ from the continuum states (band) as shown in Fig. 1 (b). The temperature-dependent energy gap $\varepsilon_{\mathrm{g}}(T)$ is defined in terms of the BCS energy parameter $\Delta$. Its temperature dependence is discussed in Appendix. The c-boson density $n_0$ at each LL is one-half the density at $\nu = 1$, which is equal to the electron density $n_{\mathrm{e}}$ fixed for the sample. Extending the theory to a general integer, we have

$$n_0 = n_{\mathrm{e}}/P\,. \tag{18}$$

The critical temperature $T_{\mathrm{c}}\,(= 1.24\hbar v_{\mathrm{F}} k_{\mathrm{B}}^{-1} n_0^{1/2})$ and the gap energy $\varepsilon_{\mathrm{g}}$ are smaller for higher $P$, making the plateau width (a measure of $\varepsilon_{\mathrm{g}}$) smaller in agreement with experiments. The c-bosons have lower energies than the conduction electrons. Hence at the extreme low temperatures the supercurrent due to the condensed c-bosons dominates the normal current due to the conduction electrons and non-condensed c-bosons, giving rise to the dip in $\rho$. The critical temperature for the condensed c-bosons is given by (see Appendix)

$$T_c = 1.24\,\hbar v_F k_B^{-1} n_0^{1/2}, \qquad n_0 \equiv N_0/A. \tag{19}$$

Then, according to Eq. (18), both the critical temperature $T_c$ and the energy gap $\varepsilon_g$ are smaller for higher $P$.

In summary we have achieved our goal of treating fractional QHE in terms of c-particles without using Laughlin's results in terms of fractional charges. Our quantum statistical theory is a finite temperature theory. The room-temperature QHE in graphene is an important topic, which is discussed separately.

## Appendix: Derivation of Eq. (10) and the superconducting energy gap

The c-bosons, each with one fluxon, will be called the fundamental (f) c-bosons. Their energies $w_{\boldsymbol{q}}^{(j)}$ are obtained from [7, 15]

$$w_{\boldsymbol{q}}^{(j)}\Psi(\boldsymbol{k},\boldsymbol{q}) = \varepsilon_{|\boldsymbol{k}+\boldsymbol{q}|}^{(j)}\Psi(\boldsymbol{k},\boldsymbol{q}) - \frac{v_0^*}{(2\pi\hbar)^2}\int' d^2k'\Psi(\boldsymbol{k}',\boldsymbol{q}), \tag{A.1}$$

where $\Psi(\boldsymbol{k},\boldsymbol{q})$ is the reduced wave function for the fc-boson; we neglected the fluxon energy. The prime on the integral means the restriction (8). The $v_0^*$ denotes the strength after the ladder diagram binding, see below. For small $q$, we may use a Taylor expansion and obtsain

$$w_{\boldsymbol{q}}^{(j)} = w_0 + \frac{2}{\pi}v_F^{(j)}q, \qquad w_0 = \frac{-2\hbar\omega_D}{\exp(1/(v_0^*\mathcal{D}_0)) - 1}, \tag{A.2}$$

where $v_F^{(j)} \equiv (2\varepsilon_F/m_j)^{1/2}$ is the Fermi velocity and $\mathcal{D}_0 \equiv \mathcal{D}(\varepsilon_F)$ the density of states per spin. Note that the energy $w_q^{(j)}$ depends *linearly* on the momentum $q$.

The system of fc-bosons undergoes a Bose-Einstein condensation (BEC) in 2D at the critical temperature [7, 15]:

$$T_c = 1.24\hbar\, v_F\, k_B^{-1}\, n_0^{1/2} \tag{A.3}$$

The interboson distance $R_0 \equiv n_0^{1/2}$ calculated from this expression is $1.24\hbar v_F(k_B T_c)^{-1}$. The boson size $r_0$ calculated from Eq. (A.3 ), using the uncertainty relation ($q_{max}r_0 \sim \hbar$) and $|w_0| \sim k_B T_c$, is $(2/\pi)\hbar v_F(k_B T_c)^{-1}$, which is a few times smaller than $R_0$. Hence, the bosons do not overlap in space, and the model of free bosons is justified. For GaAs/AlGaAs, $m^* = 0.067m_e$, $m_e$ = electron mass. For the 2D electron density $10^{11}$ cm$^{-2}$, we have $v_F = 1.36 \times 10^6$ cm s$^{-1}$. Not all electrons are bound with fluxons since the simultaneous generations of $\pm$ fc-bosons is required. The minority carrier ("hole") density controls the fc-boson density. For $n_0 = 10^{10}$ cm$^{-2}$, $T_c = 1.29$ K, which is reasonable.

In the presence of the Bose condensate below $T_c$ the unfluxed electron carries the energy $E_{\boldsymbol{k}}^{(j)} = \sqrt{\varepsilon_{\boldsymbol{k}}^{(j)2} + \Delta^2}$, where the BCS energy gap $\Delta$ is the solution of

$$1 = v_0\mathcal{D}_0\int_0^{\hbar\omega_D} d\varepsilon\frac{1}{(\varepsilon^2+\Delta^2)^{1/2}}\left\{1 + \exp[-\beta(\varepsilon^2+\Delta^2)^{1/2}]\right\}^{-1}, \qquad \beta \equiv (k_B T)^{-1}. \tag{A.4}$$

Note that the gap $\Delta$ depends on $T$. At $T_c$, there is *no* condensate and hence $\Delta$ vanishes.

Now the moving fc-boson below $T_c$ has the energy $\widetilde{w}_{\boldsymbol{q}}$ obtained from

$$\widetilde{w}_{\boldsymbol{q}}^{(j)}\Psi(\boldsymbol{k},\boldsymbol{q}) = E_{|\boldsymbol{k}+\boldsymbol{q}|}^{(j)}\Psi(\boldsymbol{k},\boldsymbol{q}) - \frac{v_0^*}{(2\pi\hbar)^2}\int' \mathrm{d}^2k'\Psi(\boldsymbol{k}',\boldsymbol{q}), \tag{A.5}$$

where $E^{(j)}$ replaced $\varepsilon^{(j)}$ in Eq. (A.1 ). We obtain

$$\widetilde{w}_{\boldsymbol{q}}^{(j)} = \widetilde{w}_0 + \frac{2}{\pi}v_{\mathrm{F}}^{(j)}q \equiv w_0 + \varepsilon_{\mathrm{g}} + \frac{2}{\pi}v_{\mathrm{F}}^{(j)}q, \tag{A.6}$$

where $\widetilde{w}_0(T)$ is determined from

$$1 = \mathcal{D}_0 v_0^* \int_0^{\hbar\omega_{\mathrm{D}}} \frac{\mathrm{d}\varepsilon}{|\widetilde{w}_0| + (\varepsilon^2+\Delta^2)^{1/2}}. \tag{A.7}$$

The energy difference,

$$\widetilde{w}_0(T) - w_0 \equiv \varepsilon_{\mathrm{g}}(T) > 0, \tag{A.8}$$

represents the *T-dependent energy gap*. The energy $\widetilde{w}_{\boldsymbol{q}}$ is negative. Otherwise, the fc-boson should break up. This limits $\varepsilon_{\mathrm{g}}(T)$ to be $|w_0|$ at 0 K. The $\varepsilon_{\mathrm{g}}$ declines to zero as the temperature approaches $T_c$ from below.

## References


[1] Laughlin R B 1983 *Phys. Rev. Lett.* **50**, 1395.

[2] Clark R G, Mallett J R, Haynes S R, Harris J J and Foxon C T 1988 *Phys. Rev. Lett.* **60**, 1747; Goldman V G and Su B 1995 *Science* **267**, 1010; Saminadayar L, Glattli D,G, Jin Y and Etienne B 1997 *Phys. Rev. Lett.* **79**, 2526; de Picciotto R, Reznikov M, Heiblum M, Umansky V, Bunin G and Mahalu D 1997 *Nature* **389**, 162.

[3] Zhang S C, Hansson T H and Kivelson S 1989 *Phys. Rev. Lett.* **62**, 82.

[4] Jain J K 1989 *Phys. Rev. Lett.* **63**, 199; 1989 *Phys. Rev.* B **40**, 8079; 1990 *ibid.* B **41**, 7653; 1992 *Surf. Sci.* **263**, 65.

[5] Ellis R K, Stirling W J and Webber B R 1996 *QCD and Collider Physics* (Cambridge UK: Cambridge University Press) pp. 1 2.

[6] Ezawa Z F 2008 *Quantum Hall Effects* (Singapore: World Scientific) 3rd Edition. This contains extensive references up to 2008.

[7] Fujita S and Okamura Y 2004 *Phys. Rev.* B **69**, 155313.

[8] Laughlin R B 1988 *Science* **242**, 525.

[9] Onsager L 1952 *Phil. Mag.* **43**, 1006.

[10] Dirac P A M 1958 *Principles of Quantum Mechanics*, 4th ed. (Oxford: Oxford Univ. Press) pp 248-252, p 267.

[11] Ehrenfest P and Oppenheimer J R 1931 *Phys. Rev.* **37**, 311; Bethe H A and Jackiw R J 1968 *Intermediate Quantum Mechanics*, 2nd ed. (New York: Benjamin), p 23; Fujita S and Morabito D L 1998 *Mod. Phys. Lett.* B **12** 1061

[12] Bardeen J, Cooper L N and Schrieffer J R 1957 *Phys. Rev.* **108**, 1175.

[13] Cooper L N 1956 *Phys. Rev.* **104**, 1189.

[14] Fujita S and Godoy S 2001 *Theory of High Temperature Superconductivity* (Dordrecht Netherlands: Kluwer) pp 133-135.

[15] Fujita S and Suzuki A 2013 *Electrical Conduction in Graphene and Nanotubes* (Weinheim Germany: Wiley-VCH) Sec. 11.3 (also see Appendix A.6).

# Thermodynamic Properties and Phase Transitions of Flexible-Semiflexible Homopolymers: A Multi-Canonical Monte Carlo Simulation

**Chahrazed Meddah**[1], **Sid Ahmed Sabeur**[2], **Amine Bouziane Hammou and Hadjira Medeghri**

Faculté de Physique, Université des Sciences et de la Technologie d'Oran, USTOMB, BP 1505 El M'naouer, Oran 31000, Algerie

E-mail: [1]chahrazedmeddah@gmail.com, [2]sidsabeur@gmail.com

**Abstract.** We focus on the simulation of flexible and semiflexible polymer systems using the multicanonical Monte Carlo method where we study their thermodynamic properties and conformational behavior. First, we investigate transition signatures of flexible homopolymer chains where monomers interact through a standard Lennard-Jones potential. In the second time, we introduce a square well potential that produces helical structures. The idea behind the use of those different potentials is to construct a phase diagram where we can characterize the liquid, solid, crystalline and helical phases.

## 1. Introduction

Several studies have been dedicated to investigate the behavior of single polymers in different models[1, 2, 3, 4, 5, 6, 7, 8] using different methods of computer simulation like molecular dynamic and Monte Carlo method. In the present paper, we intend to consider the combination of two models: flexible and semiflexible polymer chains by adding a directional potential[6] to the Lennard-Jones potential. This study is carried out by means of multicanonical Monte Carlo method[9].

The model and the simulation method are described in Sec II. The results are given in Sec III and we conclude in Sec IV.

## 2. Model and simulation method

The model is described by an off lattice homopolymer chain that contains N identical monomers. In the first time, we consider the flexible chain(the same model used in references[1, 2, 5]), where all monomers interact pairwise via a truncated and shifted Lennard-Jones potential given by

$$U_{\rm LJ}(r) = 4\epsilon_{\rm LJ}\left[\left(\frac{\sigma_{\rm LJ}}{r}\right)^{12} - \left(\frac{\sigma_{\rm LJ}}{r}\right)^{6}\right] - U_{\rm LJ}(r_c) \qquad (1)$$

$\epsilon_{\rm LJ}$ is the energy scale, r denotes the relative distance between two monomers, $r_c = 2.5\sigma_{LJ}$ is the cutoff distance, and $\sigma_{LJ} = r_0 2^{\frac{-1}{6}}$ is the distance by which the potential is zero with the minimum potential distance $r_0 = 0.7$. The interaction between nearest neighbors is given by the finite extensible nonlinear elastic FENE potential

$$U_{\mathrm{FENE}}(r) = -\frac{k}{2}R^2 \ln\left\{1 - \left[\frac{r - r_0}{R}\right]^2\right\} \tag{2}$$

$k$ is the spring constant set to 40 and $R = 0.3$.
In the case of a flexible polymer chain, the total energy is given by

$$E_{\mathrm{tot}}^{\mathrm{flexible}} = \sum_{i=1}^{N}\sum_{j=i+1}^{N} U_{\mathrm{LJ}}(r_{ij}) + \sum_{i=1}^{N-1} U_{\mathrm{FENE}}(r_{ii+1}) \tag{3}$$

In the second time, we add a directional potential[6, 11] to study helix formation in wormlike polymer chain[12]. The interaction between monomers labeled $i$ and $j$ $(i, j = 1, \ldots N, i \neq j)$ is given by

$$U_{ij}^{\mathrm{helix}} = \begin{cases} 0 & \text{for} \quad \sigma_{\mathrm{helix}} \leq r_{ij} \\ -\epsilon_{\mathrm{helix}}[\hat{u}_i \cdot \hat{r}_{ij}]^6 - \epsilon_{\mathrm{helix}}[\hat{u}_j \cdot \hat{r}_{ij}]^6 & \text{for} \quad d \leq r_{ij} \leq \sigma_{\mathrm{helix}} \\ \infty & \text{for} \quad 0 \leq r_{ij} \leq d \end{cases} \tag{4}$$

where $\hat{u}_i = (\vec{r}_i - \vec{r}_{i-1})(\vec{r}_{i+1} - \vec{r}_i)/sin(\theta)$, $\theta$ is the fixed bond angle and $\hat{r}_{ij} = (\vec{r}_i - \vec{r}_j)/|\vec{r}_i - \vec{r}_j|$. Here d is the excluded volume diameter and $\sigma_{\mathrm{helix}}$ is the attractive force range. This quantities are set respectively to $3/2a$, $\sqrt{45/8}a$, a is the fixed bond length set to 0.7 and $\theta = \pi/3$. $\epsilon_{\mathrm{helix}}$ denotes the energy scale. The total energy of semiflexible polymer chains producing helical structures is given by

$$E_{\mathrm{tot}}^{\mathrm{helix}} = \sum_{i=1}^{N-1}\sum_{j=i+2}^{N-2} U_{ij}^{\mathrm{helix}} \tag{5}$$

Finally, when we combine the two potential models described above, the total energy of the chain will be

$$E_{\mathrm{tot}}(\epsilon_{LJ}, \epsilon_{\mathrm{helix}}) = E_{\mathrm{tot}}^{\mathrm{helix}} + E_{\mathrm{tot}}^{\mathrm{flexible}} \tag{6}$$

Our study is different from the other previous works[6, 11] since we are combining the two potential models and are varying the energy scales $\epsilon_{\mathrm{helix}}$ and $\epsilon_{\mathrm{LJ}}$ which allows us to characterize the different phases. A new parameter, the ratio $\epsilon_{\mathrm{helix}}/\epsilon_{\mathrm{LJ}}$ is also used to control the simulation. To visit all possible conformations in energy space we use the multicanonical Monte Carlo method[9, 10] where the Boltzmann distribution $P_{\mathrm{can}}(E) \propto g(E)\exp(-\beta E)$ is deformed artificially in a way to produce a flat histogram and perform a random walk. Thereby, the canonical probability is multiplied by a weight factors $W(E)$ which are unknown a priori and have to be determined iteratively. The multicanonical energy distribution will be

$$P_{\mathrm{muca}}(E) \propto g(E)\exp(-\beta E)W(E) \sim H(E) = \text{constant} \tag{7}$$

$g(E)$ is the density of states and $\beta$ is the inverse temperature, $\beta = 1/k_B T$. $k_B$ is the Boltzmann constant. The iterative procedure starts by setting the weight factors $W^0(E)$ for all energies to unity and continued until the multicanonical histogram $H(E)$ is flat. In the first run, we perform a simulation at infinite temperature under canonical distribution. Each simulation $n$, $(n = 0, 1, 2, \ldots)$ which is performed with the estimate weights $W^n(E)$ yielding an estimate histogram $H^n(E)$. The estimation $W^{n+1}(E)$ is given by $W^{n+1}(E) = \frac{W^n(E)}{H^n(E)}$. After having estimated the appropriate weights $W(E)$, a long production run is performed to determine different statistical quantities which can be obtained by the following equation

$$\langle A \rangle_T = \frac{\sum_E A_E W^{-1}(E) e^{-\beta E}}{\sum_E W^{-1}(E) e^{-\beta E}} \tag{8}$$

The specific heat and structural quantities such as the square radius of gyration, the local and global helical order parameters $H_2$ and $H_4$ are calculated to characterize transition signatures and to specify the polymer structures.

$$R_g^2 = \frac{1}{N} \sum_{i=1}^{N} (\vec{r}_i - \vec{r}_{cm})^2 \tag{9}$$

where $\vec{r}_{cm}$ is the center of mass vector.

$$H_2 = \frac{1}{N-3} \sum_{i=2}^{N-1} \hat{u}_i \cdot \hat{u}_{i+1} \tag{10}$$

$$H_4 = \left( \frac{1}{N-2} \sum_{i=2}^{N-1} \hat{u}_i \right)^2 \tag{11}$$

## 3. Simulation results

In the following, we present the results of the multicanonical study in the temperature range $T \in [0, 5]$. 300 iterations are performed with $10^6$ updates at each iteration to estimate the weight factors and $10^8$ updates in the production run for polymer chain length of $N = 19$ monomers. The results presented below are obtained by averaging 10 simulations starting with different random seeds. The error bars of the energetic and structural quantities and their fluctuations turn out to be smaller than the data symbols used in the plots.

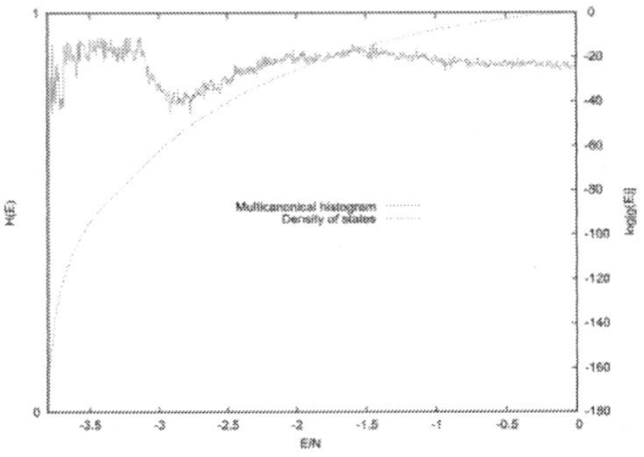


**Figure 1.** The multicanonical histogram and logarithm of the density of states versus energy per monomer obtained for polymer chain with N=19 and $\epsilon_{LJ} \gg \epsilon_{helix}$.

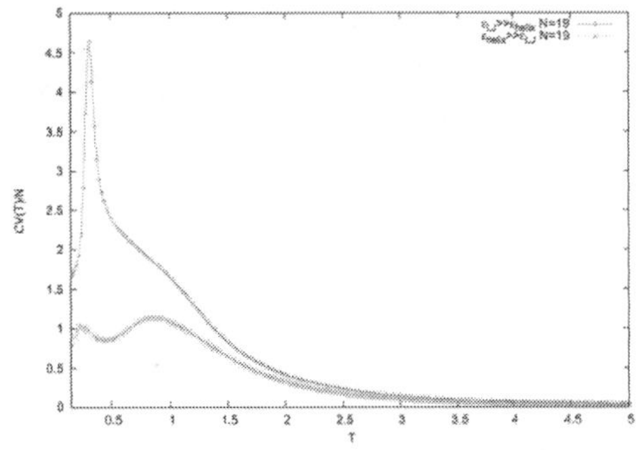


**Figure 2.** Specific heat versus temperature for chain length $N = 19$ in both cases $\epsilon_{LJ} \gg \epsilon_{helix}$ and $\epsilon_{helix} \gg \epsilon_{LJ}$.

In figure 1, we plot the histogram of the visited energies and the logarithm of the density of states when $\epsilon_{LJ} \gg \epsilon_{helix}$. Various types of displacement are used to generate conformations during the simulation. The pivot move[13] is used when $\epsilon_{helix} \gg \epsilon_{LJ}$, and in the other case addition to the pivot move, shift monomer, reptation, crankshaft and end-bridging moves[14] are all combined. The curves of specific heat in the two cases are plotted in figure 2. The transition is

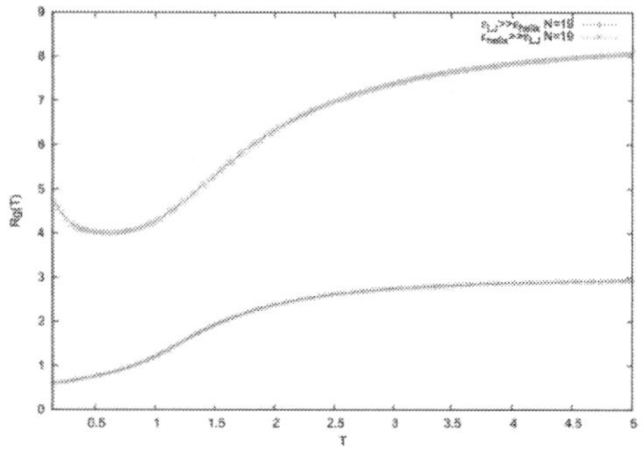

**Figure 3.** Radius of gyration of the chain of 19 monomers for $\epsilon_{\text{helix}} \gg \epsilon_{\text{LJ}}$ and $\epsilon_{\text{LJ}} \gg \epsilon_{\text{helix}}$.

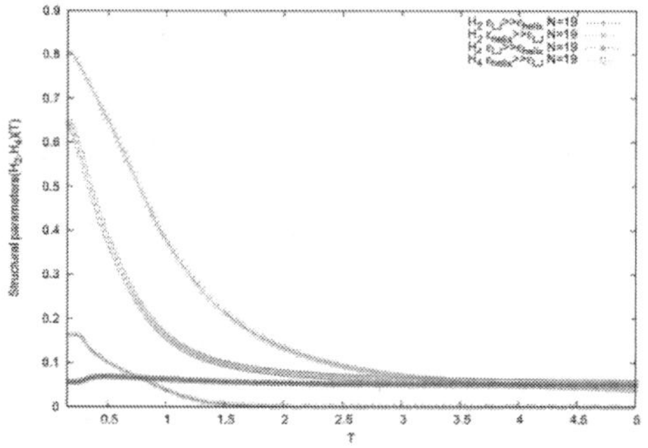

**Figure 4.** Structural parameters $H_2$ and $H_4$ versus temperature for chain length $N = 19$.

observed when $\epsilon_{\text{LJ}} \gg \epsilon_{\text{helix}}$ at T $\approx$ 0.2 and at T $\approx$ 0.8 for $\epsilon_{\text{helix}} \gg \epsilon_{\text{LJ}}$. Structural parameters are shown in figure 3 and 4. Snapshot of conformations obtained during the simulation representing different states are given in figure 5.

**Figure 5.** Representative configurations obtained during the simulation. (I):Random coil observed around $T \approx 5$. (II) and (III): Globular chains obtained for $T \approx 0.6$ and $T \approx 0.01$ when $\epsilon_{LJ} \gg \epsilon_{\text{helix}}$. (IV): Perfect helix appears for $\epsilon_{helix} \gg \epsilon_{\text{LJ}}$ at $T \approx 0.08$.

## 4. Conclusion

The aim of this paper is to classify flexible and semiflexible polymer chains through the study of their thermodynamic and structural properties by combining two different potential models. The preliminary results presented above are limited to two cases $\epsilon_{\text{helix}} \gg \epsilon_{\text{LJ}}$ and $\epsilon_{\text{LJ}} \gg \epsilon_{\text{helix}}$. At low temperatures, we have obtained helical conformations for the first case. The second case is dominated by the formation of globular structures. The behavior of the polymer chain at intermediate values of energy scales will be investigated in a future work where the phase diagram will be constructed.

## Acknowledgments

We would like to thank Professor Michael Bachmann for helpful discussions.

## References

[1] S. Schnabel, M. Bachmann and W. Janke 2009 *J. Chem. Phys.* **131** 124904
[2] S. Schnabel, W. Janke, and M. Bachmann 2011 *J. Comput. Phys.* **230** 4454
[3] S. Schnabel, T. Vogel, M. Bachmann and W. Janke 2009 *Chem. Phys. Lett* B **476** 201
[4] F. Wang and D. P. Landau 2001 *Phys. Rev. E* **64** 056101
[5] D. T. Seaton, T. Wüst and D. P. Landau 2010 *Phys. Rev. E* **81** 011802
[6] Josh P. Kemp and Zheng Yu Chen 1998 *Phys. Rev. Lett* **81** 3880

[7] M. N. Bannerman, J. E. Magee and L.Lue 2009 *Phys. Rev. E* **80** 21801
[8] Mark P. Taylor, Wolfgang Paul and Kurt Binder 2009 *Phys. Rev. E* **79** 50801
[9] B. A. Berg, T.Neuhaus 1991 *Phys. Rev. Lett. B* **267** 249
[10] B. A. Berg 1995 *J. Stat. Phys* **82**
[11] Josh P. Kemp and Jeff Z. Y. Chen 2001 *Biomacromolecules* **2** 389
[12] D. C. Rapaport 2002 *Phys Rev. E* **66** 11906
[13] N. Madras and A. D. Sokal 1988 *J. Stat. Phys* **50** 109
[14] F. A. Escobedo and J. J de Pablo 1995 *J. Chem. Phys* **102** 2636

# Unsteady compressible flows in channel with varying walls

**P Pořízková**[1], **K Kozel**[2], **J Horáček**[3]

[1] Dept. of Technical Mathematics, FME CTU, Karlovo nám. 13, CZ 121 35 Prague
[2,3] Institute of Thermomechanics AS CR, Dolejškova 5, CZ 182 00 Prague

E-mail: puncocha@marian.fsik.cvut.cz

**Abstract.** This study deals with numerical solution of a 2D and 3D unsteady flows of a compressible viscous fluid in 2D and 3D channel for low inlet airflow velocity. The unsteadiness of the flow is caused by a prescribed periodic motion of a part of the channel wall, nearly closing the channel during oscillations. The channels shape is a simplified geometry of the glottal space in the human vocal tract. Goal is numerical simulation of flow in the channels which involves attributes of real flow causing acoustic perturbations. The system of Navier-Stokes equations closed with static pressure expression for ideal gas describes the unsteady laminar flow of compressible viscous fluid. The numerical solution is implemented using the finite volume method and the predictor-corrector MacCormack scheme with artificial viscosity using a grid of quadrilateral cells. The unsteady grid of quadrilateral cells is considered in the form of conservation laws using Arbitrary Lagrangian-Eulerian method. The application of developed method for numerical simulations of flow fields in the 2D and 3D channels, acquired from a developed program, are presented for inlet velocity u=4.12 m/s, inlet Reynolds number Re=4481 and the wall motion frequency 100 Hz.

## 1. Introduction

A current challenging question is a mathematical and physical description of the mechanism for transforming the airflow energy in human vocal tract (convergent channel) into the acoustic energy representing the voice source in humans. The voice source signal travels from the glottis to the mouth, exciting the acoustic supraglottal spaces, and becomes modified by acoustic resonance properties of the vocal tract [1]. The airflow coming from the lungs causes self-oscillations of the vocal folds, and the glottis completely closes in normal phonation regimes, generating acoustic pressure fluctuations. In this study, the movement of the boundary channel is known, harmonically opening and nearly closing in the narrowest cross-section of the channel, making the investigation of the airflow field in the glottal region possible.

## 2. Governing equations

The system of Navier-Stokes equations has been used as mathematical model to describe the unsteady laminar flow of the compressible viscous fluid in a domain. The system is expressed in non-dimensional conservative form [2]:

$$\frac{\partial \mathbf{W}}{\partial t} + \frac{\partial \mathbf{F}}{\partial x} + \frac{\partial \mathbf{G}}{\partial y} + \frac{\partial \mathbf{H}}{\partial z} = \frac{1}{\mathrm{Re}} \left( \frac{\partial \mathbf{R}}{\partial x} + \frac{\partial \mathbf{S}}{\partial y} + \frac{\partial \mathbf{T}}{\partial z} \right). \tag{1}$$

$\mathbf{W} = [\rho, \rho u, \rho v, \rho w, e]^T$ is the vector of conservative variables where $\rho$ denotes density, $(u, v, w)$ is velocity vector and $e$ is the total energy per unit volume. $\mathbf{F}$, $\mathbf{G}$, $\mathbf{H}$ are the vectors of inviscid fluxes and $\mathbf{R}$, $\mathbf{S}$, $\mathbf{T}$ are the vectors of viscous fluxes. The static pressure $p$ in inviscid fluxes is expressed by the state equation in the form

$$p = (\kappa - 1)\left[e - \frac{1}{2}\rho\left(u^2 + v^2 + w^2\right)\right], \tag{2}$$

where $\kappa = 1.4$ is the ratio of specific heats.

The transformation to the non-dimensional form uses inflow parameters (marked with the infinity subscript) as reference variables (dimensional variables are marked with the hat): the speed of sound $\hat{c}_\infty = 343$ ms$^{-1}$, density $\hat{\rho}_\infty = 1.225$ kg m$^{-3}$, temperature $\hat{T}_\infty = 293.15$ K, dynamic viscosity $\hat{\eta}_\infty = 18 \cdot 10^{-6}$ Pa $\cdot$ s and a reference length $\hat{L}_r = 0.02$ m.

General Reynolds number in (1) is computed from reference variables $\mathrm{Re} = \hat{\rho}_\infty \hat{c}_\infty \hat{L}_r / \hat{\eta}_\infty$. The non-dimensional dynamic viscosity in the dissipative terms is a function of temperature in the form $\eta = (T/T_\infty)^{3/4}$.

### *2.1. Computational domains and boundary conditions*

The bounded computational domain $D_1$ used for the numerical solution of flow field in the 2D channel is shown in Figure 1. The domain is symmetric channel, the shape of which is inspired by the shape of the trachea (inlet part), vocal folds, false vocal folds and supraglottal spaces (outlet part) in human vocal tract. The upper and the lower boundaries are the channel walls. A part of the walls changes its shape between the points A and B according to given harmonic function of time and axial coordinate. The gap width is the narrowest part of the channel (in point C) and is oscillating between the minimum $g_{min} = 0.4$ mm and maximum $g_{max} = 2.8$ mm. The computational domain $D_2$ for the numerical solution of airflow in 3D is square channel $L \times H \times H = 8 \times 0.8 \times 0.8$ (160 mm $\times$ 16 mm $\times$ 16 mm).

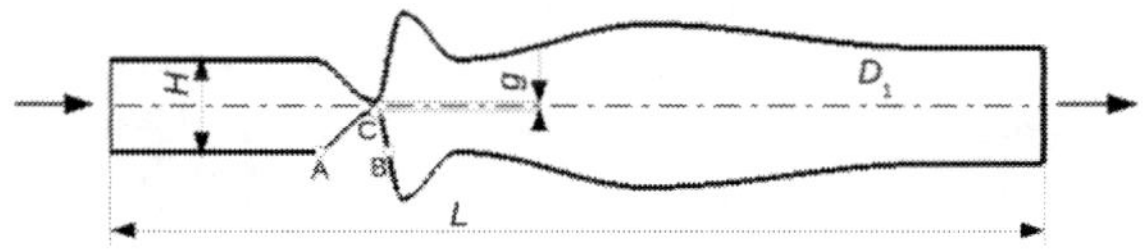


**Figure 1.** The 2D computational domain $D_1$. $L = 8$ (160 mm), $H = 0.8$ (16 mm), $g = 0.08$ (1.6 mm) - middle position.

The boundary conditions are considered in the following formulation:

(i) Upstream conditions: $u_\infty = \frac{\hat{u}_\infty}{\hat{c}_\infty}$; flow rate at the inlet is constant $H^2 \cdot u_\infty$; $\rho_\infty = 1$; $p_\infty$ is extrapolated from domain.

(ii) Downstream conditions: $p_2 = 1/\kappa$; $(\rho, \rho u, \rho v, \rho w)$ are extrapolated from domain.

(iii) Flow on the wall: $(u, v, w) = (u_{wall}, v_{wall}, w_{wall})$ - velocity of the channel walls and for temperature $T = \kappa p/\rho$ is $\frac{\partial T}{\partial \vec{n}} = 0$.

The general Reynolds number in (1) is multiply with non-dimensional value $\frac{\hat{u}_\infty}{\hat{c}_\infty} H$ represents kinematic viscosity scale and for computation of the real problem inlet Reynolds number $\mathrm{Re}_\infty = \hat{\rho}_\infty \hat{c}_\infty \frac{\hat{u}_\infty}{\hat{c}_\infty} H \hat{L}_r / \hat{\eta}_\infty$ is used.

## 3. Numerical solution

The numerical solution uses finite volume method (FVM) in cell centered form on the grid of quadrilateral cells, see e.g. [2]. In the time-changing domain, the integral form of FVM is derived using Arbitrary Lagrangian-Eulerian (ALE) formulation. The ALE method defines homomorphic mapping of the reference domain $D_{t=0}$ at initial time $t = 0$ to a domain $D_t$ at $t > 0$ [3]. The explicit predictor-corrector MacCormack (MC) scheme in the domain with a moving grid of quadrilateral cells is used. The scheme is 2nd order accurate in time and space [2]. The higher partial derivatives of velocity and temperature in viscous terms are approximated using dual volumes $V'_q$ on each face $q$ (on edge in 2D) as shown in Figure 2. The last term used in the MC scheme is the Jameson artificial dissipation $AD(W_{i,j,k})^n$ [4] for stability of computation.

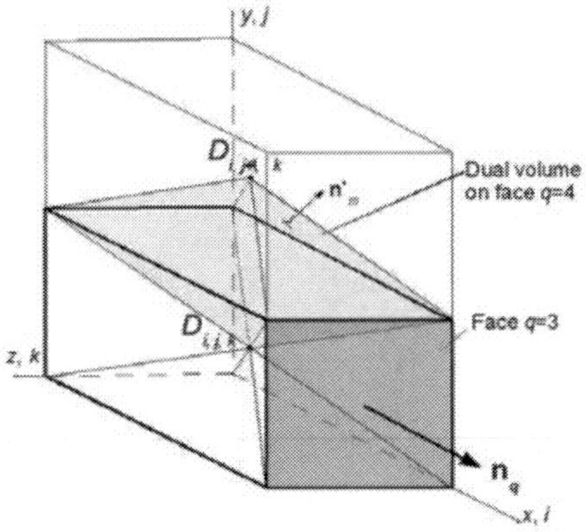


**Figure 2.** Finite volume $D_{i,j,k}$ and the dual volume $V'_q$ in 3D domain.

The grid of the 2D channel have successive refinement cells near the wall. The minimum cell size in $y$ - direction is $\Delta y_{min} \approx 1/\sqrt{\text{Re}_\infty}$ to resolve capture boundary layer effects.

## 4. Numerical results

The numerical results were obtained (using a specifically developed program) for the following input data: uniform inflow ratio velocity $\frac{\hat{u}_\infty}{\hat{c}_\infty} = 0.012$ ($\hat{u}_\infty = 4.116 \text{ ms}^{-1}$), Reynolds number $\text{Re}_\infty = 4481$ and atmospheric pressure $p_2 = 1/\kappa$ ($\hat{p}_2 = 102942$ Pa) at the outlet.

The computation in 2D channel has been carried out in two stages. First, a numerical solution is obtained, when the channel between points A and B has a rigid wall fixed in the middle position of the gap width (see Fig. 1). Then this solution is used as the initial condition for the unsteady simulation.

Figure 3 shows the unsteady flow fields computed in domain $D_1$. Simulation is captured in four time instants during one vibration period (in the fourth cycle of the wall oscillation). The highest absolute maximum velocity ratio during one vibration period is $\frac{\hat{u}_{max}}{\hat{c}_\infty} = 0.535$ ($\hat{u}_{max} = 183.5 \text{ m}s^{-1}$) at g=1.002 mm (opening phase). The application of the method for low Mach number in 3D domain $D_2$ is shown in Figure 4 with isolines of the velocity ratio.

## 5. Discussion and conclusions

The governing system (1) for flow of viscous compressible fluid based on Navier-Stokes equations for laminar flow is tested in 2D and 3D domains. In unsteady simulations (2D domain $D_1$) was possible to detect a "Coandă phenomenon" and large-scale vortices in the flow field patterns. The direction of the jet is independent on the coarseness of mesh but depends on the geometry of the channel, on the type of mesh in the domain, on the computational scheme [5] and on the governing system of flow. A similar generation of large-scale vortices, vortex convection and diffusion, jet flapping, and general flow patterns were experimentally obtained in physical models of the vocal folds by using Particle Image Velocimetry method in [6]. The method will be used for 3D simulation of unsteady flow in human vocal tract.

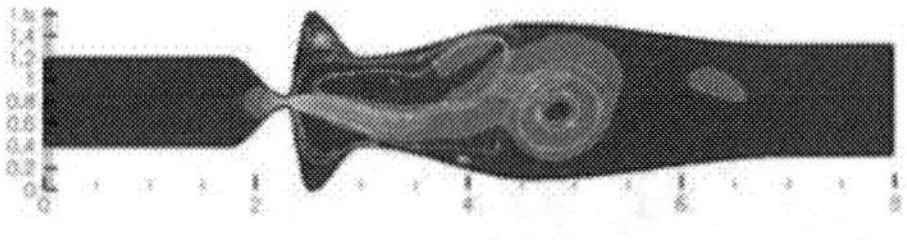

(a) $t = 30$ ms, $g = 1.6$ mm, $\hat{u}_{\max} = 55.6\text{ms}^{-1}$

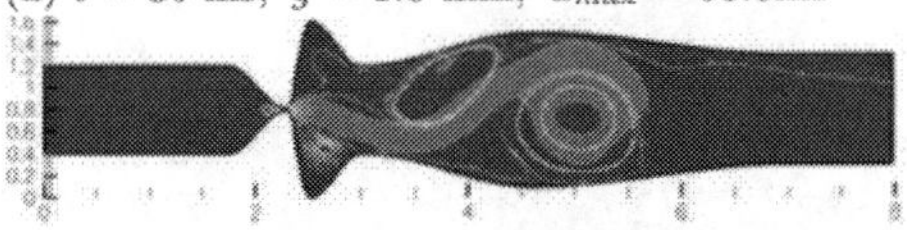

(b) $t = 32.5$ ms, $g = 0.4$ mm, $\hat{u}_{\max} = 80.9\text{ms}^{-1}$

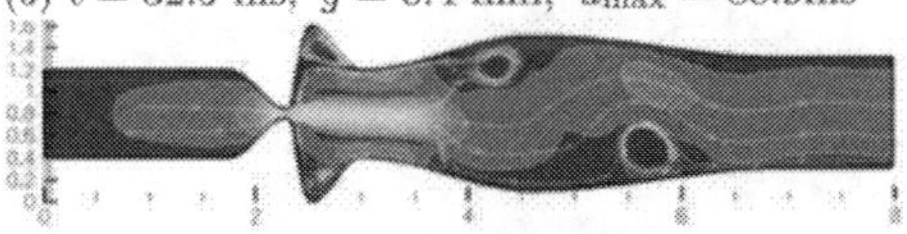

(c) $t = 35$ ms, $g = 1.6$ mm, $\hat{u}_{\max} = 126.9\text{ms}^{-1}$

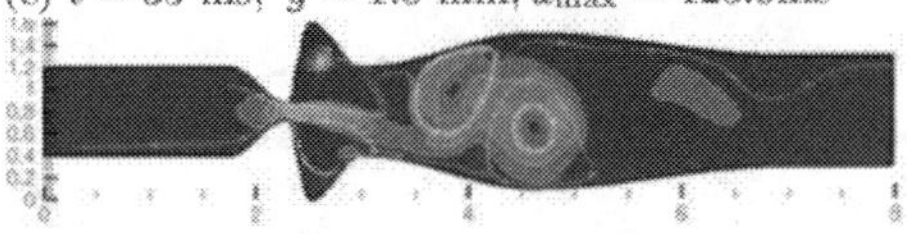

(d) $t = 37.5$ ms, $g = 2.8$ mm, $\hat{u}_{\max} = 33.3\text{ms}^{-1}$

**Figure 3.** The unsteady numerical solution of the airflow in $D_1$ - $\hat{f} = 100$ Hz, $\frac{\hat{u}_\infty}{\hat{c}_\infty} = 0.012$, $\text{Re}_\infty = 4481$, $p_2 = 1/\kappa$, $450 \times 100$ cells. Data computed during the fourth oscillation cycle. Results are mapped by iso-lines of velocity ratio and by streamlines.

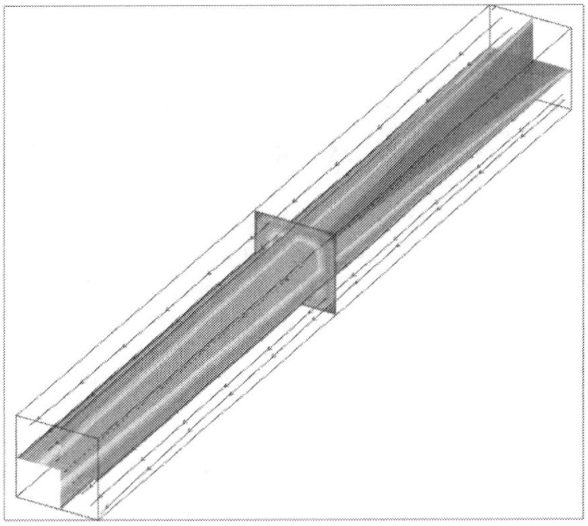

**Figure 4.** The steady numerical solution of the airflow in $D_2$ - rigid walls, $\frac{\hat{u}_\infty}{\hat{c}_\infty} = 0.012$, $\text{Re}_\infty = 4481$, $p_2 = 1/\kappa$, $100 \times 60 \times 60$ cells. Results are mapped by iso-lines of velocity ratio and by streamlines.

**Acknowledgments**

This contribution was partially supported by Research Plans MSM 6840770010, GAČR P101/11/0207, 201/08/0012 and P101/10/1329.

## References

[1] Titze, I.R., *Principles of Voice Production*. National Center for Voice and Speech, Iowa City, 2000. ISBN 0-87414-122-2.

[2] Fürst, J., Janda, M., Kozel, K., *Finite volume solution of 2D and 3D Euler and Navier- Stokes equations*. J. Neustupa, P. Penel (Eds.), Mathematical fluid mechanics, 173-194, Berlin, 2001, ISBN 3-7643-6593-5.

[3] Honzátko, R., Horáček, J., Kozel, K., Solution of inviscid incompressible flow over a vibrating profile. In: M. Beneš & M. Kimura & T. Nataki (Eds.), COE Lecture notes, **3**: 26-32, Kyushu university, 2006. ISSN 1881-4042.

[4] Jameson, A., Schmidt, W.,Turkel, E., Numerical solution of the Euler equations by the finite volume methods using Runge-Kutta time-stepping schemes, AIAA,81-125,1981.

[5] Pořízková, P., Kozel, K., Horáček, J., Numerical tests of flow in human vocal tract. In: Interaction of Dynamic Systems with Surroundings and Systems with Feedbacks. Prague: Institute of Thermomechanics, AS CR, v.v.i., 2010, p. 79-86. ISBN 978-80-87012-29-1.

[6] Horáček, J., Šidlof, P., Uruba, V., Veselý, J., Radolf, V., Bula, V., PIV Measurement of Flow-Patterns in Human Vocal Tract Model. In: Proceedings of the International Conference on Acoustic NAG/DAGA 2009, 1737-1740, Rotterdam, 2009.

# Modelling of impact of fire on safe people evacuation in tunnel

**J Glasa**[1], **L Valasek, L Halada, P Weisenpacher**
Institute of Informatics, Slovak Academy of Sciences,
Dubravska cesta 9, 84507 Bratislava, Slovakia

E-mail: jan.glasa@savba.sk

**Abstract.** In this paper, the use of Fire Dynamics Simulator (FDS) and its evacuation module, Evac for modelling fire and people evacuation in a road tunnel is illustrated. For given fire scenario and traffic situation in the tunnel, behaviour of individual evacuees as well as groups of evacuees is analyzed in order to demonstrate the impact of fire on people evacuation. Some specific features of the use of FDS+Evac for simulation of people evacuation in case of tunnel fire are also discussed.

## 1. Introduction

Current fire models and program systems based on the knowledge of CFD (computational fluid dynamics) allow to simulate fire in complex geometries incorporating a wide variety of physical phenomena related to fire and utilizing high computational power of current computers. Such systems are capable to simulate the fire spread and evaluate fire consequences on tunnel structure, facilities and people in tunnel. In this paper, Fire Dynamics Simulator (FDS, version 5.5.3, svn 7031) [5] and its evacuation module, Evac [2] are used for simulation of fire and evacuation in a short road tunnel. Several other systems for people egress simulation are available; however, most of them are able to model the impact of fire on evacuation only partially [6, 7] and only a few papers focus on the interaction of fire and evacuation simulations. In this paper, for a given traffic situation the impact of fire on behaviour of individual passengers and groups of passengers escaping from cars is illustrated in order to demonstrate some specific features of the use of FDS+Evac for simulation of evacuation in case of tunnel fire. In [8], we studied the impact of occurrence of higher capacity vehicles on the course of evacuation in tunnel. In this paper, we focus on the impact of fire on behaviour of evacuees in case of tunnel fire.

## 2. FDS+Evac

FDS (Fire Dynamics Simulator) [5] is a CFD-based field fire model capable to simulate complex physical and chemical phenomena associated with fire such as heat transfer, pyrolysis, combustion, radiation, turbulence, fluid dynamics and suppression. It was developed by NIST (National Institute of Standards and Technology, USA) in cooperation with VTT Technical Research Centre of Finland. FDS allows visualization of 3D smoke and flames spread as well as of other parameters characterizing fire in the form of 2D slices, tables, graphs and pictures. Since the first validation study elaborated by the U.S. NRC (Nuclear Regulatory Commission) [1], systematic verification and validation of FDS continue until now (see e.g. [3, 4]). Evac [2] is an evacuation module of FDS capable to utilize fire

[1] To whom any correspondence should be addressed.

parameters for modelling the impact of fire on evacuees' movement, behaviour and escape strategies. Each evacuee is treated as an independent entity (agent) with own personal properties and escape strategies. As the shape of human body is approximated by combination of three overlapping circles, motion equations of each agent include rotational movement. The model behind the movement algorithm is the social force model taking into account motive and physical forces exerted by the environment and by other agents (for more details about representation of social, motive, contact, attraction and repulsion forces, see [2]). Agents are guided to exits by a vector field of preferred walking directions calculated by FDS. Simple collision avoidance strategy is also included. The exit selection algorithm solves an optimization problem minimizing the evacuation time taking into account agent's walking time and queuing. The agents' decision making and individual and group escape strategies are influenced by familiarity with exits, visibility of exits and effects of smoke and obstacles. The interaction between fire and evacuation models and direct access of Evac to the fire information are the main strongpoints of FDS+Evac. It combines 3D fire simulation (calculated on 3D computational meshes) with 2D evacuation simulation (calculated on a single 2D computational mesh). FDS+Evac computes a form of the Purser's FED (fractional effective dose) index and the agents' intoxication and incapacitation. The impact of smoke on the agents' walking speed, exit selection and fire detection is modelled.

### 3. Fire and evacuation scenario

Let us consider a single-directional 2-lane 180 m long road tunnel with curved ceiling [8] (see figure 1). Two couples of jet fans placed by about 3 m far from each other are located about 1 m under the ceiling. Their effective diameter and length is 0.9 m and 5.2 m, respectively. Initial air flow supposed in the tunnel is modelled by fans blowing at 6.25 m/s velocity during the first 60 s of simulation providing a quasi-steady air flow with approximately 2 m/s velocity under the ceiling. The fire source is represented by 2 m x 3 m surface producing heat at 1666.667 kW/m2 heat release rate per unit area. It is placed about 1.1 m above the road. The fire is ignited at the 50th s. It linearly increases achieving 10 MW heat release rate at the 55th s. Since that time, the fire source intensity remains unchanged until the end of simulation (at the 180th s). Other flammable materials inside the tunnel are not considered. We assume the following ventilation action. At the 60th s, all fans start to work with linearly increasing velocity (from 6.25 m/s) reaching 25 m/s velocity at the 65th s. Since that time, the fans velocity is not changed. We assume 20°C ambient temperature in the tunnel at the beginning of simulation. We consider 24 cars arriving through the left tunnel portal which stop in positions shown in figure 1. Arrival times for particular cars and basic passengers' settings are also listed there. 65 passengers are considered in the scenario. We assume one 1.5 m wide emergency exit placed by about 73 m far from the left portal represented in simulation by a VENT object to which an evacuation mesh and a point lying above the middle of the exit are assigned. In FDS+Evac, the point is used to determine visibility of exit. The left tunnel portal is represented by 3 individual exits rather than by single exit. The triple of exits is selected to minimize the area from which the portal is not visible. Simulation experiments performed indicate that the portal representation as single exit would lead to the portal misrepresentation and non-realistic movement of evacuees (the problem is related to "low obstacles" implementation in the current version of FDS+Evac; for more details, see [8]). We assume that all passengers "know" (are familiar with) the left tunnel portal. Passengers escaping from cars C1-C7 know both the left portal and exit. Note that some simplifications must be adopted (for instance shortened reaction times of evacuees and alarm system in regard to times observed in real situations) to keep fire and evacuation scenarios as simple as possible and the total calculation time reasonable.

### 4. Fire Simulation

The fire and evacuation simulation was realized on 6-core PC (Intel i7-3930K, 3.26 GHz, 64 GB RAM). 4 CPU cores were used (3 cores assigned to three 3D computational meshes for fire simulation and one core assigned to single 2D evacuation mesh for evacuation simulation; all meshes had 10 cm density). Total computational time of 3-minute simulation was 95.87 hours.

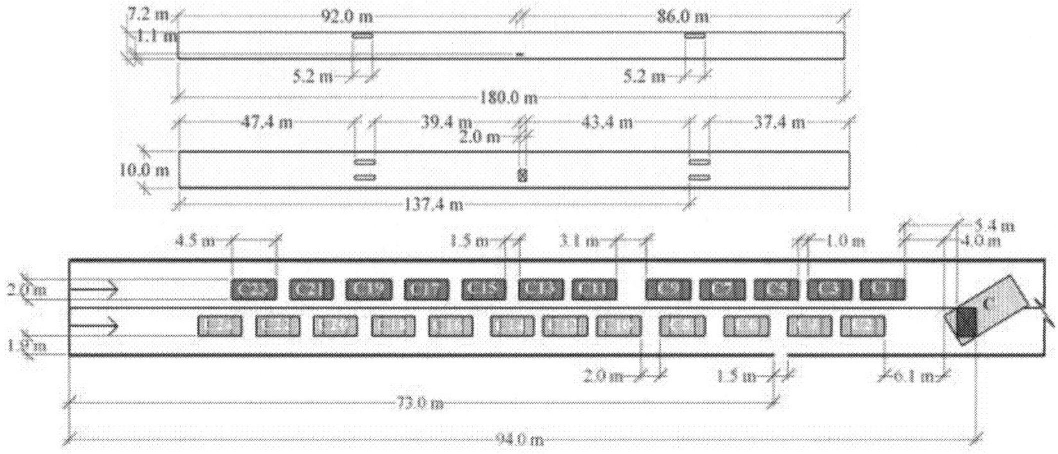


| Cars | C1 | C2 | C3 | C4 | C5 | C6 | C7 | C8 | C9 | C10 | C11 | C12 | C13 | C14 | C15 | C16 | C17 | C18 | C19 | C20 | C21 | C22 | C23 | C24 |
|---|---|---|---|---|---|---|---|---|---|---|---|---|---|---|---|---|---|---|---|---|---|---|---|---|
| AT | 53 | 54 | 55 | 56 | 57 | 58 | 59 | 60 | 61 | 62 | 63 | 64 | 65 | 66 | 67 | 68 | 69 | 70 | 71 | 72 | 73 | 74 | 75 | 76 |
| LF | A58 | A59 | M60 | F61 | A62 | M63 | A64 | M65 | M66 | M67 | M68 | M69 | M70 | M71 | E73 | M73 | M74 | M75 | F76 | M77 | M78 | M79 | E81 | M81 |
| RF | A58 | E60 | F60 | F61 | E63 | F63 | A64 | F65 | A66 | A67 | A68 | A69 | F70 | - | E74 | F73 | A74 | A75 | - | A77 | M78 | A79 | - | A81 |
| LB | - | C60 | C60 | - | - | M63 | E65 | - | M66 | - | M68 | M69 | - | - | - | E74 | M74 | M75 | - | M77 | - | M79 | - | M81 |
| RB | - | - | C60 | - | - | F63 | E65 | C65 | A66 | - | C68 | - | - | - | - | - | - | - | - | - | - | - | - | - |

Notation: AT denote arrival times of cars C1-C24; A, M, F, C and E denote adult, male, female, child and elderly, respectively; and LF, RF, LB and RB denote the left front, right front, left back and right back door, respectively.

**Figure 1.** Scheme of the road tunnel, fire source and cars location, and cars' and passengers' settings.

**Figure 2.** Fire and smoke spread at the $56^{th}$ s.

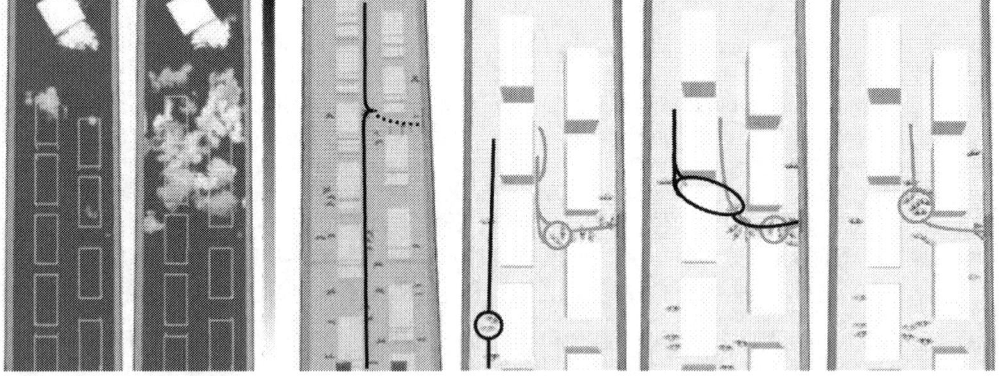

**Figure 3.** Soot visibility slices in head level at the $65^{th}$ s and $67^{th}$ s (first two figures); impact of smoke on evacuation (third figure); and examples of passengers' behaviour (last triple of figures).

The course of fire was as follows (for more details, see [8]). At first, hot gases spread upwards from the fire source reaching the ceiling at the 53rd s. Then the smoke spread under the ceiling being drifted slightly more towards the right tunnel portal because of the quasi-steady air flow under the tunnel ceiling (see figure 2). Since the 60th s, smoke spread more rapidly towards the right tunnel portal as result of ventilation action. At the 99th s, smoke was completely eliminated from the tunnel at the left of the fire source. The soot visibility slices at head level illustrate the smoke spread and increasing risk of people intoxication in the tunnel (see the first two figures in figure 3).

## 5. People Evacuation

The third figure in figure 3 illustrates the course of evacuation at the 70th s when 4 (from 36) evacuees were evacuated via the exit, 3 people were intoxicated and none of passengers was incapacitated. Trajectory of passenger from the car C1 who was forced to change escape route because of smoke is highlighted. The next figures in figure 3 illustrate escape strategies of passengers from the car C3; two couples of passengers escaping through two different emergency routes, 4 passengers escaping through the exit (however, they move relatively independently of each other) and 4 passengers behaving as a group escaping through the exit are shown. The cases differ from each other by agents' settings (familiarity with exits, preferred directions field modification, group force parameter).

## 6. Conclusion

The use of FDS+Evac for modelling people evacuation in short single-directional road tunnel in case of fire is illustrated. Detailed analysis of passengers' movement indicates that FDS+Evac is able to realistically capture the main tendencies of evacuees' behaviour and the influence of smoke on evacuees' movement, intoxication and decision making. Basic types of individual and group behaviour of evacuees modelled by FDS+Evac are illustrated. However, simulation experiments indicate that the way of implementation of emergency exits visibility must be taken into account to avoid misrepresentation of inputs for FDS+Evac simulation (for more details see also [8]).

**Acknowledgment**
This paper was partially supported by VEGA 2/0216/10 and ITMS 26240220060 research grants.

## References

[1] Hill K, Dreisbach J, Joglar F, Najafi B, McGrattan K, Peacock R and Hamins A 2007 *NUREG* **1824** U. S. Nuclear Regulatory Commission Washington USA
[2] Korhonen T and Hostikka S 2009 *Fire Dynamics Simulator with evacuation: FDS+Evac* Technical Reference and User's Guide VTT Finland
[3] McDermott R, McGrattan K B, Hostikka S and Floyd J E 2010 *Fire Dynamics Simulator (Version 5), Technical Reference Guide, Volume 2: Verification* NIST Special Publication **1018-5**, NIST Gaithersburg Maryland USA
[4] McGrattan K B, Hostikka S, Floyd J E and McDermott R 2010 *Fire Dynamics Simulator (Version 5), Technical Reference Guide, Volume 3: Validation* NIST Special Publication **1018-5**, NIST Gaithersburg Maryland USA
[5] McGrattan K, McDermott R, Hostikka S and Floyd J 2010 *Fire Dynamics Simulator (Version 5), User's Guide*, NIST Special Publication **1019-5** NIST Gaithersburg Maryland USA
[6] Ronchi E, Colonna P, Capote J, Alvear D, Berloco N and Cuesta A 2012 *Tunelling and Underground Space Technology* **30** 74-84
[7] Ronchi E, Colonna P, Gwynne S M V and Purser D A 2013 *Fire Technology* **49** 411-431
[8] Valasek L and Glasa J 2013 Simulation of the course of evacuation in tunnel fire conditions by FDS+Evac *Proc. of the Int. Conf. on Applied Mathematics and Computational Methods in Engineering* Rhodes 288-295

# Modeling of Interdisciplinary Connections in Science Courses

**Gnitetskaya Tatyana**

Departament of General Physics, The School of Natural Sciences, Far Eastem Federal University, 8, Sukhanova st., 690950, Vladivostok, Russian Federation

e-mail: gnitetskaya.tn @ dvfu.ru

**Abstract**. Graph model of interdisciplinary connections, and method of calculation quantitative parameters of model have described. The application of quantitative method to content of physics and mathematics has already done in this paper. The hierarchy of mathematical concepts used by the general physics course is established co according to two parameters - the bond strength (frequency of use) and the bond length (duration of use).

## 1. Introduction

It is very important today make a systems of inter- and intra- disciplinary connections as a basis of training course before the moment when the content of this course is began to create. These systems allow to decide some problems: 1) the intra-disciplinary connections' system helps us to understand which part of training content is actual to include into the training course but which part of training content is not so important and could be deleted; 2) we could see a level of connectedness between disciplines and optimize their content For example, concepts of derivative, integral and etc. are the main concepts in physics' description of natural phenomena, furthermore these mathematical concepts have physical meaning. The derivative - as the speed of functions' changing and an integral - as a length of way or limit of sum of lengths of an elementary segments, which the whole length of way were divided. This article is aimed to describe a graph model of inter- disciplinary connections [1] and its application to the optimization problem of the content of physics and mathematics for first years of physics learning at the university.

## 2. Graph Model of Interdisciplinary university Connection

### 2.1. Definition of interdisciplinary connection

Currently known definitions of the interdisciplinary connection, there are more than 30, represent any functions or properties of connection only qualitatively.

This approach based on suggested definition of interdisciplinary connections enable not only vividly represent these connections in disciplines but to evaluate them quantitatively [1].

According to definition illustrated above the interdisciplinary connection is understood as a structure of elements of the pedagogical system which connects the elements of the structure of interdisciplinary content of education and consists of: 1) connection object – any element of knowledge, skills and abilities belonged to a basic subject and used in the connected one (at least, in two elements of its structure); 2) connection channel – one or more elements of the educational technology, equal to subjects between a connection is established. A direction of interdisciplinary connection is given by a choice of the basic subject – from the basic subject to the connected one. The same connection cannot have a direct and reverse direction. Even with the same connection object a connection channel from subject A to subject B does not coincide with a connection channel from B to A. An illustration of the definition is given in Fig. 1.

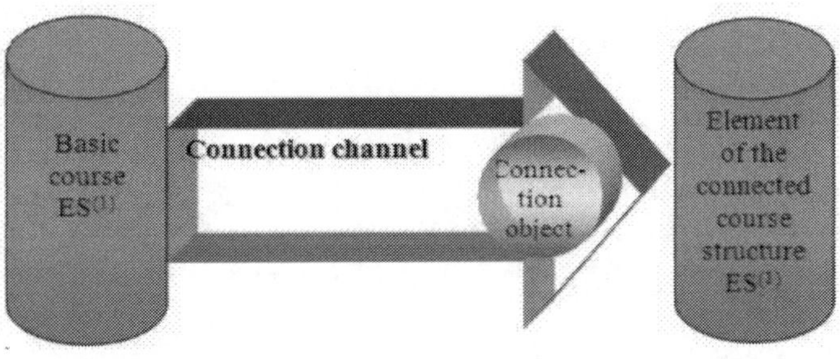


**Figure 1**. Illustration of interdisciplinary connection definition

The given interdisciplinary connection definition differs substantially from the already existing in the interpretation of the connection as an object responsible for transfer of teaching information, besides, connection structure is determined. This allows presenting the interdisciplinary connection as graph and information models that make it possible to carry out a quantitative assessment of interdisciplinary connections.

2.2. Model

Let us describe the graph model of the interdisciplinary connection by example of the interdisciplinary space created by the courses of General Physics [2] - hereinafter referred to as physics, and Mathematic Analysis [3] - hereinafter referred to as mathematics, which are obligatory to learn for the students majoring in natural sciences in their first university years. In this case mathematics is a **basic** discipline, and physics is a **connected** one. Then the **objects of the connection** are knowledge elements (mathematical notions, theorems, axioms), skills and abilities that are part of mathematics and used in physics. Interdisciplinary connections are established between the whole mathematic course (basic) and any element of physic course (connected) structure.

Let us unite, according to [1], mathematical notions, theorems, axioms, skills and abilities used in the mathematics into five relevant groups and connect the set of physics' structure elements to the set of mathematics' group elements. Let us mark every group element as $\{EG_\mu^\nu\}$ , where $\mu$ is the number of group element, $\nu$ is the number of a group. For example, the derivative could be represented as $\{EG_1^6\}$. Crossing of structure elements set with any of the set of knowledge elements divides any structure elements set into cells (see table 1). We will consider the cell filled (marked with a rhombus) if a content of physics structure element uses this mathematical notion of the set of such cells, which are the nods of the interdisciplinary connection and represent the space of the interdisciplinary contents of the physics and mathematics education.

The interdisciplinary connections implemented through the connection object, may be presented as oriented marked graphs. Each graph is a tree having its beginning in a marked (base) node $J_k(EG_\mu^\nu)$, which is the same for all interdisciplinary connections implemented through the connection object. $k$ is the number of physics' element structure where graph began. The graphs end in the nods (leaf nodes) of the subset, which are marked in case the connection object is used in the content of i $^{th}$ structure element. Our terminology coincides with that used in the book [4]. Thus, a knowledge element of the mathematics may be used in the physics course prior to learning it in the mathematics, concur in time of learning, and be learned after it was introduced in physics. The time of learning mathematical notion in the mathematics is registered by filling the m$^{th}$ cell of the physics structure element which learning falls within the same time interval, this nod is marked $J_{m_\mu}(EG_\mu^\nu)$. For derivative example, $J_{m_{6=3}}(EG_6^1)$ (see table 1).

Methodological approaches and techniques used by physics and math teachers within the time allotted according to curriculum, significantly distinguish and this determines a nonequivalence of connection channels and, finally connections themselves. There is a good reason to introduce a notion of capacity [5] of interdisciplinary connections, mentioned nonequivalence of teaching information transfer from one structural element to another. Capacity of a connection channel shall be defined by the speed (maximal) of teaching information transmission (C), i.e. amount of the teaching information (S) transferred per time unit (T): $C = S/T$ [bits/sec].

As students in a learning process carry out acquisition and application of the interdisciplinary knowledge, abilities and skills, the interdisciplinary connection is formed through time, from one element of structure to another. Therefore the capacity of interdisciplinary connection is function of time or function of not marked nod' number: $P_{i\,m}^{(ex)}(t) = p_{i\,m}^{(ex)}(J_k(EG_\mu^\nu))$.

The described graph model of the interdisciplinary connections allows introducing their quantitative characteristics.

If $T_k$ takes as the duration of study of single structural element, numbered $k$, then the duration of the i [th] interdisciplinary connection will be: $(\tau_i(EG_\mu^\nu) = \sum_{k=k_\mu}^{i} T_k$.

For example: duration of interdisciplinary connections established between mathematics and physics through derivative is $\tau_{20}(EG_6^1) = \sum_{k=1}^{20} T_k$.

This duration may be considered as an indicator of connection length or value of the connected oriented graph radius only within the scope of this task

$$L_i^{(ex)}(EG_\mu^\nu) = \sum_{k=k_\mu}^{i} p_{k_\mu i}^{(ex)}(J_k\,(EG_\mu^\nu))T_k.$$

Presence of fractures, i.e. such elements of physics structure in which a mathematical notion is not used, weakens the interdisciplinary connection. Therefore, the interdisciplinary connection strength may be an indicator of internal interconnection level: $f_i^{(ex)}(EG_\mu^\nu) = \frac{\bar{L}_i^{(ex)}(EG_\mu^\nu)}{L_i^{(ex)}(EG_\mu^\nu)}$,

where effective continuous length of interdisciplinary connection is:

$$\bar{L}_i^{(ex)}(EG_\mu^\nu) = \sum_{k=k_\mu}^{i} p_{k_\mu i}^{(ex)}\left(J_k(EG_\mu^\nu)\right)T_k - R_{k_\mu i}^{(ex)}(EG_\mu^\nu)$$

and length of fractures is: $R_{k_\mu i}^{(ex)}(EG_\mu^\nu) = \sum_k p_{k_\mu i}^{(ex)}\,(J_k(EG_\mu^\nu))T_k$

## 3. Findings

The table 1 shows the fragment of the set of mathematical notions crossing with the set of structural elements of the physics course. The structure of the chosen general physics course contains twenty structural elements which sequence is deemed traditional. Again, the application of this mathematical notion in the physics structural element is registered by the presence of a rhombus in the cell of their crossing. The graphs in the table 1 are presented for the connections established through the connection object – Derivative. It may be seen from the table that the notion of the derivative was first introduced by the physics teacher in the first structural element "Kinematics", while a grounded definition of the derivative appeared in mathematics much later and concurs with the time of studying structural physics element "Conservation law". If a mathematical notion was formed out in physics, then we consider that it was used in physics for the first time, which is why the base node of graphs of the derivative notion is referred to the first structural element. Graphs for all other mathematical notions are formed based on the same principle. Absence of cycles in the graphs depends on the learning process logics. If the interdisciplinary information is used in several elements of the connected element structure, it is impossible to transfer this information to the last of them escaping the intermediate ones. Absolute characteristics depend on the amount of structural elements in a connected discipline but it is inconvenient to use. In this connection, relative characteristics of the interdisciplinary connections have been introduced as presented in the table – relative maximal length *L* and relative connection strength *C*.

**Table 1.** Distribution of mathematical notions in the traditional general physics course structure (fragment)

| Element of structure of physics | Mechanics: Kinematics | Mechanics: Dynamics | Mechanics: Conservation laws | Mechanics: Rigid body dynamics | Mechanics: Oscillation | Mechanics: Universal gravitation | Molecular physics and thermodynamics: Thermodynamics | Molecular physics and thermodynamics: Statistic laws | Molecular physics and thermodynamics: Gases and liquids | Molecular physics and thermodynamics: Solids | Electromagnetics: Static electrical field: In vacuum | Electromagnetics: Static electrical field: In matter | Electromagnetics: Static magnetic field: In vacuum | Electromagnetics: Static magnetic field: In matter | Electromagnetics: Quasi-steady electromagnetic field | Waves: Elastic waves | Waves: Electro-magnetic interaction: In vacuum | Waves: Electro-magnetic interaction: In matter | Atomic and nuclear physics: Atomic physics | Atomic and nuclear physics: Elementary particle physics | Relative maximal length | Connection relative strength |
|---|---|---|---|---|---|---|---|---|---|---|---|---|---|---|---|---|---|---|---|---|---|---|
| Mathematical notions | 1 | 2 | 3 | 4 | 5 | 6 | 7 | 8 | 9 | 10 | 11 | 12 | 13 | 14 | 15 | 16 | 17 | 18 | 19 | 20 | L | C |
| 1 Function | ◆ | ◆ | ◆ | ◆ | ◆ | ◆ | ◆ | ◆ | ◆ | ◆ | ◆ | ◆ | ◆ | ◆ | ◆ | ◆ | ◆ | ◆ | ◆ | ◆ | 1,00 | 1,00 |
| 2 Constant | ◆ | ◆ | ◆ | ◆ | ◆ | ◆ | ◆ | ◆ | ◆ | ◆ | ◆ | ◆ | ◆ | ◆ | ◆ | ◆ | ◆ | ◆ | ◆ | ◆ | 1,00 | 1,00 |
| 3 Differential | ◆ | ◆ | ◆ | ◆ | ◆ | ◆ | ◆ | ◆ | ◆ | ◆ | ◆ | ◆ | ◆ | ◆ | ◆ | ◆ | ◆ | ◆ | ◆ | ◆ | 1,00 | 1,00 |
| : : | | | | | | | | | | | | | | | | | | | | | | |
| 6 Derivative | ◆ | ◆ | ◆ | ◆ | ◆ | ◆ | ◆ | ◆ | ◆ | ◆ | | | ◆ | ◆ | ◆ | ◆ | ◆ | ◆ | ◆ | ◆ | 1,00 | 0,87 |
| : : | | | | | | | | | | | | | | | | | | | | | | |
| 24 Definite integral | ◆ | | ◆ | ◆ | ◆ | ◆ | ◆ | ◆ | | | ◆ | ◆ | ◆ | | ◆ | | | ◆ | ◆ | | 0,95 | 0,51 |
| : : | | | | | | | | | | | | | | | | | | | | | | |
| 89 Fourier integral | | | | | | | | | | ■ | | | | | | | | ◆ | | | 0,05 | 0,05 |

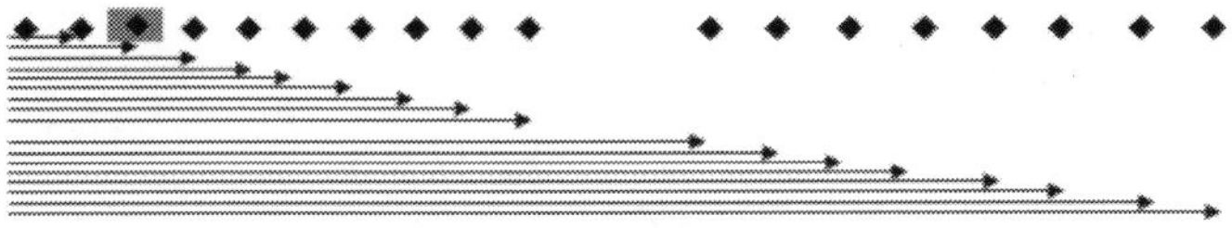

Of course the quantitative apparatus of this model was significantly simplified by bringing it to a homogeneous form. The structures which learning duration in respect to each element is identical and equal to one are accepted as "homogenous in duration".Formulae, methods of calculations and whole mathematical apparatus are given in details in the book [1].

## 4. Conclusion

Calculation of quantitative characteristics enables to establish the hierarchy of mathematical notions by their significance level in physics course. As Table 1 shows, such mathematical notions are ranked by connection length at first, and then by connection strength. The hierarchy formation permits us to abstract a fundamental kernel in the mathematical notion set, upon which we should rely when developing physics and mathematics courses for physics learning. Thus, the suggested approach based on a graph model of interdisciplinary connections supplemented by quantitative characteristics may be recommended for selection, structuring, systematization and optimization of the physical and mathematical content, which should precede the development of physics and mathematics courses in distance learning.

## References


[1] Gnitetskaya T.N.2006. *Information models of intra- and interdisciplinary connections as a basis of training physics technology*. Dr. dissert., (Moscow: MPGU Press).

[2] Saveliev I.V. 1982 *General Physics* Third ed. (Moscow: Science).

[3] Ilin V.A, Posnyak E.G. 2000 *Principles of mathematical analysis in a series of courses in higher mathematics and mathematical physics*. P.1 Fourth ed. (Moscow: Science, PHYSMATHLIT)

[4] Harary F., Palmer E.M. 1973 *Graphycal enumeration* (NY: Acad.Pr. New York and London)

[5] Lomov B.F. 196) *Man and technology* (Moscow: Soviet Radio).

# Relativistic pseudospin and spin symmetries in physical systems - recent results

**P. Alberto**[1], **A. Castro**[2], **M. Fiolhais**[1], **R. Lisboa**[3], **M. Malheiro**[4]

[1] Physics Department and Centro de Física Computacional, University of Coimbra, P-3004-516 Coimbra, Portugal

[2]Departamento de Física e Química, Universidade Estadual Paulista, 12516-410 Guaratinguetá, São Paulo, Brazil

[3]Escola de Ciências e Tecnologia, Universidade Federal do Rio Grande do Norte, 59014-615 Natal, Rio Grande do Norte, Brazil

[4]Departamento de Física, Instituto Tecnológico de Aeronáutica, Centro Técnico Aeroespacial, 12228-900 São José dos Campos, São Paulo, Brazil

E-mail: pedro.alberto@uc.pt

**Abstract.** In this paper we revise the main features of pseudospin and spin symmetries of the Dirac equation with scalar and vector potentials and mention several of its applications to physical systems. These symmetries have been extensively researched in the last 15 years, especially pseudospin symmetry, mainly in its application in understanding certain nuclear structure features of heavy nuclei. The realization of both symmetries has also been studied using several mean-field scalar and vector potentials. For many classes of potentials, these symmetries allow to have analytical solutions of the Dirac equation which otherwise would not have been possible. We report here some recent results related to anti-fermions, Coulomb and confining potentials.

## 1. Introduction

Pseudospin symmetry has been a topic in nuclear physics since the late 60's, when it was introduced to explain the near degeneracy of some single-particle levels near the Fermi surface. The subject was revived in 1997 when Ginocchio was able to relate it with a symmetry of the Dirac equation with scalar $S$ and vector $V$ mean-field potentials such that $V = -S + C$ where $C$ is a constant. However, this symmetry cannot be realized exactly in nuclei because the sum potential $V + S$ provides the binding of nucleons in nuclei. A related symmetry, the spin symmetry, was used to explain the suppression of spin-orbit splittings in states of mesons with a heavy and a light quark. In ref. [1] is reviewed the emergence of these symmetries as relativistic symmetries for the Dirac equation, as well as some of its applications. Those findings spurred much research about pseudospin symmetry in nuclei, including such topics as its very nature (how it is broken, whether is perturbative or not) [2, 3], the role of isospin [4, 5, 6] the effect of tensor potentials [7] and its realization under harmonic oscillator potentials [8]. Other theoretical studies showed the supersymmetric features of spin and pseudospin [9] and others were concerned with the realization of those symmetries for anti-fermions [10, 11, 12, 13]. More recently it was shown that pseudospin symmetry can be applied to ressonant states [14].

In this paper we will review briefly the origin of spin and pseudospin symmetries in the Dirac equation, its generators, both for general potentials and radial potentials, and their quantum numbers. Finally we report about the main conclusions of recent works, namely about the perturbative nature of spin and pseudospin symmetries, how they are realized for fermion and anti-fermion systems with Coulomb potentials [15], and also how their realization depends on the asymptotic behaviour of the radial scalar and vector potentials in the Dirac equation [16].

## 2. Spin and pseudospin symmetries in the Dirac equation

The time–independent Dirac equation for a spin 1/2 particle with mass $m$ and energy $E$, under the action of external scalar, $S$, and vector, $V$, potentials reads

$$H\psi = [\boldsymbol{\alpha}\cdot\hat{\boldsymbol{p}}\,c + \beta(mc^2+S)+V]\psi = E\psi\,, \quad \text{where} \quad \boldsymbol{\alpha}=\begin{pmatrix} 0 & \boldsymbol{\sigma} \\ \boldsymbol{\sigma} & 0\end{pmatrix}, \quad \beta=\begin{pmatrix} I & 0 \\ 0 & -I\end{pmatrix} \tag{1}$$

and $\boldsymbol{\sigma}$ are the Pauli matrices. Projecting the spinor $\psi$ into its components $\psi_\pm = P_\pm\psi$, i.e.,

$$\psi_+ = \begin{pmatrix} \phi \\ 0 \end{pmatrix} \qquad \psi_- = \begin{pmatrix} 0 \\ \chi \end{pmatrix}, \tag{2}$$

where $P_\pm = [(I\pm\beta)/2]\,\psi$, and $\phi$ and $\chi$ are respectively the upper and lower two-component spinors, and applying them to the Dirac equation (1) we get two coupled equations for $\psi_\pm$:

$$c\,\boldsymbol{\alpha}\cdot\hat{\boldsymbol{p}}\,\psi_- + (\Sigma+mc^2)\,\psi_+ = E\psi_+ \qquad c\,\boldsymbol{\alpha}\cdot\hat{\boldsymbol{p}}\,\psi_+ + (\Delta-mc^2)\,\psi_- = E\psi_-\,, \tag{3}$$

where $\Sigma = V+S$ and $\Delta = V-S$.

### 2.1. Spin symmetry

If $\Delta = 0$ ($V=S$) and after multiplying it by $c\boldsymbol{\alpha}\cdot\hat{\boldsymbol{p}}$, the second equation in (3) becomes $\hat{\boldsymbol{p}}^2/(E/c^2+m)\,\psi_+ = (E-mc^2-\Sigma)\psi_+$, which is invariant under the transformation [17]

$$\delta\psi_+ = \frac{\boldsymbol{\epsilon}\cdot\tilde{\boldsymbol{\sigma}}}{2i}\psi_+\,, \qquad \tilde{\boldsymbol{\sigma}} = \begin{pmatrix} \boldsymbol{\sigma} & 0 \\ 0 & \boldsymbol{\sigma}\end{pmatrix}. \tag{4}$$

Since $\psi_- = (c\,\boldsymbol{\alpha}\cdot\hat{\boldsymbol{p}})/(E+mc^2)\,\psi_+$, and defining $\delta\psi = \boldsymbol{\epsilon}\cdot\boldsymbol{\mathcal{S}}/(2i)\psi$, we can write the generators of this symmetry, called *spin symmetry*, as

$$\boldsymbol{\mathcal{S}} = \tilde{\boldsymbol{\sigma}}P_+ + \boldsymbol{\alpha}\cdot\hat{\boldsymbol{p}}\,\frac{\tilde{\boldsymbol{\sigma}}}{\hat{\boldsymbol{p}}^2}\,\boldsymbol{\alpha}\cdot\hat{\boldsymbol{p}}P_- = \begin{pmatrix} \boldsymbol{\sigma} & 0 \\ 0 & \boldsymbol{\sigma}\cdot\hat{\boldsymbol{p}}\,\frac{\boldsymbol{\sigma}}{\hat{\boldsymbol{p}}^2}\,\boldsymbol{\sigma}\cdot\hat{\boldsymbol{p}}\end{pmatrix}. \tag{5}$$

These generators commute with the Hamiltonian in (1) when $V=S$ and form an SU(2) algebra, i.e., $[\mathcal{S}_i,\mathcal{S}_j] = 2i\varepsilon_{ijk}\mathcal{S}_k$. The physical significance of this symmetry can be understood by looking at the second-order differential equation for $\psi_+$ when scalar and vector potentials are radial

$$\hat{\boldsymbol{p}}^2\,\psi_+ + \frac{\frac{2}{r}\,\Delta'\boldsymbol{L}\cdot\boldsymbol{S}\,\psi_+}{E-\Delta+mc^2} - \frac{\hbar^2\Delta'}{E-\Delta+mc^2}\,\frac{\partial\psi_+}{\partial r} = \frac{1}{c^2}(E-\Delta+mc^2)(E-\Sigma-mc^2)\psi_+\,, \tag{6}$$

where primes denote derivation with respect to $r$ and $\boldsymbol{S} = (\hbar/2)\,\tilde{\boldsymbol{\sigma}}$ , $\boldsymbol{L} = \boldsymbol{r}\times\hat{\boldsymbol{p}}$. From this last equation is clear that the physical significance of spin symmetry is *the disappearance of the spin-orbit coupling in a relativistic theory.* In this case there is another SU(2) symmetry connected to the orbital angular momentum, whose generators are $\boldsymbol{\mathcal{L}} = \boldsymbol{L}P_+ + \boldsymbol{\alpha}\cdot\hat{\boldsymbol{p}}\,\frac{\boldsymbol{L}}{\hat{\boldsymbol{p}}^2}\,\boldsymbol{\alpha}\cdot\hat{\boldsymbol{p}}\,P_-$. Indeed, one can check that one has $\boldsymbol{\mathcal{L}}^2\psi = \hbar^2\,\ell(\ell+1)\psi$, where $\ell$ is the orbital angular momentum quantum number of the upper component. This is true in spite of the fact that the upper and lower components of the Dirac spinor have different orbital angular momentum quantum numbers, since $\psi_+^\top = (\,i\,g_\kappa(r)/r\,\phi_{\kappa m_j}(\theta,\varphi)\quad 0\,)$ and $\psi_-^\top = (\,0\quad -f_{\tilde\kappa}(r)/r\,\phi_{\tilde\kappa m_j}(\theta,\varphi)\,)$, where $\kappa = -(\ell+1)$ if $j = \ell + \frac{1}{2}$ , $\kappa = \ell$ if $j = \ell - \frac{1}{2}$ , and $\tilde\kappa = -\kappa$. The orbital angular momentum of the lower component, $\tilde\ell$, is given by $\tilde\ell = \ell - \kappa/|\kappa|$. This means that levels with the quantum numbers $(n, l, j = l - 1/2)$ and $(n, l, j = l + 1/2)$ are degenerate (relativistic fermion levels are classified according to the quantum numbers of the upper component of their Dirac spinor). Note that the results above would still be true if $\Delta$ were just a constant.

### 2.2. *Pseudospin symmetry*

If $\Sigma = 0$ ($V = S$) or a constant, one can repeat the arguments of the previous section for the spinor $\psi_-$, whose second-order equation would be a Schroedinger-like equation. The corresponding symmetry, the *pseudospin symmetry*, has the SU(2) generators $\tilde{\boldsymbol{S}} = \tilde{\boldsymbol{\sigma}}P_- + \boldsymbol{\alpha}\cdot\hat{\boldsymbol{p}}\,\frac{\tilde{\boldsymbol{\sigma}}}{\hat{\boldsymbol{p}}^2}\,\boldsymbol{\alpha}\cdot\hat{\boldsymbol{p}}\,P_+ = \gamma^5\boldsymbol{S}$. This symmetry implies that the spin-orbit coupling for the lower component of the Dirac spinor disappears, as can be seen from the second-order radial equation for $\psi_-$

$$\hat{\boldsymbol{p}}^2\,\psi_- + \frac{\frac{2}{r}\,\Sigma'\boldsymbol{L}\cdot\boldsymbol{S}\,\psi_-}{E-\Sigma-mc^2} - \frac{\hbar^2\Sigma'}{E-\Sigma-mc^2}\frac{\partial\psi_-}{\partial r} = \frac{1}{c^2}(E-\Delta+mc^2)(E-\Sigma-mc^2)\psi_- \,. \quad (7)$$

Again, in this case, there is another SU(2) symmetry, whose generators are $\tilde{\boldsymbol{\mathcal{L}}} = \boldsymbol{L}P_- + \boldsymbol{\alpha}\cdot\hat{\boldsymbol{p}}\,\frac{\boldsymbol{L}}{\hat{\boldsymbol{p}}^2}\,\boldsymbol{\alpha}\cdot\hat{\boldsymbol{p}}\,P_+ = \gamma^5\boldsymbol{\mathcal{L}}$. In this case the orbital angular momentum of the lower component is a good quantum number, i.e., $\tilde{\boldsymbol{\mathcal{L}}}^2\psi = \hbar^2\,\tilde\ell(\tilde\ell+1)\psi$, meaning that levels with the quantum numbers $(n', l+2, j = l-1/2)$ $(n, l, j = l+1/2)$ are degenerate. In nuclei, $n' = n-1$, and it was precisely the observation of the near-degeneracy of such levels in nuclei that led to the pseudospin concept.

## 3. Spin and pseudospin symmetries for Coulomb potentials

When the scalar and vector potentials are of Coulomb type, i.e., the sum and difference potentials are of the form $\Sigma = (\alpha_\Sigma/r)\,\hbar c$ , $\Delta = (\alpha_\Delta)/r\,\hbar c$, there are analytical solutions to the respective Dirac equation. Details of the solutions can be found in refs. [9, 15]. There are two types of solutions, called $\pm$ solutions, corresponding to fermion and anti-fermion solutions in the weak coupling regime. If we expand those in terms of the $\alpha_\Delta$ and $\alpha_\Sigma$ coefficients one gets, respectively,

$$\frac{E^+_{n,\kappa}}{mc^2} = 1 - \frac{2\alpha_\Sigma^2}{\alpha_\Sigma^2+4n^2} + \frac{4\alpha_\Sigma^3\,(\kappa^2-2n|\kappa|)}{\kappa^2\,(\alpha_\Sigma^2+4n^2)^2}\,\alpha_\Delta + \mathcal{O}\left(\alpha_\Delta^2\right) \quad , \quad \frac{E^-_{n,\kappa}}{mc^2} = -1 + \frac{\alpha_\Delta^2}{2n^2} + \mathcal{O}\left(\alpha_\Delta^3\right) \ , \quad (8)$$

$$\frac{E^+_{n,\kappa}}{mc^2} = 1 - \frac{\alpha_\Sigma^2}{2n^2} + \mathcal{O}\left(\alpha_\Sigma^3\right) \quad , \quad \frac{E^-_{n,\kappa}}{mc^2} = -1 + \frac{2\alpha_\Delta^2}{\alpha_\Delta^2+4n^2} - \frac{4\alpha_\Delta^3\,(\kappa^2-2n|\kappa|)}{\kappa^2\,(\alpha_\Delta^2+4n^2)^2}\,\alpha_\Sigma + \mathcal{O}\left(\alpha_\Sigma^2\right) \quad (9)$$

where $n$ is the principal quantum number and $\kappa$ is defined above. One can see immediately that for fermion (+) solutions one cannot have bound solutions in pseudospin symmetry conditions ($\alpha_\Sigma \to 0$), but for spin symmetry ($\alpha_\Delta \to 0$) there are bound solutions. The opposite is true for anti-fermion (−) solutions, that is, there are bound solutions in pseudospin symmetry conditions but not in spin symmetry conditions. These results agree with what has been found for nuclear mean-field Woods-Saxon-like potentials, both for nucleons [1] and for anti-nucleons [10, 11, 12, 13]. These facts have been related to perturbative or non-perturbative nature of spin and pseudospin symmetries [2, 3, 5]. One may notice also from (8) and (9) that the energy for the bound states in case of spin symmetry (fermion states) and pseudospin symmetry (anti-fermion states) depends only on the principal quantum number. It is interesting to relate those to the well-known non-relativistic solutions of hidrogenic atoms. For spin symmetry, one has

$$\frac{\boldsymbol{p}^2}{2m}\,\psi_+ = (\mathcal{E}' - \Sigma')\,\psi_+ \;, \tag{10}$$

where $\mathcal{E}' = [\mathcal{E}/(2mc^2)+1]\mathcal{E}$, $\Sigma' = [\mathcal{E}/(2mc^2)+1]\Sigma$ and $\mathcal{E} = E - mc^2$. Equation (10) is just the Schroedinger equation for a hydrogenic atom of "atomic number" $Z' = -(\mathcal{E}/(2mc^2)+1)\alpha_\Sigma/\alpha$, where $\alpha$ is the fine structure constant. The energy of this "atom" is just $\mathcal{E}' = -mc^2[\mathcal{E}/(2mc^2)+1]^2\alpha_\Sigma^2/n^2$, the same as $E^+_{n,\kappa}$ in (8) when $\alpha_\Delta = 0$ . A similar reasoning can be made for the $\psi_-$ equation for pseudospin symmetry ($\alpha_\Sigma = 0$). One will get then a Schroedinger equation for a hydrogenic anti-atom of "atomic number" $Z' = -[\mathcal{E}/(2mc^2)+1]\alpha_\Delta/\alpha$ where $\mathcal{E} = -E - mc^2$. Notice that for the positive energy solutions we have $\alpha_\Sigma < 0$ while for negative energy solutions one has $\alpha_\Delta > 0$. The fact that the energy only depends on $n$ implies that there is an extra degeneracy: levels with the same $n$ and $|\kappa| \le n$ are degenerate [15]. These features of Coulomb potentials for spin and pseudospin symmetries have to do with a particular symmetry of the Dirac equation with scalar and vector Coulomb potentials (see [9, 15]).

## 4. Realization of spin and pseudospin symmetries - final considerations

As referred above, for potentials like Woods-Saxon potentials or Coulomb potentials one cannot realize pseudospin symmetry for positive energy states and spin symmetry for negative energy states. This has been observed also by many authors for a great variety of scalar and vector potentials in the Dirac equation, which have the common feature of going to zero at large distances. However, for harmonic oscillator potentials one is able to find bound solutions for both symmetries for positive and negative energy states [8]. Recently, it was shown that this behaviour is shared with general radial potentials going to infinity at large distances [16], establishing that the asymptotic behaviour of scalar and vector potentials in the Dirac equation is what determines whether spin and pseudospin symmetry can be realized for bound relativistic fermion (or anti-fermion) systems. The study of these relativistic symmetries is still very much a active subject, and its application to physical systems other than nuclei seems promising.

## References


[1] J. N. Ginocchio, Phys. Rep. **414** 165 (2005).
[2] S. Marcos, L. N. Savushkin, M. López-Quelle, and P. Ring, Phys. Rev. C **62**, 054309 (2000).
[3] S. Marcos, V.N. Fomenko, M. Lopez-Quelle, R. Niembro and L.N. Savushkin, Eur. Phys. J. A **26**, 253 (2005).
[4] P. Alberto, M. Fiolhais, M. Malheiro, A. Delfino and M. Chiapparini, Phys. Rev. Lett. **86**, 5015 (2001).
[5] P. Alberto, M. Fiolhais, M. Malheiro, A. Delfino, and M. Chiapparini, Phys. Rev. C **65**, 034307 (2002).
[6] R. Lisboa, M. Malheiro and P. Alberto, Phys. Rev. **C67**, 054305 (2003).
[7] A. S. de Castro, P. Alberto, R. Lisboa and M. Malheiro, Phys. Rev. C **73**, 054309 (2006)
[8] R. Lisboa, M. Malheiro, A. S. de Castro, P. Alberto, and M. Fiolhais, Phys. Rev. C **69**, 024319 (2004); A. S. de Castro, P. Alberto, R. Lisboa and M. Malheiro, Phys. Rev. C **73**, 054309 (2006).
[9] A. Leviatan, Phys. Rev. Lett. **92**, 202501 (2004).
[10] S.-G. Zhou, J. Meng, and P. Ring, Phys. Rev. Lett. **91**, 262501 (2003).
[11] X. T. He, S.G. Zhou, J. Meng, E.G. Zhao and W. Scheid, Eur. Phys. J. A **28**, 265 (2006).
[12] H. Liang, W- H. Long, J. Meng and N. V. Giai, Eur. Phys. J. A **44**, 119 (2010).
[13] R. Lisboa, M. Malheiro, P. Alberto, M. Fiolhais and A. S. de Castro, Phys. Rev. C **81**, 064324 (2010).
[14] B.-N. Lu, E.-G. Zhao, and S.-G. Zhou, Phys. Rev. Lett. **109**, 072501 (2012).
[15] A. S. de Castro and P. Alberto, Phys. Rev. A **86**, 032122 (2012).
[16] P. Alberto, A. S. de Castro and M. Malheiro, Phys. Rev. C **87**, 031301(R) (2013).
[17] J. S. Bell and H. Ruegg, Nucl. Phys. **B98**, 151 (1975).

# New light weakly-coupled particle searches in a neutrino detector

**A Hatzikoutelis**[1,4], **S Kotelnikov**[2], **B A Bambah**[3] **and S P Kasetti**[3]

[1]University of Tennessee Knoxville, USA
[2]Lebedev Institute of Moscow, Russia
[3]University of Hyderabad, India
[4]corresponding author.

email: a.hatzikoutelis@utk.edu

**Abstract.** Neutrino detectors at the accelerator machines of the Intensity Frontier in particle physics are becoming commonplace. As their capabilities are being understood, they seem to have the potential for studies beyond the neutrino oscillations measurements. Besides these primary neutrino physics goals, a number of exotic searches can be done with such detectors in general, and the NOvA detectors that we present here, as a particular example. Specifically, we focus on simulating signatures in NOvA experiment's Near-Detector (300 ton, 900 m from the NuMI target of Fermilab) that correspond to beam-generated new physics states from hidden sectors, dark sectors, axion-like particles, heavy or sterile neutrinos, and heavy photons. As there are no physics generators that can inherently include such states, along with the mainstream production branches, we present here the initial stages of an effort to incorporate these signatures manually in the overall simulation framework of the NOvA experiment. For this, we discuss examples and examine the potential and challenges for detecting such signatures.

## 1. Introduction

The Intensity Frontier (IF) is a world-wide initiative and the Project X [1] machines are a powerful tool consisting of a number of proton accelerators being built at the Fermi National Accelerator Laboratory facility in Illinois, USA. They will be the drivers and sources of high intensity proton, muon, neutron and neutrino beams that will be used to study rare-event phenomena. The NOvA experiment at Fermilab will use a 120 GeV proton accelerator complex at 700 kW producing the NuMI (Neutrinos at the Main Injector) neutrino beam to lead in the measurement of appearance of electron-neutrinos and anti-neutrinos [2]. These machines also have the potential to open the door into what has come to be known as the "New Physics" realms. New theoretical models and ideas of extensions of the Standard Model and the next new physics are widely discussed within the current high-energy literature. They describe exciting ways to reconcile the new discoveries at the Large Hadron Collider of the Higgs boson as they relate to the world observations of the unanswered question of the mass, the force unification, the nature of the dark matter and the matter-antimatter inconsistency of the universe. One of the more exciting and exotic candidates for these options are New Light (~GeV) Weakly-Coupled particles (NLWCp).

The motivation for various NLWCp cases comes naturally from extensions of the SM. Concepts such as Hidden or Dark Sectors in nature suggest that the constituents of the particle structure of these sectors are neutral with respect to the Standard Model charges. Models of global broken symmetries give rise to the elusive Axion (~meV) and other Axion-like-particles (ALPs) which do not bind their masses with the weak scale coupling. There are also concepts of Heavy Photons whose decay may be a source of Dark Matter in a Dark Sector and Heavy Neutrinos that may complete the puzzle of the CP violation in the lepton sector. They can all, potentially, be produced in the high intensity beams of $10^{15}$/s of protons or electrons on carbon, mercury or other high-Z material at the target or beam-dump. A brand new particle discovery will be overwhelming proof of the existence of the postulated New Sectors in nature, and will not only define the High Energy physics research in this century, it will give us the first dynamic coupling, as in the case of Axions, and a view into the nature's intentions, as in the case of Hidden Sectors, where there are no reasons to have any anthropic motivated fine tuning of parameters.

## 2. Expecting NLWCp in NOvA experiment

The theoretical motivation for expecting these discoveries at the proton accelerator fixed-target/beam-dumps is well described in IF literature [1-3]. Therefore, in this, we summarize only some of the motivations for using the neutrino detectors to search for such states. There is a significant arrangement of short- and long-baseline neutrino oscillation experiments using intense proton beams on fixed targets separated from the near (or main) neutrino detector by several hundred meters of earth or decay volumes and absorbers [1, 2]. Their primary design of creating focused and intense neutrino beams already provide significant sensitivity to nonstandard interactions and specifically to heavy and sterile neutrinos.

Various extensions to the standard model, including most grand unified theories, predict Neutral Heavy Leptons (NHLs, or heavy neutrinos) which can mix with the standard (light) neutrinos to interpret the LSND/MiniBooNE anomaly [4]. A 40-80 MeV sterile neutrino decaying into a light neutrino and photon or a 50 MeV neutrino could be produced in collisions with matter through Z- and photon-mediated processes [4]. If the mass is bigger than 140MeV, then other decay channels are also possible. For a mass region 0.25-2.0 GeV, the charged coupled decays are either through a light neutrino and two charged leptons, a lepton and two quarks, or into three neutrinos [4]. Similar decay models exist for the charged current decays in this mass ranges. NOMAD [5] sets stringent limits of $10^{-15}$ for the hypothetical single photon if it has a momentum distribution similar to that of a photon from the coherent neutral pion decay. However, using the enhanced flux $10^{15}$ POT/sec from NuMI beam and the advanced NOvA reconstruction algorithms, we can attempt detecting this heavy neutrino at the NOvA near detector.

## 3. Signatures in the NOvA Near detector

The NOvA experiment is made of two functionally identical detectors positioned off-axis by 14 mrad in front of the powerful NuMI beam of Fermilab producing neutrinos of peak energy 2GeV. The design of NOvA, of long extruded PVC tubes/cells arranged in a cross orientation, has been described in several places including reference 2. From the pair, the Near Detector (ND) of 18,000-cells and 300-tons is positioned 990 m from the NuMI beam-target. It is designed to be sensitive to muonic and electron neutrino interactions in its detectors. NLWCp signatures would appear in the detector signals as neutral current processes of the incident particle on nucleons or orbital electrons, or decays of the new particles into dilepton pairs or pion-muon combinations. The beam neutrinos have a specific energy distribution and mass (1-3 GeV in the detector's direction) but the search of NLWC particles from NuMI is model-agnostic and therefore, the energy is an open parameter. It is constrained only from geometrical and detection efficiencies. This translates to differences in kinetic energy share at their production vertex which, in turn, defines the sensitivity to production models.

In order to identify and understand the sensitivity limits of the ND to such particles, a campaign of creating specialized analysis packages has started. They can be added to the NOvA analysis framework, to supplement its capability to recognize NLWCp among neutrino interactions. The first stage of this effort includes simulations of the beam products. The beam simulation is an effort to include neutral particles with mass and kinetic energy consistent with production-channel models. It aims to create the simulation of the expected energy spectrum and angular (spatial/directional) distribution of the expected NLWCp as they travel from the target or a point along the decay volume, towards the ND. A sample of this, we show in figure 1 (right) where the angular distribution of two types of particles are shown as they were simulated by the beam-product algorithm FLUKA [6]. The two arbitrarily chosen masses are shown here as a sample of these distributions. A more generalized algorithm is still under design, one that would generate NLWCp in the form of neutral particles of any mass and determine its directional distribution probability.

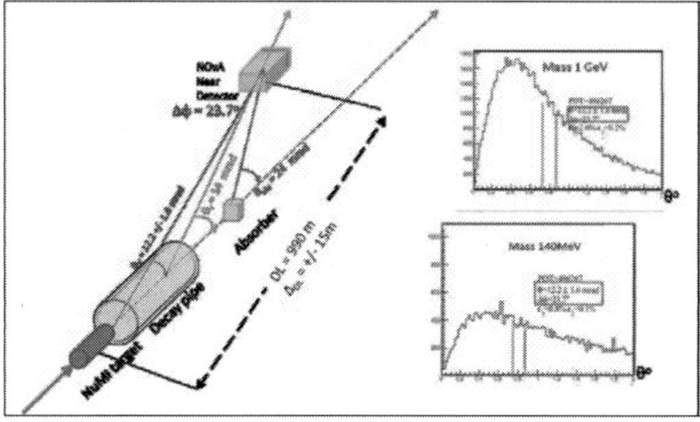


**Figure 1.** NOvA Near Detector geometry and directional acceptance (left). Angular distribution of simulated heavy neutral particle at the target. (right: top & bottom). The deviation angle-range into the detector is highlighted with two red lines.

The other part of this work is the simulations of signals. It is consistent with the most probable signatures of NLWCp in the NOvA near detector. It aims to provide an understanding of the detector efficiencies in geometry, in containment, in energy resolution. Also, the software reconstruction capabilities will have to be much wider in energy (and masses) than the baseline NOvA reconstruction is targeting [7]. The figures 2 and 3 show aspects of the detector response to simulated neutral current interactions. They come from simulated single electron and positron tracks that could be either scattering signatures or ones from the lightest dilepton decays of an NLWCp. Also, there are examples of single photons from the decays of heavy or sterile neutrinos. At this stage, the simulations are not exhaustively representing complete signatures that could challenge the analysis framework but rather the components of the detector response that affect the overall sensitivity of the detector.

Figure 2 shows a distribution of the fraction of simulated events that leave a single track in the detector as a function of their initial energy. The maximum of the distribution is about 70% and it peaks at the region of the NOvA energy range of interest. It drops significantly at higher energies as the showering becomes a big part of the signature. Of course, the single-photon simulation, where showering is more prominent, includes fewer single track events, even at the region of 1-3 GeV (highlighted in fig. 2 and 3) for which the detector is optimized. The background to the single photon signal is dominated by the asymmetric decay of neutral pions produced either in a coherent neutrino-nucleus interaction or in a neutrino-nucleon neutral-current deep-inelastic scattering or in an interaction outside the detector fiducial volume.

Even from this preliminary effort, there are signs of the sensitivity limitations. The neural nets that perform identification are trained easily for tracks but require substantially larger libraries of realistic or (better yet) real shower events. That would involve a test beam calibration which is not possible of the NOvA near detector.

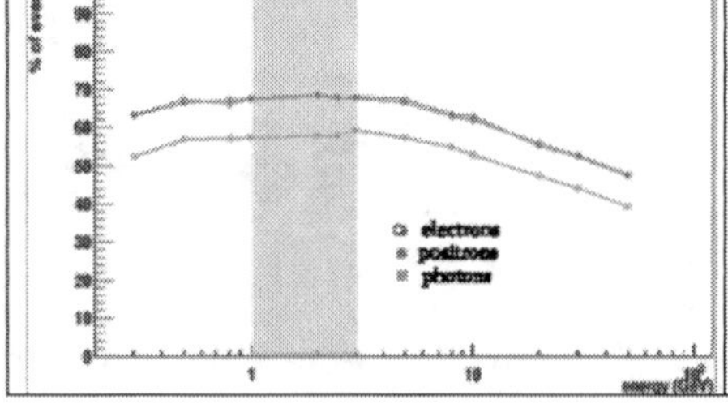


**Figure 2.** A constituent of the detector response function of a wide energy range. Only 70% of the simulated electrons or positrons leave an easily identifiable single track signature. The photons do not go above 60% and it drops for all as the track energy rises above 10GeV.

Figure 3 shows distributions of full-track containment cases over the same wide-energy range and for the three different types of particle products. As the track and/or shower length/size is a function of the energy, the drop in containment efficiency at the high energies for all types of particle signature-products is large as expected. For this aspect of the detection sensitivity there is no possible compensation from extra investment in enhancing reconstruction or particle-identification software. The detector size defines this sensitivity as its distance from the target defines the decay path sensitivity to the NLWCp lifetime, as in figure 1.

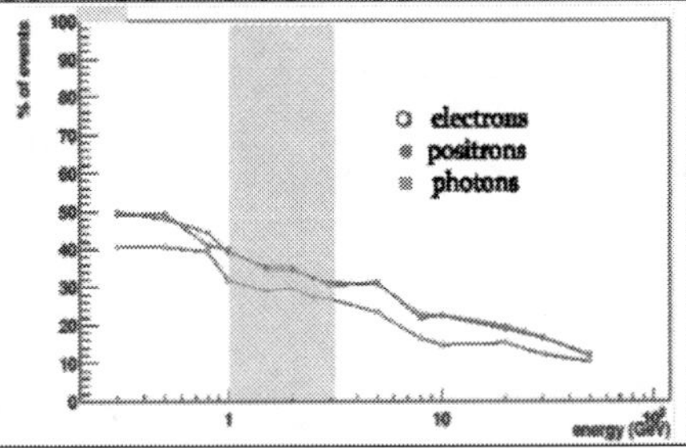


**Figure 3.** Detector response function of track containment for simulated electron, positron and single photon signatures as a function of energy. At low energies the containment efficiency is defined by the vertex position distribution and the direction of the track. For the high energies it is the track length that determines the containment probability.

## 4. Summary

The extremely exciting subject of the new light weakly-coupled particles and non-standard neutrino interactions that has been postulated in papers for years, now, is about to come to the forefront of physics research with the introduction of new powerful accelerators from the Intensity Frontier. The beginning of an effort to pave the way, before the creation of dedicated experiments, is presented here. The NOvA Near Detector is a system that can be used in the first stages of this search for NPWCp. Its limitations due to its design for neutrino detection do not prevent us from investing a wide range of mass/energies of NLWCp by studying the data it will collect, and with this, strengthen the physics program at the facility.

## References

[1] Batell B, Wester W, deNiverville P, Dharmapalan R, Hatzikoutelis A, McKeen D, Pospelov M, Ritz A and Van de Water R, 2013 "New, light, weakly-coupled particles with Project X" ed. Henderson S et al (Batavia, IL: Fermi National Accelerator Laboratory) *Project –X* chapter VII pp 142-152 http://theory.fnal.gov/people/ask/PX-book/ FERMILAB-TM-2557

[2] Hatzikoutelis A, 2013 "Neutrino oscillations at the intensity frontier: The NOvA experiment" IOP *J. of Phys.: Conf. Series*, **410** 012146

[3] deNiverville P et al 2011 *Phys. Rev.* D **84** 075020, Lin T et al 2012 *Phys. Rev.* D **85** 063503, deNiverville P et al 2012 *Phys. Rev.* D **86** 035022

[4] Aguilar-Arevalo A A et al 2009 *Phys. Rev. Lett.* **102** 101802, Gninenko S N 2011 *Phys. Rev.* D **85** 051702

[5] Kullenberg C T et al NOMAD collaboration 2012 "A search for single photon events in neutrino interactions in NOMAD" *Phys. Lett.* B **706** pp 268-275, *arXiv*:1111.3713 [hep-ex]

[6] www.fluka.org and internal NOvA technical notes (http://nova-docdb.fnal.gov/) on beam-simulations.

[7] Yang M S and Wu K L 2006 "Unsupervised possibilistic clustering" *Pattern Recognition* **39** p 521 and internal NOvA technical notes (http://nova-docdb.fnal.gov/) on offline reconstruction analysis.

# Spatial planning via extremal optimization enhanced by cell-based local search

**Epaminondas Sidiropoulos**
Faculty of Engineering, Aristotle University of Thessaloniki
54124 Thessaloniki, Greece

E-mail: nontas@topo.auth.gr

**Abstract.** A new treatment is presented for land use planning problems by means of extremal optimization in conjunction to cell-based neighborhood local search. Extremal optimization, inspired by self-organized critical models of evolution has been applied mainly to the solution of classical combinatorial optimization problems. Cell-based local search has been employed by the author elsewhere in problems of spatial resource allocation in combination with genetic algorithms and simulated annealing. In this paper it complements extremal optimization in order to enhance its capacity for a spatial optimization problem. The hybrid method thus formed is compared to methods of the literature on a specific characteristic problem. It yields better results both in terms of objective function values and in terms of compactness. The latter is an important quantity for spatial planning. The present treatment yields significant compactness values as emergent results.

## 1.Introduction

Land use planning and spatial resource allocation are important and challenging management activities, leading to mathematical optimization problems of high complexity, often solved by heuristic methods. Spatial optimization problems of this type have been treated in the context of forest planning ([1] and [2]) and land use design ([3], [4]) by utilizing concepts of simulated annealing, genetic algorithms and cellular automata; also, in the context of groundwater management ([5], [6]) and for spatial resource allocation ([7], [8], [9]), along with multi-objective treatments ([10], [11]).

A land use planning problem with a nonlinear objective function and by the use of simulated annealing has been treated in [12]. The present paper deals with a similar problem, which is typical of its kind and may be considered to possess generic characteristics. The same instance of the problem has been presented in the literature ([3], [13]) and in this way direct comparisons will be possible. The basic method followed here will be extremal optimization. It will be accompanied by a cell-based local search, thus forming a new hybrid.

Extremal optimization (EO) is a stochastic method originating in self- organized critical models of co-evolution, such as the Bak–Sneppen model [14]. The method has been employed for the solution of classical combinatorial optimization problems, such as the graph bi-section [15] and the traveling salesman problem [16]. EO has also been extended to continuous optimization problems [17] and to multi-objective ones [18]. No application of EO to a land use or spatial optimization problem is noted in the literature, at least to the present author's knowledge. This paper presents such an application, along with comparisons to alternative methods of the literature. For the specific trial problem the EO cell-based hybrid of this paper exhibits a better performance in comparison to results of the literature.

Moreover, compactness of the resulting configurations is significant, as amply documented in the literature (e.g. [19]). In this work compactness comes out as an emergent result.

## 2. Problem description and formulation

A fictive two-dimensional grid is considered divided into cells, each one of which represents a land block. Each land block will be assigned a specific land use taken from a finite set of possible land uses. Each one of the land uses entails a certain development cost depending on the location to which it is applied. The problem consists in determining an optimal distribution of land uses over the blocks of the area under study, so as to minimize the total development cost.

Following [7], let (i, j) be the coordinates of the center of the typical cell with i= 1,2,…a and j=1,2,..,b, where a and b are the lengths of the two sides of the orthogonal grid. The blocks can be numbered consecutively, so that the cell number k runs as k = 1,2,…,a•b.

Let L = {1,2,…, a•b } be the set of cells as numbered according to the above numbering scheme and let W={1,2,..,m} be the set of land uses, numbered from 1 to m:

Let p: L ⟶ W be a function that assigns a land use to each cell. This function gives rise to a set

$$C = p(L) = \{p(1),..,p(a \bullet b)\} = \{w_1,..,w_{a\bullet b}\} \qquad (1)$$

where $w_k = p(k) \in W, \quad k = 1,2,...,a \bullet b$

The set C of Equation (1) will be called the configuration of the land use mosaic as depicted in figure 1. The objective function F is a functional that maps such possible configurations to the set of positive real numbers:

F:C a ° +

The objective functional represents the development cost associated with the distribution of land uses. figure 2 shows the development cost as a function of location. Each one of the cell-blocks carries one of three different price lists for the various possible land uses.

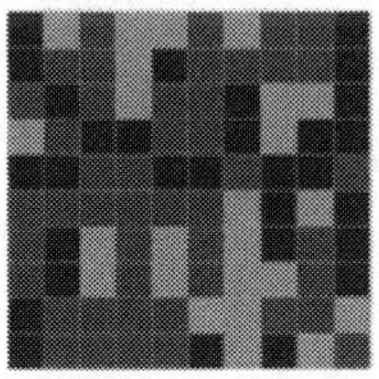

Fig. 1. Typical configuration

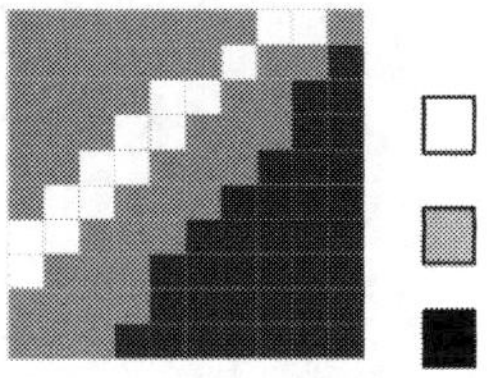

| | Lu1 | Lu2 | Lu3 |
|---|---|---|---|
| □ (white) | 750 | 400 | 1500 |
| ▒ (grey) | 2700 | 1450 | 6000 |
| ■ (black) | 5550 | 3000 | 13500 |

Fig. 2. Land use development cost in terms of location

Let PL be a price list containing the m respective development costs, such that PL(w) is development cost for land use w. Also, let $PL_k$ be the price list attached to cell k, according to figure 2. Then $PL_k(w_k)$ will be the development cost for the land use $w_k$ corresponding to cell k, according to Equation 1. Thus the local development cost will be

$$f_k = PL_k(w_k) \qquad (2)$$

The total cost will be

$$F = \sum_{k=1}^{a\bullet b} f_k \qquad (3)$$

To summarize, F depends on the configuration C as defined in Equation (1) and it forms the functional to be minimized. A constraint will be added to the formulation of the problem that prescribed fixed proportions for the numbers of the cells occupied by the corresponding land uses, namely let $L_i = \{k \in L | w_k = i\}$, i = 1,2,.., m be the subset of L consisting of cells with land use i and let $M_i = |L_i|$ be the cardinality of $L_i$.
Then the constraint requires the $M_i$'s to be constant and, of course,

$$\sum_{i=1}^{m} M_i = a \bullet b \tag{4}$$

The above constraint was not included in [12]. It is taken into account in References [3] and [13]. It will also be included in the present problem.

## 3. Method of solution

### *3.1 Extremal optimization*

The general idea of the EO method is to eliminate undesirable or low-quality parts of possible solutions, in contrast to other optimization metaheuristics, such as genetic algorithms, that work by favoring high-quality solutions. According to the EO method, the constituent parts are evaluated and each one of them is assigned a fitness that represents its state. In the present problem, these parts are the cells numbered as k= 1,2,.,a•b and the corresponding fitness for cell k is characterized by Equation (2).

The cells are ranked according to their fitness. The cell with the worst fitness is selected and the following perturbation is performed: The selected cell is kept aside and another cell is picked at random from the rest of the configuration. Then the attributes of the two cells are exchanged (figure3). Thus the proportion of land uses is preserved and the constraint (Equation 4) is not violated.

A general pseudo-code for the EO method is the following:

a. Initialize the configuration C (Equation 1). Set $C_{best} = C$, $F_{best} = F(C)$.
b. Evaluate the fitness $f_i$ of cell i, i=1, 2,...n, where n = a•b is the number of cells.
c. Find the index j such that $f_j \geq f_i \ \ \forall i$
d. Perturb C and find $C' \in N(C)$ such that cell j will be modified as above.
e. Accept C′ unconditionally.
f. If $F(C') < F_{best}$, then set $C_{best} = C'$, $F_{best} = F(C')$.
g. Repeat until a maximum number of iterations is reached.

In order to relax the stringent condition of always having to pick the worst element in step c, a probability structure can be defined over the ranks of the cells m, m= 1,2,.,a•b. If m denotes a rank (m=1,2,..,n), then m=1 will be assigned to the cell with the worst fitness and the probability is equal to

$$P(m): \ m^{-\tau}, \ \ 1 \leq m \leq a \bullet b \tag{5}$$

where τ is a parameter of the algorithm, taken as τ = 2.5.

The element to be picked will be chosen according to the probability of Equation (5), thus modifying step c. Therefore, the worst element will not necessarily be chosen. It will only have a high chance to be selected, since the probability distribution of Equation (5) favors the lower ranks.

The EO method enhanced with the stochastic selection is known as the τ-EO method [14].

It will be shown that the τ-EO method alone will not yield satisfactory results and for that reason it is complemented by a local search to be described in the next section.

### *3.2 Local neighborhood rule*

In this section perturbations will be considered at the cell level. Before that, a neighborhood structure will be defined for each cell and then a rule will determine state changes of the cells within the neighborhood, utilizing concepts of cellular automata. More specifically, a von Neumann neighborhood [5] will be defined for each cell, as shown below, in figure 4. Obviously, this definition will be suitably adjusted for the boundary cells.

A local search will be conducted following a neighborhood rule, in the sense originally defined by Fotakis [20] and further explored in [10] and [11]. The neighborhood rule aims at improving the configuration on a local scale. For the present problem the local rule will be defined as follows:

Let $f_k = PL_k(w_k)$ as in Equation 2.

Let $N_k$ denote the set of cells constituting the neighborhood of cell k and let $l_k$ denote the cardinality of $N_k$.

Then for each $l = 1,2,..,l_k$ compute

$f_l = PL_l(w_l)$ again according to equation (2).

Set $g_k = f_k + f_l$.

Exchange attributes of cells k and l : $f_{k\leftarrow l} = PL_k(w_l) \quad f_{l\leftarrow k} = PL_l(w_k)$

Set $g_l = f_{k\leftarrow l} + f_{l\leftarrow k}$ and find index $l_{min}$ such that $g_{l_{min}} = \underset{1\le l\le l_k}{Min}\{g_l\}$

If $g_{l_{min}} < g_k$ then exchange mutually the attributes of cells k and $l_{min}$ (figure 4)

Otherwise no exchange takes place.

The above process is repeated for all k = 1,2,.., ab.

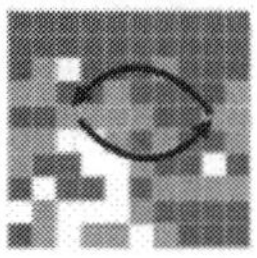

Fig. 3 Perturbation of step d

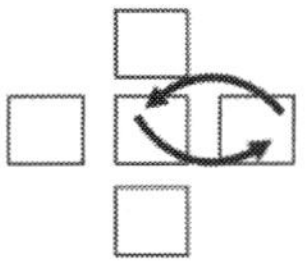

Fig. 4 Neighborhood rule

The neighborhood rule preserves by definition the proportion constraint (equation 4). The rule is applied to every generated configuration of step d in the above pseudo-code.

## 4. Results and discussion

The τ-EO method was applied first without the local neighborhood rule (NR) and then in conjunction with the rule. Three possible land uses were considered and their respective development costs are depicted in figure 2, where a 10x10 grid is shown. In both cases the initial mosaic contained 57 cells for land use 1, 29 cells for land use 2 and 14 cells for land use 3. Figure 5 shows the resulting best configuration after 200 iterations. The objective function value is 316350 and the compactness amounts to 274.

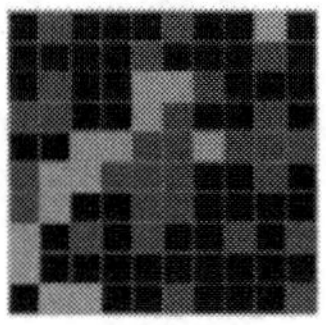

Fig. 5. Application of τ-EO alone.
F = 316350, compactness = 274

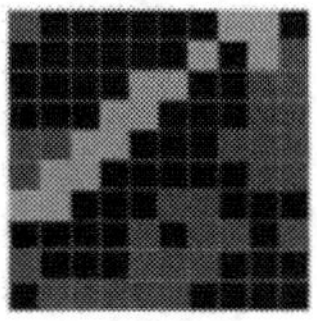

Fig. 6. Application of τ-EO with NR
F = 293500, compactness = 318

Better results were obtained by the conjunctive use of τ-EO with the neighborhood rule. After 200 iterations the objective function gave a value of 293500 with a compactness of 318 (figure 6).

These results compare directly with those obtained by Eldrandaly [13] by the use of expression programming. For exactly the same problem the corresponding results, after 1400 generations, gave a function value of 299850 with a compactness of 192. Therefore, the results of figure 6 point to a superior performance of $\tau$-EO enhanced by the neighborhood rule.

Compactness is measured by counting, for each cell, the number of cells in its neighborhood that have the same attribute. References ([3] and [14]) give precise definitions. In contrast to general practice, compactness is not included here as part of the objective function or as a separate constraint. Significant compactness values come out as emergent results. The same feature has been demonstrated in other related works of the author [6]. Apparently, the neighborhood rule contributes positively to a greater compactness, as shown in those publications.

The issue of compactness is important for spatial optimization because compactness facilitates considerably the planning activity. For this reason, further investigation of compactness from the point of view of the present paper would be beneficial to the study of this aspect.

Also, the method presented here lends itself to the treatment of more involved spatial nonlinear functions. This can be realized, if, in addition to land use planning, the allocation of commodities over the land blocks is considered, along with the respective transportation cost. Finally, some more difficulty may be added, if the objective function is non-separable with respect to individual cell contributions, such as in problems of groundwater resource allocation, as it has been presented elsewhere in the context of genetic algorithms [12].

**References**

[1] Strange N, Meilby H, and Thorsen B J 2002 *Forest Science* **48(3)** 543

[2] Heinonen T and Pukkala T 2007 *Canadian Journal of Forest Research* **37** 2188

[3] Aerts J C J H and Heuvelink G B M 2002 *Int. J. of Geographical Information Science* **16(6)** 571

[4] Stewart T, Janssen M and Herwijnen, M 2004 *Computers and Operations Research* **31** 2293

[5] Sidiropoulos E and Tolikas P 2008 *Applied Mathematical Modelling*, **32(4)** 617

[6] Sidiropoulos E 2011 Cellular automata optimization via evolutionary methods *Cellular Automata* ed. Thomas M. Li (New York: Nova), chapter 4, p. 71

[7] Sidiropoulos E and Fotakis D 2009 *WSEAS Trans. on Environment and Development* **4(5).**

[8] Sidiropoulos E and Fotakis D 2011 Spatial optimization and resource allocation in a cellular automata framework *Cellular Automata: Simplicity behind Complexity* ed A Salcido (Rijeka: INTECH Scientific Publishing), chapter 4, p.67

[9] Fotakis D and Sidiropoulos E 2012 *Annals of Operations Research,* doi 10.1007/s10479-012-1080-y

[10] Fotakis D and Sidiropoulos E 2012. *Applied Mathematics and Computation* **218 (9)** 5168

[11] Sidiropoulos, E. 2012. Multi-objective cellular automata optimization *Cellular Automata, Lecture notes in Computer Science*, vol. 7495, p. 131

[12] Sidiropoulos, E. 2011. Spatial resource allocation via simulated annealing on a cellular automaton background. *World Congress on Engineering and Technology, (Shanghai*, Oct. 28 –Nov. 2, 2011), IEEE Proceedings, vol.2, p.137.

[13] Eldrandaly K 2010 *Applied Soft Computing* **10** 694

[14] Boettcher S and Percus A G 2001. Optimization with Extremal Dynamics Phys. Rev. Lett. **86** 5211

[15] Boettcher, S. and Percus, A.G. (1999) Extremal optimization: Methods derived from co-evolution *Proceedings of the genetic and evolutionary computation conference*, p 825

[16] Chen Yu-wang, Lu Yong-zai and Chen Peng 2007 *Physica A* **385** 115

[17] Sousa, F. L. and Ramos,F.M., 2002. Function optimization using extremal dynamics. 4th International Conference on Inverse problems in Engineering, Rio de Janeiro, Brazil

[18] Chen, Min-rong, Lu, Yong-zai and Yang Gen-ke., Europ J. of Operational Research **188** 637

[19] Vanegas P, Cattrysse D and Van Orshoven J 2010 Compactness in Spatial Decision Support: A Literature Review *Lecture Notes in Computer Science 6016*, p.414 (Springer)

[20] Fotakis, D 2009 Spatial planning for decision making. Integrated land use and water resources management using evolutionary methods, *Doctoral Dissertation*, Aristotle University, Thessaloniki.

# The approximation of the normal distribution by means of chaotic expression

**M Lawnik**

Faculty of Applied Mathematics, Silesian University of Technology, ul. Kaszubska 23, 44-100 Gliwice, Poland

E-Mail: marcin.lawnik@polsl.pl

**Abstract**. The approximation of the normal distribution by means of a chaotic expression is achieved by means of Weierstrass function, where, for a certain set of parameters, the density of the derived recurrence renders good approximation of the bell curve.

## 1. Introduction

Many biological and physical processes are described by sequences of numbers derived from the normal distribution. The numbers have various applications, for example, in simulations. Hence, many numerical generators (Gaussian Random Numbers Generators, GRNG) have been created, comprising the values of the numbers corresponding to the normal distribution.

The generators are based on: transformations [1-4], methods taking advantage of the cumulative density function inversion [5], rejection methods [6-8] and recursive methods [9].

The above mentioned transformations involve the conversion of the values of the numbers derived from the uniform distribution into the values corresponding to the normal distribution. In turn, methods based on inverse distributions match the values from the normal distribution by designating the values of the inverse cumulative distribution function with given argument in the form of a number corresponding to the uniform distribution. Rejection methods, similar to transformations, convert the numbers from the uniform distribution into the normal one, but, in addition, require the fulfillment of specific assumptions concerning the derived values of the numbers, while recurrences use the numbers obtained from the normal distribution and, in the next step, generate new values by means of linear expressions.

Alternative the above mentioned procedures can be divided into exact methods [1-2,6- 9] and approximation methods [3-5]. The exact methods, upon meeting certain criteria, enable the generation of numbers that ideally match the normal distribution. On the other hand, the approximation methods generate the values of numbers that only approximate the normal distribution.

The above mentioned GRNG methods require the numbers from the uniform distribution, with the exception of recursive method. In the case of the transformations based on the central limit theorem [4] at least several values from the uniform distribution are required to derive one numerical value from the normal distribution. On the other hand, the recursive methods use the values derived from the normal distribution to obtain new numbers.

The method discussed in the paper makes it possible to derive the values approximating the normal distribution without the need to take advantage of the uniform or normal distribution.

## 2. Description of the method

The method utilizes Weierstrass function determined by the following equation[10-12]:

$$w(t) = \sum_{i=0}^{\infty} a^i \cos(b^i \pi t), \tag{1}$$

where $a$ and $b$ are the parameters fulfilling the conditions of $0 < a < 1$ and $ab > 1 + 1.5\pi$. On the grounds of the recursive method:

$$t_{k+1} = w(t_k) \tag{2}$$

a sequence of numbers is derived, the density function of which, for parameter $a$ with the values close to 1, approximates the normal density distribution. The mentioned above recursion assumes the following form:

$$t_{k+1} = \sum_{i=1}^{n} a^i \cos\left(\pi b^i t_k\right) \quad . \tag{3}$$

## 3. Conducted analyses

*3.1 Analysis of the dynamic system*

Expression (3) is a chaotic representation. Due to the fact that function $w(t)$ is nowhere differential, Lyapunov's exponent – calculated in accordance with:

$$\lambda = \lim_{n \to \infty} \frac{1}{n} \sum_{k=0}^{n-1} \ln \left| \frac{dw(t)}{dt} \right|_{t=t_k} \tag{4}$$

- has an infinitely big positive value for $n \to \infty$. For example, the dependence of the exponent on parameter $a$ for $n = 100$ terms of the sequence of Weierstrass function is shown in figure 1.

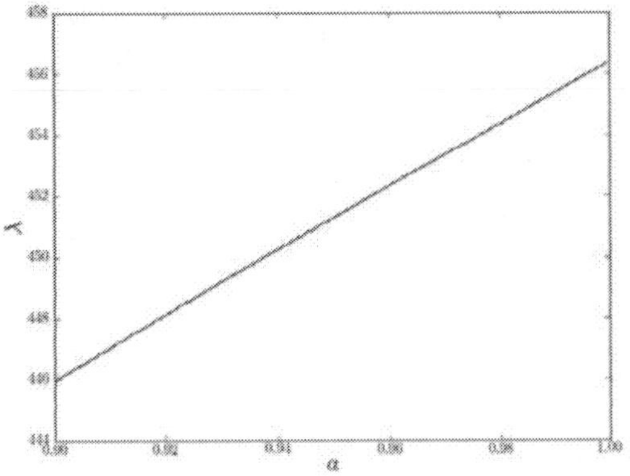


**Figure 1.** Lyapunov's exponent, $b = 100, n = 100$.

*3.2 Statistical analysis*

The density of the normal distribution is designated by the function:

$$\rho(x) = \frac{1}{\sigma\sqrt{2\pi}} \exp\left( \frac{-(x-\mu)^2}{2\sigma^2} \right) \tag{5}$$

where $\mu$ jest is the mean , whereas $\sigma$ is the standard deviation.

The above distribution is characterised, among other features, by the following measure dependencies:

- mean = median
- skewness coefficient = kurtosis = 0

Assuming that parameter $a$ is close to number 1, the mean(figure 2), kurtosis(figure 3), median(figure 4) and skewness coefficient(figure 5) were derived from equation (3). The findings of the analysis indicated that the values of these parameters are convergent with those that correspond to the normal distribution. The standard deviation for the sequence of equation (3) is represented in figure 6. Furthermore, in figure 7 the numerical density function of equation (3) is illustrated, the shape of which approximates the bell curve with the square error equal to 4.223e-05. The influence of the number of the terms of the sequence on equation (3) as well as the influence of the values of parameter $b$ is shown in figures 8 and 9.

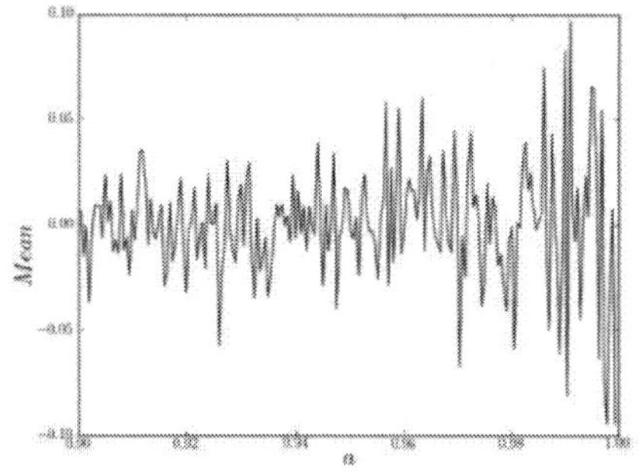


**Figure 2.** Mean, $b = 100, n = 100$.

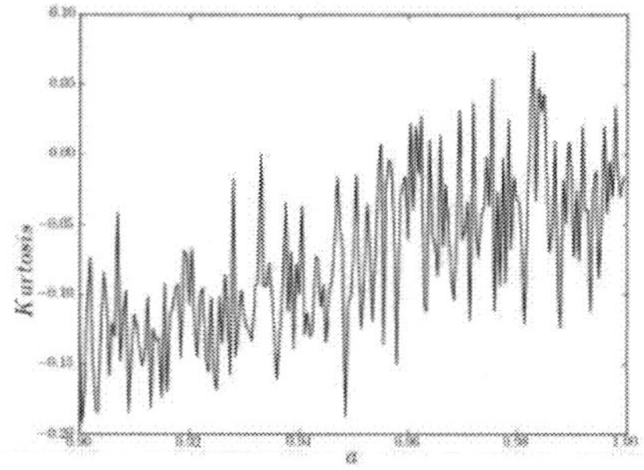


**Figure 3.** Kurtosis, $b = 100, n = 100$.

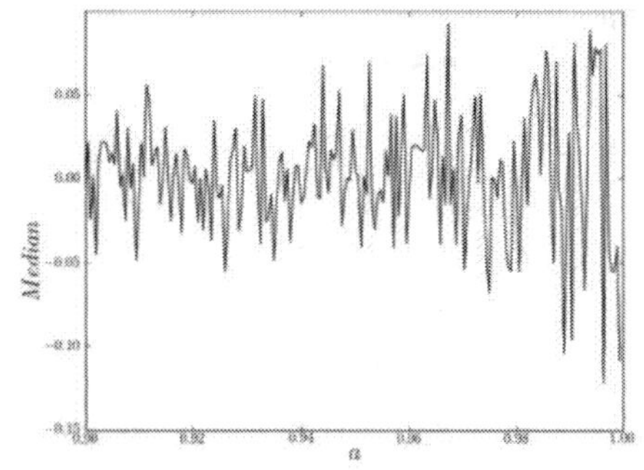


**Figure 4.** Median, $b = 100, n = 100$.

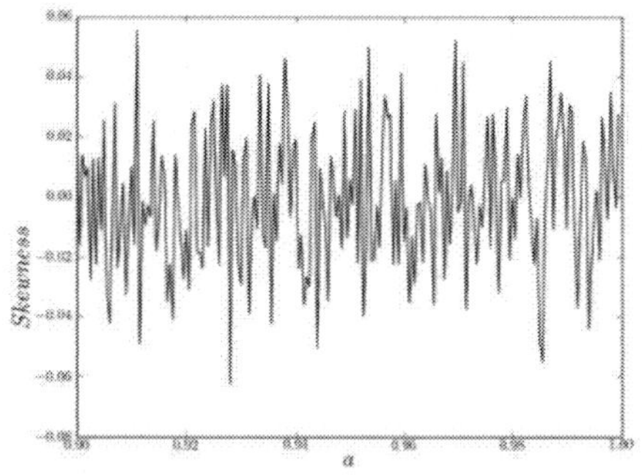


**Figure 5.** Skewness , $b = 100, n = 100$.

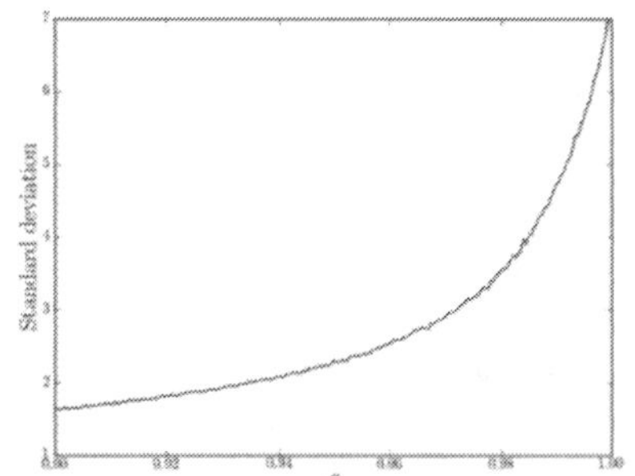


**Figure 6.** Standard deviation, $b = 100$, $n = 100$.

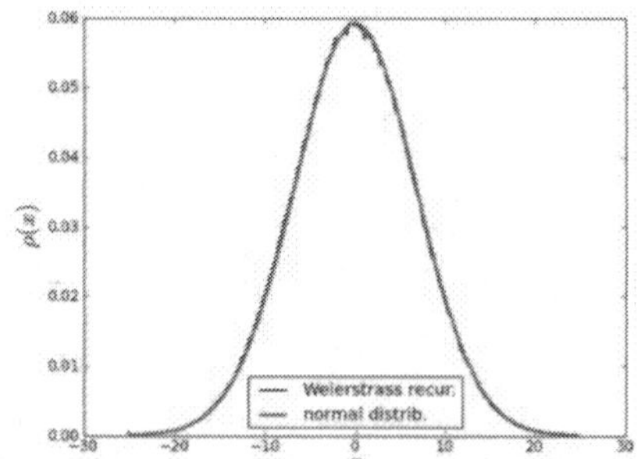


**Figure 7.** Exact and approximated density, $a = 0.999, b = 100, n = 100$.

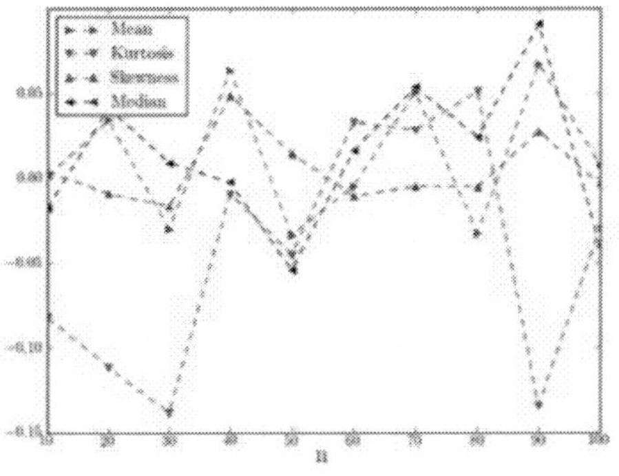


**Figure 8.** Different number of the terms of the sequence of (3), $a = 0.999, b = 100$.

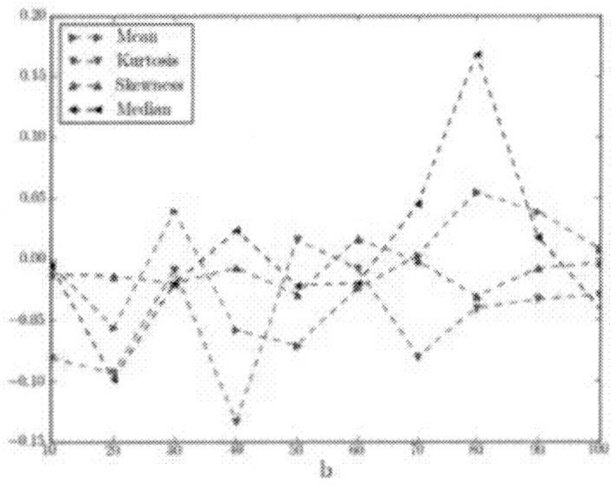


**Figure 9.** Different values of parameter $b, a = 0.999, n = 100$.

## 4. Conclusions

The method discussed in the paper enables the approximation of the normal distribution by means of a chaotic representation based on Weierstrass function. For the values of the parameter $a$ close to 1 in equation (3) the subsequent values of the dependence are derived from the normal distribution. Furthermore, the created recurrence does not require the numbers from the uniform distribution or normal distribution to generate new values.

## References

[1] Box G E P, MullerM E 1958 A note on the generation of random normal deviates *Annals Math. Stat.* **29** 610–611
[2] Marsaglia G, Tsang W W 1998 The Monty Python method for generating random variables *ACM Trans. Math. Softw.* **24** 341–350
[3] Kabal P 2000 *Generating Gaussian pseudo-random deviates* (Tech. Rep., Department of Electrical and Computer Engineering, McGill University)
[4] Muller M E 1959 A comparison of methods for generating normal deviates on digital computers *J.ACM* **6** 376–383
[5] Wichura M J 1988 Algorithm AS 241: The percentage points of the normal distribution *Appl.Statis.* **37** 477–484
[6] Brent R P 1974 Algorithm 488: A Gaussian pseudo-random number generator *Comm. ACM* **17** 704–706
[7] Bell J R 1968 Algorithm 334: Normal random deviates *Comm. ACM* **11** 498
[8] Marsaglia G, Bray T A 1964 A convenient method for generating normal variables *SIAM Rev* **6** 260–264
[9] Wallace C S 1996 Fast pseudorandom generators for normal and exponential variates *ACM Trans. Math.* Softw. **22** 119–127
[10] Weierstrass K 1886 *Abhandlungen aus der Functionenlehre* (Berlin: J. Springer) p 97
[11] Kawamoto S, Tsubata T 1997 The Weierstrass function of chaos map with exact solution *Journal of The Physical Society of Japan* **66** 2209 –10
[12] Berezowski M 2010 Phase trajectories of a certain mechanical system *Far East Journal of Dynamical Systems* **13** 85-96

# Multiparametric Data Processing on Fusion Device Stellarator L-2M

**D V Malakhov, N N Skvortsova, V D Borzosekov, E M Konchekov, B A Tedtoev and D G Vasilkov**

Department of Plasma Physics, A.M. Prokhorov General Physics Institute Russian Academy of Sciences, Vavilova Street 38, Moscow, 119991 Russia

E-mail: malakhov@fpl.gpi.ru

**Abstract**. In studying the high-temperature plasma physics, the researcher is faced with the need to process large amounts of semi-structured data. Dozens of multi-channel diagnostics, the results of which are multi-million ensembles of time samples of plasma signals, are usually installed at the same time in thermonuclear devices. These samples contain information necessary to understand the mechanisms of relation between plasma macroparameters (plasma density and temperature, plasma current, magnetic field, auxiliary heating power, etc) and plasma microparameters determined by turbulence (energy, spectral composition, amplitude density distribution, etc). The methods of analysis of plasma signals based on the processing of a single time realization of turbulence cannot be used as a basis for comparison with macroparameters and identification of new patterns of plasma containment in a magnetic trap. The main objective of this work is to obtain information about the relationship between the macro- and microparameters of the plasma based on the structuring of the acquired data and its multi-parameter processing using the modern software and hardware.

## 1. Introduction

The increasing amounts of information produced by L-2M stellarator make it necessary to automate data processing and systematization of the results. Extensive experience in the processing and analyzing the signals from different plasma diagnostics was accumulated during the existence of the device. The distributed database of the stellarator includes experimental data on more than 10,000 discharges. The main objective of this work was to develop an algorithm for searching and finding new reliable information on the relationship between the macroparameters of the plasma and the microparameters of the turbulence on the L-2M stellarator. The developed algorithm allows one to combine selecting a number of discharges with certain common macroparameters from the database with their processing using the developed software for evaluation of stable (robust) spectral characteristics of fluctuations for further physical interpretation.

The article is devoted to the evaluation of such stable characteristics of the short- and long-wavelength fluctuations of the plasma density in the L-2M stellarator. The considerable amount of experimental data obtained on the fluctuations allows one to compare the turbulence microparameters with the plasma macroparameters necessary for the study of the process of plasma confinement in a toroidal installation.

The aim of this work is to automate data processing on the L-2M stellarator, to obtain robust (stable) Fourier spectra over an ensemble of realizations with the same macroparameters (density, heat capacity, temperature, etc.), to obtain new results by averaging the spectra (selection over calculation parameters), and to build 3D spectra, their projections and multimedia images of their time evolution. The ultimate goal of the work is to develop a hardware-software system of multivariable data analysis on toroidal fusion devices.

## 2. Experimental Setup

Investigation of the relationship between the macro- and micro parameters of the plasma was carried out the L-2M stellarator. The L-2M device is an l = 2 stellarator whose parameters are described in detail in [1]. The density of the edge plasma at the radius r/a = 0.9 (where a is the separatrix radius) is $n = 1...2 \times 10^{12}$ $cm^{-3}$. The plasma is produced and heated by a 75-GHz gyrotron under the electron cyclotron resonance (ECR) conditions at the second harmonic of the electron gyrofrequency.

Turbulence microparameters were measured [2] by collective scattering diagnostics and Doppler reflectometry [3]. For each diagnostics, Table 1 provides data on its measuring region within the plasma column, the transverse wave components of the fluctuations and the obtained turbulence characteristics.

**Table 1.** Applied diagnostics, their measuring regions, wave numbers and obtained turbulence characteristics.

| Diagnostics | Wave number, sm-1 | Measuring region, r/a | Obtained characteristics |
|---|---|---|---|
| Doppler reflectometry | 1–2 | 0,8–0,9 | 1. Complex Fourier spectra and their projections<br>2. Three-dimensional spectra<br>3. Doppler velocity<br>4. Stochastic analysis |
| Collective scattering at the second harmonic of the gyrotron frequency | 24–44 | 0,3–0,4<br>0,5–0,6 | 1. Three-dimensional Fourier spectra and their projections<br>2. Noise intensity<br>3. Autocorrelation functions<br>4. Stochastic analysis |
| Small-angle scattering of the gyrotron radiation | 1–2 | 0–0,8 | 1. Fourier spectra<br>2. Noise intensity<br>3. Autocorrelation functions |

The data acquisition system of the L-2M stellarator recording the data from all its diagnostics as well as the device macroparameters and that of its gyrotron complex consists of subsystems of low- and high-speed AD converters of different standards and of distributed data stores. The L-2M database includes tens of thousands of pulses in the different operation modes of the device, which allows one to analyze the microparameters of the turbulence depending on the power of microwave heating in different modes of plasma confinement.

## 3. Data Processing

The sorted data (time samplings of the fluctuation amplitudes) from the backscattering diagnostics and scattering at the second harmonic of the gyrotron, the small-angle scattering diagnostics, and Doppler reflectometry were used in the multi-parameter complex analysis of the microparameters of the turbulence [4-8].

The sorting was based on the tables of the plasma macroparameters (plasma density, energy content, electron-cyclotron heating power). The sorting and grouping of pulses was carried out in automatic mode, so as to observe the conditions of the steady-state of the cloud of plasma macroparameters of interest (with possible errors). The sorted data was processed using the previously refined and patented algorithms of spectral analysis of fluctuations [9]. The processing resulted in obtaining spectral characteristics which are stable (robust) over an ensemble of discharges and estimating their errors.

Figure 2 shows averaged fluctuation spectra of the plasma in the stationary phase of three discharges nos. G.38515–G.38517 in sequential time windows 52–54 ms (a), 54–56 ms (b), 56–58ms (c), and 58–60 ms (d). The discharges represent the same regime with microwave heating power of 250 kW and mean plasma density of 1.8–1.9 x $10^{13}$ $cm^{-3}$. Note that the difference between full widths at half-maximum (FWHMs) and the intensities of the spectra in the appropriate windows was of 10–20%.

Estimates of the spectral components in these spectra were carried out a special method combining both the spectral and probabilistic methods of analysis. Methodology of estimation of the spectral components for the processes of structural plasma turbulence (compound Cox process) was suggested. The method is based on obtaining additional information about the number of turbulent stochastic processes though decomposition of histograms on Gaussian local-scale mixture. The decomposition was performed using the expectation-maximization (EM) algorithm and its modifications such as the stochastic EM algorithm, the median EM algorithm, and the median stochastic expectation-maximization (SEM) algorithm [11]. Figure 2 shows a smoothed robust spectrum with three allocated basic components (bands) which can be used for later comparison with the plasma macroparameters.

Such spectra can be used to determine the average energy of fluctuations over the ensemble of plasma discharges in a device with the same macroparameters and to compare the energy of the spectra in different operation regimes.

Table 2 shows the average values of the energy of long-wavelength fluctuations at different values of the transverse magnetic field for the same input microwave power of EC heating. These values indicate that the level of long-wavelength fluctuations increases with increase of the transverse field while the energy lifetime of the particles and the energy of the plasma column remain the same. The energy lifetime is one of the most important of macroparameters in fusion devices.

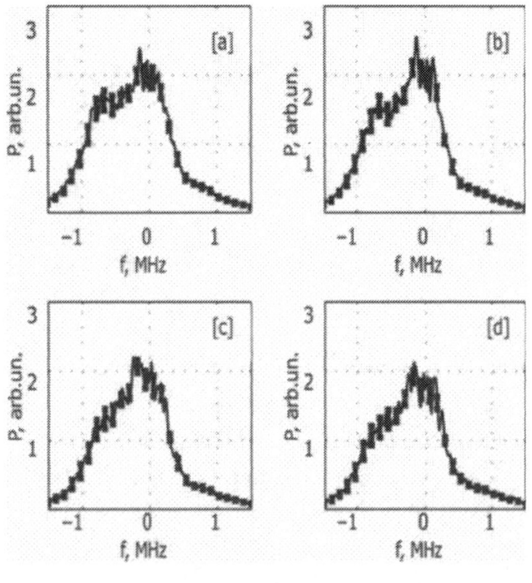


**Figure 2.** Averaged full complex fluctuation spectra in the stationary phase of three discharges in sequential time windows 52-54 ms

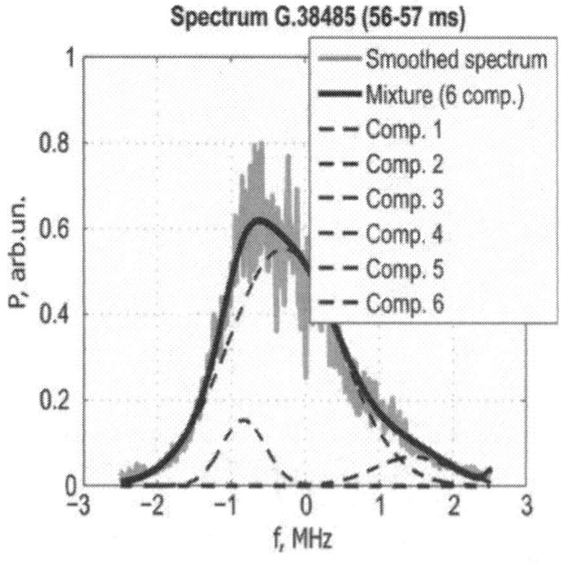


**Figure 2.** Robust turbulent spectrum of Doppler reflectometry diagnostics decomposed into 6 components. The solid line represents the average

(a), 54-56 ms (b), 56-58 ms (c), and 58-60 ms (d). The signals were obtained with the Doppler reflectometry diagnostics.

spectrum, the dotted lines show the three main spectral components.

**Table 2.** Intensity of long-wavelength fluctuations depending on the transverse magnetic field Bv

| Confinement mode | Transverse magnetic field Bv = 0 Gs, P=200 kW | Transverse magnetic field Bv = -20 Gs, P=200 kW |
|---|---|---|
| Fluctuation energy, arb. units | 5,3 ± 0,8 | 7,0±1,2 |
| Energy lifetime, ms | 2,6 ± 0,3 | 2,6±0,1 |
| Number of discharges in the ensemble with the same macro-parameters | **7 discharges** | **19 discharges** |

## 4. Conclusion

Data on the energy of the plasma fluctuations and the shape of stable spectra was obtained which allowed one to compare two microparameters of the turbulence (its fluctuation energy and energy lifetime) with the plasma confinement time in the L-2M stellarator. Analysis showed that in the L-2M stellarator, the Fourier spectra are robust and the energy of the long-wavelength fluctuations does not correlate with the plasma confinement time when the EC heating power increases [5] while there exists a correlation between the confinement time and the short-wavelength fluctuations [1].

The stable spectra of Doppler reflectometry diagnostic allowed one to determine the parameters required for optimizing the diagnosis and served as the basis for developing a new method for probabilistic analysis [10]. The analysis of the evolution of robust spectra of the diagnostics of the gyrotron radiation scattering revealed the existence of spectral bands in all recorded spatial ranges [4,8].

This work is the first step in the development of algorithms for the comparative analysis of the plasma micro- and macroparameters in toroidal fusion devices. The prototypes of the algorithms were written in MATLAB using the GPU NVIDIA Tesla K20 accelerator. In the future, a transition is planned from prototypes (m-language) toward a complete and functional application as well as obtaining patent certificates for the developed hardware-software complex for data analysis.

This work was supported in part by the Ministry of Education and Science of the Russian Federation (agreement no. 8392, "Structure creation and stochastization in plasma"), by the Russian Federation Presidential Grant (no. MK-5607.2013.2), and by Russian Foundation for Basic Research, (project no. 12-07-31267mol_a).

### References

[1] Batanov G M, Bening V E, Korolev V Yu, et al. 2003 *Pis'ma Zh. Éksp. Teor. Fiz.* **78** 974 [2003 *JETP Lett.* **78**, 502]

[2] Skvortsova N N, Akulina D K, Batanov G M, et al. 2010 *Plasma Phys. Control. Fusion* **52** 055008 11pp.

[3] Pshenichnikov A A, Kolik L V, Malykh N I, et al. 2005 *Plasma Physics Reports*, **Vol. 31**, No. 7, pp. 554–561.

[4] Skvortsova N.N., Akulina D. K., Batanov G.M., et al. Pl.Phys. Cont. Fusion 52(2010) 055008.

[5] Batanov G.M., Borzosekov V D, Kolik L V, Malakhov D V, et al. 2011 *VANT* **2** 70-75.

[6] Chernov N A, Pshenichnikov A A, Malakhov D V, Skvortsova N N 2012 *XXXIX Zvenigorod international conference on plasma physics and controlled fusion* pp M68.

[7] Malakhov D V, Borzosekov V D, Vasilkov D G 2013 *XL Zvenigorod international conference on plasma physics and controlled fusion* pp.36.

[8] Batanov G.M, Borzosekov V D, Kovriznykh L.M, et al. 2013 *Plasma Physics* Preprint .

[9] Gorshenin A K, Korolev V Yu, Malakhov D V, Skvortsova N N 2012 *Patent certificate of state registration of computer programs* 2012610645, 2012610646, 2012610923, 2011618892.

[10] Gorshenin A K, Korolev V Yu, Malakhov D V, Skvortsova N N. 2012 *Computer Research and Modeling*, , vol. 4, no. 4, pp. 793–802 (Russian).
[11] Gorshenin A K, Korolev V Yu, Malakhov D V, Skvortsova N N. 2011 *Math. Modeling* **Vol.23**. № 4. pp 83–89.

# A computational tool to design and generate crystal structures

**R C Ferreira**[1], **M B Vieira**[2] **and S O Dantas**[3] **and M Lobosco**[2]

[1] Instituto Federal Baiano, Campus Bom Jesus da Lapa, Brazil
[2] Department of Computer Science, UFJF, Juiz de Fora, Brazil
[3] Department of Physics, UFJF, Juiz de Fora, Brazil

E-mail: marcelo.lobosco@ice.ufjf.br

**Abstract.** The evolution of computers, more specifically regarding the increased storage and data processing capacity, allowed the construction of computational tools for the simulation of physical and chemical phenomena. Thus, practical experiments are being replaced, in some cases, by computational ones. In this context, we can highlight models used to simulate different phenomena on atomic scale. The construction of these simulators requires, by developers, the study and definition of accurate and reliable models. This complexity is often reflected in the construction of complex simulators, which simulate a limited group of structures. Such structures are sometimes expressed in a fixed manner using a limited set of geometric shapes. This work proposes a computational tool that aims to generate a set of crystal structures. The proposed tool consists of a) a programming language, which is used to describe the structures using for this purpose their characteristic functions and CSG (Constructive Solid Geometry) operators, and b) a compiler/interpreter that examines the source code written in the proposed language, and generates the objects accordingly. This tool enables the generation of an unrestricted number of structures, which can be incorporated in simulators such as the Monte Carlo Spin Engine, developed by our group at UFJF.

## 1. Introduction

The study of physical and chemical phenomena are always related to experimentation and observation. However, sometimes performing such experiments could not be trivial and/or be executed without major costs involved. In other situations, actual experimentation is impossible to be made, for example, in some situations involving the nanotechnology area. Thus, alternatives must be found. In this scenario, the use of software that simulates the phenomena under study can be very useful.

Simulators are built from using physical, chemical, mathematical and computational models, and allow the study of phenomena in a more practical way, since several hypotheses can be tested by the researcher by simply changing the parameters and constants used in the simulation. Also, detailed information about the experiments can be harvested and latter used for a better understanding of the phenomena been studied.

In this context, many simulators are built and deployed allowing simulations to be performed using a limited number of structures peculiar to a particular field of study, often restricting their use. The Monte Carlo Spin Engine (MCSE) [1], for example, allows the simulation of the interaction between spins organized in a three-dimensional way and formed by atoms with magnetic properties. A set of basic shapes (spheres, cylinders and cubes), composed of equally spaced atoms, can be chosen for simulation. So, it precludes the simulation of a broader set of structures. The present work deals with this limitation, presenting a new programming language that allows the user to generate crystal structure using for this purpose implicit objects. However, although the language was conceived to be used with the MCSE, it should be stressed that it generates as output a generic structure that can be used by any simulator.

This paper is organized as follows. Section 2 presents the new language. Section 3 presents some related works while Section 4 concludes the work.

## 2. The new language

The new language has been proposed with the following objectives in mind: a) to allow users to describe a Bravais lattice; b) to allow users to define a solid based on the composition of other solids previously defined; c) to perform operations that transform a solid previously defined; d) after the compilation of its description, to allow the generation of a crystal structure. The proposed language is not ambiguous, that is, if a code was not modified, a new compilation of the same code will generate the same structure again.

Using the language, four steps must be followed to create a crystal structure:

- Creation of properties. This is the part of the code where the programmer defines properties for atoms or molecules that form the structure.
- Definition of characteristics. In this part of the code the programmer defines to which Bravais lattice the structure belongs to, which vectors form the lattice and which are the positions of each atom or molecule in the base.
- Description of the solid. The programmer defines the solids or compositions that will be used. A solid is described by an expression that defines their points. A composition is a piece of code where solids are aggregated.
- Construction of a structure. In this step the programmer defines the spatial region and the solids that will be inserted on it.

The language allows the definition of variables of two types: *Number* or *Solid*. A variable of type *Number* can store any value. A *Solid* variable is used to store a solid. The language provides logical, arithmetic and relational operators (such as ==, <>, <, >, &&, ||, +, −,...) and pre-defined functions. These functions are the typical mathematical ones (such as *sin*, *cos*, *tan*, *exp*, ...), CSG operators to manipulate a solid (*union*, *intersection*, *difference* and *not* - negation), as well as functions to rotate, translate, scale and shear the solid. The *while* control structure is also available to be used by the programmer.

### 2.1. Example

Due to the lack of space to present in details the language and their functions, an illustrative example was choosen to show how complex structures can be easily constructed with the use of the proposed language. The Listing 1 presents the code that defines a spherical shell perforated by eight spheres. This structure is filled with face-centered cubic cells. Figure 1 presents the result obtained from its compilation and visualization using our compiler. A point cloud is them generated by our tool. This point cloud can be incorporated in any simulator. Figure 2 presents its incorporation in the MCSE.

**Listing 1.** Code that defines a spherical shell perforated by eight spheres.

```
cubic (face, (3.56, 0, 0), (0, 3.56, 0), (0, 0, 3.56))
base (C1 , (0, 0, 0))
base (C2 , (0.89, 0.89, 0.89))
```

```

solid MySphere (a, b, c ,r)
  pow ((x-a), 2) + pow((y-b), 2) + pow((z-c), 2) <= pow(r, 2);

composite MySphericalShell (a, b, c, r1, r2)
begin
  insert (difference(MySphere(a, b, c, r2), MySphere(a, b, c, r1)));
end

composite PerforatedShell (a ,b ,c , r1, r2)
begin
  myShell = MySphericalShell (a, b, c, r1, r2) ;
  myShell = difference (myShell , MySphere(a+r2, b+r2, c+r2, r2));
  myShell = difference (myShell , MySphere(a+r2, b+r2, c-r2, r2));
  myShell = difference (myShell , MySphere(a+r2, b-r2, c+r2, r2));
  myShell = difference (myShell , MySphere(a+r2, b-r2, c-r2, r2));
  myShell = difference (myShell , MySphere(a-r2, b+r2, c+r2, r2));
  myShell = difference (myShell , MySphere(a-r2, b+r2, c-r2, r2));
  myShell = difference (myShell , MySphere(a-r2, b-r2, c+r2, r2));
  myShell = difference (myShell , MySphere(a-r2, b-r2, c-r2, r2));
  insert (myShell);
end

lattice (-30, -30, -30, 30 , 30 ,30)

begin
  insert (PerforatedShell (0, 0, 0, 20, 30), override);
end
```

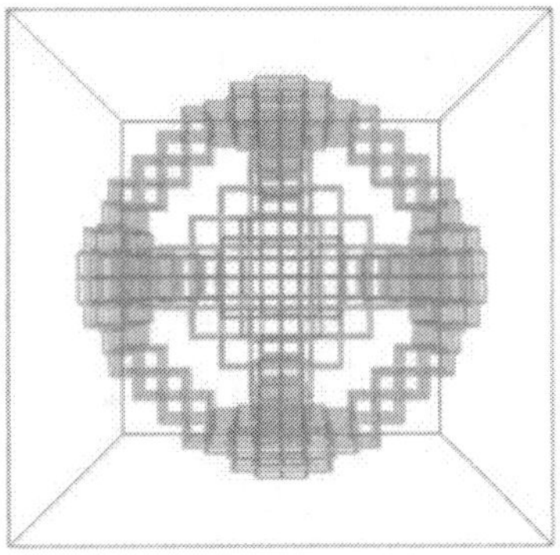

**Figure 1.** The visualization of the structure define by Listing 1.

## 3. Related Work

For the best of our knowledge, this is the first work that uses a programming language to describe crystal structures. However, lots of software [2, 3, 4, 5] are available to construct and visualize them using a graphical interface. As a general rule, in these software the user informs the Bravais lattice to be used, the atoms that compose the base and some parameters

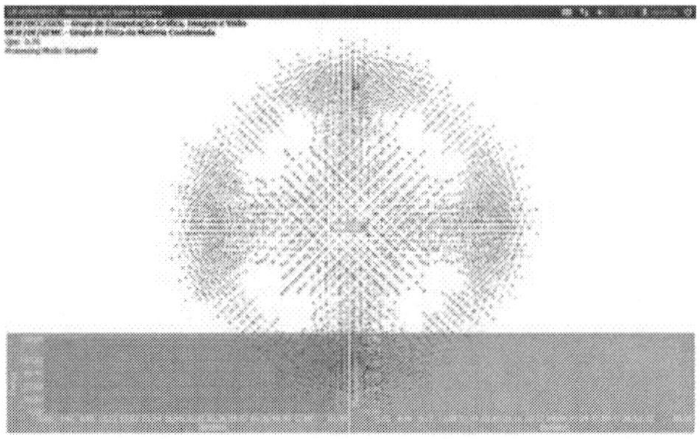

**Figure 2.** The structure generated by the language was read by the MCSE.

of the lattice. Some elements related to the crystallography, such as a crystallographic point group and the Miller indices, are present in some tools, as well as the definition of defects and impurities in the arrangements of the structure. Also some tools can simulate the X-ray diffraction phenomenon. The first version of the proposed language do not address these issues, although they can be incorporated in future versions. However, these simulators present some limitations. For example, they do not allow the user to define a spatial region where the structure must be inserted, as our proposed language does. Also, our proposed language allows the definition of monocrystalline and polycrystalline structures formed by equivalent unitary cells with distinct orientations while other tools only allow the definition of monocrystalline structures formed by only one unitary cell. Indeed, in these tools the generation of structures is based on the expansion of a unitary cell, while in our tool the generation is done filling a spatial region, defined by the user into his/her code, with unitary cells. Finally, the proposed tool allows the construction of structures of distinct shapes as well as it allows the definition of distinct properties to the elements that constitute then (atoms, ions or molecules). This characteristic was not found in other software.

## 4. Conclusion

This work presented a new computational tool to design and generate crystal structures. The tool consists of a programming language and a compiler that generates the objects accordingly. With the use of the proposed language, complex structures can be easily constructed using, for exemple, the composition and/or the transformation of objects previously defined.


## Acknowledgements

The financial support by FAPEMIG, CNPq, CAPES, and UFJF is greatly acknowledged.


## References


[1] Campos A M, Peçanha J P, Pampanelli P, de Almeida R B, Lobosco M, Vieira M B and De O Dantas S 2011 *Proceedings of the 2011 international conference on Computational science and its applications - Volume Part III* ICCSA'11 (Springer-Verlag) pp 654–667

[2] Shape software, atoms, on-line URL `http://www.shapesoftware.com`

[3] Ozawa T C and Kang S J 2004 *Journal of Applied Crystallography* **37** 679

[4] Crystalmaker software, crystalmaker, on-line URL `http://www.crystalmaker.com`

[5] Finger L W, Kroeker M and Toby B H 2007 *Journal of Applied Crystallography* **40** 188–192

# Use of Multiple GPUs to Speedup the Execution of a Three-Dimensional Computational Model of the Innate Immune System

**M P Xavier**[1], **T M do Nascimento**[2], **R W dos Santos**[3], **M Lobosco**[4]

[1,2,3,4]FISIOCOMP Lab, Federal University of Juiz de Fora, Brazil

E-mail: [1]micael.peters@ice.ufjf.br, [2]tiago.nascimento@uab.ufjf.br, [3]rwdsantos@yahoo.com, [4]marcelo.lobosco@ice.ufjf.br

**Abstract.** The development of computational systems that mimics the physiological response of organs or even the entire body is a complex task. One of the issues that makes this task extremely complex is the huge computational resources needed to execute the simulations. For this reason, the use of parallel computing is mandatory. In this work, we focus on the simulation of temporal and spatial behaviour of some human innate immune system cells and molecules in a small three-dimensional section of a tissue. To perform this simulation, we use multiple Graphics Processing Units (GPUs) in a shared-memory environment. Despite of high initialization and communication costs imposed by the use of GPUs, the techniques used to implement the HIS simulator have shown to be very effective to achieve this purpose.

## 1. Introduction

Nowadays, the scientific community makes large use of computational tools to better understand the behaviour of distinct complex systems, such as those used to mimic the functioning of the immune system. The development of such systems are essential, for example, to better understand the role of each cell and molecule in an immune response. However, such systems can demand a large amount of computational resources, in particular computation time. The use of parallel algorithms is mandatory in such cases.

This work describes the use of multiple Graphics Processing Units (GPUs), in a shared-memory environment, to reduce the computation time required to simulate the temporal and spatial dynamics of some human immune system (HIS) cells and molecules in a small three-dimensional section of tissue when this tissue is exposed to pathogens. The sequential version of the code can run for days, depending on the parameters used in simulations, such as the size of the tissue. A first attempt to reduce this huge computation time employed a single GPU [1] to perform the parallel processing. Although the parallel version achieved a respectable speedup of about 72 times over its sequential counterpart[1], a better speedup is required to deal with large tissues or more detailed simulations.

This work is organized as follows. Section 2 describes the versions of code developed in order to execute the innate HIS simulations on a multi-GPU platform. Computational results obtained experimentally are presented in Section 3. Finally, our conclusions and plans for future works are presented in the last section.

## 2. Multiple GPUs

The first version of the code used to simulate the temporal and spatial behaviour of some cells and molecules of the innate HIS in a small three-dimensional section of tissue [1] uses CUDA in a single GPU to reduce its execution time. This single-GPU version of the code is extended in this work to use multiple GPUs that can be available in a single machine in order to speedup even more the execution time. To manage these multiple GPUs, PThreads API [2] is used, so each host (or CPU) thread is responsible for invoking a CUDA kernel in a particular CUDA device. The use of multiple host threads is necessary to reduce the imbalance caused when a single thread launches all kernels in distinct GPUs. If a large number of GPUs are available, when the single host thread finishes to launch the last kernel in the last GPU, probably the first kernel launched in the first GPU has advanced a lot in its work, or even finished it. Since kernels are executed independently by each GPU, it is necessary to synchronize host threads at each time-step in order to compute populations correctly. This is done using the *cudaStreamSynchronize* directive at the end of each time iteration, so the main host thread synchronize its execution with each of the CUDA kernels launched before starting a new time-step.

To solve the system of Partial Differential Equations (PDEs) that describes the behaviour of the innate HIS cell and molecules [1], the finite difference method was employed for the spatial discretization and the explicit Euler method for the time evolution. The discretization of the chemotaxis term uses the First-Order Upwind scheme[3] that discretize the hyperbolic PDEs through the use of differences with bias in the direction given by the signal of the characteristics speeds. The upwind scheme uses an adaptive or solution-sensitive stencil to numerically simulate more precisely the direction of information propagation. With multiple GPUs, the discretized space is divided among devices, so each one will operate on a specific slice of the original space such that the whole tissue is processed by the group of GPUs. Splitting was done by dividing the $x$ dimension of a ($N_x$, $N_y$, $N_z$) mesh that describes the tissue by the number of GPUs, $N_g$, remaining a ($\frac{(N_x+N_g-1)}{N_g}$, $N_y$, $N_z$) mesh to be calculated by each device. The mesh is organized contiguously in the device memory as an unidimensional vector and the access to the points in the tissue was done linearly.

In CUDA, the execution configuration hugely impacts the performance of the application. In this work we choose a fixed block size, 128 threads, based on the memory demands of each thread. Then, a function was created to automatically generate the grid size. This function calculates the grid size taking into account that a thread computes a single point, and that the grid is unidimensional. Also, a buffer was implemented to avoid race condition among CUDA threads. Its role is quite simple: two values at times *t-1* and $t$ are stored for each point ($x$, $y$, $z$) of a given population of cells. The value at time $t$ is accessed only by the thread that is producing it, while the other one, *t-1*, is accessed by threads in neighbourhood that needs to read it. Thus, a thread at time $t$ only gets access to data produced by its neighbours at time *t-1*. These two buffer entries change their meaning at each time step to avoid data copy.

To correctly compute each point, each CUDA thread has to access its neighbours data, some of which located at distinct GPUs due to data splitting among GPUs. These data, called boundaries, are necessary simultaneously to execute the computation in two distinct GPUs. In order to deal with this issue, three distinct approaches were implemented.

In the first version, boundary data is copied between two GPUs using for this purpose the host memory space. The data copy process is divided into two steps: in the first one, boundary data is copied from the source GPU to the host memory. The final step copies data from the host memory to the destination GPU.

Although the first version of the code is quite simple, the data copy between GPU and CPU, and vice-versa, may degrade performance. To reduce the cost of memory copy operations, a second version uses the concept of unified address space among GPUs. This concept is implemented using the UVA (Unified Virtual Addressing) functions [4], so one GPU can access directly the memory positions located on a distinct one. To allow the access of a distinct GPU to its local memory space, it's necessary to call first the function *cudaEnablePeerAccess*. The access granted by a call to this function is unidirectional, so a separate symmetric call to this function is required. However, it is not possible to grant access simultaneously to two or more GPUs to a single memory space. For example, if one GPU has to access the memory space of two adjacent GPUs in a single iteration, it has to enable a peer access with both GPUs, one at a time and at each time-step, before accessing their memory positions, which can impact the performance negatively.

To solve the issues mentioned with the second version, a third version was implemented. This new version calls *cudaEnablePeerAccess* only once, during the initialization of the code, and then data can be copied among GPUs. The main difference between this version and the previous one is that in this version data must be copied explicitly, using *cudaMemcpy*, while in the previous version data is accessed directly, since the memory space is unified. Also, when compared to the first version, this version eliminates the use of the host memory space as a temporary buffer.

## 3. Experimental Results

This section presents the speedups obtained by the three distinct versions of our multi-GPU code. They were obtained in a four AMD 6272 Linux system, with 128GB of main memory. This machine has four Tesla M2050 GPUs, each with 512 CUDA cores and 5.2 GB of global memory. Linux 3.9.2, CUDA driver version 5.0, *nvcc* release 5.0 and *gcc* version 4.7.2 were used to run and compile all versions of the code.

In order to evaluate the performance gains obtained by the multi-GPU version, experiments using four distinct mesh sizes: $50 \times 50 \times 50$, $100 \times 100 \times 100$, $150 \times 150 \times 150$ and $200 \times 200 \times 200$ points. The mesh discretization generates about 1,000 points for each 1 $mm^3$ of tissue. One day of infection takes about one million of interactions to be computed. Since the execution time of each interaction is extremely regular, and the objective of this paper is to evaluate the techniques, nor the biological results, in this work we report the result for 10,000 time steps. The execution times obtained by all versions of the code were measured 5 times and the standard deviation was lower than 2.2%. At each execution, Linux *time* application was used to measure the time spent in program execution.

The relative speedups to the single-GPU version of the code are presented in Table 1, using the Equation 1 to compute the speedups:

$$S_g = \frac{t_1}{t_g} \quad (1)$$

where $t_1$ is the single-GPU execution time and $t_g$ is multi-GPU execution time with 4 GPUs. Values greater than 1.0 represent a speedup, while values less than 1.0 represent a slowdown.

Table 1 presents the obtained results. First, the second version of the code always slowdowns, regardless the size of the mesh. Our profile shows that the high cost imposed to call the function *cudaEnablePeerAccess* is responsible for this slowdown. The third version of code obtained the best results in all scenarios. Finally, as the mesh size increases, the relative speedup of the third version of code becomes better. For the largest mesh size, the speedup obtained is about 82.5% of the maximum speedup that could be obtained, a linear speedup of 4 using four GPUs.

## 4. Conclusion

This work presented an implementation of an innate HIS simulator on a multi-GPU platform. Three distinct versions of code were developed with this purpose. Surprisingly, experimental results have shown that the best results were obtained by the version that explicitly copy data between GPUs. This occurs due to the restrictions imposed by CUDA shared memory primitives

**Table 1.** Speedups obtained by the multi-GPU versions over the single-GPU one. Average execution times for 10,000 time steps are expressed in seconds.

| Mesh | Version | Runtime average (s) | Standard deviation | Speedup |
|---|---|---|---|---|
| $50 \times 50 \times 50$ | single-GPU | 30.08 | 0.5 % | - |
| | multi-GPU v.1 | 29.35 | 1.2 % | 1.0 |
| | multi-GPU v.2 | 79.51 | 0.3 % | 0.4 |
| | multi-GPU v.3 | 22.60 | 0.5 % | 1.3 |
| $100 \times 100 \times 100$ | single-GPU | 158.36 | 0.1 % | - |
| | multi-GPU v.1 | 93.35 | 2.2 % | 1.7 |
| | multi-GPU v.2 | 331.11 | 0.2 % | 0.5 |
| | multi-GPU v.3 | 66.01 | 0.5 % | 2.4 |
| $150 \times 150 \times 150$ | single-GPU | 512.45 | 0.0 % | - |
| | multi-GPU v.1 | 230.34 | 0.5 % | 2.2 |
| | multi-GPU v.2 | 843.12 | 0.2 % | 0.6 |
| | multi-GPU v.3 | 183.07 | 0.8 % | 2.8 |
| $200 \times 200 \times 200$ | single-GPU | 1224.86 | 0.0 % | - |
| | multi-GPU v.1 | 440.20 | 0.3 % | 2.8 |
| | multi-GPU v.2 | 1698.64 | 0.3 % | 0.7 |
| | multi-GPU v.3 | 373.62 | 0.7 % | 3.3 |

used by other version of the code. The best multi-GPU version was up to $3.3\times$ faster than the parallel code that uses one GPU and up to $72 \times 3.3 = 237.6$ times faster than the sequential code. As future work, we propose two new extensions. The first one will try to overlap the boundary transfers with computation. To achieve this goal, the computation of both the chemotaxis and the Laplacian operator for each point could be divided in two steps, the computation of internal points and boundary points, using for this purpose two CUDA kernels. Since the amount of boundaries points is smaller than internal points, its computation theoretically will finish first, so data copy can start early, while internal points computation is taking place. The computation and communication executions can be executed concurrently with the use of CUDA streams. The second proposed version will extend our work to deal with the situation where GPUs are spread across a cluster of computes. For this purpose, the MPI API will be used to establish communication between distinct nodes.

**Acknowledgements**
The financial support by CNPq, CAPES, UFJF, FINEP and FAPEMIG is greatly acknowledged.

**References**

[1] P. Rocha, M. Xavier, A. Pigozzo, B. M. Quintela, G. Macedo, R. Santos, and M. Lobosco, "A three-dimensional computational model of the innate immune system," in *Computational Science and Its Applications ICCSA 2012* (B. Murgante, O. Gervasi, S. Misra, N. Nedjah, A. Rocha, D. Taniar, and B. Apduhan, eds.), vol. 7333 of *Lecture Notes in Computer Science*, pp. 691–706, Springer Berlin Heidelberg, 2012.

[2] D. R. Butenhof, *Programming with POSIX threads*. Boston, MA, USA: Addison-Wesley Longman Publishing Co., Inc., 1997.

[3] M. M. Hafez and J. J. Chattot, *Innovative Methods for Numerical Solution of Partial Differential Equations*. World Scientific Publishing Company, 2002.

[4] S. Cook, *CUDA Programming: A Developer's Guide to Parallel Computing with GPUs*. San Francisco, CA, USA: Morgan Kaufmann Publishers Inc., 1st ed., 2012.

# Spectral models for orthonormal wavelets and related algorithms

**F. Gómez-Cubillo**[1], **Z. Suchanecki**[2,3], **S. Villullas**[1]

[1] Dpto de Análisis Matemático, Universidad de Valladolid, Facultad de Ciencias, 47011 Valladolid, Spain

[2] Université du Luxembourg, Campus Kirchberg, 6 Rue Coudenhove-Kalergi, L-1359 Luxembourg

[3] Institute of Mathematics and Informatics, University of Opole, 45-052 Opole, Poland

E-mail: fgcubill@am.uva.es, zdzislaw.suchanecki@uni.lu, svillu@am.uva.es

**Abstract.** This work presents new methods for constructing orthonormal wavelets from certain families of Hardy functions. Inner functions and the corresponding backward shift invariant subspaces are in the core of the structure of these families. The new algorithms focus on the local shape of the wavelets, a property making them especially useful for pattern recognition.

Keywords: Orthonormal wavelets, Hardy spaces, inner functions.

## 1. Introduction

Orthonormal bases (shortly ONBs) of $L^2(\mathbb{R})$ can be constructed by application of translations and dilations to an appropriate function. Under the name *orthonormal wavelet on* $\mathbb{R}$ is commonly understood a function $\psi \in L^2(\mathbb{R})$ such that the family $\{\psi_{m,n} := D^m T^n \psi : m, n \in \mathbb{Z}\}$ is an ONB of $L^2(\mathbb{R})$, where $T$ and $D$ denote the translation and dilation operators on $L^2(\mathbb{R})$ defined by

$$[Tf](x) := f(x-1), \quad [Df](x) := 2^{1/2} f(2x), \quad (f \in L^2(\mathbb{R})). \tag{1}$$

A way to construct wavelets and obtain efficient algorithms is given by the concept of *multiresolution analysis* (MRA). Along the literature, the transfer functions of the filters associated with an MRA have been the keystone for constructing compactly supported orthonormal wavelets (CSOWs), as well as the study of their properties. See [4] and references therein.

Almost all wavelet theory has been developed on the basis of Fourier transform, which diagonalizes the translation operator $T$. In this work we propose to consider arbitrary ONBs $\{L_i^{(0)}\}_{i\in\mathbb{I}}$ of $L^2[0,1)$ and $\{K_{\pm,j}^{(0)}\}_{j\in\mathbb{J}}$ of $L^2[\pm1,\pm2)$, and extend them to a pair of ONBs of $L^2(\mathbb{R})$ by means of translations and dilations. The resultant ONBs of $L^2(\mathbb{R})$,

$$\{L_i^{(n)} := T^n L_i^{(0)}\}_{i\in\mathbb{I}, n\in\mathbb{Z}}, \quad \{K_{s,j}^{(m)} := D^m K_{s,j}^{(0)}\}_{s=\pm, j\in\mathbb{J}, m\in\mathbb{Z}}, \tag{2}$$

lead to shift and spectral representations of the operators $T$ and $D$, respectively [1]. In Section 2 we introduce these general pairs of ONBs of $L^2(\mathbb{R})$ and use them in Theorem 1 to characterize all orthonormal wavelets $\psi(x) = \sum_{s,j,m} \tilde{\psi}_{s,j}^{(m)} K_{s,j}^{(m)}(x) \in L^2(\mathbb{R})$ (shortly $\psi = \{\tilde{\psi}_{s,j}^{(m)}\}$) by means of conditions involving the coordinates $\tilde{\psi}_{s,j}^{(m)}$ and the elements $\alpha_{i,n}^{s,j,m}$ of the change of basis matrix relating $\{L_i^{(n)}\}$ and $\{K_{s,j}^{(m)}\}$. Sections 3 and 4 deal with particular situations showing the potential of this result.

Section 3 restricts attention to orthonormal wavelets $\psi = \{\tilde{\psi}_{+,j}^{(0)}\} \in L^2(\mathbb{R})$, with compact support included in the interval $[1,2]$, and a particular choice of the ONBs $\{L_i^{(n)}\}$ and $\{K_{s,j}^{(m)}\}$ built on the classical Haar function. Proposition 2 gives simple conditions for such wavelets and ONBs under the additional assumptions $\tilde{\psi}_{+,0}^{(0)} = 0$ and $\tilde{\psi}_{+,1}^{(0)} \neq 0$.

The close relationship between orthonormal wavelets and certain families of functions belonging to the Hardy class $H^2$ is established in Section 4, where we pay attention to the situation described by Proposition 2. In particular, as Theorem 3 shows, certain families articulated around an inner function $\mathfrak{b}$ and the corresponding backward shift invariant subspace $[\mathfrak{b}H^2]^\perp$ have associated a CSOW $\psi = \{\tilde{\psi}_{+,j}^{(0)}\} \in L^2(\mathbb{R})$ with $\operatorname{supp}\psi \subseteq [1,2]$. This fact lead to algorithms focusing on the local shape of the wavelet. Rational inner functions, i.e., finite Blaschke products, with zeros $0 < |a_k| < 2^{-1/2}$, lead to optimal algorithms; see Proposition 7.

## 2. Spectral conditions for orthonormal wavelets

As we have indicated in the Introduction, suitable ONBs of $L^2(\mathbb{R})$ lead to shift and spectral representations of the translation and dilation operators, $T$ and $D$, given in (1). One begins by considering ONBs $\{L_i^{(n)}(x)\}_{i\in\mathbb{I},n\in\mathbb{Z}}$ and $\{K_{s,j}^{(m)}(x)\}_{s=\pm,j\in\mathbb{J},m\in\mathbb{Z}}$ of $L^2(\mathbb{R})$ defined by (2). For each $f \in L^2(\mathbb{R})$ one has (in $L^2$-sense) $f = \sum_{i,n} \hat{f}_i^{(n)} L_i^{(n)}$ and $f = \sum_{s,j,m} \tilde{f}_{s,j}^{(m)} K_{s,j}^{(m)}$ (shortly, $f = \{\hat{f}_i^{(n)}\} = \{\tilde{f}_{s,j}^{(m)}\}$). The change of representation between both expansions is governed by the matrix $(\alpha_{i,n}^{s,j,m})$, where $\alpha_{i,n}^{s,j,m} := \int_{\mathbb{R}} L_i^{(n)}(x)\, \overline{K_{s,j}^{(m)}(x)}\, dx$.

As the following result shows, conditions for a function $\psi = \{\hat{\psi}_i^{(n)}\} = \{\tilde{\psi}_{s,j}^{(m)}\}$ of $L^2(\mathbb{R})$ to be an orthonormal wavelet can be written in terms of the coordinates $\tilde{\psi}_{s,j}^{(m)}$ and the matrix elements $\alpha_{i,n}^{s,j,m}$. We call them "spectral conditions". As usual, $\delta_n$ denotes the *discrete Dirac delta function* defined by $\delta_n := 1$ if $n = 0$, $0$ if $n \in \mathbb{Z}\backslash\{0\}$.

**Theorem 1** *A function $\psi \in L^2(\mathbb{R})$, with $\psi = \{\tilde{\psi}_{s,j}^{(m)}\}$, is an orthonormal wavelet iff the following two conditions are satisfied:*

*(i) Orthonormality: For every $p, q \in \mathbb{Z}$,* $\displaystyle\sum_{s,j,m} \tilde{\psi}_{s,j}^{(m)} \sum_{i,n} \overline{\alpha_{i,n}^{s,j,m-p}} \sum_{r,k,l} \alpha_{i,n-q}^{r,k,l}\, \overline{\tilde{\psi}_{r,k}^{(l)}} = \delta_p\, \delta_q.$

*(ii) Completeness: For every finite set $\mathbb{F}$ of indices $(s,j)$, with $s = \pm$ and $j \in \mathbb{J}$, the matrix*[1]

$$\left( \sum_{i,n} \overline{\alpha_{i,n}^{s,j,m}} \sum_{r,k,l} \alpha_{i,n-q}^{r,k,l}\, \overline{\tilde{\psi}_{r,k}^{(l)}} \right)_{\substack{(m,q)\in\mathbb{Z}\times\mathbb{Z} \\ (s,j)\in\mathbb{F}}}$$

*has maximal range, i.e., the cardinal of $\mathbb{F}$.*

## 3. CSOWs and Haar bases

The spectral conditions of Theorem 1 are computationally useful when one deals with compactly supported orthonormal wavelets (CSOWs) and convenient ONBs $\{L_i^{(0)}(x)\}_{i\in\mathbb{I}}$ and $\{K_{\pm,j}^{(0)}(x)\}_{j\in\mathbb{J}}$. From now on we shall pay attention to CSOWs $\psi \in L^2(\mathbb{R})$ with $\operatorname{supp}\psi \subseteq [1,2]$ and the so-called

[1] Each file of this matrix corresponds with fixed values of the pair of indices $(m,q) \in \mathbb{Z} \times \mathbb{Z}$, whereas each of its columns corresponds with fixed values of the pair $(s,j) \in \mathbb{F}$, so that the matrix has an infinite number of files and the number of its columns is the cardinal of $\mathbb{F}$.

Haar ONBs $\{L_i^{(n)}\}$ and $\{K_{\pm,j}^{(m)}\}$ built as follows: Let $\varphi^H$ and $\psi^H$ be the *Haar scaling function and wavelet* given by

$$\varphi^H := \chi_{[0,1)}\,, \quad \psi^H := \chi_{[0,1/2)} - \chi_{[1/2,1)}\,,$$

where $\chi$ denotes the characteristic function, $\chi_{[a,b)}(x) := 1$ if $x \in [a,b)$, 0 otherwise. Consider the *"Haar ONBs"* $\{L_i^{(0)}(x)\}_{i\in\mathbb{N}\cup\{0\}}$ of $L^2[0,1)$ and $\{K_{\pm,j}^{(0)}(x)\}_{j\in\mathbb{N}\cup\{0\}}$ of $L^2[\pm 1, \pm 2)$ defined by

$$L_0^{(0)} := \varphi_{0,0}^H\,, \quad K_{+,0}^{(0)} := \varphi_{0,1}^H\,, \quad K_{-,0}^{(0)} := \varphi_{0,-2}^H\,, \tag{3}$$

and, for $p \in \mathbb{N}\cup\{0\}$ and $q = 0, 1, \ldots, 2^p - 1$,

$$L_{2^p+q}^{(0)} := \psi_{p,q}^H\,, \quad K_{+,2^p+q}^{(0)} := \psi_{p,2^p+q}^H\,, \quad K_{-,2^p+q}^{(0)} := \psi_{p,-2^{p+1}+q}^H\,. \tag{4}$$

For the Haar ONBs and CSOWs $\psi \in L^2(\mathbb{R})$ with $\operatorname{supp}\psi \subseteq [1,2]$ the additional assumptions $\tilde{\psi}_{+,0}^{(0)} = 0$ and $\tilde{\psi}_{+,1}^{(0)} \neq 0$ simplify significantly the situation.

**Proposition 2** *Let $\{K_{+,j}^{(0)}(x)\}_{j\in\mathbb{N}\cup\{0\}}$ be the Haar ONB of $L^2[1,2)$ given in (3) and (4), and let $\psi(x) = \sum_{j=1}^{\infty} \tilde{\psi}_{+,j}^{(0)} K_{+,j}^{(0)}(x)$ be a function of $L^2(\mathbb{R})$ with $\operatorname{supp}\psi \subseteq [1,2]$ and such that $\tilde{\psi}_{+,1}^{(0)} \neq 0$. Then, $\psi$ is an orthonormal wavelet if and only if the following set of equations is satisfied: For each $q \in \mathbb{N}\cup\{0\}$,*

$$\sum_{\substack{r=0,1,\ldots\\ t=0,1,\ldots,2^r-1}} \overline{\tilde{\psi}_{+,2^r(q+1)+t}^{(0)}}\, \tilde{\psi}_{+,2^r+t}^{(0)} = \delta_q\,. \tag{5}$$

## 4. CSOWs and inner functions

Under the assumptions of Proposition 2, Theorem 3 below shows that the set of equations (5) is closely related with certain families of Hardy functions. Inner functions and the corresponding backward shift invariant subspaces articulate the structure of these families. First of all, let us remind some facts about Hardy classes and inner functions.

Let $\mathbb{D}$ denote the open unit disc of the complex plane $\mathbb{C}$ and $\partial\mathbb{D}$ its boundary. The classical Hardy space $H^2 \equiv H^2(\mathbb{D})$ consists of complex-valued functions $\mathfrak{h} = \sum_{n=0}^{\infty} c_n z^n$, holomorphic in $\mathbb{D}$ and such that $\sum_{n=0}^{\infty} |c_n|^2 < \infty$, provided with the usual Hilbert space structure given by the scalar product $(\mathfrak{h}, \mathfrak{g}) := \sum_{n=0}^{\infty} \bar{c}_n d_n$, where $\mathfrak{h} = \sum_{n=0}^{\infty} c_n z^n$ and $\mathfrak{g} = \sum_{n=0}^{\infty} d_n z^n$ belong to $H^2$. The unilateral shift $S$ acting on $H^2$ is the linear operator "multiplication by $z$", i.e., $[S\mathfrak{h}](z) := z\,\mathfrak{h}(z)$, for $\mathfrak{h} \in H^2$ and $z \in \mathbb{D}$. The adjoint operator $S^*$, the backward shift, is given by $[S^*\mathfrak{h}](z) = (\mathfrak{h}(z) - \mathfrak{h}(0))/z$. The closed subspaces of $H^2$ invariant under $S$ are of the form

$$\mathfrak{b}H^2 := \{\mathfrak{f} \in H^2 : \mathfrak{f} = \mathfrak{b}\mathfrak{h},\ \mathfrak{h} \in H^2\}\,,$$

where $\mathfrak{b}$ is an inner function. The orthogonal complement $[\mathfrak{b}H^2]^\perp = H^2 \ominus \mathfrak{b}H^2$ is invariant under the backward shift $S^*$. See [5, 6] for details.

**Theorem 3** *Let* $\mathfrak{A} := \left\{\mathfrak{h}_l(z) = \sum_{n=0}^{\infty} c_n^{(l)} z^n \in H^2 \,:\, l \in 2\mathbb{N}-1\right\}$ *be a family of Hardy functions such that*

*(a) $\mathfrak{h}_1 = \mu_1\,\mathfrak{b}$, where $0 \neq \mu_1 \in \mathbb{C}$ and $\mathfrak{b}$ is an inner function such that $z_0 = 0$ is not a zero of $\mathfrak{b}$,*

*(b) $\mathfrak{h}_l \in [\mathfrak{b}H^2]^\perp$ for every $l \in 2\mathbb{N}+1$,*

*(c) $\sum_l ||\mathfrak{h}_l||^2 = 1$,*

*(d) the functions* $\mathfrak{h}_l$, $(l \in 2\mathbb{N}+1)$, *satisfy the set of equations: For* $q \in \mathbb{N}$,

$$\sum_{r=1}^{\infty} \sum_{u=0}^{2^{r-1}-1} (\mathfrak{h}_{2^r(q+1)+2u+1}, \mathfrak{h}_{2^r+2u+1}) = 0 \,. \tag{6}$$

*Then* $\mathfrak{A}$ *has associated a unique compactly supported orthonormal wavelet* $\psi \in L^2(\mathbb{R})$, *with* $supp\,\psi \subseteq [1,2]$, *defined by* $\psi := \sum_{j=1}^{\infty} \tilde{\psi}_{+,j}^{(0)} K_{+,j}^{(0)}$ *(in* $L^2$*-sense), where* $\{K_{+,j}^{(0)}\}_{j\in\mathbb{N}\cup\{0\}}$ *is the Haar ONB of* $L^2[1,2)$ *given in (3)-(4) and, for* $j = 2^n l$, *with* $l \in 2\mathbb{N}-1$ *and* $n \in \mathbb{N}\cup\{0\}$,

$$\tilde{\psi}_{+,j}^{(0)} = \tilde{\psi}_{+,2^n l}^{(0)} = c_n^{(l)} \,. \tag{7}$$

The set of equations (6) is satisfied under stronger orthogonality conditions for the functions $\mathfrak{h}_l$, $(l \in 2\mathbb{N}+1)$:

**Proposition 4** *Condition (d) of Theorem 3 is satisfied if*

*(d') for* $j, l \in 2\mathbb{N}+1$, *with* $j < l$ *and binary expansions* $j = 1 + (\sum_{\nu=1}^{r-1} j_\nu 2^\nu) + 2^r$ *and* $l = 1 + (\sum_{\nu=1}^{s-1} l_\nu 2^\nu) + 2^s$, *the functions* $\mathfrak{h}_j$ *and* $\mathfrak{h}_l$ *are orthogonal if* $r < s$ *and* $j_\nu = l_\nu$ *for* $0 < \nu < r$,

Each Hardy function $\mathfrak{h}_l$ in the family $\mathfrak{A}$ contributes to the wavelet $\psi$ only on a dyadic subinterval of smaller length the greater is $l$:

**Remark 5** For each odd natural number $l = 2^p + q$, $(p = 0, 1, \ldots,$ and $q = 1, 3, \ldots, 2^p - 1)$, the function $\mathfrak{h}_l = \sum_{n=0}^{\infty} c_n^{(l)} z^n$ contributes to the wavelet $\psi$ of Theorem 3 only on the subinterval $\left[1 + \frac{q}{2^p}, 1 + \frac{q+1}{2^p}\right)$ and determines the detail of $\psi$ on the right of the point $x = 1 + \frac{q}{2^p}$.

We finish with a study of the convergence of the algorithms deduced from Theorem 3.

**Proposition 6** *Under the assumptions of Theorem 3, if only a finite number of functions* $\mathfrak{h}_{l_i}$, $(1 \le i \le m)$, *are non-null, then the series in* $\sum_{j=1}^{\infty} \tilde{\psi}_{+,j}^{(0)} K_{+,j}^{(0)}$ *converges absolutely and uniformly to the wavelet* $\psi$ *except perhaps on a set* $A_\epsilon$ *of the form* $A_\epsilon = \bigcup_{i=1}^{m} \left[1 + \frac{q_i}{2^{p_i}}, 1 + \frac{q_i}{2^{p_i}} + \epsilon\right)$, *where* $\epsilon > 0$ *is arbitrarily small,* $m \in \mathbb{N}\cup\{0\}$ *and, for each* $i$, $p_i \in \mathbb{N}$ *and* $q_i \in \{0, 1, 3, \ldots, 2^{p_i} - 1\}$.

**Proposition 7** *Under the assumptions of Theorem 3, if the inner function* $\mathfrak{b}$ *is a finite Blaschke product with zeros* $a_k$ *such that* $0 < |a_k| < 2^{-1/2}$ *and only a finite number of functions* $\mathfrak{h}_l$ *are non-null, then the series* $\sum_{j=1}^{\infty} \tilde{\psi}_{+,j}^{(0)} K_{+,j}^{(0)}$ *converges absolutely and uniformly to the wavelet* $\psi$ *on the whole interval* $[1,2]$.

**Remark 8** Results similar to Theorem 3 can be given for wavelets with arbitrary support. Several families of Hardy functions with intertwined orthonormality relations must be considered in the general case. See [2] for more details. For other types of spectral algorithms see [3].

**Acknowledgments**
This work was partially supported by research projects MTM2009-12561 and MTM2012-31439 (Spain Government).

[1] Gómez-Cubillo F and Suchanecki Z 2011 *J. Fourier Anal. Appl.* **17** 191
[2] Gómez-Cubillo F and Suchanecki Z 2011 *Appl. Comput. Harmon. Anal.* **30** 273
[3] Gómez-Cubillo F, Suchanecki Z and Villullas S 2012 *Int. J. Wavelets Multiresolut. Inf. Process.* **10** 1250008
[4] Hernández E and Weiss G 1996 *A First Course on Wavelets* (Boca Raton, FL: CRC Press)
[5] Hoffman K 1962 *Banach Spaces of Analytic Functions* (Englewood Cliffs, N.J.: Prentice-Hall)
[6] Koosis P 1998 *Introduction to* $H^p$ *Spaces* (Cambridge: Cambridge University Press)

# A discrete geometric formulation for eddy-current problems in fusion devices

**Paolo Bettini[1,2], Maurizio Furno Palumbo[1] Ruben Specogna[3]**

[1] Università di Padova, Dipartimento di Ingegneria Industriale (DII), 35131 Padova, Italy
[2] Consorzio RFX, EURATOM-ENEA Association, 35127 Padova, Italy
[3] Università di Udine, Dipartimento di Ingegneria Elettrica, Gestionale e Meccanica (DIEGM), 33100 Udine, Italy

E-mail: furnopalumbo@igi.cnr.it

**Abstract.**
All thermonuclear controlled fusion devices under construction or design have such high performances to require a special care in the dimensioning of various components, specifically from the electromagnetic point of view. To this purpose, it is fundamental to develop models which are both accurate (i.e. able to describe the physical phenomena) and predictive (i.e. useful not only to explain what happens in running experiments, but also to reliably extrapolate to other range of parameters).

The dynamics of fusion plasmas is often conveniently described by Magneto-Hydro-Dynamics (MHD) equations, which predict that some unstable evolution modes may exist. On the other hand, the complexity of the intrinsically 3D model of the interactions between a realistic unstable plasma, the surrounding passive structures (important to guarantee a good MHD stability) and the active conductors (coils) require the numerical solution of challenging electromagnetic problems.

In this work a discrete geometric formulation for eddy-current problems in the frequency domain is developed; the magnetic fields produced by a typical active coil system is calculated in the presence of 3D conductive structures.

## 1. Introduction

The dynamics of fusion plasmas is often conveniently described by Magneto-Hydro-Dynamics (MHD) equations, which predict that some unstable evolution modes may exist [1], [2]. The MHD modes are usually categorised using the toroidal mode number $n$, i.e. the harmonic index of a Fourier decomposition of plasma perturbations along the toroidal direction.

While some MHD instabilities develop on very fast timescales (few milliseconds) and are only marginally influenced by the interaction with the structures surrounding the plasma, one particular category, known as Resistive Wall Modes (RWMs), critically depends on the detailed characteristics of the conductive wall and in particular on its non-uniformities and its typical penetration time for electro-magnetic perturbations [3]. Close-fitting passive conducting structures are an efficient way to prevent the growth of magnetohydrodynamic (MHD) instabilities, but are not suitable for a steady state fusion reactor, because of the finite diffusion time of any material shell.

Given the relatively slow growth rate of RWMs, feedback control by means of active coils is possible and is at present one of the most active research fields. RWMs can develop in both tokamak and Reversed Field Pinch (RFP) devices and for both they represent one of the most important performance limiting MHD instabilities [4], [5]. In general, given the characteristics of both the target instability and the active control method, successful experiments and modelling take greatly advantage of a careful knowledge of the impact of 3-dimensional effects in the mode dynamics [6]. In fact, when the plasma is in presence of closed loop control actions, eddy currents in 3D conducting structures and plasma perturbations will evolve in a coupled way. The comprehension of their interaction represents a very complex problem both on the experimental side and the modelling one [7]. The controller design itself needs an accurate evaluation of all the effects in the coil actions due to the non uniformities of the passive boundary [8], [9].

In this paper we focus on the problem of carefully computing the interactions between active conductors (coils) and passive structures surrounding the plasma.

## 2. Discrete Geometric Formulation

A discrete geometric formulation for eddy-current problems in the frequency domain is presented, which is based on the circulation of the magnetic vector potential over hexahedral grids.

The 3-D domain of interest $\mathcal{D}$ is covered by a mesh of generic hexahedra, whose incidences are encoded in the *cell complex* $\mathcal{K}$ represented by the standard incidence matrices $\mathbf{G}$, $\mathbf{C}$ and $\mathbf{D}$ [10]. A dual barycentric complex $\tilde{\mathcal{K}}$ is obtained from $\mathcal{K}$ by using the *barycentric subdivision*; its incidence matrices are $\tilde{\mathbf{G}} = \mathbf{D}^T$, $\tilde{\mathbf{C}} = \mathbf{C}^T$ and $\tilde{\mathbf{D}} = -\mathbf{G}^T$.

Three subdomains of $\mathcal{D}$ are identified: the passive conductive region $\mathcal{D}_c$ (including all conductive structures surrounding the plasma, see Fig.1), the non-conductive region $\mathcal{D}_a$ (vacuum or air –outside the vacuum vessel), and the source region $\mathcal{D}_s$ (active coils used to control the plasma instabilities).

When modeling stranded coils, it is useful to introduce integral sources, which avoid to cover $\mathcal{D}_s$ with a fine mesh. With this aim, the circulations of the magnetic vector potential $\mathbf{A}$ along primal edges $e \in \mathcal{D}$ can be expresses as $\mathbf{A} = \mathbf{A}_s + \mathbf{A}_r$, where $\mathbf{A}_r$ are the circulations of the magnetic vector potential due to eddy currents in $\mathcal{D}_c$ and $\mathbf{A}_s$ are the circulations of the magnetic vector potential produced by the sources in $\mathcal{D}_s$.

By combining the discrete Ampère's law and Faraday's law with the discrete counterpart of the constitutive laws for the flux density $\mathbf{B}$ and the current density $\mathbf{J}$, a symmetric complex linear system of equations is obtained [9],

$$\begin{aligned} &(\mathbf{C}^T \boldsymbol{\nu}\, \mathbf{C})\, \mathbf{A}_r = \mathbf{0}, && \forall e \in \mathcal{D}_a \bigcup \mathcal{D}_s \\ &(\mathbf{C}^T \boldsymbol{\nu}\, \mathbf{C} + i\omega\, \boldsymbol{\sigma})\, \mathbf{A}_r = -i\omega\, \boldsymbol{\sigma}\, \mathbf{A}_s, && \forall e \in \mathcal{D}_c \end{aligned} \tag{1}$$

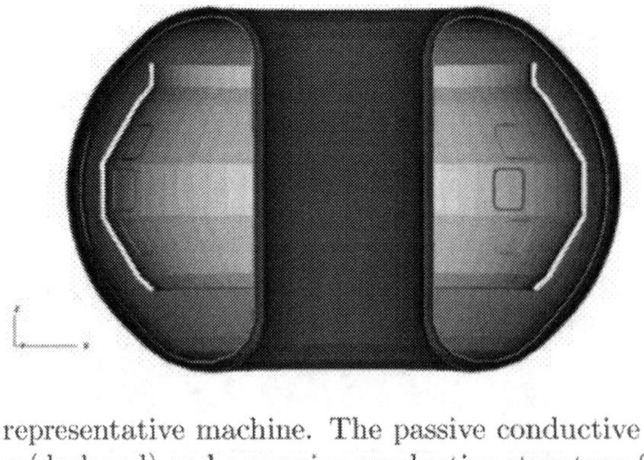

**Figure 1.** Sketch of a representative machine. The passive conductive region $\mathcal{D}_c$ includes two closed toroidal structures (dark red) and a massive conductive structure (gray) facing the plasma region. A set of typical saddle coils devoted to the active control of RWMs are also shown.

where $\omega$ is the angular frequency, $\boldsymbol{\nu}$ and $\boldsymbol{\sigma}$ are square matrices that require metric notions, material properties, and some hypothesis on the fields in order to be computed.

On the RHS, $\mathbf{A}_s$ denotes the circulations of the magnetic vector potential along $e \in \mathcal{D}_c$ produced by the sources in $\mathcal{D}_s$, only; each entry of $\mathbf{A}_s$ can be computed with standard closed formulas.

The sparse and symmetric linear system (1) is solved with a standard *tree-cotree gauge*, using a state-of-the-art direct solver (PARDISO, included in the Intel MKL library).

## 3. Numerical results

The proposed approach has been applied to calculate the magnetic field produced by a set of saddle coils, individually fed by sinusoidal currents, in the presence of three main conductive structures of a representative machine (see Fig. 1). It must be noted that for this particular geometry it would be possible to model only a part of the geometry due to its toroidal symmetry. Nonetheless, we developed a tool considering the most general application, since most fusion reactors we aim to study present no symmetries at all.

The module of the current density induced on the most relevant structure (facing the plasma) by a sinusoidal current feeding a single saddle coil is shown in Fig. 2. Due to linearity, any combination of currents feeding the RWM coils can be calculated by superposition.

The geometric elements of the mesh are summarized in Table 1, together with degrees of freedom (DOFs) and non zero entries (NNZs) of the system matrix. The numerical solution of the problem by means of a numerical code (CAFE), takes less than 10 minutes[1], including pre-processing (assembling of the system matrix and its RHS) and post-processing (calculation of the magnetic field components by Biot-Savart'law).

The shielding effect of the conductive structure is clearly visible in Fig. 3 where the amplitude of the magnetic flux density is presented (for different frequencies in the range $0 - 100\,Hz$) as a function of the distance along the radial direction from the centre of the coil towards plasma.

## 4. Conclusions

The development of methods for the active control of MHD instabilities and for the correction of error fields is mandatory in view of fusion reactors. A discrete geometric formulation for eddy-current problems in the frequency domain is introduced to calculate the effects on active control systems of 3D conductive structures surrounding the plasma in a representative machine.

[1] CAFE (Computer Aided Fusion Engineering), research code developed by first and third authors, runs on a workstation equipped with two 8-core processors (Xeon E5-2680 2.7GHz 20MB) and 256GB DDR3-1600 RAM.

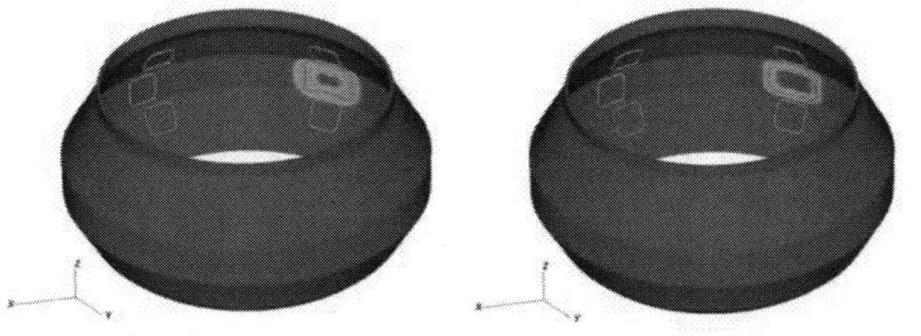

(a) (b)

**Figure 2.** Module of the current density induced on the conductive structure facing the plasma by a sinusoidal current feeding a single saddle coil: (a) $f = 10Hz$, (b) $f = 100Hz$.

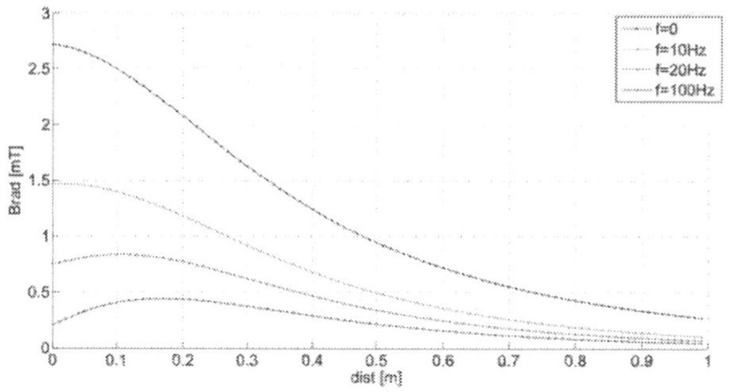


**Figure 3.** Amplitude of the magnetic flux density as a function of the distance along the radial direction from the centre of the coil towards plasma.


## 5. Acknowledgment

This work was partially supported by the European Communities under the Contract of Association between EURATOM and ENEA/Consorzio RFX, and by the Italian MIUR under PRIN grant 2010SPS9B3. The views and opinions expressed herein do not necessarily reflect those of the European Commission.


## 6. References


[1] J. P. Freidberg. *Plasma Physics and Fusion Energy*. Cambridge University Press, 2007.

[2] G. Bateman. *G. Bateman, MHD instabilities*. MIT Press, 1978.

[3] M. S. Chu and M. Okabayashi. Stabilization of the external kink and the resistive wall mode. *Plasma Phys. Control. Fusion*, 52:123001, 2010.

[4] F. Troyon, R. Gruber, H. Saurenmann, S. Semenzato, and S. Succi. MHD-limits to plasma confinement. *Plasma Phys. Control. Fusion*, 26:209, 1984.

[5] B. Alper. RFP stability with a resistive shell in HBTX1C. *Plasma Phys. Control. Fusion*, 31:205, 1989.

[6] F. Villone, Y. Q. Liu, R. Paccagnella, T. Bolzonella, and G. Rubinacci. Effects of three-dimensional electromagnetic structures on resistive-wall-mode stability of reversed field pinches. *Phys. Rev. Lett.*, 100:255005, 2008.

[7] M. Baruzzo, T. Bolzonella, S.C. Guo, Y. Q. Liu, G. Marchiori, R. Paccagnella, A. Soppelsa, F. Villone, and Z.R. Wang. 3D effects on RWM physics in RFX-mod. *Nucl. Fusion*, 51:083037, 2011.

[8] G. Marchiori, M. Baruzzo, T. Bolzonella, Y.Q. Liu, A. Soppelsa, and F. Villone. *Plasma Phys. Control. Fusion*, 52:023020, 2012.

[9] P. Bettini, L. Marrelli, and R. Specogna. Calculation of 3D magnetic fields produced by MHD active control systems in fusion devices. *IEEE Trans. on Magnetics*, DOI: 10.1109/TMAG.2013.2279141, in press.

[10] E. Tonti. *The Mathematical Structure of Classical and Relativistic Physics*. Birkhäuser Basel, 2013.


**Table 1.** Geometric elements (nodes, hexahedra, edges) of the mesh. Corresponding degrees of freedom (DOFs) and non zero entries (NNZs) of the system matrix.

| | |
|---|---|
| Nodes | 819,789 |
| Hexahedra | 807,408 |
| Edges | 2,447,132 |
| DOFs (with gauge) | 1,752,331 |
| NNzs (with gauge) | 22,379,984 |

# Numerical modelling of electromagnetic loads on fusion device structures

**Paolo Bettini**[1,2], **Maurizio Furno Palumbo**[1] **Ruben Specogna**[3]

[1] Università di Padova, Dipartimento di Ingegneria Industriale (DII), 35131 Padova, Italy
[2] Consorzio RFX, EURATOM-ENEA Association, 35127 Padova, Italy
[3] Università di Udine, Dipartimento di Ingegneria Elettrica, Gestionale e Meccanica (DIEGM), 33100 Udine, Italy

E-mail: `furnopalumbo@igi.cnr.it`

**Abstract.** In magnetic confinement fusion devices, during abnormal operations (disruptions) the plasma begins to move rapidly towards the vessel wall in a vertical displacement event (VDE), producing plasma current asymmetries, vessel eddy currents and open field line halo currents, each of which can exert potentially damaging forces upon the vessel and in-vessel components. This paper presents a methodology to estimate electromagnetic loads, on three-dimensional conductive structures surrounding the plasma, which arise from the interaction of halo-currents associated to VDEs with a magnetic field of the order of some Tesla needed for plasma confinement. Lorentz forces, calculated by complementary formulations, are used as constraining loads in a linear static structural analysis carried out on a detailed model of the mechanical structures of a representative machine.

## 1. Introduction

Disruptions represent one of the main concerns for tokamak operations, especially in view of fusion reactors or experimental test reactors under design [1]. In elongated plasmas, vertical control can be lost generating what are called *Vertical Displacement Events* (VDEs), that induce plasma current asymmetries, eddy currents and halo currents, each of which can exert potentially damaging forces upon the tokamak components. In particular, when the plasma hits the first wall (FW), enormous currents are injected in the passive conductive structures of the machine, in a region submerged in the magnetic field needed for plasma confinement giving rise to appalling forces. Aim of this paper is to describe a methodology used for approaching the problem from the engineering point of view. The proposed procedure, based on multiple physical models, is summarized in the following steps.

(i) Development of a detailed 3D model of the fusion device under study (including all conductive structures surrounding the plasma) and identification of the mesh elements facing the plasma;

(ii) Evaluation of the currents exchanged between plasma and FW (BCs for step *iii*) by means of a suitable plasma model (full MHD or simplified);

(iii) Evaluation of the resistive distribution of the current density **J** in each mesh element by means of a 3D electromagnetic code (see Sec. 2);

(iv) Evaluation of Lorentz force ($\mathbf{f} = \mathbf{J}\times\mathbf{B}$) in each mesh element, assuming a known distribution of magnetic flux density $\mathbf{B}$, to provide the electromagnetic loads (nodal forces) for step $v$;

(v) Solution of a linear static structural problem (see Sec. 3).

## 2. Stationary current conduction problem

An efficient numerical code (CAFE[1]) is used to solve the stationary current conduction problem in a three-dimensional, non-trivial domain. In fusion devices, the geometry is so complicated that complementarity is used to provide a robust error estimator for mesh refinements and to have a reliable control on the accuracy of the solution [2]. An irrotational electric field is obtained by a geometric formulation based on the electric scalar potential $\mathbf{V}$, whereas a formulation based on the electric vector potential $\mathbf{T}$ is used to obtain a solenoidal current density.

Let us consider a conductive region $K$ immersed in a perfectly insulating medium. $K$ is covered by a hexahedral mesh, whose incidences are encoded in the *cell complex* $\mathcal{K}$ represented by the standard incidence matrices $\mathbf{G}$, $\mathbf{C}$ and $\mathbf{D}$ [3], [4], [5]. A dual barycentric complex $\tilde{\mathcal{K}}$ is obtained from $\mathcal{K}$ by using the *barycentric subdivision*: $\tilde{\mathbf{G}} = \mathbf{D}^T$, $\tilde{\mathbf{C}} = \mathbf{C}^T$ and $\tilde{\mathbf{D}} = -\mathbf{G}^T$ represent the incidence matrices of $\tilde{\mathcal{K}}$.

In the first formulation, based on the electric scalar potential $\mathbf{V}$, fixing the potential on one arbitrary node to zero, the other potentials are found by solving the symmetric and positive definite linear system $\mathbf{G}^T\mathbf{S}\mathbf{G}\mathbf{V} = \tilde{\mathbf{I}}_s$, where $\mathbf{S}$ is the Ohm's constitutive matrix and $\tilde{\mathbf{I}}_s$ is the current through the boundary dual faces.

The second formulation, based on a electric vector potential $\mathbf{T}$, requires a complicated topological pre-processing due to the fact that the domain under study is topologically non-trivial[2] [6]. The *electric current* $\mathbf{I}$ can be expressed in terms of $\mathbf{T}$ as

$$\mathbf{I} = \mathbf{I}_s + \mathbf{C}\mathbf{T} + \sum_{i=1}^{N} I_g^i \mathbf{\Pi}^i, \tag{1}$$

where the 2-cocycles $\mathbf{\Pi}^i$ are the *thick links* [7] and $I_g^i$ are the *independent currents*. By combining the discrete Ohm's law $\tilde{\mathbf{U}} = \mathbf{R}\mathbf{I}$ and Faraday's law $\mathbf{C}^T\tilde{\mathbf{U}} = \mathbf{0}$ with (1) and taking into account a set of *non-local Faraday's laws*, the final algebraic linear system of equations is obtained [8]

$$\begin{aligned} &\mathbf{C}^T\mathbf{R}\mathbf{C}\mathbf{T} + \textstyle\sum_{j=1}^{N} \left(\mathbf{C}^T\mathbf{R}\mathbf{\Pi}^j\right) I_g^j = -\mathbf{C}^T\mathbf{R}\mathbf{I}_s \\ &\left(\mathbf{\Pi}^{iT}\mathbf{R}\mathbf{C}\right)\mathbf{T} + \textstyle\sum_{j=1}^{N} \left(\mathbf{\Pi}^{iT}\mathbf{R}\mathbf{\Pi}^j\right) I_g^j = -\mathbf{\Pi}^{iT}\mathbf{R}\mathbf{I}_s. \end{aligned} \tag{2}$$

The coefficients of the source current $\mathbf{I}_s$ on mesh faces belonging to $\partial\mathcal{K}$ are set to the plasma halo current imposed through them. The sparse and symmetric linear system (2) can be solved using an *ungauged* technique with iterative linear solvers, whereas the standard *tree-cotree gauge* is applied for direct solvers.

## 3. Static Structural Analysis

The components involved in the computational domain represent the conductive structures surrounding the plasma. For each of them an accurate geometric model is carried out by a 3D CAD software and a multiblock structured hexahedral mesh is generated by a commercial software. A reference machine with an ITER-like model (including vessel, port extensions, blanket modules and divertor) is considered, discretized with more than 1 million elements (all hexahedra, see Fig. 1). First components facing the plasma are the Blanket Modules (BMs).

[1] CAFE (Computer Aided Fusion Engineering) is a research code developed by first and third authors.

[2] $K$ is a compact connected subset of the 3D Euclidean space which possesses topological handles and cavities.

Lower area of the domain is occupied by the Divertor. The conductive structure surrounding the BMs and the Divertor represents the Vacuum Vessel (VV) made of two shells welded by means of the stiffening ribs. All these structural elements have been assumed as made of stainless steel (AISI 316LN, at a reference temperature $t = 150C$).

In order to set up the FE-model we have considered each contact area welded to one another, so non linear contacts or gaps are not investigated. No symmetries are taken into account: we developed this procedure considering the most general application, since most fusion reactors present no symmetries at all. Besides the structure is constrained with Single Point Constraints (SPC), created on the lower nodes of the VV lower port extensions by fixing them on the ground for preventing the six degrees of freedom.

Finally, we have set up a load step where constraints, the gravitational force and nodal electromagnetic forces computed by CAFE, are applied in order to perform a linear static analysis for computing static elastic stress. The problem, in the linear form $\mathbf{F} = \mathbf{KU}$, where $\mathbf{F}$ is the nodal forces vector, $\mathbf{K}$ is the stiffness matrix, and $\mathbf{U}$ the vector representing the structure degrees of freedom, is solved by Altair RadiOss Bulk© code using a direct method for sparse matrices.

## 4. Numerical results

The proposed procedure has been applied to two test cases. First, axisymmetric BCs are assumed (step $i$) to asses the procedure in the presence of almost symmetric electromagnetic loads, corresponding to an upward symmetric VDE: the results exhibit the expected symmetry in terms of current density flowing in the conductive structures $J$ (Fig. 2(a)), force density $f$ (Fig. 2(c)) and equivalent (*Von Mises*) stress (Fig. 3(a)). Then, a representative asymmetric case, corresponding to an upward asymmetric VDE is considered: the currents exchanged between plasma and FW (BCs calculated at step $i$) are provided by a simplified plasma model, coherent with experimental observation associated with the so called *kink mode* of the plasma [1]. The results exhibit the expected asymmetry for current density $J$ flowing in the structures (Fig. 2(b)), force density $f$ (Fig. 2(d)) and equivalent (*Von Mises*) stress (Fig. 3(b)).

## 5. Conclusions

A procedure, based on multiple physical models, to calculate the electromagnetic loads induced by halo currents on fusion devices has been proposed and applied to representative test cases.

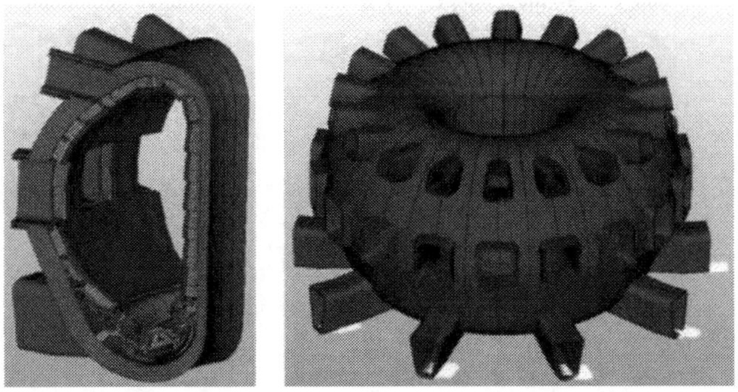

(a) (b)

**Figure 1.** Representative fusion device, discretized with 1,088,055 hexahedra. (a) Detail of the 3D model, including vessel, ports, blanket modules and divertor. (b) Whole model, with Single Point Constraints (SPC) created on the lower nodes of the VV lower port extensions.

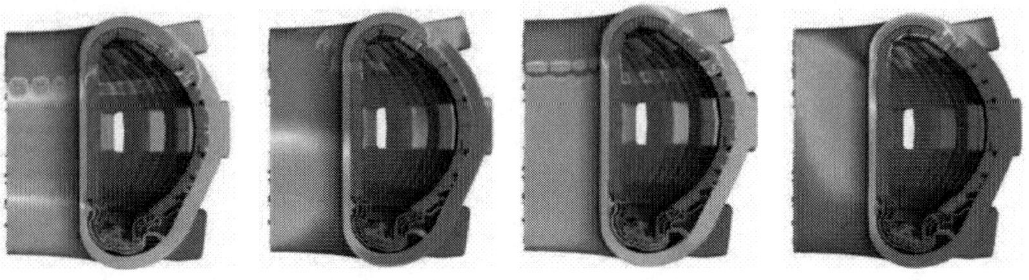

(a) (b) (c) (d)

**Figure 2.** Current density $J$ (color map saturation (red): $0.1A/mm^2$): (a) symmetric, (b) asymmetric. Force density $f$ (saturation: $10^6N/m^3$): (c) symmetric, (d) asymmetric.

(a) (b)

**Figure 3.** Equivalent (*Von Mises*) stress (order of magnitude $10MPa$) for upward VDES: (a) symmetric, (b) asymmetric.

## 6. Acknowledgment

This work was partially supported by the European Communities under the Contract of Association between EURATOM and ENEA/Consorzio RFX, and by the Italian MIUR under PRIN grant 2010SPS9B3. The views and opinions expressed herein do not necessarily reflect those of the European Commission.

## 7. References

[1] P. Bachmann, M. Sugihara, R. Roccella, G. Sannazzaro, Y. Gribov, V. Riccardo, T.C. Hender, S.N. Gerasimov, G. Pautasso, A. Belov, E. Lamzin, M. Roccella, and JET EFDA Contributors. Specification of asymmetric VDE loads of the ITER tokamak. *Fusion Engineering and Design*, 86:1915–1919, 2011.

[2] J.A.H. Rikabi, C.F. Bryant, and E.M. Freeman. An error-based approach to complementary formulations of static field solutions. *Int. J. Numer. Meth. Eng.*, 26(9):1963–1987, 1988.

[3] E. Tonti. *The Mathematical Structure of Classical and Relativistic Physics.* Birkhäuser Basel, 2013.

[4] A. Bossavit. *Computational Electromagnetism.* Academic Press, 1998.

[5] P. Dłotko, R. Specogna, and F. Trevisan. Automatic generation of cuts on large-sized meshes for $t$-$\omega$ geometric eddy-current formulation. *Comput. Meth. Appl. Mech. Eng.*, 198:3765–3781, 2009.

[6] P.W. Gross and P. R. Kotiuga. *Electromagnetic Theory and Computation: A Topological Approach.* Cambridge University Press, 2004.

[7] R. Specogna. Complementary geometric formulations for electrostatics. *Int. J. Numer. Meth. Eng.*, 86:1041–1068, 2011.

[8] P. Bettini and R. Specogna. Lazy cohomology generators enable the use of complementarity for computing halo current resistive distribution in fusion reactors. *IEEE Transactions on Magnetics*, DOI: 10.1109/TMAG.2013.2281506, in press.

# A Higher-Order Neural Network Design for Improving Segmentation Performance in Medical Image Series

**Eşref Selvi**[1] **M. Alper Selver**[2] **Cüneyt Güzeliş**[3] **Oğuz Dicle**[4]

[1]Dokuz Eylül University, Institute of Natural and Applied Sciences, Buca, Izmir, Turkey.
[2]Dokuz Eylül University, Dept. of Electrical and Electronics Engineering, Buca, Izmir, Turkey.
[3]Izmir University of Economics, Dept. of Electrical and Electronics Engineering, Izmir, Turkey.
[4]Dokuz Eylül University, Faculty of Medicine, Dept. of Radiology, Balçova, Izmir, Turkey.

E-mail: `esref.selvi@gmail.com`, `alper.selver@deu.edu.tr`

**Abstract.** Segmentation of anatomical structures from medical image series is an ongoing field of research. Although, organs of interest are three-dimensional in nature, slice-by-slice approaches are widely used in clinical applications because of their ease of integration with the current manual segmentation scheme. To be able to use slice-by-slice techniques effectively, adjacent slice information, which represents likelihood of a region to be the structure of interest, plays critical role. Recent studies focus on using distance transform directly as a feature or to increase the feature values at the vicinity of the search area. This study presents a novel approach by constructing a higher order neural network, the input layer of which receives features together with their multiplications with the distance transform. This allows higher-order interactions between features through the non-linearity introduced by the multiplication. The application of the proposed method to 9 CT datasets for segmentation of the liver shows higher performance than well-known higher order classification neural networks.

## 1. Introduction

Medical imaging devices (i.e. Computed Tomography (CT), Magnetic Resonance Imaging (MRI) etc.) can acquire image slices with very small slice thickness (ST) (i.e. inter-slice distance). Due to the trade-off between efficiency (i.e. computational time) and performance (i.e. accuracy of the segmentation result) extracting a body part of interest (i.e. organ, tissue, tumor etc.) via segmentation from high number of image slices is still an emerging field [1].

Current studies on organ segmentation (i.e. automatic, semi-automatic) area usually are slice-by-slice approaches. Therefore, inter-slice information is very important. The distance transform (DT) can be used to represent inter-slice info as a feature or a multiplier for image prior to extraction of other features [2]. This corresponds to High Order Neural Network (HONN) structures, which are designed to overcome the performance drawbacks of NN architectures at classifying complicated data with high order non-linearity. HONN is suggested by [3] and its learning capabilities are analysed in [3], which shows that the higher order inputs and weights combinations gives higher generalization performance as they need a smaller training set [4].

This study presents a novel approach by constructing an HONN, the input layer of which receives multiplicative pairs of each feature with the DT. The proposed method is applied to

9 Computed Tomography Angiography (CTA) datasets acquired for the measurement of liver volumes of transplantation donors prior to living-donated surgery [2]. The results show higher accuracy, sensitivity and specificity compared to other well known HONN structures such as pi-sigma and productive unit NN.

## 2. Datasets

CTA datasets, which were obtained by using a Philips Secura CT and a Philips Mx8000 CTA with contrast agent injection at portal phase. The CTA datasets consist of DICOM images, which have 512x512 resolution and 3.2 mm slice thickness. Each data set include 70 to 110 images. The datasets selected for the study are chosen to have challenging difficulties for better illustration of the performance comparison of the different NN designs used in this study. The liver can be adjacent to other organs as seen. According to partial volume effects and their similar texture, the boundary between other organs and liver can become hard to segment. There are numerous variations about the appearances of these organs through different data sets due to countless variations in human anatomy, injection of contrast media, different image characteristics etc.

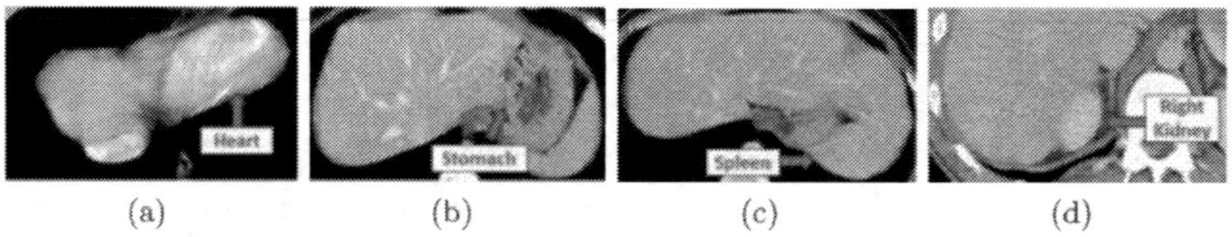


Figure 1: Examples of challenging problems in liver segmentation (a) unclear boundary between liver and heart, (b) unclear liver boundary due to stomach, (c) liver and spleen become adjacent due to atypical shape of the liver that results in visually indistinguishable border between them, (d) the liver is completely connected to right kidney.

## 3. Methodology

In this paper, the fully automatic three stage (i.e. pre-processing, classification, post-processing) procedure proposed in [2] is used as the main algorithm. The pre-processing part removes the fat tissue, bone structure from images by using a combination of adaptive thresholding and morphological processing. The classification stage takes the output of pre-process as input, and extracts three features that are mean, Standard Deviation (SD), and Distance Transform (DT). The post-processing stage receives classification output and applies non-linear filtering to finish the segmentation process. Since this study deals with the classification stage only, pre-process and post-process operations are the same for all approaches.

In total, 5 different types of input space structures (Table 1) are constructed and tested in this study. The first one uses the algorithm in [2], results of which are known to be clinically acceptable for evaluating living liver transplantation donors. The remaining five systems are constructed by using multiplicative and additive combinations of features at the input of the classifiers. The details of these five NN structures are given below:

If we define mean feature vector with, $\mathbf{m}$, SD feature vector with, $\mathbf{s}$, and DT feature vector with, $\mathbf{d}$, extracted from image, $\mathbf{I}$, then the input space, $\mathbf{x}$, is constructed using $\mathbf{x}$=[$\mathbf{s}$;$\mathbf{m}$;$\mathbf{d}$]. Then the output of the classifier is equal to:

$$o^i = f(\sum_{i=1}^{n} \mathbf{w} * \mathbf{x}^i)(1)$$

Table 1: NN input space architectures used in this study

| NN type | Information | Source | Feature 1 | Feature 2 | Feature 3 | Feature 4 | Feature 5 |
|---|---|---|---|---|---|---|---|
| type 1 | 0,51 | I | M | SD | DT | | |
| type 2 | 0,51 | I*DT | M | SD | | | |
| type 3 | 0,45 | I | M*DT | SD*DT | | | |
| type 4 | 0,49 | I | M | SD | M*DT | SD*DT | DT |
| type 5 | 0,49 | I*DT | M | SD | DT | | |

where, **f** is the activation function, **w** is the weight vector, and **o** is the output of the classifiers.

The SD and the mean features are calculated in a 9 x 9 window size centred for a given pixel. The size of the kernel is determined after extensive experimentation. The mean feature is calculated to represent the regions of low frequency, such as homogeneous areas (i.e. parenchyma), in the slice to segment. The SD feature is used to represent regions of high frequency, such as edges and organ boundaries. These features are calculated as

$$\bar{m}_{ij} = \frac{1}{N}(\sum_{i-4}^{i+4}\sum_{i-4}^{i+4} m_{ij}), \quad \sigma_{ij} = \frac{1}{N}(\sum_{i-4}^{i+4}\sum_{i-4}^{i+4} x_{ij}) - \bar{m}$$

where $\bar{m}_{ij}$ is the mean values of the pixel at the point (i,j) and $\sigma_{i,j}$ is the SD of the pixel located at (i,j). N is the total number of the pixels in window and is equal to 81 in this study.

Finally, the DT feature is used to represent the previously segmented image. By creating a distance map from the center of the segmented liver at slice N, the DT shows a probabilistic map of liver position at slice N+1.

In type 2 and type 5, features extracted after multiplying the image by DT. This process reduces the brightness of the pixels according to their locations, where the pixels that are far away from the previously segmented liver become darker and their gray values are reduced. In these approaches, the effect of DT is projected onto mean and SD features. The only difference between them is that the type 5 uses also DT as an input. Therefore, the feature vector of type 2 is **x**=[**s**;**m**] and and the feature vector of type 5 is **x**=[**s**; **m**; **d**], where **s** and **m** are extracted from DT*I (i.e. the DT is multiplied by the current image to segment).

In the type 3, the SD and the mean features of the image is multiplied by the DT feature of the previous image.Type 3 is similar to pi-sigma with a single difference. At the input, all combinations of input feature multiplications except one are used. The multiplication of mean with SD feature is not used since the multiplication of these complementary features suppresses the discriminating information coming from both of them. Therefore, the input feature vectors of type 3 is **x**=[**s*d**; **m*d**]. The output is calculated by using the equation in (1).

In type 4, the SD, the mean and the DT features are used in addition to type 3 features. Thus, this network structure is a mixture of PUNN and pi-sigma network types. The similarity to pi-sigma is due to the reason mentioned in type 3 and the similarity to PUNN is due to the features, which are also used as input besides their multiplications. The input vector of this system becomes **x**=[**m**; **s**; **s*d**; **m*d**; **d**] and the output is calculated as in (1).

## 4. Results

The HONN systems described in the previous section are applied to 9 challenging data sets. The results obtained with different systems are compared image (i.e. algorithm result) by image (i.e. ground truth created by manual delineation of expert physician). The 2D error metrics used to evaluate the results are false positive (FP), false negative (FN), true positive (TP), true negative (TN), accuracy (CC), selectivity (SE), specificity (SP), positive predictive value

(PPV), and negative predictive value (NPV). The results, which show average values of each error metric among all data sets, are presented in Table 2. All HONN structures used in the study (i.e. type 2-5) perform better or equally well when compared to original NN system (i.e. type 1). The results also show that multiplying other features with DT feature gives better results than multiplying image with DT feature before extracting other features. Thus, projecting the DT information onto the other features outperform the projection onto the image with a slight difference. On the other hand, using DT feature itself together with other features extracted from image multiplied with DT is also useful (i.e. type 5) and gives better results than the other systems (i.e. type 2 and type 3). However, according to the results (i.e. type 4), using another feature may not further increase the performance.

Table 2: The results of calculated 2D error metrics

| NN type | FP | FN | TP | TN | CC | SE | SP | PPV | NPV |
|---|---|---|---|---|---|---|---|---|---|
| type 1 | 0,51 | 31,15 | 99,49 | 68,85 | 84,17 | 59,10 | 99,27 | 99,49 | 68,85 |
| type 2 | 0,51 | 23,04 | 99,49 | 76,96 | 88,23 | 56,38 | 99,34 | 99,49 | 76,96 |
| type 3 | 0,45 | 23,03 | 99,55 | 76,97 | 88,26 | 56,39 | 99,41 | 99,55 | 76,97 |
| type 4 | 0,49 | 23,16 | 99,51 | 76,84 | 88,18 | 56,43 | 99,37 | 99,51 | 76,84 |
| type 5 | 0,47 | **20,44** | 99,53 | **79,56** | **89,55** | 55,58 | 99,42 | 99,53 | **79,56** |

## 5. Conclusions

In this study, the performance of the liver segmentation method in [2] is increased by modifying the classification stage using HONN systems. The simulations show that well-known successful HONN architectures such as pi-sigma and PUNN, which uses multiplicative combinations of inputs, achieve improved performance compared to the direct use of the same features. Moreover, it is shown that even a higher performance can be achieved using 'physically meaningful' combinations of features instead of using all combinations as in pi-sigma and PUNN. Developed HONN designs based on this idea (type 5) qualifies 'physically meaningful' use of multiplicative feature combinations by projecting the most discriminating feature (i.e. DT), which defines a probabilistic map at the vicinity of liver by using adjacent slice information, onto the other features (i.e. mean, SD). By using only the selected combinations in the developed HONN design, the local minima problem of HONN systems is reduced and thus, introduced non-linearity by increasing input space dimension is advantageously used. The results show that proposed HONN system outperforms both well-known HONN systems (i.e. pi-sigma, PUNN) and previously reported system in [2], which is shown to produce results at clinically acceptable levels.

## 6. Acknowledgments

This study is supported by TUBITAK EEEAG under grant number 112E032.

## 7. References

[1] S.D Olabarriaga and A.W.M Smeulders, "Interaction in the segmentation of medical images: A survey " *Medical Image Analysis*, pp. 127-142, 2001.

[2] Selver, M. Alper and Kocaoğlu, Aykut and Demir, Güleser K. and Doğan, Hatice and Dicle, Oğuz and Güzeliş, Cüneyt, "Patient oriented and robust automatic liver segmentation for pre-evaluation of liver transplantation.", *Comp. in Bio. and Med.*, Vol.38, pp.765-784, 2008.

[3] Dembo, A. and Farotimi, O. and Kailath, T., "High-order absolutely stable neural networks" *Circuits and Systems, IEEE Transactions on*, Reading, vol 6, num 1, pp. 267-272, 1991.

[4] Villalobos, L. and Merat, F.L., "Learning capability assessment and feature space optimization for higher-order neural networks" *Neural Networks, IEEE Transactions on*, vol 38, num 1, pp. 57-65. 1995

# Numerical estimates for the regularization of nonautonomous ill-posed problems

**Matthew A. Fury**

Assistant Professor of Mathematics, Penn State Abington, Division of Science & Engineering, 1600 Woodland Road, Abington, PA 19001, USA

E-mail: maf44@psu.edu

**Abstract.** The regularization of ill-posed problems has become a useful tool in studying initial value problems that do not adhere to certain desired properties such as continuous dependence of solutions on initial data. Because direct computation of the solution becomes difficult in this situation, many authors have alternatively approximated the solution by the solution of a closely-defined well-posed problem. In this paper, we demonstrate this process of regularization for the backward heat equation with a time-dependent diffusion coefficient, among other nonautonomous ill-posed problems. In the process, we provide two different approximate well-posed models and numerically compare convergence rates of their solutions to a known solution of the original ill-posed problem.

## 1. Introduction

In this paper, motivated by the recent work of Trong and Tuan [9, 10], we illustrate regularization for certain nonautonomous ill-posed problems and demonstrate an array of numerical estimates for the regularization of the backward heat equation in $L^2[0,\pi]$ with a time-dependent diffusion coefficient, e.g.

$$\begin{aligned} \frac{\partial u}{\partial t} &= -e^t \frac{\partial^2 u}{\partial x^2}, \quad 0 < x < \pi, \ 0 < t < 1 \\ u(0,t) &= u(\pi,t) = 0, \quad 0 < t < 1 \\ u(x,0) &= \phi(x), \quad 0 < x < \pi. \end{aligned} \tag{1}$$

Because solutions $u$ may not exist or, if they do exist, generally do not depend continuously on initial data, numerical calculations are often difficult. The regularization of the problem offers an alternate method of obtaining information about such solutions $u$ through comparison to the solution of an approximate well-posed problem. For example, for $\epsilon > 0$, one may consider an approximate solution $v_\epsilon$ where the model in (1) is replaced by Lattes and Lions's quasi-reversibility method $\frac{\partial v}{\partial t} = -e^t \frac{\partial^2 v}{\partial x^2} - \epsilon \frac{\partial^4 v}{\partial x^4}$ (cf. [5]).

Regularization for ill-posed problems, particularly the abstract Cauchy problem $\frac{du}{dt} = Au(t), \quad 0 \le t < T, \quad u(0) = \phi$, where $A$ is an operator in a Banach space $X$, e.g. $A = -\Delta$ in $X = L^p(\mathbb{R})$, has been established in various settings by authors including Lattes and Lions [5], Miller [7], Showalter [8], and more recently Mel'nikova and Filinkov [6], Ames and Hughes [1], and Huang and Zheng [3, 4]. Very recently, error estimates for regularization of the backward heat equation have been scrutinized by Trong and Tuan (cf. [9, 10]). In [9], using the quasi-reversibility method, they determine a table of numerical estimates for $\|u(x,0.5) - v_\epsilon(x,0.5)\|_{L^2(0,\pi)}$ depending on different values of $\epsilon$ limiting to 0.

The tables that we provide in this paper extend the work of Trong and Tuan to the *nonautonomous* equation (1) and also encompass the difference $\|u(x,t) - v_\epsilon(x,t)\|_2$ for multiple values of $t$ in $[0,1]$. Further, our paper illustrates two methods of regularization for other higher order nonautonomous ill-posed problems. The results of this paper find that the quasi-reversibility method yields a much faster convergence of the well-posed solution $v_\epsilon$ to the solution $u$ of the ill-posed problem, as compared to a second method following Showalter [8].

## 2. Regularization

In this section, we provide background theory which explains the way in which problem (1), among other higher-order equations, may be regularized by the methods we propose. Consider the nonautonomous ill-posed problem

$$\frac{du}{dt} = a(t)A^k u(t), \quad 0 \le t < T, \quad u(0) = \phi \tag{2}$$

in a Hilbert space $H$ where $A$ is a positive, self-adjoint operator in $H$, $k$ is a positive integer, and $a \in C([0,T] : \mathbb{R}^+)$. The regularization of this problem is defined as follows.

**Definition 1.** [4, Definition 3.1] A family $\{R_\epsilon(t) \mid \epsilon > 0,\ t \in [0,T]\}$ of bounded linear operators on $X$ is called a *family of regularizing operators for the problem* (2) if for each solution $u(t)$ of (2) with initial data $\phi \in H$, and for any $\delta > 0$, there exists $\epsilon(\delta) > 0$ such that

(i) $\epsilon(\delta) \to 0$ as $\delta \to 0$,

(ii) $\|u(t) - R_{\epsilon(\delta)}(t)\phi_\delta\| \to 0$ as $\delta \to 0$ for $0 \le t \le T$ whenever $\|\phi - \phi_\delta\| \le \delta$.

In order to define a regularizing family, we consider an approximate well-posed problem

$$\frac{dv}{dt} = f_\epsilon(t,A)v(t), \quad 0 \le t < T, \quad v(0) = \phi \tag{3}$$

where $\epsilon > 0$ and $f_\epsilon(t,A)$ is defined by either of two examples $f_\epsilon(t,A) = a(t)A^k - \epsilon A^{k+1}$ or $f_\epsilon(t,A) = a(t)A^k(I + \epsilon A^k)^{-1}$, the first being a generalization of Lattes and Lions's quasi-reversibility method [5] and the second motivated by work of Showalter [8]. In [2], Fury and Hughes show under certain stabilizing conditions on a known solution $u(t)$ of (2), e.g. $\|u(T)\| \le M'$, that $\|u(t) - v_\epsilon(t)\| \le C\epsilon^{1-\frac{t}{T}}M^{\frac{t}{T}}, 0 \le t < T$ where $v_\epsilon(t) = e^{\int_0^t f_\epsilon(\tau,A)d\tau}\phi$ is the unique solution of (3) and $C$ and $M$ are constants independent of $\epsilon$ ([2, Theorem 3.9, Example 4.1, Example 4.2]).

Hence, with the aid of this inequality, we may define $R_\epsilon(t) = e^{\int_0^t f_\epsilon(\tau,A)d\tau}$ as a regularizing operator in which case $v_\epsilon(t) = R_\epsilon(t)\phi$. Regularization is then established by the calculation $\|u(t) - R_\epsilon(t)\phi_\delta\| \le \|u(t) - v_\epsilon(t)\| + \|R_\epsilon(t)\phi - R_\epsilon(t)\phi_\delta\| \to 0$ as $\delta \to 0$ where the first quantity tends to 0 if $\epsilon \to 0$ as $\delta \to 0$, while the second quantity on the right tends to 0 as $\delta \to 0$ because (3) is well-posed (and so satisfies continuous dependence on initial data). Choosing $\epsilon$ in terms of $\delta$ typically depends on the growth order of $\|R_\epsilon(t)\|$. Generally, this is not difficult and will be illustrated explicitly in the following section.

## 3. Numerical Estimates

Following recent work of Trong and Tuan [9, 10], we illustrate the process of regularization discussed in Section 2 for concrete partial differential equations represented by (2) and provide tables of numerical estimates for the regularization.

**Example 1.** Let us consider, in $H = (L^2[0,\pi], \|\cdot\|_2)$ where $\|\phi\|_2 = (\int_0^\pi |\phi(x)|^2 dx)^{1/2}$ for $\phi \in L^2[0,\pi]$, the following nonautonomous partial differential equation by setting $A = -\Delta$ and $a(t) = e^t$ in (2):

$$\begin{aligned} \frac{\partial u}{\partial t} &= (-1)^k e^t \frac{\partial^{2k} u}{\partial x^{2k}}, \quad 0 < x < \pi,\ 0 < t < 1 \\ u(0,t) &= u(\pi,t) = 0, \quad 0 < t < 1 \\ u(x,0) &= e \sin x, \quad 0 < x < \pi. \end{aligned} \tag{4}$$

The problem (4) is ill-posed with exact solution $u(x,t) = e^{e^t} \sin x$. In order to illustrate regularization for (4) we will consider a perturbed initial data $\phi_n(x) = e \sin x + \frac{1}{n}\sin(nx)$ which converges to $u(x,0) = e \sin x$ as $n \to \infty$. The approximate well-posed problem (3) according to the first approximation $f_\epsilon(t,A) = e^t A^k - \epsilon A^{k+1}$ becomes

$$\begin{aligned} \frac{\partial v}{\partial t} &= (-1)^k \left( e^t \frac{\partial^{2k} v}{\partial x^{2k}} + \epsilon \frac{\partial^{2k+2} v}{\partial x^{2k+2}} \right), \quad 0 < x < \pi,\ 0 < t < 1 \\ v(0,t) &= v(\pi,t) = 0, \quad 0 < t < 1 \\ v(x,0) &= e \sin x + \frac{1}{n}\sin(nx), \quad 0 < x < \pi \end{aligned} \tag{5}$$

with solution $v_\epsilon^n(x,t) = e^{e^t - \epsilon t} \sin x + \frac{1}{n} e^{(e^t-1)n^{2k} - \epsilon t n^{2k+2}} \sin(nx)$. For $0 < t \leq 1$, assuming $n > 1$, it is easy to show that $\|u(x,t) - v_\epsilon^n(x,t)\|_2 = \sqrt{\frac{\pi}{2}} \sqrt{\left[e^{e^t}(1 - e^{-\epsilon t})\right]^2 + \left[\frac{1}{n} e^{(e^t-1)n^{2k} - \epsilon t n^{2k+2}}\right]^2}$. Because of the negative leading term $-\epsilon t n^{2k+2}$ in the exponent of the last exponential, $\epsilon$ may be chosen as $\epsilon = \frac{e}{n^2}$, in which case we have that $\epsilon \to 0$ as $n \to \infty$ and $\|u(x,t) - v_\epsilon^n(x,t)\|_2 = \sqrt{\frac{\pi}{2}} \sqrt{\left[e^{e^t}(1 - e^{-\frac{et}{n^2}})\right]^2 + \left[\frac{1}{n} e^{(e^t - 1 - et)n^{2k}}\right]^2} \to 0$ as $n \to \infty$ since $e^t - 1 - et < 0$ for $0 < t \leq 1$. Note for the case that $t = 0$, we simply have $\|u(x,0) - v_\epsilon^n(x,0)\|_2 = \|e \sin x - \phi_n(x)\|_2 = \frac{1}{n}\sqrt{\frac{\pi}{2}} \to 0$ as $n \to \infty$. Hence, regularization is complete.

Note that when $k = 1$, (4) becomes the backward heat equation with time-dependent diffusion coefficient $e^t$. Table 1 shows numerical estimates for $\|u(x,t) - v_\epsilon^n(x,t)\|_2$ in this specific case ($k = 1$) based on different values of $n$ and $t$ and assuming the choice $\epsilon = \frac{e}{n^2}$.

**Table 1.**

| $\|u(x,t) - v_\epsilon^n(x,t)\|_2$ | | | | | |
|---|---|---|---|---|---|
| | | $n$ | | | |
| | | 1000 | 10000 | 100000 | 1000000 |
| $t$ | 0 | 0.001253314 | 0.000125331 | 1.25331E-05 | 1.25331E-06 |
| | 0.1 | 1.02878E-06 | 1.02878E-08 | 1.02878E-10 | 1.02861E-12 |
| | 0.2 | 2.31117E-06 | 2.31118E-08 | 2.31117E-10 | 2.31126E-12 |
| | 0.3 | 3.94196E-06 | 3.94196E-08 | 3.94196E-10 | 3.94183E-12 |
| | 0.4 | 6.05767E-06 | 6.05767E-08 | 6.05767E-10 | 6.05789E-12 |
| | 0.5 | 8.85839E-06 | 8.85839E-08 | 8.85839E-10 | 8.85836E-12 |
| | 0.6 | 1.26427E-05 | 1.26428E-07 | 1.26428E-09 | 1.26424E-11 |
| | 0.7 | 1.78654E-05 | 1.78655E-07 | 1.78655E-09 | 1.78656E-11 |
| | 0.8 | 2.52339E-05 | 2.52339E-07 | 2.52339E-09 | 2.52335E-11 |
| | 0.9 | 3.58747E-05 | 3.58748E-07 | 3.58748E-09 | 3.58753E-11 |
| | 1 | 5.16284E-05 | 5.16285E-07 | 5.16285E-09 | 5.16282E-11 |

**Table 2.**

| $\|u(x,t) - w_\epsilon^n(x,t)\|_2$ | | | | | |
|---|---|---|---|---|---|
| | | $n$ | | | |
| | | 1000 | 10000 | 100000 | 1000000 |
| $t$ | 0 | 0.001253314 | 0.000125331 | 1.25331E-05 | 1.25331E-06 |
| | 0.1 | 0.129960995 | 0.10662892 | 0.09040324 | 0.078463391 |
| | 0.2 | 0.301472976 | 0.248217982 | 0.210952415 | 0.183414183 |
| | 0.3 | 0.530421901 | 0.438378357 | 0.373542488 | 0.325406902 |
| | 0.4 | 0.839826665 | 0.696954498 | 0.59558013 | 0.519929335 |
| | 0.5 | 1.26361519 | 1.053350114 | 0.902955844 | 0.790084105 |
| | 0.6 | 1.852663844 | 1.551900529 | 1.334863161 | 1.170949288 |
| | 0.7 | 2.684706206 | 2.26073153 | 1.951776304 | 1.71681859 |
| | 0.8 | 3.880948905 | 3.286694025 | 2.848953561 | 2.513494718 |
| | 0.9 | 5.634602185 | 4.801148727 | 4.179848201 | 3.699671806 |
| | 1 | 8.261138225 | 7.085642684 | 6.197758763 | 5.505105575 |

Alternatively, we may consider the well-posed problem according to the second approximation

$f_\epsilon(t, A) = e^t A^k (I + \epsilon A^k)^{-1}$. Here, problem (3) becomes

$$\frac{\partial w}{\partial t} = (-1)^k \left( e^t \frac{\partial^{2k} w}{\partial x^{2k}} - \epsilon \frac{\partial^{2k+1} w}{\partial x^{2k} \partial t} \right), \quad 0 < x < \pi,\ 0 < t < 1 \tag{6}$$
$$w(0,t) = w(\pi,t) = 0, \quad 0 < t < 1$$
$$w(x,0) = e \sin x + \frac{1}{n} \sin(nx), \quad 0 < x < \pi$$

with solution $w_\epsilon^n(x,t) = e^{\frac{e^t+\epsilon}{1+\epsilon}} \sin x + \frac{1}{n} e^{\frac{(e^t-1)n^{2k}}{1+\epsilon n^{2k}}} \sin(nx)$. Note, for $0 < t \leq 1$, again assuming $n > 1$, it may be shown that $\|u(x,t) - w_\epsilon^n(x,t)\|_2 = \sqrt{\frac{\pi}{2}} \sqrt{\left[ e^{e^t} - e^{\frac{e^t+\epsilon}{1+\epsilon}} \right]^2 + \left[ \frac{1}{n} e^{\frac{(e^t-1)n^{2k}}{1+\epsilon n^{2k}}} \right]^2}$. This time, a reasonable choice is to choose $\epsilon$ so that $\frac{1}{n} e^{\frac{(e^t-1)n^{2k}}{1+\epsilon n^{2k}}}$ grows like $\frac{1}{\sqrt{n}}$. Equating these two quantities, and setting $t = 1$, we arrive at $\epsilon = \frac{(e-1)n^{2k} - ln\sqrt{n}}{n^{2k}\, ln\sqrt{n}}$. Hence, $\epsilon \to 0$ as $n \to \infty$ and

$$\|u(x,t) - w_\epsilon^n(x,t)\|_2 = \sqrt{\frac{\pi}{2}} \sqrt{\left[ e^{e^t} - e^{\frac{(e^t n^{2k} - 1) ln\sqrt{n} + (e-1)n^{2k}}{(n^{2k}-1) ln\sqrt{n} + (e-1)n^{2k}}} \right]^2 + n^{\frac{e^t-1}{e-1} - 2}} \to 0 \quad \text{as} \quad n \to \infty.$$

Again for the case that $t = 0$, we have $\|u(x,0) - w_\epsilon^n(x,0)\|_2 = \|e \sin x - \phi_n(x)\|_2 = \frac{1}{n}\sqrt{\frac{\pi}{2}} \to 0$ as $n \to \infty$. Regularization is thus accomplished in the case of the second approximation. Returning again to regularization of the backward heat equation, Table 2 shows estimates for the difference $\|u(x,t) - w_\epsilon^n(x,t)\|_2$ in the case that $k = 1$, based on different values of $n$ and $t$ and assuming $\epsilon = \frac{(e-1)n^2 - ln\sqrt{n}}{n^2\, ln\sqrt{n}}$.

In comparing Table 1 and Table 2, we note that the convergence to 0 in the case of the second approximation is considerably slower than that in the first approximation.

## Acknowledgments

The author would like to thank Rhonda J. Hughes for her support, and also Stephen Hill, Ryan Jacobs, Jacquelyn Nguyen, and Bhumika Patel for their contributions to the tables.

## References

[1] Ames K A and Hughes R J 2005 Structural stability for ill-posed problems in Banach space *Semigroup Forum* **70**, 127–45

[2] Fury M and Hughes R J 2010 Continuous dependence of solutions for ill-posed evolution problems *Electron. J. Differ. Equ.* **Conf. 19** 99–121

[3] Huang Y and Zheng Q 2004 Regularization for ill-posed Cauchy problems associated with generators of analytic semigroups *J. Differ. Equ.* **203** 38–54

[4] Huang Y and Zheng Q 2005 Regularization for a class of ill-posed Cauchy problems *Proc. Am. Math. Soc.* **133(10)** 3005–12

[5] Lattes R and Lions J L 1969 *The Method of Quasireversibility, Applications to Partial Differential Equations* (New York: Amer. Elsevier)

[6] Melnikova I V and Filinkov A I 2001 *Abstract Cauchy Problems: Three Approaches* (London: Chapman & Hall)

[7] Miller K 1973 Stabilized quasi-reversibility and other nearly-best-possible methods for non-well-posed problems, in: *Symposium on Non-Well-Posed Problems and Logarithmic Convexity (Heriot-Watt Univ. Edinburgh) Springer Lecture Notes in Mathematics* vol 316 (Berlin: Springer) pp 161–76

[8] Showalter R E 1974 The final value problem for evolution equations *J. Math. Anal. Appl.* **47** 563–72

[9] Trong D D and Tuan N H 2006 Regularization and error estimates for nonhomogeneous backward heat problems *Electron. J. Differ. Equ.* **2006(4)**

[10] Trong D D and Tuan N H 2008 A nonhomogeneous backward heat problem: regularization and error estimates *Electron. J. Differ. Equ.* **2008(33)**

# Multiscale principal component analysis

**A A Akinduko and A N Gorban**

Mathematics Department, University of Leicester, Leicestershire, LE1 7RH, UK

E-mail: aaa78@le.ac.uk and ag153@le.ac.uk

**Abstract**. Principal component analysis (PCA) is an important tool in exploring data. The conventional approach to PCA leads to a solution which favours the structures with large variances. This is sensitive to outliers and could obfuscate interesting underlying structures. One of the equivalent definitions of PCA is that it seeks the subspaces that maximize the sum of squared pairwise distances between data projections. This definition opens up more flexibility in the analysis of principal components which is useful in enhancing PCA. In this paper we introduce scales into PCA by maximizing only the sum of pairwise distances between projections for pairs of datapoints with distances within a chosen interval of values $[l,u]$. The resulting principal component decompositions in Multiscale PCA depend on point $(l,u)$ on the plane and for each point we define projectors onto principal components. Cluster analysis of these projectors reveals the structures in the data at various scales. Each structure is described by the eigenvectors at the medoid point of the cluster which represent the structure. We also use the distortion of projections as a criterion for choosing an appropriate scale especially for data with outliers. This method was tested on both artificial distribution of data and real data. For data with multiscale structures, the method was able to reveal the different structures of the data and also to reduce the effect of outliers in the principal component analysis.

## 1. Introduction

In 1901, Pearson proposed approximating high dimensional data with lines and planes and hence invented the Principal Component Analysis (PCA). At present time, PCA is a powerful analysis tool with many interesting applications which include: dimension reduction, blind source separation, data visualization, image compression, and with relevance in many applied disciplines such as quantitative finance, biology, pharmaceutics, taxonomy, healthcare and many more [1-3].

However, despite the many applications of PCA, it is not without its drawbacks. An example of such drawbacks is that PCA is based on the covariance matrix which is sensitive to outliers. In this paper, outliers are defined as data elements with large distance from the other data elements in a data sample. Even though outliers can be filtered before performing PCA on the dataset, however in some contexts, identifying outliers could be cumbersome. In addition to the above, datasets are usually noisy (here we define noises as data elements with rather small variance) and the presence of noise in data analysis can further obfuscate the underlying structure(s) of the data being investigated [5].

One of the definitions of PCA is that PCA finds subspaces (lines, planes or higher dimensional subspaces) that maximize the sum of point-to-point squared distances between the orthogonal projections of data points to them. In this paper, scale was introduced to enhance the performance of PCA on datasets. That is, we will use in the definition of multiscale PCA maximization of the sum of point-to-point squared distances between the orthogonal projections of data points for the pairs of points with distances in some intervals. The result of this is PCA decomposition of the data which depend on the scale chosen. A further study of these PCA decompositions reveals some underlying structures which could have been obfuscated by other structures such has the presence of outliers or repeated patterns as shown later. We also proposed a criterion for determining the appropriate scale for computing the principal components for data with outliers.

## 2. Definitions and Mathematical Background

### *2.1. Weighted PCA*

Consider the problem of finding the principal component using weighted pairwise distances of projected data. This problem is stated below.

$$\sum_{i<j} w_{ij}[dist^2(P_L\mathbf{x}_i, P_L\mathbf{x}_j)]$$

$$D_X = \sum_{i<j} w_{ij} \| P_L(\mathbf{x}_i - \mathbf{x}_j)\|^2 \to \max \tag{1}$$

$$\text{Subject to: } (\mathbf{v}_\alpha, \mathbf{v}_\beta) = \delta_{\alpha\beta}.$$

Where $w_{ij} = w_{ji}$ is the non-negative weight assigned to the distance between element $i$ and $j$ and $w_{ij} = 0$, for $i = j$.

Let $L^w = [L^w{}_{ij}] = \left(\delta_{ij}(\sum_{j=1}^{n} w_{ij}) - w_{ij}\right)$, where $w_{ij} = 0$, for $i = j$. Let us introduce the matrix $\widetilde{M}_{\alpha\beta} = (X_\alpha{}^T L^w X_\beta)$. In these notations, the problem given by (15) is reduced to

$$\max_{v_1,\dots,v_k} \sum_{\alpha=1}^{k} \mathbf{v}_\alpha^T \widetilde{M} \mathbf{v}_\alpha$$
$$\text{Subject to } (\mathbf{v}_\alpha, \mathbf{v}_\beta) = \delta_{\alpha\beta} \quad \alpha, \beta = 1,2,..,k, \tag{2}$$

where $\widetilde{M}$ is a symmetric positive semi-definite matrix, and the eigenvectors corresponding to the sorted eigenvalues of the matrix $\widetilde{M}$ is a maximizer of the constrained maximization problem (2). In the case of degenerated eigenvalues, the set $\mathbf{e}_1,\dots,\mathbf{e}_m$ is not uniquely defined.

## 3. Multiscale PCA (MPCA)

In this section, we introduce the Multiscale PCA (MPCA) algorithm to enhance the robustness of the PCA especially in revealing hidden structure(s) that may be present in dataset but which the conventional approach might not reveal. MPCA compute principal components by maximizing the sum of pairwise distances between data projection for only pairs of datapoints for which the distance is within the chosen scale:

$$\begin{cases} w_{ij} = 1 & l \le \| \mathbf{x}_i - \mathbf{x}_j \|_2 \le u \\ w_{ij} = 0 & otherwise. \end{cases} \tag{3}$$

In the scale interval, $(l,u)$, $l$ is the lower limit of the scale and $u$ is the upper limit. Let $d^{\min}$ be the minimum pairwise distance greater than zero and $d^{\max}$ be the maximum pairwise distance in the data. We select the pairs $(l,u)$ from a triangle $\Delta=\{(l,u)\colon d^{\min} \le l < u \le d^{\max}\}$.

With this control over the pairwise distances, we are able compute PCA at various scales and the outcome of this is scale dependent PCA which can reveal interesting underlying structure(s) that may be present in data. For example, reducing the upper limit of the scale while keeping the lower limit at 0 translate to computing PCA by considering smaller distances and excluding very large distances. This has the effect of minimizing without explicit exclusion the contribution of certain influential data elements in the analysis of the principal components.

### *3.1. The Multiscale PCA Algorithm*

Here we discuss the Multiscale PCA Algorithm.

1. Given the data sample.
2. Centralize the data by subtracting the mean of the variables from each observation.
3. Find the dissimilarity matrix by computing the Euclidean distance.
4. Choose an appropriate scale between 0 and the maximum distance. For easy analysis, a scale between 0 and 1 could be chosen and then multiplied by the maximum distance. For this paper when using scale between 0 and 1 we call it standard scale.
5. Calculated the binary weight as given in equation (27)
6. Calculate the matrix $L^w$ as given below
7. Calculate the matrix $A = Y^T L^w Y$, where $Y$ is the centralized data matrix.
8. Find the sorted eigenvalues of the matrix $A$ in descending order of magnitude and project the data onto their corresponding eigenvectors. This will be the principal components at the selected scale.

## 4. Clustering Analysis on the Interval of Scales

To further study these structures, we consider clustering analysis on the interval of scales and we introduce the Ratio of Distortion in this section. We represent the PCA structure of the data at any point $(l,u)$ by the sum of the projectors corresponding to MPCA at that point. This will be denoted by

$\rho_k = \sum_{i=1}^{k} \mathbf{e}_i \otimes \mathbf{e}_i$, $k = 1,2,,,m-1$. The full description of the principal components decompositions of data $X$ is given by an ordered set ("cortege") of projectors $\rho_1, \rho_2, \ldots, \rho_{m-1}, \rho_m = 1$ (compare to [4]) If we arrange the $e_i$ $i = 1,2,\ldots,k$ as columns of matrix $E$, then $\rho_k X = EE^T X$ and $\rho_k = EE^T$. For $k = m,\ EE^T = \mathbf{I}$.

MPCA lead to scale dependent PCA structures and with these PCA structures represented as defined above, we can study the structures in our data further by analyzing these projectors.

The PCA structures associated with two different points on the plane is said to be similar if their corresponding projectors $\rho_k$ are similar. We guess that in some cases there are clear internal structures in the data which depend on scales. Performing MPCA on the data leads to a continuum of PCA structures depending on scales used and to reveal the structures in the data, we join scales with similar PCA structures and separate scales with dissimilar PCA structures. This leads to the idea of clustering of scales. We represent the distance between two points on the scale by the distance between their corresponding PCA structures. Clustering analysis of the scales group similar PCA structures together and this reveals some structures in the data. We describe each cluster by the projector corresponding to the medoid point of the cluster. We denote the projector $\rho_k$ at a point $\chi_p$ by $\rho_{\chi_p}$. For any pair of points $\chi_p, \chi_q$ in the space of scales we can compute the distance between the associated projectors

$\rho_{\chi_p}, \rho_{\chi_q}$ for a given $k$ using the matrix norm. This clustering is illustrated on a simple example of data with hierarchical microstructures (Fig. 1-3).

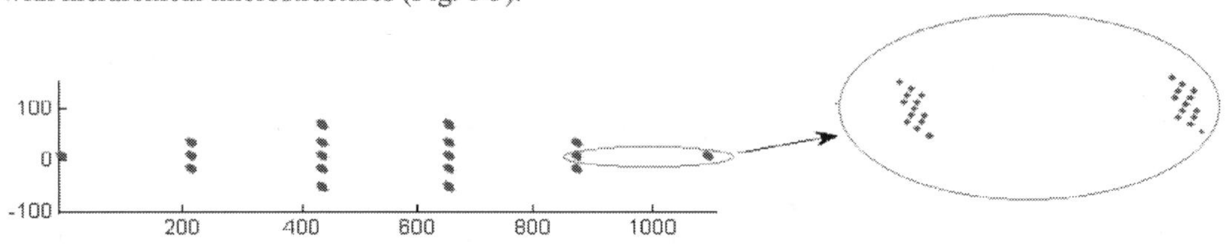


**Figure 1.** Scatter plot of data with hierarchical microstructure.

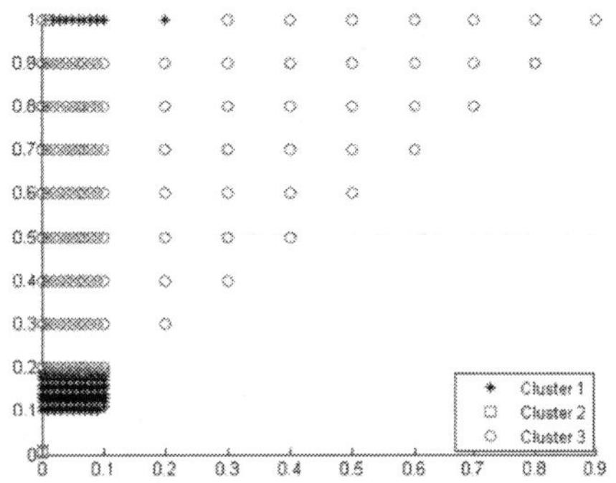


**Figure 2.** Cluster of scales on the plane for data from Fig. 1.

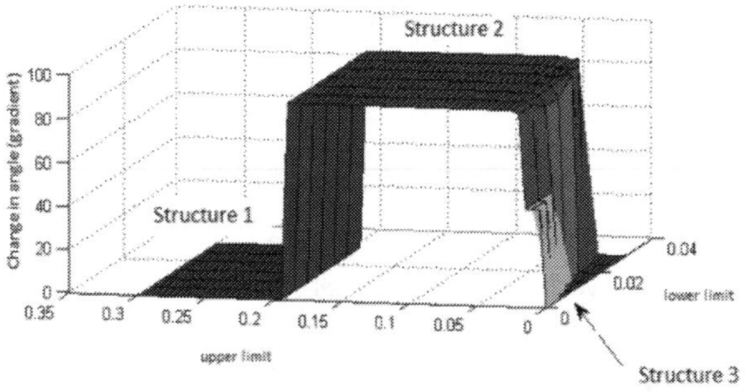


**Figure 3.** The angle between the first principal component using PCA and the first principal component using MPCA at a given scale.

## 5. Conclusion

MPCA is developed to solve the problem of revealing hidden geometric structures in data. The result of MPCA on data leads to a continuum of PCA structures of the data which is dependent on the intervals chosen. For data with clear multiscale structures, the cluster analysis of the scales reveals some underlying structures in the data which conventional PCA cannot reveal due to the fact that such structures are obfuscated by other structures of higher variance.

## References

[1] Jolliffe I T 2002 Principal Component Analysis Second Edition (New York: Springer)

[2] Gorban A N and Zinovyev A Y 2009. Principal Graphs and Manifolds *Handbook of Research on Machine Learning Applications and Trends: Algorithms, Methods and Techniques* (Hershey, PA: IGI Global) pp 28-59.

[3] Burges C J C 2010 Geometric Methods for Feature Extraction and Dimensional Reduction - A Guided Tour *Data Mining and Knowledge Discovery Handbook* (New York: Springer) ed O Maimon and L Rokach . 2nd Edition ISBN 978-0-387-09822-7 pp 53-82.

[4] Arnold R and Jupp P E 2013 Statistics of orthogonal axial frames, *Biometrika* pp 1-16

[5] Koren Y and Carmel L 2004 Robust linear dimensionality reduction, Visualization and Computer Graphics, *IEEE Transactions* vol.10, no.4, pp.459-70.

[6] Akinduko A A, Gorban A N 2013 arXiv:1307.8339 [stat.ME]

# Is it possible to predict long-term success with k-NN? Case study of four market indices (FTSE100, DAX, HANGSENG, NASDAQ)

**Y Shi, A N Gorban, TY Yang**

Department of Mathematics, University of Leicester, LE1 7RH, UK

ys98@leicester.ac.uk , ag153@leicester.ac.uk, yty725@sina.com

**Abstract**. This case study tests the possibility of prediction for 'success' (or 'winner') components of four stock & shares market indices in a time period of three years from 02-Jul-2009 to 29-Jun-2012.We compare their performance ain two time frames: initial frame three months at the beginning (02/06/2009-30/09/2009) and the final three month frame (02/04/2012-29/06/2012).To label the components, average price ratio between two time frames in descending order is computed. The average price ratio is defined as the ratio between the mean prices of the beginning and final time period. The 'winner' components are referred to the top one third of total components in the same order as average price ratio it means the mean price of final time period is relatively higher than the beginning time period. The 'loser' components are referred to the last one third of total components in the same order as they have higher mean prices of beginning time period. We analyse, is there any information about the winner-looser separation in the initial fragments of the daily closing prices log-returns time series.The Leave-One-Out Cross-Validation with k-NN algorithm is applied on the daily log-return of components using a distance and proximity in the experiment. By looking at the error analysis, it shows that for HANGSENG and DAX index, there are clear signs of possibility to evaluate the probability of long-term success. The correlation distance matrix histograms and 2-D/3-D elastic maps generated from ViDaExpert show that the 'winner' components are closer to each other and 'winner'/'loser' components are separable on elastic maps for HANGSENG and DAX index while for the negative possibility indices, there is no sign of separation.

Keywords: possibility of prediction, long-term success, Leave-One-Out Cross-Validation, k-NN, ViDaExpert and elastic maps

## 1. Introduction

It is important to study the predictability of the time series separately from constructing the specific predictors because in creation of each model we assume some additional hypotheses about the model structure.

Our case study is aimed to find possibility of the prediction of 'success' within three years' time interval from 02-JUL-2009 to 29-JUN-2012 for four selected stock and share market indices. We compare their performance in two time frames: initial frame three months at the beginning

(02/07/2009-30/09/2009) and the final three month frame (02/04/2012-29/06/2012).The idea of the main experiment is based on backward analysis. The backward analysis can be defined as an analysis to determine properties of the inputs of a program from properties or contexts of outputs. This case study is aimed to construct experiments on the data to test if it is possible to predict the long-term success. The possibility of long-term 'success' of the selected indices is tested from the results of the experiment in the three years' time interval. For each stock market index, the closing prices of all components are collected from Yahoo! Finance website. After data pre-processing step, the remaining components are labeled with 'winner' companies or 'loser' companies (or simply just 'winner', 'loser') by using the '1/3 average price' approach. For this approach, we compute the average price ratio which is defined as the mean price of the end period divided by the mean price of beginning period. The companies are then sorted by the descending order of this computed ratio. We label the first 33.3% of companies as 'winner'and label the last 33.3% of the companies as 'loser'. Then the log-return prices are computed on this data. The k-NN algorithm with Leave-One-Out Cross-Validation with two distance measurements is used as the indicator to test the possibility of prediction.

We use Leave-One-Out Cross-Validation for k-NN classifier [1] to test the possibility of prediction of 'success' components for each market index. The data is collected from data and cleaned. Then we did experiment of Cross-Validation for k-NN using two different forms of distance measurements. Then we analyze the total error and separate error and use two methods to visualize our result. We investigate that there is a possibility of predictions for long-term success for HANGSENG and DAX indices in the result section. We summarize our result and conclude in the last part.

## 2. Results and analysis

The experiment is computed using MATLAB. It begins with data pre-processing step. The companies that do not have enough amounts of closing prices are deleted from the company list. The next step is to apply the'1/3 Average Price' method for the remaining companies, labeled them with 'winner' if the corresponding average prices are the largest 1/3 of the sequence of average prices, and 'loser' if the average prices are the last 1/3 of the sorted sequence. The result data is generated by joining the 'winner' and 'loser' companies in matrix form vertically. In this matrix each column represents each date and each row represents each company. The daily log-return matrix is computed from the joint matrix. This log-return matrix is then used for Leave-One-Out Cross-Validation of 1-NN algorithm. The 1-NN (k=1 case) is admissible as the error is up bounded by twice of Bayes error rate. We use 1-NN for our experiments.

### *2.1. Analysis of total error and separate error*

The error of Leave-One-Out Cross-Validation for 1-NN is performed for different time periods. (i.e. from 3 months to 18 months) The total error is referred to the number of misclassified points for both 'winner' and 'loser' companies. The errors generated using different functions of measurement are almost identical. The proximity can generate a bit smaller error than the distance function for several months. The total error for time period of 3 months is the minimum for three indices. For indices FTSE, DAX and NASDAQ, the errors are mostly around 50%.This means the prices are random and there is no sign of possibility of prediction. The error percentage is less than 50% for HANGSENG index. This shows this market is not completely random. To analyze the characteristics for 'winner' and 'loser' companies, the separate error analysis is applied.The sum of 'winner' error number and 'loser' error number should be the same as the total error number.

For indices FTSE and NASDAQ, the error percentages are approximately 50% for both 'winner' and 'loser' companies. Hence the separate error analysis shows that there is no sign of possibility of prediction. For DAX index, the errors are ranged from 30% to 60%, it can be considered as a border case and the 'loser' companies in general have slightly lower errors than 'winner' companies. For HANGSENG index, both errors are below 50% and the errors of 'winner' companies are much smaller than the errors of 'loser' companies. Hence this means for HANGSENG index, there are some conclusions about predictability for the 'winner' companies.

### *2.2. Visualization using histograms of correlation distance matrix*

From result of experiment using different time frames, the minimum error rate occurs when the time interval is within 3 months for FTSE, DAX and HANGSENG index. The sign of possibility of prediction in this time period is most dominant among all. The histograms of correlation distances can be used to study the distribution of correlation distance. For each index, a set of four histogram plots are generated. They represent the distributions of in-class (winner/winner or loser/loser) correlation distances and cross-class (winner/loser) correlation distances. The in-class histograms show that the distribution of distances between 'winner' themselves and 'loser' themselves are almost identical. The distributions of in-class distances and cross-class distances are quite similar as well. This makes 'winner' and 'loser' companies are not easily separated. Since k-NN is structure sensitive algorithm, if the points are mixed together, it generates errors. Hence if 'winner' and 'loser' companies are mixed together, this market has no sign of possibility of prediction. The histogram figures for FTSE and NASDAQ indices show no sign of possibility of prediction. For DAX index, the histogram shows some possibility of prediction but this sign is not very clear. From the in-class distributions, the 'winner' and 'loser' seems to be separated, but this separation is not clear enough since the distribution peak of winner/winner distances is slightly shifted to the left hand side of the distribution peak of loser/loser distances.

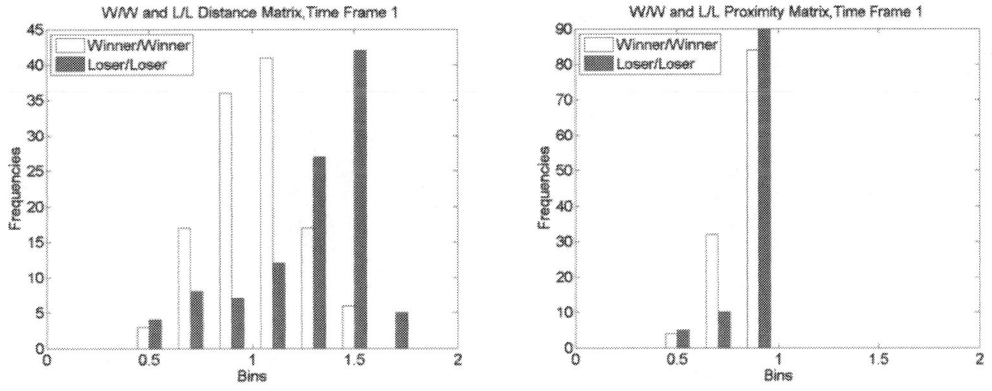


(a) (b)

Figure 2.1 Histograms of different expressions of correlation distances for HANGSENG index when using first 3 months closing prices for experiment. (a) Using Distance between winner/winner and loser/loser companies (b) Using Proximity between winner/winner and loser/loser companies

The histograms Figure 2.1 show the distributions of correlation distances for HANGSENG index. For this index, it has a clear sign of possibility of prediction. The distribution of winner/winner distances is on the left hand side of the distribution of loser/loser distances. Hence the 'winner' companies are more compactthan 'loser' companies and there is a good separation between 'winner' and 'loser' companies.

### *2.3. Visualization of using 2-D/3-D elastic maps*

The principal graphs and manifolds can be used to visualize the companies using idea of metaphor of elastic membrane and plate to construct the principal manifold approximations of varies topologies. The software 'ViDaExpert' enables us to visualize multidimensional data with the idea of using principal object to reduce the dimension of this data [2].

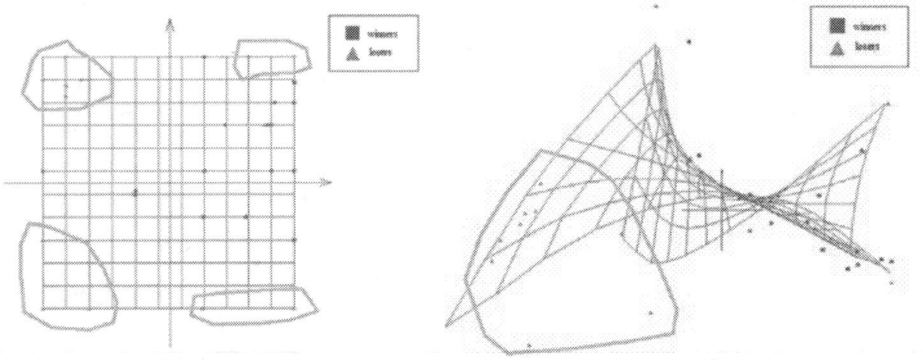

(a) (b)

Figure 2.2. Visualization of components (companies) of HANGSENG index (log-returns) using elastic maps: (a) 2D - Elastic Map (b) 3D - Principal Manifold Graph The Hand-made green lines show the "clusters" of loser companies

For HANGSENG index with first 3-month time frame, the elastic maps Figure 2.2 are generated. From (a), it shows that the winners and losers are almost separated nicely. The losers generate 4 "clusters" within the internal coordinates. These "clusters" are located on each corner of the coordinate plane. From (b), it shows the winners are closer to each other than the loser companies as the red points are more compact to each other. However, the map boundary is not linear since some loser companies that are very close to the winner companies and there exists around six points that is seems to be very close to the red points. These green points can be considered as outliers. As the error of k-NN is dominated by the structure of data, this visualization supports the LOOCV error analysis.

## 3. Conclusions

In this case study, the 1-NN classifier combined with Leave-One-Out Cross Validation is used as an indicator to test the possibility of long-term success. The error analysis and two visualization methods show that for HANGSENG index, there is a possibility of predicting long-term success as the winner companies are closer to each other. The DAX index may be possible to predict but the sign is not clear. For FTSE and NASDAQ indices, there is no sign of possibility of prediction. Our results support some previous observations about better predictability of emerging markets [3]. For more details we refer to [4].

**References**

[1] Fernández-Rodríguez F, Sosvilla-Rivero S, Andrada-Félix J 1999 *International Journal of Forecasting* **15** 383
[2] Gorban A N, Zinovyev A 2010 *Int. J. Neural Syst.* **20** 219
[3] Harvey C R 1995 Review of Financial studies **8** 773
[4] Shi Y, Gorban A N and Yang T Y 2013 arXiv: 1307.8308

# Finite Element Analysis of Piping Feature in Sheet-extrusion Process

**W Phanitwong**[1] **and S Thipprakmas**[2,3]
[1]Graduate Student, Department of Tool and Materials Engineering, King Mongkut's University of Technology Thonburi, Prachautid, Thungkru, Bangkok 10140, Thailand
[2]Department of Tool and Materials Engineering, King Mongkut's University of Technology Thonburi, Prachautid, Thungkru, Bangkok 10140, Thailand

E-mail: [1]wiriyakorn.wp@gmail.com, [2]sutasn.thi@kmutt.ac.th

**Abstract.** Considering the advantages of the press process as compared to the machining process, the making good utilization of work piece materials and the saving of production time, the Press machine associated with press form tooling is increasingly served for complicated shape parts production. The sheet-extrusion is the most important metal-forming process, which controls the material flow to fulfil the extruded shape. By using this process, the complicated extruded-shape parts could be manufactured without the machining process. Owing to the piping defect, however, the complicated extruded-shape parts are limited in the sheet-extrusion process. In this study, the finite element method (FEM) is used to investigate the piping feature. In addition, a relationship between extruded shape and piping defect is also examined. The laboratory experiments are performed to validate the accuracy of FEM simulation results. Based on the stress distribution and velocity analysis, the relationship between extruded shape and piping defect could be clearly identified. Furthermore, The FEM simulation results show good agreement with the experimental results with reference to the extruded shape and piping defect.

## 1. Introduction

In recent years, the complicated shape parts are increasingly required in many industrial fields such as automobile industry, aerospace industry, electronics industry, and housing-utensil industry. To fabricate these parts, the material removable process is required resulting in the increases in work piece material loss and production time. The press process with press form tooling that making good utilization of work piece materials and saving of production time is considered to serve for complicated shape parts production without the material removable process. In this study, the sheet-extrusion being a process for producing raised designs in sheet metal is investigated. This process can be made by applying high pressure on sheet of metal with the desired pattern of hollowed male or female dies. The pressed metal sheet is thinning and is flown into a hollowed male of female dies, producing a pattern of an extruded shape raised out of the surface of a pressed metal sheet. In the past researches, however, most researchers studied the conventional extrusion process. Atzema and Huétink used the arbitrary Lagrangian Eulerian method and a remeshing technique in the finite element method to simulate forward–backward extrusion [1]. A similar process was also simulated by

[3] Corresponding author

Kuzman et al. that the simultaneous and divided material flows were studied [2]. Kiuchi et al. performed an analysis of the steady three-dimensional extrusion processes, and obtained the values of the average extrusion pressure and metal flow characteristics [3]. However, there are a few researches carried out sheet-extrusion process. Zheng et al. [4] studied the numerical analysis of the sheet metal extrusion process. Chen et al. [5] investigated the flow-through phenomenon in the press forging of magnesium-alloy sheets. Hirota studied the fabrication of micro-billit by sheet extrusion process [6]. However, the occurrence of piping feature and a relationship between extruded shape and piping defect in the sheet-extrusion process have rarely been investigated. In this study, therefore, the finite element method (FEM) is used as a simulation tool to investigate the piping feature and the relationship between extruded shape and piping defect. The laboratory experiments are also carried out to validate the FEM simulation results. Based on the stress distribution and velocity analysis, the relationship between extruded shape and piping defect could be clearly identified. Furthermore, The FEM simulation results show good agreement with the experimental results with reference to the extruded shape and piping defect.

## 2. FEM simulation and experimental procedures

Figure 1 show the FEM simulation model for the sheet extrusion process. Only half of the simulation model was used to reduce the calculation time. The commercial analytical code for the three-dimensional implicit, quasi-static finite element method (DEFORM-3D) with automatic remeshing generation was used as the FEM simulation tool. The work piece material was set as an elasto-plastic type, and the punch, die and blank holder were assigned as the rigid type. The process conditions investigated were listed in Table 1. In order to validate the accuracy of FEM simulation results, the experiments were performed with the same FEM simulation conditions. A 150 kN hydraulic press machine was used. The pressed parts were sectioned and further processed by subsequent mounting. Optical microscopy was used to observe and capture extruded shape images.

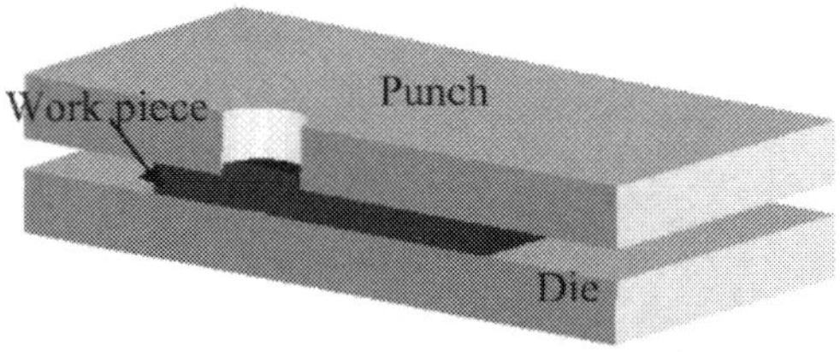


**Figure 1.** FEM simulation model.

**Table 1.** Experimental and FEM simulation conditions.

| | |
|---|---|
| Work piece material | A1100-O ($\sigma_B = 88$ MPa, $\lambda = 46\%$), Thickness ($t$): 7 mm |
| Initial work piece shape | Width 66 mm × Length 85mm |
| Extruded shape | Extruded diameter ($\varnothing_E$) : 20, 30 and 40 mm |

## 3. Results and discussion

### *3.1. Velocity analysis and stress distribution during extrusion phase*

Figure 2 shows the velocity analysis and stress distribution during extrusion phase. With the punch stroke of 0.1 mm, the work piece was pressed and the material was forced to flow into the hole as shown in Figure 2(a-1). With this feature, as shown in Figure 2(a-2), the compressive stress was generated in the pressed work piece, whereas the tensile stress was slightly generated on the extruded area. With this characteristic, the extruded shape could be raised by continuously pressing punch on the work piece. As the punch stroke increased to be 2.2 mm, the pressed work piece was thinning and the pressed material was strongly forced to flow into the hole, as shown in Figure 2(b-1). Therefore, the extruded shape could be formed. It was observed that the compressive stress increased in the pressed work piece and the tensile stress also increased on the upper surface of extruded area, as shown in Figure 2(b-2). However, it is notable that the velocity on the centre of hole is larger than that on the edge of hole. This resulted in the piping defect generated on the lower surface of extruded zone, as shown in Figure 2(b).

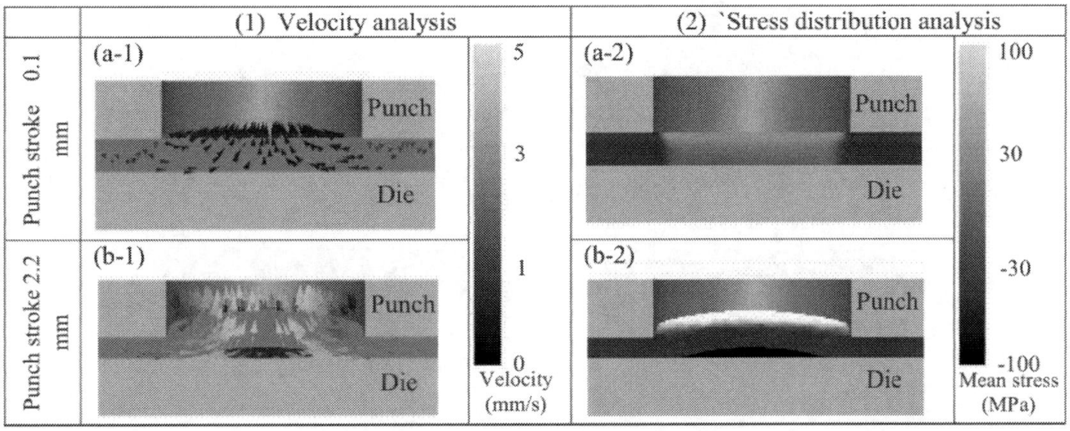


**Figure 2.** Illustration of stress and velocity distribution analyses during extrusion process ($\varnothing_E$ 30 mm).

*3.2. Relationship between extruded diameter and piping shape*
As shown in figure 3, the extruded diameter affected on the velocity on extruded area. As the results, the larger extruded diameter applied, the faster velocity generated. With these different levels of velocity, the different levels of piping diameters were formed. Figure 4 shows the relationship between extruded diameter and piping diameter analysed by FEM. The results illustrated that the piping diameter increased as the extruded diameter increased. It is also observed that as the thickness after pressing decreased the piping diameter increased, as shown in Figure 4. In addition, the extruded diameter also affected on the piping height. Figure 5 shows the relationship between extruded diameter and piping height analysed by FEM. The results illustrated that the piping height increased as the extruded diameter increased. It is also observed that as the thickness after pressing decreased the concave height increased.

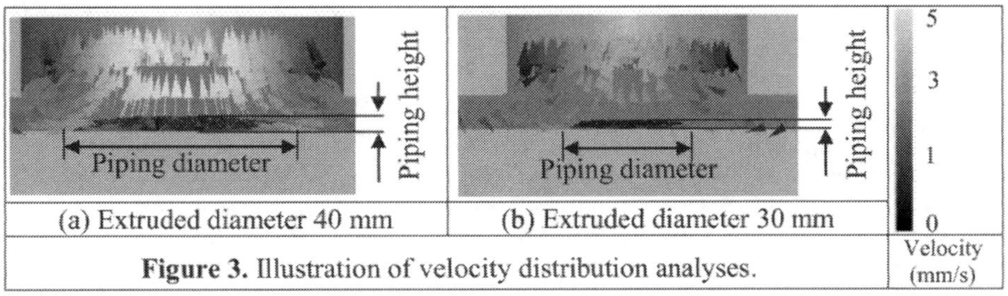


**Figure 3.** Illustration of velocity distribution analyses.

*3.3. Comparison of FEM simulation results and experimental results*
To validate the FEM simulation results, laboratory experiments were performed. Figure 6 compares the extruded shape obtained by the experiments and the FEM simulation, where good agreement is shown. The FEM simulation results agreed well with the experimental results, in which the error in the piping diameter and piping height compared with that obtained by the experimental results, were approximately 2%, respectively.

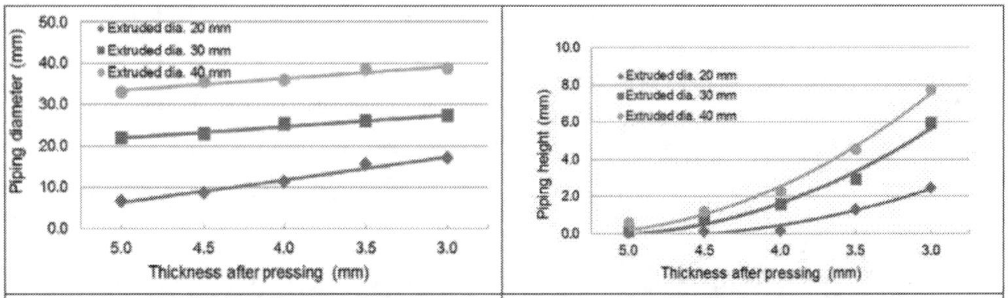


(a) Piping diameter (b) Piping height

**Figure 4.** Relationship between extruded diameter and piping shape.

(a) FEM simulation result (b) Experimental result

**Figure 5.** Comparison of the FEM simulation and the experiment results ($\varnothing_E$ 30mm).

## 4. Conclusions

In this study, the sheet-extrusion characteristic was clearly characterized based on a velocity analysis and a stress distribution analysis. The relationship between extruded shape and piping shape, including piping and piping height, was investigated using the FEM and laboratory experiments. The research illustrated the extruded diameter affected on the velocity on extruded zone, where the piping defect was generated. The extruded diameter directly affected on the piping shape. Specifically, as the extruded diameter increased the piping diameter and height increased. In addition, as the thickness of initial work piece after pressing decreased, the same tendency of relationship between extruded diameter and piping shape could be obtained. Namely, the thickness after pressing decreased the piping diameter and height increased. Laboratory experiments were performed to validate the accuracy of the FEM simulation results. Based on the extruded shape, the FEM simulation results showed good agreement with the experimental results in terms of piping diameter and height.

## 5. References

[1] Atzema E.H and Huétink 1995 Simulation of Materials Processing, Theory, Methods and Apllications. 383
[2] Kuzman K, Pfeifer E, Bay N 1996 J. Mater. Process. Technol. 60 141
[3] Kiuchi M, Yanagimoto J, Mendoza V 1995 Simulation of Materials Processing, Theory, Methods and Applications. 847
[4] Zheng P.F, Chan L.C, Lee T.C 2005 Finite Elem Anal Des. 42 189
[5] Chen F.K, Huang T.B, Wang S.J 2007 J. Mater. Process. Technol. 187-188 770
[6] Hirota K 2007 J. Mater. Process. Technol. 191 283

## Acknowledgments

The present research is partially supported by a grant from Thai-German Institute (TGI) under grant no. S2-CD-03/2554 and the Higher Education Research Promotion and National Research University Project of Thailand, Office of the Higher Education Commission, under grant no. 56000519. The authors also thank Mr. Arkarapon Sontamino and Mr. Pakkawat Komolruji for their assistance to this study.

# Mathematical modeling of mass generation via confinement of relativistic particles

**Constantinos G. Vayenas**

School of Engineering, University of Patras, GR 26504 Patras, Greece & Division of Natural Sciences, Academy of Athens, 28 Panepistimiou Avenue,10679 Athens, Greece

E-mail: cgvayenas@upatras.gr

**Abstract.** A simple mathematical model is presented for the generation of mass via the spontaneous confinement of elementary particles, e.g. electrons, neutrinos or quarks, in bound rotational states with relativistic particle velocities. In this simple mechanism, which is operative both in chemical and in physical systems, the new mass which is created corresponds to the kinetic energy of the trapped particles. It is shown that the ratio of the new mass created to the initial mass is very small for chemical systems but can be very high in the case of gravitational confinement of neutrinos in rotational states with relativistic particle velocities.

## 1. Introduction

The mechanism of mass generation is a central problem of interest in modern particle physics. Among the various proposed mechanisms, the Higgs boson mechanism [1] has attracted significant attention during the last few decades [2]. It postulates that matter obtains mass by interacting with a field, known as Higgs field [2].

In this work we describe a simple mechanism of mass generation via the confinement of elementary particles, e.g. electrons, neutrinos or quarks, in bound rotational states. It is shown that the ratio of new mass created to initial mass can be quite high for relativistic particle velocities inside the bound state.

## 2. Mathematical modeling

We examine the changes induced in the total energy (Hamiltonian) of a system as well as to its relativistic energy and rest mass for two cases: First when a H atom is formed via the Coulombic attraction of a proton and an electron and second when a rotational state corresponding to a baryon (e.g. a neutron) is formed via the relativistic gravitational attraction of three neutrinos [3, 4, 5].

The total energy, i.e. the Hamiltonian, $\mathcal{H}$, of a system consisting of $N$ particles (i=1,2,..., N) is the sum of the total relativistic energy, $E$, and of the potential energy of the system, $V$:

$$\mathcal{H} = E + V. \tag{1}$$

The relativistic energy, $E$, of the system is the sum of its rest energy, $\sum m_{o,i}c^2$, plus its kinetic energy $T$, which equals $\sum(\gamma_i - 1)m_{o,i}c^2$, i.e.

$$E = \sum_{i=1}^{N} m_{o,i}c^2 + \sum_{i=1}^{N}(\gamma_i - 1)m_{o,i}c^2. \tag{2}$$

### 2.1. H atom

Thus in the case of a H atom, using the Bohr model and neglecting the (rather small) kinetic energy of the proton, we can write with good accuracy:

$$E = m_p c^2 + m_e c^2 + (\gamma_e - 1) m_e c^2 \tag{3}$$

$$\mathcal{H} = m_p c^2 + m_e c^2 + (\gamma_e - 1) m_e c^2 + V_e, \tag{4}$$

where $V_e$ is the electrostatic energy of the electron. Since, $\mathrm{v} = \alpha c$, thus $\mathrm{v} \ll c$, the kinetic energy term reduces to $(1/2) m_e \mathrm{v}^2 = (1/2)\alpha^2 m_e c^2$, where $\alpha$ $(= e^2/\epsilon c\hbar = 1/137.036)$ is the fine structure constant. For the circular orbit of the Bohr model it is $V_e = -2T$, thus one obtains:

$$E = m_p c^2 + m_e c^2 + (1/2)\alpha^2 m_e c^2 \tag{5}$$

$$\mathcal{H} = m_p c^2 + m_e c^2 - (1/2)\alpha^2 m_e c^2. \tag{6}$$

In the initial state (a proton and an electron at rest and "infinite" distance), both $E$ and $\mathcal{H}$ equal $(m_p + m_e)c^2$, thus upon formation of the bound H atom state it is

$$\Delta E = (1/2)\alpha^2 m_e c^2 (= T) \tag{7}$$

$$\Delta \mathcal{H} = -(1/2)\alpha^2 m_e c^2 = -13.6 eV. \tag{8}$$

Thus while the system Hamiltonian has decreased by 13.6 $eV$, and therefore the formation of the H atom occurs spontaneously and is exoergic, i.e. energy is being released, at the same time the relativistic energy of the system, $\Delta E$, has increased by the Bohr energy $(1/2)\alpha^2 m_e c^2$, which actually equals $T$, i.e. the kinetic energy of the electron. Consequently for a laboratory observer who cannot detect the circular motion, the apparent *rest* energy of the system has increased by $(1/2)\alpha^2 m_e c^2$. This implies that the rest mass of the system has increased by $(1/2)\alpha^2 m_e$ and that therefore accounting also for the values of $m_e$ and $m_p$, the ratio of the masses in the final and initial state is 1.0000000145.

Thus the rest energy increase is equal to the kinetic energy of the electron (or of the proton-electron pair) in the confined H atom state (Fig. 1, left).

### 2.2. Neutrino confinement

Similar is the situation when a composite particle is formed via the gravitational confinement of a number (e.g. two or three [3, 4]) of light particles such as neutrinos [3, 4]. The bound rotational states resulting from the confinement of three highly relativistic neutrinos have been found recently to have, surprisingly, many of the properties of baryons, including their masses [3, 4, 5].

Table 1 summarizes the recent Bohr-type model for the gravitational confinement of three relativistic neutrinos and compares it with the Bohr model for the H atom [3, 4, 5]. The only difference is that the elestrostatic attraction of the H atom model is replaced in the three neutrino model by the gravitational attraction, accounting for special relativity and for the equivalence principle [3, 4, 5]. The justification for using Newton's universal gravitational law rather than general relativity has been discussed [5].

For the initial state, i.e. three non-interacting neutrinos at rest, it is $E_o = \mathcal{H}_o = 3m_o c^2$ where $m_o$ is the rest mass of each neutrino. Upon formation of the bound state it is

$$E = 3m_o c^2 + 3(\gamma - 1) m_o c^2 = 3\gamma m_o c^2 \tag{9}$$

$$\mathcal{H} = 3\gamma m_o c^2 + V_G = 3\gamma m_o c^2 - 5\gamma m_o c^2 = -2\gamma m_o c^2, \tag{10}$$

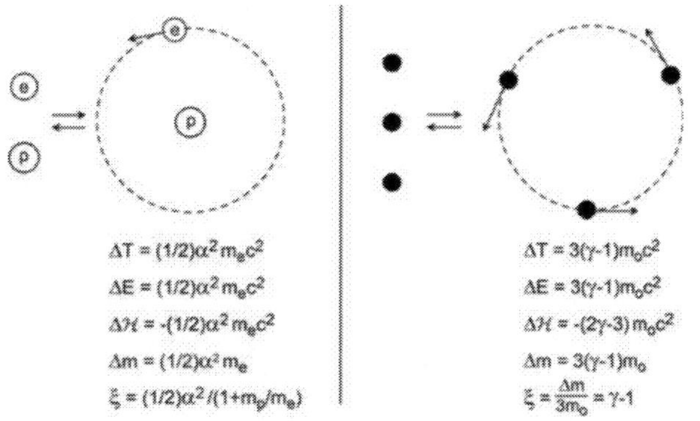


**Figure 1.** Apparent mass generation mechanism during formation of a H atom from a proton and an electron (left) and of a neutral baryon (e.g. a neutron) from three neutrinos (right).

where in (10) we have expressed the gravitational potential energy, $V_G$, using refs. [3] and [4].

Thus again it is $\Delta\mathcal{H} < 0$, i.e. the reaction is exoergic and occurs spontaneously, but at the same time it is $\Delta E > 0$, i.e. the relativistic energy of the system increases. Consequently, for a laboratory observer, the apparent rest energy increase is $\Delta E = 3(\gamma - 1)m_o c^2$ [3, 4, 5] and thus again the rest energy increase equals the kinetic energy of the three rotating particles (Fig. 1, right). However in this case, the ratio of the masses of the final and initial states is very high. It equals $\gamma_n = m_n/3m_o (= 7.163 \cdot 10^9)$ where $\gamma_n$ is the Lorentz factor corresponding to the rotational speed of the three neutrinos and $m_n$ is the neutron mass. This is computed from the following expression obtained by solving the two equations of the Bohr-type rotating neutrino model of Table 1 [3, 4, 5]:

$$\gamma_n = 3^{1/12} m_{Pl}^{1/3} / m_o^{1/3} (= 7.163 \cdot 10^9). \tag{11}$$

In this expression $m_{Pl} (= (\hbar c/G)^{1/2})$ is the Planck mass and $m_o (= 0.051 \pm 0.01\ eV/c^2)$ [6] is the neutrino mass. Exact agreement of the computed neutron mass with experiment is obtained for $m_o = 0.043723\ eV/c^2$ [3, 4, 5]. Thus in this case apparent rest energy and apparent rest mass are generated simply by the action of gravity, which causes particle confinement in a high kinetic energy, thus high $\gamma$ value state. The two Bohr models for electrostatic and for gravitational confinement are compared in Table 1 and also in Figure 1. In both cases the rest energy increase and concomitant rest mass increase corresponds to the kinetic energy of the confined particles. It should be noted that this very simple mass generation mechanism does not attempt to compete with the Higgs mechanism, since it is obvious that some initial mass is necessary for the present mechanism to become operative.

One can define a parameter, $\xi$, as the ratio of new mass created to initial mass, i.e.

$$\xi = \frac{new\ mass\ created}{initial\ mass}. \tag{12}$$

Thus in the case of the H atom it is:

$$\xi = \frac{(1/2) m_e \alpha^2}{m_p + m_e} = 1.45 \cdot 10^{-8}, \tag{13}$$

while for the case of the neutron it is:

$$\xi = \frac{3(\gamma_n - 1) m_o}{3 m_o} = \gamma_n - 1 \approx 7.163 \cdot 10^9. \tag{14}$$

**Table 1.** Comparison of the Bohr model for the H atom and of the Bohr type model for the neutron [4, 5]

| **Bohr model for the H atom** | **Bohr model for the neutron** |
|---|---|
| Electron as *particle*<br>$m_e \frac{v^2}{R} = \frac{e^2}{\varepsilon R^2}$<br>↑ Newton's 2nd law ↑ Coulomb law | Neutrino as *particle*<br>$\gamma m_o \frac{v^2}{R} = \frac{G m_o^2 \gamma^6}{\sqrt{3} R^2}$<br>↑ Relativistic equation of motion for circular motion ↑ Newton's gravitational law accounting for special relativity ($m_i = \gamma^3 m_o$) and for equivalence principle ($m_g = m_i$) |
| Electron as *wave*<br>$\frac{\hbar}{m_e v} = R$<br>De Broglie (for n=1) | Neutrino as *wave*<br>$\frac{\hbar}{m_o v} \approx \frac{\hbar}{m_o c} = R$<br>De Broglie (for n=1) |
| Results<br>$v = \alpha c \; ; \; \gamma \approx 1$<br>$R = \frac{\hbar}{\alpha m_e c} = 0.5292 \cdot 10^{-10} m$<br>$V_e = -27.2$ eV<br>$T = 13.6$ eV<br>$\mathcal{H} = T + V_e = -13.6$ eV<br>$\Delta(RE) = \Delta mc^2 = T = (1/2)\alpha^2 m_e c^2 = 13.6$ eV | Results (with $m_o$=0.043723 eV/$c^2$)<br>$v \approx c \; ; \; \gamma = 3^{1/12} (\hbar c / G)^{1/6} / m_o^{1/3} (= 7.163 \cdot 10^9)$<br>$R = \frac{\hbar}{\gamma m_o c} = 0.630 \cdot 10^{-15} m$<br>$V_g = -(5/3) m_n c^2 = -1565.9$ MeV<br>$T = m_n c^2 = 939.565$ MeV<br>$\mathcal{H} = T + V_G = -(2/3) m_n c^2 = -626.4$ MeV<br>$\Delta(RE) = \Delta mc^2 = T = m_n c^2 = 939.565$ MeV<br>Per particle: $T_p = 313.2$ MeV; $\mathcal{H}_p = -208.8$ MeV |

This process of mass generation via particle confinement is simple and different from other proposed mass generation schemes [1, 2, 7, 8, 9]. It is similar in its conclusions with the QCD results of Dürr et al. [7] who showed that even if the quark masses vanished, the baryon mass would not change much, a phenomenon sometimes called "mass without mass" [8, 9].

## References


[1] Higgs B (1964) Broken symmetries and the masses of gauge bosons. Phys Rev Lett 13:508-509

[2] Yao WM et al (2006) Searches for Higgs bosons, in Review of Particle Physics. J Phys G: Nucl Part Phys 33:1

[3] Vayenas CG & Souentie S (2012) arXiv:1106.1525v4 [physics.gen-ph]

[4] Vayenas CG & Souentie S (2012), Gravity, special relativity and the strong force: A Bohr-Einstein-de-Broglie model for the formation of hadrons. Springer, ISBN 978-1-4614-3935-6

[5] Vayenas CG, Souentie S & Fokas A (2013) A Bohr-type model with gravity as the attractive force. arXiv:1306.5979v1 [physics.gen-ph]

[6] Mohapatra RN et al (2007) Rep Prog Phys 70:1757-1867

[7] Dürr S et al (2008) Ab Initio Determination of Light Hadron Masses. Science 322(5905):1224-1227

[8] Kronfeld AS (2008) The Weight of the World is Quantum Chromodynamics. Science 322:1198-1199

[9] Wilczek F (2000) Mass without Mass II: The Medium is the Mass-age. Phys. Today 53:13-14

# Fractal descriptors for discrimination of microscopy images of plant leaves

**N R Silva[1,2], J B Florindo[2], M C Gómez[3], R M Kolb[4] and O M Bruno[2,1]**

[1] Institute of Mathematics and Computer Science, University of São Paulo (USP), Avenida Trabalhador são-carlense, 400 13566-590 São Carlos, São Paulo, Brazil
[2] Scientific Computing Group, São Carlos Institute of Physics, University of São Paulo (USP), cx 369 13560-970 São Carlos, São Paulo, Brazil - www.scg.ifsc.usp.br
[3] Department of Physics, Faculty of Biochemistry and Biological Sciences, National University of Littoral, S3000ZAA Santa Fe, Argentina
[4] Department of Biological Sciences, Faculty of Sciences and Letters, Universidade Estadual Paulista Júlio de Mesquita Filho, UNESP. Av. Dom Antônio, 2100, 19806-900, Assis, Brazil

E-mail: nubiars@icmc.usp.br, jbflorindo@gmail.com, mcgpna@gmail.com, rosanakolb@hotmail.com, bruno@ifsc.usp.br

**Abstract.** This study proposes the application of fractal descriptors method to the discrimination of microscopy images of plant leaves. Fractal descriptors have demonstrated to be a powerful discriminative method in image analysis, mainly for the discrimination of natural objects. In fact, these descriptors express the spatial arrangement of pixels inside the texture under different scales and such arrangements are directly related to physical properties inherent to the material depicted in the image. Here, we employ the Bouligand-Minkowski descriptors. These are obtained by the dilation of a surface mapping the gray-level texture. The classification of the microscopy images is performed by the well-known Support Vector Machine (SVM) method and we compare the success rate with other literature texture analysis methods. The proposed method achieved a correctness rate of 89%, while the second best solution, the Co-occurrence descriptors, yielded only 78%. This clear advantage of fractal descriptors demonstrates the potential of such approach in the analysis of the plant microscopy images.

## 1. Introduction

There is a strong need for the development of systems to predict plant species, given the difficulty in recognizing species as it requires in-depth knowledge about the species. Moreover, in most cases, it is necessary to obtain flowers, fruits or seeds for the species identification. However, these features are not available all the time for analysis [1]. Because of this weakness, it becomes necessary robust computing systems capable of recognizing species by analyzing only the plant leaves. Recent studies have shown that the leaves can also be an important element for classifying plant species, by obtaining features such as shape, texture and color for the characterization of each sample.

In [2], Casanova et al. extracted texture features of windows cropped from leaves obtained by a scanner. For this purpose they developed a system for feature extraction using Gabor wavelet filters. The feature vector used for representing each image was composed of the results obtained by computing the energy of each filtered image, varying the rotation and scale parameters of Gabor method. A variety of studies have investigated the potential of the complexity analysis of texture tissues and leaf anatomical characteristics by extracting the canonical fractal dimension and multiscale fractal dimension [3, 4, 5, 6, 7, 8]. Structures such as the cuticle, epidermis and full thickness of the leaves were analyzed through their color and texture characteristics [7, 9, 10]. On another studies, the vein is used for analysis of leaf species. In this case, the leaf image is segmented to extract the venation and some features of this venation are extracted for classification [11, 12, 9].

In this study we aim at exploring the use of information contained in the anatomical structure of cross-section median leaf veins obtained by microscopy for identification of species. The fractal dimension through the Bouligand-Minkowski volumetric method was extracted, generating texture descriptors for each histological image. Such descriptors represent the spatial arrangement of pixels within each texture on different scales, and such arrangements are directly related to the inherent properties of the material depicted in the image. The descriptors obtained by this method are generated from the expansion of a surface texture mapping of gray levels of each pixel in the texture. Subsequently, the classification of descriptors that represent the images is made by using the Support Vector Machine (SVM) classifier, with the best classification performance with about 89% accuracy, using 5-fold cross-validation whereas the second best approach compared, co-occurrence matrix, achieved a 78% accuracy.

This work is divided into 5 sections. The following section discusses the volumetric Bouligand-Minkowski fractal dimension for the extraction of texture features. Section 3 describes the experiments with 10 plant species ans the results of such experiments and the conclusions are stated on the Section 4.

## 2. Volumetric Bouligand-Minkowski Fractal Descriptors

Bouligand-Minkowski is one of the most used methods to estimate the fractal dimension of real-world objects. Here, we describe the approach to estimate the dimension of objects represented in a gray-level texture image.

Let $I : [1 : M] \times [1 : N] \rightarrow \Re$ be the digital image representing the texture. Firstly, this image is mapped onto a three-dimensional surface $S$, by means of the following definition:

$$S = \{(i, j, k)|(i, j) \in [1 : M] \times [1 : N], k = I(i, j)\}.$$

In the following, each surface point with co-ordinates $(x, y, z)$ is dilated by a sphere with radius $r$. This radius is varied within a pre-defined range. For each value of $r$, the volume of the dilated structure $V(r)$ may be computed by:

$$V(r) = \sum \chi_{\mathfrak{D}(r)}[(i, j, k)],$$

where $(i, j, k) \in S$ is any point in the surface, $\chi$ is the characteristic function and $\mathfrak{D}(r)$ corresponds to the following set:

$$\mathfrak{D}(r) = \{(x, y, z)|[(x - P_x)^2 + (y - P_y)^2 + (z - P_z)^2]^{1/2} \leq r\},$$

where $(P_x, P_y, P_z) \in S$.

The fractal dimension $D_{BM}$ is given by:

$$D_{BM} = 3 - \lim_{r \to 0} \frac{\log(V(r))}{\log(r)}.$$

Although the fractal dimension is a meaningful measure of an object, it is not sufficient to describe more complex patterns like those in the plant leaves analyzed here. Therefore, we employ a generalization of the fractal dimension, named fractal descriptors [4]. In this approach, instead of only using the global value of the dimension, a set of features are computed from the dimension taken at different scales. In practice, the set of descriptors $u$ are obtained from Equation 2:

$$u : \log(r) \rightarrow \log(V(r)).$$

### 3. Experiments and Results

The volumetric Bouligand-Minkowski fractal dimension was applied to 127 vein images of 10 species of Brazillian flora: *Anacardium humile*, *Annona crassiflora*, *Aristolochia galeata*, *Arrabidaea brachypoda*, *Aspidosperma subincanum*, *Baccharis salzmannii*, *Banisteriopsis stellaris*, *Bauhinia pulchella*, *Bauhinia ungulata* and *Byrsonima laxiflora*. Medial segments of leaves (n=4, four individuals per species) were collected at Brazilian Cerrado (tropical savanna ecoregion), being processed by histological routine to obtain cross-sections, as shown at Figure 1. The images were obtained in a trinocular microscope Axio Lab A1, 10X lens coupled to a digital camera AxioCam ICc 1.

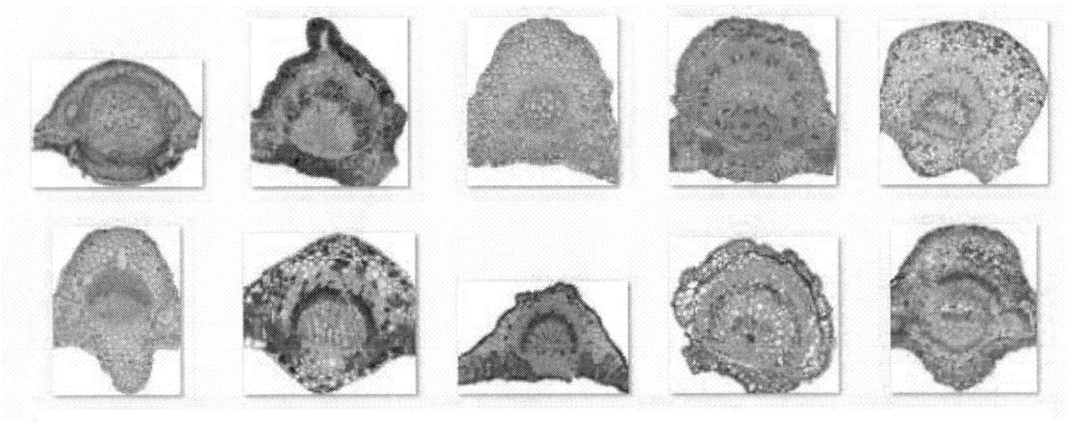

**Figure 1.** Cross-section of foliar veins of each specie. First line, from left to right: *Anacardium humile*, *Annona crassiflora*, *Aristolochia galeata*, *Arrabidaea brachypoda* and *Aspidosperma subincanum*. Second line, from left to right: *Baccharis salzmannii*, *Banisteriopsis stellaris*, *Bauhinia pulchella*, *Bauhinia ungulata* and *Byrsonima laxiflora*.

The leaves are classified by a Support Vector Machine (SVM) classifier, using a 5-fold cross-validation scheme. The fractal descriptors are compared to two other texture descriptors widely used in the literature: Gray-Level Co-occurrence Matrix (GLCM) and Gabor-wavelets descriptors. Table 1 shows the success rates in the classification of the species.

**Table 1.** Results: compare classification performance of ten different species.

| Method | Success Rate (RMSE) |
|---|---|
| Gabor descriptors | 77.1($\pm$ 0.2) |
| GLCM descriptors | 77.9 ($\pm$ 0.2) |
| Fractal descriptors | 89.0($\pm$ 0.1) |

## 4. Conclusions

The fractal descriptors provided an accurate and reliable result in the discrimination of the plant species. Such performance complies with the capabilities of these descriptors described in the literature. The measures of fractal dimension at different scales gives a detailed picture of the complexity of micro and macro-patterns inside the object analyzed and this is tightly related to anatomical and physical properties of the plant leaf. The results also confirmed the potential of fractal descriptors as a powerful method for analyzing plants using only their leaves. This is a remarkable finding, as the leaves are almost always available and the computational technique is simple and inexpensive.

## 5. Acknowledgments

Núbia R. Silva, João B. Florindo, Rosana M. Kolb and Odemir M. Bruno are grateful for the support from The State of São Paulo Research Foundation (FAPESP) with respective grants Nos. 2011/21467-9, 2012/19143-3, 2011/01523-1, 2011/23112-3. Bruno also acknowledges the financial support of the National Council for Scientific and Technological Development (CNPq), grant Nos. 308449/2010-0 and 473893/2010-0) and Rosana M. Kolb acknowledges PROPe UNESP (edict 14/2012/Programa Renove).

## References

[1] Ash A, Ellis B, Hickey L J, Johnson K, Wilf P and Wing S 1999 *Manual of Leaf Architecture - morphological description and categorization of dicotyledonous and net-veined monocotyledonous angiosperms by Leaf Architecture Working Group* Smithsonian Institution 10th St. & Constitution Ave., N.W. Washington, DC 20560-0121

[2] Casanova D, Sá Junior J J d M and Bruno O M 2009 *International Journal of Imaging Systems and Technology* **19** 236–243 ISSN 1098-1098

[3] Mugnai S, Pandolfi C, Azzarello E, Masi E and Mancuso S 2008 *Plant Systematics and Evolution* **270** 95–108 ISSN 0378-2697

[4] Bruno O M, de Oliveira Plotze R, Falvo M and de Castro M 2008 *Information Sciences* **178** 2722 – 2733 ISSN 0020-0255

[5] Backes A R, S Junior J J M, Kolb R M and Bruno O M 2009 Plant species identification using multi-scale fractal dimension applied to images of adaxial surface epidermis *Computer Analysis of Images and Patterns* (*Lecture Notes in Computer Science* vol 5702) ed Jiang X and Petkov N (Springer Berlin Heidelberg) pp 680–688 ISBN 978-3-642-03766-5

[6] Florindo J B, Backes A R and Bruno O M 2010 Leaves shape classification using curvature and fractal dimension *Image and Signal Processing* (*Lecture Notes in Computer Science* vol 6134) ed Elmoataz A, Lezoray O, Nouboud F, Mammass D and Meunier J (Springer Berlin Heidelberg) pp 456–462 ISBN 978-3-642-13680-1

[7] Sa Junior J J d M, Backes A R, Rossatto D R, Kolb R M and Bruno O M 2011 *Botany* **89** 467–479

[8] Rossatto D R, Casanova D, Kolb R and Bruno O 2011 *Plant Systematics and Evolution* **291** 103–116 ISSN 0378-2697

[9] xiang Du J, Zhai C M and Wang Q P 2013 *Neurocomputing* **116** 150 – 156 ISSN 0925-2312

[10] Sa Junior J J d M, Rossatto D R, Kolb R M and Bruno O M 2013 *Ecological Informatics* **15** 34 – 43 ISSN 1574-9541

[11] Fu H and Chi Z 2006 *Vision, Image and Signal Processing, IEE Proceedings* - **153** 881–892 ISSN 1350-245X

[12] Larese M G, Namas R, Craviotto R M, Arango M R, Gallo C and Granitto P M 2013 *Pattern Recognition* – ISSN 0031-3203

# Satellite classification and segmentation using non-additive entropy

**Lucas Assirati**[1], **Alexandre Souto Martinez**[2], **Odemir Martinez Bruno**[1]

[1]Instituto de Física de São Carlos (IFSC), Universidade de São Paulo, Av. Trabalhador São Carlense, 400, 13560-970 São Carlos, SP, Brazil

[2]Faculdade de Filosofia, Ciências e Letras de Ribeirão Preto (FFCLRP), Universidade de São Paulo (USP), Avenida Bandeirantes, 3900, 14040-901 Ribeirão Preto, SP, Brazil

Instituto Nacional de Cincias e Tecnologia de Sistemas Complexos

E-mail: assirati@usp.br, asmartinez@ffclrp.usp.br, bruno@ifsc.usp.br

**Abstract.** Here we compare the Boltzmann-Gibbs-Shannon (standard) with the Tsallis entropy on the pattern recognition and segmentation of colored images obtained by satellites, via "Google Earth". By segmentation we mean particionate an image to locate regions of interest. Here, we discriminate and define an image partition classes according to a training basis. This training basis consists of three pattern classes: aquatic, urban and vegetation regions. Our numerical experiments demonstrate that the Tsallis entropy, used as a feature vector composed of distinct entropic indexes $q$ outperforms the standard entropy. There are several applications of our proposed methodology, once satellite images can be used to monitor migration form rural to urban regions, agricultural activities, oil spreading on the ocean etc.

## 1. Introduction

Image pattern recognition is a common issue in medicine, biology, geography etc, in short, in domains that produce huge data in images format. Entropy, in its origins is interpreted as a disorder measure. Nevertheless, nowadays it is interpreted as the lack of information. Thus, it has been used as a methodology to measure the information content of a signal or an image. In image analysis, the greater the entropy is, the more irregular and patternless a given image is. The additive property of the standard entropy allows its use in several situations just by summing up image characteristics. Among the non-additive entropies, we study the Tsallis entropy, which has been proposed to extent the scope of application of classical statistical physics. Here, we compare the additive Boltzmann-Gibbs-Shannon (standard) [1] and non-additive Tsallis entropy [2] when dealing with colored satellite images.

We start defining the standard entropy for black and white images and we simply extend its use to colored images, justified by its additive property. Next, we consider the Tsallis entropy for black and white images and extend it to colored images. Due to non-additiveness, we call attention to some characteristics that help to qualify these images more efficiently that the standard entropy.

## 2. Non-additive entropy

Firstly, consider an black and white image with $L_x \times L_y$ pixels. The integers $i \in [1, L_x]$ and $j \in [1, L_y]$ run along the $\hat{x}$ and $\hat{y}$ directions, respectively. Let the integer $\tilde{p}_{i,j} \in [0, 255]$ represent the image gray levels intensity of pixel $(i, j)$. The histograms $\tilde{p}(x)$ of a gray levels image are obtained by counting the number of pixels with a given intensity $\tilde{p}_{i,j}$. To properly use the entropic indexes, one must consider normalized quantities: $p(x) = \tilde{p}(x)/(L_x \times L_y)$, so that normalization condition $\sum_{x=0}^{255} p(x) = 1$ is satisfied. The standard entropy of this image is: $H = -\sum_{x=0}^{255} p(x) \ln p(x) = \sum_{x=0}^{255} p(x) \ln(1/p(x))$.

For colored images, a given pixel has three components: red ($k = 1$), green ($k = 2$) and blue ($k = 3$), and the integer intensity concerning each one of these colors are written as $\tilde{p}_{i,j,k} \in [0, 255]$, so that $k = 1, 2, 3$. This leads to different histograms for each color: $p_k(x)$, and hence different entropies for each color: $H_k$, with $k = 1, 2, 3$.

For two images $A$ and $B$, for a given color, the entropy of the composed image, is the entropy of one image plus the other $H_k(A + B) = H_k(A) + H_k(B)$. This is the additivity property of the standard entropy, which leads to:

$$H_1 = \sum_{x=0}^{255} p_1(x) \ln\left(\frac{1}{p_1(x)}\right) ; H_2 = \sum_{x=0}^{255} p_2(x) \ln\left(\frac{1}{p_2(x)}\right) ; H_3 = \sum_{x=0}^{255} p_3(x) \ln\left(\frac{1}{p_3(x)}\right) . \quad (1)$$

Secondly, consider an black and white image mentioned before. The Tsallis entropy is for it generalizes the standart entropy[3]: $S_q = \sum_{x=0}^{255} p(x) \ln_q(1/p(x))$, where the generalized logarithmic function is $\ln_q(x) = (x^{q-1} - 1)/(q - 1)$, so that, as $q \to 1$, one retrieves the standard logarithm, consequently the standard entropy.

To build a feature vector, one simply uses $n$ different entropic values: $\vec{S}_{bw} = (S_{bw,q_1}, S_{bw,q_2}, \ldots, S_{bw,q_n})$, so that $n = 1$ and $q = 1$, one retrieves the standard entropy image qualifier. Notice the richness introduced by this qualifier. If $n = 1$, we have already an infinity range of entropy indexes to address. This richness is amplified for $n > 1$, considering instances of : $q < 1$, $q = 1$ and $q > 1$ [4].

Since $\ln_q(x_1 x_2) = \ln_q(x_1) + \ln_q(x_2) + (1 - q)\ln_q(x_1)\ln_q(x_2)$, see Ref. [5], $S_{bw,q}$ is non-additive leading to interesting results when composing two images $A$ and $B$. The entropy of the composed image is $S_{bw,q}(A + B) = S_{bw,q}(A) + S_{bw,q}(B) + (1 - q)S_{bw,q}(A)S_{bw,q}(B)$, which, for $q \neq 1$ is not simply summation of two entropic values. This property leads to different entropic values depending on how one partitions a given image. The final image entropy is not simply to summation of the entropy of all its partitions, but it depends on the sizes of these partitions.

For colored images, we proceed as before, we calculate the entropy of each color component, in principle with different entropy indices values: $q^{(r)}$, $q^{(g)}$ and $q^{(b)}$. For sake of simplicity, we consider the same entropic index for all the color components. For color $k$ the entropy is:

$$S_q(k) = \sum_{x=0}^{255} p_k(x) \ln_q\left(\frac{1}{p_k(x)}\right) , \quad (2)$$

so that so that $k = 1, 2, 3$ retrieves Eq. (1), for $q = 1$.

## 3. Metodology

Considering pattern recognition in images, the main objective is to classify a given sample according to a set of classes from a database. In supervised learning, the classes are predetermined. These classes can be conceived of as a finite set, previously arrived by a human. In practice, a certain segment of data is labelled with these classifications. The classifier task is search for patterns and classify a sample as one of the database classes.

To perform this classification, classifiers usually uses a feature vector that comes from a method of data extraction. Here, we use the multi-$q$ analyses method [6] [7] that composes a feature vector using certain $q$-entropy values: $\vec{S}_q = (S_{q_1}(1); S_{q_1}(2); S_{q_1}(3); \ldots ; S_{q_n}(1); S_{q_n}(2); S_{q_n}(3))$.

The reason to use the multi-$q$ analysis is that a feature vector gives us more and richer information than a single value of entropy. The correct choice of $q$ indexes emphasize characteristics and provide better classifications.

The following steps describe image treatment, training and validation:

- Using Google Earth software, capture images from several locations;
- each image must be segmented in $16 \times 16$ pixels partitions;
- for each partition the colors Red, Green and Blue are written in a tridimensional array;
- for each array and for each color, histograms are built and the Tsallis entropies (Eq. 2) are calculated, for $q \in [0, 2]$ in steps of $1/10$;
- the feature vector is created and the classifiers $k$-nearest neighbors (KNN), Support Vector Machine (SVM) and Best-First Decision Tree (BFTree) are applied;
- an output image are delivered with the segmented partitions highlighted (aquatic region = yellow, urban region = cyan, vegetation regions = magenta) according with the classification of KNN classifier.

**Figure 1.** Images obtained from Google Earth, from different regions. (a) Urban, (b) Aquatic, (c) Vegetation

Table 1 presents the hit rate percentage of each classifier evaluated for the 3 methods: multi-$q$ analysis, multi-$q$ analysis with attribute selection and standard entropy analysis. Since the use of a feature vector gives us more information than a single entropy value it also gives some redundant information. In this context, the feature selection is important to eliminates those redundancies.

**Table 1.** Several classifiers are used (SVM, KNN and BFTree) to compare the performance of the generalized entropy with respect to the standard one in pattern recognition.

*attribute selection

| Method (# of features) / Classifier | Multi-$q$ (60) | Multi-$q$ * (8) | BGS (3) |
|---|---|---|---|
| SVM | 69.60 % | 68.96 % | 65.60 % |
| KNN (1 neighbour) | 70.80 % | 69.76 % | 63.04 % |
| KNN (3 neighbours) | 72.32 % | 72.96 % | 64.00 % |
| KNN (5 neighbours) | 72.80 % | 73.28 % | 64.16 % |
| KNN (7 neighbours) | 74.88 % | 72.96 % | 68.48 % |
| BFTree | 72.16 % | 72.80 % | 67.36 % |

Figure 2 depicts image highlights produced by KNN method, evaluated in a region that contains the three types of pattern classes: aquatic, urban and vegetation regions.

(a) Original Image (b) KNN with 1 neigbour classifier (c) KNN with 3 neigbour classifier

**Figure 2.** Segmentation obtained by Multi-q method and highlights provided by KNN classifier. The yellow color indicates an aquatic region, the cyan color indicates an urban region and the magenta color indicates a vegetation region.

## 4. Conclusion

Our study indicates that the Tsallis non-additive entropy can be successfully used in the construction of a feature vector, concerning colored satellite images. This entropy generalizes the Boltzmann-Gibbs one, which can be retrieved with $q = 1$. For $q \neq 1$, the image retrieval success is better that the standard case ($q = 1$), once the entropic parameter $q$ allows thorougher image exploration.

## References

[1] Shannon C E 1948 *The Bell System Technical Journal* **27** 379–423–623–656
[2] Tsallis C 1988 *Journal of Statistical Physics* **52** 479–487
[3] Tsallis C 2011 *Entropy* **13** 1765–1804
[4] Barbieri A L, de Arruda G F, Rodrigues F A, Bruno O M and da Fontoura Costa L 2011 *Physica A: Statistical Mechanics and its Applications* **390** 512–518
[5] Borges E P 2004 *Physica A: Statistical Mechanics and its Applications* **340** 95–101
[6] Fabbri R, Gonçalves W N, Lopes F J and Bruno O M 2012 *Physica A: Statistical Mechanics and its Applications* **391** 4487–4496
[7] Fabbri R, Bastos I N, Neto F D M, Lopes F J P, Gonçalves W N and Bruno O M 2013 *arXiv* **abs/1305.2876**

# Reconstructing f(R) modified gravity with dark energy parametrization

**Masaaki Morita**[1] **and Hirotaka Takahashi**[2]

[1] Okinawa National College of Technology, 905 Henoko, Nago, Okinawa 905-2192, Japan
[2] Nagaoka University of Technology, 1603-1 Kamitomioka, Nagaoka, Niigata 940-2188, Japan

E-mail: morita@okinawa-ct.ac.jp, hirotaka@kjs.nagaokaut.ac.jp

**Abstract.** We demonstrate the reconstruction of $f(R)$ modified gravity theory with late-time accelerated cosmic expansion. A second-order differential equation for Lagrangian density is obtained from the field equation, and is solved as a function of the cosmic scale factor in two cases. First we begin with the case of a wCDM cosmological model, in which a dark-energy equation-of-state parameter $w$ is constant, for simplicity. Next we extend the method to a case in which the parameter $w$ is epoch-dependent and is expressed as the Chevallier-Polarski-Linder parametrization. Thus we can represent Lagrangian density of $f(R)$ modified gravity theory in terms of dark energy parameters.

## 1. Introduction

The discovery of accelerated expansion in the late-time universe by observations of distant type Ia supernovae [1, 2, 3] is manifestly one of the most remarkable advances in modern cosmology. Combining it with observational data of other astrophysical objects such as the cosmic microwave background, the universe acceleration is being established, but the theoretical solution to the cause of the acceleration is yet to be seen although theoretical possibilities have been explored with a large number of dark energy models [4, 5]. A '$f(R)$ modified gravity' dark energy model, in which some function of the scalar curvature $R$ is added to the Einstein-Hilbert action as a correction, is one of such theoretically possible models [6]. It is of great interest for theorists to study this model because it is, in a sense, most straightforward extension of Einstein's general relativity, and also will provide a clarification of the link between the universe acceleration and gravitational theories. (We should bear in mind the role of Starobinsky's $R^2$-model in inflationary cosmology [7].) There is, however, no guiding principle for determining the form of $f(R)$; only severe constraints have been obtained from local-scale experiments (e.g. solar system tests) [8] and cosmological observations [9, 10, 11]. Therefore, we can only *design* the form of $f(R)$ so that it satisfies those observational constraints.

In this article, we demonstrate the reconstruction of $f(R)$ modified gravity theory for a given accelerated cosmic expansion as an illustration of designing $f(R)$. As models of the cosmic expansion, we employ a wCDM cosmological model, in which a dark-energy equation-of-state parameter $w$ is constant, and the Chevallier-Polarski-Linder (CPL) parametrization [12, 13], which has been frequently used in dark energy phenomenology. This work extends the previous works that treat the reconstruction of $f(R)$ gravity with mainly a $\Lambda$CDM model [14, 15, 16].

## 2. Basic equations in $f(R)$ modified gravity

We consider $f(R)$ modified gravity theory with the action

$$\mathcal{S} = \frac{1}{16\pi G} \int f(R)\sqrt{-g}\; \mathrm{d}^4x \; + \mathcal{S}_{\mathrm{m}} \,, \tag{1}$$

where $G$ is Newton's gravitational constant, $R$ is the four-dimensional scalar curvature, $g$ is the determinant of the space-time metric $g_{\mu\nu}$, and $\mathcal{S}_{\mathrm{m}}$ denotes the action of matter field described as a pressureless fluid. The action principle yields the field equations

$$F(R)R_{\mu\nu} - \frac{1}{2}f(R)g_{\mu\nu} - \nabla_\mu\nabla_\nu F(R) + g_{\mu\nu}\Box F(R) = 8\pi G T^{(\mathrm{m})}_{\mu\nu} \,, \tag{2}$$

where $F(R) := \partial_R f(R)$, $\nabla$ represents the four-dimensional covariant derivative, $\Box := g^{\mu\nu}\nabla_\mu\nabla_\nu$, and $T^{(\mathrm{m})}_{\mu\nu}$ is the energy-momentum tensor of pressureless matter. Hereafter we write the function $f(R)$ as $R - \xi(R)$, where $\xi(R)$ plays the role of an effective dark energy component, and then $F(R) = 1 - \partial_R\xi(R)$.

In a homogeneous and isotropic universe model with the scale factor $a(t)$, a spatial constant curvature $K$, and the energy density $\rho_{\mathrm{M}}(t)$ of pressureless matter, the $(0,0)$-component of the field equations (2) reads

$$H^2 + \frac{K}{a^2} = \frac{8\pi G}{3}\rho_{\mathrm{M}} + \frac{1}{6}\xi(R) + H\partial_t\left(\partial_R\xi(R)\right) - \partial_R\xi(R)\left(\partial_t H + H^2\right) \,, \tag{3}$$

where $H := \partial_t a/a$ is the Hubble parameter, and $R = 6\partial_t H + 12H^2 + 6K/a^2$. Note that this equation is reduced to the Friedmann equation in the standard FLRW universe model if we set $\xi(R) = 2\Lambda$ with a positive cosmological constant $\Lambda$. Replacing the time variable $t$ with $a$, and the variable $R$ with $a$, and rearranging Eq. (3), we obtain

$$\partial_a^2\xi - \left(\frac{1}{a} + \frac{\partial_a H}{H} + \frac{\partial_a^2 R}{\partial_a R}\right)\partial_a\xi + \frac{\partial_a R}{6aH^2}\xi = \frac{\partial_a R}{aH^2}\left(H^2 + \frac{K}{a^2} - \frac{8\pi G}{3}\rho_{\mathrm{M}}\right) . \tag{4}$$

## 3. Reconstruction of $f(R)$ gravity with dark energy parametrization

We assume that the cosmic expansion is that of the universe with a spatial constant curvature $k$, matter density $\rho_{\mathrm{m}}(t)$, and dark energy density $\rho_{\mathrm{d}}(t)$, obeying Einstein's general relativity. Then the Friedmann equation is written as

$$H^2 + \frac{k}{a^2} = \frac{8\pi G}{3}\left(\rho_{\mathrm{m}} + \rho_{\mathrm{d}}\right) . \tag{5}$$

Introducing the dark-energy equation-of-state parameter $w := P_{\mathrm{d}}/\rho_{\mathrm{d}}$ with the dark energy pressure $P_{\mathrm{d}}$, the integration of the continuity equation gives the dark energy density as

$$\rho_{\mathrm{d}}(a) = \rho_{\mathrm{d}}(a_0)\,\mathcal{E}(a)\,; \qquad \mathcal{E}(a) := \exp\left[-3\int_{a_0}^{a}\frac{1+w(\tilde{a})}{\tilde{a}}\,\mathrm{d}\tilde{a}\right] , \tag{6}$$

where the subscript '0' denotes the present value. Then the Friedmann equation (5) is rewritten as $H^2 = H_0^2\left(\Omega_{\mathrm{m}0}\,a^{-3} + \Omega_{\mathrm{d}0}\mathcal{E} + \Omega_{k0}\,a^{-2}\right)$, where $\Omega_{\mathrm{m}0} := 8\pi G\rho_{\mathrm{m}}(a_0)/(3H_0^2)$, $\Omega_{\mathrm{d}0} := 8\pi G\rho_{\mathrm{d}}(a_0)/(3H_0^2)$, $\Omega_{k0} := -k/H_0^2$ and $a_0 := 1$, and the scalar curvature as

$$R = 3H_0^2\left[\Omega_{\mathrm{m}0}\,a^{-3} + (1-3w)\Omega_{\mathrm{d}0}\mathcal{E} + 2\left(\Omega_{k0} - \Omega_{K0}\right)a^{-2}\right] , \tag{7}$$

where $\Omega_{K0} := -K/H_0^2$. Rewriting the right-hand side of Eq. (4), we arrive at the following key equation for the reconstruction:

$$\partial_a^2\xi - \left(\frac{1}{a} + \frac{\partial_a H}{H} + \frac{\partial_a^2 R}{\partial_a R}\right)\partial_a\xi + \frac{\partial_a R}{6aH^2}\xi = \frac{\partial_a R}{aH^2}H_0^2\Big[(\Omega_{k0} - \Omega_{K0})\,a^{-2} + (\Omega_{m0} - \Omega_{M0})\,a^{-3} + \Omega_{d0}\mathcal{E}\Big]\,, \tag{8}$$

where $\Omega_{M0} := 8\pi G\rho_M(a_0)/(3H_0^2)$. Hereafter, we assume that $\Omega_{m0} = \Omega_{M0}$ and $\Omega_{k0} = \Omega_{K0} = 0$ for simplicity.

### 3.1. Case of the wCDM model

First we demonstrate the reconstruction in a wCDM model, in which the parameter $w$ is constant, and then $\mathcal{E}(a) = a^{-3(1+w)}$. The key equation (8) is reduced to

$$\partial_a^2\xi - \frac{1}{a}C_1(a)\partial_a\xi - \frac{3}{2a^2}C_2(a)\xi = -\frac{9}{a^2}C_2(a)H_0^2\Omega_{d0}\,a^{-3(1+w)}\,, \tag{9}$$

where

$$C_1(a) := -\frac{9}{2} - \frac{3w}{2}\frac{\Omega_{d0}}{\Omega_{m0}}a^{-3w}\left[\left(1 + \frac{\Omega_{d0}}{\Omega_{m0}}a^{-3w}\right)^{-1} + 2(1-3w)(1+w) \times\left(1 + (1-3w)(1+w)\frac{\Omega_{d0}}{\Omega_{m0}}a^{-3w}\right)^{-1}\right]\,, \tag{10}$$

$$C_2(a) := 1 - (2w + 3w^2)\frac{\Omega_{d0}}{\Omega_{m0}}a^{-3w}\left(1 + \frac{\Omega_{d0}}{\Omega_{m0}}a^{-3w}\right)^{-1}\,. \tag{11}$$

To find homogeneous solutions of Eq. (9) in an approximate form, we make an ansatz $\xi(a) \propto a^n$, and then find that $n = \left(1 + C_1 \pm \sqrt{(1+C_1)^2 + 6C_2}\right)/2 =: n_\pm$, which asymptotically becomes $n_\pm \approx \left(-7 \pm \sqrt{73}\right)/4$ in matter-dominated era ($a^{-3w} \ll 1$), and $n_\pm \approx \left(-7 - 9w \pm \sqrt{73 + 78w + 9w^2}\right)/4$ in dark-energy-dominated era ($a^{-3w} \gg 1$). Since $R \approx 3H_0^2\Omega_{m0}\,a^{-3}$ and $R \approx 3(1-3w)H_0^2\Omega_{d0}\,a^{-3(1+w)}$ in these respective eras, the homogeneous solutions lead to $\xi(R) \propto R^{(7+\sqrt{73})/12} \approx R^{1.3}$ in matter-dominated era, and $\xi(R) \propto R^{(7+9w-\sqrt{73+78w+9w^2})/(12+12w)} \approx R^{-3.6}$ (for $w \approx -0.9$) in dark-energy-dominated era. These have been found to be observationally inconsistent [9, 10], and thus we take only the particular solution of Eq. (9) into account, omitting the homogeneous solutions. Using the method of a Green function, we obtain

$$\xi \approx \frac{6H_0^2\Omega_{d0}\,a^{-3(1+w)}}{2 - 5w - 6w^2} \approx \frac{6H_0^2\Omega_{d0}}{2 - 5w - 6w^2}\left(\frac{R}{3H_0^2\Omega_{m0}}\right)^{1+w} \quad \text{in matter-dominated era,} \tag{12}$$

$$\xi \approx 3(1-3w)H_0^2\Omega_{d0}\,a^{-3(1+w)} \approx R \quad \text{in dark-energy-dominated era.} \tag{13}$$

### 3.2. Case of the CPL parametrization

Next we proceed to a more general case, in which the parameter $w$ is expressed as $w = w_0 + w_1(1-a)$ with constants $w_0$ and $w_1$, and then $\mathcal{E}(a) = a^{-3(1+w_0+w_1)}\exp[-3w_1(1-a)]$. This phenomenological model is known as the Chevallier-Polarski-Linder (CPL) parametrization. Here we focus on finding the form of $\xi(R)$ only in matter-dominated era because the universe is considered to have been matter dominant during most of the epoch after photon decoupling, except for the most recent epoch. Then the key equation (8) becomes

$$\partial_a^2\xi + \frac{9}{2a}\partial_a\xi - \frac{3}{2a^2}\xi = -\frac{9H_0^2\Omega_{\mathrm{d}0}}{a^2}a^{-3(1+w_0+w_1)}\exp[-3w_1(1-a)]\,, \tag{14}$$

whose particular solution is

$$\xi = \frac{-18H_0^2\Omega_{\mathrm{d}0}\,\mathrm{e}^{-3w_1}}{\sqrt{73}}\left[a^{n_+}\int^a \tilde{a}^{-(n_++4+3w_0+3w_1)}\mathrm{e}^{3w_1\tilde{a}}\,\mathrm{d}\tilde{a} - a^{n_-}\int^a \tilde{a}^{-(n_-+4+3w_0+3w_1)}\mathrm{e}^{3w_1\tilde{a}}\,\mathrm{d}\tilde{a}\right], \tag{15}$$

where $n_\pm = (-7\pm\sqrt{73})/4$. Utilizing the Taylor expansion for $\mathrm{e}^{3w_1 a}$, we obtain

$$\xi = \sum_\ell \frac{-18H_0^2\Omega_{\mathrm{d}0}\,\mathrm{e}^{-3w_1}a^{-3(1+w_0+w_1)}}{2\left[\ell-3(1+w_0+w_1)\right]^2+7\left[\ell-3(1+w_0+w_1)\right]-3}\frac{(3w_1a)^\ell}{\ell!} \tag{16}$$

$$\approx \sum_\ell \frac{-18H_0^2\Omega_{\mathrm{d}0}\,\mathrm{e}^{-3w_1}(3w_1)^\ell}{\ell!\left\{2\left[\ell-3(1+w_0+w_1)\right]^2+7\left[\ell-3(1+w_0+w_1)\right]-3\right\}}\left(\frac{R}{3H_0^2\Omega_{\mathrm{m}0}}\right)^{1+w_0+w_1-\ell/3}. \tag{17}$$

## 4. Summary and outlook

In this article, the reconstruction of Lagrangian density of $f(R)$ gravity has been demonstrated for a given cosmic expansion in the context of dark energy cosmology. We have adopted a wCDM model, in which dark-energy equation-of-state parameter $w$ is constant, and the CPL parametrization, as phenomenological models of accelerated cosmic expansion. In these cases, Lagrangian density of $f(R)$ gravity has been expressed in terms of dark energy parameters. We hope that the results obtained would be useful in clarifying the relation between dark energy parameters and local gravity constraints, and also in evaluating the effect of gravity modification on cosmological perturbations in the framework of $f(R)$ gravity theory.

## Acknowledgments

This work was partially supported by Nagaoka University of Technology grant for collaborative research with Colleges of Technology.

## References

[1] Perlmutter S *et al* 1999 *Astrophys. J.* **517** 565
[2] Riess A G *et al* 1998 *Astron. J.* **116** 1009
[3] Riess A G *et al* 1999 *Astron. J.* **117** 707
[4] Copeland E J, Sami M and Tsujikawa S 2006 *Int. J. Mod. Phys.* D **15** 1753
[5] Amendola L and Tsujikawa S 2010 *Dark Energy: Theory and Observations* (Cambridge: Cambridge University Press)
[6] De Felice A and Tsujikawa S 2010 *Living Rev. Relativity* **13** 3
[7] Starobinsky A A 1980 *Phys. Lett.* B **91** 99
[8] Chiba T, Smith T L and Erickcek A L 2007 *Phys. Rev.* D **75** 124014
[9] Song Y-S, Hu W and Sawicki I 2007 *Phys. Rev.* D **75** 044004
[10] Tsujikawa S 2008 *Phys. Rev.* D **77** 023507
[11] Tsujikawa S, Gannouji R, Moraes B and Polarski D 2009 *Phys. Rev.* D **80** 084044
[12] Chevallier M and Polarski D 2001 *Int. J. Mod. Phys.* D **10** 213
[13] Linder E V 2003 *Phys. Rev. Lett.* **90** 091301
[14] Capozziello S, Cardone V F and Troisi A 2005 *Phys. Rev.* D **71** 043503
[15] Nojiri S, Odintsov S D and Saez-Gomez D 2009 *Phys. Lett.* B **681** 74
[16] Dunsby P K S, Elizalde E, Goswami R, Odintsov S and Saez-Gomez D 2010 *Phys. Rev.* D **82** 023519

# Extensions of Natural Hamiltonians

**G Rastelli**

Independent researcher, cna Ortolano 7, Ronsecco, Italia

E-mail: giorast.giorast@alice.it

**Abstract.** Given an n-dimensional natural Hamiltonian L on a Riemannian or pseudo-Riemannian manifold, we call "extension" of L the n+1 dimensional Hamiltonian $H = \frac{1}{2}p_u^2 + \alpha(u)L + \beta(u)$ with new canonically conjugated coordinates $(u, p_u)$. For suitable L, the functions $\alpha$ and $\beta$ can be chosen depending on any natural number m such that H admits an extra polynomial first integral in the momenta of degree m, explicitly determined in the form of the m-th power of a differential operator applied to a certain function of coordinates and momenta. In particular, if L is maximally superintegrable (MS) then H is MS also. Therefore, the extension procedure allows the creation of new superintegrable systems from old ones. For m=2, the extra first integral generated by the extension procedure determines a second-order symmetry operator of a Laplace-Beltrami quantization of H, modified by taking in account the curvature of the configuration manifold. The extension procedure can be applied to several Hamiltonian systems, including the three-body Calogero and Wolfes systems (without harmonic term), the Tremblay-Turbiner-Winternitz system and n-dimensional anisotropic harmonic oscillators. We propose here a short review of the known results of the theory and some previews of new ones.

## 1. Introduction

Natural Hamiltonian systems are the mathematical models of those physical systems for which the energy is constant, for example harmonic oscillators or the Kepler system. Often, as in the previous two examples, more quantities are constants of the motion (or *first integrals*): angular momentum, Laplace-Runge-Lentz vector, etc. Usually, these constants are expressed by quadratic polynomials in the momenta or, for quantum systems, by second-order differential operators. Hamiltonian systems with constants of the motion of degree higher than two are less common, nevertheless, some of them are of great interest, as for instance the three-body Jacobi-Calogero and Wolfes systems. These systems represent the dynamics of three point-masses on a line under forces determined by the potential functions

$$V_{JC} = k\left((x-y)^{-2} + (x-z)^{-2} + (y-z)^{-2}\right),$$
$$V_W = k\left((x+y-2z)^{-2} + (x+z-2y)^{-2} + (y+z-2x)^{-2}\right),$$

respectively (we do not consider here the harmonic oscillator terms) and they have essentially the same dynamics [1]. Both the resulting natural Hamiltonians in $\mathbb{E}^3$ admit one linear and one quadratic in the momenta constants of the motion, making the systems Liouville-integrable and solvable by separation of variables (see [1] and references therein). Other two independent constants of the motion do exist, one quadratic, due to the multiseparability of the Hamiltonian, and one cubic. The systems are then maximally superintegrable (MS), having a number of functionally independent constants of the motion equal to twice the degrees of freedom, minus one (for quantum systems, the same number of algebraically independent symmetry operators). MS systems are of the greatest importance in mathematical physics, harmonic oscillators and Kepler are MS and this makes them to satisfy Bertrand's theorem. Indeed, maximal superintegrability manifests itself, for classical systems, in the fact that all finite orbits of MS systems are closed while, for quantum systems, in the fact that the energy levels are totally degenerate [6]. In recent years, several techniques made possible the construction of classical and quantum Hamiltonian systems, MS and not, with first integrals of arbitrarily high degree [1, 13, 7, 3, 10] whose study, still in development, produced remarkable results in special functions, quantum algebras, canonical quantization theories [7, 8, 12, 11, 9]. In this note it is shortly introduced the work on the topic done by Claudia Chanu, Luca Degiovanni and the author (in short CDR) in several joint articles. In few words, the extension procedure (Theorem 1) adds one degree of freedom to some suitable Hamiltonian $L$ in such a way an extra non trivial first integral, polynomial of degree $m \in \mathbb{N}$, of the new Hamiltonian do exist.

**Main results**

The following Theorem, stated in [2], defines and characterizes what we intend for "extensions" in the particular case of natural Hamiltonians on cotangent bundles of Riemannian manifolds; for a more general definition, see [4]. Given an $n$-dimensional natural Hamiltonian $L$ on the cotangent bundle of a (pseudo)-Riemannian manifold $Q$, let be

$$H = \frac{1}{2}p_u^2 + \alpha(u)L + \beta(u) \tag{1}$$

and

$$U = p_u + \gamma(u)X_L, \tag{2}$$

where $X_L$ is the Hamiltonian vector field of $L$, then

**Theorem 1.** *Let $Q$ be a $n$-dimensional (pseudo-)Riemannian manifold with metric tensor $\mathbf{g}$. The natural Hamiltonian $L = \frac{1}{2}g^{ij}p_ip_j + V(q^i)$ on $M = T^*Q$ with canonical coordinates $(p_i, q^i)$ admits an extension $H$ in the form (1) with a first integral $F = U^m(G)$ with $U$ given by (2) and $G(q^i)$, if and only if the following conditions hold:*

*i) the functions $G$ and $V$ satisfy*

$$\mathbf{H}(G) + mc\mathbf{g}G = \mathbf{0}, \quad m \in \mathbb{N} - \{0\}, c \in \mathbb{R}, \tag{3}$$

$$\nabla V \cdot \nabla G - 2m(cV + L_0)G = 0, \quad L_0 \in \mathbb{R}, \tag{4}$$

*where $\mathbf{H}(G)_{ij} = \nabla_i\nabla_j G$ is the Hessian tensor of $G$.*

*ii) for $c = 0$ the extended Hamiltonian $H$ and the first integral $U^mG$ are*

$$H = \frac{1}{2}p_u^2 + mA(L + V_0) + mL_0A^2(u + u_0)^2, \quad U^mG = (p_u - A(u + u_0)X_L)^m G, \tag{5}$$

*for $c \neq 0$ the extended Hamiltonian $H$ and the first integral $U^mG$ are*

$$H = \frac{1}{2}p_u^2 + \frac{m(cL + L_0)}{S_\kappa^2(cu + u_0)} + W_0, \quad U^mG = \left(p_u + \frac{C_\kappa(cu + u_0)}{S_\kappa(cu + u_0)}X_L\right)^m G, \tag{6}$$

*with $\kappa$, $u_0, V_0, W_0, A \in \mathbb{R}$, $A \neq 0$ and*

$$S_\kappa(x) = \begin{cases} \frac{\sin\sqrt{\kappa}x}{\sqrt{\kappa}} & \kappa > 0, \\ x & \kappa = 0, \\ \frac{\sinh\sqrt{|\kappa|}x}{\sqrt{|\kappa|}} & \kappa < 0. \end{cases} \qquad C_\kappa(x) = \begin{cases} \cos\sqrt{\kappa}x & \kappa > 0, \\ 1 & \kappa = 0, \\ \cosh\sqrt{|\kappa|}x & \kappa < 0. \end{cases}$$

In [2] it is proved that $U^m(G)$ is functionally independent from $H$, $L$ and any other first integral of $L$ in $T^*Q$. The integrability conditions of (3) are discussed in [2] and it is found that their complete integrability requires $mc = \kappa$, where $\kappa$ is the constant curvature of $Q$. Then, the function $G(q^i)$ can depend linearly on up to $n+1$ parameters and the maximal number of parameters is attained on constant curvature manifolds only. However, non complete solutions can be found in non-constant curvature manifolds (CDR to appear). From equation (4), the expressions of the admissible potentials $V$ can be computed. Several examples are given in [2, 4, 5].

The particular form of $U$ makes possible to explicit any $U^m(G)$ by expanding the $m$-th power of a binomial, obtaining [4]

$$U^m G = P_m G + D_m X_L G, \tag{7}$$

with

$$P_m = \sum_{k=0}^{[m/2]} \binom{m}{2k} \gamma^{2k} p_u^{m-2k} \left(-2m(cL + L_0)\right)^k,$$

$$D_m = \sum_{k=0}^{[(m-1)/2]} \binom{m}{2k+1} \gamma^{2k+1} p_u^{m-2k-1} \left(-2m(cL + L_0)\right)^k, \quad m > 1,$$

where $[\cdot]$ denotes the integer part and $D_1 = \gamma$. We remark that first integrals of high degree obtained in other ways than by the extension procedure [7, 10] can be explicitly expressed only thanks to the fact that the dynamical equations are in these cases separated in some coordinate system.

As a first example of the extension procedure we consider the one-dimensional Hamiltonian [2]

$$L = \frac{1}{2}p_v^2 + V(v).$$

The geodesic term of the extended Hamiltonian $H$ is

$$\frac{1}{2}\left(p_u^2 + \frac{mc}{S_\kappa^2(cu + u_0)} p_v^2\right),$$

where $\kappa$ is here the constant curvature of the extended configuration manifold. The solutions of equations (3) and (4) are

$$G(v) = b_1 S_{mc}(v + v_0), \quad V(v) = b_2 \frac{1}{C_{mc}^2(v + v_0)},$$

where $b_i \in \mathbb{R}$. When $c = 1$ and $\kappa = 0, 1, -1$, the configuration manifold of $H$ is the Euclidean plane, the Sphere $\mathbb{S}_2$ and the pseudosphere $\mathbb{H}_2$ respectively, while for $c = -1$ and $\kappa = 0, 1, -1$, the Minkowski plane, the deSitter and anti-deSitter manifolds, respectively. After a rescaling of the coordinate $v$, the parameter $m$ in $H$ passes into $L$ and

$$V(v) = b_2 \frac{1}{C_{mc}^2(mv + v_0)}.$$

This makes evident that the extension procedure introduces some discrete symmetry into $H$, in this case a dihedral symmetry of order $2m$, somehow connected with the extra first integral $U^m(G)$. In the Euclidean plane (i.e. $\kappa = 0$) with $c = m^{-1}$ and $m = 3$, $V$ is associated with the Jacobi-Calogero potential or, equivalently, with the Wolfes potential [1, 3]. Indeed, in cylindrical coordinates of $\mathbb{E}^3$, $(u,v,w)$, with axis $w$ parallel to $(1,1,1)$ w.r. to Cartesian coordinates $(x,y,z)$, we have

$$V_{JC} = \frac{k}{u^2\sin^2(3v)}, \quad V_W = \frac{k}{u^2\cos^2(3v)}.$$

The procedure of extension provides two new functionally independent first integrals to the extended Hamiltonian $H$: $H$ itself and $U^mG$. This fact is particularly relevant when the Hamiltonian $L$ is MS. In this case, $H$ is MS too, admitting $2n+1 = 2(n+1)-1$ functionally independent first integrals. In [5] this property of extended Hamiltonians is studied in several cases. For example, let us consider

$$L = \frac{1}{2}p_1^2 + \frac{\zeta}{S_\chi^2(x_1)}\left(\frac{1}{2}p_2^2 + F(x_2)\right), \quad \chi,\zeta \in \mathbb{R},$$

that is a particular case of the generalized Tremblay-Turbiner-Winternitz system (TTW) [13, 7, 10] for $\zeta = 1$ and

$$F(x_2) = \frac{\alpha_1}{\cos^2\lambda x_2} + \frac{\alpha_2}{\sin^2\lambda x_2},$$

defined on constant-curvature manifolds of curvature $\chi$. This system is MS for any rational parameter $\lambda$ and admits polynomial first integrals of degree related to $\lambda$ [13, 7, 10]. In [5] it is shown that $L$ always admits extensions of the form

$$H = \frac{1}{2}p_u^2 + \frac{\chi}{S_\kappa^2(\frac{\chi}{m}u)}L,$$

with $G = a_0C_\chi(x_1) + (a_1S_\zeta(x_2) + a_2C_\zeta(x_2))\,S_\chi(x_1)$, creating in this way new MS systems. Similarly, harmonic oscillators in $\mathbb{E}^n$, isotropic or not, can be extended into harmonic oscillators in $\mathbb{E}^{n+1}$ [5]. The assumption $G(q^i)$ can be generalized to $G(q^i,p_i)$. This leads to important generalizations, introduced in [4], that will be developed in a paper in preparation (CDR). The extension procedure can be in this case applied with $m$ substituted by any positive rational $\mu$ after a suitable definition of $U^\mu G$, so that the generalized TTW system of above, with $\alpha_1\alpha_2 = 0$, can be written as an extension for any rational $\lambda$.

To classical extended Hamiltonians and their first integrals can be associated quantum Hamiltonians and symmetry operators by some procedure of quantization, usually in form of Laplace-Beltrami operators. When the curvature of $Q$ is not constant, the quantization requires additional terms (quantum corrections) in order to keep integrability or superintegrability . The quantum correction is then determined by the scalar curvature and by the Weyl tensor [8, 12]. At least in the case $m=2$, the simultaneous quantization of $H$ and $U^2G$ is possible, allowing the preservation of maximal superintegrability of MS classical extended systems to the quantum limit. This will be shown in a paper in preparation (CDR).

## References


[1] Chanu C, Degiovanni L and Rastelli G 2008, *J. Math. Phys.* **49**, 112901
[2] Chanu C, Degiovanni L and Rastelli G 2011, *SIGMA* **7** 038
[3] Chanu C, Degiovanni L and Rastelli G 2011, *J. Math. Phys.* **52** 032903
[4] Chanu C, Degiovanni L and Rastelli G 2012, *J. Phys.: Conf. Ser.* **343**, 012101
[5] Chanu C, Degiovanni L and Rastelli G 2012, *SIGMA* **8**, 070
[6] Jauch J M and Hill E L 1940, *Phys. Rev* **57**, 641
[7] Kalnins E G, Kress J M and Miller W Jr 2010, *SIGMA* **6** 066
[8] Kalnins E G, Kress J M and Miller W Jr 2013, *J. Phys.* A: *Math. Theor.* **46** 022002
[9] Levesque D, Winternitz P and Post S, 2012 *J. Phys.* A: *Math. Theor.* **45** 465204
[10] Maciejewski A J, Przybylska M and Yoshida H 2010, *J. Phys.* A: *Math. Theor.* **43** 382001
[11] Rañada M F 2012 *J. Phys.* A: *Math. Theor.* **45** 465203
[12] Riglioni D, 2013 *J. Phys.* A:*Math. Theor.* **46** 265207
[13] Tremblay F, Turbiner A V and Winternitz P 2009 *J. Phys.* A **42** n. 24, 242001

# Modelling and analysing oriented fibrous structures

**M Rantala[1], M Lassas[1], J Sampo[2], J Takalo[3], J Timonen[3] and S Siltanen[1]**

[1] Department of Mathematics and Statistics, University of Helsinki, P.O. Box 68, FI-00014 University of Helsinki, Finland

[2] Department of Mathematics and Physics, Lappeenranta University of Technology, P.O. Box 20, FI-53851 Lappeenranta, Finland

[3] Department of Physics, University of Jyväskylä, P.O. Box 35, FI-40014 University of Jyväskylä, Finland

E-mail: maaria.rantala@iki.fi

**Abstract.** A mathematical model for fibrous structures using a direction dependent scaling law is presented. The orientation of fibrous nets (e.g. paper) is analysed with a method based on the curvelet transform. The curvelet-based orientation analysis has been tested successfully on real data from paper samples: the major directions of fibrefibre orientation can apparently be recovered. Similar results are achieved in tests on data simulated by the new model, allowing a comparison with ground truth.

## 1. Introduction

The quality of paper and also other fibre-based products depends essentially on how wood fibres are distributed in a more or less random network of predominantly planar orientation. For this reason, it would be important to be able to measure and control orientation and other properties of the fibre network already during the manufacturing process.

The formation of paper has traditionally been inspected visually and later by analysing in different ways its optical transmission image [1]. However, there are artefacts in optical transmission images due to strong scattering of visible electromagnetic radiation in paper-like fibrous structures. Also, it is not certain as yet from which part of the paper structure an optical transmission image contains information. Therefore, it is of utmost importance to properly calibrate the information gained from optical images. There are two especially suitable methods of calibration: x-ray tomography and simulated networks with precisely known properties.

A method to transform an optical transmission image to one that fairly closely resembles that of x-ray transmission has already been formulated [2]. In this work we present a mathematical model for simulating simple fibre nets using a direction dependent scaling law. The orientation of simulated data is analysed with a method based on the curvelet transform [3, 4], which has been successfully tested in analysing real data from paper samples [5]. The aim of the work is to design an algorithm that simulates a fibrous system from a given orientation distribution.

## 2. Theory

### *2.1. Analysing orientation with curvelets*

The orientation of complex patterns has traditionally been analysed by applying the Fourier transform [6] or gradient-based methods like the structure tensors [7, 8]. However, more sophisticated methods, especially the wavelet transform, [9] have become popular in the last two decades. Furthermore, in recent years transforms like the curvelet, contourlet and shearlet transform have been developed and proved to be well-suited for some applications [3, 4, 10, 11].

The curvelet transform is tightly localised in both space and frequency domain, and has in addition an angle parameter that makes it an optimal tool for orientation analysis. The mother curvelet is defined at each scale $0 < a < a_0$ in the frequency domain as

$$\widehat{\gamma}_{a00}(r\cos(\omega), r\sin(\omega)) = a^{\frac{3}{4}} W(ar) V(\omega/\sqrt{a}),$$

where $r \leq 0$ and $0 \leq \omega \leq 2\pi$, the radial window $W$ is a non-negative, infinitely smooth real-valued function supported inside the interval $(\frac{1}{2}, 2)$, and the angular window $V$ is a non-negative, infinitely smooth real-valued function supported on the interval $[-1, 1]$ (see Figure 1). The whole curvelet is achieved as

$$\gamma_{ab\theta}(x) = \gamma_{a00}(R_{-\theta}(x - b)),$$

where $0 \leq \theta \leq 2\pi$ is a rotation parameter, $b \in \mathbb{R}^2$ is a translation parameter, and $R_\theta$ is the matrix of planar counter-clockwise rotation by the angle $\theta$. The curvelet transform is then defined by

$$\Gamma_f(a, b, \theta) := \langle \gamma_{ab\theta}, f \rangle = \int_{\mathbb{R}^2} f(x) \gamma_{ab\theta}(x)\, dx,$$

for all $0 < a < a_0$, $b \in \mathbb{R}^2$ and $0 \leq \theta \leq 2\pi$.

Curvelets follow the parabolic scaling law in the aspect ratio of the area that contains most of their energy. If we take a piece of smooth curve (corresponding to an edge of a fibre) with a length of about $a$, then the whole piece will fit into a rectangle with the side lengths $a$ and $\sqrt{a}$. We can think of $\gamma_{ab\theta}$ as a sensor that tries to detect if there is a fibre with orientation $\theta$ in the neighbourhood of $b$. If $f$ denotes a fibrous image, then the inner product $\langle \gamma_{ab\theta}, f \rangle$ presents the response of sensor $\gamma_{ab\theta}$. A small value of parameter $a$ means we focus into a part of a fibre, while its larger values can embed a whole fibre. If there is no fibre with the orientation angle $\theta$ located at point $b$, the value of $|\langle \gamma_{ab\theta}, f \rangle|$ is very small.

**Figure 1.** Schematic illustration of curvelet functions, left: curvelet functions of one angle $\theta$ in frequency domain, right: the same curvelet functions in spatial domain.

### *2.2. The H model for fibrous systems*

The statistics of physical parameters (e.g. density) of fibrous structures measured on lines depend on the line direction. If we assume that the physical parameters on a line follow the fractional

Brownian motion statistics, then the Hurst index models the order of the fractional Brownian motion [12]. We create a random model for objects of controllable lengths, where a greater value of the Hurst index means larger features in the object.

Assume now that $Y(x)$ is a stationary two-dimensional Gaussian random field with zero mean, which satisfies a non-standard scaling law for the covariance operator. Here, a Gaussian random field $Y(x)$ is stationary if the covariance function $K_Y(x,y) = \mathbb{E}\,(Y(x)Y(y))$ satisfies $K_Y(x+a, y+a) = K_Y(x,y)$. We consider two one-dimensional stationary Gaussian processes $p_1$ and $p_2$ having covariances

$$\mathbb{E}\,(p_1(t)p_1(s)) \sim 1 - |t-s|^{2H_1} \quad \text{and} \quad \mathbb{E}\,(p_2(t)p_2(s)) \sim 1 - |t-s|^{2H_2} \quad \text{as } t \to s,$$

and zero mean, $\mathbb{E}\,p_j = 0$. This means that $p_1$ and $p_2$ are independent fractional Brownian motions with Hurst indices $H_1$ and $H_2$. The random field $Y(x)$ is obtained by taking oriented fibres having one end point at $(y_1, y_2)$, i.e. the functions $(x_1, x_2) \mapsto K_{y_1,y_2}(x_1, x_2) = k_1(x_1 - y_1)k_2(x_2 - y_2)$, and summing such functions with all possible $(y_1, y_2)$ together with random weights

$$Y(x_1, x_2) = \int_{\mathbb{R}^2} k_1(x_1 - y_1)k_2(x_2 - y_2)W(y_1, y_2)\,dy_1 dy_2,$$

where $W(y_1, y_2)$ is a 2D Gaussian white noise. The Fourier transform of the function $\mathcal{K}(x_1, x_2) = k_1(x_1)k_2(x_2)$ satisfies

$$\widehat{\mathcal{K}}(\xi_1, \xi_2) = \widehat{k}_1(\xi_1)\widehat{k}_2(\xi_2) \sim \quad c_1c_2(1+|\xi_1|)^{-1/2-H_1}(1+|\xi_2|)^{-1/2-H_2} \quad \text{as } \xi \to \infty.$$

## 3. Results and discussion

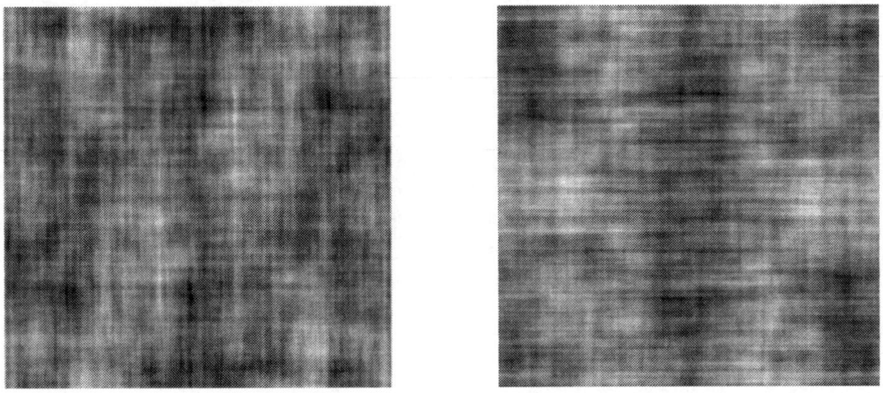

**Figure 2.** Left: fibre net with Hurst indices $H_1 = 0.3$ and $H_2 = 0.5$, right: fibre net with Hurst indices $H_1 = 0.5$ and $H_2 = 0.3$.

The H model gives us fibre nets with two perpendicular orientations. Two samples of fibre nets simulated with the model are shown in Figure 2. A common example of this type of fibre net is newsprint. The simulated fibre nets can then be transformed to change the orientation of the fibres. Examples of the transformed fibre nets are shown in Figure 3. Note that the vertical fibres are turned to the same orientation in both transformations, but the horizontal fibres are transformed differently in rotation and shearing. The orientation analysis of these fibre nets is

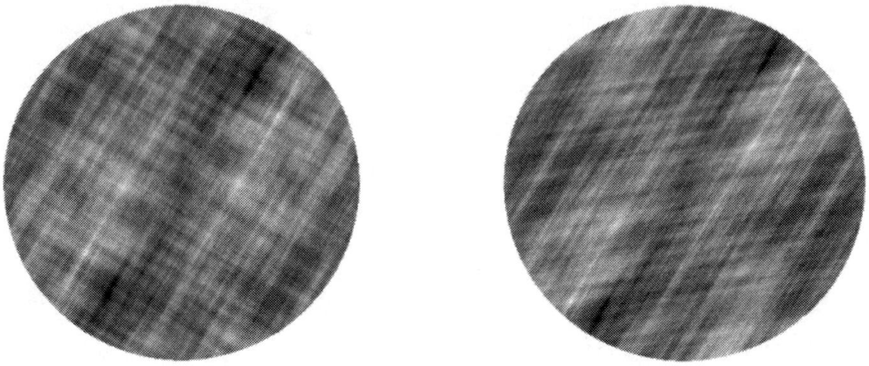

**Figure 3.** Left: the left fibre net of Figure 2 rotated 30° clockwise, right: the same fibre net [illegible]

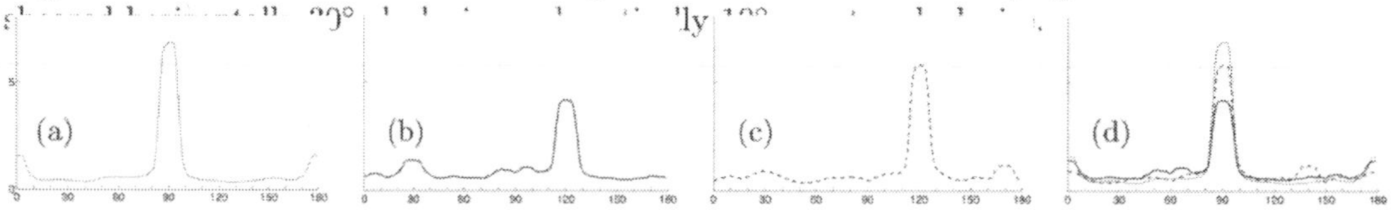


**Figure 4.** (a): orientation analysis of the left fibre net of Figure 2, (b): orientation analysis of the left rotated fibre net of Figure 3, (c): orientation analysis of the right sheared fibre net of Figure 3, (d): orientation curves (a)-(c) shifted to match the strongest orientation.

shown in Figure 4. The strongest orientations are found in the expected positions, although the transformations vary the orientation strengths.

The numerical evidence shows that we get the desired orientations in the fibre nets simulated with the H model. A further goal is to analyse the orientations in real fibre data and then simulate data with the exact same orientations.

## 4. References

[1] Niskanen K 1998 *Papermaking Science and Technology* (Jyväskylä, Fapet)

[2] Takalo J, Timonen J, Sampo J, Siltanen S and Lassas M 2011 Evaluation of the areal material distribution of paper from its optical transmission image *Europ. Phys. J. - Appl. Phys.* **55** 20701

[3] Candès E J and Donoho D L 2005 Continuous curvelet transform. I: Resolution of the wavefront set *Appl. Comput. Harmon. Anal.* **19** (2) 162–97

[4] Candès E J and Donoho D L 2005 Continuous curvelet transform. II: Discretization and frames *Appl. Comput. Harmon. Anal.* **19** (2) 198–222

[5] Sampo J, Takalo J, Siltanen S, Lassas M, Miettinen A and Timonen J 2013 Curvelet-based method for orientation estimation of particles *Submitted*

[6] Enomae T, Han Y-H and Isogai A 2006 *Nord. Pulp Pap. Res. J.* **21** (2) 253–9

[7] Brox T, van den Boomgaard R, Lauze F, van de Weijer J, Weickert J, Mrzek P and Kornprobst P 2006 *Visualization and Processing of Tensor Fields* ed J Weickert, H Hagen (Berlin, Springer-Verlag) pp 17–47

[8] Krause M, Hausherr J M, Burgeth B, Herrmann C and Krenkel W 2010 *J. Mater. Sci.* (45) 888–96

[9] Daubechies I 1992 *Ten lectures on wavelets* (Philadelphia, SIAM)

[10] Do M N and Vetterli M 2005 *IEEE T. Image Process.* **14** (12) 2091–106

[11] Guo K and Labate D 2007 *SIAM J. Math. Anal.* **39** (1) 298–318

[12] Kallenberg O 1997 *Foundations of Modern Probability* (New York, Springer-Verlag)

# Towards an accurate representation of the continuum in calculations of electron, positron and laser field interactions with molecules

**Zdeněk Mašín, Jimena D. Gorfinkiel**

The Open University, Department of Physical Sciences, Walton Hall, Milton Keynes MK7 6AA, UK

E-mail: zdenek.masin@open.ac.uk

**Abstract.** Gaussian-type orbitals (GTOs) are the most common choice of basis functions in calculations of electronic structure of molecules, i.e. for the description of bound electrons. The main advantage of this approach is the analytic form of the multicentre molecular integrals. For the same reason GTOs have been adopted as basis functions for the description of the unbound particle in many ab-initio calculations of electron, positron and laser fields interacting with molecules. However, the accurate description of the unbound particle using GTOs may become very difficult and in some cases numerically unstable. We describe an approach for the representation of the continuum in which the unbound particle is described using a mixed GTO and B-spline basis set in a manner which exploits the best features of these functions. Analytical expressions for the GTO-only molecular integrals allow us to accurately represent the part of the wavefunction close to the target, while the B-splines enable us to represent accurately the wavefunction's tail, corresponding to the unbound particle, over a wide range of kinetic energies. The main challenge posed by this approach is the accurate and rapid numerical evaluation of a large number of mixed BTO/GTO molecular integrals. We propose a scheme for this integral calculation in which the overlap integrals between GTOs and B-spline functions play a central role and demonstrate that they can be calculated rapidly and accurately.

## 1. Introduction

Ab-initio methods for studying molecular processes involving an unbound particle, such as the R-matrix method [1], rely on the use of Gaussian-type orbitals (GTOs) for the description of both the bound electrons of the molecule and the unbound particle (continuum). The central object of the R-matrix method is the R-matrix sphere whose radius must be chosen large enough to fully contain charge density of the molecule. The use of GTOs for the description of the continuum (i.e. the unbound particle) becomes inaccurate or fails completely when large R-matrix radii are needed or when high incident electron energies are required [2].

The origin of the problems in the representation of the continuum using center-of-mass GTOs lies in the need to use functions with very diffuse exponents to guarantee that the resulting continuum orbitals span the whole R-matrix sphere and are non-zero on the boundary. Consequently, the orthogonal continuum orbitals (given as linear combinations of the GTOs) can contain large coefficients and interchanging signs. The combination of those two effects may result in loss of precision of the transformed integrals with obvious consequences for precision of the eigenvalues of the Hamiltonian.

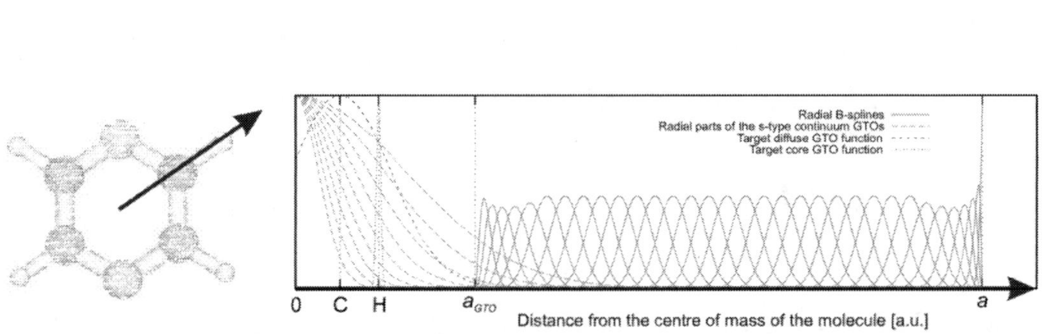


**Figure 1.** On the left: the target molecule (pyrazine in this case). The black arrow defines the radial coordinate for the figure on the right hand side. On the right: the R-matrix sphere has radius $r = a$ and coincides with the endpoint of the last radial B-spline. The radius for which the continuum GTOs have been optimized is denoted as $a_{GTO}$. The continuum radial GTOs and B-splines are shown in green and red. The blue curve corresponds to a cross section through a diffuse target GTO centered on the carbon atom. A compact core GTO function centered on the hydrogen is shown in pink.

We propose a new (hybrid) approach for the representation of the continuum which aims to improve significantly its quality and allow large R-matrix radii ($\approx 30 - 50\ a_0$) to be used. The method builds on the successful use of B-splines in calculations of unbound processes involving atoms [3, 4] and on the recent breakthrough [5, 6, 7] in calculations of two-electron molecular integrals.

The method requires the use of basis functions of two types: standard GTOs and functions whose radial parts are the B-spline functions (see [4] for a detailed description of B-splines). Consequently, the main challenge posed by this approach is the calculation of the required types of molecular integrals in the mixed GTO/B-spline orbital basis. This work presents results of calculations of the overlap and kinetic energy integrals in the mixed basis and proposes a framework for calculation of the rest of the mixed one- and two-electron molecular integrals.

## 2. The choice of the continuum basis

Figure 1 illustrates concisely our approach for the representation of the continuum. The R-matrix radius $a$ is chosen large enough to fully contain the molecule's atomic basis functions (more precisely the molecular orbitals). This radius is determined by the spatial extent of the most diffuse target GTOs. However, the atomic core and polarization functions usually have a smaller spatial extent and can be fitted inside a sphere with the radius $a_{GTO} < a$. This sphere will always contain only GTOs and therefore we call it the *GTO sphere*. Based on this observation we construct the continuum basis from two types of functions:

- **GTOs centered on the center of mass** (CMS) of the molecule and optimized for use with the radius $a_{GTO}$. These functions are generally nonzero inside the whole R-matrix sphere: $r \in [0, a]$.
- **B-spline orbitals (BTOs) centered on the CMS** which span *only* the radial range: $r \in [a_{GTO}, a]$. (The exact form of these functions is defined in Appendix A).

Consequently, the only basis functions which are non-zero in the region outside the GTO sphere are the BTOs, continuum GTOs and some target (usually diffuse) GTOs. This *hybrid approach for representation of the continuum* has several important properties:

- The quality of representation of the continuum inside the GTO sphere is given by the choice of the continuum GTOs. We can control the quality of the continuum inside the GTO sphere by changing its radius (and/or the corresponding continuum GTOs) [8].

- The product of a BTO and those target GTOs fully enclosed by the GTO sphere is identically zero. The product of a BTO and the remaining GTOs is generally non-zero.

These properties may be used to achieve significant savings in the computational cost of calculating the mixed GTO/BTO molecular integrals. The most computationally demanding part of the calculation are the mixed BTO/GTO two-electron integrals:

$$\langle \phi_1 | r_{12}^{-1} | \phi_2 \rangle = \int d\mathbf{r_1} \int d\mathbf{r_2} a(\mathbf{r_1}) b(\mathbf{r_1}) r_{12}^{-1} c(\mathbf{r_2}) d(\mathbf{r_2}), \tag{1}$$

$$\phi_1(\mathbf{r_1}) = a(\mathbf{r_1}) b(\mathbf{r_1}),\ \phi_2(\mathbf{r_2}) = c(\mathbf{r_2}) d(\mathbf{r_2}), \tag{2}$$

where at least one of $a(\mathbf{r_1}), b(\mathbf{r_1})$ and/or $c(\mathbf{r_2}), d(\mathbf{r_2})$ is a BTO and the rest of the functions are GTOs. Utilizing the properties listed above these integrals can be split into three classes:

(i) Identically zero two-electron integrals. This class contains integrals in which at least one of the densities ($\phi_1(\mathbf{r_1})$ and/or $\phi_2(\mathbf{r_2})$) contains the product of a BTO and a GTO fully enclosed by the GTO sphere.

(ii) The densities $\phi_1(\mathbf{r_1})$ and $\phi_2(\mathbf{r_2})$ are spatially separated. This situation arises when e.g. the density $\phi_1(\mathbf{r_1})$ involves at least one GTO which is fully enclosed by the GTO sphere while the density $\phi_2(\mathbf{r_2})$ involves the product of a BTO with a GTO that is non-zero outside the GTO sphere.

(iii) The densities $\phi_1(\mathbf{r_1})$ and $\phi_2(\mathbf{r_2})$ spatially overlap, i.e. they both involve products of functions that are non-zero outside the GTO sphere.

Depending on the choice of $a_{GTO}$ we can determine which target functions are fully enclosed by the GTO sphere and therefore assign each two-electron integral into one of the three classes. Class (ii) integrals can be calculated using the multipole expansion of the Coulomb interaction [9]. The most difficult case are the class (iii) integrals. To calculate them we suggest using the method based on resolution of the Coulomb operator into single-particle functions [7] centered on the CMS:

$$r_{12}^{-1} \approx \sum_{k=1}^{n} \psi_k(\mathbf{r_1}) \psi_k(\mathbf{r_2}) \rightarrow \langle \phi_1 | r_{12}^{-1} | \phi_2 \rangle \approx \sum_{k=1}^{n} \langle \phi_1 | \psi_k \rangle \langle \phi_2 | \psi_k \rangle. \tag{3}$$

The same approach can be used to calculate the mixed one electron nuclear attraction integrals. We can see that the calculation of the mixed GTO/BTO overlap integrals plays a crucial role in the suggested approach.

## 3. Interaction-free electron scattering

As the first step towards implementing the proposed approach for the representation of the continuum we calculated the mixed GTO/BTO overlap and kinetic energy integrals utilizing the CMS-centered partial wave expansion of the GTO. Formulae for these integrals can be found in Appendix A. In order to test the approach on a realistic system we solved, using the R-matrix method, the one-electron free potential scattering problem. In order to solve this problem we also needed the GTO/GTO and BTO/BTO overlap and kinetic energy integrals whose evaluation is trivial.

To mimic the description of the continuum electron in the R-matrix suite the set of orbitals used were those of the continuum and the Hartree-Fock (HF) molecular orbitals of two selected molecules: water and pyrazine. In order to obtain a set of orthogonal orbitals we utilized the standard procedure used in the molecular R-matrix calculations: the HF (orthogonal) orbitals were kept unchanged throughout the calculation. Orbitals based on the CMS continuum GTOs and BTOs were first Schmidt orthogonalized to the target orbitals and then symmetrically

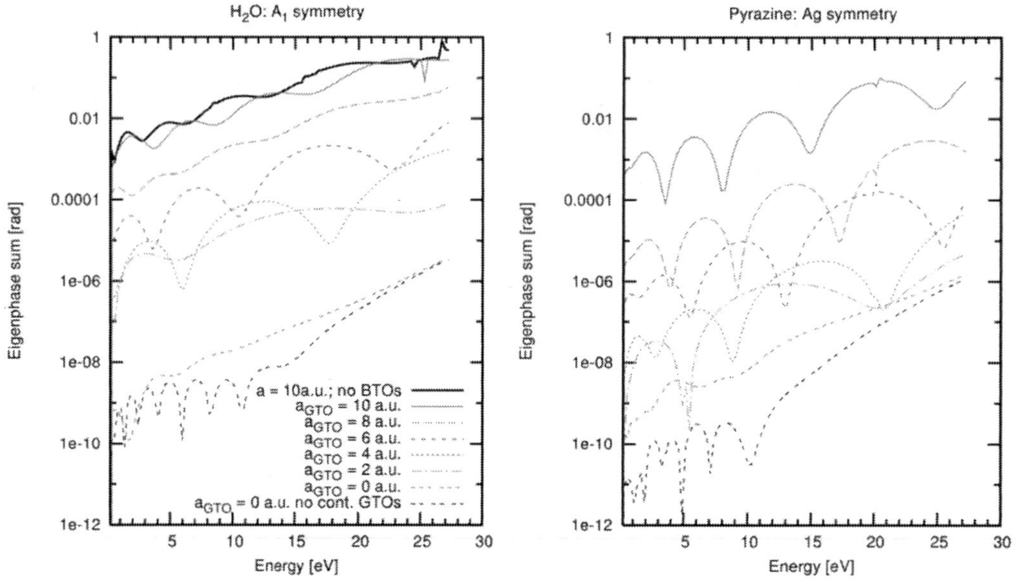


**Figure 2.** Eigenphase sums for interaction-free scattering from water ($C_{2v}$ symmetry) and pyrazine ($D_{2h}$ symmetry) for the totally symmetric contribution. The exact eigenphase sum is zero. The colour lines are for results when both the GTOs and BTOs were used; the starting radius for the BTOs ($a_{GTO}$) is indicated in the Figure.

orthogonalized among themselves. Finally, all continuum orbitals with eigenvalues of the continuum-continuum overlap matrix smaller than $10^{-7}$ were deleted. In all calculations we used the same continuum GTO basis optimized for a radius $a_{GTO} = 10a_0$. The BTO radial basis consisted of B-splines of order 8 with break points separated by 1 $a_0$ and spanning the radial distance of $29a_0$. This set of radial B-splines was kept fixed and only translated along the radial coordinate according to the value of $a_{GTO}$. For water (pyrazine) an R-matrix radius $a$ of $10a_0$ ($18a_0$) was sufficient to enclose all target orbitals.

Figure 2 shows the results obtained when the target orbitals were those of water (6 orbitals of $A_1$ symmetry) and pyrazine (14 orbitals of $A_g$ symmetry). We show only the results obtained for scattering in the totally symmetric irreducible representation; results for the rest of the symmetries were similar. For water we show the eigenphase sum obtained when only GTOs (and $a = 10a_0$) were used for the continuum as well as the result obtained when only the BTOs ($a_{GTO} = 0$ and $a = 28a_0$) were used for the continuum. In both cases (water and pyrazine) we observe a rapid and systematic improvement of the quality of representation of the continuum as $a_{GTO}$ decreases, i.e. as the BTOs are pushed closer towards the CMS. Clearly, for $a_{GTO} = 6a_0$ we already obtain eigenphase sums which are for all energies smaller than the value of the smallest eigenphase sum obtained for $a_{GTO} = 10a_0$.

The relative precision of the overlap and the kinetic energy integrals was never worse than $10^{-10}$ for pyrazine and $10^{-13}$ for water. As expected based on the properties of the partial wave expansion of the GTO, the relative precision of the integrals generally decreases as the distance of the center of the GTO from the CMS is increased and as the angular momentum and exponent of the GTO are increased. The only numerical problems (loss of precision) were detected during the integral transformation step; we are currently investigating possible resolutions of this problem.

The calculation of the mixed GTO/BTO integrals utilized the point group symmetries of the target molecules and in both cases took $\approx 9.7$s on an Intel Xeon 2.50 GHz workstation.

## 4. Conclusions

We propose and test a methodology for a hybrid GTO/BTO representation of the continuum that can be used for calculations of electron, positron and light interactions with molecules. This approach was found to be much more accurate than our currently employed GTO-only approach. We showed that the mixed GTO/BTO overlap and kinetic energy integrals can be calculated accurately and efficiently. Somewhat surprisingly we found that the most serious numerical problem is the loss of precision that may occur on the level of the integral transformation. The numerical stability of the mixed BTO/GTO primitive overlap and kinetic energy integral calculation can be controlled for and does not appear to cause significant problems.

## Appendix A. Overlap and kinetic energy integrals between GTOs and BTOs

We use GTOs in the standard form:

$$\chi_{\alpha LM}(\mathbf{r}-\mathbf{R}) = N_{\alpha L}^{GTO} S_{LM}(\mathbf{r}-\mathbf{R}) \exp[-\alpha|\mathbf{r}-\mathbf{R}|^2],\ N_{\alpha L}^{GTO} = \sqrt{\frac{2(2\alpha)^{L+3/2}}{\Gamma[L+3/2]}\frac{2L+1}{4\pi}}, \tag{A.1}$$

where $\mathbf{R}$ is the radius vector pointing in the direction of the centre of the GTO, $\alpha$ is the exponent of the GTO and $L, M$ describe the angular dependence of the GTO through the normalized real solid harmonics $S_{LM}(\mathbf{r}-\mathbf{R})$ centered at $\mathbf{r}=\mathbf{R}$. We follow the standard conventions [10] for the definition of the real solid harmonics. The normalization factor $N_{\alpha L}^{GTO}$ is chosen to reproduce the unit self-overlap of the GTO. The B-spline orbital centered on the center of mass of the molecule (BTO) is defined by the equation:

$$\mathcal{B}_{L_iM_i}^{i}(\mathbf{r}) = N_i^{BTO}\frac{B_i(r)}{r}X_{L_iM_i}(\hat{\mathbf{r}}),\ N_i^{BTO} = \left[\int_{t_i}^{t_{i+k}} dr (B_i(r))^2\right], \tag{A.2}$$

where $X_{L_iM_i}(\hat{\mathbf{r}})$ is the normalized real spherical harmonic [11], $r$ is the distance from the origin, $N_i^{BTO}$ is the normalization factor chosen to reproduce the unit self-overlap of the BTO and $B_i(r)$ is the B-spline function in coordinate $r$ (piecewise polynomial with compact support); $t_i, \ldots, t_{i+k}$ is the set of $k+1$ knots, where $k > 0$ is the order of the B-splines (see [4] for details).

The overlap integral between a GTO (not centered on the CMS) and a BTO is given *exactly* by a finite sum of terms originating in the partial wave expansion of the GTO [12] and has the form:

$$\langle \chi_{\alpha LM}|\mathcal{B}_{L_iM_i}^{i}\rangle = \frac{4\pi}{\sqrt{\alpha R(2L+1)}} N_i^{BTO} N_{\alpha L}^{GTO} \sum_{j=1}^{n_{LM}} d_j^{LM} (-1)^{\mu_j + l'_j + m'_j} \mathcal{I}_{\lambda_j\mu_j}\mathcal{I}_{\lambda'_j\mu'_j} R^{l'_j} X_{l'_jm'_j}(\hat{\mathbf{R}}) \times$$
$$\times \sum_{l=l_{min}}^{l_{max}} R_{i,\lambda_j,l}(\alpha,R) \sum_{m=-l}^{l} \langle L_iM_i|\lambda_j\mu_j|lm\rangle_R X_{lm}(\hat{\mathbf{R}}), \tag{A.3}$$

where $d_j^{LM}$ are coefficients in the translation formula [13] for the real unnormalized solid harmonics $z_{LM}(\mathbf{r}_1+\mathbf{r}_2)$:

$$z_{LM}(\mathbf{r}_1+\mathbf{r}_2) = \sum_{j=1}^{n_{LM}} d_j^{LM} z_{l_jm_j}(\mathbf{r}_1) z_{l'_jm'_j}(\mathbf{r}_2), \tag{A.4}$$

$$z_{LM}(\mathbf{r}) = (-1)^M \mathcal{I}_{LM} r^L X_{LM}(\hat{\mathbf{r}}),\ \mathcal{I}_{LM} = \sqrt{\frac{2\pi(1+\delta_{M,0})}{2L+1}\frac{(L+|M|)!}{(L-|M|)!}}, \tag{A.5}$$

$\langle L_i M_i | \lambda_j \mu_j | lm \rangle_R$ are the Gaunt coefficients for the real spherical harmonics and

$$R_{i,\lambda_j,l}(\alpha, R) = \int_{t_i}^{t_{i+k}} dr B_i(r) r^{\lambda_j+1/2} \exp[-\alpha(r-R)^2] \tilde{I}_{l+1/2}(2\alpha r R) \tag{A.6}$$

are the radial integrals in which $\tilde{I}_{l+1/2}(2\alpha r R) = \exp[-2\alpha r R] I_{l+1/2}(2\alpha r R)$ stands for the exponentially scaled modified Bessel function. These radial integrals are obtained using an adaptive numerical quadrature. In order to speed up the calculation we use recurrent relations for $R_{i,\lambda_j,l}(\alpha, R)$ which (for fixed $i, \alpha$ and $R$) couple the $\lambda$ and $l$ parameters:

$$R_{i,\lambda,l-1}(\alpha, R) = R_{i,\lambda,l+1}(\alpha, R) + \frac{l+1/2}{\alpha R} R_{i,\lambda-1,l}(\alpha, R). \tag{A.7}$$

These recurrent relations can be derived using the well-known downward (numerically stable) recurrent relation for the modified Bessel functions $I_{l+1/2}(x)$. The formula for the kinetic energy integral $\langle \chi_{\alpha LM} | \Delta | \mathcal{B}^i_{L_i M_i} \rangle$ retains the same form as (A.3) with exception of the radial integral which has the form:

$$\begin{aligned} R^{\Delta}_{i,\lambda_j,l}(\alpha, R) &= \int_{t_i}^{t_{i+k}} dr B_i''(r) r^{\lambda_j+1/2} \exp[-\alpha(r-R)^2] \tilde{I}_{l+1/2}(2\alpha r R) - \\ &- L_i(L_i+1) R_{i,\lambda_j-2,l}(\alpha, R), \end{aligned} \tag{A.8}$$

where $B_i''(r)$ stands for the second derivative of the B-spline $B_i(r)$. The relative precision of the individual terms entering (A.3) is not lower than $\approx 10^{-14}$ and the only loss of precision may occur as a result of subtraction of competing partial-waves when the overlap (or the kinetic energy) integral is close to zero due to symmetry. However, the occurrence of these instabilities can be mitigated by the choice of the breakpoint sequence for the radial B-splines which destroys the spatial symmetry of the integrand $\chi_{\alpha LM}(\mathbf{r}-\mathbf{R}) \mathcal{B}^i_{L_i M_i}(\mathbf{r})$. Other means by which the calculation can be stabilized are being investigated.

## References


[1] Carr J M, Galiatsatos P G, Gorfinkiel J D, Harvey A G, Lysaght M A, Madden D, Mašín Z, Plummer M, Tennyson J and Varambhia H N 2012 *The European Physical Journal D* **66** 58

[2] Mašín Z and Gorfinkiel J D 2011 *The Journal of Chemical Physics* **135** 144308

[3] Moore L, Lysaght M, Nikolopoulos L, Parker J, van der Hart H and Taylor K 2011 *Journal of Modern Optics* **58** 1132–1140

[4] Bachau H, Cormier E, Decleva P, Hansen J and Martin F 2001 *Reports on progress in physics* **64** 1815–1943

[5] Hoggan P E 2009 *Self-Organization of Molecular Systems: From Molecules and Clusters to Nanotubes and Proteins* ed Russo N, Antonchenko V Y and Kryachko E S (Dordrecht: Springer) pp 199–219

[6] Varganov S A, Gilbert A T B, Deplazes E and Gill P M W 2008 *The Journal of Chemical Physics* **128** 201104

[7] Limpanuparb T 2011 *Applications of Resolutions of the Coulomb Operator in Quantum Chemistry* Ph.D. thesis Australian National University

[8] Faure A, Gorfinkiel J D, Morgan L A and Tennyson J 2002 *Computer Physics Communications* **144** 224–241

[9] Morgan L A, Gillan C J, Tennyson J and Chen X 1997 *Journal of Physics B: Atomic, Molecular and Optical Physics* **30** 4087–4096

[10] Helgaker T, Olsen J and Jorgensen P 2000 *Molecular Electronic-Structure Theory* (Wiley) ISBN 9781118531471

[11] Homeier H H H and Steinborn E O 1996 *Journal of Molecular Structure: THEOCHEM* **368** 31–37

[12] Kaufmann K and Baumeister W 1989 *Journal of Physics B-Atomic Molecular and Optical Physics* **22** 1–12

[13] Rico J F, Lopez R, Ema I and Ramirez G 2013 *International Journal of Quantum Chemistry* **113** 15441548

# Velocity selective trapping of atoms in a frequency-modulated standing laser wave: wave function and stochastic trajectory approaches

**Victor Argonov**

Laboratory of Geophysical Hydrodynamics, Pacific Oceanological Institute of the Russian Academy of Sciences, 43, Baltiiskaya Street, Vladivostok, Russia 690041

E-mail: argonov@poi.dvo.ru

**Abstract.**
The wave function of a moderately cold atom in a stationary near-resonant standing light wave delocalizes very fast due to wave packet splitting. However, frequency modulation of the field may suppress packet splitting for some atoms having specific velocities in a narrow range. These atoms remain localized in a small space for a long time. We show that modulated field can not only trap, but also cool the atoms. We perform a numerical experiment with a large atomic ensebmble having wide initial velocity and energy distribution. During the experiment, most of atoms leave the wave while trapped atoms have narrow energy distribution

## 1. Introduction

Laser cooling and trapping of atoms and ions is a rapidly developing field of modern physics. Cold particles in a laser field are a common physical substrate used in numerous fundamental and applied issues such as Bose-Einstein condensates, quantum chaos, single-atom laser, quantum computer, etc. A significant number of methods of atomic cooling in a laser field were developed in the recent decades (the Doppler cooling [1, 2], the Sisyphus cooling [3, 4], the velocity selective coherent population trapping (VSCPT) [5], dynamical localization and trapping [6], etc. [7]).

In this paper we suggest a method of coherent laser cooling in the absence of spontaneous emission. When an atom moves in a near-resonant standing light wave, two periodic optical potentials form in the space [8]. When the atom crosses a standing wave node, it may undergo the Landau-Zener transition between these two potentials. Such transitions cause splitting of the wave packets [9] and rapid delocalization of the wave function [10]. In this paper we show that frequency modulation of the field may suppress the splitting of wave packets for atoms that have velocities in the specific narrow range determined by the field modulation parameters. We suppose that in a real experiment, this may significantly decrease the energy distribution of moderately cold atoms.

## 2. Equations of motion

Let us consider a two-level atom (with the transition frequency $\omega_a$ and mass $m_a$) moving in a strong standing laser wave with the modulated frequency $\omega_f[t]$. In absence of spontaneous emission (the atomic excited state must have long lifetime, or some experimental methods must suppress the decoherence) the atomic motion may be described by the Hamiltonian [11]

$$\hat{H} = \frac{\hat{P}^2}{2m_a} + \frac{1}{2}\hbar(\omega_a - \omega_f[t])\hat{\sigma}_z - \hbar\Omega\,(\hat{\sigma}_- + \hat{\sigma}_+)\cos k_f\hat{X}, \tag{1}$$

where $\hat{\sigma}_{\pm,z}$ are the operators of transitions between the atomic excited and ground states (the Pauli matrices), $\hat{X}$ and $\hat{P}$ are the operators of the atomic coordinate and momentum, and $\Omega$ is the Rabi frequency.

Let us use the following dimensionless normalized quantities: momentum $p \equiv P/\hbar k_f$, time $\tau \equiv \Omega t$, position $x \equiv k_f X$, mass $m \equiv m_a\Omega/\hbar k_f^2$ and detuning $\Delta[\tau] \equiv (\omega_f[\tau] - \omega_a)/\Omega$. Let us suppose that the field modulation is harmonic,

$$\Delta[\tau] = \Delta_0 + \Delta_1 \cos[\zeta\tau + \phi], \quad \zeta \ll 1, \quad \Delta_0 \lesssim \Delta_1 \ll 1. \tag{2}$$

Now let us obtain the equations for the probability amplitudes to find an atom with the normalized momentum $p$ in the excited or ground state, $a[p,\tau]$ and $b[p,\tau]$, correspondently:

$$\begin{aligned} i\dot{a}[p,\tau] &= \left(\frac{p^2}{2m} - \frac{\Delta[\tau]}{2}\right)a[p] - \frac{1}{2}(b[p-1] + b[p+1]), \\ i\dot{b}[p,\tau] &= \left(\frac{p^2}{2m} + \frac{\Delta[\tau]}{2}\right)b[p] - \frac{1}{2}(a[p-1] + a[p+1]). \end{aligned} \tag{3}$$

Here the dot designates the differentiation with respect to $\tau$. For every value of $p$, there is its own pair (3).

## 3. Wavefunction approach: numerical manifestations of velocity selective trapping

Let us choose the values of the parameters and initial conditions in order to perform the numerical simulation. The average initial atomic momentum $\langle p[0]\rangle$ will be a variable condition for the purpose of this paper. All other conditions will be fixed: normalized mass $m = 10^5$ (by order of magnitude this corresponds to the experiments with Cs [12] and Rb [13] atoms, but for a stronger field $\Omega \sim 10^{9-10}$ Hz), field parameters $\Delta_0 = -0.02$, $\Delta_1 = 0.047$, $\zeta = 0.00508$, $\phi = 0$. The initial wave packet has a Gaussian form with $\langle x[0]\rangle = 0$ and the initial probability to find the atom in the excited state 0.5. The standard deviation of the atomic momentum is $\sigma_p[0] = 5\sqrt{2}$. Therefore, in accordance with the Heisenberg relation, the standard deviation of the initial coordinate is $\sigma_x[0] = 1/(2\sigma_p[0]) = 0.1/\sqrt{2}$ (it is much less than the normalized optical wavelength $2\pi$). In numerical experiments, we use these initial conditions to simulate the system of 8000 equations (3) with $-1000 \leq p \leq 1000$. For larger values of $|p|$, we put $a[p,\tau] = b[p,\tau] = 0$ due to the energy restrictions. Obtaining the solution in the momentum space we perform the Fourier transform and get the wave function in the coordinate space.

Let us study the effect of field modulation on atomic delocalization. In [9, 10], the atomic motion was studied in absence of modulation. At $\Delta = 0$ and $|\Delta| \gtrsim 1$ the atomic motion is simple. Atoms move in constant spatially periodic potentials. Slow atoms are trapped in potential wells and fast atoms move ballistically through the wave. At $0 < |\Delta| \ll 1$ the atomic motion is more complex. The slowest atoms ($|\langle p[0]\rangle| < \sqrt{2m}$) are trapped in potential wells. Faster atoms ($\sqrt{2m} \leq |\langle p[0]\rangle| < 2\sqrt{m}$) perform a kind of random walk. Their wave packets split each time they cross standing-wave nodes, and this causes fast delocalization of the wave functions. The fastest atoms ($|\langle p[0]\rangle| > 2\sqrt{m}$) move ballistically through the wave. Their wave packets split, but all products move in the same direction, so the overall delocalization is slow.

In Fig. 1 we calculate the variance of the atomic position $\sigma_x^2$ after a relatively long time span of coherent evolution $\tau = 5000$ as a function of the initial atomic momentum $\langle p[0]\rangle$. For the

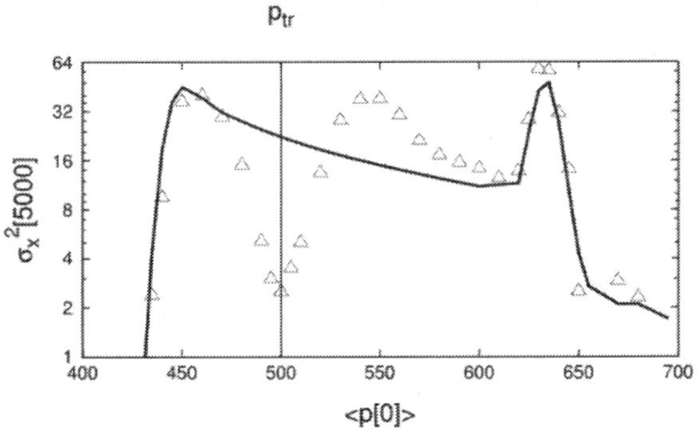


**Figure 1.** The variance of atomic position $\sigma_x^2$ at $\tau = 5000$ as a function of initial atomic momentum $\langle p[0] \rangle$: curve — constant field $\Delta = -0.02$, triangles — modulated field $\Delta = -0.02 + 0.047 \cos[0.00508\tau]$

constant field (solid curve) this function shows fast delocalization of all atoms in the range of $\sqrt{2m} \simeq 440 \lesssim \langle p[0] \rangle \lesssim 2\sqrt{m} \simeq 640$ (cold atoms with velocities of the order of 1 m/s). Local peak at $\langle p[0] \rangle \simeq 630$ is produced by moderately fast atoms having an uncertain scenario of either random walking or flying ballistically.

Now let us "switch on" the field modulation and see the changes. In Fig. 1 the analogous function of $\sigma_x^2$ is shown with triangles. This function has a more complex structure. In particular, it has a prominent additional minimum at $\langle p[0] \rangle = p_{tr} \simeq 500$. These atoms are not trapped in potential wells in a strict sense (their energy is too high, see the theory in the next sections), but some mechanism significantly suppresses the delocalization of their wave functions (note that both functions are shown in a logarithmic scale).

## 4. Semiclassical approach: explanation of the effect

In order to explain the effect of velocity selective trapping, let us mention some analytical results from [9, 10, 11]. Near resonance, the atomic motion can be described in terms of two potentials

$$U^- = -\sqrt{\cos^2[x] + \frac{\Delta^2[\tau]}{4}}, \quad U^+ = \sqrt{\cos^2[x] + \frac{\Delta^2[\tau]}{4}}. \tag{4}$$

An atom moves in one of these potentials when it is far from standing-wave nodes ($x = \pi/2 + n$). The semiclassical energy and equations of motion are [11]

$$E \equiv \frac{p^2}{2m} + U[x, \tau], \quad \dot{x} = \frac{p}{m}, \quad \dot{p} = -\text{grad}[U]. \tag{5}$$

If initial energy $E[0] \lesssim 0$ (for $x[0] = 0$, this corresponds to $|p[0]| \lesssim \sqrt{2m}$), then an atom cannot reach any standing-wave node. It is trapped in the bottom the of the first potential well near $x = 0$. If initial energy is in the range of $0 \lesssim E \lesssim 1$ (for $x[0] = 0$, this corresponds to $\sqrt{2m} \lesssim |p[0]| \lesssim 2\sqrt{m}$), then an atom may either perform a random walk or being trapped (if $p[0] = p_{\text{tr}}$). Faster atoms with $E \gtrsim 1$ move ballistically through the wave in a constant direction.

When an atom crosses the node, the potential may change the sign (atom undergoes the Landau-Zener tunneling between potentials $U^{\pm}$) with the probability

$$W_{LZ} \approx \exp \frac{-\Delta^2 m\pi}{4\langle p_{\text{node}} \rangle}, \tag{6}$$

where $\langle p_{\text{node}} \rangle$ is an average momentum of an atom when it crosses the node. At $0 < |\Delta| \ll 1$ the tunneling causes splitting of wave packets (observed in numerical experiments [9, 10, 11]).

and fast delocalization of the wave function. However, at $\Delta = 0$ potentials coincide at nodes, so the probability of tunneling is equal to 1 and wave packets do not split. Delocalization of wave function is very slow and the potential takes the simplest form $U = \pm\cos[x]$.

Velocity-selective trapping of atom occurs if the field modulation is synchronized with the atomic mechanical motion. It is possible to choose such modulation parameters and atomic momentum (see analytical estimations in [11]) that $\Delta[\tau]$ takes zero values each time an atom crosses the node. With our parameters such synchronization occurs at $\langle p[0]\rangle = p_{\rm tr} \simeq 500$. Therefore, packet splitting and delocalization are suppressed.

Note that for trapped atoms, equations (5) stay correct during entire evolution (even during node crossings), and the potential takes simplest form. Trapping occurs, if atom either does not crosses nodes at all, or node crossings take place when $\Delta[\tau] = 0$. Therefore, the term $\Delta^2/4$ in (4) is always neglible, and the atom moves in a constant potential with a constant energy

$$U \simeq -\cos[x], \quad \tilde{E} \equiv \frac{p^2}{2m} - \cos[x]. \tag{7}$$

Initial conditions used in this paper correspond the negative sign of $U$ (conserved for trapped atoms).

## 5. Stochastic trajectory approach: modeling of atomic cooling process

In order to show that reported effect is not only trapping of atoms but also their cooling, we must simulate the dynamics of an atomic ensemble having wide initial velocity (and energy) distribution and show that the distribution goes narrow during the evolution. Such simulation with quantum equations requires a huge computational time. Therefore, we develop an alternative simplified model of atomic motion based on the following principles.

1. Atom is a dot-like particle having a particular trajectory.
2. Between standing-wave nodes, an atom moves in an effective potential $U^{\mp}[x,\tau]$ (4) having constant sign. Such motion is governed by semiclassical equations (5).
3. At initial time moment, the potential $U^{\mp}$ has negative sign. Any time when an atom crosses a node, the potential changes its sign with the probability (6).

In Fig. 2, we simulate the dynamics of an atomic ensemble (several thousands of atoms) with comparatively wide initial momentum distribution moving in positive direction with average velocity $\langle p[0]\rangle = 550$ (Fig. 2a). Corresponding energy distribution is shown in Fig. 2b (we calculate simplified energy $\tilde{E}$ (7), but it is equal to general energy $E$ (5) at initial time moment).

In order to show that velocity selective trapping really cools atoms, let us consider a small part of laser wave in a range

$$-\frac{3}{2}\pi < x < \frac{3}{2}\pi. \tag{8}$$

At the beginning of the experiment, all the atoms have $x \simeq 0$. During the evolution, trapped atoms ($p[0] \lesssim 440$, $E[0] \lesssim 0$ and $p[0] \simeq p_{\rm tr} = 500$, $E[0] \simeq 0.25$) stay in the range (8) while most of other atoms leave it (due to ballistic flight or random walk). Trapped atoms have wide momentum distribution because their momenums oscillate in a wide range. However, their energy distributuon is very narrow. In Figs. 2c, d, there is a prominent peak near $\tilde{E} = 0.25$, and it is very narrow in comparison with initial energy distribution. This is because the majority of atoms with other initial values of energy leaved the wave. Note that simplified energy $\tilde{E}$ is conserved only for trapped atoms. Other atoms can change it during the evolution (see, for example, spontaneous peak at $\tilde{E} \simeq -0.6$, Fig. 2c). However, the number of such atoms in area (8) decays fastly, so they do not change the overall picture.

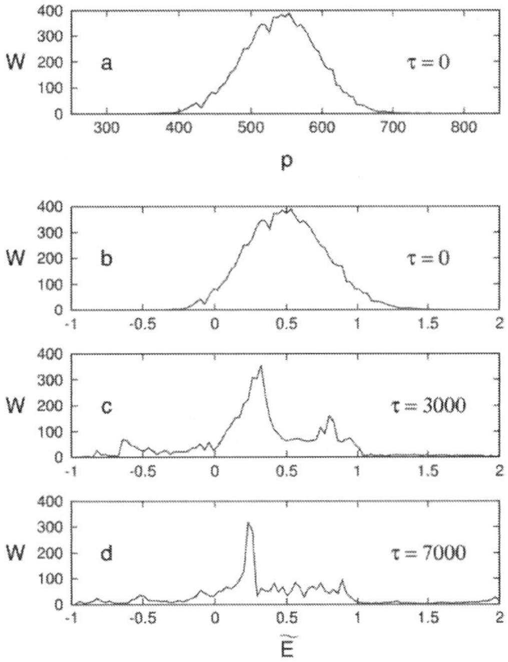


**Figure 2.** Cooling of an atomic cloud due to velocity selective trapping (statistics of atoms having positions in a range $-3\pi/2 < x < 3\pi/2$). Probability density $W$ to find an atom with a given momentum or energy is shown in arbitrary units.

## 6. Conclusion

Intensive coherent light produces significant mechanical action on cold atoms having velocities of the order of 1 m/s. There is a wide range of field parameters at which atom performs a kind of random walk accompanied by wave packets splitting and fast delocalization of wave function. However, a specific field modulation can suppress wave packet splitting for atoms with precisely selected velocities. These atoms oscillate in a small space, and their wave functions are almost completely localized. This effect cannot cool atoms in the sense of achieving zero velocity, but it can significantly decrease their mechanical energy distribution (see Fig. 2).

Note that the effect is purely coherent, it takes place only in absence of dissipation. However, we believe that this is just a quantitative technical limitation that may be overcome by an appropriate choice of atoms and hi-Q cavities.

This work has been supported by the Grant of the Russian Foundation for Basic Research 12-02-31161.

[1] Hansch T W and Schawlow A L 1975 *Opt. Comm.* **13** 68
[2] Wineland D J, Drullinger R E and Walls F L 1978 *Phys. Rev. Lett.* **40** 1639
[3] Dalibard K and Cohen-Tannouudji C 1989 *J. Opt. Soc. Am. B* **6** 2023
[4] Bezverbny A V, Niz'ev V G and Tumaikin A M 2004 *Quant. Electronics* **34** 685
[5] Aspect A, Arimondo E, Kaiser R, Vansteenkiste N and Cohen-Tannoudji C 1988 *Phys Rev Lett.* **61** 826
[6] Raizen M and Steck D A 2011 Scholarpedia **6** 10468
[7] Phillips W 1997 *Laser Cooling and Trapping of Neutral Atoms* (Nobel Lecture)
[8] Kazantsev A P, Surdutovich G I and Yakovlev V P 1990 *Mechanical action of light on atoms* (Singapore; London: World Scientific)
[9] Prants S V 2009 *J. Exp. Theor. Phys.* **109** 751
[10] Argonov V Yu 2009 *J. Exp. Theor. Phys. Lett.* **90** 739
[11] Argonov V Yu 2013 *J. Exp. Theor. Phys. Lett.* **98** 655
[12] Ammann H, Gray R, Shvarchuck I and Christensen N 2009 *Phys. Rev. Lett.* **80** 4111
[13] Hensinger W K, Heckenberg N R, Milburn G J, et al 2003 *J. Opt. B: Quantum Semiclass. Opt.* **5** R83

# Bayesian estimation of realized stochastic volatility model by Hybrid Monte Carlo algorithm

**Tetsuya Takaishi**

Hiroshima University of Economics, Hiroshima 731-0192, JAPAN

E-mail: tt-taka@hue.ac.jp

**Abstract.** The hybrid Monte Carlo algorithm (HMCA) is applied for Bayesian parameter estimation of the realized stochastic volatility (RSV) model. Using the 2nd order minimum norm integrator (2MNI) for the molecular dynamics (MD) simulation in the HMCA, we find that the 2MNI is more efficient than the conventional leapfrog integrator. We also find that the autocorrelation time of the volatility variables sampled by the HMCA is very short. Thus it is concluded that the HMCA with the 2MNI is an efficient algorithm for parameter estimations of the RSV model.

## 1. Introduction

In empirical finance, volatility of asset returns plays an important role to manage financial risk. Since volatility is not directly observed in financial markets, one needs to estimate volatility from observed quantities in the markets. A typical way to estimate volatility is to use parametric models which mimic properties of asset returns. One of such models is the stochastic volatility (SV) model[1] which allows the volatility to be a stochastic process. Recently an SV-type model that further utilizes daily realized volatilities (RV) [2, 3] was proposed[4]. This model is called the RSV model. An advantage of the RSV model is that it uses the RV as additional information to the model and thus can estimate the daily volatility more accurately.

To estimate parameters of SV-type models one usually use the Markov Chain Monte Carlo (MCMC) method based on the Bayesian approach[5, 6, 7]. In the MCMC method of the SV model the most time-consuming part is the volatility update. The number of the volatility variables to be updated increases with the data size of time series. Generally a local update algorithm such as the Metropolis algorithm is not efficient for the update of volatility variables.

In this study we use an alternative MCMC method, the HMCA[8] to estimate the parameters of the RSV model. The main advantage of the HMCA is that it is a global update algorithm that can update volatility variables simultaneously. The HMCA has been tested for the SV model[9, 10, 11] and it is found that the HMCA can de-correlate MCMC samples of volatility variables faster than the Metropolis algorithm. In this study we investigate the performance of the HMCA for the RSV model. We also examine the efficiency of an improved integrator called the 2nd order minimum norm integrator (2MNI) used for the MD simulation in the HMCA.

## 2. Realized Stochastic Volatility Model

The basic realized stochastic volatility (RSV) model by Takahashi *et al.*[4] is given by

$$y_t = \exp(h_t/2)\epsilon_t, \quad \epsilon_t \sim N(0,1), \tag{1}$$

$$\ln RV_t = \xi + h_t + u_t, \quad u_t \sim N(0, \sigma_u^2), \tag{2}$$
$$h_{t+1} = \mu + \phi(h_t - \mu) + \eta_t, \quad \eta_t \sim N(0, \sigma_\eta^2), \tag{3}$$

where $y_t$ and $RV_t$ are a daily return and a daily realized volatility at time $t$ respectively, and $h_t$ is a latent volatility defined by $\ln \sigma_t^2$. This model contains 5 parameters ($\theta = \phi, \mu, \xi, \sigma_\eta^2, \sigma_u^2$) that have to be determined so that the model matches the underlying returns and realized volatilities. We apply the Bayesian inference for parameter estimations and perform the Bayesian inference by the MCMC method. For the MCMC simulation of volatility variables we use the HMCA. On the other hand, we update model parameters by the standard MCMC method previously used in parameter estimations of the SV model[5, 6, 7, 9].

## 3. Hybrid Monte Carlo Algorithm

Originally the HMCA was invented and developed for the MCMC simulations of the lattice Quantum Chromo Dynamics (QCD) calculations[8] where non-linear interactions between gauge variables are major obstacles to perform the MCMC method[14]. The basic idea of the HMCA is a combination of the MD simulation and the Metropolis accept/reject test. Candidates for next volatility variables in Markov chain are obtained by solving the Hamilton's equations of motion (HEOM) in fictitious time $\tau$, $\frac{dh_i}{d\tau} = \frac{\partial H}{\partial p_i}$, $\frac{dp_i}{d\tau} = -\frac{\partial H}{\partial h_i}$, where $p_i$ is a conjugate momentum to $h_i$. The Hamiltonian $H$ is defined by $H(p,h) = \frac{1}{2}\sum_i^n p_i^2 - \ln f(h,\theta)$, where $f(h,\theta)$ is the conditional posterior density of the RSV model[4]. Since in general HEOM can not be solved analytically, we integrate them numerically through the MD simulation. The conventional integrator for the MD simulation in the HMCA is the 2nd order leapfrog integrator (2LFI)[8] given by

$$h_i(\tau + \delta\tau/2) = h_i(\tau) + \frac{\delta\tau}{2} p_i(\tau), \quad p_i(\tau + \delta\tau) = p_i(\tau) - \delta\tau \frac{\partial H}{\partial h_i}, \tag{4}$$
$$h_i(\tau + \delta\tau) = h_i(\tau + \delta\tau/2) + \frac{\delta\tau}{2} p_i(\tau + \delta\tau),$$

where $\delta\tau$ stands for the step size in the MD simulation. In the operator representation the 2LFI can be expressed as $e^{\lambda\delta\tau T/2} e^{\delta\tau V} e^{\lambda\delta\tau T/2}$, where $T$ and $V$ are the linear operators corresponding to the kinetic term and potential term of the Hamiltonian respectively[15, 16] Eqs.(5) are repeatedly applied and, volatility and momentum variables are integrated up to a certain time $l$ called the total integration length. The numerical integration introduces the integration errors that violate the conservation of the Hamiltonian. Let $h'$ and $p'$ be volatilities and momenta after integration. Then we define the difference of the Hamiltonian between after and before integration as $\Delta H = H(p', h') - H(p, h)$. Finally new candidates for volatilities $h_t'$ are accepted according to the Metropolis probability, $\min\{1, \exp(-\Delta H)\}$.

The magnitude of $\Delta H$ depends on the integrator we choose. It is possible to use higher order integrators (HOI) that can decrease the magnitude of $\Delta H$ at the same step size and increase the Metropolis acceptance. On the other hand HOI are usually computationally expensive. Therefore the total performance of HOI is not known a priori and depends on the model we use[9, 10]. In Ref.[11] it is shown that an improved integrator called the 2MNI*et al.*[12] is superior to the 2LFI for the lattice QCD simulations. The 2MNI is expressed as $e^{\lambda\delta\tau T} e^{\delta\tau V/2} e^{(1-2\lambda)\delta\tau T} e^{\delta\tau V/2} e^{\lambda\delta\tau T}$ where $\lambda$ is a tunable parameter. $\lambda$ is set to 0.193183327. This value is approximately the optimum value that minimizes the error function of 2MNI[12]. In this study we also use the 2MNI for the HMCA and examine its efficiency for the HMCA of the RSV model.

## 4. Simulation Study

We generated artificially 4000 data with RSV parameters $(\phi, \mu, \xi, \sigma_\eta^2, \sigma_u^2) = (0.93, -1.0, 0.3, 0.1, 0.2)$ and performed the Bayesian estimation of the RSV model for this artificial data.

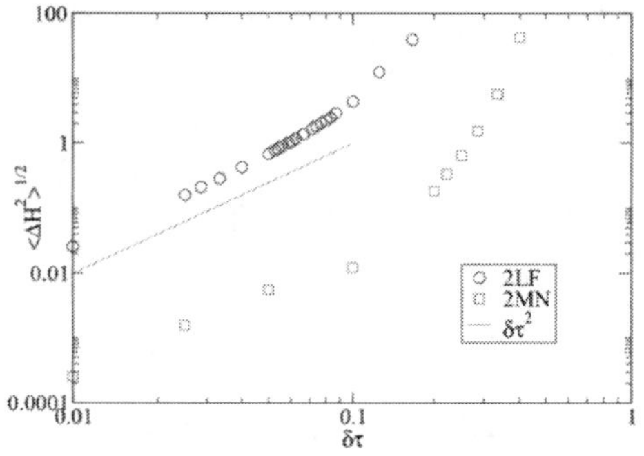


**Figure 1.** $\langle \Delta H^2 \rangle^{1/2}$ as a function of $\delta\tau$. The solid line shows a line proportional to $\delta\tau^2$.

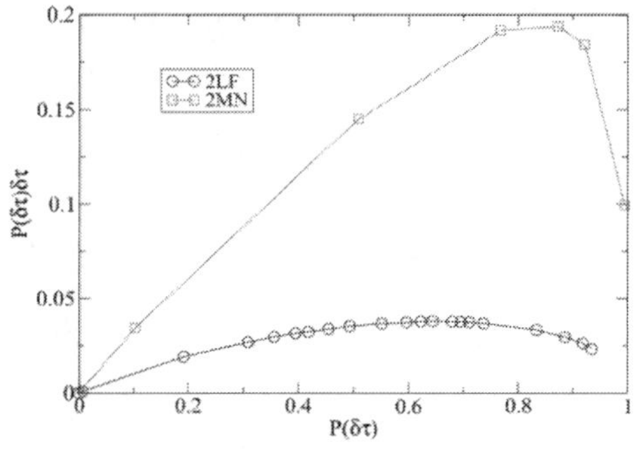


**Figure 2.** The efficiency function $P(\delta\tau)\delta\tau$ versus $P(\delta\tau)$.

**Table 1.** Results obtained by MCMC simulations. S.D. stands for standard deviation.

| | $\phi$ | $\mu$ | $\xi$ | $\sigma_\eta^2$ | $\sigma_u^2$ | $h_{10}$ |
|---|---|---|---|---|---|---|
| input | 0.93 | -1.0 | 0.3 | 0.1 | 0.2 | |
| average | 0.926 | -0.97 | 0.31 | 0.097 | 0.203 | -9.77 |
| S.D. | 0.007 | 0.10 | 0.03 | 0.006 | 0.010 | 0.23 |
| $2\tau_{int}$ | 8.8(20) | 96(40) | 1130(480) | 64(18) | 38(10) | 21(6) |

Fig. 1 shows the root mean square (RMS) of $\Delta H$ as a function of $\delta\tau$. Here the total integration length is set to 2. At small $\delta\tau$ the RMS of $\Delta H$ for both integrators is proportional to $\delta\tau^2$ as theoretically expected[19, 17]. As $\delta\tau$ increases higher order terms contribute the RMS of $\Delta H$ and the RMS of $\Delta H$ increases more rapidly with $\delta\tau$. The efficiency of the integrators can be measured by the efficiency function defined by $P(\delta\tau)\delta\tau$ where $P(\delta\tau)$ stands for the acceptance at $\delta\tau$. Fig.2 shows the efficiency function as a function of $P(\delta\tau)$. It is found that the efficiency function takes a maximum at a certain optimum acceptance $P_{opt}$. For the 2LFI, $P_{opt}$ is found to be around 0.65 which is consistent with the theoretical value 0.61[17] for the 2nd order integrators. On the other hand for the 2MNI, $P_{opt}$ is around $0.8 \sim 0.9$ which is rather close to the optimum value for the higher order integrators[9]. This can be explained by that higher order error terms of the 2MNI dominate the RMS of $\Delta H$ already at lower $\langle \Delta H^2 \rangle^{1/2}$. From fig.2 we see that the efficiency function of the 2MNI at the optimum is about 5 times bigger than that of the 2LFI. Since the computational cost of the 2MNI is 2 times bigger than that of the 2LFI[11], in total the 2MNI is about 2.5 times more efficient than the 2LFI.

Table 1 shows the results obtained by the MCMC simulations. Here the HMCA with the 2MNI is performed with $\delta\tau = 0.222$ which corresponds to the optimum acceptance. We find that the results obtained by the MCMC simulations are consistent with the input values, which means that our MCMC algorithm worked correctly. The autocorrelation time (ACT), $2\tau_{int}$ is defined by $2\tau_{int} = 1 + 2\sum_{t=1}^{\infty} ACF(t)$ where $ACF(t)$ stands for the autocorrelation function. In table 1 we only show the results of $h_{10}$ as a representative one for the volatility variables. $2\tau_{int}$ of $h_{10}$ is around 21 which is very small compared to about 200 of the Metropolis algorithm[10].

## 5. Empirical Study

As an empirical study we also performed the Bayesian inference of the RSV model for the stock price data of Panasonic Co. sampled from 3 July 2006 to 30 Dec. 2009 and estimated the model parameters of the RSV model. The RV is constructed by using 1min-sampling high-frequency returns. We use the 2MNI for the HMCA and set the step size to 0.1 and the total integration length to 1. We discarded the first 5000 MCMC samples and then accumulated 50000 samples for analysis.

Table 2 shows the results estimated from the MCMC simulation. The ACT of $h_{10}$ is found to be short, around 37. Thus MCMC samples of volatility variables are well de-correlated by the HMCA. The parameter $\xi$ corresponds to the bias correction of the RV[4]. The raw RV is distorted by bias such as microstructure noise and non-trading effects. To correct such bias Hansen and Lunde[20] introduced an adjustment factor $c$ which rescales the average of the RV to the variance of the daily returns. Using $c$, the raw $RV$ is corrected to $cRV$. If $\xi$ explains the same bias as the adjustment factor $c$ the relation $\xi = -\log(c)$ should hold. For the RV with 1min sampling frequency we obtain $c = 1.244$, thus $-\log(c) = -0.2183$. On the other hand we obtain $\xi = -0.101$ which is considerably different from $-\log(c)$. Therefore this indicates that $\xi$ may also include other type of bias in the RV.

**Table 2.** Results obtained by MCMC simulations for Panasonic Co.

| | $\phi$ | $\mu$ | $\xi$ | $\sigma_\eta^2$ | $\sigma_u^2$ | $h_{10}$ |
|---|---|---|---|---|---|---|
| average | 0.958 | -7.63 | -0.101 | 0.0440 | 0.0176 | -7.21 |
| S.D. | 0.013 | 0.14 | 0.049 | 0.0036 | 0.0031 | 0.12 |
| $2\tau_{int}$ | 39(6) | 24(0) | 178(42) | 65(11) | 112(19) | 37(11) |

## 6. Conclusion

We have performed the Bayesian estimation of the RSV model by the HMCA. We find that the HMCA with the 2MNI is more efficient than the conventional 2LFI. We also find that the ACT of the volatility variables by the HMCA is short, which means that the HMCA can sample well de-correlated volatility variables. Finally we conclude that the HMCA with the 2MN integrator is efficient for the Bayesian parameter estimation of the RSV model.

## Acknowledgement

Numerical calculations in this work were carried out at the Yukawa Institute Computer Facility and the facilities of the Institute of Statistical Mathematics. This work was supported by JSPS KAKENHI Grant Number 25330047.

## References

[1] Taylor S J 1986 *Modelling Financial Time Series* John Wiley & Sons, New Jersy
[2] Andersen T G and Bollerslev T 1998 *International Economic Review* **39** 885-905
[3] Andersen T G, Bollerslev T, Diebold F X and Labys P 2001 *J. Am. Statist. Assoc.*A **96** 42-55
[4] Takahashi M, Omori Y and Watanabe T 2009 *Compt. Stat. & Data anal.* **53** 2404-2425
[5] Kim S, Shephard N and Chib S 1998 *Review of Economic Studies* **65** 361-393
[6] Shephard N and Pitt M K 1997 *Biometrika* **84** 653-667
[7] Watanabe T and Omori Y 2004 *Biometrika* **91** 246-248
[8] Duane S, Kennedy A D, Pendleton B J and Roweth D 1987 *Phys. Lett. B* **195** 216-222
[9] Takaishi T 2008 *Lecture Notes in Computer Science* **5226** 929-936
[10] Takaishi T 2009 *Journal of Circuits, Systems, and Computers* **18** 1381-1396
[11] Takaishi T 2013 *J. Phys.: Conf. Ser.* **423** 012021
[12] Omelyan I P, Mryglod I M and Folk R 2003 *Comput. Phys. Commun.* **151** 272
[13] Takaishi T and de Forcrand Ph 2008 *Phys. Rev. E* **73** 036706
[14] Montvay I and Münster G 1994 *Quantum Fields on a Lattice* Cambridge University Press
[15] Forest E and Ruth R D 1990 *Physica D* **43** 105-117
[16] Yoshida H 1990 *Phys. Lett. A* **150** 262-268
[17] Takaishi T 2000 *Comput. Phys. Commun.* **133** 6-17
[18] Takaishi T 2002 *Phys. Lett. B* **540** 159-165
[19] Gupta S, Irbäck, Karsch F and Petersson B 1990 *Phys. Lett. B* **242** 437
[20] Hansen P R and Lunde A 2005 *Journal of Applied Econometrics* **20** 873-889

# Appropriateness of dynamical systems for the comparison of different embedding methods via calculation of the maximum Lyapunov exponent

**M Franchi and L Ricci**

Dipartimento di Fisica, Università di Trento, I-38123 Trento, Italy

E-mail: franchi@science.unitn.it

**Abstract.** The embedding of time series provides a valuable, and sometimes indispensable, tool in order to analyze the dynamical properties of a chaotic system. To this purpose, the choice of the embedding dimension and lag is decisive. The scientific literature describes several methods for selecting the most appropriate parameter pairs. Unfortunately, no conclusive criterion to decide which method – and thus which embedding pair – is the best has been so far devised. A widely employed quantity to compare different methods is the maximum Lyapunov exponent (MLE) because, for chaotic systems that have explicit analytic representations, MLE can be numerically evaluated independently of the embedding dimension and lag. Within this framework, we investigated the dependence on the calculated MLE on the embedding dimension and lag in the case of three dynamical systems that are also widespreadly used as reference systems, namely the Lorenz, Rössler and Mackey–Glass attractors. By also taking into account the statistical fluctuations of the calculated MLE, we propose a new method to assess which systems provide suitable test benches for the comparison of different embedding methods via MLE calculation. For example we found that, despite of its popularity in this scientific context, the Rössler attractor is not a reliable workbench to test the validity of an embedding method.

## 1. Introduction

Embedding experimental data provides an indispensable tool in order to analyse the dynamical properties of a chaotic system. Given a time series $x_i$, $i = 0, \ldots n$, an embedding vector $i$ is defined as $\mathbf{X}_i = (x_i, x_{i+L}, \ldots, x_{i+(m-1)L})$, $m \in \mathbb{N}$ is the embedding dimension and $L \in \mathbb{N}$ is the lag. Provided that proper values of $m$ and $L$ are chosen, a dynamics $F : \mathbf{X}_i \to \mathbf{X}_{i+1}$ can be defined, which is representative of the original chaotic system; this is the essential statement of Takens-Mañé embedding theorem [1]. Regrettably, Takens-Mañè theorem says nothing on how to choose the two parameters $(m, L)$.

Because the choice of the embedding parameters is crucial to the subsequent analysis, many embedding methods were developed. The most cited are: Gao–Zheng's method [2], Schuster's method [3], the method of characteristic length (CL), global false neighbours and the autocorrelation function (GFNN–A) and global false neighbours and mutual information (GFNN–MI) [4]. Each motivates its choice of $(m, L)$ by addressing one or more particular features of the dynamics of the system.

One of the basic features of chaos is the peculiar dependence on initial conditions. A quantitative measure of the response sensitivity to small changes in initial conditions is provided by the Lyapunov exponents of the dynamical system. Lyapunov exponents are asymptotic measures that characterize the average rate of growth (or, conversely, "shrinking") of small perturbations in the phase-space. In the case of a chaotic orbit, at least one Lyapunov exponent is positive: initially nearby orbits exponentially diverge from each other.

Cellucci et al. compared different embedding methods by considering the difference between the maximum – and therefore most significant – Lyapunov exponent (MLE) $\lambda_\alpha(m, L)$ calculated for the embedded system and the MLE $\chi_\alpha$ calculated by using the algorithm developed by Benettin et al. [5]: the best embedding method gives an embedding pair $(m, L)$ that minimizes this difference. In their paper, Cellucci et al. used different chaotic systems, and it was tacitly assumed that all chaotic systems were equally effective as test benches. However, a chaotic system, for which the calculated MLE is essentially embedding–independent, is *de facto* useless to test an embedding method. Conversely, a stronger dependence is linked to a higher appropriateness of the dynamical system for testing and comparing different embedding methods via MLE calculation. In this paper, we consider the dependence of $\lambda_\alpha$ on the embedding pairs $(m, L)$ for three chaotic systems, namely the Lorenz, Rössler and Mackey–Glass attractors.

The paper is organized as follows: Section 2 presents a short overview of the concepts regarding the embedding, as well as Gao–Zheng's method to determine MLE; the reference dynamical systems are the topic of Section 3; the results are discussed in Section 4.

## 2. Theoretical background

For each system $\alpha$, we calculated a reference maximum Lyapunov exponent $\chi_\alpha$ by applying the so–called *standard method* [6]. This method, developed by Benettin *et al.* [7, 8], evaluates the maximum Lyapunov exponent $\chi_\alpha$ directly from the differential equations that characterize the dynamical system. Consequently, $\chi_\alpha$ is independent of the embedding pair $(m, L)$.

As mentioned at the beginning of Section 1, an embedding procedure with dimension $m$ and lag $L$ maps a sample time sequence $\{x_i\}$ of a scalar, real variable $x$, into a sequence of *embedding* vector $\{\mathbf{X}_i\}$, with $\mathbf{X}_i \in \mathbb{R}^m$. The MLE $\lambda_\alpha(m, L)$ can be then evaluated by using the method proposed by Gao and Zheng [2]. Let us consider all pairs $\mathbf{X}_i$, $\mathbf{X}_j$ that satisfy the following two conditions [5]: 1) $\|\mathbf{X}_i - \mathbf{X}_j\| \leqslant r$, with $r$ corresponding to the 0.1 percentile of the distribution of all distances $\|\mathbf{X}_i - \mathbf{X}_j\|$; 2) $|i - j| \geqslant c_0$, with $c_0$ being the location of the first minimum of the autocorrelation function. Let us also define the quantity $\Lambda(k)$ as follows:

$$\Lambda(k) = \frac{1}{N_{ref}} \sum_{i,j} \log \frac{\|\mathbf{X}_{i+k} - \mathbf{X}_{j+k}\|}{\|\mathbf{X}_i - \mathbf{X}_j\|},$$

where $k$ is an non–negative, integer delay, $N_{ref}$ is the number of pairs and $\|\ldots\|$ represents the Euclidean norm. Then, for small values of $k$, the dependence of $\Lambda$ on $k$ turns out to be linearly growing and independent of the boundary conditions. In this linear regime, the MLE can be estimated by the slope of $\Lambda(k)$, i.e. $[\Lambda(k_2) - \Lambda(k_1)]/([k_2 - k_1]\delta t)$, where $\delta t$ is the time step of the sample time sequence.

## 3. Reference dynamical systems

Three of the most common continuous dynamical systems that are considered in the literature are [2, 9]: the Lorenz attractor, the Rössler attractor and the Mackey–Glass attractor. In the present work, to compute the evolution of these dynamical systems, a fourth–order Runge–Kutta algorithm was used. The system of differential equations of the Lorenz attractor is [10]:

$$\begin{aligned} \frac{dx}{dt} &= \sigma(y-x) \\ \frac{dy}{dt} &= x(r-z)-y \\ \frac{dz}{dt} &= xy-bz \end{aligned} \tag{1}$$

To simulate this system, we used $\sigma = 10$, $r = 45.92$, $b = 8/3$ and a sampling interval $\delta t$ equal to 0.03.

The Rössler attractor is governed by the following system of differential equations [11]:

$$\begin{aligned} \frac{dx}{dt} &= -y-z \\ \frac{dy}{dt} &= x+ay \\ \frac{dz}{dt} &= b+z(x-c) \end{aligned} \tag{2}$$

To simulate this system, we used $a = 0.15$, $b = 0.2$, $c = 10$ and a sampling interval $\delta t$ equal to 0.125.

The Mackey–Glass attractor is governed by the following equation [12]:

$$\frac{dx(t)}{dt} = a\frac{x(t-\tau)}{1+[x(t-\tau)]^c} - bx(t), \quad a, b, c > 0\,. \tag{3}$$

To simulate this system, we used $a = 0.2$, $b = 0.1$, $c = 10$, $\tau = 30$ and a sampling interval $\delta t$ equal to 1.5.

For each dynamical system, only the $x$–coordinate was considered in the embedding process.

## 4. Results

Considering the dynamical systems $\alpha$ in Section 3, we calculated the reference MLE $\chi_\alpha$ provided by the standard method and the MLE $\lambda_\alpha(m,\, L)$ estimated by using Gao–Zheng's method with the embedding parameters $(m,\, L)$.

"Good" $(m,\, L)$ pairs are ideally characterized by equal values of $\lambda_\alpha(m,\, L)$ and $\chi_\alpha$. However, because $\chi_\alpha$ and $\lambda_\alpha(m,\, L)$ were calculated on finite samples, a perfect assessment of the potential equality of the two quantities is hindered by statistical fluctuations. To tackle this issue, given a dynamical system $\alpha$ and for each point $(m,\, L)$ of the lattice $1 \leqslant m \leqslant 9$, $0 \leqslant L \leqslant 18$, we carried out a number of 10 independent calculations of $\lambda_{i,\alpha}(m,\, L)$ $(i = 1, \ldots 10)$. Then, on each 10–fold set, the average $\lambda_\alpha(m,\, L)$ and the standard deviation $\sigma_\alpha(m,\, L)$ was calculated. Finally, we calculated the mean $\Sigma_\alpha$ of the set of $\sigma_\alpha(m,\, L)$; the quantity $\Sigma_\alpha$ is therefore representative of the dynamical system $\alpha$ only.

We propose to deem an embedding pair $(m,\, L)$ to be "good" if $\lambda_\alpha(m,\, L)$ is in the interval $[\chi_\alpha - \Sigma_\alpha,\, \chi_\alpha + \Sigma_\alpha]$. For the three dynamical systems under consideration here, Tab. 1 reports $\chi_\alpha$, $\Sigma_\alpha$ and finally the statistics of the "good" $(m,\, L)$ pairs. To compare the dynamical system we considered the percentage of "good" pairs: this parameter gives a quantitative value of the difficulty to select a $(m,\, L)$ pair compatible with the respective standard value $\chi_\alpha$. Rössler and Mackey–Glass attractors have a value of 81% and 68%, respectively. This means that it is not difficult to randomly select a "good" pair. On the other hand, the Lorenz attractor has a value of 25% only. Our conclusion is that the Rössler and Mackey–Glass attractors, at least with the boundary conditions described in Section 3, are not as useful to compare embedding methods as the Lorenz attractor is.

**Table 1.** Statistic of the "good" embedding pairs for each of the three dynamical system of Section 3. The standard value $\chi_\alpha$ and confidence interval $\Sigma_\alpha$ are also reported.

| Dynamical system ($\alpha$) | $\chi_\alpha$ | $\Sigma_\alpha$ | # of "good" pairs | total # of pairs | % "good" pairs |
|---|---|---|---|---|---|
| Lorenz | 2.09 | 0.10 | 36 | 144 | 25 |
| Rössler | 0.129 | 0.009 | 117 | 144 | 81 |
| Mackey–Glass | 0.0066 | 0.0003 | 98 | 144 | 68 |

## References


[1] F. Takens, in: D. A. Rand, L. S. Young (Eds.), Dynamical Systems and Turbolence (Warwick 1980), Vol. 898, Springer Verlag, Berlin, 1980, pp. 366–380.

[2] J. Gao, Z. Zheng, Direct dynamical test for deterministic chaos and optimal embedding of a chaotic time–series, Phys. Rev. E 49 (1994) 3807–3814.

[3] W. Liebert, K. Pawelzik, H. G. Schuster, Optimal embedding of chaotic attractors from topological considerations, Europhys. Lett. 14 (6) (1991) 521–526.

[4] H. D. I. Abarbanel, Analysis of Observed Chaotic Data, Springer Verlag, 1996.

[5] C. J. Cellucci, A. M. Albano, P. E. Rapp, Comparative study of embedding methods, Phys. Rev. E 67 (2003) 066210.

[6] C. Skokos, The Lyapunov Characteristic Exponents and Their Computation, Lect. Notes Phys. 790 (2010) 63–135.

[7] G. Benettin, L. Galgani, A. Giorgilli, J. Strelcyn, Lyapunov characteristic exponents for smooth dynamical systems and for Hamiltonian systems; a method for computing all of them. Part 1: Theory, Meccanica 15 (1980) 9–20.

[8] G. Benettin, L. Galgani, A. Giorgilli, J. Strelcyn, Lyapunov characteristic exponents for smooth dynamical systems and for Hamiltonian systems; a method for computing all of them. Part 2: Numerical application, Meccanica 15 (1980) 21–30.

[9] A. Wolf, J. B. Swift, H. L. Swinney, J. A. Vastano, Determining Lyapunov exponents from a time–series, Physica D 16 (3) (1985) 285–317.

[10] E. N. Lorenz, Deterministic Nonperiodic Flow, J. Atmos. Sci. 20 (1963) 130–141.

[11] O. E. Rössler, An equation for continuous chaos, Phys. Lett. 57A (1976) 397–398.

[12] M. C. Mackey, L. Glass, Oscillation and Chaos in Physiological Control Systems, Science 197 (1977) 287–289.

# A method for system identification of a structure supported by nonlinear springs using evolutionary computing

**Krassimire Karabeliov**

**Matthias Baeßler**

**Pablo Cuéllar**

**Werner Rücker**

BAM, Federal Institute for Materials Research and Testing, Unter den Eichen 87, 12205 Berlin, Germany

krassimire.karabeliov@bam.de

**Abstract.** A mechanical structure supported by nonlinear springs subjected to an external load is considered. If all mechanical parameters of the system were known, the displacement of the system subjected to this load could be easily calculated. If not all of the parameters are known, but the load and the displacement are measured at one location, an inverse problem exists. In the presented problem the nonlinear springs are unknown and have to be determined. At first glance a problem needs to be solved, which is underdetermined due to the number of unknown variables. However, evolutionary computing can be applied to solve this inverse, nonlinear and multimodal problem. Sometimes a prior knowledge exists on certain system properties, which is difficult to implement into analytical or numerical solver. This knowledge can play a decisive role in identifying the system properties and it can be easily included as boundary condition when applying evolutionary algorithm. This article examines how and under what conditions the spring resistances can be identified. The procedure is exemplified at a mechanical system of a pile foundation.

## 1. Introduction

The pile foundation with predominantly axially loaded driven piles is currently one of the most used foundations for offshore wind turbines in Germany (see fig. 1). The current state of scientific knowledge suggests that this type of foundation might be at risk due to cyclic loads caused by wind and waves. As part of on-going research projects, the Federal Institute for Materials Research and Testing is currently studying the behavior of cyclic axially loaded piles and developing an appropriate monitoring method for this type of foundation.

An axially loaded pile can be modeled numerically with the FE method as shown in Figure 2. The pile is considered there as a linear elastic body and the soil-pile interaction is represented by the so-called t-z springs (springs with a non-linear stiffness characteristic). According to [1] for the calculation of the displacements of the piles founded in sand a bilinear load vs. displacement function can be adopted for the t-z springs. The quality of this assumption is not of interest for the general approach and it is used only as an example for the studies described below. The crucial thing, however, is that the definition of bilinear springs with only one unknown (the maximal resistance t of the bilinear spring) is fixed, since the yield limit for displacements $z_{pl}$ is assumed as constant and equal to 2.5 mm.

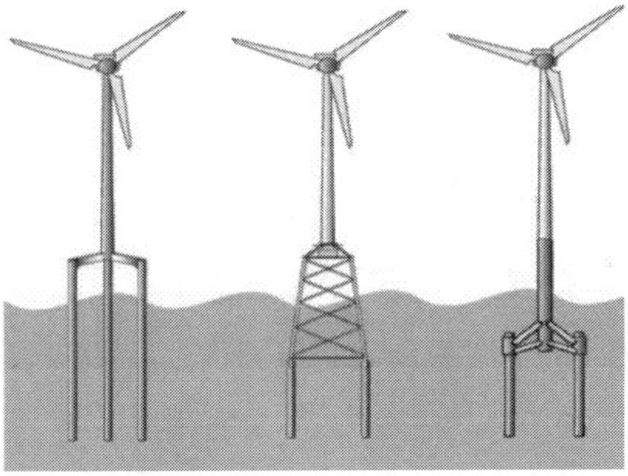

Figure 1: Offshore foundations with axially loaded piles (Tripile, Jacket and Tripod)

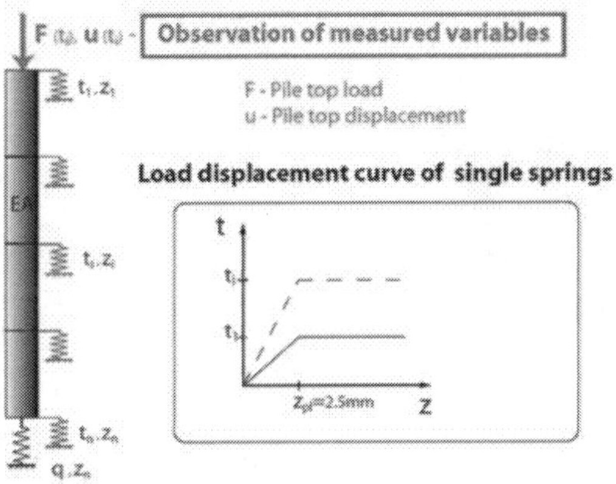


Figure 2: FE model und definition of the t-z spring according to ISO 19902

Measurements below the ground surface, and especially in the offshore environment, are extremely costly and time-consuming. Therefore, it is important to reduce the use of measurement to an absolute minimum. On the other hand, the measurements should be sufficient to obtain the system identification. This is a classical model-update problem where a parameter identification has to be performed. Based on mechanical considerations of the structure it can be shown that the measurement of the pile forces and of the displacements at the pile head could meet these two opposed requirements, and so despite a high number of unknown parameters (in this case, the number of springs).

This article proves the thesis above and provides an approach for the system identification of such a model using multi population evolution strategies [2]. Moreover, it also presents possibilities and limitations of this approach. This strategy permits the solution of problems that can be replaced by similar FE models.

## 2. Solution strategy

The problem of the system identification can be considered as classic black box model. The pile and the pile-soil interaction represent the unknown system. The pile loads are the input parameters in our unknown system. The system reacts to it and its output values are the displacements at the pile head. As mentioned before, the black box can be replaced by a FE model with nonlinear springs. The maximal resistances of these springs are the free parameters of the model that have to be determined (fig. 2). The number of free parameters depends consequently on the FEM discretization. Therefore the specific task to be performed is the determination of the resistance of the non-linear springs, so that the best possible match between the measured and the calculated load displacement curves at the pile head can be achieved.

A system can be referred to as identified if the substitute model shows the same reaction caused by the same action. However, the solution space can be multimodal. In this case, it means that a complete match can be achieved between the calculated and the measured progress of the load displacement curve for different sets of spring load capacities. A different possible method for providing the correlation between the problem variables and the problem solution might be the Monte Carlo simulation. This approach was discussed in [3]. The disadvantage of this method is that the representation of the dependencies between the input and the output parameters is difficult.

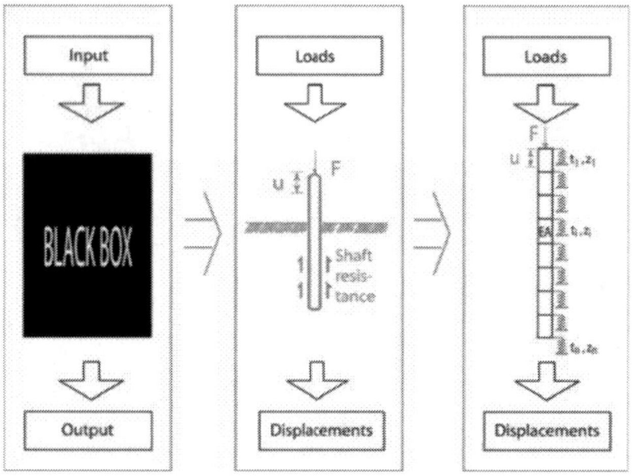


Figure 3: Consideration of a pile foundation as classic black box model

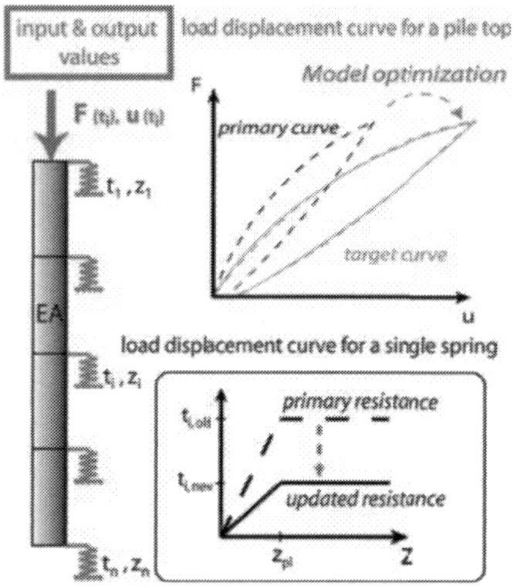


Figure 4: Model optimization

### 3. Example

A FE model, as shown in figure 2, with a given set of 35 springs is used here as target system to be identified. The full load displacement curve of the system is calculated for the given spring resistance values (as can be seen in figure 5 and $z_{pl}$=2.5 mm (figure 2). These values are the target values, which the method has to find by optimization with evolutionary computing out of the given load-displacement curve. Free parameters in this optimization problem are the 35 values for the spring resistances and the one value of the plastic threshold $z_{pl}$. In order to avoid the possibility that the algorithm finds always the same local optimum, different arbitrary start values can be used (fig. 5).

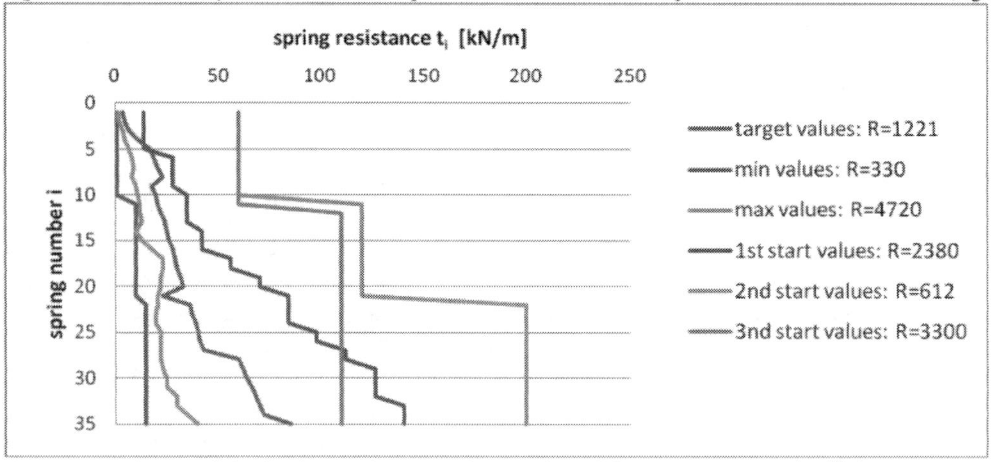


Figure 5: Boundary condition for the search space and start values of the spring resistance ($R = \sum t_i$)

The solution to the problem is multimodal. However, we can use a prior knowledge to constrain the solution space. In this case, the following boundary conditions can be specified:

1. For each spring the maximum resistance can be constrained within an expected range based on a preliminary ground investigation for example
2. The ground constitutes a continuum. So for homogeneous soils, the difference between two consecutive springs should in general not be arbitrary. Based on this assumption, one may constrain it to 30% and disregard any solution outside this range of variation

The difference between two solution results without using both boundary conditions can be very significant. In figure 6 the results of two optimizations using both boundary conditions and in figure 7 only with the first boundary condition are shown. In the latter case the difference between two solution results is very large, while in former cae it is relatively small. The results are not unimodal, but they can be adopted as a good engineering solution or as a good approximation.

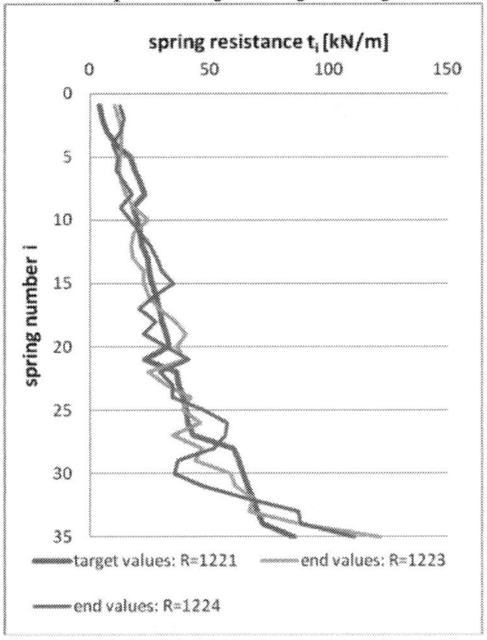


Figure 6: Results of the parameter identification by using both of the boundary conditions ($R = \sum t_i$)

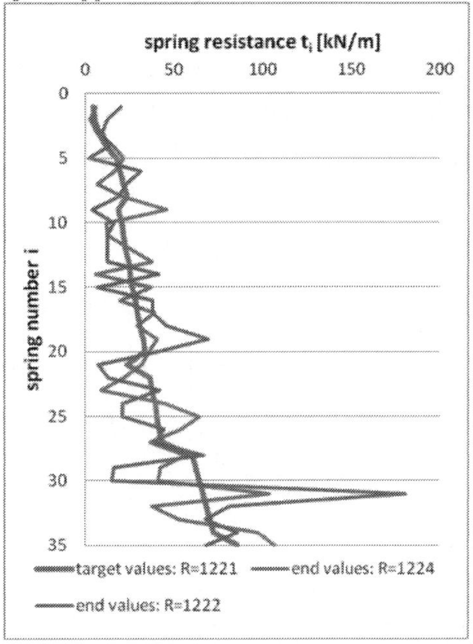


Figure 7: Results of the parameter identification by using only the first boundary condition ($R = \sum t_i$)

## 4. Summary and Outlook

This article shows that the identification of such static systems with non-linear springs is possible. The most important requirements for the application to real cases are:

- To provide a mathematical description of the spring characteristic as precisely as possible, since the number of free parameters is limited.
- High requirements on the measurement of the observation variables, as small measurement errors can lead to large scattering of the results by the inverse parameter identification.

As a summary, the present study has investigated the conditions to be met by the definition of a spring curve (its constitutive law) and how it should be developed in order to present a well-conditioned optimization problem for parameter identification.

## References

[1] ISO-19902 (2007). International Standard: Petroleum and natural gas industries - Fixed steel offshore structures

[2] Rechenberg, I., *Evolutionsstrategie' 94*. Vol. Band 1. 1994: frommann-holzboog.

[3] Stopper, J. (2010). Stochastische Untersuchungen zur Kausalität zwischen dem Last-Verschiebungs- und dem Bodenverhalten von axial belasteten Gründungspfählen. TU Berlin - Fachgebiet Statik und Dynamik, BAM - 7.2 Ingenieurbau. **Diplomarbeit**.

# An alternative model, explaining the VLS, created by the gravitational interaction of two matter populations, one composed of positive mass and the other of negative mass.

**Jean-Pierre Petit** **Gilles d'Agostini**

Lambda Laboratory, 8 Bd F.Buisson, Rochefort 17300, France

contact@lambda-laboratory.fr

**Abstract**. Negative masses arise naturally from dynamic groups, as shown in 1970 by the French mathematician Jean-Marie Souriau. We recall this, based on the coadjoint action of the Poincaré group on its momentum. Negative populations include negative energy photons. If we admit that positive and negative masses cannot interact through virtual photons, they only interact through gravitational force. Two particles whose masses display the same sign attract each other through Newton law. Two particles of opposite signs repel each-other through «anti-Newton» law. Then the two populations tend to separate. If the mass density of the negative material is much larger, it first forms clusters by gravitational instability. Then the positive matter is repelled in the remnant place, shaped like adjoining bubbles, which looks like VLS. To illustrate this scheme, we provide 2D simulation results. In addition this model provides a new insight on the galaxies' birth mechanism.

## 1. Introduction

The French mathematician Jean-Marie Souriau writes the momentum of the Poincaré group in the following matrix notation and calls "passage" the vector $f$ and "whirstling" the vector $l$. Then the coadjoint action of the group has the following simple form :

$$g=\begin{pmatrix} L & C \\ 0 & 1 \end{pmatrix} \quad C=\begin{pmatrix} \Delta t \\ \Delta x \\ \Delta y \\ \Delta z \end{pmatrix} \quad J=\begin{pmatrix} 0 & -l_z & l_y & f_x & E \\ l_z & 0 & -l_x & f_y & p_x \\ -l_y & l_x & 0 & f_z & p_y \\ -f_x & -f_y & -f_z & 0 & p_z \\ -E & -p_x & -p_y & -p_z & 0 \end{pmatrix} \quad P=\begin{pmatrix} E \\ p_x \\ p_y \\ p_z \end{pmatrix} \quad J=\begin{pmatrix} M & P \\ {}^tP & 0 \end{pmatrix}$$

$$J' = g\,J\,{}^tg \qquad M' = L\,M\,{}^tL + C\,{}^tP\,L - L\,P\,{}^tC \qquad P' = L\,P$$

If one chooses a coordinate system relative to the particle, the passage vector $f$ becomes zero, and the $l$ vector becomes the kinetic momentum associated to the mass-point particle. In his book [1], Souriau, developing geometrical method for quantization shows that this vector is quantized and becomes the spin (but this extension to quantum world is out of the scope of the present paper). The Poincaré group is built around the Lorentz group, which owns four connex components, forming two subsets : the orthochron one ($L_o$ matrixes ) and the antichron one ($L_a$ matrixes). It inherits this property and can be written :

$$g=\begin{pmatrix} L & C \\ 0 & 1 \end{pmatrix}=\begin{pmatrix} \lambda L_o & C \\ 0 & 1 \end{pmatrix} \quad with \ \lambda=\pm 1 \quad \Rightarrow \ P=\lambda L_o P$$

The ($\lambda$ = - 1) terms implies *T*-symmetry and subsequently the inversion of the energy *E*. Then these movements refer to matter whose energy and mass are negative [1]. In addition, if we consider the complete Poincaré group we must take into account negative energy photons. In addition, Souriau shows that an electric charge can be added to the particle, considering its movement as inscribed in a 5D Kaluza ( t , x , y , z , $\zeta$ ) space with signature ( + - - - - ). Moreover he showed [2] that the charge conjugation ( $q \rightarrow -q$ ) goes with the inversion of the fifth dimension ($\zeta \rightarrow -\zeta$ ), the Kaluza dimension, which can be expressed through the following matrix representation of the extended group :

$$\begin{pmatrix} \mu & 0 & \mu\theta \\ 0 & \lambda L_o & C \\ 0 & 0 & 1 \end{pmatrix} \quad \begin{matrix} with \ \lambda=\pm 1 \\ and \ \mu=\pm 1 \end{matrix} \quad \begin{matrix} M' = L_o M' L_o + \lambda C' P L_o - \lambda L_o P' C \\ P' = \lambda L_o P \qquad q' = \lambda \mu q \end{matrix}$$

According to this vision, among others things, classical antimatter corresponds to a peculiar movement of the charged mass point in a five dimensions space. This shows that the $\zeta$-symmetry ($\mu$ = -1 ; $\zeta \rightarrow -\zeta$) goes hand in hand with a *C*-symmetry ($q \rightarrow -q$) which is a geometrical interpretation of the matter-antimatter symmetry through an extension of the Poincaré dynamic group. As derived from such dynamic groups, built with real components, the representation of matter becomes somewhat different. Given the *T*-symmetry, the *CPT*-symmetric and *PT*-symmetric of a particle refer to a negative energy and negative mass objects. A negative world arises, where the matter antimatter duality holds.

**2. Could positive and negative energy particles co-exist in the same space ?**

Negative mass problem was investigated by H.Bondi [3] and W.B.Bonnor [4]. In 1967 A.Sakharov proposed a twin universe model, with a time inversion [5]. Adding the Souriau's 1970 theorem we built a new dynamical system, ruled by gravitationally interacting masses, of opposite signs, represented as two interacting field equations [6]. The solution of such system is a pair of non-independent metrics g+ and g-. Newtonian approximation provides Poisson equation and interaction laws. Particles whose masses own the same signs mutually attract through Newton's law. When these signs are opposite they repel each other according to "anti-Newton's law". What would be an Universe made of positive and negative masses? The latter would emit negative energy photons. Neither our eyes, nor our optical devices react to such negative energy light. Therefore we cannot observe such structures with our telescopes. Negative mass structures would be perfectly invisible to us. As shown in reference [7] exact (bimetric) solutions, Schwarzschild type, of the system presented in [6] can be built. If, for example, we look at the external Schwarzschild metric, in its classical expression (m is a length, not a mass) :

$$ds^2 = (1-\frac{2m}{r})c^2\,dt^2 - \frac{dr^2}{(1-\frac{2m}{r})} - r^2(d\theta^2 + \sin^2\theta\, d\varphi^2) \qquad M=\frac{mc^2}{G}$$

m is nothing but an integration constant, which can be chosen positive or negative. The equations of geodesics, (where $\varphi$ is the polar angle and $u = 1/r$) is the following (right : null geodesics) :

$$\varphi = \varphi_o + \int_{u_o}^{u} \frac{du}{\sqrt{\frac{c^2 a^2 - 1}{h^2} + \frac{2m}{h^2} u - u^2 + 2mu^3}} \qquad \varphi = \varphi_o + \int_{u_o}^{u} \frac{du}{\sqrt{\frac{c^2 a^2}{h^2} - u^2 + 2mu^3}}$$

and similar for the "internal" Schwarzschild solutions. On the following figures photon's trajectories around a positive and negative mass $M$. Notice that a positive energy photon can experience negative lensing due to "antigravitational effect", when encountering a cluster of negative mass, for it does not interact with negative material through electromagnetic force, so that it cannot collide with any negative mass particle.

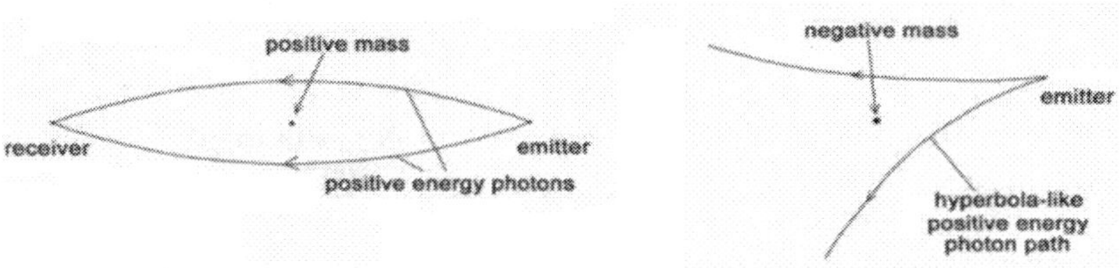


**Figure 1.** : Positive and negative gravitational lensing effect

As a conclusion, although invisible, negative material clusters would induce negative gravitational lensing effect, reducing the apparent magnitude of large redshift galaxies, which fits observational data [8]. Combining Euler and Poisson equations we get Jeans' like joint gravitational instability equations [9] :

$$\Delta\,\delta n + \frac{\delta n}{L_j^2} - \frac{m}{kT}\frac{\partial^2 \delta n}{\partial t^2} = 0 \quad (Jeans) \quad \rightarrow \quad \begin{cases} \Delta\,\delta n^+ + \dfrac{1}{L_j^{+2}}\left(\delta n^+ + \dfrac{m^-}{m^+}\delta n^-\right) - \dfrac{1}{n^+}\dfrac{\partial^2 \delta n^+}{\partial t^2} = 0 \\ \Delta\,\delta n^- - \dfrac{1}{(L_j^-)^2}\left(\delta n^- + \dfrac{m^+}{m^-}\delta n^+\right) - \dfrac{1}{n^-}\dfrac{\partial^2 \delta n^-}{\partial t^2} = 0 \end{cases}$$

A spherically symmetric non-linear steady state exact solution can be built, corresponding to a Maxwellian velocity distribution function in both systems. Hereafter: when the absolute values of the densities $\rho^+$ and $\rho^-$ are equal at infinite, and absolute values of temperatures $T^+$ and $T^-$ are constant (in time and space), we get the following figures:

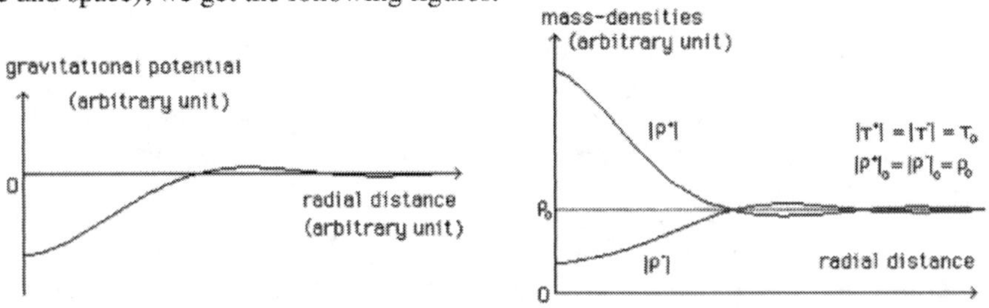


**Figure 2.** : Spherically symmetric cluster nested in negative material [9]

This is similar, with two interacting populations to the model presented in 1942 by S. Chandrasekhar [10]. Now, assume that the absolute value of the negative mass density is 50 times higher than the one of the positive mass. Then its associated Jeans times are such as:

$$t_J^- = \frac{1}{\sqrt{4\pi G|\rho^-|}} \ll t_J^+ = \frac{1}{\sqrt{4\pi G\rho^+}}$$

Hereafter the result of 2D numerical simulations [7]. Negative species forms clusters in Figure 3(a) and repels the positive mass in the remnant space in Figure 3(b). In Figure 3(c) the two species, are superimposed.

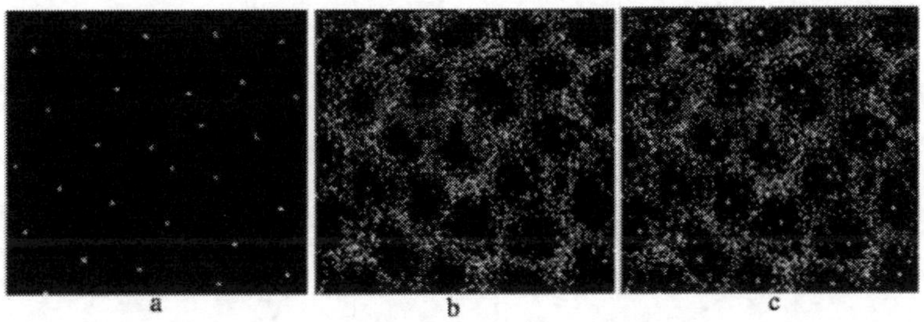


**Figure 3.** : Results of 2D numerical simulations [7]

This global pattern is stable in time. The cellular structure keeps the negative clusters in place. These negative clusters behave like anchors with respects to the positive matter structure which looks like joined soap bubbles. In 3D it would be similar to VLS.

## 3. Some insight on the galaxies' birth problem.

To get condensed matter objects, initially formed by gravitational instability, the problem is to evacuate the heat by emission of radiation. In proto stars, their radiatizve cooling is due their spherical surface. The larger the proto star the longer its cooling time. 3D gravitational instability creates spherical negative clusters, with huge mass and dimensions. Then, if the physics of negative matter is similar to ours, the associated cooling time could be larger than the age of the Universe. Even if we fundamentally cannot have optical information on such structures, as they emit negative energy photons, we can try to have some idea about what is going on there. These big clusters would emit long wavelength radiations (reddish, infrared). They would resemble to proto stars that would never ignite. Subsequently the negative world would not have stars, no heavy atoms, planets nor ... life.

On the contrary, when negative clusters first form, for their Jeans time is smaller, they repel and compress the positive matter, forming soap bubbles like pattern. This is optimum for energy dissipation by a radiative process and triggers, promotes galaxies' formation. Later, negative matter could also play a role in galaxies' confinement. In addition, this invisible and negative energy content could be the searched so-called "dark energy".

## References

[1] J.M.Souriau : Structure des systèmes dynamiques. Dunod Ed. France, 1970 and Structure of Dynamical Systems. Boston, Birkhaüser Ed. 1997

[2] J.M.Souriau : Géométrie et Relativité, Hermann Ed. 1964, France.

[3] H. Bondi: Negative mass in General Relativity : Negative mass in General Relativity. Rev. of Mod. Phys., Vol 29, N°3, july1957

[4] W.B.Bonnor : Negative mass and general relativity. General Relativity and Gravitation Vol.21, N°11, 1989

[5] A.D. Sakharov (1980). Cosmological Model of the Universe with a Time Vector Inversion. ZhETF (Tr. JETP 52, 349-351) (79): 689–693

[6] J.P. Petit (July 1994). The Missing Mass Problem. Il Nuovo Cimento B, 109: 697–710

[7] J.P. Petit (1995). Twin Universe Cosmology. Astrophysics and Space Science (226): 273–307.

[8] P.J.E.Peebles : Principles of Physical Cosmology. Princeton Series in Physics. 1993

[9] J.P. Petit; P. Midy, F. Landsheat (June 2001). Twin matter against dark matter. International Meeting on Atrophysics and Cosmology. "Where is the matter ?", Marseille, France.

[10] S.Chandrasekhar. Principles of Stellar Dynamics. Dover Pub. New York, 1942

# Parallel iterative solution of the Hermite Collocation equations on GPUs II

**N. Vilanakis and E. Mathioudakis** [1]

Department of Sciences, Technical University of Crete, University Campus, 73100 Chania, Crete, Greece

E-mail: nivilanakis@isc.tuc.gr, manolis@science.tuc.gr

**Abstract.** Hermite Collocation is a high order finite element method for Boundary Value Problems modelling applications in several fields of science and engineering. Application of this integration free numerical solver for the solution of linear BVPs results in a large and sparse general system of algebraic equations, suggesting the usage of an efficient iterative solver especially for realistic simulations. In part I of this work an efficient parallel algorithm of the Schur complement method coupled with Bi-Conjugate Gradient Stabilized (BiCGSTAB) iterative solver has been designed for multicore computing architectures with a Graphics Processing Unit (GPU). In the present work the proposed algorithm has been extended for high performance computing environments consisting of multiprocessor machines with multiple GPUs. Since this is a distributed GPU and shared CPU memory parallel architecture, a hybrid memory treatment is needed for the development of the parallel algorithm. The realization of the algorithm took place on a multiprocessor machine HP SL390 with Tesla M2070 GPUs using the OpenMP and OpenACC standards. Execution time measurements reveal the efficiency of the parallel implementation.

## 1. Introduction

The Collocation discretization method, based on Hermite bi-cubic elements, is a well-known high accurate finite element technique for solving elliptic boundary value problems (BVPs) [1, 2, 3]. Applying the method to problems on square domains, and by using uniform $n_s \times n_s$ discretization and the appropriate *red-black* numbering of unknowns and equations [4] the resulting red-black collocation linear system $A\boldsymbol{x} = \mathbf{b}$ is in 2-cyclic normal form [5], namely,

$$\begin{bmatrix} D_R & H_B \\ H_R & D_B \end{bmatrix} \begin{bmatrix} \boldsymbol{x}_R \\ \boldsymbol{x}_B \end{bmatrix} = \begin{bmatrix} \mathbf{b}_R \\ \mathbf{b}_B \end{bmatrix} \tag{1}$$

where the matrices $D_R$ and $D_B$ are block diagonal and nonsingular [6]. The large size of the collocation system, especially for fine discretizations, and the demanding storage requirements refer immediately to iterative methods for its efficient solution [7, 8, 9] on parallel computing environments [10, 3, 11, 12]. The preconditioned BiCGSTAB [13] iterative method is an efficient solver for the collocation linear system (1) for high performance parallel architectures [11, 10, 12].

[1] This work was supported by EU (European Social Fund ESF) and Greek funds through the operational program *Education and Lifelong Learning* of the National Strategic Reference Framework (NSRF) - Research Funding Program: THALIS.

In [14] a new Schur Complement type iterative solver was proposed for multicore machines with a Graphic Processing Unit (GPU). The method approximates iteratively only the black coloured unknowns only, while the red ones are directly calculated from the black ones. Since the method is based on the application of a two sided precondition technique into the linear system (1), the iterative procedure of the algorithm for the Schur complement linear system includes an unpreconditioned solver. The BiCGSTAB method is chosen for this purpose evaluating all the black unknowns. Taking into account that this is the most computationally intense part of the solution process, and more specifically the two matrix-vector multiplications involving the Schur complement matrix for every iterative step of BiCGSTAB algorithm, a significant part of the parallel calculations for these matrix-vector multiplications are chosen to be performed on the GPU in order to achieve performance acceleration. The algorithm is based on the shared model for the CPU and GPU cores. But if more than one GPU are available, which means that each one stores the available data in its local memory, and due to the specific structure of matrices $H_R$ and $H_B$, there is data dependency for every matrix-vector multiplication. The following section presents this extended hybrid algorithm based on the shared-distributed memory model required for the case of multiple GPUs.

## 2. The parallel algorithm

The parallel algorithm for shared memory architectures in [14] is based on a specific mapping of unknowns into the CPU/GPU computational threads permitting uniform load balancing for the computation among cores, minimizing the communication cost between threads and shared memory and also eliminating the idle core threads during the parallel procedures. Following the same mapping of unknowns in case of $N$ number of GPUs available and for even $k = \frac{n_s}{N}$ the parallel procedures of $\boldsymbol{t} = H_R\boldsymbol{z}$ and $\boldsymbol{q} = H_B\boldsymbol{s}$ GPU matrix vector multiplications have to use GPU and CPU core threads in order to avoid transferring multiple data copies from CPU memory to each GPU local memory. The core threads for every GPU $j = 1, \ldots, N$ calculate the vector parts $\boldsymbol{t}_l$ and $\boldsymbol{q}_l$ with $l = (j-1)k+1, \ldots, jk$ being the number of subvectors of length $2n_s$. For the evaluation on the $jth$ GPU of subvectors $\boldsymbol{t}_{(j-1)k+1}, \boldsymbol{t}_{(j-1)k+2}, \boldsymbol{t}_{jk-1}, \boldsymbol{t}_{jk}, \boldsymbol{q}_{(j-1)k+1}$ and $\boldsymbol{q}_{jk}$ the subvectors $\boldsymbol{z}_{(j-1)k}, \boldsymbol{z}_{jk+1}, \boldsymbol{s}_{(j-1)k-1}, \boldsymbol{s}_{(j-1)k}, \boldsymbol{s}_{jk+1}$ and $\boldsymbol{s}_{jk+2}$ are needed, which have been stored on the $j-1$ and $j+1$ GPU's local memories. For this reason these operations are performed by CPU core threads, since the required data vectors are also stored on the CPU's shared memory.

The above are implemented in the parallel algorithm that follows and describes the $\boldsymbol{t} = H_R\boldsymbol{z}$ GPU/CPU matrix vector multiplication.

```
!$OMP PARALLEL
        k = n_s/N
        call acc_device_num(j,acc_device_nvidia)
!$ACC KERNELS COPYIN(z((j − 1)k + 1 : jk)) COPYOUT(t((j − 1)k + 1 : jk))
!$ACC LOOP INDEPENDENT
        do   l = (j − 1)k + 1   to   jk
            The j^th GPU computes   t_l
        enddo
!$ACC END KERNELS
!$OMP SECTION
            The j^th CPU core computes   t_(j−1)k+1 , t_(j−1)k+2 , t_jk−1 , t_jk
!$OMP END SECTION
!$OMP END PARALLEL
```

The GPU/CPU matrix vector multiplication $\boldsymbol{q} = H_B\boldsymbol{s}$ can be described with the following analogous parallel algorithm

```
!$OMP PARALLEL
        k = n_s/N
        call acc_device_num(j,acc_device_nvidia)
!$ACC KERNELS COPYIN(s((j-1)k+1 : jk)) COPYOUT(q((j-1)k+1 : jk))
!$ACC LOOP INDEPENDENT
        do    l = (j-1)k+1  to  jk
            The j^th GPU computes  q_l
        enddo
!$ACC END KERNELS
!$OMP SECTION
            The j^th CPU core computes  q_(j-1)k+1 , q_jk
!$OMP END SECTION
!$OMP END PARALLEL
```

We point out that, for the efficient implementation of the above algorithm parts at least $N$ CPU core threads have to be available. The CPU threads for $j = 1, \ldots, N$ manage the GPU processes. More specifically, each one loads the required data from the CPU shared memory and transfers it to the corresponding GPU local memory. When all GPU calculations are performed the same CPU thread transfers the data from GPU's memory back to the CPU's memory. This is expressed in the above algorithms by the outer OMP PARALLEL procedure. The OpenACC subroutine acc_device_num assigns each GPU card to a CPU core thread created by the OMP parallel region.

## 3. Realization on a Shared-memory parallel computer with GPUs

The shared memory machine HP SL390s G7 consists of a 6-core Xeon X5660@2.8GHz type processor with 12 MB Level 3 cache memory. The total memory is 24 GB and the operating system is Oracle Linux version 6.2. This machine is also equipped with two Fermi edition Tesla M2070 GPUs [15] connected via PCI-e gen2 slots. Each GPU has 6GB of memory and 448 cores on 14 multiprocessors. The application is developed in double precision Fortran code using OpenMP [16, 17] and OpenACC [18] standards with PGI's compilers [19] version 12.9. The basic linear algebra operations subroutines from scientific libraries BLAS and LAPACK [20] are considered.

For the implementation of the above parallel algorithm the test Dirichlet modified Helmholtz problem, which accepts the following exact solution

$$u(x,y) = 10\,\phi(x)\,\phi(y)\,,\quad \phi(x) = e^{-100(x-0.1)^2}(x^2-x),$$

with parameter $\lambda = 1$ was solved.

For the algorithm's performance investigation a single CPU core thread is used for the CPU implementation, while for the CPU/GPU implementations the CPU core threads where as many as the number of GPU cards. Several problem sizes are solved for $n_s = 256$ up to 2048 finite elements in each spatial direction. As every Hermite collocation finite element has 16 degrees of freedom the total degrees of freedom for every problem size is $16n_s^2$. For example in the case of the finest problem the total degrees of freedom are more than 67 millions.

Table 1 presents the total computation time in seconds and the speedup measurements using only the CPU , the CPU and one GPU card and finally cores from CPU and the two GPUs for all problem sizes.

**Table 1.** Speedup and time execution measurements in seconds.

| $n_s$ | CPU time | CPU + GPU time | CPU + GPU speedup | CPU + 2GPUs time | CPU + 2GPUs speedup |
|---|---|---|---|---|---|
| 256 | 12.24 | 11.18 | 1.09 | 8.58 | 1.42 |
| 512 | 88.83 | 71.25 | 1.25 | 54.61 | 1.63 |
| 1024 | 750.35 | 549.82 | 1.37 | 399.42 | 1.88 |
| 2048 | 9176.02 | 6770.11 | 1.36 | 5209.01 | 1.76 |

As expected the size of the problem and the GPU number affect the algorithm performance. An acceleration performance of almost 50% is observed using all available GPU cores for fine discretization problems.

## Acknowledgment

The present research work has been co-financed by the European Union (European Social Fund ESF) and Greek national funds through the Operational Program *Education and Lifelong Learning* of the National Strategic Reference Framework (NSRF) - Research Funding Program: THALIS. Investing in knowledge society through the European Social Fund.

## References

[1] Christara C C 1996 *Advances in Engineering Software* **27** 71–89
[2] Houstis C E, Houstis E N and Rice J 1997 *Par. Comp.* **5** 141163
[3] Mathioudakis E, Papadopoulou E and Saridakis Y 1996 *Parallel Algorithms and Applications* **8** 141–154
[4] Mathioudakis E, Papadopoulou E and Saridakis Y 2004 *Computers and Maths with Appl.* **48** 951–970
[5] Varga R 2000 *Matrix Iterative Analysis* (New York: Springer Verlag)
[6] Papatheodorou T 1983 *Math. Comp.* **(41),164** 511–525
[7] Saad Y 2003 *Iterative methods for sparse linear systems* (SIAM)
[8] Dongarra J, Duff I, Sorensen D and van der Vorst H 1998 *Numerical Linear Algebra for high-performance computers* (Phil.: SIAM)
[9] Mathioudakis E, Papadopoulou E and Saridakis Y 2006 *WSEAS Trans. on Mathematics* **(5),7** 811–816
[10] Brill S H and Pinder G F 2002 *Parallel Computing* **28** 399–414
[11] Mathioudakis E, Papadopoulou E and Saridakis Y 2003 *Numerical Mathematics and advanced applications - ENUMATH 2001, Springer* 957–966
[12] Mathioudakis E and Papadopoulou E 2007 *Int. J. App. Maths and comp. sciences* **(4),3** 179–184
[13] van der Vorst H A 1992 *SIAM J. Sci.Stat.Comp.* **13** 631–644
[14] Mathioudakis E, Vilanakis N, Papadopoulou E and Saridakis Y Parallel iterative solution of the hermite collocation equations on gpus *Proc. of the World Congress on Engeneering 2013 (WCE2013, Imperial College - London, U.K.), Best Paper Award of The 2013 International Conference of Parallel and Distributed Computing* vol 2 pp 1281–1286 URL `http://www.iaeng.org/publication/WCE2013/WCE2013_pp1281-1286.pdf`
[15] *http://www.nvidia.com/object/tesla-servers.html*
[16] Rohit C 2001 *Parallel programming with OpenMP* (M. K.)
[17] *http://www.openmp.org*
[18] *http://www.openacc.org*
[19] *http://www.pgroup.com*
[20] *http://www.netlib.org*

# Donor binding energy under magnetic field in cylindrical nanotube with two GaAs/GaAlAs quantum wells

**J D Gonzalez[1,2], F J Rondano[1], J E González-Cujia[3]**

[1] Grupo Teoría de la Materia Condensada, Universidad del Magdalena, Santa Marta, Colombia.
[2] Departamento de Física, Universidad Nacional de Colombia, Bogotá, Colombia.
[3] Escuela de Física, Universidad Industrial de Santander, Bucaramanga, Colombia

E-mail: jedgonzalezac@unal.edu.co

**Abstract.** A systematic study of the binding energy of the ground state of a hydrogenic off-axis donor in a cylindrical quantum wire containing two quantum wells in a section of the tube layer is calculated in the presence of a uniform magnetic field applied parallel to the wire axis with different potential shape. We express the wave function as a product of combinations of $s$ and $p$ subband wave functions and an envelope function that depends only on the electron–ion separation . By using the variational principle we derive a differential equation for the envelope function, which we solve numerically. Two peaks in the curves for the dependence of the ground-state binding energies on the donor distance from the axis are presented and it is shown that the increasing the magnetic field increasing the binding energy while the impurity is located in the QW1, whereas the opposite occurs when the impurity is located in the QW2.

## 1. Introduction

The study of the behavior of the on-axis hydrogenic impurity center in quantum-well wires (QWWs) that represent a quasi-one-dimensional system has attracted considerable attention in the multilayer structures including strained layers are useful to form nanotubes and nanocoils [1,2]. By using lattice-mismatched epitaxial layers that rolled up when freed from the substrate due to the built-in strain and a micro-tube including two quantum wells (QWs) located at positions with different types of strain was fabricated and its optical properties before and after the fabrication process were investigated by photoluminescence (PL) spectroscopy. Micro-tubes containing two GaAs/AlGaAs quantum wells (QWs) in a section of the tube layer, and also the optical properties of the embedded QWs have been experimentally obtained [3]. In our previous work we show that the low lying $1s$ and $2p_{x,y}$ subbands of the free electron in nanotube, become almost degenerated as the width of the repulsive core grows and the mixing of these subbands in the presence of the off-axis donor no longer should be depreciated [4].The purpose of this paper is to analyze the ground state energy of the off-axis neutral donor and the density of the impurity states in cylindrical nanotube with two GaAs/AlGaAs quantum wells with and without the presence of a magnetic field oriented along the growth axis.

## 2. Theory

Neglecting the differences between material parameters, the dielectric constant $\varepsilon$=12.53 and the electron effective mass $m^*$=0.067$m_0$ of the well and those of the barrier, the Hamiltonians of the electron, $H_0$ and the donor, $H$ in a heterostructure with cylindrical symmetry in the presence of an applied uniform magnetic field $\vec{B} = B\hat{z}$ in the effective-mass approximation, can be written as:

$$H = H_0 - \frac{2}{|\mathrm{r} - \vec{\xi}|} \; ; H_0 = -\nabla^2 + V(\rho) + \frac{1}{4}\gamma^2\rho^2 - i\gamma\frac{\partial}{\partial\varphi} \tag{1}$$

Where $\vec{r}$ and $\vec{\xi}$ are used to designate the electron and ion positions, respectively and we have introduced the effective Bohr radius $a_0^* = \varepsilon\hbar/m^*e^2$ as the unit of length, the effective Rydberg $Ry_0^* = e^2/2a_0^*\varepsilon$ as the unit of energy and $\gamma = e\hbar B/2m^*cRy^*$ as the dimensionless unit of the magnetic-field strength corresponding to the first Landau level.

To calculate the donor ground state energy, we choose a trial function as the product of a linear combination of the waves functions of the electron ground, $s$ and the first excited, $2p_{x,y}$ states, with an unknown isotropic function $\Phi(|\vec{r} - \vec{\xi}|)$ could be interpreted as a variational wave function:

$$\psi(\vec{r}) = \left[\alpha f_{1s} + \sqrt{1-\alpha^2} f_{2p}\right]\Phi(|\vec{r} - \vec{\xi}|), \tag{2}$$

where $\alpha$ is a variational parameter $(-1 < \alpha < 1)$ which gives the grade of the mixing of the subbands [4]. One can see that in the particular case as $\alpha = 1$, there are not subband mixing and the trial function (2) becomes similar to the simple Bastard-type trial function. In general case, the mixing provided a decrease of the parameter $\alpha$. Greater the mixing the smaller is the parameter $\alpha$. In our calculations the parameter $\alpha$ decrease at to 0.8. Due to the inclusion of two quantum wells in the nanotube, appear the strong mixing subbands mostly in the central barrier. In this way, we must include the effect in our calculations.

For our model the wave equation for the free electron $H_o f(\vec{r}) = E_0 f(\vec{r})$ is separable and the electron wave function can be written as follows:

$$f(\vec{r}) = e^{ikz}e^{im\varphi}g(\rho), \tag{3}$$

where $m = 0, \pm1, \pm2, \ldots$ is the angular momentum in $z$-direction, $k$ is the wave number corresponding to a free motion in $z$-direction $(-\pi < k < \pi)$ and the function $g(\rho)$ is the solution of the one-dimensional boundary value problem:

$$\begin{gathered} g''(\rho) + \frac{1}{\rho}g'(\rho) + U(\rho)g(\rho) = 0; \; g'(0) = 0; \; g(\infty) = 0 \\ U(\rho) = E_0 - \gamma m - k^2 - V(\rho) - \frac{\gamma^2\rho^2}{4} - \frac{m^2}{\rho^2} \end{gathered} \tag{4}$$

The differential equation (4) is solved numerically using the trigonometric sweep method [5]. For modelling a nanotube containing two $GaAs/Ga_{1-x}Al_xAs$ quantum wells (QWs), we use the confinement potential given by the following expression:

$$V(\rho) = V_1\theta(-\rho,-R_1,W) + V_2\theta(\rho,R_2,W) - V_3\theta(\rho,R_3+W,W) + V_4\theta(\rho,R_4,W), \quad (5)$$

where:

$$\theta(z,z_0,W) = \begin{cases} 0 \; ; & z < z_0 - W \\ \left[(z-z_0)/W^2 - 1\right]^2 ; & z - W \le z < z_0 \\ 1; & z \ge z_0 \end{cases} \quad (6)$$

is soft-edge version of the Heaviside function, $R_i$, $V_i$ represent the radii and heights of the barriers, respectively; $W$ is a parameter related to the width of the transition region (see Fig. 1).

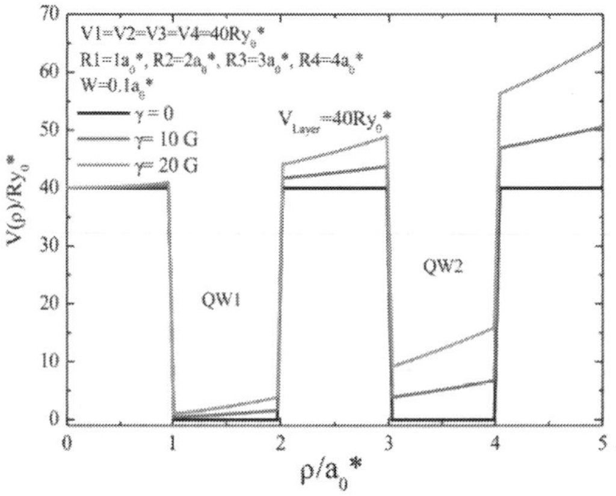


Fig.1 Confinement potential shape containing two GaAs/GaAlAs quantum wells (QWs), different magnetic field applied and potential concentric barrier $V_{Layer}$. $V_{eff}$ include the confinement potential plus the magnetic field induced potential due to the paramagnetic term in the eq. (1).

## 3. Results and Discussion

The peak position coincides approximately with the midpoint between the interior and exterior radii for QW1 and QW2, $(R_1 + R_2)/2$ and $(R_3 + R_4)/2$ respectively where the confinement is the largest. In addition, we display the variation of the binding energies as a function of the distance from the donor location to the axis for different values of the magnetic field parallel to the axis. Increasing the magnetic field decreases the cyclotron radius relative to the wire radius for the electron. Inconsequence, the binding energies of the donor located close to the axis increases, while for donors removed far from the axis the distance between the electron and the donor under the magnetic field grows and the binding energy decreases. This is why the order of the curves of the binding energies dependencies for different magnetic fields presented in Fig. 4 is inverted as the donor is removed from the axis. It is interesting that the crossovers of all curves related to the inversion of their order occur approximately at the same point, corresponding to a distance $2.2a_0^*$ from the axis of the nanotube. In this position of the donor binding energy is almost insensitive to the external magnetic field, while the binding energies of donors arranged closer or further from the axis increases or decreases respectively.

## 4. Conclusions

In this work using the effective mass-approximation within the variational approach and different values of the magnetic field applied parallel to the axis, we calculate the binding energy for the lowest state of the off-axis $D^0$ donors in a nanotube with two square wells, taking into account the variation of the potential shape $\boldsymbol{W = 0.1a_0^*, 0.2, 0.5a_0^*}$, square, soft and parabolic barrier, respectively. We found that under external magnetic field applied parallel to the axis the binding energies of donors located close to the axis increases and located far from the axis decreases.

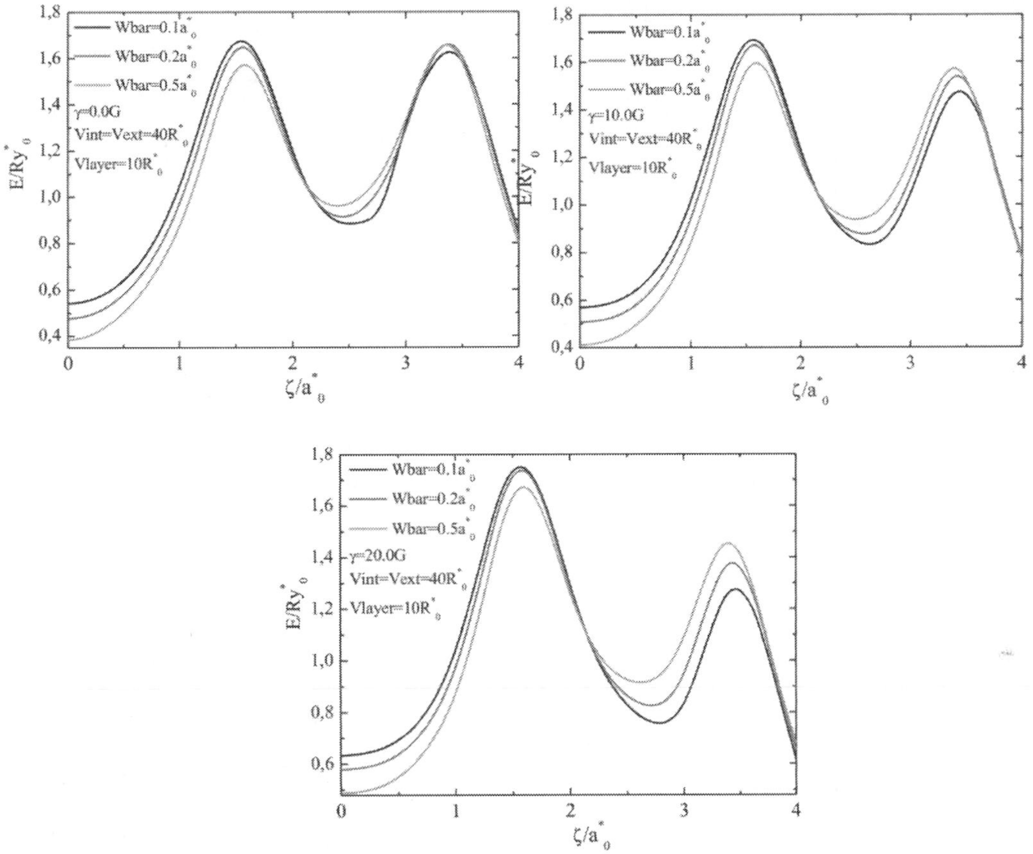


Fig. 4 $D^0$ ground state binding energies as a function of the donor displacement $\xi_\rho$ from the axis of the cylindrical nanotube with different values of the magnetic field and potential shape.

**Acknowledgement**
This work was partially financed by the Universidad del Magdalena and the Colombian Agency COLCIENCIAS through doctoral scholarships 567.

**References**

[1] Villamil P and Porras-Montenegro N, 1998 *J. Phys.Condens. Matter* **10**, 10599.
Branis S V , Li G, and Bajaj K K, 1993 *Phys. Rev. B* **47**, 1316.
Cen J, Lee S M, and Bajaj K K, 1993 *J. Appl. Phys*. **73**, 2848.
[2] Sierra-Ortega J, Mikhailov I D, and Betancur F J, 2004, *Physica B* **348**, 66.
[3] Kubota K, Vaccaroa P O, Ohtania N, Hiroseb Y, Hosodab M, Aida T, 2002 *Physica E* **13,** 313.
[4] Betancur F J, Orozco E A, Gonzalez J D and Mikhailov I D, 2005 *Phys. Stat. Sol*. (b), **242**, 1833.
[5] Betancur F J, I D Mikhailov and L E Oliveira, 1998 *J. Appl. Phys. D* **31**, 3391.

# Ligth-hole exciton in cylindrical microtube with two quantum wells

**J D González[1,2], J E Gonzalez[3], J Barba-Ortega[2]**

[1] Grupo Teoría de la Materia Condensada, Universidad del Magdalena, Santa Marta, Colombia.
[2] Departamento de Física, Facultad de Ciencias, Universidad Nacional de Colombia, Bogotá.
[3] Escuela de Física, Universidad Industrial de Santander, Bucaramanga, Colombia

E-mail: jedgonzalezac@unal.edu.co

**Abstract.** We report on microtube with double quantum well and large radius of curvature. Method for calculating the ground-state energy of light hole exciton and density of states confined in a square potential model that consist of a narrow well, which is produced by a symmetrical structure. The exciton trial function is taken as a product of the ground state wave functions of both the unbound electron and hole in the heterostructure, with an arbitrary correlation function that depends only on electron-hole separation. A renormalized Schrödinger equation for the correlation function is derived and coincides with the corresponding equation for a hydrogen atom in an effective and space-isotropic homogeneous. The binding energy of the ground state to an exciton in this heterostructure, the contribution to the energy given by the sublevels and the density of states is determined as a function of the width of the well, the aluminum concentration and confinement potential profile is obtained by solving the equation calculated by the variational model proposed.

## 1. Introduction

The progress in nanoscale technology has made possible the fabrication of low-dimensional heterostructures with controlled thickness and relatively sharp interfaces, where the excitons remain present even at room temperature because of the quantum confinement increases highly the electron–hole attraction [1]. A great deal of attention has been devoted to experimental studies of excitons in heterostructures based on III–V semiconductor [2–8] particularly quantum wells (QWs), superlattices (SLs), quantum-well wires (QWWs), quantum dots (QDs), quantum rings (QRs) and more recently microtube [9]. The multilayer structures including strained layers are useful to form nanotubes and nanocoils [10]. By using lattice-mismatched epitaxial layers that rolled up when freed from the substrate [11-13] due to the built-in strain and a micro-tube including two GaAs/GaAlAs quantum wells (QWs) [14,15] located at positions with different types of strain was fabricated and its optical properties before and after the fabrication process were investigated by photoluminescence (PL) spectroscopy [16]. The purpose of this paper is to analyze the density of the states of a light-hole exciton as a function of the width of quantum wells taking into account the ground state energy and the contribution given by sublevels in a microtube with two GaAs/AlGaAs quantum wells.

## 2. Theory

In this work, we consider a single exciton in semiconductor heterostructures within the effective mass approximation by using an spherical hole mass approximation and constant physical parameters throughout the heterostructures [10]. In this case the characteristic energy of an exciton can consequently be defined through the effective Rydberg $R_y^* = e^2\ /2\varepsilon\, a_0^*$, and its characteristic length through the effective Bohr radius, $a_0^* = \varepsilon\hbar/\mu e^2$, with $\mu = m_e^*\, m_h^*/m_e^* + m_h^*$ being the reduced exciton effective mass. Thus the dimensionless Hamiltonian for the electron-hole pair confined in a heterostructure in the presence of an external field may be written as:

$$H(\boldsymbol{r}_e, \boldsymbol{r}_h, \tau) = H_e(\boldsymbol{r}_e) + H_h(\boldsymbol{r}_h) + V(\boldsymbol{r}_e, \boldsymbol{r}_h) - 2\tau/r_{eh} \quad (\tau = 0{,}1) \tag{3}$$

$$H_e = -\eta_e \nabla_e^2 + V_e(\mathbf{r}_e);\ H_h = -\eta_h \nabla_h^2 + V_h(\mathbf{r}_h) \tag{4}$$

Where $H_e$ and $H_h$ describe the free motion of electron and hole confined in the heterostructure respectively, $V(\boldsymbol{r}_e, \boldsymbol{r}_h)$ is a potential confinement, $-2/r_{eh}$ is the energy of the electron-hole interaction , $\boldsymbol{r}_e$ and $\boldsymbol{r}_e$ are electron and hole position vectors, and $r_{eh} = |\boldsymbol{r}_e - \boldsymbol{r}_h|$ is the electron-hole separation. The parameter $\tau$ in Eq.(1) is equal to 1 for exciton and to 0 for uncorrelated electron-hole pair, and the parameters $\eta_e = \mu/m_e^*$ and $\eta_h = \mu/m_h^*$ in Eq.(4) are characteristics of the corresponding dimensionless effective masses of the hole in units of the reduced exciton effective mass $\mu$ respectively. For an uncoupled electron-hole pair ($\tau = 0$), the Hamiltonian is separable, on the contrary for exciton ($\tau = 1$), and the Schrödinger equation can only be solved by using some approximation methods. Assuming that the center of mass motion is affected by the electron-hole interaction to a considerably smaller degree than the relative motion, we choose the ground state exciton trial function in the form:

$$\Psi(\boldsymbol{r}_e, \boldsymbol{r}_e) = f_0(\boldsymbol{r}_e, \boldsymbol{r}_h)\Phi(r_{eh});\ r_{eh} = \sqrt{(x_e - x_h)^2 + (y_e - y_h)^2 + (z_e - z_h)^2} \tag{5}$$

where $f_0(\boldsymbol{r}_e, \boldsymbol{r}_h)$ is the product of the one-particle electron and hole ground state wave functions, whereas $\Phi(r_{eh})$ is a variational function that describes the intrinsic properties of the exciton and depends only on the electron-hole separation. The ground state energy of the exciton is found by minimizing the functional, after some algebraic manipulations we can obtain the following variational problem:

$$F[\Phi] = \int_0^\infty \left\{ J_0(r) \left[ \left( \frac{d\Phi(r)}{dr} \right)^2 + \left( E_0 - E_{ex} - \frac{2}{r} \right) \right] \Phi^2(r) \right\} dr \to min \tag{6}$$

with $E_0 = E_e + E_h$ being the ground state energy of the uncoupled electron-hole pair, $r = r_{eh}$ and

$$J_0(r) = \int d\boldsymbol{r}_e \int f_0^2(\boldsymbol{r}_e, \boldsymbol{r}_h)\delta(|\boldsymbol{r}_e - \boldsymbol{r}_h| - r) d\boldsymbol{r}_h \tag{7}$$

The minimization of the functional (6) with respect to $\Phi$ and its first derivative yields the Euler-Lagrange equation

$$\frac{1}{J_0(r)} \frac{d}{dr} J_0(r) \frac{d\Phi(r)}{dr} - \frac{2}{r}\Phi(r) = [E_{ex} - E_0]\Phi(r) \tag{8}$$

The binding energy of the exciton, $E_b = E_0 - E_{ex}$, is then obtained by solving numerically Eq. (8).

Taking into account the dimensions of the microtube we can consider two independent possible states of exciton inside each one separately. On the other hand, the scales for the longitudinal and transversal motions are very different and the radii are huge one can take the advantage of theadiabatic approximation and calculate the energies of the exciton as a sum of the energies of the exciton $E_x(w_i)$ confined inside the QWs of the widths $w_i$ and the rotational kinetic energies $m^2/R_i^2$ corresponding to slow exciton motiuon along the circular guidelines:

$$E_i(m) = \Delta_i + \frac{m^2}{R_i^2};\ \Delta_i = E_g + E_x(w_i) \tag{9}$$

Here $E_g$ is the gap between the valence and the conduction bands for the material of the QW. All energies are taken to be reffered to the top of the valence band equal to zero. In order to calculate the density of the energy states ( DOS) with respect from the valence band to excitonic state transition we assume that the resulting DOS es an envelope curves of all individual transitions

$$\rho(E) = \sum_{i=1,2}\sum_{m=0,10} f\left(E - E_i(m)\right); f(x) = \frac{1}{\sigma\sqrt{(2\pi)}} exp(-x^2/2\sigma^2) \tag{10}$$

## 3. Results and Discussion

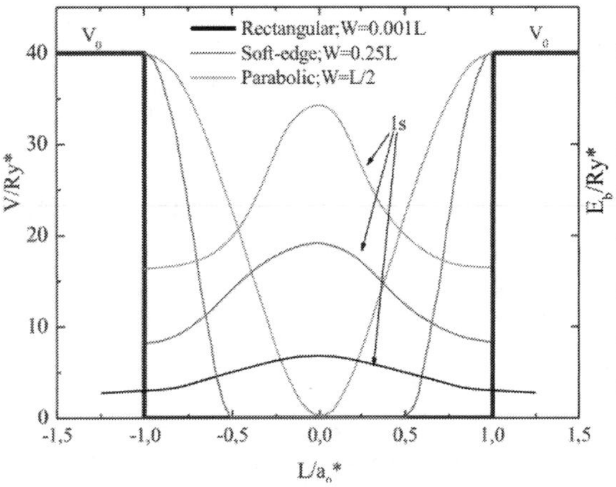


Figure 1. Potential shape with different widths of the transition region

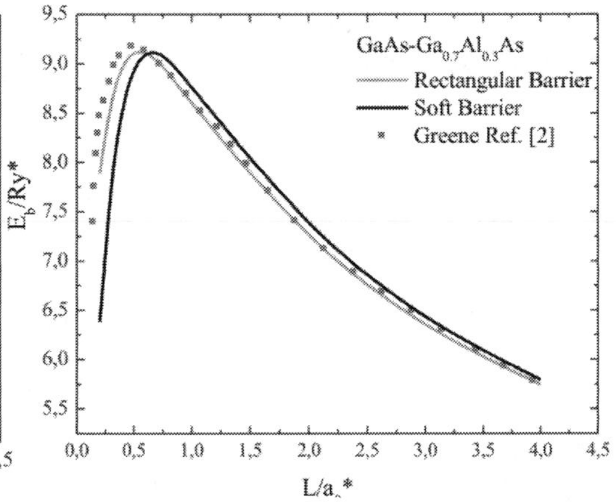


Figure 2. Ligth-hole exciton energies in cylindrical microtube as a function of the width. Curves show our results in the solid lines and the results obtained by Greene, Bajaj and Phelps in the dashed lines (Ref. 2).

In Fig. 2 We compare the binding energies of the ligth-hole exciton as a function of the width $L$ of the cylindrical microtube GaAs/Ga$_{0.7}$Al$_{0.3}$As for different potential shapes Figure 1, with those previously obtained by Greene, Bajaj and Phelps [2]. In this calculation we intentionally chose the material parameters that roughly correspond to those from Ref. 2. The excellent agreement observed in all cases suggests that we can obtain novel curves for different configurations of the heterostructure.

Figure 3. Shows that the peaks position of the density of states of light hole exciton are inverted due to peaks with longer wavelengths have less energy according to the equation $\lambda = \frac{1242}{E_{exc}}$. As we can see the behavior of excitonic states density is similar to that obtained by photoluminescence curve ref. [9], high coincidence of peaks corresponding to wavelengths in both curves is shown, by taking the well width $0.2a_0^*$ and $0.3a_0^*$. Moreover, the difference in the height of the peaks in Figure 2. is due to the choice in the location of wells QW2 and QW1, for radii $R_1$ and $R_2$ in the microtube. The higher peaks in the QW's is located in the position corresponding to the value of R with smaller radius of curvature.

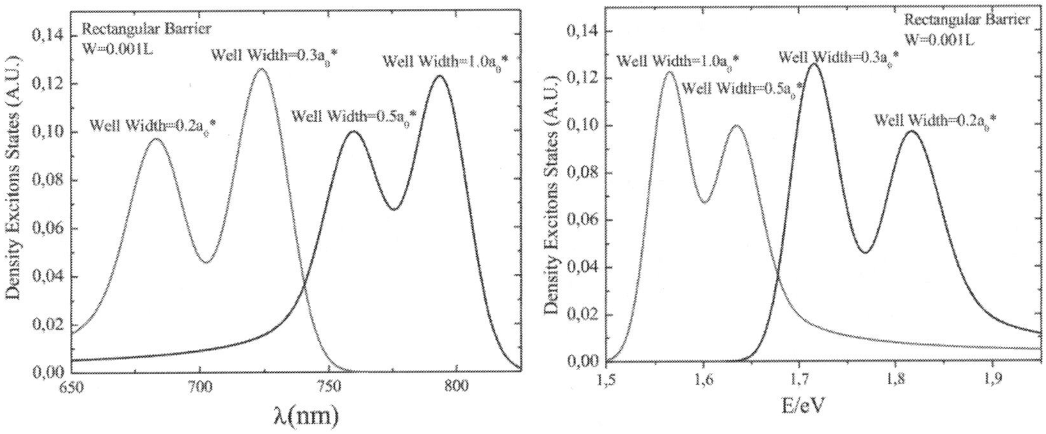


Figure 3. Curves of light-hole density states as a function of a). exciton energy and b). Wavelength in a microtube with two rectangular quantum well.

## 3. Conclusion

The executed numerical simulations show the binding energy behavior for the lowest state of the light-hole in a microtube with two wells with rectangular and soft barriers, taking into account the sublevels due to the rotational kinetic energy. We found that ours calculus are in a good agree with experimental case. Also we present new curves that show novel density of states peaks and different configuration of a microtube can be analyzed.

**Acknowledgement**

This work was partially financed by the Universidad Del Magdalena and the Colombian Agency COLCIENCIAS through doctoral scholarships 567.

## 4. References

[1] Miller D A 1988 in: H. Hang, L. Banyai (Eds.), *Optical Switching in Low-Dimensional Systems, Plenum*, New York, pp. 1–8.
[2] Greene R L , Bajaj K K , Phelps D E 1984 *Phys. Rev. B* **29** 1807.
[3] Cen J , Bajaj K K , 1992 *Phys. Rev. B* **46** 15280; Cen J , Bajaj K K , 1994 *Phys. Rev. B* **50** **g**10947.
[4] Gang Li, Branis S V , Bajaj K K , 1995 *J. Appl. Phys.* **77** 1097.
[5] Goff S Le, Stebe B, 1993 *Phys. Rev. B* **47** 1383.
[6] Garm T, J. 1996 *Condens. Matter.* **8** 5725.
[7] Ferreyra J M, Proetto C R , 1998 *Phys. Rev. B* **57** 9061; Bolcatto P G , Proetto C R , 1999 *Phys. Rev. B* ***59*** *12487.*
[8] Glutsch S, Bechstedt F, Wegscheider W, Schedelbeck F, 1997 *Phys. Rev. B* **56** 4108.
[9] Kubota K, Vaccaroa P O, Ohtania N, Hiroseb Y, Hosodab M, Aida T, 2002 *Physica E* **13** 313–316.
[10] Prinz V Ya, Seleznev V A, Samoylov V A, Gutakovsky A K, 1996 *Microelectron. Eng.* **30** 439.
[11] Lao Y F, 2010 *Applied Physics Letters* **97**, 091104.
[12] Rodriguez R D , *Nanoscale Research Letters*, 2012 **7**, 594.
[13] Balhorn F, Mansfeld S, Krohn A, Topp J, Hansen W, Heitmann D, Mendach S, Phys Rev Lett, **104** (3):037205. (2010).
[14] Miranda G L, Mora-Ramos M E, Duque C A, 2013 *Physica B* **409** 78–82.
[15] Syrbu N, Dorogan A, Dragutan N, Vieru T, Ursaki V, 2011 *Physica E* **44**, 202–206.
[16] Nakayama M, Hirao T, Hasegawa T, 2010 *Physica E* **42**, 2644–2647.

# Implementation of interactive virtual simulation of physical systems

**H Sanchez, J J Escobar, J D Gonzalez, J Beltran**

Research group in didactic and pedagogical on higher education, Universidad del Magdalena, Santa Marta, Colombia.

E-mail: jedgonzalezac@unal.edu.co

**Abstract.** Considering the limited availability of laboratories for physics teaching and the difficulties this causes in the learning of school students in Santa Marta Colombia, we have developed software in order to generate greater student interaction with the phenomena physical and improve their understanding. Thereby, this system has been proposed in an architecture Model/View- View- Model (MVVM), sharing the benefits of MVC. Basically, this pattern consists of 3 parts: The Model, that is responsible for business logic related. The View, which is the part with which we are most familiar and the user sees. Its role is to display data to the user and allowing manipulation of the data of the application. The ViewModel, which is the middle part of the Model and the View (analogous to the Controller in the MVC pattern), as well as being responsible for implementing the behavior of the view to respond to user actions and expose data model in a way that is easy to use links to data in the view. .NET Framework 4.0 and editing package Silverlight 4 and 5 are the main requirements needed for the deployment of physical simulations that are hosted in the web application and a web browser (Internet Explorer, Mozilla Firefox or Chrome). The implementation of this innovative application in educational institutions has shown that students improved their contextualization of physical phenomena.

## 1. Introduction

The exponential growth of Virtual Manipulation and ownership by the students offers an advantage in the purpose to improve their learning of physics; we therefore propose the handling of a web environment that offers a friendly tool for excellent results in appropriation of physical concepts. Contemporary learning theories, such as constructivism, emphasize the importance of providing appropriate instructional supports and guidance in order to facilitate meaningful learning and maximize students' learning outcomes [1]. Many education researchers further recommend that learning environments should provide appropriate instructional scaffolding to help learners acquire knowledge [2,3]. They also reported that educational practices such as chalk and talk lessons were viewed as boring, and students preferred active participation using strategies that involved more opportunities for interaction and discussions. It is probably this perceived difficulty and the way the subject is presented which leads to boredom, disengagement and eventually to poor learning outcomes. In this way, we propose the use of laboratories with virtual environments that work as computer-based simulations.

## 2. Design simulation

The topics for the creation of software that include the application protocol in the form of manuals. Given the difficulties encountered in learning physics, we have developed software under Web environment that can generate more interaction with physical systems and improve their understanding. This has been proposed in an architecture model Vista / Vista / Model (MVVM). MVVM pattern is an evolution of MVC (Model View Controller) that seeks to facilitate parallel work between someone in charge of designing the user interface and another person responsible for generating the code that will sustain. In other words, it is a design pattern to decouple application code from user interface code that is not UI. With MVVM, defines the user interface declaratively (eg, using XAML or Extensible Application Markup Language) and uses the data binding markup to link to other layers that contain data and user commands. The data link infrastructure provides a weak link that keeps synchronized the user interface and related data, and redirects all user inputs to the appropriate commands. Figure 5 shows the structure of MVVM pattern.

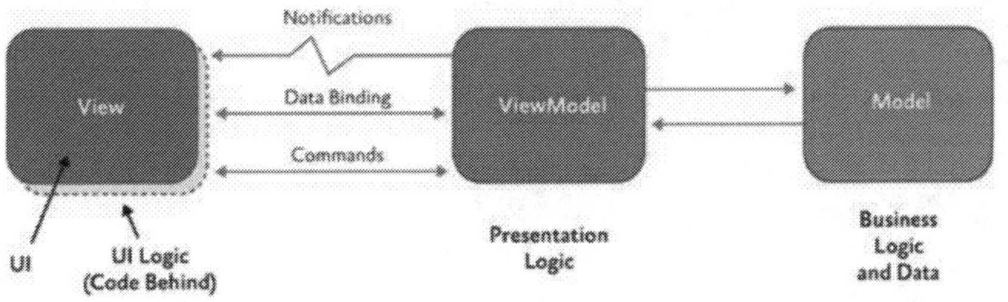


Figure. 1 Structure of the Model MVVM (Model View / View / Model). Image taken from Microsoft.

Basically, this pattern consists of 3 parts: The view or View that is expressed in XAML (UI), the View-Model or ViewModel containing the language we use to develop the presentation logic (VB.NET, C # or C + +) and Model or Model that will be in charge of data access and business logic.
. NET Framework 4.0 and editing package Silverlight 4 and 5 are the main requirements needed for the deployment of physical simulations that are hosted in the web application and a web browser (Internet Explorer, Mozilla Firefox or Chrome).

## 3. Description of virtual laboratory

The implementation of this innovative application in educational institutions has shown that students improved their contextualization of physical phenomena. Currently we have developed 25 virtual labs, considering different topics of physics as: optics, mechanics, electromagnetism and waves. Figure 1 and 2 show two of the mentioned labs.

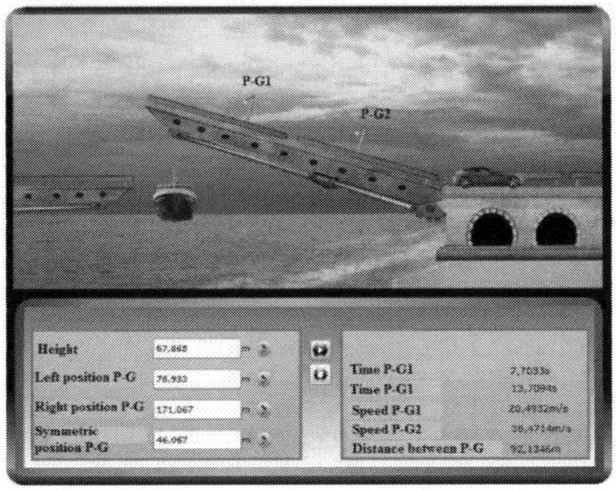


Fig 1. Interface computational simulation of energy conservation lab.

In the Figure 1. We take as input parameters in this simulation, the lifting height of the bridge rail and the distance between the Photo-Gate (P-G). To start the simulation, the times taken for the car to pass each P-G are being processed, as well as the respective speeds. The Figure 2, show simulation of Second Law of Newton, in which one we can find the same structure by using input and output parameters. The friendly environment interface with the users is one of ours objectives due to we want to obtain the attention of the student. The output information obtained for each simulation is used by the student to determine physical phenomena related to lab by obtaining the graphics and values related with the problem.

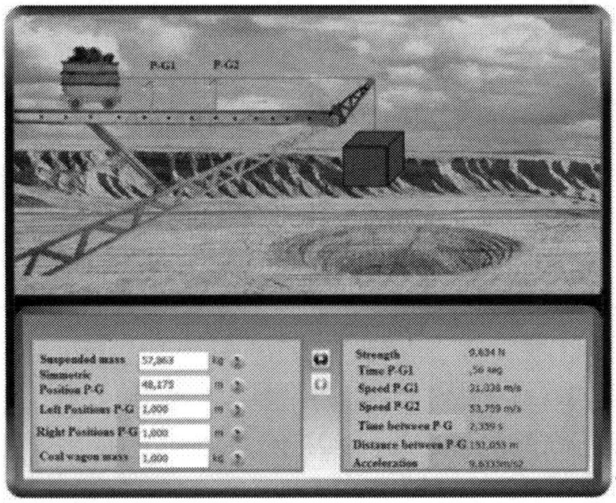


Fig 2. Interface computational simulation of Second Law of Newton lab.

## 4. Conclusions

It should be noted that the present proposal was considerate in the context of a normal course in physics. In which, we develop a new tool for physics teaching by using the architecture Model/View-View- Model (MVVM) to generate greater student interaction with the phenomena physical and improve their understanding. We have developed 25 new Virtual Labs in different topics of physics, whit friendly environments and a simple way to study first principles of physics and engineering.

### Acknowledgement

This work was partially financed by the Universidad del Magdalena and the Colombian Agency COLCIENCIAS through doctoral scholarships 567.

### References

[1] Vygotsky, L S 1978. Mind in society: *Development of higher psychological processes. Cambridge*, MA: Harvard University Press.

[2] Collins, A , Brown, J S, & Holum, A. 1991. *Cognitive apprenticeship: making thinking visible. American Educator*, 6(11), 38–46.

[3] Palincsar, A S, & Brown, A L 1984. *Reciprocal teaching of comprehension-fostering and comprehension-monitoring activities. Cognition and Instruction*, 1(2), 117–175.

[4] Angell, C, Guttersrud, O, Henriksen, E K, & Isnes, A 2004. *Physics: frightful, but fun. Pupils' and teachers' views of physics and physics teaching. Science Education*, 88(5), 683–706.

[5] Marulanda J I, Gómez L A, 2006. *Revista Colombiana de Física*, VOL. **38**, 115-117.

[7] Fonseca M, Hurtado A, Lombana C, Ocaña O, 2006. *Revista Colombiana de Física*, VOL. **38**, No.**2** .
[8] Bates T, 2001. "*Afrontar el reto tecnológico en los centros universitarios e institutos*". *Y Cómo gestionar el cambio tecnológico*". Barcelona. *Gedisa*; pág. 25-26.
[9] Epper, R; Bates A, 2004. *Enseñar al profesorado cómo utilizar la tecnología*. Barcelona, *Editorial UOC*.

# Time-dependent generalization of the Nosé-Hoover thermostatting technique for molecular dynamics simulations

**Edward D Davis**[1] **and Ashraf A Zaher**[2]

[1] Department of Physics, Faculty of Science, Kuwait University, Safat 13060, Kuwait

[2] Department of Electrical and Computer Engineering, College of Arts and Sciences, American University of Kuwait, PO Box 3323, Safat 13034, Kuwait

E-mail: davis@kuc01.kuniv.edu.kw

**Abstract.** The Nosé-Hoover scheme demonstrates that molecular dynamics simulations can be used to calculate the properties of systems at constant temperature (i.e. canonical ensemble averages). There is interest in deterministic generalizations of Nosé-Hoover dynamics which are ergodic even for simple systems like the harmonic oscillator. Prompted by parallels with studies of the Duffing oscillator within control theory, we have investigated a non-autonomous version of the Nosé-Hoover oscillator in which the temperature is replaced by a weakly time-dependent function. This function is chosen so that its average over time coincides with the temperature desired. Calculations are facilitated by graphical programming with a MATLAB-Simulink platform. A time series analysis of our simple non-autonomous system yields the position and momentum distributions expected for the harmonic oscillator.

## 1. Introduction

Use of Nosé-Hoover dynamics [1, 2] is an ingenuous and simple way of computing canonical ensemble averages within molecular dynamics simulations. For a single one-dimensional harmonic oscillator (of unit mass and angular frequency), the Nosé-Hoover equations of motion are

$$\dot{q} = p, \quad \dot{p} = -q - \frac{p_\eta}{Q} p, \quad \dot{\eta} = \frac{p_\eta}{Q}, \quad \dot{p}_\eta = p^2 - T \tag{1}$$

where $q$ ($p$) is the oscillator coordinate (momentum), $\eta$ ($p_\eta$) is the coordinate (momentum) of the thermostat and $T$ is the temperature in units such that the Boltzmann constant $k_B = 1$. (For simplicity, we shall, henceforth, set the thermostat mass $Q = 1$.) Ergodicity of the dynamics is essential to the success of the Nosé-Hoover method, i.e. the average over a trajectory must be equal to the average over the full phase space (which includes the thermostat variables). Unfortunately, the system corresponding to equation (1) is not ergodic (see [3] and references therein).

The example of the Nosé-Hoover thermostat and its limitations has spurred the development of several other deterministic thermostatting schemes [4-9]. The success of *deterministic* thermostats is, nevertheless, a little surprising. A more plausible approach to the coupling between a system and a thermostat (or heat-bath) is to treat it as *stochastic*. Elimination of the heat-bath degrees of freedom (via Mori-Zwanzig projection [10]) then gives rise to equations of motion for the system's degrees of freedom which are generalized Langevin equations. These equations of motion feature: (i) the replacement, in principle, of the forces within the system by effective forces which include small time-dependent interactions mediated by the heat-bath; (ii) non-Markovian frictional forces which account for the loss of energy by the system to the heat-bath, and; (iii) random forces (with vanishing average) which are responsible for the transfer of energy from the heat-bath to the system. An explicit realization of generalized Langevin equations (obtained by a careful and transparent derivation) can be found in [11] (see also [12]).

It would be nice to establish a bridge between deterministic thermostats on the one hand and the stochastic description of heat-baths on the other. A pragmatic response to this dichotomy has been to construct hybrid models in which a previously deterministic thermostat is subjected to random forces [13]. Significant improvements in sampling have been reported.

A different perspective entails using the physics of the stochastic description (such as the qualitative features of the generalized Langevin equation discussed above) as a guide in writing down the equations of motion for deterministic thermostats. It is in this spirit that we have experimented with a time-dependent generalization of the Nosé-Hoover thermostat for the harmonic oscillator. It is also the rationale behind the configurational thermostats of [14].

## 2. A time-dependent Nosé-Hoover thermostat

Within the Nosé-Hoover scheme for the harmonic oscillator [see equation (1)], just one interaction, namely the term $-p_\eta p$ in the equation for $\dot{p}$, causes the transfer of energy to and from the heat-bath. As we noted above, stochastic models possess independent mechanisms for these two transfers. In casting about for an additional device for energy-transfer, our aim is to retain as far as possible the simplicity of the Nosé-Hoover model. An option compatible with these considerations is to replace the temperature $T$ in the equation for $\dot{p}_\eta$ by a weakly time-dependent temperature:

$$T \Rightarrow \tau(t) = T[1 + g(t)] \qquad (2)$$

It is natural to suppose that changes in the energy of the system would be associated with fluctuations in the temperature $\tau$. Experience with the Duffing oscillator [15] suggests that it will be possible to control the dynamics of the present system with an appropriately chosen "driving term" $g(t)$. Provided the average of $g$ over time vanishes, the average of $p^2$ over a trajectory will tend to the desired limit of $T$ for large times. This assertion can be proved, under the assumption that $p_\eta$ is bounded (which is consistent with the results of our numerical simulations), by integrating the modified equation for $\dot{p}_\eta$. In this study, we set $g(t) = \varepsilon \sin(\omega_T t)$ and explore the effect of different choices of the positive parameters $\varepsilon\,(< 1)$ and $\omega_T\,(\ll \omega = 1)$.

Many numerical techniques have been reported in the literature to solve equation (1) and its variants, among them a second order operator splitting technique [3], the second order velocity Verlet integrator method [4], and the fourth order Hamming's predictor-corrector algorithm [16].

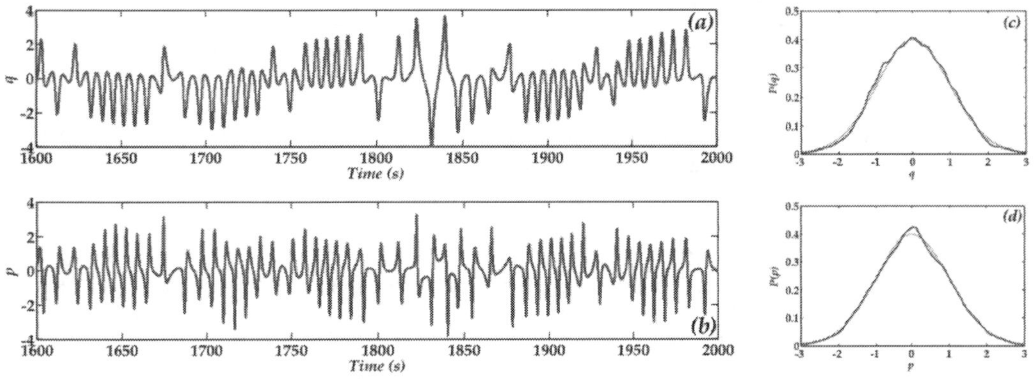


**Figure 1.** The displacement and momentum of a harmonic oscillator coupled to our time dependent Nosé-Hoover thermostat are shown in (*a*) and (*b*), respectively [$q(0) = p(0) = p_\eta(0) = 1$]. The corresponding distribution functions (solid line) are plotted in (*c*) and (*d*), respectively, superimposed on the exact equilibrium distributions (dotted line).

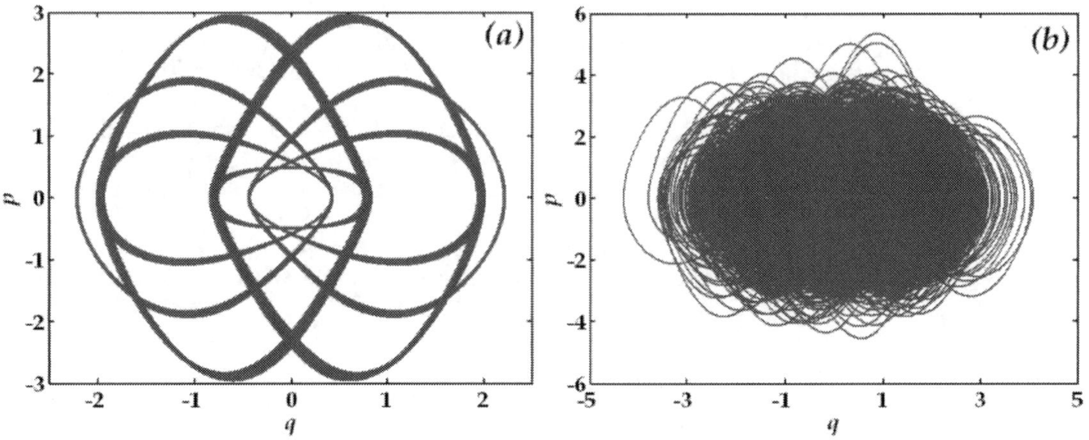


**Figure 2.** Phase space trajectories for the original and the modified Nosé-Hoover models in (*a*) and (*b*), respectively. The initial conditions are $q(0) = 2.2$ and $p(0) = p_\eta(0) = 0$ as in [3].

In this paper, we use Heun's second order integration algorithm with an integration step of 0.005 s, error tolerance of $10^{-14}$ and a total simulation time of 10,000 s. This algorithm is a two-stage Runge-Kutta method that can be easily implemented within a MATLAB-Simulink environment. To ensure accuracy, a MATLAB script and a graphical Simulink model were run in parallel to reproduce the results reported in [4].

The typical choice of $\tau(t) = 1 + 0.5\sin(0.1t)$ for the modified Nosé-Hoover thermostat results in figure 1, which shows the system is unstable but bounded. In addition, the distribution functions are seen to be very close to the Gaussian distributions expected for an ergodic system. Decreasing $\omega_T$ below 0.1, while keeping $\varepsilon = 0.5$, one finds non-ergodic behaviour with negligible effect on the dominant frequency of the system. If, instead, one decreases $\varepsilon$ below 0.5, with $\omega_T = 0.5$, the dynamics is quite similar to that of the original Nosé-Hoover system [equation (1)], indicating loss of sensitivity to the time-dependent perturbation in equation (2). The robustness of the (apparent) ergodic behaviour of the modified Nosé-Hoover model for other choices of initial conditions is illustrated in figure 2 for the same case reported in [3] (which deals with the original Nosé-Hoover model).

Phase space trajectories for the original and the modified Nosé-Hoover models are plotted in figure 3, along with the power spectra of the displacements. Our results agree with those reported in the literature for the dominant frequency when $Q = 1$ (small thermostat mass). The power spectrum of the modified system is non-trivial and qualitatively similar to that found for hyperchaotic high-order oscillators, whereas the power spectrum of the unmodified Nosé-Hoover oscillator has a single *super*-dominant frequency (reflecting the torus-like shape of its strange attractor). It is reasonable to infer that the perturbation due to the forcing function *seems to be sufficient to produce an ergodic system with just one thermostat*. By contrast, two or more *chained* thermostats were required in [4].

## 3. Discussion

The work reported here is at a preliminary stage. Our results are encouraging. A harmonic oscillator coupled to a weakly *time-dependent* Nosé-Hoover thermostat does seem to be ergodic, which is, perhaps, not unexpected: the Hamiltonian-like function $H = \frac{1}{2}(p^2 + q^2 + p_\eta^2) + \eta T$ loses its status as a constant of the motion when $T$ is replaced by $\tau(t)$ in equation (2). (One finds that $\dot{H} = \eta T\dot{g}$.) However, our modification of the Nosé-Hoover thermostat does have one potentially serious drawback. The equilibrium distribution function $f_{eq} \propto \exp[-(p^2 + q^2 + p_\eta^2)/2T]$ does not belong to the kernel of the Liouvillian operator $L$ for our modified Nosé-Hoover model. Instead,

$$Lf_{eq} = g(t) p_\eta f_{eq} \tag{3}$$

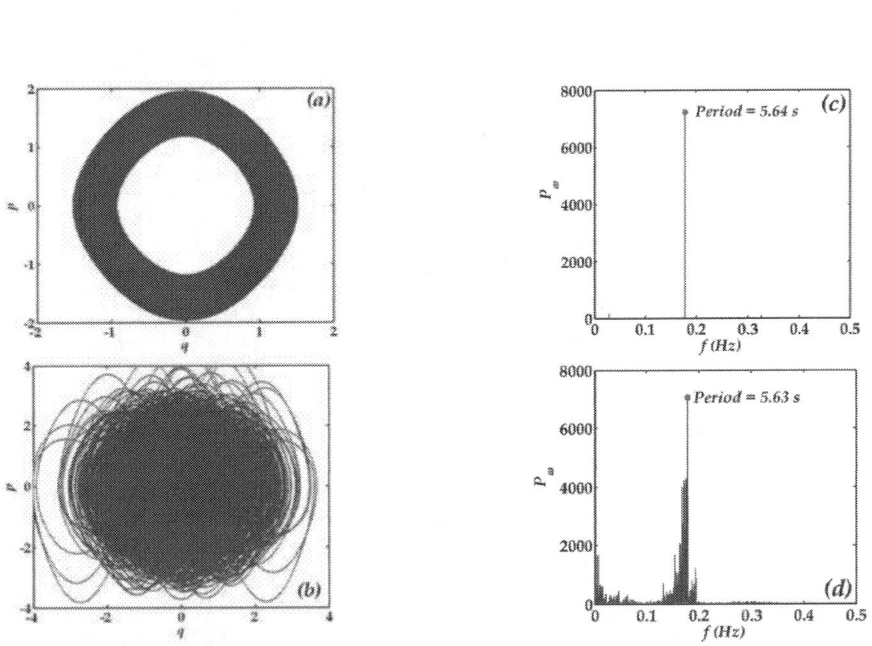


**Figure 3.** Phase space trajectories of the harmonic oscillator coupled to the original and to the time-dependent Nosé-Hoover thermostat are plotted in (*a*) and (*b*), respectively. The corresponding power spectra of the displacement *q*(*t*) are shown in (*c*) and (*d*), respectively. These indicate that, while the original Nosé-Hoover system is regular, the modified system appears to be not only ergodic but also chaotic.

[Actually, the result in equation (3) is independent of the choice of potential $V(q)$.] Nevertheless, the distributions we have generated to date (examples of which are given in figure 1) are not manifestly inconsistent with the exact equilibrium distributions, which may be a consequence of the fact that the right-hand side of equation (3) is, by design, a small oscillating function of time with vanishing average. Perhaps, it is sufficient if $Lf_{eq}$ is zero on average, i.e.

$$\lim_{t\to\infty}\frac{1}{t}\int_0^t Lf_{eq}dt' = 0 \tag{4}$$

Clearly, this issue merits further careful investigation.

## 4. References

[1] Nosé S 1984 *J. Chem. Phys.* **81** 511-9
[2] Hoover WG 1985 *Phys. Rev.* A **31** 1695-7
[3] Legoll F, Luskin M and Moeckel R 2007 *Arch. Rational Mech. Anal.* **184** 449-63
[4] Martyna GJ, Klein ML and Tuckerman ME 1992 *J. Chem. Phys.* **97** 2635-43
[5] Bulgac A and Kusnezov D 1990 *Phys. Rev.* A **42** 5045-8
[6] Liu Y and Tuckerman ME 2000 *J. Chem. Phys.* **112** 1685-1700
[7] Bond S, Leimkuhler B and Laird B 1999 *J. Comput. Phys.* **151** 11434
[8] Leimkuhler B and Sweet S 2004 *J. Chem. Phys.* **121** 108-17
[9] Leimkuhler B and Sweet S 2005 *SIAM J. Applied Dynamical Systems* **4** 187-216
[10] Grabert H 1982 *Projection Operator Techniques in Nonequilibrium Statistical Mechanics* (Berlin: Springer)
[11] Kantorovich L 2008 *Phys. Rev.* B **78** 094304
[12] Dürr D, Goldstein S and Lebowitz JL 1981 *Comm. Math. Phys.* **78** 507-30
[13] Leimkuhler B, Noorizadeh E and Theil F 2009 *J. Stat. Phys.* **135** 261-77
[14] Samoletov AA, Dettmann CP and Chaplain MAJ 2007 *J. Stat. Phys.* **128** 1321-36
[15] Kovacic I and Brennan MJ 2011 *The Duffing Equation: Nonlinear Oscillators and Their Behavior* (New York: John Wiley & Sons)
[16] Posch HA, Hoover WG and Vesley FJ 1986 *Phys. Rev.* A **33** 4253-65

# Boson Model of Quantized EM-Field and Nature of Photons

**Dmitri Yerchuck**[a], **Alla Dovlatova**[b], **Andrey Alexandrov**[b]

[a] Heat-Mass Transfer Institute of National Academy of Sciences of RB, Brovka Str.15, Minsk, 220072

[b] M.V.Lomonosov Moscow State University, Moscow, 119899

E-mail: dpy@tut.by

**Abstract.** The fundamental result, obtained by Dirac, that the dynamical system, which consists of the ensemble of identical bosons is equivalent to the dynamical system, which consists of the ensemble of oscillators, was used to show, that the presence of scalar charge function $\rho(\vec{r},t)$ to be peer force characteristic of electromagnetic (EM) field along with vector force characteristics $\vec{E}(\vec{r},t)$, $\vec{H}(\vec{r},t)$, that was established earlier, agrees with charge neutrality of photons. The simplest analogue in its mathematical description in the physics of condensed matter is the chain of bosonic (spin S = 1) carbon atoms in *trans*-polyacetylene. It has been shown, that neutral photons are topological relativistic solitons with nonzero spin value, which is equal to $\frac{1}{2}$ instead of prevalent viewpoint, that the photons possess by spin $S = 1$. Naturally, they have nonzero size, that is, they cannot be considered to be point objects. At the same time, the main excitations in so called "doped" rest massless "boson-atomic" structure of EM-field will be charged spinless EM-solitons. It seems to be reasonable to suggest, that "doping" can be effective by propagation of EM-field in the medium like to rain-clouds, although detailed mechanism has to be additionally studied.

The representation of photons to be the result of spin-charge separation effect in rest massless "boson-atomic" structure of EM-field makes substantially more clear the nature of corpuscular-wave dualism.

## 1. Introduction

Earlier symmetry studies of Maxwell equations [1] gave new insight on the nature of electromagnetic (EM) field. The existence of dual symmetry of Maxwell equations, determined by dual transformations, established by Rainich [2] and additional dual symmetry, established in [1], determines the mathematical structure of the quantities incoming in Maxwell equations. They are quaternions and, consequently, Maxwellian EM-field has in general case quaternion structure. It consists of four independent field constituents, which differ each other by the parities under space inversion and time reversal. The presence along with 3-vector quaternion characteristics the independent scalar quaternion characterics of EM-field [1] allowed to describe EM-field instead of unobservable vector and scalar potentials by observable electric field 4-vector-function with the components $E_\alpha(\vec{r},t) = \{E_x(\vec{r},t), E_y(\vec{r},t), E_z(\vec{r},t), i\frac{c\rho_e(\vec{r},t)}{\lambda}\}$ and (or in the case of free EM-field) by means of magnetic field 4-vector-function $H_\mu(\vec{r},t) = \{H_x(\vec{r},t), H_y(\vec{r},t), H_z(\vec{r},t), i\frac{c\rho_m(\vec{r},t)}{\lambda}\}$, where $ic\rho_e(\vec{r},t)$, $ic\rho_m(\vec{r},t)$ are the $j_4(\vec{r},t)$-components of 4-current density, corresponding to contribution of electric and magnetic components of charge densities correspondingly, $\lambda$ is conductivity, which for the case of EM-field propagation in vacuum is $\lambda_v = \frac{1}{120\pi}\ (Ohm)^{-1}$. The most actual for the tasks of modern theories of interaction of EM-field with matter, taking into account their practical appliation to nanoelectronics, spintronics andto a number of the other nanotechnology branches represents the quantum description of EM-field. The well known now simple formula $E = h\nu$, $h = const > 0$, proposed right at the beginning of the 20th century by Planck [4] and by Einstein [5] became epoch-making and a real symbol of the substantial progress in the science. The interpretation, given by Einstein, indicates straight on real existence of light quanta of frequency $\nu$ with the total energy $E$, which in its turn has led to a new understanding of the nature of the electromagnetic field. In fact it was the indication on oscillatory, that is discrete in space structure of EM-field, which in the same time is the set of physical objects strongly connected to some periodic with time period $T = 1/\nu$ process, being to be intrinsic for given objects, at that Lorentz invariant product $ET$ is equal to $h$. Similar interpretation of De Broglie's relationship [6] leads to conclusion on quantum nature of the charges. Although there are the great achievements in quantum theory, the great challenge to give an adequate description of light quanta, which were called by Lewis photons [7], still has not brought satisfactory results in fact up to now. Even Einstein himself recognizes [8] in one's time, that the whole fifty years of conscious brooding have not brought me nearer to the answer to the question what are light quanta, and now, more than half a century later, theoretical physics still needs progress to present a satisfactory explanation of the photon nature and to propose its structure. We will show, that the quaternion structure of EM-field and the existence of the charge, being to be inherent scalar force characteristic of EM-field, allow to understand the nature of photons. Let us remark the following. On the one hand, the photons seem to be charge neutral particles, which is confirmed in a number of observations, including astronomic observations, and experiments on light absorption, transmission, reflection, Rayleigh and Raman scatterings and so on. On the other hand, the photons seem to be charged particles, which is confirmed, for example, by observations of zigzag-like light propagation between adjacent rainclouds during storm, since zigzag-like light propagation (forked lightning) in the same conditions (that is by the absence of lightning between the same rainclouds) does not takes place. The fact, that the appearance of forked lightning is connected with charge particles, was confirmed experimentally long time ago. Consequently, it is reasonable to suggest, that quantum structure of EM-field is responsible for lightning development and propagation, since the velocity of the process of lightning development and propagation is too much, in order to refer given effect the only to dielectric medium, in which any electrically conducting cannels are absent. In its turn, it allows to suggest, that by EM-field propagation in rainclouds the charged photons, that is, the photons of the other kind in comparison with usual ones, can be created. It will be proved in given work. It is the aim of the work presented.

## 2. Results and Discussion

The fundamental result, obtained by Dirac, that dynamical system, which consists of the ensemble of identical bosons, is equivalent to dynamical system, which consists of the ensemble of oscillators, will be used to show, that the existence of the charge agrees with charge neutrality of photons. The simplest analogue in the physics of condensed matter of the system of interacting S = 1 bosons is carbon. So we come to the model of linearly polarized EM-field to be the chain of bosons, which is like in its mathematical description to the chain of carbon atoms in *trans*-polyacetylene (t-PA), at that both in "atomic" and "electronic" structure. One-dimensionality of the task can be argued in the following way. Since for the description of EM-field instead of unobservable vector and scalar potentials the 4-vector of electrical and/or magnetic field strength can be used, then, to describe linearly polarized EM-field in Euclidian space $R_3$ it is sufficient to specify the propagation direction, that is vector $\vec{k}$ and to define $\vec{E}$. Given vectors determine the plane, in which a frame of reference with $z$-axis along the propagation direction and orthogonal to it $x$-axis can be set. Taking into account the homogeneity of Minkowski space $R_4$ and homogeneity of free EM-field in it, free EM-field can be modelled by the set of noninteracting (or weak interacting) between themselves "boson-atomic" chains, similarly to those ones in many carbon-based systems. What concerned the "atomic" structure, we have to include the contribution of vacuum fluctuations, which presents in oscillator task and which is absent in the case of boson task. The presence of charge, being to be scalar characteristic of EM-field, gives the possibility to model "electronic" structure of equivalent boson chain like to t-PA electronic structure, that is consisting of "$\sigma$-subsystem" and "$\pi$-subsystem". It becomes to be understandable, if to take into account, that charge space distribution is directly connected with $\vec{E}$ space distribution. In other words the presence of $E_z$-component will determine the appearance of EM-field charge "$\sigma$-subsystem", while $E_x$-component will determine the appearance of EM-field charge "$\pi$-subsystem", at that, like to t-PA, its distribution in space $R_3$ will be twice degenerated. Given conclusion can be argued on the basis quaternion structure of EM-field in the following way. $E_x$-polar component by EM-field propagation every other half-period alters its sign, at the same time $E_x$-axial component does not alters its sign, which is equivalent to appearance of alternating single-double interbosonic "$\pi$-bonds" in EM-field charge "$\pi$-subsystem", at that two configuration - single-double and double-single are topologically equivalent. Consequently we come to conclusion, that the interaction between equivalent to oscillators "bosonic atoms" can be described in the frames of !D Fermi gas model in zero-th order approximation or in the frames of 1D Fermi liquid model in the first order approximation. Mathematical description in zero-th order approximation will be similar to well known Su-Schrieffer-Heeger (SSH) model of organic conductors with some corrections, concerning two branch of quasiparticles, given in [3]. Fermi liquid model for 1D case will represent the generalization of SSH-model and it will be considered in details in applicability to EM-field description. The concept of quantum 1D Fermi liquid for description of 1D correlated electronic systems was recovered in [15]. It has topological soliton origin in distinction from well known spinon-holon spin-charge separation effect in Luttinger liquids. Analogous model will be developed in given paper for description of propagation of quantized EM-field. We will start from Hamiltonian

$$\hat{\mathcal{H}}(u) = \hat{\mathcal{H}}_0(u) + \hat{\mathcal{H}}_{\pi,t}(u) + \hat{\mathcal{H}}_{\pi,u}(u). \tag{1}$$

Like to works [13], [14] we will consider Born-Oppenheimer approximation. The first term in (1) is

$$\hat{\mathcal{H}}_0(u) = \sum_m \sum_s (E_m^k \hat{a}_{m,s}^+ \hat{a}_{m,s} + K u_m^2 \hat{a}_{m,s}^+ \hat{a}_{m,s}) \tag{2}$$

is the operator of kinetic energy of "boson atom" motion (the first term), and the operator of the $\sigma$-bonding energy the second term, $K$ is effective $\sigma$-bonds spring constant, $u_m$ is configuration coordinate for $m$-th "boson atom", which corresponds to translation of $m$-th "boson atom" along the symmetry axis $z$ of the chain, $m = \overline{1,N}$, $N$ is number of "boson atoms" in the chain, $m = \overline{1,N}$, $\hat{a}_{m,s}^+$, $\hat{a}_{m,s}$ are creation and annihilation operators of creation or annihilation of quasipartile with spin projection $s$ on the $m$-th chain site in "$\sigma$-subsystem". The second term in (1) is

$$\hat{\mathcal{H}}_{\pi,t}(u) = \sum_m \sum_s [t_0(\hat{c}_{m+1,s}^+ \hat{c}_{m,s} + \hat{c}_{m,s}^+ \hat{c}_{m+1,s})]. \tag{3}$$

It is the resonance interaction (hopping interaction in tight-binding model approximation) of quasiparticles in "$\pi$-subsystem" of all charge system, which is considered to be Fermi liquid, and in which the only constant term in Taylor series expansion of resonance integral about the dimerized state is taking into account. Here $\hat{a}_{m,s}^+$, $\hat{a}_{m,s}$ are creation and annihilation operators of creation or annihilation of quasipartile with spin projection $s$ on the $m$-th chain site in "$\pi$-subsystem". The third term in (1) is

$$\hat{\mathcal{H}}_{\pi,u}(u) = \sum_{m}\sum_{s}[(-1)^m 2\alpha_1 u(\hat{c}^+_{m+1,s}\hat{c}_{m,s}+ \hat{c}^+_{m,s}\hat{c}_{m+1,s}) + (-1)^m 2\alpha_2 u\hat{c}^+_{m,s}\hat{c}^+_{m+1,s}\hat{c}_{m+1,s}\hat{c}_{m,s}]. \tag{4}$$

It represents correspondingly the terms, which are proportional to linear terms in Taylor series expansion about the dimerized state of the resonance interaction of quasiparticles in "$\pi$-subsystem" of charge system and potential energy of the pairwise interaction of quasiparticles in "$\pi$-subsystem" between themselves. It is taken into account, that in Born-Oppenheimer approximation in perfectly dimerized chain the coordinates $\{u_m\}$, $m = \overline{1,N}$, can be represented in the form $\{u_m\} = \{(-1)^m u\}$, where $u$ is displacement amplitude, corresponding to minimum of ground state energy [13]. In result of solving of the task formulated with Hamiltonian (1) two values for the energy of quasiparticles were obtained, indicating on the possibility of formation of the quasiparticles in $c$-band and $v$-band of two kinds. They are

$$E_k^{(c)}(u) = \frac{Q^2\Delta_k^2 - \epsilon_k^2}{\sqrt{\epsilon_k^2 + Q^2\Delta_k^2}}, \quad E_k^{(v)}(u) = \frac{\epsilon_k^2 - Q^2\Delta_k^2}{\sqrt{\epsilon_k^2 + Q^2\Delta_k^2}} \tag{5}$$

and

$$E_k^{(c)}(u) = \sqrt{\epsilon_k^2 + Q^2\Delta_k^2}, \quad E_k^{(v)}(u) = -\sqrt{\epsilon_k^2 + Q^2\Delta_k^2}, \tag{6}$$

where $\Delta_k = 4\alpha_1 u \sin ka$, $\epsilon_k = 2t_0 \cos ka$, the value for factor $Q$ satysfies the equation

$$[1 + \frac{\alpha_2}{2\alpha_1}\sum_k\sum_s \frac{Q\Delta_k \sin ka}{\sqrt{\epsilon_k^2 + Q^2\Delta_k^2}}(n_{k,s}^{(c)} - n_{k,s}^{(v)})] = Q, \tag{7}$$

where $n_{k,s}^{(c)}$ is eigenvalue of density operator of quasiparticles' number in $c$-band, $n_{k,s}^{(v)}$ is eigenvalue of density operator of quasiparticles' number in $v$-band. It is evident, that at $Q = 1$ in (5), (6) we have the case, equivalent to SSH-model. It can be realized, if $\frac{\alpha_2}{\alpha_1}\sum_k\sum_s \frac{1}{2}\frac{\Delta_k}{\sqrt{\epsilon_k^2+\Delta_k^2}} \sin ka(n_{k,s}^{(c)} - n_{k,s}^{(v)})] \to 0$, which, in its turn, is realized, if $\alpha_2 \to 0$. Consequently, it seems to be interesting to consider the opposite case, when $|\frac{\alpha_2}{\alpha_1}\sum_k\sum_s \frac{1}{2}\frac{\Delta_k}{\sqrt{\epsilon_k^2+\Delta_k^2}} \sin ka(n_{k,s}^{(c)} - n_{k,s}^{(v)})]| \gg 1$. Passing on to continuum limit, in which $\sum_k\sum_s \to 2\frac{Na}{\pi}\int_0^{\frac{\pi}{2a}}$, and assuming $n_{k,s}^{(v)} = 1$, $n_{k,s}^{(c)} = 0$, we have integral equation for determination of factor $Q$

$$\frac{2Nua\alpha_2}{\alpha_1\pi t_0}\int_0^{\frac{\pi}{2a}} \frac{\sin^2 ka}{\sqrt{1 - \sin^2 ka[1 - (\frac{2uQ}{t_0})^2]}}dk = 1. \tag{8}$$

In the case $|\frac{2uQ}{t_0}| < 1$ and in the case $|\frac{2uQ}{t_0}| > 1$ $Q$ was evaluated from the relation (8) appoximately. In the case $\frac{2uQ}{t_0} = 1$ parameter $Q$ was calculated exactly. The quasiparticles of the second kind at $Q = 1$ are quite similar quasiparticles in its mathematical description, that those ones, which were obtained in [13], [14]. It was found, that SSH-like solution is unapplicable for the description of standard processes, passing near equilibrium state by any parameters. The quasiparticles, described by SSH-like solution, can be created the only in strongly nonequilibrium state with inverse population of the levels in $c$- and $v$-bands. At the same time, the first solution can be realized both in near equilibrium and in strongly nonequilibrium states of the "$\pi$-subsystem" of boson-"atomic" chain, which is considered to be quantum Fermi liquid. The continuum limit for the ground state energy of boson-"atomic" chain with SSH-like quasiparticles will coincide in its mathematical form with known solution [14], if to replace $\Delta_k Q \to \Delta_k$. Let us calculate the ground state energy $E_0^{[u]}(u)$ of the boson-"atomic" chain with quasiparticles' branch, which is stable near equilibrium. Taking into account, that in ground state $n^c_{k,s} = 0$, $n^v_{k,s} = 1$, in the continuum limit we have

$$E_0^{[u]}(u) = -\frac{2Na}{\pi}\int_0^{\frac{\pi}{2a}} \frac{(Q\Delta_k)^2 - \epsilon_k^2}{\sqrt{(Q\Delta_k)^2 + \epsilon_k^2}} dk + 2NKu^2, \tag{9}$$

then, calculating the integral, using the complete elliptic integral of the first kind $F(\frac{\pi}{2}, 1 - z^2)$ and the complete elliptic integral of the second kind $E(\frac{\pi}{2}, 1 - z^2)$, we obtain

$$\begin{aligned} E_0^{[u]}(u) = \frac{4Nt_0}{\pi}\{F(\frac{\pi}{2}, 1 - z^2) + \\ \frac{1 + z^2}{1 - z^2}[E(\frac{\pi}{2}, 1 - z^2) - F(\frac{\pi}{2}, 1 - z^2)]\} + 2NKu^2, \end{aligned} \tag{10}$$

where $z^2 = \frac{2Q\alpha_1 u}{t_0}$. Approximation of (10) at $z \ll 1$ gives

$$\begin{aligned} E_0^{[u]}(u) = N\{\frac{4t_0}{\pi} - \frac{6}{\pi}\ln\frac{2t_0}{Q\alpha_1 u}\frac{4(Q\alpha_1)^2u^2}{t_0} + \\ \frac{28(Q\alpha_1)^2u^2}{\pi t_0} + ...\} + 2NKu^2. \end{aligned} \tag{11}$$

It is seen from (11), that the energy of quasiparticles, described by the first solution, has the form of Coleman-Weinberg potential with two minima at the values of dimerization coordinate $u_0$ and $-u_0$ like to the energy of quasiparticles, described by SSH-solution for t-PA [14].

Therefore, all qualitative conclusions of the model proposed in [14] are holding in Fermi-liquid consideration of "$\pi$-subsystem" of the chain (instead of Fermi-gas consideration) for the quasiparticles, corresponding to the second-branch-solution. It is evident, that the mechanism of the phenomenon of spin-charge separation in Fermi-liquid is soliton mechanism, analogous to the mechanism proposed by Jackiw and Rebbi [11] on the basis of field theory positions and to the mechanism proposed by Luther and Emery [10], but it is quite different from Anderson spinon-holon mechanism [12].

Thus, the results obtained allow to propose the reasonable explanation of the existence in the fields with charges of chargeless particles - solitons with nonzero spin value, which in the case of EM-field is equal to $\frac{1}{2}$ instead prevalent viewpoint, that photons possess by spin $S = 1$. The photons in quantized EM-field are main excitations in oscillator structure, which is equivalent to spin S = 1 boson-"atomic" structure, like mathematically to well known spin S = 1 boson matter structure - carbon atomic backbone structure in many conjugated polymer chains. The photons have two kind nature. The photons of the first kind represent themselves neutral EM-solitons of SSH-soliton family. They are main excitations in so-called "undoped " structure of EM-field, including free EM-field in vacuum. Naturally, they have nonzero size, that is they cannot be considered to be point objects. It seems to be evident, that like to Fermi-gas SSH-model, the main excitations in "doped" boson-"atomic" structure of EM-field will be charged spinless EM-solitons, which also can be referred to SSH-soliton family. It is reasonable to suggest, that "doping" can be effective in the medium like to rain-clouds, although detailed mechanism has to be additionally studied. The representation of photons to be the result of spin-charge separation effect and their assignment with main excitations in ground state of "$\pi$-subsystem" in rest masless boson-"atomic" EM-field structure - chargeless spin $\frac{1}{2}$ topological solitons - makes substantially more clear the nature of corpuscular-wave dualism. Really, like to matter atomic structure, the quantized EM-field represents itself the discrete boson-"atomic" structure, the separate bosons in which produce the structure like to atomic structure in condensed matter. The main difference between EM-field "atomic" structure and atomic structure in condensed matter consists in that the "atoms" in EM-field structure have zeroth rest mass. The origin of wave in given structure is determined by the mechanism, quite analogous to Bloch wave formation in solid state of condensed matter. It is harmonic trigonometric functions for Maxwellian EM-field, which determine wave character of quantized Maxwellian EM-field. At the same time there are simultaneously the corpuscules, propagating along given EM-field boson-"atomic" chain structure, that is, chargeless spin $\frac{1}{2}$ topological relativistic solitons - photons, formed in usual conditions (or spinless charged solitons in so-called "doped" EM-field structure). It becomes now to be understandable, that the display of the corpuscular or wave nature of EM-field will be dependent on experimental conditions.

Let us remark, that the model proposed is in fact the development of an idea of multiphoton "molecules". The idea of multiphoton "molecules" arose at the earliest stage of quantum physics around 1910, and it belongs to Debye [16]. Debye presented in above cited work a simple, ondulatory method to obtain the radiation law. Let us remark, that the derivation of the complete radiation formula from the corpuscular concept of the photon "molecules" has been obtained the only relatively recently, on the frontier of the millennia, in 2000, in [18]. The authors of [18] have obtained the complete radiation formula on the base of Debye photon "molecules" idea, using the description of 'multiphoton states' in multiphoton "molecules", which was proposed in [17]: light of definite direction and frequency $\nu$ presents itself in units "molecules" of 0, 1, 2, ..., n, ...photons, with energy $nh\nu$, that is, with zero binding energy, and they contribute independently to the energy density. The authors of [18] accentuate, that given assumption is rational, it is in the spirit of the atomistic Democritean viewpoint, and it means that all the photons of the same frequency are identical.

Especially interesting, that we came quite independently to boson chain description of EM-field, the photons in which, being to be noninteracting between themselves topological solitons in the atomistic 1D-chain structure of the linearly polarized light, can really be represented by 1D "molecules" of 0, 1, 2, ..., n, ...quasiparticles, with energy $nh\nu$, that is, with zero binding energy, in full accordance with the model, proposed by [16], [17], [18]. On the existance of the works [16], [17], [18] we have got to know quite recently. However given independence seems to be testifying on the correctness of Debye idea and its development in subsequent works, including our work.

Let us give the experimental confirmations of the model proposed.

Very strong argument in favour of the model proposed is the well known experimentally observed phenomenon of electron-positron annihilation. It is well theoretically described, see, for example [19], and experimentally confirmed that by direct annihilation of electron-positron pair two photons are produced. At the same time the explanation for the case of the relative velocity of annihilating particles near to zero, how can be produced from two particles with spin value 1/2 also two particles, however with spin value 1, is in fact absent. It is evident, that the boson model presented explains given disagreement between experimental data on electron-positron annihilation and the theoretical viewpoint on the photon to be the spin-1 particle in a natural way - from two particles with spin values 1/2, that is from electron-positron pair can be produced two photons the only with spin values 1/2.

The second argument is the existence of the dependence of a resonance microwave absorption rate on the spin value of absorbing centers, which was found in electron spin resonance (ESR) spectroscopy for the first time in [20], [21]. It was established, that spin 1-centers absorb microwave power with the rate exceeding the the absorption rate by spin 1/2-centers precisely two times more. From give result it follows the reasonable suggestion, that if the photons were possessing by spin 1 they couldn't be absorbed in ESR conditions by the centers with the spin value 1/2 (like to the absence of any ESR-absorption by the centers with zeroth spin value).

The third argument is the following. In the concept of elementary patticles existing at present "the Higgs boson emerges and disappears by borrowing energy from and returning energy to the particles in the electroweak interaction, respectively. The returning of energy from the Higgs scalar boson to the particles is through the absorption of the Higgs scalar boson by the particles. When a massless particle in the electroweak interaction absorbs the Higgs scalar boson, the Higgs scalar boson becomes the longitudinal component of the massless particle, resulting in the massive particle and the disappearance of the Higgs scalar boson" [22]. According to [22] "The observed Higgs boson at the LHC is a remnant of the Higgs boson. At the beginning of the universe, all particles in the electroweak interaction were massless. The Higgs boson appeared by borrowing energy symmetrically from all particles in the electroweak interaction. The Higgs boson coupled with all massless particles, including leptons, quarks, and gauge bosons, in the electroweak interaction. All massless particles except photon absorbed the Higgs boson to become massive particles. The asymmetrical returning of energy from the Higgs boson by the absorption of the Higgs boson is called the symmetrical breaking of the electroweak interaction in the Standard Model. In the cases of such massive particles, including leptons, quarks, and weak gauge bosons, the Higgs boson disappeared. In the case of massless photon the un-absorbed Higgs boson became the remnant of the Higgs boson. Being specific to the electroweak interaction, the remnant of the Higgs boson could not return the borrowed energy to any other massless particles". So it is established that the photons occupy a peculiar place regarding the interaction with Higgs bosons. From here follows also the reazonable suggestion, that the photons cannot be considered to be the gauge bosons of electromagnetism, possessing by spin 1.

## 3. Conclusions

New model of photons is proposed. The photons in quantized EM-field are main excitations in oscillator structure of EM-field, which is equivalent to spin S = 1 rest massless boson-"atomic" structure, like mathematically to well known spin S = 1 boson matter structure - carbon atomic backbone chain structure in many conjugated polymers. They have two kind nature. The photons of the first kind represent themselves neutral chargeless topological EM-solitons of SSH-soliton family. The photons of the second kind represent themselves charged spinless EM-solitons, which also can be referred to SSH-soliton family. The origin of waves in given structure is determined by the mechanism, quite analogous to Bloch wave formation in solid state of condensed matter. At the same time, the existence simultaneously of the corpuscules, propagating along given EM-field boson-"atomic" chain structure, that is, chargeless spin $\frac{1}{2}$ topological relativistic solitons - photons, formed in usual conditions (or spinless charged solitons in so-called "doped" EM-field structure) clarifies the nature of corpuscular-wave dualism.

[1] Alla Dovlatova and Dmitri Yerchuck, Journal of Physics: Conference Series, **343** (2011) 012133, DOI:10.1088/1742-6596/343/1/012133

[2] Rainich G Y, Trans.Am.Math.Soc.,**27** (1925) 106-136
[3] D.Yerchuck, A.Dovlatova, Journal of Physical Chemistry C, **116**, N 1 (2012) 63-80, DOI:10.1021/jp205549b
[4] Planck M, Ann.d.Phys., **4** (1901) 553-563
[5] Einstein A, Ann.d.Phys., **17**, (1905) 132-148
[6] De Broglie L.V., Comp.Rend., **177** (1923) 507-510
[7] Lewis G.N, Nature, **118** (1926) 874-875
[8] Speziali P, Ed., Albert Einstein-Michele Besso Correspondence (1903-1955), Herman, Paris, 1972, 453 pp
[9] Scully M O, Zubairy M S, Quantum Optics, Cambridge University Press, 1997, 650 pp
[10] Emery V J and Luther A, Phys.Rev.Lett., **33** (1974) 589-592
[11] Jackiw R and Rebbi C, Phys.Rev.D, **13** (1976) 3398-3409
[12] Anderson P W, Science, **235** (1987) 1196-1198
[13] Su W.P., Schrieffer J.R, and Heeger A.J, Phys.Rev.Lett., **42** (1979) 1698-1701
[14] Su W.P., Schrieffer J.R, and Heeger A.J, Phys.Rev.B, **22** (1980) 2099-2111
[15] Alla Dovlatova, Dmitri Yerchuck, Felix Borovik, arXiv:1112.3339 [cond-mat.str-el], 2013
[16] Debye P, Ann.d.Phys., **33** N 4 (1910) 1427-1434
[17] Wolfke M, Phys.Zs., **15** (1914) 308-311, Ibid.,**22** (1921) 375-379 (1921)
[18] Boya Luis J, Duck Ian M, and Sudarshan E C G, Deduction of Planck's Formula from Multiphoton States, arXiv:quant-ph/0010010v1 2 Oct 2000
[19] Berestecky V.B, Lifshitz E.M, Pitaevsky L.P, Quantum Electrodynamics, M., Nauka, 1989, 728 pp
[20] Fedoruk G G, Rutkovskii I.Z, Yerchak D.P, Stelmakh V.F, Zhurnal Experimentalnoi i Teoreticheskoi Fiziki, **80**, N 5 (1981) 2004-2009, Fedoruk G G, Rutkovskii I.Z, Erchak D.P, Stel'makh V.F, Sov.J.Exp.Theor.Physics, USA, **53**, N 5 (1981) 1042-1044
[21] Yerchak D.P, Rutkovskii I.Z, Stelmakh V.F, Fedoruk G G, "Nonsteady State Radiation Defects' EPR in Silicon", in "Questions of Atomic Science and Technique. Radiation Damage Physics and Radiation Materialkeeping", Kharkov, 1982, 2/21, 79-80
[22] Ding-Yu Chung, Ray Hefferlin, Journal of Modern Physics, **4** (2013) 21-26, DOI:10.4236/jmp.2013.44A004

# Dynamic gesture classification using skeleton model on RGB-D data

**Y. Tamura[1], T. Umetani[1], N. Kashima[2] and H. Nakamura[3]**

[1] Department of Intelligence and Informatics, Konan University, Okamoto, Higashinada, Kobe, 658-8501, Japan

[2] Department of Energy Engineering and Science, Nagoya University, Frocho, Chikusa, Nagoya, 464-8603, Japan

[3] Department of Helical Plasma Research, National Institute for Fusion Science, Oroshi, Toki, 509-5292, Japan

E-mail: tamura@konan-u.ac.jp

**Abstract.** This study aims to subjectively detect and classify similar gestures using a red-green-blue-depth camera. Human gesture recognition is one of the crucial components for realizing natural user interfaces (NUIs) using computers and machines. The quality of the NUI highly depends on the robustness of the achieved gesture recognition. We, therefore, propose a gesture classification method using singular spectrum transformation. Using this method, we can robustly classify gestures and behavior.

## 1. Introduction

Dynamic gesture classification is an important and growing area in fields such as human interface, robotics, and virtual reality. One obstacle encountered in automatically detecting gestures is that the raw data include noise; second, there are some representations to show one's decision by gestures. To overcome these problems, we propose a gesture classification method using singular spectrum transformation (SST) [1].

This method has been applied to different fields, including earth sciences, physics, and financial research. In human interface research, this method has been applied to the segmentation of human motions [2]. We propose a method to classify gestures using similar factors involving the results of SST.

## 2. System configuration

Our system consists of a Microsoft Kinect sensor and a PC. The sensor has a red-green-blue and a depth sensor; such sensors detect the position of not only an object but also human joints. The Kinect sensor can detect three degree-of-freedom (DOF) positions of 20 joints approximately 30 times per second. We used only six joints for gesture detection: elbows, shoulders, and knees. This translates to 18 DOF position data being sent from the Kinect sensor to the PC. The key reason for using only six joints is to reduce the required processing time; furthermore, most joint positions (e.g., the spine) are not related to gestures.

## 3. Method

Figure 1 shows the procedure governing our proposed system.

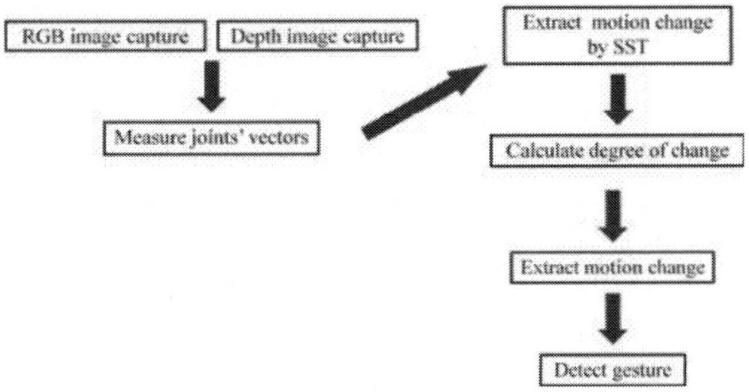


**Figure 1.** Procedure to detect gestures and behavior

In this study, we use SST for gesture classification. SST is a nonparametric decomposition method suitable for the decomposition of unsteady datasets. Consider one-dimensional time series dataset $x(1), x(2), x(3), \cdots, x(t)$. A vector $\mathbf{j}$ having $w$ components is defined as

$$\mathbf{j}(t-1) = [x(t-w) \;\; x(t-w+1) \cdots x(t-1)]^T. \tag{1}$$

The Hankel matrix (i.e., trajectory matrix) $\mathbf{H}(t)$ having n columns is described as

$$\mathbf{H}(t) = [\mathbf{j}(t-n) \;\; \mathbf{j}(t-n+1) \cdots \mathbf{j}(t-1)]. \tag{2}$$

This matrix can be decomposed using singular value decomposition as

$$\mathbf{H}(t) = \mathbf{U}(t)\mathbf{W}(t)\mathbf{V}(t)^T. \tag{3}$$

We apply this method to the time series of joint vector data in both the past dataset and the future dataset. The past dataset is composed of data from time $(t-n-w-1)$ to $(t-1)$, whereas the future dataset is composed of data from $(t+s)$ to $(t+s+w+n)$, where $s$ is the time shift and $wandn$ are parameters mentioned in eq. 1 and 2, respectively. Comparing results of both datasets, we can extract the motion change point.

For detecting motion change, we introduce "the degree of change" factor. A hyperplane $\mathbf{A_p}(t)$ is defined using n largest left-singular vectors $(\mathbf{u_{p1}}, \mathbf{u_{p2}}, \cdots \mathbf{u_{pn}})$ from past matrix $\mathbf{U_p}(t)$ by eq. (3). The largest left-singular vector $\mathbf{u_f}(t)$ is calculated from future matrix $\mathbf{U_f}(t)$. If the tendency of the past and the future dataset is similar, the inner product between vector $\mathbf{u_f}(t)$ and the vector, which is the projected $\mathbf{u_f}(t)$ to hyperplane $\mathbf{A_p}(t)$, is approximately 0. In the opposite case, the factor becomes 1. We define the degree of change by the following equations:

$$S(t) = 1 - \mathbf{u_f}(t)\mathbf{P}(t) \tag{4}$$

$$\mathbf{P}(t) = \mathbf{A_p}(t)\mathbf{A_p}(t)^T, \tag{5}$$

where all components of matrix $\mathbf{A_p}$ are orthonormal bases.

Next, we estimate similarity using the Hillinger distance between a control gesture and the other gesture as follows:

$$d(i,j)^2 = \int (\sqrt{S_i(t)} - \sqrt{S_j(t)})^2 dt. \tag{6}$$

| | |
|---|---|
| Case 1 | Right-hand gesture - Stand still |
| Case 2 | Right-hand gesture - Walk |
| Case 3 | Left-hand gesture - Stand still |
| Case 4 | Right-hand up - Stand still |
| Case 5 | Right-hand gesture - Stand still |

**Table 1.** Gestures for estimation

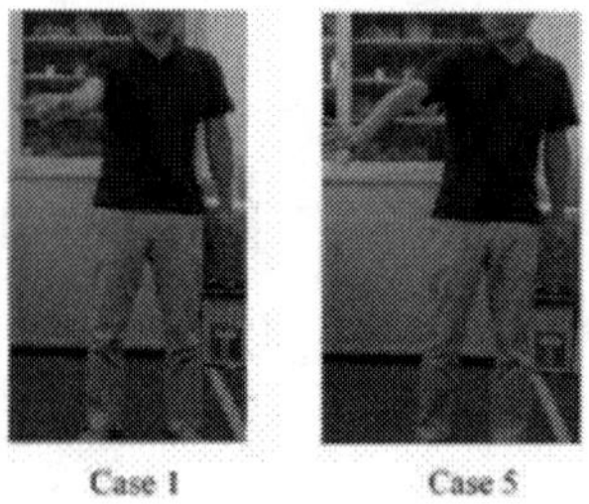


**Figure 2.** Gestures using the right hand and arm corresponding to cases 1 and 5 of Table 1

## 4. Results

Table 1 shows gestures for estimation of the proposed method. The user stands in front of the Kinect sensor approximately 3 m away. Except case 2, the user gestures while standing still; however, in case 2, the user gestures by walking. In cases 1 and 5, the user gestures to indicate the right direction; however, these gestures differ slightly, as shown in fig. 2. The parameters used in eq. 1 and 2 are $w = 30, n = 30$, and $s = -15$. Parameter $w$ needs to be set sufficiently long to include all gesture signals; therefore, $w$ is set to 30, indicating that the time range of the matrix is approximately 1 s. In this experiment, we used only $\mathbf{u_{p1}}, \mathbf{u_{p2}}$ to obtain hyperplane $\mathbf{A_p}$. Figure 3 shows results of case 1. The left-hand image shows raw data of the elbow joint direction, and the right-hand image shows the SST results. In this case, the user gestured three times, and results show the gestures successfully extracted from the raw data.

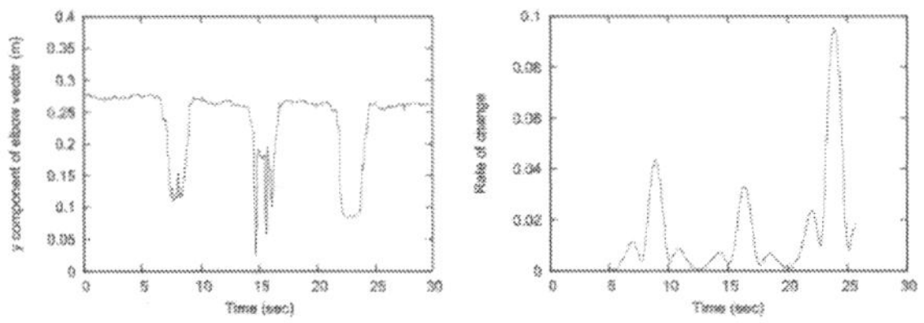


**Figure 3.** Successful extraction of meaningful gestures: on the left, the x component of raw data of the right elbow to right wrist; on the right, results of extraction using SST

Figure 4 shows results of case 2. In this case, the user gestured four times while walking. In general, it is difficult to extract distinct gesture signals and walking signals; however, our system successfully extracts the gesture signals. Moreover, as shown in the left-hand column of fig. 4, if this method is applied to the left-hand movement (i.e., no gesture), no signal is detected.

After extracting the gesture area, the next step is to detect what the extracted gesture denotes (i.e., classify the gesture). From eq. 6, we determine the gesture type. Figure 5 shows the classification results. The distance is calculated between the right-hand gesture (i.e., case 1 is the control gesture) and the other gestures. Results indicate that our method can successfully determine the joint, which has moved, and the gesture by the distance factor.

## 5. Conclusion

We applied SST to extract meaningful gestures and behavior from raw data obtained via a Microsoft Kinect system. The raw data are generally noisy, and therefore, it is difficult to extract useful data. SST removes continuous noisy data and extracts objective data with only a few specified parameters. Furthermore, we successfully classified the meaning of the gesture using the Hillinger distance.

**Acknowledgement**
This study is supported by KAKENHI (22500114).

## References

[1] J. B. Elsner and A. A. Tsonis 1996 *Plenum Press*
[2] H. Nakanishi, S. Kanata, H. Hattori, T. Sawaragi and Y. Horiguchi 2011 *Journal of Advanced Computational Intelligence and Intelligent Informatics* **15** 8

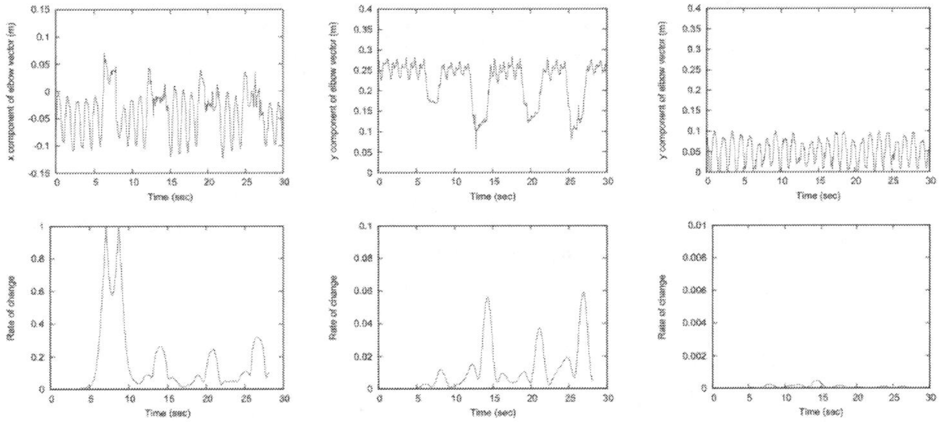


**Figure 4.** Results of extraction in case 2 (gesturing with walking) in which the upper half shows raw data, and the lower half shows SST results: the left-hand and middle columns present the x and y vector components, respectively, from the right elbow to the right wrist; the right-hand column presents the y vector component of the left hand

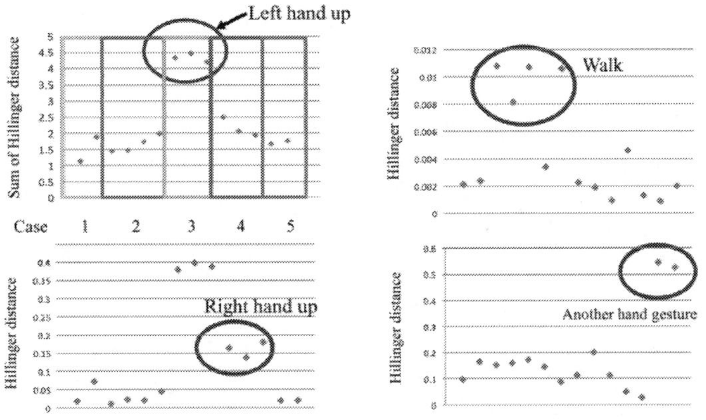


**Figure 5.** Classification results using the Hillinger distance: the upper-left graph shows the sum of all distances of all components (i.e., six joints); the upper-right graph shows the knee vector; the lower-left graph shows the right hand vector; and the lower-right graph shows the right shoulder vector

# Statistical analysis of road-vehicle-driver interaction as an enabler to designing behavioural models

**T Chakravarty[1], A Chowdhury[1], A Ghose[1] and C Bhaumik[1], P. Balamuralidhar[2]**
[1]TCS Innovation Labs, BIPL, Salt Lake, Sector-V, Kolkata, India
[2]TCS Innovation Labs, Abhilash, Plot-96, EPIP, Whitefield, Bangalore

E-mail: tapas.chakravarty@tcs.com

**Abstract.** Telematics form an important technology enabler for intelligent transportation systems. By deploying on-board diagnostic devices, the signatures of vehicle vibration along with its location and time are recorded. Detailed analyses of the collected signatures offer deep insights into the state of the objects under study. Towards that objective, we carried out experiments by deploying telematics device in one of the office bus that ferries employees to office and back. Data is being collected from 3-axis accelerometer, GPS, speed and the time for all the journeys. In this paper, we present initial results of the above exercise by applying statistical methods to derive information through systematic analysis of the data collected over four months. It is demonstrated that the higher order derivative of the measured Z axis acceleration samples display the properties Weibull distribution when the time axis is replaced by the amplitude of such processed acceleration data. Such an observation offers us a method to predict future behaviour where deviations from prediction are classified as context-based aberrations or progressive degradation of the system. In addition we capture the relationship between speed of the vehicle and median of the jerk energy samples using regression analysis. Such results offer an opportunity to develop a robust method to model road-vehicle interaction thereby enabling us to predict such like driving behaviour and condition based maintenance etc.

## 1. Introduction

Telematics is defined as a technology that relates to the collection of moving vehicle data and transporting these over telecom network to a central server for further analysis. Apart from a sensor enabled dedicated device, some emerging applications demand use of Smartphones also [1]. Insurers analyse these data in order to identify a given driving behaviour in terms of potential risk induced. The analysis must lead to quantitative figures that are used for underwriting purpose. On a different context, fleet managers may use such telematics application to identify safe driving behaviour on daily basis; in fact, fleet managers may like to identify aberrations (on a particular trip) with respect to a baseline behaviour where the baseline behaviour is classified based upon a-priori knowledge; often captured as part of blind profiling. Analysis of the data being collected on daily basis leads to the ability to create predictive models about the three-way interaction between road, vehicle and driver. This paper presents some initial results of such exercise being carried out on an office bus. In the following sections, the test setup is explained along with the observations and computed inference.

[1] To whom any correspondence should be addressed.

## 2. Experiment and observation

In this section we discuss the setup and method used for acquisition and analysis of the physical signatures related to the road-vehicle interaction. Towards such experimentation, we deployed 'ConnectPort® X5 R' (a telematics gateway device for Vehicle Area Network from M/s Digi International Inc.) in the office bus meant for the associates transporting to TCS Siruseri (India) campus. This device collects location (GPS), speed, time and accelerometer data. A python script is implemented to collect the 3-axis raw acceleration data, pre-process it before sending the processed data to a device cloud over cellular network. The bus travels 5-days a week and a large data set has been collected over last four months. It is also to be noticed that the route and the timings remain identical for the given bus.

### *2.1 Data pre-processing*

The typical trip duration is approximately 90 minutes. To capture meaningful signatures of the road-vehicle interaction, it is needed to measure the vibrations at sampling rate of 20Hz or higher. Thus the raw data size for a single trip alone becomes significantly large thereby resulting in high cellular charges. To mitigate this issue, we pre-processed the raw data by computing jerk energy (JE) for every 20 acceleration samples [2] and sending only the JE values to device cloud.

Let us assume that $a_1$, $a_2$, .$a_n$ be the consecutive discrete acceleration samples at time $t_1$, $t_2$,..$t_n$ where $\Delta t = t_n - t_{n-1}$ for uniform sampling rate. Then jerk, i.e. rate of change of acceleration (m/sec$^3$) is defined as:

$$J_i = {a_{i+1} - a_i}/{\Delta t} \quad \forall 1 \le i \le n-1 \tag{1}$$

Thus, 'jerk energy' is $JE_s = J_{s1}^2 + J_{s2}^2 + \ldots . J_{s19}^2$ (2)

where, 's' represents a time-window of 1 sec. The JE computation takes place inside the telematics device itself. Subsequently, the computed JE values are uploaded. These JE data are being collected on regular basis. Further analysis is carried out for trip level data set as well as aggregated (over week & month) data; in order to identify any behavioral patterns that are unique to the given interaction between the three entities namely rough road, vehicle and the driver. In our analysis, we utilized the statistical tool called MINE (maximal information-based nonparametric exploration) [3,4] to evaluate relationship between the known parameters and measured variables like JE. In the tool MINE, 'maximal information coefficient (MIC)' is a measure of 'two variable dependence'. Its value ranges between 0 (no dependency) and 1(dependent). In the collected data, we have 19 JE values that correspond to one location and one speed attribute. Intuitively it is suspected that the jerk energies corresponding to vertical vibration (+Z axis) and the average speed must be related. Statistical measures like mean, median and maximum of 19 jerk energy samples are calculated to get one representation signature in order to pair with the corresponding average speed. It was found that value of maximum information coefficient (MIC) was highest for the pair of 'median JE' and 'average speed'. A representative result is presented in table 1.

**Table 1**: Measure of relationship between average speed (X) and JE (Y) for 4th April, 2013

| X variable | Y variable | MIC (strength) | MIC-p^2 (nonlinearity) | MAS (non-monotonicity) |
|---|---|---|---|---|
| speed | Median of JE | 0.74 | 0.16 | 0.06 |

The MIC value leads us to establish functional relationship between speed and JE. Based on high MIC score for the collected real data, we fitted regression equation for average speed and JE. Figure 1 is an example scatter plot for jerk energy vs average speed of one trip. Included in the scatter plot, is the fitted curve representing the mathematical relationship between the above two variables. From figure 1, it is clear that even at very low speeds, the measured JE is about 200 whereas the system noise ( in terms of JE) in the telematics device is measured to be approximately 50. Thus, there exists an offset value in JE that is entirely a particular signature of the vehicle itself (boarders alighting is one possible cause). This result is used later, when we attempt to measure vehicle degradation. In figure 1, we observe multiple JE values for any small band of speed variation. This is attributed to different road roughness as well as driving behaviour. After deriving suitable regression lines for each day of the three months; April to June, we derived a monthly equation of predicting JE based on the values of vehicle speed. We found out that most of the fitted regression lines were quadratic in nature. After collecting all the coefficients (a, b, c), we took Trimmed Mean for each of them to obtain the monthly coefficients. This is given as:

$$v = 238.53 - 4.42 * u + 0.56 * u^2 \quad \& \hat{\sigma} = 151.1 \qquad (3)$$

*2.2 Hazard rate*

Analysis of JE data shows that ' x = 1/ JE' follows Weibull distribution with shape parameter k >1. From the observed data, we can derive the scale parameter λ and shape parameter **k**. Additional observations on different vehicles inform us that these two parameters vary primarily on driving behavior and are weakly-persistent on vehicle condition. For Weibull distribution the failure rate *h(x)* (or hazard rate) is given as

$$h(x) = \frac{k}{\lambda}\left(\frac{x}{\lambda}\right)^{k-1} \qquad (4)$$

From the acquired data acquired for a given trip, k and λ and h(x) are computed. Traditionally, Weibull distribution is used to measure mean-time-to-failure. But we have replaced time't' by 'x' = inverse of jerk energy; thus h(x) represents the probability of the vehicle system failing when subjected to a small jerk, having survived large initial 'jerks'. We characterize the above described vehicle at x= 1/300 where 300 is the nominal JE value for a vehicle at low speeds from the aboveregression model. Figure 2 shows the variation of scale & shape parameters from measured +Z axis accelerometer data on consecutive days

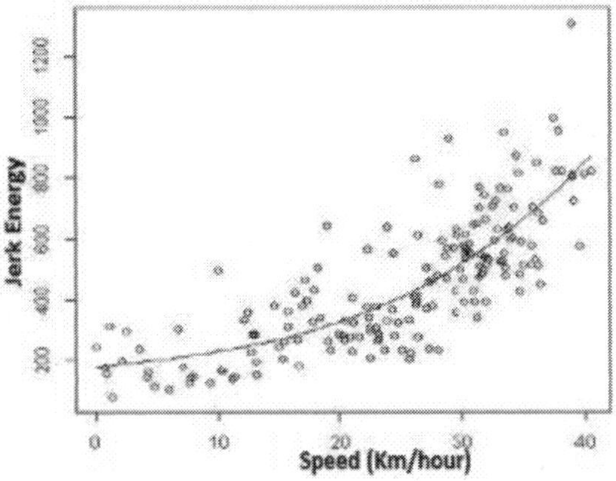


**Figure 1**: Regression equation between u (speed in Km/hr) and JE ($m^2/sec^6$)

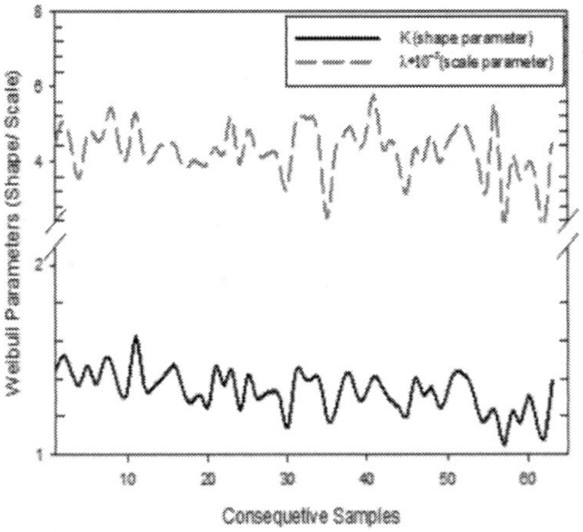


**Figure 2**: Variation of Weibull parameters computed from 3 months data

### 3. Results

The setup described as above is used to calculate hazard rate (HR) for each trip: journey to office and back. In figure 3, the measured hazard rate for the bus is presented for two months. The mean HR (of monthly samples) & the upper/lower variability levels can potentially display a degradation of the suspension-damper system of the vehicle. In the previous work, Chakravarty et al [5] presented the method to model complex road-vehicle interaction. The given method is also used to simulate HR for a large vehicle when interacting with different road types; from very good to bad. The results are displayed in figure 4. Here, HR values are computed for speeds of 20km/hr and 30km/hr. These results are indicative only. When HR is measured for a real drive, such computed HR can be compared with these indications to estimate the road type. It is known that the ISO proposed geometric mean of spectral densities for different road grades increases exponentially with road type [6]. From the values of hazard function it is clear that road roughness attribute is correctly captured by hazard function. Now, higher HR value also denotes vehicle suspension degradation. When a vehicle continues to traverse same road on a daily basis, an out-of-range value of HR can be directly associated with the nature of driving for that trip. Similarly, it is possible to classify potentially hazardous driving by comparing the measured HR with the peers on the same road.

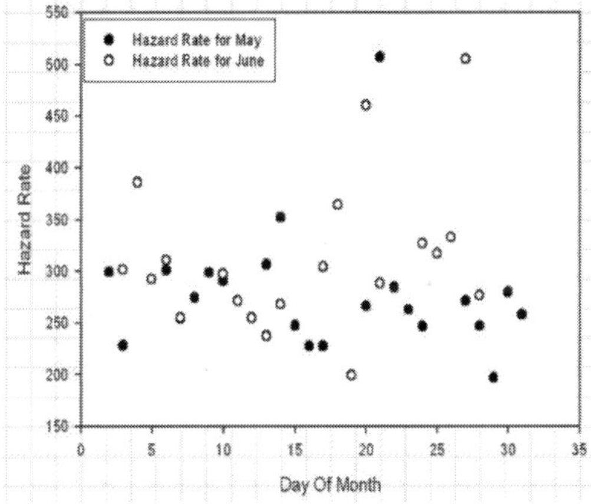


**Figure 3**: Computed hazard rate for the bus for two months

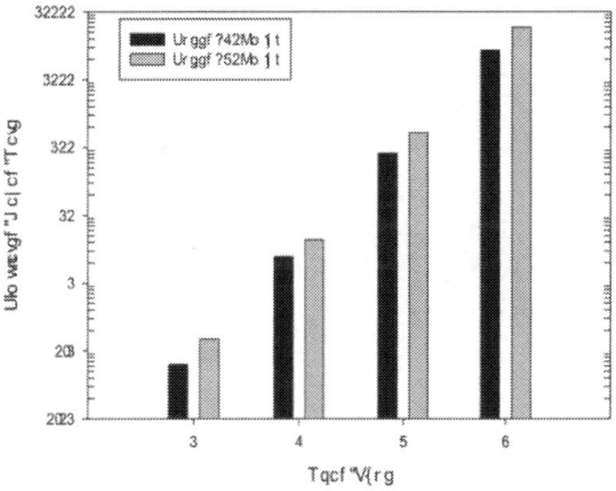

**Figure 4**: Simulated hazard rate for interaction of large car with different road types. (1- very good road, 2- good road, 3-average road, 4- bad road)

## 5. Concluding remarks

In this paper, methods of analysing the physical signatures obtained from complex road-vehicle-driver interaction are presented. It is shown that the vertical vibration signatures of a moving vehicle offer strong insights into the nature of such interaction. It is possible to build predictive models of such interaction so that behavioural aberrations can be identified even for a smaller data sample

## Reference

[1] Ghose A, Biswas P, Bhaumik C, Sharma M, Pal A, Jha A 2012 *IEEE Int. Conf. on Pervasive Computing and Communications Workshops (PERCOM Workshops)*, vol., no., pp.489,491

[2] Bilal S, Pal A, Srinivasarengan K, Balamuralidhar P, 2012 *IEEE 6th Int. Conf. on Sensing Technology (ICST)*, Kolkata p 8-12

[3] http://www.exploredata.net/

[4] http://www.sciencemag.org/content/suppl/2011/12/14/334.6062.1518.DC1

[5] Chakravarty T, Srinivasarengan K , Roy S, Bilal S, Balamuralidhar P, 2013 *IOP Journal of Physics: Conference Series, Vol. 410(2013) 012065*

[6] Wong J, 2001 *Theory of Ground Vehicles* (Wiley-Interscience)

# $\alpha$-decay half-life calculations of superheavy nuclei using artificial neural networks

**Tuncay Bayram**[1,*], **Serkan Akkoyun**[2], **S. Okan Kara**[3]

[1] Department of Physics, Sinop University, Sinop, Turkey
[2] Vocational School of Health Services, Cumhuriyet University, Sivas, Turkey
[3] Bor Vocational School, Nigde University, Nigde, Turkey

E-mail: *tbayram@sinop.edu.tr

**Abstract.** Investigations of superheavy elements (SHE) have received much attention in the last two decades, due to the successful syntheses of SHE. In particular, $\alpha$-decay of SHEs has a great importance because most synthesized SHE have $\alpha$-decay and the experimentalists have evaluated the theoretical predictions of the $\alpha$-decay half-life during the experimental design. Because of this, the correct prediction of $\alpha$-decay half-life is important to investigate superheavy nuclei as well as heavy nuclei. In this work, artificial neural networks (ANN) have been employed on experimental $\alpha$-decay half-lives of superheavy nuclei. Statistical modeling of $\alpha$-decay half-life of superheavy nuclei have been found as to be successful.

## 1. Introduction

One of the hottest research subjects in nuclear physics is superheavy nuclei beyond Fermium ($Z = 100$). Better understanding of structures and properties of superheavy nuclei can provide quite significant knowledge for the area of existing nuclei and chemical elements. Furthermore, this can make a change in our conception of the material world's boundaries [1]. Many nuclear research centers have started preparing experiments for the synthesising SHE since 1970. Due to the successful synthesizing of SHEs in the laboratory, some of them has been stimulated, such as $Z = 107-112$ at GSI [2, 3, 4], $Z = 110-113$ at RIKEN [5, 6] and $Z = 113-118$ at Dubna [7, 8, 9]. From theoretical side, some properties of superheavy nuclei, such as binding energy, $\alpha$-decay energy and $\alpha$-decay half-lives have been intensively studied by using various nuclear models, such as the macroscopic-microscopic models, Skyrme-Hartree-Fock and relativistic mean field (RMF) models. The authors refer to read [10, 11] and references therein. In particular, $\alpha$-decay half-lives of SHE are important because most synthesized SHE have $\alpha$-decay. The theory of $\alpha$-decay was firstly interpreted as a consequence of quantum penetration of $\alpha$ particle by Gamow [12] in 1928. Since then various models for description of $\alpha$-decay have been proposed. On the other hand some simple semi-empirical formulas have been introduced for the estimations of the $\alpha$-decay half-lives because experimentalists evaluate the values of the $\alpha$-decay half-life during the experimental design (See [13] and references therein). It should be noted, however, that phenomenological formulae cannot replace theoretical studies of $\alpha$-decay which provide a microscopic description of $\alpha$-decay process.

In recent years, artificial neural networks (ANN) have been employed in nuclear physics research ([14] and references therein) as in the many field of sciences. Discriminating between neutron and gamma-rays [15], developing nuclear mass systematic [16, 17] and obtaining nuclear charge radii [18] have been done by using ANN. In this work, feed-forward ANN have been used in order to obtain $\alpha$-decay half-lives of superheavy nuclei. The fundamental task of ANN is to give outputs through computation of the inputs. The method does not need any relationship between input and output data. The main purpose of the present study is to show success of ANN in describing of the unknown $\alpha$-decay half-lives of superheavy nuclei.

## 2. Artificial Neural Networks (ANN)

ANN are nonlinear, nonalgorithmic and parallel processing systems that mimic the human brain functionality. They consist of several processing units called neurons which have adaptive synaptic weights [19]. ANN are known as very powerful tools that are used when standard techniques fail to estimate the correlation between the variables. ANN have several advantages, such as requiring less formal statistical training, ability to detect complex highly non-linear relationships between input and output variables and ability to detect all possible interactions between predictor variables.

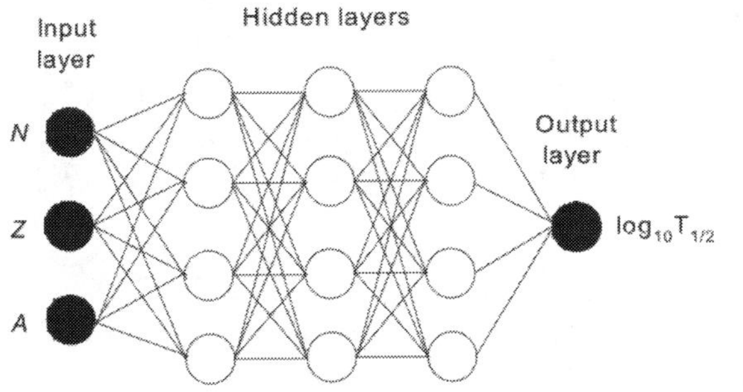


**Figure 1.** Used ANN topology (3-4-4-4-1) in the present study.

The layered feed-forward ANN consist of input, hidden and output layers. In this study, one input layer with three neurons ($p = 3$), three hidden layers with four ($h$) neurons in each and one output layer with one neuron ($r = 1$) ANN topology has been used for accurate prediction of the $\alpha$-decay half-lives of superheavy nuclei (Figure 1). The number of total adjustable weights, which can be calculated by $p \times h_1 + h_1 \times h_2 + ... + h_i \times r$ (i=1 to maximum number of hidden layers), has been 48 with no bias. Three ANN inputs are neutron $N$, proton $Z$ and mass $A$ numbers and the desired output is logarithm of the $\alpha$-decay half-lives ($\log T_{1/2}$) for superheavy nuclei. The neurons in the input layer collect the data from environment and transmits via weighted connections to the neurons of the hidden layers which are needed to approximate any non-linear function. In this work, the chosen type of hidden neuron activation function is hyperbolic tangent. In the present study, neural network software NeuroSolutions v6.02 has been used.

ANN are composed of two stages, training and test stages. In the training stage in which Levenberg-Marquardt back-propagation algorithm was used, the weights are adjusted to construct ANN. Because of this, these weights play a key role for solving the problem considered. This stage continues until the acceptable error level which is calculated by the difference between desired and ANN outputs. After the construction, ANN have been tested on the data in the test stage. The data used in this stage has never been seen by ANN before. In this work, all data has been partitioned into two separate sets: one for the supervised training of ANN and the rest (belonging to the Z=107 nuclei) for the test.

## 3. Results and discussions

In the present study, the available experimental data for $\alpha$-decay half-lives of 154 superheavy nuclei starting from $Z = 100$ to 118 has been obtained from [7, 8, 9, 20, 21]. These values except those of Bh ($Z = 107$) nuclei have been used as output of ANN in units of second while related proton $Z$, neutron $N$ and mass $A$ numbers have been used as inputs. However, ANN have been found as unsuccessful for understanding connections between these variables. The $\alpha$-decay half-lives $T_{1/2}$ performed in the Viola-Seaborg phenomenological formula [22] is given by

$$\log T_{1/2}(Z, N) = (aZ + b)Q^{1/2} + cZ + d + h_{log} \tag{1}$$

where $Z$ is proton number, $N$ is neutron number and $Q$ is the $\alpha$-decay energy of a parent nucleus. In this equation, $T_{1/2}$ and $Q$ are in units of second and MeV, respectively. The quantities $a$, $b$, $c$, $d$ and $h_{log}$ are adjustable parameters. By considering equation (1), we have used the logarithm of the experimental $T_{1/2}$ values as output of ANN instead of the $T_{1/2}$ values. In this case, ANN has been found as to be successful in detecting relationships between inputs ($Z$, $N$ and $A$) and the output ($\log T_{1/2}$). In Figure 2, ANN training results (open symbols) for the logarithm of the $T_{1/2}$ values for some superheavy nuclei ($Z = 100, 103, 105, 108, 110, 114$ and 117) are shown as in comparison with experimental values (filled symbols). The data error sizes are smaller than those of symbols. The maximum deviation between ANN training results and experimental ones is 1.001. The root mean square error (RMSE) value of ANN training results is 0.636 for 147 nuclei. As can be seen in Figure 2, ANN training results are close to the available experimental data for the $\log T_{1/2}$ values of superheavy nuclei. Also, ANN can be found as to be successful in understanding the general tendency of experimental data points for isotopic chains of superheavy nuclei.

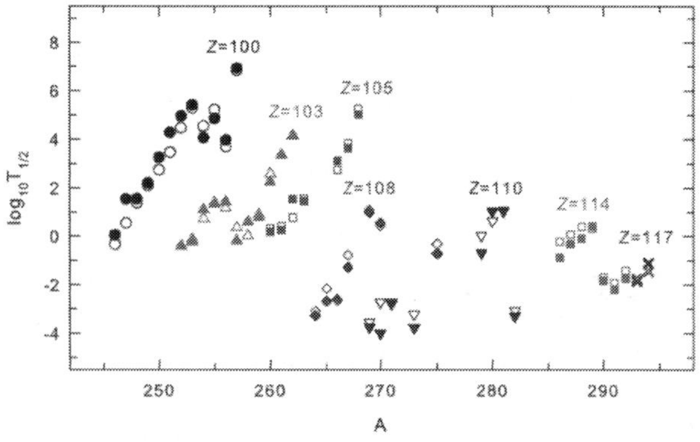


**Figure 2.** The logarithm of $T_{1/2}$ values for some superheavy nuclei as a function of mass number $A$. The experimental values [7, 9, 20, 21] are represented by filled symbols while ANN training results are denoted by open symbols. Units of the half-lives are in second.

Properties of superheavy nuclei, such as binding energy, $\alpha$-decay energy and $\alpha$-decay half-life are strongly influenced by the shell effects. Various theoretical models predict different proton and neutron magic numbers in superheavy region. The macroscopic-microscopic models predict that $^{270}Hs_{162}$ ($Z = 108$) is a doubly magic nucleus and this prediction has been favored by experiments ([10] and references therein). One can expect that the fission barrier near this region is high because of the large shell correction energy. This leads to the fact that the spontaneous fission half-life will be increased. This indication is clearly visible in Figure 2. Generally, ANN results for the $\log T_{1/2}$ values for an isotopic chain of superheavy nuclei is increasing around neutron number $N = 162$ as in agreement with the experimental $\log T_{1/2}$ values.

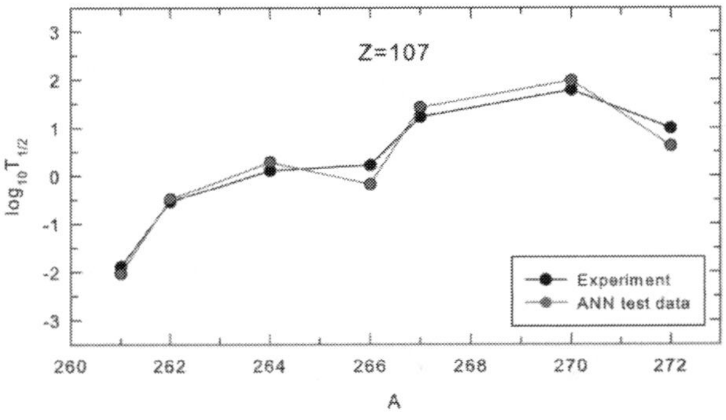


**Figure 3.** ANN test results for the $\log T_{1/2}$ of superheavy Bh ($Z = 107$) nuclei in comparison with the available experimental data. Units are in second and the experimental data error size are smaller than those of symbols.

After training of ANN, we performed the test of ANN. The training had been practised on the data without taking the experimental $\log T_{1/2}$ values of Bh ($Z = 107$) isotopes into account. Later, we have obtained ANN results for these isotopes which have never been seen by ANN. In Figure 3, ANN test results for the $\log T_{1/2}$ values of superheavy Bh isotopes are shown as in comparison with the available experimental data [8, 20, 21]. The maximum difference between the logarithm of the experimental and ANN $T_{1/2}$ values is seen at mass number $A = 266$ and its value is 0.399. The RMSE value of ANN test results is 0.246. As can be seen in Figure 3, ANN results are close to the experimental data points. This indicates that ANN has predictive power for the $\log T_{1/2}$ of superheavy nuclei. On the other hand, ANN method cannot be replaced by microscopic or macroscopic-microscopic approaches which provide deep descriptions of $\alpha$-decay process. It should be noted, however, that ANN gives a practical value, for an easy prediction of not yet measured values of $T_{1/2}$.

## 4. Conclusions

ANN has been employed to investigate $T_{1/2}$ of superheavy nuclei starting from $Z = 100$ to 118. ANN results have been found to be in agreement with the experimental data. ANN method has also been performed for estimating $T_{1/2}$ of Bh ($Z = 107$) isotopes without introducing their original data to the network. The results show that in the region $Z = 100 - 118$, ANN is a practical tool for obtaining unmeasured $T_{1/2}$ values and it can be useful for describing systematics of $\alpha$-decay half-lives.

## References

[1] Oganessian Y 2013 *Nuclear Physics News* **23** 15
[2] Hofmann S 1998 *Rep. Prog. Phys.* **61** 639
[3] Hofmann S and Münzenberg 2000 *Rev. Mod. Phys.* **72** 733
[4] Hofmann S *et al* 2004 *Nucl. Phys.* A **734** 93
[5] Morita K *et al* 2004 *J. Phys. Soc. Jpn.* **73** 2593
[6] Morita K *et al* 2007 *J. Phys. Soc. Jpn.* **76** 045001
[7] Oganessian Y *et al* 2006 *Phys. Rev.* C **74** 044602
[8] Oganessian Y 2007 *J. Phys. G: Nucl. Part. Phys.* **34** R165
[9] Oganessian Y *et al* 2010 *Phys. Rev. Lett* **104** 142502
[10] Dong T and Ren Z 2010 *Phys. Rev.* C **82** 034320
[11] Bayram T and Yilmaz A H 2013 *Mod. Phys. Lett.* A **28** 1350068
[12] Gamow G 1928 *Z. Phys* A **51** 204
[13] Zhang Di-Da *et al* 2010 *Phys. Rev.* C **81** 044319
[14] Medhat M E 2012 *Ann. Nucl. Energy* **45** 73
[15] Akkoyun S 2013 *Ann. Nucl. Energy* **55** 297
[16] Athanassopoulos S, Mavrommatis E, Gernoth K A and Clark J W 2004 *Nucl. Phys.* A **743** 222

[17] Bayram T, Akkoyun S and Kara S O 2014 *Ann. Nucl. Energy* **63** 172
[18] Akkoyun S, Bayram T, Kara S O and Sinan A 2013 . *J. Phys.* G **40** 055106
[19] Haykin S 1999 *Neural Networks: A Comprehensive Foundation* (International: Prentice-Hall)
[20] Audi G, Bersillon O, Blachot J and Wapstra A H 2003 *Nucl. Phys.* A **729** 3
[21] Gupta M and Burrows T W 2005 *Nucl. Data Sheets* **106** 251
[22] Viola V E and Seaborg G T 1966 *J. Inorg. Nucl. Chem.* **28** 741

# A study on structural evolution of $^{142-164}$Nd isotopes

**Seyit O. Kara**[1,*], **Tuncay Bayram**[2], **Serkan Akkoyun**[3]

[1] Bor Vocational School, Nigde University, Nigde, Turkey
[2] Department of Physics, Sinop University, Sinop, Turkey
[3] Vocational School of Health Services, Cumhuriyet University, Sivas, Turkey

E-mail: *seyitokankara@gmail.com

**Abstract.** Constrained Hartree-Fock-Bogoliubov theory with SLy5 Skyrme force has been applied for even-even $^{142-164}$Nd isotopes to investigate the structural evolution of Nd isotopic chain. In this work, ground-state energies and charge radii of Nd isotopes have been carried out as in good agreement with the experimental data. The systematic investigation of ground-state shape evolution between spherical U(5) and axially deformed SU(3) for $^{142-164}$Nd has been studied by using potential energy curves.

## 1. Introduction

Nuclear models can describe main features of nuclei along the nuclear chart while a full microscopic description of the nuclear many-body system is often needed to describe the full variety of nuclear behaviours. In particular, the size and shape of nuclei are very important to describe their behaviours and different regions of nuclidic chart can be characterized by using macroscopic concept of nuclear shape (spherical nuclei, axially deformed nuclei, gamma unstable nuclei, etc). Nuclei can exhibit a change of their shapes along the isotopic or isotonic chains based on changes of their neutron or proton numbers. This refers to Quantum Phase Transition (QPT) in shape of nuclei. The Interacting Boson Model (IBM) [1] has three phases, with the U(5), SU(3) and O(6) dynamical symmetries corresponding to the breaking of U(6) into its sub-algebras. U(5), SU(3) and O(6) symmetries are associated with spherical, axially deformed and $\gamma$-unstable shapes of nuclei, respectively. More recently, Iachello has proposed the critical-point symmetries X(5) and E(5) which corresponds to the first order [2] and second order [3] phase transition, respectively. The symmetry X(5) corresponds to the critical-point between the U(5) and SU(3) dynamical symmetry limits while the E(5) symmetry characterizes the phase transition region between the U(5) and O(6) dynamical symmetries. Experimental evidences of the X(5) and E(5) critical-point symmetries have been found in the spectrum of $^{152}$Sm [4] and $^{134}$Ba [5], respectively. Because of these progresses, theoretical and experimental investigations of nuclei with possible critical-point symmetry have been a hot topic in recent decade.

Bohr collective Hamiltonian and the IBM are used for investigating shape phase transitions in nuclei. On the other hand, Hartree-Fock-Bogoliubov (HFB) method [6, 7] and relativistic mean field (RMF) model [8, 9] successfully have been applied to search possible critical-point nuclei [10, 11, 12, 13]. Generally, structural evolution of an isotopic chain has been used for determining possible critical-point nuclei by using calculated potential energy curves (PECs) of nuclei in these studies. Many nuclei have been pointed out as possible critical-point nuclei with E(5) and X(5) symmetry within the framework of RMF theory and HFB method. A brief summary on studies for determining of possible critical-point nuclei within the mean field models can be found in [11, 13] and references therein.

Various quantities such as two-neutron separation energies $S_{2n}$, various electric quadrupole transition rates (B(E2;$2_1^+ \rightarrow 0_1^+$) and B(E0;$0_2^+ \rightarrow 0_1^+$)), isomer shifts, isotope shift and the characteristic ratio $R_{4/2} = E(4_1^+)/E(2_1^+)$ of nuclei are used to search critical-point nuclei. The $N = 90$ isotones of Nd, Sm, Gd and Dy are associated with X(5) symmetry. After the first experimental identification for X(5) behavior of $^{152}$Sm, $^{150}$Nd was identified as exhibiting X(5) behavior [14]. Additionally, $^{148}$Nd has been suggested as an example of nucleus with X(5) critical-point symmetry via an analytic solution using $\beta^2$ potential [15]. Furthermore, the PECs of Nd isotopic chain has been carried out within the framework of axially deformed RMF theory [10]. In this study, authors have concluded that the PECs for nuclei suggested as good examples of X(5) exhibit a bump by using chains of isotopes involving nuclei suggested to be good examples of the X(5) critical-point symmetry. In the present study, the ground-state properties of even-even $^{142-164}$Nd isotopes have been carried out by using HFB method. The structural evolution of $^{142-164}$Nd isotopes has been discussed by examining their PECs obtained by imposing a constraint on the mass quadrupole moment for Nd isotopes. In addition, some signatures of X(5) symmetry such as two-proton separation energies, the differences between squares of ground-state charge radii and proton-neutron interaction strength $\delta V_{pn}$ of $^{142-164}$Nd isotopes have been investigated.

## 2. HFB method and details of calculations

In HFB method, a two-body Hamiltonian of a system of fermions in terms of a set of annihilation and creation operators ($c$, $c^\dagger$) has the form:

$$H = \sum_{n_1 n_2} e_{n_1 n_2} c_{n_1}^\dagger c_{n_2} + \frac{1}{4} \sum_{n_1 n_2 n_3 n_4} \bar{v}_{n_1 n_2 n_3 n_4} c_{n_1}^\dagger c_{n_2}^\dagger c_{n_4} c_{n_3}\,, \tag{1}$$

where $\bar{v}_{n_1 n_2 n_3 n_4} = \langle n_1 n_2 \mid V \mid n_3 n_4 - n_4 n_3 \rangle$ are anti-symmetrized matrix elements of the two-body effective nucleon-nucleon interaction. In terms of Skyrme forces, the total Hamiltonian is the sum of the mean-field and pairing energy densities. In the present study, a parametric form of total HFB energy with the Skyrme force SLy5 [16] has been used as in Ref. [17]. Tensor effect has been taken into account. The code HFBTHO (v1.66p) [17] has been used to reproduce ground-state properties of even−even $^{142-164}$Nd isotopes which utilizes an axially symmetric deformed harmonic oscillator potential. The PECs of even−even isotopic chain of Nd have been obtained by utilizing quadrupole moment constraint as prescribed in Ref. [17]. The number of oscillator shells taken into account was 16 and good convergence were obtained in the calculations.

## 3. Results and discussions

Several nuclear models which can be mainly classified into three categories (macroscopic, macroscopic-microscopic and microscopic models) have been successfully employed to carry out masses of nuclei through the nuclear chart. A brief summary on the studies for mass predictions with nuclear models can be found in [18] and references therein. As a microscopic model, HFB method with Skyrme type forces [16] can be applied in the whole mass region along the nuclear chart to reproduce ground-state energies, sizes and deformations of nuclei [19, 20]. Calculated ground-state binding energies (BEs) of even-even $^{142-164}$Nd obtained from this work are listed in Table 1. The predictions of Finite Range Droplet Model (FRDM) [21] and RMF model with NL3* functional [18] and experimental data [22] are also listed for comparison. As can be understood from Table 1, the predictions of FRDM whose parameters are fitted by using many experimental masses of nuclei are closer to the experimental data rather than those of HFB and RMF. On the other hand, it can be say that HFB and RMF predictions for BEs of Nd isotopes are reliable. Two-neutron separation energy $S_{2n}$ which can be derived easily from the relation

**Table 1.** Ground-state BEs of $^{142-164}$Nd isotopes (in units of MeV) obtained from this work by using HFB method with SLy5 Skyrme force. The predictions of FRDM and RMF model with NL3* functional and experimental data are also listed for comparison.

| Nuclei | FRDM [21] | RMF [18] | This work | Experiment [22] |
|---|---|---|---|---|
| $^{142}$Nd | 1185.67 | 1188.83 | 1182.84 | 1185.14 |
| $^{144}$Nd | 1199.16 | 1200.97 | 1194.26 | 1199.01 |
| $^{146}$Nd | 1212.00 | 1210.69 | 1207.42 | 1212.40 |
| $^{148}$Nd | 1225.33 | 1223.38 | 1220.66 | 1225.03 |
| $^{150}$Nd | 1238.46 | 1235.15 | 1233.24 | 1237.45 |
| $^{152}$Nd | 1250.93 | 1247.88 | 1246.17 | 1250.06 |
| $^{154}$Nd | 1262.45 | 1258.96 | 1257.24 | 1261.74 |
| $^{156}$Nd | 1273.02 | 1269.10 | 1267.17 | 1272.71 |
| $^{158}$Nd | 1282.88 | 1278.82 | 1276.88 | 1282.80 |
| $^{160}$Nd | 1292.14 | 1287.89 | 1286.48 | 1291.84 |
| $^{162}$Nd | 1300.81 | 1295.60 | 1293.72 | |
| $^{164}$Nd | 1308.31 | 1302.27 | 1300.56 | |

$$S_{2n} = BE(N, Z) - BE(N - 2, Z) \quad (2)$$

is important quantity to expose the shell structure of nuclei. In this equation (2), BE is the total binding energy of the nuclei with $N$ neutrons and $Z$ protons. Figure 1 shows the calculated $S_{2n}$ values of even-even $^{142-164}$Nd isotopes obtained by using HFB method with the SLy5 parameters. The available experimental data and the predictions of FRDM and RMF model are also shown for comparison. As can be seen in Figure 1, except the neutron number $N = 84$, our calculations and the predictions of RMF model are in good agreement with the experimental data. An abrupt decrease in $S_{2n}$ at magic neutron number $N = 82$ corresponds to the closed shell.

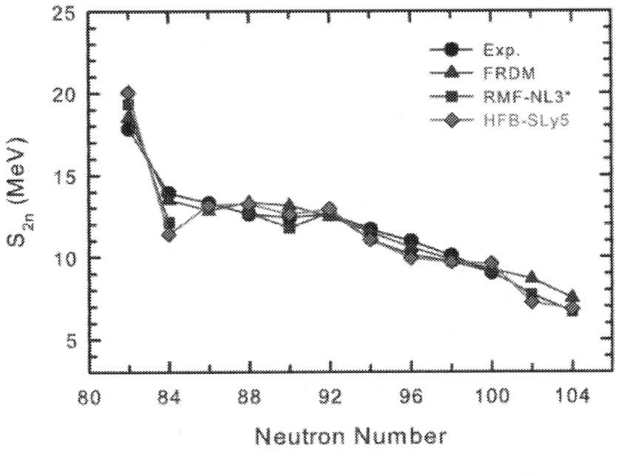


**Figure 1.** Comparison of the calculated two-neutron separation energies ($S_{2n}$) of $^{142-164}$Nd with the RMF model, FRDM and experimental results.

The concept of critical point symmetry includes analytic expressions for observables such as transition rates B(E2) and excitation energies. These observables of a nucleus have to be calculated to indicate that it is a critical-point nucleus quantitatively and this is not possible on the mean-field level. On the other hand, relativistic and non-relativistic mean field models have been employed in calculations of potential energy curves (PECs) as functions of the quadrupole deformation parameter $\beta$ for isotopic chains in which the occurrence of critical point symmetries had been predicted [10, 12]. The resulting PECs display shape transitions from spherical to deformed configurations. In these studies relatively flat PECs over an extended range of the deformation parameter have been obtained in nuclei showing the E(5) symmetry, while in nuclei showing the X(5) symmetry, PECs with a bump have been found. It should be noted, however, that these assumptions are reliable qualitatively.

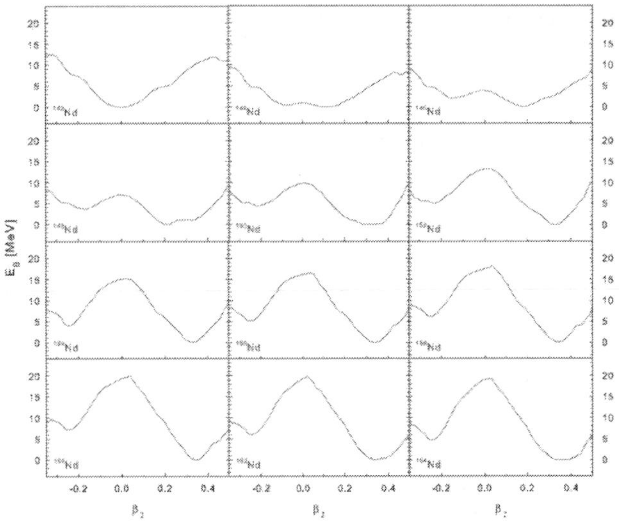


**Figure 2.** The PECs for $^{142-164}$Nd obtained from HFB method with the Skyrme force SLy5.

In Figure 2, the PECs of even-even $^{142-164}$Nd as a function of quadrupole moment deformation parameter $\beta$ obtained from HFB calculations with the Skyrme force SLy5 are shown. The PECs as function of $\beta$ deformation has been obtained by imposing a constraint on the mass quadrupole moment for Nd nuclei. The particle number is only approximately restored by the Lipkin-Nogami procedure. The BE of each Nd isotope for the ground-state is taken as reference in these PECs. As can be clearly seen from Figure 2, the PECs exhibit a gradual transition between the spherical $^{142}$Nd and well deformed prolate $^{164}$Nd. In particular, a wide flat minimum on the prolate side and additional 4 MeV excitation energy on the oblate side of the PEC of $^{150}$Nd are found and the potential barrier between two minima is seen as being approximately 10 MeV. Because of the assumption of Ref. [10] that the PECs found for nuclei suggested as good examples of X(5) exhibit a bump, $^{164}$Nd could be indicated as a possible critical-point nucleus with X(5) symmetry. Furthermore, it should be noted that the similarity between the projections on the prolate and oblate axes of the original X(5) potential considered by Iachello [2] and the calculated PEC of $^{150}$Nd shown in Figure 2 is notable.

In the present study, the differences between squares of ground-state charge radii of Nd isotopes with neutron number $N$ and those of the reference nucleus ($N = 82$) have been calculated for better understanding the structural evolution of Nd isotopes with increasing neutron number. Calculations obtained from HFB method with SLy5 parameters and available experimental data [23] are shown in Figure 3. As can be seen in Figure 3, after three steps rising in neutron numbers $N \leq 88$ an abrupt change starts at $N \leq 90$ ($A = 150$) and then the nuclei remain in deformed shape. This is a signature of phase-transitional behavior of $^{150}$Nd.

In addition, first order phase transitional behavior in the equilibrium deformation has further highlighted the key role of the proton-neutron (p-n) interaction. Direct correlation between

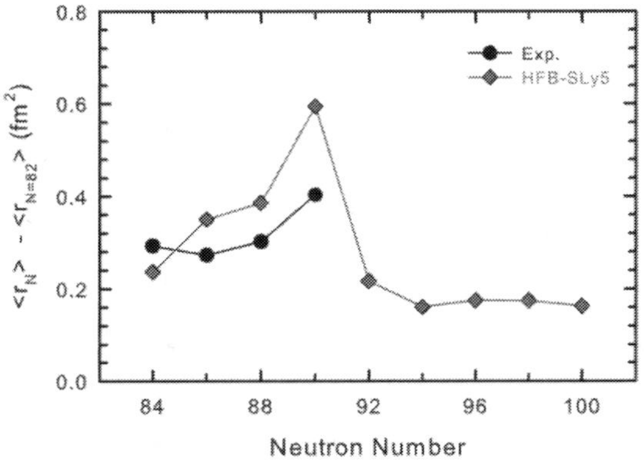


**Figure 3.** Calculated differences between squares of ground-state charge radii: $\langle r_N^2 \rangle - \langle r_{N=82}^2 \rangle$ as functions of neutron number in Nd isotopes.

observed growth rates of collectivity and empirical p-n interaction strengths have been found by Cakirli and Casten [24]. Because of this, we have calculated p-n interaction strengths of $^{142-164}$Nd. The average p-n interaction strength for an even-even nucleus can be calculated by using the formula [25]

$$| \delta V_{pn}^{ee}(Z,N) |= \frac{1}{4}[(B_{Z,N} - B_{Z,N-2}) - (B_{Z-2,N} - B_{Z-2,N-2})]. \tag{3}$$

The calculated and available experimental data [22] for $\delta V_{pn}$ of even-even $^{142-164}$Nd are shown in Figure 4. As can be seen from this figure, the calculated values of $\delta V_{pn}$ for $^{142-164}$Nd are in agreement with the experimental data. An abrupt increase in $\delta V_{pn}$ starts at neutron number $N = 90$ $(A = 150)$ which means that collectivity of nuclei grows in this isotopic chain and this may be considered as a signature of shape phase transitional character of $^{150}$Nd from spherical to axially deformed shape.

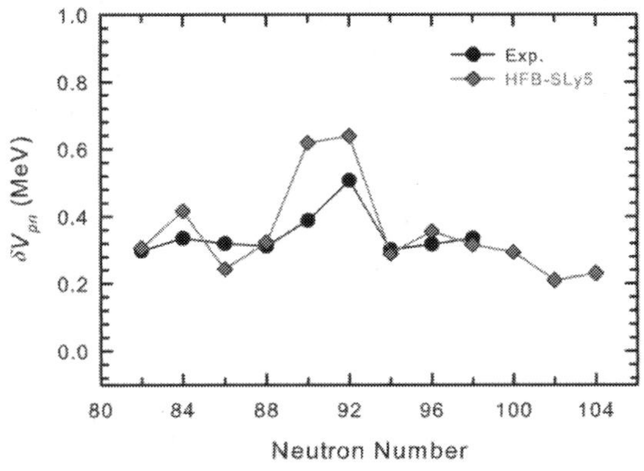


**Figure 4.** The calculated and available experimental $\delta V_{pn}$ for $^{142-164}$Nd.

## 4. Conclusions

Axially deformed HFB calculations with the SLy5 Skyrme force have been carried out to investigate structural evolution of ground-state of even-even $^{142-164}$Nd. The calculated ground-state BEs and $S_{2n}$ of Nd isotopes have been found in good agreement with the experimental data.

By examining the PECs of Nd isotopic chain, $^{150}$Nd has been indicated as a possible critical-point nucleus with X(5) symmetry that is favored by the experimental data. Furthermore, the same conclusion of the transition has been supported by investigating differences in the ground-state charge radii of Nd isotopes.

## References


[1] Iachello F and Arima A 1987 *The Interacting Boson Model* (New York, Cambridge Univ. Press)
[2] Iachello F 2001 *Phys. Rev. Lett.* **87** 052502
[3] Iachello F 2000 *Phys. Rev. Lett.* **85** 3580
[4] Casten R F and Zamfir N V 2001 *Phys. Rev. Lett.* **87** 052503
[5] Casten R F and Zamfir N V 2000 *Phys. Rev. Lett.* **85** 3584
[6] Decharge J and Gogny D 1980 *Phys. Rev.* C **21** 1568
[7] Ring P and Schuck P 1980 *The Nuclear Many-Body Problem* (New York, Springer-Verlag).
[8] Serot B D and Walecka J D 1986 *Adv. Nucl. Phys.* **16**, 1
[9] Meng J, Toki H, Zhou S G, Zhang S Q, Long W H and Geng L S 2006 *Prog. Part. Nucl. Phys.* **57** 470
[10] Fossion R, Bonatsos D and Lalazissis G A 2006 *Phys. Rev.* C **73** 044310
[11] Bayram T and Akkoyun S 2013 *Phys. Scr.* **87** 065201
[12] Rodrıguez-Guzman R and Sarriguren P 2007 *Phys. Rev.* C **76** 064303
[13] Bayram T 2012 *Mod. Phys. Lett.* A **27** 1250162
[14] Krücken R *et al* 2002 *Phys. Rev. Lett.* **88** 232501
[15] Bonatsos D, Lenis D, Minkov N, Raychev P P and Terziev P A 2004 *Phys. Rev.* C **69** 014302
[16] Chabanat E, Bonche P, Haensel P, Meyer J and Schaeffer R 1998 *Nucl. Phys.* A **635** 231
[17] Stoitsov M V, Dobaczewski J, Nazarewicz W and Ring P 2005 *Comp. Phys. Commun.* **167** 43
[18] Bayram T and Yilmaz A H 2013 *Mod. Phys. Lett.* A **28** 1350068
[19] Dobaczewski J, Stoitsov M V and Nazarewicz W 2004 *AIP Conf. Proc.* **726** 51-56
[20] Goriely S, Chamel N and Pearson J M 2010 *Phys. Rev.* C **82** 035804
[21] Möller P, Nix J R and Kratz K -L 1997 *At. Data Nucl. Data Tables* **66** 131
[22] Audi G, Wapstra A H and Thibault C 2003 *Nucl. Phys.* A **729** 337
[23] Angeli I and Marinova K P 2013 *At. Data Nucl. Data Tables* **99** 69
[24] Cakirli R B and Casten R F 2006 *Phys. Rev. Lett.* **96** 132501
[25] Cakirli R B, Blaum K and Casten R F 2010 *Phys. Rev.* C **82** 061304(R)

# Modelling the effects of solar activity onto the Greek national electric grid

**I P ZOIS**[1]
PPCR, R&D Department, 3, Kapodistriou Street,
Agia Paraskevi, GR-153 42, Athens, Greece

E-mail: i.zois@exeter.oxon.org

**Abstract.** We study both the short term and long term effects of solar activity on the large transformers (150kV and 400kV) of the Greek national electric grid. We use data analysis and various analytic and statistical methods and models. Contrary to the common belief in PPC Greece, we see that there are considerable both short term (immediate) and long term effects of solar activity onto large transformers in a mid-latitude country like Greece. Our results can be summarized as follows: For the short term effects: During 1989-2010 there were 43 "stormy days" (namely days with for example Ap ≥ 100) and we had 19 failures occurring during a stormy day plus or minus 3 days and 51 failures occurring during a stormy day plus or minus 7 days. All these failures can be directly related to Geomagnetically Induced Currents (GIC's). Explicit cases are presented. For the long term effects we have two main results:
The maximum number of transformer failures occur 3-4 years after the maximum of solar activity.
There is statistical correlation between solar activity expressed using various newly defined long term solar activity indices and the annual number of transformer failures. These new long term solar activity indices were defined using both local (from geomagnetic stations in Greece) and global (planetary averages) geomagnetic data. Applying both linear and non-linear statistical regression we compute the regression equations and the corresponding coefficients of determination.

## Introduction

Explosions powered by the Sun's magnetic field (flare and coronal mass ejection or CME's) are among the principal causes of space weather. These electromagnetic storms can affect our technological infrastructure in space, interfere with communications and GPS signals, and couple through the geomagnetically induced currents (GIC's) into the large-scale high-voltage electric grid. Despite the known impact of large space weather events on the electrical power grid (see [1]) – including the 1989 Hydro-Québec blackout – relatively few studies of the general correlation are available; case studies of individual events (see [2]) and compilations of events for comparison with the solar cycle generally focus on large storms and large impacts.

There is a recognized hazard of catastrophic outages that may be caused by geomagnetic superstorms larger than what we have experienced in recent decades. Such superstorms may cause trillions of dollars of damage (see again [1]), although it is acknowledged that such estimates are rather uncertain

1 To whom any correspondence should be addressed.

The main cause of GIC's is the interaction of the geomagnetic field with the magnetic field carried within CME's and the surrounding background magnetized solar wind that is modulated by them. With speeds of 400–2500 km/s, it takes some (1–4) days for CME's to propagate from the Sun to the Earth, with a typical transit time of (2–3) days. Furthermore, the magnitude of GIC's depends on the location and time of day (through the geomagnetic position relative to the Sun-Earth line) at impact, on the structure of the magnetic field within the CME as that field interacts with the magnetic field of the Earth during the CME's passage (thereby inducing electric fields in the Earth dependent on the direction and the rate of change of the CME magnetic field), on the ground conductivities in a wide area around any particular site for depths from ground level down to in excess of 100 km, and on the evolving architecture of the electric power grid into which the induced electric field couples (see [3]).

Within the electric power system, GIC's can cause transformers to operate in their nonlinear saturation range during half of the AC cycle. The consequences of half-cycle saturation include distortions of the voltage pattern (reflected in the existence of harmonics to the primary frequency), heating within the transformers, or voltage-to-current phase shifts expressed as reactive power consumption in the system. The detection of harmonics or of strong GIC's may cause protective systems to trip, taking one or more transformers off line to protect them from severe damage. The implementation of such protective measures changes the grid's overall configuration as well as the regional balance between power generation and use which, in turn, can lead to power-quality variations in the form of voltage and frequency swings (causing, for example, the 1989 Hydro-Québec blackout). Moreover, the detection of GIC's may cause system operators to change the operational standards to protect the overall system from damage, for example by changing the transfer limits for power that may be transported between segments of a grid (from where surplus power is more economically available to regions where the demand is highest) to create a buffer interval to keep GIC's from pushing transformers into their nonlinear range. Strong GIC's can result in dissipative heating within the transformers which may lead to their failure, either within minutes or because of cumulative damage done over the lifetime of the transformer (see [4]).

Our first task was to gather large (i.e. 150 kV up to 400 kV) transformer failure data from PPC (Public Power Corporation) throughout Greece. Reliable data were digitized for the period 1989-2010. We got rid of obviously irrelevant failures (like those caused by human error, earthquakes, falling of objects, short circuits caused by animals, extreme weather conditions, lightnings etc). This gave us the first set of data, namely annual transformer failure numbers (ATFN for short). If one plots on the same time graph both ATFN (red curve, scaled and smoothed) and solar activity (blue curve, say as expressed by the annual number of sunspots) one gets the following graph:

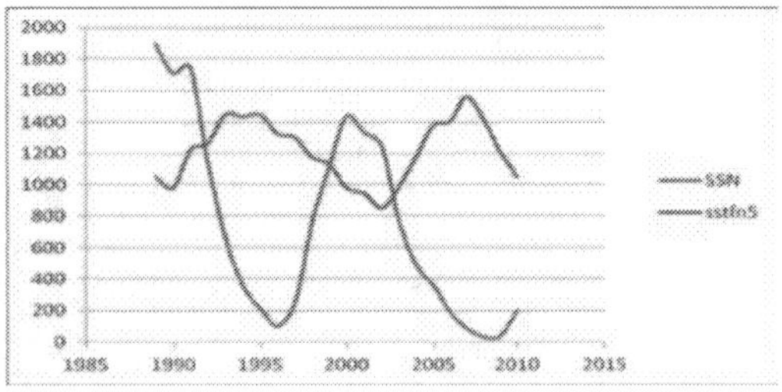


This rather revealing graph indicates that the annual number of failures generally follows the solar activity pattern, it has a periodicity similar to the 11 year solar cycle but it has a "phase delay" of about 3-4 years. This can be explained by the fact that GIC's in Greece are existing but smaller compared to say Scandinavian countries and cumulative effects dominate: Failures occur after some threshold is crossed and since GIC's in Greece are generally smaller, this threshold is crossed later.

For example in Finland the maxima of these 2 curves almost coincide or have only 1 (max 2) years time difference.

The next thing we did was to see the geomagnetic activity during days when some failure was recorded. This revealed that during 1989-2010 there were 43 "stormy days" (namely days with for example $Ap \geq 100$) and we had 19 failures occurring during a stormy day plus or minus 3 days and 51 failures occurring during a stormy day plus or minus 7 days. From the detailed reports (type of damage) all these failures can be directly related to Geomagnetically Induced Currents (GIC's). Explicit cases were analysed.

The third thing we did was to try to apply regression and correlation between ATFN data and earth affecting solar activity (abbreviated to EASA) data. EASA is being monitored by a variety of indices (like the K-index, the A-index, the Dst index etc). These indices are short term indices, measured every 3 hours and broadly speaking they are related to the variations of the tangential components of the geomagnetic filed caused by its interaction with the magnetic field of the plasma coming from the sun. During a geomagnetic storm these variations can be quite large. We defined various new long term (on an annual basis) EASA indices (like annual Ap mean, annual Dst mean, annual number of days with $Ap \geq 40$, annual number of days with $Dst \leq -40$, annual mean of the daily K-index sum etc using both planetary means and local geomagnetic data) and then we applied statistical regression between these 2 sets of data, namely various annual EASA indices (considered as the independent variable X) and ATFN (considered as the dependent variable Y). We tried various functional models (polynomial, exponential, hyperbolic, logarithmic, Box-Cox transforms of both variables), we used the least squares method and computed the resulting regression equations along with the coefficient of determination. Not surprisingly (bearing in mind the Stone-Weierstrass theorem from functional analysis), the highest coefficient of determination (best fit) was obtained using polynomials (our basic ansatz was that X and Y are related by a continuous function). Some of our results can be summarized in the following table:

| Index | Function | Equation | $R^2$ |
|---|---|---|---|
| SSN | *linear* | $Y = -0.01X + 48.09$ | 0.27 |
| | *ln* | $Y = -4.12 \ln X + 65.92$ | 0.19 |
| | *poly* | $Y = 36.872 + 0.2032X - 0.0011X^2 + 2E\text{-}06X^3 - 2E\text{-}09X^4 - 9E\text{-}13X^5 - 2E\text{-}16X^6$ | 0.35 |
| | *hyp* | $Y = 75.09 X^{-0.11}$ | 0.20 |
| | *exp* | $Y = 47.01 e^{-0.0002X}$ | 0.26 |
| | | | |
| Ap mean | *linear* | $Y = -0.6416X + 49.16$ | 0.078 |
| | *ln* | $Y = -7.49 \ln X + 59.39$ | 0.082 |
| | *poly* | $Y = 353.07 - 195.63X + 47.664X^2 - 5.7355X^3 + 0.3622X^4 - 0.0115X^5 + 0.0001X^6$ | 0.123 |
| | *hyp* | $Y = 62.54 X^{-0.19}$ | 0.081 |
| | *exp* | $Y = 47.84 e^{-0.0153X}$ | 0.069 |
| | | | |
| Dst mean | *linear* | $Y = -0.3375X + 46.539$ | 0.05 |
| | *ln* | $Y = -4.2932 \ln X + 52.471$ | 0.04 |
| | *poly* | $Y = 39.998 + 10.631X - 3.9763X^2 + 0.5504X^3 - 0.0348X^4 + 0.001X^5 - 0.00001X^6$ | 0.23 |
| | *hyp* | $Y = 52.831 X^{-0.1099}$ | 0.04 |
| | *exp* | $Y = 45.888 e^{-0.089X}$ | 0.04 |
| | | | |
| Annual no of days with $Ap \geq 40$ | *linear* | $Y = -0.2007X + 44.33$ | 0.06 |
| | *ln* | --- | |
| | *poly* | $Y = 43.942 + 5.0937X - 1.2764X^2 + 0.1086X^3 - 0.0042X^4 + 8E\text{-}05X^5 - 5E\text{-}07X^6$ | 0.14 |
| | *hyp* | --- | |
| | *exp* | $Y = 42.451 e^{-0.046X}$ | 0.05 |
| | | | |
| Annual no of days with $Dst \leq -40$ | *linear* | $Y = -0.1101X + 44.87$ | 0.08 |
| | *ln* | --- | |
| | *poly* | $Y = 50.453 - 4.9074X + 0.5092X^2 - 0.0192X^3 + 0.0003X^4 - 2E\text{-}06X^5 + 7E\text{-}09X^6$ | 0.22 |
| | *hyp* | --- | |
| | *exp* | $Y = 43.215 e^{-0.024X}$ | 0.07 |

The final and perhaps the most interesting thing we did was this: We tried to push polynomial regression to the limit, namely find the largest polynomial degree for which the least squares method gave a unique solution and try other more elaborate algebras instead of ordinary polynomials, which are also dense in the algebra of continuous functions and hence separate points (see [5], [6], [7] and [8]). This requires some use of advanced mathematics like the K-Theory long exact sequence. Our results can be summarized in the following table.

| **Long term solar activity index** | **Polynomials** | | | |
|---|---|---|---|---|
| | **(ordinary) Raw** | | **Orthogonal** | |
| | **Max allowable degree** | **$R^2$** | **Max allowable degree** | **$R^2$** |
| *SSN* | 12 | 0.60 | 17 | 0.81 |
| *Annual Ap mean* | 10 | 0.22 | 12 | 0.30 |
| *Annual Dst mean* | 10 | 0.33 | 16 | 0.52 |
| *Annual no of days with Ap ≥ 40* | 10 | 0.43 | 13 | 0.58 |
| *Annual no of days with Dst ≤ - 40* | 11 | 0.71 | 15 | 0.88 |
| *Annual mean of daily K-index sum (Penteli)* | 7 | 0.26 | 16 | 0.58 |

The annual number of days with Dst ≤ - 40 long term index gives a rather impressive $R^2$ value of 0.88 using orthogonal polynomials (of 15<sup>th</sup> degree).

**References**

**[1]** Space Studies Board, *Severe space weather events – understanding societal and economic impacts*. National Academy Press, Washington, DC, USA, 2008.

[2] Kappenman J *Geomagnetic storms and their impacts on the US power grid (Tech. Rep. Meta-R-319)*, Metatech Corp, Goleta, CA, 2010.

[3] United Kingdom Royal Academy of Engineering: "Extreme Space Weather: impacts on engineered systems and infrastructure", 2013

[4] Coetzee G and Gaunt CT Transformer failures in regions incorrectly considered to have low GIC-risk (2007)

[5] Jeffreys H "Weierstrass's Theorem on Approximations by Polynomials" in methods of Mathematical Physics, 3rd edition, Cambridge University Press 1988.

[6] Zois I P "18 Lectures on K-Theory", *arxiv.org 1008.1346, (2010)*

[7] Kaw A and Kalu E. "Numerical Methods with Applications" (2008) (Chapter 6 deals with linear and non-linear regression).

[8] Kennedy J Jr and Gentle J E *Statistical Computing*, Marcel Dekker 1980.

# Route prediction model of infectious diseases for 2018 Winter Olympics in Korea

**Eungyeong Kim**[1], **Seok Lee**[1], **Young Tae Byun**[1], **Jae Hun Kim**[1], **Hyuk-jae Lee**[2], **Taikjin Lee**[1]
[1] Sensor System Research Center, Korea Institute of Science and Technology (KIST), Seoul, 136-791, Korea
[2] Nano and Electronic Physics, Kookmin University, Seoul, 136-702, Korea

E-mail: eungyeong@kist.re.kr

**Abstract.** There are many types of respiratory infectious diseases caused by germs, virus, mycetes and parasites. Researchers recently have tried to develop mathematical models to predict the epidemic of infectious diseases. However, with the development of ground transportation system in modern society, the spread of infectious diseases became faster and more complicated in terms of the speed and the pathways. The route of infectious diseases during Vancouver Olympics was predicted based on the Susceptible-Infectious-Recovered (SIR) model. In this model only the air traffic as an essential factor for the intercity migration of infectious diseases was involved. Here, we propose a multi-city transmission model to predict the infection route during 2018 Winter Olympics in Korea based on the pre-existing SIR model. Various types of transportation system such as a train, a car, a bus, and an airplane for the interpersonal contact in both inter- and intra-city are considered. Simulation is performed with assumptions and scenarios based on realistic factors including demographic, transportation and diseases data in Korea. Finally, we analyze an economic profit and loss caused by the variation of the number of tourists during the Olympics.

## 1. Introduction

In order to predict for the infection route most countries of world use mathematical models of infectious diseases. Longini et al. [1] and Carrat et al. [2] have explained the transmission of infectious diseases such as measles and SARS by a mathematical modelling. Glenn et al. [3] predicted the transmission of infectious disease using the SIR model based on pandemic influenza A data. In [4], Susceptible-Infectious-Recovered (SIR) Model was applied for the prediction of the transmission of infectious diseases in Vancouver Olympics considering an airplane as a transportation factor. However, with the development of the modern transportation system, the spreading speed and route of infectious diseases became faster and more complicated. Therefore, transportation factor including only an airplane in [4] is very limited.

In this study, we propose a multi-city transmission model to predict the infection route during 2018 Winter Olympics in Korea based on the pre-existing SIR Model [5]. Various types of transportation system such as a train, a car, a bus, and an airplane for the interpersonal contact in both inter- and intra-city are considered. Simulation is performed with assumptions and scenarios based on realistic factors including demographic data from Statistics Korea [6], daily traffic data available from the Korean Statistical Information Service (KOSIS) [7] and diseases data form the Korea Centers for Disease Control and Prevention (KCDC) [8]. Although this work does not include all the possible

[1] To whom any correspondence should be addressed.

factors related to infectious diseases, it can be a guideline to predict the infection route and minimize an epidemic disaster through this basic modeling.

## 2. Multi-city transmission model

We represented the transmission of pandemic influenza A in Korea based on pre-existing SIR Model [5]. Here individuals are classified as susceptible ($S(t)$), infectious ($I(t)$), and recovered ($R(t)$) and we assume that the population and the traffic flow are homogeneous. Our multi-city transmission model includes modes of transportation in order to consider the interpersonal contact over a broad range. Therefore, our model sufficiently predicts the transmission of infectious diseases because of including interpersonal contact in a broad range and based on analysing the traffic data. Our model is the system with nonlinear differential equations:

$$\begin{aligned}
\frac{dS_i[t]}{dt} &= -\alpha\left[\frac{I_i[t]}{N_i[t]} + \sum_{j\neq i}\left(\frac{m_{ij}}{N_i[t]} + \frac{m_{ji}}{N_j[t]}\right)\frac{I_j[t]}{N_j[t]}\right]S_i[t] \\
\frac{dI_i[t]}{dt} &= -qI_i[t]\theta[t - t_{ii}] - \beta I_i[t] + \alpha\left[\frac{I_i[t]}{N_i[t]} + \sum_{j\neq i}\left(\frac{m_{ij}}{N_i[t]} + \frac{m_{ji}}{N_j[t]}\right)\frac{I_j[t]}{N_j[t]}\right]S_i[t] \qquad (1) \\
\frac{dR_i[t]}{dt} &= qI_i[t]\theta[t - t_{ii}] + \beta I_i[t]
\end{aligned}$$

where $\alpha$ represents the rate of infection (number of adequate contacts per unit time [contacts/day]), $\beta$ is the rate at which infected people recover (1/day). $m_{ij}$ is the number of persons moving from city $i$ to city $j$. $m$ includes transportation factors by train, passenger car, bus, and airplane. $(m_{ij}/N_i[t]) \times (I_j/N_i[t])$ is the probability that a person that migrated from city $i$ to city $j$ can contact with an infected person in city $j$. Therefore $(m_{ij}/N_i[t]) \times (I_j/N_i[t]) \times S_i[t]$ describes the number of susceptible persons in the city $I$, that contact with infected persons in city j through the transportation. $qI_i[t]$ is the number of persons in city $i$ that inoculated according to the rate of antiviral drug ($q$) to infected persons at time $t$. $\beta I_i[t]$ is the number of infected persons in city $i$ that recovered with recovery rate $\beta$. $\theta\ [t - t_{ii}]$ is the Heaviside unit step function as

$$\theta[t - t_{ii}] = \begin{cases} 1 & \text{for } t > t_{ii} \\ 0 & \text{for } t < t_{ii} \end{cases} \qquad (2)$$

The timing of the antiviral drug is $t_{ii}$, which can be set differently for each city.

## 3. Results

The simulation calculated the spreading course of infectious diseases during the outbreak of pandemic influenza A prior to the 2018 Winter Olympics in Korea using Mathematica 6.0. The simulation covers January 6, 2018 to April 24, 2018 (t=1~100day) and the Winter Olympics is held from February 9, 2018 to February 25, 2018 (t=25~42day). The following suppositions have been set up in order to predict the spreading course of infectious diseases. Pandemic influenza A occurs from February 5 to February 14 (t=21~30day) at the Incheon International Airport and 25 people (0.05%) out of Incheon International Airport population are carriers. And the foreign tourists enter the country through Incheon International Airport during this period. The simulation has selected five major cities with a large flow in population using the population data of [6]. They are Seoul (10,223,243 people), Incheon International Airport (51,224 people), Gangneung (219,245 people), Pyeongchang (43,706 people) and Wonju (320,563 people). It was assumed that Pandemic influenza A spreads using only the transportations methods of trains, passenger cars, buses and airplanes. Seoul has the largest transportation system and Incheon International Airport, Gangneung, Wonju and Pyeongchang are followed in order [7]. However, Pyeongchang, the hosting city of Winter Olympics, was assumed that it is the second after Seoul in terms of the traffic flow during the Winter Olympics.

**Table 1.** Pandemic influenza A occurs during t=21~30day at the Incheon International Airport, proportion of infectious persons by each city during the Olympics according to the three types of scenario. ($I$= 0.05%, $q$= 0.03 (3%)).

| | Incheon Airport | Seoul | Wonju | Pyeongchang | Gangneung |
|---|---|---|---|---|---|
| Scenario 1 | 23 % | 25 % | 16 % | 26 % | 16 % |
| Scenario 2 | 16 % | 17 % | 10 % | 17 % | 10 % |
| Scenario 3 | 22 % | 23 % | 15 % | 23 % | 15 % |

There are three types of scenario. First, the antivirus drug is not administered to all cities during the Winter Olympics. Second, the antivirus drug is administered to Pyeongchang which is the hosting area of Olympics on February 9th (t=25day) which is the first day of Winter Olympics. Third, the antivirus drug is administered to Pyeongchang on February 19th (t=35day) which is a day in the middle of Winter Olympics to compare how the proportion of in Winter Olympics according to the three types of scenario. Here, $q$ was assumed as 3%, $\alpha$ was assumed as 17.535% and $\beta$ was assumed as 6.25%. As a result, the fact that proportion of infectious persons getting decreased could be found if the antivirus drug is quickly administered in all cities. In other words, the proportion of infectious persons has reduced considerably in scenario 2 than scenario 3. And the proportion of infectious persons in scenario 2 has reduced considerably in Pyeongchang which is the antivirus drug administered area. Next were in the order of Seoul with a lot of traffic, Incheon International airport and Gangneung or Wonju with little traffic.

Figure 1 has compared the following with same conditions as Table 1. (a) proportion of infectious persons in all cities during the Winter Olympics, (b) antivirus treatment cost of infectious persons (the price of Tamiflu is $22 per person per dosage), (c) rate of visiting foreign tourists and (d) cost of economic loss due to decrease of foreign tourists. Here, the total number of foreign tourists are 195,000 and $421 million is spent [9]. As a result of simulation, (a-c) reduces the number of infectious persons and treatment cost as the administration of antivirus drug gets earlier while the fact that visiting foreign tourist also getting increased has been verified. And (d) has verified the fact that economic loss is great as the consumption of foreign tourist gets reduced as the treatment of infectious persons gets delayed.

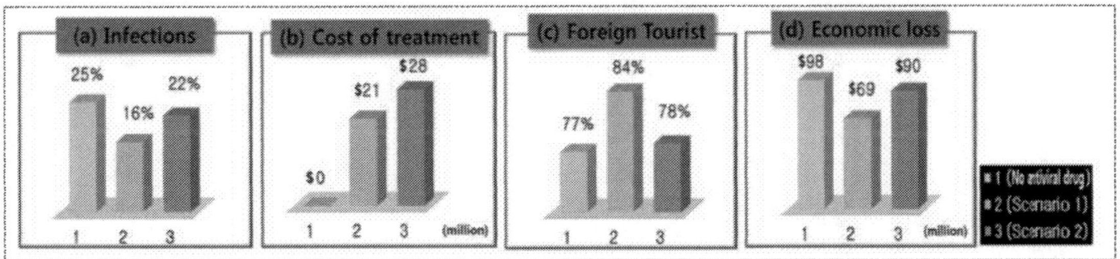


**Figure 1.** Pandemic influenza A occurs during t=21~30day at the Incheon International Airport, (a) proportion of infectious persons, (b) antivirus treatment cost of infectious persons, (c) rate of visiting foreign tourists and (d) cost of economic loss due to decrease of foreign tourists.

## 4. Discussion

This paper has proposed a Multi-city transmission model for tracking and predicting the spreading course of pandemic influenza A by extending the SIR model while simulating the epidemics of contagious disease during the outbreak of pandemic influenza A during the period of the Winter Olympics of Korea in 2018. As a result of the calculation, the infectious disease spreads out being influenced by the transportations and the time and position for the administration of the antivirus with high priority can be determined by the analysis based on the amount of the traffic. Also, the alternative for attracting most number of tourists could be found through an economic analysis followed by increase or decrease of tourists.

The simulation is limited in that the modes of transport only included buses, trains, passenger cars, and airplanes and only 5 cities are considered. Therefore, the future study will be researched by adding more cities and various types of transportation.

This subject is supported by Korea Ministry of Environment (MOE) as "Global Top Environmental Technology Research and Business Development Program" (GT-11-F-02-002-1).

**References**

[1] Longini, I. M. Jr et al. 2005 *Science* 309 1083.
[2] Carrat F., Luong J, Lao H. et al. 2006 *BMG Med* 4 1.
[3] Glenn R. Fulford & Belinda Barnes. 2009 *Taylor & Francis Group. NY: CRC Press* 95.
[4] Hyman, J.M. & LaForce, T. 2003 *H. Banks, C. Castillo-Chavez (Eds.), Bioterrorism, SIAM* 215.
[5] Kermack, W. O. & McKendrick, A. G. 1991 *Bull. Math. Biol.* 53, 33.
[6] [http://kostat.go.kr/
[7] http://www.kosis.kr/
[8] http://www.cdc.go.kr/
[9] Report of Korea Institute for Industrial Economics & Trade 2008 1.

# Water Self-Diffusion in the Roots of the Plant Effected by Stress Factor under Long-Continued NMR-Experiment

V A Tyurin[1,4], T A Serebrennikova[2] and L P Khokhlova[3]

[1]Associate professor of Department of Radiophysics, Institute of Physics, Kazan (Volga Region) Federal University, Kremlevskaja str., 18, Kazan 420008, Tatarstan, Russian Federation
E-mail: vtyurin0@gmail.com
[2]Associate professor of Department of physics, Kazan State Energy University, Krasnoselskaja str., 51, Kazan 420066, Tatarstan, Russian Federation
[3]Professor of Department of Physiology and Biochemistry of Plant, Institute of Fundamental Medicine and Biology, Kazan (Volga Region) Federal University, Kremlevskaja str., 18, Kazan 420008, Tatarstan, Russian Federation

**Abstract.** The water self-diffusion in the root segments of winter wheat plantlet after their cutting off from the intact plants during continued «adaptive aging» was investigated by PGSE technique. The diffusion decay of the stimulated echo relative amplitude under the diffusion times 10 ms, 155 ms, 350 ms may be described by the sum of three exponents. Strongly marked correlated dependences of diffusion coefficients and population of the diffusion decay components on the sample lifetime have been revealed. These dependences may be caused by the change of diffusion direction of water molecules running through the tonoplast aquaporin channels in the different cell compartments. This change of direction seems to be the important mechanism of the cell water homeostasis maintenance and it is considered as the part of the general stress roots reaction.

Kay words: NMR, roots, self-diffusion of water molecules, membranes, vacuole, cytoplasm.

Abbreviations: PGSE – pulse-field-gradient spin-echo technique; DD – diffusion decay of the spin-echo amplitude; SDC – self-diffusion coefficient.

## Introduction

The processes of self-diffusion of water in segments of roots cut off from the whole plant; carry important information about metabolism of root cells and the effect of stress mechanisms on the living system [1]. To research these processes the PGSE technique is used [2-4]. Cutting off the roots as mechanical stress as well as other types of phytostress includes the whole complex of metabolic and physiological changes referred to as nonspecific adaptative syndrome, observed in the cells [5]. On the

[4] To whom any correspondence should be addressed.

first stages of the stressor's effect these changes have the lifetime character, revealing itself in the adaptative reorganization of structural-functional characteristic of cells/ tissues and the most complicated hierarchy of regulatory interrelations. In case the effect of damage factor continues, spare capacities inevitably exhaust and «defatigation phase» comes. This phase is characterized by intensive destruction, nonreversible metabolic disturbance and the death of cells. In respect to cut off roots this phenomena is known as «adaptative aging» or «survival» [1], and the understanding of its nature lies within the modern concept of stress and general principles of responsiveness of living system [5,6]. At the same time biophysical aspects of prolonged stress response of roots, particularly its relationship with the process of self-diffusion of water, need in the additional study and precise interpretation.

Thereby the aim of the present work is the research of dependence of self-diffusion of water molecules on «adaptative aging» time of vegetable patterns in the conditions of prolonged mechanical stress.

## 1. Methods and objects of research

### 1.1 Diffusion decay

Primary information about self-diffusion in PGSE technique is obtained due to DD analysis. Diffusion decay is known to be logarithm dependence with normalized amplitude of spin echo $M(k^2)=\ln\left[A(k^2)/A(0)\right]$ on the square wave-vector $k^2=\gamma^2\delta^2g^2$ and diffusion time $t_d$. Here $\gamma$ – gyromagnetic ratio of proton, $\delta$ - duration and g – amplitude pulse of the gradient of magnetic field. DD have been registered by means of diffusiometer with computer process control and resonance frequency of protons equal to value 60 MHz. Peak amplitude of gradient pulse was equal ≈ 30 T/m. Measurements have been made by «stimulated echo» technique [6] with diffusion time ~ 10, 155 and 350 ms. The registration time of one DD was determined by accumulation number (from 4 to 6) and was equal ~ 25 - 40 min. Measurements were taken at 23 $^0$C degree. The measurement error was equal ~ 10% in each case.

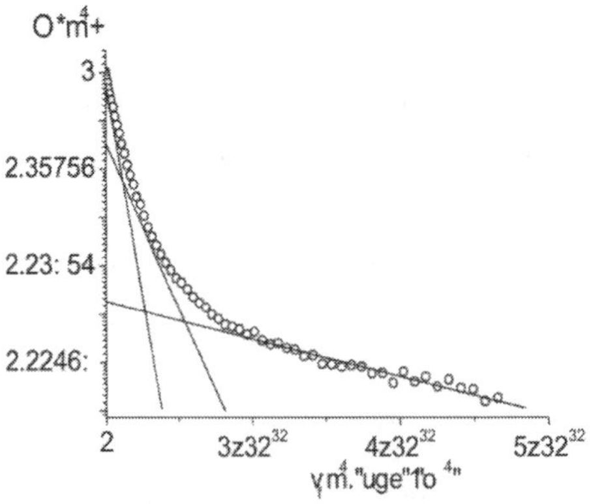


Figure 1. Diffusion decay in the segments of roots with $t_d$ = 10 ms.

Figure 1 shows DD in the research samples at $t_d$ =10 ms. DD had the same nonexponential shape also at $t_d$=155 и 350 ms. In each case DD may be described by sum of three exponents

$$A(k^2)\Big/A(0)=p_1\exp\left(-D_1t_dk^2\right)+p_2\exp\left(-D_2t_dk^2\right)+p_3\exp\left(-D_3t_dk^2\right).$$

Here $p_i$ и $D_i$ (i = 1, 2, 3) – populations and SDC ($D_1 < D_2 < D_3$) respectively. Differentiation of DD components was performed on computer by algorithm step-by-step subtraction with «ORIGIN 7.5». Population and SDC of the components were identified while approximating by linear functions.

### 1.2 Objects of research

Objects of research were primary roots of 7 - daily plantlets of winter wheat (Triticum aestivum) of Mironovskaja 808 brand, handled in the water culture with luminance 100 W/m$^2$ and 12- hour photoperiod. The above-ground part of fasciculate plants was cut off, but 2-3 sm of spear. This material was incubated in double distilled water pending for 3 h. After that the segment of 8 мм lengthwise matching tension zone and the origin of differentiation zone were cut out of each root. Parallel oriented segments of roots, 60-70 in number, were shaped in the pack, which was wrapped up by thin fluoroplastic film and put in hermetic glass ampoule. The sample was disposed in the NMR probe perpendicularly to vector pulse gradient of the magnetic field. Due to the said position we could observe the self-diffusion in direction, perpendicularly to longitudinal axis of root. In each case measurements were completed before defatigation phase, thus and so further alongside with the term «adaptative aging» we'll use the term «lifetime».

## 2. Results and discussion

In this report we discuss only the dependences, obtained at $t_d$ = 350 ms, the plots of them shown on figure 2. Mean-square displacement, to which the molecule of water is subjected in the direction of the gradient at $t_d$ = 350 ms, according to Einstein equation $(< r^2 >)^{1/2} = 2D_s t_d$ is ranged ≈ from 19 to 33 micron, what is comparable with the cell size of root tension zone.

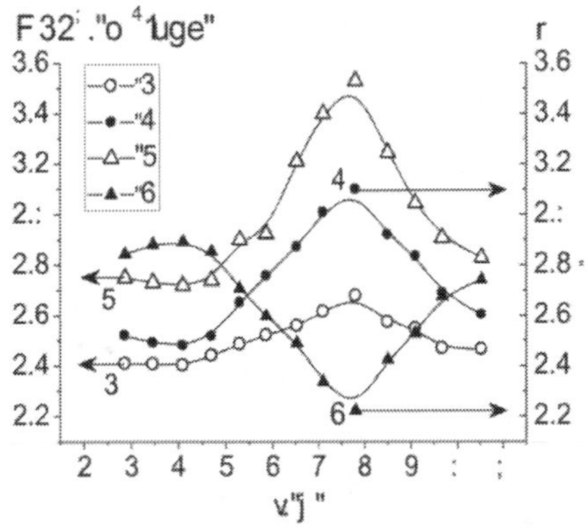


Figure 2. Dependences of $D_2$ (1), $p_2$ (2) и $D_3$ (3), $p_3$ (4) on a lifetime sample and td = 350 мс.

According to existing opinions [7-9], molecules of apoplast and simplast vacuole water contribute to component DD with biggest SDC $D_3$. Protons simplast cytoplasm water contributes to the component characterized by $D_2$. As for the component, characterized by $D_1$, the definite opinion is not known at the time being [3]. The parameters $D_1$ and $p_1$ of the slow component do not depend upon lifetime and therefore are not shown in figure 2.

The present results testify practically determined change of translational mobility of water molecules, starting in the cellular space of roots' segments after they were cut off. This process it should be considered as the part of the total cell reaction against the stress caused by the stop of the directed water stream and the replacement the external water medium by the air medium. It is experimentally determined [1], that biggest stimulation of a number of functional systems: breath and oxidative association with phosphorus, ion pump, and also polarization and restoration of barrier properties of plasmatic membrane takes place in the cells of roots in 4 – 6 hours after their cutting off. Since maximums of diffusion characteristics ($D_2$, $D_3$ и $p_2$) are being observed in the same time range (figure 2), there are good reasons to expect close interconnection between translational mobility of water molecules and functional stressful responses of the roots.

The response to water deficiency is the compression of protoplast [10] and, as consequence, the reduction of vacuole volume, which normally takes up to 80 – 90% of the cell volume in the tension zone [11]. As a result, the osmotic pressure in vacuole increases, and the water stream running through tonoplast and plasmalemma makes its way in direction to the cytoplasm. Therefore population $p_3$ decreases, while $p_2$ и $D_2$ of cytoplasmic water increase due to more mobile water molecules of vacuole. SDC $D_3$ also increases as a result of probable augmentation of the relative portion of apoplast water molecules.

## Conclusion

The research has shown that diffusion decay in segments of roots of plantlet winter wheat under hours-long effect of mechanical stress is of irregular shape. The analysis of DD by its approximation by the sum of the three exponents reveals the response to the stress in a form of a distinct extreme dependence of self-diffusion coefficients and populations of intermediate and rapidly decaying component on the lifetime of the sample. For explanation of results thus obtained the model based on the change of the water diffusion direction from vacuole to cytoplasm and inversely through aquaporin water canals of tonoplast has been proposed. It is hypothesized, that periodical change of diffusion direction prevents the critical deaquation and, eo ipso, provides for preservation of the cells water homeostasis being subjected to stresses, by maintenance of coordinated variations of volume and water content of vacuole and cytosolic compartments.

## References

[1] Pakhomova G M 1995 *Cytology* **37** 66
[2] Steiskal E O, Tanner G E 1965 *J. Chem. Phys.* **42** 288
[3] Velikanov G A, Volobueva O V and Khokhlova L P 2001 *Physiol. of plant* **48** 375
[4] Anisimov A V, Ionenko I F and Romanov A V 2004 *Biophysics* **49** 891
[5] Braun A D, Mogenok T P 1987 *Nonspecific adaptative syndrome of cellular system* (Leningrad: Science)
[6] Selje G 1972 *At level of whole organism* (Moscow: Science)
[7] Ionenko I F, Anisimov A V and Romanov A V 2003 *Physiol. of plant* **50** 88
[8] Volobueva O V, Velikanov G A and Balushka F 2004 *Physiol. of plant* **51** (5), 751
[9] Ionenko I F, Anisimov A V 2007 *Physiol. of plant* **54** (2), 253
[11] Gorshkova G A 2007 *Vegetable cell wall like dynamic system* (Moscow: Science)
[10] Genkel P A 1982 *Physiology of heat - and drought-resistance of plant* (Moscow: Science)

2nd International Conference on Mathematical Modeling in Physical Sciences 2013 IOP Publishing
Journal of Physics: Conference Series **490** (2014) 012110 doi:10.1088/1742-6596/490/1/012110

# Optimal capacity and buffer size estimation under Generalized Markov Fluids Models and QoS parameters

**José Bavio and Beatriz Marrón**
Universdidad Nacional del Sur, Av. Alem 1253, Bahía Blanca, Buenos Aires, Argentina
E-mail: jmbavio@yahoo.com.ar - beatriz.marron@uns.edu.ar

**Abstract.**
Quality of service (QoS) for internet traffic management requires good traffic models and good estimation of sharing network resource.

A link of a network processes all traffic and it is designed with certain capacity $C$ and buffer size $B$.

A Generalized Markov Fluid model (GMFM), introduced by Marrón (2011), is assumed for the sources because describes in a versatile way the traffic, allows estimation based on traffic traces, and also consistent effective bandwidth estimation can be done.

QoS, interpreted as buffer overflow probability, can be estimated for GMFM through the effective bandwidth estimation and solving the optimization problem presented in Courcoubetis (2002), the so call inf-sup formulas.

In this work we implement a code to solve the inf-sup problem and other optimization related with it, that allow us to do traffic engineering in links of data networks to calculate both, minimum capacity required when QoS and buffer size are given or minimum buffer size required when QoS and capacity are given.

## 1. Introducción

Modeling different digital sources is a wide topic. Markov models have been studied for many kinds of sources like ON/OFF sources or video sources but they have some limitations when the data rate transfers can take too many value. In this work a Generalized Markov Fluid model, introduced in [1] is used to describe the traffic of each source.

For this model, a sources is a data network assumes the state $Z_s$ at time $s$, where $Z$ is a continuous time, homogeneous and irreducible Markov chain, with finite state space $K = \{1, \cdots, k\}$, invariant distribution $\pi$ and infinitesimal generator $\mathbb{Q}^Z$. Let us consider $f_1, f_2, \cdots, f_k$, $k$ laws of probability with known and disjoint support. When the chain $Z$ reaches states $i$, at time $s$, the speed with which the sources transfer data it is drawn, independently of the chain $Z$, by the law $f_i$. This is, the random variable $Y_s|Z_s = i$, is distributed according the probability law $f_i$ for $i = 1, \cdots, k$.

This new model can be interpreted as follows: the state of the chain indicate some type of activity in the transfer data such as email, chat, conversation, video conferences, etc., and the rate for each state is drawn according to a probability law, within a range of reasonable values for such activity.

The process $Y$ takes the drawn value as long as the chain $Z$ remains in that state, if the modulating chain change state, a new value for $Y$ is drawn. Let us note that the process $Y_s$ is observable and, since the supports of the $k$ laws of probability are known and disjoint, so is the process $Z_s$.

The Markov flow modulated by the chain $Z_s$ that represent the work load received from the source that delivers information with speed $Y_s$ is

$$X_t = \int_0^t Y_s \, ds. \tag{1}$$

Given an expected $QoS$, interpreted as the probability of buffer overflow, the actual resources that should be reserved lie between the mean rate and the peak rate of the connection. These resources are generally referred to as the Effective Bandwidth (EB) of the traffic sources, was proposed by Kelly in [2] and is defined as follows.

Let $X_t$ be a process with stationary increments, representing the amount of work arriving from a source in the interval $[0, t]$, then the EB of the source is

$$\alpha(s,t) = \frac{1}{st} \log E e^{sX_t} \qquad 0 < s, t < \infty, \tag{2}$$

where the parameters $s$ and $t$ characterize a link's operating point and depends on the context of the stream. Specifically, the space parameter $s$ indicate the degree of multiplexing and the time parameter $t$ corresponds to the most probable duration of the buffer busy period prior to overflow.

## 2. Effective bandwidth

We can compute the EB for the GMFM by Kesidis,Walrand, Chang formula [3] by the following theorem,

**Theorem 1** *Let $\{X_t\}_{t\geq 0}$ be a GMFM modulated by a continuous time, homogeneous and irreducible Markov chain $Z$ with invariant distribution $\pi$ and infinitesimal generator $\mathbb{Q}^Z$. Let us consider the random variables $Y_i$ with density function $f_i$, mean $\mu_i$ and variance $\sigma_i^2$ for $i = 1, \cdots, k$ and the $k$ dimensional diagonal matrix $\mathbb{H}$, whose nonzero elements are the first moments $\mu_i$ of each distribution, then*

$$\alpha(s,t) = \frac{1}{st} \log \left\{ \pi \exp \left[ \left( \mathbb{Q}^Z + s\mathbb{H} \right) t \right] \mathbf{1} \right. , \tag{3}$$

*where* $\mathbf{1}$ *is a column vector with all entries equal to 1.*

The importance of this theorem is that it provides an expression for the EB that depends on the infinitesimal generator of the modulating chain, its invariant distribution and a matrix containing information of the transfer rate, and all these elements can be estimated with traffic traces. In [1] a consistent and asymptotically Gaussian estimator is obtained from traffic traces.

## 3. Operational point

Each node or link of the network has two major design parameters, capacity and buffer size, and one QoS parameter buffer overflow probability (BOP).

The EB is relationally with the probability of overflow by the called $\inf - \sup$ formula $\lim_{N\to\infty} \frac{1}{N} \log P(Q_N > B) = -\gamma$, with

$$-\gamma = \inf_{t\geq 0} \sup_{s\geq 0} \left\{ (b + ct)s - st\alpha(s,t) \right\}, \tag{4}$$

where $Q_N$ is the stationary workload in tail, $c$ is the capacity of the node, $b$ is the buffer size, $N$ is the number of sources in the link and $\alpha(s,t)$ is its EB of the link [4]. The values $t^*$ and $s^*$ in which the inf – sup is reach is called the link operating point.

The parametric approach it is possible with a model for the traffic, and our aim is to estimate $\gamma, t^*$ and $s^*$ using an estimator of $\alpha(s,t)$.

In [1] is proved that if $\alpha^n(s,t)$ is an estimate of $\alpha(s,t)$, then replacing $\alpha^n(s,t)$ in (4) we obtain $(s_n^*, t_n^*)$ that are consistent estimators of $(s^*, t^*)$.

Finally we can estimate $\gamma$ with $\gamma_n$ using $\alpha_n$ and solving the inf – sup equation.

## 4. Algorithm and Code

In the paper [5] there is an algorithm to solve (4). First call $F(s,t) = (b+ct)s - st\alpha(s,t)$. The algorithm involves two optimization procedures:

(i) find for a fixed time, the maximum $F^*(t) = \max_s F(s,t)$,

(ii) find the minimum of $M = \min_t F^*(t)$.

The first step can be numerically solved in an efficient manner by taking into account that the logarithmic moment generating function $st\alpha(s,t)$ is convex in $s$, whereas $s(b+ct)$ is linear in $s$. Due to this, $F_t(s) = F(s,t)$ is an unimodal function of $s$ and the maximizer is unique. To find the maximum one can start from an initial "uncertainly" interval $[s_a, s_b]$ that contains the maximum and decrease it using a golden section search until its length is less than some small value $\epsilon$.

Unlike the function $F_s(t)$, there is no general property for $F^*(t)$ that we can exploit in order to perform minimization $\min_t F^*(t)$ efficiently. For this reason, the minimization is solved by linearly searching the values of $t$ in the interval $[0, \rho]$. The value of $\rho$ is determined empirically and depends on the buffer size. This value indicates the time scale at buffer overflow occurs.

### *4.1. Code*

We present here an example of effective bandwidth estimation solving the inf – sup problem.

First of all we generate a trace from a GMFM constructing a Markov chain of 13 states where each state correspond to one of the transfer speed range:

```
transf_range=[0 64 128 256 512 1024 2048 3072 4096 5120 6144 7168 8292 10240];
```

this is, when the chain is in state 1, the transfer rate of the trace lies between $[0;64]$ bits per second, in state 2 between $[64;128]$, and so on.

The higher transfer rate available in the transmission channel, in our case the state 13, is expected to be the usual state and is also usual it jump from one state to neighboring states. With these considerations infinitesimal generator is designed to simulate the model. The speed is choose randomly within it range with a truncated gaussian centered.

As we know when the trace is within a range, the modulating chain is in certain state, to estimate the infinitesimal generator we count the jumps, from one state to another in a in increasing window time. We also estimate the average rate of each state by averaging the states that the trace visits in the corresponding range in an increasing window time.

With the infinitesimal generator estimate ($Q_{estim}$) and the average rate estimated (*gammas*) we can calculate the effective bandwidth from 3:

```
pi_est=expm(Q_estim*100);
alpha=@(s,t) 1/(s*t)*log(pi_est(1,:)*expm((Q_estim+diag(gammas)/1000*s)*t)*ones(13,1));
```

Now we can solve the inf – sup problem: For fix $t_0$ you find the maximum $s_{\text{star}}$, and evaluate the function $F$.

```
F=@(s,t) s*(B+C*t)-log(pi(1,:)*expm((Q+h*s)*t)*ones(13,1)); %The last part
%coincides with the alpha
g=(sqrt(5)-1)/2; epsilon=0.01;  %Golden Ratio and Error threshold
%The following while structure performs the golden search
while s_min-s_max>epsilon
gamma=g*(s_max-s_min); sl=s_max-gamma; sr=gamma+s_min;
Fsl=subs(F,[s,t],[sl,tfijo]); Fsr=subs(F,[s,t],[sr,tfijo]);
    if Fsl>Fsr ;    s_max=sr;
    elseif Fsl<Fsr ; s_min=sl;
    elseif Fsl==Fsr
        s_min=sl; s_max=sr;     end;    end; s_star=(s_min+s_max)/2
```

The value $s_{\text{star}}$ is the maximum in the variable $s$ when $t = t_0$. This gives three numbers $[s_{\text{star}}, t_0, F(s_{\text{star}}, t_0)]$.

We have to repeat the procedure until a minimum in is found. This may be a long loop, but always possible to reach. Finally we have found the link operating point, the BOP and the parameters of $s^*$ and $t^*$.

The above results can be extended to the design of a network link that requires certain quality of service. The minimum buffer size of a link can be calculated for a given capacity of the link, the traffic and the maximum loss probability desired. In a similar way, having the same information as before, but defined the desired buffer size, it is possible to calculate minimum link capacity required to ensure certain probability of loss. In Figure 1 we show a trace, the estimation of the operational point and the estimation of the probability of loss.

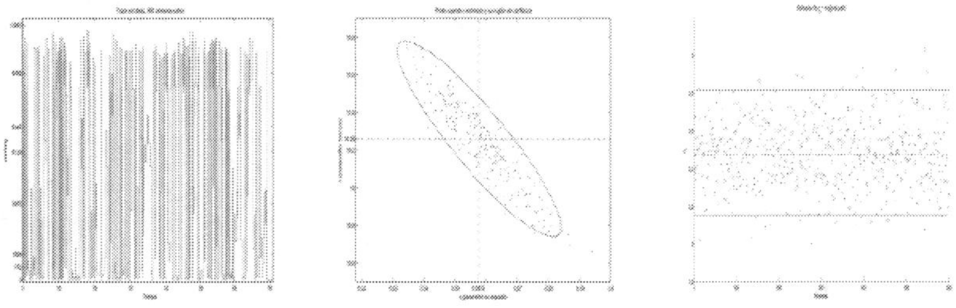

(a) (b) (c)

**Figure 1.** (a) GMFM trace. (b)Operational point estimation. c) Probability of loss estimation.

## 5. Conclusions and Remarks

In this work we want to give an insight of the traffic engineering process given QoS parameters, and presenting an application to Generalized Markov Fluids exploiting the availability of "good" effective bandwidth estimators for this source class. Knowledge on this subject allows better (QoS) networks design and resource utilization, two important features in telecommunications.

Code presented in section 4, is developed in Matlab environment an, due to space requirements, it is an extract to illustrate some key parts of the algorithm. For a fully compilable version contact the authors at *jmbavio@yahoo.com.ar*.

## References

[1] Marrón B 2012 *Estadística de Procesos Estocásticos aplicados a Redes de Datos y Telecomunicación* Ph.D. thesis Departamento de Matemática, Universidad Nacional del Sur

[2] Kelly F 1996 Notes on effective bandwidths

[3] Kesidis G, Walrand J and shang Chang C 1993 Effective bandwidths for multiclass markov fluids and other atm sources

[4] Courcoubetis C and Weber R 1995 *Journal of Applied Probability* **33** 886–903

[5] Courcoubetis C and Siris V A 2002 *Perform. Eval.* **48** 5–23

# Numerical solution of unsteady generalized Newtonian and Oldroyd-B fluids flow by dual time-stepping method

**R Keslerová, K Kozel**

Czech Technical University in Prague, Faculty of Mechanical Engineering, Dep. of Technical Mathematics, Czech Republic

E-mail: Radka.Keslerova@fs.cvut.cz, Karel.Kozel@fs.cvut.cz

**Abstract.** This work deals with the numerical solution of viscous and viscoelastic fluids flow. The governing system of equations is based on the system of balance laws for mass and momentum for incompressible laminar fluids. Different models for the stress tensor are considered. For viscous fluids flow Newtonian model is used. For the describing of the behaviour of the mixture of viscous and viscoelastic fluids Oldroyd-B model is used. Numerical solution of the described models is based on cell-centered finite volume method in conjunction with artificial compressibility method. For time integration an explicit multistage Runge–Kutta scheme is used. In the case of unsteady computation dual-time stepping method is considered. The principle of dual-time stepping method is following. The artificial time is introduced and the artificial compressibility method in the artificial time is applied.

## 1. Mathematical model

The governing system of equations is the system of balance laws of mass and momentum for incompressible fluids. This system is completed by the equation for a viscoelastic part of stress tensor, [1]:

$$\text{div}\ \boldsymbol{u} = 0 \tag{1}$$

$$\rho\frac{\partial \boldsymbol{u}}{\partial t} + \rho(\boldsymbol{u}.\nabla)\boldsymbol{u} = -\nabla P + \text{div}\ \mathbf{T_s} + \text{div}\ \mathbf{T_e} \tag{2}$$

$$\frac{\partial \mathbf{T}_e}{\partial t} + (\boldsymbol{u}.\nabla)\mathbf{T}_e = \frac{2\mu_e}{\lambda_1}\mathbf{D} - \frac{1}{\lambda_1}\mathbf{T}_e + (\mathbf{W}\mathbf{T}_e - \mathbf{T}_e\mathbf{W}) + (\mathbf{D}\mathbf{T}_e + \mathbf{T}_e\mathbf{D}) \tag{3}$$

where $P$ is the pressure, $\rho$ is the constant density, $\boldsymbol{u}$ is the velocity vector. The symbols $\mathbf{T_s}$ and $\mathbf{T_e}$ represent the Newtonian and viscoelastic parts of the stress tensor and

$$\mathbf{T}_s = 2\mu_s\mathbf{D}, \qquad \mathbf{T}_e + \lambda_1\frac{\delta \mathbf{T}_e}{\delta t} = 2\mu_e\mathbf{D} \tag{4}$$

where $\mathbf{D}$ is symmetric part of the velocity gradient $\mathbf{D} = \frac{1}{2}(\nabla\boldsymbol{u} + \nabla\boldsymbol{u}^T)$ and $\mathbf{W}$ is antisymmetric part of the velocity gradient $\mathbf{W} = \frac{1}{2}(\nabla\boldsymbol{u} - \nabla\boldsymbol{u}^T)$.

Both models could be generalized. In this case the viscosity $\mu$ is defined by viscosity function according to the cross model (for more details see [2])

$$\mu(\dot{\gamma}) = \mu_\infty + \frac{\mu_0 - \mu_\infty}{(1 + (\lambda\dot{\gamma})^b)^a}, \quad \dot{\gamma} = 2\sqrt{\frac{1}{2}\text{tr } \mathbf{D}^2} \tag{5}$$

where $\mu_0 = 1.6 \cdot 10^{-1} Pa.s, \mu_\infty = 3.6 \cdot 10^{-3} Pa.s, a = 1.23, b = 0.64, \lambda = 8.2s$.

## 2. Numerical solution

### *2.1. Steady case*

Numerical solution of the described models is based on cell-centered finite volume method using explicit Runge–Kutta time integration. Steady state solution is achieved for $t \to \infty$. In this case the artificial compressibility method can be applied. It means that the continuity equation is completed by the time derivative of the pressure (for more details see e.g. [3], [4], [5]). The system of equations (including the modified continuity equation) could be rewritten in the vector form.

$$\tilde{R}_\beta W_t + F_x^c + G_y^c = F_x^v + G_y^v, \quad \tilde{R}_\beta = \text{diag}(\frac{1}{\beta^2}, 1, 1), \quad \beta \in \mathbf{R}^+ \tag{6}$$

where $W$ is vector of unknowns, $F^c, G^c$ are inviscid fluxes and $F^v, G^v$ are viscous fluxes.

Eq. 6 is discretized in space by the finite volume method and the arising system of ODEs is integrated in time by the explicit multistage Runge-Kutta scheme [6], [7].

*Steady boundary conditions* The flow is modelled in a bounded computational domain where a boundary is divided into three mutually disjoint parts: an inlet, an outlet and a solid wall. At the inlet Dirichlet boundary condition for velocity vector is used. At the outlet the pressure value is given. The homogeneous Dirichlet boundary condition for the velocity vector is used on the wall.

### *2.2. Unsteady case*

For the unsteady computation the dual-time stepping method is used. The principle of dual-time stepping method is following. The artificial time $\tau$ is introduced and the artificial compressibility method in the artificial time is applied. The system of Navier-Stokes equations is extended to unsteady flows by adding artificial time derivatives $\partial W/\partial \tau$ to all equations, for more details see [8]

$$\tilde{R}_\beta W_\tau + \tilde{R} W_t + F_x^c + G_y^c = F_x^v + G_y^v, \quad \tilde{R} = \text{diag}(0, 1, 1), \quad \tilde{R}_\beta = \text{diag}(\frac{1}{\beta^2}, 1, 1). \tag{7}$$

The derivatives with respect to the real time $t$ are discretized using a three-point backward formula, it defines the form of unsteady residual

$$\tilde{R}_\beta \frac{W^{l+1} - W^l}{\Delta\tau} = -\tilde{R}\frac{3W^{l+1} - 4W^n + W^{n-1}}{2\Delta t} - \text{Res}(W)^l = -\overline{\text{Res}}(W)^{l+1}, \tag{8}$$

where $\Delta t = t^{n+1} - t^n$ and $\text{Res}(W)$ is the steady residual. The symbol $\overline{\text{Res}}(W)$ denotes unsteady residual.

*Unsteady boundary condition* In the inlet, in the solid wall and in one of the outlet part the steady boundary conditions are prescribed. In the branch going up the pressure value is prescribed by the function

$$p_{out} = \frac{1}{4}\left(1 + \frac{1}{2}\sin(\omega t)\right), \tag{9}$$

where $\omega$ is the angular velocity defined as $\omega = 2\pi f$, where $f$ is the frequency.

## 3. Numerical results

### *3.1. Steady numerical results*

In this section the steady numerical results of two dimensional incompressible laminar viscous and viscoelastic flows for generalized Newtonian and Oldroyd-B fluids flow are presented. The following model parameters are used: $\mu_e = 0.001 Pa.s$, $\mu_s = 0.009 Pa.s$, $\lambda_1 = 0.06s$, $U_0 = 0.1 m.s^{-1}$, $L_0 = 0.01m$, $\rho = 1000 kg.m^{-3}$.

In Fig. 1 velocity isolines for tested fluids cases are presented.

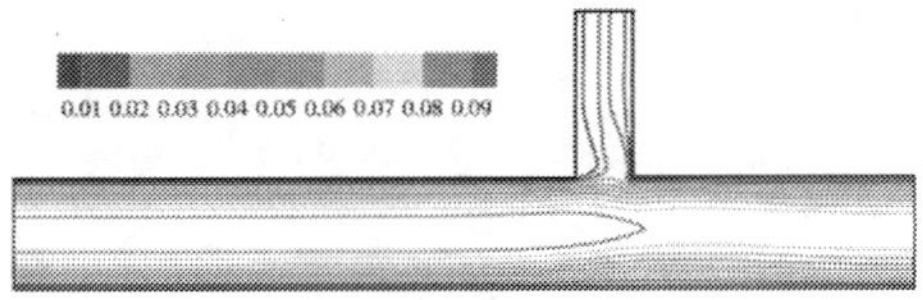


(a) Newtonian

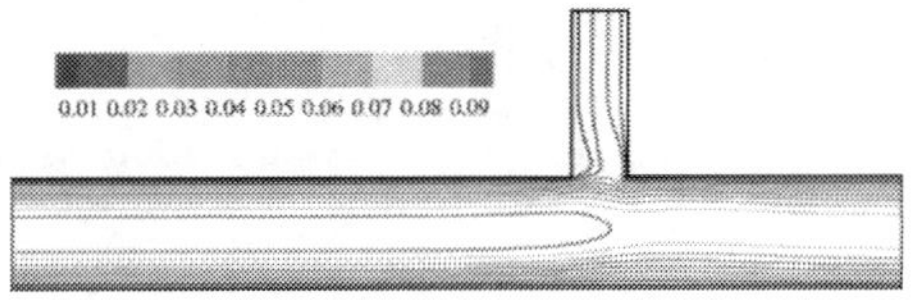


(b) generalized Newtonian

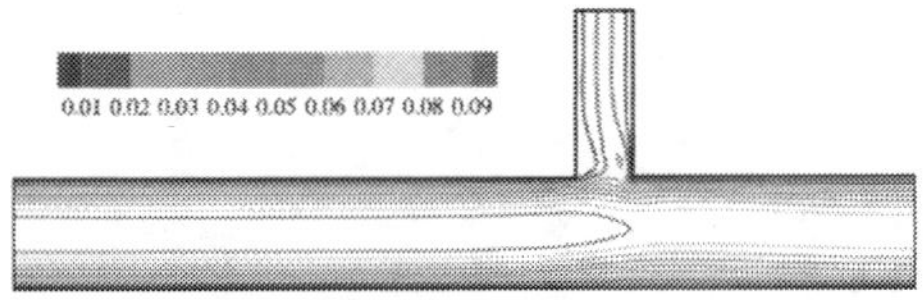


(c) Oldroyd-B

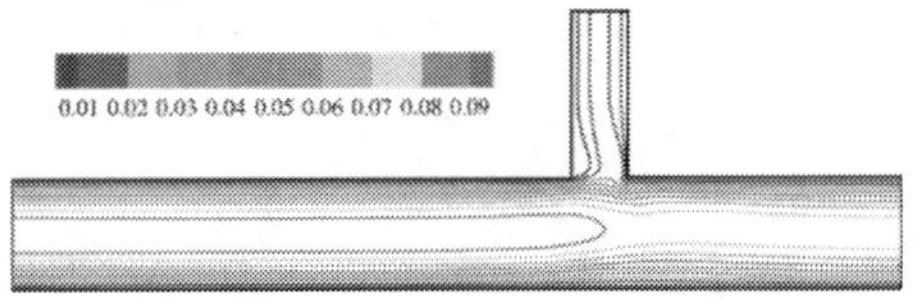


(d) generalized Oldroyd-B

**Figure 1.** Velocity isolines of steady flows for generalized Newtonian and Oldroyd-B fluids.

### *3.2. Unsteady numerical results*

The used method is the the dual-time stepping method with artificial compressibility coefficient $\beta = 1.0 m.s^{-1}$. The frequency $f = 2H$ is used. In Figs. 2 and 3 graphs of velocity as the function of time and the velocity distribution are shown. First pictures show graphs of velocity as the function of the time. The square symbols mark positions in time of the snapshots shown in next three pictures during one period. As initial data the numerical solution of steady fully developed flow of generalized Newtonian fluid was used (see Figs. 1).

## 4. Conclusions

In this paper a finite volume solver for incompressible laminar viscous flows in the branching channel for two dimmensional case was described. Newtonian model was generalized for generalized Newtonian fluids flow. Power-law model with different values of power-law index were used. Three different values of this coefficient for corresponding Newtonian, shear thickening and shear thinning fluids flow were tested. The explicit Runge-Kutta method was considered for time integrating. The convergence history confirms robustness of the applied method. The numerical results obtained by this method were presented and compared.

Two unsteady approaches were considered, the artificial compressibility method and the dual-time stepping method. Both methods were applied for generalized Newtonian fluids with initial data obtained by steady numerical computation. In this computations two different values of frequency were used.

## Acknowledgments

This work was partly supported by the grant GACR 201/09/0917 and SGS13/174/OHK2/3T/12.

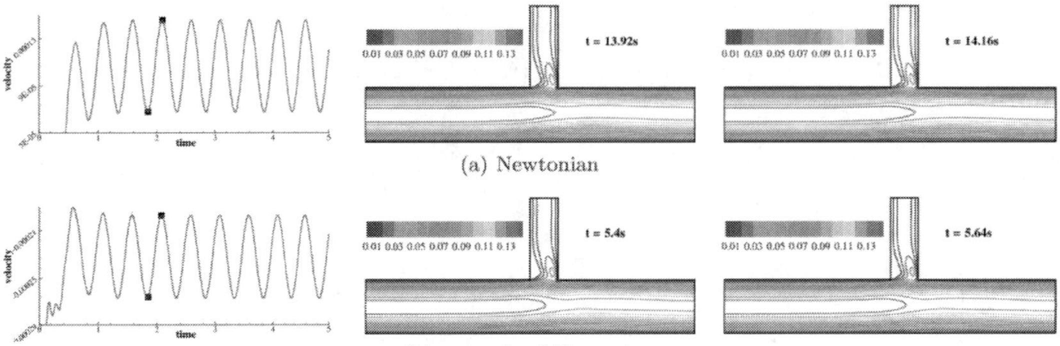


(a) Newtonian

(b) generalized Newtonian

**Figure 2.** The graphs of the velocity as the function of time and velocity isolines of unsteady flow for Newtonian and generalized Newtonian fluids.

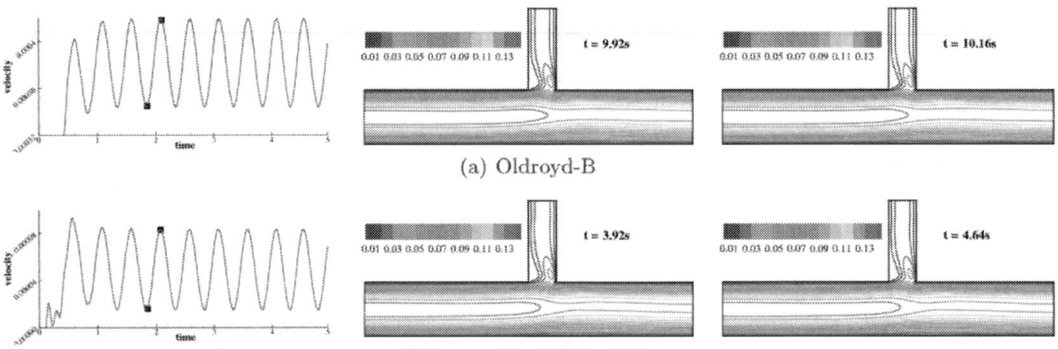


(a) Oldroyd-B

(b) generalized Oldroyd-B

**Figure 3.** The graphs of the velocity as the function of time and velocity isolines of unsteady flow for Oldroyd-B and generalized Oldroyd-B fluids.

## References

[1] Bodnar T and Sequeira A 2010 *Advances in Mathematical Fluid Mechanics* ed R. Ranacher, A. Sequeira pp 83–104.
[2] Vimmr J and Jonášová A 2010 *Mathematics and Computers in Simulation* vol 80 pp 1324–1336
[3] Chorin AJ 1967 *Journal of Computational Physics* vol 135 pp 118–125
[4] Dvořák R and Kozel K 1996 *Mathematical Modelling in Aerodynamics* (CTU, Prague, Czech Republic)
[5] Louda P and Kozel K and Příhoda J and Beneš L and Kopáček T 2011 *Journal of Computers & Fluids* vol 46 pp 318–324
[6] Keslerová R and Kozel K 2010 *Journal of Mathematics and Computers in Simulation* vol 80 pp 1783–1794
[7] LeVeque R 2004 *Finite-Volume Methods for Hyperbolic Problems* (Cambridge University Press)
[8] Honzátko R 2007 *Numerical simulations of incompressible flows with dynamical and aeroelastic effects* (PhD thesis, CTU Prague, Czech Republic)

# ALOPEX stochastic optimization for pumping management in fresh water coastal aquifers

**P.N. Stratis[1], Y.G. Saridakis[2], M.S. Zakynthinaki[3] and E.P. Papadopoulou[4]**

Applied Mathematics & Computers Laboratory, Department of Sciences
Technical University of Crete,
Chania 73100, Greece

E-mail: [1]`pstratis@science.tuc.gr`, [2]`yiannis@science.tuc.gr`, [3]`marzak@science.tuc.gr`, [4]`elena@science.tuc.gr`

**Abstract.** Saltwater intrusion in freshwater aquifers is a problem of increasing significance in areas nearby the coastline. Apart from natural disastrous phenomena, such as earthquakes or floods, intense pumping human activities over the aquifer areas may change the chemical composition of the freshwater aquifer. Working towards the direction of real time management of freshwater pumping from coastal aquifers, we have considered the deployment of the stochastic optimization *Algorithm of Pattern Extraction* (ALOPEX), coupled with several penalty strategies that produce convenient management policies. The present study, which further extents recently derived results, considers the analytical solution of a classical model for underground flow and the ALOPEX stochastic optimization technique to produce an efficient approach for pumping management over coastal aquifers. Numerical experimentation also includes a case study at Vathi area on the Greek island of Kalymnos, to compare with known results in the literature as well as to demonstrate different management strategies.

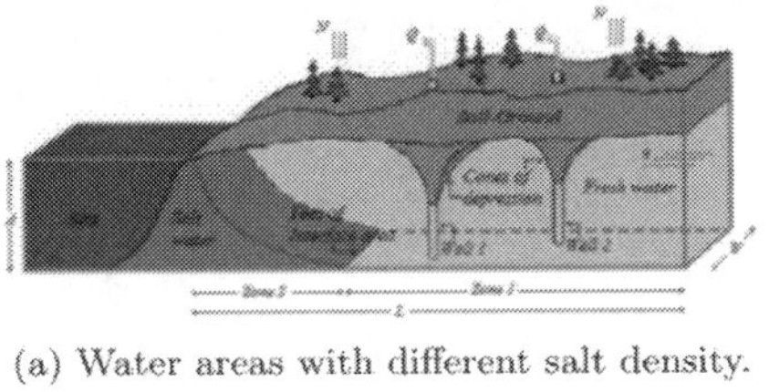

(a) Water areas with different salt density.

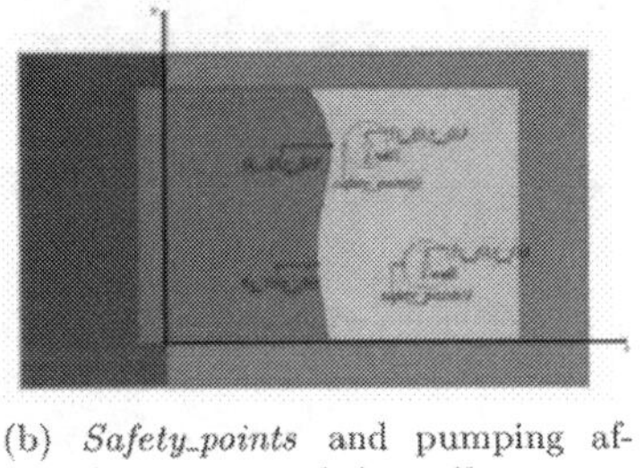

(b) *Safety_points* and pumping affected areas around the wells.

Figure 1: Coastal freshwater aquifer.

## 1. ALOPEX II optimization

In the present study a well-known existing mathematical model, presented in [2] and [3], is numerically simulated and the management of freshwater pumping is investigated and optimized by means of stochastic optimization. Extending our results, recently included in [4], a new, well established (e.g. [2]), profit function is used and, although it is a linear one, the convergence of the stochastic optimization ALOPEX II algorithm ([1], [5]) is achieved in a few iterations. Furthermore, the penalties management setup, proposed in [4], is being qualitatively improved to a management procedure that effectively controls well pumping rates taking into consideration wells' positions and their pumping capabilities.

### 1.1. *Toe constraint formulation*

The problem at hand is the determination of the optimal values for the pumping rates $Q(i)$ , $i = 1, \dots, n$, with $n$ being the number of wells, without contaminating with saltwater any one of the wells pumping fresh water over the aquifer. This problem is formulated as follows:

$$\begin{aligned} &\text{maximize} \quad \text{Profit}(Q_1, \cdots, Q_N) = \frac{\sum_{i=1}^{n} Q(i)}{\sum_{i=1}^{n} Q_{local}^{max}(i)} \leq 1 \\ &\text{under the constrains} \\ &x_{well}(i) - x_{toe}(i) \geq D \ , \ 0 \leq Q_{local}^{min}(i) \leq Q(i) \leq Q_{local}^{max}(i) \ , \ \sum_{i=1}^{n} Q(i) \leq Q_{aquifer}^{total}, \end{aligned} \tag{1}$$

where $x_{toe}(i)$ denotes the x-coordinate of the saltwater interface opposite to the $i$-well and $D = x_{well}(i) - safety\_point(i) > 0$ is a safety distance (see also figure 1b). The new values of pumping rates $Q(i)$ are determined at every ALOPEX II iteration step, using the following iteration scheme:

$$\begin{aligned} &Q^{(k)}(i) = Q^{(k-1)}(i) + c\big[Q^{(k-1)}(i) - Q^{(k-2)}(i)\big]\big[Profit^{(k-1)} - Profit^{(k-2)}\big] + noise^{(k)}(i) \\ &\text{where c} \in \mathbb{R}\text{:constant, } noise^{(k)}\text{:Gaussian noise .} \end{aligned} \tag{2}$$

### 1.2. *Penalties management*

In order to protect the aquifer wells from saltwater intrusion, appropriately defined pumping penalties are activated and enforced in each iteration step of the optimization algorithm.

More specifically, $Q_{min}$ and $Q_{max}$ are the penalties controlling respectively the minimum and the maximum volume of water that can be pumped by each well of the aquifer, in order to enforce the second and third constrain of (1) in every iteration (refer to [4] for detailed information). The *x_movement* penalty, also presented in [4], is responsible for an early attempt of protecting the integrity of each well, by lowering its pumping rate, when saltwater front reaches the *safety_points* (see subsection 2.4 for a demonstration of this penalty).

For the successful enforcement of the first constrain of (1), the previous penalty is coupled with a new penalty that implements the pumping influence of the $i$-th well to the $j$-th well inside the aquifer, in every iteration. For this purpose we introduce the *risk_to_each_other* $n \times n$ real matrix, every element of which is calculated by the following scheme:

$$\begin{aligned} &\text{if } i \neq j \text{ and } x_{well}(i) >= x_{well}(j) : \\ &risk\_to\_each\_other^{(k)}(i,j) = \frac{Q_{local}^{max}(i) - Q^{(k)}(i)}{max\{Q_{local}^{max}\}} * \frac{safety\_point(i) - x_{toe}^{(k)}(i)}{L} * \frac{\sqrt{(x_{well}(i) - x_{well}(j))^2 + (y_{well}(i) - y_{well}(j))^2}}{\sqrt{L^2 + W^2}} \\ &\text{else : } risk\_to\_each\_other^{(k)}(i,j) = 0, \text{ for } i,j \in \{1,2,..,n\}, \text{ at } k\text{-th iteration.} \end{aligned} \tag{3}$$

Then, *risk_penalties* are activated for every aquifer well and applied using the following scheme:
if $risk\_to\_each\_other^{(k)}(i,j) = 0 : risk\_penalty = 1$,
if $risk\_to\_each\_other^{(k)}(i,j) \in (0, 0.020) : risk\_penalty = 0.98$,
if $risk\_to\_each\_other^{(k)}(i,j) \in [0.020, 0.045) : risk\_penalty = 0.96$,
if $risk\_to\_each\_other^{(k)}(i,j) \in [0.45, 0.070) : risk\_penalty = 0.92$,
if $risk\_to\_each\_other^{(k)}(i,j) \in [0.070, 1] : risk\_penalty = 0.88$.

## 2. Numerical simulations: Vathi aquifer, Kalymnos, Greece

As in [4] (cf. also[2]), we examine the case of a homogeneous aquifer of almost rectangular geometry, located at the Vathi area on the Greek island Kalymnos, with the following properties: $L = 7000m$, $W = 3000m$, $K = 100m/day$, $q = 1.23m^2/day$, $d = 25m$, $N = 30mm/year$, $safety\_distance = 400m$, $Q_{aquifer}^{total} = 20000m^3/day$, $(Q_{local}^{min}, Q_{local}^{max}) = (200, 1500)m^3/day$. Inside the aquifer area there exist 5 pumping wells: $x_{well} = (3932, 2657, 4873, 3353, 4632)m$, $y_{well} = (975, 1572, 1586, 2200, 2470)m$.

The critical well in this area is the well no 2 (see figure 2a), the first one at the left side of the aquifer. This is the closest to the saltwater interface area and so, it is the most likely to be saltwater intruded. In order to prevent this intrusion, every time the saltwater front reaches the *safety_points* in front of every aquifer well, the pumping penalties presented before are activated to manage the pumping. The objective is to maximize the water pumping from all the aquifer wells, given the constraints. We examine a number of subcases, presenting different pumping setups.

### 2.1. Vathi aquifer: Main case

The ALOPEX II optimization method, in a typical run of 300 iterations, is called to determine the optimal pumping rates for all aquifer wells, using the following penalty management setup: $(Q_{min}, Q_{max}) = (1.20, 0.95)$, *x_movement penalty*=0.95 and *risk_penalties*=(1.00,0.98,0.96,0.92,0.88). Although the initial values of the pumping rates correspond to a low value of the profit function, the optimization method is able to converge to an optimal solution within a few iterations. The optimal pumping rates achieved during this procedure are the following:
$Q^{opt} = (762.90, 346.21, 1240.82, 1378.89, 206.12)m^3/day$, with $\sum_{i=1}^{5} Q^{opt}(i) = 3934.95m^3/day$.
Numbers of penalties activation: $(Q_{min}, Q_{max}, x\_movement, risk) = (85, 79, 230, 174)$.
All wells are kept safe from saltwater intrusion (see figure 2a).

### 2.2. Subcase 1: Optimal pumping rates artificially increased

In order to examine the sensitivity of the solution presented in the main case above, we studied the influence of a 2% increase of the optimal pumping rate of all aquifer wells obtained earlier. The modified new pumping scheme is as follows:
$Q^{mod} = (778.16, 353.13, 1265.64, 1406.48, 210.24)m^3/day$, with $\sum_{i=1}^{5} Q^{mod}(i) = 4013.65m^3/day$.
Inspecting figure 2b it becomes obvious that such an increase results to saltwater contamination of well 2. This leads us to the conclusion that the ALOPEX optimization procedure and the penalty management system used in the main case produced near optimal results in the sense that small global perturbations may lead to violation of the constrains.

### 2.3. Subcase 2: Rain factor N increased

We consider, as an example, the case of a 20% increase on the rain factor $N$ over the aquifer area. The new rain conditions of the aquifer are assumed to be $N = 1.20 * 30 = 36mm/year$. ALOPEX II is called to calculate the new optimal pumping rates. We can see a 7.72% increase of the total volume of pumping water, while all the wells are kept safe from saltwater intrusion. The penalties management setup was the one proposed at the main case (see 2.1) and the new optimal pumping rates are the following:
$Q^{opt} = (1449.21, 240.14, 946.72, 1380.36, 222.27)m^3/day$, with $\sum_{i=1}^{5} Q^{opt}(i) = 4238.69m^3/day$.
Numbers of penalties activation: $(Q_{min}, Q_{max}, x\_movement, risk) = (60, 114, 215, 148)$.

### 2.4. Subcase 3: X_movement penalties turned off

The *x_movement* penalties are activated when the saltwater front reaches the *safety_points* in front of every well (see figure 1b). They are the "first line of defense" against the saltwater advancement to the interior of the aquifer. In this way, when a well is in danger, the first counter-reaction is to reduce the volume of pumping water from the specific well only. This means that the *x_movement* penalty has only a local effect in the pumping plan of the aquifer. It offers small changes in the total volume of pumping water and as a result, this penalty can only be used for fine tuning of the optimization procedure. On the other hand, lack of this penalty will not destroy the pumping management of the aquifer, but it will result in an erratic behavior of the profit function in the optimization procedure and it is possible that some local extrema will not be examined (see figure 2c). For the optimization run we used the following penalty management setup:
$(Q_{min}, Q_{max}) = (1.20, 0.95)$, *x_movement penalty*=**1.00** and *risk_penalties*=(1.00,0.98,0.96,0.92,0.88).
The optimal values obtained for the pumping rates are the following:
$Q^{opt} = (830.22, 843.38, 732.72, 713.35, 736.74)m^3/day$, with $\sum_{i=1}^{5} Q^{opt}(i) = 3856.41m^3/day$.
Numbers of penalties activation: $(Q_{min}, Q_{max}, x\_movement, risk) = (165, 177, 0, 187)$.

### 2.5. Subcase 4: Risk_to_each_other penalties turned off

In order to demonstrate the necessity of the *risk_to_each_other* penalty set and its important role in the optimization process, we considered the case of turning off these penalties. The penalty management setup is as follows:
$(Q_{min}, Q_{max}) = (1.20, 0.95)$, *x_movement penalty*=0.95 and *risk_penalties*=(**1.00,1.00,1.00,1.00,1.00**).
The pumping rate values obtained after a few iterations were the following:
$Q = (724.34, 279.59, 1495.30, 292.26, 1449.60)m^3/day$ with $\sum_{i=1}^{5} Q(i) = 4241.09m^3/day$
The results are disastrous for the freshwater integrity of the aquifer wells no 2 and 4 (see figure 2d). In a few iterations saltwater contaminates both wells, in spite the fact their pumping rates were dropped close to the minimum acceptable values, due to frequent activation of the *x_movement* penalty. Wells

no 3 and 5 have almost reached their maximum pumping capabilities, without risking their freshwater integrity, keeping safe at the same time well no 1. The need for a robust pumping management is in this case more obvious than ever.

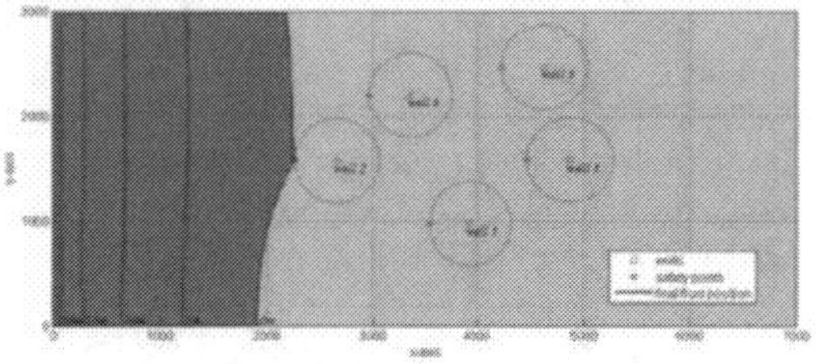

(a) Main case: Optimal results achieved by ALOPEX II algorithm with the penalty system fully activated. All wells are kept safe from saltwater intrusion.

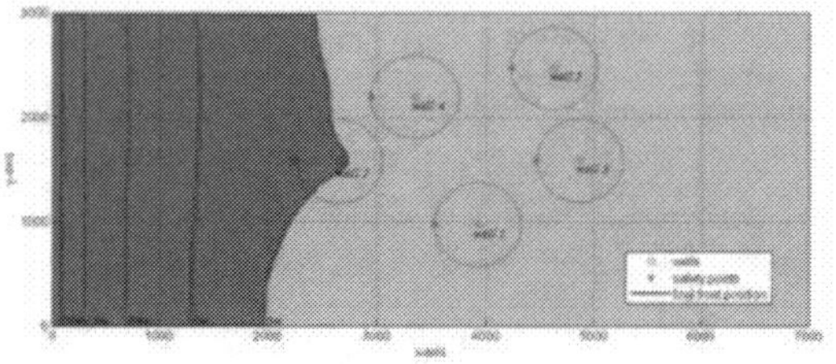

(b) Subcase 1: Main case's optimal pumping rates artificially increased by 2%. Well no 2 is intruded.

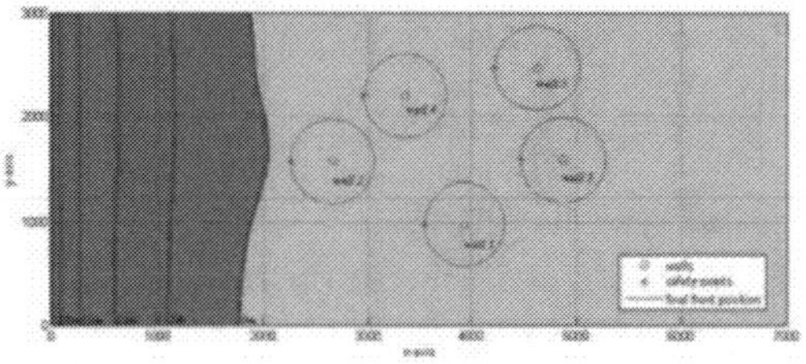

(c) Subcase 3: Optimal results achieved by ALOPEX II algorithm. The *x_movement* penalty is turned-off. Main case's optimal results cannot be achieved due to the lack of fine-tuning of the optimization procedure.

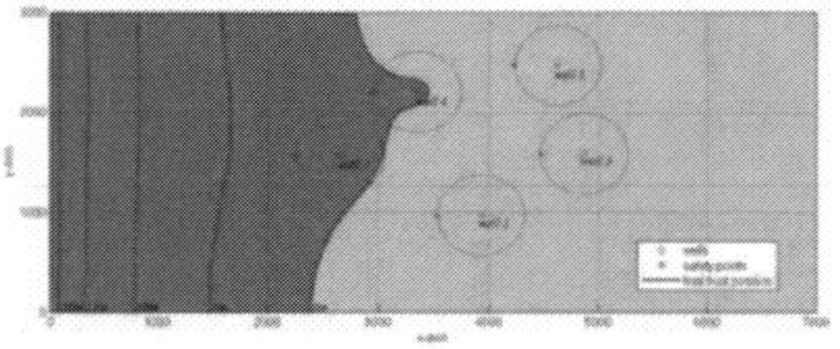

(d) Subcase 4: Results obtained after a few iterations of ALOPEX II algorithm with the *risk_to_each_other* penalties turned-off. The two frontal wells of the aquifer were intruded.

Figure 2: Contour plots of the interface area, at different elevation heights (5m step) from the bottom to the top of the aquifer, after completion of the optimization procedure for the case of Vathi aquifer in the Greek island of Kalymnos.

## 3. Conclusions

The present work effectively implements the stochastic optimization ALOPEX method, coupled with a constrain management system, to the problem of prevention saltwater intrusion in freshwater aquifers. Simulations are presented for the aquifer of Vathi area, on the Greek island of Kalymnos. A study on the sensitivity of the optimization process confirmed the efficiency and applicability of the optimization method, as well as the importance of the penalties imposed.

## Acknowledgments

The present research work has been co-financed by the European Union (European Social Fund ESF) and Greek national funds through the Operational Program Education and Lifelong Learning of the National Strategic Reference Framework (NSRF) - Research Funding Program: THALIS . Investing in knowledge society through the European Social Fund (MIS 379416). This work has also been supported by a research funding grand of Technical University of Crete.

## References

[1] Harth E and Tzanakou E 1974 ALOPEX: a stochastic method for determining visual receptive fields *Vision Research* **14** 1475

[2] Mantoglou A 2003 Pumping management of coastal aquifers using analytical models of saltwater intrusion *Water Recourses Research* **39(12)** 1335

[3] Strack O D L 1989 *Groundwater Mechanics* (Prentice Hall)

[4] Stratis P, Saridakis Y, Zakynthinaki M and Papadopoulou E 2013 *Stochastic optimization for a model problem of saltwater intrusion in coastal aquifers* (submitted)

[5] Zakynthinaki M and Saridakis Y G 2003 Stochastic optimization for a tip-tilt adaptive correcting system *Comp. Phys. Commun.* **150(3)** 274

# Construction of the most suitable unbiased estimate distortions of radiation processes from remote sensing data

**Iskakova Ayman**

Mechanic and Mathematic Department, L.N. Gumilyov Eurasian Nathional University, Mirzoyan st, 2, Astana, Kazakhstan

E-mail: ayman.astana@gmail.com

**Abstract.** Proposed and studied in a multivariate discrete probability model of distribution of processes distortion of radiation from remote sensing data. Introduce the concept of the most appropriate evaluation of the set of unbiased, which has good asymptotic properties.

## 1. Introduction

One of the characteristics assigned to the Astana Branch of JSC "NTSKIT" Center of Space-Based Monitoring is the regular reception and recording of the input data stream remote sensing satellites IRS-1C, IRS-1D, IRS-P6, RADARSAT, AQUA in the zone of radio visibility.

International Satellite System based on geostationary spacecraft is designed to meet the challenges of the global meteorological support and for consumers in the European , Asian and African regions. The management and maintenance work on the program is implemented by Eumetsat.

The remote sensing (RS) is an observation of the surface of the Earth aviation and space vehicles equipped with different types of imaging equipment. Wavelength range, which is receiving apparatus, is constitutes from a fraction of micrometer (visible optical radiation) to m (radio waves). The methods can be passive, that is to use a natural reflection or secondary heat radiation of objects on the Earth's surface due to solar activity, and active, these are using stimulated emission objects initiated by artificial light source directed action. The datas of RS, which is received from the spacecraft, is characterized by a large degree of depending on the transparency of the atmosphere. So on spacecraft equipment the multichannel passive and active types are used, which have recording of electromagnetic radiation in different ranges.

RS can be represented as a process by which information is gathered about an object, area or phenomenon without direct contact with him. Part of RS are entering in the digital form that allows you to use them for processing modern computer technology. Pictures on photo can be converted into a digital format for a bitmap using special scanning devices (scanners) . Digital image in the form of a raster is a matrix of numbers.

RS contain a series of random, systemic and systematic distortions due to the influence of the atmosphere, the Earth's curvature, motion shooting the vehicle with respect to its surface at the time of the shooting, the physical characteristics of sensors used and communication channels.

To eliminate the mentioned quite numerous distortions, taking into account their specific uses several types of correction: radiation, radiometric, and geometric calibration.

Objectives of the study of: a) submission of a probabilistic model processes distortion RS; b) the construction of unbiased estimates of the probability distribution of the model under investigation. All major results are new and are as follows: a) proposed and studied by multivariate discrete probability model for the distribution of sums of unobservable independent identically distributed random matrices; b) it is shown that the best unbiased estimate of the probability of this distribution exist only if the uniqueness of the partition of the observed matrix of all possible components. c) if some element of the implementation of the sample has more than one partition on the possible components, then we construct a set of unbiased estimators studied for distribution; d) introduce the concept of the most appropriate estimate of the set of unbiased, which has good asymptotic properties.

The results and methods presented in the paper can be used in research specialists in Mathematical Statistics and in Space Monitoring .

Telemetry data transmitted from the Board of the satellite at a frequency of 1675.928 MHz. To send telemetry (see [1]) using directional antenna S-band high-gain, and at the stage of withdrawal satellite into orbit - directional antenna S-range frequencies. Command information is taken at a frequency of 2098.0 MHz using a directional antenna S-band high-gain.

On the energy characteristics of the radio transmission of data in HR with the IRS series of Meteosat satellites (see [2], p. 205) affect the EIIM, loss of distribution losses in the antenna pointing ($\pm 1$), gain receiving antenna with a mirror diameter of 4 m , losses in the feeder lines, return loss, received power $A+2+3+4+5+6$, modulation loss, the effective received signal $G+8$, the noise temperature of antenna elevation angle $30^\circ$, the noise temperature, system noise temperature $A0+11$, noise spectral density $198.6+A2$, data transmission speed 166 kbit/second, the total noise power of the $A3+A14$ and so on. In other words, the distortion of the influence factor of 23, ie, $d = 23$.

## 2. Multivariate discrete probability distribution processes remote sensing

Consider a probabilistic model of the processes of energy characteristics of radio satellite IRS Meteosat. In [3], a probabilistic-statistical model of weather prediction.

Assume that the true image can be represented as a matrix $\mathbf{l}_0 = \|l_{0_{i,j}}\|_{m\times q}$, which imposed distortion, consisting of four factors (matrices) of losses $\mathbf{u} = \|u_{i,j}\|_{m\times q}$, taking values from the set of $\mathbf{l}_1, \mathbf{l}_2, \dots,, \mathbf{l}_d$.

Obviously, the factors (the matrix), the loss $\mathbf{l}_1, \mathbf{l}_2, \dots,, \mathbf{l}_d$ are realizations of random matrices $\mathbf{L}_1, \mathbf{L}_2, \dots,, \mathbf{L}_d$, which appear relevant probabilities $\mathbf{p} = (p_1, \dots, p_d)$, where

$$\sum_{\alpha=1}^{d} p_\alpha = 1.$$

Assume that $V_{\mathbf{u}}$ is the number of possible combinations $r_{1_{v_{\mathbf{u}}}}\mathbf{L}_1, \dots, r_{d_{v_{\mathbf{u}}}}\mathbf{L}_d$, which together form a matrix of $\mathbf{u}$, where $r_{1_{v_{\mathbf{u}}}}, \dots, r_{d_{v_{\mathbf{u}}}}$ determine the possible number of balls taken out, which marked the relevant matrices $\mathbf{L}_1, \mathbf{L}_2, \dots,, \mathbf{L}_d$. In other words, from [4] it follows that $V_{\mathbf{u}}$ is the number of partitions on the part of the matrix $\mathbf{u}$ on $\mathbf{L}_1, \mathbf{L}_2, \dots,, \mathbf{L}_d$.

**Theorem 1.** *The probability that the distortion takes the value u is determined by the formula*

$$P(\mathbf{U} = \mathbf{u}) = \sum_{v_{\mathbf{u}}=1}^{V_{\mathbf{u}}} n! \prod_{\alpha=1}^{d} \frac{p_\alpha^{r_{\alpha_{v_{\mathbf{u}}}}}}{r_{\alpha_{v_{\mathbf{u}}}}!}. \tag{1}$$

## 3. Unbiased estimation of the probability distribution of the proposed model

In practice, as a rule, elements of the vector $\mathbf{p} = (p_1, \ldots, p_d)$ are not known. It is also not known matrix $\mathbf{L}_1$, $\mathbf{L}_2$, ..., $\mathbf{L}_d$. Consequently, formula (1) does not find the actual application.

Assume that there are photos in the number of $k$ particular locality with the distortions $\mathbf{x} = \{\mathbf{x}_1, ..., \mathbf{x}_k\}$. In other words, a number of evidence-x can be interpreted as a realization of a sample of k, whose elements are subject to distribution (1). We denote $\mathbf{r}_{v_\beta}$ vector $(r_{1_{v_\beta}}, \ldots, r_{d_{v_\beta}})$, which defines $v_\beta$ - th solution of equation

$$\begin{cases} \sum\limits_{\alpha=1}^{d} L_\alpha r_{\alpha_{v_\beta}} = \mathbf{u}, \\ \sum\limits_{\alpha=1}^{d} r_{\alpha_{v_\beta}} = n, \end{cases} \tag{2}$$

, where $v_\beta = 1, \ldots, V_\beta, V_\beta$ - the number of partitions of the matrix $\mathbf{x}_\beta$ on the matrices $\mathbf{L}_1$, $\mathbf{L}_2$, ..., $\mathbf{L}_d$. Using the system of equations (3), the matrices $\mathbf{L}_1$, $\mathbf{L}_2$, ..., $\mathbf{L}_d$, and the actual data $\mathbf{x}$, we define for each $\beta = 1, \ldots, k$ the number of partitions $V_\beta$ matrix $\mathbf{x}_\beta$ at $\mathbf{L}_1$, $\mathbf{L}_2$, ..., $\mathbf{L}_d$, and vectors $\mathbf{r}_{1_\beta}$, ..., $\mathbf{r}_{V_\beta}$.

Suppose that for each $j = 1, \ldots, \mu$, where

$$\mu = \prod_{\beta=1}^{k} V_\beta,$$

there is a vector $\mathbf{z}_j = (z_{1_j}, \ldots, z_{d_j})$, defined as

$$\mathbf{z}_j = \sum_{\beta=1}^{k} \mathbf{r}_{v_\beta}, \tag{3}$$

and the indices on the right and left side are linked one-to-one correspondence, which is not unique.

Thus, from the above lemma that if some element of the implementation of the sample $\mathbf{x} = (\mathbf{x}_1, \ldots, \mathbf{x}_k)$ of the distribution (1) has more than one partition on the submitted part, it is impossible, using the theorem Rao - Blackwell-Kolmogorov construct an unbiased estimate with minimum variance for the probability distribution (1).

**Theorem 2.** *The elements of* $W(\mathbf{u}, \mathbf{z}) = W(\mathbf{u}, \mathbf{z}_1), \ldots, W(\mathbf{u}, \mathbf{z}_\mu)$ *is an unbiased estimate for the probability* $P(\mathbf{U} = \mathbf{u})$ *distribution* (1) *that for* $j = 1, \ldots, \mu$ *is defined as*

$$W(\mathbf{u}, \mathbf{z}_j) = \frac{\sum\limits_{v_\mathbf{u}=1}^{V_\mathbf{u}} \prod\limits_{\alpha=1}^{d} \binom{z_{\alpha_j}}{r_{\alpha_{v_\mathbf{u}}}}}{\binom{nk}{n}}, \tag{4}$$

*where* $V_\mathbf{u}$ - *number of partitions on the part of the matrix* $\mathbf{u}$ $\mathbf{L}_1$, $\mathbf{L}_2$, ..., $\mathbf{L}_d$; *for each partition* $r_{1_{v_\mathbf{u}}}, \ldots, r_{d_{v_\mathbf{u}}}$ *determine the possible number of matrices* $\mathbf{L}_1$, $\mathbf{L}_2$, ..., $\mathbf{L}_d$; $k \geq 1$ *and* $z_{\alpha_j} \geq r_{\alpha_{v_\mathbf{u}}}$, *when* $\alpha = 1, \ldots, d$, $v_\mathbf{u} = 1, \ldots, V_\mathbf{u}$.

## 4. The most suitable unbiased estimates for the probability distribution of the proposed model and their properties

Thus, we have a lot of unbiased estimates of the probability of distortion.

**Definition 1.** *Decision* $\mathbf{z}_g$, *based on observation, is the most appropriate set of* $\mathbf{z} = \{\mathbf{z}_1, \ldots, \mathbf{z}_m\}$, *if*

$$\prod_{\beta=1}^{k} W(\mathbf{x}_\beta, \mathbf{z}_g) = \max_{j=1,\ldots,\mu} \prod_{\beta=1}^{k} W(\mathbf{x}_\beta, \mathbf{z}_j), \tag{5}$$

*where for* $\beta = 1, \ldots, k$ elements of $W(\mathbf{x}_\beta, \mathbf{z}) = \{W(\mathbf{x}_\beta, \mathbf{z}_1), \ldots, W(\mathbf{x}_\beta, \mathbf{z}_\mu)\}$ *is an unbiased estimate for the probability* $P(\mathbf{U} = \mathbf{u})$ *distribution* (1) *defined in* (5).

**Definition 2.** *Unbiased estimate of* $W(\mathbf{x}_\beta, \mathbf{z}_g)$ *for the probability* $P(\mathbf{U} = \mathbf{u})$ *distribution* (1) *is the most suitable from the entire set of unbiased estimates of* $W(\mathbf{x}_\beta, \mathbf{z}) = \{W(\mathbf{x}_\beta, \mathbf{z}_1), \ldots, W(\mathbf{x}_\beta, \mathbf{z}_\mu)\}$ *defined in* (5), *if* $\mathbf{z}_g$ *- the most appropriate solution, based on observation.*

**Theorem 3.** *The most suitable unbiased estimate of* $W(\mathbf{x}_\beta, \mathbf{z}_g)$ *for the probability* $P(\mathbf{U} = \mathbf{u})$ *model* (1) *is consistent, asymptotically normal and asymptotically efficient.*

## References

Analysis conducted in this paper, allows us to formulate the following main results.

- Proposed and studied a new probability distribution of probabilistic model of the processes of energy characteristics of radio satellite Meteost.
- Developed an algorithm for computing the probability and define the generating function for the distribution of the proposed model.
- The set of unbiased estimates for the probability distribution of the proposed model and the variance of these estimates.
- Introduced a new concept of the most appropriate evaluation of the set of unbiased estimates, with asymptotic properties.

## References

[1] Lisov I. Satellites. // Space News, 01, 1996.
[2] Garbuk S.V., Gershenzon VE Space remote sensing. Moscow, 1997, 297 p.
[3] Iskakova A.S. Identifying the most suitable unbiased estimate for a weather forecast. // Siberian Journal of Industrial Mathematics. 2002.Tom V, 1 (9). P.79-84.
[4] Andrews G.E. The theory of partitions, Encyclopedia of Mathematics and Its Applications (Rota, et.). G.-C. Addison - Wesley, Reading. 1976. Vol.2. 256p
[5] Cramer G., Methods of Mathematical Statistics. - M. 1975. - 648 p.
[6] Nikulin M.S., Smirnov T.I., Voinov V.G. Multivariate distributions induced by an urn scheme, linear diophantine equations, unbiased estimating end testing. // Rapports Internes de l'Unit Math. Applique'es Bordeaux. 1998. No. 98 023 .- 11 p.

# Two-Dimensonal Decaying Turbulence in Different Shaped Containers

**Gabor Toth**[1] **and Gabor Hazi**[2]

[1]Eotvos University, Faculty of Science,
Department of Physics of Complex System, H-1117 Budapest, Pazmany Peter setany 1/A, Hungary
[2]Hungarian Academy of Sciences,
Centre for Energy Research, Budapest H-1525, Konkoly Th. 29-33, Hungary

E-mail: tooth@caesar.elte.hu, hazi.gabor@energia.mta.hu

**Abstract.** Several interesting phenomena have been observed simulating two-dimensional decaying turbulence in bounded domains. In this paper an overview is given about our observations obtained by simulating freely decaying turbulence in different regular polygon shaped containers with no-slip walls. For these simulations the lattice Boltzmann method has been used as a numerical approach. The initial Reynolds number based on the container dimension was in the order of 1000. The initial condition was the same in each simulation, therefore, we were able to compare the effect of geometrical constraints on the evolution of relevant physical quantities such as the kinetic energy and the enstrophy.

## 1. Introduction

Two-dimensional turbulence in bounded domains plays an important role in geosciences, meteorology and oceanography. One of the most impressive example has been taken from the nature; the vortex formation in coastal currents, a relevant phenomenon in geophysical fluid dynamics. In the last two decades numerical computational studies of decaying two-dimensional turbulence in a container have revealed some interesting results. It was pointed out, for instance, that the decay process of two-dimensional turbulence in a container differs significantly from the one which takes place in a double-periodic domain [1]. Clercx et al. [2] studied the evolution of 2D turbulence in square and circle shaped containers. They demonstrated that the presence of walls acts on the evolution of enstrophy due to the interactions between walls and vortices. In contrast to the evolution of 2D turbulence in a double periodic domain, the enstrophy is not a monotone decaying function of time in a container, but there can be local maxima in the enstrophy as a consequence of interactions between walls and vortices. The final state of the evolution is also different in a container and in a double periodic domain. While a simple dipole is formed at the end of the decay in a double periodic domain, in a container the flow spontaneously spin-up forming a large vortex in the center with small satellites around this structure.

In this paper we present simulation results of 2D decaying turbulence in regular polygon shaped (circle, octagon, hexagon, square) containers. Our aim is to learn more about the spontaneous spin-up process and the effect of the walls in 2D turbulence. Therefore we have peformed several simulations starting them from the same initial condition, i.e. the same

Reynolds number and the same initial vortex arrangement were used for each geometry. Then the flow was left to evolve and the relevant physical quantities were recorded. Finally, the simulation results were compared with each other as it is presented in the paper. The outline of the paper is follows. In Section 2 our numerical approach, the lattice Boltzmann method (LBM) is briefly introduced. In Section 3 the problem description is outlined and in Section 4 our results are presented. Finally, the conclusions are drawn in Section 5.

## 2. Lattice Boltzmann method

This section is devoted to briefly summarize some basic information on the lattice Boltzmann method [3],[4], which had been applied first for the simulation of two-dimensional turbulence in the pioneering work of Benzi and Succi [5].

Using the lattice Boltzmann method one solves a discrete kinetic equation for the one-particle velocity distribution functions $f_i$-s:

$$f_i(\mathbf{x} + \mathbf{c}_i\delta_t, t + \delta_t) - f_i(\mathbf{x}, t) = \Omega_i, \tag{1}$$

where $\mathbf{c_i}$ is the lattice vector, $\delta_t$ is the time step and $\Omega_i$ is the collision operator. In this paper we use the simplest form of the latter i.e. the BGK operator [6], [7]:

$$\Omega_i = -\frac{1}{\tau}\left[f_i(\mathbf{x}, t) - f_i^{eq}(\mathbf{x}, t)\right], \tag{2}$$

where $\tau$ is the relaxation time and $f_i^{eq}$ is a local equilibrium distribution function, which can take the following form:

$$f_i^{eq} = w_i\rho\left[1 + \frac{c_{i\alpha}u_\alpha}{c_s^2} + \frac{u_\alpha u_\beta}{2c_s^4}\left(c_{i\alpha}c_{i\beta} - c_s^2\delta_{\alpha\beta}\right)\right], \tag{3}$$

where $w_i$ is the lattice weight, $u$ is the hydrodynamic velocity and $c_s$ is the speed of sound (repeated Greek indices imply summation).

Solving Eq. (1) one can obtain the macroscopic quantities by taking the suitable moments of the distribution functions:

$$\rho = \sum_i f_i\,, \quad \rho u_\alpha = \sum_i c_{i\alpha} f_i. \tag{4}$$

The pressure can be obtained through the equation of state of an ideal gas, i.e. $p = \rho c_s^2$.

For a specific model, the lattice vector and the lattice weights need to be selected. In our analysis we will use the parameters of a two-dimensional nine-velocity model (D2Q9).

In this work Eq. (1) was solved by a simple streaming and collision procedure and to model no-slip walls we used the bounce-back rules [9], i.e. distribution functions encountering a solid surface were reflected back in the direction they came from.

## 3. Initial field and problem description

We consider a circle, an octagon, a hexagon and a square container with rigid no-slip walls and divide the domain into lattices using the resolution N x N = 1024 x 1024. The initial vorticity field contains forty-eight equal-size shielded Gaussian vortices [10]:

$$\omega = -\omega_0\left(\frac{r^2}{r_0^2} - 1\right)e^{-\frac{r^2}{r_0^2}}, \tag{5}$$

placed on a regular lattice in the center of the domain, with some small perturbations in their individual positions $(x_k, y_k)$ in order to break the symmetry of the problem. The corresponding velocity field of a vortex is given by

$$u_x = \frac{1}{2}\omega_0(y - y_k)e^{-\frac{r^2}{r_0^2}}, \quad u_y = -\frac{1}{2}\omega_0(x - x_k)e^{-\frac{r^2}{r_0^2}}, \tag{6}$$

where $r_0$ is the initial vortex radius. Half of the vortices have positive and the rest have negative circulation.

## 4. Results

In order to characterize the evolution process we define the kinetic energy $E(t)$ and enstrophy $\Omega(t)$ as follows [11]:

$$E(t) = \frac{1}{2}\int_\Omega |\mathbf{u}|^2 dA, \tag{7}$$

$$Z(t) = \frac{1}{2}\int_\Omega |\omega|^2 dA. \tag{8}$$

These quantities were continuously calculated and recorded during each simulation and here we briefly present how they evolve in the various geometry.

First the decay of the kinetic energy is shown in Fig. 1 for each geometry and for two different initial Reynolds numbers. As one can see the trend of the decay is almost the same in each case, it depends strongly on the initial Reynolds number, but apparently, the shape of the geometry does not influence the evolution of the kinetic energy. This observation might be surprising, since there is inverse energy cascade in 2D turbulence, which means that the energy is dissipated in large, integral scales. So one might expect that the shape of the container have some influence on the decaying process. However, the evolution of the flow is very similar in each geometry. In the initial stage, vortices interact with each other forming larger and larger structures, while they also interact with the wall producing small but strong counter-rotating vortices. This fact can be seen in Fig. 2 and Fig. 3, where snapshots of vorticity are shown at the same time instant for the different containers. In Fig. 2 the difference is almost negligible, and the size of the relevant vortices is very similar in the later stage, too as one can see in Fig. 3.

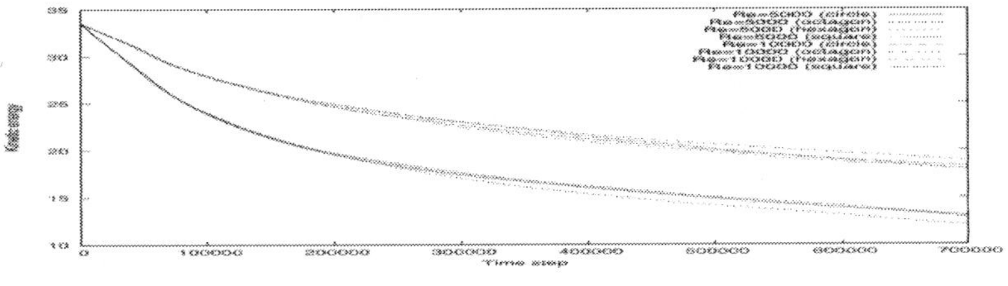


**Figure 1.** Evolution of kinetic energy between $t = 0$ and $t = 700000$ step at $Re = 5000$ and $Re = 10000$ Reynolds number

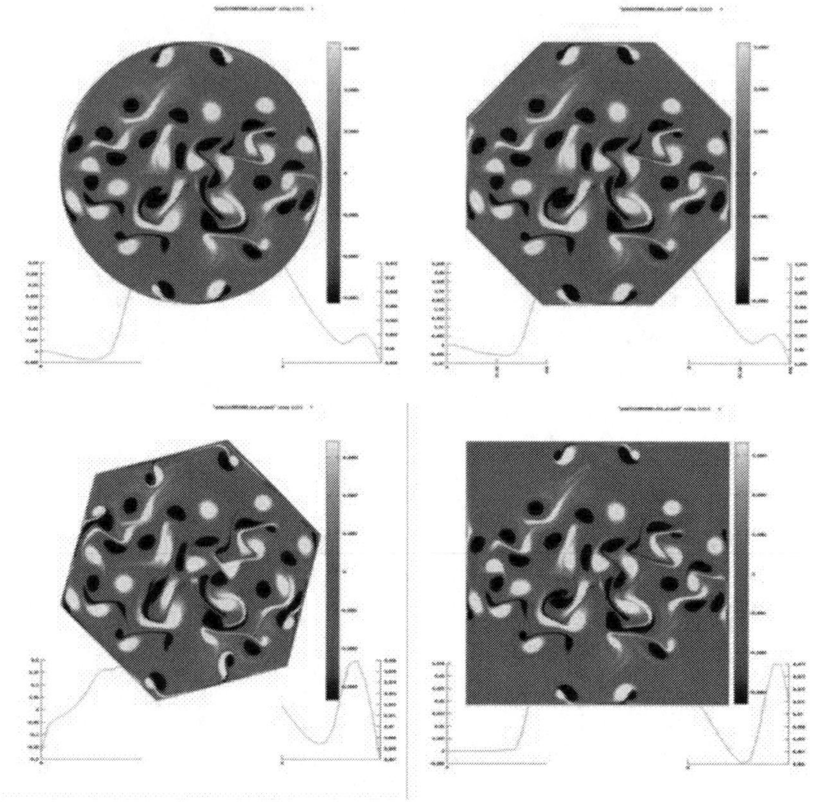

**Figure 2.** Vorticity fields at at t = 65000 time step in four different geometries.

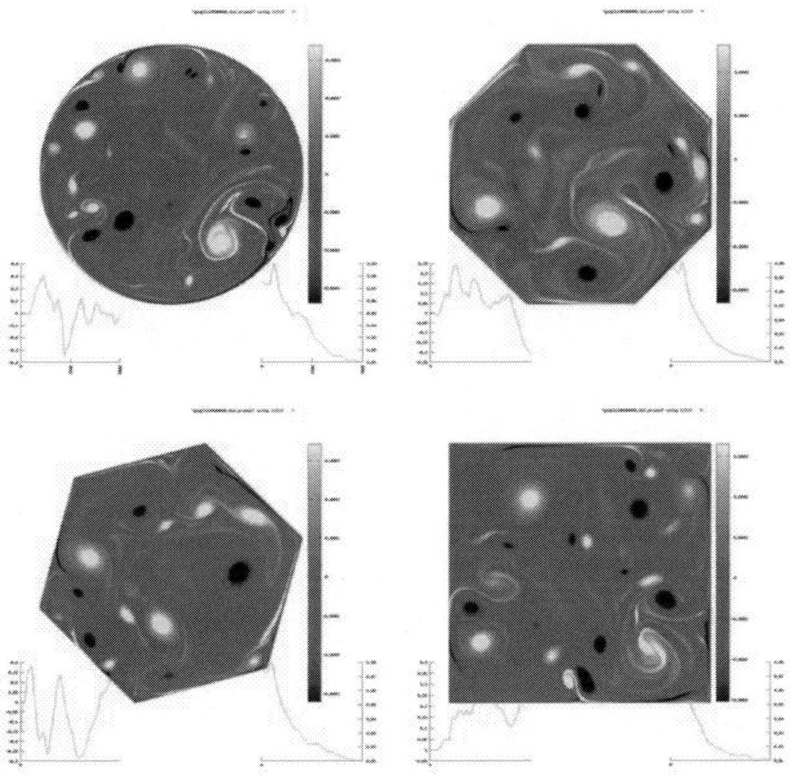

**Figure 3.** Vorticity fields at t = 500000 time step in four different geometries.

So even if the evolution is chaotic and differs localy in the various containers, the global quantities like the kinetic energy erodes in a very similar manner. The evolution of enstrophy follows similar behaviour as one can see in Fig. 4 for each geometry. However, here the situation is somewhat different. In 2D turbulence there is a direct enstrophy cascade, so enstrophy is generated in large scales and dissipated in small scales. The first interations between the walls and the vortices can be clearly recognized in Fig. 4 as a local maxima in the enstrophy around 65000 steps. It is worth emphasizing that the enstrophy is a monotone decreasing function of time in a double periodic domain, since there is no physical mechanism to produce entstrophy in such geometry. Here, however, enstrophy can be produced by wall shear stress and although our containers have different shapes, we cannot see significant differences in the vorticity production, the enstrophy functions follow more or less the same trend. The reason of this agreement is obvious, the corners of the shapes do not have significant effect on the initial evolution as we have already seen in Fig. 2. The first interaction between the walls and the vortices take place in a very similar manner, therefore the enstrophy functions do not differ from each other.

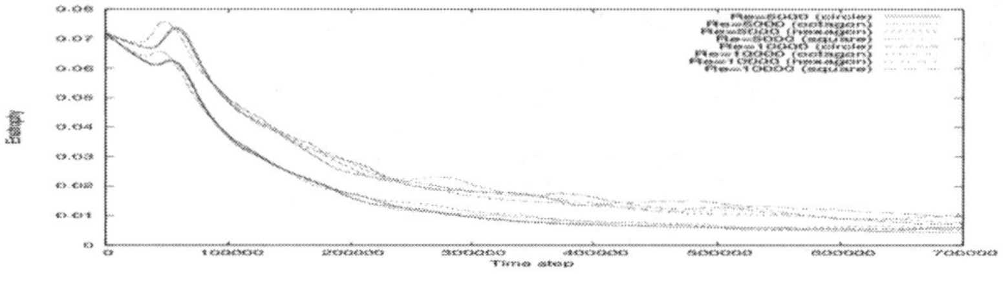


**Figure 4.** Evolution of enstrophy between $t = 0$ and $t = 700000$ step at $Re = 5000$ and $Re = 10000$ Reynolds number

## 5. Conclusion

Studying the decay of 2D turbulence in four different shaped containers We have shown that the relevant physical quantities such as the kinetic energy and enstrophy evolve almost independently on the geometry of the containers. Enstrophy has a local maxima when the initial vortices reach the walls, then it decays with the same trend in each container. The decay of the kineric energy also follows the same trend independently from the shape of the container.

## 6. References

[1] Li, S., Montgomery, D. 1996, *Phys. Lett. A*, **218**, 281
[2] Clercx, H. J. H., Nielsen, A. H., Torres, D. J., Coutsias,E.A. 2001, *Eur. J. Mech. B - Fluids*, **20**, 557
[3] Benzi R., Succi S., Vergassola M., 1992, *Phys. Rep.*, **222**, 145
[4] Chen, S., Doolen, G. D. 1998, *Ann. Rev. Fluid Mech.*, **30**, 329
[5] Benzi R., Succi S., 1990, *J. Phys. A: Math. Gen.*, **23**, L1
[6] Bhatnagar, P. L., Gross, E. P., Krook, M. 1954, *Phys. Rev.*, **94**, 511
[7] Qian, Y.H., d'Humieres, D., Lallemand, P., 1992, *Europhys. Lett.*, **17**, 479
[8] Karlin I. V., Succi S., 1998, *Phys. Rev. E*, **58**, R4053
[9] Rohde, M., Kandhai, D., Derksen, J. J., van den Akker, H. E., 2003, *Phys. Rev. E*, **67**, 066703
[10] Toth, G., Hazi, G., 2010, *Phys. Fluids*, **22**, 053101
[11] R.H. Kraichnan, D. Montgomery, 1980, *Rep. Prog. Phys*, **43**, 547

# A new paradigm on battery powered embedded system design based on User-Experience-Oriented method

**Zhuoran Wang**[1] **and Yue Wu**[2,*]

[1]Solution Dept of iStarChip Semiconductor Technology (Shanghai) Co., Ltd., Shanghai, P.R.C

[2] Department of Chemical and Life Science Engineering, Virginia Commonwealth University, Richmond, VA 23284 USA

**Abstract.** The battery sustainable time has been an active research topic recently for the development of battery powered embedded products such as tablets and smart phones, which are determined by the battery capacity and power consumption. Despite numerous efforts on the improvement of battery capacity in the field of material engineering, the power consumption also plays an important role and easier to ameliorate in delivering a desirable user-experience, especially considering the moderate advancement on batteries for decades. In this study, a new Top-Down modelling method, User-Experience-Oriented Battery Powered Embedded System Design Paradigm, is proposed to estimate the target average power consumption, to guide the hardware and software design, and eventually to approach the theoretical lowest power consumption that the application is still able to provide the full functionality. Starting from the 10-hour sustainable time standard, average working current is defined with battery design capacity and set as a target. Then an implementation is illustrated from both hardware perspective, which is summarized as Auto-Gating power management, and from software perspective, which introduces a new algorithm, SleepVote, to guide the system task design and scheduling.

## 1. Introduction

The concept, user experience, has been attracting increasing attention since the revolutionary commercial portable device introduced by Apple Ltd in 2007. As nobody wants to charge their devices every several hours, the 10-hour sustainable time user experience has been brought up and become the facto standard (or reference) for smart phones and tablets since the successful introduction of the iPad in 2010.

Generally, there are two ways to increase battery sustainable time. One is to use a big battery with high capacity, which is widely used by android device vendors. The idea is easy but proved to be a big challenge for production design and is against the desire for thin and tiny commercial portable device with a reasonably long sustainable time. Numerous efforts have been made to increase the energy density since the commercialization of the first Li-ion battery by Sony in 1991[1]. However, the improvement is not satisfied enough up to now [2].The second method of increasing battery sustainable time is to reduce the total system power-consumption, which involves hardware and software integration. This method takes battery as a constant factor in the system and focuses on theories and

* Corresponding author: wuy@vcu.edu

approaches for a) evaluating the lowest system power-consumption with which the device can still provide strictly defined functionality and b) the practical implementation. Compare with the first approach, reducing the total system power-consumption can increase the battery sustainable time significantly without changing the device volume, extend the battery life time, and provide a more accurate state of charge indication in full temperature range (0~45°C).

This paper will apply the second method and reason a practical embedded system design paradigm based Top-Down User-Experience-Oriented modeling, which will be illustrated from both software and hardware perspectives.

## 2. 10-hour User Experience

The 10-hour User Experience describes a standard from human feelings, which could be ambiguous and arbitrary. To make it a real standard, a precise, practical and reasonable engineering definition is required.

C-rate is the charge or discharge current, which equals in Amperes to the rated capacity in Ah numerically. For example, the C-rate is 1100 mA in the case of an 1100 mAh battery, while the 0.2 C and 0.1 C are 220mA and 110mA respectively[1]. With C-rate, the 10hour User Experience can be interpreted as the remaining run-times ($t_r$) of a battery under 0.1 C discharge rate in average. The real value of 0.1 C current is determined by the design capacity of the battery. The bigger the battery is, the higher the 0.1 C current value will be. Typically, a discharge current which is less than or equals 0.2 C is regarded as low discharge rate. By using low discharge C-rate, 5-10hours battery sustainable time is expected, meanwhile the battery life-time is extended and the state of charge indication accuracy is also improved[1][2]. By using the non-replaceable Li-ion Polymer batteries, devices can be designed even thinner and lighter than usual, and keep available for at least one year.

## 3. General Model and Working Pattern

The 10-Hour User-Experience set the target power-consumption. To reach the goal, the easy-to-get idea Sleep As Soon As Possible is modelled and interoperated into guidelines first. Building on system modularization, new paradigm creates several models for different system levels and environments. As a general modeling guideline, the client of service provider applies Sleep As Soon As Possible strategy to the target component based on the idle signal(s) provided by services.

### 3.1. Component

In this paper, the component is a black box providing strictly defined services for client(s). The component always provides idle signal and the method for power-control at the same time. From the power-consumption perspective, the component is passive, providing idle information to the outside and waiting for power-management control.

Idle signal is provided by the component and used to indicate the life-time of self-determined service(s) in the component. Self-determined services is a kind of services that are requested, controlled or triggered by the client(s) and the life time is determined once started, for example, the UART (Universal Asynchronous Receiver / Transmitter) transmission service. By contrast, the life time of the none-self-determined services is not guaranteed. It provides information for client(s) but typically triggered or controlled by others, for example, the UART receiving service. The component should provide idle signal as the busy flag of self-determined services and provide clue(s) to client(s) for power-management if necessary.

The power-management is performed at client side. The target object is not only energy but also including power-related supplies, such as clock source in hardware, CPU time in software and etc. To make it short, power-supply will include all power-related supplies throughout this paper unless special statement is made. For those components consisting of self-determined services, once the idle signal is asserted, the power supply will be cut off immediately, which is the most straightforward strategy. This strategy works for some cases but not all, since repowering may take time, power and/or the worst, the functionality. The power can be switched frequently if the cost is small or negligible, while sophisticated algorithm should be carefully applied to prioritize the functionality delivering if the cost is large. For those components containing none-self-determined services, the general strategy is the same, but clues are taken into consideration, a) if the life-time of none-self-determined service is determined, the clue will enable the power-supply and hence the idle signal will be used to indicate the service lifetime; b) if the life-time is undetermined, short-term power supply will be applied for each clue or the power of the target component will be manually controlled by application.

### 3.2. Power-Consumption Component Categories

Table 1 lists several different power-consumption component categories based on the information complexity of idle signal(s), style of power-management and the power-supply types.

**Table 1.** Power-Consumption Components Categories

| | Idle | Style | Supply | Example |
|---|---|---|---|---|
| Basic | 1 | On/Off | Single | Simple Software Tasks, divider resistance in A/D conversion circuit, and etc. |
| Composite | n | Multi-Levels | Single | Microcontrollers Clock Management System |
| Multi-Layer | n | Multi-Levels | Multiple | Microcontrollers Sleep Modes, A/D Convertor, and etc. |

Basic power-consumption component is the smallest unit for power-management. It provides one idle signal and on/off switch for specified power-supply, i.e. most of the software tasks are basic power-consumption components, and the divider resistance in A/D conversion circuit can also be considered as a basic power-consumption component as it is passive and always idle. Composite power-consumption component is a set of components with the same power-supply type, and, hence the set of internal idle signals is able to provide either a simple wired-and idle signal or detailed working status. It's possible to line out all power-switches of internal components but the flexibility introduces significant complexities at the same time, so typically, control templates are designed in certain order, which are referred as power-management levels. The clock management system in microcontrollers is the typical application for composite power-consumption component. A set of components with the same type of power-supply is defined as a power-layer. If a component supports more than one type of power-supply-managements, it is a multi-layer power-consumption component. Analog IPs in IC design always involve clock management and power-management at the same time. User can gate the clock to reduce the dynamic power-consumption when idle and shut-down the energy-supply when it's disabled or not used for certain time to reduce the leakage current.

### 3.3. Burst Working Pattern

If certain power-management strategy based on idle signal(s) applied to a target component, this part of system behaves in a burst working pattern in some level. With a given component, power-consumption is the function of time under the specified strategy, represented as $f(t)$. The burst working pattern is the shape of $f(t)$ in a selected time-window. So the average power consumption is represented as

$$g(t) = \frac{\int f(t)dt}{t} = \frac{E(x_0)\cdot t_0 + E(x_2)\cdot t_1 + \cdots + E(x_n)\cdot t_n \ (n\epsilon N^+)}{t} \qquad (1)$$

where $E(x_n)$ is the power-consumption at level $x_n$ in a component and $t_n$ is the duration time of the specified level. $t_n$ is fixed once the system design is settled. With a given component, the $g(t)$ will be a bounded function for both self-determined and none self-determined services if the service working time is limited. So, we can further get

$$g(t) = \frac{\int f(t)dt}{t} = \frac{E(x_0)\cdot t_0 + E(x_2)\cdot t_1 + \cdots + E(x_n)\cdot t_n \ (n\epsilon N^+)}{t} = E(x_0)\cdot r_0 + E(x_1)\cdot r_1 + \cdots + E(x_n)\cdot r_n \ (n\epsilon N^+) \quad (2)$$

where $r_n$ is the time ratio for the corresponding power-consumption level. Equation 2 elaborates the essential method for evaluating system minimal power-consumption while delivering strictly defined functionalities and guiding the system design to approach the evaluation result. The system power-consumption can be reduced by dividing big components into necessary sub-components with different power consumptions, increasing the power-management levels and increasing the ratio for low-power consumption levels. Equation 2 also indicates that it is possible to balance the power-consumption by increasing system performance to reduce the time ratio.

With a given component, if the service working time is undetermined, the service should be evaluated together with the object, which triggers and/or controls the none-self-determined service(s), and, hence new integrated-component is generated. If the life-time of the combination is determined, Equation 2 can be applied. If the combination is still undermined, the iteration should be repeated again and again until self-determined service is generated. For some cases, users become the final objects who control the undetermined components such as your smart phones. In those cases the general system power-consumption is controlled by human behaviours.

## 4. Practical Implementation

In real application, hardware modules provide idle signals to implement auto-clock gating power-management to reduce the dynamic part of power-consumption. Meanwhile, different depths of sleep models are designed to gate unused power-domain with sophisticated strategy and avoid the repowering overhead to reduce the static part of power-consumption caused by leakage current.

In order to enter sleep mode as soon as possible, event-driven techniques is applied to all tasks in software design, so that the task life-time is determined, and self-determined services are always triggered by events. If all tasks complete their services and no event is triggered, idle state will be identified, in which the sleep mode should be entered. To make sure the deepest sleep mode is entered, and system can still wake up by specified events, all sleep sensitive tasks are required to vote just before any sleep mode entered and as the vote result, the shallowest sleep mode will be used. This algorithm is called SleepVote.

## 5. Conclusion

The new paradigm proposed in this paper provides identical modelling method and guidelines for software and hardware implementation at the same time. By employing it, the theoretical minimal working power-consumption can be evaluated soon after the system is modelled and/or designed with the guideline of burst working pattern. The result can be compared with the target average power-consumption defined by 10-hour user experience battery sustainable time standard to help the decision making for product design with convincible data.

## References


[1] H.J Bergveld, W.S. Kruijt, P.H.L Notten, Battery Management Systems, Design by Modelling, Philips Research Book Series, Kluwer Academic Publishers, Boston(2002)

[2] Valer Pop, Henk Jan Bergveld, Dmitry Danilov, Paul P.L Regtien and Peter H.L Notten, Battery Management System, Accurate State-of-Charge Indication for Battery-Powered Application, Philips Research Book Series Volume 9, Springer Science Business Media B.V(2008).

[3] Elaine Rich, Automata, Computability and Complexity Theory and Applications, PEARSON Prentice Hall, ISBN 0-13-228806-0

[4] Ken Pugh, Prefactoring, 1005 Gravenstein Highway North, Sebastopol, CA95472, O'Reilly Media, Inc. ISBN 0-596-00874-0

# Modeling of explosive electron emission and electron beam dynamics in high-current devices

**S.V. Anishchenko and A.A. Gurinovich**

Research Institute for Nuclear Problems, Bobruiskaya 11 str, Minsk 220030, Belarus

E-mail: sanishchenko@mail.ru

**Abstract.** Based on a detailed analysis of explosive electron emission in high-current electronic devices, we formulate a system of equations that describes the expansion of the cathode plasma and the generation of high-current electron beams. The system underlies the numerical algorithm for the hybrid code which enables simulating the charged particles' dynamics in high-current vircators with open resonators. Using the Gabor-Morlet transform, we perform the time-frequency analysis of vircator radiation.

## 1. Introduction

A lot of algorithms and computer codes have been developed that permit studying different phenomena in high-current electronic devices. The explosive electron emission is particularly interesting, because during this process plasma changes from nonideal, having strong particle interaction, to practically collisionless, emitting a high-density electron beam in the presence of an external electric field. The beam acts as an active medium in high-power microwave generators – e.g. vircators [4] (Fig. 1).

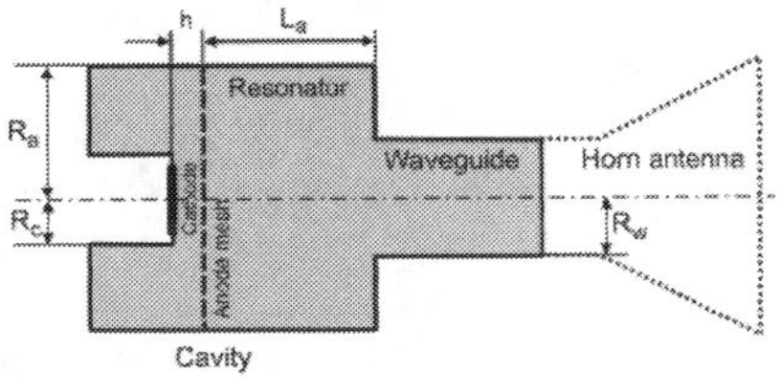


**Figure 1.** Vircator. Schematic diagram

Vircators, or the sources of electromagnetic radiation with an oscillating virtual cathode, have been the subject of theoretical and experimental studies for more than 30 years. The interest in vircators is natural: the devices based on an oscillating virtual cathode can produce high-power signals. Already in the first experiments with vircators, the output power was reported to achieve more than 100 MW [4]. The radiation in the source is generated owing to the development of Pierce instability of the electron beam injected through a semi-transparent anode mesh of a high-current diode into an equipotential cavity (drift tube), where the oscillating virtual cathode –

the electron bunch that reflects a fraction of electrons backward to the anode – is formed. The accelerated motion of charged particles is accompanied by the emission of electromagnetic waves.

The present paper suggests a variant of a hybrid code in which the beam dynamic is computed in the kinetic and plasma dynamics – in the hydrodynamic approximations. The second section gives a detailed analysis of explosive emission processes leading to the formation of cathode plasma in a vircator, substantiates the applicability of a hydrodynamic approach to the cathode plasma description. Section 3 provides the time-frequency analysis of radiation of a vircator with an open resonance cavity.

## 2. Explosive electron emission

A high-current diode is started by applying the voltage $U$ to the interelectrode gap of a high-current diode (Fig. 1) and thus inducing the field-emission current from the cathode surface. This current is a flow of electrons tunneling from metal to vacuum under the action of the electric field [4]. Moving in a metal, the electrons heat up the cathode surface. Because of the surface defects of the conductor (microprotrusion tips), dielectric inclusions, oxide and other inorganic films, layers of the adsorbed gas, emerging intergrain boundaries, crater edges, and pores cracks) the microscopic electric field near the cathode is nonuniform. Particularly, the value of the field in the vicinity of microprotrusion tips is increased compared to the average value, and so the tips are rapidly heated and explode when the specific energy density amounts to $\sim 10^4$ J/g. When the electric field strength $U/h$ attains values of the order of 1 MV/cm, the explosion delay is about 1 ns. Microexplosions result in plasma formation and intense thermionic emission from the expanding plasma surface. The plasma initiated on the cathode surface leads to further heating of the cathode areas and charging the dielectric films, and thus favors the formation of new emitting regions.

The properties of the cathode plasma change significantly during the expansion. Being a nonideal plasma with strong interparticle interaction at the moment of hydrodynamic expansion of the tip, it turns into ideal plasma as its density gradually decreases and the collision rate reduces.

Explosive electron emission is considered by the example of a graphite cathode. Graphite cathodes are extensively used in high-current electronics for their durability and low vacuum breakdown field strength. In our simulation we shall neglect the phenomena related to the transition of plasma from a nonideal to a rarefied state, occurring at a distance of about 100 $\mu$m form the cathode surface [2, 4]. There are three reasons why such a simplification is needed. First, computations of microprocesses on the cathode increase the computation time. Second, a detailed information about the cathode state is often missing. Third, most experimental data about the cathode plasma refer mainly to that expansion stage when the plasma is already ideal.

It was experimentally determined that microexplosions initiate the emission from the cathode surface of a plasma flux of velocity $v \approx 2 \cdot 10^6$ cm/sec. The temperature of the electronic component of the flux is $T_e \approx 5 \cdot 10^4$ K [4], while the temperature $T_i$ of its ionic component is by a factor of five smaller.

Knowing the experimentally measured cathode mass loss per 1 coulomb ($K \approx 1.4 \cdot 10^{-6}$ g/C [7]), the total diode current $I$, and the area $S$ of the emission region, we can readily determine the ion concentration $n_{0i} = n_{0e} = KI/m_i S v$ and the pressure $P_0 = n_i k(T_e + T_i)$ of the cathode plasma. Let $I = 20$ kA and $S = 30$ cm$^2$, then $n_{0i} = 3.0 \cdot 10^{15}$ cm$^{-3}$ and $P_0 \approx 1$ kPa.

The main parameter characterizing the transport and relaxation properties of cathode plasma is the electron-ion collision rate $\nu_{ei} = 5.5\Lambda_e n_{0i} T^{-3/2} \approx 10^{10}$ sec$^{-1}$ [8]. Because the value of $\nu_{ei}$ is small, the local thermodynamic equilibrium between electrons and ions is set up in a time interval as long as $m_i/m_e\nu_{ei} \simeq 2$ mcs. The electron heat conductivity can be neglected because the entire process of explosive emission lasts for about a microsecond, and so the characteristic distance over which the temperature can be expected to equalize is as small as 0.5 cm. Hence, a two-component plasma expands adiabatically, and this motion can be described by the following hydrodynamical equation:

$$\frac{d\vec{V}}{dt} = -\frac{\nabla P}{\rho}, \tag{1}$$

written in the Lagrangian form. The pressure $P$ relates to the density $\rho$ as $P = P_0(\rho/\rho_0)^{5/3}$ ($p_0$ and $\rho_0$ are the plasma pressure and density at the cathode surface, respectively).

Note also that carbon plasma in the interelectrode gap remains completely ionized due to the small rate of charged particle recombination. For this reason, its Langmuir frequency is $\omega_e = \sqrt{e^2 n_{0i}/4\pi m_e} \approx 500$ GHz, being much greater than the characteristic frequency of beam oscillations $1 - 10$ GHz [4]. As a consequence, the electric field hardly penetrates the plasma. That is why when simulating the processes occurring in the diode, we can consider the plasma and cathode potentials to be equal.

We solve hydrodynamical equations by means of the particle-in-cell method (PIC), widely used in hydrodynamics and plasma physics simulations. The code for the solution of hydrodynamical equations is based on a somewhat modified version of the algorithm [10] developed for magnetohydrodynamical applications. Our algorithm differs from that described in [10] by the absence of the magnetic fields that play only a minor role in the absence of the pinch effect.

In a high-current diode, the beam is generated through thermal electron emission from the expanding cathode plasma, whose electron temperature is of the order of $10^4$ K. Because the thermionic current density $j_T = en_e\sqrt{kT_e/2\pi m_e} \approx 10 \div 20$ kA/cm$^2$ is one-two orders of magnitude greater than the limiting charge current density in the diode, $j_{CL} = \frac{mc^3}{e}\frac{2(e\phi/mc^2)^{3/2}}{9\pi d^2} <$ 1 kA/cm$^2$ ($\phi \sim 500$ kV, $d \sim 1$ cm), the emission capability of the cathode plasma can be considered unlimited; the cathode can be considered as an electric-field-screened spatial charge of the beam.

The motion of electrons can be studied in the framework of collisionless plasmadynamics [3].

## 3. Analysis of radiation from the vircator

On the basis of the described theory, we developed a computer code for simulation of vircators with open resonant cavity (Fig. 1). As has been shown in [5, 6] a noticeable stabilization of the radiation frequency in such vircators can be achieved when one of the the resonator's eigenfrequencies $\omega_0$ is approximately the same as the frequency $\omega_b$ of the electrostatic oscillations of the beam. The latter depends on the cathode-anode gap $h$ in a most straightforward way.

This naturally raises the question of how the cathode plasma expansion affects the output characteristics of vircator radiation. To find the answer, we performed a numerical simulation of a vircator, in which charged particles interact with the transverse mode $TM_{010}$ of the electric field of a cylindrical resonator. The vircator parameters were as follows: the cathode-anode gap $h = 1.6$ cm, the cathode radius $R_c = 3.2$ cm, the resonator radius $R_a = 7.5$ cm, and the resonator length $L_a = 6.0$ cm. The mode $TM_{010}$ had the eigenfrequency $f_0 = 3.4$ GHz and the damping constant $\lambda_0 = 0.025\omega_0$, which corresponds to the resonator Q-factor equal to 20. At modes different from $TM_{010}$, the resonator Q-factor was considered much lower.

Numerical calculations show that a virtual cathode is formed and the oscillations of the transverse electromagnetic field are excited at the frequency $\sim$ 3.6 GHz when the current exceeds the threshold current ($I_0 \approx 9$ kA ). When the current further grows by a factor of 2.5, the generation frequency increases monotonically. This is readily seen by the Gabor-Morlet wavelet transform of the oscillation amplitude $q_0$ of the field oscillator [3, 11]:

$$q_0(t, s) = \int_{-\infty}^{+\infty} q_0(\eta)\pi^{-1/4} e^{i\omega\eta} e^{-\frac{\eta^2}{2s^2}}, \tag{2}$$

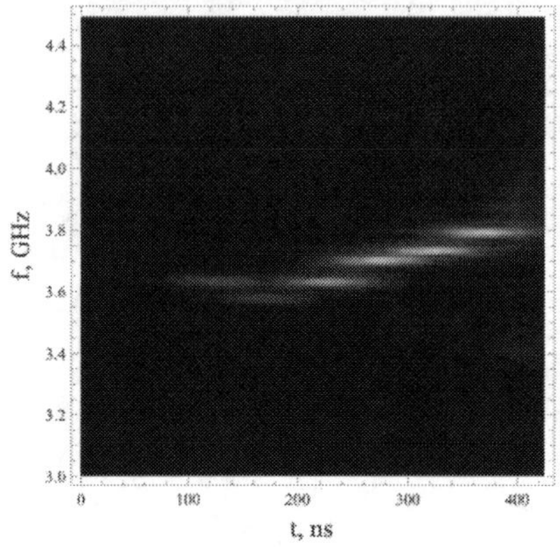


**Figure 2.** Time-frequency diagram of radiation from the vircator.

where $s$ is set equal to 6.7 ns.

One can infer from Fig. 2 that the cathode plasma expansion leads to a 5.5% shift of the basic generation frequency in 300 ns (from 3.6 GHz to 3.8 GHz).

## 4. Conclusion

A thorough analysis of the explosive emission processes in high-current electronic devices was given. It allowed us to formulate the physical model that describes the expansion of the cathode plasma and the generation of high-current electron beams used for excitation of transverse electromagnetic oscillations in open-cavity resonators. On the basis of the physical model, we developed a numerical algorithm and a computer code for simulating charged particle dynamics in high-currency devices. The code was used for time-frequency analysis of radiation generated by an open-resonator vircator with a low Q-factor ($Q \sim 20$). It was shown that when the vircator current is increased by a factor of 2.5 compared to the threshold current the basic frequency of radiation grows by $5 - 6\%$.

## References

[1] Verboncoeur J P 2005 *Plasma Phys. Control. Fusion.* **47** 231.
[2] Welsh D R et al 2009 *Phys. Plasmas.* **16** 123102
[3] Baryshevsky V, Gurinovich A, Molchanov P, Anishchenko S, and Gurnevich E 2013 *IEEE Trans. Plasma Sci.* DOI: 10.1109/TPS.2013.2275022
[4] Mesyatz G A 2004 *Pulsed Power* (Moscow: Nauka)
[5] Benford J, Price D, Sze H, and Bromley D 1987 *J. Appl. Phys.* **61** 2098
[6] C. Möller, M. Elfsberg, A. Larsson, and Nyholm S E 2010 *IEEE Trans. Plasma Sci.* **38** 024705
[7] Gunin A V et al 1999 *JTP Lett.* **25** 84
[8] Artsimovich L A and Sagdeev R Z 1979 *Plasma Physics For Physicists* (Moscow: Atomizdat)
[9] Parker R K, Anderson R E, and Duncan C V 1974. *J. Appl. Phys.* **45** 2463
[10] Brunel F et al 1981 *J. Comput. Phys.* **43** 268
[11] Anfinogentov V G, Koronovsky A A, and Hramov A E 2000 *Izvestiya RAN. Ser. Fiz.* **64** 2383

# Parameters estimation from the diffusion MRI signal using a macroscopic model

[1]H. T. Nguyen, [2]J. R. Li, [3]D. S. Grebenkov, [1]D. Le Bihan, [1]C. Poupon

[1] Neurospin - CEA Saclay, Gif-sur-Yvette, France

[2] INRIA Saclay-Equipe DEFI, CMAP, Ecole Polytechnique, Palaiseau, France

[3] LPMC, CNRS – Ecole Polytechnique, Palaiseau, France

E-mail: hang.tuan.nguyen@cmap.polytechnique.fr, jingrebecca.li@inria.fr

**Abstract.** Diffusion magnetic resonance imaging (dMRI) probes the diffusion characteristics of a sample via the application of magnetic field gradient pulses. The dMRI signal from a heterogeneous sample includes the water proton magnetization from all spatial positions in a voxel. If the voxel consists of different diffusion compartments with weak exchange, while the duration of the diffusion-encoding gradient pulses is short compared to the diffusion time (the narrow pulse approximation), the dMRI signal can be approximated by the Karger model. A new macroscopic ODE model for the dMRI signal was recently derived mathematically from the microscopic multiple compartments Bloch-Torrey partial differential equation (PDE) without the narrow pulse restriction. We illustrate by numerical simulations that this ODE model accurately approximates the dMRI signal in a domain containing spherical cells of various sizes, and show preliminary results on solving the inverse problem to estimate the cellular volume fraction and surface area.

## 1. Introduction

A microscopic model for the diffusion magnetic resonance imaging (dMRI) signal is the Bloch-Torrey partial differential equation (PDE) [1, 2] on multiple compartments [3]. In general, for tissue geometries containing cells with permeable membranes, analytical solutions of this PDE are not known even for domains containing simple-shaped cells. One needs thus to resort to numerical simulations as well as to macroscopic models. An existing macroscopic model is the Karger model [4] which consists of a system of ordinary differential equations (ODEs) and takes into account a certain form of exchange between compartments. The validity of the Karger model and the "physical" meaning of the various parameters of this model was discussed in [5]. One important limitation of the Karger model is that the duration of the diffusion-encoding gradient pulses must be short compared to the diffusion time (the narrow-pulse approximation, see [6]). Here we conduct a numerical study of a recently formulated macroscopic ODE model [7] that overcomes this limitation, in a 3D geometry containing spheres of various sizes, and solve the inverse problem to identify unknown parameters of the ODE model.

## 2. Microscopic multiple compartments Bloch-Torrey PDE

The starting point of the macroscopic model is the microscopic multiple compartment Bloch-Torrey PDE, which is a generalization of the original Bloch-Torrey PDE [1] to heterogeneous media. Inside a representative volume $C$, we define many spherical compartments, $\Omega^{s_j}$, $j = 1, \cdots, P$, where each $\Omega^{s_j}$ is a sphere. The complementary set, $\Omega^e = C \backslash \left( \bigcup_j \Omega^{s_j} \right)$, is an extracellular compartment. For the macroscopic model to be discussed later, one can also group all cells of a certain type into a single compartment. The complex transverse water proton magnetization $M^l(\mathbf{r}, t|\mathbf{g})$ in each compartment $\Omega^l$ satisfies the Bloch-Torrey PDE [1]:

$$\frac{\partial M^l(\mathbf{r}, t|\mathbf{g})}{\partial t} = -If(t)(\gamma \mathbf{g} \cdot \mathbf{r})\, M^l(\mathbf{r}, t|\mathbf{g}) + \nabla \cdot (D^0 \nabla M^l(\mathbf{r}, t|\mathbf{g})), \quad \forall\, l, \tag{1}$$

where we denote the amplitude and direction of the diffusion-encoding gradient by $\mathbf{g} = (g_1, g_2, g_3)$ (normalized direction $\mathbf{q} = \mathbf{g}/|\mathbf{g}|$) and its time profile by $f(t)$, $I$ the imaginary unit, $\gamma = 2.67513 \times 10^8 \,\mathrm{rad\,s^{-1}T^{-1}}$ the gyromagnetic ratio of the water proton, and $D^0$ the intrinsic water diffusion coefficient. For the pulsed gradient spin echo (PGSE) sequence [8], with two rectangular pulses of duration $\delta$, separated by a time interval $\Delta - \delta$, the profile $f(t)$ is

$$f(t) = \begin{cases} 1, & 0 \le t \le \delta, \\ -1, & \Delta < t \le \Delta + \delta, \\ 0, & \text{otherwise}, \end{cases} \tag{2}$$

where the starting time of the first gradient pulse is $t = 0$ and $\Delta > TE/2$.

We supplement the PDE in (1) with interface conditions at the interface $\Gamma^{ln}$ where two compartments $\Omega^l$ and $\Omega^n$ come in contact. One interface condition is the continuity of flux:

$$D^0 \left( \nabla M^l(\mathbf{a}, t|\mathbf{g}) \cdot \mathbf{n}^l(\mathbf{a}) \right) = -D^0 \left( \nabla M^n(\mathbf{a}, t|\mathbf{g}) \cdot \mathbf{n}^n(\mathbf{a}) \right), \quad \mathbf{a} \in \Gamma^{ln}, \tag{3}$$

where $\mathbf{n}^l(\mathbf{a})$ and $\mathbf{n}^n(\mathbf{a})$ are the *outward*-point normals to $\Omega^l$ and $\Omega^n$ at $\mathbf{a}$, so in fact $\mathbf{n}^l(\mathbf{a}) = -\mathbf{n}^n(\mathbf{a})$. The second interface condition

$$D^0 \left( \nabla M^l(\mathbf{a}, t|\mathbf{g}) \cdot \mathbf{n}^l(\mathbf{a}) \right) = \kappa \left( M^n(\mathbf{a}, t|\mathbf{g}) - M^l(\mathbf{a}, t|\mathbf{g}) \right), \quad \mathbf{a} \in \Gamma^{ln}. \tag{4}$$

incorporates a permeability coefficient $\kappa$ across $\Gamma^{ln}$. Finally, if the excitation of the magnetization in the imaging voxel is uniform, the following initial condition is added: $M(\mathbf{r}, 0|\mathbf{g}) = 1, \quad \mathbf{r} \in \Omega^l, \quad \forall\, l$. Then, same as [9], we assume that the computational domain $C = [-L_1/2, L_1/2] \times [-L_2/2, L_2/2] \times [-L_3/2, L_3/2]$ is extended by periodic copies of itself. The dMRI signal measured in experiments (without the imaging gradients and $T_2$ effects) is proportional to

$$S_{PDE}(b) := \sum_l \int_{\mathbf{r}\in\Omega^l} M^l(\mathbf{r}, TE|\mathbf{g})\, d\mathbf{r}, \tag{5}$$

where the $b$-value is defined as $b(\mathbf{g}) = \gamma^2|\mathbf{g}^2|\delta^2\,(\Delta - \delta/3)$. In a homogeneous medium, the signal attenuation is $\exp(-D^0 b)$.

## 3. Macroscopic ODE model

By an asymptotic analysis of the cell membrane permeability coefficient $\kappa$, a macroscopic (homogenized) model for the multiple compartment Bloch-Torrey PDE was obtained in [7], consisting of a system of ODEs:

$$\frac{dS^m_{ODE}(b,t)}{dt} = -\left(c(t)\gamma^2\mathbf{g}^T\overline{D}^m\mathbf{g} + \sum_{l=1,\, l\neq m}^{P} \frac{1}{\tau^{lm}_{ODE}}\right) S^m_{ODE}(b,t) + \sum_{l=1,\, l\neq m}^{P} \frac{1}{\tau^{ml}_{ODE}} S^l_{ODE}(b,t)$$
$$m = 1, \cdots, P$$

where

$$\frac{1}{\tau^{ml}_{ODE}} := \kappa\frac{|\Gamma^{ml}|}{|\Omega^l|} \qquad \Longrightarrow \frac{\tau^{lm}_{ODE}}{\tau^{ml}_{ODE}} = \frac{|\Omega^m|}{|\Omega^l|} = \frac{v^m}{v^l}, \tag{6}$$

and $|\Gamma^{ml}|$ is the surface area of the interface between $\Omega^m$ and $\Omega^l$; $|\Omega^l|$ and $v^l$ are respectively the volume and volume fraction of compartment $\Omega^l$ ($\sum_{l=1}^{P} v^l = 1$). For the pulsed gradient spin echo (PGSE) sequence, it was shown in [7]:

$$c(t) = \begin{cases} t^2, & 0 \leq t \leq \delta, \\ \delta^2, & \delta < t \leq \Delta, \\ (t - \Delta - \delta)^2, & \Delta < t \leq \Delta + \delta. \end{cases} \tag{7}$$

In the narrow pulse regime, $\delta \ll \Delta$, the ODE model reduces to the Karger model [4]:

$$\frac{dS^m_{KAR}(b,t)}{dt} = -\left(\delta^2\gamma^2\mathbf{g}^T\overline{D}^m\mathbf{g} + \sum_{l=1,\, l\neq m}^{P} \frac{1}{\tau^{lm}}\right) S^m_{KAR}(b,t) + \sum_{l=1,\, l\neq m}^{P} \frac{1}{\tau^{ml}} S^l_{KAR}(b,t)$$
$$m = 1, \cdots, P.$$

Both the new ODE model and the Karger model rely on a system of coupled ODEs, subject to initial conditions: $S^l(b,0) = v^l$, $l = 1, \cdots, P$. The total dMRI signals for the ODE model and the Karger model are

$$S_{ODE}(b) = \sum_{m=1}^{P} S^m_{ODE}(b, TE), \qquad S_{KAR}(b) = \sum_{m=1}^{P} S^m_{KAR}(b, TE). \tag{8}$$

The effective diffusion tensors, $\overline{D}^m$, $m = 1, \cdots, P$, in the compartments can be obtained after solving three steady-state Laplace PDEs (see [7] for details).

## 4. Numerical results

We solved the multiple compartment Bloch-Torrey PDE in the computational domain $C = [-5\mu m, 5\mu m]^3$ containing the cellular configuration shown in Fig. 1a, with 76 spherical cells of radii in the range $0.6 - 2.55\mu m$, using the finite elements method described in [10]. We set the intrinsic diffusion coefficient in all the compartments to $D^0 = 3 \times 10^{-3} mm^2/s$. To obtain the ODE and the Karger model signals we combined the 76 spheres to form one compartment $\Omega^s$. The extra-cellular space forms a second compartment $\Omega^e$. The corresponding volume fractions are $v^s = 0.65$ and $v^e = 0.35$. The surface of the spherical compartment is $|\Gamma^{es}| = 1198\mu m^2$. The effective diffusion coefficient of $\Omega^s$ is $\overline{D}^s = 0$ because the spheres are compact. We computed the effective diffusion tensor of $\Omega^e$ (see [7] for details) to be $\overline{D}^e = \text{diag}\,(2.20, 2.25, 2.24) \times 10^{-3} mm^2/s$. For simplicity, we set $TE = \delta + \Delta$. The ODE signal is numerically solved by the Matlab routine *ode*45 with absolute tolerance of $10^{-8}$. The Karger solution is obtained explicitly by the eigen-decomposition of a $2 \times 2$ matrix [4].

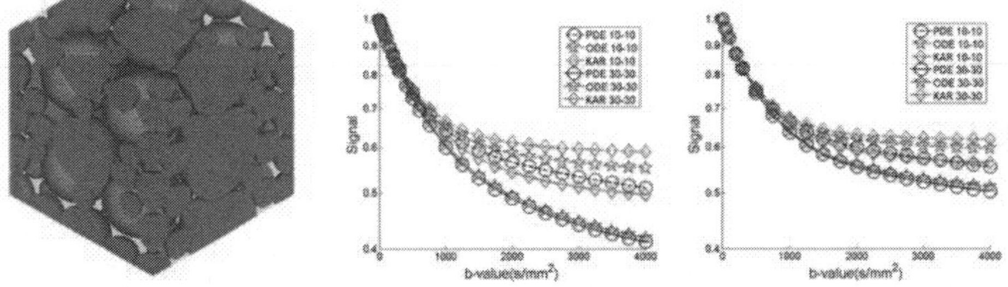


(a) Finite elements mesh of $C$ (b) Signals: $\kappa = 10^{-5}$m/s (c) Signals: $\kappa = 5 \times 10^{-6}$m/s

Figure 1: a) Finite element mesh of a domain $C = [-5\mu m, 5\mu m]^3$ containing spheres of many sizes. b) and c) DMRI signals with the PGSE sequence in the direction $\mathbf{g} = (1, 0, 0)$ at two diffusion times, $\delta = \Delta = 10$ms and $\delta = \Delta = 30$ms, and two permeabilities $\kappa = 10^{-5}$m/s and $\kappa = 5 \times 10^{-6}$m/s.

Figure 1b, 1c shows the dMRI signals at 20 $b$-values between 0 and $4000 s/mm^2$ for two PGSE sequences, in the gradient direction $\mathbf{g} = (1, 0, 0)$. The signals are labeled: "PDE" (the multiple compartment Bloch-Torrey PDE), "ODE" and "KAR", respectively. We can see that the ODE signal is significantly closer to the PDE signal than the Karger signal in these "non-narrow pulse" examples ($\delta$ is not small compared to $\Delta$). We also see that the ODE signal is closer to the PDE signal as diffusion time increases. Both the ODE model and the Karger model operate with an "effective diffusion coefficient" (the slope of the logarithm of the signal versus the b-value) that is independent of the diffusion time. Thus, if the diffusion time is not long enough to get the "effective diffusion coefficient" independent of the diffusion time, then the ODE and the Karger models will definitely not be good approximations to the full PDE model. Note that this necessary condition is not sufficient to guarantee that the macroscopic models are accurate. We see that the slope of the logarithm of the signal curve is the same at $\delta = \Delta = 10$ms and at $\delta = \Delta = 30$ms, but the ODE signal does not give a very good approximation to the PDE signal at the $\delta = \Delta = 10$ms, but it give an excellent approximation at $\delta = \Delta = 30$ms.

Now we consider the "PDE" signals in Fig. 1b and Fig. 1c as data (without noise) and solve the parameters estimation problem with the ODE and Karger models, i.e., determine the parameters of the macroscopic models from the signal. In particular, we assume that the volume of the computational domain $|C|$ and the permeability $\kappa$ are known, and the effective diffusion tensor of the spherical compartment $\overline{D}^s$ is set at 0. Thus, we consider three unknown parameters: 1) the effective extra-cellular diffusion coefficient $\mathbf{q}^T\overline{D}^e\mathbf{q}$, 2) the volume fraction of spherical compartment $v^s$, and 3) the surface area $|\Gamma^{es}|$. The Matlab routine *lsqnonlin* (tolerance of $10^{-8}$) was used to find the least-square fit of the "PDE" signal to the ODE and the Karger models. The initial guess of $\mathbf{q}^T\overline{D}^e\mathbf{q}$ is $D^0/2$, of $v^s$ is 0.5 and of $|\Gamma^{es}|$ is a random number in $[0.1, 10]$ times the true value. Using the parameters found by the Matlab routine, we computed the resulting $S_{ODE}(b_i)$ and $S_{KAR}(b_i)$ at the same 20 b-values shown in Fig. 1b and Fig. 1c. From these values we define the fitting error for each model in the following way:

$$E^{fit}_{ODE} \equiv \sqrt{\frac{1}{n_b}\sum_{i=1}^{n_b}\left|\frac{S_{PDE}(b_i) - S_{ODE}(b_i)}{S_{PDE}(b_i)}\right|^2}, \quad E^{fit}_{KAR} \equiv \sqrt{\frac{1}{n_b}\sum_{i=1}^{n_b}\left|\frac{S_{PDE}(b_i) - S_{KAR}(b_i)}{S_{PDE}(b_i)}\right|^2},$$

where $n_b$ is the number of $b$-values ($n_b = 20$ in our example). The above expressions give a measure of the average relative error of the signal fit at all the b-values. Table 1 contains parameters estimation results for two membrane permeabilities, $\kappa = 10^{-5}$m/s and $\kappa = 5 \times 10^{-6}$m/s, and at two diffusion times, $\delta = \Delta = 10$ms and $\delta = \Delta = 30$ms; the table also includes the fitting errors.

We see that the estimation of all three parameters is better using the ODE model than the Karger model in these "non-narrow pulse" cases. In addition, because both macroscopic models are better approximations of the PDE model at the longer diffusion time ($\delta = \Delta = 30$ms) than the shorter diffusion time ($\delta = \Delta = 10$ms), the parameters estimation is also more accurate at the higher diffusion time for these examples.

## 5. Conclusion

We illustrated by numerical simulations that a recently formulated ODE model for the dMRI signal accurately approximates the dMRI signal from the full PDE model, and strongly outperforms the Karger model when the duration of the magnetic field gradient pulses is not small compared to the diffusion time. We estimated macroscopic geometrical quantities using the full PDE signal as the data without noise, and showed that the parameters estimation results using the new ODE model are promising. Future work on macroscopic parameter estimation in more physically realistic settings is in progress.

## References

[1] H. Torrey, "Bloch equations with diffusion terms," *Physical Review Online Archive (Prola)*, vol. 104, no. 3, pp. 563–565, 1956.

| $\kappa = 1 \times 10^{-5}$m/s | | $\kappa = 5 \times 10^{-6}$m/s | |
|---|---|---|---|
| $\delta = \Delta = 10$ms | $\delta = \Delta = 30$ms | $\delta = \Delta = 10$ms | $\delta = \Delta = 30$ms |
| ODE (Karger) | ODE (Karger) | ODE (Karger) | ODE (Karger) |
| $E^{fit}_{ODE}$ $\left(E^{fit}_{KAR}\right)$ | | | |
| 0.2% (0.3%) | 0.03% (0.1%) | 0.2% (0.3%) | 0.04% (0.1%) |
| $v^s$ (true value = 0.65) | | | |
| 0.70 (0.73) | 0.65 (0.72) | 0.70 (0.73) | 0.66 (0.70) |
| $\lvert\Gamma^{es}\rvert$ ($\mu$m$^2$) (true value = 1198) | | | |
| 2200 (4467) | 1234 (2554) | 3130 (6623) | 1397 (2792) |
| $\mathbf{q}^T D^e \mathbf{q}$ (mm$^2$/s) (true value = 0.0022) | | | |
| 0.0026 (0.0030) | 0.0023 (0.0028) | 0.0026 (0.0030) | 0.0023 (0.0026) |

Table 1: Parameter estimation results for the domain in Fig. 1a. In each entry, the result is shown for the ODE model first, followed by the value for the Karger model in the parentheses.

[2] W. S. Price, A. V. Barzykin, K. Hayamizu, and M. Tachiya, "A model for diffusive transport through a spherical interface probed by pulsed-field gradient nmr," *Biophysical Journal*, vol. 74, pp. 2259–2271, May 1998.

[3] D. S. Grebenkov, "Pulsed-gradient spin-echo monitoring of restricted diffusion in multilayered structures," *Journal of Magnetic Resonance*, vol. 205, pp. 181–195, Aug. 2010.

[4] J. Karger, H. Pfeifer, and W. Heinik, "Principles and application of self-diffusion measurements by nuclear magnetic resonance," *Advances in magnetic resonance*, vol. 12, pp. 1–89, 1988.

[5] E. Fieremans, D. S. Novikov, J. H. Jensen, and J. A. Helpern, "Monte carlo study of a two-compartment exchange model of diffusion," *NMR in Biomedicine*, vol. 23, no. 7, pp. 711–724, 2010.

[6] D. Grebenkov, "Nmr survey of reflected brownian motion," *Reviews of Modern Physics*, vol. 79, no. 3, pp. 1077–1137, 2007.

[7] J. Coatléven, H. Haddar, and J.-R. Li, "A new macroscopic model including membrane exchange for diffusion mri." See http://www.cmap.polytechnique.fr/ jingrebeccali/preprints.html, 2013.

[8] E. O. Stejskal and J. E. Tanner, "Spin diffusion measurements: Spin echoes in the presence of a time-dependent field gradient," *The Journal of Chemical Physics*, vol. 42, no. 1, pp. 288–292, 1965.

[9] J. Xu, M. Does, and J. Gore, "Numerical study of water diffusion in biological tissues using an improved finite difference method.," *Physics in medicine and biology*, vol. 52, p. N111, Apr. 2007.

[10] D. V. Nguyen, D. Grebenkov, and J.-R. Li, "An efficient finite-elements code to simulate diffusion mri signals in complex tissue geometries." Submitted, see http://www.cmap.polytechnique.fr/ jingrebeccali/preprints.html, 2013.

# Mathematical modeling on the force experienced by the wall during the deflection of granular flows

**Yu Hui Deng and Yi Zhao**

Applied Mathematics Center, Harbin Institute of Technology Shenzhen Graduate School, Shenzhen University Town, XiLi Nanshan, Shenzhen, P.R. China

E-mail: 1.yuhuideng@hitsz.edu.cn 2.zhao.yi@hitsz.edu.cn

**Abstract.** We develop a mathematical model to describe the system where a dilute granular flow is deflected by an oblique smooth wall. Both particles in the granular flow and the oblique wall experience inelastic collisions with other particles. An analytical solution for the mean force experienced by the wall is obtained through the derivation based on the mathematical model and probability theory. Our theory shows that large angle between the wall and the granular flow may decrease the force experienced by the wall. This is opposite to one's intuition that large deflected angle implies large force. It is probable particle collisions and scattering after the deflection that prevent the wall from experiencing a further strong impact, thus the wall experience less force.

## 1. Introduction

Granular flows are commercially important in industrial applications such as pharmaceutical industry, agriculture, and energy production. However, some kinds of granular flows such as rock falls may pose significant natural disaster. In this case, oblique deflectors such as deflecting dams are applied to divert particle flows.

Despite various applications of granular flows and great efforts which have been devoted to the study of them, current understanding in the mechanics of granular flows is limited, and many fundamental difficulties still remain, so there is a great interest in explaining their properties. Behringer [1] gave a general introduction to the focus issue on granular materials, and talked about what we have learned since Jaeger's article [2] had been published. Pouliquen [3] studied experimentally and deduced a new model of steady flows of particles down a roughened plane. More recently, By using discrete element numerical simulations, Wassgren *et al.* [4] investigated a two-dimensional dilute granular flow around an immersed cylinder. Hogg *et al.* [5] investigated more thoroughly the steady deflection of granular flows by obstacles and established a theoretical model of the interaction that may predict quantitatively the flow following deflection.

The purpose of this paper is to investigate the force experienced by the oblique wall, which is very important when diverting granular flows. By mathematical modeling and probability theory we will derive an analytical formula for the mean force. Our theory shows an interesting phenomenon that large angle between the wall and the granular flow may decrease the force experienced by the wall, which may be opposite to one's intuition. The interpretation for this will be given in this paper.

## 2. Theoretical approach and results

We consider a two dimensional system where a dilute horizontal granular flow is deflected by an oblique smooth wall with a deflected angle $\theta$, which is defined as the acute angle between the wall and the moving direction of the flow. We assume that the flow consists of one type of smooth sphere with the same radius $a$ and mass $m$. The particles enter the system with different time interval. Different spatial distribution along the boundary can be specified for the initial positions of particles. Particles are removed from the system after interactions with other particles and the wall. In this paper, we choose Gaussian distribution with standard deviation $\sigma$ for the initial positions of particles which are independent and assume that the mean location of the flow is zero due to its unimportance. For simplicity, we can also choose the same input time interval $\tau$ as the rate of particle entering. The temperature in system is relatively low which means we can suppose that all particles have the same initial horizontal velocity $v_0$ which, we assume, is much larger than any initial thermal energy of the particles.Both collisions between particles and collisions between particles and the wall are inelastic with constant restitution coefficients $e$ and $e_w$, respectively. $e$ (or $e_w$ ) is defined as the ratio of the relative velocities of the particle to another particle (or the wall) in the direction along their line of centers immediately after and before the particle-particle (or the particle-wall) collision. For simplicity, we neglect the gravity because including gravity does not qualitatively effect the interesting results we present here.

We nondimensionalize lengths by $a$, then the system can be described by five dimensionless parameters, the deflected angle $\theta$, the particle-particle restitution coefficient $e$, the particle-wall restitution coefficient $e_w$, $V = \frac{v_0\tau}{a}$ and $S = \frac{\sigma}{a}$. In fact, small value of $V$ (or $S$) implies that the particles are close to each other in horizontal ( or perpendicular) direction . In this paper, we aim to investigate dimensionless mean force experienced by the oblique wall $f_{mean} = \frac{F_{mean}}{mv_0/\tau}$.

Since we consider a dilute system where multiple-particle collisions are negligible and pairwise particle collisions dominate, we suppose that there is only one collision between particles, and after the particle-particle collision, one or both of the two particles could hit the oblique wall again only once. With this simplification we can apply statistic average to determine the total average impact on the oblique wall by analyzing possible interactions between two particles. We randomly choose a particle from the system, and denote it by $B_{n+1}$. According to the simplicities, we know that it may collide with anyone of previous particles $B_n$, $B_{n-1}$, $\ldots B_1$ only once. First let us consider a simple case where $B_{n+1}$ may only collide with its previous neighbor particle $B_n$. Let $p_1^{(n+1)}$, $p_2^{(n+1)}$ and $p_0^{(1)}$ denote the following probabilities,

$$\begin{aligned} p_1^{(n+1)} &= P\big(B_{n+1} \text{ collides with } B_n\big), \quad p_2^{(n+1)} = P\big(B_{n+1} \text{ does not collide with } B_n\big), \\ p_0^{(1)} &= P\big(B_{n+1} \text{ collides with } B_n \mid B_n \text{ does not collide with } B_{n-1}\big). \end{aligned}$$

By the definition of $p_1^{(n+1)}$, $p_2^{(n+1)}$, $p_0^{(1)}$ and notice that $B_{n+1}$ may collide with $B_n$ if and only if $B_n$ does not collide with $B_{n-1}$, then we can obtain,

$$p_1^{(n+1)} = p_0^{(1)}(1 - p_1^{(n)}) \tag{1}$$

Similarly, we know that $B_{n+1}$ will not collide with $B_n$ if and only if $B_n$ does collide with $B_{n-1}$ or $B_{n+1}$ misses the rebounding particle $B_n$ when $B_n$ does not collide with $B_{n-1}$, thus we can also write $p_2^{(n+1)}$ as a function of $p_1^{(n)}$ and $p_2^{(n)}$,

$$p_2^{(n+1)} = 1 - p_1^{(n+1)} = p_1^{(n)} + (1 - p_0^{(1)})p_2^{(n)}. \tag{2}$$

Combining (1) and (2), we can write the results in a matrix form,

$$\begin{pmatrix} p_1^{(n+1)} \\ p_2^{(n+1)} \end{pmatrix} = \begin{pmatrix} 0 & p_0^{(1)} \\ 1 & 1 - p_0^{(1)} \end{pmatrix} \begin{pmatrix} p_1^{(n)} \\ p_2^{(n)} \end{pmatrix} \triangleq \mathbf{M}^{(1)} \begin{pmatrix} p_1^{(n)} \\ p_2^{(n)} \end{pmatrix} \tag{3}$$

In fact, if $B_{n+1}$ may only collide with its previous neighbor particle $B_n$, all the possible states of $B_{n+1}$ must only depends on the states of $B_n$. In other words, the states of a single particle in the system have Markov property and the matrix $\mathbf{M^{(1)}}$ in (3) is the transition probability matrix for a single step. When a steady state has been achieved, there will be of no difference between the states of $B_{n+1}$ and the states of $B_n$, then by letting $n \to +\infty$ in (3), we can obtain the probability for a single particle experiencing collisions as follows,

$$p^{(1)} = \lim_{n\to+\infty} p_1^{(n+1)} = \frac{p_0^{(1)}}{1+p_0^{(1)}} \tag{4}$$

Now let us consider a little bit more complicated case where $B_{n+1}$ may collide with $B_n$ or $B_{n-1}$. Let $p_1^{(n+1)}$, $p_2^{(n+1)}$, $p_3^{(n+1)}$, $p_0^{(1)}$ and $p_0^{(2)}$ respectively denote the following probabilities,

$$\begin{aligned}
p_1^{(n+1)} &= P(B_{n+1}\ \textit{collides with}\ B_n),\quad p_2^{(n+1)} = P(B_{n+1}\ \textit{collides with}\ B_{n-1})\\
p_3^{(n+1)} &= P(B_{n+1}\ \textit{does not collide with}\ B_n\ \textit{and}\ B_{n-1}),\\
p_0^{(1)} &= P(B_{n+1}\ \textit{collides with}\ B_n \mid B_n\ \textit{does not collide with}\ B_{n-1}\ \textit{and}\ B_{n-2}),\\
p_0^{(2)} &= P(B_{n+1}\ \textit{collides with}\ B_{n-1} \mid B_{n+1}\ \textit{does not collide with}\ B_n,\\
&\qquad \textit{and}\ B_{n-1}\ \textit{does not collide with}\ B_n,\ B_{n-2}\ \textit{and}\ B_{n-3}).
\end{aligned}$$

Similar to the previous case but with a little more calculation, we can also describe the system in a matrix form as follows,

$$\begin{pmatrix} p_1^{(n+1)} \\ p_2^{(n+1)} \\ p_3^{(n+1)} \end{pmatrix} = \begin{pmatrix} 0 & 0 & p_{13}^{(n)} \\ 0 & p_{22}^{(n)} & p_{23}^{(n)} \\ 1 & 1-p_{22}^{(n)} & 1-p_{13}^{(n)}-p_{23}^{(n)} \end{pmatrix} \begin{pmatrix} p_1^{(n)} \\ p_2^{(n)} \\ p_3^{(n)} \end{pmatrix} \triangleq \mathbf{M^{(2)}} \begin{pmatrix} p_1^{(n)} \\ p_2^{(n)} \\ p_3^{(n)} \end{pmatrix} \tag{5}$$

$$p_{13}^{(n)} = p_0^{(1)}(1-p_{23}^{(n)}),\ p_{22}^{(n)}p_2^{(n)} = p_0^{(2)}(p_2^{(n)}-p_{22}^{(n-1)}p_2^{(n-1)}),\ p_{23}^{(n)}p_3^{(n)} = \frac{(1-p_0^{(1)})p_0^{(2)}}{1-p_0^{(1)}p_0^{(2)}}p_3^{(n-1)}(1-p_{13}^{(n)}-p_{23}^{(n)}).$$

The difference between the two cases is that the matrix $\mathbf{M^{(2)}}$ is no longer a constant matrix but a nonconstant matrix with iterative parameters. In fact, if $B_{n+1}$ may collide with $B_n$ or $B_{n-1}$, all the possible states of $B_{n+1}$ not only depend on the states of $B_n$ but also depend on the states of $B_{n-1}$. When a steady state has been achieved, by letting $n \to +\infty$ in (5), we can obtain

$$\begin{pmatrix} p_1 \\ p_2 \\ p_3 \end{pmatrix} = \begin{pmatrix} 0 & 0 & p_{13} \\ 0 & p_{22} & p_{23} \\ 1 & 1-p_{22} & 1-p_{13}-p_{23} \end{pmatrix} \begin{pmatrix} p_1 \\ p_2 \\ p_3 \end{pmatrix} \tag{6}$$

$$p_{22} = \frac{p_0^{(2)}}{1+p_0^{(2)}},\ p_{13} = \frac{p_0^{(1)}(1-p_0^{(1)}p_0^{(2)})}{1-p_0^{(1)}p_0^{(2)}+(1-p_0^{(1)})^2p_0^{(2)}},\ p_{23} = \frac{(1-p_0^{(1)})^2p_0^{(2)}}{1-p_0^{(1)}p_0^{(2)}+(1-p_0^{(1)})^2p_0^{(2)}}.$$

Solving the linear system (6) gives the probability for a single particle experiencing collisions in this case,

$$p^{(2)} = p_1+p_2 = \frac{p_0^{(1)}+B}{1+p_0^{(1)}+B},\ B = p_{23}(1-p_0^{(1)}+p_0^{(2)}) = \frac{p_0^{(2)}(1-p_0^{(1)})^2(1-p_0^{(1)}+p_0^{(2)})}{1-p_0^{(1)}p_0^{(2)}+(1-p_0^{(1)})^2p_0^{(2)}} \tag{7}$$

Following this procedure we can work out the probabilities for a single particle experiencing collisions with more previous particles. However, by comparing (4) and (7), we find that including the interaction with more previous particles has a weak effect for the collision probability since $B$ is relatively small in a dilute system. Therefore, we choose the leading part of $p = \lim_{n\to+\infty} p^{(n)}$, i.e $p^{(1)}$ in (4), as the approximate collision probability for a single particle in the system. The force experienced by the wall comes from two aspects. One is the impact, denoted by $I$, which is caused by the direct particle-wall collision without any collision with previous particles. The other is the impact, denoted by $J$, which is caused by the particle-wall collision after the particle-particle collision. Thus by letting $n \to +\infty$, we can write the total average force $F_{mean}$ on the wall as follows due to the arbitrariness of $B_{n+1}$ in the system.

$$\begin{aligned} F_{mean} &= \lim_{n\to+\infty} \Big[\big(1-p_1^{(n+1)}\big)\cdot \mathbf{E}\big(I\big|B_{n+1}\ does\ not\ collide\ with\ B_n\big) + \big(1-p_1^{(n)}\big) \\ &\quad \cdot\mathbf{E}\big(J\big|B_n\ does\ not\ collide\ with\ B_{n-1}\big)\Big] = (1-p^{(1)})(\tilde{I}+\tilde{J}) = \frac{1}{1+p_0^{(1)}}(\tilde{I}+\tilde{J}). (8) \end{aligned}$$

Therefore, by using probability theory and performing integration, we will obtain the analytical formula of dimensionless mean force $f_{mean}$ as follows,

$$\begin{aligned} &\frac{(1+e_w)\sin\theta}{4+2\mathrm{erf}(D_+)-4\mathrm{erf}(D_-)}\Bigg\{4+\Big((1+e)-e_w(1-e)\Big)\Big(\mathrm{erf}(H_+)-\mathrm{erf}(H_-)\Big)+\Big((1-e)-e_w(1+e)\Big) \\ &\Big(\mathrm{erf}(D_+)-\mathrm{erf}(D_-)+M\Big)+\Big(1+e\Big)\Big(1+e_w\Big)\Bigg[\Big(\frac{S^2\sin^2\theta}{2}+\frac{V^2\cos^2\theta}{4}\Big)\Big(\mathrm{erf}(D_+)-\mathrm{erf}(D_-) \\ &+\mathrm{erf}(H_-)-\mathrm{erf}(H_+)+M\Big)+\frac{VS\sin\theta\cos\theta}{\sqrt{\pi}}\Big(e^{-D_-^2}-e^{-D_+^2}+e^{-H_+^2}-e^{-H_-^2}+N\Big)+\frac{S^2\sin^2\theta}{\sqrt{\pi}} \\ &\Big(D_-e^{-D_-^2}-D_+e^{-D_+^2}+H_+e^{-H_+^2}-H_-e^{-H_-^2}+R\Big)\Bigg]\Bigg\}, \end{aligned}$$

where $\mathrm{erf}(x) = \frac{2}{\sqrt{\pi}}\int_0^x e^{t^2/2}\mathrm{d}t$, and

$$M = \begin{cases} 0 \\ \mathrm{erf}(L_-)-\mathrm{erf}(L_+), \end{cases} \quad N = \begin{cases} 0 \\ e^{-L_+^2}-e^{-L_-^2}, \end{cases} \quad R = \begin{cases} 0 & \text{if } 0 \leqslant e_w \leqslant \frac{1-e}{1+e} \\ L_+e^{-L_+^2}-L_-e^{-L_-^2} & \text{if } \frac{1-e}{1+e} < e_w \leqslant 1 \end{cases},$$

$$D_\pm = \frac{\pm 1 - \frac{V\cos\theta}{2}}{S\sin\theta}, \quad H_\pm = \frac{\pm\sqrt{\frac{(1+e)-e_w(1-e)}{(1+e)(1+e_w)}} - \frac{V\cos\theta}{2}}{S\sin\theta}, \quad L_\pm = \frac{\pm\sqrt{\frac{e_w(1+e)-(1-e)}{(1+e)(1+e_w)}} - \frac{V\cos\theta}{2}}{S\sin\theta}$$

One can show that the mean force $f_{mean}$ is a non-monotonic function of the deflected angle $\theta$. Large angle $\theta$ may decrease the force experienced by the wall. This is opposite to one's intuition that large deflected angle implies a large impact perpendicular to the wall, and consequently large force. However, it is the particle-particle collisions and the scattering after the deflection that prevent the wall from experiencing a further strong impact, thus the wall experience less force.

## 3. Acknowledgements

The work described in this paper was fully supported by the China Postdoctoral Science Foundation (2011M500647).

## References

[1] Behringer R, 1999, *Chaos.* **9**, 509.
[2] Jaeger H.M, Nagel S.R and Behringer R.P, 1996, *Rev. Mod. Phys.* **68**, 1259–1273.
[3] Pouliquen O, 1999, *Phys. Fluids.* **11**, 542.
[4] Wassgren C.R, Cordova J.A, Zenit R and Karion A, 2003, *Physics of Fluids.* **15**, 3318.
[5] Hogg A.J and Hkonardttir K.M, 2005, *Physics of Fluid.* **17**, 077101.

# Analogy between electrochemical oscillations and quantum physical processes

**J Grzanna**[1] **and H J Lewerenz**[1,2]
[1]Helmholtz Zentrum Berlin for Materials and Energy
Hahn-Meitner Platz 1, D-14109 Berlin, Germany
[2]Joint Center for Artificial Photosynthesis, California Institute of Technology
1200 E. California Blvd, Pasadena, CA 91125, USA

E-mail: grzanna@helmholtz-berlin.de

**Abstract.** In (photo) electrochemical oscillations, the discretization of phase oscillators leads to a sequence of time dependent oscillator density functions which describe the passing of the oscillators through their minimum at each cycle. Two consecutive oscillator density functions are connected by a Markov process represented by a linear integral equation of second order which is homogeneous in the case of sustained oscillations. The kernel of the integral equation is a normalized Greens Function and represents the probability density for the periods of the oscillators. Both together, the oscillator density function and the two-dimensional probability density for the periods of the oscillators define a random walk. The relation of the model to the holographic principle is discussed briefly. Further, a detailed analysis of a kernel of the integral equation leads to a frequency distribution $g$ for the period length. Additionally, it is possible to determine the energy $E$ in dependence on the period length from the electrochemical process. The product $g\ E$ shows qualitatively the same behaviour as the radiation of a black body, indicating that the discretization of phase oscillators, when represented by phase space analysis, show an analogy to quantum processes.

## 1. Introduction

The oscillatory behaviour of silicon electrodes has been reported about two decades ago [1]. Particularly, electrochemical oscillations on Si photo-electrodes have been extensively studied [2]. Several models were developed based on self-oscillating domains [3], the so-called current bursts [4], and oxide-induced interfacial stress [5-9] relating to the observation of nano-dimensioned pores, fluctuating with the phase of the oscillating (photo)current of Si electrodes immersed in dilute ammonium fluoride solutions. The latter model was presented by a special phase space analysis based on a discretization of phase oscillators indicating that two types of oxides (with different nanopore densities) can explain sustained current oscillations. In section 2 the relation of the model to a random walk and to the holographic principle is discussed. In section 3 the frequency distribution $g$ and the energy $E$ in dependence on the period length is determined in the case of sustained and damped oscillations. The product $g\ E$ is compared with the radiation of a black body.

## 2. Oscillatory process and mathematical model

The basic process at the electrode surface domain $\Omega \subseteq R^2$ is characterised by a cyclic oxidation of a small silicon layer (with a thickness of ca. 10 nm) and the following removal of the oxide by etching. The oxide thickness evolves not uniformly at the electrode due to cracks and nanopores at the oxide which result from stress due to the lattice mismatch between silicon and its oxide. The oxide thickness is given at the time $t$ at each point $(x,y)$ of the electrode surface by

$$d(t,x,y) = \frac{\cos(\varphi(t,x,y)+\pi)+1}{2}(d^{\max} - d^{\min}) + d^{\min} \quad \text{for } i2\pi \le \varphi(t,x,y) \le (i+1)2\pi,\ i = 0,1,2,... \tag{1}$$

at which the phase $\varphi$ propagates nonlinear monotonically increasing. In [9], the so called phase oscillator $\varphi$ is discretized in the phase space $\Phi = \{(x,y,\varphi) \mid (x,y) \in \Omega, \varphi \in R_+\}$ by registering its value only at the so-called snap-shot screens $S_i = \{(x,y,i\,2\pi) \mid (x,y) \in \Omega\}$ that are planes with a constant phase difference of $2\pi$. The result is the time dependent oscillator density function $p_i(t)$ which defines for the time $t$ the differential number of oscillators passing their minimum $d^{\min}$ at the $i$-th time. In the following, we assume that all phase oscillators pass $i\,2\pi$ during the time interval $[t_i^{\min}, t_i^{\max}]$ at the $i$-th cycle. Then two consecutive oscillator density functions are connected by a Markov process represented by a linear integral equation of second order (figure 1 top left)

$$p_i(t) = \int_{t_{i-1}^{\min}}^{t_{i-1}^{\max}} p_{i-1}(s)\, q_{i,s}(t-s)\, ds \qquad \text{for } i = 1,2,3,... \text{ and } t \in [t_i^{\min}, t_i^{\max}] \tag{2}$$

Eq. (2) is homogeneous in the case of sustained oscillations i.e. $f(t)$ is zero in the eq. $p_i(t) = p_{i-1}(t-\overline{T}) + f(t)$ ($\overline{T}$ - macroscopic oscillation period). The kernel $q_{i,s}(t-s)$ of the integral equation is a normalized Greens Function [8] and represents the probability density for the periods $t-s$ of the oscillators in dependence on the starting times $s$ of the oscillators during the $i$-th cycle.

Both together, the oscillator density function $p_{i-1}(s)$ and the two-dimensional probability density $q_{i,s}(t-s)$ for the periods of the oscillators define a random walk

$$P_i = P_{i-1} + Q_{i,s} \qquad \text{for } i = 1,2,3,... \text{ and } s \in [t_{i-1}^{\min}, t_{i-1}^{\max}] \tag{3}$$

for which $P_i$ and $P_{i-1}$ denote random variables and $Q_{i,s}$ the random process. $p_i(t)$ defines the probability density for $P_i$ and $q_{i,s}(t-s)$ for $Q_{i,s}$. The random variable $P_i$ assigns to each event (oscillation process starting with a bare silicon electrode) the time $t$ at which the phase oscillator $\varphi(t,x,y)$ equals $i\,2\pi$. The $s$ dependent random variable $Q_{i,s}$ assigns to each event the period $t-s$ at which the phase oscillator equals $(i-1)\,2\pi$ at the time $s$ and $i\,2\pi$ at the time $t$.

The given description of the oscillatory process reveals a relation to the holographic principle as presented in [10]. We describe in [5-9] via phase oscillators a process in the space volume (evolution of a finite set of pores in the grown oxide) by a process at the two-dimensional snap shoot screen $S_i$ (unique fingerprint of the pore evolution at the minimum oxide thickness in terms of strings of intersection between the phase and $S_i$). $S_i$ is a holographic screen located at an equipotential surface corresponding to a constant potential at the minimum oxide thickness. Hence all important information of the process can be displayed at the holographic screen and the space volume has one holographic direction perpendicular to the electrode surface directed into the bulk.

## 3. Mathematical model and the black body radiation

We determine the frequency distribution $g$ for the period length $T$ from the probability density $q_i(t,s)\ (= q_{i,s}(t-s))$ for the periods $t-s$ of the oscillators. In the case of sustained oscillations we choose for $q$ a normalized Greens Function [8] and introduce additionally a damping parameter $v$ which defines the transition from sustained $(v=0)$ to damped $(v>0)$ oscillations (figure 1 top left) i.e.

$$q_i(t,s)=\begin{cases}\dfrac{2(t-\overline{T})}{(L+v)s}, & for\ \ \overline{T}\le t\le \overline{T}+\dfrac{L+v}{L}s\\ \dfrac{2(\overline{T}+L+v-t)}{(L+v)(L+v-\dfrac{L+v}{L}s)}, & for\ \ \overline{T}+\dfrac{L+v}{L}s\le t\le \overline{T}+L+v\end{cases} \tag{4}$$

for $s\in[t_{i-1}^{\min},t_{i-1}^{\max}]$, $L=t_{i-1}^{\max}-t_{i-1}^{\min}$ and $\overline{T}$ denotes the macroscopic oscillation period. The coordinate transformation $T=t-s$ transforms $q_i$ to $g$ which means $g_{\overline{T},L}^{v}(T,s)=q_i(T+s,s)$ defines the frequency distribution for all period lengths *T* which create one macroscopic oscillation in dependence on $v$, $\overline{T}$, *L* and *s*. To eliminate the dependence on *s*, we integrate and get $g_{\overline{T},L}^{v}(T)=L^{-1}\int_0^L g_{\overline{T},L}^{v}(T,s)\,ds$ (figure 1 top right). In the next step, the dependence on *L* is eliminated by integration (figure 1 bottom left) with the result $g_{\overline{T}}^{v}(T)=\overline{T}^{-1}\int_0^{\overline{T}} g_{\overline{T},L}^{v}(T)\,dL$. Let the time $\overline{T}_{\max}$ denote the temporal analogue to the maximum wavelength which corresponds to the size of a black body. For an arbitrarily positive integer *k* we consider the mode $\overline{T}_{\max}k^{-1}$. Then the equations (2) and (4) define oscillations with a sequence of macroscopic periods $\overline{T}_i=(\overline{T}_{\max}k^{-1}+iv_k2^{-1})$ which have (arranged one after another) to match $\overline{T}_{\max}$ i.e. $\sum_{i=1}^{\mu_k}\overline{T}_i=\overline{T}_{\max}$. Hence, the integer $\mu_k$ equals *k* in the case of sustained oscillations $(v_k=0)$ and $v_k=4(\mu_k{}^2+\mu_k)^{-1}(1-\mu_k k^{-1})\overline{T}_{\max}$ for $0<\mu_k<k$ in the case of damped oscillations. Finally, the frequency distribution *g* for the periods *T* is defined by

$$g(T)=\sum_{k=1}^{m}\sum_{i=1}^{\mu_k} g_{\overline{T}_i}^{v_k}(T) \tag{5}$$

Oscillations with macroscopic periods shorter than $\overline{T}_{\max}m^{-1}$ are excluded to avoid an unrealistic unlimited velocity of the oscillation process. In the model, the current resulting from one macroscopic oscillation with period $\overline{T}$ is defined by $I_i^{\overline{T}}(t)=\int_{t_i^{\min}}^{t_i^{\max}} p_i(s)I_{elem}^{\overline{T}}(t-s)\,ds$ where $I_{elem}^{\overline{T}}$ denotes the elementary current peak [7] resulting from the oxide growth. The related (per oscillation cycle *i*) quantized energy is given by $E_{\overline{T}}=U\int_{t_i^{\min}}^{t_i^{\max}+\Delta t} I_i^{\overline{T}}(t)\,dt$ where *U* denotes the potential at the boundary between the silicon and its oxide and $\Delta t$ the oxide growth time. Measurements [11] indicate that during each of the $\mu_k$ oscillations the same energy quant $E_{\overline{T}_{\max}k^{-1}}$ is emitted. Hence, the energy density is defined by

$$\rho(T)=\sum_{k=1}^{m}E_{\overline{T}_{\max}k^{-1}}\sum_{i=1}^{\mu_k} g_{\overline{T}_i}^{v_k}(T) \tag{6}$$

Defining $E(T)=\rho(T)(g(T))^{-1}$, equation (6) leads to the relation $\rho(T)=E(T)g(T)$. Measurements [11] show, that for larger modes $\overline{T}_{\max}k^{-1}$ sustained oscillation appear i.e. $v_k=0$ for $k=1,2,...,n$. For shorter modes the oscillations are more and more damped i.e. $0<v_{n+1}<v_{n+2}<...<v_m$. The energy $\mu_k E_{\overline{T}_{\max}k^{-1}}$ (resulting from $\mu_k$ oscillations during the time $\overline{T}_{\max}$) depends on the mode $\overline{T}_{\max}k^{-1}$ i.e. on *k*. In the case of sustained oscillations, the resulting energy increases progressively for shorter modes reflected by the graph $(\overline{T}_{\max}k^{-1},kE_{\overline{T}_{\max}k^{-1}})$ for $k=1,2,...,n$. Moreover, during sustained oscillations measurements [11] of $E_{\overline{T}}$ indicate additionally a progressive increase of $E_{\overline{T}}$ for decreasing $\overline{T}$. In the

case of damped oscillations, the resulting energy $\mu_k E_{\bar{T}_{max}k^{-1}}$ decreases because the number $\mu_k$ of periods decreases for $k = n+1,...,m$. Also, $E_{\bar{T}_{max}k^{-1}}$ decreases because the oxide growth time falls more and more below a time limit for a sufficiently large oxide growth. This behaviour of the energy is reflected by $E(T) = c(T^2(e^{dT^{-1}} - 1))^{-1}$ quantitatively. Finally, a suitable and with measurements [11] consistent choice of the model parameters $n$, $m$, $n > \mu_{n+1} > \mu_{n+2} > ... > \mu_m > 0$, and $E_{\bar{T}_{max}k^{-1}}$ for $k = 1,2,...,m$ leads to a fit of $g(T)E(T)$ to the radiation of a black body (figure 1 bottom right).

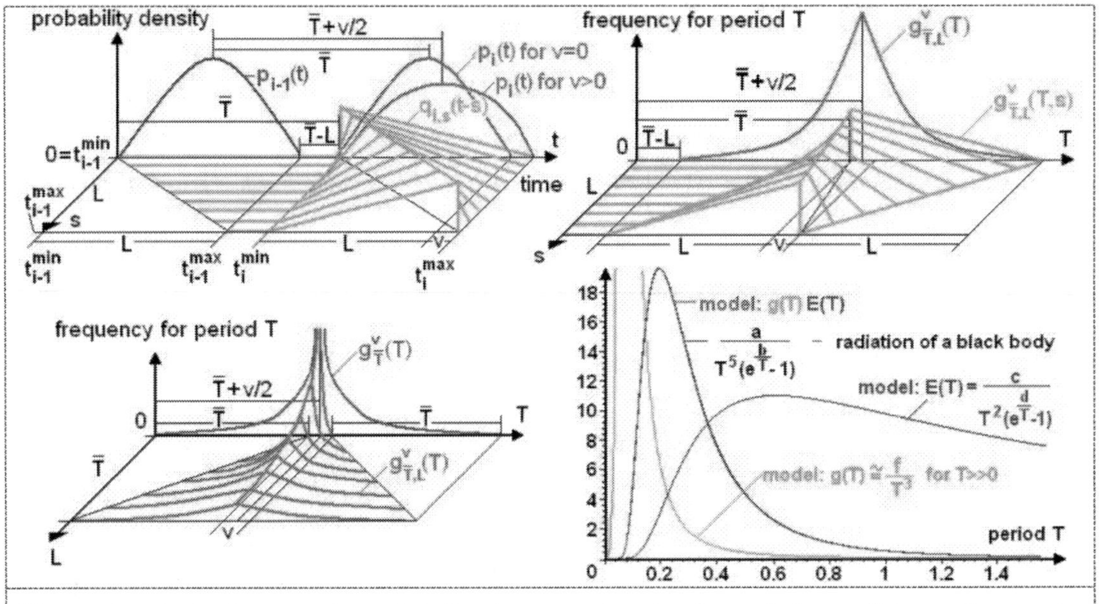

**Figure 1.** Relation between the probability density $q_{i,s}(t\text{-}s)$ for the periods $t$-$s$ of the oscillators, the various frequency distributions $g$ for the period lengths $T$ and the radiation of a black body compared with the model. The magnitude of the coordinate axes is chosen arbitrarily and $a,b,...,f$ are constants.

## 4. Conclusions

The mathematical concept of the discretization of phase oscillators leads to a linear integral equation of second order representing a Markov process and a random walk. The relation of the model to the holographic principle is shown. Moreover an analysis of the frequency distribution for the periods of the phase oscillators and of the energy of the (photo) electrochemical system leads to the graph of the radiation of a black body. Particularly the energy is quantized in the model.

## References


[1] Gerischer H, Lübke M 1988 *Ber. Bunsenges. Phys. Chem.* **92** 573
[2] Aggour M, Giersig M and Lewerenz H J 1995 *J. Electroanal. Chem.* **383** 67
[3] Ozanam F, Blanchard N and Chazalviel J N 1993 Electrochim. Acta **38** 1627
[4] Foca E, Carstensen J and Föll H 2007 J. Electroanal. Chem. **603** 175
[5] Grzanna J, Jungblut H and Lewerenz H J 2000 J. Electroanal. Chem. **486** 181+190
[6] Grzanna J, Jungblut H and Lewerenz H J 2007 Phys. Stat. Sol. (a) **204** 1245
[7] Grzanna J, Notz T and Lewerenz H J 2009 Phys. Stat. Sol. (c) **6** 1639
[8] Grzanna J, Notz T, Stempel T and Lewerenz H J 2011 Phys. Stat. Sol. (c) **8** 1734
[9] Grzanna J and Lewerenz H J 2013 Journal of Physics: Conference Series **410** 012160
[10] Verlinde E 2011 Journal of High Energy Physics **4** 29
[11] Aggour M 1994 Dissertation Technical University Berlin Department of Physics

# Waves in periodic media: Fourier analysis shortcuts and physical insights, case of 2D phononic crystals

**S Dupont[1,2,3], J Gazalet[1,2,3], JC Kastelik[1,2,3]**
[1]Univ Lille Nord de France, F-59000 Lille, France
[2]UVHC, IEMN-DOAE, F-59313 Valenciennes, France
[3]CNRS, UMR 8520, F-59650 Villeneuve d'Ascq, France

E-mail: samuel.dupont@univ-valenciennes.fr

**Abstract.** Phononic crystal is a structured media with periodic modulation of its physical properties that influences the propagation of elastic waves and leads to a peculiar behaviour, for instance the phononic band gap effect by which elastic waves cannot propagate in certain frequency ranges. The formulation of the problem leads to a second order partial differential equation with periodic coefficients; different methods exist to determine the structure of the eigenmodes propagating in the material, both in the real or Fourier domain. Brillouin explains the periodicity of the band structure as a direct result of the discretization of the crystal in the real domain. Extending the Brillouin vision, we introduce digital signal processing tools developed in the frame of distribution functions theory. These tools associate physical meaning to mathematical expressions and reveal the correspondence between real and Fourier domains whatever is the physical domain under consideration. We present an illustrative practical example concerning two dimensions phononic crystals and highlight the appreciable shortcuts brought by the method and the benefits for physical interpretation.

## 1. Introduction

Man-made periodic media have emerged several decades ago [1, 2] and have drawn much attention: they offer a mean to control the propagation of waves; yet one dimension photonic periodic structures like fibre Bragg gratings, multilayer dielectric mirrors or Distributed Feed-Back laser have reached industrial maturity and allow the effective control of photons. Such devices for phonon would be highly desirable; in fact, any kind of wave could be subject to such practical developments: the existence of band structures in Fourier domain is a universal feature of media periodicity regardless the nature of propagating wave. Wave propagation in periodic media is governed by a second order partial differential equation with periodic coefficients. Usually, the general form of the solutions, the so called Bloch modes, are determined either in the real (direct) domain or in the Fourier (reciprocal) domain. Brillouin has exploited the two domain correspondence in order to explain the behaviour of the band structure [3]. In this paper we go on with the Brillouin conception of the direct and reciprocal domains correspondence, we derive an expression of the Bloch theorem in the Fourier domain. Along the way we show that the general form of the solutions (Bloch modes) comes out naturally as a consequence of media periodicity. The cornerstone of the proposed method is to introduce the Dirac delta comb to express the periodic media parameters in the equation system. The property of Dirac delta comb have been extensively studied in the frame of distribution functions theory and are well known, we take advantage of namely: the replicating property, the shifting property and the sifting property along with their associated Fourier transform properties. These properties are intensively used in digital signal processing; the reader who isn't familiar with these notions can find a detailed summary in [4] that gives a survey of classical methods to establish wave properties in periodic media. In section two, we present the proposed method and show how straightforward it is to establish the wave equation for elastic waves in two-dimension periodic media and Dirac comb properties are briefly recalled. Then, in section three, we highlight the benefits of the proposed method for physical interpretation. We end up with some concluding remarks.

## 2. Elastic waves in 2D periodic media

In this section, we establish the expression of periodic media parameters as a function of a Dirac comb in direct domain and we derive the expression of the reciprocal domain wave equation.

### *2.1. Expression of the elastic wave equation in direct and reciprocal domains*

Starting from the dynamic equation of motion for the displacements $u_i$ and the expression of the stress tensor $\sigma_{ij}$, the well-known acoustic wave equation in a homogeneous media is:

$$\rho(\boldsymbol{r})\frac{\partial^2 u_i(\boldsymbol{r})}{\partial t^2}=\sum_j\left(\frac{\partial}{\partial x_j}\left[\sum_{kl}c_{ijkl}(\boldsymbol{r})\frac{\partial u_k(\boldsymbol{r})}{\partial x_l}\right]\right) \tag{1}$$

With $\rho(\boldsymbol{r})$ the mass density and $c_{ijkl}(\boldsymbol{r})$ the stiffness tensor of the considered media.
Assuming an harmonic dependence of the form $e^{i(\omega t-\boldsymbol{k}.\boldsymbol{r})}$ and recalling that a simple product in real domain converts to a convolution product (represented by $\otimes$) in reciprocal domain, the Fourier transform of the wave equation (1) writes :

$$\rho(\boldsymbol{k})\otimes\left(-\omega^2 u_i(\boldsymbol{k})\right)=\sum_j\left(-ik_j\left[\sum_{kl}c_{ijkl}(\boldsymbol{k})\otimes\left(-ik_l u_k(\boldsymbol{k})\right)\right]\right) \tag{2}$$

Now consider the shifting property that reads: $u(\mathbf{r})\otimes\delta(\mathbf{r}-\mathbf{R})=u(\mathbf{r}-\mathbf{R})$ and let $o(\boldsymbol{r})$ be the restriction of the mass density of a periodic media to one period. Thus, thanks to the replicating property, $\rho(\boldsymbol{r})$ and its Fourier transform can be expressed as:

$$\rho(\boldsymbol{r})=o(\boldsymbol{r})\otimes\sum_{\boldsymbol{R}\in R}\delta(\boldsymbol{r}-\boldsymbol{R}) \qquad \Leftrightarrow \qquad \rho(\boldsymbol{k})=o(\boldsymbol{k})V_{RL}\sum_{\boldsymbol{G}\in G}\delta(\boldsymbol{k}-\boldsymbol{G}) \tag{3}$$

where $R$ and $G$ are the direct and reciprocal lattices, $V_{RL}$ stands for the reciprocal lattice unit cell volume and the symbol $\Leftrightarrow$ stands for the Fourier transform operation. Hereafter, we refer to this kind of formulations as the Real (or Fourier) Domain Close Form Expression: RD-CFE (FD-CFE). Similar expressions can be written for $c_{ijkl}(\boldsymbol{r})$ the stiffness tensor defining $\gamma_{ijkl}(\boldsymbol{r})$ as its restriction to a period. Substituting the FD-CFE for mass density and stiffness tensor in equation (2), the Fourier domain wave equation becomes:

$$\begin{aligned}&\left(\sum_G o(\boldsymbol{G})\delta(\boldsymbol{k}-\boldsymbol{G})\right)\otimes\left(-\omega^2 u_i(\boldsymbol{k})\right)\\&=\sum_j\left(\left(-ik_j\right)\left[\sum_{kl}\left(\sum_G\gamma_{ijkl}(\boldsymbol{G})\delta(\boldsymbol{k}-\boldsymbol{G})\right)\otimes\left(-ik_l u_k(\boldsymbol{k})\right)\right]\right)\end{aligned} \tag{4}$$

We recall that the convolution product with a Dirac comb shifts the functions so that equation (4) gives us:

$$\omega^2\left(\sum_G o(\boldsymbol{G})u_i(\boldsymbol{k}-\boldsymbol{G})\right)=\sum_j\left(k_j\left[\sum_{kl}\sum_G\gamma_{ijkl}(\boldsymbol{G})(k_l-G_l)u_k(\boldsymbol{k}-\boldsymbol{G})\right]\right) \tag{5}$$

Equation (5) is the Fourier domain expression of the elastic wave equation. In the above development, since no assumption was made concerning the nature of the solution, the variable k is considered as continuous (not discrete).

*2.2. Case of a transverse elastic wave in a two dimension lattice*

To avoid the unnecessary heaviness of the tensor product in equation (5) in the discussion that follows and without loss of generality about the method presented, we restrict this section to the 2-D periodic media. Moreover, we limit ourselves to a special case where the coupling between the longitudinal and transversal waves does no longer exist. This situation appears in all crystal symmetries excluding trigonal and triclinic. More specifically, we consider a media with 2D periodicity in the (x, y) plane and uniformity along the z direction where a transverse acoustic wave propagates in the $(x, y)$ plane with a displacement polarised along the $z$ axis. In this situation equation (5) leads to scalar wavefunctions and writes:

$$\omega^2\left(\sum_G o(\boldsymbol{G})u_3(\boldsymbol{k}-\boldsymbol{G})\right) = k_1\left[\sum_l\sum_G \gamma_{313l}(\boldsymbol{G})(k_l - G_l)u_3(\boldsymbol{k}-\boldsymbol{G})\right] + k_2\left[\sum_l\sum_G \gamma_{323l}(\boldsymbol{G})(k_l - G_l)u_3(\boldsymbol{k}-\boldsymbol{G})\right] \quad (6)$$

Introducing Voigt notation, we substitute the fourth rank tensor expression of the unit cell elastic constants by the equivalence: $\gamma_{31kl} = \gamma_{5m}$ ; $\gamma_{32kl} = \gamma_{4m}$. Again, for the sake of simplicity we further exclude monoclinic crystals thus: $\gamma_{55} = \gamma_{44}$ ; $\gamma_{45} = \gamma_{54} = 0$ and equation (6) reduces to:

$$\omega^2\left(\sum_G o(\boldsymbol{G})u_3(\boldsymbol{k}-\boldsymbol{G})\right) = \sum_G \gamma_{44}(\boldsymbol{G})\boldsymbol{k}.(\boldsymbol{k}-\boldsymbol{G})u_3(\boldsymbol{k}-\boldsymbol{G}) \quad (7)$$

*2.3. Correspondence with traditional methods*

Equation (7) is the Fourier domain expression of the transverse elastic wave equation in a 2D periodic media; we have implicitly assumed a harmonic dependence of the wave in the form of $e^{-i\boldsymbol{k}.\boldsymbol{r}}$. The correspondence with traditional methods which suppose as a starting point that the solutions are Bloch functions (i.e. a superposition of harmonics of the form $e^{-i(\boldsymbol{k}+\boldsymbol{G}'').\boldsymbol{r}}$ [5]) is simply found by substituting ***k*** by ***k+G”*** and changing the dummy variable ***G*** to ***G’=G”- G*** in (7).

## 3. The two domain correspondence

We discuss how the proposed method is a helpful tool for physical insights in the domain of wave propagation in periodic media.

*3.1. Physical content of the formalism*

In this section, we show how the periodicity of the band structure in Fourier domain as well as the general expression of Bloch functions are implicit results of equation (7).

*3.1.1. Periodicity of the band structure in Fourier domain.* A close inspection of equation (7) reveals that when we consider a fixed wave vector ***k***, we are in fact handling a subsystem of coupled linear equations that contains all shifted values of ***k*** by reciprocal lattice vectors, as expressed by the sum over all reciprocal vectors $\boldsymbol{G} \in G$. This justifies the periodicity of the band structure in Fourier domain since any ***k’= k+nG*** leads to the same subsystem (permutation of its rows) with the same solutions. Also this observation justifies the restriction of ***k*** over its principal values i.e. the first Brillouin zone.

*3.1.2. Bloch theorem as a consequence of band structure periodicity.* Having established the periodicity of the band structure, it appears that if we consider a travelling mode with a fixed eigen-frequency $\omega$. It will contain all the discrete Fourier components of the form $\boldsymbol{k_{0'}} = \boldsymbol{k_0}\text{-}\boldsymbol{G}$. We can write:

$$u_{3,k_0}(\boldsymbol{k}) = \sum_{G \in G} u_3(\boldsymbol{k_0} - \boldsymbol{G})\,\delta(\boldsymbol{k} - (\boldsymbol{k_0} - \boldsymbol{G})) \tag{8}$$

The inverse Fourier transform of equation (8) gives us:

$$u_{3,k_0}(\boldsymbol{r}) = \frac{1}{2\pi}\left(\sum_{G \in G} u_3(\boldsymbol{k_0} - \boldsymbol{G})\, e^{i\boldsymbol{Gr}}\right) e^{-i\boldsymbol{kr}} \tag{9}$$

The term between parentheses is a periodic function in real domain; we recognise the general expression of Bloch function. So Bloch theorem comes out naturally as a consequence of media periodicity, our unique assumption. Indeed, in equation (7) we consider the Fourier transform $u_3(\boldsymbol{k})$ of the wavefunction as a continuous function of k (and not a discrete one). Consequently, we have not anticipated any assumption concerning the wavefunction periodicity. It is the discrete nature of the periodic media parameters Fourier components $\rho(\boldsymbol{k})$ as expressed by expression (3) as well as $c_{ijkl}(\boldsymbol{k})$ that introduce the summation term over the reciprocal lattice vectors in equation (7). As mentioned above this term generates a selective coupling of such Fourier components whose wavevectors are of the form ($\boldsymbol{k}'=\boldsymbol{k}\text{-}\boldsymbol{G}$). In turn, this selective coupling explains why the initial "infinite equation system" splits into an infinite set of "infinite but numerable subsystems" each of these subsystems belongs to a fixed k-value in the first Brillouin zone. Finally, the discretization of the Fourier components of $u_3(\boldsymbol{k})$ is a consequence of the summation term in equation (7). We recall that this summation term has been introduced by the convolution product with $\rho(\boldsymbol{k})$ and $c_{ijkl}(\boldsymbol{k})$ which are weighted Dirac combs.

*3.2. Comments*

We have presented a formalism that is devoted to derive the Fourier domain mathematical expression of wave equations in periodic media. This approach takes full advantage of digital signal processing techniques specifically, the replicating property i.e. the convolution with Dirac combs. Its derivation within the framework of distribution functions theory enables one to express the periodic coefficients in terms of continuous Fourier transform rather than Fourier series. This seemingly minor point appears to be crucial. On the one hand, the mathematical tools closely match the periodic media both in the real and the Fourier domains [4]. On the other hand, it leads to appreciable shortcuts: the problem reduces to a two steps procedure: having established RD and FD-CFE of the periodic coefficients, the Fourier domain formulation of the wave equation may be written by simple inspection.

Finally, we have illustrated the clearness and the straightforwardness of the formalism using the practical case of elastic waves in 2D periodic media. These features are direct results respectively of: First, from a conceptual point of view, thanks to the replicating property, the method clearly dissociates the contribution of the lattice as depicted by the Dirac comb from the contribution of the unit cell as depicted by the restriction of the physical parameters. Second, from a mathematical point of view, the concept of reciprocal lattice and its correspondence with the direct lattice is an already solved problem in the field of distribution functions theory as emphasized by the Fourier transform pair RD-CFE ⇔ FD-CFE.


[1] Yablonovitch E, 1987 *Phys. Rev. Lett.* **58** 2059

[2] Kushwaha MS, Halevi P, Dobrzynski L, Djafari-Rouhani B, 1993 *Phys. Rev. Lett.* **71** 2022

[3] Brillouin L, 1953 *Wave Propagation in Periodic Structures: Electric Filters and Crystal Lattices* (New York: Dover Publication)

[4] Gazalet J, Dupont S, Kastelik JC, Rolland Q, 2013 *Wave Mot.* **50** 619

[5] Wu TT, Huang ZG, Lin S, 2004 *Phys. Rev. B* **69** 094301

# Towards the Dressing Phase in the $AdS_3/CFT_2$ Duality

**M Beccaria**[1] **and G Macorini**[2]

[1] Dipartimento di Matematica e Fisica "Ennio De Giorgi", Universita' del Salento & INFN, Via Arnesano, 73100 Lecce, Italy

[2] Niels Bohr International Academy and Discovery Center, Niels Bohr institute, Blegdamsvej 17 DK-2100 Copenhagen, Denmark

E-mail: Matteo.Beccaria@le.infn.it, macorini@nbi.ku.dk

**Abstract.** In this paper we report some recent progresses and open problems in the determination of the first quantum correction at strong coupling of the dressing phases appearing in the Bethe Ansatz equation which have been conjectured to describe the spectrum of string theory on the $AdS_3 \times S^3 \times T^4$ background.

## 1. Introduction

Besides the two most studied examples of integrable theories in the AdS/CFT duality ($AdS_5 \times S^5$ and $AdS_4 \times CP^3$), other backgrounds allow for (classically) integrable string theories; recently the background $AdS_3 \times S^3 \times T^4$ got much attention. A first set of Bethe Ansatz (BA) equations describing the asymptotic spectrum of the theory has been conjectured in [1] followed by a second proposal in [2]. These equation, as in the $AdS_{5,4}$ cases, contain a "dressing factor" (two different phases appear in the BA for $AdS_3 \times S^3 \times T^4$), which is not completely fixed by the symmetry of the problem; at strong coupling, the leading term of the dressing phase is required to be the AFS phase [3], to match the classical finite gap limit of the model. The aim of this paper is to report some progresses [4, 5] in understanding the first quantum correction of the dressing phases at strong coupling, i.e. the determination of the $c^{(1)}_{r,s}$ coefficients appearing in the expansion of the scattering phase of elementary magnons with momenta $p_j$ and $p_k$:

$$\vartheta(p_j, p_k) = 2\sum c_{r,s}(\lambda)\left(\frac{\lambda}{16\,\pi^2}\right)^{\frac{r+s-1}{2}}\left[q_r(p_j)\,q_s(p_k) - q_s(p_j)\,q_r(p_k)\right]$$

where $q_n(p)$ is the elementary magnon $n$-th charge,

$$c_{r,s}(\lambda) = c^{(0)}_{r,s} + \frac{1}{\sqrt{\lambda}}\,c^{(1)}_{r,s} + \dots$$

and the $c^{(0)}_{r,s}$ are the classical AFS phase.

## 2. Algebraic Curve determination of the $c_{r,s}^{(1)}$

The 1-loop correction to the dressing phase and the $c_{r,s}^{(1)}$ coefficients can be computed, in principle in a very general way, without refering to any particular solution, in the Algebraic Curve (AC) framework, following the approach of [6]. We will need anyway an independent calculation of the 1-loop energy, done for the particular solution of a rigid circular string using the World-Sheet (WS) method: we need this second result both to understand regularization issues and to check the independence of the result from the missing modes in the AC, realtive to the $T^4$ component of the background.

In the semiclassical quantisation of the AC, the first correction to the phase is encoded in a set of potentials $V_I$ correcting the classical quasimomenta that characterize the curve: these potentials compute the effect of the quantum fluctuations around the classical solution; for each quasi-momentum $p_I$ the correction $V_I$ is obtained summing over all the fluctuations connecting the sheets of the AC. The total phase corrections to the Bethe equations are obtained by evaluating $\mathcal{V} = V_I - V_J$.For the middle node 2 equation (the other BA equations are not corrected) $\mathcal{V}(x) \equiv V_2(x) - V_3(x)$

$$\mathcal{V}(x) = \int_{-1}^{1} \frac{dy}{2\pi} \left[ \left(G_2(y) + \overline{G}_{\overline{2}}(y)\right)' \frac{\widehat{\alpha}(x)}{x-y} + \left(\overline{G}_2(y) + G_{\overline{2}}(y)\right)' \frac{\widehat{\alpha}(1/x)}{1/x-y} \right]. \tag{1}$$

where $x$ is the spectral parameter,

$$\widehat{\alpha}(x) = \frac{4\pi}{\sqrt{\lambda}} \frac{x^2}{x^2-1}, \tag{2}$$

and $G$ the discrete resolvent defined in terms of the Bethe roots $x_{a,k}$ as:

$$G_a(x) = \sum_{k=1}^{K_a} \frac{\widehat{\alpha}(x_{a,k})}{x - x_{a,k}}, \qquad \overline{G}(x) = G(1/x) \tag{3}$$

Expanding the resolvent in terms of the conserved charges $Q_r$, we get:

$$\mathcal{V}_2(x) = \frac{\widehat{\alpha}(x)}{2\pi} \sum_{r=2}^{\infty}\sum_{s=1}^{\infty} \widehat{c}_{r,s} \frac{Q_r}{x^s}, \qquad \widehat{c}_{r,s} = -4\left(1 - \frac{1}{2}\delta_{s,1}\right) \frac{1-(-1)^{r+s}}{2} \frac{r-1}{r+s-2}.$$

This result can't be interpreted as a phase, since the coefficient are not antisymmetric in the $r, s$ indeces; moreover, if we apply the AC method to the case of the $SU(2)$ circular string, we get for the 1-loop contribution of the dressing phase to the energy ($\mathcal{J}$ is the angular momentum of the string, $m$ the winding number):

$$\delta E_1^{AC} = \underbrace{\frac{1}{\sqrt{m^2+\mathcal{J}^2}}\left(m^2 + \mathcal{J}^2 \log \frac{\mathcal{J}^2}{m^2+\mathcal{J}^2}\right)}_{\delta E_1^{AdS_3}} - \underbrace{\frac{m^2\left(2\mathcal{J}\left(\mathcal{J} - \sqrt{\mathcal{J}^2+m^2}\right) + m^2\right)}{2\left(\mathcal{J}^2+m^2\right)^{3/2}}}_{\Delta E_1},$$

While, computing the 1-loop correction to the energy of the circular string in the WS framework, and isolating the dressing part as in [7], the result is

$$\delta E_1^{AdS_3} = \frac{1}{\sqrt{\mathcal{J}^2+m^2}}\left(m^2 + \mathcal{J}^2 \log \frac{\mathcal{J}^2}{m^2+\mathcal{J}^2}\right) \tag{4}$$

where $\Delta E_1$ is a discrepancy between the two approaches.

Both problems, the non-antisymmetry of the coefficients and the mismatch $\Delta E_1$, can be traced back to a regularization issue in the sum over the frequencies of the quantum fluctuations: in the WS approach the natural cut-off is a common mode number $N$, while in the AC approach the cut-off is a common radius for the contour integral in the spectral plane defining the potentials; this difference is translated in a reordering of the terms in the sum over the frequences to compute the 1-lopp energy $E_1 \sim \sum_{n\in Z} (\omega_n^B - \omega_n^F)$, and is the origin of the $\Delta E_1$ term. We can repeat the calculation of $\mathcal{V}$ enforcing the regularization of the AC curve to reproduce the WS result, solving both problems at once: indeed, with the new prescription, we get rid of the $\Delta E_1$ term obtaining full agreement with WS prediction and the the results for the new potential $\widehat{\mathcal{V}}$ and the coefficients change in:

$$\widehat{\mathcal{V}}(x) = \frac{\widehat{\alpha}(x)}{2\pi} \sum_{r=1}^{\infty}\sum_{s=1}^{\infty} \frac{c_{r,s}^{(1)}\, Q_r - \bar{c}_{r,s}^{(1)}\, \overline{Q}_r}{x^s},$$

$$c_{r,s}^{(1)} = 2\,\frac{1-(-1)^{r+s}}{2}\,\frac{s-r}{r+s-2}\,, \qquad \bar{c}_{r,s}^{(1)} = -2\,\frac{1-(-1)^{r+s}}{2}\,\frac{r+s-2}{s-r} \tag{5}$$

Using the new regularization, we can now compute the scattering phases between magnons, following [6]: We identify the dressing phase contribution in the BA as:

$$e^{i\,\widehat{\mathcal{V}}(x_{2,i})} = \prod_{k\neq j}^{K_2} e^{i\,\vartheta(x_{2,j},x_{2,k})} \prod_{k}^{K_{\bar{2}}} e^{-i\,\widetilde{\vartheta}(x_{2,j},x_{\bar{2},k})}. \tag{6}$$

and the result for the two phases is

$$\vartheta(x,y) = -\frac{\widehat{\alpha}(x)\,\widehat{\alpha}(y)}{2\pi\,(x-y)^2}\left[2\,\log\left(\frac{x+1}{x-1}\,\frac{y-1}{y+1}\right) + 2\,\frac{(x-y)\,(x^2+y^2-2)}{(x^2-1)(y^2-1)}\right], \tag{7}$$

$$\widetilde{\vartheta}(x,y) = -\frac{\widehat{\alpha}(x)\,\widehat{\alpha}(y)}{2\pi\,(1-x\,y)^2}\left[2\,\log\left(\frac{x+1}{x-1}\,\frac{y-1}{y+1}\right) - 2\,\frac{(x-y)\,(x^2y^2-1)}{(x^2-1)\,(y^2-1)}\right]. \tag{8}$$

## 3. Folded string and open problems

The choice of regularization used in the previous section allows to separate the original potential $\mathcal{V} = \mathcal{V}_{phase} + \delta\mathcal{V}$, where only the first part can be consistently interpreted as a phase, while the second is understood as a regularization effect. Nevertheless the prescription is based on the agreement between the AC and WS approaches in the particular case of the $SU(2)$ curcular string. If we consider as a second example the $SL(2)$ folded string solution, we can repeat the same steps: computing the 1-loop dressing contribution to the energy for the folded string, the AC and WS give different result, and the mismatch is again due to the different regularizations. While we can remove the discrepancy and get a consistent result on the string theory side, if we compare the string prediction with the BA, using the phase derived in the previous section we still have a discrepancy.

$$\text{WS} \equiv \text{AC-reg. mismatch :} \qquad E_1^{\text{dressing}} = \frac{\coth^{-1}\left(\sqrt{\mathcal{J}^2+1}\right)}{2\mathcal{J}^3\sqrt{\mathcal{J}^2+1}}\,\mathcal{S}^2 + \mathcal{O}(\mathcal{S}^3),$$

$$\text{BA with } c_{r,s} \text{ coeff. :} \qquad E_1^{\text{dressing}} = \left[\frac{\coth^{-1}\left(\sqrt{\mathcal{J}^2+1}\right)}{2\mathcal{J}^3\sqrt{\mathcal{J}^2+1}} + \frac{1}{2\,\mathcal{J}^4\,\sqrt{\mathcal{J}^2+1}}\right]\mathcal{S}^2 + \mathcal{O}(\mathcal{S}^3).$$

where the first line is the string theory result for a folded string with semiclassical spin $\mathcal{S}$ and angular momentum $\mathcal{J}$, the second line is the result obtained from the BA equations, assuming the coefficients $c_{r,s}$ for the dressing phase.

This disagreement calls for a deeper understanding of the role of $\delta\mathcal{V}$; while the regularization/antisymmetrization prescription works perfectly in the $SU(2)$ sector, the comparison between string theory and BA predictions for more general cases is still problematic.

Another open issue is related to the fact that the phases in eq.(7) don't satisfy the crossing relations found in [2]: to get crossing respecting phases the coefficients $c_{r,s}$ with $r = 1$ should be modified with an additional factor 1/2 [8, 9]. The relative phases for the magnon scattering can be found starting from the original potential in eq.(1), and making the resulting phase antisymmetric by hand $\theta^{AC,asym}(x,y) = 1/2[\theta^{AC}(x,y) - \theta^{AC}(y,x)]$. But this phase, while crossing symmetric, gives a contribution to the 1-loop energy not in agreement with the string theory WS prediction, even in the case of the circular string solution.

## References


[1] A. Babichenko, J. Stefanski, B., and K. Zarembo, *Integrability and the AdS(3)/CFT(2) correspondence*, *JHEP* **1003** (2010) 058

[2] R. Borsato, O. Ohlsson Sax, A. Sfondrini, B. Stefanski, Jr. and A. Torrielli, *The all-loop integrable spin-chain for strings on $AdS_3 \times S^3 \times T^4$: the massive sector* arXiv:1303.5995 [hep-th].

[3] G. Arutyunov, S. Frolov, and M. Staudacher, *Bethe ansatz for quantum strings*, *JHEP* **0410** (2004) 016

[4] M. Beccaria, F. Levkovich-Maslyuk, G. Macorini and A. A. Tseytlin, *Quantum corrections to spinning superstrings in $AdS_3 x S^3 x M^4$: determining the dressing phase* JHEP **1304** (2013) 006

[5] M. Beccaria and G. Macorini, *Quantum corrections to short folded superstring in $AdS_3 x S^3 x M^4$* JHEP **1303** (2013) 040

[6] N. Gromov and P. Vieira, *Constructing the AdS/CFT dressing factor*, *Nucl.Phys.* **B790** (2008) 72–88,

[7] N. Beisert and A. A. Tseytlin, *On quantum corrections to spinning strings and Bethe equations*, *Phys.Lett.* **B629** (2005) 102–110,

[8] R. Borsato, O. O. Sax, A. Sfondrini, B. Stefanski and A. Torrielli, *Dressing phases of AdS3/CFT2* arXiv:1306.2512 [hep-th].

[9] M. C. Abbott, *The AdS3 x S3 x S3 x S1 Hernandez-Lopez Phases: a Semiclassical Derivation* arXiv:1306.5106 [hep-th].

# Non-abelian T-duality, generalised geometry and holography.

**N. T. Macpherson**

Department of Physics, Swansea University, Singleton Park, Swansea, SA2 8PP, UK

E-mail: pymacpherson@swansea.ac.uk

**Abstract.** Recent progress which relates non-abelian T-duality of $\mathcal{N}=1$ SuGra solutions to the powerful techniques of Generalised geometry is reviewed. It is shown that $SU(3)$ structure solutions are mapped to $SU(2)$ structures and the transformation rule of the corresponding pure spinors is presented. This constitutes an important step on the road towards the utility of the duality within holography, showing for example, how smeared sources must transform and so how to add flavour to the T-duals.

## 1. Introduction

Duality transformations play an important role in a wide range of physical systems from the 2d Ising Model to many dualities of superstring theory. Within the context of the gauge-gravity correspondence, duality transformation have in recent years proven to be a very useful way of generating new supergravity solutions dual to theories with phenomenologically desirable dynamics. An important example [2] is how U-duality can be used to map a deformation of the Maldacena-Nunez solution [3], which give a holographic description of a QFT with an irrelevant operator insertion, to the Baryonic branch of Klebanov-Strassler [4] which is well behaved asymptotically.

Here non-abelian T-duality is considered as a SuGra solution generator. Performing T-dualities along non-abelian isometries is not a new idea, indeed it was first proposed over 20 years ago [5]. However, in order for it to have any real utility as a solution generating technique with applications to holography one first needed to know how non trivial RR fluxes transform under the duality. This was realised far more recently in [6] and expanded upon in [7].

Recently attention has been focused on generating new SuGra duals of $\mathcal{N}=1$ gauge theories in 4-d [8] by dualising along the $SU(2)$ isometries of well established conifold solutions such as Klebanov-Witten [9]. Such solutions have a well understood generalised geometric description in terms of pure spinors [10]. This note reviews the results of [1], in which the techniques of non-abelian T-duality and generalised geometry are combined, constituting an important step towards constructing useful holographic duals of $\mathcal{N}=1$ QFT's in 4-d via the duality.

## 2. Non-abelian T-duality along $SU(2)$ isometries

The NS sector of a SuGra solution with an $SU(2)$ isometry can be generically expressed in terms of $SU(2)$ left invariant 1-forms, $L^i=-iTr(g^{-1}dg)$ , as:

$$ds^2=G_{\mu\nu}dX^\mu dX^\nu+2G_{\mu i}dX^\mu L^i+g_{ij}L^iL^j,\quad B=B_{\mu\nu}dX^\mu\wedge dX^\nu+B_{\mu i}dX^\mu\wedge L^i+\frac{1}{2}b_{ij}L^i\wedge L^j \quad (1)$$

where $\mu = 1, ...7$. The duality acts on the non linear sigma model corresponding to this solution

$$S = \int d^2\sigma Q_{\mu\nu}\partial_+ X^\mu \partial_- X^\nu + Q_{\mu i}\partial_+ X^\mu L^i_- + Q_{i\mu} L^i_+ \partial_- X^\mu + E_{ij} L^i_+ L^j_-, \tag{2}$$

where $Q = G + B$ and similarly for $E$. The dual sigma model is obtained by gauging the $SU(2)$ isometry by promoting $\partial_\pm$ to a covariant derivative in the pull back of the left invariant 1-forms, that is

$$\partial_\pm g \to D_\pm g = \partial_\pm - A_\pm g. \tag{3}$$

A Lagrange multiplier $-iTr(vF_{+-})$ should also be added to ensure the gauge field is non dynamical.

Upon integrating the Lagrange multiplier by parts the gauge fields may be solved for and and the T-dual sigma model obtained. Gauge fixing must be imposed as the dual theory depends on Euler angles of $SU(2)$ in addition to 3 dual coordinates $v_i$. The most simple fixing is $g = I$ in which only the $v_i$ are kept and one obtains the dual Lagrangian

$$\tilde{S} = \int d^2\sigma Q_{\mu\nu}\partial_+ X^\mu \partial_+ X^\nu + (\partial_+ v_i + \partial_+ X^\mu Q_{\mu i})(E_{ij} + f_{ij}{}^k v_k)(\partial_- v_j - Q_{j\mu}\partial_- X^\mu), \tag{4}$$

from which the dual metric and NS 2-form may be extracted. In addition the dilaton obtains a correction at the quantum level

$$e^{-2\hat{\Phi}} = det(E + fv)e^{-2\Phi}. \tag{5}$$

The transformation on the RR sector may be ascertained in a different fashion. After dualisation, left and right movers naturally couple to 2 distinct sets of vielbeins. However because these vielbeins describe the same geometry they must be related by a Lorentz transformation $\Lambda$. This may be used to define an action on spinors parametrised by a matrix $\Omega$ satisfying $\Omega^{-1}\Gamma^a\Omega = \Lambda^a{}_b\Gamma^b$. For $SU(2)$ isometries this is solved by

$$\Omega = \Gamma^{(11)}\frac{\Gamma^{123} + \sum_{a=1}^{3}\zeta^a\Gamma_a}{\sqrt{1+\zeta^2}} \tag{6}$$

where $\zeta^a$ depends on the original geometry and the gauge fixing condition and can be found in [8]. Dualising along odd dimensional isometrics maps type-IIA $\leftrightarrow$ type-IIB and the RR sectors of the original and dual theories are the related by

$$e^{\Phi_{IIA}}\not{F}_{IIA} = e^{\Phi_{IIB}}\not{F}_{IIB}.\Omega^{-1} \tag{7}$$

where $F_{IIA/IIB}$ are RR polyforms and the slash denotes contraction with 10-d gamma matrices once the polyforms have been mapped to bispinors under the Clifford map.

## 3. Generalised geometry and non-abelian T-duality

As in [1] the focus of this section shall be geometries that can be expressed as

$$ds^2_{str} = e^{2A}dx_{1,3} + ds^2_6, \tag{8}$$

which preserve $\mathcal{N} = 1$ and have non-trivial RR fluxes turned on. The MW 10-d killing spinors of such a solution my be written in terms of a 4+6 split

$$\epsilon^1 = \zeta_+ \otimes \eta^1_+ + \zeta_- \otimes \eta^1_-, \quad \epsilon^2 = \zeta_+ \otimes \eta^2_\mp + \zeta_- \otimes \eta^2_\pm, \tag{9}$$

where $\pm$ denotes chirality, the upper signs in $\epsilon^2$ corresponding to type-IIA, the lower to IIB and a basis such that $(\eta_+^{1,2})^* = \eta_-^{1,2}$ is chosen. From the internal spinors (The $\eta$'s) it is possible to define 2 $Cliff(6,6)$ spinors

$$\Psi_\pm = \eta_+^1 \otimes (\eta_\pm^2)^\dagger \tag{10}$$

which map to polyforms under the Clifford map. The condition for $\mathcal{N}=1$ SUSY to hold may be caste in terms of differential constraints on these pure spinors

$$(d - H\wedge)(e^{2A-\Phi}\Psi_\mp) = e^{2A-\Phi} dA \wedge \bar{\Psi}_\mp + i\frac{e^{3A}}{8} \star_6 F_{IIA/IIB}, \quad (d - H\wedge)(e^{2A-\Phi}\Psi_\pm) = 0 \tag{11}$$

where once more the upper signs should be taken in type-IIA.

A 6-d internal space which preserves $\mathcal{N}=1$ SUSY supports either an $SU(3)$ or an $SU(2)$-structure [10]. If $\eta_+^1$ is parallel to $\eta_+^2$, then the structure is $SU(3)$ and the pure spinors may be expressed as

$$\Psi_+ = \frac{e^A}{8} e^{i\chi} e^{-iJ}, \quad \Psi_- = \frac{e^A}{8}\Omega_{hol}. \tag{12}$$

where $\chi$ denotes a possibly point dependent phase. There exists a basis such that the projections of the background are canonical and $J$ and $\Omega_{hol}$ may be expressed in terms of 6-d vielbeins as

$$J = e^{12} + e^{34} + e^{56}, \quad \Omega_{hol} = (e^1 + ie^2) \wedge (e^3 + ie^4) \wedge (e^5 + ie^6). \tag{13}$$

When $\eta_+^1$ is not parallel to $\eta_+^2$ each spinor defines an independent $SU(3)$-structure and the largest common subgroup is $SU(2)$. A special case is when the spinors bisect perpendicularly which gives an orthogonal $SU(2)$ characterised by pure spinors

$$\Psi_+ = \frac{e^A}{8} e^{-iv\wedge w} \wedge \omega, \quad \Psi_- = ie^{i\xi}\frac{e^A}{8}(v + iw)e^{-ij} \tag{14}$$

were $\omega$ is a 2-form and $v, w$ are 1-forms and $\xi$ is another phase.

The action of non-abelian T-duality on the 10-d spinors must change the chirality of $\epsilon_2$. When the original geometry is type-IIB the transformation is

$$\hat{\eta}_+^1 = \eta_+^1, \quad \hat{\eta}_+^2 = \Omega\eta_-^2, \quad \hat{\eta}_-^{1,2} = (\tilde{\eta}_+^{1,2})^*, \tag{15}$$

while performing the duality on a type-IIA background requires a transformation by $\Omega^{-1}$. This makes it abundantly clear that non-abelian T-duality is rotating the spinors with respect to each other and so one must map $SU(3) \to SU(2)$-structure under the duality. And as most known solutions of this type are $SU(3)$ structures, the duality provides a way of generating many more $SU(2)$-structure solutions, as long as the dualisation is performed on isometries that preserve SUSY.

The pure spinors transform under a non-abelian T-duality in much the same way as RR fluxes,

$$\Psi_{IIA\pm} = \Psi_{IIB\mp}\Omega^{-1}. \tag{16}$$

As well as giving an explicit check of whether SUSY is preserved this also gives an immediate definition of the calibration conditions for space-time filling Dn-branes in the dual geometry.

$$\sqrt{g}d^{n+1}\xi\Big|_{\Sigma_{Dn}} = \Psi_{cal}\Big|_{\Sigma_{Dn}}, \quad \Psi_{cal} = -8e^{3A-\Phi}Im(\Psi_\mp) \tag{17}$$

where the upper sign is in IIA [11]. This enables one to find SUSY brane embedding in the T-dual geometry which is important for adding flavour among other things. In addition smeared sources which enter a SuGra solution as a violation of the Bianchi identity of the RR polyforms

$$(d - H\wedge)F = \Xi \wedge e^B \tag{18}$$

transform in precisely the same way as the RR sector. Together these facts allow one to generate new flavoured $SU(2)$-structure solution from flavoured $SU(3)$-structure solutions directly.

In [1] a first example of the utility of the techniques described above can be found. There KW [9] is dualised along one of its $SU(2)$ isometries. The resulting solution supports an orthogonal $SU(2)$-structure. The dualisation is repeated on KW with massless flavours [12] which generates sources for D8, D6 and D4 branes that are all SUSY. This constitutes a solid proof of concept.

## 4. Outlook

Holographic applications of non-abelian T-duality clearly require a greater understanding of the dual QFT's being generated. The large number of fluxes generated means that the duality probably maps to quivers, but this needs to be made more precise. The largest obstacle at this time is the periodicity of the dual coordinates. There is currently no canonical way to fix these and so many global properties of the dual manifolds are unknown. However one might reasonably hope that the generalised geometric picture may yet shed some light on this conundrum.

$SU(2)$-structures are not that common in the literature so non-abelian T-duality could give some useful information about their general construction. Rarest of all is the dynamic $SU(2)$-structure where the angle between the internal spinors varies through out the geometry, it would be interesting to see if they can be generated with the duality.

Finally to what extent can generalised geometry help in the understanding non-abelian T-duals of other types types of solutions? For instance in [13] $G_2$-structure solutions dual to confining $\mathcal{N} = 1$ Chern-Simons theories are considered. Does the duality map $G_2 \to SU(3)$-structure in 7-d? This question is currently being addressed.

## Acknowledgments

I would like to thank A. Barranco, J. Gaillard, C. Nunez and D. C. Thompson for their collaboration on [1] which this note is mostly based on. I am supported by an SCFT studentship.

## References

[1] A. Barranco, J. Gaillard, N. T. Macpherson, C. Nunez and D. C. Thompson, JHEP **1308** (2013) 018 [arXiv:1305.7229 [hep-th]].
[2] J. Maldacena and D. Martelli, JHEP **1001** (2010) 104 [arXiv:0906.0591 [hep-th]].
[3] J. M. Maldacena and C. Nunez, Phys. Rev. Lett. **86** (2001) 588 [hep-th/0008001]. R. Casero, C. Nunez and A. Paredes, Phys. Rev. D **73** (2006) 086005 [hep-th/0602027].
[4] A. Butti, M. Grana, R. Minasian, M. Petrini and A. Zaffaroni, JHEP **0503** (2005) 069 [hep-th/0412187]. I. R. Klebanov and M. J. Strassler, JHEP **0008** (2000) 052 [hep-th/0007191].
[5] X. C. de la Ossa and F. Quevedo, Nucl. Phys. B **403** (1993) 377 [hep-th/9210021].
[6] K. Sfetsos and D. C. Thompson, Nucl. Phys. B **846** (2011) 21 [arXiv:1012.1320 [hep-th]].
[7] Y. Lozano, E. .O Colgain, K. Sfetsos and D. C. Thompson, JHEP **1106** (2011) 106 [arXiv:1104.5196 [hep-th]]. G. Itsios, Y. Lozano, E. .O Colgain and K. Sfetsos, JHEP **1208** (2012) 132 [arXiv:1205.2274 [hep-th]].
[8] G. Itsios, C. Nunez, K. Sfetsos and D. C. Thompson, Phys. Lett. B **721** (2013) 342 [arXiv:1212.4840]. G. Itsios, C. Nunez, K. Sfetsos and D. C. Thompson, Nucl. Phys. B **873** (2013) 1 [arXiv:1301.6755 [hep-th]].
[9] I. R. Klebanov and E. Witten, Nucl. Phys. B **536** (1998) 199 [hep-th/9807080].
[10] M. Grana, R. Minasian, M. Petrini and A. Tomasiello, JHEP **0511** (2005) 020 [hep-th/0505212].
[11] L. Martucci and P. Smyth, JHEP **0511** (2005) 048 [hep-th/0507099].
[12] F. Benini, F. Canoura, S. Cremonesi, C. Nunez and A. V. Ramallo, JHEP **0702** (2007) 090 [hep-th/0612118].
[13] N. T. Macpherson, JHEP **1206** (2012) 136 [arXiv:1204.4222 [hep-th]]. N. T. Macpherson, JHEP **1304** (2013) 076 [arXiv:1301.5178 [hep-th]]. N. T. Macpherson, PoS Corfu **2012** (2013) 119.

# Soft breaking in Hořava Gravitational theory

**Alvaro Restuccia**[1], **Jorge Bellorín**[2], **Adrián Sotomayor**[3]

[1] Departamento de Física, Universidad de Antofagasta, Antofagasta, Chile
[2] Departamento de Física, Universidad Simón Bolívar, Caracas, Venezuela
[3] Departamento de Matemáticas, Universidad de Antofagasta, Antofagasta, Chile

E-mail: arestu@usb.ve,jbellorin@usb.ve,asotomayor@uantof.cl

**Abstract.** We consider anisotropic gravity at the conformal point of the kinetic term of the Hořava Gravity. This value is protected under quantum corrections by the constraints of the theory. The symmetry of the $z = 3$ interacting terms, $z$ being the time scale, which ensure the renormalizability of the theory is softly broken by $z = 1$ interacting terms. We show that the resulting theory has the same number of physical degrees of freedom as General Relativity. Moreover the linearized theory around the Minkowski solution is exactly equivalent to General Relativity. We also study the spherically symmetric solutions to the fields equations. We find a close expression for the solutions. It terms out that the solutions are structurally stable from the right and unstable from the left around the value of the coupling constant corresponding to the Schwarzchild solution of General Relativity.

## 1. Introduction

The main idea of the Hořava Gravity [1] theory is to consider an anisotropic behaviour of space and time. The scaling being of the form

$$t \sim [b]^z$$
$$x \sim [b]^1.$$

The scaling breaks the relativistic symmetry, with the hope of recovering it at low energies or equivalently at large distances. The main benefit would be, to obtain a renormalizable theory, and this goal may be achieved by considering a suitable scaling $z$. If at low energy one recovers General Relativity the Hořava theory could be regarded as the $UV$ completion of it.

The Hořava proposal considered the foliation-preserving diffeomorphism of the 3-dimensional space as a gauge symmetry of the theory and the hope is to have an enhancement of it to space-time diffeomorphisms at low energies. The action of the non-projectable Hořava theory written in terms of the $ADM$ variables $g_{ij}, N, N_i$ is

$$S = \int dt d^3x \sqrt{g} N \left( G^{ijkl} K_{ij} K_{kl} - \nu \right) \quad (1)$$

where

$$K_{ij} = \frac{1}{2N} \left( \dot{g}_{ij} - 2\nabla_{(iN_j)} \right) \quad (2)$$

$$G^{ijkl} = \frac{1}{2} \left( g^{ik} g^{jl} + g^{il} g^{jk} \right) - \lambda g^{ij} g^{kl} \quad (3)$$

and the potential density $\nu = \nu(g_{ij}, a_i, \ldots)$ is the most general combination of the spatial metric, its curvature tensor, the vector $a_i \equiv \frac{N_{,i}}{N}$ and covariant spatial derivatives of these objects transforming as a scalar under spatial diffeomorphisms. In order to have power-counting renormalizativity the potential must include terms up to sixth order in spatial derivatives. The potential may include a cosmological constant but we do not consider it here.

The potential including the sixth order terms is given by

$$\nu = -R - \alpha a_i a^i + wC_{ij}C^{ij} \tag{4}$$

where $\alpha$ and $w$ are coupling constants. The second term on the right hand member was introduced in [2]. $C^{ij}$ is the Cotton tensor

$$C^{ij} = \frac{1}{\sqrt{g}}\epsilon^{kl}\left({}^{i}\nabla_k R_l^j\right). \tag{5}$$

The sixth order term $\sqrt{g}C_{ij}C^{ij}$ was introduced in Hořava [1] and it is the unique interacting term of that order arising from the detailed balance principle proposed by Hořava. It transforms homogeneously, with weight $\frac{3}{2}$, under conformal transformations of the metric. The lower order differential terms in the potential violate the detailed balance principle but the action is still renormalizable. In this sense the inclusion of these terms is a soft-breaking mechanism.

Most of the analysis of the theory has been performed for $\lambda \neq \frac{1}{3}$ [3, 4, 5, 6, 7, 8, 9, 10, 11]. In that case the theory contains an additional degree of freedom compared to GR. In this work we consider $\lambda = \frac{1}{3}$. It is the particular case where the tensor $G^{ijkl}$ is not invertible. It turns out that this case is quite interesting because the number of physical degrees of freedom of the theory is the same as in GR.

## 2. The hamiltonian

From the action (1) we may obtain the hamiltonian of the theory [12]. Its expression is

$$\int d^3x \left(\frac{N}{\sqrt{g}}\pi^{ij}\pi_{ij} + \sqrt{g}N\nu + N_i\mathcal{H}^i + \sigma\phi + \mu\pi\right) + E_{ADM} \tag{6}$$

where

$$\mathcal{H}^i \equiv -2\nabla_j\pi^{ij} + \phi\partial^i N. \tag{7}$$

$\phi$ is the conjugate momentum associated to $N$ and $\sigma, \mu$ are Lagrange multipliers. The primary constraint

$$\pi = 0 \tag{8}$$

is a particular property of the Hořava theory formulated for $\lambda = \frac{1}{3}$. This constraint ends up being a second class one and it is preserved on the quantum formulation of the theory. In that sense the $\lambda = \frac{1}{3}$ value is protected from quantum corrections of the theory. The constraint $\phi = 0$ arises as a consequence of the absence of $\dot{N}$ terms in the hamiltonian.

$E_{ADM}$ denotes the $ADM$ energy

$$E_{ADM} = \oint d\Sigma_i \left(\partial_j g_{ij} - \partial_i g_{jj}\right). \tag{9}$$

It has to be included in the hamiltonian, following [13], in order to obtain the equations of motion compatible with the boundary conditions from the most general variations of the hamiltonian.

The preservation of the primary constraints give rise to two new constraints, the hamiltonian constraint

$$\mathcal{H} \equiv \frac{1}{\sqrt{g}}\pi^{ij}\pi_{ij} - \sqrt{g}R + \alpha\sqrt{g}\left(2\nabla_i a^i + a_i a^i\right) + w\sqrt{g}C_{ij}C^{ij} = 0 \tag{10}$$

and

$$C \equiv \frac{3N}{2\sqrt{g}}\pi^{ij}\pi_{ij} + \frac{1}{2}\sqrt{g}NR - \sqrt{g}N\left(2\nabla_i a^i + \left(2-\frac{\alpha}{2}\right)a_i a^i\right) + \frac{3w}{2}\sqrt{g}NC_{ij}C^{ij} = 0. \tag{11}$$

The conservation of these new constraints provides two elliptic partial differential equations for the lagrange multipliers $\sigma$ and $\mu$. The Dirac algorithm ends up at this stage.

From (10) and (11) we obtain

$$\left(1-\frac{\alpha}{2}\right)\nabla^2 N = N\left(\frac{1}{g}\pi^{ij}\pi_{ij} + wC_{ij}C^{ij}\right). \tag{12}$$

The right hand member is positive provided $w > 0$. If, in addition, $\alpha < 2$ then the results in [11] imply the existence and uniqueness of a globally defined solution $N$, moreover it is guaranteed that $N \geq 0$ over all the foliation.

## 3. The energy of the theory

The hamiltonian (6) may also be rewritten as a sum of constraints plus boundary terms

$$H = \int d^3x\left(H\mathcal{H} + N_i\mathcal{H}^i + \sigma\phi + \mu\pi\right) + E_{ADM} - 2\alpha\Phi_N \tag{13}$$

where

$$\Phi_N \equiv \int d\Sigma_i \partial_i N. \tag{14}$$

The role of the boundary terms ensure the existence of the Gateaux derivative of the Hamiltonian under variations of $N$ that behave as $\delta N = \mathcal{O}(r^{-1})$ at infinity. The energy of the theory is given by the value of the Hamiltonian on the constrained submanifold and it is thus equal to $[E_{ADM} - 2\alpha\Phi_N]$.

The positive mass theorem in [14, 15] may be used to analyze the positivity of $E_{ADM} - 2\alpha\Phi_N$. A strong result was obtained in [12] where we show that if $-\frac{2}{3} \leq \alpha \leq 0$ then $R \geq 0$ and using the positivity mass theorem one gets $E_{ADM} - 2\alpha\Phi_N \geq 0$. Besides, a new result follows by considering a conformal transformation on the three dimensional leaves of the foliation, see also [16]. If we take $g_{ij} \to \widetilde{g}_{ij} = g_{ij}e^{\alpha N}$ then $\widetilde{E}_{ADM} = E_{ADM} - 2\alpha\Phi_n$ and

$$\begin{aligned}\widetilde{R} &= e^{-\alpha N}\left[R - 2\alpha\nabla_i\nabla^i N - \frac{\alpha^2}{2}(\nabla N)^2\right] \\ &= e^{-\alpha N}\left[R - \frac{4\alpha}{(2-\alpha)}N\left(\frac{1}{g}\pi^{ij}\pi_{ij} + wC^{ij}C_{ij}\right) - \frac{\alpha^2}{2}(\nabla N)^2\right].\end{aligned}$$

Hence if $R \geq \frac{4\alpha}{(2-\alpha)}N(\frac{1}{g}\pi^{ij}\pi_{ij} + wC^{ij}C_{ij} + \frac{\alpha^2}{2}(\nabla N)^2)$ then $E_{ADM} - 2\alpha\Phi_N \geq 0$.

## 4. The physical degrees of freedom

From the analysis of the constraints it follows that the number of degrees of freedom of the theory is the same as in General Relativity. Moreover a perturbative analysis around a Minkowski background

$$g_{ij} = \delta_{ij} + h_{ij}\ ,\ \ \pi^{ij} = p_{ij}\ ,\ \ N = 1 + n \tag{15}$$

yields

$$h^T = n = 0 \tag{16}$$

and

$$H = \int d^3x\left(p_{ij}^{TT}p_{ij}^{TT} + \frac{1}{4}\partial_i h_{jk}^{TT}\partial_i h_{jk}^{TT}\right) \tag{17}$$

which is exactly the hamiltonian of linearized General Relativity.

## 5. The static, spherical symmetric solutions

It turns out that the relevant coupling in the analysis of the static, spherically symmetric solutions is $\alpha$, the solutions depend only on $\alpha$. When $\alpha = 0$, one obtains the Schwarzchild solution of General Relativity. In terms of the spherical coordinates, the metric has the form

$$ds^2 = -N^2 dt^2 + \frac{dr^2}{f} + r^2 d\Omega^2 \tag{18}$$

where $N = N(r)$ and $f = f(r)$.

The exact solution for $\alpha < 2$ has the expression

$$\begin{aligned} r &= \beta G M e^{\frac{\chi}{\beta}} \quad \text{where}\, \chi > 0 \quad \text{and}\, \beta = \sqrt{1 - \frac{\alpha}{2}}, \\ N &= e^{-\frac{\chi}{\beta}}, \\ f &= \left(\cosh\chi - \frac{\sinh\chi}{\beta}\right)^2. \end{aligned}$$

For $0 \leq \alpha < 2$ the metric presents an horizon (for relativistic matter) at $\hat{\chi}$ where $\tanh\hat{\chi} = \beta$ and a essential singularity at $\chi \to \infty$.

For $\alpha < 0$ the solution does not present an horizon, it has a naked essential singularity at $\chi \to \infty$.

The solution for $\alpha > 2$ has a different expression but it always presents an horizon and an essential singularity.

The asymptotic behaviour of the solution is

$$\begin{aligned} N &= 1 - 2\tfrac{GM}{r} \\ f &= 1 - 2\tfrac{GM}{r} \end{aligned} \tag{19}$$

as in Schwarzchild solution for General Relativity.

It is remarkable that the exact solution at $\alpha = 0$, corresponding to the Schwarzchild solution, is stable from the left and unstable from the right when we vary the coupling constant $\alpha$. The solution corresponding to $\alpha > 0$ approachs the Schwarzchild solution, in a suitable norm, when $\alpha \to 0$. In distinction when $\alpha < 0$, the solution is qualitatively different to the one at $\alpha = 0$.

Perturbation analysis may be performed for $\alpha > 0$ around $\alpha = 0$, for any $r$. The explicit solution in the asymptotic regime $r \to \infty$ agrees with the already reported result in [17].

## 6. Conclusions

The Hořava theory for $\lambda = \frac{1}{3}$ whose hamiltonian formalism is given by (6) is a good candidate for a quantum renormalizable gravity theory. The value $\lambda = \frac{1}{3}$ is protected by the constraints of the theory and does not acquires quantum corrections. The theory has the same number of physical degrees of freedom as General Relativity. Moreover at the linearized level around Minkowski space-time it exactly agrees with General Relativity. The static spherically symmetric solution can be obtained explicitly in a closed form. The solutions for $\alpha > 0$ converge smoothly to the Schwarzchild solution, corresponding to $\alpha = 0$. When $\alpha < 0$ the solutions are qualitatively different from the one at $\alpha = 0$. They do not have an horizon, only a naked essential singularity.

**Acknowledgments** A. R. and A. S. are partially supported by Project Fondecyt 1121103, Chile.

## References


[1] P. Hořava, Phys. Rev. D 79, 084008 (2009).
[2] D. Blas, O. Pujolas, and S. Sibiryakov, Phys. Rev. Lett. 104, 181302 (2010).
[3] C. Charmousis, G. Niz, A. Padilla, and P. M. Saffin, J. High Energy Phys. 08 (2009) 070; X. Gao, Y. Wang, R. Brandenberger, and A. Riotto, Phys. Rev. D 81, 083508 (2010); J.-O. Gong, S. Koh, and M. Sasaki, Phys. Rev. D 81, 084053 (2010); A. Kehagias and K. Sfetsos, Phys. Lett. B 678, 123 (2009); Y.W. Kim, H.W. Lee, and Y. S. Myung, Phys. Lett. B 682, 246 (2009); M.-i. Park, Classical Quantum Gravity 28, 015004 (2011); Gen. Relativ. Gravit. 43, 2979 (2011).
[4] D. Blas, O. Pujolas, and S. Sibiryakov, J. High Energy Phys. 10 (2009) 029.
[5] M. Li and Y. Pang, J. High Energy Phys. 08 (2009) 015.
[6] A. Papazoglou and T. P. Sotiriou, Phys. Lett. B 685, 197 (2010); D. Blas, O. Pujolas, and S. Sibiryakov, Phys. Lett. B 688, 350 (2010); I. Kimpton and A. Padilla, J. High Energy Phys. 07 (2010) 014; A. Cerioni and R. H. Brandenberger, arXiv:1008.3589; R.-G. Cai, B. Hu, and H.-B. Zhang, Phys. Rev. D 83, 084009 (2011).
[7] J. Bellorin and A. Restuccia, Int. J. Mod. Phys. D 21, 1250029-1 (2012).
[8] Bellorin and A. Restuccia, Phys. Rev. D 84, 104037 (2011).
[9] J. Bellorin and A. Restuccia, Phys. Rev. D 83, 044003 (2011).
[10] W. Donnelly and T. Jacobson, Phys. Rev. D 84, 104019 (2011).
[11] J. Bellorin, A. Restuccia and A. Sotomayor, Phys. Rev. D 85, 124060 (2012).
[12] J. Bellorin, A. Restuccia and A. Sotomayor, Phys. Rev. D 87, 084020 (2013).
[13] T. Regge and C. Teitelboim, Ann. Phys. (N.Y.) 88, 286 (1974).
[14] R. Schoen and S-T Yau, Commun. Math. Phys. 79, 47 (1981); 79, 231 (1981).
[15] E. Witten, Commun. Math. Phys. 80, 381 (1981).
[16] D. Garfinkle and T. Jacobson, Phys. Rev. Lett. 107, 191102 (2011).
[17] E. Kiritsis, Phys. Rev. D 81:044009, (2010).

# Vector model for mapping of visual space to subjective 4-D sphere

**Dalius Matuzevicius**[1,2] **and Henrikas Vaitkevicius**[1]

[1] Department of General Psychology, Faculty of Philosophy, Vilnius University, Universiteto str. 9/1, Vilnius, Lithuania

[2] Department of Electronic Systems, Faculty of Electronics, Vilnius Gediminas Technical University, Naugarduko str. 41-413, Vilnius, Lithuania

E-mail: dalius.matuzevicius@gmail.com, henrikas.vaitkevicius@ff.vu.lt

**Abstract.** Here we present a mathematical model of binocular vision that maps a visible physical world to a subjective perception of it. The subjective space is a set of 4-D vectors whose components are outputs of four monocular neurons from each of the two eyes. Monocular neurons have one of the four types of concentric receptive fields with Gabor-like weighting coefficients. Next this vector representation of binocular vision is implemented as a pool of neurons where each of them is selective to the object's particular location in a 3-D visual space. Formally each point of the visual space is being projected onto a 4-D sphere. Proposed model allows determination of subjective distances in depth and direction, provides computational means for determination of Panum's area and explains diplopia and allelotropia.

## 1. Introduction

Perception of 3-D visual space through processing of two different projections of the scene onto each retina is addressed in many papers [1–8]. Current widely accepted models of stereo vision interprets the coding of either disparity or visual direction; moreover, no biologically plausible models combine processing of both. Such properties of stereo vision as diplopia, allelotropia, and stereoacuity phenomena still are not computationally explained. Well-known energy model suggested explanation for physiological mechanisms of disparity detectors' construction [5–8]. However the disparity describes only one spatial coordinate of the object. Mechanisms of direction encoding are not considered, suggesting the opinion about different means of coding disparity and direction.

We present a mathematical model of binocular vision and show that coding mechanisms of spatial coordinates (depth and direction) could be the same. Using proposed model we quantitatively describe perceived depth and direction as a function of disparity and parallax, Panum's fusional area, diplopia and the exponential decrease of depth sensitivity while object moves away from horopter.

## 2. Model structure

Schematic diagram of a binocular vision system is presented in Fig. 1 where the neural structure (model) for binocular computations is the key element.

The model has two distinct parts: *local linear part* (structured array of local analyzers – LAs, Fig. 1b) and the *global nonlinear part* (global analyzer – GA, Fig. 1c). LAs (detailed in

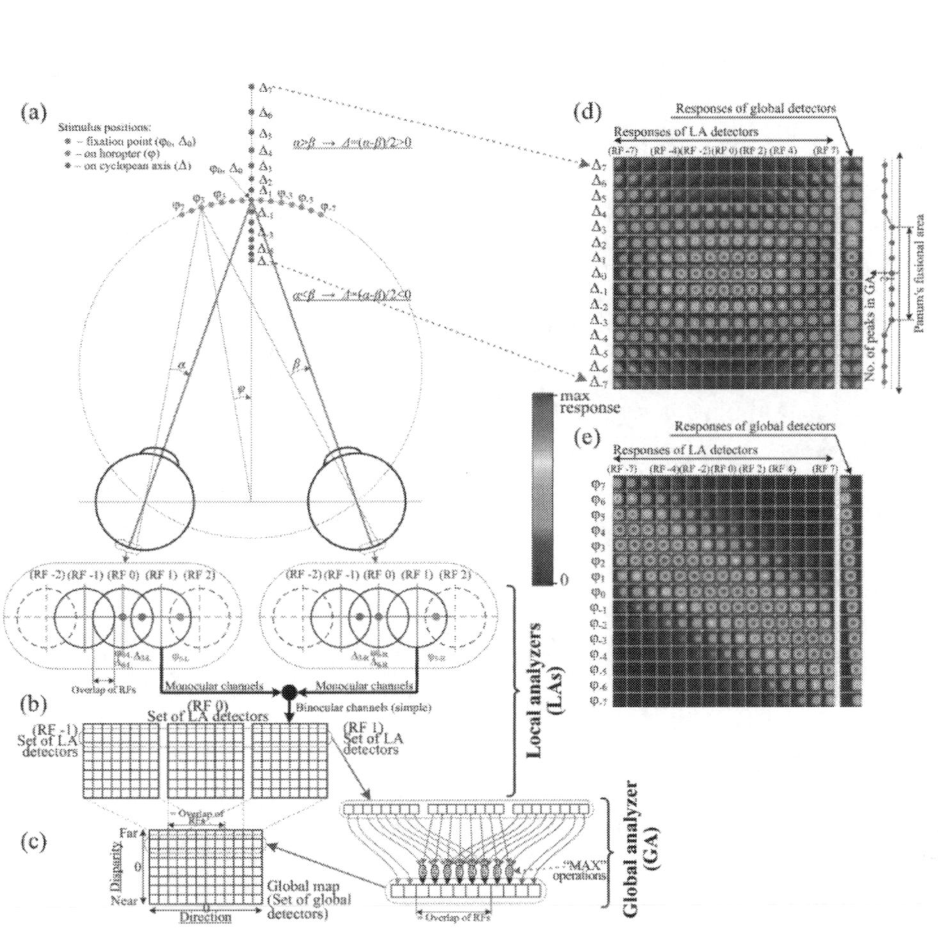


**Figure 1.** Schematic diagram of stereo vision system: (a) positions of stimulus in space with projections onto receptive fields of retinas; (b) array of local analyzers; (c) Structure of global analyzer and transduction of signals from local detectors to global detectors; (d) responses of local and global detectors and identification of Panum's area when stimulus moves on cyclopean axis; (e) responses of local and global detectors when stimulus moves on horopter.

Fig. 2a) perform linear mapping of each point from the local visual space onto a 4-D sphere. Such procedure assigns local spatial coordinates to the object presented in a particular small area of 3-D visual space. Meanwhile, the global analyzer combines responses of LAs by means of nonlinear operations (here – MAX operation) and provides the global spatial coordinates (horizontal parallax and disparity) of a presented binocular feature.

Local analyzers (LAs) also have two distinguishable parts: *monocular* and *binocular* (Fig. 2a). Monocular part is the group of four monocular neurons situated in each eye and being connected to the same crowd of photoreceptors but having different types of Gabor-like receptive fields (RFs) [9]. So monocular outputs is a weighted sum of photoreceptor outputs and can be treated as components of the 4-D vector whose direction encodes position of the light centroid over the current RF. It is noteworthy that LA has corresponding RFs, i.e. its RFs are in the corresponding areas of retinas [10]. RF structures (weight functions) of four monocular neurons are presented in Fig. 2b and are obtained according to Eq. (1). Given the distribution of the light over corresponding RFs $I(\alpha, \gamma)$ and $I(\beta, \gamma)$, Eq. (2) defines outputs of monocular neurons $M_{i\mathrm{L}}$ and

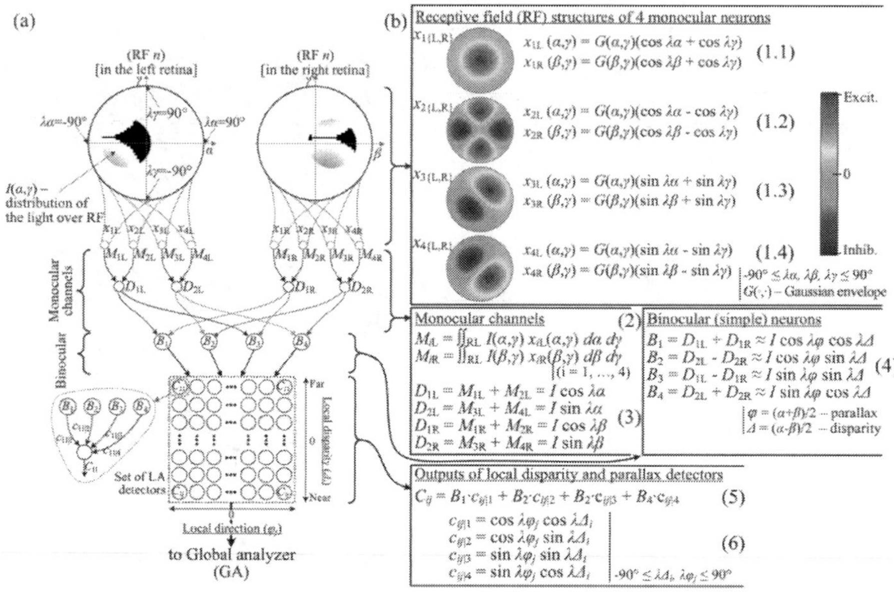


**Figure 2.** Structure of local analyzer (LA).

$M_{i\text{R}}$, $(i = 1 \ldots 4)$.

Neurons of binocular part (Fig. 2a) resemble class of "simple" neurons. There are four independent neurons that linearly sum up signals received from monocular neurons (Fig. 2a, Eq. (4)) and a pool of linearly dependent neurons that act as detectors of disparity and parallax (Fig. 2a, Eq. (5)). Four independent neurons can be treated as components of the 4-D vector $B(\varphi_c, \Delta_c) = \{B_1, B_2, B_3, B_4\}$ whose direction encodes parallax and disparity of a binocular feature. Simple binocular neurons-detectors receive signals from neurons $B_1 - B_4$ and are tuned to the specific spatial coordinates, i.e. they are selective to the specific orientation of vector $B$, or – to particular location of the object in a visual space spanned by the current LA. When projections of the object are presented on RFs of that LA, only one detector is maximally excited. Finding the row and column numbers of maximally excited detector means identification of local depth and direction respectively. So the mission of local analizer is to assign local coordinates to the object that is presented in a such part of visual space which is being projected onto left and right RFs of that LA.

Global analyzer (GA) (Fig. 1c) is the internal map of external visual field. Here global spatial coordinates of the featured object are derived from the local ones. Neurons of GA resemble class of "complex" neurons. Neurons in GA are ordered in such way that neurons located in the middle column are responding to zeroth direction (when stimulus is on horopter), and neurons in the middle row are selective to zeroth depth (when stimulus is on cyclopean axis). Side neurons are signalling left or right direction, and top/bottom -row neurons are tuned to crossed/uncrossed disparities. Each GA neuron receives signals from neurons of several adjacent LAs and passes the maximal one, i.e. connections are created only between LAs and GA neurons which are selective to the same location in visual space. Depth and direction of the object in visual space are identified by the row and column numbers of a maximally excited global neuron-detector. Responses of LAs and GA detectors when stimulus is moving on horopter or cyclopean axis are presented in Fig. 1d,e.

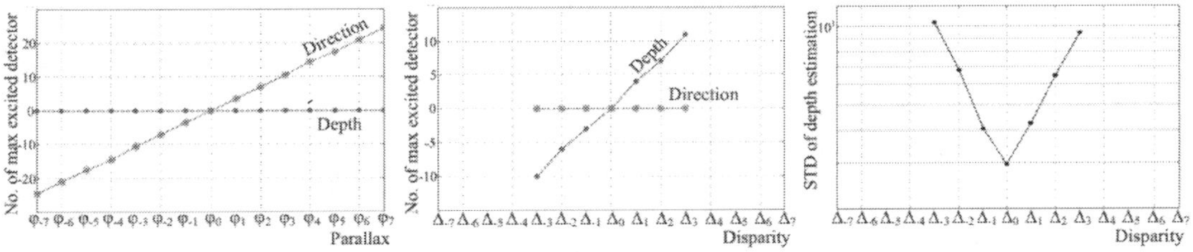


**Figure 3.** Identification of depth, direction, and exponential decrease of depth sensitivity.

## 3. Results and discussion

The proposed neural model of binocular vision is able to identify direction (parallax) and depth (disparity) of the object, and identified direction and depth are linear functions of physical parallax (Fig. 3a) and disparity (Fig. 3b) of that object. In these graphs numbers of maximally excited detectors were obtained from GA responses in Fig. 1d,e. Identified direction and depth are constant while stimulus is located on horopter and cyclopean axis respectively.

Events in GA when stimulus is moving along the horopter and cyclopean axis are significantly different. Only one detector is maximally excited while the stimulus is moving along the horopter. When stimulus gets far from horopter two detectors encoding different parallaxes appear maximally excited, and here diplopia occurs. Area in visual space in which presented stimulus causes only one peak in GA may be called as the Panum's region of the model.

Identified parallax remains constant until stimulus is presented within Panum's area and the identified depth is linear function of the physical disparity. That is in agreement with experimental data [11]. Moreover the identified parallaxes of diplopic stimuli differ from parallaxes of the same monocular stimuli, when each of them is presented alone, i.e. there is resemblance to the phenomena of allelotropia [12].

It is known that the accuracy of depth perception is decreasing exponentially while the stimulus is moving away from the horopter [13, 14]. Our model demonstrated such pattern (Fig. 3c) when the noise was introduced, and this argues that there is no necessity to have two (course and fine) systems for disparity determination in order to explain exponential decrease.

## Acknowledgments

B. Postdoctoral fellowship is being funded by European Union Structural Funds project "Postdoctoral Fellowship Implementation in Lithuania"

## References

[1] Barlow H B, Blakemore and Pettigrew J D 1967 *J. Physiol.* **193** 327–42
[2] Pettigrew J D, Nikara T and Bishop P O 1968 *Exp. Brain. Res.* **6** 353-72
[3] Poggio G F, Gonzalez F and Krause F 1988 *J. Neurosci.* **8** 4531–50
[4] Marr D and Poggio T 1979 *Proc. R. Soc. Lond.* B **204** 301–28
[5] Freeman R D and Ohzawa I 1990 *Vision Res.* **30** 1661–76
[6] Ohzawa I 1998 *Curr. Opin. Neurobiol.* **8** 509–15
[7] Ohzawa I, DeAngelis G C and Freeman R D 1997 *J. Neurophysiol.* **77** 2879–909
[8] Fleet D, Wagner H and Heeger D J 1996 *Vision Res.* **36** 1839–57
[9] Fomin S V, Sokolov E N and Vaitkevicius H H 1979 *Artificial sensory system: a problem of sensory simulating* (Nauka: Moscow) p 189
[10] Kulikowski J J and Vidyasagar T R 1986 *Exp. Brain Res.* **64** 5–18
[11] Lawson R B, Cowan E, Gibbs T D and Whitmore C G 1974 *J. Exp. Psychol.* **103** 1142–6
[12] Rose D and Blake R 1988 *J. Opt. Soc. Am.* A **5** 1512–21
[13] Farell B, Li S and McKee S P 2004 *J. Vision* **4** 488–99
[14] McKee S P, Verghese P and Farell B 2005 *J. Vision* **5** 783-92

# Algorithms for automatic segmentation of bovine embryos produced *in vitro*

**D H Melo [1], M Z Nascimento [2], D L Oliveira [3], L A Neves [4] and K Annes [5]**

[1,3] Center of Mathematics, Computing and Cognition, Federal University of ABC (UFABC), Santo André, SP, Brazil
[2] Faculty of Computing, Federal University of Uberlândia (UFU), Uberlândia, MG, Brazil
[4] Institute of Biosciences, Letters and Science, Department of Computer Science and Statistics, São Paulo State University (UNESP), São José do Rio Preto, SP, Brazil
[5] Laboratory of Cellular and Molecular Biology of the Center of Natural Sciences and Humanities, Federal University of ABC (UFABC), Santo André, SP, Brazil

E-mail: douglas.melo@ufabc.edu.br

**Abstract.** In vitro production has been employed in bovine embryos and quantification of lipids is fundamental to understand the metabolism of these embryos. This paper presents a unsupervised segmentation method for histological images of bovine embryos. In this method, the anisotropic filter was used in the differents RGB components. After pre-processing step, the thresholding technique based on maximum entropy was applied to separate lipid droplets in the histological slides in different stages: early cleavage, morula and blastocyst. In the post-processing step, false positives are removed using the connected components technique that identify regions with excess of dye near pellucid zone. The proposed segmentation method was applied in 30 histological images of bovine embryos. Experiments were performed with the images and statistical measures of sensitivity, specificity and accuracy were calculated based on reference images (gold standard). The value of accuracy of the proposed method was 96% with standard deviation of 3%.

## 1. Introduction

In vitro production (IVP) has been employed to improved features genetics of bovine embryos [1]. However, the increase in the levels of cytoplasmic lipid droplets in the IVP of bovine embryos are relationship reduction of embryos after cryopreservation [2]. Therefore quantify the lipids is fundamental to understand the metabolism of these embryos.

Some methods are applied to evaluate and measure the features of lipids in the investigation of IVP of bovine embryos. Among the techniques applied, the Sudan black B staining is a method that consists of a lipophilic dye, which reacts with lipids. Regions with lipids are represented by darker colours compared to the rest of the embryo without lipid droplets. After Sudan black B staining of bovine embryos, a microscope equipped with digital camera is used to acquire images [3].

Some computational techniques have been applied to quantification of lipid droplets in images of bovine embryos [3, 4]. However, the accuracy rate is low and these methods are supervised. Thus, quantification of lipid by Sudan black B staining becomes one of the difficulties for the specialists.

This paper presents a unsupervised segmentation method for histological images of bovine embryos. In this method, the anisotropic filter was used in the RGB colour components. After pre-processing step, the thresholding technique based on maximum entropy was applied to separate lipid droplets in the histological slides in different stages: early cleavage, morula and blastocyst. In the post-processing step, false positives are removed using the connected components technique that identify regions with excess of dye near pellucid zone. The proposed segmentation method was applied in 30 histological images of bovine embryos.

## 2. Materials and Methods

The method was organized in three steps: (1) Pre-processing, (2) Segmentation and (3) Post-Processing. Each step is described in detail in the following subsections.

### *2.1. Image Database*

The database image used in the present study was made up of 30 histological slides in different stages: early cleavage, morula and blastocyst. The images were acquired in the Laboratory of Cellular and Molecular Biology of the Federal University of ABC. The microscopic images were acquired using an inverted light microscope Olympus IX71 with digital camera coupled (Lumenera Innity 1-1). Each slide was photographed at a magnification of 100x, using 10x objective lens and ocular lens and image was saved in .png using the RGB model, 2048x1536 pixels, with 24 quantization bits.

### *2.2. Pre-processing*

In the mounting process of each slice, fragments, dyes and other parts of the embryo can leak out of the zona pellucida. The algorithm considers these problems by calculating the center of the image and considering only the embryo centered. In this step, the pixels of the image were compared with the coordinate of the center and the values were used to calculate the hypotenuse. Regions with value 22% greater than the image area were removed.

Anisotropic diffusion filter can be used to smooth the outer regions of the edge giving a blur effect and highlight the edges with a sharp effect keeping unchanged the inner regions of the edges [5]. In this work, we applied this filter using the equation 1:

$$I(s,t+1) = I(s,t) + \frac{\lambda}{|\eta_s|} \sum_{p \in \eta_s} g(|\nabla I_{s,p}(t)|) \nabla I_{s,p}(t) \tag{1}$$

where $I(s,t)$ is the image, $s$ is the pixel position, $t$ is the number of interactions, $\eta_s$ is the set of neighbors space, $\lambda$ is the diffusion speed, $I_{s,p}(t)$ is the gradient magnitude of the image $I$ at the point $s$ in the direction $(s,p)$ in iteration: $t\nabla I_{s,p}(t) = I(p,t) - I(s,t), p \in \eta_s$. In this paper, we used parameter $g$ defined by equation 2 .

$$g_2(x) = exp\left[\frac{-x^2}{2\sigma^2}\right] \quad . \tag{2}$$

In equation 1, $t = 20$, $\lambda = 0.2$ and in equation 2, $\sigma = 15$.

### *2.3. Segmentation*

In this step, a global thresholding was used, considering the maximum entropy technique. This technique maximize entropy of the image dividing the histogram into two probability distributions [6]. Each distribution represents one of the classes: background and object. The entropy $H_b(T)$ and $H_w(T)$ associated to the pixels of the background and objects, respectively, are expressed by

$$H_b(T) = -\sum_{i=0}^{T} p_i \; log \; p_i \tag{3}$$

$$H_w(T) = -\sum_{i=T+1}^{L-1} p_i \; log \; p_i \tag{4}$$

where $n_i$ is the number of pixels with gray scale $i$, $p_i$ is the probability of gray scale $i$ found in the image, $n$ is the number of pixels of the image and $L$ is the number of gray scales of image.

The gray scale is analyzed in order to find a threshold value $T$, such that $T = H_b(T) + H_w(T)$ is maximum. This technique defines a good separation between the object and the background of the image, $T = arg \; max[H_b(T) + H_w(T)]$

### 2.4. *Post-Processing*

The connected components technique was used to identify all segmented regions and remove the regions that do not represent lipids: the 4-adjacency technique was used to remove regions false-positives [7].

### 2.5. *Quantitative Evaluation*

The quantitative evaluation was performed by calculating the overlap between the image regions segmented by the proposed method and the regions of a binary reference demarcated by a specialist. This calculation was performed using statistical measurements of sensitivity ($SE$), specificity ($ES$) and accuracy ($ACC$) ([8]).

## 3. Results and Discussion

The proposed segmentation method was applied to an image bank with 30 samples and measures of sensitivity (SE), specificity (SP) and accuracy (ACC) were calculated. Figure 1 respectively shows an image of embryo which was manually segmented by a specialist (gold standard) and with the proposed method. Table 1 shows the measures considering sensitivity (SE), specificity (SP) and accuracy (ACC).

The best ACC results (proportion of pixels defined as correctly segmented, both true positives and true negatives) were obtained with the R colour component for the Early cleavage, Morula and Blastocyst stages. Also, the proposed method provided segmented images with high rates of SE and SP. In this study, sensitivity relates to the method's ability to identify positive areas or belong to a specific stage such as Early cleavage (SE of 94%), Morula (SE of 96% ) and Blastocyst (SE of 92%). These values were obtained with the B colour component. The method's ability to identify negative results (specificity) also is important: Early cleavage with 87%, Morula with 78% and Blastocyst with 88%, where these values were obtained with R colour component. These measures suggest that textural features from bovine embryos images are sufficient to separate lipids on different stages. Segmentation methods have been proposed for investigation on different histological images but are not yet applied to bovine embryos. This is a limitation for comparisons of our results. So, the proposal presented here provides a significant contribution to studies focusing on IVP of bovine embryos.

## 4. Conclusion

This paper presented an unsupervised method for the automatic segmentation of bovine embryos images considering different stages: early cleavage, morula and blastocyst. The proposed method is effective for the segmentation of embryos images, considering the quantitative results calculated. Also, the method offers the advantage of making automatically the segmentation.

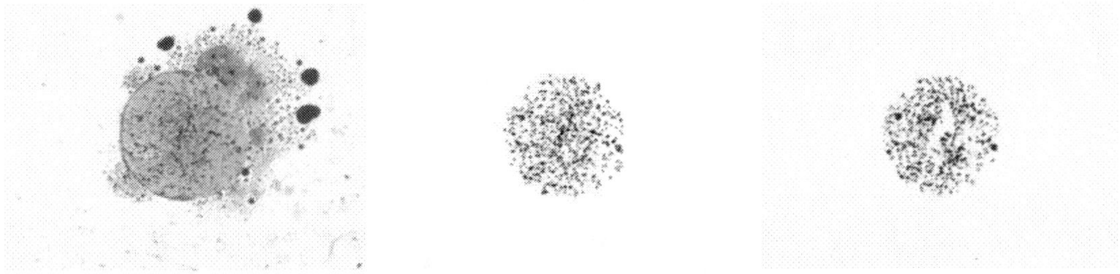

(a) (b) (c)

Figure 1: Sample of an embryo is shown (a), the segmentation provided by a specialist is shown in (b) and the result obtained with the proposed method is illustrated in (c).

Table 1: Measures of sensitivity (SE), specificity (SP) and accuracy (ACC) obtained for each stage and colour component.

| Stages | Colour Channel | ACC | SE | SP |
|---|---|---|---|---|
| Early cleavage | Red | $0.92 \pm 0.04$ | $0.92 \pm 0.04$ | $0.87 \pm 0.12$ |
| | Green | $0.91 \pm 0.03$ | $0.91 \pm 0.04$ | $0.83 \pm 0.22$ |
| | Blue | $0.93 \pm 0.04$ | $0.94 \pm 0.04$ | $0.68 \pm 0.32$ |
| Morula | Red | $0.93 \pm 0.04$ | $0.93 \pm 0.04$ | $0.78 \pm 0.30$ |
| | Green | $0.94 \pm 0.04$ | $0.95 \pm 0.04$ | $0.66 \pm 0.33$ |
| | Blue | $0.96 \pm 0.04$ | $0.96 \pm 0.04$ | $0.56 \pm 0.32$ |
| Blastocyst | Red | $0.92 \pm 0.04$ | $0.92 \pm 0.04$ | $0.88 \pm 0.14$ |
| | Green | $0.92 \pm 0.04$ | $0.92 \pm 0.04$ | $0.84 \pm 0.22$ |
| | Blue | $0.92 \pm 0.04$ | $0.92 \pm 0.04$ | $0.83 \pm 0.24$ |

In future studies, performance tests will be performed and the results compared to others segmentation methods.

## References

[1] Goovaerts I G F, Leroy J L M R, Langbeen A, Jorssen E P A, Bosmans E and Bols P E J 2012 Unravelling the needs of singly in vitro-produced bovine embryos from cumulus cell co-culture to semi-defined, oil-free culture conditions. *Reprod Fertil Dev*, **24** 1084-1092.

[2] Mucci N, Aller J, Kaiser G, Hozbor F, Cabodevila J and Alberio R 2006 Effect of estrous cow serum during bovine embryo culture on blastocyst development and cryotolerance after slow freezing or vitrification. *Theriogenology*, **65** 1551-1562.

[3] Sudano M, Paschoal D, Rascado T, Magalhaes L C, Crocomo L, de Lima-Neto J and Landim-Alvarenga F 2011 Lipid content and apoptosis of in vitro-produced bovine embryos as determinants of susceptibility to vitrification. *Theriogenology*, **75** 1211-1220

[4] Sudano M, Santos V, Tata A, Ferreira C, Paschoal D, Machado R, Buratini J, Eberlin M and Landim-Alvarenga F 2012 Phosphatidylcholine and sphingomyelin profiles vary in bos taurus indicus and bos taurus taurus in vitro- and in vivo-produced blastocysts. *Biology of reproduction*, **87** 1-11.

[5] Perona P and Malik J 1990 Scale-space and edge detection using anisotropic diffusion. *Pattern Analysis and Machine Intelligence, IEEE Transactions on*, **12** 629-639.

[6] Pun T 1980 A new method for grey-level picture thresholding using the entropy of the histogram. *Signal Processing*, **2** 223-237.

[7] Gonzalez R and Woods R 2007 Digital Image Processing. Pearson.

[8] Sonka M 2000 Handbook of medical imaging: medical image processing and analysis, volume 2. Society of Photo Optical.

# Numerical simulations of a minimal model for the fluid dynamics of dense bacterial suspensions

**Sebastian Heidenreich**[1], **Sabine H. L. Klapp**[2] **and Markus Bär**[1]

[1]Department of Mathematical Modelling and Data Analysis, Physikalisch-Technische Bundesanstalt, Abbestr. 2-12, 10587 Berlin, Germany

[2]Institute for Theoretical Physics, Technische Universität Berlin, Hardenbergstr. 36, 10623 Berlin, Germany

E-mail: sebastian.heidenreich@ptb.de

**Abstract.** Collective behavior is a fascinating phenomenon and ubiquitous in nature. A large variety of complex dynamic structures from swarming to turbulence arise in active particle systems. In recent investigations a set of minimal continuum equations was proposed to model mesoscale bacterial turbulence. Numerical solutions are validated with experimental data of *Bacillus subtilis* bacteria. In this short paper we present a recently used pseudo-spectral operator splitting method that directly solves the nonlinear equations in the turbulent regime. In two and three spatial dimensions we show the resulting typical velocity and vorticity fields as well as energy spectra to highlight the strong difference between turbulence in ordinary fluids and in bacterial suspensions.

## 1. Introduction

Swimming bacteria, fish schools, moving algae or artificial microswimmers are intriguing examples of self-propelled particle systems. Such active particles are capable to convert ambient or stored energy into directed motion and often show coordinated collective behavior. The related phenomena range from large scale vortex structures, clustering, giant number fluctuations to mesoscale turbulence [1, 2, 3, 4]. Mesoscale turbulence was observed in dense suspensions of Bacillus subtilis and is characterized by self-sustained complex flow pattern. Recently, a phenomenological equation was proposed to model self-sustained turbulence [5]. The model combines properties from the Navier-Stokes equation with a pattern-forming instability mechanism closely related to the one in the Swift-Hohenberg equation. This instability drives the flow even when no external forces are applied. Surprisingly, this simple model could explain turbulent structures which are found in experiments of Bacillus subtilis suspensions in two and three dimensions quantitatively [3, 6]. In this article we will briefly introduce the phenomenological equation and describe in detail the numerical algorithm employed for simulations. Furthermore, we show snap-shots of two and three dimensional self-sustained turbulence and discuss the difference to turbulence of Navier-Stokes type on the basis of energy spectra.

## 2. Governing equations and numerical scheme

Dense bacterial suspensions of Bacillus subtilis or E-coli show an irregular flow behavior at mesoscales.

Recently, a continuum equation based on the effective velocity $\mathbf{v}$ was introduced to model the dynamics of the dense suspensions of the bacteria [5] ,i.e.,

$$(\partial_t + \lambda_0 \mathbf{v} \cdot \nabla)\mathbf{v} = -\nabla p + \lambda_1 \nabla v^2 - \beta(v^2 - v_0^2)\mathbf{v} + \Gamma_0 \Delta \mathbf{v} - \Gamma_2 \Delta^2 \mathbf{v} \qquad (1)$$

$$\nabla \cdot \mathbf{v} = 0. \tag{2}$$

Here, $\lambda_i, \beta, \Gamma_i, v_0^2$ are phenomenological coefficients, $p$ is the hydrodynamical pressure and $\Delta$ specifies the Laplace operator. For the parameter values $\lambda_0 = 1$, $\lambda_1 = \beta = \Gamma_2 = 0$ and $\Gamma_0 > 0$ the Navier-Stokes equation is obtained.

A similar theory for flocking of animals was introduced by Toner and Tu. The constant density limit of the Toner-Tu theory is obtained when $\Gamma_0 > 0, \Gamma_2 = 0$. However, a positive $\Gamma_0$ stabilizes a polar state and does not exhibit the characteristic vortex structures observed for swimming bacteria. When negative values are allowed the disordered state is exhibiting a Turing like irregular, turbulence like flow pattern appear in simulations even in the absence of external forces that usually causes turbulence in fluid flows. In the following we will explain how the nonlinear partial differential equaitons (1) are solved. Furthermore, we discuss some specific features of bacterial turbulence.

To solve the equations for the bulk flow, we used a pseudo-spectral algorithm with periodic boundary conditions and an operator splitting to treat the time integration of the linear part exactly. The nonlinear part is integrated by a simple Euler scheme. In particular, the algorithm consists of the following steps:

- A Fast Fourier transformation of Eq. (1); treatment of the convection term by forward and backward transformation, i.e., $F[\mathbf{v} \cdot F^{-1}[-i\mathbf{k}F[\mathbf{v}]]]$
- Application of anti-aliasing 2/3 rule to suppress oscillation effects due truncated Fourier series of nonlinear terms
- Splitting of the resulting terms into linear and nonlinear parts
- Exact integration of the linear part (see [7])
- Explicit Euler integration of the remaining equation
- Pressure correction routine to ensure incompressibility
- Inverse Fourier transform; store results

In the next section we will discuss results for simulations of the continuum equation Eq. (1) in two and three dimensions.

## 3. Results

For simulations, we used a grid resolution of $256 \times 256$ points in 2D and $256^3$ in 3D, respectively. Initially we generated a random flow field. Eq. (1) are scaled according to Refs. [3, 6]. We used time steps of $\Delta t = 0.1$ for 2D and $\Delta t = 0.05$ for 3D. After a relaxation time of $t = 10$ time units we saved velocity fields for generating snap-shots and for calculating spectra. The final simulation time was between $t = 500$ and 2000 time units which corresponds to a maximum time of about one minute.

Fig. 1 shows representative snap-shots for simulations in two and three dimensions. In the three dimensional case the figure depicts a two dimensional slice from the middle of the simulation box. In both snap-shots the flow pattern are irregular and show vortex structures. Remarkably, the flow is different from that found for inertial turbulence described by the Navier-Stokes equation. The reason is the emergence of vortices of a typical size in contrast to decaying vortices that occur for inertial turbulence found in the classical Navier-Stokes equation.

The key is that the typical vortex size in bacterial turbulence introduces a characteristic length scale into the system. In contrast simple fluids without boundaries are scale invariant. The energy spectra of the two dimensional bacterial turbulence pattern for different values of the parameter $\lambda_0$ are shown in Fig.2 (calculated according to [5]). For large values of $\lambda_0$ changes of the energy spectra are small.

The spectra of bacterial turbulence differs qualitatively from that obtained for inertial turbulence (see Fig. 2b) for a sketch). Two dimensional initial turbulence is characterized by an upper energy and a lower enstrophy cascade. For bacterial turbulence, the energy injection happens at small scales by the individual swimmer; followed by an upward cascade up to a specific maximum. From the maximum to larger scales we observe a downward cascade. It is the maximum in the energy spectrum that corresponds

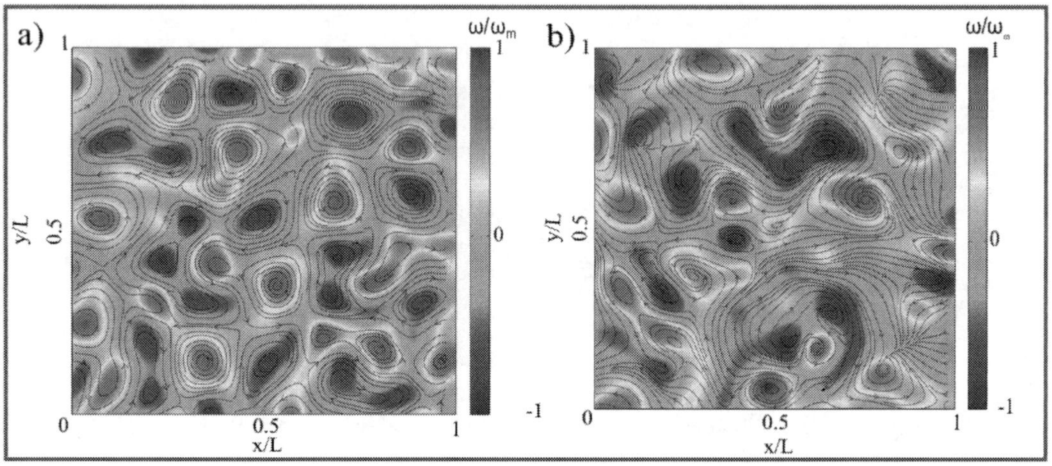


**Figure 1.** Snapshots of bacterial turbulent states. The color bar encodes vorticity strength and the box length corresponds to $200\mu m$ and $300\mu m$ for 2D and 3D, respectively. a) Two dimensional simulations at parameter values $\Gamma_0 = -1, \Gamma_2 = 1\lambda_0 = 5, \lambda_1 = 2, L = 48 * \pi, \beta = 1, v_0^2 = 0.2$. b) Two dimensional slice of three dimensional simulations. The used parameter values are $\Gamma_0 = -0.9, \Gamma_2 = 0.9\lambda_0 = 1.7, \lambda_1 = 0.35, L = 24 * \pi, \beta = 0.1, v_0^2 = 0.1$.

to the typical length scales of vortices. The typical length scale is encoded in the phenomenological parameters $\Gamma_0$ and $\Gamma_2$, i.e.,

$$\Lambda = 2\pi\sqrt{\Gamma_2/(-\Gamma_0)}. \tag{3}$$

Typically values for Bacillus subtillis suspensions are $30\mu m$ to $50\mu m$. Unfortunately, there is still no microscopic derivation of the phenomenological equations and the microscopic meaning of the length scale as well as the mechanism that selects the scale remains unknown.

For the three-dimensional case, calculations are challenging and we have simulated the flow pattern, but not yet spectra. The stream lines of the two-dimensional slice of a three-dimensional flow Fig. 1b) show spirals rather than vortices. Here, bacteria are swimming in cycles with a velocity component orthogonal to the plane. The resulting collective motion looks like a twister. Fig. 3b, c) show a three dimensional plot of that flow structure. Fig. 3b) shows a cut perpendicular to a twister and Fig. 3c) the overall shape. In the turbulence regime such twister like structures spontaneously emerge and disappear all the time. The overall flow structure in the simulation box shows isotropic turbulence seen in the iso-energy surface plot in Fig. 3a).

In this article, we introduced phenomenological equations to model the collective flow of active particle suspensions. We presented the numerical scheme and discussed results from two and three dimensional simulation of the bulk flow. Bacterial turbulence is different from inertial turbulence of the Navier-Stokes type due to a typical vortex length. A fingerprint of that vortex size can be found in energy spectra which is also different from inertial turbulence. Finally, we discussed a new type of flow structure that we have found in three dimensional bacterial turbulence; collective twister motion.

In further work we plan to investigate the influence of confinement to the turbulent flow structures. There is experimental evidence that in the presence of boundaries turbulent structures are suppressed in favor to a single vortex [9]. Furthermore, we plan to study rheological properties of the continuum model and compare it to more complex descriptions where higher order parameters are included [8].

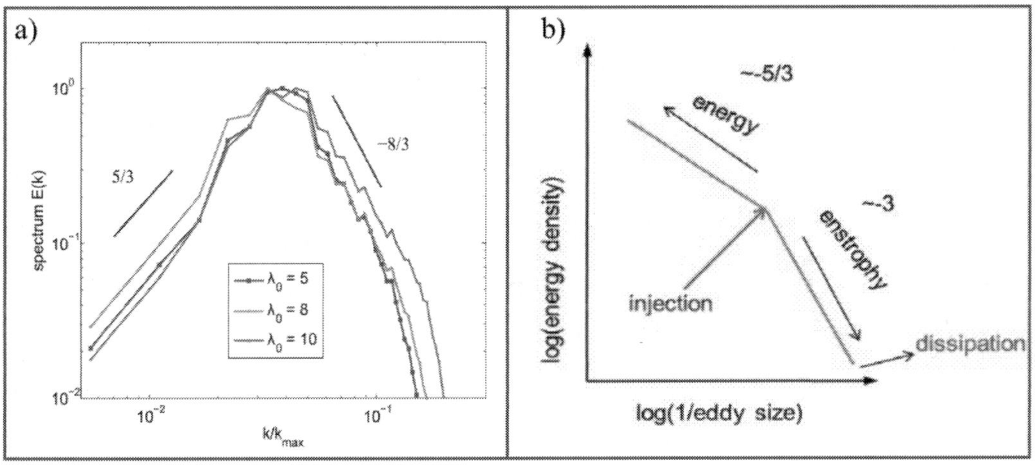


**Figure 2.** a) Energy spectra of two dimensional bacterial turbulence simulations for different parameter values $\lambda_0$. The remaining parameters are the same as in Fig. 1. Energy is injected at the length scale of microswimmer ( $k_{max}$) b) Sketch of a spectrum of inertial turbulence.

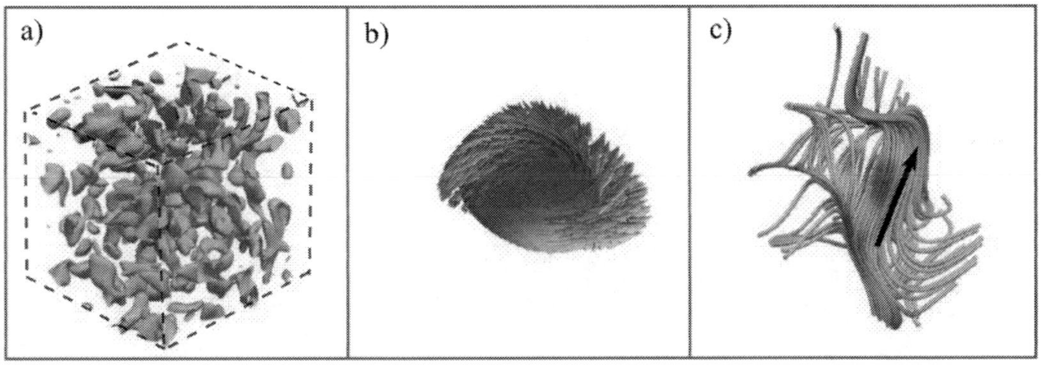


**Figure 3.** Illustration of three dimensional mesoscale turbulence using Paraview. Parameter values are the same as in Fig. 1. a) Iso-energy surface plot. b) Velocity field of a three dimensional vortex. c) Tube graph of velocity stream lines. The twister like structure shows complex 3D collective bacterial motion.

## Acknowledgment

The authors are grateful to Jörn Dunkel and Holger Stark for helpful discussion. This work was supported by the Deutsche Forschungsgemeinschaft (DFG) via GRK 1558 (M.B.) and SFB910 (S. K.).

## References

[1] Sumino Y, Nagai K H, Shitaka Y, Tanaka D, Yoshikawa K, Chaté H and Oiwa K 2012 *Nature* **483** 448

[2] Schaller V, Weber C, Semmrich C, Frey E, Bausch A R 2010 *Nature* **467** 73

[3] Wensink H H, Dunkel J, Heidenreich S, Drescher K, Goldstein R E, Löwen H and Yeomans J M 2012 *Proc. Nat. Acad. Sci. USA* **109** 14308

[4] Peruani F, Starruß J, Jakovljevic V, Sogaard-Anderson L, Deutsch A, and Bär M 2012 *Phys. Rev. Lett.* **108** 098102

[5] Dunkel J, Heidenreich S, Bär M and Goldstein R E 2013 *New J. Phys.* **15** 045016
[6] Dunkel J, Heidenreich S, Drescher K, Wensink H H, Bär M and Goldstein R E 2013 *Phys. Rev. Lett.* **110** 228102
[7] Pedrosa J, Hoyuelos M, and Martel C, *Eur. Phys. J. B* **66** 525
[8] Heidenreich S , Hess S, Klapp S H L 2011 *Phys. Rev. E* **83** 011907
[9] Wioland H, Woodhouse F G, J. Dunkel J, Kessler J O, Goldstein R E 2013 *Phys. Rev. Lett.* **110** 268102

# Classification of mammographic lesion based in Completed Local Binary Pattern and using multiresolution representation

**Y A S Duarte[1], M Z Nascimento[2] and D L L Oliveira[1]**

[1]Center of Mathematics, Computing and Cognition, Federal University of ABC (UFABC), Santo Andre, São Paulo, Brazil

[2]Faculty of Computing (FACOM), Federal University of Uberlândia (UFU), Uberlândia, Minas Gerais, Brazil

E-mail: yan_asd@hotmail.com

**Abstract.** This paper presents a comparison of two methods for features extraction of mammograms based in completed local binary pattern (CLBP) and wavelet transform. In first part, CLBP was applied in digitized mammograms. In second part, we applied CLBP in the sub-bands obtained from the wavelet multi-resolution representation of the mammographies. In this study, we evaluated the CLBP in the image in the spatial domain and in the sub-bands obtained with wavelet transform. Then, the statistical technique of variance analysis (ANOVA) was used to reduce the number of features. Finally, the classifier Support Vector Machine (SVM) was applied in the samples. The proposed methods were tested on 720 mammographies which 240 was diagnosed as normal samples, 240 as benign lesion and 240 as malign lesion. The images were obtained randomly of the Digital Database for Screening Mammography (DDSM). The system effectiveness was evaluated using the area under the ROC curve (AUC). The experiments demonstrate that the textural feature extraction of the multi-resolution representation was more relevant with value of AUC=1.0. In our experiments, CLBP in the spatial domain resulted in value of AUC=0.89. The proposed method demonstrated promising results in the classification of different classes of mammographic lesions.

## 1. Introduction

The breast cancer is the second disease that most cause obits among women in all the world. In Brazil, the Instituto Nacional do Câncer (National Institute of Cancer - INCA) estimated more than 52,000 of new cases and more than 12,000 deaths occur due to this disease.

An early diagnostic is essential to increase the chance of a successfully treatment. The most common method used in diagnostic is the analysis of the screen/film mammography by a specialist. However, this method has some limitations. For example, the experience level of the specialist and the human vision limitation. A possible solution to assist the specialist in the breast cancer diagnostic is the used of a computer aided diagnosis (CAD) system. This kind of system can produces consistent results for the classification of disease. Several CAD systems have been employed in the breast lesions classification using the digitalized image of mammography [1, 2, 3]. However, there have been few investigations that utilize more recent techniques for textural feature extraction in mammographies and the obtained results can still be improved.

In this paper we present a method for extraction and classification of regions of interest (ROI) of mammograms. The method was based in the wavelet multi-resolution technique [4], completed local binary pattern (CLBP) operators [5], analysis of variance (ANOVA) [6] and Support Vector Machine (SVM) classifier. We used a subset of mammography cases from a publicly available database, the Digital Database for Screening Mammography.

## 2. Materials and methods

The method was organized in four steps: (1) calculation of Wavelet coefficients from the mammographies; (2) extraction of CLBP from wavelet coefficients and the original mammographies; (3) application of ANOVA statistical method and (4) classification of images using SVM. Fig. 1 presents the flowchart of the proposed method.

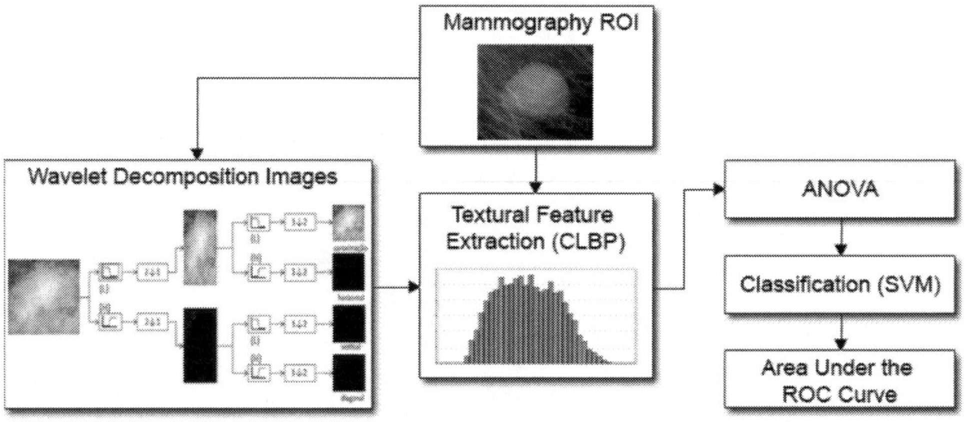


Figure 1: Flowchart of the proposed system.

### 2.1. Image database

In this experiments we used 720 mammographic images where 240 were diagnosed as normal samples, 240 as benign lesion and 240 as malign lesion. Each ROI was obtained of each mammograms of CC and MLO views get from the the database named Digital Database for Screening Mammography (DDSM). The dataset is composed of 360 mammographic ROI's in cranio-caudal (CC) view and 360 ROI's in mediolateral oblique (MLO) view randomly selected from the DDSM. Therefore, the ROI's were cuttings of size 512 x 512 pixels.

### 2.2. Wavelet coefficients

A multi-resolution technique allows an image data analysis that results in decomposition details at different resolution levels. Wavelet transform is a technique that represents an image in frequency and time domains [4]. The task is to decompose the original image into sub-bands that preserve high and low frequency information. The most used approach for discrete images is the wavelet transform with quadrature mirrors filter (QMF) banks. This process is made recursively by high-pass and low-pass filtering the image. Four sub-images are generated in this process: the approximation, which represents the original image with a smaller resolution, and the details, which represent the horizontal, vertical and diagonal directions, respectively. However, only the wavelet detail sub-band coefficients were used. For multi-resolution analysis, we used wavelet function Symlet 8 with 2 levels of decomposition.

### 2.3. Completed Local Binary Pattern

Completed local binary pattern provides a robust and discriminated way for the textural features extraction [5]. This technique has three distinct operators: (1) CLBPs that considers the signal component information; (2) CLBPm that considers the magnitude component information and (3) CLBPc that considers the intensity of the central pixel. This method consist in compare each point of a matrix with its neighbors and get information of signal, magnitude and the analyzed point (central). This procedure results in two histogram, one for signal and other for magnitude, and one matrix that contains the intensity of each analyzed pixel. In our experiments, we used the CLBP as much in the wavelet details sub-bands as in the digitalized image in order to compare the results. The parameters neighbors and radius were set to 8 and 1 respectively.

### 2.4. Attribute reduction and classification

ANOVA were applied to improve the performance of the classifier and eliminate outliers of data. This technique is a statistical model that compares the means of two or more experiments. This evaluation is performed to the extent that the observed differences between the means are significant for the comparison of two estimates. The ANOVA method was applied in the attributes obtained from the CLBP technique used in the ROI and we also investigated the CLBP operators obtained from the details wavelet sub-bands. This was done to determine the statistical significance of these values. Finally, the SVM classifier was applied. SVM is a machine learning technique based in the statistic learning theory. This theory find a hyperplane with maximum separation between the classes and assumes that the data is linearly separable [7]. In our experiments the SVM was used with linear kernel. For the training and the validation step, the 10-fold cross validation technique was applied with 90% of the data for training and the remaining 10% for test.

## 3. Results and discussion

The AUC results obtained of evaluation of each group lesion based in the CLBP are presented in Table 1.

Table 1: AUC results for each lesion group with CLBP and wavelet detail sub-band coefficients.

| Classes | View | CLBP | CLBP on Coefficients |
|---|---|---|---|
| Benign versus Malign | CC | $0.71 \pm 0.17\%$ | $1.0 \pm 0.0\%$ |
| | MLO | $0.71 \pm 0.16\%$ | $1.0 \pm 0.0\%$ |
| Benign versus Normal | CC | $0.78 \pm 0.14\%$ | $1.0 \pm 0.0\%$ |
| | MLO | $0.78 \pm 0.15\%$ | $1.0 \pm 0.0\%$ |
| Normal versus Malign | CC | $0.89 \pm 0.11\%$ | $1.0 \pm 0.0\%$ |
| | MLO | $0.87 \pm 0.12$ % | $1.0 \pm 0.0$ % |

Fig. 2(a), Fig. 2(b) and Fig. 2(c) show, respectively, ROC curves with the values of AUC more relevant for the classes of images.

The results showed that the technique CLBP applied in the wavelet detail sub-band coefficients were very effective reaching an AUC value of 1.0 in all the classes. The CLPB applied in the mammographic images without the use of wavelet transformed has similarly results in CC and MLO views of each group. In this case, the value of AUC was 0.71 for group of benign versus malign lesions. In the benign versus normal group AUC was 0.78. At last, the Normal versus Malign group reached an AUC value of 0.89. Using CLBP without wavelet detail sub-band coefficients the method provided results different groups evaluated. This experiments shows that the use of operators CLBP and wavelet improve AUC for different classes of mammographic images.

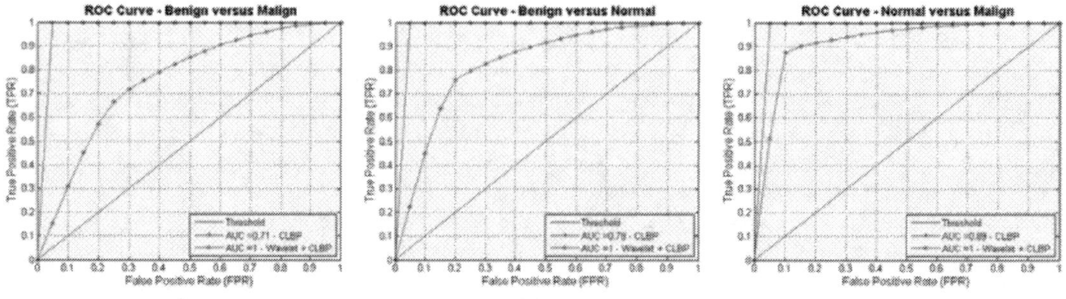


(a) (b) (c)

Figure 2: ROC curves of classes of images: Benign versus Malign (a), Benign versus Normal (b) and Normal versus Malign (c).

## 4. Conclusion

In this study we presented a method based in the wavelet multi-resolution technique, completed local binary pattern (CLBP) operators and analysis of variance (ANOVA) for features extraction. In general, the aggregation of CLBP and wavelet has reached better results in the classification of different kinds of abnormalities. Based on these relevant values of AUC, we can state that the proposed technique using this textural descriptors is an efficient tool for analysis of breast tissue on mammography. For further works we pretend to investigate the proposed method in another mammography database and analyse others techniques of texture feature extraction.

## References

[1] Llado X, Oliver A, Freixenet J, Marti R, Mari J 2009 A textural approach for mass false positive reduction in mammography *Comput Med Imaging Graph* **33** 415-422.

[2] Dantas R D, Nascimento M Z, Jacomini R S, Pereira D C and Ramos R P 2012 Fusion of two-view information: SVD based modeling for computerized classification of breast lesions on mammograms *Mammography - Recent Advances* **1** 261-278.

[3] Ramirez-Villegas J F, Ramirez-Moreno D F 2012 Wavelet packet energy, tsallis entropy and statistical parameterization for support vector-based and neural-based classification of mammographic regions *Neurocomputing* **77** 82-100.

[4] Mallat S 1999 *A Wavelet Tour of Signal Processing* Access Online via Elsevier.

[5] Guo Z, Zhang D and Zhang D 2010 A completed modeling of local binary pattern operator for texture classification *Image Processing, IEEE Transactions on* **19** 1657-1663.

[6] Jacomini R S, Nascimento M Z, Dantas R D, Ramos R P 2013. Classification of mass in two views mammograms: use of analysis of variance (ANOVA) for reduction of the features *Recent Patents on Medical Imaging* **3** 80-88.

[7] Osta H, Qahwaji R, Ipson S 2008 Comparisons of feature selection methods using discrete wavelet transform and support vector machines for mammogram images *In System Signals and Devices* 1-6

# An analytic reconstruction method for PET based on cubic splines

**George A Kastis**[1], **Dimitra Kyriakopoulou**[1] **and Athanasios S Fokas**[1,2]

[1] Research Center of Mathematics, Academy of Athens, Soranou Efessiou 4, Athens 11527, Greece

[2] Department of Applied Mathematics and Theoretical Physics, University of Cambridge, Cambridge, CB30WA, UK

E-mail: gkastis@academyofathens.gr

**Abstract.** PET imaging is an important nuclear medicine modality that measures *in vivo* distribution of imaging agents labeled with positron-emitting radionuclides. Image reconstruction is an essential component in tomographic medical imaging. In this study, we present the mathematical formulation and an improved numerical implementation of an analytic, 2D, reconstruction method called SRT, Spline Reconstruction Technique. This technique is based on the numerical evaluation of the Hilbert transform of the sinogram via an approximation in terms of 'custom made' cubic splines. It also imposes sinogram thresholding which restricts reconstruction only within object pixels. Furthermore, by utilizing certain symmetries it achieves a reconstruction time similar to that of FBP. We have implemented SRT in the software library called STIR and have evaluated this method using simulated PET data. We present reconstructed images from several phantoms. Sinograms have been generated at various Poison noise levels and 20 realizations of noise have been created at each level. In addition to visual comparisons of the reconstructed images, the contrast has been determined as a function of noise level. Further analysis includes the creation of line profiles when necessary, to determine resolution. Numerical simulations suggest that the SRT algorithm produces fast and accurate reconstructions at realistic noise levels. The contrast is over 95% in all phantoms examined and is independent of noise level.

## 1. Introduction

In classical tomography, one needs to reconstruct a two-dimensional (2D) function $f(x_1, x_2)$ from the following projections:

$$\hat{f}(\rho,\theta) = \int_{-\infty}^{\infty} f\left(\tau\cos\theta - \rho\sin\theta, \tau\sin\theta + \rho\cos\theta\right) d\tau, \quad -\infty < \rho < \infty, \quad 0 \le \theta < 2\pi, \qquad (1)$$

where $\rho = x_2\cos\theta - x_1\sin\theta$ and $\tau = x_2\sin\theta + x_1\cos\theta$. The above line integral is known as the 2D Radon transform. Its inversion was first constructed by Radon [1], and it is equivalent to the filtered backprojection (FBP) reconstruction formula [2, 3, 4].

Filtered backprojection (FBP) is the predominant analytic reconstruction method today. The main advantages of FBP are speed and simplicity. FBP assumes a simple Radon model where the data consist of line integrals along the radioactivity distribution, ignoring the randomness of the gamma-ray counting process. Therefore, in FBP it is difficult to incorporate complex physical phenomena such as attenuation and scatter. Noise issues are treated by selecting appropriate filtering parameters, such as the roll-off and cutoff frequencies of the reconstruction filter (usually at the expense of spatial resolution). Another disadvantage of FBP is the streak artifacts that are particular prominent near hot regions of the object.

In this study, we present an improved numerical implementation of an analytic, two-dimensional, reconstruction method called SRT (Spline Reconstruction Technique), which was presented earlier in the literature [5, 6]. This technique involves the numerical evaluation of the Hilbert transform of the sinogram via an approximation in terms of cubic splines. In this new version we have corrected singularity issues and optimized reconstruction time by employing mathematical symmetries and imposing sinogram thresholding to restrict reconstruction within object pixels. We have implemented SRT in the Open Source software library called STIR (Software for Tomographic Image Reconstruction), and have evaluated this method using simulated data from a clinical PET system.

## 2. Materials and methods

### *2.1. Spline Reconstruction Technique (SRT)*

Associated with Eq. (1) there exists the following inverse problem: Given $\hat{f}(\rho,\theta)$ for all $0 \le \theta < 2\pi$ and $-\infty < \rho < \infty$, determine the corresponding function $f(x_1,x_2)$. The relevant formula, called the *Inverse Radon transform*, can be expressed in the following form [9]:

$$f(x_1,x_2) = -\frac{1}{2\pi}\int_0^{2\pi}\left[\frac{\partial F(\rho,\theta)}{\partial \rho}\right]_{\rho=x_2\cos\theta - x_1\sin\theta} d\theta, \qquad -\infty < x_1, x_2 < \infty, \tag{2}$$

where $F(\rho,\theta)$ denotes half the Hilbert transform of $\hat{f}(\rho,\theta)$ with respect to $\rho$, *i.e.*

$$F(\rho,\theta) \equiv \frac{1}{2\pi}\oint_{-\infty}^{\infty}\frac{\hat{f}(r,\theta)}{r-\rho}dr, \qquad -\infty < \rho < \infty, \qquad 0 \le \theta < 2\pi, \tag{3}$$

and $\oint$ denotes the principal value integral.

For the numerical calculation of the Hilbert transform of $\hat{f}(\rho,\theta)$ we assume that $\hat{f}(\rho,\theta)$ has support in the interval $-1 \le \rho \le 1$ with $\hat{f}(-1,\theta) = \hat{f}(1,\theta) = 0$, and that $\hat{f}(\rho,\theta)$ is given for every $\theta$ at the $n$ points $\{\rho_i\}_1^n$. In the interval $\rho_i \le \rho \le \rho_{i+1}$, we approximate $\hat{f}(\rho,\theta)$ by cubic splines:

$$\hat{f}(\rho,\theta) = a_i(\theta) + b_i(\theta)\rho + c_i(\theta)\rho^2 + d_i(\theta)\rho^3, \ \ \rho_i \le \rho \le \rho_{i+1}, \ \ 0 \le \theta < 2\pi, \ \ i = 1,\dots,n, \tag{4}$$

where $\{a_i(\theta), b_i(\theta), c_i(\theta), d_i(\theta)\}_1^n$ are constants given in terms of $\hat{f}_i$, $\hat{f}_{i+1}$, $\hat{f}''_i$, $\hat{f}''_{i+1}$, $\rho_i$ and $\rho_{i+1}$. Note that $\hat{f}_i$ denotes the value of $\hat{f}$ at $\rho_i$, and $\hat{f}''_i$ denotes the value of the second derivative of $\hat{f}(\rho,\theta)$ with respect to $\rho$ evaluated at $\rho_i$. By substituting Eq. (4) in Eq. (3) we can derive the following expression for the partial derivative of $F(\rho,\theta)$ with respect to $\rho$:

$$\begin{aligned}\frac{\partial F(\rho,\theta)}{\partial \rho} &= \frac{1}{2\pi}\Bigg\{C(\theta) + \frac{1}{2}(\hat{f}''_n - \hat{f}''_1)\rho + D_{n-1}(\rho,\theta)\ln|\rho-\rho_n| - D_1(\rho,\theta)\ln|\rho-\rho_1| + \\ &\quad \sum_{i=1}^{n-2}[D_i(\rho,\theta) - D_{i+1}(\rho,\theta)]\ln|\rho-\rho_{i+1}|\Bigg\},\end{aligned} \tag{5}$$

where $-1 \le \rho \le 1$ and $0 \le \theta < 2\pi$. $C(\theta)$ and $\{D_i(\rho,\theta)\}_1^n$ are defined as follows:

$$C(\theta) = \sum_{i=1}^{n-1}[2c_i(\theta)\Delta_i + \frac{3}{2}d_i(\theta)(\rho_{i+1}^2 - \rho_i^2)], \quad \Delta_i = \rho_{i+1} - \rho_i \tag{6a}$$

$$D_i(\rho,\theta) = b_i(\theta) + 2c_i(\theta)\rho + 3d_i(\theta)\rho^2, \quad \rho_i \le \rho \le \rho_{i+1}, \quad i = 1,\dots,n. \tag{6b}$$

The quantities $\left\{\widehat{f}_i''(\theta)\right\}_1^n$ can be computed in terms of $\widehat{f}_i(\theta)$ by solving the following $n$ linear equations:

$$D_i(\rho_{i+1},\theta) = D_{i+1}(\rho_{i+1},\theta), \quad i=1,\dots,n-2, \quad 0\le\theta<2\pi, \tag{7a}$$

$$D_1(\rho_1,\theta) = D_{n-1}(\rho_n,\theta)=0. \tag{7b}$$

We note that the points $\{\rho_{i+1}\}_{i=1}^{n-1}$ are *removable* logarithmic singularities. This is a direct consequence of Eqs. (7*a*).

Therefore, the *Inverse Radon transform* of a function $\widehat{f}(\rho,\theta)$, using our splines approach, can be written as:

$$f(x_1,x_2) = -\frac{1}{4\pi^2}\int_0^{2\pi}\left\{C(\theta)+\frac{1}{2}(\widehat{f}_n''-\widehat{f}_1'')\rho + D_{n-1}(\rho,\theta)\ln|\rho-\rho_n| - D_1(\rho,\theta)\ln|\rho-\rho_1| + \sum_{i=1}^{n-2}[D_i(\rho,\theta)-D_{i+1}(\rho,\theta)]\ln|\rho-\rho_{i+1}|\right\}d\theta, \tag{8}$$

where $C(\theta)$ and $D_i(\rho,\theta)$ are given by Eqs. (6*a*) and (6*b*), respectively.

We note that in the construction of the so-called 'natural' splines, one requires continuity of the first derivatives (the set of equations in Eq. (7*a*)), as well as the conditions $\widehat{f}_1''=\widehat{f}_n''=0$. The former requirement implies that there *cannot* be logarithmic singularities at the interior points $\rho=\rho_i, \quad i=2,\dots,n-1$. In order to eliminate the logarithmic singularities at the end points $\rho_1=-1$ and $\rho_n=1$, we impose the set of equations in (7*b*) (instead of $\widehat{f}_1''=\widehat{f}_n''=0$). In this way we construct a set of splines 'custom made' for the evaluation of the Hilbert transform.

In order to numerically evaluate Eq. (8) for $f(x_1,x_2)$ from the given sinogram $\widehat{f}(\rho,\theta)$ and the known detector locations $\rho$ and projection angles $\theta$, we first evaluate the second derivatives of the sinogram $\widehat{f}''(\rho_i,\theta)$ by solving the system of linear equations in Eqs. (7*a*) and (7*b*). Then, we calculate the term $C(\theta)$ for all $\theta$'s and $\rho_i$'s. Finally, for any $x_1$, $x_2$, we compute $\rho$ for every $\theta$, and then $f(x_1,x_2)$ using Eq. (8).

The reconstruction time of this algorithm can be reduced by employing object specific information that is 'hidden' in the sinogram. In this respect we consider the important case that the boundary of the object is convex. In this case, a pixel which is outside the boundary spanned by an object and hence has zero value, can be singled out from the sinogram by first identifying the detector locations for all angles $\theta$ that receive contribution from this pixel; then, for every $(x_1,x_2)$, if there is even one $\theta$ such that $\widehat{f}(\rho,\theta)=0$, it follows that $f(x_1,x_2)$ must be zero.

Using the above condition we can restrict the reconstruction process only to pixels within the object boundary and exclude all zero pixels outside the object. In this way, in addition to improving considerably the reconstruction time, we can also obtain a 'clean' reconstruction without any streak artifacts outside the object. Note that for real data, the condition $\widehat{f}(\rho,\theta)=0$ must be replaced by $\widehat{f}(\rho,\theta)\le$ *threshold*, since in the presence of system noise, pixels outside the object's boundary in the sinogram can have values greater than zero. The *threshold* can be determined manually by examining the sinogram values outside the object boundary, or automatically using thresholding selecting techniques.

We have implemented the above algorithm in STIR [7], which is an object-oriented library using C++. For this purpose, we have utilized STIR's built-in classes and also have created a new class to accommodate our algorithm. The speed of the algorithm was optimized by using certain mathematical symmetries which exist in the standard case of constant detector spacing, which is indeed the case for the Discovery ST PET scanner employed in this study.

### 2.2. Simulation studies

We have used STIR to simulate the GE Discovery ST PET scanner [8]. In order to evaluate the performance of the SRT algorithm we have employed an image quality (IQ) phantom, a Jaszczak phantom and a slice of the digital 3D Hoffman phantom.

The IQ phantom simulates the human torso and it consists of two circular cold regions (with diameter of 38 mm and 32 mm) and four circular hot regions (with diameters of 25 mm, 19 mm, 15 mm, and 12 mm) inside a larger warm region that simulates background. The radioactivity ratio between hot regions and the surrounding background is 4:1 for the three hot regions. The Jaszczak phantom is separated into six sections. Each section has cold lesions of different size uniformly arranged to form an equilateral triangle (with circle diameters of 27 mm, 18 mm, 16 mm, 13 mm, 11 mm, 9 mm) inside a hot region 30.67 cm in diameter. The center-to-center distance between lesions of equal diameter is twice their diameter. The Hoffman phantom simulates the radioactivity distribution of a cerebral PET study and it consists of three distinct radioactive regions: Gray Matter (GM), White Matter (WM), and Cerebrovascular Fluid (CSF).

After placing the three phantoms in the center of the scanner, 2D projection data were generated in STIR using a ray tracing technique with 10 rays per detector. Scatter and attenuation were not modeled. These sinograms provide the noiseless PET measurements. For each noiseless sinogram, 20 Poisson noise realizations have been generated at 3 different levels (NL1-NL3), where NL3 corresponds to the highest noise level. No filtering or smoothing has been applied to the SRT reconstructed images post reconstruction.

The contrast for the hot and cold regions were calculated according to [8]. For the IQ phantom, the contrast for the three smallest hot lesions was determined in each noise level by drawing an appropriate size circular ROI to determine the average counts measured in hot lesion and background, respectively. These values were averaged over all noise realizations. Similar procedure was followed for the Jaszczak phantom, where the contrast for the 18 mm, 16 mm and 13 mm cold lesions was determined by calculating the mean activity in all lesions of the section. Furthermore, a line profile was drawn through the smallest lesions of the Jaszczak phantom at NL3, in order to determine the capability of SRT to distinguish two cold lesions placed close to each other. The line profile has been normalized to its maximum value.

## 3. Results

The reconstruction time for SRT was 2.1 sec per sinogram, executed on a PC with Intel® Core™i7-920 Processor. We note that no parallel programming or other accelerating techniques have been employed.

Visual comparisons of the reconstructed images with no noise, as well as with noise are shown in Fig. 1. The noisy images presented are representative reconstructions of one realization at the specific noise level. Note that the SRT algorithm can generate negative values in the reconstructed images. In all images presented, the all-black color corresponds to zero values, whereas the white color represents the maximum value of the distribution. No streak artifact are present outside the object boundary.

The contrast for the three smallest hot spheres of the IQ phantom (19 mm, 15 mm, and 12 mm) as a function of noise level, is presented in Fig. 2A. Furthermore, the contrast for three cold sections for the Jaszczak phantom (with diameters of 16 mm, 13 mm, 11 mm) is illustrated in Fig. 2B. The SRT algorithm exhibits high contrast (over 95%) in all cold and hot lesions independently of noise level. Especially, SRT delivers almost 100% contrast for the cold regions.

## 4. Conclusion

In this work, we have presented the mathematical formulation and numerical implementation of a new, analytic, 2D, image reconstruction method for parallel beam geometry. Simulated images

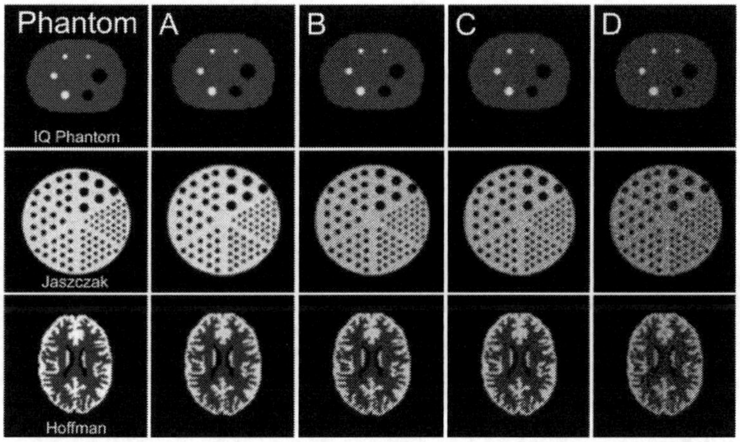


**Figure 1.** The original phantoms and respesentative images at various noise levels (The noise level increases moving from left to right): (A) No Noise, (B) NL1, (C) NL 2, and (D) NL3.

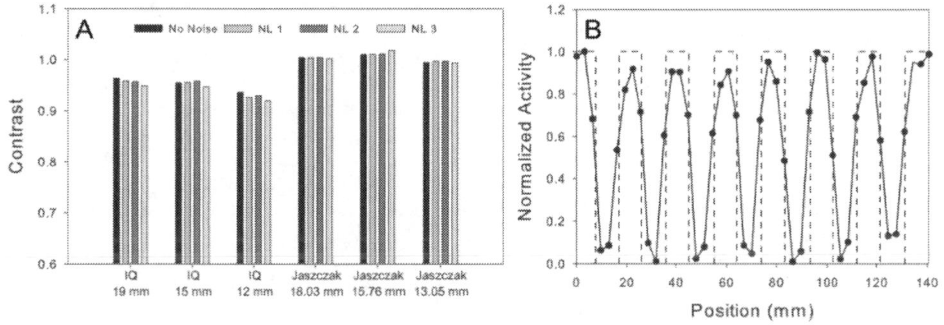


**Figure 2.** A) Contrast for the hot lesions of the IQ phantom and the cold lesions of the Jaszczak phantom. B) Line profile across the smallest cold lesions of the Jaszczak phantom at NL3.

have been reconstructed for various noise levels and have been evaluated in terms of cold and hot lesion contrast. Overall, the SRT algorithm provided high contrast images with no streak artifacts. Furthermore, unlike other analytic reconstruction algorithms, the reconstruction time of SRT is comparable with that of FBP.

## References

[1] Radon J 1917 Uber die Bestimmung von Funktionen durch ihre Integralwerte langs gewisser Mannigfaltigkeiten *Ber. Verh. Sachs. Akad. Wiss.* **67** 262
[2] Barrett H H 1984 *Progress in Optics XXI* ed E Wolf (Amsterdam: Elsevier) p 217
[3] Kak A C and Slaney M 1988 *Principles of computed tomographic imaging* (New York: IEEE Press)
[4] Natterer F 1986 *The mathematics of computerized tomography* (New York: John Wiley)
[5] Fokas A S, Iserles A and Marinakis V 2006 *J. R. Soc. Interface* **3** 45
[6] Kastis G A, Gaitanis A, Fernandez Y, Kontaxakis G *et al.* 2010 *IEEE Nuclear Science Symposium Conf. Record (Knoxville)* p 3282
[7] Thielemans K, Tsoumpas C, Mustafovic S, Beisel T *et al.* 2012 *Phys. Med. Biol.* **57** 867
[8] Bettinardi V, Danna M, Savi A, Lecchi M, Castiglioni I *et al.* 2004 *Eur. J. Nucl. Med. Mol. Imaging* **31** 867
[9] Fokas AS and Novikov RG 1991 *C. R. Acad. Sci. Paris* **313 75**

# Competing computational approaches to reaction-diffusion equations in clusters of cells

**Sabrina Stella**[1], **Roberto Chignola** [2] **and Edoardo Milotti**[1,3]

[1]Department of Physics, University of Trieste, Italy
[2]Department of Biotechnology, University of Verona, Italy
[3]Istituto Nazionale di Fisica Nucleare, Sezione di Trieste, Italy

E-mail: sabrina.stella@ts.infn.it

**Abstract.** We have developed a numerical model that simulates the growth of small avascular solid tumors. At its core lies a set of partial differential equations that describe diffusion processes as well as transport and reaction mechanisms of a selected number of nutrients. Although the model relies on a restricted subset of molecular pathways, it compares well with experiments, and its emergent properties have recently led us to uncover a metabolic scaling law that stresses the common mechanisms that drive tumor growth. Now we plan to expand the biochemical model at the basis of the simulator to extend its reach. However, the introduction of additional molecular pathways requires an extensive revision of the reaction-diffusion part of the C++ code to make it more modular and to boost performance. To this end, we developed a novel computational abstract model where the individual molecular species represent the basic computational building blocks. Using a simple two-dimensional toy model to benchmark the new code, we find that the new implementation produces a more modular code without affecting performance. Preliminary results also show that a factor 2 speedup can be achieved with OpenMP multithreading, and other very preliminary results indicate that at least an order-of-magnitude speedup can be obtained using an NVidia Fermi GPU with CUDA code.

## 1. Introduction

Biophysical models of cancer are widely believed to be important tools in the understanding of the disease, and currently there are two main – and often competing – approaches: analytical models [1], and numerical models [2, 3]. We think that both approaches are useful and important, and we have recently given an overview of a synergetic approach that combines both methods [4]. While the difficulties of the analytical methods are well understood, the actual feasibility of computational modeling depends on a plethora of details, and here we discuss an interesting numerical implementation of the reaction-diffusion part of our program for the simulation of the growth of small avascular solid tumors [5]. The program includes the definition of single cells with their phenotype, and of their metabolic activity, growth, and proliferation. The simulations produce 3-D lattice free cell aggregates with nearly spherical shape which can eventually contain more than one million cells [6]. The diffusion and the interaction of a limited, but accurately chosen set of molecular species, within cells and the environment is described by a system of partial differential equation, and the numerical integration of this system is one of the important aspect of the simulation process. The diffusion problem is naturally discretized over the network of cells' centers. In this context it is natural to implement an object-oriented code where single cells are the basic computational objects, and the masses of metabolites inside cells are object data. Here we describe a new abstract model with a reversed logic, where the molecular species are the basic objects while cells are a mere spatial scaffolding; this approach makes the code more reliable, thanks to a better encapsulation, and more easily expandable.

## 2. Material and Methods

We explain the inner workings of the new code with the aid of the 2-dimensional toy model which is illustrated in fig.1. It consists of a grid of cells (represented by circles) in which the diffusion

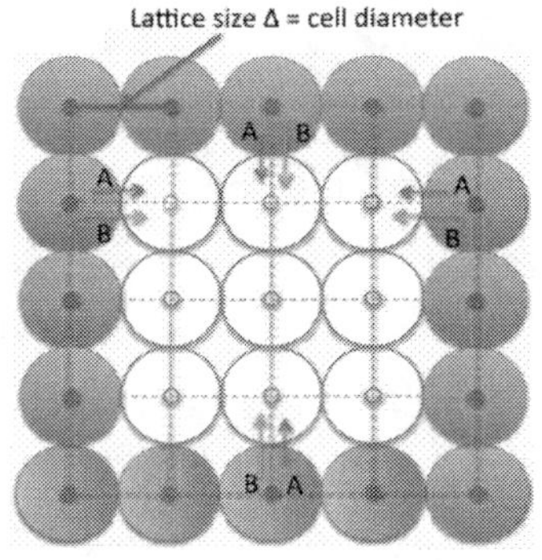


$$\frac{\partial \rho^{(A)}}{\partial t} = D_A \nabla^2 \rho^{(A)} - \frac{V_{max}^{(A)} \rho^{(A)}}{K^{(A)} + \rho^{(A)}}$$

$$\frac{\partial \rho^{(B)}}{\partial t} = D_B \nabla^2 \rho^{(B)} - \frac{V_{max}^{(B)} \rho^{(B)} \rho^{(A)}}{K^{(B)} + \rho^{(B)} \rho^{(A)}}$$

**Figure 1.** Geometry of the toy model. Cells are arranged in a square two-dimensional lattice. Two different chemical species diffuse throughout the cell aggregate, and the initial concentration values are set to 1 (arbitrary units) along the edge of the lattice and vanish in the remaining cells.

of two different molecular species, A and B, takes place. The initial concentration values are set to 1 (arbitrary unit) along the edge and remain constant over the time, whereas the inner cells has a vanishing concentration value. The molecular species can diffuse throughout the lattice and, at the same time, they take part to reaction processes; the evolution of their concentrations in the lattice can be described by second order partial differential equations

$$\frac{\partial \rho^{(k)}(x,y,t)}{\partial t} = D^{(k)} \nabla^2 \rho^{(k)}(x,y,t) - f^{(k)}[\boldsymbol{\rho}(x,y,t)] \tag{1}$$

where the two terms on the r.h.s. are respectively the diffusion and the reaction terms, and $\boldsymbol{\rho} = (\rho^{(A)}, \rho^{(B)})$ is the vector of the concentrations of two interacting chemicals. The parameters $D^{(k)}$ $(k = A, B)$ are the diffusion constants, whereas the functions $f^{(k)}$ corresponds to a single and double substrate Michaelis-Menten terms respectively for the molecular species A and B as illustrated in fig.1. The equations are solved numerically by discretizing the spatial domain with a step size equal to the diameter of the cells; a $N \times N$ sets of ordinary differential equation is obtained, where $N$ is the lattice size. A stable solution can be obtained using the implicit Euler method [5] which then yields a system of coupled nonlinear equations. These equations are then solved with the Newton-Raphson method; the iterations are stopped when a predetermined accuracy value is finally achieved over the whole lattice.

## 3. C++ implementation and test runs

The new abstract model has been implemented using two classes that we call `System` and `Chemical`. The class `System` stores information about the environment, in this case the square lattice, and it manages the collective diffusion of all the initially defined molecular species. The class `Chemical` represents a single diffusing molecular species and it stores its spatial concentration map in a single STL vector where the $n$-th component corresponds to the $n$-th cell. `Chemical` manages the time evolution of the concentration using the method `Interact()` that calculates the concentration of the molecular species in the whole lattice at time $t + \Delta t$ from the value known at time $t$. `Interact()` is specific to each different molecular species, and we use the constructor to define a pointer to external functions that will be successively dereferenced in the body of `Interact()`. In particular two external functions are used to determine the numerical integration method (which might eventually be different from the straightforward implicit Euler method) and to identify the specific metabolic function. Using these classes the toy model is implemented according to the following pseudocode:

1 create the environment where diffusion takes place: define lattice size, lattice spacing, and time steps;

2 define the biochemical system: create `Chemical` objects representing the molecular species A and B;

3 iterate the diffusion steps by calling the `System::CollectiveDiffusion()` method.

Figure 2 shows a concentration map for the molecular species A (roughly corresponding to glucose), obtained with the toy model, with a $100 \times 100$ lattice, with lattice spacing (i.e., cell diameter) 5 $\mu$m, and with the parameters specified in the cell legend.

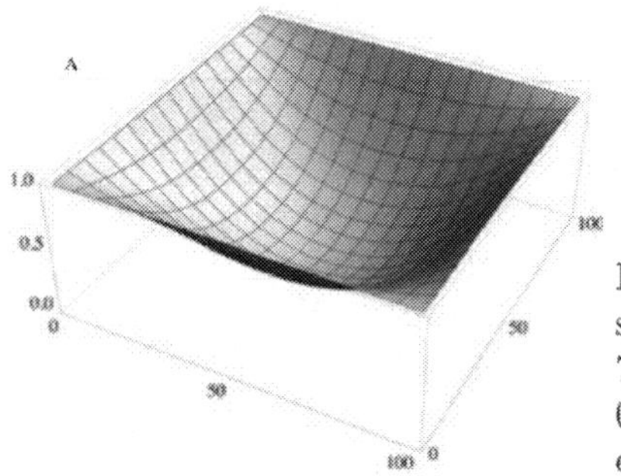


**Figure 2.** Concentration map of the molecular species A for simulated time of 3600 s, $D = 7 \cdot 10^{-6}\text{cm}^2 s^{-1}$, $V_{max} = 1.2 \cdot 10^{-4}\text{g cm}^{-3} s^{-1}$; $K = 0.27 \cdot 10^{-3}\text{g cm}^{-3}$. The general shape follows the expected behavior.

Both to check the correctness of the results and to compare the respective performances and structural properties, we implemented the toy model also with a conventional scheme in which the cells are the basic objects (hereafter we call this the *cell-based* model). In the *cell-based* model the concentration values of A and B are data of the `Cell` class, and the corresponding parameters are data of an additional class `CellType` that represents the cell's phenotype. Finally a `CellSystem` class is the equivalent of `System` in previous abstract scheme. The *chemical-based* approach (that is the previously describe one) has a strikingly better encapsulation, since the introduction of a new molecular species does not require any change in the already defined classes but only the introduction of a new metabolic function in a header file. No other additional code is required. On the contrary, in the *cell-based* model the introduction of a new molecular species requires the modification of the whole set of classes.Table 1 shows that the new scheme does not decrease the performance either. Another set of runs with variable lattice size shows that for both abstract models the total execution time $T$ has a faster-than-linear growth with respect to the total number of lattice sites $N$, $T \approx 2.15 \cdot 10^{-5} N + 2.16 \cdot 10^{-7} N^{1.5}$, where the $N^{1.5}$ term is due to the global convergence condition in the Newton algorithm, since the number of loops required to reach convergence is proportional to the number of lattice sites and to the linear lattice size, i.e., to $N \times \sqrt{N} = N^{1.5}$. These considerations show that the total execution time is dominated by the numerical solution algorithm, and not by memory transfers (more efficient in the *chemical-based* version).

**Table 1.** Execution times [s] of the diffusion process using the cell-based and chemical-based method, for different simulated diffusion times in a 100 × 100 lattice. The simulations were run on a quad core MacBook Pro using the compilers Intel C++ XE 13.0 and GCC 4.7.

| | Intel XE 13.0 | | GCC 4.7 | | GCC 4.7 (opt -O ) | |
|---|---|---|---|---|---|---|
| Diff.time (s) | Cell-b | Chem-b | Cell-b | Chem-b | Cell-b | Chem-b |
| 3600 | 0.43 | 0.47 | 1.25 | 1.24 | 0.44 | 0.46 |
| 5400 | 0.55 | 0.58 | 1.56 | 1.56 | 0.56 | 0.58 |
| 7200 | 0.67 | 0.70 | 1.90 | 1.91 | 0.67 | 0.70 |
| 9000 | 0.78 | 0.83 | 2.25 | 2.25 | 0.78 | 0.82 |

## 4. Code parallelization

The new code has high intrinsic parallelism: using straightforward OpenMP preprocessor directives we obtain a speed-up of about 1.95x in a simulation with 4 different molecular species on a 100x100 lattice, however, this can be pushed much further using a GPU, and the code was slightly modified to exploit the power of a Quadro 4000 NVidia GPU. In this preliminary implementation that does not use any optimization procedure, such as exploiting texture or shared memory [7] we obtain a speedup of about 5x for a lattice size of 1000 × 1000 and total simulation time equal to 3600 s. A comparison of the execution times for the diffusion-reaction processes implemented on CPU and GPU is shown in table 2.

**Table 2.** Execution times [s] of the diffusion-reaction process implemented on GPU and CPU with or without optimization flag for different lattice size and simulation time = 3600s.

| Lattice size | GPU | CPU (no opt) | speed-up | CPU (-O3) | speed-up |
|---|---|---|---|---|---|
| 96 × 96 | 0.73 | 1.2 | 1.6 | 0.46 | <1 |
| 296 × 296 | 3.04 | 20.8 | 6.8 | 7.8 | 2.6 |
| 600 × 600 | 13.53 | 144.8 | 10.6 | 55.2 | 4.8 |
| 1000 × 1000 | 46.02 | 624.8 | 13.5 | 239.5 | 5.2 |

## 5. Conclusions

The new computational scheme shall be used in the numerical simulator of tumor spheroids. It is more easier extendible, allowing for an easier introduction of new chemicals, and more robust, providing a better encapsulation of the code, without loss in terms of performance. Moreover, the code running on GPU can be further optimized exploiting its hierarchical memory, to significantly increase the execution speed.

## References


[1] Araujo R and McElwain D 2004 *Bulletin of mathematical biology* **66** 1039–1091

[2] Chignola R, FABBRO A D, Farina M and Milotti E 2011 *Journal of Bioinformatics and Computational Biology* **9** 559–577

[3] Chignola R and Milotti E 2012 *AIP Advances* **2** 011204–011204

[4] Milotti E, Vyshemirsky V, Sega M, Stella S, Dogo F and Chignola R 2013 *IEEE/ACM TCBB*

[5] Milotti E, Del Fabbro A and Chignola R 2009 *Computer Physics Communications* **180** 2166–2174

[6] Milotti E and Chignola R 2010 *PLoS One* **5** e13942

[7] Jr F M, Izsákb F, Mészárosc R and Lagzic I 2011 *Chemometrics and Intelligent Laboratory Systems* **108** 76 – 85

# An ensemble of dissimilarity based classifiers for Mackerel gender determination

**A. Blanco, R.Rodriguez, I. Martinez-Maranon**
AZTI-Tecnalia Astondo Bidea, Edificio 609 - Parque Tecnológico de Bizkaia - 48160 Derio (Bizkaia)
E-mail: ablanco@azti.es

**Abstract.**
Mackerel is an infravalored fish captured by European fishing vessels. A manner to add value to this specie can be achieved by trying to classify it attending to its sex. Colour measurements were performed on Mackerel females and males (fresh and defrozen) extracted gonads to obtain differences between sexes. Several linear and non linear classifiers such as Support Vector Machines (SVM), $k$ Nearest Neighbors ($k$-NN) or Diagonal Linear Discriminant Analysis (DLDA) can been applied to this problem. However, theyare usually based on Euclidean distances that fail to reflect accurately the sample proximities. Classifiers based on non-Euclidean dissimilarities misclassify a different set of patterns. We combine different kind of dissimilarity based classifiers. The diversity is induced considering a set of complementary dissimilarities for each model. The experimental results suggest that our algorithm helps to improve classifiers based on a single dissimilarity.

## 1. Introduction
Mackerel is a pelagic specie plentiful in the Northeast atlantic waters. The idea of distinguishing and selecting fish species is justified by the high value of roes compared with the price of the fish. The method chosen is based on the color difference between the gonads of male and female fish consists in a sensor that allows to characterize the spectrum obtained after the incidence of electromangetic radiation into the gonads

A variety of machine learning techniques have been proposed such as SVM, $k$ NN or DLDA. However they rely on the use of the Euclidean distance that fails often to reflect accurately the proximities among samples. The classifiers have been extended to work with non-Euclidean dissimilarities. In spite of this, the resulting algorithms misclassify a different set of patterns. Combining non-optimal classifiers can help to reduce particularly the variance of the predictor. To achieve this goal, different versions of the classifier are usually built by sampling the patterns or the features. Nevertheless, this kind of resampling techniques reduce the size of the training set. We build the diversity of classifiers considering three models and a set of complementary dissimilarities for each model. Finally, the classifiers are aggregated using a voting strategy.

## 2. Combination of dissimilarity based classifiers
An important step in the design of a classifier is the choice of a proper dissimilarity that reflects the proximities among the objects.

SVM and DLDA are not able to work directly from a dissimilarity matrix. The SVM algorithm is extended to work from a dissimilarity matrix by defining a kernel of dissimilarities. DLDA is adapted following a different approach by embedding the patterns in a Euclidean space.

The DLDA is a variant of the Linear Discriminant Analysis with diagonal and constant covariance matrices. A vectorial representation of the data should be obtained.

Our method builds the diversity of classifiers considering three different kind of models such as SVM, $k$-NN and DLDA and several dissimilarities.

Our combination algorithm proceeds as follows: A set of complementary dissimilarities are computed. For the SVM algorithm, the kernel of dissimilarities is computed and the optimization problem is solved in the usual way. $k$-NN is able to work directly from a dissimilarity matrix. DLDA algorithm, the dissimilarities should be embedded in an Euclidean space via a Multidimensional Scaling algorithm. The ensemble of classifiers is aggregated by a standard voting strategy.

## 3. Results and Discussion

Mackerel in their best maturity stages from Basque fishing ports was used to the test the ensemble. Colour measurements was performed on females and males extracted gonads. The reflectance spectra were recollected and twelve colour parameters were analyzed to characterize differences between both sexes.

Two datasets were considered. The first consisted on fresh mackerel samples. 1006 samples (from 2007-2010). The second dataset consisted on 1439 samples of defrozen Mackerel.

The dissimilarities have been computed without normalizing not to increase the correlation among them.The algorithm chosen to train the SVM is C-SVM. The $C$ regularization parameter has been set up by ten fold-crossvalidation. We have considered non-linear kernels. The number of neighbors for $k$-NN algorithm is estimated by cross-validation.

**Table 1.** Empirical results for the best single classifier for each technique.

| Technique | Datasets | Error % | False negative % | False positive % |
|---|---|---|---|---|
| SVM RBF (Correlation) | Fresh | 7.05% | 5.16% | 1.88% |
| SVM RBF (Manhattan) | Defrozen | 2.51% | 1.47% | 1.04% |
| $K$-NN (Manhattan) | Fresh | 8.84% | 5.16% | 3.67% |
| $K$-NN(Chi-squared) | Defrozen | 2.51% | 0.86% | 1.65% |
| DLDA (Correlation) | Fresh | 8.74% | 2.98% | 5.76% |
| DLDA (Correlation) | Defrozen | 3.67% | 0.19% | 3.47% |
| | Fresh | 6.16% | 3.280% | 2.88% |
| **Combination** | Defrozen | 1.78% | 0.19% | 1.58% |

The algorithms have been evaluated considering the global errors and the false negative and positive errors. Both have been estimated by ten-fold cross-validation

Table 1 shows the experimental results for the best single classifier for each technique and the proposed technique and the following conclusions can be drawn:

- The dissimilarity that minimizes the error depends strongly on the classifier and on the particular dataset considered. No dissimilarity outperforms the others for a wide range of models and datasets. Hence the choice of a proper dissimilarity is not an easy task for human experts.
- The combination strategy proposed outperforms the misclassification errors of the best single classifiers. In particular, the ensemble of classifiers improves significantly the SVM

algorithms for the two datasets considered. We also report that our method improves the best $k$-NN classifier for Fresh and Defrozen Mackerel. Finally, DLDA is also improved for both datasets.

## 4. Conclusions

A new classification scheme that consider different dissimilarities and different classifiers have been proposed. The algorithm has been applied to the Mackerel gender determination in fresh and defrozen species Our method helps to reduce the misclassification errors of the best single classifier for both datasets. As future research trends, a new classification scheme will be developed to take into account different classes of kernels a nd new combination strategies will be applied.

[1] Blanco A, Manuel Martn-Merino M , De Las Rivas J 2008, *Innnovations in Hybrid Intelligent Systems.* 393 – 400.
[2] Dudoit S,Fridlyand J, and Speed T 2002, *Journal of the American Statistical Association,* B **97** 77–87.
[3] Furey T, Cristianini N, Duffy N, Bednarski D, Schummer M, and Haussler D 2000 *Bioinformatics.* B **16** 906–914.
[4] Kuncheva L I 2004. Pattern Classifiers.
[5] Lucio P 1997 *Biology and Behaviour.* B **21** 1–23 .
[6] Pekalska E , Paclick P and Duin R 2001, of Machine Learning Research. B **2** 175–211, .
[7] Valentini G and Dietterich T 2004 *Journal of Machine Learning Research.* B **5** 725–775.
[8] Vapnik V 1998 *Statistical Learning Theory* .

# Delaunay Tetrahedralization of the Heart Based on Integration of Open Source Codes

**E Pavarino[1], L A Neves[1], J M Machado[1], M F de Godoy[2], Y Shiyou[3], J C Momente[1], G F D Zafalon[1], A R Pinto[1], C R Valêncio[1] and M Z do Nascimento[4]**

[1] Department of Computer Science and Statistics (DCCE), São Paulo State University (UNESP), São José do Rio Preto, Brazil.

[2] Interdisc. Center for the Study of Chaos and Complexity (NUTTECC), Faculty of Medicine of São José do Rio Preto (FAMERP), São José do Rio Preto, SP, Brazil.

[3] College of Electrical Engineering Zhejiang University Hangzhou, People Republic of China.

[4] Faculty of Computer Science (FACOM), Federal University of Uberlandia (UFU), Uberlandia, MG, Brazil.

E-mail: `leandro@ibilce.unesp.br`

**Abstract.** The Finite Element Method (FEM) is a way of numerical solution applied in different areas, as simulations used in studies to improve cardiac ablation procedures. For this purpose, the meshes should have the same size and histological features of the focused structures. Some methods and tools used to generate tetrahedral meshes are limited mainly by the use conditions. In this paper, the integration of Open Source Softwares is presented as an alternative to solid modeling and automatic mesh generation. To demonstrate its efficiency, the cardiac structures were considered as a first application context: atriums, ventricles, valves, arteries and pericardium. The proposed method is feasible to obtain refined meshes in an acceptable time and with the required quality for simulations using FEM.

## 1. Introduction

Studies using the Finite Element Method (FEM) may be used to improve cardiac ablation procedure, such as radiofrequency cardiac ablation, and reduce possible complications. The radiofrequency cardiac ablation is being used for over 10 years to treat tachycardia, atrial fibrillation and atrial flutter [1]. However, the esophageal injury is a common damage: the consequence for the patient is death due to the internal bleeding. The focus of the problem is monitoring the temperatures of the tissues and FEM simulations have contributed significantly to the improvement of this technique [2, 3, 4]. For such simulations, the finite element meshes should consider the anatomical and histological features of the target structures and its quality has to be acceptable. Some methods or tools used to generate meshes of biological structures are still limited by constructing non-detailed models or due to limitation in the use condition [2, 4, 5].

Therewith, strategies to integrate open source software as an alternative to solid modeling and automatic mesh generation are demonstrated in this paper. Some related ideas were previously presented in [6]. The novelty proposed here is the pericardium covering the internal layers, with detailed anatomical features. This was motivated by the importance that internal and external cardiac structures represent for studies of cardiac ablation. This approach brings many advantages for the community interested in FEM simulations and cardiac ablation, providing means to perform the entire pre-processing step (solid modeling and mesh generation) of complex geometrical problems.

## 2. Methodology

The choices of cardiac structures were motivated by the complex geometrical domain and by the clinical relevance that these structures represent for investigating esophageal injury. The model defined in our study was composed of three main parts: right internal portion (venous), left internal portion (arterial) and external portion (pericardium). Each internal portion has an atrium, a ventricle, an artery and valves. The dimensions of these structures are presented in Table 1 [7], which were defined in cardiac diastole. The pericardium was represented without thickness, because it is a very thin tissue which surrounds the heart. Also, an echocardiographic image was used as another reference during the construction process of the 3D model.

The Blender package was chosen as the solid modeler for this study. This package is a multiplatform integrated system of tools and contains resources to export and import objects in different formats using Python scripts. There are different methods for generating 3D meshes. We chose the Delaunay algorithm [8] for being one of the most popular and one of the most efficient algorithms [9], available in the TetGen package [8]. Both packages are under public licenses.

To ensure the integration of Blender with TetGen, two construction strategies must be respected: (1) definition of faces and (2) density control of vertices. The first strategy consists of choosing the appropriate type of face to discretize a solid. Regular or non-complex regions of the domain must be discretized with quadrilateral or rectangular faces (eg atriums and ventricles). Critical or complex regions must be discretized with triangular faces (eg bifurcations, regions of connections). The second strategy defines the number of nodes in a region. The increasing number of vertices allows a better representation of curved regions, smoothing the direction transition and respecting the features defined (eg arteries). Regions represented with triangular faces, quadrilateral or rectangular faces, and with increased density of vertices are shown in Figure 1(a). A domain discretized in Blender is stored in its native file format: *.blend*. This file is composed by datablocks (file structures which store data used to discretize a solid). It could be of object type, mesh type, scene type and other types. The final model is composed by 17 datablocks of object, mesh and material types (Figure 1(b)).

The exportation of the model was accomplished by a script written in Python. This script writes the information of all datablocks in a single *.poly* file, required as input by TetGen. The *.poly* file is composed by four lists: lists of nodes, faces, volume holes and region attributes (such as color markers). For a better visualization, a distinct color maker was assigned for each structure of the model, totalizing 11 markers.

The outputs of the mesh generation process are three files: *.node*, *.face* and *.ele* files,

**Table 1.** Morphometry applied in the hearth model (diastolic dimensions of an adult heart [7]).

| Structure | Dimension (cm) | Longest Axis (cm) |
|---|---|---|
| Left Atrium | 2.1 – 3.0 | 4.6 |
| Right Atrium | 1.9 – 4.4 | 5.3 |
| Left Ventricle | 3.1 – 3.8 | 4.3 |
| Right Ventricle | 2.1 – 3.7 | 4.2 |
| Aorta, Pulmonary Artery and Bicuspid Valve | 2.5 | 2.5 |
| Tricuspid Valve | 3.0 | 3.0 |
| Aortic Valve | 2.4 | 2.4 |
| Pulmonary Valve | 2.3 | 2.3 |
| Interventricular Septum | 0.8 – 1.2 | 4.0 |
| Pericardium | — | 10.6 |

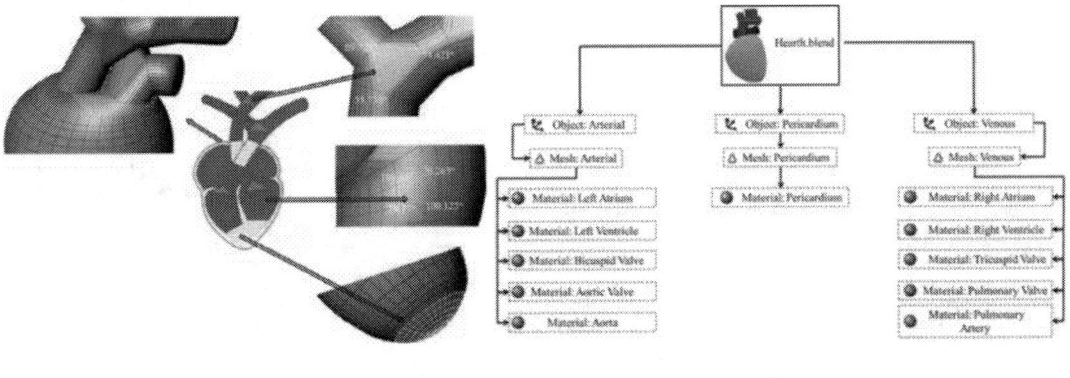


(a) (b)

**Figure 1.** Visualization of the location of different types of faces (a) and hierarchy of datablocks model (b).

containing information, about vertices, faces and tetrahedral elements of the resulting mesh, respectively.

## 3. Results and Discussions

FEM simulations require highly refined meshes of simple or complex domains, and its generation in an acceptable time. The methodology proposed in this paper performs the requirements. The complex geometry of the cardiac structures allowed testing the feasibility of the combined use of anatomical features, free and open source packages and the application of the defined strategies (Figure 2). The *.poly* file stores data of 4,363 vertices, 8,718 faces and 11 color markers. This file is shown in Figure 2(b). A first mesh was generated in 9.235 seconds with 189,125 tetrahedral elements and is shown if Figure 2(c).

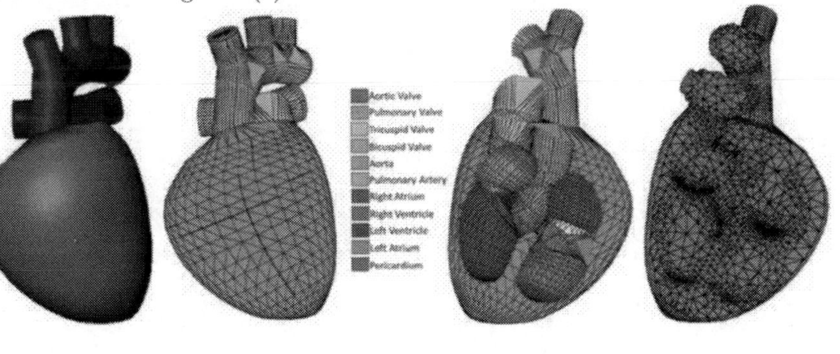


(a) (b) (c)

**Figure 2.** Resulting model: (a) *.blend* file with assigned boudary markers, (b) *.poly* file with assigned color markers and (c) .ele file with tetrahedral mesh.

Meshes were generated through successive refinements of the initial mesh (Figure 2(c)): this strategy allowed estimating the processing time growth due to the number of tetrahedral elements (Figure 3(a)). The most refined mesh was generated in 7.72 minutes with 17,645,744 elements. The quality of the mesh is an important aspect in proposals designed to generate meshes through FEM simulations. Adopting the criterion of measurement of dihedral angles, the results are relevants. This can be verified in a histogram constructed with the total of dihedral angles present in ranges of angles, as well as identification of the smallest and largest values involved. A highly refined mesh is shown in Figure 3(b) and the corresponding histogram is presetend in Figure 3(c). In this histogram, the tetrahedral elements were constructed with dihedral angles whose values are between 5 and 170 degrees, with 32.88 percent belonging between 80 and 110 degrees, which are close to values set on a regular tetrahedron [10].

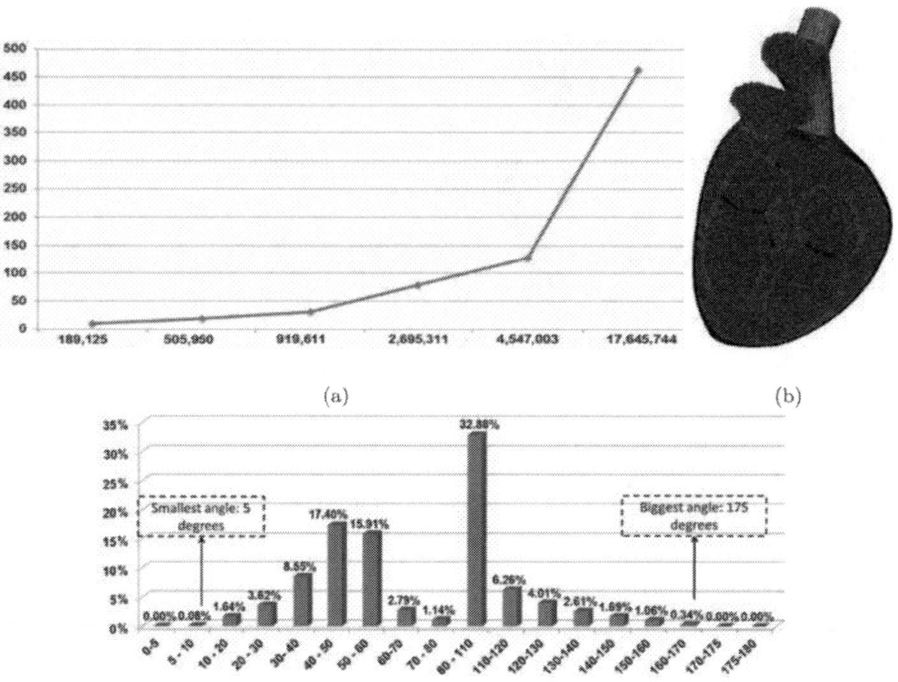


(a) (b)

(c)

**Figure 3.** Processing time (seconds) (y axis) versus number of elements (x axis) is shown in (a), refined mesh with 4,547,003 elements is shown in (b) and total of dihedral angles (y axis) inside prescribed angle intervals in degrees (x axis) is shown in (c).

## 4. Conclusions

The proposal presented in this paper considers strategies for solids construction and automatic mesh generation through integration of open source packages. We have proved the efficiency of this integration. Highly refined and detailed meshes (atriums, ventricles, valves, arteries and pericardium) were generated in acceptable times and with the required quality for simulations using FEM: the values of dihedral angles are compatible with the specialized literature. Thus, our proposal provides a significant contribution to mesh generation for studies using FEM which can be applied in different areas, such as studies to improve cardiac ablation procedures.

## 5. Acknowledgments

This work was financially supported by PROPe/UNESP (Pró-Reitoria de Pesquisa/UNESP) and Fundação de Amparo à Pesquisa do Estado de São Paulo (FAPESP).

## 6. References

[1] Sosa E and Scanavacca M 2005 *Journal of cardiovascular electrophysiology* **16** 249–250
[2] Haemmerich D and Webster J G 2005 *Biomedical engineering online* **4** 42
[3] Rodríguez I, Lequerica J L, Berjano E J, Herrero M and Hornero F 2007 *Physiological measurement* **28** 453
[4] Yang W, Fung T C, Chian K S, Chong C K *et al.* 2007 *World Journal of Gastroenterology* **13** 1352
[5] Sun Z and Chaichana T 2010 *Korean Journal of Radiology* **11** 95–106
[6] Pavarino E, Neves L, Machado J, de Godoy M, Shiyou Y, Momente J C, Zafalon G F, Pinto A and Valêncio C R 2013 *International journal of biomedical imaging* **2013**
[7] Bonow R O, Mann D L, Zipes D P and Libby P 2005 *Braunwald's Heart Disease: A Textbook of Cardiovascular Medicine, 2-Volume Set* (Saunders)
[8] Si H and TetGen A 2006 *Weierstrass Institute for Applied Analysis and Stochastic, Berlin, Germany*
[9] Lizier M, Shepherd J, Nonato L, Comba J and Silva C 2008 Comparing techniques for tetrahedral mesh generation *Inaugural International Conference of the Engineering Mechanics Institute*
[10] Labelle F and Shewchuk J R 2007 Isosurface stuffing: fast tetrahedral meshes with good dihedral angles *ACM Transactions on Graphics (TOG)* vol 26 (ACM) p 57

# Radiotherapy treatments using Tsallis entropy statistical approach

**Rodríguez-Pérez, D.**
UNED, Departamento de Física Matemática y de Fluidos, 28040 Madrid, Spain
E-mail: daniel@dfmf.uned.es

**Sotolongo-Grau, O.**
Fundació ACE, Institut Català de Neurociències Aplicades, 08029 Barcelona, Spain
E-mail: osotolongo@fundacioace.com

**Sotolongo-Costa, O.**
University of Havana, Cátedra de Sistemas Complejos Henri Poincaré, Havana 10400, Cuba
E-mail: osotolongo@fisica.uh.cu

**Antoranz, J. C.**
UNED, Departamento de Física Matemática y de Fluidos, 28040 Madrid, Spain
University of Havana, Cátedra de Sistemas Complejos Henri Poincaré, Havana 10400, Cuba
E-mail: jcantoranz@dfmf.uned.es

**Abstract.** Several radiobiological models mimic the biologic effect of one single radiation dose on a living tissue. However, the actual fractionated radiotherapy requires accounting for a new magnitude, i.e., time. Here, we explore the biological consequences posed by the mathematical prolongation of a previous single radiation model to fractionated treatment. The survival fraction is obtained, together with the equivalent physical dose, in terms of a time dependent factor (similar to a repair coefficient) describing the tissue trend to recovering its radioresistance. The model describes how dose fractions add up to obtain the equivalent dose and how the repair coefficient poses a limit to reach an equivalent dose equal to the critical one that would completely annihilate the tumor. On the other hand, the surrounding healthy tissue is a limiting factor to treatment planning. This tissue has its own repair coefficient and thus should limit the equivalent dose of a treatment. Depending on the repair coefficient and the critical dose of each tissue, unexpected results (failure to fully remove the tumor) can be obtained. To illustrate these results and predictions, some realistic example calculations will be performed using parameter values within actual clinical ranges. In conclusion, the model warns about treatment limitations and proposes ways to overcome them.

## 1. Introduction

A recently developed model [1, 2] of radiobiology shows that Tsallis entropy definition [3] could be applied to describe radiotherapy treatments. As shown in [4] nonextensive Tsallis entropy has been successful in describing a vast class of natural systems, some of them as complex as the development of living systems [5, 6].

This new radiobiological model, or *maxent* model, describe the survival fraction as a functional of the radiation absorbed dose, taking advantage of Tsallis entropy expression, and a minimum number of statistical and biologically motivated hypotheses.

*Maxent* model takes into account that a critical dose, $D_0$, exists that annihilates every single cell in a tissue. Thus, the radiation dose, $d$, can be written as a dimensionless dose, $x = d/D_0$. This means that the support of the cell death probability density function, $p(x)$, must be $\Omega = [0;1]$. Under these conditions Tsallis entropy functional becomes,

$$S_q = \frac{1}{q-1}\left[1 - \int_0^1 p^q(x)\,dx\right], \tag{1}$$

with $q$ the nonextensivity index. The survival fraction of cells will be given by $f(x) = \int_x^1 p(x)\,dx$, that is the complement of the fraction of cells killed by radiation. The probability normalization condition implied by this interpretation of $p(x)$ must be complemented by the existence of a finite $q$-mean value as was shown in [1]. Both conditions allow the calculation of the probability density maximizing (1), using Lagrange multipliers, resulting in

$$p(x) = \gamma(1-x)^{\gamma-1}, \tag{2}$$

where $\gamma = \frac{q-2}{q-1}$. Hence, the survival fraction predicted by the model is

$$f(x) = (1-x)^\gamma, \tag{3}$$

valid for $x \in \Omega$ and requiring $\gamma > 1$.

This model shows remarkable agreement with experimental data [1, 7], even at high doses where previous models are less accurate. Furthermore, analysis of the model shows that interaction is universal for a given tissue, and is characterized by a single exponent, independent on the radiation type, energy or dose rate; on the other hand, the cutoff dose, above which every single cell dies, does depend on the characteristics of the radiation. Previous models can be derived as particular limiting cases of the mathematical expression, which can also be easily plotted and interpreted [1].

Expression (3), understood as survival probability, lacks the extensivity property, i.e., for $n$ events following (3) the total survival probability should be found as a composition of the survival probabilities of the successive events. However, there is not a straightforward composition rule for those probabilities. For instance, let two doses, $x_A$ and $x_B$ be applied; the resulting composition probability can take two possible limit values:

- If the total dose is assumed additive: $f_{AB} = (1 - x_A - x_B)^\gamma$, that is, the individual probabilities under $A$ and $B$ events could not be treated as independent probabilities, $f_{AB} \neq f_A f_B$.
- If probabilities are multiplicative, $f = (1-x_A)^\gamma (1-x_B)^\gamma$ , doses would not fulfill the superposition principle for the equivalent physical dose, $x_{AB} \neq x_A + x_B$.

A mathematically consistent way to crossover between both limits, is explained and analysed in what follows.

## 2. Composition rules

If the survival probabilities are deemed independent, the total probability for two events $A$ and $B$ is $f_{AB} = f_A f_B$. This condition and equation 3 lead to the composition rule

$$\begin{gathered} f_{AB} = f_A f_B \\ x_{AB} = x_A \oplus x_B = x_A + x_B - x_A x_B \end{gathered} \tag{4}$$

Analogously, if the dose should be additive, $x_{AB} = x_A + x_B$, the same equation leads to the composition rule

$$\begin{aligned} x_{AB} &= x_A + x_B \\ f_{AB} = f_A \otimes f_B &= \left({f_A}^{1/\gamma} + {f_B}^{1/\gamma} - 1\right)^{\gamma} \end{aligned} \tag{5}$$

In clinical treatments, that both limits are not clearly distinct. Tissue recovering capabilities make two treatments separate enough in time independent from each other as the system "loses" memory of its previous state. In radiobiological terms, the next radiotherapy session is applied after late effects of the previous one occur. However, if both doses were applied simultaneously they must be summed up, i.e., they will be composed additively. Thus, (4) and (5) are limit cases corresponding to $t = 0$ and $t \gg T$, where $t$ is the time between successive doses and $T$ some biological characteristic time (tissue repair time, for instance).

The description of a real fractionated process, requires to generalize sum $\oplus$ and product $\otimes$ operators, in a form that allows a smooth transition between these two limits. This was done in [2] through the introduction of a coefficient $\epsilon \in [0, 1]$ such that for $\epsilon = 1$ the expression describes completely correlated events, while $\epsilon = 0$ means they are fully independent. Thus $\epsilon$ plays the role of a "repair coefficient" between doses. The final expression encompassing both limits reads:

$$x_{AB} = x_A \oplus x_B = x_A + x_B \left(\frac{1-x_A}{1-\epsilon x_A}\right) \tag{6}$$

$$f_{AB} = f_A \otimes f_B = f_A \left[\frac{{f_B}^{1/\gamma} - \epsilon\left(1-{f_A}^{1/\gamma}\right)}{1-\epsilon\left(1-{f_A}^{1/\gamma}\right)}\right]^{\gamma} \tag{7}$$

from which is readily checked that both limit definitions (4) and (5) are recovered for $\epsilon = 0$ and $\epsilon = 1$ respectively.

## 3. Biological and physical implications

### *3.1. Isoeffect relationship*

One of the central concepts in radiotherapy is isoeffect relationships. An oncologist usually seeks treatments that produce the same outcome on the target tissue, while causing the least damage on the surrounding healthy tissues.

Equations (7) can be generalized for the case of a multifractionated treatment to render a survival fraction $F_n$:

$$F_n = (1 - X)^{\gamma} = \left(1 - \bigoplus_{i=1}^{n} x_i\right)^{\gamma} \tag{8}$$

where $n$ is the number of fractions $x_i$ and $X$ is the effective dose.

All fractionated treatments with the same value of $X$, will provide the same value for the survival fraction: this is the isoeffect criterion for the fractionated therapy.

The oncologist will seek among the different pairs of values $(n, x_i)$ in equation (8) providing the same value of $F$ for a given tissue while keeping small collateral damages to healthy tissues.

In order to check the model reliability, it has been fitted to data from [8, 9, 10] using the same weighted least squares algorithm as in [1] (see Figure 1). Those data sets are considered as a reliable source of clinical parameters (as the $\alpha/\beta$ relation of LQ model [11]). The obtained $\epsilon$ values are far from the limit cases ($\epsilon = 0$ and $\epsilon = 1$) showing a survival fraction behavior far from the pure $q$-algebraic limits.

Once the involved coefficients for a treatment ($\epsilon$ and $D_0$) are known the dosage can be tuned to obtain the desired effective dose by changing $n$ and $d$. Notice that $\gamma$ does not play any role in this composition.

To illustrate the practical use of the theory exposed above, let us consider the following example. A normal tissue $H$, characterized by typical coefficients $\gamma = 10.0$ and $D_0 = 40.0$ Gy,

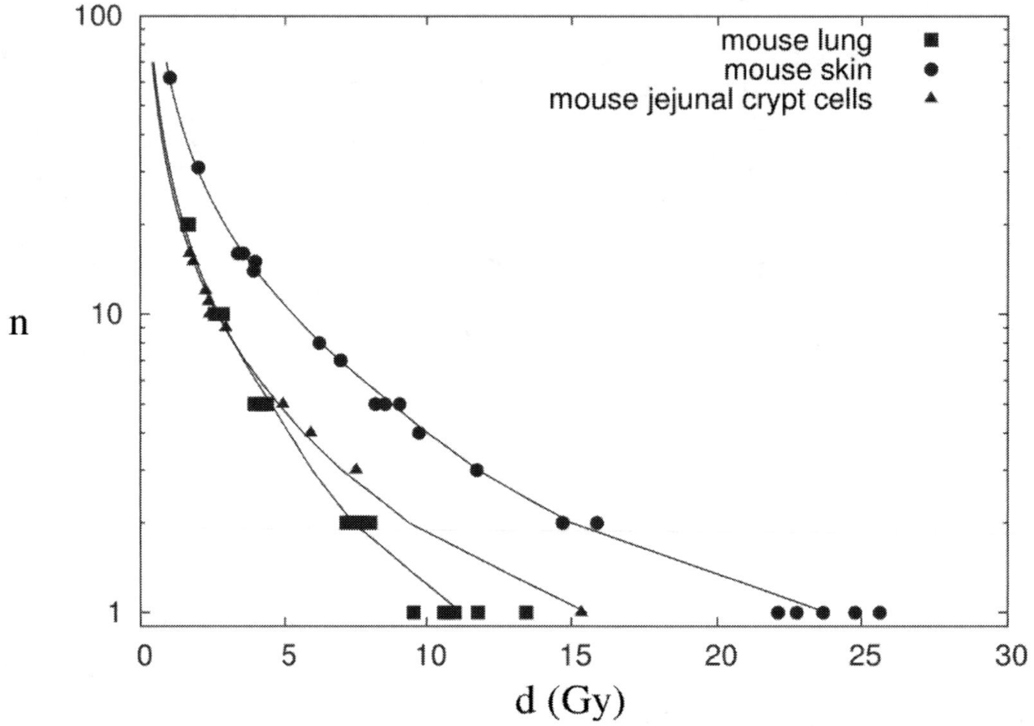


**Figure 1.** Isoeffect relationship data reported for mouse lung by [8] ($\epsilon = 0.50$, $D_0 = 11.3$ Gy), mouse skin by [9] ($\epsilon = 0.58$, $D_0 = 24.0$ Gy) and mouse jejunal crypt cells by [10] ($\epsilon = 0.62$, $D_0 = 16.1$ Gy), fitted to (9).

surrounded by a more resilient tumoral tissue $T$ with $\gamma = 15.0$ and $D_0 = 80.0$ Gy. $H$ cannot receive more than 36.0 Gy, or equivalently, $X = 0.9$. Following [11] the isoeffect curves for incomplete repair can be computed using the LQ model approximation; we will assume a typical repair half time of 3 hours for both tissues in this example. Different treatment schedules, characterized by a time ($\Delta t$) between sessions, render different survival curves. From them the $\epsilon$ values can be found in terms of $\Delta t$. This is shown in Figure 2.

### 3.2. *Critical dosage*

Assuming the same physical dose per fraction, $x_i = x$, as is the case in many radiotherapy protocols, we obtain from (6) the equivalent dose for $n$ fractions:

$$X_i = X_{i-1} + x\left(\frac{1 - X_{i-1}}{1 - \epsilon X_{i-1}}\right) \tag{9}$$

This is the expression of a recursive map describing the evolution of the effective dose in a treatment. The stability analysis of this map shows that, for every $\epsilon$ there is a critical value of $x$,

$$x_c = 1 - \epsilon, \tag{10}$$

that divides the plane $(\epsilon, x)$ in two different regions.

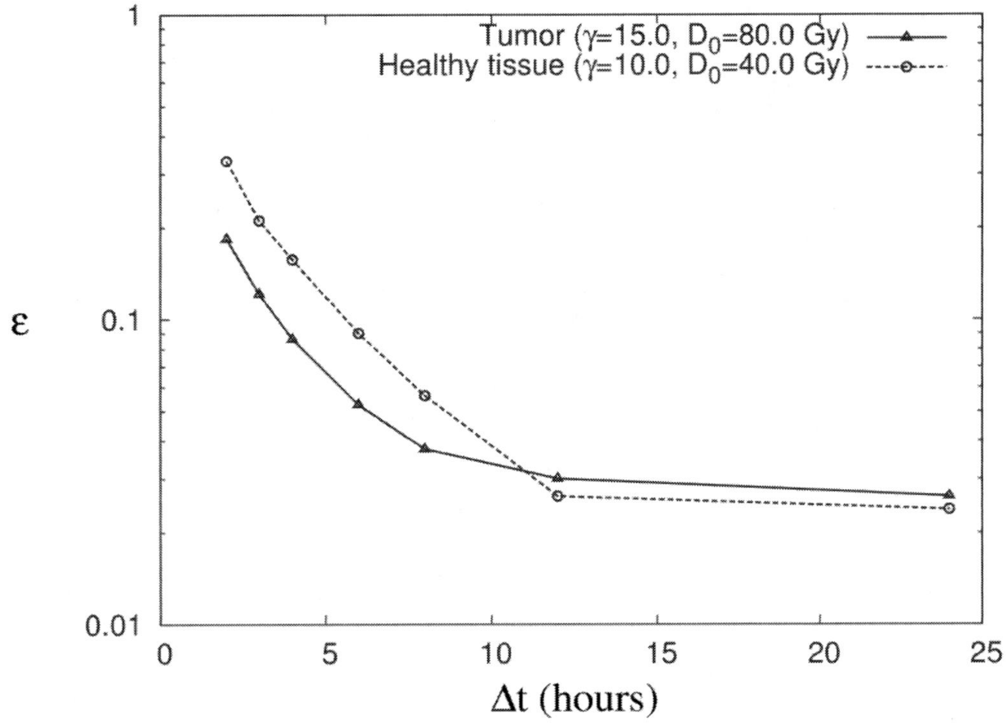


**Figure 2.** Session-coupling values found following the LQ model of incomplete repairment for an hypothetical tumor and healthy tissue as function of treatment time schedule.

- For $x < x_c$, there will always be a surviving portion of the tissue since always $X_n < 1$, for every $n$.
- For $x > x_c$, after enough fractions $X_n$ will reach the critical value 1, thus every single cell of tissue has been removed by the treatment.

Then, for some values of $\epsilon$ it is possible to find a threshold value of $n$, that kills every cell, for a given therapy protocol. However, for some $\epsilon$ values, this $n$ does not exist. This is shown in the inset of Figure 3.

## 4. Discussion

The session-coupling parameter $\epsilon$ is a cornerstone on isoeffect relationships. Notice that unlike $\gamma$, that characterizes the tissue primary response to radiation, $\epsilon$ characterizes the tissue trend to recover its previous radioresistance.

$\epsilon$ acts as a relaxation term: after radiation damage occurs ($\epsilon = 1$) tissue begins to recover; then $\epsilon$ gradually decreases, until the tissue recovers its initial recovery capacity ($\epsilon = 0$). This can be also interpreted as if formerly applied radiation decrease the total annihilation dose (equal, initially, to $D_0$). The more coupled a session is to the previous one, the larger this change (scaled by $\epsilon$) in the the critical dose of the next session.

As it was shown in [1], all these nonextensive properties of tissue response to radiation are more noticeable the closer to the critical values, i.e., for higher doses. This has as a consequence that, for a given effect, doses predicted by traditional modes, are larger than those predicted

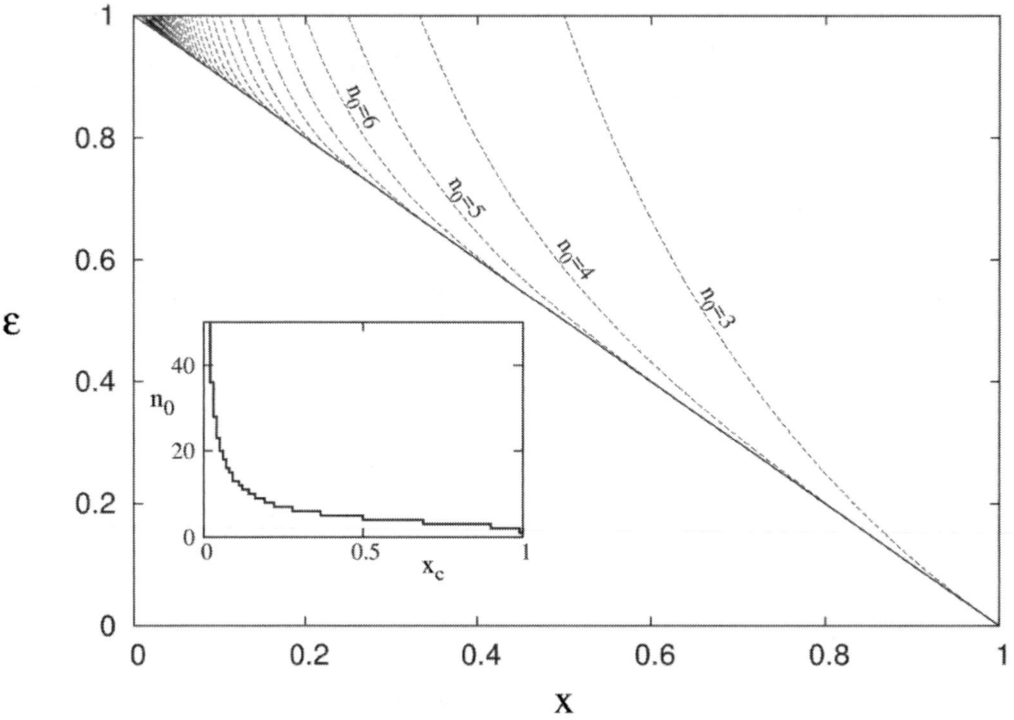


**Figure 3.** The larger plot represents $n_0$ isolines as a function of $x$ and $\epsilon$ (dashed lines) above $x_c(\epsilon)$ (solid line); below this line, killing all tissue cells is impossible. The small one represents critical values $n_0$ in terms of $x_c$.

by our model, because traditional models do not take into account effect enhancement close to the critical point. For the same total dose, a lower dose per fraction will enhance nonextensive properties in fractionated therapies. For high doses per fraction, few fractions will be required, and then $n$ will not depend on $\epsilon$. However, for small doses per fraction, more sessions will be required to cause the desired effect ($F_n = 0$); furthermore, $\epsilon$ will determine whether this achievement will be possible at all at a given dose per fraction $x$ (equation (10)).

With the model introduced above, a fractionated radiotherapy can be planned for a tissue described by $\gamma$, varying $x$ according to $\epsilon$, in order to achieve the desired therapy effects with minimal dose.

## 5. Conclusions

The *ansatz* of maximum Tsallis entropy led us to a simple nonextensive expression for the single dose survival fraction. The mathematical constraints, required to consistently define treatment compositions, naturally introduce a new parameter relating radiation sessions. The new *session-coupling* parameter takes values between 0 and 1 that describe the cross effects of independent to simultaneous doses, respectively. This has been checked estimating its value from experimental data available from multisession treatments, showing that it is relevant to describe them within the non-extensive framework.

An interesting interpretation of this new parameter is that it modulates the critical dosage (the total annihilation dose that kills every tumor cell) for the following session in terms of the previous treatment history. As the session-coupling parameter plays the role of a "repair coefficient" between doses its dependence on time and type of tissue will require experimental characterization, as do the critical dose (radiation and tissue dependent) and the $\gamma$ exponent (tissue dependent).

The session-coupling parameter completes the non-extensive radiotherapy model allowing the design of isoeffective treatments, i.e., that have the same effect on tumoral tissue, while minimizing the damage caused to healthy tissues more effectively.

## Acknowledgments

Authors acknowledge the financial support from the Spanish Ministerio de Economía y Competitividad under project FIS2012-37408. Authors want to thank Juan Antonio Santos, MD for his priceless help in the understanding of radiobiological processes and models.

## References

[1] O. Sotolongo-Grau, D. Rodríguez-Pérez, J. C. Antoranz, and O. Sotolongo-Costa. Tissue radiation response with maximum Tsallis entropy. *Physical Review Letters*, 105(15):158105, 2010.

[2] O. Sotolongo-Grau, D. Rodríguez-Pérez, O. Sotolongo-Costa, J.C. Antoranz. Tsallis entropy approach to radiotherapy treatments. *Physica A: Statistical Mechanics and its Applications*, 392(9):2007–2015, 2013.

[3] C. Tsallis. Nonextensive statistics: Theoretical, experimental and computational evidences and connections. *Brazilian Journal of Physics*, 29:1–35, 1999.

[4] H. Swinney and C. Tsallis. Anomalous distributions, nonlinear dynamics, and nonextensivity. *Physica D: Nonlinear Phenomena*, 193(1-4):1 – 2, 2004.

[5] Jean Paul Rieu, Arpita Upadhyaya, James A. Glazier, Noriyuki Bob Ouchi, and Yasuji Sawada. Diffusion and deformations of single Hydra cells in cellular aggregates. *Biophysical Journal*, 79(4):1903 – 1914, 2000.

[6] Arpita Upadhyaya, Jean-Paul Rieu, James A. Glazier, and Yasuji Sawada. Anomalous diffusion and non-gaussian velocity distribution of Hydra cells in cellular aggregates. *Physica A: Statistical Mechanics and its Applications*, 293(3–4):549 – 558, 2001.

[7] O. Sotolongo-Grau, D. Rodriguez-Perez, J.C. Antoranz, and O. Sotolongo-Costa. Non-extensive radiobiology. In A. Mohammad-Djafari, J-F. Bercher, and P. Bessiere, editors, *Bayesian inference and maximum entropy methods in science and engineering (Proceedings of the 30th International Workshop on Bayesian Inference and Maximum Entropy Methods in Science and Engineering, 4-9 July 2010, Chamonix, France)*, volume 1305 of *AIP Conference Proceedings*, pages 219–226. AIP, 2010.

[8] C. S. Parkins, J. F. Fowler, R. L. Maughan, and M. J. Roper. Repair in mouse lung for up to 20 fractions of X rays or neutrons. *British Journal of Radiology*, 58(687):225–41, 1985.

[9] B. G. Douglas and J. F. Fowler. The effect of multiple small doses of X rays on skin reactions in the mouse and a basic interpretation. *Radiation Research*, 66(2):401–426, 1976.

[10] H. D. Thames, R. Withers, K. A. Mason, and B. O. Reid. Dose-survival characteristics of mouse jejunal crypt cells. *International Journal of Radiation Oncology*Biology*Physics*, 7(11):1591 – 1597, 1981.

[11] A.J. van der Kogel and C.C.R. Arnout. Calculation of isoeffect relationships. In G.G. Steel, editor, *Basic Clinical Radiobiology for Radiation Oncologists*, pages 72–80. Edward Arnold Publishers, London, 1993.

# Runge-Kutta and Hermite Collocation for a biological invasion problem modeled by a generalized Fisher equation

**IE Athanasakis , EP Papadopoulou and YG Saridakis**

Applied Math & Computers Lab, Technical University of Crete, 73100 Chania, Crete, Greece

E-mail: yiannis@science.tuc.gr

**Abstract.** Fisher's equation has been widely used to model the biological invasion of single-species communities in homogeneous one dimensional habitats. In this study we develop high order numerical methods to accurately capture the spatiotemporal dynamics of the generalized Fisher equation, a nonlinear reaction-diffusion equation characterized by density dependent non-linear diffusion. Working towards this direction we consider strong stability preserving Runge-Kutta (RK) temporal discretization schemes coupled with the Hermite cubic Collocation (HC) spatial discretization method. We investigate their convergence and stability properties to reveal efficient HC-RK pairs for the numerical treatment of the generalized Fisher equation. The Hadamard product is used to characterize the collocation discretized non linear equation terms as a first step for the treatment of generalized systems of relevant equations. Numerical experimentation is included to demonstrate the performance of the methods.

## 1. Introduction

One way to incorporate density-dependent species motility, such as small scale migrations, to the Fisher-Kolmogorov ([1], [2]) classical biological invasion model, is by replacing the constant diffusion coefficient $D$ by a density-dependent $D(u)$ one (cf. [3], [4], [5] and the references therein). Assuming that the diffusivity depends linearly on density (cf. [6]), namely $D(u) = \lambda_0 u + \lambda_1$, the generalized Fisher's equation takes the form:

$$u_t = [(\lambda_0 u + \lambda_1)u_x]_x + \lambda_2 u - \lambda_3 u^2 \tag{1}$$

where $u \equiv u(x,t)$, $x \in [a,b]$, $t \in [0,T]$ and $\lambda_i \in \mathbf{R}$, for $i = 0,1,2,3$, with $\lambda_2\lambda_3 > 0$, while Neumann boundary conditions $u_x(a,t) = 0$ and $u_x(b,t) = 0$ are imposed at the boundaries and an initial density distribution $u(x,0) = f(x)$ is assumed. As exact solutions have been recently derived (cf. [6]), equation (1) can play the role of a model problem for investigating the behavior of numerical techniques.

Working towards the development of high order numerical schemes, in Section 2 we implement the Hermite Collocation (HC) method to discretize in space and we use the Hadamard product to characterize the non linear equation terms. In Section 3, explicit (to avoid solving nonlinear systems) Strong Stability Preserving (SSP) Runge-Kutta time discretization schemes are coupled with the HC method. Their efficiency, convergence and ability to maintain strong stability are investigated in Section 4 through numerical experimentation.

## 2. Hermite Collocation (HC) Spatial Discretization Method

Assuming sufficiently smooth solutions of equation (1), a uniform partition of $[a,b]$ into $N$ subintervals, with spacing $h=(b-a)/N$ and nodes $x_j := a+jh$ , $j=1,\ldots,N+1$, the Hermite Collocation method seeks $O(h^4)$ approximations in the form:

$$U(x,t)=\sum_{j=1}^{N+1}\left[\alpha_{2j-1}(t)\phi_{2j-1}(x)+\alpha_{2j}(t)\phi_{2j}(x)\right] \tag{2}$$

where $\phi_{2j-1}(x)$ and $\phi_{2j}(x)$ are the Hermite cubic basis functions, defined by

$$\phi_{2j-1}(x)=\begin{cases}\phi\left(\frac{x_j-x}{h}\right) & , \quad x\in I_{j-1}\\ \phi\left(\frac{x-x_j}{h}\right) & , \quad x\in I_j\\ 0 & , \quad \text{otherwise}\end{cases} \quad , \quad \phi_{2j}(x)=\begin{cases}-h\psi\left(\frac{x_j-x}{h}\right) & , \quad x\in I_{j-1}\\ h\psi\left(\frac{x-x_j}{h}\right) & , \quad x\in I_j\\ 0 & , \quad \text{otherwise}\end{cases} \tag{3}$$

with $I_j=[x_j,x_{j+1}]$ , $j=1,\ldots,N$, and $\phi(s)=(1-s)^2(1+2s)$ , $\psi(s)=s(1-s)^2$ for $s\in[0,1]$.

Substitution of the approximate solution (2) into equation (1) yields a residual

$$\mathcal{R}(x,t):=U_t-\left[(\lambda_0U+\lambda_1)U_x\right]_x+\lambda_2U-\lambda_3U^2 \ . \tag{4}$$

The Collocation method at the Gauss points (cf. [7]) produces a system of algebraic equations, for the evaluation of the unknown parameters $\alpha_i\equiv\alpha_i(t)$ , $i=1\ldots,2(n+1)$, by forcing the residual $\mathcal{R}(x,t)$ to vanish at the two Gaussian interior collocation points per subinterval. In doing so, the two elemental collocation equations that correspond to the element $I_j$ , $j=1,\ldots,N$ may be written in the matrix form:

$$\begin{aligned}C_j^{(0)}\begin{bmatrix}\dot\alpha_{2j-1}\\ \dot\alpha_{2j}\\ \dot\alpha_{2j+1}\\ \dot\alpha_{2j+2}\end{bmatrix}=\lambda_0&\left[\left(C_j^{(1)}\begin{bmatrix}\alpha_{2j-1}\\ \alpha_{2j}\\ \alpha_{2j+1}\\ \alpha_{2j+2}\end{bmatrix}\right)\circ\left(C_j^{(1)}\begin{bmatrix}\alpha_{2j-1}\\ \alpha_{2j}\\ \alpha_{2j+1}\\ \alpha_{2j+2}\end{bmatrix}\right)+\left(C_j^{(0)}\begin{bmatrix}\alpha_{2j-1}\\ \alpha_{2j}\\ \alpha_{2j+1}\\ \alpha_{2j+2}\end{bmatrix}\right)\circ\left(C_j^{(2)}\begin{bmatrix}\alpha_{2j-1}\\ \alpha_{2j}\\ \alpha_{2j+1}\\ \alpha_{2j+2}\end{bmatrix}\right)\right]\\ &+\lambda_1C_j^{(2)}\begin{bmatrix}\alpha_{2j-1}\\ \alpha_{2j}\\ \alpha_{2j+1}\\ \alpha_{2j+2}\end{bmatrix}+\lambda_2C_j^{(0)}\begin{bmatrix}\alpha_{2j-1}\\ \alpha_{2j}\\ \alpha_{2j+1}\\ \alpha_{2j+2}\end{bmatrix}-\lambda_3\left(C_j^{(0)}\begin{bmatrix}\alpha_{2j-1}\\ \alpha_{2j}\\ \alpha_{2j+1}\\ \alpha_{2j+2}\end{bmatrix}\right)\circ\left(C_j^{(0)}\begin{bmatrix}\alpha_{2j-1}\\ \alpha_{2j}\\ \alpha_{2j+1}\\ \alpha_{2j+2}\end{bmatrix}\right)\end{aligned} \tag{5}$$

where $\dot\alpha_i\equiv\frac{d\alpha_i}{dt}$ and

$$C_j^{(k)}=\frac{1}{h^k}\begin{bmatrix}s_1^{(k)} & hs_2^{(k)} & s_3^{(k)} & hs_4^{(k)}\\ s_3^{(k)} & hs_4^{(k)} & s_1^{(k)} & hs_2^{(k)}\end{bmatrix}$$

with

$$s_1=\begin{cases}\frac{9+4\sqrt3}{18}\\ -1\\ -2\sqrt3\end{cases} \quad s_2=\begin{cases}\frac{3+\sqrt3}{36}\\ \frac{\sqrt3}{6}\\ -1-\sqrt3\end{cases} \quad s_3=\begin{cases}\frac{9-4\sqrt3}{18}\\ 1\\ 2\sqrt3\end{cases} \quad s_4=\begin{cases}\frac{3-\sqrt3}{36} & , \quad for\ k=0\\ -\frac{\sqrt3}{6} & , \quad for\ k=1\\ -1+\sqrt3 & , \quad for\ k=2\end{cases} .$$

In the above equations the symbol $\circ$ denotes the Hadamard matrix product. Apparently, the collocation discretized non-linear terms are formed as the Hadamard matrix product of the collocation discretized corresponding linear terms.

Furthermore, the additional two needed equations are obtained by forcing the approximate solution $U(x,t)$, to satisfy the boundary conditions. This directly implies that

$$\alpha_2=0 \ , \ \alpha_{2N+2}=0 \quad \text{hence also} \quad \dot\alpha_2=0 \ , \ \dot\alpha_{2N+2}=0$$

Careful assembling now all the interior and boundary collocation equations, described above, we obtain the ODE system:

$$\dot{\mathbf{a}} = C(\mathbf{a}) \tag{6}$$

$$C(\mathbf{a}, t) \ := \ C^{(0)^{-1}} \left[\lambda_0 \left(C^{(1)}\mathbf{a} \circ C^{(1)}\mathbf{a} + C^{(0)}\mathbf{a} \circ C^{(2)}\mathbf{a}\right) + \lambda_1 C^{(2)}\mathbf{a} + \lambda_2 C^{(0)}\mathbf{a} - \lambda_3 C^{(0)}\mathbf{a} \circ C^{(0)}\mathbf{a}\right]$$

where $C^{(k)}$ , $k = 0, 1, 2$ denotes the one dimensional Collocation matrix corresponding to the $k$-th derivative, $\mathbf{a} \equiv \mathbf{a}(t) = [\alpha_1 \ \alpha_3 \ \ldots \ \alpha_{2N+1}]^T$ denotes the vector of unknowns and $\dot{\mathbf{a}} := \frac{d}{dt}\ \mathbf{a} = [\dot\alpha_1 \ \dot\alpha_3 \ \ldots \ \dot\alpha_{2N+1}]^T$. We note that the linear independence of the Hermite cubic basic functions implies the non-singularity of the matrix $C^{(0)}$.

## 3. Strong Stability Preserving Runge-Kutta Time Discretization Schemes

High order strong stability preserving (SSP) Runge-Kutta methods were developed (cf. [8], [9], [10]) for the time discretization of the semi-discrete system of ODEs obtained from the spatial discretization of PDEs by a Total Variation Diminishing (TVD) finite difference or finite element method. The essence of the SSP class of time discretization methods lies on their ability to maintain strong stability in any given norm $||\cdot||$, that is to say

$$||\mathbf{a}^{n+1}|| \leq ||\mathbf{a}^n|| \quad , \quad \mathbf{a}^n := \mathbf{a}(t^n) \ \text{ with } \ t^n := n\Delta t \ , \ n = 0, 1, \ldots \tag{7}$$

assuming that forward Euler is strongly stable and providing suitable restrictions of the time stepping, while fully preserving the order of the error of the spatial discretization.

General $m$ stage and order $p$ SSP schemes are denoted by SSP(m,p) or SSPRK(m,p). In this work we consider two optimal (cf. [11], [12]) SSP(3,3) and SSP(4,3) schemes for the solution of the semi-discrete ODE system in (6). They can be written in the form:

$\underline{SSP(4,3)}$

$$\begin{aligned}
\mathbf{a}^{(1)} &= \mathbf{a}^n + \tfrac{1}{2}\Delta t C(\mathbf{a}^n)\\
\mathbf{a}^{(2)} &= \mathbf{a}^{(1)} + \tfrac{1}{2}\Delta t C(\mathbf{a}^{(1)})\\
\mathbf{a}^{(3)} &= \tfrac{2}{3}\mathbf{a}^n + \tfrac{1}{3}\mathbf{a}^{(2)} + \tfrac{1}{6}\Delta t C(\mathbf{a}^{(2)})\\
\mathbf{a}^{n+1} &= \mathbf{a}^{(3)} + \tfrac{1}{2}\Delta t C(\mathbf{a}^{(3)})
\end{aligned}$$

$\underline{SSP(3,3)}$

$$\begin{aligned}
\mathbf{a}^{(1)} &= \mathbf{a}^n + \Delta t C(\mathbf{a}^n)\\
\mathbf{a}^{(2)} &= \tfrac{3}{4}\mathbf{a}^n + \tfrac{1}{4}\mathbf{a}^{(1)} + \tfrac{1}{4}\Delta t C(\mathbf{a}^{(1)})\\
\mathbf{a}^{n+1} &= \tfrac{1}{3}\mathbf{a}^n + \tfrac{2}{3}\mathbf{a}^{(2)} + \tfrac{2}{3}\Delta t C(\mathbf{a}^{(2)})
\end{aligned}$$

We note that the classical RK4 is also considered and used in our experiments for comparison purposes.

## 4. Numerical Results

The model problem used to investigate the performance of the HC-RK methods is given by

$$\begin{aligned}
&u_t = [(1-u)u_x]_x + 2u - 2u^2 \quad , \qquad -5\pi/2 \leq x \leq 5\pi/2, \quad 0 \leq t \leq T\\
&u_x(\frac{-5\pi}{2}, t) = 0, \quad u_x(\frac{5\pi}{2}, t) = 0 \quad , \qquad u(x,0) = \frac{1}{3}\left[2 + \sin(-x)\right]
\end{aligned} \tag{8}$$

and admits the analytical solution $u(x,t) = \frac{1}{3}\left[\frac{e^{-t}(3e^{2t}+1+2\sin(-x))}{e^t+e^{-t}}\right]$. The spatial absolute error used in all experiments is defined as $\mathcal{E}^n := ||U(x,t^n) - u(x,t^n)||_2$ and the necessary time stepping restrictions imposed are $\Delta t \leq \frac{1}{5}h^2$ for SSP(4,3), $\Delta t \leq \frac{1}{10}h^2$ for SSP(3,3) and $\Delta t \leq 0.3h^3$ for RK4. Under these restrictions all time discretization schemes were strongly stable, as it is depicted in Figure 1b for SSP(4,3) while, at the same time, the $O(h^4)$ order of convergence of the HC method is preserved (see Table I). The maximum $\mathcal{E}^1 = \max_n\{\mathcal{E}^n\}$ error norm and the computational time needed to reach the time level $t = 1$ are also included in Table I to demonstrate the efficiency of the methods.

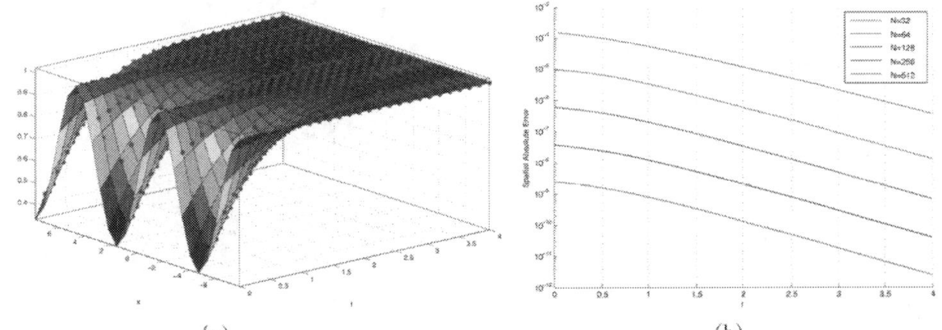

(a) (b)

**Figure 1.** a) Exact (solid) and HC-SSPRK(4,3) approximate (point) solutions for $N = 64$
b) Spatial absolute error as a function of time for the HC-SSPRK(4,3)

TABLE I Computational Performance of HC-RK schemes

| | *Spatial Absolute Error at* $t = \Delta t$ | | | *Spatial Order of Convergence* | | | *Time (sec)* | | |
|---|---|---|---|---|---|---|---|---|---|
| N | SSP(4,3) | SSP(3,3) | RK4 | SSP(4,3) | SSP(3,3) | RK4 | SSP(4,3) | SSP(3,3) | RK4 |
| 32 | 1.53e-04 | 1.56e-04 | 1.53e-04 | - | - | - | 0.01 | 0.03 | 0.02 |
| 64 | 9.85e-06 | 9.89e-06 | 9.89e-06 | 3.95 | 3.98 | 3.95 | 0.05 | 0.05 | 0.15 |
| 128 | 6.20e-07 | 6.20e-07 | 6.21e-07 | 3.99 | 3.99 | 3.99 | 0.14 | 0.20 | 1.60 |
| 256 | 3.88e-08 | 3.89e-08 | 3.89e-08 | 4.00 | 4.00 | 4.00 | 0.72 | 1.07 | 19.78 |
| 512 | 2.56e-09 | 2.56e-09 | 2.56e-09 | 3.93 | 3.93 | 3.93 | 4.28 | 6.48 | 269.91 |

## 5. Conclusion

In this work, the fourth order $HC$ is coupled to third order $SSPRK$ schemes for the treatment of a generalized Fisher equation. Numerical evidence is presented for the method's efficient and stable behaviour (Table I) as well as the method's ability to maintain strong stability (Fig. 1b).

## Acknowledgement

The present research work has been co-financed by the European Union (European Social Fund ESF) and Greek national funds through the Operational Program Education and Lifelong Learning of the National Strategic Reference Framework (NSRF) - Research Funding Program: THALIS. Investing in knowledge society through the European Social Fund.

## References

[1] Fisher RA *The wave of advance of advantageous genes*, Ann. Eugen., **7**, 255-369, 1937
[2] Kolmogorov AN, Petrovsky IG and Piskunov NS *Investigation of the equation of diffusion combined with increasing of the substance and its application to a biology problem*, Bull. Moscow State Univ. Ser. A: Math. and Mech., **1(6)**, 1-25, 1937
[3] Hengeveld R *Dynamics of Biological Invasions*, Chapman and Hall, London, 1989
[4] Murray JD *Mathematical Biology*, Springer, Berlin, 1989
[5] Petrovskii SV and Li BL *Exactly Solvable Models of Biological Invasion*, Taylor & Francis, 2010
[6] Cherniha R and Dutka V *Exact and Numerical Solutions of the Generalized Fisher Equation*, Reports on Mathematical Physics, **47**, 393-412, 2001
[7] de Boor C and Swartz B *Collocation at Gaussian points*, SIAM Num. Anal., vol. 10, pp. 582-606, 1973
[8] Shu CW *Total-variation-diminishing time discretizations*, SIAM J. Sci. Stat. Comput., **9**, 1073-1084, 1988
[9] Shu CW and Osher S *Efficient implementation of essentially non-oscillatory shock-capturing schemes*, J. Comput. Phys., **77**, 439-471, 1988
[10] Gottlieb S, Shu CW and Tadmor E *Strong Stability-Preserving High-Order Time Discretization Methods*, SIAM Num. Anal., **43**, 89-112, 2001
[11] Gottlieb S and Shu CW *Total variation diminishing Runge-Kutta schemes*, Mat. Comp., **67**, 73-85, 1998
[12] Ruuth S and Spiteri R *Two barriers on strong-stability-preserving time discretization methods*, J. Scientific Computation, **17**, 211-220, 2002

# Investigations on the electrical current-voltage response in protein light receptors

**Eleonora Alfinito**
Dipartimento di Ingegneria dell'Innovazione,
Università del Salento, via Monteroni, I-73100 Lecce-Italy
CNISM, via della Vasca Navale, 84, I-00146 Roma -Italy
E-mail: eleonora.alfinito@unisalento.it

**Jeremy Pousset and Lino Reggiani**
Dipartimento di Matematica e Fisica "Ennio De Giorgi"
Università del Salento,via Monteroni, I-73100 Lecce- Italy

**Abstract.** We report a theoretical/computational approach for modeling the current-voltage characteristics of sensing proteins. The modeling is applied to a couple of transmembrane proteins, bacteriorhodopsin and proteorhodopsin, sensitive to visible light and promising biomaterials for the development of a new generation of photo-transducers. The agreement between theory and experiments sheds new light on the microscopic interpretation of charge transfer in proteins and biological materials in general.

## 1. Introduction

A recent frontier in physics is to investigate organic and living matter, trying to formulate models able to capture their main features. This investigation has the double mark of being in itself interesting due to the lack of knowledge in the subject and of exploiting wide potentialities in the production of a new generation of electronic devices. As a matter of fact, some specific tasks of sensing protein can be reproduced in artificial devices to improve their performances. It is the case of nano-biosensors which aim to mimic *in vitro* the peculiar sensing abilities shown *in vivo* by proteins and organic materials. Accordingly, much attention has been devoted to transmembrane proteins (TMPs), which interface external and internal cell environments. Specifically, TMPs are proteins consisting of one or several helices crossing the cell membrane and able to modify their structure when capturing a specific molecule or a small protein (ligand). Among available experiments, here we report on those performed on two different seven-helix TMPs, bacteriorhodopsin (bR) and proteorhodopsin (pR), proteins able to transform energy coming by light in energy useful for the cell survival. Therefore, they appear very interesting to work out bio-solar-cells and/or bio-diodes. To this purpose, recent measurements on the electrical properties of bR and pR have revealed the possibility to use them *in vitro* like organic transistors. In particular, they are almost insulators at the lowest values of the applied bias, and their conductance suddenly increases for orders of magnitude, above some threshold value where they enter a sharp super-Ohmic regime. Furthermore, in the presence of green light, i.e.

after a conformational change, the conductance exhibits a further increase above the dark value, thus confirming the hypothesis of a structure-mediated charge transfer.

To provide a microscopic model of the main features described above, we developed a numerical approach based on the analogy between a single protein and an impedance network. The charge transport is obtained by introducing a procedure of stochastic changes of the network impedances. The investigations are performed on the bR, whose current-voltage (I-V) responses were carried out on a wide range of applied voltages [1], then extended to pR [2], on which experiments are only at the preliminary stage of low applied bias.

## 2. Experiments and Theory

Former experiments were carried out on bR, probably the best known protein in the seven-helix TMP family. Samples of bR in its natural membrane were used in a metal-insulator-metal structure [3, 1] and their I-V characteristics measured in dark [1] and also in green light [3]. The samples exhibited a small conductivity (less than 1 pS/cm) at low applied field. This conductivity was found to increase as the protein sample is irradiated with green light and/or when the electrical field overcomes a critical value. In particular, green light produces a higher current with respect to dark conditions, as a consequence of the modification of the chromophore-opsin complex [4]. The super-linear I-V characteristics point toward a tunneling charge-transfer mechanism [3, 1, 4]. In particular, a transition between a direct tunneling (DT) regime and a Fowler-Nordheim (FN) tunneling regime is revealed (see Figure 1). First measurements of I-V characteristics in pR were reported in Ref. [2]. The experiment was performed on a thin-film device with a contact distance of 50 $\mu m$, just about $10^4$ times longer than the bR nanolayers used in the previous experiments [3, 1]. The film was obtained by dehydrating pR in a buffer solution. The protein activity after dehydration was tested by absorbtion measurements that confirmed the preservation of the structure and the function after the treatment. Accordingly, the experiments performed on bR and pR evidence the possibility to obtain, *in vitro* like so *in vivo*, a significant photocurrent, a result of great interest for optoelectronic applications.

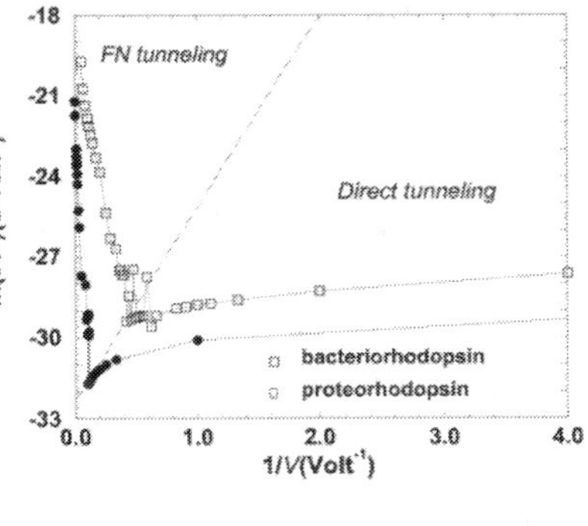


**Figure 1.** Current-voltage characteristics in pR and bR. Symbols refer to numerical calculations, lines are guide to the eyes. The dashed line indicate the DT and FN tunneling regions.

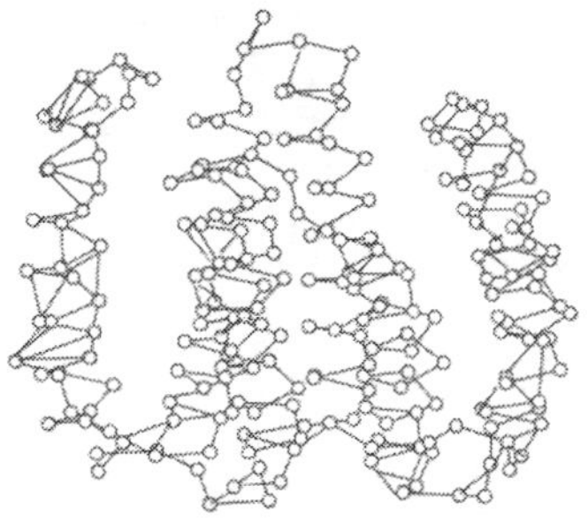

**Figure 2.** Draw of the protein graph. Amino acid positions are projected on the plane sheet and identified by circles; links are drawn by assuming a small interaction radius.

Following the indication of experiments [3, 1], we assume that charge transfer through the amino acids of the single protein is the main mechanism of transport. Therefore, a microscopic interpretation of experiments is given in terms of the single protein electrical properties, taken as intrinsic characteristics of the protein morphology. To this purpose we make use of the impedance network protein analogous (INPA) technique already detailed in [4]. In short, each protein is associated with a topological network, consisting of nodes and links, as pictured in Figure 2. Each node corresponds to a single amino acid, and its position is taken as that of the related $C_\alpha$ atom [5]. Each couple of nodes is connected with a link when their distance is less than an assigned interaction radius, $R_c$ (usually in the range $3.8 - 30$ Å), so that the network captures the protein topology. The graph turns into an impedance network when an elemental impedance is associated with each link. In the present case of a static response only resistances are considered. Accordingly, the elemental resistance between the $i$-th and $j$-th nodes, say, $R_{i,j}$ is taken as [4]: $R_{i,j} = \frac{l_{i,j}}{\mathcal{A}_{i,j}} \rho$, where $\mathcal{A}_{i,j} = \pi(R_c^2 - l_{i,j}^2/4)$, is the cross-sectional area between the spheres centered on the $i, j$ nodes, respectively, $l_{i,j}$ is the distance between these nodes, $\rho$ is the resistivity. By positioning the input and output electrical contacts on the first and last node, respectively, for a given applied bias the network is solved within a linear Kirchhoff scheme. By construction, this network produces a parameter-dependent static I-V characteristic. To account for the super-linear current at increasing applied voltages, a tunneling mechanism of charge transfer is included by using a stochastic approach within a Monte Carlo scheme[4]. In particular, the resistivity value of each link is chosen between a low value $\rho_{min}$, taken to fit the current at the highest voltages, and a high value $\rho(V)$, which depends on the voltage drop between network nodes as:

$$\rho(V) = \rho_{MAX} \quad (eV \leq \Phi), \qquad \rho(V) = \rho_{MAX}(\frac{\Phi}{eV}) + \rho_{min}(1 - \frac{\Phi}{eV}) \quad (eV \geq \Phi) \qquad (1)$$

where $\rho_{MAX}$ is the maximum resistivity value taken to fit the I-V characteristic at the lowest voltages (Ohmic response) and $\Phi$ is the height of the tunneling barrier between nodes. The

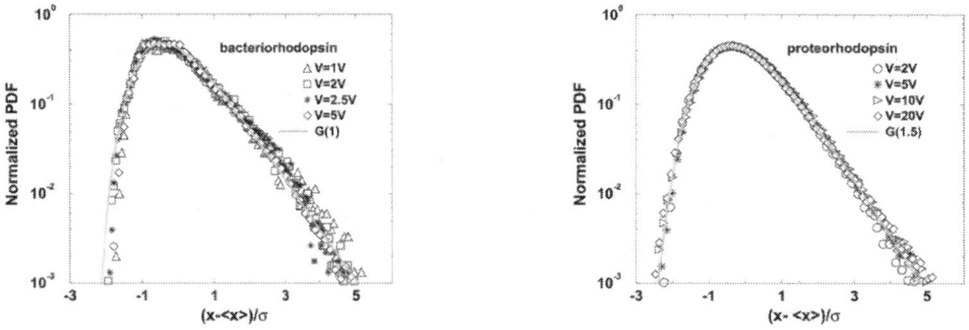


**Figure 3.** Normalized PDF of conductance fluctuations for bR and pR, at different bias values, x=ln($g$), $\sigma^2$ is the variance [7] of x.

transmission probability of each tunneling process is given by:

$$\mathcal{P}_{i,j}^{DT} = \exp\left[-\alpha\sqrt{(\Phi - \frac{1}{2}eV_{i,j})}\right] \quad (eV_{i,j} < \Phi), \quad \mathcal{P}_{ij}^{FN} = \exp\left[-\alpha \frac{\Phi}{eV_{i,j}}\sqrt{\frac{\Phi}{2}}\right] \quad (eV_{i,j} \geq \Phi) \quad (2)$$

where $V_{i,j}$ is the potential drop between the couple of $i, j$ amino acids, $\alpha = \frac{2l_{i,j}\sqrt{2m}}{\hbar}$, and $m$ is the electron effective mass, here taken the same of the bare value.

## 3. Results

The investigation is carried out through a comparative analysis of pR and bR properties. The INPA model is calibrated on the experiments performed on bR monolayers [1]. In particular, starting from the experimental I-V characteristic [1] the model parameters used to get the best agreement are: $R_c = 6$ Å, $\Phi = 3.5 \times 10^{-20}$ J, $\rho_{MAX} = 4 \times 10^{13}$ $\Omega$ Å, and $\rho_{min} = 4 \times 10^{5}$ $\Omega$ Å [4]. Figure 1 reports the I-V characteristics calculated for a single protein of bR and pR, on a wide bias range. In the common range of applied voltages, calculations agree with experiment when properly rescaled to take into account the sample size. Figure 1 reports the data on scale axes appropriate to evidence the transition between DT and FN tunneling regimes, the latter being evidenced by a sharp increase (over ten orders of magnitude) at high voltages of the dependent variable. The average chord-conductance, $\langle g \rangle = \langle I \rangle / V$, is calculated over a large ensemble of configurations, and is taken as a measurable global-quantity and order parameter in recent investigations [7]. As a further step, the conductance fluctuations which emerge as a result of the same numerical simulations are estimated. Indeed, these fluctuations are essential to determine the average value of the global quantity $\langle g \rangle$ since they are particularly high close to the transition between the DT and FN regimes, as generally expected near to a phase transition. Indeed, the probability distribution function (PDF) of these fluctuations, normalized to its first and second moment, is found to be well described by the generalized Gumbel distribution, $G(a)$,

$$G(a) = \frac{\theta(a)a^a}{\Gamma(a)} exp\{-aw - ae^{-w}\} \quad (3)$$

where $a$ is a positive number, $\sigma^2$ the variance, $w = \theta(a)(z + \nu(a))$ with $z = \frac{ln(g) - <ln(g)>}{\sigma}$. Furthermore, $\nu(a) = \frac{1}{\theta(a)}(\ln(a) - \psi(a))$, with $\Gamma(a)$, $\psi(a)$ and $\theta^2(a)$ indicating the Gamma, digamma and trigamma function, respectively. Finally, $G(a)$ is a normalized PDF with zero mean and unitary variance; it is a skewed function, recently used to describe finite-size systems in both critical and not critical conditions [6]. The numerical value of the shaping parameter $a$ is related to the system internal dynamic [6, 7]. In the present case, the lack of a Gaussian behaviour is due to the existence of a low and upper bound on the conductance values [7]. The values of $a$ which, in a wide range of applied bias, better describe the PDFs for bR and pR, are $a = 1$ and $a = 1.5$, respectively (see Figure 3). Integer values of $a$ allow to interpret the $G(a)$ as the distribution of the $a$-th minimum/maximum of the independent variables, i.e. $a = 1$ should be correlated with the existence of a single dominant conductance value. Otherwise, the interpretation of on integer $a$ value remains an open question.

## 4. Conclusions

Microscopic investigations of the I-V and conductance characteristics of the two light receptors bR and pR reproduce available experiments and predict the pR behaviour on a bias range not yet explored. From experiments, the protein activity after dehydration was tested by absorption measurements that confirmed the preservation of the structure and the function after the treatment. From theory, the PDF of the fluctuations of the conductance, $\langle g \rangle$, taken as order parameter are found to follow the generalized Gumbel distribution $G(a)$, for both proteins. These results confirm the INPA approach as a powerful numerical method to investigate protein electrical properties on the basis of their tertiary structure.

## References

[1] I Casuso, L Fumagalli, J Samitier, *et al.*, 2007 *Phys. Rev. E* **76** 041919
[2] H Melikyan, B -E Khishigbadrakh, A Babajanyan, *et al.*, 2011 *Thin Solid Films* **519** 3425
[3] Y Jin, N Friedman, M Sheves, *et al.*, 2006 *Proc. Natl. Acad. Sci.* **103** 8601 .
[4] E Alfinito, J -F Millithaler and L Reggiani 2011 *Phys. Rev. E* **83** 042902
[5] H M Berman, J Westbrook, Z Feng, *et al.* Nucleic Acids Research *2000* **28** 235
[6] A Noullez, and J -P Pinton, 2002 *Eur. Phys. J.* **B 28** 231 ; S C Chapman, G Rowlands and N W Watkins, 2005 *J. Phys. A: Math. Gen.* **38** 2289
[7] E Alfinito and L Reggiani, 2013 *J. Phys. C: Solid State Phys.* **25** 375103

# Computational diagnosis of canine lymphoma

**E M Mirkes[1], I Alexandrakis[2], K Slater[3], R Tuli[2] and A N Gorban[1]**

[1] Department of Mathematics, University of Leicester, Leicester, LE1 7RH, UK
[2] Avacta Animal Health, Unit 706, Avenue E, Thorp Arch Estate Wetherby, LS23 7GA
[3] PetScreen Ltd, Biocity, Pennyfoot Street, Nottingham, NG1 1GF, UK

E-mail: ag153@le.ac.uk

**Abstract**. One out of four dogs will develop cancer in their lifetime and 20% of those will be lymphoma cases. PetScreen developed a lymphoma blood test using serum samples collected from several veterinary practices. The samples were fractionated and analysed by mass spectrometry. Two protein peaks, with the highest diagnostic power, were selected and further identified as acute phase proteins, C-Reactive Protein and Haptoglobin. Data mining methods were then applied to the collected data for the development of an online computer-assisted veterinary diagnostic tool. The generated software can be used as a diagnostic, monitoring and screening tool. Initially, the diagnosis of lymphoma was formulated as a classification problem and then later refined as a lymphoma risk estimation. Three methods, decision trees, kNN and probability density evaluation, were used for classification and risk estimation and several pre-processing approaches were implemented to create the diagnostic system. For the differential diagnosis the best solution gave a sensitivity and specificity of 83.5% and 77%, respectively (using three input features, CRP, Haptoglobin and standard clinical symptom). For the screening task, the decision tree method provided the best result, with sensitivity and specificity of 81.4% and >99%, respectively (using the same input features). Furthermore, the development and application of new techniques for the generation of risk maps allowed their user-friendly visualization.

## 1. Introduction

Lymphoma (Lymphosarcoma, LSA) is one of the most common cancers seen in dogs. One in four dogs will develop cancer in their lifetime. It accounts for approximately 20% of all canine tumours [1].

The PetScreen Canine Lymphoma Blood Test employs advanced technology to detect lymphoma biomarkers present in a dog's serum [1]. Concentration of two acute phase proteins is evaluated: Haptoglobin (Hapt) and C-Reactive Protein (CRP) Detection of these biomarkers indicates a high likelihood that the dog has lymphoma [1], [2]. The growth of the database enables new methods of data mining. We analyse usability of attributes for the lymphoma diagnostic test. We divide the database in two cohorts and formulate two different tasks: (i) *differential diagnostic in clinically suspected cases* and (ii) *screening*. The isolation of the clinically suspected cohort is necessary for formulation of the task of differential diagnostics and selection of the appropriate methods.

Three methods are used. The first is the method of *decision trees* [3-5]. The second is *k nearest neighbours method* (kNN) in several versions [6,7]. The third is the method of *probability density function estimation* (PDFE) [8,9]. We use them for direct estimation of the lymphoma risk.

We present the case study for both tasks: for the diagnostics task we have tested 2,432,000 variants of the kNN method, 248,400 variants of decision tree algorithms and 280 variants of PDFE method; for the screening task we have tested 48,640 variants of kNN, 4,968 variants of decision trees and 280 variants of PDFE.

The versions differ by impurity criteria, kernel functions, number of nearest neighbours, weights and other parameters. They are compared by the standard goodness of fit, the entropic criteria and the generalization ability.

The best results are implemented in web-accessed software for the diagnosis of canine lymphoma.

If we solve the classification problem then for each value of input features we can obtain only two answers: dog with lymphoma or dog without lymphoma. Formulation of the task as a problem of risk estimation provides obtaining the real value between 0 and 1. Analysis of risk map provides a new ability to formulate hypotheses.

The obtained results provide the creation of a reliable diagnostic and screening system for canine lymphoma.

## 2. Data description

Table 1. Relative information gain for 'Lymphoma'

| Tested feature | RIG | RIG under given Lymphadenopathy | |
|---|---|---|---|
| | | Y | N |
| Lymphadenopathy | 28.92% | - | - |
| CRP binned | 24.38% | 15.00% | 23.52% |
| Hapt binned | 7.02% | 1.76% | 14.32% |
| Age binned | 6.07% | 1.62% | 9.39% |
| Sex | 0.95% | 3.79% | 22.84% |

**Categorical features**. Lymphoma ('Y' or 'N'), Sex ('M' or 'F'), and a most important clinical symptom, Lymphadenopathy ('Y' or 'N')

**Real features.** Age, CRP (concentration), Hapt (concentration).

**Analysis of importance**. Table 1 contains values of relative information gain (RIG) [3] for target feature Lymphoma of any input feature. RIG is calculated for whole database and for two samples: (Y) with Lymphadenopathy='Y' and (N) with Lymphadenopathy='N'.

## 3. Problem refinement

The results of the database analysis show that there are two different cohorts of data in the database.

The first cohort is titled 'clinically suspected' and contains records collected by PetScreen from dogs undergoing differential diagnosis. All these records correspond to dogs which have been referred to differential diagnostics by veterinary practitioners on the base of one or more clinical symptoms. It is not possible to find a posteriori these symptoms for each instance and we have to introduce a new synthetic attribute: 'clinically suspected'. The second cohort is titled 'healthy' and contains records obtained from healthy dogs courtesy of the Pet Blood Bank. These two cohorts have very different statistics of the attributes. By expert estimations, the prior probability of lymphoma is located between 2% and 5% in the canine population. The number of records of dogs with lymphoma is 97 or 32% of all the records in the initial database. All these cases have been clinically suspected and form 42% of the clinically suspected cases. This imbalance entails the use of specific methods to solve screening tasks. The 'clinically suspected' feature was added to the database to identify the two cohorts. The values of feature 'clinically suspected' were defined by using additional information of veterinary cards. The existence of the two cohorts allows the formulating of two different tasks: differential diagnostics and screening.

**Differential diagnostics**. The differential diagnostic task can be formulated as a problem of lymphoma diagnosis for dogs with some clinical symptoms of lymphoma. A diagnostic task is a usual classification problem and all classification methods can be used. We use three types of classification methods: kNN, decision tree and the method based on probability distribution function estimation. The first two methods have an additional parameter – weighting of the positive class.

**Screening**. The screening task can be formulated as an estimate of lymphoma risk for any dog. To solve this task we used all the database records. The experts' estimation of prior probability of lymphoma is between 2% and 5% however the fraction of dogs with lymphoma records in the database is 32%. To compensate for this imbalance all methods take into account the prior probability of lymphoma. For this task, the weighting of classes is defined by prior probability.

We use the following notations: $p$ is the prior probability of lymphoma, $N_L$ is the number of dogss with lymphoma, $N_{CS}$ is the number of all clinically suspected dogs and $N_H$ is the number of healthy dog. The weight of the class of dogs with lymphoma is equal to $p$. The weight of one dog with lymphoma is equal to $w_L = p/N_L$. Indeed, this is the weight of any record of the clinically suspected cohort. The weight of each record of a healthy dog is calculated as $w_H = (1\text{-}w_L N_{CS})/N_H$. These two weights are used to calculate the risk of lymphoma in the screening task.

## 4. Risk map analysis

Visualization of data and probability distributions may use various screens, from the coordinate planes and PCA to non-linear principal graphs and manifolds [10]. In this work, we use visualisation of risk of lymphoma on the plane of two real attributes, CRP and Hapt. For example, we use risk maps to generate hypotheses about impact of input features. All maps below use the legend which is presented in any figure. The brown colour indicates the low level of risk: the greater intensity of brown colour indicates the less risk of lymphoma. The white colour indicates the median value of risk (50%). Blue colour indicates the high level of risk: the greater intensity of indicates the greater risk value.

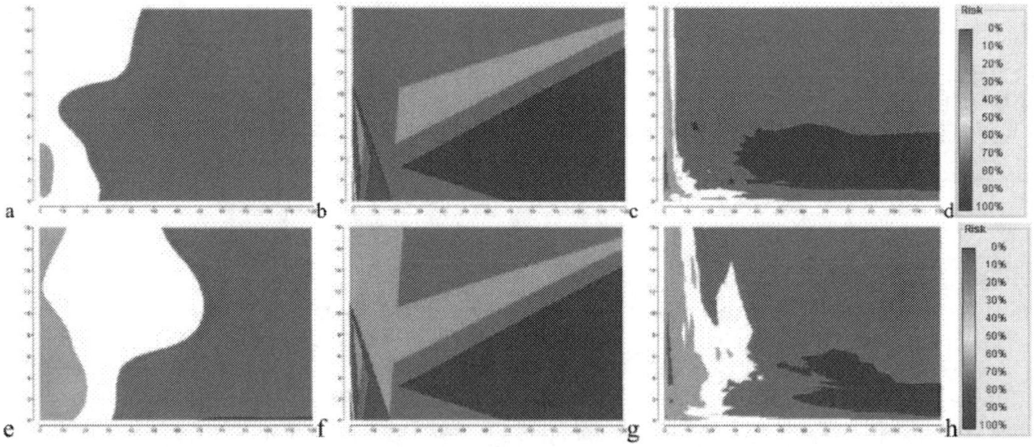


Figure 1. The maps of lymphoma risk for male and female dogs: a) PDFE map for male, b) decision tree map for male, c) KNN map for male, e)PDFE map for female, f) decision tree map for female, g) KNN map for female, d) and h) are legend.

Let us consider the risk of lymphoma in relation to sex for clinically suspected dogs. The probabilities are represented (Fig 1). The maps show that there are some areas where the risk of lymphoma is greater for male dogs. For the best decision tree and kNN the qualitatively same results were obtained (Fig. 2). In this area, the risk of lymphoma may depend on the steroid hormones. This hypothesis needs additional verification.

## 5. Results

During the case-study the best solution for each task is selected with the minimal number of errors.

**Differential diagnostic**. The best result is obtained by a decision tree which uses three input features: the concentrations of CRP and Hapt, and Lymphadenopathy. The concentrations of CRP and

Hapt are used in linear combinations. DKM is used as a splitting criterion. The sensitivity of this method is 83.5%, specificity is 77%. ROC integral for this method is 0.879.

If 'Lymphadenopathy' is unknown then we use decision tree which only uses CRP and Hapt. In the best version, these input features are used in linear combinations after logarithmic transformation. Information gain is used as the splitting criterion. The sensitivity of this method is 81.5%, specificity is 76%. ROC integral for this method is 0.810.

**Screening**. The best result is obtained by the decision tree which uses three input features: the concentrations of CRP and Hapt, and Lymphadenopathy. CRP and Hapt are used separately. DKM is used as the splitting criterion. The sensitivity of this method is 81.4%, specificity is >99% (no false negative results in one-leave-out cross-validation).

If 'Lymphadenopathy' is unknown then we use another decision tree, which uses CRP and Hapt only. These input features are used in linear combinations after logarithmic transformation. Gini gain is used as the splitting criteria. The sensitivity is 65%, specificity is 83%.

To evaluate the quality of the achieved results, we compare them to some current human cancer screening tests. The tests that rely upon single biomarkers demonstrate often the worse performance. For example, the CA-125 screen for human ovarian cancer provides sensitivity approximately 53% and specificity 98%, and the male PSA test gives sensitivity approximately 85% and specificity 35%. Supplementation of CA-125 by several other biomarkers increases sensitivity of at least 75% for early stage disease and specificity of 99.7%. For PSA marker, using age-specific reference ranges improved test specificity and sensitivity, but did not improve the overall accuracy of PSA testing.

Visualisation of risk maps provides a friendly tool for explanatory data analysis and affords an opportunity to generate hypotheses about impact of input feature on the final diagnosis. For more details and the additional bibliography we refer to [11].

## References

[1] Ratcliffe L, Mian S, Slater K, King H, Napolitano M, Aucoin D and Mobasheri A 2009 Proteomic identification and profiling of canine lymphoma patients, *Veterinary and Comparative Oncology* **7(2)** pp 92–105

[2] Alexandrakis I 2012 The use of CART algorithms to combine serum acute phase protein levels as a diagnostic aid in canine lymphoma *Proc. of 15th Congress of the Int. Society for Animal Clinical Pathology, 14th Conf. of the European Society of Veterinary Clinical Pathology, (Ljubljana, Slovenia, 3rd-7th July, 2012)* (Ljubljana: Veterinary Faculty) p 65

[3] Rokach L and Maimon O 2010 Decision trees *Data Mining and Knowledge Discovery Handbook* ed O Maimon and L Rokach (Berlin: Springer) pp 165–192

[4] Quinlan J R 1987 Simplifying decision trees *Int. J. of Man-Machine Studies* **27** pp 221–234

[5] Gelfand S B, Ravishankar C S and Delp E J 1991 An iterative growing and pruning algorithm for classification tree design *IEEE Transaction on Pattern Analysis and Machine Intelligence* **13(2)** pp 163–174

[6] Clarkson K L 2005 Nearest-neighbor searching and metric space dimensions *Nearest-Neighbor Methods in Learning and Vision: Theory and Practice* (The MIT Press) pp 15–59.

[7] Hastie T and Tibshirani T 1996 Discriminant adaptive nearest neighbor classification *IEEE Transactions on Pattern Analysis and Machine Intelligence* **18 (6)** pp 607–616

[8] Scott D W 1992 *Multivariate Density Estimation: Theory, Practice and Visualization* (New York: Wiley)

[9] Buhmann M D 2003 *Radial Basis Functions: Theory and Implementations* (Cambridge University Press)

[10] Gorban A N, Kégl B, Wunsch D C and Zinovyev A (eds) 2007 *Principal Manifolds for Data Visualisation and Dimension Reduction* (*LNCSE*) vol 58 (Berlin – Heidelberg – New York: Springer)

[11] Mirkes E M, Alexandrakis I, Slater K, Tuli R and Gorban A N 2023 arXiv:1305.4942 [q-bio.QM]

# A modified Seeded Region Growing algorithm for vessel segmentation in breast MRI images for investigating the nature of potential lesions

**D Glotsos[1], K Vassiou[3,4], S Kostopoulos[1], El Lavdas[2], I Kalatzis[1], P Asvestas[1], D L Arvanitis[3], I V Fezoulidis[4] and D Cavouras[1]**

[1]Department of Medical Instruments Technology, Technological Education Institute of Athens, Greece
[2]Department of Medical Radiologic Technology, Technological Educational Institute of Athens, Athens, Greece
[3]Department of Anatomy, Medical School, University of Thessaly, Viopolis, Larissa, Greece
[4]Department of Radiology, University Hospital of Larissa, Viopolis, Larissa, Greece

E-mail: dimglo@teiath.gr

**Abstract.** The role of Magnetic Resonance Imaging (MRI) as an alternative protocol for screening of breast cancer has been intensively investigated during the past decade. Preliminary research results have indicated that gadolinium-agent administrative MRI scans may reveal the nature of breast lesions by analyzing the contrast-agent's uptake time. In this study, we attempt to deduce the same conclusion, however, from a different perspective by investigating, using image processing, the vascular network of the breast at two different time intervals following the administration of gadolinium. Twenty cases obtained from a 3.0-T MRI system (SIGNA HDx; GE Healthcare) were included in the study. A new modification of the Seeded Region Growing (SRG) algorithm was used to segment vessels from surrounding background. Delineated vessels were investigated by means of their topology, morphology and texture. Results have shown that it is possible to estimate the nature of the lesions with approximately 94.4% accuracy, thus, it may be claimed that the breast vascular network does encodes useful, patterned, information, which can be used for characterizing breast lesions.

## 1. Introduction

In the presence of abnormal growth, the need for oxygen and nutrients rapidly increases. Mechanisms for stimulating vascular growth are activated and new vessels are generated to support the increased metabolic demands of the tumour. This effect is captured by proper MRI scan protocols, such as T1-weighted scans, and, thus, diagnostic assessments are feasible [1-4]. These assessments are made on the basis of the so-called Time Intensity Curves (TIC) or Kuhl Curves [5, 6], which are divided into three categories: a/ type I characterized by a continuous increase in the measured signal (contrast-agents uptake) over time; these curves are indicative of benign cases, b/ type II characterized by signal increment upon a plateau value, raising concerns that malignant formations might be present, and c/ type III characterized by a rapid signal increase following by a rapid decrease, which has been shown as typical pattern in the presence of malignancy. Although angiogenesis is an important diagnostic lead strongly correlating with malignancy, it is also appears in non-malignant cases, such as benign lesions, inflammations etc. The latter complicates assessments. Extensive clinical reviews have shown a great sensitivity (around 90%) of the method with controversial specificity (from 37% to 100%) [7] that increases the overall cost of the exam and distress of the patients, since biopsies are obtained needlessly. The latter has initiated a number of studies aiming in standardizing the procedures and protocols for breast MRI scans assessments. Yabuuchi et al [8] have proposed a method that combines DCE and DWI (Diffusion Weighted Imaging) leading to increased sensitivity, specificity, positive predictive value, negative predictive value, and accuracy. Sivarajan et al [9] have concluded that the introduction of morphologic and perfusion parameters in the interpretation of breast DCE-MRI improves both sensitivity and specificity. Macura et al [10] concluded that intensity enhancement measured in various time intervals after the contrast agent administration (thus, the TICs) is the most crucial. In this study, we attempt to investigate whether the quantification of the vascular network of the breast might lead to the same diagnostic conclusions as the investigation of the TICs. For this reason, a structured sequence of image processing step was developed with basis a modified Seeded Region Growing (SRG) algorithm in order to effectively delineate blood vessels.

## 2. Material and Methods

### *2.1. Material*

Clinical material comprised 20 breast MRI images from 20 female patients, 18 with confirmed cancer. The MRI examinations were carried out on a 3T system (Signa, GE Healthcare) and the array spatial sensitivity encoding technique (ASSET) or parallel imaging technique was used. Two MIP (maximum intensity projection) images were obtained from each patient, one 1.5 minute after the administration of the contrast agent (pre-contrast) and one after 1.5-3 min (post-contrast). The contrast agent administration consisted of injection of 0.1 mmol/kg of gadopenate dimeglumine (Omniscan, Magnevist, Multihance) followed by a 10-mL saline solution flush, over a period of 5-8 sec. Detailed information regarding the data and protocols used can be found in [11].

### *2.2. Methods*

The methodology consisted of a structured sequence of image processing algorithms:

**Step 1:** Background removal by subtracting from the original image a morphological opened image using a 7x7 disk component to remove structures that had similar grey level values with vessels (i.e. dense breast regions and masses) [12]. **Step 2:** Median filtering (3x3) to reduce noise [13]. **Step 3:** Otsu's thresholding [14] for an initial estimation of seed points used by subsequent step of the SRG algorithm. **Step 4:** Application of the SRG algorithm, which was constructed using a hysteresis criterion (see section *2.2.1*), for the finalization of the segmentation (white pixels in the binary image represented vessel regions). **Step 5:** Based on the segmented image, the total area of vessels was calculated as the sum of white pixels for both the pre- and post contrast image of each patient. By dividing this area with the total number of pixels of the breast (Whole Breast Area - WBA) the percentage of vessels within the breast (Vascular Coverage - VC) was computed:

$$VC_{pre/post-contrast} = \frac{Segmented\ Vessel\ Area_{pre/post-contrast}}{WBA} \tag{1}$$

WBA was calculated as the sum of all segmented pixels found after applying thresholding to the morphologically processed image (using a 7x7 opening filter). **Step 6:** The ratio of VC of the pre- and the post-contrast image was computed as an index of the Vascular Growth (VG). Values of VG<1 meant that the vascular coverage appeared increased in the post-contrast image.

$$VG = \frac{VC_{pre-contrast}}{VC_{post-contrast}} \tag{2}$$

#### *2.2.1. Seeded Region Growing (SRG)*

The SRG algorithm aims in partitioning the image into meaningful regions on the basis of a set of initial predefined points, the so called seeds. The seeds can be manually set or automatically labelled. Given the seeds, the algorithm searches the seeds' immediate (connected) neighbours to find those sharing some common similarity. The seeds are also referred to as allocated pixels, whereas the rest of the pixels as non-allocated. Considering a set of initial seeds $S_1, S_2, S_3, \ldots, S_n$, $n$ is the number of total seeds, then non-allocated pixels found in the immediate vicinity of $S$ are [15]:

$$U = \left\{ x \notin \bigcup_{i=1}^{n} S_i \,\middle|\, N(x) \cap \bigcup_{i=1}^{n} S_i \neq 0 \right\} \tag{3}$$

where $N(x)$ is the set of non-allocated immediate neighbours of pixel $x$. Usually the immediate neighbourhood is considered rectangular with 8 neighbour-pixels surrounding $x$, which is found at the center of this region. To estimate whether the neighbours will be allocated to the 'seed-class', a similarity criterion should be defined. The simplest criterion $\delta(x)$ is

$$\delta(x) = \left| g(x) - \underset{y \in S_i(x)}{mean}\left[ g(y) \right] \right| \tag{4}$$

where $g(x)$ is the grey value of pixel $x$. When this criterion is minimized then pixel $x$ is classified as a seed-class pixel. In this study a modified version of the SRG algorithm was developed and applied. The modified version uses a different criterion for investigating the immediate neighbourhood of the seed-class pixels. The proposed criterion separates the immediate neighbourhood into strong pixels belonging to the seed region (Region 1), strong pixels belonging to the background region (Region 2) and weak pixels, which are the non-allocated candidate pixels (Region 3), according to:

$$REGION\ 1: \delta(x) = \left| g(x) - \underset{y \in S_i(x)}{mean}\left[ g(y) \right] \right| < 0.2\left( \max\{S\} - \min\{S\} \right) \tag{5}$$

$$REGION\ 2: \delta(x) = \left| g(x) - \underset{y \in S_i(x)}{mean}\left[ g(y) \right] \right| > 0.8\left( \max\{S\} - \min\{S\} \right) \tag{6}$$

$$REGION\ 3: \delta(x) = \left| g(x) - \underset{y \in S_i(x)}{mean}\left[ g(y) \right] \right| \tag{7}$$

## 3. Results and Discussion

Figure 1 illustrates an example of the application of the proposed structured sequence of image processing steps to a post-contrast image of a female patient with breast cancer at the center of the right breast with type III TIC curve. Figure 1(a) is the original image. Figure 1(b) is the background removed median filtered image, where it is clearly visible that the intensity and grey level distribution of pixels belonging to the mass at the center of the right breast are greatly weakened. In this way, vessel segmentation is facilitated. Non-informative texture other than vessels was smoothed, since this kind of texture did not constitute useful information. Figure 1(c) depicts the result of an Otsu-based thresholding, which was used as a gross estimation of the initial seeds for the subsequent SRG. This segmentation result was incomplete since many vessels were lost. The result was greatly enhanced (Fig 1(d)) following the application of the modified SRG. More vessels were detected as compared with thresholding, and corrupted vessels were completed. However, there were still vessel regions undetected. The latter did not significantly affect the computation of vascularity, since it is a reasonable to assume that the magnitude of under-segmentation was approximately the same in both the pre- and post-contrast images. Thus, the ratio of VC of the pre- and post-contrast images eliminated this under-segmentation error. Figure 1(e) illustrates the superposition of the segmented vessel pixels to the original image (Fig 1(a)), clearly indicating a segmentation underestimation. It is logical to assume that this underestimation in the first and second time interval image is the same, thus, the computation of the VG factor is not biased. Figure 1(f), which is a scatter diagram of the VG factor for each case included in the study, illustrates the Vascular Growth factor for all 18 cancer cases tested. For all cases with breast cancer except one (for the 94.4% of the cases), the vascular network appeared more prominent in the post-contrast image ($VG \ll 1$) confirming that abnormal formations do require an increase metabolic, and, thus, blood circulation coverage. Moreover, this increment of vascularity was found as high as 60% for the more advanced malignancies. Only one case was found as of no change (VG≈1). In the two control images (normal cases), no change of the appearing vascular network was computed (VG≈1) between the pre- and post-contrast images. Thus, results indicate that it is possible to estimate the presence of cancer in breast MRI images by estimating the extent and the area of the vascular network.

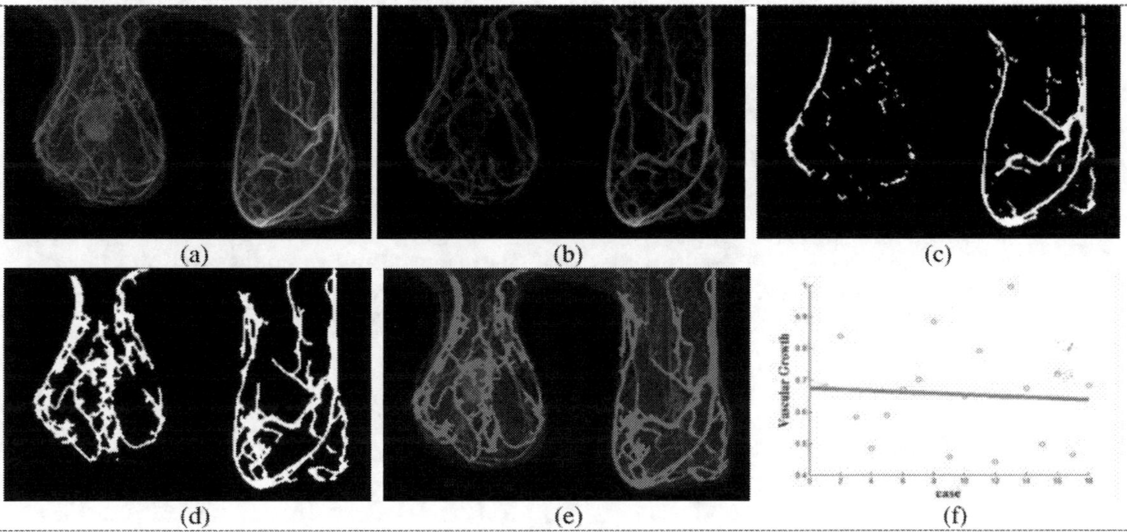


**Figure 1.** (a) Original Image, (b) Background Corrected Image, (c) Median Filtered Image, (d) Threshold Image, (e) Final SRG Segmented Image, (f) Vascular Growth

The sensitivity of the proposed method (94.4%) is in line with results presented in literature [7] (around 90%), although the proposed methodology investigates the problem from a different point of view focusing on the vascular network of the breast, rather than investigating the TOC curves [7].

**Acknowledgments**
This research has been co-financed by the European Union (European Social Fund – ESF) and Greek national funds through the Operational Program "Education and Lifelong Learning" of the National Strategic Reference Framework (NSRF) - Research Funding Program: **ARCHIMEDES III**. Investing in knowledge society through the European Social Fund.

## 4. References

[1] Yabuuchi, et al. 2010 *Eur J Radiol* **75**(1) e126-32
[2] Kuhl 2007 *Magn Reson Imaging Clin N Am* **15**(3) 315-20, vi
[3] Kuhl 2000 *Eur Radiol* **10**(1) 46-58
[4] Yuan, et al. 2013 *Eur J Radiol* **82**(9) 1506-11
[5] Mann, et al. 2008 *Eur Radiol* **18**(7) 1307-18
[6] Kuhl and Schild 2000 *J Magn Reson Imaging* **12**(6) 965-74
[7] Lee 2004 *Radiol Clin North Am* **42**(5) 919-34
[8] Yabuuchi, et al. 2008 *J Magn Reson Imaging* **28**(5) 1157-65
[9] Lee, et al. 2012 *Eur J Radiol* **81**(4) 784-8
[10] Macura, et al. 2006 *Radiographics* **26**(6) 1719-34; 1719
[11] Vassiou, et al. 2013 *Acta Radiol*
[12] Wong, Kamyar, and Molloi 2007 *Med Phys* **34**(10) 4003-15
[13] Nayebifar and Abrishami Moghaddam 2013 *Comput Biol Med* **43**(5) 541-8
[14] Sukkaew, et al. 2007 *J Med Assoc Thai* **90**(9) 1780-92
[15] Adams R 1994 *IEEE Trans on Pattern Analysis and Machine Intelligence* **16**(6)

# Multimodality imaging and state-of-art GPU technology in discriminating benign from malignant breast lesions on real time decision support system

**S Kostopoulos[1], K Sidiropoulos[2], D Glotsos[1], N Dimitropoulos[3], I Kalatzis[1], P Asvestas[1] and D Cavouras[1]**

[1]Department of Biomedical Engineering, Technological Educational Institute of Athens, Greece

[2]School of Engineering and Design, Brunel University West London, Uxbridge, Middlesex, UB8 3PH,United Kingdom

[3]Delta Digital Diagnostic Center, Semitelou 6, Athens, 11528, Greece

E-mail: skostopoulos@teiath.gr

**Abstract.** The aim of this study was to design a pattern recognition system for assisting the diagnosis of breast lesions, using image information from Ultrasound (US) and Digital Mammography (DM) imaging modalities. State-of-art computer technology was employed based on commercial Graphics Processing Unit (GPU) cards and parallel programming. An experienced radiologist outlined breast lesions on both US and DM images from 59 patients employing a custom designed computer software application. Textural features were extracted from each lesion and were used to design the pattern recognition system. Several classifiers were tested for highest performance in discriminating benign from malignant lesions. Classifiers were also combined into ensemble schemes for further improvement of the system's classification accuracy. Following the pattern recognition system optimization, the final system was designed employing the Probabilistic Neural Network classifier (PNN) on the GPU card (GeForce 580GTX) using CUDA programming framework and C++ programming language. The use of such state-of-art technology renders the system capable of redesigning itself on site once additional verified US and DM data are collected. Mixture of US and DM features optimized performance with over 90% accuracy in correctly classifying the lesions.

## 1. Introduction

Breast cancer is the most frequent malignant tumour among women. Diagnostic mammography (DM) is the leading imaging protocol for screening of breast tumours [1]. Among the most important signs indicating the presence of malignant lesions in mammograms is the existence of masses, microcalcifications (mCs) and clusters of mCs [2]. However, in diagnostically difficult cases, such as in early stage cancer, the differentiation between normal and abnormal mammograms is not straightforward due to the complexity of the composition of breast tissues, which might lead to misinterpretations for up to 30% of the cases [3, 4]. Moreover, DM is among the projection radiological protocols with the highest dose profiles, although modern systems have reduced the dose profile. The above reasons have evoked attempts to propose different imaging modalities that would a/ replace and/or improve diagnostic mammography and b/ reduce overall dose. Such imaging modalities are the Magnetic Resonance Imaging (MRI) and the Ultrasonography (US) [5, 6]. MRI offers a competitive alternative to DM, however, the cost and the complexity of the exam is still one major drawback. US, on the other hand, offers a non-ionizing alternative with low cost and complementary information to DM [7]. In this study we attempt to investigate the nature of this complementary information of US with respect to DM by means of pattern recognition system design for discriminating benign from malignant cancer cases.

## 2. Material and Methods

### *2.1. Material*

Clinical material comprised 59 patients that underwent both DM and US examinations, which were conducted by the same experienced radiologist (N.D.). Twenty nine (29) cases were diagnosed as malignant whereas thirty (30) as benign. All patients were examined at the Department of Radiology of the Delta Digital diagnostic center, Athens, Greece.

### *2.2. Methods*

An experienced radiologist outlined breast lesions on both US and DM images from 59 patients employing a custom designed computer software application (Figure 1). Textural features were extracted from each segmented lesion (Figure 2) and were used to design the pattern recognition system. These features comprised four (4) from the lesion's grey-tone histogram, twenty-six (26) from the lesion's co-occurrence matrix [8] and 10 from the lesion's run-length matrix [9]. Thus, eighty (80) features were computed for each case (patient), forty (40) from the segmented DM lesion and forty (40) from the segmented US lesion.

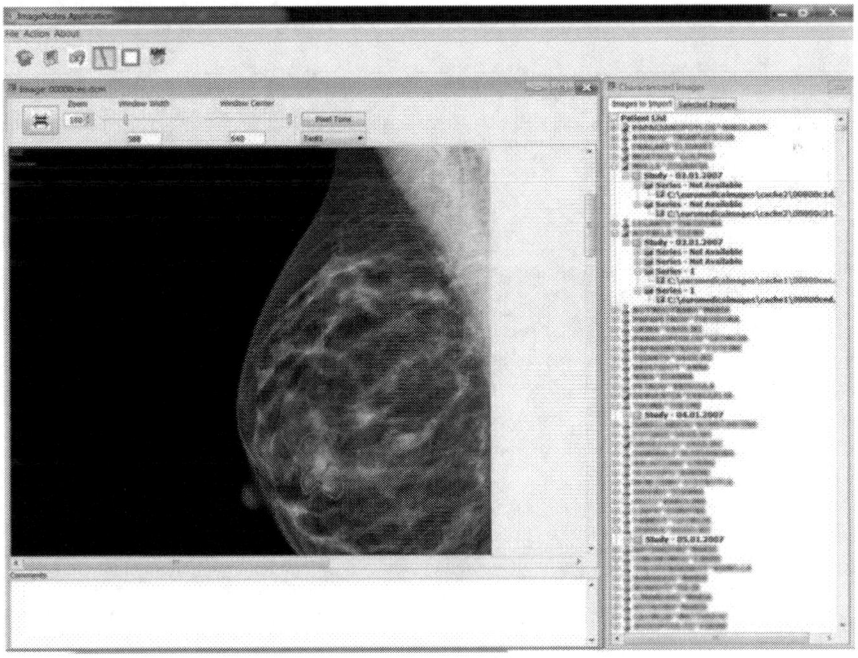

**Figure 1.** Custom designed software for breast lesion delineation

Three classifiers were tested (Support Vector Machines [10], k-Nearest Neighbor [11] and Probabilistic Neural Network [12]) for highest performance and were also combined into ensemble schemes for further improvement of system classification accuracy. The exhaustive search and the Leave One Out method [11] were used for system performance evaluation. That meant that for each possible feature for the total of 80 features used, a different classifier should be designed and tested giving about 3 billion different architectures if one was to search for combinations up to 7 features (more feature combinations were not tested to avoid over-fitting [11]). In order to make this computational problem practical and feasible, when optimizing the system (selecting the best classifier configuration) the accuracy was not the only criterion. Another, equally important criterion was the complexity and simplicity of the classifier configuration scheme. The only classifier tested that combined both criteria (accuracy improvement and design simplicity) was the Probabilistic Neural Network classifier (PNN), which was then designed on the GPU card (GeForce 580GTX) using CUDA programming framework and C++ programming language [13].

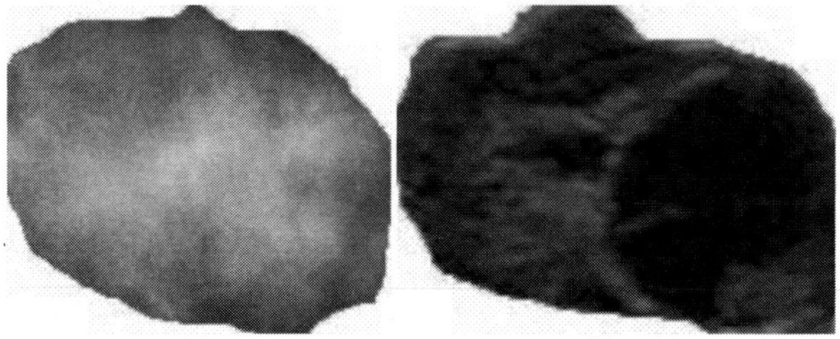

(a) (b)

**Figure 2.** Segmented regions from a patient with benign lesion; (a) DM and (b) US images

To investigate whether the combination of features from both DM and US improves the overall accuracy of the system in distinguishing benign from malignant cases, the pattern recognition system was designed under three different scenarios: 1. using only DM features, 2. using only US features, 3. using both DM and US features.

## 3. Results and Discussion

Table 1 summarizes the results of the pattern recognition system for the three scenarios. Using only DM best accuracy (83%) was obtained with 8 features. Using only US best performance was 84.7% with 6 features. Using both DM and US overall accuracy was boosted to 93.2% with 7 features, 5 from the DM lesion and 2 from the US lesion. It has to be mentioned that there are no dependenies between the features of the two modalities since the physicians are treated separately. It was possible to examine so many combinations of features exhaustively only because the PNN classifier was applied on the GPU card. More specifically, it took about 2 hours on the 580GTX, in contrast to more than 24 hours required on the CPU, which in fact failed to complete after 24 hours when 7 features combinations were asked.

**Table 1.** Classification accuracies using the PNN classifier in discriminating benign from malignant breast tumours.

| | | DM | US | DM+US |
|---|---|---|---|---|
| Accuracies (%) | Benign | 80.0 | 90.0 | 93.1 |
| | Malignant | 86.2 | 79.3 | 93.3 |
| | Overall | 83.1 (8 features) | 84.7 (6 features) | 93.2 (7 features) |

The five DM features comprised the correlation[r], that is a measure of grey-level dependencies, the difference entropy[r], that is a measure of anisotropy in grey-level randomness, the short run emphasis[a], that indicates the existence of small structures within the lesion, the run percentage[a], that is related to percentage of pixels that have the same grey-levels and the run percentage$^{r}$, where $^{a}$ and $^{r}$ stand for average and range over the four directions (0°, 45°, 90°, 135°). On the other hand, the two US features included the kurtosis and the run percentage$^{a}$. These results imply four main issues: a/ The texture of lesions in both DM and US images contain useful information regarding the malignancy of suspected tumours and may be used to discriminate benign from cancerous formations. b/ It is well known that morphology play a crucial role in the characterization of lesions, however, this study suggests that also the spatial homogeneity encoded by the textural features used, might be a strong, and perhaps complementary indicator for diagnosis. c/ The feature run percentage appears as important for both US and DM. This feature takes lower values for smooth textures, whereas its value increases for unsystematic distributions. Table 2 present the average feature values concerning benign and malignant for both modalities.

**Table 2.** The mean values and the standard deviation of the Run Percentage feature for different modalities.

| Run Percentage/Modality | DM | US |
|---|---|---|
| Benign | 0.48±0.06 | 0.22±0.10 |
| Malignant | 0.45±0.11 | 0.37±0.17 |

Thus, it may be speculated that benign tumours in DM present a coarser texture (many small runs, i.e. small homogeneous structures such as microcalcifications) as compared to malignant tumours, which are characterized by larger, smoother structures (large runs, i.e. masses). On the other hand, in US a different behaviour is observed regarding the values of the Run Percentage feature. Figure 2 visualizes such behaviour, since it is quite obvious that the benign lesion in the DM image is more homogeneous in terms of texture as compared to the benign lesion in the US image. Benign cases in US presented smaller values compared to malignant cases, indicating that the texture is coarser for malignant cases in US. d/ The combination of information from both DM and US boosts up the performance in discriminating the nature of the tumours, thus, it may be claimed that US gives complementary information to DM and should be further investigated not only as an additional protocol for screening of breast cancer, but as an essential prerequisite to DM that should be performed along with DM.

**Acknowledgments**
This research has been co-financed by the European Union (European Social Fund – ESF) and Greek national funds through the Operational Program "Education and Lifelong Learning" of the National Strategic Reference Framework (NSRF) - Research Funding Program: **ARCHIMEDES III**. Investing in knowledge society through the European Social Fund.

## 4. References

[1] del Junco, et al. 2008 *J Natl Cancer Inst* **100**(5) 333-46
[2] Carney, et al. 2010 *Radiology* **255**(2) 354-61
[3] Buist, et al. 2011 *Radiology* **259**(1) 72-84
[4] Jorgensen 2010 *Maturitas* **65**(1) 1-2
[5] Houssami and Ciatto 2011 *Prev Med* **53**(3) 123-6
[6] Lee, et al. 2009 *Acad Radiol* **16**(11) 1323-8
[7] Laine, et al. 1996 *European Journal of Ultrasound* **3**(1) 9-14
[8] Haralick, Shanmugam, and Dinstein 1973 *IEEE Trans Sys Man Cyb* **3** 610-621
[9] Galloway 1975 *Computer Graphics and Image Processing* **4** 172-179
[10] Burges 1998 *Data Mining and Knowledge Discovery* **2**(2) 121-167

[11] Jain, Duin, and Mao 2000 *IEEE Transactions on Pattern Analysis and Machine Intelligence* **22**(1) 4-37
[12] Specht 1990 *Neural Networks* **3** 109-118
[13] Sidiropoulos, et al. 2013 *Int J Comput Assist Radiol Surg* **8**(4) 547-60

# A Parallel 2D Numerical Simulation of Tumor Cells Necrosis by Local Hyperthermia

**R F Reis**[1], **F S Loureiro**[2] **and M Lobosco**[3]
[1,2,3]Department of Computer Science, Federal University of Juiz de Fora, Brazil
E-mail: [1]ruyfreis@gmail.com, [2]felipe.loureiro@ufjf.edu.br, [3]marcelo.lobosco@ufjf.edu.br

**Abstract.** Hyperthermia has been widely used in cancer treatment to destroy tumors. The main idea of the hyperthermia is to heat a specific region like a tumor so that above a threshold temperature the tumor cells are destroyed. This can be accomplished by many heat supply techniques and the use of magnetic nanoparticles that generate heat when an alternating magnetic field is applied has emerged as a promise technique. In the present paper, the Pennes bioheat transfer equation is adopted to model the thermal tumor ablation in the context of magnetic nanoparticles. Numerical simulations are carried out considering different injection sites for the nanoparticles in an attempt to achieve better hyperthermia conditions. Explicit finite difference method is employed to solve the equations. However, a large amount of computation is required for this purpose. Therefore, this work also presents an initial attempt to improve performance using OpenMP, a parallel programming API. Experimental results were quite encouraging: speedups around 35 were obtained on a 64-core machine.

## 1. Introduction

The tissue temperature variation caused by the magnetic nanoparticle hyperthermia process can be mathematically modeled by means of the Pennes equation [1]. Although there are other mathematical bioheat transfer models, the Pennes one is the most widely adopted in modeling due to its simplicity and good approximation [2]. Normally, heat generation in hyperthermia process is given by a quantity namely specific absorption rate (SAR) that are added to the bioheat equation as a source term. The SAR distribution and its parameters in the context of magnetic nanoparticles were determined by a previous study of temperature elevations in a rat hind limb [3].

The objective of the current study is to analyze different ways of employing injection sites inside the tumor. The tissue temperature in the hyperthermia process is written as a function of a steady-state temperature since before the heating the tumor tissue is at a higher temperature than the normal tissue. Furthermore, due to the large amount of floating point operations required to implement the numerical method, the current work also presents the speedup achieved by the parallel version of the code, developed using OpenMP[4].

## 2. Bioheat transfer model

The transient bioheat model used is the well-known Pennes equation given by [1]:

$$\begin{cases} \rho c \dfrac{\partial T_1(\vec{x},t)}{\partial t} = \nabla . k \nabla T_1(\vec{x},t) + \omega_b \rho_b c_b (T_a - T_1(\vec{x},t)) + Q_m(\vec{x}) + Q_r(\vec{x},t) & \text{in } \Omega, t > 0 \\ \alpha T_1(\vec{x},t) + \beta \nabla T_1(\vec{x},t) . \vec{n} = f(\vec{x},t) & \text{on } \Gamma, t > 0 \\ T_1(\vec{x},0) = T_2(\vec{x}) & \text{in } \Omega \end{cases} \tag{1}$$

where $\rho$, $c$ and $k$ are, respectively, the density, the specific heat and thermal conductivity of the tissue; $c_b$, $\rho_b$ and $\omega_b$ are, respectively, the specific heat of the blood, the density of the blood and blood perfusion rate; $Q_m$ metabolic heat generation; $T_a$ is the arterial blood temperature and $T_1$ the tissue temperature; $Q_r$ external spatial heating.

The initial temperature, called $T_2(\vec{x})$, over the whole tissue with a tumor can be calculated from the steady-state bioheat equation considering $Q_r = 0$, that are obtained taking $\frac{\partial T(\vec{x},t)}{\partial t} = 0$.

Once the tissue temperature $T_2(\vec{x})$ is known, one can proceed with the solution of equation 1. However, to simplify the analysis of the model we use a transient equation that represents the relative temperature $\overline{T}(\vec{x},t) = T_1(\vec{x},t) - T_2(\vec{x})$ [5]. Hence, we obtain:

$$\begin{cases} \rho c \dfrac{\partial \overline{T}(\vec{x},t)}{\partial t} = \nabla . k \nabla \overline{T}(\vec{x},t) - \omega_b \rho_b c_b \overline{T}(\vec{x},t) + Q_r(\vec{x},t) & \text{in } \Omega, t > 0 \\ \overline{T}(\vec{x},0) = 0 & \text{in } \Omega \end{cases} \tag{2}$$

## 3. Numerical scheme

The numerical method applied to solve equation 2 is the finite difference method [6]. In this way, a uniform Grid over the closed domain $\Omega \bigcup \Gamma \subseteq \mathbb{R}^2$ with coordinates $x_i = ih_x$ and $y_i = jh_y$ is constructed with $h_x$ and $h_y$ being the grid spacing in each direction. The discretization used in the time domain is the forward difference with a time step size $h_t$ such that $t_n = nh_t$ while the second order central difference is employed for the spatial discratization. Hence, the discrete counterpart of equation 2 reads:

$$T_{i,j}^{n+1} = \frac{h_t}{\rho_{i,j} c_{i,j}} \left[ -\frac{(q_{i+1/2,j}^n - q_{i-1/2,j}^n)}{h_x} - \frac{(q_{i,j+1/2}^n - q_{i,j-1/2}^n)}{h_y} - \omega_{bi,j} \rho_{bi,j} c_{bi,j} T_{i,j}^n + Q_{ri,j}^n \right] + T_{i,j}^n \tag{3}$$

where $q_{i+1/2,j}^n = -k_{i+1/2,j}^n \frac{\partial T_{i,j}^n}{\partial x} \approx -k_{i+1/2,j}^n \frac{T_{i+1,j}^n - T_{i,j}^n}{h_x}$ (all the other fluxes are approximated in a similar manner). When thermal conductivity is a smooth function, the values of $k_{i+1/2,j}$ can be computed directly at that point. However, if $k(\vec{x})$ is discontinuous as in the case of piecewise homogeneous media (e.g. tissue with tumor), the harmonic mean of the thermal conductivity expressed by $k_{i\pm1/2,j} = \frac{2k_{i,j}k_{i\pm1,j}}{k_{i,j}+k_{i\pm1,j}}$ is adopted to ensure the flux continuity.

## 4. Numerical simulations and parallelization strategy

All simulations performed in this study consider that above $T = 43^oC$ the cell is destroyed causing necrosis [7]. Actually, the temperature is computed in terms of $\overline{T}(\vec{x},t) = T_1(\vec{x},t) - T_2(\vec{x})$ with $T_2(\vec{x})$ being slightly higher than $37^oC$ due to the tumor.

In hyperthermia treatment the heat generation from the applied injection sites can be described by the so-called SAR. According to Salloum et. al [3] the SAR around the injection site due to magnetic nanoparticles can be approximated by a Gaussian distribution expressed by $SAR = A.e^{-r^2/r_0^2}$ where $A$ represents the maximum value; $r$ is the spatial distance from the injection site; and $r_0$ is how far from the injection site the heating affects the tissue. The external spatial heating $Q_r$ accounts for the $n$ injection sites which is represented by $Qr(\vec{x},t) = \sum_{i=1}^{n} A_i.e^{-r(\vec{x})_i^2/r_{0,i}^2}$

In the model analyzed here, we consider a square domain with length $0.1m$ and a square tumor with length $0.02m$ located at the center of the domain. The thermal conductivity considered for normal tissue and tumor are $k = 0.5W/m^oC$ and $k = 0.55W/m^oC$, respectively; the blood density $\rho_b = 1000.0Kg/m^3$; the blood specific heat $c_b = 4200.0J/Kg^oC$; the blood perfusion rate for normal tissue and tumor are, respectively, $\omega_b = 5.10^{-4}s^{-1}$ and $\omega_b = 1.25.10^{-3}s^{-1}$.This study was divided into 2 cases, the first assumes a single injection site and the second one 4 injection sites. At both cases we consider a hyperthermia with 50 min of duration. In the first case the Gaussian distribution SAR parameters are assumed to be $A = 1.3.10^6W$ and $r_0 = 3.1.10^{-3}m$[7] while in the second case $A = 0.325.10^6W$ and $r_0 = 3.1.10^{-3}m$ for each injection site, in other words, each injection site has a quarter power of the first case.

Figure 1 depicts the temperature distribution in both cases, and Figure 2 the temperature time histories at two distinct points, one located in the health tissue at ($x = 0.030$,$y = 0.050$) and another one inside the tumor at ($x = 0.045$,$y = 0.050$).

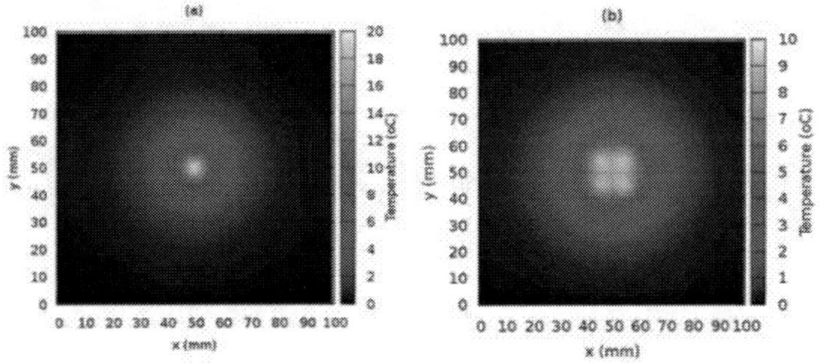


**Figure 1.** Temperature contour plot at $t = 50$min: (a) one injection site; and (b) four injection sites

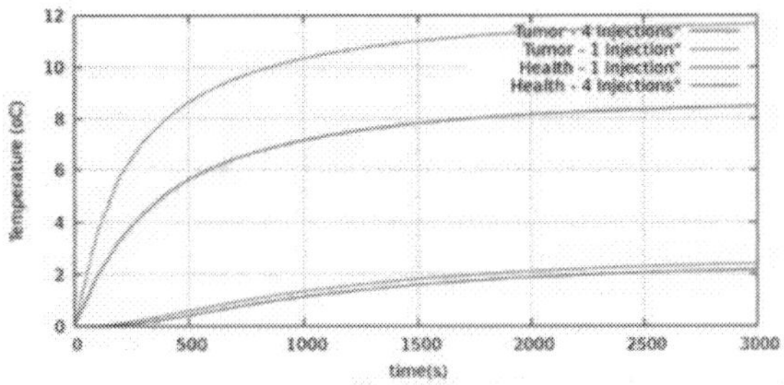


**Figure 2.** Temperature time-histories inside and outside the tumor in both cases

It is worth mentioning the effectiveness of the tumor necrosis at both cases. However, in the former we get 78.75% of tumor necrosis and 0.125% of health tissue necrosis, while in the latter we get 91.75% of tumor necrosis and 0.260% of health tissue necrosis. The point is that with four injection sites more tumor cells are affected compared to one injection site which means that the temperature is more distributed at the latter case.

However, this sequential version of the code takes almost 10 hours to execute a simulation. In order to reduce this huge execution time, a parallel version of the code was implemented using OpenMP[4]. Due to dependencies among distinct time-steps, only the computation of the spatial discretization was implemented in parallel. Supposing that the domain is discretized into $N_x \times N_y$ grid points, each thread is responsible for processing $(N_x \times N_y)/n$ points, where $n$ is the number of threads. Although the loop that implements the spatial computation is nested into the temporal loop, threads are created only once, before the temporal loop, and then associated statically with the spatial data it must compute, so thread creation overhead is paid only once.

Our experiments were performed on a SMP Linux (3.9.2-200) computer, consisting of 4 AMD Opteron 6272 CPU and 128 GB of RAM. Each CPU has 16 cores, so a total of 64 cores are available. It is worthwhile to note that in this CPU two cores share a single FPU. This architectural characteristic hurts performance. In particular, depending on the way threads are scheduled, two distinct threads can dispute the same FPU, even when the number of threads are below to the total number of FPUs available. The speedup presented in Figure 3 were obtained using the average value from three executions whose standard deviation was below to 0.4957. As one can observe, the results are quite encouraging: speedups up to 35 were achieved, which means that the execution time drops from 10 hours to about 16 minutes.

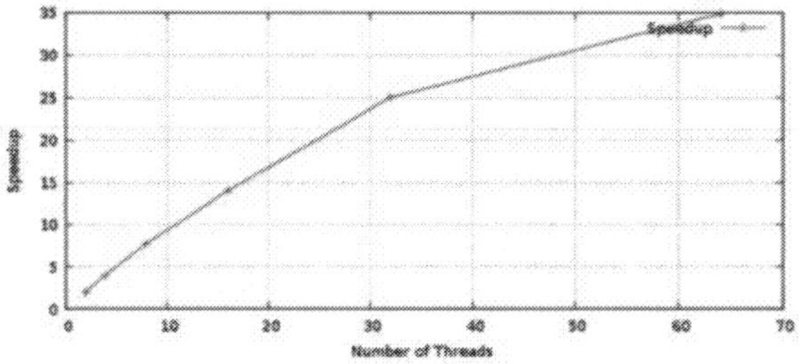


**Figure 3.** Speedups on a 64-core computer

## Conclusions

A numerical simulation of tumor cells necrosis using hyperthermia by applying magnetic nanoparticles has been presented. It has been observed that the number of injection sites and their locations inside the tumor play an important role to achieve the desired temperature distribution (recall that hyperthermia normally involves heating tissue above $T = 43^oC$). Moreover, speedups up to 35 were achieved. As a future work, we plan to handle more complex models in 3D using for this purpose GPUs[8].

## Acknowledgements

The financial support by CNPq, CAPES, UFJF and FAPEMIG is greatly acknowledged.

## References

[1] Pennes H H 1948 *Journal of Applied Phisiology* **1** 93–122
[2] Minkowycz W J and Sparrow E M 2009 *Advances in Numerical Heat Transfer* 1st ed vol 3 (CRC Press)
[3] Salloum M, Ma R and Zhu L 2008 *International Journal of Hyperthermia* **24** 589–601
[4] Chandra R, Dagum L, Kohr D, McDonald D M J and Menon R 2001 *Parallel Programming in OpenMP* (Morgan Kaufmann Publishers)
[5] Liu J and Xu L 1999 *Biomedical Engineering, IEEE Transactions on* **46** 1037–1043
[6] RJ L 2007 *Finite Difference Methods for Ordinary and Partial Differential Equations, Steady State and Time Dependent Problems* 1st ed (SIAMI)
[7] Salloum M, Ma R and Zhu L 2009 *International Journal of Hyperthermia* **25** 309–321 pMID: 19670098
[8] Sanders J and Kandrot E 2010 *CUDA by Example* 1st ed (Addison-Wesley Professional)

# A Novel Method for the Image Quality assessment of PET Scanners by Monte Carlo simulations: Effect of the scintillator

**G E Karpetas[1], C M Michail[2], G P Fountos[2], N I Kalyvas[2], I G Valais[2], I S Kandarakis[2], G S Panayiotakis[1]**

[1]Department of Medical Physics, Faculty of Medicine, University of Patras, 265 00 Patras, Greece
[2]Department of Medical Instruments Technology, Technological Educational Institute (TEI) of Athens, Ag. Spyridonos, 12210, Athens, Greece

E-mail: kandarakis@teiath.gr

**Abstract.** The aim of the present study was to propose a comprehensive method for PET scanners image quality assessment, by the simulation of a thin layer chromatography (TLC) flood source with a previous validated Monte-Carlo (MC) model. The model was developed by using the GATE MC package and reconstructed images were obtained using the STIR software, with cluster computing. The PET scanner simulated was the GE Discovery-ST. The TLC source was immersed in 18F-FDG bath solution (1MBq) in order to assess image quality. The influence of different scintillating crystals on PET scanner's image quality, in terms of the MTF, the NNPS and the DQE, was investigated. Images were reconstructed by the commonly used FBP2D, FPB3DRP and the OSMAPOSL (15 subsets, 3 iterations) reprojection algorithms. The PET scanner configuration, incorporating LuAP crystals, provided the optimum MTF values in both 2D and 3DFBP whereas the corresponding configuration with BGO crystals was found with the higher MTF values after OSMAPOSL. The scanner incorporating BGO crystals were also found with the lowest noise levels and the highest DQE values after all image reconstruction algorithms. The plane source can be also useful for the experimental image quality assessment of PET and SPECT scanners in clinical practice.

## 1. Introduction

Positron Emission Tomography (PET) scanners have been an essential imaging modality for various medicine disciplines. Their imaging performance is affected by various factors all of which contribute to the decreased PET resolution and image degradation [1]. The Detective Quantum Efficiency (DQE) is a figure of merit which is accepted as the best single objective indicator of overall image quality, for comparison between various imaging detector technologies [2]. DQE combines spatial resolution through the Modulation Transfer Function (MTF) and image noise through the Normalized Noise Power Spectrum (NNPS), to provide a measure of the Signal to Noise ratio (SNR) transfer through the imaging system, as a function of spatial frequency. Several studies have been carried out concerning PET image quality. However, in the majority of them, resolution was assessed from point sources in terms of the Point Spread Function (PSF) and consequently from the Full width at Half Maximum

(FWHM) [3]. Only in a few studies resolution was assessed in terms of the MTF calculated from the PSF method [4]-[5]. MTF can be also assessed from a line source through the estimation of the Line Spread Function (LSF). The use of the LSF method for determining the MTF in tomographic imagers was initially introduced by Boone [6] who applied this method for CT scanners evaluation. Fountos *et al.* [7] recently introduced a similar method for Single Photon Emission Computed Tomography (SPECT) scanners to obtain the MTF through the LSF method. The concept of DQE was previously approached by various methods for PET scanners [8]. According to our knowledge the overall imaging quality performance has never been assessed in terms of the spatial frequency dependent DQE for PET scanners. The aim of this study was to extend a previous validated Monte Carlo (MC) model [9] for the complete characterization and the further improvement of the image quality in PET scanners. This novel image quality method was tested by investigating the effect of crystal material on PET scanner performance. To this aim, DQE was estimated by the simulation of a Thin Layer Chromatography (TLC) plane source filled with 18F-FDG. The Geant4 Application for Tomographic Emission (GATE) was used in combination with the Software for Tomographic Image Reconstruction (STIR) [10]. Various GATE studies have been published on implementing STIR reconstruction for large commercial PET scanners [1], [10]-[12]. None of these, however, simulated the GE Discovery-ST scanner in order to fully characterise the image quality of PET in terms of DQE. Furthermore, the influence of the replacement of the BGO crystals with various crystal materials on the image quality of the scanner was also investigated. The simulation of the plane source phantom provides an accurate model that is useful to fully characterize the performance of nuclear medicine imaging systems.

## 2. Materials and methods

### *2.1. Geometry of the modeled PET scanner*

The GE Discovery ST PET/CT incorporates BGO [13], [14] crystals with dimensions of 6.3x6.3x30 mm [9]. The crystals were assembled into blocks of 6x6 crystals. Each block was coupled to a Photomultiplier Tube (PMT) consisting of four square channels, and is assembled in modules consisting of 8 blocks (2x4) each. The detector ring is finally comprised of 35 modules, i.e. 280 crystal blocks, or 24 rings of 420 crystals. The scanner is designed to acquire images in both 2D and 3D modes. The system was also simulated by replacing the BGO crystal arrays with $YAlO_3$:Ce (YAP) [14], LuAP:Ce, LuYAP:Ce-70, LuYAP:Ce-80 [14]-[16], $Lu_2SiO_5$:Ce (LSO) [17]-[19], and $Gd_2SiO_5$:Ce (GSO) [20], [21] crystals.

### *2.2. Preparation of the MTF test object*

The plane source was implemented by a layer of Silica gel on Aluminium (Al) foil substrates (Figure 1). The dimensions of the TLC plate were 5x10 $cm^2$ and it was immersed in 18F-FDG bath solution (1 MBq). The MTF test object was simulated within a phantom, consisting of two semi-cylindrical polyethylene blocks with 20 cm diameter and 70 cm length. Plane source images were acquired from STIR, after reconstruction with the commonly used FBP2D, the FPB3DRP and the MLE-OSMAPOSL reprojection algorithms [10].

### *2.3. Modulation transfer function (MTF) and Normalized noise power spectrum (NNPS)*

Image quality was assessed in terms of the MTF that was obtained, by using the LSF method [7]. To estimate noise, NNPS data were obtained at the same activity concentration as that of the MTF from the coronal images [22]:

### 2.4. *Detective Quantum Efficiency (DQE)*

Detective Quantum Efficiency (DQE) is calculated by comparing the SNR at the detector output with that at the detector input as a function of spatial frequency [2]. DQE is dependent on radiation exposure, spatial frequency, MTF, NNPS and detector material. High DQE values indicate that less radiation is needed to achieve identical image quality. The DQE of the PET scanner was calculated from the MTF, NNPS, and the incoming SNR squared (defined as the plane source phantom activity (counts/mm$^2$) incident on the detectors) by using (1) [22]:

$$DQE(u) = MTF^2(u) / (SNR_{in}^2 \cdot NNPS(u)) \tag{1}$$

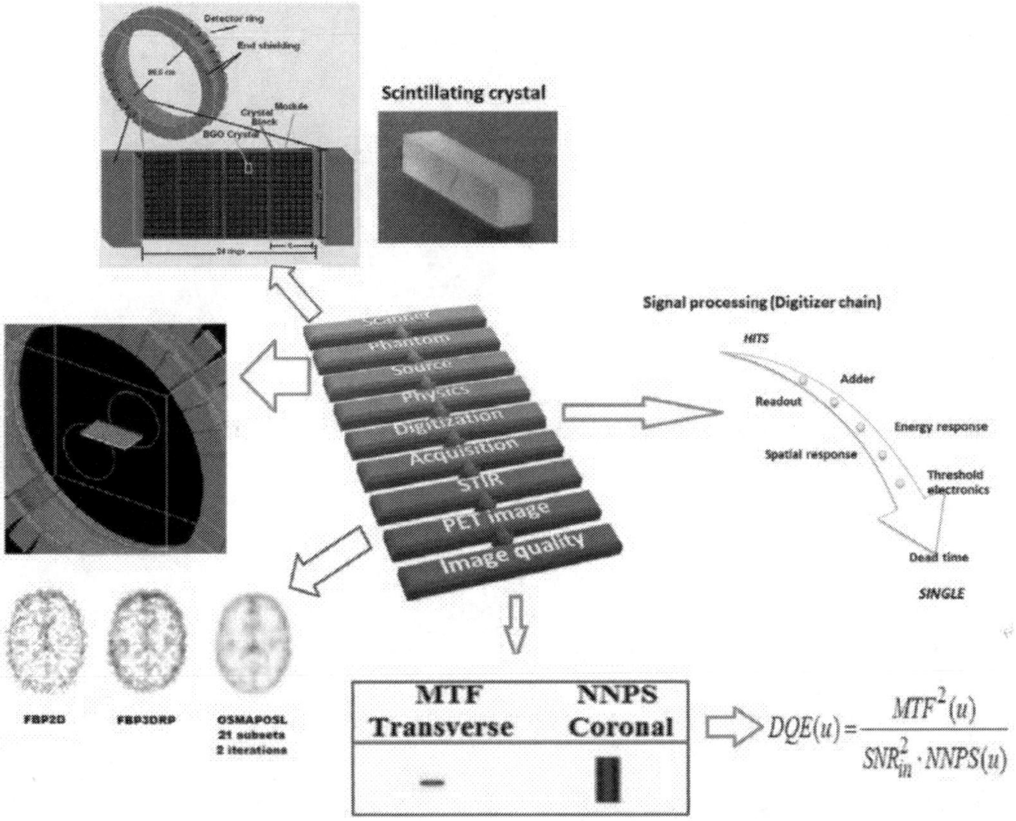


**Figure 1**. Block scheme of the DQE procedure.

## 3. Results and discussion

### *3.1. DQE*

Figures 2 and 3 show DQE results obtained from the FBP2D, FBP3DRP and OSMAPOSL reconstruction algorithms, for the PET scanner configurations incorporating successively all the crystals under investigation. The curves correspond to the combined MTF and NNPS results from both the transverse and coronal reconstructed slices. The behavior of the DQE was influenced by both MTF and NNPS. DQE values of the PET configuration incorporating BGO crystals were found higher after FBP3DRP and OSMAPOSL image reconstruction algorithms examined in this study. DQE values of the LuYAP where found higher after FBP2D.

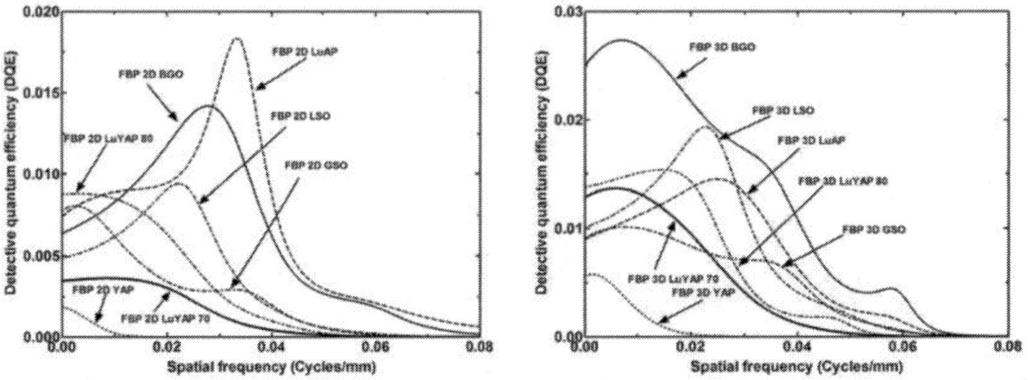


**Figure 2**. Comparison between the DQEs obtained from the plane source reconstructed images with the FBP2D (left) and the FBP3DRP (right) for various crystals.

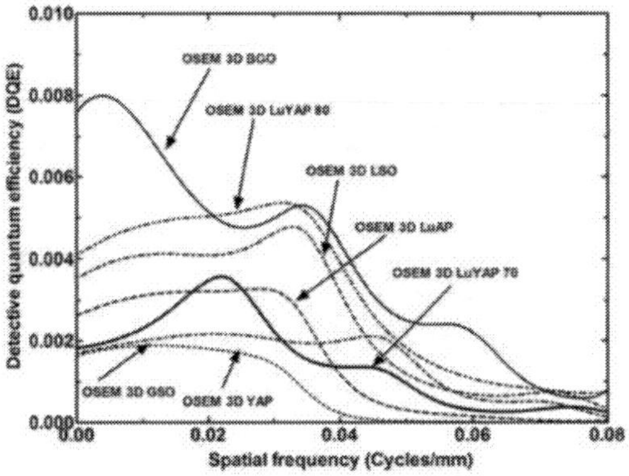


**Figure 3**. Comparison between the DQEs obtained from the plane source reconstructed images with the OSMAPOSL (15 subsets, 3 iterations) for various crystals.

## 4. Conclusion

A novel image quality method for PET scanners was implemented with GATE Monte Carlo methods. PET images were obtained from STIR through the simulation of a novel thin 18F-FDG plane source. Image quality was assessed from the reconstructed images, by the estimation of the spatial frequency dependent DQE of the plane source reconstructed images. The influence of the detector material on the image quality of a PET scanner was also investigated. The method modelled and simulated in this study was used for the image quality assessment and optimisation, but it can be also useful for the further development of PET and SPECT scanners though GATE simulations.

### Acknowledgments

This research has been co-funded by the European Union (European Social Fund) and Greek national resources under the framework of the "Archimedes III: Funding of Research Groups in TEI of Athens" project of the "Education & Lifelong Learning" Operational Programme.

STIR is based on work performed by the PARAPET project, which was supported by the European Esprit LTR project PARAPET (EP23493) and the Swiss Federal Office for Education and Science under grant 96.193.

## 5. References


[1] Schmidtlein C R, Kirov A S, Nehmeh S A, Erdi Y E, Humm J L, Amols H L, Bidaut L M, Ganin A, Stearns C W, McDaniel D L and Hamacher K A 2006 *Med. Phys.* **33(1)** 198.

[2] Darambara D 2006 *Nucl. Instrum. Meth. Phys. Res. A.* **569** 153.

[3] Lartizien C, Kuntner C, Goertzen A, Evans A and Reilhac A 2007 *Phys. Med. Biol.* **52** 4845.

[4] Stickel J R and Cherry S R 2005 *Phys. Med. Biol.* **50** 179.

[5] Liang H, Yang Y, Yang K, Wu Y, Boone J M and Cherry S R 2007 *Phys. Med. Biol.* **52** 3881.

[6] Boone J M 2001 *Med. Phys.***28** 356.

[7] Fountos G P, Michail C M, Zanglis A, Samartzis A. Martini N, Koukou V, Kalatzis I and Kandarakis I S 2012 *Med. Phys.* **39(3)** 1561.

[8] Herbert D, Saveliev V, Belcari N, D'Ascenzo N, Del Guerra A and Golovin A 2006 *IEEE Trans. Nucl. Sci.* **53(1)** 389.

[9] Karpetas G E, Michail C M, Fountos G P, Valsamaki P N, Kandarakis I S and Panayiotakis G S 2013 *Hell. J. Nucl. Med.* **16(2)** 111.

[10] Thielemans K, Tsoumpas C, Mustafovic S, Beisel T, Aguiar P, Dikaios N and Jacobson M W 2012 *Phys. Med. Biol.* **57** 867.

[11] Polycarpou I, Marsden P K and Tsoumpas C 2011 *J. of Physics* **317** 012001.

[12] Delso G, Martinez M J, Torres I, Ladebeck R, Michel C, Nekolla S and Ziegler S I 2009 *Med. Phys.* **36(9)** 4126.

[13] Valais I, Michail C, David S, Costaridou L, Nomicos C D, Panayiotakis G S and Kandarakis I 2008 *Physica Medica* **24** 122.

[14] Van Eijk C W E 2002 *Phys. Med. Biol.* **47** R85.

[15] Valais I G, Michail C M, David S L, Konstantinidis A, Cavouras D A, Kandarakis I S and Panayiotakis G S 2008 *IEEE Trans. Nucl. Sci.* **55(2)** 785.

[16] Valais I G, Michail C M, David S L, Liaparinos P F, Fountos G P, Paschalis T V, Kandarakis I S and Panayiotakis G S 2010 *IEEE Trans. Nucl. Sci.* **57(1)** 3.

[17] Michail C, David S, Liaparinos P, Valais I, Nikolopoulos D, Kalivas N, Toutountzis A, Sianoudis I, Cavouras D, Dimitropoulos N, Nomicos C D, Kourkoutas K, Kandarakis I and Panayiotakis G S 2007 *Nucl. Instrum. Meth. Phys. Res. A* **580** 558.

[18] Michail C M, Fountos G P, David S L, Valais I G, Toutountzis A E, Kalyvas N E, Kandarakis I S and Panayiotakis G S 2009 *Meas. Sci. Technol.* **20** 104008.

[19] Michail C, Toutountzis A, David S, Kalivas N, Valais I, Kandarakis I and Panayiotakis G S 2009 *App. Phys. B* **95** 131.

[20] Valais I, David S, Michail C, Konstantinidis A, Kandarakis I and Panayiotakis G S 2007 *Nucl. Instrum. Meth. Phys. Res. A* **581** 99.

[21] Valais I, Kandarakis I, Nikolopoulos D, Michail C, David S, Loudos G, Cavouras D and Panayiotakis G S 2007 *IEEE Trans. Nucl. Sci.* **54(1)** 11.

[22] Michail C M, Spyropoulou V A, Fountos G P, Kalyvas N E, Valais I G, Kandarakis I S and Panayiotakis G S 2011 *IEEE Trans. Nucl. Sci.* **58(1)** 314.

# Applying robust control theory to solve problems in bio-medical sciences: study of an apoptotic model

**M Trujillo** [1], **V M Becerra and S Hadjiloucas**
School of Systems Engineering, University of Reading, RG6 6AY, UK

E-mail: s.hadjiloucas@reading.ac.uk

**Abstract.** Biological models of an apoptotic process are studied using models describing a system of differential equations derived from reaction kinetics information. The mathematical model is re-formulated in a state-space robust control theory framework where parametric and dynamic uncertainty can be modelled to account for variations naturally occurring in biological processes. We propose to handle the nonlinearities using neural networks.

## 1. Introduction

Robust control theory has been considered as a tool in the biomedical community for model validation, optimal experiment design for hypothesis testing, systems oriented drug design, target identification, or multi-target drug dosage optimization. Unfortunately practical application of this method is limited to models that can be linearized without substantial loss of quantitative or qualitative features. Since most problems arising in biomedical applications do not fall in this category, we present a different perspective where traditional parameter sensitivity analysis is translated into sensitivity of biological features that can be measured experimentally. In this study, biological models of an apoptotic process are studied using models describing a system of differential equations derived from reaction kinetics information. The mathematical model is re-formulated in a state-space robust control theory framework where parametric and dynamic uncertainty can be modelled to account for variations naturally occurring in biological processes. Since the standard robust control theory framework is valid for linear systems only, we re-formulate the problem as piecewise linear allowing for specific non-linearities. We propose to handle the nonlinearities using neural networks. The approach overcomes the limitations imposed by forced linearization and extends the applicability of robust control theory to problems encountered within the Bio-medical community.

## 2. Apoptosis model

Apoptosis is the programed death that regulates the disposal of cells that have been damaged or are no longer required by an organism. Mis-regulation of this process can lead to pathological conditions such as cancer when cells do not die or Alzheimer's condition where cells die more than needed. Many models of apoptosis are available in the literature and choosing an adequate model can be a difficult process. The current study focuses on the EARM (Extrinsic Apoptosis Reaction Model) described in [1] because in the process of creating, evaluating, calibrating and evolving this model, as

[1] To whom any correspondence should be addressed.

this is now broadly accepted in the literature as a comprehensive one, where the associated experimental parameters are well calibrated. Furthermore, the model incorporates the latest hypotheses associated to the different reaction mechanisms postulated in the literature.

The model of the extrinsic apoptotic pathway published in [1] is formed by four sub-circuits that interact. These describe the entire pathway leading to the apoptotic process. The first circuit starts by the addition of tumor necrosis factor (TNF). The TNF-related apoptosis-inducing ligand (TRAIL), binds to its receptor forming the death-inducing signalling complexes (DISC) and activating pro caspase 8 (pro-C8). The second circuit represents the activation of pro-C8 produces caspase 8 (C8). C8 directly activates caspase-3 (C3) which is responsible for the cleavage of Poly (ADP-ribose) polymerase (PARP), an essential cell substrate for cell survival. The third circuit is a feed-forward loop where C8 promotes the cleavage of Bid to form tBid which activates Bax. Activated Bax promotes the formation of pores in the mitochondria to translocate Smac and cytochrome c (CyC) to the cytosol. Smac binds to X-linked IAP (XIAP) preventing it from inhibiting C3. CyC binds to Apaf and then attracts caspase 9 (C9) to form the apoptosome which in turn produces more C3. The fourth circuit is a feedback loop where the production of C3 activates caspase 6 (C6) to produce more C8.

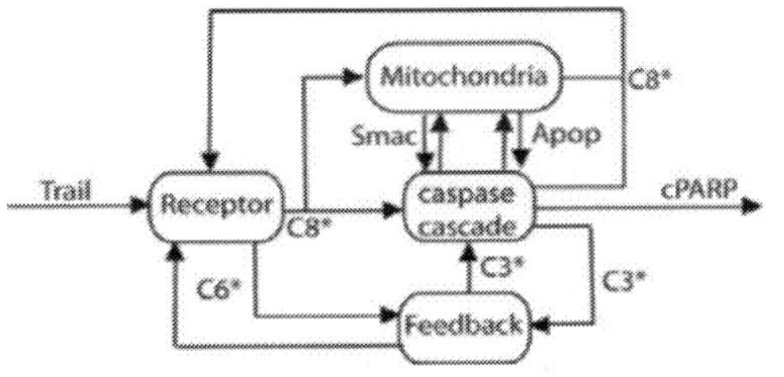


**Figure 1**. Simplified block diagram of the apoptosis model divided in four sub-circuits. The input is the death inducing ligand TRAIL and the output is the amount of cleaved PARP. The four blocks are the receptor complex, mitochondria, caspace cascade and feedback.

The model has 18 species with non-zero initial conditions and 40 additional species that represent modified forms of the initial 18 species (cleaved and short-lived intermediate species). When these species have moved across membranes, they are annotated differently. All species interact via 28 biochemical reactions that use 70 rate reactions including forward, reverse and catalytic rates for each reaction. The biochemical reactions were then translated into 58 ordinary differential equations.

It has been proposed and experimentally tested by the authors of the model that the apoptotic network works as a variable delay snap-action switch described by the following function:

$$c(t) = f - f\left[1 + e^{\frac{(t-T_d)}{4T_s}}\right]^{-1} \tag{1}$$

where $c$ is the amount of substrate cleaved at time $t$, $f$ is the fraction cleaved at the end of the reaction, $T_d$ is the delay between TRAIL addition and half the total substrate cleaved (cPARP) and $T_s$ is the switching time between initial and total effector caspase cleavage. This concept is illustrated with the response of cPARP for an initial concentration of TRAIL (Figure 2a).

The authors of the model found that $T_d$ depends on TRAIL dose, but $T_s$ and $f$ are independent of TRAIL dose. The model parameters were calibrated to follow experimental values of $T_d$, $T_s$ and $f$. The cell is destined to die when 10% of the total amount of PARP is cleaved therefore, the value of cPARP = 0.1 PARP is the threshold necessary for cell death (Figure 2b). Experiments with TRIAL doses from 2 to 1,000 ng/ml corresponded to a variation in $T_d$ from 140 to 660 minutes. We assume that reducing the TRIAL dose enough to prevent reaching the cPARP threshold will create a cell survival steady state for long enough to compare the set of parameter values key for differentiating death and survival states. In figure 2b four species are shown as representative of the different stages of the apoptotic process. The initiator caspase C8* is the only concentration that has a slow increment during the delay period Td followed by a sudden increment after reaching the switching time $T_s$.

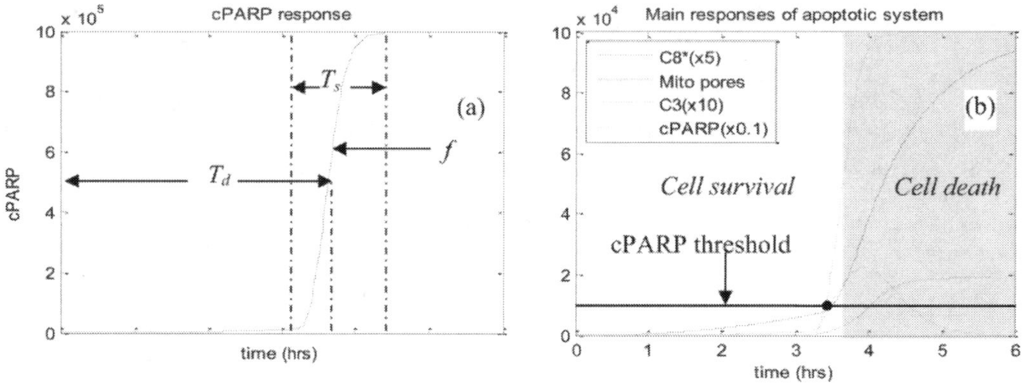


**Figure 2**. a) cPARP trajectory in response to ligand treatment as described in (1), b) main responses of apoptotic system species; initiator caspaces, mitochondria pores, effector caspaces and cleaved PARP.

## 3. Nonlinear robust control approach

The most obvious application of robust control for the apoptosis network is to specify desirable behaviour, translate it as performance functions, define boundaries for uncertain parameters and dynamics and design a controller that achieves the required performance for all the possible uncertain scenarios. However, the apoptosis pathway is not yet fully understood, many contradicting models describe different possible hypotheses and experimental data cannot be yet generated to test them. The model we selected has gone through various iterations of modelling, experimental data fitting and calibration but since the experimental measurements of variables in individual cells destroy the cell, it is not possible yet to monitor the same cell through all its stages in the process from reception of the apoptosis-inducing ligand to cell death. Hence we focus in the use of robust control to provide further understanding of the process, before we could design controllers to intervene at the cellular level.

### *3.1. General robust control problem*

Almost any dynamical system can be put in the general form described in figure 3a. In this general form $P$ is a plant consisting of the nominal dynamics of our system augmented with performance functions defined by the user to force the system to satisfy design requirements, $K$ is the controller to be designed to satisfy performance and stability requirements for all the systems resulting from the combination of the plant $P$ with the uncertain block $\Delta$ which contains the allowed variations in parameter space and un-modelled dynamics that may affect the system. $w$ is a vector of the external inputs to the system, $z$ are the outputs to be controlled, $u$ are the inputs to the controller, $y$ are the measurements available from the system, $w_p$ and $z_p$ are the output and input to the uncertainty block.

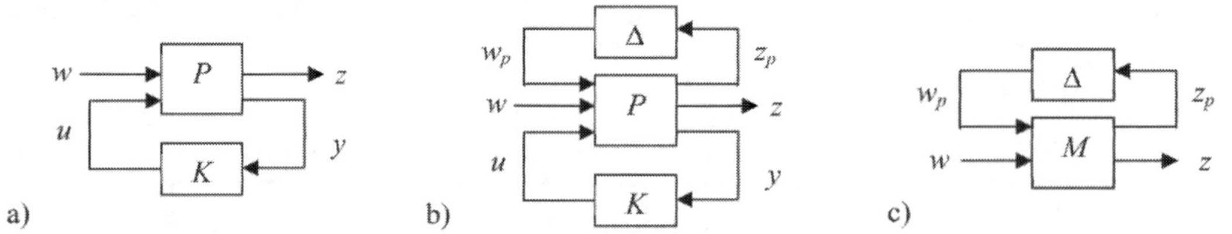


**Figure 3**. a) General plant $P$ and controller $K$, b) Robust control problem for uncertainty $\Delta$, c) Plant and controller integrated in $M$ for the analysis of robust performance

The mathematical representation of the perturbed plant $P$ is given by

$$P:\begin{cases}\dot{x}=f(x)+g_1(x)w+g_2(x)u;\ \ x(t_0)=x_0\\ y=x\\ z=h_1(x)+k_{12}(x)u\end{cases} \tag{2}$$

where $x$ is the state vector of the nominal system, $f(x)$ are the dynamics of the system as described in the nominal model, $g_1, g_2, h_1, h_2, k_{12}$ and $k_{21}$ are functions of $x$. The nonlinear $H_\infty$ problem is to find a controller action $u$ that results in an $\mathcal{L}_2$-gain of the system P from $w$ to $z$ less or equal to a given $\gamma$ while maintaining system stability. This is an optimisation problem where the allowed disturbance has to be maximized while the control energy has to be minimized. Solutions to this problem come from theories such as differential games, disipativity, viscosity among others. The final result is a Hamilton-Jacobi-Isaacs equation (HJIE) of the form:

$$V_x(x)f(x)+\frac{1}{2}V_x(x)\left[\frac{1}{\gamma^2}g_1(x)g_1^T(x)-g_2(x)g_2^T(x)\right]V_x^T(x)+\frac{1}{2}h_1^T(x)h_1(x)=0, V(0)=0 \tag{3}$$

Finding a storage function $V_x(x)$ that satisfies the HJI in most cases is not straightforward. Once a function $V_x(x)$ satisfying equation 3 is found, the minimum energy control and the maximum disturbance allowed for the system in figure 3 a) are given by:

$$u(x)=\alpha(x)=-g_2^T(x)V_x^T(x), \qquad w(x)=\frac{1}{\gamma^2}g_1^T(x)V_x^T(x) \tag{4}$$

For the system given in figure 3 b) the perturbed plant $P_\Delta$ becomes

$$P_\Delta:\begin{cases}\dot{x}=f(x)+\Delta f(x,\theta,t)+g_1(x)w+[g_2(x)+\Delta g_2(x,\theta,t)]u;\ \ x(t_0)=x_0\\ y=[h_2(x)+\Delta h_2(x,\theta,t)]+k_{21}(x)w\\ z=h_1(x)+k_{12}(x)u\end{cases} \tag{5}$$

where $\Delta f, \Delta h_2$ and $\Delta g_2$are unknown functions that belong to the set of admissible uncertainties and $\theta$ are the system parameters that may vary over time in a defined range [2].

For the purpose of this paper, we are not concerned with the design of a controller since we consider the apoptosis network as a whole system with an integrated controller. Figure 3b shows the general framework used in robust control theory for analysis rather than synthesis where $P$ and $K$ form a single block $M$. This framework allows us to ask how big can the perturbation be and still permit the system to achieve the desired performance. Then, by manipulating the weighting functions (performance criteria) and the allowed uncertainties we can perform sensitivity analysis to measure the impact of various parameters in the system, test different hypothesis for model validation/invalidation and identify underlying mechanisms and interactions between species of the system.

### *3.2. Performance functions, modeling uncertainty and solving the HJIE with Neural networks*

The performance functions are used to modify the nominal system to include desired trajectories. In the case of the apoptosis network, it was suggested in [3] that four features of the time trajectories from cytosolic Smac and cPARP could define the difference between a normal and a pathological response. These are: 1) $t_{PARP}$ which is the time it takes 50% of PARP to be cleaved, 2) $t_{MOMP}$ which is the time it takes for half the Smac to translocate from the mitochondria, 3) $t_{switch}$ which is the time between the start and finish of PARP cleavage and 4) $f_{PARP}$ which is the amount of PARP cleaved after the experiment or simulation. It is possible then to identify values of interest for these features and define performance functions accordingly. By defining structural perturbations on various species we can identify which once affect the most each feature. To study the effects that variations in the system have on the defined features of interest we have to model such variations as uncertainties $\Delta$ and incorporate them in the system *M*. Parametric uncertainty is the variation in parameter space that we want to explore and dynamic uncertainty accounts for the unknown dynamics taking place in the system. When a mixture of both uncertainty types has to be analyzed then $\Delta$ becomes a structured uncertainty block that can be separated from the plant as in figure 3b. A solution for the HJIE from equation 3 can be found indirectly by using iterative processes. The HJI is broken in a sequence of linear partial differential equations and the aim becomes finding a possible storage function $V_x(x)$. In each iteration step, the performance of an initial stabilizing controller is improved until a satisfactory solution to the HJI is found through several iterations. Even though finding $V_x(x)$ for the linear partial differential equations at each iteration is less complicated than finding it for the HJIE, nevertheless, it is still a difficult task that needs to be solved. One possible way is by approximating $V_x(x)$ with a neural network at each iteration. A more complete description of the process can be found in [4].

## 4. Conclusion

So far, because of its implementation complexity, robust control theory has failed to gain popularity as a tool in the bio-medical field. Non-linear robust control theory is still under development and soft computing alternatives are constantly introduced to overcome the complexity of solving problems that cannot be tackled by linearization. The HJIE approach should be seen as an extension to linear $\mathcal{H}\infty$-control theory, providing a unifying approach in treating disturbances using a nonlinear counterpart of the Riccati equations which are conventionally used to solve linear problems. Different soft computing methodologies such as neural networks and fuzzy systems theory are worthwhile exploring to overcome the difficulties of solving nonlinear HJI equations. A particular advantage in the general formulation developed for the apoptotic process described in section 3.1 is that it clarifies the parameters that need to be estimated (for example the different uncertainties introduced in the model and their inter-relationship to other biochemical pathways).

Furthermore, because of the nice stability and stabilizability properties of passive systems, such as (global) asymptotic stabilizability by pure-gain output-feedback it is considerably desirable in pharmacological or radiological intervention to render the associated biochemical non-linear systems passive. In non-linear control theory, it is well understood that if the state space of a system is reachable from the origin $x=0$, by an appropriate choice of an input $u(t)$ so that its output is finite, then it can be rendered dissipative. A further goal of the proposed formulation is, therefore, to identify the control points in biochemical reactions that would render interconnected pathways passive through the control of the supply rate (in our case the cPARP threshold in Fig. 2). Since most biochemical pathways can be seen as interconnected dissipative sub-systems of processes described by storage functions, the approach should have a wide domain of applications across Systems Biology. The present work, therefore, serves also as an intellectual scaffolding upon which other Systems-Biology problems can be formulated where seemingly non-linear processes can be controlled though the control of the supply rate to an identified storage function of the system.

**References**

[1] Albeck J G, Burke J M, Spencer S L, Lauffenburger D A and Sorger P K 2008 *PLoS Biology*, **6**(12): e299.

[2] Abu-Khalaf M, Huang J and Lewis F L 2006 *Nonlinear H2/H∞ Constrained Feedback Control* (Springer-Verlag London)

[3] Gaudet S, Spencer S L, Chen W W, Sorger P K, 2012, *PLoS Computational Biology*, 1-15

[4] Aliyu M D S 2011 *Nonlinear H∞-control, Hamiltonian systems and Hamilton-Jacobi equation* (CRC Press, Taylor &Francis, Boca Raton FL, USA)

# Fractional Order and non-linear System Identification algorithms for Biomedical applications

**S Hadjiloucas**[1] **and R K H Galvão**[2]
[1]Cybernetics, School of Systems Engineering, University of Reading, RG6 6AY, UK
[2]Divisão de Engenharia Eletrônica, Instituto Tecnológico de Aeronáutica, São José dos Campos, SP, 12228-900 Brazil.

E-mail: s.hadjiloucas@reading.ac.uk

**Abstract.** We discuss the modelling of dielectric responses of amorphous biological samples. Such samples are commonly encountered in impedance spectroscopy studies as well as in UV, IR, optical and THz transient spectroscopy experiments and in pump-probe studies. In many occasions, the samples may display quenched absorption bands. A systems identification framework may be developed to provide parsimonious representations of such responses. To achieve this, it is appropriate to augment the standard models found in the identification literature to incorporate fractional order dynamics. Extensions of models using the forward shift operator, state space models as well as their non-linear Hammerstein-Wiener counterpart models are highlighted. We also discuss the need to extend the theory of electromagnetically excited networks which can account for fractional order behaviour in the non-linear regime by incorporating nonlinear elements to account for the observed non-linearities. The proposed approach leads to the development of a range of new chemometrics tools for biomedical data analysis and classification.

## 1. Introduction

Many broadband spectroscopic investigations of biomedical interest are performed using femtosecond transient pulse systems, as well as using continuous wave and Fourier domain mode-locked lasers. These spectrometers ensure a persistent excitation of all the modes of the sample across each spectral bin. In these experiments, one often observes Jonscher-like responses that display a fractional order behavior as a function of excitation frequency [1-8]. Furthermore, Havriliak-Negami or Fröhlich mixture [9, 10] type responses are also more faithfully described using fractional order calculus [11, 12]. The use of parametric models to the dielectric spectra of biological tissues has been often discussed in the medical physics community. For example permittivity and conductivity measurements of blood, bone, kidney, liver, spleen, tendon, muscle, skin, fat and brain (Grey and White matter) as well as heart muscle have been characterized and Cole-Cole models have been fitted to those datasets over a large range of frequencies (10 Hz to 100 GHz) by separating the datasets into four different regions and fitting a model for each region [13].

This work provides an overview of modeling approaches that are of relevance to the signal processing of this type of broadband spectra. System identification techniques may be used as a chemometric tool to infer the composition of the sample and observe subtle differences in composition

[1] To whom any correspondence should be addressed.

between samples on the basis of studying the input and output waveforms from the associated experiments. Modeling techniques enable very parsimonious expressions of the dynamics of the excited samples. On the basis of the generated models, guidelines for optimal excitation using tailored chirped or PRBS sequences can be subsequently developed ensuring optimal signal-to-noise ratio in the experiments. Optimal excitation can be conveniently implemented experimentally using well-established pulse shaping techniques.

The work also highlights the utility of using random *RLC* networks to model the dynamics of multiple excitation/de-excitation processes. These networks can emulate complex diffusive charge transport processes and provide alternatives to existing percolation and spectral density approaches, as well as Archie's law and Kohlrausch-Williams-Watts models. As a consequence, they can also form a basis to extract chemometric information for classification tasks.

## 2. Fractional order calculus and system identification algorithms for the study of the interaction of excitatory waves with biological tissue

In the following sections we discuss generic methodologies for studying the interaction of excitatory waves with biological tissue. The approach considers an emerging area in applied mathematics, that of fractional order calculus and discusses its application within a wider signal processing framework applicable within a biomedical context. We also propose to use 3-dimensional networks containing a different composition of resistive, capacitive and inductive elements to model the dynamic response of biological tissue to excitatory waves. This approach is justified on the basis that the associated dynamics of such networks are of fractional order.

### *2.1. Fractional order models using the forward shift operator*

In linear spectroscopies, the output of a deterministic linear system that describes a biological or chemical process at time instance $k$ can be computed by filtering the excitation input $u(k)$ through a linear filter $G(q)$ which has the dynamic of the system under study. In this formulation, $q$ denotes the forward shift operator $q^{-1}x(k)=x(k-1)$ which may be associated with its time domain counterpart $z=e^{j\omega}$ in the frequency domain [14-16]. Fractional order linear system identification can be based on parametric methods which would need to describe the true process behavior exactly with a finite number of parameters based on fractional order differential or difference equation models, or non-parametric methods where an infinite number of parameters would be needed to describe the process exactly. Normally, parametric methods will be most appropriate when a relatively low number of parameters need to be determined. Non-parametric methods, however, are generally more flexible, as less structure is imposed on the model in an *a priori* basis.

Usually, the parameters described in parametric models can be extracted in the time, frequency or wavelet domains using least squares, nonlinear optimization or repeated least squares, whereas non-parametric methods would require impulse or transient response analysis or correlation analysis in the time domain, or frequency response, Fourier and spectral analysis in the frequency domain; these techniques are also extendable in the wavelet domain. A general linear fractional order model based on the forward shift operator is shown below:

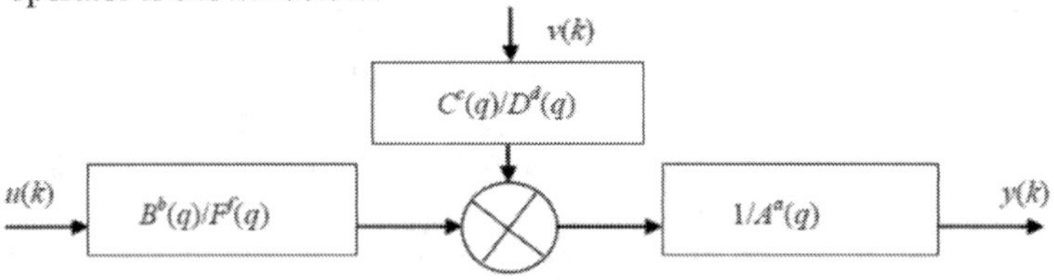


**Figure 1.** Linear fractional order model in generic form.

This generic structure forms the basis for all linear fractional order models that are based on the unit forward shift operator. The fractional order autoregressive with exogenous input (FO-ARX), autoregressive moving average with exogenous input (FO-ARMAX), output error (FO-OE), Box-Jenkins (FO-BJ), autoregressive-autoregressive with exogenous input (FO-ARARX) and finite impulse response (FO-FIR) models are special cases of the above structure. The identification of the above models is a direct extension from integral order calculus but it also requires an additional step for the estimation of the associated exponents.

### 2.2. State space fractional order models

The dynamics of an unknown medium may also be modeled using an input-output state space representation where a subspace multiple input-multiple output (MIMO) model may be fitted [16-20].

The fractional state space representation for the system is given from:

$$D^{a}\mathbf{x}(t) = A\mathbf{x}(t) + B\mathbf{u}(t) \tag{1a}$$

$$\mathbf{y}(t) = C\mathbf{x}(t) + D\mathbf{u}(t) \tag{1b}$$

where $\mathbf{x} \in \mathbb{R}^{n}$ is the state vector, $\mathbf{u} \in \mathbb{R}^{m}$ is the input vector, $\mathbf{y} \in \mathbb{R}^{p}$ is the output vector and $A \in \mathbb{R}^{n \times n}$, $B \in \mathbb{R}^{n \times m}$, $C \in \mathbb{R}^{p \times n}$ and $D \in \mathbb{R}^{p \times m}$ are the state, input, output and feed-through system constant matrices to be determined. As discussed in [21], such fractional order system is stable if $0 < \alpha < 2$ and $|\arg(\lambda_k)| > \alpha\pi / 2$ and $-\pi < \arg(\lambda_k) \leq \pi$ where $\lambda_k$ corresponds to the $k^{\text{th}}$ eigenvalue of $A$. For time-domain simulations of a system, a recursive distribution of poles and zeros is obtained to approximate the frequency behavior of $s^{a}$ over an interval $[\omega_A, \omega_B]$. Because the asymptotic behavior at the low and high limits of the above frequency interval can have a static error between the fractional order model and its approximation, it is common to minimize this using an integrator operating outside that interval. Given a large number of inputs and outputs related to the unknown dielectric system, the goal of the subspace algorithms is to determine the order of the system, $A, B, C, D$ (to a similarity transformation). For simplicity, one may only consider the deterministic case where there is no noise in the measured inputs or process (estimated state) although the proposed approach is also valid to the stochastic case where an explicit augmented model in innovation form can be considered. In that case, an additional step is required to obtain an estimate of covariance matrices of the noise sequences. Simulations using the associated formulations are mentioned elsewhere [22].

### 2.3. Combining fractional order calculus models with non-linear system identification models

Advances in instrumentation and measurement methods have also resulted in a new surge of interest to extend chemometrics algorithms to describe the non-linear phenomena observed in biological sciences. Such problems, arise in 2-photon microscopy, four-wave mixing studies, saturation absorption spectroscopies, Raman spectroscopies, fluorescence lifetime methods or other non-linear spectroscopies using high power laser sources e.g. synchrotrons, free electron lasers etc. Non-linear models need also to be assumed in femtosecond pulse pump-probe experiments where the pump pulse introduces a static non-linearity in the sample which is then observed by the probe pulse at various phase delays. Hammerstein and Wiener models (Figure 2) have been extensively used by the systems identification communities for such purposes. In previous works our group has used integral order calculus with polynomial fitting for the derivation of a static non-linearity [23]. Neural networks, however, are also appropriate to describe the non-linearity in the identification process. Since solute interactions can involve non-linear processes as well as forces acting at a distance, it is quite appropriate to extend such framework to account for processes displaying fractional order dynamics.

Taking into consideration the generic framework shown above, it is appropriate to extend existing Hammerstein-Wiener algorithms to account for fractional order dynamics.

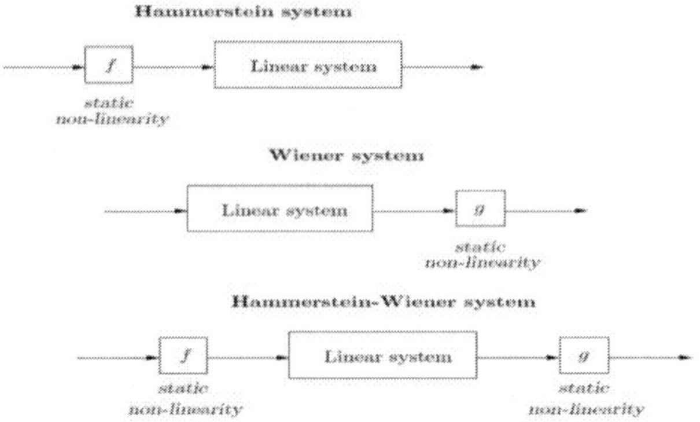


**Figure 2**. Pulse excited systems described by a linear dynamic model preceded and/or followed by static non-linearities.

### *2.4. Simulations with randomly connected RLC networks*

In previous works [22, 24] we have discussed the possibility to simulate dynamic processes encountered in spectroscopy experiments using 2-dimensional and 3-dimensional random *RLC* networks. Our work reconfirmed results in the existing dielectrics literature [9, 25-28] and has paved the way for generating parsimonious representation of the excitation and de-excitation dynamics as well as place experimental results within a chemometrics context. For example, it is possible to distinguish between batches of samples prepared in powder form and compressed into pellets as is customary in the pharmaceutical industries by assuming each element in the composition is associated with a capacitive or resistive behavior. Such framework is also of particular relevance to the interpretation of results from laser-induced fluorescence lifetime experiments. By converting the state space responses of these networks into input/output transfer function formulations, it is possible to link the overall admittance observed for the system to few resistive and capacitive components [29]. Classification can then be conveniently performed using very parsimonious representations [30-32]. This work may also be extended to *RLC* networks [33] to account for resonant structures, as would be the case for molecular optics experiments where the electric field resonantly excites the studied molecules. Current work aims to further extend these simulators to Fröhlich mixture type responses as encountered when polarized structures such as membranes are probed, since the observed responses can be highly non-linear. This should be possible by incorporating in the existing networks memristive, memcapacitive and memductive elements.

## 3. Discussion

Although systematic studies of dielectric responses of tissue across the electromagnetic spectrum are regularly being performed using acoustic, nuclear (including functional) magnetic resonance, electron spin resonance, positron emission tomography as well as several other dielectric spectroscopic modalities, in the hope that malignant or diseased tissue can be identified and localized at early stages so that it can be isolated and targeted by therapeutic pharmaceutical modalities in a systematic or in a non-invasive manner through radiation therapies, there is a diverse range of tools currently available to perform parametric and non-parametric identification of such datasets. None of these, however, can fully capture the dynamics of the physical processes associated with the complex observed responses in most samples. This is due to the compositional complexity of the associated tissues. Fractional order modelling approaches can accommodate more complex responses in a manner that is not possible using standard integro-differential equation models. The associated fractional order expressions can capture the dynamics of an arbitrarily complex medium with a greater fidelity (a much smaller residual error in a least squares sense) so small differences in the model structure or parameters can be associated to small changes in the condition of the tissue. The proposed approach should be further explored in a systematic manner within a biomedical dataset classification context e.g. using support vector machine methodologies [34], taking advantage of the parsimonious fractional order expressions so that the generalization ability of the inference engines are not compromised.

## References


[1] Nowick A S, Vaysleyb A V and Kuskovsky I 1998 *Phys. Rev. B*, **58** 8398-8406
[2] Dyre J C 1991 *J. Non-Crystal. Solids,* **135** 219-226
[3] Wieczorek W, Zalewska A, Siekierski M and Przyluski J 1996 *Solid State Ionics,* **86-88** 357-362
[4] Nowick A S and Lim B S 2001 *Phys. Rev. B* **63** 184115
[5] Sidebottom D L 1999 *Phys. Rev. Lett.*, **83** 983-986
[6] Jonscher A K 1977 *Nature* (London) **267** 673-679
[7] Jonscher A K *Dielectric Relaxation in Solids* (Chelsea Dielectrics Press, London, 1983).
[8] Fröhlich H *Theory of Dielectrics* (Oxford University Press, London 1949).
[9] Macdonald JR 2000 *Solid State Ionics* **133** 79-97
[10] Qi J and Sihvola A 2011 *IEEE Trans. Dielectr. Electr. Insul.* **18** 149-154
[11] Hartley T T and Lorenzo C F 2003 *Signal Process.* **83** 2287-2300
[12] Podlubny I 2002 *Fractional Calculus & Applied Analysis* **5** 367–386.
[13] Gabriel S, Lau R W and Gabriel C 1996 *Phys. Med. Biol.* **41** 2271-2293
[14] Hadjiloucas S, Galvão R K H, Bowen J W, Martini R, Brucherseifer M, Pellemans H P M, Bolivar P H, Kurz H, Digby J, Parkhurst G M, Chamberlain J M 2003 *Journal of the Optical Society of America B* **20** 391-401
[15] Hadjiloucas S, Galvão R K H, Becerra VM, Bowen J W, Martini R, Brucherseifer M, Pellemans H P M, Haring-Bolívar P, Kurz H, and Chamberlain J M 2004 *IEEE Trans. Microw. Theory,* **52** 2409-2419.
[16] Galvão R K H, Hadjiloucas S, Becerra V M and Bowen J W 2005 *Meas. Sci. Technol.* **16** 1037-1053
[17] van Overschee P and De Moor B 1994 *Automatica* **30** 75–93
[18] Katayama T 2005, *Subspace methods for system identification*, Springer.
[19] Favoreel W, De Moor B and Van Overschee P 2000 *J. Process Contr.* **10** 149-155.
[20] Viberg M 1995 *Automatica* **31** 1835-1851
[21] Matignon D 1998 Stability properties for generalized fractinal differential systems, In *ESAIM: Proc. Fractional Differential Systems: Models, Methods and Applications* **5** 145-158
[22] Hadjiloucas S, Walker G C, Bowen J W and Galvão R K H 2011 *J. Phys. Conf. Ser.* **310** 012002
[23] Galvão R K H, Izhac A, Hadjiloucas S, Becerra V M and Bowen J W, 2007 *IEEE Commun. Lett.* **11** 513-515
[24] Galvão R K H, Hadjiloucas S, Kienitz K H, Paiva H M and Afonso, R J M 2013 *IEEE Trans. Circuits and Systems I* **60** 624-637
[25] Panteny S, Stevens R and Bowen CR 2005 *Ferroelectrics* **319** 199-208.
[26] Almond D P, Bowen C R and Rees D A S 2006 *J. Phys. D Appl. Phys.* **39** 1295-1304.
[27] Bouamrane R and Almond D P 2003 *J. Phys. Condens. Matt.* **15** 4089-4100.
[28] McCullen N J, Almond D P, Budd C J and Hunt G W 2009 *J. Phys. D Appl. Phys.* **42** 064001.
[29] Galvão R K H, Kienitz K H, Hadjiloucas S, Walker G C, Bowen J W, Soares S F C, and Araújo M C U, 2013 *IEEE Trans. Dielectr. Electr. Insul.* **20** 995-1008.
[30] Araújo M C U, Saldanha T C B, Galvão R K H, Yoneyama T, Chame H C and Visani V, 2001 *Chemom. Intell. Lab. Syst.***57** 65-73.
[31] Galvão R K H, Araújo M C U, Fragoso W D, Silva E C, José G E, Soares S F C and Paiva H M, 2008 *Chemom. Intell. Lab. Syst.* **92** 83-91
[32] Paiva H M, Soares S F C, Galvão R K H and Araújo M C U 2012, *Chemom. Intell. Lab. Syst.* **118** 260-266
[33] Paiva H M, Duarte M A Q, Galvão R K H and Hadjiloucas S 2013 *J. Phys. Conf. Ser.* Dielectrics Conference Reading UK (*in press*)
[34] Goethals I, Pelckmans K, Suykens A K and De Moor B 2005 Identification of MIMO Hammerstein models using least squares support vector machines *Automatica* **41** 1263-1272

# On the application of optimal wavelet filter banks for ECG signal classification

**S Hadjiloucas[1*], N Jannah[1] F. Hwang[1] and R K H Galvão[2]**
[1]School of Systems Engineering the University of Reading
[2]Divisão de Engenharia Eletrônica, Instituto Tecnológico de Aeronáutica, São José dos Campos, SP, 12228-900 Brazil

E-mail: s.hadjiloucas@reading.ac.uk

**Abstract.** This paper discusses ECG signal classification after parametrizing the ECG waveforms in the wavelet domain. Signal decomposition using perfect reconstruction quadrature mirror filter banks can provide a very parsimonious representation of ECG signals. In the current work, the filter parameters are adjusted by a numerical optimization algorithm in order to minimize a cost function associated to the filter cut-off sharpness. The goal consists of achieving a better compromise between frequency selectivity and time resolution at each decomposition level than standard orthogonal filter banks such as those of the Daubechies and Coiflet families. Our aim is to optimally decompose the signals in the wavelet domain so that they can be subsequently used as inputs for training to a neural network classifier.

## Introduction

Signal decomposition using perfect reconstruction quadrature mirror filter banks can provide a very parsimonious representation of ECG signals. In previous works [1] we have shown that optimal wavelets can be used for the post-processing of ECG signals so that classifiers can operate directly in the wavelet domain as opposed to the time or frequency domains. Our approach extends the wavelet parametrization approach proposed by Sherlock and Monro [2] to ensure that the derived wavelets have at least two vanishing moments. In the current work, the filter parameters are adjusted by a numerical optimization algorithm in order to minimize a cost function associated to the filter cut-off sharpness. The goal consists of achieving a better compromise between frequency selectivity and time resolution at each decomposition level than standard orthogonal filter banks such as those of the Daubechies and Coiflet families.

## Wavelet filter bank parametrization

In the signal decomposition using the DWT, both a low pass (LPF) and a high pass (HPF) filter bank are used to generate time domain responses, these are convolved with the time domain ECG signal. Convolving the response function of the chosen filter (corresponding to a particular mother wavelet) with the signal provides an output which has different energy at different scales. Approximation coefficients relate to the low frequency components of the signal whereas detail coefficients relate to the higher frequency components in the signal. Wavelet decomposition using the DWT provides essentially a multi-resolution representation of the input signal. The user normally retains coefficients

[1*] To whom any correspondence should be addressed.

up to a particular scale whereas more detailed decompositions become redundant as their incorporation have a negligible effect on the signal. The convolution operation may be conveniently performed in the frequency domain where it is implemented through a simple multiplication process.

In this filter bank, the low-pass filtering result undergoes successive filtering iterations with the number of iterations $N_{it}$ chosen by the analyst. The final result of the decomposition of data vector **x** is a vector resulting from the concatenation of row vectors $\mathbf{c}(N_{it})$ (termed approximation coefficient at the largest scale level) and $\mathbf{d}(s)$ (termed detail coefficients at the $s^{\text{th}}$ scale level, $s = 1,\ldots, N_{it}$) in the following manner:

$$\mathbf{t} = [\mathbf{c}(N_{it}) \mid \mathbf{d}(N_{it}) \mid \mathbf{d}(N_{it} - 1) \mid \ldots \mid \mathbf{d}(1)] \tag{1}$$

with coefficients in larger scales (e.g. $\mathbf{d}(N_{it}),\mathbf{d}(N_{it} - 1),\mathbf{d}(N_{it} - 2),\ldots$) associated with broad features in the data vector, and coefficients in smaller scales (e.g. $\mathbf{d}(1),\mathbf{d}(2),\mathbf{d}(3),\ldots$) associated with narrower features such as sharp peaks. The filter bank transform can be regarded as a change in variables from $\Re^J$ to $\Re^J$ performed according to the following operation,

$$t_j = \sum_{n=0}^{J-1} x_n v_j(n), \quad j = 0, 1, \ldots, J-1 \tag{2}$$

where $t_j$ is a transformed variable and $v_j(n) \in \Re$ is a transform weight. It proves convenient to write the transform in matrix form as:

$$\mathbf{t}_{1\times J} = \mathbf{x}_{1\times J}\mathbf{V}_{J\times J} \tag{3}$$

where $\mathbf{x} = [x_0\ x_1\ \ldots\ x_{J-1}]$ is the row vector of original variables, **t** is the row vector of new (transformed) variables and **V** is the matrix of weights. Choosing **V** to be unitary (that is, $\mathbf{V}^{\mathrm{T}}\mathbf{V} = \mathbf{I}$), the transform is said to be orthogonal and it, therefore, consists of a simple rotation in the coordinate axes (with the new axes directions determined by the columns of **V**).

Let $\{h_0, h_1, \ldots, h_{2N-1}\}$ and $\{g_0, g_1, \ldots, g_{2N-1}\}$ be the impulse responses of the low-pass and high-pass filters respectively. Assuming that filtering is carried out by circular convolution, the procedure for generating the approximation coefficients from the data vector **x** is illustrated in Table 1. The convolution consists of flipping the filtering sequence and moving it alongside the data vector. For each position of the filtering sequence with respect to the data vector, the scalar product of the two is calculated (with missing points in the filtering sequence replaced with zeros). For instance, if $N = 2$, the third row in Table 1 shows that $c_1' = x_1h_3 + x_2h_2 + x_3h_1 + x_4h_0$. Dyadic down-sampling is then performed to $c_{2i}'$ to generate coefficients $c_i$. The detail coefficients $d_i$ are obtained in a similar manner by using the high-pass filtering sequence.

**Table 1.** Convolution procedure for low-pass filtering showing results before and after dyadic down-sampling.

| $x_0$ | $x_1$ | $\cdots$ | $x_{2N-1}$ | $x_{2N}$ | $\cdots$ | $x_{J-1}$ | $x_0$ | $x_1$ | $\ldots$ | $x_{2N-2}$ | Before | After |
|---|---|---|---|---|---|---|---|---|---|---|---|---|
| $h_{2N-1}$ | $h_{2N-2}$ | $\cdots$ | $h_0$ | | | | $\cdots$ | | | | $c_0'$ | |
| | $h_{2N-1}$ | $\cdots$ | $h_1$ | $h_0$ | | | | $\cdots$ | | | $c_1'$ | $c_0$ |
| | | $\cdots$ | | $\vdots$ | | | | $\cdots$ | | | $\vdots$ | $\vdots$ |
| | | | | | | $h_{2N-1}$ | $h_{2N-2}$ | $h_{2N-3}$ | $\cdots$ | | $c_{J-2}'$ | |
| | | | | | | | $h_{2N-1}$ | $h_{2N-2}$ | $\cdots$ | $h_0$ | $c_{J-1}'$ | $c_{J/2-1}$ |

If the approximation **c** and detail **d** coefficients are stacked in vector $\mathbf{t} = [\mathbf{c} \mid \mathbf{d}]$, the wavelet transform can be expressed in the matrix form with the transformation matrix given by:

$$\mathbf{V}=\left[\begin{array}{ccccc|ccccc}
0 & 0 & \cdots & h_{2N-4} & h_{2N-2} & 0 & 0 & \cdots & g_{2N-4} & g_{2N-2} \\
h_{2N-1} & 0 & \cdots & h_{2N-5} & h_{2N-3} & g_{2N-1} & 0 & \cdots & g_{2N-5} & g_{2N-3} \\
h_{2N-2} & 0 & \cdots & h_{2N-6} & h_{2N-4} & g_{2N-2} & 0 & \cdots & g_{2N-6} & g_{2N-4} \\
h_{2N-3} & h_{2N-1} & \cdots & h_{2N-7} & h_{2N-5} & g_{2N-3} & g_{2N-1} & \cdots & g_{2N-7} & g_{2N-5} \\
\vdots & \vdots & \vdots & \vdots & \vdots & \vdots & \vdots & \vdots & \vdots & \vdots \\
h_0 & h_2 & \cdots & 0 & 0 & g_0 & g_2 & \cdots & 0 & 0 \\
0 & h_1 & \cdots & 0 & 0 & 0 & g_1 & \cdots & 0 & 0 \\
0 & h_0 & \cdots & 0 & 0 & 0 & g_0 & \cdots & 0 & 0 \\
\vdots & \vdots & \vdots & \vdots & \vdots & \vdots & \vdots & \vdots & \vdots & \vdots \\
0 & 0 & \cdots & h_{2N-2} & 0 & 0 & 0 & \cdots & g_{2N-2} & 0 \\
0 & 0 & \cdots & h_{2N-3} & h_{2N-1} & 0 & 0 & \cdots & g_{2N-3} & g_{2N-1}
\end{array}\right] \tag{4}$$

A requirement for the transform to be orthogonal (i.e., $\mathbf{V}^{\mathrm{T}}\mathbf{V}=\mathbf{I}$) is that the sum of the squares of each column must be equal to one and the scalar product of different columns must be equal to zero [3]. Therefore, for a filter bank that utilizes low-pass and high-pass filters, the following conditions ensure orthogonality of the transform so that no information is lost in the decomposition process [4]:

$$\sum_{n=0}^{2N-1-2l} h_n h_{n+2l} = \begin{cases} 1, & l=0 \\ 0, & 0<l<N \end{cases} \tag{5a}$$

$$g_n = (-1)^{n+1} h_{2N-1-n},\ n=0,1,\ldots,2N-1 \tag{5b}$$

Under these conditions, the filter bank is said to enjoy a perfect reconstruction (PR) property, because **x** can be reconstructed from **t** which means that there is no loss of information in the decomposition process. Although other non-orthogonal filter bank transforms can also enjoy a PR property, provided that they are associated to a non-singular matrix **V**, the analysis in the present work is restricted to orthogonal transforms. In fact, the orthogonality of the transform (with the consequent PR property) ensures that no information that may be potentially useful for classification purposes is lost in the decomposition process. Moreover, convenient parameterisation schemes may then be employed to cast the transform filters into forms amenable to optimization.

**Parametrization and optimization approach.**

Let:

$h_0^{(1)} = \cos(a_1)$,

$h_0^{(N+1)} = \cos(a_{N+1})h_0^{(N)}$

$h_{2i}^{(N+1)} = \cos(a_{N+1})h_{2i}^{(N)} - \cos(a_{N+1})h_{2i-1}^{(N)},\quad i=1,2\ldots N-1$

$h_{2N}^{(N+1)} = -\sin(a_{N+1})h_{2N-1}^{(N)}$

$h_1^{(1)} = \sin(a_1)$

$h_1^{(N+1)} = \sin(a_{N+1})h_0^{(N)}$

$h_{2i+1}^{(N+1)} = \sin(a_{N+1})h_{2i}^{(N)} + \cos(a_{N+1})h_{2i-1}^{(N)},\quad i=1,2\ldots N-1$

$h_{2N+1}^{(N+1)} = \cos(a_{N+1})h_{2N-1}^{(N)}$

As stated in [5, 6], in order to ensure two vanishing moments for the resulting transform,

$$a_N = \frac{\pi}{4} - \sum_{i=1}^{N-1} a_i \tag{6}$$

$$a_{N-1} = \frac{1}{2}\arcsin\left\{-\frac{1}{2} - \sum_{k=1}^{N-2}\left[\sin\sum_{i=1}^{k} 2a_i\right]\right\} - \sum_{i=1}^{N-2} a_i \tag{7}$$

In what follows, an optimization process is proposed that maximizes the selectivity of the pair of high-pass/low-pass orthonormal wavelet filters with a given length.

As discussed in [5, 6] the expression in (7) has a real value solution if the set of angles $a_i$ where $1 \le i \le N-2$ satisfy a set of constrains that define a non-convex region in $R^{N-2}$. Additional constrains are imposed to ensure this, by invoking a new parameter $\chi_i$ so that:

$$\chi_i = \sin \sum_{k=1}^{i} 2a_k, \quad 1 \le i \le N-2 \tag{8}$$

$$-\frac{3}{2} \le \sum_{i=1}^{N-2} \chi_i \le \frac{1}{2}, \quad -1 \le \chi_1, \chi_2 \ldots \chi_{N-2} \le 1 \tag{9}$$

Rearranging (8) and (9) we have:

$$a_1 = \frac{1}{2} \arcsin(\chi_1) \tag{10}$$

$$a_i = \frac{1}{2} \arcsin(\chi_i) - \sum_{k=1}^{i-1} 2a_k, \quad 2 \le i \le N-2 \tag{11}$$

The cost $J$ is defined from the frequency response of the low pass filter as:

$$J = \frac{\int_{0.25\omega_s}^{0.5\omega_s} \left| H^{(N)} e^{j\omega T} \right| d\omega}{\int_{0}^{0.5\omega_s} \left| H^{(N)} e^{j\omega T} \right| d\omega} \tag{12}$$

with $\omega_s = 2\pi / T$ being the sampling frequency. As discussed in [6], optimization of the cost function with respect to the $\chi_i$ parameters is accomplished by using sequential quadratic programming (SQP) which uses local quadratic approximations of the cost function and local linear approximations of the restrictions. An interesting variance to the above algorithm where additional constrains are imposed to ensure a third vanishing moment can be found in [7].

**Example of a signal decomposition process**

The signal in Fig. 1 represents the first 1000 points out of a 3600 data points record from patient number 100 (lead 1) from the MIT database. The patient does not have a pathogenic condition and his record (among others from that database) is normally used as a training set to different classifiers to discriminate from other pathogenic patient records. A typical decomposition of the signal to approximation and detail coefficients at the first decomposition level is shown as an inset to that figure. Normally, a much smaller filter tap is generated by the user as shown in the Figure 1b. Figure 2 depicts the angular parameter alpha associated with a standard db6 filter bank as well as for the filter bank generated on the basis of the proposed procedure. Figure 2b shows the difference in the function of these filters in the frequency domain. The introduction of a vanishing moment in the random parameter filter shown in Fig. 2b ensures that its gain drops to zero at high frequencies.

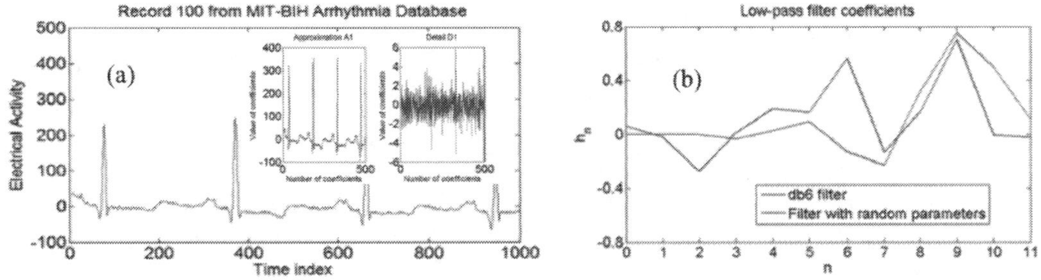


**Figure 1.** a) Typical signal from the MIT-BIH database with corresponding reconstruction on the basis of approximation and wavelet coefficients at the first decomposition level and b) comparison of filter coefficients impulse response function assuming 12 taps.

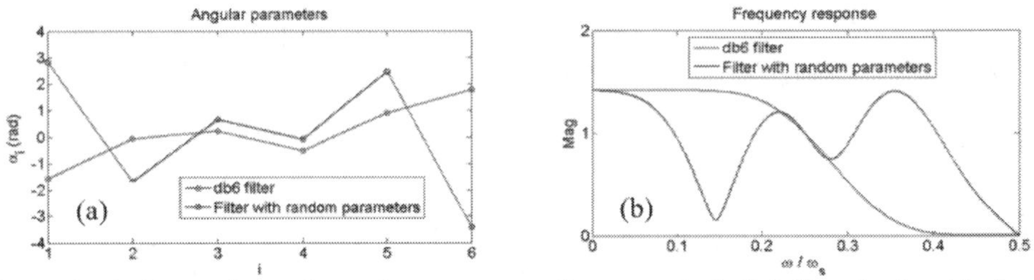


**Figure 2.** a) Comparison of angular parameters for a standard db6 filter bank with those of a filter bank generated using the proposed procedure and b) normalized frequency response for the two filter banks depicting the difference in their function.

## Conclusion

The current work advances previous ECG wavelet decompositions by placing recent work on wavelet parametrizations [5, 7] within a biomedical applications context. A test waveform was decomposed to qualitatively describe the new parametrization process being implemented. This decomposition was contrasted to that of a standard db6 parametrization. We intend to further explore this approach as a feature reduction method before presenting ECG signals from the MIT-BIH database to different classifiers (e.g. a successive projections algorithm [8]). Further work intends to optimize the number of wavelet coefficients as well as number of decomposition levels presented to the classifier. Contrary to previous works in the ECG literature, this needs to be performed in a more systematic manner (e.g., using Kohonen maps), evaluating the performance of the classifier output when it is presented with normal beats, premature ventricular contraction beats, paced beats, left and right bundle branch block beats, atrial premature contraction beats, ventricular flutter wave and ventricular escape beats.

## Acknowledgment

The authors acknowledge CNPq for the provision of post-doctoral and research fellowships and FAPESP for the provision of grants 2011/13777-8 and 2011/17610-0 that enabled this work to be performed. N Jannah acknowledges scholarship support from the Ministry of Higher Education, Kingdom of Saudi Arabia.

## References

[1] Froese T, Hadjiloucas S, Galvão R K H, Becerra V M and Coelho C J 2006 Comparison of extrasystolic ECG signal classifiers using Discrete Wavelet Transforms. *Pattern Recognition Letters* **27**(5) 393-407.

[2] Sherlock B G, and Monro D M 1998 On the space of orthonormal wavelet. *IEEE Transactions on Signal Processing* **46** 1716-1720.

[3] Vaidyanathan P P 1993 *Multirate Systems and Filter Banks*. Prentice Hall, New Jersey.

[4] Strang G, and Nguyen T, 1996 *Wavelets and Filter Banks*. Wellesley-Cambridge Press Massachusetts.

[5] Paiva H M, Marins M N, Galvão R K H and Paiva J P L M 2009 On the space of orthonormal wavelets: additional constrains to ensure two vanishing moments. *IEEE Signal Processing Letters*, **16** 101-104.

[6] Paiva H M and Galvão R K H 2012 Optimized orthonormal wavelet filters with improved frequency separation. *Digital Signal Processing* **22** 622-627.

[7] Uzinski J C, Paiva H M, Alvarado F V, Duarte M A Q and Galvão R K H 2013 Additional Constraints to Ensure Three Vanishing Moments for Orthonormal Wavelet Filter Banks *Anais do Congresso de Matemática Aplicada e Computacional* CMAC Centro-Oeste 2013, 16-19.

[8] R. K. H. Galvão, M. C. U. Araújo, W. D. Fragoso, E. C. Silva, G. E. José, S. F. C. Soares, H. M. Paiva, 2008 A variable elimination method to improve the parsimony of MLR models using the successive projections algorithm *Chemom. Intell. Lab. Syst.*, **92**, 83-91.

# Fokas method for a multi-domain linear reaction-diffusion equation with discontinuous diffusivity

**M Asvestas, AG Sifalakis, EP Papadopoulou and YG Saridakis**

Applied Math & Computers Lab, Technical University of Crete, Chania 73100, Greece

E-mail: yiannis@science.tuc.gr

**Abstract.** Motivated by proliferation-diffusion mathematical models for studying highly diffusive brain tumors, that also take into account the heterogeneity of the brain tissue, in the present work we consider a multi-domain linear reaction-diffusion equation with a discontinuous diffusion coefficient. For the solution of the problem at hand we implement Fokas transform method by directly following, and extending in this way, our recent work for a white-gray-white matter brain model pertaining to high grade gliomas. Fokas's novel approach for the solution of linear PDE problems, yields novel integral representations of the solution in the complex plane that, for appropriately chosen integration contours, decay exponentially fast and converge uniformly at the boundaries. Combining these method-inherent advantages with simple numerical quadrature rules, we produce an efficient method, with fast decaying error properties, for the solution of the discontinuous reaction-diffusion problem.

## 1. Introduction

Reaction-diffusion linear PDEs have been the core biomathematical model for studying highly invasive and aggressive forms of brain tumors for many years now (e.g. [1] and the references therein). The incorporation (cf. [2], [3]) of brain's tissue heterogeneity (white-gray matter) into the basic model led to the differential equation:

$$\frac{\partial \overline{c}}{\partial t} = \nabla \cdot (\overline{D}(\overline{x}) \nabla \overline{c}) + \rho \overline{c}, \tag{1}$$

where $\overline{c}(\overline{x}, \overline{t})$ denotes the tumor cell density at location $\overline{x}$ and time $\overline{t}$, $\rho$ denotes the net proliferation rate, and $\overline{D}(\overline{x})$ is the diffusion coefficient representing the active motility of malignant cells satisfying:

$$\overline{D}(\overline{x}) = \begin{cases} D_w, & \overline{x} \text{ in white matter} \\ D_g, & \overline{x} \text{ in gray matter} \end{cases}, \tag{2}$$

where $D_w$ and $D_g$ are scalars and $D_w > D_g$. The model formulation had zero flux boundary conditions, which impose no migration of cells beyond the brain boundaries, and an initial condition $\overline{c}(\overline{x}, 0) = \overline{f}(\overline{x})$, where $\overline{f}(\overline{x})$ is the initial spatial distribution of malignant cells.

Using the classical transformation $c(x,t) = e^t u(x,t)$, the dimensionless initial-boundary value problem (cf. [2], [4]), in $1D$, may be written as

$$\begin{cases} u_t = (Du_x)_x, & x \in [a,b], \quad t \geq 0 \\ u_x(a,t) = 0 \quad \text{and} \quad u_x(b,t) = 0 \\ u(x,0) = f(x) := \sum \delta(x - \xi_i), & \xi_i \in (a,b) \end{cases}, \tag{3}$$

where $\delta(x)$ denotes Dirac delta function.

Due to brain tissue heterogeneity, the domain $[a, b]$ is considered partitioned into $n+1$ regions $R_j := (w_{j-1}, w_j)$, with $a \equiv w_0 < w_1 < w_2 < \ldots < w_n < w_{n+1} \equiv b$, and if, for some $j$, $R_j$ is white matter region, then $R_{j-1}$ and $R_{j+1}$ are grey matter regions. Thus, the dimensionless diffusion coefficient $D(x)$ is defined to be:

$$D(x) = \gamma_j \ , \ x \in R_j \ , \ j = 1, \ldots, n+1 \ \text{ with } \gamma_j := \begin{cases} D_g/D_w, & \text{when } R_j \text{ is grey matter} \\ 1, & \text{when } R_j \text{ is white matter} \end{cases} . \tag{4}$$

Furthermore, notice that the parabolic nature of the problem directly implies continuity of both $u$ and $Du_x$ across each interface point $w_j$. Hence:

$$u(w_j, t) \ := \ \lim_{x \to w_j^+} u(x,t) = \lim_{x \to w_j^-} u(x,t), \quad \forall j = 1, 2, \ldots, n \tag{5}$$

$$Du_x(w_j, t) \ := \ \lim_{x \to w_j^+} D(x) u_x(x,t) = \lim_{x \to w_j^-} D(x) u_x(x,t), \quad \forall j = 1, 2, \ldots, n. \tag{6}$$

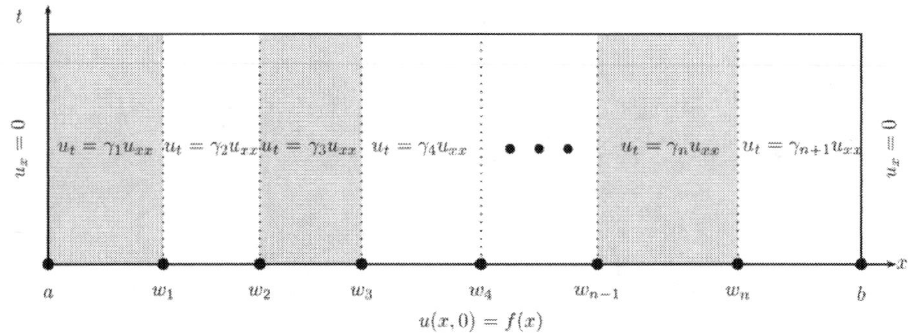


**Figure 1.** The multi-domain initial-boundary value problem

With the multi-domain problem depicted in Figure 1 above, we proceed, in Section 2, with the development of Fokas method (cf. [5], [6]) for its solution and the numerical investigation of its behaviour in Section 3.

## 2. Fokas Method

Let $u^{(j)}(x,t)$ denote the solution of the multi-domain problem defined over $\overline{R_j} := [w_{j-1}, w_j]$. Namely,

$$u^{(j)}(x,t) := \begin{cases} u(x,t), & x \in R_j \\ \lim_{x \to w_{j-1}^+} u(x,t), & x = w_{j-1} \\ \lim_{x \to w_j^-} u(x,t), & x = w_j \end{cases} , \quad j = 1, \ldots, n+1, \tag{7}$$

and, naturally,

$$u_x^{(j)}(w_{j-1}, t) := \lim_{x \to w_{j-1}^+} u_x(x,t) \quad \text{and} \quad u_x^{(j)}(w_j, t) := \lim_{x \to w_j^-} u_x(x,t) \ . \tag{8}$$

Apparently then,

$$u_t^{(j)} = (\gamma_j u_x^{(j)})_x = \gamma_j u_{xx}^{(j)} \ , \tag{9}$$

while, recalling the constrains (5)-(6), there also holds:

$$u^{(j)}(w_j, t) = u^{(j+1)}(w_j, t) \quad \text{and} \quad \gamma_j u_x^{(j)}(w_j, t) = \gamma_{j+1} u_x^{(j+1)}(w_j, t) \ . \tag{10}$$

Working similarly as in [4], we write equation (9) in divergence form

$$(e^{-ikx+\gamma_j k^2 t}u^{(j)})_t - (e^{-ikx+\gamma_j k^2 t}\gamma_j(u_x^{(j)} + iku^{(j)}))_x = 0\ ,\ k \in \mathbb{C} \text{ arbitrary}, \tag{11}$$

and by integrating over $A_j := \{(x,t) : x \in \overline{R_j},\ 0 \le t \le T\}$, using Green's Theorem and Fourier transforms we obtain the *"global relation"* in the form:

$$\begin{aligned} e^{\gamma_j k^2 t}\widehat{u}^{(j)}(k,t) &= \widehat{f}^{(j)}(k) - \gamma_j e^{-ikw_{j-1}}[\widetilde{u}_x^{(j)}(w_{j-1},\gamma_j k^2) + ik\widetilde{u}^{(j)}(w_{j-1},\gamma_j k^2)] \\ &+ \gamma_j e^{-ikw_j}[\widetilde{u}_x^{(j)}(w_j,\gamma_j k^2) + ik\widetilde{u}^{(j)}(w_j,\gamma_j k^2)], \end{aligned} \tag{12}$$

where

$$\widehat{f}^{(j)}(k) = \int_{w_{j-1}}^{w_j} e^{-ikx} f^{(j)}(x)dx, \quad \widehat{u}^{(j)}(k,t) = \int_{w_{j-1}}^{w_j} e^{-ikx}u^{(j)}(x,t)dx, \tag{13}$$

and

$$\widetilde{u}^{(j)}(x,\gamma_j k^2) := \int_0^T e^{\gamma_j k^2 t}u^{(j)}(x,t)dt, \quad \widetilde{u}_x^{(j)}(x,\gamma_j k^2) := \int_0^T e^{\gamma_j k^2 t}u_x^{(j)}(x,t)dt. \tag{14}$$

Inverting the windowed Fourier transform $\widehat{u}^{(j)}(k,t)$, setting $k^2 \leftarrow \gamma_j k^2$, $c_j \leftarrow \gamma_j^{-1/2}$, applying conditions (10), and after some algebraic manipulations we find that the solution, for all $j = 1, 2, \ldots, n$, is given by

$$\begin{aligned} u^{(j)}(x,t) &= \frac{c_j}{2\pi}\int_{-\infty}^{+\infty} e^{ic_j kx - k^2 t}\widehat{f}^{(j)}(c_j k)dk \\ &- \frac{1}{2\pi c_j}\int_{-\infty}^{+\infty} e^{ic_j k(x-w_{j-1})-k^2 t}[\frac{\gamma_{j-1}}{\gamma_j}\widetilde{u}_x^{(j-1)}(w_{j-1},k^2) + ic_j k\widetilde{u}^{(j-1)}(w_{j-1},k^2)]dk \\ &+ \frac{1}{2\pi c_j}\int_{-\infty}^{+\infty} e^{ic_j k(x-w_j)-k^2 t}[\widetilde{u}_x^{(j)}(w_j,k^2) + ic_j k\widetilde{u}^{(j)}(w_j,k^2)]dk\ , \end{aligned} \tag{15}$$

where $\widetilde{u}_x^{(j)}(w_j,k^2)$ and $\widetilde{u}^{(j)}(w_j,k^2)$ are given by the following $2(n+1) \times 2(n+1)$ linear system:

$$G\widetilde{\mathbf{u}} = \widehat{\mathbf{f}}, \tag{16}$$

where:

$$G = \begin{bmatrix} A_1^{(1)} & A_3^{(1)} & A_4^{(1)} & 0 & 0 & \cdots & 0 & 0 & 0 & 0 & 0 \\ A_5^{(1)} & A_7^{(1)} & A_8^{(1)} & 0 & 0 & \cdots & 0 & 0 & 0 & 0 & 0 \\ 0 & A_1^{(2)} & A_2^{(2)} & A_3^{(2)} & A_4^{(2)} & \cdots & 0 & 0 & 0 & 0 & 0 \\ 0 & A_5^{(2)} & A_6^{(2)} & A_7^{(2)} & A_8^{(2)} & \cdots & 0 & 0 & 0 & 0 & 0 \\ \vdots & \vdots & \vdots & \vdots & \vdots & \ddots & \vdots & \vdots & \vdots & \vdots & \vdots \\ 0 & 0 & 0 & 0 & 0 & \cdots & A_1^{(n)} & A_2^{(n)} & A_3^{(n)} & A_4^{(n)} & 0 \\ 0 & 0 & 0 & 0 & 0 & \cdots & A_5^{(n)} & A_6^{(n)} & A_7^{(n)} & A_8^{(n)} & 0 \\ 0 & 0 & 0 & 0 & 0 & \cdots & 0 & 0 & A_1^{(n+1)} & A_2^{(n+1)} & A_3^{(n+1)} \\ 0 & 0 & 0 & 0 & 0 & \cdots & 0 & 0 & A_5^{(n+1)} & A_6^{(n+1)} & A_7^{(n+1)} \end{bmatrix},$$

| $i$ | $A_i^{(j)}$ |
|---|---|
| 1 | $ic_j\gamma_j ke^{-ic_j kw_{j-1}}$ |
| 2 | $\gamma_{j-1}e^{-ic_j kw_{j-1}}$ |
| 3 | $-ic_j\gamma_j ke^{-ic_j kw_j}$ |
| 4 | $-\gamma_j e^{-ic_j kw_j}$ |
| 5 | $-ic_j\gamma_j ke^{ic_j kw_{j-1}}$ |
| 6 | $\gamma_{j-1}e^{ic_j kw_{j-1}}$ |
| 7 | $ic_j\gamma_j ke^{ic_j kw_j}$ |
| 8 | $-\gamma_j e^{ic_j kw_j}$ |

$$,\ \widetilde{\mathbf{u}} = \begin{bmatrix} \widetilde{u}^{(1)}(w_0,k^2) \\ \widetilde{u}^{(1)}(w_1,k^2) \\ \widetilde{u}_x^{(1)}(w_1,k^2) \\ \vdots \\ \widetilde{u}^{(n)}(w_n,k^2) \\ \widetilde{u}_x^{(n)}(w_n,k^2) \\ \widetilde{u}^{(n+1)}(w_{n+1},k^2) \end{bmatrix} \text{ and } \widehat{\mathbf{f}} = \begin{bmatrix} \widehat{f}^{(1)}(c_1 k) \\ \widehat{f}^{(1)}(-c_1 k) \\ \widehat{f}^{(2)}(c_2 k) \\ \widehat{f}^{(2)}(-c_2 k) \\ \vdots \\ \widehat{f}^{(n)}(c_n k) \\ \widehat{f}^{(n)}(-c_n k) \\ \widehat{f}^{(n+1)}(c_{n+1} k) \\ \widehat{f}^{(n+1)}(-c_{n+1} k) \end{bmatrix}.$$

Finally we point out that the analytic nature of the integrand functions in relation (15) above allows the determination of appropriate integration contours in the complex plane (cf. [7], [4]) to achieve fast decay of the integrands hence stability of the whole calculation.

## 3. Numerical Solution

Following our work in [4], to numerically evaluate the integrals in relation (15) above, we apply the trapezoid rule on appropriate hyperbolic contours (cf. [8]) defined by mapping the points $\theta$ on the real line to the points $\pm k(\theta)$ of the complex plane, where $k_\theta \equiv k(\theta) := i\sin(\beta - i\theta)$.

The numerical model used in our experiments is described by:

$$[a, w_1, w_2, w_3, w_4, w_5, b] = [-4, -2, -1.5, 0, 3, 4, 5], \quad \gamma_1 = D_g/D_w = 0.2, \xi_1 = -3 \;.\xi_2 = 2.5 \quad (17)$$

The relative error used in our calculations is given by $E_N := \|u_{N_{i+1}} - u_{N_i\|_\infty} / \|u_{N_{i+1}}\|_\infty$, where $N$ denotes the number of quadrature points, and is depicted in Figure 2 that follows.

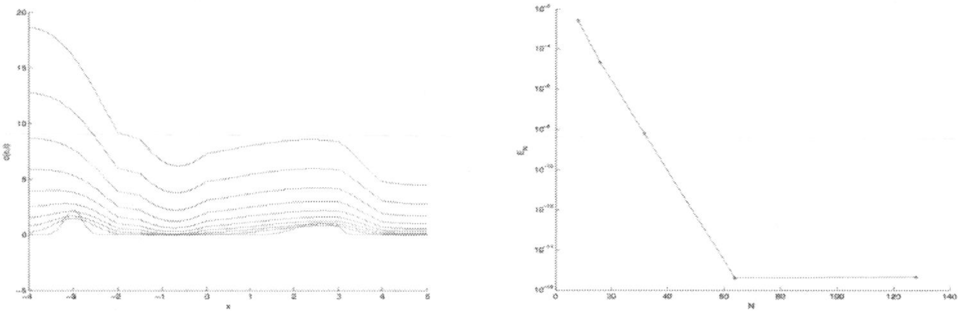

**Figure 2.** Time evolution of cell density $c(x, t)$ and the corresponding relative error $E_N$

## 4. Conclusion

The Fokas transform method, combined with numerical integration on hyperbolic contours, is applied to the solution of a multi-domain brain tumor invasion model. The analytical solution is produced in integral form at any space-time point and evaluated by a fast convergent quadrature.


## Acknowledgement

The present research work has been co-financed by the European Union (European Social Fund ESF) and Greek national funds through the Operational Program Education and Lifelong Learning of the National Strategic Reference Framework (NSRF) - Research Funding Program: THALIS. Investing in knowledge society through the European Social Fund.


## References


[1] Murray JD *Mathematical Biology*, Springer-Verlag, 2002

[2] Swanson KR *Mathematical modeling of the growth and control of tumors*, PHD Thesis,University of Washington, 1999.

[3] Swanson KR, Alvord EC Jr and Murray JD *A quantitive model for differential motility of gliomas in grey and white matter*, Cell Proliferation, 33, 317-329, 2000.

[4] Mantzavinos D, Papadomanolaki MG, Saridakis YG and Sifalakis AG *A novel transform approach for a brain tumor invasion model with heterogeneous diffusion in 1+1 dimensions*, (submitted) 2013

[5] Fokas AS *A unified transform method for solving linear and certain nonlinear PDEs*, Proc.R.Soc. A, 453, 1411-1443, 1997.

[6] Fokas AS *A new transform method for evolution PDEs*, IMA J. Appl. Math.,67(6), 559-590, 2002.

[7] Flyer N and Fokas AS *A hybrid analytical-numerical method for solving evolution partial differential equations I: The half-line*, Proc. R. Soc. A, 464, 1823-1849, 2008.

[8] Weideman JAC and Trefethen LN *Parabolic and hyperbolic contours for computing the bromwich integral*, Math. Comp., 76(259), 13411356, 2007

# A pattern recognition system for prostate mass spectra discrimination based on the CUDA parallel programming model

**Spiros Kostopoulos[1], Dimitris Glotsos[1], Konstantinos Sidiropoulos[2], Pantelis Asvestas[1], Dionisis Cavouras[1] and Ioannis Kalatzis[1]**
[1] Department of Medical Instruments Technology, Technological Education Institute of Athens, Greece
[2] School of Engineering and Design, Brunel University West London, Uxbridge, Middlesex, UB8 3PH, UK

E-mail: skostopoulos@teiath.gr

**Abstract.** The aim of the present study was to implement a pattern recognition system for the discrimination of healthy from malignant prostate tumors from proteomic Mass Spectroscopy (MS) samples and to identify m/z intervals of potential biomarkers associated with prostate cancer. One hundred and six MS-spectra were studied in total. Sixty three spectra corresponded to healthy cases (PSA<1) and forty three spectra were cancerous (PSA>10). The MS-spectra are publicly available from the NCI Clinical Proteomics Database. The pre-processing comprised the steps: denoising, normalization, peak extraction and peak alignment. Due to the enormous number of features that rose from MS-spectra as informative peaks, and in order to secure optimum system design, the classification task was performed by programming in parallel the multiprocessors of an nVIDIA GPU card, using the CUDA framework. The proposed system achieved 98.1% accuracy. The identified m/z intervals displayed significant statistical differences between the two classes and were found to possess adequate discriminatory power in characterizing prostate samples, when employed in the design of the classification system. Those intervals should be further investigated since they might lead to the identification of potential new biomarkers for prostate cancer.

## 1. Introduction

The prostate specific antigen (PSA) has been successfully introduced in daily clinical practice as a diagnostic serum biomarker for prostate cancer detection. Although PSA is sensitive, its specificity is not adequate so as to be used as a single screening method for prostate cancer detection [1]. Due to absence of accurate biomarkers that detect, quantify, and reliable distinguish patients with different prostate cancer stages, many early stage prostate cancer patients are treated as having an aggressive state of prostate cancer [2]. Thus the need of multiple biomarkers arises for improving early detection, diagnosis, and prognosis of prostate cancer [3]. In proteomic clinical research, the Mass spectrometry (MS) technology has been employed for the identification of new biomarkers and offers an interesting insight to biological samples derived from tissue, serum or urine [4]. The MS spectrum contains tens of thousands pair wise data points accounting for m/z location and the corresponding intensities [5].

Several studies have used signal analysis and pattern recognition methods to identify candidate biomarkers for prostate cancer [6-8]. The major concern in those approaches is the dimensionality of the feature space that exists in all high-throughput methods [4], and must be reduced although there is an increased risk of plausible loss of candidate biomarkers. In the present study, we propose a pattern-recognition (PR) system to discriminate between healthy and malignant prostate MS-spectra. The implementation and evaluation of the PR-system were performed by means of a graphics processing unit (GPU) framework using the parallel architecture of the GPU card and the Compute Unified Device Architecture (CUDA) programming environment [9] in order to attain optimal classifier design, minimizing the loss of candidate biomarkers.

## 2. Material and Methods

Mass spectrometry prostate cancer spectra were obtained from the National Cancer Institute Clinical Proteomics Database (JNCIdataset 7-3-02). In the present work, 106 MS signals were collected; 63 spectra accounting for patients with no evidence of disease (PSA<1) and 43 spectra corresponding to prostate cancer patients with PSA>10. Subsequently, all the MS-spectra were pre-processed for smoothing, noise estimation and peak detection [10].

Peaks that were found to differ less than 1% of relative mass [6, 11] were classified as of corresponding to the same mz-interval. Only common mz-intervals amongst at least 30% of the available class-spectra (either PSA<1 or PSA>10) were further considered. Eighty one mz-intervals were obtained from the pre-processing steps and were further considered for the classification stage. The k-Nearest Neighbour (kNN) [12], the Support Vectors Machines (SVM) [13], and the Probabilistic Neural Network (PNN) [14] classifiers were tested. For feature selection purposes, an optimal and two suboptimal methods were tested. In the first sub-optimal method the best features combination was identified by means of the sequential backward selection (SBS) technique [15]. The precision of the design was tested by means of the leave-one-out (LOO) evaluation method. In the second suboptimal method, the Wilcoxon statistical test was employed and only the mz-intervals that presented statistical significant differences ($p<0.001$) were further examined by the exhaustive search method [15]. The optimal feature selection method (exhaustive search) exhaustively forms all possible mz-interval combinations, designing the classifier by each combination, and testing its precision by means of the LOO method [15]. In order to avoid overfitting, features combinations of up to fourteen mz-intervals were examined. The total number of combinations, thus resulting, may be estimated by equation (1), where $n$ is the number of features (81 mz-intervals) and $k$ is the number of feature combinations ($k$=1,2,...,14). This is an enormous number of combinations ($fc>2\cdot 10^{15}$) and, one has to consider that, for each combination, the classifier's performance has to be evaluated by the LOO method, which requires that the classifier is redesigned as many times as is the number of data in both classes.

$$fc=\sum_{k=1}^{14}\frac{n!}{(n-k)!k!} \tag{1}$$

To deal with the high-dimensionality problem of the PR-system design, most of the system design was transferred onto the processors of a GPU (GeForce GTX 580), attached to the computer, by adopting parallel processing programming using the CUDA toolkit v4.0 and the C/C++ programming language [9]. Figure 1 shows the parallelization of the training process, where in each GPU thread an mz-interval combination was fed, the PNN classifier was designed, trained and was evaluated for overall accuracy by means of LOO method. Resulted overall accuracies from GPU were transferred to host PC memory for the identification of the mz-interval combination that achieved the highest classification accuracy. The PNN with Gaussian kernel was selected for the parallelization implementation as it proved fast in execution and of high discriminatory ability based on its performance in the suboptimal designs. The discriminant function of the PNN for class $j$ is given in:

$$g_j(x)=\frac{1}{(2\pi)^{d/2}\sigma^d N_j}\sum_{i=1}^{N_j}\exp\left(-\frac{(x-x_{ji})^T(x-x_{ji})}{2\sigma^2}\right) \tag{2}$$

where $\sigma$ is the spread of the Gaussian activation function, that was experimentally determined for best performance to 0.2, $N_j$ is the number of MS-Spectra of class $j$, $d$ is the number of mz-intervals, $x_{ji}$ is the feature vector of the *i-th* MS-spectrum of class $j$, and $x$ is the feature vector of the unknown MS-spectrum. The latter is classified to the class with the highest discriminant function value.

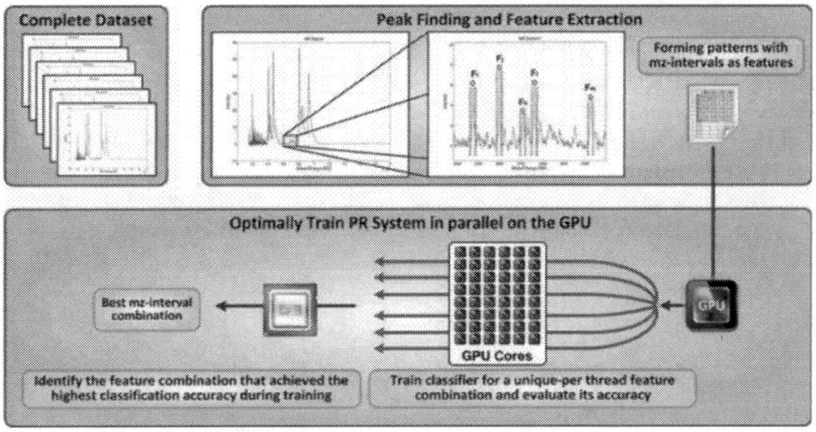


**Figure 1.** Block diagram of the PR-system design and evaluation in GPU.

### 3. Results and Discussion

Table 1 shows the classification accuracies achieved following the two suboptimal feature selection methods and three different classifiers. The highest accuracies for PNN, kNN and SVM were 91.5%, 90.6% and 90.6% respectively. When the optimal features selection method was employed 98.1% overall classification accuracy was achieved with two cancerous cases were mistaken classified as healthy. The specificity of the system when suboptimal methods were employed range from 87.3% to 95.2%. These values are in agreement with other works [16]. Although, it has to be noticed that in the present work only healthy and malignant cases were accounted, that, a priori, are better separable. The PNN classifier, the exhaustive search, for testing all possible mz-intervals combinations, and the LOO evaluation method on the 512 cores of the GPU card were utilised for final system design. The best PR-system design was achieved by using seven mz-intervals combination (see Figure 2).

**Table 1.** Classification accuracies achieved employing the Sequential Backward Selection and the Wilcoxon statistical test as feature selection methods.

| | Sequential Backward Selection | | | Wilcoxon statistical test | | |
|---|---|---|---|---|---|---|
| Classifier | PSA<1 | PSA>10 | Overall | PSA<1 | PSA>10 | Overall |
| PNN | 95.2% (60/63) | 86.1% (37/43) | 91.5% | 92.1% (58/63) | 86.1% (37/43) | 89.6% |
| kNN | 92.1% (58/63) | 88.4% (38/43) | 90.6% | 90.1% (57/63) | 88.4% (38/43) | 89.6% |
| SVM | 93.7% (59/63) | 86.1% (37/43) | 90.6% | 87.3% (55/63) | 72.1% (31/43) | 81.1% |

Figure 2 shows the boxplot diagrams of the seven mz-intervals that were combined to the optimum PR-system design and resulted in the highest classification accuracy. Five out of seven mz-intervals were found to present statistical significant differences (*p<0.001*) between the two classes, of PSA<1 and PSA>10. It is also noticeable that the mz-intervals contain not only high but also low intensity peaks. Specifically, for three of the mz-intervals '3459.023-3475.5088', '4724.3401-4747.4607' and

'5339.7242-5365.6699' the spectra belonging to cancerous class have low intensity peaks. This finding may be indicative of peptide absence in those mz regions for the cancerous cases.
Figure 3 demonstrates the transformed scatter diagrams, where the multi-dimensional mz-intervals were transformed into two-dimensional spaces (dimensionality is equal to the number of classes) by the quadratic least squares mapping process [17]. Diamonds and stars correspond to PSA<1 and PSA>10 cases, respectively. As it can be observed, a fine class separability was attained based on the mz-intervals combination that provided the highest classification accuracy. This good separability of the classes was succeeded only when the GPU technology employed, so as the search amongst a huge number of mz-intervals combinations was plausible.

The identified mz-intervals displayed significant statistical differences between the classes and were found to possess adequate discriminatory power in characterizing prostate samples, when employed in the design of the classification system. Those intervals should be further investigated since they might lead to the identification of potential new biomarkers for prostate cancer.

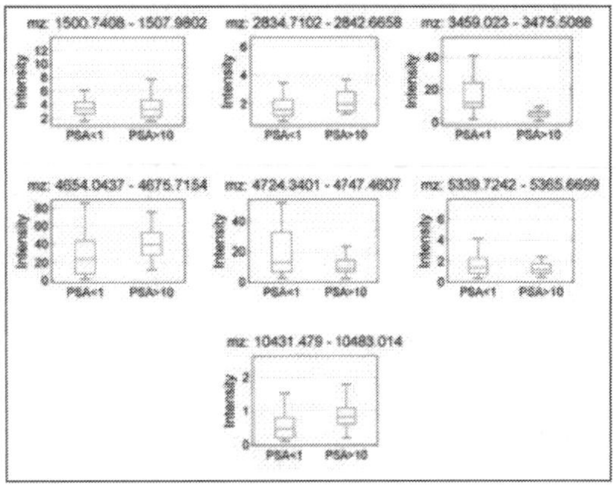


**Figure 2.** Boxplots of the seven mz-intervals that formed the best feature combination in the optimal system design.

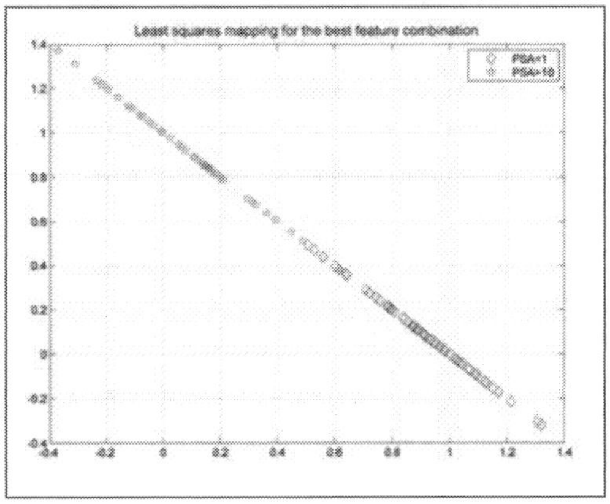


**Figure 3.** Least square mapping using seven mz-intervals combination.

**Acknowledgments**
This research has been co-financed by the European Union (European Social Fund – ESF) and Greek national funds through the Operational Program "Education and Lifelong Learning" of the National Strategic Reference Framework (NSRF) - Research Funding Program: ARCHIMEDES III.
Investing in knowledge society through the European Social Fund.

## 4. References

[1] Feuerstein I, et al. 2005 *Journal of Proteome Research* **4** 2320-6.
[2] Wright ME, et al. 2005 *Molecular and Cellular Proteomics* **4** 545-54.
[3] Zhang X, et al. 2007 *Mass Spectrometry Reviews* **26** 403-31.
[4] Roy P, et al. 2011 *Briefings in Bioinformatics* **12** 176-86.
[5] Chen S, et al. 2007 *Computational Statistics and Data Analysis* **52** 211-20.
[6] Petricoin Iii EF, et al. 2002 *Journal of the National Cancer Institute* **94** 1576-8.
[7] Qu Y, et al. 2002 *Clinical Chemistry* **48** 1835-43.
[8] Lilien RH, et al. 2003 *Journal of Computational Biology* **10** 925-46.
[9] Sidiropoulos K, et al. 2012 *Computers in Biology and Medicine* **42** 376-86.
[10] Bougioukos P, et al. 2010 *Comput Meth Prog Bio* **99** 147-53.
[11] Wang X. 2006 *Proteomics* **6** 2095-100.

[12] Gose E, et al. 1996 *Pattern Recognition and Image Analysis* (New Jersey: Prentice Hall PTR)
[13] Burges CJC. 1998 *Data Mining and Knowledge Discovery* **2** 121-67.
[14] Specht DF. 1990 *Neural Networks* **3** 109-18.
[15] Theodoridis S, et al. 2003 *Pattern Recognition* (San Diego: Elsevier)
[16] Vu TN, et al. 2008 *LNCS* **4901** 17-24.
[17] Ahmed N, et.al. 1975 *Orthogonal Transforms for Digital Signal Processing* (Springer-Verlag)

# Network inference of AP pattern formation system in *D.melanogaster* by structural equation modeling

**S Aburatani and H Toh**

Computational Biology Research Center, AIST, AIST Tokyo Waterfront Bio-IT Research Building, 2-4-7 Aomi, Koto-ku, Tokyo, 135-0064, Japan

E-mail: s.aburatani@aist.go.jp

**Abstract.** Within the field of systems biology, revealing the control systems functioning during embryogenesis is an important task. To clarify the mechanisms controlling sequential events, the relationships between various factors and the expression of specific genes should be determined. In this study, we applied a method based on Structural Equation Modeling (SEM), combined with factor analysis. SEM can include the latent variables within the constructed model and infer the relationships among the latent and observed variables, as a network model. We improved a method for the construction of initial models for the SEM calculation, and applied our approach to estimate the regulatory network for Antero-Posterior (AP) pattern formation in *D. melanogaster* embryogenesis. In this new approach, we combined cross-correlation and partial correlation to summarize the temporal information and to extract the direct interactions from the gene expression profiles. In the inferred model, 18 transcription factor genes were regulated by not only the expression of other genes, but also the estimated factors. Since each factor regulated the same type of genes, these factors were considered to be involved in maternal effects or spatial morphogen distributions. The interpretation of the inferred network model allowed us to reveal the regulatory mechanism for the patterning along the head to tail axis in *D. melanogaster*.

## 1. Introduction

Clarifying the process that generates an organism's body plan is one of the fascinating themes, and the process of *D. melanogaster* embryo formation is a suitable model for studying embryogenesis. During *D. melanogaster* embryogenesis, anterior-posterior (AP) patterning is generated by serial transcriptional regulation, beginning with maternally supplied transcripts, through gap genes and pair-rule genes, and finally segment polarity genes [1]. The expression of these genes generates gradients of their proteins in the embryo, and thus these gradients determine the expression of the following genes [2]. The regulation of developmental genes has been extensively studied, but many aspects remain unexplained and we are far from a complete understanding of pattern formation.

To obtain better insights into the transcriptional regulatory mechanism for pattern formation, a gene regulatory network is useful [3, 4]. Since the underlying mechanism of transcriptional regulation is the localized expression of transcription factor genes at specific times and places, revealing the regulatory networks between these genes would provide a schema of developmental regulation in embryogenesis. Furthermore, the influences from several types of cellular components should be considered as regulatory factors of gene expression. Thus, we have developed a new approach, based on Structural Equation Modeling (SEM), to investigate the regulatory relationships between different cellular components [5, 6].

In this study, we developed a new method to assume an initial model for SEM calculation, and improved our SEM approach for inferring complicated transcriptional regulation in *D. melanogaster* embryogenesis. We applied our approach to 18 transcription factor genes with regulated expression for AP pattern formation. Using our methods, the regulatory factors for AP pattern formation were estimated by factor analysis, and the regulatory relationships from the regulators to the genes were estimated by SEM.

## 2. Materials & Methods

### 2.1. Expression data and selected genes

For the construction of the initial model, we utilized the expression profiles of 12,868 genes measured during 28 time points, covering the entire 24-h period during *D. melanogaster* embryo development [7]. All expression data in *D. melanogaster* embryo cells were downloaded from GEO Database (http://www.ncbi.nlm.nih.gov/geo/). Among the empirically identified transcription factor genes for AP pattern formation [8], 18 genes displayed sufficient expression profiles, and these were utilized to infer the regulatory network.

### 2.2. Construction of the initial model

To construct an initial model, we developed a new method that combines cross correlation and partial correlation. Causal relationships between genes were detected by cross correlation, given by

$$r_{xy} = \frac{\sum_{i=1}^{N}(x_i - \bar{x})(y_i - \bar{y})}{\left\{\sum_{i=1}^{N}(x_i - \bar{x})^2\right\}^{1/2}\left\{\sum_{i=1}^{N}(y_{i+d} - \bar{y})^2\right\}^{1/2}} \tag{1}$$

where $x_i$ and $y_i$ are two time series data with $N$ time points, and $d$ is the time-lag between variables $x$ and $y$. In this case, each gene pair was calculated with $d$ = -5, ... , +5. The eleven absolute values of cross correlations were compared, and the highest absolute value was arranged in a fundamental cross correlation matrix. The corresponding time-lag value $d$ was arranged as a matrix element in a discrete matrix.

The direct interactions were estimated by the partial correlation coefficients calculated from the cross correlation matrix, as follows:

$$r_{ij|rest} = \frac{r^{ij}}{(r^{ii})^{1/2}(r^{jj})^{1/2}} \tag{2}$$

where $r_{ij|rest}$ is the partial correlation coefficient between variables $i$ and $j$, given the rest, and $r^{ij}$ is the ($i$, $j$) element in the reverse of the cross correlation coefficient matrix. To detect the significant value of the partial correlation coefficient, we use the statistic

$$t_{ij} = \frac{r_{ij|rest}(n-q-2)^{1/2}}{(1-r_{ij|rest}^2)^{1/2}} \tag{3}$$

where $q$ is the number of fixed variables and $n$ is the number of samples. The statistic $t_{ij}$ is distributed according to the $t$ distribution. Thus, the relationship between the expression profiles can be tested by the t-test. In this study, we selected the significant gene pairs at $p < 0.01$.

Finally, we combine partial correlation and cross correlation. The regulatory directions, which were estimated by the sign of the time-lag values, were added to the extracted gene pairs by PCC. The procedure developed for constructing the initial model is displayed in Figure 1.

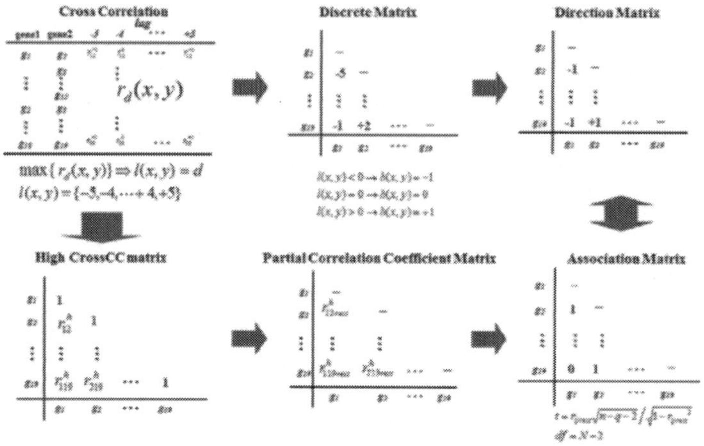


**Figure 1.** Procedure for initial model construction.

*2.3. Factor Analysis*

To determine the optimal number of factors for inclusion in the network model as latent variables, we performed a factor analysis. In the factor analysis, the covariance matrix between the observed variables $\Sigma$ is structurized by parameters, as follows:

$$\Sigma = Var[X] = \Lambda\Phi\Lambda' + \Psi^2 \tag{4}$$

where $\Psi^2$ is the covariance matrix of error terms, $\Lambda$ is the factor loading matrix of latent variables, and $\Phi$ is the covariance matrix among factors. From this structurized matrix, the values of the matrix and the variances of the error terms are estimated. In this study, the Kaiser criterion states and the scree plot were utilized to estimate the number of factors. The number of latent variables was suggested by a principal factor method with promax rotation, which is a general method for rotating factors to fit a hypothesized structure of latent variables.

*2.4. Structural Equation Modeling (SEM)*

In this study, the regulatory model is defined as follows:

$$\begin{bmatrix} f \\ v \end{bmatrix} = \begin{bmatrix} \mathrm{B} & \Gamma \\ \Theta & \mathrm{K} \end{bmatrix} \begin{bmatrix} f \\ v \end{bmatrix} + \begin{bmatrix} \delta \\ \varepsilon \end{bmatrix} \tag{5}$$

Here, $f$ is a vector of $p$ latent variables, and $v$ is a vector of $q$ observed variables. The matrix B is a $p$x$p$ matrix representing the relationships between latent variables; $\Gamma$ is a $p \times q$ matrix representing the causal relationships between observed and latent variables; $\Theta$ is a $q \times p$ matrix representing the effectiveness of latent variables to the observed variable; and K is a $q \times q$ matrix representing the regulatory relationships between observed variables. In the SEM analysis, the parameter estimation was performed by comparing the actual covariance matrix S, calculated from the measured data, with the estimated covariance matrix $\Sigma(\theta)$ of the constructed model. The maximum likelihood method was utilized as a fitting function to estimate the model parameters. To optimize the network model, we developed an iteration algorithm by using fitting scores and modification index (MI) scores, which measure how much the chi-square statistic is expected to decrease if a particular parameter setting is constrained. The SEM software package SPSS AMOS 17.0 (IBM, USA) was used.

## 3. Results & Discussion

By our combination of cross correlation and partial correlation, we estimated 35 causal relationships between 18 genes. Furthermore, the application of factor analysis provided information about unobserved effective factors, and 4 factors were estimated as regulators for the 18 genes. Thus, the optimal model was inferred from an initial model with the 18 genes and the detected factors. The whole network is displayed in Figure 2a. In Figure 2b, the order of the 18 genes is consistent with the biological knowledge of them, even though the order was disturbed in the initial model. The inclusion of latent variables into an inferred model is considered to be suitable for representing a biological phenomenon. Furthermore, Figure 2c shows that the known temporal and spatial gene expressions during *D. melanogaster* development are well reflected by the inferred networks between genes.

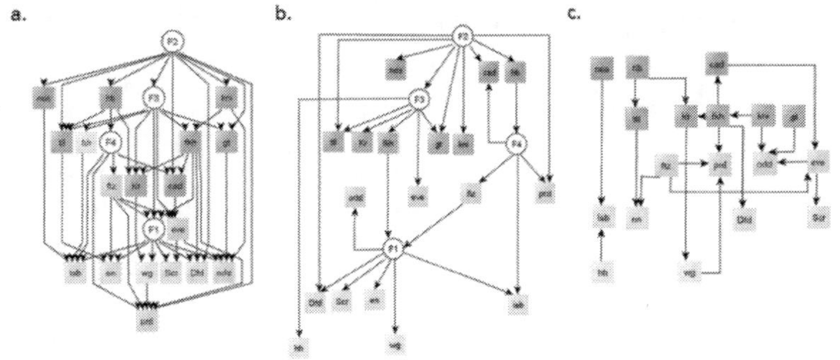


**Figure 2.** Inferred networks. (a) Whole network model. (b) Relationships between factors and genes. (c) Relationships between genes. Orange: Maternal, Pink; Gap, Green: Pair-rule, Blue; Segment, Yellow; Signal.

Since F2 was a positive regulator of all maternal genes and some gap genes, F2 was considered as Bicoid protein density in the early embryo. Furthermore, F3 regulated many gap genes positively, but negatively regulated one pair-rule gene, therefore F3 was considered to be an inducer of the gap genes' process by repressing pair-rule gene expression.

## 4. Conclusion

In this study, the spatial and temporal controls that function in the developmental process were identified by our inferred network. In the inferred network, not only the effects of transcription factor proteins, but also the protein densities were suggested as latent variables. The network inference by SEM is applicable to clarify the control of gene expression by intracellular factors.

## References

[1] Pankratz M J and Jäckle H 1993 *Blastoderm segmentation. In The Development of Drosophila melanogaster.* (Cold Spring Harbor: Cold Spring Harbor Laboratory Press) 467-516

[2] Rivera-Pomar R and Jäckle H 1996 From gradients to stripes in Drosophila embryogenesis: Filling in the gaps *Trends Genet.* **12** 478–483

[3] Friedman N, Linial M, Nachman I and Pe'er D 2000 Using Bayesian networks to analyze expression data *J. Comput. Biol.* **7** 601-620

[4] Akutsu T, Miyano S and Kuhara S 2000 Inferring qualitative relations in genetic networks and metabolic pathways *J. Comput. Biol.* **7** 331-343

[5] Aburatani S 2011 Application of Structure Equation Modeling for inferring a serial transcriptional regulation in yeas *Gene. Regul. Syst. Biol.* **5** 75–588

[6] Aburatani S 2011 Network Inference of pal-1 Lineage-Specific Regulation in the C. elegans Embryo by Structural Equation Modeling *Bioinformation* **8** 652–657

[7] Hooper S D, Boué S, Krause R, Jensen L J, Mason C E, Ghanim M, White K P, Furlong E E and Bork P 2007 Identification of tightly regulated groups of genes during Drosophila melanogaster embryogenesis *Mol. Sys. Biol.* **3** 1-11

[8] MacArthur S *et al.* B 2009 Developmental roles of 21 Drosophila transcription factors are determined by quantitative differences in binding to an overlapping set of thousands of genomic regions *Genome Biol.* **10** R80

# Simulating external compressions of the breast with the Surface Evolver

**Marcelo Zanchetta do Nascimento**[1], **Valério Ramos Batista**[2]

[1] *FACOM-UFU, Av. João Naves de Ávila 2121, Bl.A, 38400-902 Uberlândia-MG, Brazil*
[2] *CMCC-UFABC, R. Sta. Adélia 166, Bl.B, 09210-170 St. André-SP, Brazil*

E-mail: nascimento@facom.ufu.br

**Abstract.** In this paper we introduce a computational modelling that reproduces the breast compression processes used to obtain the mammogram. The main result is a programme in which one can track the first steps of virtual mammography. On the one hand, our modelling enables addition of structures that represent different tissues, muscles and glands in the breast. On the other hand, we shall validate and implement it by means of laboratory tests with phantoms. To the best of our knowledge, these two characteristics do confer originality to our research. This is because their interrelation seems not to be properly established elsewhere yet. We conclude that our model reproduces the same shapes and measurements really taken by the volunteer's breasts.

The 2nd author dedicates this work to his wife Clarice.

## 1. Introduction

The obit rate caused by cancer around the globe is tracked by the World Health Organization. Their statistics indicate that 7.6 million of the obits in 2008 were due to cancer. This ratio has been increasing and some estimates show that it can reach 13.1 million cases until 2030. Lung cancer in women is still the leading obituary kind, and breast cancer follows right next. See `http://www.who.int/mediacentre/factsheets/fs297/en` for details.

Medical imaging methods help detect breast cancer. The most used are resonance, ultrasonography and mammography. However, screen/film mammography is the primary imaging modality due to low cost and accessibility. It has proved to be an effective aid in the early detection, and most of the restraint in obit rate is solely thanks to this method. But it shares a disadvantage with the others: in the imaging procedure, forces applied to the breast change its shape considerably. For the surgery, tumour location is highly uncertain, and so a great portion of the breast has to be removed [1].

Along the past decade there has been an effort to develop softwares devoted to anatomic modelling of the breast and of deformations by mammography procedures. Among others, we cite [1, 2, 3, 4, 5] and [6]. Some are aesthetically impressive, like [7]. For a model at rest, this latter is indeed very complete, but its functionality has not been presented yet. The problem will reside in the excessive computational complexity at managing all different elements that take part in the model. The source codes tend to be untreatable in such cases.

In order to bypass these difficulties, perhaps a simpler approach like the ITK software of [1] turns out to be more accessible. However, the human breast is composed by different tissues, muscles and glands, as highly taken into account in [7]. Since ITK works with three dimensional meshes, it is a hard task to separate all those components into specific clusters. Moreover, this produces a "Lego effect" that demands too much orchestration to control the belonging of each mesh element to each tissue, muscle or gland.

Which simplifications can be made without going into considerable errors? Regarding numeric modelling of the breast deformation during mammography, we still do not have enough satisfactory answers. In this case, the dilemma between complexity and simplification motivated us to work with the *Surface Evolver* [8]. Its computational representation of any element is made by surface *layers* that enclose it. Evolver allows to add as many elements as needed, together with their individual attributes like mass, geometric restrictions, contact and elastic bending energies, etc. See http://www.susqu.edu/brakke for details.

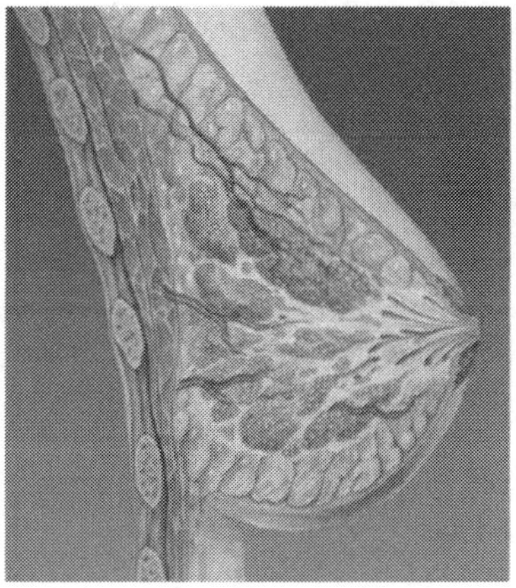

Figure 1: Side view of the breast[1].

Figure 2: Surgery position with components.

In the long-term, our research aims at a full and detailed reproduction of the mammography procedure with Evolver. By starting with the mammographies that show tumour nodules, we rewind the virtual procedure. This way, nodules will be located with precision for the surgery. For the time being, our work is devoted to simulations of external compressions of the breast. In order to track tumour nodules, we shall either consider the complexity of internal parts, or equate their trajectory by performing experiments with transparent breast phantoms.

## 2. Methodology

In order to take mammographies in our virtual environment, we have subdivided the whole process in 6 main steps for each breast. We call them SRG (*surgery*), STU (*stand-up*), LAT (*lay-on-table*), CRC (*cranio-caudal*), LET (*lean-on-table*) and MLO (*medio-lateral-oblique*).

Of course, at each step the mass of the breast remains constant, while its shape changes drastically. Except for CRC and MLO, changes in the volume and average density are much slighter than in the area. There is enough literature with average values, but for our purposes they are inadequate, because measurements vary a lot from woman to woman.

It is possible to apply electrodes to draw a computer model and then get magnitudes numerically. However, this procedure is not practical and requires expensive equipment.

But we check our data in Evolver by means of a virtual tape-measure that we call TMR. Our procedure is to enter the initial values and then use the TMR to fine-tune them. For the time being, the fine-tuning is still interactive. We shall make it automatic in future.

Figure 3 left shows a standing woman with her spine straight. The dotted line indicates a tape measure that passes *under* her breast. With the open right hand upon her belly we locate our coordinates: $Ox$ points as the thumb, $Oy$ points as the other fingers, and the palm is raised along $Oz$. Figure 3 right shows the oval curve obtained after having taken the volunteer's measurements. Important values are indicated there, and we have fixed all magnitudes in the cgs-system. Notice that the overall perimeter is 86cm.

[1] Taken from http://en.wikipedia.org/wiki/Portal:Medicine/Selected_picture/26

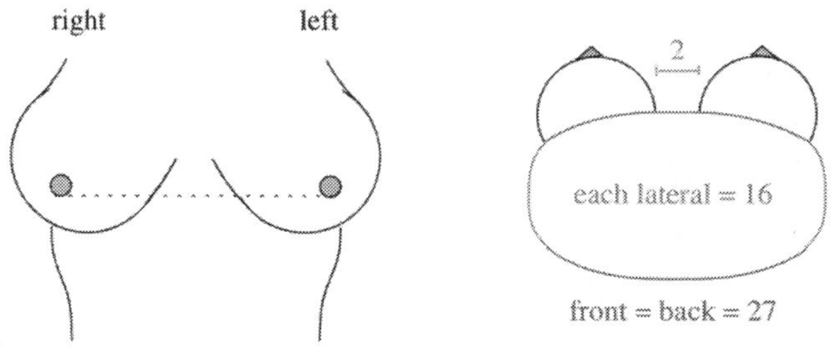


Figure 3: Measuring strategic stretches of the thorax.

See the link *Slides-ICMSQUARE* of our home page `http://www.facom.ufu.br/~nascimento` for other important measurements. There you will also find a detailed description of the methods and the modelling used in our work.

Regarding CRC and MLO measurements, they are taken directly from the mammographies. Our simulations with Evolver always start at SRG, because it is the most symmetric position. In practice, not only a surgical drape holds the base of the breast, but also an auxiliary surgeon and sometimes an extra support fitted under the woman's armpit.

Our Evolver datafiles follow the sequence SRG $\rightarrow$ STU and then either LAT $\rightarrow$ CRC or LET $\rightarrow$ MLO. From STU to LAT, early simulations showed that the virtual breast *does not* necessarily reach a symmetric configuration. This is because Evolver seeks to attain some pre-defined values and stabilises right thereafter. The authors believe that the breasts' shape naturally minimises a certain energy whenever they rest at a given position. In the case of cell membranes, this energy is related to the *Willmore Functional* (for instance, see [9]). This functional is defined as $\int_M H^2 dS$, where $M$, $H$ and $dS$ stand for the surface of the body, its mean curvature and the area element of integration, respectively. By asking Evolver to minimise this integral, we have obtained more symmetric answers. That is why the Willmore Functional is included in our datafiles. We conjecture that this energy is minimised whenever the breast accommodates in a rest position.

## 3. Results

The main result is our programme, in which the user can track the first steps of the virtual mammography, as depicted in Figures 4 and 5.

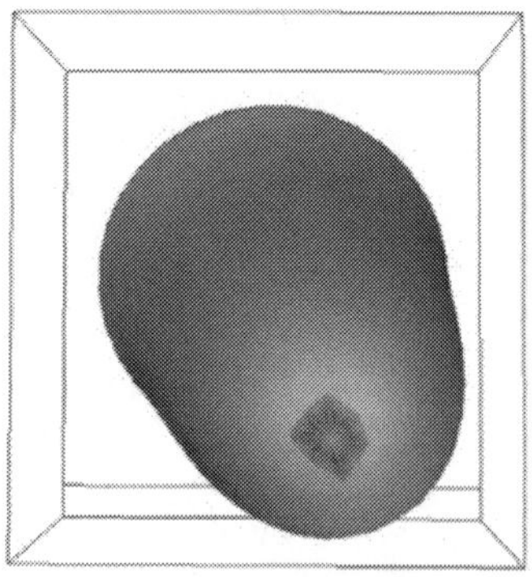

Figure 4: STU enhanced with Geomview.

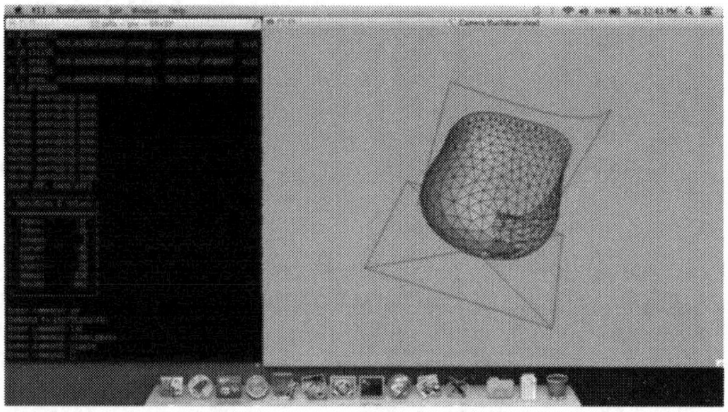

Figure 5: LAT with measurements.

In order to run our programme, we recommend Linux Operating System Ubuntu 11.10 or later, Evolver 2.50 and Geomview 1.9.4. For the time being, we have implemented SRG, STU and LAT, all of them for the left breast only. You can either run it in our Virtual Machine, available on our home page in the link *Softwares*, or access our link *Slides-ICMSQUARE* for an overview. The Virtual Machine contains all source codes, and therefore all details of our modelling.

Regarding its *validation*, the link Softwares also contains three sample Videos of tests with a transparent phantom. The phantom dimensons are compatible with the volunteer's, and addition of virtual nodules to our model will be implemented in a forthcoming work.

It is important to notice that the breast volume *decreases* from SRG to STU, and again from STU to LAT. Indeed, when the woman is lain down on the operating table, the breast structure is uniformly distributed on its base. This is when we have the least pressure due to gravity.

The breast *mass* is constant and *mean density = mass/volume*. When the woman stands up, the breast mass concentrates at the bottom. Pressure due to gravity gets higher, and the same happens to the mean density. This makes the breast volume smaller. In the volunteer's case, our simulations with Evolver indicate that it decreases from 700cm$^3$ to 680cm$^3$ approximately.

Further on, when the woman lays her breast on the table, not only gravity by also the contact force acts. The volume decreases again. In the volunteer's case, it drops from 680cm$^3$ to nearly 660cm$^3$. Because of that, in this paper we have used the word *compression* as *the action of pressing something into a smaller space*, rather than *pressing or squeezing together*. This second meaning applies to our next paper, devoted to simulating the virtual CC and MLO mammographies.

## 4. Conclusions

Our model reproduces the same shapes and measurements really taken by the volunteer's breasts. We have just presented the first results of a long-term research. These initial results are already of high complexity. For instance, nipple location, changes in shape, values and dimensions follow many geometric and physical laws. There are equated and implemented in the programmes, but technical details will be postponed to future publications.

## 5. Acknowledgements

We thank Dr Ana Cláudia Veronesi for details about gynaecological surgery, and her husband Alejandro Montepeloso for translating technical terms. This research is supported by FAPESP.

## References

[1] Pathmanathan P, Gavaghan D J, Whiteley J P, Chapman S J and Brady J M 2008 Predicting tumor location by modeling the deformation of the breast *IEEE Transactions on Bio-medical Engineering* **55** 2471–80

[2] Bliznakova K, Bliznakov Z, Bravou V, Kolitsi Z and Pallikarakis N 2003 A three-dimensional breast software phantom for mammography simulation *Physics in Medicine and Biology* **48** 3699–721

[3] Nishikawa R M 2007 Current status and future directions of computer-aided diagnosis in mammography *Computerized Medical Imaging and Graphics* **31** 224–35

[4] Tang J, Rangayyan R M, Xu J J, El-Naqa I and Yang Y 2009 Computer-aided detection and diagnosis of breast cancer with mammography: recent advances *IEEE Transactions on Information Technology in Biomedicine* **13** 236–51

[5] Hsu C M L, Palmeri M L, Segars W P, Veress A I and Dobbins J T 2011 An analysis of the mechanical parameters used for finite element compression of a high-resolution 3D breast phantom *Medical Physics* **38** 5756–70

[6] Kuhlmann M, Fear E C, Ramirez-Serrano A and Federico S 2013 Mechanical model of the breast for the prediction of deformation during imaging *Medical Engineering & Physics* **35** 470–8

[7] Bhatti S N and Sridhar-Keralapura M 2012 A novel breast software phantom for biomechanical modeling of elastography *Medical Physics* **39** 1748–68

[8] Brakke K A 1992 The Surface Evolver *Experimental Mathematics* **1** 141–65

[9] Koiter W T 1966 *On the nonlinear theory of thin elastic shells. Proc. of the Koninklijke Nederlandse Akademie van Wetenschappen* vol B69 (Amsterdam: North-Holland) pp 1–54

# Just three water molecules can trigger the undesired nonenzymatic reactions of aspartic acid residues: new insight from a quantum-chemical study

**O Takahashi**

Tohoku Pharmaceutical University, 4-4-1 Komatsushima, Aoba-ku, Sendai 981-8558, Japan

E-mail: ohgi@tohoku-pharm.ac.jp

**Abstract.** Aspartic acid (Asp) residues in peptides and proteins (L-Asp) can undergo spontaneous, nonenzymatic reactions under physiological conditions by which abnormal L-β-Asp, D-Asp, and/or D-β-Asp residues are formed. These altered Asp residues may affect the three-dimensional structures of the peptides and proteins and hence their properties and functions. In fact, the altered Asp residues are relevant to age-related diseases such as cataract and Alzheimer's disease. Most of the above reactions of the L-Asp residue proceed via a cyclic succinimide intermediate. In this paper, I propose a detailed mechanism of cyclization of an Asp residue (forming a precursor of the succinimide) by the B3LYP/6-31+G(d,p) density functional theory calculations carried out for a small Asp-containing model compound complexed with three water molecules which act as general acid–base catalysts in proton transfers. In the proposed mechanism, the amide group on the C-terminal side of the Asp residue is first converted to the tautomeric iminol form. Then, successive reorientation of a water molecule and conformational change occur followed by the nucleophilic attack of the iminol nitrogen atom on the carboxyl carbon atom of the Asp side chain to form a five-membered ring. A satisfactory agreement was obtained between the calculated and experimental energetics.

## 1. Introduction

The aspartic acid (Asp) residues (L-Asp residues) in peptides and proteins tend to undergo nonenzymatic alteration to β-Asp, D-Asp, and/or D-β-Asp residues [1–3]. These reactions which occur under physiological conditions can affect the three-dimensional structures of the peptides and proteins and hence their properties and functions. In fact, the altered Asp residues are relevant to age-related diseases such as cataract and Alzheimer's disease [3].

Most of the above "undesired" reactions of the L-Asp residue are believed to proceed via a cyclic succinimide intermediate as shown in figure 1 [1–4]. However, there have been no quantum-chemical calculations reported which reasonably explain the experimental findings about the formation of the succinimide intermediate, especially from a viewpoint of energetics. In this paper, I present a plausible computational model of the cyclization of an Asp residue to a *gem*-diol species (a possible precursor of the succinimide). In the proposed model, three water molecules play important roles as general acid–base catalysts in proton transfers. The results strongly suggest that water molecules are crucial for the succinimide formation in peptides and proteins.

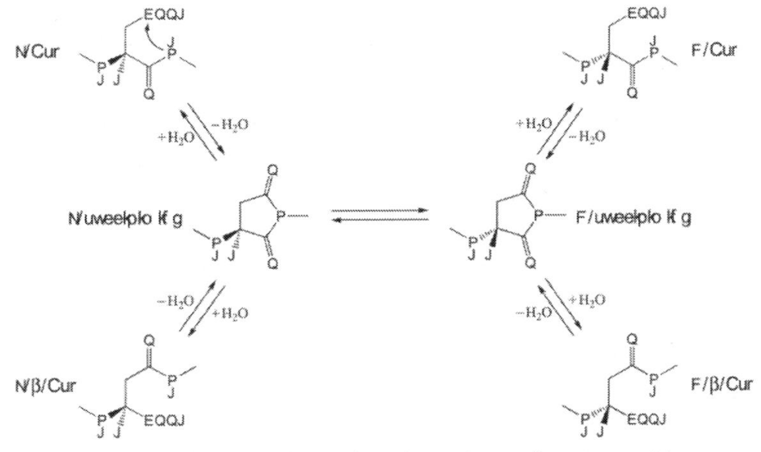

**Figure 1.** Succinimide-mediated reactions of an Asp residue.

## 2. Method

All calculations were performed using Spartan '08 (Wavefunction, Inc., Irvine, CA, USA). Figure 2 shows the model compound used in the present study. The side-chain carboxyl group was taken as protonated, since only the protonated form is considered to undergo the nucleophilic attack by the nitrogen atom of the following residue [2,4]. Three water molecules were placed to form the reactant complex so that they make a bridge between the oxygen and NH hydrogen atoms of the amide group on the C-terminal side of the Asp residue.

Energy-minimum and transition state (TS) geometries were located by the density functional theory (DFT) calculation with the B3LYP functional and the 6-31+G(d,p) basis set. The relative energies were corrected for the zero-point energy (ZPE).

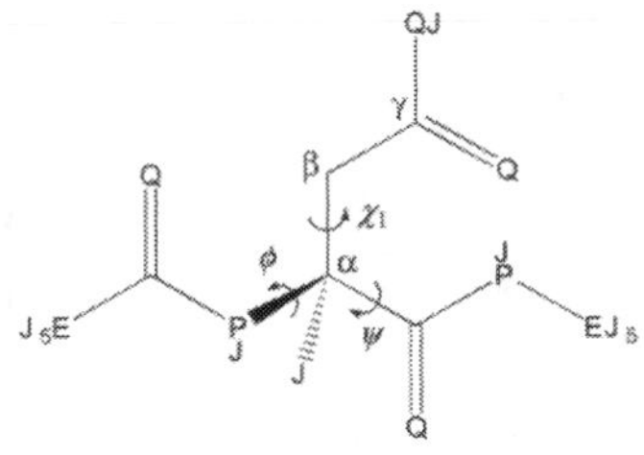

**Figure 2.** Model compound (Ace–Asp–Nme) used in the present study. Dihedral angles $\phi$, $\psi$, and $\chi_1$ characterize the main-chain and side-chain conformations.

## 3. Results and discussion

As shown in figure 3, a four-step reaction pathway was found starting from the reactant complex **AM** (amide form). The geometry of **AM** is shown in figure 4.

The first step is iminolization (i.e., conversion to the tautomeric iminol form) of the amide group on the C-terminal side to form **IM1**. This step may be schematically designated as in figure 5; namely, the three water molecules enable a quadruple proton transfer by which the conversion of the amide group to the iminol form occurs. By iminolization, the nucleophilicity of the nitrogen atom is expected to be greatly reinforced [4]. The TS for the first step is denoted $\mathbf{TS_{AI}}$. The corresponding activation barrier was calculated to be 21.2 kcal mol$^{-1}$, which is possible for a biological reaction. **IM1** is 16.2 kcal mol$^{-1}$ higher in energy than **AM**. The geometries of $\mathbf{TS_{AI}}$ and **IM1** are shown in figure 6. The iminol OH group in **IM1** is *cis* with respect to the C=N bond.

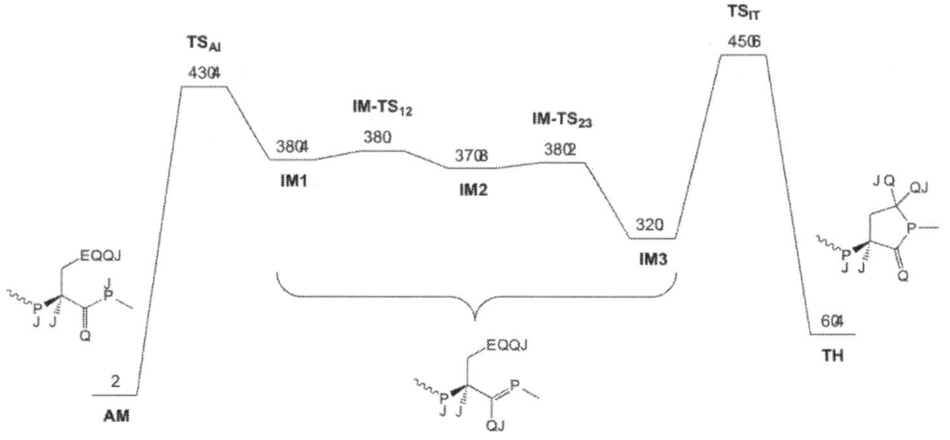


**Figure 3.** Energy diagram. ZPE-corrected energies relative to the reactant complex **AM** are shown in kcal mol$^{-1}$.

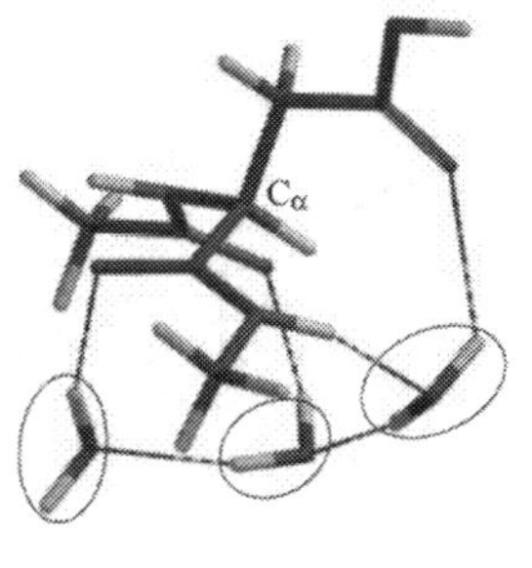


**Figure 4.** The geometry of **AM** ($\phi = -120°$, $\psi = 158°$, $\chi_1 = -158°$). The three catalytic water molecules are circled.

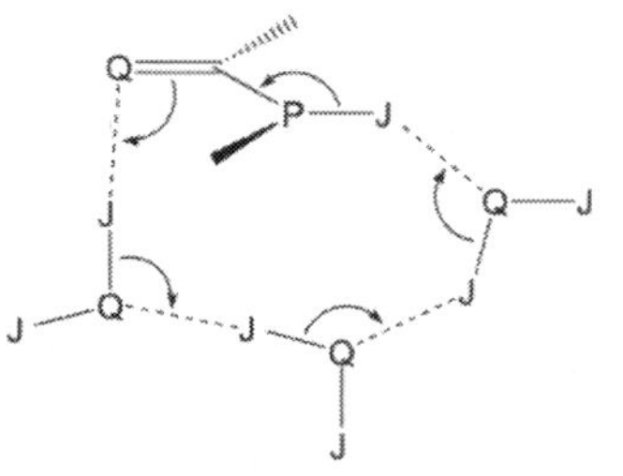

**Figure 5.** A schematic representation of the three-water-mediated iminolization of the amide group.

The second step is reorientation of a water molecule leading to **IM2** (figure 6) The corresponding TS is **IM-TS$_{12}$** (geometry not shown). More specifically, the hydrogen bond involving the iminol nitrogen in **IM1** is broken and a new hydrogen-bond bridge is formed between the iminol OH group and the side-chain C=O group. This step can occur with an extreme ease, the energy barrier being only 0.6 kcal mol$^{-1}$. **IM2** is lower in energy than **IM1** by 0.6 kcal mol$^{-1}$.

The third step is a conformational change to **IM3** (figure 6) through the transition state **IM-TS$_{23}$** (geometry not shown) by which the iminol OH group becomes *trans*. The barrier for this step is also very small (0.4 kcal mol$^{-1}$), and the energy of **IM3** is lower than that of **IM2** by 4.8 kcal mol$^{-1}$.

No direct cyclization pathways from **IM1** and **IM2** leading to a five-membered ring could be found. Instead, a cyclization pathway from **IM3** was successfully found (the fourth step). The *gem*-diol product of this step, denoted **TH**, is an example of the tetrahedral intermediates of the nucleophilic substitution reactions at carbonyl carbon atoms. The energy of **TH** relative to **AM** is 4.2 kcal mol$^{-1}$. The TS connecting **IM3** and **TH** is denoted **TS$_{IT}$**. The geometries of **TS$_{IT}$** and **TH** are also shown in figure 6. It should be noted that the three water molecules again act as catalysts in this step, mediating the proton transfer from the iminol OH group to the side-chain C=O group. The energy of **TS$_{IT}$** is 23.4 kcal mol$^{-1}$ with respect to **AM**, and the cyclization step is rate-determining in the overall reaction shown in figure 3. Interestingly, this value is close to the experimental activation energy (21.7 kcal mol$^{-1}$) estimated for the succinimide formation from the Asp residue in the hexapeptide Val-Tyr-Pro-Asp-Gly-Ala [1]. The succinimide is formed by dehydration of **TH**. This process has not been examined in the present study, because other water molecules (near the *gem*-diol group) should catalyze it. Nevertheless, the cyclization step may be rate-determining in the succinimide formation from the Asp residue.

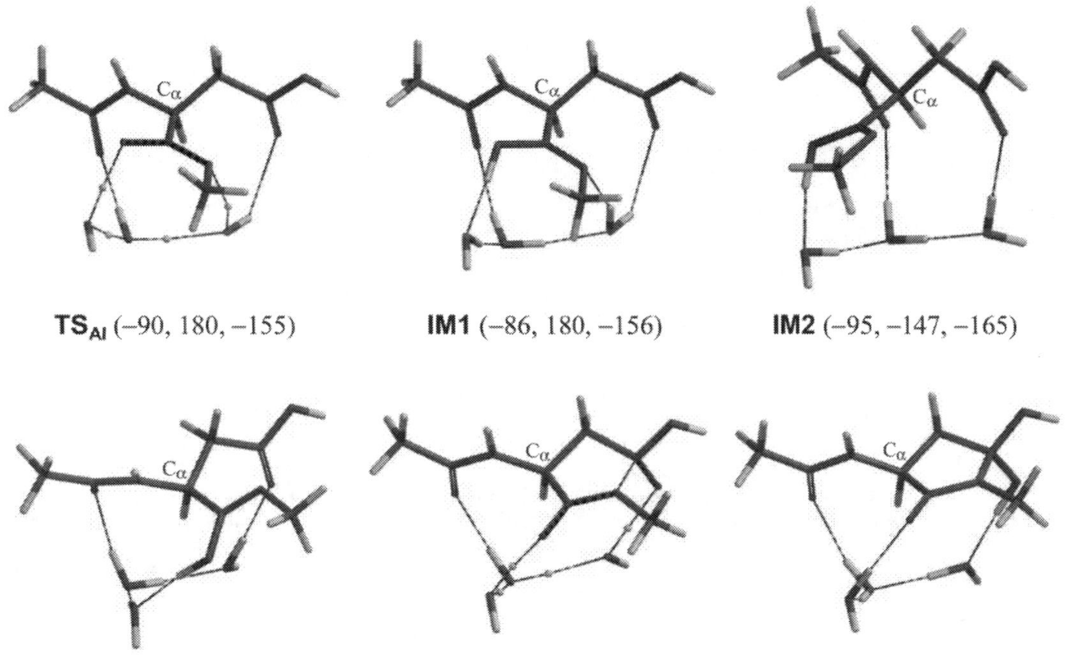


**TS**$_{AI}$ (−90, 180, −155) **IM1** (−86, 180, −156) **IM2** (−95, −147, −165)

**IM3** (−148, −115, −173) **TS**$_{IT}$ (−121, −141, 158) **TH** (−128, −144, 151)

**Figure 6**. The geometries of **TS**$_{AI}$, **IM1**, **IM2**, **IM3**, **TS**$_{IT}$, and **TH**. For each geometry, the $\phi$, $\psi$, and $\chi_1$ values (in degrees) are given in parentheses in this order.

## 4. Conclusion

A reaction model has been proposed for the cyclization reaction of Asp residues leading to the tetrahedral intermediate in the succinimide formation. Key steps are (1) iminolization of the amide group on the C-terminal side, and (2) cyclization from the iminol intermediate. Both steps are assisted by three water molecules, which act as general acid–base catalysts in proton transfers. The "undesired" reactions of the Asp residues in proteins and peptides may be triggered by just three water molecules.

## References

[1] Geiger T and Clarke S 1987 *J. Biol. Chem.* **262** 785
[2] Stephenson R C and Clarke S 1989 *J. Biol. Chem.* **264** 6164
[3] Ritz-Timme S and Collins M J 2002 *Aging Res. Rev.* **1** 43
[4] Takahashi O and Oda A 2012 Amide–Iminol Tautomerization of the C-Terminal Peptide Groups of Aspartic Acid Residues. Two-Water-Assisted Mechanism, Cyclization from the Iminol Tautomer Leading to the Tetrahedral Intermediate of Succinimide Formation, and Implication to Peptide Group Hydrogen Exchange *Tyrosine and Aspartic Acid: Properties, Sources and Health Benefits* ed Jones J E and Morano D M (New York: Nova Science Publishers) chapter 7 pp 131–147

# Generalized Backprojection Operator: Fast Calculation

**Eduardo X Miqueles**[1], **Elias S Helou**[2] **and Alvaro R De Pierro**[2]
[1] Brazilian Synchrotron Light Source/CNPEM - Campinas, SP/Brazil
[2] Universidade de São Paulo/ICMC - São Carlos, SP/Brazil

E-mail: edu.miqueles@gmail.com

**Abstract.** The inverse Radon transform and his straightforward implementation, known as *filtered backprojection* (also known as FBP), has become a powerful algorithm for solving a tomographic inverse problem. It has a wide range of applications, including geophysics, medicine and synchrotrons, and from *kilo* to *centi* to *micro* scale respectively. Such a classical inversion has a major computational disadvantage: increasing slowness proportionally to the data size. An ordinary implementation of this algorithm relies on a simple integral that has to be done pixelwise. Many accelerating techniques were proposed in the literature so as to make this part of the inversion as fast as possible. One the most promising strategies is converting the backprojection as a convolution operator (at log-polar coordinates). The generalized backprojector has many applications, for instance in the analytical inversion of single-photon emission tomography or x-ray fluorescence tomography. Our aim in this paper is to show how these ideas can be used for other inversion methods, the iterative ones; which deal much better with noise.

## 1. Introduction

Most of the hot-spot tomographic problems, viz.: CT (*computed tomography*), SPECT (*single-photon emission computed tomography*), PET (*positron emission tomography*), ERT (*exponential radon transform*) and XFCT (x-rays fluorescence computed tomography), are related to the inversion of following integral operator:

$$g(t,\theta) = \int_{\Omega(t,\theta)} f(\boldsymbol{x})\omega(\boldsymbol{x},\theta)\mathrm{d}\boldsymbol{x} \equiv \mathscr{R}_\omega f(t,\theta) \tag{1}$$

with $\Omega(t,\theta) = \{\boldsymbol{x} \in \mathbb{R}^2 : \boldsymbol{x}\cdot\boldsymbol{\xi}_\theta = t\}$ a straight line parameterized by $(t,\theta) \in [-1,1] \times [0,2\pi]$. Here $\xi_\theta = (\cos\theta, \sin\theta)^T$ is the normal vector to $\Omega$ and $\boldsymbol{x}$ lies in the unit ball $\|\boldsymbol{x}\|_\infty \leqslant \frac{\sqrt{2}}{2}$. The above integral operator is called *the generalized Radon transform* and the common task is: find $f$ given the function $g$ over the set $\{\Omega(t,\theta)\}$. Let us denote $U$ as the *feature space*, lying all the map functions of the form $z = z(\boldsymbol{x})$, and denote $V$ as the *Radon space*, gathering all the sinogram functions $g = g(t,\theta)$. Some typical weight functions $\omega$ are summarized on table 1. Operator $\mathscr{D}$ is the *divergent beam transform*, meaning that $\mathscr{D}h(\boldsymbol{x},\theta)$ (for $h \in U$) is the integral of $h$ over the line $\mathsf{span}\{\boldsymbol{x} + s\boldsymbol{\xi}_\theta^\perp ; s \in \mathbb{R}_+\}$. Analytical inversion formulas for $\mathscr{R}_\omega$ at each case of table 1 are presented in [4, 6, 7, 8, 9]. Note that, for a fixed angle $\theta$, $\mathscr{R}_\omega f(t,\theta)$ is the classical Radon transform of the function $f(\cdot)\omega(\cdot,\theta) \in U$, i.e., $\mathscr{R}_\omega f(t,\theta) = \mathscr{R}[f\omega(\cdot,\theta)](t,\theta)$.

**Discretization** When dealing with a discretized version of $z = z(\boldsymbol{x}) \in U$, we consider that $\boldsymbol{x}$ belongs to a pixel grid $\{\boldsymbol{x}_0, \ldots, \boldsymbol{x}_{P-1}\}$; usually with $P = s^2$. Therefore, we have $z$ as a

**Table 1.** Some typical weight functions for the generalized Radon transform (1).

| Name | Weight function $\omega(\boldsymbol{x}, \theta)$ | Parameters |
|---|---|---|
| CT | 1 | – |
| ERT | $e^{-(\boldsymbol{x}\cdot\boldsymbol{\xi}^{\perp})\mu}$ | $\mu \in \mathbb{R}_+$ |
| PET | $e^{-\mathscr{R}\mu(\boldsymbol{x}\cdot\boldsymbol{\xi},\theta)}$ | $\mu \in U$ |
| SPECT | $e^{-\mathscr{D}\mu(\boldsymbol{x},\theta)}$ | $\mu \in U$ |
| XFCT | $e^{-\mathscr{D}\lambda(\boldsymbol{x},\theta+\pi)} \int_\Gamma \mathrm{d}\gamma e^{-\mathscr{D}\mu(\boldsymbol{x},\theta+\gamma)}$ | $\lambda, \mu \in U$, $\Gamma \subset [0, \pi]$ |

linear combination of basis functions $\{\phi_0(\boldsymbol{x}), \ldots, \phi_{P-1}(\boldsymbol{x})\}$ (each per pixel); hence it suffices to represent $z$ as a vector $\boldsymbol{z} \in \mathbb{R}^P$ with coefficients $\boldsymbol{z}_i = z(\boldsymbol{x}_i)$. The same applies to $g = g(t, \theta) \in V$ with $(t, \theta)$ lying in the grid $\{-1 + j\frac{2}{\mathrm{r}}\}_{0\leqslant j\leqslant \mathrm{r}}^{\mathrm{r}} \times \{k\frac{2\pi}{\mathrm{v}}\}_{0\leqslant k\leqslant \mathrm{v}}$; yielding a matrix $\boldsymbol{g} \in \mathbb{R}^{\mathrm{vr}}$ with entries given by $\boldsymbol{g}_n = g(t_{j(n)}, \theta_{k(n)})$. Here, $n = j + k\mathrm{r} \in \{0, 1, \ldots, \mathrm{D} - 1\}$ (therefore $k_n = [n/r]$). We denote $D = \mathrm{rv}$.

## 2. The backprojection operator

It was shown in [10] that the adjoint operator of $\mathscr{R}_\omega$ in (1), is the one acting as $\mathscr{B}_\omega \colon g \in V \mapsto b \in U$, defined through

$$b(\boldsymbol{x}) = \mathscr{B}_\omega g(\boldsymbol{x}) = \int_{[0,2\pi]} g(\boldsymbol{x}\cdot\boldsymbol{\xi}_\theta, \theta)\omega(\boldsymbol{x}, \theta)\mathrm{d}\theta \approx \frac{2\pi}{\mathrm{v}} \sum_{k=0}^{\mathrm{v}} g(\boldsymbol{x}\cdot\boldsymbol{\xi}_{\theta_k}, \theta)\omega(\boldsymbol{x}, \theta_k) \tag{2}$$

The approximation above comes from a finite set of measurements $\{\theta_0, \ldots, \theta_{\mathrm{v}}\}$. The following fact, is a well-known procedure for discretized computed tomography, i.e., $\omega \equiv 1$. We generalize it for $\omega \neq 1$.

**Lemma 2.1** *Equations $g = \mathscr{R}_\omega f$ are equivalent to $\boldsymbol{g} = \boldsymbol{A}_\omega \boldsymbol{f}$ in the grid $\{t_j\} \times \{\theta_k\} \times \{\boldsymbol{x}_i\}$*

*Proof*: Indeed, writing $f(\boldsymbol{x})$ as a linear combination of the basis functions $\{\phi_0, \ldots, \phi_{P-1}\}$ and taking the Radon transform we get

$$\boldsymbol{g}_{j+k\mathrm{r}} = \mathscr{R}\left\{\sum_{i=0}^{P-1} \boldsymbol{f}_i\phi_i(\cdot)\omega(\cdot, \theta_k)\right\}(t_j, \theta_k) = \sum_{i=0}^{P-1} \boldsymbol{f}_i\mathscr{R}\left\{\omega(\cdot, \theta_k)\phi_i(\cdot)\right\}(t_j, \theta_k) = \sum_{i=0}^{P-1} \boldsymbol{f}_i\boldsymbol{w}_{n,i}\boldsymbol{a}_{n,i} \tag{3}$$

with $n = j + k\mathrm{r}$, $\boldsymbol{a}_{n,i} = \mathscr{R}\phi_i(t_j, \theta_k)$ and $\boldsymbol{w}_{n,i} = \frac{\mathscr{R}\{\omega(\cdot,\theta_k)\phi_i(\cdot)\}(t_j,\theta_k)}{\mathscr{R}\{\phi_i(\cdot)\}(t_j,\theta_k)}$ if $\mathscr{R}\left\{\omega(\cdot, \theta_k)\phi_i(\cdot)\right\}(t_j, \theta_k) \neq 0$ or $\boldsymbol{w}_{n,i} = 0$ otherwise. Notice that this is well defined if, say, $\phi_i(\boldsymbol{x}) \geqslant 0$. It follows from (3) that $\boldsymbol{g} = \boldsymbol{A}_\omega \boldsymbol{f}$, being $\boldsymbol{a}_{n,i}\boldsymbol{w}_{n,i}$ the entries of $\boldsymbol{A}_\omega \in \mathbb{R}^{D\times P}$. ■

Matrix $\boldsymbol{A}_\omega$ of Lemma 2.1 is a Hadamard product of matrices $(\boldsymbol{a}_{ni})$ and $(\boldsymbol{w}_{k_n,i})$ (i.e., element-wise matrix product, see [11, 12]). Here, matrix $\boldsymbol{A} = (\boldsymbol{a}_{ni})$ is the one giving the intersection length of a ray $\Omega_n$ and a pixel $\boldsymbol{x}_i$. As a Hadamard product, denoted by symbol $\odot$, we can see $\boldsymbol{A}_\omega$ as

$$\boldsymbol{A}_\omega = \begin{bmatrix} \ddots & & \\ & \boldsymbol{A} & \\ & & \ddots \end{bmatrix} \odot \begin{bmatrix} \boldsymbol{e}\boldsymbol{w}_1^T \\ \vdots \\ \boldsymbol{e}\boldsymbol{w}_{\mathrm{v}}^T \end{bmatrix} \Leftrightarrow \boldsymbol{A}_\omega = \boldsymbol{A} \odot \boldsymbol{M}, \quad \boldsymbol{M} = (\boldsymbol{W} \otimes \boldsymbol{e}). \tag{4}$$

with $\boldsymbol{e} \in \mathbb{R}^{\mathrm{r}}$ a vector of ones and $\boldsymbol{w}_k^T \in \mathbb{R}^{1\times P}$ the $k$th row of matrix $\boldsymbol{W} = (\boldsymbol{w}_{k,i}) \in \mathbb{R}^{\mathrm{v}\times P}$. Symbol $\otimes$ stands for Kronecker operation [12]. The product $\odot$ satisfies the property (see [12])

$(\boldsymbol{A} \odot \boldsymbol{M})^T = (\boldsymbol{A}^T \odot \boldsymbol{M}^T)$, giving us, for all $\boldsymbol{g} \in U$

$$\begin{aligned} [\boldsymbol{A}_\omega^T \boldsymbol{g}]_i = \left[(\boldsymbol{A}^T \odot \boldsymbol{M}^T)\boldsymbol{g}\right]_i &= [(\boldsymbol{a}_1 \odot \boldsymbol{m}_1)\boldsymbol{g}_1 + \ldots + (\boldsymbol{a}_D \odot \boldsymbol{m}_D)\boldsymbol{g}_D]_i & (5)\\ &= (\boldsymbol{a}_{1,i} \odot \boldsymbol{m}_{1,i})\boldsymbol{g}_1 + \ldots + (\boldsymbol{a}_{D,i} \odot \boldsymbol{m}_{D,i})\boldsymbol{g}_D & (6)\\ &= \underbrace{(\boldsymbol{a}_{1,i} \ldots \boldsymbol{a}_{D,i})^T}_{\text{backprojection}} \underbrace{\operatorname{diag}(\boldsymbol{m}_{1,i}, \ldots, \boldsymbol{m}_{D,i})\,\boldsymbol{g}}_{\text{weighting of } g} & (7) \end{aligned}$$

Equation (7) is the discrete analogous of (2), i.e, the generalized backprojection of sinogram $\boldsymbol{g}$, evaluated at pixel $\boldsymbol{x}_i$. Note that if $\omega \equiv 1$, matrix $\boldsymbol{M}$ becomes a matrix of ones and the above diagonal matrix becomes the identity, where we retrieve the classical backprojection algorithm. For $\omega \neq 1$ the generalized backprojection of sinogram $\boldsymbol{g} \in U$ becomes the usual backprojection of a weighted projection. Such a weighting step could be done by means of a Kronecker product. Denoting $\boldsymbol{c}_i \in \mathbb{R}^P$ as the $i$th canonical vector, we get from (4):

$$\operatorname{diag}(\boldsymbol{m}_{1,i}, \ldots, \boldsymbol{m}_{D,i})\,\boldsymbol{g} = \boldsymbol{g} \odot \mathbf{vec}\left(\boldsymbol{W}\boldsymbol{c}_i \otimes \boldsymbol{e} \otimes \boldsymbol{e}^T\right), \quad \boldsymbol{e} \in \mathbb{R}^{\mathrm{r}}. \tag{8}$$

## 3. Fast computation

From (7) and (8) we realize that, for computing the generalized backprojection operator $\mathscr{B}_\omega$, we only need to compute the classical backprojection $\mathscr{B}$ preceded by a weighting step of the sinogram. The same happens in a continuous setting; in fact, we just have to observe that $g(\boldsymbol{x} \cdot \boldsymbol{\xi}_\theta, \theta)\omega(\boldsymbol{x}, \theta)$ is a pointwise product of two sinograms, the first being the input sinogram $g$ and the second being $\omega$; which is dependent only on $\theta$. In the discrete angle set $\{\theta_0, \ldots, \theta_{\mathrm{v}}\}$, this pointwise action is just a Hadamard product of both sinograms, plus an interpolation from $\boldsymbol{x} \cdot \boldsymbol{\xi}$ to $t$.

Let $\mathbb{B}$ denote the algorithm to compute the classical backprojection operator $\mathscr{B}$, i.e., $\mathscr{B}g(\boldsymbol{x}) \approx \mathbb{B}[\boldsymbol{g}](\boldsymbol{x})$. In a generalized/vectorial notation, we are able to write $\boldsymbol{b}(\boldsymbol{x}_i) = \mathscr{B}_\omega(\boldsymbol{x}_i) \approx \mathbb{B}\left[\boldsymbol{g} \odot \mathbf{vec}\left(\boldsymbol{W}\boldsymbol{c}_i \otimes \boldsymbol{e} \otimes \boldsymbol{e}^T\right)\right]$. Or, using matrices:

$$\boldsymbol{b}(\boldsymbol{x}_i) \approx \mathbb{B}\left( \underbrace{\begin{bmatrix} g(t_1,\theta_1) & \ldots & g(t_{\mathrm{r}},\theta_1) \\ g(t_1,\theta_2) & \ldots & g(t_{\mathrm{r}},\theta_2) \\ \vdots & & \vdots \\ g(t_1,\theta_{\mathrm{v}}) & \ldots & g(t_{\mathrm{r}},\theta_{\mathrm{v}}) \end{bmatrix}}_{\text{input sinogram } g} \boxdot \underbrace{\begin{bmatrix} \omega(\boldsymbol{x}_i,\theta_1) \\ \omega(\boldsymbol{x}_i,\theta_2) \\ \vdots \\ \omega(\boldsymbol{x}_i,\theta_{\mathrm{v}}) \end{bmatrix}}_{\text{weight vector } \boldsymbol{W}\boldsymbol{c}_i} \right) \tag{9}$$

with $\boxdot$ a row-wise hadamard product between matrix/vector with the same number of rows. It is easy to verify that $\boxdot$ is closely related to the $\{\odot, \otimes\}$ operations. The fast computation of $\mathscr{B}_\omega$ is dependent of a fast computation of $\mathscr{B}$, represented by algorithm $\mathbb{B}$. The following statement give us a powerful tool for calculating $\mathbb{B}$. The proof can be found in [3].

**Theorem 3.1** $\mathbb{B}\boldsymbol{g}$ *can be computed as the convolution of* $\boldsymbol{g}$ *with an appropriate point-spread function, both in log-polar coordinates.*

The above Theorem allow us to compute the backprojection with $O(s^2 \log s)$ complexity using the Fast Fourier transform, in log-polar coordinates, for each pixel. This is in contrast with a naïve implementation of the integral in (2) which requires approximately $O(s^3)$ computations using a quadrature rule for each pixel $\boldsymbol{x}$. It is not the only existing method with low complexity $O(s^2 \log s)$, see [1, 2].

## 4. Application: iterative methods

We propose an immediate application of Sections 2 and 3 on iterative methods for solving (1). From Lemma 2.1 we have to find the solution $\boldsymbol{f}$ of the linear system $\boldsymbol{A}_\omega \boldsymbol{f} = \boldsymbol{g}$. Iterative techniques can be used to solve this inverse problem [5, 10], and the backprojection operator $\mathscr{B}$ plays an important role. For instance,

$$f_i^{k+1} = f_i^k + \frac{1}{\|\boldsymbol{W}\boldsymbol{c}_i\|_1} \mathbb{B}\mathbb{F}\left\{\boldsymbol{g} - (\boldsymbol{A} \odot \boldsymbol{M})\boldsymbol{f}^k\right\}\Big|_i, \quad \boldsymbol{f}^{k+1} = \left[\boldsymbol{f}^k \odot \mathbb{B}\left\{\boldsymbol{g} \oplus (\boldsymbol{A} \odot \boldsymbol{M})\boldsymbol{f}^k\right\}\right] \oplus [\mathbb{B}\boldsymbol{e}] \qquad (10)$$

are the continuous iterative method [10] and the expectation maximization method [5]. It should be noted that $\mathbb{F}$ is a low-pass filtering operator while $\mathbb{B}\mathbb{F}$ can be regarded as the FBP inversion method. Symbol $\oplus$ stands for the element-wise division of vectors and $\boldsymbol{e} = (1) \in \mathbb{R}^D$.

Figure 1 shows an application with data from a micro-tomography beam line at the Brazilian Synchrotron Light Source. The reconstruction using the filtered backprojection algorithm is shown in (c). For a noisy data $\boldsymbol{g} \in \mathbb{R}^{6184960}$ (since $6184960 = 3020$ angles $\times 2048$ rays) depicted in (a), it is clear that an analytic method is not the best choice. To use an iterative technique (EM or RAMLA), one has to implement a very fast backprojection operator, whose action is shown in (b). Therefore, Theorem 3.1 seems to be the best option for a fast implementation of $\mathbb{B}$ and a straightforward fast iterative technique. Fast generalized backprojection techniques could also be easily implemented for XFCT measurements.

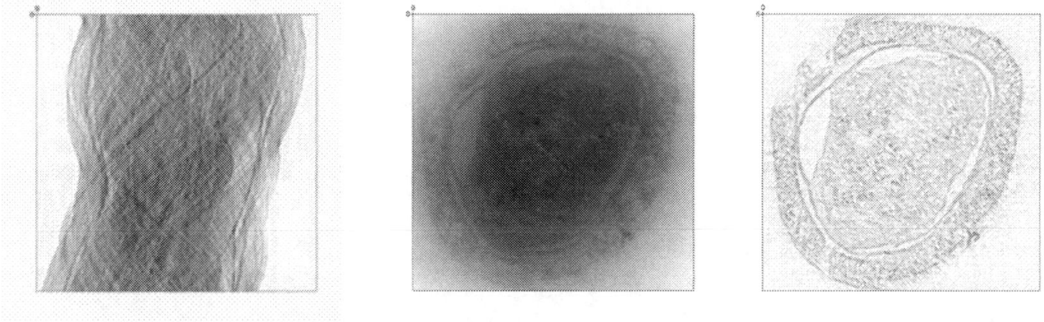

**Figure 1.** Left: Input sinogram $\boldsymbol{g}$. Center: Action of algorithm $\mathbb{B}[\boldsymbol{g}]$. Right: Filtering of $\mathbb{B}[\boldsymbol{g}]$

## 5. Conclusion

Fast backprojection can be implemented efficiently using log-polar coordinates. In this manuscript, we demonstrate that the generalized backprojection operator can also be made fast, using appropriate matrix products. Such an strategy has great impact at x-rays fluorescence tomography and single-photon emission tomography.

## Acknowledgments

We thank *Francis P. O'Dowd* and *Hugo H. Slepicka*, from the Brazilian Synchrotron Light Source for providing the data. Work by Alvaro R. De Pierro was supported by Cnpq grant no. 304820/2006-7.

## References

[1] Brandt A, Mann J, Brodski M and Galun M, 1999 A fast and accurate multilevel inversion of the radon transform, *SIAM, J.Appl. Math*, **60 (2)**, 437-62

[2] Basu S and Bresler Y, 2000, $O(N^2 \log_2 N)$ filtered backprojection reconstruction algorithm for tomography, *IEEE, Trans.Med.Imaging*, **9 (10)**, 1760-73.

[3] Andersson F, 2005 Fast inversion of the radon transform using log-polar coordinates and partial back-projections, *SIAM, J.Appl. Math*, **65 (3)**, 818.

[4] Natterer F and Wubbeling F, 2001 Mathematical methods in image reconstruction *SIAM*.

[5] Helou E S and De Pierro A R, 2005 Convergence results for scaled gradient algorithms in positron emission tomography, *Inverse Problems*, **21(6)**, 1905-14.
[6] Tretiak O and Metz C E, 1980 The exponential radon transform, *SIAM J. Appl. Math.*, **39**, 341.
[7] Metz C E and Pan X, 1995, A unified analysis of exact methods of inverting the 2-D exponential radon transform, with implications for noise control in spect, *IEEE Trans.Medical Imaging*, **14 (4)**, 643.
[8] Fokas A S, 2008 A unified approach to boundary value problems, *SIAM, CBMS-NSF Regional Conference Series in Applied Mathematics.*
[9] Miqueles E X and De Pierro A R, 2011 On the inversion of the xfct radon transform, *Studies in Applied Mathematics*, **127(4)**, 394.
[10] Miqueles E X, De Pierro A R, 2013 On the iterative inversion of generalized attenuated radon transforms, *Journal of Inverse and Ill-posed Problems*, **21(5)**, 695.
[11] Horn R A, Johnson C R, 1985 Matrix analysis, *Cambridge University Press, New York.*
[12] Van Loan C F, 2000 The ubiquitous kronocker product, *Journal of Computational and Applied Mathematics*, **123**, 85.

# Numerical aspects of grazing incidence XRF

**Eduardo Miqueles**[1], **Carlos Pérez**[1] **and Vanessa I.T. Suaréz**[2]

[1] Brazilian Synchrotron Light Source/CNPEM - Campinas, SP/Brazil

[2] Universidade Federal de Viçosa/Dept. of Physics - Viçosa, MG/Brazil

E-mail: edu.miqueles@gmail.com

**Abstract.** In this article, we give a different mathematical approach for background aspects of grazing incidence x-ray fluorescence, GIXRF for short. Our contribution comes from an applied point of view, in order to have a computer program to simulate the fluorescence intensity from a stacking of thin layer films. A typical ill-posed inverse problem is formulated. Our aim is to reconstruct the fluorescence intensity for a variety of grazing angle measurements. We rederive some classical equations pointing out the numerical aspects of the inversion procedure and giving new directions for direct in inverse algorithms.

## 1. Introduction

Grazing incidence x-ray fluorescence - GIXRF - experiments are used to study the propagation of electromagnetic waves in a stratified media. The fluorescence yield by wave penetration at glancing incidence above the critical angle, provides information about the upper layers. The theory for GIXRF is beyond the scope of this article. For a complete description of the main equations and physical properties we refer to [1, 2, 3, 4]. A typical layered media, representing a thin-film with 5 layers, is shown in figure 1. Interfaces separating each layer are marked at the left, starting from 1 to 6. The eletromagnetic source is placed at vacuum, and is represented by the asterisk in figures 1 and 2. In this short example, the incoming wave strikes the first interface with a grazing angle $\theta$. Assuming $\theta$ above the critical angle, the electromagnetic wave penetrates in the medium, were reflected and transmitted waves propagates according to Snell's law. If $\theta$ varies within a short interval, containing the critical angle, we have a family of waves striking the first interface, propagating through layers, and coming back to the surface, were the fluorescence intensity is finally measured, as depicted in figure 2. In general, we denote $\Sigma_j$ as the $j$th interface separating layer $j$ and $j+1$. Each layer has a complex refraction index $n_j$ and the pair $(a_j, b_j) \in \mathbb{C}$ at each interface denote the amplitude of the incident and reflected wave at $\Sigma_j$.

Each layer is a composition of many chemical elements, say copper, cobalt, zinc, gold, among others. According to the scattering theory [1, 3], the total fluorescence intensity for a given element $m$ is given by

$$F_m(\theta, \boldsymbol{u}) = \sum_{j=1}^{N} c_{mj}(\boldsymbol{u}) \int_0^{h_j} f_j(z, \theta, \boldsymbol{u}) e^{-d_{mj}(\boldsymbol{u})z} \mathrm{d}z, \quad f_j(z, \theta, \boldsymbol{u}) = |a_j(\theta, \boldsymbol{u}) e^{ikzp_j(\theta)} + b_j(\theta, \boldsymbol{u}) e^{-ikzp_j(\theta)}|^2, \qquad (1)$$

with $c_{mj}(\boldsymbol{u})$ and $d_{mj}(\boldsymbol{u})$ appropriate physical constants, dependent on the parameter $\boldsymbol{u} \in \mathbb{R}^N$. Each entry of the vector $\boldsymbol{f} = (f_j) \in \mathbb{R}^N$ represents the incident fluorescent intensity at the layer

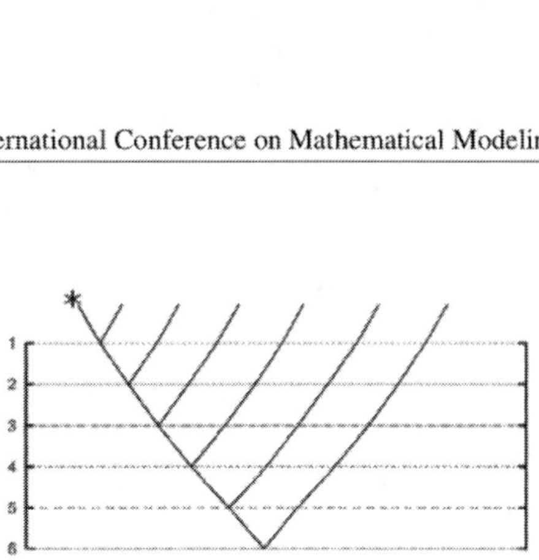


**Figure 1.** Single x-ray incidence

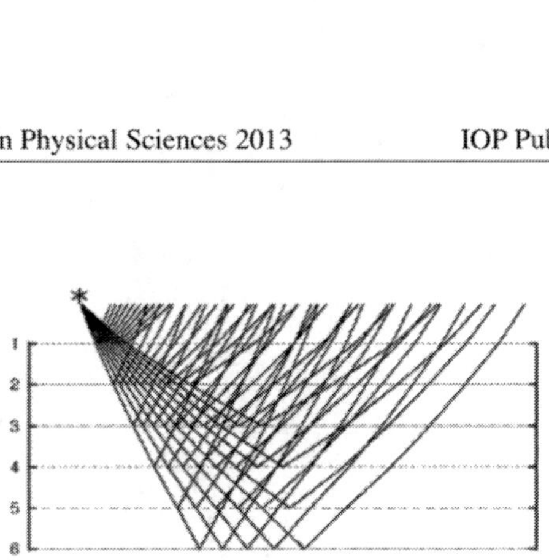


**Figure 2.** Total x-rays incidence

$j$, whereas $\boldsymbol{h} = (h_j) \in \mathbb{R}^N$ is the thickness of the $j$th layer and $|\cdot|$ stands for the complex modulus. Function $\theta \in \Theta \mapsto F_m(\theta, \boldsymbol{u}) \in \mathbb{R}$ is a real-valued function of the angle $\theta$. In pratice, a measure of total fuorescence $F_m$ is obtained, say $\{q_1, q_2, \ldots, q_v\}$ satisfying $F_m(\theta_k) = q_k$ (at least hypothetically), $\theta_k \in \Theta$. The mathematical challenge can be stated as the following non-linear fitting problem:

**Problem [♠]:** Find $\boldsymbol{u} \in \mathbb{R}^N$ such that $q_k - F_m(\theta_k, \boldsymbol{u})$ is minimized.

There are many optimization techniques to solve the above non-linear least squares problem, for instance the Gauss-Newton or the Levenberg-Marquardt method. Like most of the iterative approaches, we find a sequence $\{\boldsymbol{u}^k\}$ converging to the approximated solution, and the computation of each iterate strongly depends on the value $F_m(\theta, \boldsymbol{u}^k)$, for a given $\theta$. The parameter $\boldsymbol{u}$ could be the vector of layer thickness $\boldsymbol{u} = \boldsymbol{h}$, or the concentration of a given element within each layer.

## 2. The direct problem

We aim at calculating the function value $F_m(\theta, \boldsymbol{u})$, for a fixed pair $(\theta, \boldsymbol{u})$, through (1). Let us denote the inner integral in (1) by $\ell_{jm}(\theta, \boldsymbol{u})$. It is easy to note that $\ell_{jm}$ resembles a Laplace transform of function $f_j$ and it is easily calculated using a quadrature formula. Before that, we have to evaluate function $f_j$, which is dependent on the amplitude values $a_j$ and $b_j$. From the matrix formalism [1, 3, 4, 5], these amplitudes are given by the Fresnel equations and the principle of reversibility:

$$\begin{cases} \nu_j a_j + \nu_j r_j b_j - a_{j-1} t_j = 0 \\ \frac{r_j}{\nu_j} a_j + \frac{1}{\nu_j} b_j - b_{j-1} t_j = 0 \end{cases} \Leftrightarrow \begin{bmatrix} a_{j-1} \\ b_{j-1} \end{bmatrix} = \frac{1}{t_j} \begin{bmatrix} \nu_j & 0 \\ 0 & \frac{1}{\nu_j} \end{bmatrix} \begin{bmatrix} 1 & r_j \\ r_j & 1 \end{bmatrix} \begin{bmatrix} a_j \\ b_j \end{bmatrix}, \quad \nu_j \equiv \exp\left(i \frac{2\pi}{\lambda} p_j h_j\right) \qquad (2)$$

where $t_j$, $r_j$ are the transmission and reflection coefficients[1] at $\Sigma_j$, respectively. As stablished in the literature, the matrix relation in (2) is used to find $(a_j, b_j)$ in terms of $(a_{j-1}, b_{j-1})$ in a recursive way, see [3, 5, 6]. Now, we take a different approach. Let us write down the Fresnel equations in (2), with index $j$ varying from 1 to $N+1$. To make easier the equations, we suppose $N = 2$; the generalization will go straightforward. In the absence of reflected waves at the substrate, i.e., $b_{N+1} = 0$, we will have:

$$\begin{bmatrix} \nu_1 & 0 & 0 & 0 & r_1\nu_1 & 0 \\ -t_2 & \nu_2 & 0 & 0 & 0 & r_2\nu_2 \\ 0 & -t_3 & \nu_3 & 0 & 0 & 0 \\ \frac{r_1}{\nu_1} & 0 & 0 & -t_1 & \frac{1}{\nu_1} & 0 \\ 0 & \frac{r_2}{\nu_2} & 0 & 0 & -t_2 & \frac{1}{\nu_2} \\ 0 & 0 & \frac{r_3}{\nu_3} & 0 & 0 & -t_3 \end{bmatrix} \begin{bmatrix} a_1 \\ a_2 \\ a_3 \\ b_0 \\ b_1 \\ b_2 \end{bmatrix} = \begin{bmatrix} a_0 t_1 \\ 0 \\ 0 \\ -b_3 \frac{1}{\nu_3} \\ 0 \\ 0 \end{bmatrix} \Leftrightarrow \underbrace{\begin{bmatrix} \mathbf{H}_U & \mathbf{Q} \\ \mathbf{D} & \mathbf{H}_L \end{bmatrix}}_{\text{matrix } \mathbf{L}} \underbrace{\begin{bmatrix} \mathbf{a} \\ \mathbf{b} \end{bmatrix}}_{\text{unknown}} = \begin{bmatrix} a_0 t_1 \boldsymbol{\rho}_1 \\ \mathbf{0} \end{bmatrix} \qquad (3)$$

[1] For a grazing incidence angle $\theta$ at the surface, the coefficients $(t_j, r_j)$ at $\Sigma_j$ - see [3] - are given by $t_j = \frac{2p_{j-1}}{p_{j-1}+p_j}$ and $r_j = \frac{p_{j-1}-p_j}{p_{j-1}+p_j}$, with $p_j = \sqrt{\theta^2 - 2\delta_j - 2i\beta_j}$

with $\mathbf{a} = (a_1, a_2, \ldots, a_{N+1})^T$ and $\mathbf{b} = (b_0, b_1, \ldots, b_N)^T$. In the multidimensional case (right hand side of (3)), $\mathbf{D} = \mathrm{diag}[(r_j \nu_j^{-1})] \in \mathbb{C}^{(N+1)\times(N+1)}$ is a diagonal matrix while $\mathbf{H}_U, \mathbf{H}_L \in \mathbb{C}^{(N+1)\times(N+1)}$ are the so called Hessenberg matrices, upper and lower respectively. Also, $\boldsymbol{\rho}_j \in \mathbb{C}^{N+1}$ is the $j$th canonical vector. A similar approach, not exactly in the same matrix context, was also stablished in [2]. It should be noted that (3) gives us the following results:

$$\begin{cases} \mathbf{H}_U \mathbf{a} + \mathbf{Q} b = a_0 t_1 \rho_1 \\ \mathbf{D} \mathbf{a} + \mathbf{H}_L \mathbf{b} = \mathbf{0} \end{cases} \Rightarrow \begin{cases} \mathbf{a} = -\mathbf{D}^{-1} \mathbf{H}_L \mathbf{b} \\ \left[\mathbf{Q} - \mathbf{H}_U \mathbf{D}^{-1} \mathbf{H}_L\right] \mathbf{b} = a_0 t_1 \boldsymbol{\rho}_1 \end{cases} \tag{4}$$

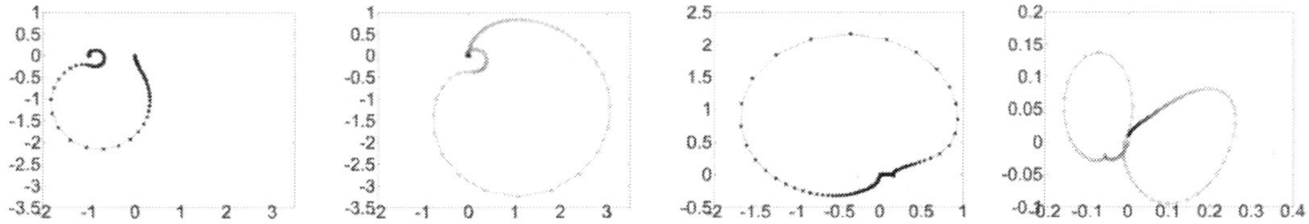


**Figure 3.** From left to right: Plot of $\{a_1(\theta), a_2(\theta), b_0(\theta), b_1(\theta)\}$ in the complex plane, with $0 \leqslant \theta \leqslant 1°$.

To find the solution $\{\boldsymbol{a}, \boldsymbol{b}\}$, we first solve a linear system to find the vector of reflected amplitudes $\mathbf{b}$, followed by a matrix product to compute the transmitted amplitudes $\mathbf{a}$, as given in (4). It is easy to note that the matrix $\mathbf{M} \equiv \mathbf{Q} - \mathbf{H}_U \mathbf{D}^{-1} \mathbf{H}_L$ is in the tridiagonal form. Therefore, solving $\mathbf{M}\mathbf{b} = a_0 t_1 \mathbf{p}_1$ requires about $O(N)$ operations using a fast algorithm. This is in contrast with the $O(N^3)$ operations requested by the complete Gaussian elimination, see [8, 7]. Another non-optimal approach, involves computing the analytical inverse of matrix $\mathbf{L}$ using the Aitken-block diagonalization (Banachiewicz inversion formula, see [9]). Such an inversion, although explicit, involves at least three tridiagonal inversion; being expensive than solving the system with matrix $\mathbf{M}$. We remark that our strategy has the same computational cost as the one depicted in [3, 5], where $N$ two-dimensional linear systems must be solved to find $\mathbf{a}$ and $\mathbf{b}$.

Example: In order to verify the matrix formalism in (4), we consider an example extracted from [1]. Let us consider a 2-layered thin film, deposited on a thick Silicium substrate; first layer being Cobalt and second Gold. Layers have thickness around $h_1 = 3.4 \times 10^{-9}$nm and $h_2 = 2 \times 10^{-}9$nm respectively. Complex refractive indexes are given by $n_j = 1 - \delta_j - i\beta_j$ with $\boldsymbol{\delta} = (5.6 \times 10^{-6}, 10.5 \times 10^{-6}, 1.6 \times 10^{-6})$ and $\boldsymbol{\beta} = (19.8 \times 10^{-8}, 129.5 \times 10^{-8}, 0.84 \times 10^{-8})$. Figure 3 shows the complex solution $\boldsymbol{a} = (a_1, a_2) \in \mathbb{C}^2$ and $\boldsymbol{b} = (b_0, b_1) \in \mathbb{C}^2$ of equation (4), for a variety of glancing angles. The intensity function $f_0(z, \theta) = |e^{-ikzp_0} + b_0(\theta) e^{ikzp_0}|^2$ is shown in figures 4 and 5 for fixed values of $z$ and $\theta$. We considered the initial amplitude $a_0 = 1$. The results obtained here are slightly different from [1] since we are not considering mixtures within each layer. ■

## 3. The inverse problem

As previously stated, we want to solve a fitting problem [♠]. Let the measurement data be given by vector $\boldsymbol{q} \in \mathbb{R}^v$. Functional $\boldsymbol{u} \in \mathbb{R}^N \mapsto r(\boldsymbol{u}) \in \mathbb{R}^v$ (with $v \geqslant N$) has to be minimized using a standard optimization algorithm. Many differential techniques can be used in this problem since $\boldsymbol{r}$ is a differentiable function of $\boldsymbol{u}$. In fact,

$$\boldsymbol{r}(\boldsymbol{u}) = \boldsymbol{q} - \boldsymbol{F}_m(\boldsymbol{u}), \quad \boldsymbol{F}_m(\boldsymbol{u}) = (F_m(\theta_s, \boldsymbol{u})) \in \mathbb{R}^v, \quad F_m(\theta, \boldsymbol{u}) = \boldsymbol{c}_m^T(\boldsymbol{u}) \boldsymbol{\ell}_m(\theta, \boldsymbol{u}) \tag{5}$$

with $m$ standing for a given element under investigation. Here, $\boldsymbol{c}_m^T$ is the $m$th row of the matrix $\boldsymbol{C} = (c_{mj})$ whereas $\boldsymbol{\ell}_m$ is the $m$th column of a matrix $\boldsymbol{L} = (\ell_{jm})$. To implement an optimization

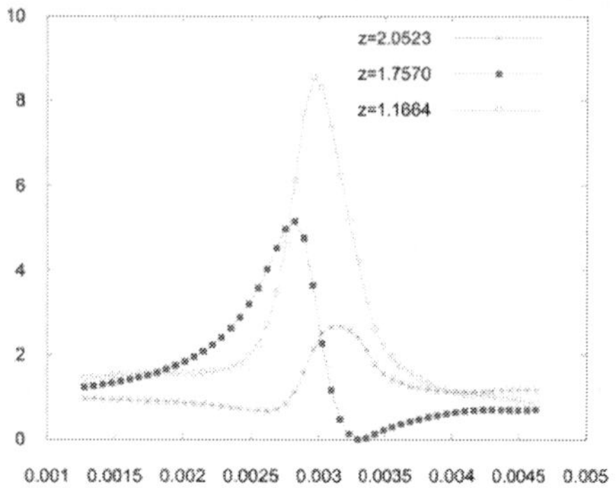


**Figure 4.** Intensity curve $f_0(z,\cdot)$ at depth $z \in \{2.0523, 1.7570, 1.1664\}$ (unity in nanometer range)

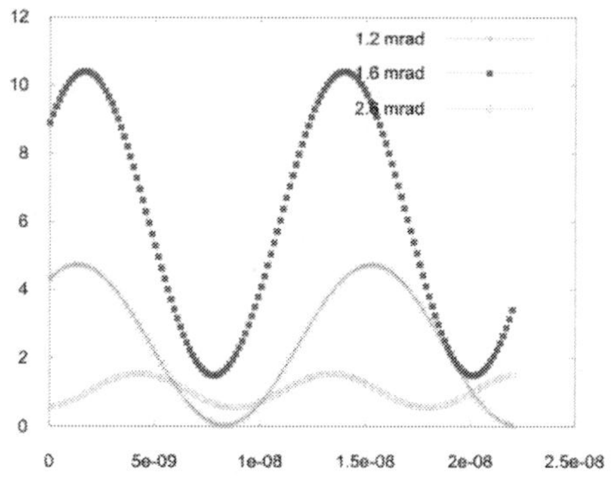


**Figure 5.** Intensity curve $f_0(\cdot,\theta)$ at glancing angles $\theta \in \{0.0012752, 0.0016107, 0.0026174\}$.

method, one must have the Jacobian of $\boldsymbol{r}$, which is written as $\boldsymbol{J}_{\boldsymbol{r}}(\boldsymbol{u})^T = [\nabla \boldsymbol{r}_1(\boldsymbol{u}) \dots \nabla \boldsymbol{r}_v(\boldsymbol{u})]$ with $\boldsymbol{r}_s(\boldsymbol{u}) = \boldsymbol{q}_s - F_m(\theta_s, \boldsymbol{u})$. The gradient $\nabla \boldsymbol{r}_s$ depends on the derivatives $\partial_l F_m(\theta_s, \boldsymbol{u}) = \partial_l(\boldsymbol{c}_m)^T \boldsymbol{\ell}_m + \boldsymbol{c}_m^T \partial_l(\boldsymbol{\ell}_m)$, with $\partial_l$ denoting $\equiv \frac{\partial}{\partial \boldsymbol{u}_l}$. Hence, to obtain $\partial_l F_m$, first we need alternatives to obtain the partial derivatives of $\boldsymbol{\ell}_m$, and therefore, the derivatives of $\boldsymbol{a}(\theta, \boldsymbol{u})$ and $\boldsymbol{b}(\theta, \boldsymbol{u})$. This is our new problem:

**Problem [★]:** Find $\partial_l \boldsymbol{a}(\theta, \boldsymbol{u}) \in \mathbb{C}^{N+1}$ and $\partial_l \boldsymbol{b}(\theta, \boldsymbol{u}) \in \mathbb{C}^{N+1}$ for $l = 1, \dots, v$

Ommiting the angle dependence for simplicity, equation (4) give us a direct formula to compute $\boldsymbol{a}$ and $\boldsymbol{b}$ through simple matrix product, where each matrix also depends on the parameter $u$. We use this fact to obtain the derivatives. In fact, using $\alpha \equiv a_0 t_1$ and $\boldsymbol{b}(\boldsymbol{u}) = \alpha \mathbf{M}(\boldsymbol{u})^{-1} \boldsymbol{\rho}_1$; and from trivial matrix calculus $\partial_l[\mathbf{M}^{-1}] = -\mathbf{M}^{-1}(\partial_l \mathbf{M})\mathbf{M}^{-1}$:

$$(\partial_l \boldsymbol{b}) = -\mathbf{M}^{-1}(\partial_l \mathbf{M})\mathbf{M}^{-1}(\alpha \boldsymbol{\rho}_1) \quad \Rightarrow \quad \begin{cases} \mathbf{M}(\partial_l \boldsymbol{b}) = -(\partial_l \mathbf{M})\boldsymbol{b} & \text{(a)} \\ \mathbf{M}\boldsymbol{b} = \alpha \boldsymbol{\rho}_1 & \text{(b)} \end{cases} \tag{6}$$

This means that $\partial_l \boldsymbol{b}$ is obtained solving the tridiagonal linear system (6.a), while (6.b) was previously solved. To find $\partial_l \boldsymbol{b}$ we use (4) again. Indeed, from $\mathbf{D}\boldsymbol{a} + \mathbf{H}_L \boldsymbol{b} = 0$ and matrix calculus we obtain $-\mathbf{D}(\partial_l \boldsymbol{a}) = (\partial_l \mathbf{D})\boldsymbol{a} + (\partial_l \mathbf{H}_L)\boldsymbol{b} + \mathbf{H}_L(\partial_l \boldsymbol{b})$. It only remais to find matrices $\{\partial_l \mathbf{D}, \partial_l \mathbf{H}_L, \partial_l \mathbf{H}_U\}$. After some observation, it is easy to realize from (3) that

$$\mathbf{H}_L = -\text{diag}(\boldsymbol{t}) + \sum_{k=1}^{N} \frac{\boldsymbol{\rho}_k \boldsymbol{\rho}_{k+1}^T}{\nu_k}, \quad \mathbf{H}_U = \sum_{k=1}^{N+1} \nu_k \boldsymbol{\rho}_k \boldsymbol{\rho}_k^T - t_{k+1} \boldsymbol{\rho}_{k+1} \boldsymbol{\rho}_k, \quad \mathbf{Q} = \sum_{k=1}^{N} r_k \nu_k \boldsymbol{\rho}_k \boldsymbol{\rho}_{k+1}^T, \quad \mathbf{D} = \sum_{k=1}^{N} \frac{r_k}{\nu_k} \boldsymbol{\rho}_k \boldsymbol{\rho}_k^T \tag{7}$$

where now follows immediately $-\partial_l \mathbf{M} = (\partial_l \mathbf{H}_U)\mathbf{D}^{-1}\mathbf{H}_L + \mathbf{H}_U \partial_l(\mathbf{D}^{-1})\mathbf{H}_L + \mathbf{H}_U \mathbf{D}^{-1} \partial_l(\mathbf{H}_L)$ with

$$\partial_l \mathbf{H}_L = \partial_l \left(\frac{1}{\nu_l}\right) \boldsymbol{\rho}_l \boldsymbol{\rho}_{l+1}^T, \quad \partial_l \mathbf{H}_U = \partial_l(\nu_l) \boldsymbol{\rho}_l \boldsymbol{\rho}_l^T, \quad \partial_l \mathbf{Q} = \partial_l(\nu_l) r_l \boldsymbol{\rho}_l \boldsymbol{\rho}_{l+1}^T, \quad \partial_l \mathbf{D} = \partial_l \left(\frac{r_l}{\nu_l}\right). \tag{8}$$

## 4. Future work

Our matrix formalism is new and very different from the one pointed out in [3, 5, 1]. A fast computation of the amplitude coefficients $\boldsymbol{a}, \boldsymbol{b} \in \mathbb{R}^{N+1}$ and their partial derivatives $\partial_l \boldsymbol{a}, \partial_l \boldsymbol{b} \in \mathbb{R}^{N+1}$ was presented. A complete algorithmic description, using standard differential optimization techniques, depends on the calculation of these vectors.

## References


[1] de Boer D K G, 1991 Glancing-incidence x-ray fluorescence of layered materials, *Physical Review B*, **42(2)**, pp. 498.

[2] Pérez R D, Sánchez H J, Rubio M, Pérez C A, 1999 Mathematical model for evaluation of surface analysis data by total reflection xrf, *X-Ray Spectrometry*, **28**, pp.342.

[3] Sánchez H J, Pérez C A, Pérez R D, Rubio M, 1996 Surface analysis by total-reflection x-ray fluorescence, *Radiat.Phys.Chem.* **48**, pp.325.

[4] Parrat L G, 1994 Surface studies of solids by total reflection of x-rays, *Physical Review*, **95(2)**, pp 359.

[5] Król A, Sher C J and Kao Y H, 1988 X-ray fluorescence of layered synthetic materials with interfacial roughness, *Phyical Review B*, **38(13)**, 8579-92.

[6] Gerrard A and Burch J M, 1975, Introduction to Matrix Methods in Optics, *Dover*, John Wiley & Sons.

[7] Lewis J W, 1982 Inversion of tridiagonal matrices, *Numerische Mathematik*, **38(3)**, pp.333.

[8] Usmani R A, 1994 Inversion of jacobi's tridiagonal matrix, *Computer Math. Applic.*, **27(8)**, pp.59.

[9] Tian Y, Takane Y, 2009 The inverse of any two-by-two nonsingular partitioned matrix and three matrix inverse completion problems, *Computers & Mathematics with Applications*, **57(8)**, pp. 12941304.

# Performance of a new atomistic geometrical model of the B-DNA configuration for DNA-radiation interaction simulations

**M A Bernal[1], D Sikansi[1], F Cavalcante[1], S Incerti[2], C Champion[2], V Ivanchenko[3], Z Francis[4,5] and M Karamitros [2,6]**

[1]Instituto de Física Gleb Wataghin, Universidade Estadual de Campinas, SP, Brazil
[2]Université Bordeaux 1, CNRS/IN2P3, Centre d'Etudes Nucléaires de Bordeaux-Gradignan, CENBG, Chemin du Solarium, BP 120, 33175 Gradignan, France.
[3]Ecoanalytica, 119899 Moscow, Russia
Geant4 Associates International Ltd, United Kingdom
[4]Université Saint Joseph, Science Faculty, Department of Physics, Beirut, Lebanon.
[5]The Open University, Faculty of Science, Department of Physical Sciences, Walton Hall MK7 6AA, Milton Keynes, United Kingdom.
[6]CNRS, INCIA, UMR 5287, 33400 Talence, France

E-mail: mbernalrod@gmail.com

**Abstract.** We have recently developed an atomistic model of the B-DNA configuration, up to the 30-nm chromatin fiber. This model is intended to be used in Monte Carlo simulations of the DNA-radiation interaction, specifically in conjunction with the Geant4-DNA extension of the Geant4 Monte Carlo toolkit. In this work, 11449 parallel chromatin fibers have been arranged within a cube mimicking a cell nucleus containing about $6.5\times10^9$ base pairs. Each atom in the model is represented by a sphere with the corresponding van der Waals radius. Direct single, double and total DNA strand break yields due to the impact of protons and alpha particles with LET ranging from 4.57 to 207.1 keV/$\mu$m have been determined. Also, the corresponding site-hit probabilities have been calculated.

## 1. Introduction

The knowledge on the biological effects induced by ionizing radiations on living beings is of paramount importance for areas such as medical radiation therapy and diagnostic, among others. Monte Carlo simulations of the radiation-DNA interaction process have been successfully coupled to biophysical models to study this matter [1, 2]. The use of ion beams in radiation therapy is particularly complex from the radiobiological point of view due to the stronger dependency of their relative biological effectiveness as a function of the particle energy when compared to that of photon and electron beams. In a recent work [3], the development of an atomic-resolution B-DNA geometrical model has been addressed and an accompanying open source subroutine to determine the closest atom containing an arbitrary point in space has been released. This model accounts for various organization levels of the DNA, from the nucleotide pair up to the 30-nm chromatin fiber, and it should be integrated in the GEANT4-DNA extension of the Geant4 Monte Carlo simulation toolkit to determine direct and indirect strand break yields in the near future.

In this work, a cell nucleus model has been built by arranging segments of the chromatin fiber described in our previous work [3]. This development is a long step forward with respect to a recently published work [4] in which a cell nucleus model was constructed by using the non-atomistic DNA model described in Ref. [5]. The performance of this model was tested by determining the direct total, single and double strand break yields due to the impact of protons and alpha particles within a wide linear energy transfer (LET) range. Site-hit probabilities have been also determined.

## 2. Material and methods

The details of the B-DNA atomistic model used in this work can be found elsewhere [3]. A simple geometrical model of a cell nucleus was developed by aligning 11449 3.36-$\mu$m long chromatin segments within a cube. This nucleus contains about 6.52 Gbp, which is approximately the number of bp in a human cell, and it defines the region of interest (ROI) in our simulations. The ROI was placed at 2.6 $\mu$m depth in a semi-infinite water phantom, with the fibers axis normal to the surface. This system was uniformly irradiated with 0.5, 1, 5, 7, and 10 MeV protons, and 2, 5, 7 and 10 MeV alpha particles. The corresponding Monte Carlo simulations were carried out with the GEANT4-DNA sub-package [8]. The LET of the incident particles at the middle of the ROI ranges from 4.57 to 207.1 keV/$\mu$m. Direct total (TSB), single (SSB) and double (DSB) strand break yields were determined by the use of two approaches. In the first one, a SSB is recorded if an inelastic collision with an energy deposition greater than 8 eV occurs within one van der Waals radius around any of the atoms conforming the phosphodiester group. The union volume of these atoms defines the target volume. In the second one, a SSB is accounted for in the same situation but using two van der Waals radii and with an additional probability going from 0 to 1 when the deposited energy ranges from 5 to 35 eV. The latter method was implemented in order to compare our results with those reported by Friedland et al. [7]. In any case, a DSB is counted if two SSB occur on opposite strands and within 10 bp. A fixed number of histories were assigned to each simulation batch, according to the incident particle type and energy.

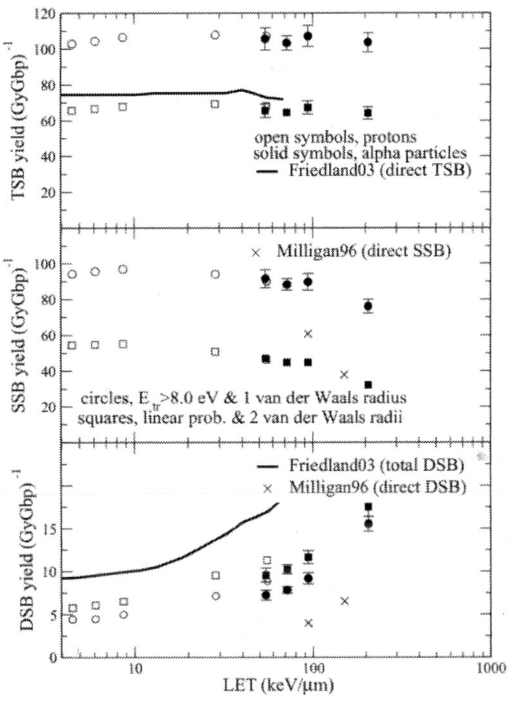


**Figure 1.** Direct total (top graph), single (middle graph) and double (bottom graph) strand break yields due to the impact of protons (○ & □) and alpha particles (● & ■). Also, the corresponding damage yields for alpha particles (◇, experimental) [6] and protons (——, simulation) [7] are shown. Uncertainties in our results of the order of the symbol size or smaller are not shown.

## 3. Results

The radiation doses absorbed into the ROI for all particle-energy combinations were in the 12-15 Gy range. Figure 1 shows the direct TSB, SSB and DSB yields determined by following the two approaches described in the former section. The first one leads to TSB yields significantly higher than those reported for protons by the Friedland's group [7], mainly because practically all deposited energies are high enough to provoke a strand break, despite that only one van der Waals radius around each atom was used to define the target volume. However, our results are close to those reported in the mentioned reference when the second approach is used, which is similar to the one used by these authors. Differences should be attributed to slight differences in the van der Waals radii used and in the geometrical model *per se.* It is good to remark that our results also describe the quasi-constant direct TSB yield (see the discussion on this behavior in Ref. [9]) and the slight decrement shown for LET greater than about 40 keV/$\mu$m.

It can be observed in the mid graph of Fig. 1 that the SSB yield decreases as the primary particle LET increases. This fact has been previously reported by experimental works (e.g. see Ref. [10]). In addition, SSB yields due to alpha particle impact, determined by Milligan et al. [6] in solutions with high free radical scavenger concentrations, are also shown. It should be pointed out that these experimental results were obtained with up to three-fold uncertainties, so they should only serve as a coarse reference.

From the analysis of the bottom graph of Fig. 1, it can be seen that, unlike the TSB results, DSB yields obtained with the second approach are higher than those determined with the first one. This means that the increase of the atom size has a greater influence on the DSB yields than the reduction of the probability to produce a SSB according to the deposited energy. Unfortunately, only the total DSB yields determined by Friedland and colleagues are shown since they did not report the corresponding direct damage probabilities. According to the consistency of the TSB yields and the DSB yield results, it could be induced that direct and indirect damages have similar contribution to the total DSB yield, at least for protons in this LET range. It should be kept in mind that by using two van der Waals radii non-scavengable damage is accounted as part of the direct damage. Also, it can be seen that protons are more effective to induce DSB than alpha particles at the same LET. This was attributed to a higher capacity of protons to form clustered depositions when compared to alpha particles with the same LET (see discussions on this matter in Refs. [10] and [11]). DSB yields determined by Milligan et al. in the conditions described above are also shown. The experimental method used to determine these yields strongly underestimates the countii

Figure 2 displays the probabilities for any primary or secondary particle to hit a phosphodiester group (known as site-hit probability) as a function of the primary particle LET at the half depth of the ROI. The horizontal line represents the site-hit probability (SHP) determined through the division of the total volume occupied by the targets by that of ROI. This SHP should be retrieved if the energy deposition events were uniformly distributed within the ROI [3]. The SHPs obtained during simulations agree with that predicted by the volumes ratio for LET values up to about 100 keV/$\mu$m. Above this LET value, this SHP tends to increase appreciably over the geometrically predicted value. Thus, it seems that ionizing particles with very high LET produce SHPs greater than that induced by uniformly distributed ionization events.

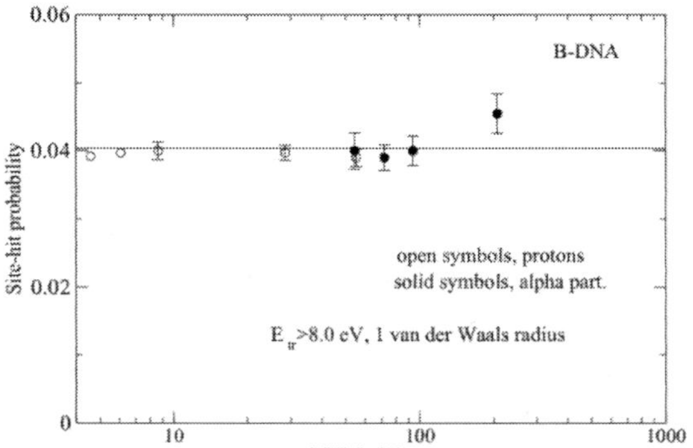


**Figure 2.** Site-hit probabilities due to the impact of protons (○) and alpha particles (●). Uncertainty of the order of the symbol size or smaller are not shown. The line (——) represents the geometrically determined SHP.

## 4. Conclusions

A new B-DNA atomistic model has been preliminarily tested in this work by determining direct DNA damage yields and site-hit probabilities due to the impact of protons and alpha particles for a wide LET range. The direct TSB yields found with this model for protons are consistent with those reported by Friedland et al., who used a similar atomistic geometrical model. Our SSB yields decrease with the increment of the incident particle LET, as described in other published works. In addition, these values seem to be consistent with the corresponding experimental determinations reported by Milligan et al. However, the results for direct DSB yields overestimate those reported by these authors, probably due to the difficulties they had to count very short DNA fragments. In the near future, this geometrical model will be tested by also having into account the indirect damage induced by chemical species.

## 5. Acknowledgements

M Bernal wishes to thank the FAPESP foundation in Brazil for financing his research activities through the FAPESP 2011/51594-2 project.

## References

[1] Pomplun E 1991 *Int. J. Radiat. Biol.* **59** 625–642

[2] Friedland W, Dingfelder M, Kundrat P and Jacob P 2011 *Mutat Res Fundam Mol Mech Mutagen* 28–40

[3] Bernal M A, Sikansi D, Cavalcante F, Incerti S, Champion C, Ivanchenko V and Francis Z 2013 *Computer Physics Communications* –

[4] Incerti S, Champion C, Tran H N, Karamitros M, Bernal M, Francis Z, Ivanchenko V and Mantero A 2013 *Nuclear Instruments and Methods in Physics Research Section B: Beam Interactions with Materials and Atoms* **306** 158–164

[5] Bernal M A and Liendo J A 2009 *Medical Physics* **36** 620–625

[6] Milligan J R, Aguilera J A, Wu C C L, Ng J Y Y and Ward J F 1996 *Radiation Research* **145** 442–448

[7] Friedland W, Jacob P, Bernhardt P, Paretzke H G and Dingfelder M 2003 *Radiation Research* **159** 401–410

[8] Incerti S, Ivanchenko A, Karamitros M, Mantero A, Moretto P, Tran H N, Mascialino B, Champion C, Ivanchenko V N, Bernal M A, Francis Z, Villagrasa C, Baldacchino G, Guèye P, Capra R, Nieminen P and Zacharatou C 2010 *Medical Physics* **37** 4692–4708

[9] Bernal M A, deAlmeida C E, Sampaio C, Incerti S, Champion C and Nieminen P 2011 *Medical Physics* **38** 4147–4153

[10] Leloup C, Garty G, Assaf G, Cristovão A, Breskin A, Chechik R, Shchemelinin S, Paz-Elizur T, Livneh Z, Schulte R W, Bashkirov V, Milligan J R and Grosswendt B 2005 *International Journal of Radiation Biology* **81** 41–54

[11] Francis Z, Incerti S, Ivanchenko V, Champion C, Karamitros M, Bernal M A and Bitar Z E 2012 *Physics in Medicine and Biology* **57** 209–224

# Automatic classification of prostate stromal tissue in histological images using Haralick descriptors and Local Binary Patterns

**D L L Oliveira[1], M Z Nascimento[2], L A Neves[3], V R Batista[1], M F Godoy[4], R S Jacomini[5], Y A S Duarte[1], P F F Arruda[6] and D S Neto[7]**

[1] Mathematics, Computer Science and Cognition Centre, Federal University of ABC (UFABC), Santo André, SP, Brazil.
[2] Faculty of Computer Science (FACOM), Federal University of Uberlândia (UFU), Uberlândia, MG, Brazil.
[3] Institute of Biosciences, Literature and Exact Sciences, Department of Computer Science and Statistics, São Paulo State University (UNESP), São José do Rio Preto, SP, Brazil.
[4] Interdisciplinary Center for the Study of Chaos and Complexity (NUTTECC), Faculty of Medicine of São José do Rio Preto (FAMERP), São José do Rio Preto, SP, Brazil.
[5] Faculdade de Tecnologia Termomecnica, Centro Educacional Fundação Salvador Arena, São Bernardo, SP, Brazil.
[6] Surgery Department of Renal Transplantation, Regional Medical Faculty Foundation (FUNFARME), São José do Rio Preto, SP, Brazil.
[7] Department of Pathology of Base Hospital, Regional Medical Faculty Foundation (FUNFARME), São José do Rio Preto, SP, Brazil.

E-mail: domingos.latorre@gmail.com

**Abstract.** In this paper we presente a classification system that uses a combination of texture features from stromal regions: Haralick features and Local Binary Patterns (LBP) in wavelet domain. The system has five steps for classification of the tissues. First, the stromal regions were detected and extracted using segmentation techniques based on thresholding and RGB colour space. Second, the Wavelet decomposition was applied in the extracted regions to obtain the Wavelet coefficients. Third, the Haralick and LBP features were extracted from the coefficients. Fourth, relevant features were selected using the ANOVA statistical method. The classication (fifth step) was performed with Radial Basis Function (RBF) networks. The system was tested in 105 prostate images, which were divided into three groups of 35 images: normal, hyperplastic and cancerous. The system performance was evaluated using the area under the ROC curve and resulted in 0.98 for normal versus cancer, 0.95 for hyperplasia versus cancer and 0.96 for normal versus hyperplasia. Our results suggest that texture features can be used as discriminators for stromal tissues prostate images. Furthermore, the system was effective to classify prostate images, specially the hyperplastic class which is the most difficult type in diagnosis and prognosis.

## 1. Introduction

Automatic classification of prostate stromal tissue in histology images is still a challenge due to the large variation in shape and colour of the histological components caused by prostate cancer.

Computer-aided diagnosis (CAD) has been developed and can produce consistent results for the detection of diseases. Several CAD systems have been proposed for the analysis of prostate histological images [1, 2]. However, none of these methods used stromal tissue regions in their analysis, and they only considered two classes of tissue: cancerous and normal.

In contrast with these approaches, in this paper we propose a classification method of prostate images using regions of interest (ROI) of the stromal tissue. The method was based on Wavelet decomposition [3], combinations of Haralick descriptors [4], Local Binary Patterns (LBP) [5], ANOVA statistical method and Radial Basis Function (RBF). The method has been evaluated considering histological images of cancer, hyperplastic and normal classes.

## 2. Materials and methods

The method was organized in five steps: (1) detection and definition of stromal regions, (2) calculation of Wavelet coefficients, (3) extraction of Haralick and LBP features, (4) aplication of ANOVA statistical method and (5) classification of images using Radial Basis Function (RBF) networks. Fig. 1 shows the flowchart of the method.

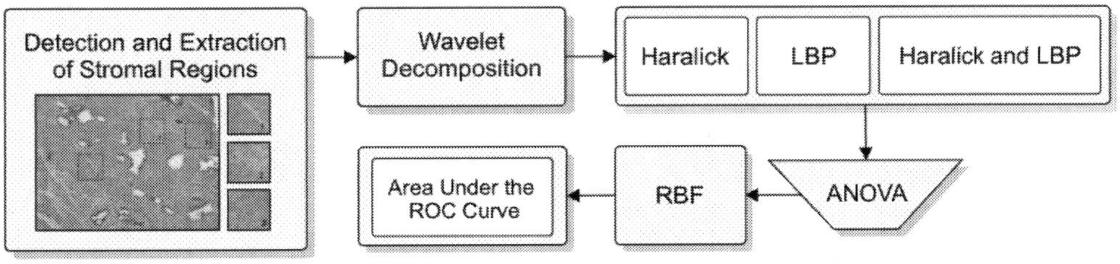


Figure 1: Flowchart of the proposed classification system.

### 2.1. *Image bank*

The image bank used in our study consists of histological slides from the prostates of 35 patients from 50 to 75 years old (mean age: 64.4 ±5.9 years). The patients were treated at the Urology Service of the Base Hospital in São José do Rio Preto, state of São Paulo, Brazil, between 2007 and 2008. This study was approved by the Human Research Ethics Committee of the São José do Rio Preto School of Medicine (Brazil). The sample consists of 105 images, which were divided into three groups of 35 images: normal, hyperplastic and cancerous.

### 2.2. *Detection and extraction of stromal regions*

Stromal regions were segmented by removing cuboidal cells and glandular lumens areas in the image. These regions were segmented using methods described by Oliveira *et al.* [6] and [7], which render two binary images $I_{ncc}$ and $I_{lg}$. The stromal regions were segmented by the equation $I_{est}(x, y) = \neg(I_{ncc}(x, y)|I_{lg}(x, y))$, where the value of each pixel in the image $I_{est}$ in the coordinates $x$ and $y$ was computed by logic operations using the binary images $I_{ncc}$ and $I_{lg}$.

### 2.3. *Calculation of wavelet coefficients*

A multiscale approach allows image data analysis considering different resolutions. One of the techniques used to achieve the multiresolution representation is the Wavelet Transform (WT). In this technique the spatial domain is transformed into frequency and time domains by the formula

$$Wf(a,b) = \int_{-\infty}^{\infty} f(t)\psi_{a,b}(t)dt. \quad (1)$$

In Eq. 1 $Wf(a,b)$ is the the inner product of a signal $f(t)$ with a Wavelet function $\psi$ scaled and translated in the time axis. Namely, $a$ is the scale (dilation) and $b$ is the translation (shift in time) of the Wavelet $\psi_{a,b}(t) = \frac{1}{\sqrt{a}}\psi(\frac{x-b}{a})$. The process of obtaining the wavelet coefficients was performed by Wavelet Daubechies 8 with 3 levels of decomposition. The application generated 4 sub-images of wavelet coefficients at each level.

### *2.4. Extraction of Haralick and LBP features*

Haralick descriptors [4] are second-order statistical measures obtained by calculating the co-occurrence matrices. The matrices are constructed for each sub-image with $N$ Wavelets coefficients. The corresponding descriptors were extracted with angles $\theta$ (0°, 45°, 90° and 135°) and distance $d = 1$. The Local Binary Patterns (LBP) operator proved to be robust at extracting image textures [5]. The operator assigned a label to each pixel of an image by thresholding the 3×3 neighbourhood of each pixel with the centre pixel value and giving the result as a binary number. LBP encode the gray-scale pattern of $N$ neighbouring pixels.

### *2.5. Dimensionality reduction and classification*

To increase the performance of the classifier, the ANOVA statistical method was applied to reduce the vector of features [8]. Afterwards, the RBF classifier was applied. This classifier consists of a multilayer neural model that is able to learn complex pattern and solve non-linearly separable problems. The cross-validation technique was used in the training and validation steps.

## 3. Results and discussion

The AUC results for the vectors constructed with Haralick, Haralick LBP and LBP are presented in Table 1, considering each combination of tissue class and RGB colour hues.

Table 1: AUC results for tissue combinations and colour channels using Haralick, LBP and Haralick LBP.

| Tissue Combination | Colour Channel | Haralick | LBP | Haralick LBP |
|---|---|---|---|---|
| Normal versus Cancerous | Red | 0.85 | 0.89 | 0.86 |
| | Green | **0.88** | **0.98** | **0.90** |
| | Blue | 0.87 | 0.95 | 0.89 |
| Normal versus Hyperplastic | Red | 0.67 | 0.90 | 0.69 |
| | Green | **0.71** | **0.95** | **0.84** |
| | Blue | 0.70 | 0.95 | 0.80 |
| Hyperplastic versus Cancerous | Red | 0.73 | **0.95** | 0.80 |
| | Green | 0.76 | 0.94 | **0.78** |
| | Blue | **0.76** | 0.93 | 0.74 |

The ROC curves of the best AUC results for the combinations of classes Normal versus Cancerous, Normal versus Hyperplastic and Hyperplastic versus Cancerous are shown in Fig. 2 (a), Fig. 2 (b) and Fig. 2 (c), respectively.

The best results were obtained with LBP features, which gave the most relevant AUC (0.98) for the class Normal versus Cancerous. This occurred due to the abnormalities caused by cancer in areas of stromal tissue, since cancer changes the patterns of form and colour intensity. For

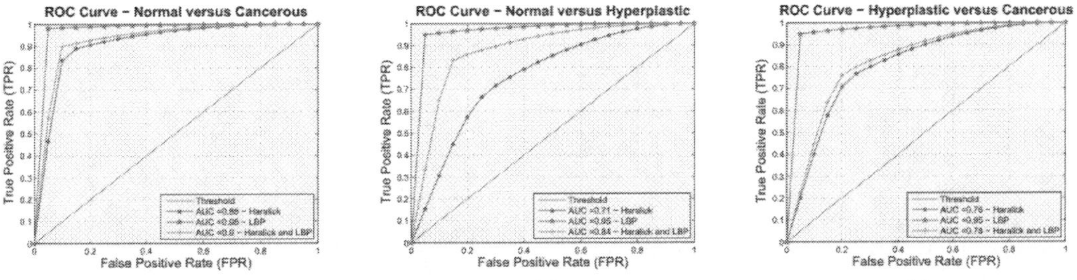


(a) (b) (c)

Figure 2: ROC curves for combinations of Normal versus Cancerous (a), Normal versus Hyperplastic (b) and Hyperplastic versus Cancerous (c).

Normal versus Hyperplastic and Hyperplastic versus Cancerous classes the best results were: 0.95 and 0.95 of AUC, respectively. These results suggest that textural features from the stromal tissue are sufficient to separate the Hyperplastic class: an important contribution to studies focused on the diagnosis of prostate cancer. Regarding the analysis of colour hues (Table 1), the green colour hues showed better results in 8 of the 9 combinations. This occurs because of the contrast information in intensity between the different components such as nuclei and cytoplasm. This contrast is related to the properties of the dyes hematoxylin and eosin. Hematoxylin has affinity with nucleic acids in the cell nucleus and eosin stains the cytoplasm with pink colour.

## 4. Conclusion

In this paper an automatic classification method was proposed to analyse prostate stromal tissue of normal, hyperplastic and cancer images. The main features of this method are: (1) the use of texture features of wavelet coefficients, (2) relevante to classification results between images of normal and cancer classes, (3) effectiveness in classification of the Hyperplastic class, which is the most difficult type in diagnosis and prognosis and (4) the use of small ROIs (spatial resolution of 64×64 pixels). In future works, new features of stromal tissue and glandular components such as lumen and cuboidal cells will be investigated.

## References

[1] Doyle S, Madabhushi A, Feldman M and Tomaszeweski J 2006 *Medical Image Computing and Computer-Assisted Intervention–MICCAI 2006* (Springer) pp 504–511

[2] Nguyen K, Sabata B and Jain A K 2012 *Pattern Recognition Letters* **33** 951–961

[3] Mallat S 1999 *A wavelet tour of signal processing* (Access Online via Elsevier)

[4] Haralick R M, Shanmugam K and Dinstein I H 1973 *Systems, Man and Cybernetics, IEEE Transactions on* 610–621

[5] Wang L and He D C 1990 *Pattern Recognition* **23** 905–910

[6] de Oliveira D L L, do Nascimento M Z, Neves L A, de Godoy M F, de Arruda P F F and de Santi Neto D 2013 *Expert Systems with Applications* **40** 7331–7340

[7] de Oliveira D L L, do Nascimento M Z, Neves L A, de Godoy M F, de Arruda P F F and de Santi Neto D 2012 *Anais do XXIII Congresso Brasileiro em Engenharia Biomdica – XXIII CBEB* (XXIII CBEB) pp 1265–1269

[8] Jacomini R, Nascimento M, Dantas R and Ramos R 2013 *Recent Patents on Medical Imaging* **3** 80–88

# Loop Quantum Cosmological Perturbations

**M Fernández-Méndez**[1], **G A Mena Marugán**[1] **and J Olmedo**[2]

[1] Instituto de Estructura de la Materia, IEM-CSIC, Serrano 121, 28006 Madrid, Spain
[2] Instituto de Física, Facultad de Ciencias, Iguá 4225, esq. Mataojo, Montevideo, Uruguay

E-mail: m.fernandez.m@csic.es

**Abstract.** In the context of loop quantum cosmology, we provide a complete quantization of an inhomogeneous inflationary model consisting of a perturbed Friedmann-Lemaître-Robertson-Walker universe filled with a minimally coupled massive scalar field. We focus on the case of flat, compact spatial sections. After fixing the local gauge freedom, the kinematical Hilbert space is constructed by combining the representation used in loop quantum cosmology for the homogeneous sector with a preferred Fock quantization of the inhomogeneities. We characterize the physical states annihilated by the quantum Hamiltonian constraint and discuss the evolution of the inhomogeneities in the presence of an emergent relational time.

## 1. Introduction

Loop Quantum Cosmology (LQC) [1] employs the polymer quantization of Loop Quantum Gravity (LQG) [2] in symmetry-reduced systems. Its successful application to the Friedmann-Lemaître-Robertson-Walker (FLRW) model, in which the big-bang is replaced with a bounce, led to the study of more complicated situations, such as the Bianchi or Gowdy models. In the latter, the combination of a polymer quantization of the gravitational homogeneous degrees of freedom and the standard Fock quantization of the inhomogeneities allowed one to deal with systems with an infinite number of degrees of freedom [3]. In this approximation, usually called *hybrid quantization*, the effects of quantum geometry are included only in the zero modes. Nevertheless, it provides a rigorous way of constructing a kinematical Hilbert space in which the constraints can be represented.

The inclusion of inhomogeneities in LQC is an inevitable step in order to make contact with cosmological observations. In this work, we study the case of a (scalarly) perturbed FLRW model with a massive scalar field as matter content. As it is well known, this simple model can nonetheless undergo inflation, thus providing a natural arena for the emergence of primordial inhomogeneities which is compatible with observations [4].

## 2. Classical parametrization

The parametrization of the classical system that we use is essentially that of Halliwell and Hawking [5]. The Arnowitt-Deser-Misner variables are expanded in the eigenbasis of the Laplace-Beltrami operator of a reference metric (the three-torus in the case of a flat FLRW model [6] or the three-sphere for a closed scenario [7]). We only analyze scalar harmonics, although the generalization to vector and tensor ones is straightforward. The zero modes are then identified with the homogeneous variables of an FLRW model, while the inhomogeneities are considered perturbations of them. Hence, the action is truncated at quadratic order in the inhomogeneities.

A Legendre transform leads to a Hamiltonian that is a linear combination of constraints: a global Hamiltonian constraint with quadratic corrections, which appears multiplied by the homogeneous lapse, and two local constraints linear in the perturbations, whose corresponding Lagrange multipliers are the modes of the perturbations of the lapse and shift. These linear constraints are removed with a suitable gauge fixing. In particular, we have considered the so-called longitudinal gauge and the gauge in which the spatial sections have constant curvature [6, 7]. The reduced system is parametrized by the homogeneous variables, the modes of the matter-field perturbation and their canonically conjugate momenta. However, these variables may not have canonical Dirac brackets (this is the case e.g. in the longitudinal gauge). In the truncation of the action at quadratic order in perturbations, a new set of canonical variables can be introduced if the homogeneous variables are corrected with quadratic terms [6, 7].

The Fock quantization of a field theory in a (classical) curved spacetime suffers from an infinite ambiguity in the choice of representation for the canonical commutation relations. It has been shown that in many cosmological scenarios this ambiguity can be eluded by imposing (a) the invariance of the vacuum state under the spatial symmetries of the field equation and (b) the unitary implementation of the field dynamics in the quantum theory. Furthermore, these two requirements select a preferred scaling of the field and a particular canonically conjugate momentum, apart from picking out a unique class of unitarily equivalent Fock representations for them [8]. A representative of this class can be constructed by choosing the annihilation- and creation-like variables that would be natural if the field were massless.

We incorporate these criteria in our quantization so as to recover the privileged quantum field theory in the regimes in which the background behaves classically. In the case under discussion, the matter field perturbation must be scaled with the scale factor of the background metric [9]. With this scaling, the field equation becomes a Klein-Gordon equation with background-dependent mass and subdominant ultraviolet corrections. The privileged momentum coincides—up to subdominant ultraviolet corrections—with the time-derivative of the field. To extend this change of variables to a canonical transformation in the complete phase space (at the considered perturbative order), the homogeneous variables need to be corrected with terms quadratic in the perturbations [9].

It is worth mentioning that an alternative quantization with the same requirements of symmetric vacuum and unitary dynamics can be performed in terms of a gauge-invariant canonical pair [6, 7]. Since the canonical transformation that relates these variables with the scaled field perturbation and its privileged momentum is non-local, while the uniqueness results on the Fock representation commented above apply only to local transformations, the two quantizations might—in principle—be inequivalent. Nevertheless, we have proven that the symplectomorphism is unitarily implementable in the classical background [6, 7], and thus the two quantizations are equivalent in that regime.

### 3. Quantization of the model

In LQG, the fundamental variables are the holonomies of the Ashtekar-Barbero connection and the fluxes of a densitized triad. In a homogeneous and isotropic model, it suffices to consider their counterparts $N_\mu = \exp(i\mu c/2)$ ($\mu$ is proportional to the length of the holonomy) and $p$. Following the so-called *improved dynamics* scheme [10], we consider the $p$-dependent length $\bar{\mu} = \sqrt{\Delta/p}$, where $\Delta$ is the minimum non-zero eigenvalue of the area operator in LQG. These variables are represented in the space of square-integrable functions in the Bohr compactification of the real line, $\mathbb{R}_\mathrm{B}$, with respect to its Haar measure $d\mu_\mathrm{B}$. Alternatively, we can choose an orthonormal basis $\{|v\rangle\}_{v\in\mathbb{R}}$ such that

$$\hat{p}|v\rangle = \mathrm{sgn}(v)(2\pi\gamma G\hbar\sqrt{\Delta}|v|)^{2/3}|v\rangle, \quad \hat{N}_{\bar{\mu}}|v\rangle = |v+1\rangle. \tag{1}$$

The constants $\gamma$, $G$, and $\hbar$ are the Immirzi, Newton, and reduced Planck constants, respectively.

The homogeneous part of the scalar field, $\phi$, is quantized à la Schrödinger, whereas for its perturbation, $\delta\phi$, we adopt the preferred Fock representation described above. We denote the corresponding Fock space by $\mathcal{F}$. As anticipated, the total kinematical Hilbert space is just the product of the three pieces, $L^2(\mathbb{R}_{\rm B}, d\mu_{\rm B}) \otimes L^2(\mathbb{R}, d\phi) \otimes \mathcal{F}$. This is the space where the Hamiltonian constraint is to be represented. The structure of the constraint is $C = C_0 + \sum C_2^{(n)}$, where $n$ labels the modes of the Laplace-Beltrami operator. We identify the zeroth-order part of the constraint, $C_0$, with that of the unperturbed system. In the case of a flat universe, the prescription we adopt is

$$\hat{C}_0 = \left[\widehat{\frac{1}{\sqrt{|p|}}}\right]^{3/2} \left[-\frac{6}{\gamma^2}\hat{\Omega}^2 + 8\pi G\left(\hat{\pi}_\phi^2 + m^2|\hat{p}|^3\hat{\phi}^2\right)\right] \left[\widehat{\frac{1}{\sqrt{|p|}}}\right]^{3/2}, \tag{2}$$

where we have introduced the regularized operators

$$\left[\widehat{\frac{1}{\sqrt{|p|}}}\right] = \frac{3}{4\pi\gamma G\hbar\sqrt{\Delta}}\widehat{\mathrm{sign}(p)}\sqrt{|\hat{p}|}\left(\hat{N}_{-\bar{\mu}}\sqrt{|\hat{p}|}\hat{N}_{\bar{\mu}} - \hat{N}_{\bar{\mu}}\sqrt{|\hat{p}|}\hat{N}_{-\bar{\mu}}\right), \tag{3}$$

$$\hat{\Omega} = \frac{1}{4i\sqrt{\Delta}}\hat{V}^{1/2}\left[\widehat{\mathrm{sign}(v)}\left(\hat{N}_{2\bar{\mu}} - \hat{N}_{-2\bar{\mu}}\right) + \left(\hat{N}_{2\bar{\mu}} - \hat{N}_{-2\bar{\mu}}\right)\widehat{\mathrm{sign}(v)}\right]\hat{V}^{1/2}. \tag{4}$$

In the case of a closed universe, the constraint can be led to a similar form by means of a unitary transformation [7]. In both cases, the homogeneous constraint is a second-order difference operator and the states with support on the semilattices $\mathcal{L}_\varepsilon^\pm = \{v = \pm(\varepsilon + 4n) \,|\, \varepsilon \in (0, 4],\ n \in \mathbb{N}\}$ form superselection sectors.

The quadratic part of the constraint contains both homogeneous and inhomogeneous variables. Although at present we do not have at our disposal a systematic method to regularize it, we adopt a prescription that respects the superselection sectors of the unperturbed theory [6, 7].

Once the constraint is represented in the kinematical Hilbert space, one has to find the (generalized) states that it annihilates. At this point, the use of a homogeneous scalar field as a relational time is frequently invoked in LQC. The solutions of the Hamiltonian constraint satisfy an equation of the form

$$-\hbar^2\partial_\phi^2\Psi = \left(\hat{\mathcal{H}}_0^2 + {}^{(0)}\hat{\Theta}_2 - i\hbar\,{}^{(1)}\hat{\Theta}_2\partial_\phi\right)\Psi, \tag{5}$$

where $\hat{\mathcal{H}}_0$ is a zeroth-order operator (perturbatively), while ${}^{(0)}\hat{\Theta}_2$ and ${}^{(1)}\hat{\Theta}_2$ are operators quadratic in the perturbations and that contain no derivatives with respect to $\phi$. If $\phi$ can be used as a relational time, this is an evolution equation that characterizes the physical states.

We can adopt the ansatz

$$\Psi(\phi, v, \delta\phi) = \frac{\chi_0(\phi, v)\psi(\phi, \delta\phi)}{\langle\hat{\mathcal{H}}_0(\phi)\rangle_{\chi_0}}, \tag{6}$$

where $\chi_0$ is a positive-frequency background state (such that $-i\hbar\partial_\phi\chi_0 = \hat{\mathcal{H}}_0\chi_0$) and $\langle\bullet\rangle_{\chi_0}$ denotes the expectation value at $\chi_0$ with respect to the inner product in the unperturbed theory. Then, if we neglect the contributions that mix different background states and assume that the characteristic 'time' scale of the inhomogeneities is much smaller than that of the unperturbed background, we arrive at the first-order differential equation

$$-i\hbar\partial_\phi\psi = \frac{1}{2}\frac{\langle{}^{(0)}\hat{\Theta}_2 + {}^{(1)}\hat{\Theta}_2\hat{\mathcal{H}}_0\rangle_{\chi_0}}{\langle\hat{\mathcal{H}}_0\rangle_{\chi_0}}\psi. \tag{7}$$

This equation can be interpreted as corresponding to a quantum field theory in an effective curved spacetime, given by the background state.

Actually, there is in fact no need of adopting a relational time to characterize the physical states. Since the constraint is a difference operator that does not mix the semilattices $\mathcal{L}_\varepsilon^\pm$, its solutions can be characterized in terms of their initial data on the minimum volume section of each superselection sector, $|v| = \varepsilon$. Assuming a perturbative expansion for the physical states, $\Psi = \Psi_0 + \Psi_2 + \ldots$, and imposing the constraint, one can see that $\Psi_0$ must be a solution of the unperturbed constraint, whereas $\Psi_2$ must satisfy a similar equation with a source term arising from the action of the inhomogeneities on $\Psi_0$. If we endow the space of initial data with a suitable inner product, we obtain the physical Hilbert space $L^2(\mathbb{R}, d\phi) \otimes \mathcal{F}$.

## 4. Summary and outlook

We have sketched the complete quantization of an FLRW model with scalar perturbations. The gauge is fixed up to (global) time reparametrizations at the classical level, although a description in terms of gauge-invariants is available and leads to consistent results. The parametrization of the inhomogeneous sector is chosen so that unitary dynamics can be recovered in the limit of quantum field theory in a curved spacetime. In order to maintain canonical Dirac brackets, the homogeneous variables need suitable corrections that are quadratic in the perturbations.

The kinematical Hilbert space has been obtained as the product of the LQC one and a Fock space, corresponding to the homogeneous background and the matter-field perturbation, respectively. However, the constraint operator mixes both spaces in a non-trivial way. Nevertheless, we have proposed two different characterization of the physical states. The first one employs the homogeneous scalar field as a relational time, which allows us to interpret the constraint equation as an evolution equation. The other one determines the states in terms of their initial value on the minimum volume section of each superselection sector.

Of course, the goal is now to confront the predictions of this model with observations. A first approach to do this is provided by the effective dynamics of semiclassical states [1]. In the case of the hybrid quantization of the Gowdy models, the analysis of the effective dynamics showed very interesting features, namely the persistence of the bounce and the statistical amplification of the inhomogeneities through it [11]. These results hold qualitatively in the perturbed FLRW model, and will be the object of a forthcoming publication.

## Acknowledgments

This work was supported by the Project No. FIS2011-30145-C03-02 from Spain and by Pedeciba (Uruguay).

## References

[1] Ashtekar A and Singh P 2011 *Class. Quant. Grav.* **28** 213001.
[2] Giesel K and Sahlmann H 2011 *Proc. Sci.* QGQGS 002
[3] Martín-Benito M, Garay L J and Mena Marugán G A 2008 *Phys. Rev.* D **78** 083516
[4] Ade P A R *et al.* (Planck Collaboration) 2013 Planck 2013 results. XXII. Constraints on inflation *Preprint* astro-ph/1303.5082
[5] Halliwell J J and Hawking S W 1985 *Phys. Rev.* D **31** 1777
[6] Fernández-Méndez M, Mena Marugán G A and Olmedo J 2013 *Phys. Rev.* D **88** 044013
[7] Fernández-Méndez M, Mena Marugán G A and Olmedo J 2012 *Phys. Rev.* D **86** 024003
[8] Cortez J, Mena Marugán G A, Olmedo J, and Velhinho J M 2012 *Phys. Rev.* D **86** 104003
[9] Fernández-Méndez M, Mena Marugán G A, Olmedo J and Velhinho J M 2012 *Phys. Rev.* D **85** 103525
[10] Ashtekar A, Pawłowski T, and Singh P 2006 *Phys. Rev.* D **74** 084003
[11] Brizuela D, Mena Marugán G A, Pawłowski T 2011 *Phys. Rev.* D **84** 124017

# Experimental perspectives in (low-energy) photon-photon scattering

**Federico Della Valle[1], Aldo Ejlli[2], Ugo Gastaldi[3], Giuseppe Messineo[2], Edoardo Milotti[1], Ruggero Pengo[3], Livio Piemontese[2], Giuseppe Ruoso[3], Stefano Sabotig[1], and Guido Zavattini[2]**

[1] Dipartimento di Fisica dell'Università di Trieste and I.N.F.N. – Sezione di Trieste, Via Valerio, 2 – I-34127 Trieste, Italy

[2] Dipartimento di Fisica dell'Università di Ferrara and I.N.F.N. – Sezione di Ferrara, Via Saragat 1, Edificio C, I-44122 Ferrara, Italy

[3] I.N.F.N. – Lab. Naz. di Legnaro, viale dellUniversità 2, I-35020 Legnaro (Padova), Italy

E-mail: milotti@ts.infn.it

**Abstract.** The possibility of photon-photon scattering is a striking difference between classical and quantum electrodynamics. This genuinely quantum feature is made possible by the fluctuations of charged fields, and it makes quantum vacuum a nonlinear optical medium. Photon-photon scattering is thus a delicate probe into the structure of quantum electrodynamics and any departure from the expected behavior would be a powerful signal of "new physics". To date this process has never been observed – except as a radiative correction to other processes – and several experiments are trying to detect it at very low energy, in the scattering of real photons in powerful light beams off the virtual photons of intense magnetic fields. Here we briefly review the experimental state-of-the-art, with special emphasis on the PVLAS experiment, and we describe a new proposal to observe photon-photon scattering in the range 1 – 2 MeV.

## 1. Introduction

Classical electrodynamics is a linear theory, the superposition principle plays an important and ubiquitous role in all its applications, and it usually comes as a surprise when one first learns that this is not so in quantum electrodynamics, although the departure from linearity is exceedingly small. The nonlinearity is associated with the quantum fluctuations of charged fermion fields, and therefore with the very nature of quantum vacuum, and a detailed examination of its nature leads to a wealth of interesting and deep results. Figure 1 shows the lowest-order Feynman diagram for photon-photon interaction: there is no tree-level diagram, and only an even number of photon vertices is allowed around the loop, because of Furry's theorem. This was studied early in the development of quantum field theory, and its nonlinear nature was first captured in the low-energy Lagrangian density correction $\mathcal{L}_{\rm EH}$ introduced by Euler and Heisenberg [1] and by Weisskopf [2] (EHW) – written here in standard S.I. units:

$$\mathcal{L}_{\rm EH} = \frac{A_e}{\mu_0}\left[\left(\frac{E^2}{c^2} - B^2\right)^2 + 7\left(\frac{\mathbf{E}}{c}\cdot\mathbf{B}\right)^2\right], \tag{1}$$

**Figure 1.** The lowest order Feynman diagram for photon-photon scattering.

where the $A_e$ coefficient is defined by

$$A_e = \frac{2}{45\mu_0} \frac{\alpha^2 \lambda_e^3}{m_e c^2} = 1.32 \cdot 10^{-24}\ \mathrm{T}^{-2}. \qquad (2)$$

The two quadratic terms in parentheses in eq. (1) correspond to the invariants $\mathcal{F} = F_{\mu\nu}F^{\mu\nu}$ and to $\mathcal{G} = \epsilon_{\mu\nu\sigma\tau}F^{\mu\nu}F^{\sigma\tau}$, so that the EHW Lagrangian is a linear combination of the squares of the admissible next-to-lowest-order invariants. As far as the linear terms are concerned, the lowest order, classical Lagrangian is indeed proportional to $\mathcal{F}$, while $\mathcal{G}$ is forbidden by parity conservation. A careful study of the generic Lagrangian density where the coefficients of the linear combination are free to vary leads to a rich low-energy phenomenology [3, 4], and using the optical theorem we can connect the observed vacuum birefringence with the photon-photon scattering amplitude, and finally with the photon-photon cross section [5, 6].

## 2. The PVLAS experiment

Although the EHW Lagrangian has been known for a long time – the original calculations date back to 1936 – the effects produced by the nonlinear correction are exceedingly small with near-visible photons, and the first experimental proposal was put forward by Emilio Zavattini and collaborators only at the end of the 1970's [7]. The experiment aims at measuring the scattering of photons from a laser source off the virtual photons of an intense, static, uniform dipolar magnetic field. The direction of the magnetic field lies in a plane perpendicular to the laser beam direction, and it influences the laser beam polarization, much as a uniaxial birefringent optical medium would do. It can be shown [8] that with an external field $B_{ext}$ the difference between the refractive indexes for beam polarization parallel and perpendicular to the field is

$$\Delta n^{(EHW)} = n_{\|}^{(EHW)} - n_{\perp}^{(EHW)} = 3A_e B_{ext}^2, \qquad (3)$$

so that linearly-polarized light propagating in the field region acquires a small ellipticity, which depends on the $A_e$ coefficient, the result of the QED calculation. Even with strong magnetic fields, this difference is extremely small: a 2.5 T magnet (roughly the saturation field of iron) gives $\Delta n^{(EHW)} \approx 10^{-23}$. This means that even tiny phase changes $\Delta\phi$ require very long path lengths $L = \lambda\Delta\phi/2\pi\Delta n^{(EHW)}$. Long path lengths are commonly obtained by folding the light path: in PVLAS this is done by means of a high-finesse Fabry-Perot cavity. In addition, we must also remember that photodetectors are only sensitive to intensity, and therefore in a crossed-polarizer arrangement the transmitted intensity changes by a factor $1 - \cos\Delta\phi \approx \Delta\phi^2$, and apparently we have to revise our estimates to make the path much longer. However, this can be corrected by modulating the effect, so that at the modulation frequency it becomes proportional to the derivative of the cosine and therefore proportional to $\Delta\phi$. This is usually achieved by pulsing or rotating the magnetic field, either electrically, as in the LAS experiment

**Figure 2.** The new PVLAS experimental setup in Ferrara (Italy). The optical table sits in a clean room and has both passive and active noise damping. The magnetic field region is defined by two 1 m-long 2.5 T permanent magnets (dark gray cylinders on the left). The laser beam is produced by a 1 W Nd-YAG infrared laser (hidden by the support structure), it is linearly polarized at the far end of the table, and its ellipticity is modulated by a PEM. The outgoing beam is detected at the near end. The structure above the optical table is used to support the magnets and the motor that rotates them: these parts are mechanically isolated from the optics.

[9, 10], or mechanically, as in PVLAS [8]. The light intensity is detected by a photodiode and the amplitude of the spectral line at twice the modulation frequency carries the physical information [8]. Unfortunately, it is not possible to modulate the magnetic field faster than a few Hz: thus the spectral line at the modulation frequency is drowned in the low-frequency $1/f$ noise peak. This can be fixed, by further modulating the initial polarization at a frequency of a few kHz – for instance by means of a photo elastic modulator (PEM) as in PVLAS – and thus moving the physical signal to higher frequency, far from the $1/f$ noise peak due to mechanical vibrations and to electronic noise. The final sensitivity of the experiment is determined by the noise level close to the PEM modulation frequency, and the latest tests performed with the new PVLAS apparatus yield a noise figure for the measured ellipticity close to $3 \times 10^{-7}$ radians/$\sqrt{\text{Hz}}$. Further improvements shall be needed to bring this down to the level of $10^{-8}$ radians/$\sqrt{\text{Hz}}$, required to successfully carry out a measurement in about a month of data taking time. More details can be found in the original proposal [7], and in recent papers, such as [8]. At the moment the PVLAS collaboration is setting up a new improved apparatus (see figure 2). The first tests are encouraging, and the collaboration hopes to collect data by the end of 2013. PVLAS is one of several competing experiments: two of them (OSQAR [11] and Q&A [12]) use the same basic technique, BMV [13] utilizes a strong, pulsed magnetic field, to take advantage of the quadratic dependence of the effect on field intensity – see eq. (3) – while an experiment at ELI could utilize the very high beam intensity [14]. Differences and comparisons among experiments can be found in the ref. [8].

## 3. Higher-energy experiments

Even though the birefringence of vacuum has not yet been observed, it is already possible to estimate an upper limit to the photon-photon scattering cross section in the 1 eV region, and in PVLAS we find [8] $\sigma_{\gamma\gamma} < 1.2 \times 10^{-32}\mu$b with 1064 nm photons. The fast growth $\sigma_u \propto (\hbar\omega)^6$ at energies below $\sim 0.7$ MeV (see figure 3), brings the total cross section to nearly 2 $\mu$b at the

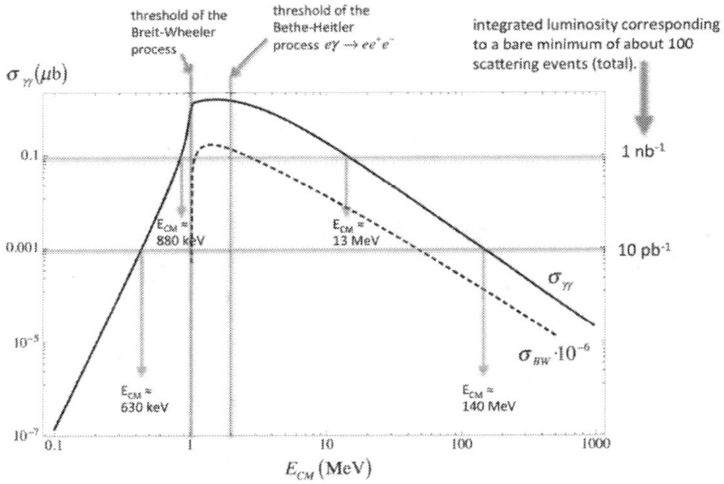


**Figure 3.** Unpolarized photon-photon cross section $\sigma_{\gamma\gamma}$ vs. CM energy, from the electronic contribution to the fermion loop of figure 1. The initial fast growth brings the cross section to about 2 $\mu$b close to the peak. $\sigma_{BW}$ is the cross section of the background Breit-Wheeler process.

peak, and it is possible to contemplate a different experimental setup with colliding gamma rays [4, 15]: a photon-photon collider facility with center of mass energy around 1-2 MeV could soon make this true, if the IRIDE project – now in the preliminary Technical Design Report phase – shall be a approved as wished and construction shall start in a new laboratory near Rome [16].

## References


[1] Heisenberg W and Euler H 1936 *Zeitschrift für Physik* **98** 714–732

[2] Weisskopf V 1936 *Dansk. Vidensk. Seisk., Mat.-Fys. Meddr 14 No* **6**

[3] Denisov V I, Krivchenkov I V and Kravtsov N V 2004 *Phys. Rev. D* **69** 066008

[4] Milotti E, Della Valle F, Zavattini G, Messineo G, Gastaldi U, Pengo R, Ruoso G, Babusci D, Curceanu C, Iliescu M *et al.* 2012 *International Journal of Quantum Information* **10**

[5] Haïssinski J, Dagoret S, Urban M and Zomer F 2006 *Physica Scripta* **74** 678

[6] Bregant M, Cantatore G, Carusotto S, Cimino R, Della Valle F, Di Domenico G, Gastaldi U, Karuza M, Lozza V, Milotti E *et al.* 2008 *Phys. Rev. D* **78** 032006

[7] Iacopini E and Zavattini E 1979 *Phys. Lett. B* **85** 151–154

[8] Della Valle F, Gastaldi U, Messineo G, Milotti E, Pengo R, Piemontese L, Ruoso G and Zavattini G 2013 *New Journal of Physics* **15** 053026

[9] Semertzidis Y, Cameron R, Cantatore G, Melissinos A, Rogers J, Halama H, Prodell A, Nezrick F, Rizzo C and Zavattini E 1990 *Phys. Rev. Lett.* **64** 2988–2991

[10] Cameron R, Cantatore G, Melissinos A, Ruoso G, Semertzidis Y, Halama H, Lazarus D, Prodell A, Nezrick F, Rizzo C *et al.* 1993 *Phys. Rev. D* **47** 3707

[11] Pugnat P, Ballou R, Schott M, Husek T, Sulc M, Deferne G, Duvillaret L, Finger Jr M, Finger M, Flekova L *et al.* 2013 Results and Perspectives for Laboratory Search of Weakly Interacting Sub-eV Particles with the OSQAR Experiment (Preprint hep-ex:1306.0443)

[12] Chen S J, Mei H H and Ni W T 2007 *Modern Physics Letters A* **22** 2815–2831

[13] Cadène A, Berceau P, Fouché M, Battesti R and Rizzo C 2013 Vacuum magnetic linear birefringence using pulsed fields: the BMV experiment (Preprint physics.optics:1302.5389)

[14] Bernard D, Moulin F, Amiranoff F, Braun A, Chambaret J, Darpentigny G, Grillon G, Ranc S and Perrone F 2000 *The European Physical Journal D-Atomic, Molecular, Optical and Plasma Physics* **10** 141–145

[15] Becker W, McIver J and Schlicher R 1988 *Phys. Rev. A* **38** 4891–4894

[16] Alesini D *et al.* 2013 IRIDE White Book, An Interdisciplinary Research Infrastructure based on Dual Electron linacs&lasers (Preprint physics.ins-det:1307.7967)

# Breakdown of the equivalence between gravitational mass and energy for a composite quantum body

**Andrei G. Lebed**

Department of Physics, University of Arizona, 1118 E. 4-th Street, Tucson, AZ 85721, USA and L.D. Landau Institute for Theoretical Physics, RAS, 2 Kosygina Street, Moscow, Russia

E-mail: lebed@physics.arizona.edu

**Abstract.** The simplest quantum composite body, a hydrogen atom, is considered in the presence of a weak external gravitational field. We define an operator for the passive gravitational mass of the atom in the post-Newtonian approximation of the general relativity and show that it does not commute with its energy operator. Nevertheless, the equivalence between the expectation values of the mass and energy is shown to survive at a macroscopic level for stationary quantum states. Breakdown of the equivalence between passive gravitational mass and energy at a microscopic level for stationary quantum states can be experimentally detected by studying unusual electromagnetic radiation, emitted by the atoms, supported by and moving in the Earth's gravitational field with constant velocity, using spacecraft or satellite

## 1. Introduction

The notion of gravitational mass for a composite body is known to be non-trivial in general relativity and related to several paradoxes. The role of the classical virial theorem in establishing of the equivalence between averaged over time gravitational mass and energy is discussed in detail in Refs.[1,2] for different types of classical composite bodies. In particular, for electrostatically bound two bodies, it is shown that the gravitational field is coupled to a combination $3K + 2U$, where $K$ is the kinetic energy and $U$ the potential energy. Since the classical virial theorem states that the following time average is equal to zero, $\langle 2K + U \rangle_t = 0$, we can conclude that averaged over time gravitational mass is proportional to the total amount of energy, $E$ [1,2]:

$$\langle m^g \rangle_t = \langle 3K + 2U \rangle_t / c^2 = \langle K + U \rangle_t / c^2 = E/c^2. \quad (1)$$

## 2. Goal

The main goal of our paper is to study the quantum problem concerning the passive gravitational mass of a composite body. As the simplest example, we consider a hydrogen atom in the Earth's gravitational field, where we take into account only kinetic and Coulomb potential energies of an electron in a curved spacetime. We claim three main results in the paper (see also Refs. [3,4]). Our first result is that the equivalence between passive gravitational mass and energy in the absence of gravitational field survives, at the macroscopic level, in the quantum case. More strictly speaking, we show that the expectation value of the mass is equal to $E/c^2$ for stationary quantum states due to the quantum virial theorem [5]. Our second result is a breakdown of the equivalence between passive gravitational mass and energy, at the microscopic level, for stationary quantum states due to the fact that passive gravitational mass operator does not commute with energy operator. As a result, there exist a non-zero probability that a measurement of passive gravitational mass can yield a value, which is different from $E/c^2$, given by the Einstein's equation. Our third result is a suggestion of a realistic experiment to detect this inequivalence by measurement of electromagnetic radiation, emitted by a macroscopic ensemble of hydrogen atoms, supported by and moving in the Earth's gravitational field, using spacecraft or satellite.

## 3. Gravitational mass in classical physics

Here, we derive the Lagrangian and Hamiltonian of a hydrogen atom in the Earth's gravitational field, taking into account couplings of kinetic and potential Coulomb energies of an electron with a weak gravitational field. Note that we keep only terms of the order of $1/c^2$ and disregard all tidal effects. Therefore, we can write the interval, using a weak field approximation [6]:

$$ds^2 = -(1 + 2\phi/c^2)(cdt)^2 + (1 - 2\phi/c^2)(dx^2 + dy^2 + dz^2), \ \phi = -GM/R, \tag{2}$$

where $G$ is the gravitational constant, $c$ is the velocity of light, $M$ is the Earth mass, $R$ is the distance between center of the Earth and center of mass of a hydrogen atom (i.e., a proton). Then, in the local proper spacetime coordinates, which in the first approximation in $\phi/c^2$ are

$$x' = (1 - \phi/c^2)x, \ y' = (1 - \phi/c^2)y, z' = (1 - \phi/c^2)z, \ t' = (1 + \phi/c^2)t, \tag{3}$$

the classical Lagrangian and action of an electron have the following standard forms:

$$L' = -m_e c^2 + \frac{1}{2} m_e (\mathbf{v}')^2 + \frac{e^2}{r'} , \quad S' = \int L' dt', \tag{4}$$

where $m_e$ is the bare electron mass, $e$ and $\mathbf{v}'$ are the electron charge and velocity, respectively; $r'$ is a distance between electron and proton. It is possible to show that the Lagrangian (4) can be rewritten in coordinates $(x, y, z, t)$ as

$$L = -m_e c^2 + \frac{1}{2} m_e \mathbf{v}^2 + \frac{e^2}{r} - m_e \phi - \left(3m_e \frac{\mathbf{v}^2}{2} - 2\frac{e^2}{r}\right) \frac{\phi}{c^2}, \tag{5}$$

which corresponds to the following Hamiltonian:

$$H = m_e c^2 + \frac{\mathbf{p}^2}{2m_e} - \frac{e^2}{r} + m_e \phi + \left(3\frac{\mathbf{p}^2}{2m_e} - 2\frac{e^2}{r}\right) \frac{\phi}{c^2}. \tag{6}$$

## 4. Gravitational mass in quantum physics

The Hamiltonian (6) can be quantized by substituting a momentum operator, $\hat{\mathbf{p}} = -i\hbar\partial/\partial\mathbf{r}$, instead of canonical momentum, $\mathbf{p}$:

$$\hat{H} = m_e c^2 + \frac{\hat{\mathbf{p}}^2}{2m_e} - \frac{e^2}{r} + \hat{m}_e^g \phi , \tag{7}$$

where we introduce passive gravitational mass operator of an electron to be proportional to its weight operator in a weak gravitational field (2),

$$\hat{m}_e^g = m_e + \left(\frac{\hat{\mathbf{p}}^2}{2m_e} - \frac{e^2}{r}\right) \frac{1}{c^2} + \left(2\frac{\hat{\mathbf{p}}^2}{2m_e} - \frac{e^2}{r}\right) \frac{1}{c^2} . \tag{8}$$

Note that the Hamiltonian (7),(8) contains the virial contribution (i.e., the last term) to the mass operator and, therefore, does not commute with electron energy operator, taken in the absence of the field. Below, we discuss some consequences of Eqs.(7),(8). Suppose that we have a macroscopic ensemble of hydrogen atoms with each of them being in a ground state with energy $E_1$. Then the expectation value of the electron mass operator (8) per atom is

$$< \hat{m}_e^g >= m_e + \frac{E_1}{c^2} + \left\langle 2\frac{\hat{\mathbf{p}}^2}{2m_e} - \frac{e^2}{r} \right\rangle \frac{1}{c^2} = m_e + \frac{E_1}{c^2}, \tag{9}$$

where the third term in Eq.(9) is zero in accordance with the quantum virial theorem [5]. Therefore, we conclude that the equivalence between passive gravitational mass and energy survives at a macroscopic level for stationary quantum states. Let us discuss how Eqs.(7),(8) break the equivalence between passive gravitational mass and energy at a microscopic level. Here, we illustrate the above mentioned inequivalence, using the following thought experiment. Suppose that at $t = 0$ we create a ground state wave function of a hydrogen atom, corresponding to the absence of gravitational field,

$$\Psi_1(r,t) = \Psi_1(r) \exp(-iE_1 t/\hbar) \ . \tag{10}$$

In gravitational field (2), wave function (10) is not anymore a ground state. It is possible to show that a general solution of the Schrödinger equation, corresponding to the Hamiltonian (7),(8), can be written as

$$\Psi(r,t) = (1-\phi/c^2)^{3/2} \sum_{n=1}^{\infty} a_n \Psi_n[(1-\phi/c^2)r] \exp[-im_e c^2(1+\phi/c^2)t/\hbar] \exp[-iE_n(1+\phi/c^2)t/\hbar]. \tag{11}$$

We wish to draw attention to the fact that wave function (11) is a series of eigenfunctions of passive gravitational mass operator (8). $\Psi_n(r)$ is a normalized wave function of an electron in a hydrogen atom in the absence of gravitational field, corresponding to energy $E_n$.

In accordance with quantum mechanics, probability that, at $t > 0$, an electron occupies excited state with energy $m_e c^2(1+\phi/c^2) + E_n(1+\phi/c^2)$ is

$$P_n = |a_n|^2, \ a_n = \int \Psi_1^*(r)\Psi_n[(1-\phi/c^2)r]d^3\mathbf{r} = -(\phi/c^2)\int \Psi_1^*(r) r \Psi_n'(r) d^3\mathbf{r},$$
$$\int \Psi_1^*(r) r \Psi_n'(r) d^3\mathbf{r} = V_{n,1}/(\hbar\omega_{n,1}), \ \hbar\omega_{n,1} = E_n - E_1, \ n \neq 1,$$
$$V_{n,1} = \int \Psi_1^*(r)\hat{V}(\mathbf{r})\Psi_n(r)d^3\mathbf{r}, \ \ \hat{V}(\mathbf{r}) = 2\frac{\hat{\mathbf{p}}^2}{2m_e} - \frac{e^2}{r}. \tag{12}$$

Let us discuss Eq.(12). We stress that it directly demonstrates that there is a finite probability,

$$P_n = |a_n|^2 = (\phi/c^2)^2 \ [V_{n,1}/(E_n - E_1)]^2 \ , \ n \neq 1, \tag{13}$$

that, at $t > 0$, an electron occupies n-th ($n \neq 1$) energy level, which breaks the expected Einstein's equation, $m_e^g = m_e + E_1/c^2$. In fact, this means that quantum measurement of passive gravitational mass in a quantum state with a definite energy (10) gives the following quantized values:

$$m_e^g(n) = m_e + E_n/c^2 \ . \tag{14}$$

## 5. Suggested experiment

Here, we describe a realistic experiment: We consider a hydrogen atom being in its ground state at $t = 0$, located at distance $R'$ from the center of the Earth and having the wave function:

$$\tilde{\Psi}_1(r,t) = (1-\phi'/c^2)^{3/2}\Psi_1[(1-\phi'/c^2)r]\exp[-im_e c^2(1+\phi'/c^2)t/\hbar]\exp[-iE_1(1+\phi'/c^2)t/\hbar]. \quad (15)$$

The atom is supported by in the Earth's gravitational field and moving from the Earth at constant velocity, $u \ll \alpha c$ (where $\alpha$ is the fine structure constant), by spacecraft or satellite. It is possible to show that the electron wave function and the time dependent perturbation for the Hamiltonian (7)-(8), in the coordinate system related to center of mass of the atom, can be expressed as

$$\tilde{\Psi}(r,t) = (1-\phi'/c^2)^{3/2}\sum_{n=1}^{\infty}\tilde{a}_n(t)\Psi_n[(1-\phi'/c^2)r]\exp[-im_e c^2(1+\phi'/c^2)t/\hbar] \\ \exp[-iE_n(1+\phi'/c^2)t/\hbar], \quad (16)$$

$$\hat{U}(\mathbf{r},t) = \frac{\phi(R'+ut)-\phi(R')}{c^2}\left(3\frac{\hat{\mathbf{p}}^2}{2m_e} - 2\frac{e^2}{r}\right). \quad (17)$$

We wish to draw attention that in a spacecraft (satellite), which moves with constant velocity, gravitational force, which acts on each individual hydrogen atom, is compensated by some non-gravitational forces. This causes very small changes of a hydrogen atom energy levels and is not important for our calculations. Therefore, the atoms do not feel directly gravitational acceleration, $\mathbf{g}$, but feel, instead, gravitational potential, $\phi(R") = \phi(R'+ut)$.

It is important that, if excited levels of a hydrogen atom were strictly stationary, then probability to find the passive gravitational mass to be quantized with $n \neq 1$ (14) would be

$$\tilde{P}_n = (V_{n,1}/\hbar\omega_{n,1})^2[\phi(R')/c^2]^2 \;, n \neq 1,\; |\phi(R'+ut)| \ll |\phi(R')|, u \ll \omega_{n,1}R, \quad (18)$$

which coincides with Eq. (13). In reality, the excited levels spontaneously decay with time and, therefore, it is possible to observe the quantization law (14) indirectly by measuring electromagnetic radiation from a macroscopic ensemble of the atoms. In this case, Eq.(18) gives a probability that a hydrogen atom emits a photon with frequency $\omega_{n,1} = (E_n - E_1)/\hbar$ during the time of the experiment. To estimate the probabilities (18) we use the following numerical values of the Earth's mass, $M \simeq 6 \times 10^{24} kg$, and its radius, $R_0 \simeq 6.36 \times 10^6 m$. It is important that, although the probabilities (18) are small, the number of photons, $N$, emitted by macroscopic ensemble of the atoms, can be large. For instance, for 1000 moles of hydrogen atoms, $N$ is estimated as

$$N(n \to 1) = 2.95 \times 10^8\,[V_{n,1}/(E_n - E_1)]^2, \quad N(2 \to 1) = 0.9 \times 10^8, \quad (19)$$

which can be experimentally detected. [Here, $N(n \to 1)$ stands for a number of photons, emitted with frequency $\omega_{n,1} = (E_n - E_1)/\hbar$.]

## Acknowledgments

We are thankful to N.N. Bagmet (Lebed), V.A. Belinski, Steven Carlip, Li-Zhi Fang, Douglas Singleton, and V.E. Zakharov for useful discussions. This work was supported by the NSF under Grant DMR-1104512.

## References

[1] Nordtvedt K 1994 *Class. Quantum Grav.* **11**, A119
[2] Carlip S 1998 *Am. J. Phys.* **66**, 409
[3] Lebed A G 2011 *Preprint* arXiv:1111.5365v1
[4] Lebed A G 2012 *Preprint* arXiv:1205.3134v1
[5] Park D 2005 *Introduction to the Quantum Theory, 3-rd Edition* (Dover Publications, New York)
[6] Misner C W, Thorne K S and Wheeler J A 1973 *Gravitation* (W.H. Freeman and Co, New York)

# Spinning black hole in the puncture method: Numerical experiments

**Tim Dietrich, Bernd Brügmann**
Friedrich-Schiller-Universität Jena, 07743 Jena, Germany
E-mail: tim.dietrich@uni-jena.de, bernd.bruegmann@uni-jena.de

**Abstract.** The strong-field region inside a black hole needs special attention during numerical simulation. One approach for handling the problem is the moving puncture method, which has become an important tool in numerical relativity since it allows long term simulations of binary black holes. An essential component of this method is the choice of the '1+log'-slicing condition. We present an investigation of this slicing condition in rotating black hole spacetimes. We discuss how the results of the stationary Schwarzschild '1+log'-trumpet change when spin is added. This modification enables a simple and cheap algorithm for determining the spin of a non-moving black hole for this particular slicing condition. Applicability of the algorithm is verified in simulations of single black hole, binary neutron star and mixed binary simulations.

## 1. Introduction

The moving puncture method is a successful approach for the simulation of black holes, see [1, 2] for pedagogic explanations and references. Numerically the method can be applied to multiple, spinning black holes in quasi-circular or eccentric orbits, but analytical results are only known for the Schwarzschild spacetime. Without spin and linear momentum a single black hole settles down to a stationary state similar to a 'trumpet', containing one asymptotically flat end and one with finite Schwarzschild radius, e.g. [3, 4, 5, 6]. In this article we begin an investigation of punctures in axisymmetry by performing numerical experiments of rotating black holes. We will show that the spinning puncture settles down to a stationary state similar to the spherical symmetric case and generalize results obtained for the Schwarzschild spacetime. A short description about possible modification of the ansatzes of [3] will be given, which offers the possibility to read off the spin value directly without additional computational costs.

## 2. Spinning Puncture

### 2.1. Numerical setup

We use the BAM-code [7] for fully general relativistic evolutions with the moving puncture method [8, 9]. Unless otherwise stated the total mass of the system $M$ is set to unity and the spin axis is equivalent to the z-axis. We use the BSSN [10, 11] evolution scheme for the numerical evolution. As a particular gauge choice, we apply the commonly used 1+log-slicing condition [12] together with the $\tilde{\Gamma}$-driver shift [13]. This special choice, sometimes called moving puncture coordinates, is well established in numerical relativity and widely used.

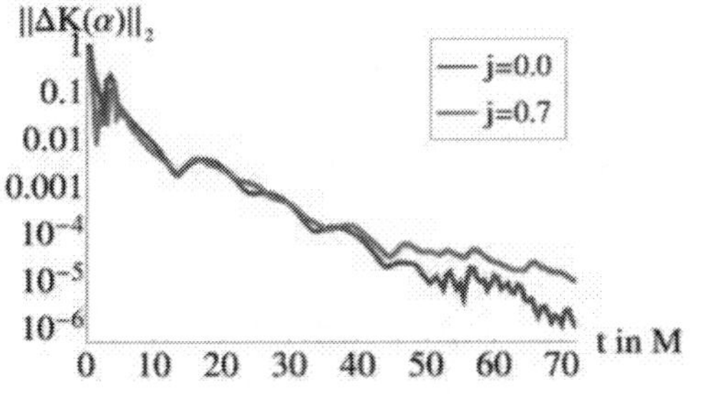


**Figure 1.** $||\Delta K(\alpha)||_2$ for $j = 0.0$ and $j = 0.7$ with $t_1 - t_2 = 0.5M$.

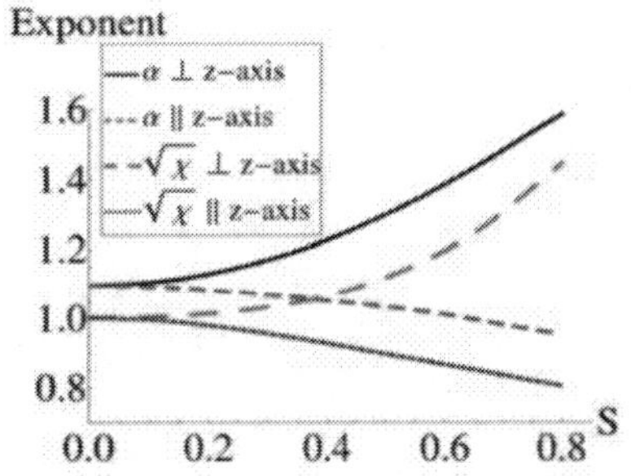


**Figure 2.** Exponents for the lapse and the conformal factor along different axes.

*2.2. Stationarity*

As a starting point, we want to demonstrate that for spinning punctures a stationary state is reached similar to the one in spherical symmetry. For this purpose, we consider the trace of the extrinsic curvature $K$ as a function of the lapse $\alpha$, because $K(\alpha)$ is independent of the spatial coordinates as long as we use the same slicing condition for all simulations. When a stationary state is reached, the difference of $K(\alpha)$ between two hypersurfaces, i.e. $||\Delta K(\alpha)||_2 = \sqrt{\int |K(\alpha)_{t_1} - K(\alpha)_{t_2}|^2}$, becomes zero for $t \to \infty$. In Fig. 1 we compare directly the spinning and the non-spinning configuration. We know that the non-spinning puncture reaches a stationary state and the same seems to be true for the spinning black hole. Additionally, one can show that the time derivatives of the evolution variables become zero. We checked this for the lapse $\alpha$ by computing the right hand side of the 1+log-slicing condition, e.g. for a black hole with $j = 0.4$ the derivative after $100M$ is smaller than $5 \cdot 10^{-5}$.

*2.3. Influences on the BSSN-evolution variables*

The stationary state in spherical symmetry can be computed semi-analytically. The lapse behaves near the puncture like $\alpha \sim r^{1.0916}$, see [6]. We find an exponent of $\gamma = 1.091$ when we use our numerical data and apply the fitting function $\alpha = \alpha_0 + \alpha_1 r^\gamma(1 + \alpha_2 r + \alpha_3 r^2)$. We can do the same for the conformal factor $\psi$ and get for $\sqrt{\chi} = \psi^{-2}$ with $\sqrt{\chi} = c_0 + c_1 r^\gamma(1 + c_2 r + c_3 r^2)$ an exponent of $\gamma = 1.002$, while $\gamma = 1.000$ is the analytical result. One should mention that we are using the radial coordinate of our numerical grid $r$, which does not necessarily agree with the isotropic radius for which the exponents were derived in [6], but because of the particular gauge choices both descriptions become approximately the same near the puncture. Thus, we will use the coordinate radius to get a better insight in the numerical simulation, not necessarily in the geometry of the spacetime. According to Fig. 2 the exponents of $\alpha$ and $\sqrt{\chi}$ change differently, depending on the axis between the spin and the axis used for the fit.

Remarkable is that even when the exponent is decreasing the puncture method works and can handle the non-differentiable functions. The important point is that although the regularity of the functions is reduced, compared with spherical symmetry, the moving puncture method does not fail. This effect could explain why stationary spinning punctures are less accurate than non-spinning punctures in numerical simulations [14]. Additionally, the angular dependence of the exponent $\gamma$ complicates possible attempts to factor out the singularity analytically.

Independent of the non-integer powers of $r$ for $\alpha$ and $\sqrt{\chi}$, we can use a Taylor expansion to model the behavior near the black hole, see [3] for the description in spherical symmetry. Considering axisymmetry we have to introduce an additional angular dependence, where we

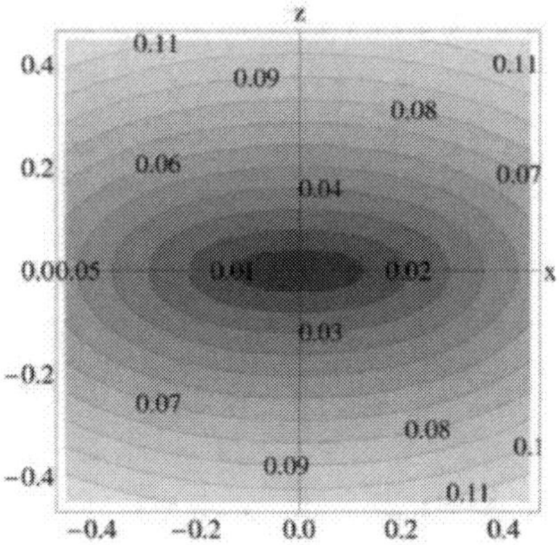


**Figure 3.** Level surface of $\alpha$ for a black hole with $j = 0.8$. Although the contour lines might suggest that the puncture is stretched to a one dimensional line instead of a single point numerical experiments showed no indication that the puncture has an extension $> 10^{-3}M$. Thus, also a spinning puncture can be treated as a point.

define $\varphi$ as the angle between the axis of rotation $S^i$ and $n^i$ as the outward-pointing unit radial vector. This dependence is described reasonably well with spherical harmonics, e.g. for the trace of the extrinsic curvature $K = \sum_{i=0}^{\infty} K_i r^i$, where we choose $K_i = \sum_{l=0}^{\infty} K_{il} Y_{2l,0}(\varphi, 0)$ to encode axisymmetry in the components $K_i$. The same holds for the other BSSN scalars. The shift vector $\beta^i$ in axial symmetry can be described by $\beta^i = (b_1 r + b_2 r^2) n^i + (c_1 r + c_2 r^2)\bar{\epsilon}^{ijk} n_j S_k$, where we introduced the additional term $(c_1 r + c_2 r^2)\bar{\epsilon}^{ijk} n_j S_k$. A closed description of the extrinsic curvature $\tilde{A}_{ij}$ and the conformal metric $\tilde{g}_{ij}$ is more complicated and is explained up to some point in [15].

## 3. Spin-determination without integration

### 3.1. *The new algorithm*

An important issue during the numerical evolution of black holes is obtaining a good estimate of the black hole spin. Using the properties of the stationary state described in sec. 2.2, it is simple to find the spin of a non-moving black hole evolved with the puncture gauge. For increasing spin the absolute value of $K$ at the puncture, $K_0$, becomes smaller. The behavior is well approximated by a quadratic function. Inverting this formula leads to

$$S = \sqrt{1.41789 - 4.71218 K_0}, \tag{1}$$

which correlates an evolution variable of the BSSN scheme and properties of the underlying spacetime. Thus, we are able to read off the amplitude of the spin of a non-moving puncture simply by measuring $K_0$. To get the entire information of the black hole spin we need the spin axis. This can be found with the lapse $\alpha$. Our experiments showed that the coordinate distance between the origin and a contour line is smallest when we look along the rotation axis, see Fig. 3. The direction of rotation can be determined with the shift vector $\beta^i$, because of the additional term $(c_1 r + c_2 r^2)\bar{\epsilon}^{ijk} n_j S_k$. Empirical studies showed $c_1 > 0$, so we can find the sign of $S_k$ by regarding $\beta^i$ along one axis.

### 3.2. *Numerical tests and measurement uncertainties*

The first test for the validity of our new algorithm are single black hole simulations. Table 1 shows some test cases in which the spin was calculated using BAM's curvature flow apparent horizon finder and the new method. For further tests we include results for the final remnant of a mixed binary (BHNS) and a neutron star binary (NSNS) merger. Comparing with the apparent horizon finder, the results agree within the estimated uncertainty.

To define the measurement uncertainty we investigate every part of our algorithm individually. The first step is to determine the absolute spin value with the help of the interpolation function (1). The error comes from the estimate of $K_0$ and can be evaluated by using different resolutions. Using the corresponding uncertainty, we can compute the error according to equation (1), including also the standard error of the regression. The second step of our algorithm defines the angular uncertainty. Considering a box with $N^3$ grid points the angular error is approximately $2/N$, in our case $\lesssim 0.02rad$. On the other hand using an interpolation to find the level surface one achieves a much lower uncertainty of around $\sim 10^{-3}rad$. Because the last step of our algorithm uses only the qualitative behavior of $\beta^i$, it comes without an additional error. We can sum up that our method determines the spin with an uncertainty of $\Delta S \lesssim 0.02$ and an angular resolution of at least $0.02rad$ for $j \gtrsim 0.2$. For smaller spins the change of $S(K)$ in equation (1) is small, which leads to higher uncertainties. Furthermore, equation (1) was defined as an adequate function for $j \in [0, 0.7]$, nevertheless using higher resolution for higher spins the method is applicable as well.

**Table 1.** Single black hole simulations (left) and results for the final remnant for a binary merger (right). The first line refers to the new algorithm, the second line to the apparent horizon finder.

| BH-mass | $S_x$ | $S_y$ | $S_z$ | j | system | $M_{ADM}^{init}$ | $K_0$ | $j$ |
|---|---|---|---|---|---|---|---|---|
| 1 | 0.10 | 0.66 | 0.21 | 0.70±0.01 | BHNS | $8.328M_\odot$ | $\sim 0.0315M_\odot$ | $0.42 \pm 0.05$ |
| | 0.11 | 0.66 | 0.20 | 0.70 | | | | 0.43 |
| 1 | 0.09 | -0.50 | 0.21 | 0.55±0.01 | NSNS | $2.998M_\odot$ | $\sim 0.059M_\odot$ | $0.76 \pm 0.04$ |
| | 0.10 | -0.50 | 0.20 | 0.55 | | | | 0.76 |
| 1.92 | 0.04 | -0.32 | 0.15 | 0.10±0.07 | NSNS | $3.439M_\odot$ | $\sim 0.057M_\odot$ | $0.71 \pm 0.05$ |
| | 0.10 | -0.50 | 0.20 | 0.15 | | | | 0.70 |

## 4. Conclusion

We presented numerical results for rotating black holes and generalized the ansatzes of [3]. For spinning punctures a simple $r^{\gamma(\varphi)}$-model for the evolution variables is possible, where in contrast to spherical symmetry the exponent depends on the angle $\varphi$. This complicates attempts to factor out the singularity analytically. In the future we plan to use these results to find an analytical description for a rotating black hole using the moving puncture gauge.

Additionally, a fast algorithm was introduced to determine the spin of a non-moving puncture. Adding linear momentum to a spinning black hole to obtain a binary orbit breaks the axisymmetry, but a generalization of the numerical method appears possible.

## References

[1] Alcubierre M 2008 *Introduction to 3+1 Numerical Relativity* (Oxford University Press)
[2] Baumgarte T and Shapiro S 2010 *Numerical Relativity: Solving Einsteins Equations on the Computer* (Cambridge University Press)
[3] Hannam M, Husa S, Pollney D, Brügmann B and O Murchadha N 2007, *Phys. Rev. Lett.* **99** 241102
[4] Brown J 2008 *Phys. Rev. D* **77** 044018
[5] Hannam M, Husa S, Ohme F and O Murchadha N 2008 *Phys. Rev. D* **78** 064020
[6] Brügmann B 2009 *Gen. Rel. Grav.* **41** p.2131-51
[7] Brügmann B, Gonzales J, Hannam M, Husa S, Sperhake U and Tichy W 2008 *Phys. Rev. D* **77** 024027
[8] Campanelli M, Lousto C O, Marronetti P and Zlochower Y 2006 *Phys. Rev. Lett.* **96** 111101
[9] Baker J G, Centrella J, Choi D I, Koppitz M and van Meter J 2006 *Phys. Rev. D* **73** 104002
[10] Shibata M and Nakamura T 1995 *Phys. Rev. D* **52** p. 5428-44
[11] Baumgarte T and Shapiro S 1998 *Phys. Rev. D* **59** 024007
[12] Bona C, Seidel E and Stela J 1995, *Phys. Rev. Lett.* **75** p. 600-603
[13] Alcubierre M, Brügmann B, Diener P, Koppitz M, Pollney D, Seidel E and Takahashi R 2003, *Phys. Rev. D* **67** 084023
[14] Lousto C and Zlochower Y 2009 *Phys. Rev. D* **79** 064018
[15] Dietrich T 2012 *Black Hole Data in Axial symmetry*, Master Thesis at University of Jena

# The Post-Newtonian Hill Problem

**A. F. Steklain Lisbôa and P. S. Letelier**
Departamento de Matematica, Universidade Tecnologica Federal do Parana, 80230-901, Curitiba, Parana, Brazil
Instituto de Fısica Gleb Wataghin, Universidade Estadual de Campinas, 13083-970, Campinas, São Paulo, Brazil
Departamento de Matematica Aplicada, Instituto de Matematica, Estatıstica e Computação Cientıfica, Universidade Estadual de Campinas, 13083-970, Campinas, São Paulo, Brazil
E-mail: steklain@utfpr.edu.br

**Abstract.** A relativistic approach of the Hill Problem is presented by using an approximate binary system metric obtained from the first post-Newtonian expansion (1PN). We employ Poincare maps and Lyapunov exponents to study the stability of bounded orbits of the system for different mass arrangements, and compare with the classical problem based on Newtonian dynamics. We find that for larger masses the system become totally stable, a striking behaviour that is not predicted by the Newtonian dynamics.

## 1. Introduction
The Hill Problem was formulated in the nineteenth century by Hill [1, 2], in order to study the Moon-Earth-Sun system. The Hill Problem is still applied in solar system models where bodies in nearly circular orbits are perturbed by other far away massive bodies, and it is very useful in the study of stellar dynamics. In many systems the Hill Problem can be taken as a first approximation and can easily accommodate necessary modifications (see for instance Heggie [3]). The interaction of a Keplerian binary system with a normally incident circularly polarized gravitational wave can be represented by a Hill system, as shown by Chicone *et al.* [4]. The classical Hill Problem was proved to be non-integrable by Meletlidou et al. [5], and it is chaotic, as shown by Simó and Stuchi [6].

## 2. The Post-Newtonian Hill Problem
Before we study the post-Newtonian Hill problem we must situate the restricted three-body problem in the context of general relativity. We consider the spacetime under influence of the binary system composed by BH1 and BH2. The motion of the third, small mass body, is given by a timelike geodesic of this spacetime.

The exact metric for a binary system is not known. Due to the lack of symmetries of a such mass distribution the search for an exact solution in this case it does look unworkable. There are several approximate metrics that can represent different regions of the spacetime with limited accuracy. The approximate metric proposed by Alvi [7] is a composition of several different metrics, each one valid for a different region of the spacetime. The delimiters $r_1^{in}$, $r_2^{in}$ and $r^{out}$ are given by $r_1^{in} = \sqrt{Gm_1R/c^2}$, $r_2^{in} = \sqrt{Gm_2R/c^2}$ and $r^{out} = \lambda_c/2\pi$, where $R$ is the separation distance between BH1 and BH2 and $\lambda_c$ is the characteristic wavelength of gravitational radiation emitted by the binary system.

In this work we consider the motion of the small mass body only in the Region III (the zone between BH1 and BH2). We could have considered as well the Region II (close to BH2 but outside the apparent horizon). The choice of Region III is justifiable as $r_2^{in}$ is very small for typical systems, and the orbits we are considering are located in the Buffer Zone between the two regions. The analysis of the other regions shall be considered in future works. In Region II we have $r_1 > r_1^{in}$, $r_2 > r_2^{in}$ and $r < r^{out}$, where $r_1 = \sqrt{(x-x_1)^2+(y-y_1)^2+(z-z_1)^2}$, $r_2 = \sqrt{(x-x_2)^2+(y-y_2)^2+(z-z_2)^2}$ and $r = \sqrt{x^2+y^2+z^2}$ are the distances of the small mass body from BH1, BH2 and the center of mass, respectively. This region can be described using post-Newtonian expansions [8]. We carry out the post-Newtonian expansion of the binary system only up to first order (1PN), so we consider the motion of the two massive bodies BH1 and BH2 in Newtonian orbits (0PN). According to the statements of the Hill Problem, we make the additional assumption that these bodies are in circular orbits around their center of mass with angular velocity $\omega$, given by $\omega = \sqrt{G(m_1+m_2)/R^3}$.

We use this metric to obtain the equations of motion from the geodesic equations. We use units such that $R = 1, \omega = 1$, and we change the origin and the scale of the problem with the transformation $x \to \mu^{1/3}x + 1$, $y \to \mu^{1/3}y$, where $\mu = m_2/(m_1+m_2)$ is the mass ratio. The 1PN Hill Problem is then obtained by considering $m_1 \gg m_2$ such that $\mu \approx m_2/m_1$ is small and keeping terms up to the $\mu^{2/3}$ order of approximation. We have,

$$\begin{aligned}
\frac{d^2x}{dt^2} &= 2\frac{dy}{dt} + \left(3 - \frac{1}{r^3}\right)x + \frac{1}{c^{*2}}\left[-\frac{5}{\mu^{1/3}} + \left(-12 + \frac{3}{r}\right)x + 2\frac{dy}{dt}\right], \\
\frac{d^2y}{dt^2} &= -2\frac{dx}{dt} - \frac{y}{r^3} + \frac{1}{c^{*2}}\left[\left(5 + \frac{3}{r}\right)y - 2\frac{dy}{dt}\right],
\end{aligned}$$

where $c^*$ is given by $c/\mu^{1/3}$, in the new units of the problem. These are the Hill equations, in the first post-Newtonian approximation. The terms on the right-hand side of each equation with no dependency on $c^*$ correspond to the Newtonian Hill problem [1, 9].

## 3. Stability of Orbits

In this work we consider the variation of the masses such that the mass ratio $\mu$ is fixed. We use $r_S^* = 2Gm_1/c^2\mu^{1/3}$ as a parameter. A variation of $r_S^*$ is associated with the variation of $m_1$, and the dynamics of the system is changed. We label the different systems with the corresponding value of $r_S^*$, as in our previous work [10]. For values less than $r_S^* = 10^{-4}$ the 1PN expansion is still valid, but for larger values more terms in our series expansions ought to be considered.

### *3.1. Poincaré Sections*

We study surfaces of section (Poincaré sections) by evaluating the orbits for different values of the Jacobi constant and registering the crossings of the hypersurface $y = 0$ with $\dot{y} > 0$.

The results for $C_J = -2.17$ (typical value for a bounded system) are shown on the Figure 1 for several values of $r_S^*$. The first figure, in upper left, shows a section for $r_S^* = 5.10^{-12}$. The other figures, in upper right, lower left and lower right, show the sections obtained for $r_S^* = 5.10^{-6}$, $r_S^* = 10^{-5}$ and $r_S^* = 5.10^{-5}$, respectively. We see that, as $r_S^*$ increases, less Kolmogorov-Arnold-Moser (KAM) tori are destroyed, indicating an increasing stability. In particular, the system with $r_S^* = 5.10^{-5}$ seems to be completely stable. Besides, the dependence on initial conditions, the main characteristic of chaotic systems, still must be analyzed by appropriate tools, like Lyapunov exponents. This analysis can decide if a system is more unstable than other [11].

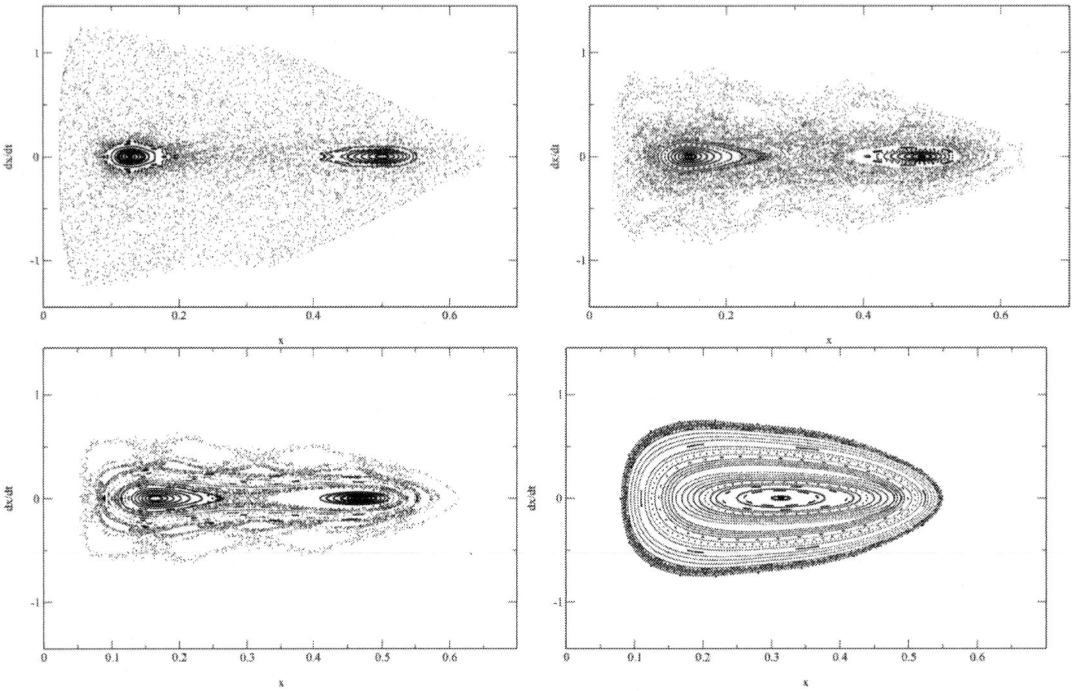


**Figure 1.** Poincaré sections for different values of the parameter $r_S^*$. From lrft to right and top to bottom: $r_S^* = 5.10^{-12}$, $r_S^* = 5.10^{-6}$, $r_S^* = 10^{-5}$ and $r_S^* = 5.10^{-5}$. The rate of destruction of KAM tori decreases and the system becomes more stable.

### *3.2. Lyapunov Exponents*

To analyze quantitatively the orbits stability we study the Lyapunov exponents for the systems described above. We get the largest $\lambda$ by applying the technique suggested by Benettin *et al.* [12] and the algorithm of Wolf *et al.* [13].

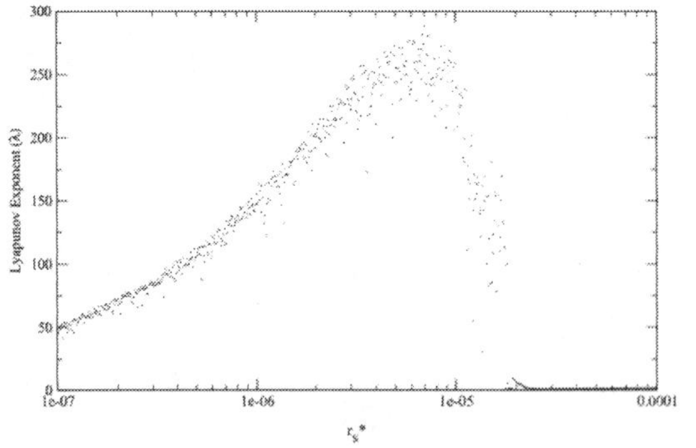


**Figure 2.** Lyapunov exponents for different values of $r_S^*$ for the 1PN Hill Problem.

The Lyapunov exponents are not absolute, but dependent on the choice of the time scale. We recall that we have fixed the time scale for each system by requiring $\omega = 1$, so the different

Lyapunov exponents must be normalized to a fixed time unit in order to be comparable. Each coefficient was computed until convergence is reached. To achieve this precision the system of equations was integrated for at least one hundred thousand periods. The calculations are performed with the aid of the Burlisch-Stoer method with step control, that works well for non-stiff systems, and the error due to the integration is proportional to the tolerance imposed ($10^{-10}$). For the Newtonian system we have $\lambda = 0.141$ that is independent of the mass of the bodies.

The Lyapunov exponents obtained for the relativistic systems are shown in Figure 2. We can see in this figure, that the value of the Lyapunov exponents increases until it reaches a maximum, approximately at $r_S^* = 8.10^{-6}$ and begins to decrease, reaching approximately zero for values of $r_S^*$ larger than $5.10^{-5}$. These results confirm the analysis of the Poincaré sections shown in the last section.

## 4. Conclusions

The dynamics of the relativistic system is also different from the dynamics of systems based on pseudo-Newtonian potentials, in particular the Paczyński-Wiita potential [10, 14]. The Paczyński-Wiita potential is obtained from a slight modification of the Newtonian potential by changing $1/r$ for $1/(r - r_S)$. It correctly reproduces the locations of innermost stable circular orbit and the marginally bound orbit, and it is widely used in numerical simulations of black hole accretion. However, it does not correspond to the actual post-Newtonian approximation. The singularity present on the Paczyński-Wiita potential is different from the general relativistic system, so each system present a different dynamics when the masses are increased. Namely the relativistic system becomes more unstable in a rate greater than in the case of the pseudo-Newtonian system. Moreover, the pseudo-Newtonian the systems becomes more unstable as $r_S^*$ increases. On the other hand, the relativistic system achieves a maximum and then starts to decrease, until it reaches a almost stable state.

The behaviour of the relativistic system must be analysed more carefully. The decrease in the stability can be due to the fact that we carried the post-Newtonian expansion only up to first order. By considering more terms we can confirm if our results indicate the true behaviour of the relativistic system. However, this is a non trivial task, as these expansions need long calculations to be performed.

## Acknowledgments

*In memoriam* of PSL. Partial financial support by FAPESP and Fundação Araucária.

## References

[1] Hill G 1878 *Am. J. Math.* **1** 129
[2] Szebehely V 1967 *Theory of Orbits in the Restricted Problem of Three Bodies* (Waltham: Academic Press)
[3] Heggie D C 2001 Escape in Hill's problem *The Restless Universe* ed Steves B A and Maciejewski A J pp 109 128
[4] Chicone C, Mashhoon B and Retzloff D 1996 *Ann. Inst. H. Poincaré, Phys. Théor.* **64** 87
[5] Meletlidou E, Ichtiaroglou S and Winterberg F 2001 *Celest. Mech. Dynam. Astron.* **80** 145
[6] Simo C and Stuchi T 2000 *Physica D* **140** 1
[7] Alvi K 1999 *Phys. Rev. D* **61** 124013
[8] Weinberg S 1972 *Gravitation and Cosmology : Principles and Applications of the General Theory of Relativity* (Wiley)
[9] Murray C and Dermott S 1999 *Solar System Dynamics* (Cambridge: Cambridge Univ. Press)
[10] Steklain A and Letelier P 2006 *Phys. Lett. A* **352** 398
[11] Eckmann J and Ruelle D 1985 *Rev. Mod. Phys.* **57** 617
[12] Benettin G, Galgani L, Giorgilli A and Strelcyn J 1980 *Meccanica* **15** 9
[13] Wolf A, Swift J, Swinney H and Vastano J 1985 *Phys. D* **16** 285
[14] Paczynski B and Wiita P 1980 *Astron. Astrophys.* **88** 23

# Lorentz-symmetry violating gauge field as regulator and origin of dynamical flavour oscillations

**J. Alexandre**

King's College London, Department of Physics, WC2R 2LS, UK

E-mail: jean.alexandre@kcl.ac.uk

**Abstract.** We show how a mass mixing matrix can be generated dynamically, for two massless fermion flavours coupled to a Lorentz invariance violating (LIV) gauge field. The LIV features play the role of a regulator for the gap equations, and the non-analytic dependence of the dynamical masses, as functions of the gauge coupling, allows to consider the limit where the LIV gauge field eventually decouples from the fermions. Lorentz invariance is then recovered, to describe the oscillation between two free fermion flavours, and we check that the finite dynamical masses are the only effects of the original LIV theory.

## 1. Introduction

The generation of quark, lepton and vector boson masses, as described in the Standard Model due to their coupling with the Higgs boson, seems to have been confirmed by the latest experimental results at the Large Hadron Collider [1], with the discovery of a Higgs-like (scalar) particle. However, the origin of neutrino masses is still not well established, although the seesaw mechanism seems the most elegant and simple for such a purpose [2].

The possibility, therefore, of generating neutrino masses dynamically is still at play. A scenario has been proposed in [3], in which flavour oscillations can arise dynamically from the flavour-mixing interaction of two massless bare fermions with an Abelian gauge field, which has a Lorentz-Invariance-Violating (LIV) propagator. Lorentz symmetry violation is achieved by higher order space derivatives, which are suppressed by a large mass scale $M$. This mass scale allows the dynamical generation of fermion masses, as was shown in [4] with the Schwinger-Dyson approach. Another role of this mass scale is to lead to a finite gap equation, and therefore to regulate the model.

Moreover, LIV U(1) gauge models of the form suggested in [4] have been shown to arise in the low-energy limit of some consistent quantum gravity theories [5], for instance when the U(1) gauge theory is embedded in a stringy space time foam model, with the foamy structures being provided by (point-like) D-brane space-time defects ("D-particles"). In such microscopic models, the gauge field is one of the physical excitations on brane world universes interacting with the D-particles. It was observed in [5] that the LIV Lagrangian of [4] can be obtained from a Born-Infeld-type Lagrangian of the U(1) gauge field in the D-particle background, upon an expansion in derivatives. Lorentz Violation arises locally in such models as a result of the recoil of the D-particle defects during their interaction with open strings representing the U(1) excitations. Finally, such a microscopic model has been used for the study of decoherence in neutrino oscillations [6].

An important point is the following structure of the fermion dynamical mass [4]

$$m_{dyn} \simeq M \exp(-a/e^2) , \tag{1}$$

where $a$ is a positive constant and $e$ is the gauge coupling. From the expression (1), one can see that it is possible to take the simultaneous limits $M \to \infty$ and $e \to 0$, in such a way that the dynamical mass (1) remains finite, corresponding to a *physical* fermion mass. In the previous limit, the non-physical gauge field decouples from the theory, and we also note that the gauge dependence of the dynamical mass is avoided.

An essential feature of the mechanism described here is the following. Although LIV features are suppressed by the large mass scale $M$, so that the corresponding effects are negligible at the classical level, quantum corrections completely change this picture, and lead to finite effects. In our present study, the finite effect is the dynamical generation of fermion masses, which is present even after setting the LIV-suppressing mass scale $M$ to infinity and the coupling $e$ to 0. Also, after this limit is taken, relativistic dispersion relations for fermions are recovered, such that the dynamical masses are the only finite effect from the original LIV model.

## 2. Model

The LIV Lagrangian considered in [3] is

$$\mathcal{L} = -\frac{1}{4}F_{\mu\nu}(1 - \frac{\Delta}{M^2})F^{\mu\nu} + \bar{\Psi}(i\partial\!\!\!/ - \tau A\!\!\!/)\Psi, \tag{2}$$

where $F_{\mu\nu}$ is the Abelian field strength for the gauge field $A^\mu$ and $\Delta = -\partial_i\partial^i$ is the Laplacian. The mass scale $M$ suppresses the LIV derivative operator $\Delta$, and can be thought of as the Plank mass, but which will eventually be set to infinity. $\Psi$ is a massless fermion doublet $\Psi = (\psi_1, \psi_2)$ and the flavour mixing matrix $\tau$ features the gauge couplings $(e_1, e_2, \epsilon)$ as

$$\tau = \begin{pmatrix} e_1 & -i\epsilon \\ i\epsilon & e_2 \end{pmatrix} = \frac{e_1 + e_2}{2}\mathbf{1} + \frac{e_1 - e_2}{2}\sigma_3 + \epsilon\sigma_2 , \tag{3}$$

where $\sigma_i$ are the usual Pauli matrices and $\mathbf{1}$ is the $2 \times 2$ identity matrix. The fermions $\psi_1$ and $\psi_2$ in eq.(2) are Dirac, but the structure of the gap equations which is discussed bellow remains the same in the case of Majorana fermions, hence the corresponding dynamical masses are independent of the nature of fermions.

In order to study the possibility of generating masses dynamically, we assume the dressed fermion mass matrix

$$\mathbf{M} = \begin{pmatrix} m_1 & \mu \\ \mu & m_2 \end{pmatrix} = \frac{m_1 + m_2}{2}\mathbf{1} + \frac{m_1 - m_2}{2}\sigma_3 + \mu\sigma_1 , \tag{4}$$

with eigenvalues $\lambda_{m\pm} = (m_1 + m_2)/2 \pm \sqrt{(m_1 - m_2)^2 + 4\mu^2}/2$. We allow for the presence of $\mathbf{M}$ in the gap equations, obtained from the self-consistent Schwinger-Dyson equation for the fermion propagator, which has the usual structure and is not modified by the LIV term in the Lagrangian (2). If we neglect corrections to the wave functions, the vertices and the gauge propagator, the Schwinger-Dyson equation reads for our model

$$G^{-1} - S^{-1} = \int_p D_{\mu\nu}\, \tau\gamma^\mu\, G\, \tau\gamma^\nu , \tag{5}$$

where $S, G, D_{\mu\nu}$ are the bare fermion propagator, the dressed fermion propagator and the gauge propagator respectively, given by (we denote $p_\mu = (\omega, \vec{p})$)

$$\begin{aligned} S &= i\not{p}/p^2 \qquad (6)\\ G &= i\frac{p^2 + \not{p}(m_1+m_2) + m_1 m_2 - \mu^2}{(p^2-m_1^2)(p^2-m_2^2) - 2\mu^2(p^2+m_1m_2) + \mu^4}\left[(\not{p} - \frac{m_1+m_2}{2})\mathbf{1} + \frac{m_1-m_2}{2}\sigma_3 + \mu\sigma_1\right]\\ D_{\mu\nu} &= -\frac{i}{1+\vec{p}^2/M^2}\left(\frac{\eta_{\mu\nu}}{\omega^2-\vec{p}^2} + \zeta\frac{p_\mu p_\nu}{(\omega^2-\vec{p}^2)^2}\right), \end{aligned}$$

where $\zeta$ is a gauge fixing parameter. The loop integral (5) is finite as a consequence of the LIV term $\vec{p}^2/M^2$ in the denominator of the gauge propagator.

## 3. Solutions

The Schwinger-Dyson equation (5) leads to 4 self-consistent gap equations, which must be satisfied by three unknowns $m_1, m_2, \mu$. This is possible provided constraints are satisfied, and several possibilities are derived in [3], which are summarized here:

**(a)** $\{m_1 = m_2 = 0,\ \mu \neq 0 \text{ and } e_1e_2 > \epsilon^2\}$ or $\{m_1 = -m_2 \neq 0 \text{ and } e_1 = e_2 > \epsilon\}$: Mixing but no oscillation because the mass eigenvalues are opposite. In both cases we have

$$m^2 + \mu^2 = M^2 \exp\left(\frac{-16\pi^2}{(4+\zeta)(e_1e_2 - \epsilon^2)}\right), \qquad (7)$$

where $m = |m_1| = |m_2|$, and only the sum $m^2 + \mu^2$ is constrained;

**(b)** $\{m_1 = m_2 \neq 0,\ \mu^2 = m_1m_2 \text{ and } e_1 = e_2,\ \epsilon = 0\}$: Mixing and flavour oscillations. In this situation, the mass matrix has eigenvalues

$$\lambda_+ = 2m = M\exp\left(-\frac{8\pi^2}{(4+\zeta)e^2}\right), \quad \lambda_- = 0, \qquad (8)$$

where $m = m_1 = m_2$ and $e = e_1 = e_2$. The mixing angle is $\pm\pi/4$, depending on the sign of $\mu$;

**(c)** $\{\mu = 0 \text{ and } \epsilon = 0\}$: No mixing. In this case, the two eigenvalues of the mass matrix are

$$m_i = M\exp\left(\frac{-8\pi^2}{(4+\zeta)e_i^2}\right), \quad i = 1, 2, \qquad (9)$$

so the mass matrix is diagonal in flavour space, with masses $m_i$.

The latter case is relevant to Majorana mass eigenstates. Indeed, one defines the Majorana fields $\nu_k$ in terms of the left-handed Dirac fields $\nu_k^L$ as $\nu_k = \nu_k^L + (\nu_k^L)^C$, such that the Lagrangian for free Majorana fields reads

$$\mathcal{L} = \frac{1}{2}\sum_k \left(\overline{\nu}_k i\not{\partial}\nu_k - m_k\overline{\nu}_k\nu_k\right) \qquad (10)$$

where the extra overall factor 1/2 compared to the Dirac case does not play a role in our discussion, as it can be absorbed in a redefinition of masses.

## 4. Discussion

One can consider the coupling of a doublet of (mass eigenstate) Majorana fields to the regulator U(1) gauge field $A_\mu$ in the case **(c)**. The fact that a Majorana field contains both chiralities allows for a straightforward extension of the Dirac case discussed in previous sections to the current situation. In this way, we are able to generate dynamically different mass eigenvalues (9) for the two species, without mixing, as implied by the corresponding solution. This is a consistent way of discussing the dynamical appearance of a Majorana mass for left-handed neutrinos of the standard model. The non-trivial mixing of flavour eigenstates is then obtained by coupling neutrinos to the physical $SU(2)_L$ gauge fields of the standard model. Thus it is because Majorana neutrinos are mass eigenstates that the solution *without mixing* is relevant to the scenario with Majorana neutrinos.

In order to recover Lorentz invariance in the different solutions found, we finally take the simultaneous limits

$$M \to \infty \quad \text{and} \quad e_1,\ e_2,\ \epsilon \to 0, \tag{11}$$

in such a way that the dynamical masses are *finite*, and we denote the corresponding "renormalized" mass matrix by $\mathbf{M}_R$. This mass matrix can be fixed by experimental data, which in the interesting cases **(b)** and **(c)** leads to the couplings $e_1, e_2$ as functions of $M$, for the limit (11) to be defined unambiguously. This procedure is independent of the gauge parameter $\zeta$, and the resulting fermion mass is set to any desired value. In this limit, the gauge field decouples from fermions, and the only finite effect from Lorentz violation in the original model is the presence of finite dynamical masses for fermions. Indeed, the fermion dispersion relations are relativistic in the limit (11), since the fermion self-energy is then [3]

$$\Sigma(\omega,\vec{p}) \quad \to \quad -\frac{1}{4}(\omega\gamma^0 - \vec{p}\cdot\vec{\gamma})\mathbf{1} - \mathbf{M}_R\ , \tag{12}$$

such that time and space derivatives are dressed with the same (non-perturbative) corrections.

We also note that, because of flavour mixing interactions of fermions with the vector field, that latter may become massive [7], but this feature does not occur here in the relevant cases **(b)** or **(c)**, where $\epsilon = 0$.

Finally, the case **(b)** might seem restrictive for Dirac fermions, since the mixing angle is necessarily equal to $\pm\pi/4$. This is a result of the high symmetry of the $2\times 2$ mass matrix, and the extension to the three-flavour case is planned, in order to allow for more flexibility and phenomenological studies.

[1] G. Aad *et al.* [ATLAS Collaboration], Phys. Lett. B **716** (2012) 1 [arXiv:1207.7214 [hep-ex]]; S. Chatrchyan *et al.* [CMS Collaboration], Phys. Lett. B **716** (2012) 30 [arXiv:1207.7235 [hep-ex]].
[2] For a review see: R. N. Mohapatra, S. Antusch, K. S. Babu, G. Barenboim, M. -C. Chen, A. de Gouvea, P. de Holanda and B. Dutta *et al.*, Rept. Prog. Phys. **70** (2007) 1757 [hep-ph/0510213] and references therein.
[3] J. Alexandre, J. Leite and N. E. Mavromatos, arXiv:1304.7706 [hep-ph].
[4] J. Alexandre, arXiv:1009.5834 [hep-ph]; J. Alexandre and A. Vergou, Phys. Rev. D **83** (2011) 125008 [arXiv:1103.2701 [hep-th]].
[5] N. E. Mavromatos, Phys. Rev. D **83** (2011) 025018 [arXiv:1011.3528 [hep-ph]].
[6] J. Alexandre, K. Farakos, N. E. Mavromatos and P. Pasipoularides, Phys. Rev. D **77** (2008) 105001 [arXiv:0712.1779 [hep-ph]].
[7] R. Jackiw, K. Johnson and , Phys. Rev. D **8** (1973) 2386; J. M. Cornwall, R. E. Norton and , Phys. Rev. D **8** (1973) 3338.

# A molecular dynamics study of structure and thermal properties of carbon dioxide in sodium chloride aqueous solution

**Shigeki Matsunaga**

Nagaoka National College of Technology,
Nishikatakai 888, Nagaoka, 940-8532, Japan

E-mail: matsu@nagaoka-ct.ac.jp

**Abstract**. The structure and thermal conductivity of the sodium chloride (NaCl) aqueous solution in similar concentration to the seawater with saturated carbon dioxide ($CO_2$) have been investigated by molecular dynamics simulation. The effects of pressure have been investigated under various pressures corresponding to the depth of the sea from 40m to 10000m. The *negative* pressure dependence of the thermal conductivity has been detected in the depth of more than 8000m, whereas that of NaCl aqueous solution without $CO_2$ shows the *positive* pressure dependence.

## 1. Introduction

Recent years, much attention has been paid to the glowing ratio of carbon dioxide ($CO_2$) in the atmosphere, because it is considered to be a main reason of the greenhouse effect of the earth environment [1]. $CO_2$ gas is easily soluble to the seawater, and besides, the extra $CO_2$ and higher temperature of the atmosphere may enhance the dissolution of $CO_2$ into the seawater, or the sodium chloride (NaCl) aqueous solution as a simplified expression. Although many molecular dynamics (MD) studies have been performed in the $CO_2$ and NaCl aqueous solution, most of them focus on the solubility of $CO_2$, the stability of the solution, or the dilute $CO_2$ solution [2-4]. There is no MD study of the thermal conductivity of $CO_2$ and NaCl aqueous solution, to the best of our knowledge. Under these circumstances, in this study, we wish to perform molecular dynamics simulation in order to investigate the structure and the thermal properties of NaCl aqueous solution with saturated $CO_2$ under various pressures corresponding to the depth of the sea, which would be one of the fundamental data for the ocean environment.

## 2. Simulation Procedure

In this study, we perform the equilibrium (EQ) and non-equilibrium (NE) MD simulation. The structure is obtained by EQMD. On the other hand, the thermal conductivity is obtained by NEMD, which will be discussed in the following section. The essential numerical procedure of EQMD simulation in this study is same as our previous works of molten salts [5]. In MD for $CO_2$ and NaCl aqueous solution, the rigid body models are used for water (TIP4P)[6] and $CO_2$ molecules. The interactions between constituent molecules of TIP4P water, $CO_2$, $Na^+$ and $Cl^-$ ions, except between $Na^+$-$Cl^-$, TIP4P-$Na^+$, and TIP4P-$Cl^-$, are expressed as the charged L-J type potentials, as,

$$\phi_{ij}(r) = \frac{z_i z_j e^2}{r} + \frac{A}{r^{12}} - \frac{B}{r^6} \tag{1}$$

The interactions between $Na^+$-$Cl^-$, TIP4P-$Na^+$, and TIP4P-$Cl^-$ are expressed as [7],

$$\phi_{ij}(r) = \frac{z_i z_j e^2}{r} + \frac{C}{r^9} - \frac{D}{r^6} \tag{2}$$

In (1) and (2), $i$ and $j$ stand for the constituent atoms; $e$ is the elementary charge. The used charges for the constituent species $z_i$, and the interaction parameters are taken from the literature; TIP4P-TIP4P [6]; $CO_2$-$CO_2$, TIP4P-$CO_2$, and $CO_2$-$Cl^-$ (OPLS) [8]; $CH_4$-$Na^+$ [9]; TIP4P-$Na^+$, TIP4P-$Cl^-$, $Na^+$-$Na^+$, $Cl^-$-$Cl^-$, and $Na^+$-$Cl^-$ [7]. The Ewald method is used for the calculation of the Coulomb interaction. For the structure calculation, MD is performed in *NTP* constant condition [10,11] under the pressures from 0.5MP to 100MP corresponding to the depth of the sea from 40m to 10000m, respectively. MD is performed for 50000steps with 0.1fs one time step in 1.1% NaCl aqueous solution at the saturated concentration of $CO_2$ [12]. MD cell contains about 10000 molecules (30000 atoms) for the structure calculation, about 5000 molecules (15000 atoms) for the thermal conductivity. The main part of MD is performed using SIGRESS ME package (Fujitsu) at the supercomputing facilities in Kyushu University. The snap shot after 50000 steps under 100MP at 275K is shown in figure 1.

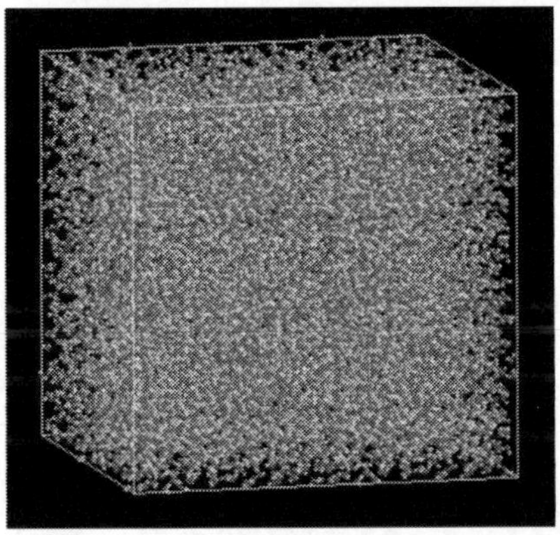

**Table 1.** The number of particles used in MD.

| | water | $Na^+$ | $Cl^-$ | $CO_2$ |
|---|---|---|---|---|
| 0.5MP | 10000 | 112 | 112 | 26 |
| 100MP | 10000 | 112 | 112 | 812 |

Figure.1.(left) The snapshot of $10000H_2O$ (O: purple), $112Na^+$ (red), $112Cl^-$ (green) and $812CO_2$ (C: light brown) at 275K, 100MP.

## 3. Results and Discussion

The pair distribution functions, $g_{ij}(r)$, have been obtained from the MD results which are shown in figure 2-3 for 0.5MP and 100MP. The slight enhancement of the first peak heights of $g_{ij}(r)$ of $CO_2$-$H_2O$, $Na^+$-$H_2O$, $Cl^-$-$H_2O$ and $H_2O$-$H_2O$ can be observed at 100MP, which may correspond to the cage formation by $H_2O$ molecules.

The interference functions for neutron, $I_{ij}(q)$, are obtained from $g_{ij}(r)$, which is expressed as [13],

$$I_{ij}(q) = c_i b_i c_j b_j / (\textstyle\sum_k c_k b_k)^2 \int_0^\infty 4\pi r^2 \rho_0 (g_{ij}(r) - 1) \frac{\sin(rq)}{qr} dr \tag{3}$$

where $c_i$ is the atomic fraction of the $i$-type atoms; $\rho_0$ is the average number density; $b_i$ is the neutron scattering amplitude of the $i$-type atoms. The obtained $I_{ij}(q)$'s using $g_{ij}(r)$'s are shown in figure 4-5 for the immediate comparison with experiment. In $I_{C\text{-water}}(q)$ a sharp peak can be observed at 2[Å$^{-1}$], which is enhanced more in figure 5. The peak heights of $I_{Na\text{-water}}(q)$ and $I_{Cl\text{-water}}(q)$ decrease, and their width have been broadened in figure 5. These facts can be explained that as the molecule fraction of $CO_2$ increases, the structure formation of water molecules around $CO_2$ is enhanced.

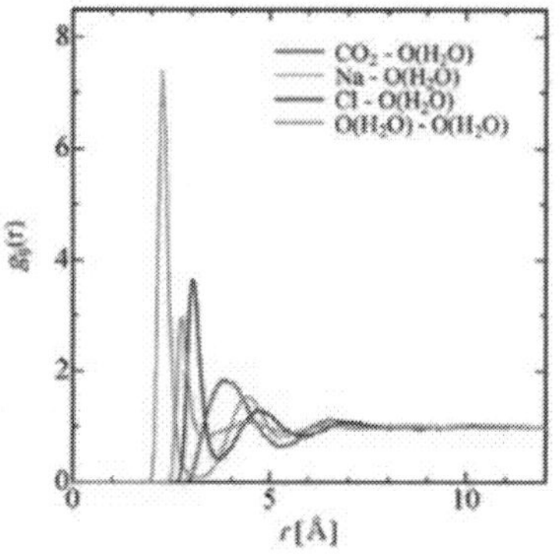

Figure.2 (left) $g_{ij}(r)$ at 0.5MP, 283K

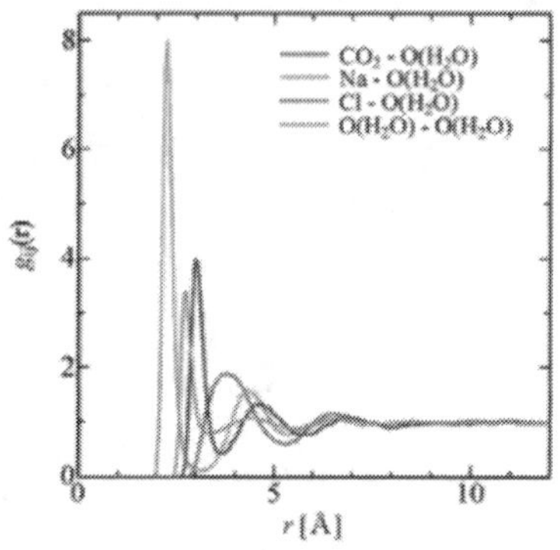

Figure.3 (right) $g_{ij}(r)$ at 100MP, 275K

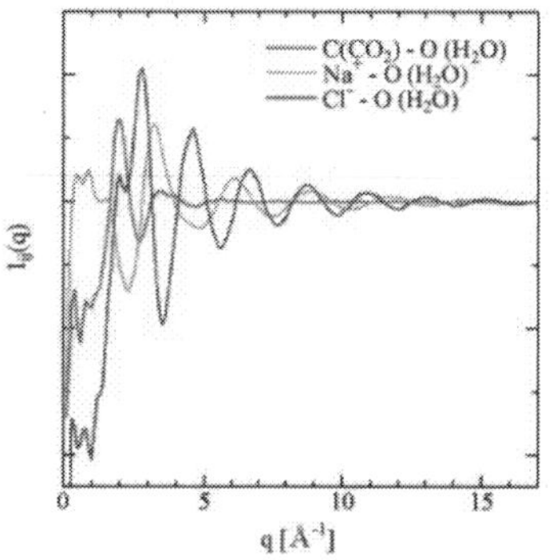

Figure.4 (left) $I_{ij}(r)$ at 0.5MP, 283K

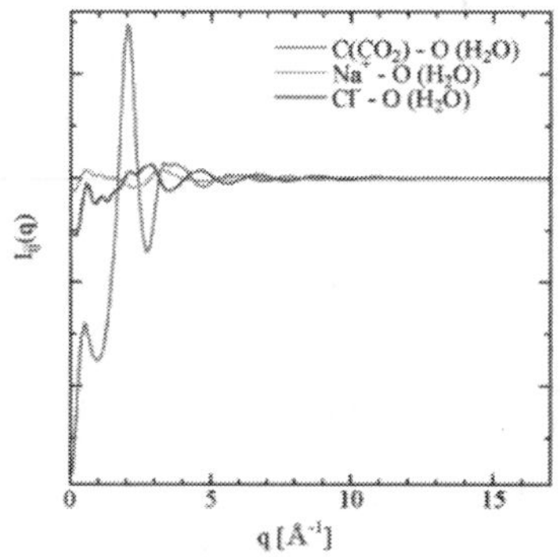

Figure.5 (right) $I_{ij}(r)$ at 100MP, 275K

To investigate the thermal properties of 1.1% NaCl aqueous solution with saturated $CO_2$, we have calculated the thermal conductivity. To the best of the author's knowledge, this is the first report of the thermal conductivity of $CO_2$ and NaCl aqueous solution obtained by MD. The calculation procedure is briefly described as follows. There are several methods to calculate the thermal conductivity by MD; (1) the direct method; (2) the equilibrium method; (3) the perturbation method or NEMD method. In this study, we adopt the NEMD method, which can avoid the margin of error caused by integrating the energy flux autocorrelation function, the averaging of energy flux over time is performed, instead [14]. The perturbation is applied to the system in the thermal equilibrium at time $t = 0$. According to the Green-Kubo formula, the thermal conductivity $\lambda$ is expressed as [15],

$$\lambda = \frac{1}{Vk_BT^2}\int_0^{\infty}\langle J_x(\tau)J_x(0)\rangle d\tau \tag{4}$$

where $J_x(\tau)$ is the $x$ component of the perturbation current; $V$ is the volume of the system; $k_B$ is the Boltzmann constant; $T$ is the temperature of the system. The relation between the applied perturbation $F_{ext}$ and the average of perturbation current is expressed as,

$$\langle J_x\rangle_t = \frac{F_{ext}}{k_BT}\int_0^{t}\langle J_x(\tau)J_x(0)\rangle d\tau \tag{5}$$

Then, the relation between the thermal conductivity and the perturbation is expressed as [14],

$$\lambda = \frac{1}{VF_{ext}T}\lim_{t\to\infty}\langle J_x\rangle_t \tag{6}$$

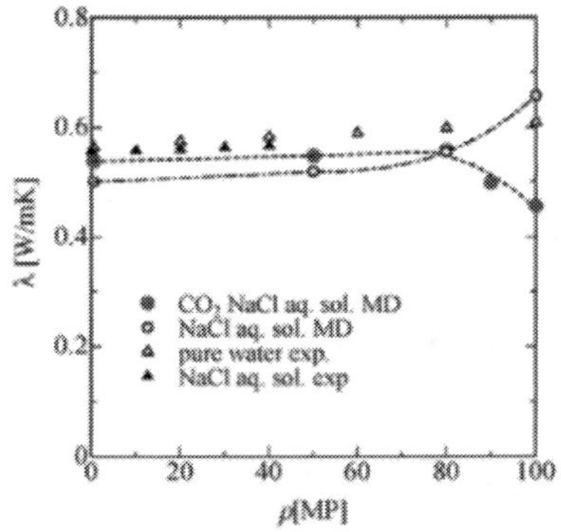


Figure 6. The pressure dependence of the thermal conductivity. The lines are the guides to the eye.

In the case of 'low' molecules, which consist of a few atoms, the thermal conductivity is also obtainable on the condition that the contribution from the atom in the same molecule is omitted from the perturbation current [16]. The obtained results of pressure dependence of the thermal conductivity are shown in figure 6. The experimental thermal conductivity data of pure water and 1% NaCl aqueous solution are also shown in figure 6 [17,18]. The pressure dependence for the saturated concentration of $CO_2$ in 1.1% NaCl aqueous solution is quite noteworthy, because the thermal conductivity decreases above 80MP. It is a marked contrast with the other thermal conductivity data without $CO_2$, which increase as the pressure increases. This fact also implies the structure formation of the water molecules around $CO_2$ by pressure.

**Acknowledgment**

The author thanks Professor S. Tamaki for his helpful comments and encouragement on this study. This study was supported by the Grant-in-Aid for Scientific Research of the Ministry of Education and Culture. Part of the results in this study was obtained using the supercomputing facilities at Research Institute for Information Technology, Kyushu University.

**References**

[1] Gruber N 2011 *Phil. Trns. Royal Soc. A:Math. Phys. Eng. Sci.* **369** 180
[2] Sabil K, Witkamp G –J and Peters C 2009 *Fluid Phase Equilibria* **284** 38
[3] Martin C and Patroescu C 2006 *Revista De Chimie* **57** 669
[4] Huang T 2012 *PhD Thesis* Swinburne University of Technology
[5] Matsunaga S 2012 *Mol. Sim.* **38** 384; Matsunaga S 2013 *Trans. Mat. Res. Soc. Jpn.* **38** 113
[6] Jorgensen W L, Chandrasekhar J, Madura J D, Impey R W and Klein M L 1983 *J.Chem. Phys.***79** 926.
[7] Chandrasekhar J, Spellmeyer D C and Jorgensen W L 1984 *J. Am. Chem. Soc.* **106** 903.
[8] Zhengwei P, Ewig C S, Hwang M-J, Waldman M and Hagler A T 1997 *J. Phys. Chem.* A **101** 7243
[9] Mayo S L, Olafson B D and Goddard III W A 1990 *J. Phys. Chem.* **94** 8897
[10] Nosé S 1984 *Mol. Phys.* **52** 255; Nosé S 1984 *J. Chem. Phys.* **81** 511.
[11] Andersen H C 1980 *J. Chem. Phys.* **72** 2384.
[12] Z Duan and R Sun 2003 *Chem. Geo.* **193** 257
[13] Waseda Y 1980 *The Structure of Non-Crystaline Materials* (New York: McGraw-Hill)
[14] Yoshida M, Harada A, Takeuchi A, Matsunaga K and Matsubara H 2004 *Molecular Simulation* **30** 953
[15] Hansen J –P and McDonald I R 1986 *Theory of Simple Liquids* (London: Academic Press)
[16] Petravic J 2005 *J. Chem. Phys.* **123** 174503
[17] Abdulagatov I M, Magomedov U B 2003/2004 *High Temp. – High Press.* **35/36** 149
[18] Nagasaka Y, Okaga H, Suzuki J and Nagashima A 1983 Ber. Bunsenges. Phys. Chem. **87** 859

# Quantum and classical phenomena in photoionization of carbon nanostructures

**A V Verkhovtsev[1,2,3], A V Korol[1,4] and A V Solov'yov[1,5]**

[1] Frankfurt Institute for Advanced Studies, Ruth-Moufang-Str. 1, 60438 Frankfurt am Main, Germany

[2] St. Petersburg State Polytechnic University, Politekhnicheskaya ul. 29, 195251 St. Petersburg, Russia

[3] A.F. Ioffe Physical-Technical Institute, Politekhnicheskaya ul. 26, 194021 St. Petersburg, Russia

[4] Department of Physics, St. Petersburg State Maritime Technical University, Leninskii prospekt 101, 198262 St. Petersburg, Russia

E-mail: verkhovtsev@fias.uni-frankfurt.de

**Abstract.** We present the results of investigation of electron excitations in various carbon-based nanoscale systems in the process of photoionization. As a case study, we consider a number of highly symmetric fullerenes, namely $C_{20}$, $C_{60}$ and $C_{80}$, as well as aromatic hydrocarbons benzene ($C_6H_6$) and coronene ($C_{24}H_{12}$). The calculations are performed within the *ab initio* TDDFT framework and model approach, based on the plasmon resonance approximation. Analysis of the results demonstrates that the main contribution to the photoionization spectra of nanoscale carbon systems is due to collective excitations of delocalized electrons, known as plasmons. Results of the model-based calculations are in close agreement with those of the more accurate quantum-chemical calculations and correspond also to the existing experimental data.

Photoionization of atomic clusters, fullerenes as well as other nanoscale systems, represents a complex phenomenon and involves a number of features which can be studied by means of various theoretical methods. Being quantum by its nature, the photoionization process can be described within the *ab initio* framework based on the time-dependent density-functional theory (TDDFT) [1]. It deals with the time-dependent Schrödinger equation and allows one to obtain information on the excited-state properties of a complex many-electron system. However, it is well established that photoionization of nanoscale carbon systems, fullerenes in particular, as well as metallic clusters and nanoparticles, takes place through plasmons – collective excitations of delocalized valence electrons, which are induced by an external electric field. The plasmon excitations correspond to oscillations of the electron density with respect to the positively charged ions [2, 3] and are described in the classical physics terms [2–4].

In this paper, we elucidate the contributions of classical and quantum phenomena appearing in the ionization process of various carbon nanosystems. By comparing the TDDFT results with those based on the plasmon resonance approximation (see [3–5] and references therein), we map well-resolved features of the photoionization spectra of fullerenes and polycyclic aromatic hydrocarbons (PAHs) to different types of single-particle and collective electron excitations having the different physical nature. It is demonstrated that the results of the *ab initio* and

[5] On leave from A.F. Ioffe Physical-Technical Institute, Politekhnicheskaya ul. 26, 194021 St. Petersburg, Russia

model-based calculations are in close mutual agreement as well as with available experimental results on photoionization of carbon nanostructures.

In most cases, the ionization spectra, calculated within the *ab initio* framework, can be obtained in a broad range of excitation energies only for small molecules or clusters consisting of a few atoms. For larger system, such as, for instance, fullerenes, a vast majority of contemporary software packages for *ab initio* calculations can describe accurately only a limited number of low-lying excited states located below or just above the ionization threshold. A detailed structure of the spectrum at higher excitation energies, where the plasmon excitations dominate the ionization spectrum, could be hardly revealed due to significant computational costs. However, it has become possible recently to calculate optical spectra of complex molecular systems in a broad excitation energy range using the TDDFT methods (see [6,7] and references therein) in the linear regime within the dipole approximation. The method introduced is based on the frequency representation of the response function and allows one to calculate the full photoionization spectrum of complex molecular systems in a broad energy range without repeating time-consuming operations for different excitation frequencies.

An effective tool for evaluation of the contribution of plasmon excitations to the ionization spectra is based on the plasmon resonance approximation. The advantage of this approach is that it provides a clear physical explanation of the resonant-like structures in the photoionization [3,8] and inelastic scattering cross sections [5, 9–11] on the basis of excitation of plasmons by the photon or electron impact. Within this approach it is assumed that a contribution from single-particle excitations, which are of the quantum nature, is small as compared to the collective modes [5, 12] and, therefore, can be neglected.

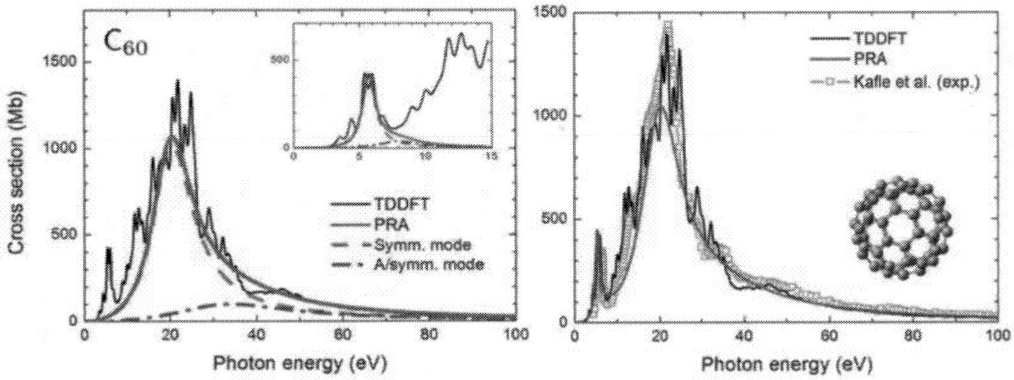


**Figure 1.** The photoionization cross section of $C_{60}$ calculated within the TDDFT method (black line) and the plasmon resonance approximation (green line). Contribution of the symmetric and antisymmetric modes of the surface plasmon is shown by the dashed red and dash-dotted blue lines, respectively. Theoretical curves are compared to the experimental data of Kafle *et al.* [16].

In order to evaluate the contribution of plasmon excitations to the photoionization spectra of carbon nanosystems, we utilize the following model approach. The systems are described by a homogeneous electron density distribution over the shell of a finite width, $\Delta R = R_2 - R_1$, where $R_{1,2}$ are the inner and the outer radii of a system, respectively [10, 13–15]. In all case studies considered, except the coronene molecule, carbon atoms are almost equally-spaced from the geometrical center of the system. Therefore, in order to describe fullerenes and the benzene molecule within the model approach, the shell width $\Delta R$ was chosen to be equal to 1.5 Å. The chosen value corresponds to the typical size of a carbon atom [14]. In order to model correctly geometry of the coronene molecule, we used the value $\Delta R = 3.4$ Å.

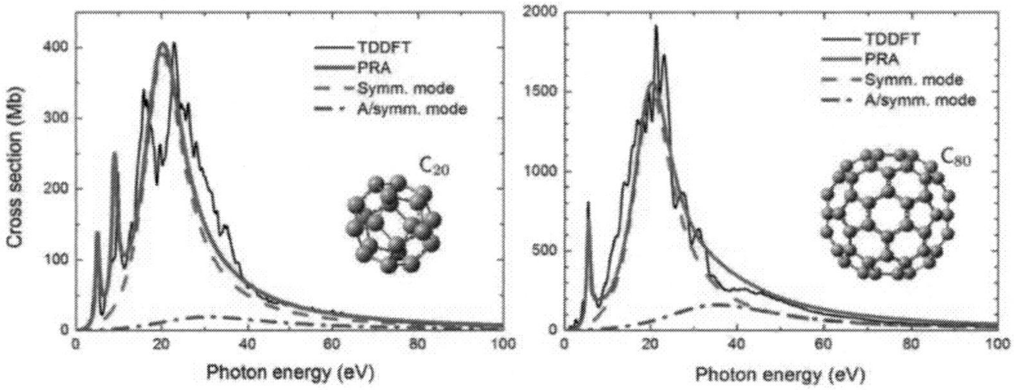


**Figure 2.** The photoionization cross section of the $C_{20}$ and $C_{80}$ fullerenes calculated within the TDDFT method (black line) and the plasmon resonance approximation (green line). Contributions of the symmetric and antisymmetric modes of the plasmons are shown by the dashed red and dash-dotted blue lines, respectively.

It is well established (see, e.g. [17]) that a cloud of delocalized electrons in fullerenes and aromatic hydrocarbons is formed due to the contribution of four $2s^22p^2$-electrons from each carbon atoms. Therefore, we assume that the number of delocalized electrons $N = 80$, 240 and 320 in case of the $C_{20}$, $C_{60}$ and $C_{80}$ fullerenes, respectively, while 24 and 96 electrons contribute to the plasmon excitations in benzene and coronene, respectively.

Due to the interaction with a uniform external field, the electron density variation occurs on the inner and outer surfaces of the spherical shell. This variation leads to the formation of the surface plasmon, which has two normal modes, the symmetric and antisymmetric ones [8, 13, 14]. It was shown [3, 4, 8], that only the surface plasmon can occur in the system interacting with a uniform external electric field, as it happens in the photoionization process. When a system interacts with a non-uniform electric field created, for instance, in collision with charged particles, the volume plasmon [12] can also occur due to a local compression of the electron density in the shell interior [11].

Figure 1 shows the photoionization spectrum of $C_{60}$ calculated within the *ab initio* and classical approaches in the photon energy region up to 100 eV. The thin solid (black) line represents the results of TDDFT calculations within the LDA approach, and the thick solid (green) one represents the contribution from the plasmon excitations. The main resonant structure presented in the left panel of figure 1 is formed due to collective oscillations of both $\sigma$ and $\pi$ delocalized electrons of the system, while a prominent peak in the low-energy region of the spectrum (shown in the inset) is attributed to the collective excitation of only $\pi$-electrons. The $\sigma$- and $\pi$-electrons occupy, respectively, $\sigma$- and $\pi$-orbitals of a fullerene, which are formed due to the $sp^2$-hybridization of carbon atomic orbitals [17]. The dashed (red) and dash-dotted (blue) lines show, respectively, contributions from the symmetric and antisymmetric modes of the plasmons to the cross section (see [19] for further details). In the right panel of figure 1, the theoretical curves are compared to the results of recent experimental measurements of photoabsorption of $C_{60}$ [16]. The analysis performed shows that the plasmon resonance approximation gives an adequate description of the experimental results and more precise TDDFT calculations. In particular, the model approach describes the main features of the spectrum, such as height and position of the plasmon resonances [18]. The spectrum calculated within the TDDFT approach reveals a more detailed structure which is formed atop the plasmon resonances and represents a series of individual peaks. It has been demonstrated recently [19] that the peculiarities arising in the photoionization spectrum of $C_{60}$ atop the dominating plasmon excitations have the quantum origin and can be assigned to the particular discrete single-electron transitions and to ionization of the innermost valence molecular orbitals of the system.

The photoionization spectra of $C_{20}$ and $C_{80}$ fullerenes are presented in figure 2, while figure 3 demonstrates spectra of aromatic hydrocarbons: the simplest one – benzene ($C_6H_6$) (left panel), and a more complicated compound – coronene ($C_{24}H_{12}$) which consists of six benzene rings (right panel). Similar to the spectrum of the $C_{60}$ molecule, the main contribution to the ionization spectra of other fullerenes and PAHs comes from the collective excitation of both types ($\sigma$ and $\pi$) of delocalized electrons, which is described by a prominent broad peak located at about 15-25 eV. The spectra are also characterized by a single or several much narrower peaks in the low-energy part of the spectrum, which are due to the collective excitation of only $\pi$-electrons.

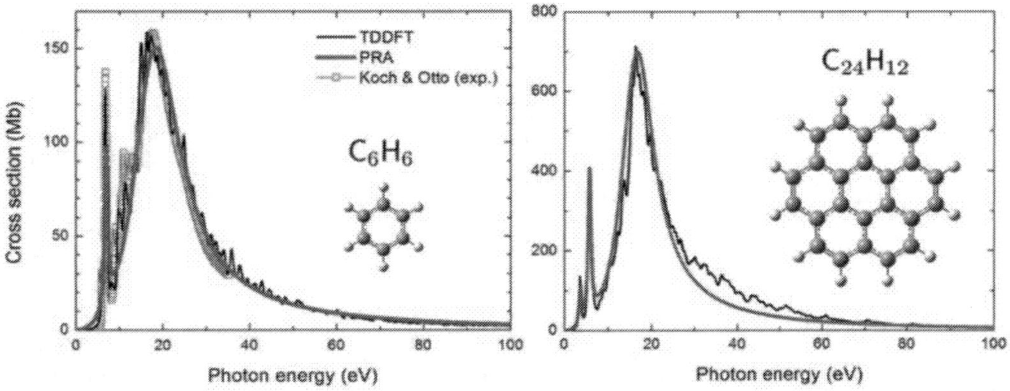


**Figure 3.** The photoionization cross section of the $C_{20}$ and $C_{80}$ fullerenes calculated within the TDDFT method (thin black line) and the plasmon resonance approximation (thick green line). In case of benzene theoretical curves are compared to the experimental data [20].

## References

[1] Runge E and Gross E K U 1984 *Phys. Rev. Lett.* **52** 997
[2] Lundqvist S and March N H 1983 *Theory of the Inhomogeneous Electron Gas* (New York: Plenum Press)
[3] Connerade J-P and Solov'yov A V 2002 *Phys. Rev.* A **66** 013207
[4] Verkhovtsev A V, Korol A V, and Solov'yov A V 2012 *Eur. Phys. J.* D **66** 253
[5] Solov'yov A V 2005 *Int. J. Mod. Phys.* B **19** 4143
[6] Walker B, Saitta A M, Gebauer R, and Baroni S 2006 *Phys. Rev. Lett.* **96** 113001
[7] Malcioğlu O B, Gebauer R, Rocca D, and Baroni 2011 *Comp. Phys. Commun.* **182** 1744
[8] Korol A V and Solov'yov A V 2007 *Phys. Rev. Lett.* **98** 179601
[9] Solov'yov A V 2001 *Electron scattering on metal clusters and fullerenes*, in *Atomic Clusters and Nanoparticles* eds C Guet, P Hobza, F Spiegelman et al. (Berlin-Heidelberg-New York: Springer-Verlag) pp 401-435
[10] Verkhovtsev A V, Korol A V, Solov'yov A V, et al 2012 *J. Phys. B: At. Mol. Opt. Phys.* **45** 141002
[11] Bolognesi P, Ruocco A, Avaldi L, et al 2012 *Eur. Phys. J.* D **66** 254
[12] Gerchikov L G, Ipatov A N, Polozkov R G, and Solov'yov A V 2000 *Phys. Rev.* A **62** 043201
[13] Lambin Ph, Lucas A A, and Vigneron J-P 1992 *Phys. Rev.* B **46** 1794
[14] Östling D, Apell P, and Rosen A 1993 *Europhys. Lett.* **21** 539
[15] Lo S, Korol A V, and Solov'yov A V 2007 *J. Phys. B: At. Mol. Opt. Phys.* **40** 3973
[16] Kafle B P, Katayanagi H, Prodhan M, et al. 2008 *J. Phys. Soc. Jpn.* **77** 014302
[17] Martins J L, Troullier N, Weaver J H 1991 *Chem. Phys. Lett.* **180** 457
[18] Verkhovtsev A V, Korol A V, Solov'yov A V 2013 *J. Phys.: Conf. Ser.* **438** 012011
[19] Verkhovtsev A V, Korol A V, Solov'yov A V 2013 submitted (see also arXiv:1307.3514 [physics.atm-clus])
[20] Koch E E and Otto A 1972 *Chem. Phys. Lett.* **12** 476

# An analysis of the quantum effect on the thermodynamic and transport properties of cryogenic hydrogen using molecular dynamics method

**Hiroki Nagashima[1], Shin-ichi Tsuda[2], Nobuyuki Tsuboi[3], Mitsuo Koshi[4], A. Koichi Hayashi[5], and Takashi Tokumasu[1]**

[1]Tohoku University, 2-1-1 Katahira, Aoba-ku, Sendai, Miyagi, Japan
[2]Shinshu University, 4-17-1 Wakasato, Nagano, Nagano, Japan
[3]Kyushu Institute of Technology, 1-1 Sensui-cho, Tobata-ku, Kitakyushu, Fukuoka, Japan
[4]Yokohama National University, 79-7 Tokiwadai, Hodogaya, Yokohama, Kanagawa, Japan
[5]Aoyama Gakuin University, 5-10-1 Fuchinobe, Chuo-ku, Sagamihara, Kanagawa, Japan

E-mail: nagashima@nanoint.ifs.tohoku.ac.jp

**Abstract**. In this paper, we have analysed an effect of quantum nature of the hydrogen molecule on its thermodynamic and transport properties using molecular dynamics (MD) method based on the path integral method. We performed *NVE* constant MD simulation and the quantum effect on the molecular mechanism was analysed. The simulation results were compared with experimental data. As a result, we clarified that the quantum nature makes the virial pressure larger than in classical mechanics and taking account the quantum nature makes smaller intermolecular interaction energy and larger repulsive force than classical representation. Besides, we have confirmed that the path-integral-based MD method well reproduces the thermal conductivity and quantum effect on the transport properties is also large.

## 1. Introduction

Analyses of nanosacle thermal flow properties of hydrogen using molecular dynamics (MD) method are important for the safe and efficient use of hydrogen. However, in the case of hydrogen molecules, their quantum nature, that is, the uncertainty or wave nature of the nucleus resulting from its small mass cannot be neglected. Because the molecular motion of hydrogen cannot be treated by classical mechanics, it is difficult to reproduce nanoscale thermal flow properties of hydrogen using the classical MD method [1]. Recently, some quantum MD methods those are able to reproduce the time evolution of the molecules whose quantum effect cannot be neglected were proposed. However, because this quantum nature affects not only microscopic thermal flow properties but also macroscopic thermodynamic properties, it is needed to comprehend the quantum effect on the both sides of property in detail. Therefore, in this study, we investigated the effect of the quantum nature on the thermodynamic properties of liquid hydrogen and its molecular mechanism using classical MD method and a MD method based on the path integral method.

## 2. Simulation method

In this paper, we applied the molecular dynamics methods which are based on the path integral method. Using the path integral method, the quantum partition function becomes the same form as the classical partition function which consists of $L$ particles ring polymer [2]. In short, the quantum partition function is described as the Trotter product discretization,

$$Q(\beta,N)=\lim_{L\to\infty}\frac{1}{N!}\left(\frac{mL}{2\pi\beta\hbar^2}\right)^{\frac{3NL}{2}}\int\cdots\int\prod_{i=1}^{N}\prod_{j=1}^{L}dr_i^{(j)}\exp\left[-\beta\left(\sum_{i=1}^{N}\sum_{j=1}^{L}\frac{mL}{2\beta^2\hbar^2}\left(r_i^{(j)}-r_i^{(j+1)}\right)^2+\frac{1}{L}\phi\left(r_i^{(j)}\right)\right)\right], \tag{1}$$

where $m$ is the mass of molecule, $N$ is the number of molecules, $L$ is the number of discretized points, that is, the number of beads composed a ring polymer, $\beta$ (= $1/k_BT$) is the inverse temperature, $\boldsymbol{r}_i^{(j)}$ is the $j$th bead position of molecular $i$, $\phi$is the interaction of beads. The centroid MD (CMD) [3] is one of the new quantum MD method which is able to reproduce the time evolution of the molecules whose quantum effect cannot be neglected. The CMD method can reproduce the thermodynamic characteristics of liquid hydrogen as compared with other methods [4]. The centroid position and momentum are defined as

$$\bar{\boldsymbol{r}}_i=\frac{1}{L}\sum_{j=1}^{L}\boldsymbol{r}_i^{(j)},\ \bar{\boldsymbol{p}}_i=\frac{1}{L}\sum_{j=1}^{L}\boldsymbol{p}_i^{(j)}. \tag{2}$$

Where $\boldsymbol{p}_i^{(j)}$ is the $j$th bead momentum of molecular $i$. Using the centroid information, the centroid distribution function becomes classical like form,

$$\rho_c(\bar{\boldsymbol{r}},\bar{\boldsymbol{p}})=\rho_c(\bar{\boldsymbol{r}})\rho_c(\bar{\boldsymbol{p}})=\exp\left(-\beta\frac{\bar{\boldsymbol{p}}^2}{2m}\right)\exp(-\beta\phi_c(\bar{\boldsymbol{r}})), \tag{3}$$

where $\phi_c(\boldsymbol{r}^c)$ is the centroid potential of mean force. The equation of motion of hydrogen molecule also can be expressed by a classical like form [3],

$$m_i\ddot{\boldsymbol{r}}_i^c=-\frac{\partial\phi_c\left(\boldsymbol{r}_1^c,\cdots,\boldsymbol{r}_N^c\right)}{\partial\boldsymbol{r}_i^c}=-\frac{\int\cdots\int\prod_{i=1}^{N}\prod_{j=1}^{L}dr_i^{(j)}\delta\left(\boldsymbol{r}_i^c-\bar{\boldsymbol{r}}_i\right)\frac{\partial\phi(\bar{\boldsymbol{r}}_1,\cdots,\bar{\boldsymbol{r}}_N)}{\partial\boldsymbol{r}_i^0}\exp\left(-\beta S\left[\boldsymbol{r}_i^{(j)}\right]\right)}{\int\cdots\int\prod_{i=1}^{N}\prod_{j=1}^{L}dr_i^{(j)}\delta\left(\boldsymbol{r}_i^c-\bar{\boldsymbol{r}}_i\right)\exp\left(-\beta S\left[\boldsymbol{r}_i^{(j)}\right]\right)}, \tag{4}$$

Moreover, centroid correlation function $\langle\hat{B}_c(0)\hat{A}_c(t)\rangle$ can be exactly related to the Kubo-transformed quantum correlation function,

$$\left\langle\hat{B}_c(0)\hat{A}_c(t)\right\rangle=\frac{1}{Z}\int\int\frac{d\bar{r}d\bar{p}}{2\pi\hbar}\rho(\bar{r},\bar{p})B_cA_c(t)=\frac{1}{\beta Z}\int_0^\beta d\lambda\mathrm{Tr}\left[e^{-(\beta-\lambda)\hat{H}}\hat{B}e^{-\lambda\hat{H}}e^{iHt/\hbar}\hat{A}e^{-iHt/\hbar}\right], \tag{5}$$

where the operator $\hat{B}_c$ and $\hat{A}_c$ are restricted in a linear function of the position and/or momentum [5]. Because of this restriction, it is being pointed out that in the case of nonlinear operator, for example heat flux operator [6], CMD is difficult to apply for the calculation. Therefore, in this study, we applied the maximum entropy numerical analytic continuation (MEAC) method [7] with imaginary-time path integral MD method for the calculation of the thermal conductivity of liquid hydrogen. The quantum correlation function $C(t)$ is related to its spectral density, $I(\omega)$, by a Fourier transform,

$$C(t)=\int_{-\infty}^{\infty}d\omega e^{-i\omega t}I(\omega). \tag{6}$$

The imaginary time correlation function $G(\tau)$ can be written as

$$G(\tau)=\int_{-\infty}^{\infty}d\omega e^{-\omega\tau}I(\omega). \tag{7}$$

by substituting $-i\tau$ into $t$ of equation (6). The same spectral density $I(\omega)$ can be obtained by inverse Laplace transform of the imaginary time correlation function of equation (7). Because the inversion is an ill-posed problem, however, the MEAC method is applied for the inversion. The quantum correlation function can be obtained by the inverse Fourier transform of equation (6) using the $I(\omega)$ which derived by inversion of equation (7). The imaginary time heat flux correlation function can be written using heat flux operator $\hat{\boldsymbol{J}}$ [6] as

$$G_J(\tau)=\frac{1}{Z}\mathrm{Tr}\left[e^{-\beta H}e^{\tau H}\hat{\boldsymbol{J}}e^{-\tau H}\cdot\hat{\boldsymbol{J}}\right]. \tag{8}$$

In the case of the CMD simulation, the number of molecules and the number of beads were 256 and 64 respectively. The REference System Propagator Algorithm (RESPA) [8] was used for the numerical integration method. The time steps for the centroids motion was 0.2 fs, and the number of Trotter expansion was set at 10, that is, the time step for the beads motion of 0.02 fs. Silvera-Goldman (SG) [9] was applied for the bead interaction. We have performed *NVE* constant classical MD and CMD simulations at constant density ($\rho$ = 80 kg/m$^3$) by changing temperature. The pressure formulation using the path integral representation can be obtained from the equation (1) [10] as

$$P = \frac{N}{V\beta} + \frac{1}{3V}\sum_{i=1}^{N} \boldsymbol{r}_i^c \cdot \boldsymbol{F}_i^c \cdot \qquad (10)$$

Here $\boldsymbol{F}_i^c$ is the force which acts on the centroid and $V$ is the volume of the system. In this study, the quantum effect on the molecular mechanism of pressure is analysed.

In the case of the MEAC simulation, first we have performed path integral MD simulation [11] to evaluate the imaginary time heat flux auto correlation function. The number of molecules and the number of beads were 256 and 48, respectively. The time step was 0.5 fs. At every 10 steps, we calculated the imaginary time heat flux correlation function of equation (8) and we repeated this calculation 1000 times. We considered this data as a set and we calculated 1000 data sets. These data sets were used for the inverse Laplace transform of equation (7). We applied the same procedure as described in Ref. [7] to obtain the spectral density using the MEAC method. The simulation conditions were saturation points at 14 K and 30 K. We calculated the thermal conductivity of liquid hydrogen using the spectral density and clarified the quantum effect on the transport properties.

## 3. Results and discussion

Temperature dependence of the pressure formula of classical MD and CMD at constant density $\rho$ = 80 kg/m$^3$ is shown in figure 1. In this figure, the total pressure of the CMD is consistent with experimental data [12] in whole temperature region. By contrast, there are disagreements between the classical MD results and experimental data in the low temperature region. Moreover, from the comparison of the each term of the pressure formula, it is found that although the kinetic pressure of each method agrees well with each other in whole temperature region, the virial pressure of the CMD is larger than classical case and the difference of the virial pressure is equivalent to the total pressure difference. From these results it is clarified that taking account of the quantum nature produces the virial pressure larger than classical case. To clarify the quantum nature on the virial term in detail, intermolecular potential of hydrogen is derived using force matching method [13]. In this method, the force which acts on the centroid is expressed by classical like force using least squares approximation. The fitted results are shown in figure 2. As shown in this figure, the potential which acts on the centroid depends on the temperature. Additionally, in the case of the lower temperature, the potential well is shallower than and the well positions are much more far away from the original SG potential well position. These results suggest that taking into account the quantum nature makes weaker intermolecular interaction energy and broader repulsive force than classical representation. This is the

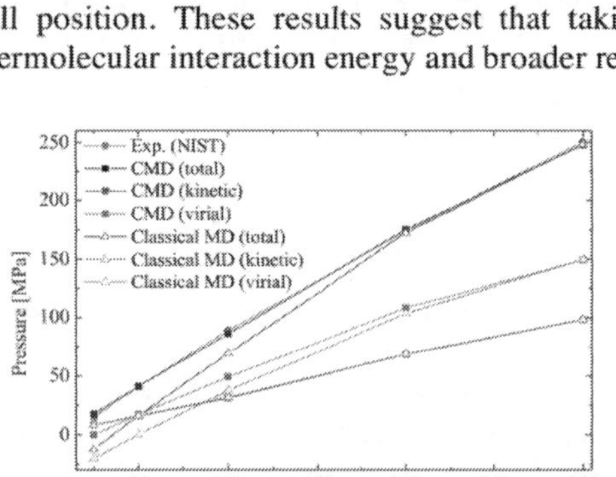


**Figure 1.** Comparison of temperature dependence of pressure at constant density (80 kg/m$^3$)

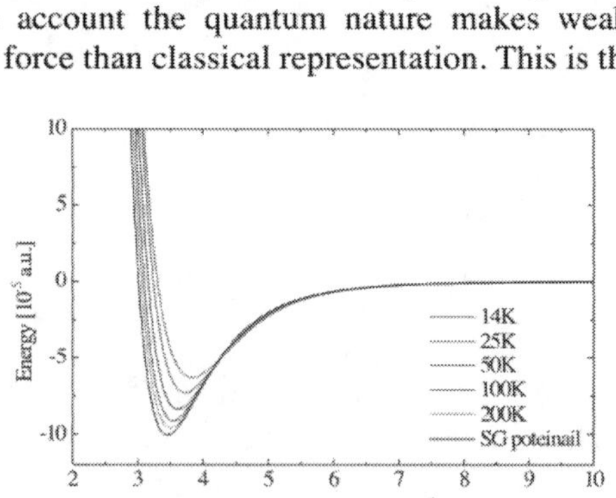


**Figure 2.** Comparison of mean intermolecular potential between hydrogen molecules.

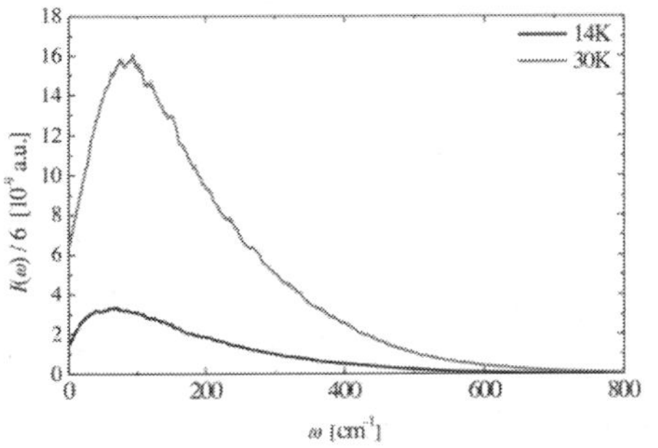


**Figure 3.** Comparison of spectral density of heat flux correlation function at 14 K and 30 K.

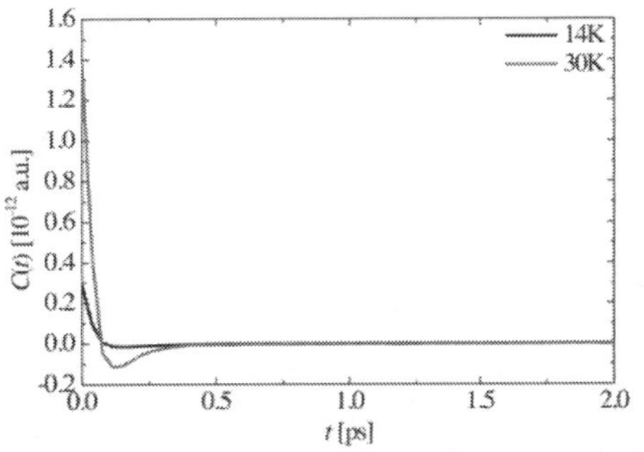


**Figure 4.** The quantum heat flux correlation functions at 14 K and 30 K.

reason why the virial pressure increase.

Next, the comparison of the spectral density of heat flux correlation function is shown in figure 3 and its quantum correlation function is shown in figure 4. The thermal conductivity can be obtained from the heat flux correlation through the Green-Kubo method. The thermal conductivity result of each temperature is 0.105 W/m/K and 0.0650 W/m/K, respectively. These results are in good agreement with the experimental data (0.0767 W/m/K and 0.0937 W/m/K) [12] as compared with those of the classical method results (0.239 W/m/K and 0.110 W/m/K) especially in low temperature region. From these results we found that the quantum effect of hydrogen molecule on its transport properties is large. However, it is need to more detailed analysis of quantum effect on the thermal conductivity, for example the quantum effect on the molecular mechanism of energy transport.

## 4. Concluding remarks

In this paper, we have analyzed the quantum effect of the hydrogen molecule on its thermodynamic and transport properties using molecular dynamics (MD) method based on the path integral method. We calcified that the quantum effect is large and taking into account the quantum nature makes weaker intermolecular interaction energy and broader repulsive force than classical representation.

**Acknowledgments**

This study has been supported by the Ministry of Education, Science, Sports, and Culture, Grant-in-Aid for Scientific Research (B) (No. 23360380) and (No. 24360065), and the Collaborative Research Project of the Institute of Fluid Science, Tohoku University.

## References

[1] Nagashima H, Tokumasu T, Tsuda S, Tsuboi N, Koshi M, and Hayashi A K, 2012 *Mole. Sim.*, **38** 404.
[2] Chandler D and Wolynes P G, 1981 *J. Chem. Phys.*, **74** 4078.
[3] Cao J and Voth G A, 1993 *J. Chem. Phys.*, **100**, 5093.
[4] Hone T D, Rossky P J, and Voth G A, 2006 *J.* Chem. Phys., **124** 154103.
[5] Jang S and Voth G A, 1999, *J.* Chem. Phys., **111** 2357.
[6] Liu J, Alder B J, and Miller, W H, 2011 *J.* Chem. Phys., **135** 114105.
[7] Rabani E, Reichman D R, Krilov G, and Berne B J, 2002 *Proc. Natl. Acad. Sci. U.S.A.* **99** 1129.
[8] Martyna G J, Tuckerman M E, Tobias D J, and Klein M L, 1996 *Mole. Phys.*, **87** 1117.
[9] Silvera I F and Goldman V V, 1978 *J. Chem. Phys.*, **69** 4209.
[10] Martyna G J, Hughes A, and Tuckerman M E, 1999, *J. Chem. Phys.* **110** 3275.
[11] Habershon S, Braams B J, and Manolopoulos D E, 2007 *J. Chem. Phys.* **127** 174108.
[12] NIST database: http://webbook.nist.gov/chemistry/.
[13] Izvekov S, Parrinello M, Burnham C J, and Voth G A, 2004 *J. Chem. Phys.*, **120** 10896.

# Mathematical modeling of nanomachining with atomic force microscope cantilevers

Win-Jin Chang, Haw-Long Lee, and Yu-Ching Yang

Department of Mechanical Engineering
Kun Shan University, Tainan 71003, Taiwan

E-mail: changwj@mail.ksu.edu.tw

**Abstract.** This article theoretically analyzes the cutting depth and material removal rate of an atomic force microscope (AFM) cantilever during nanomachining. An analytical expression for the vibration frequency and displacement of the cantilever has been obtained by using the modified couple stress theory. The theory includes one additional material length scale parameter revealing the micro-scale effect. According to the analysis, the results show that the effect of size-dependent on the vibration behavior of the AFM cantilever is obvious. The maximum displacement of nanomachining with the AFM cantilever represents the cutting depth. The area under the displacement-time curve is related to the material removal rate. When the excitation frequency is closer to the nature frequency of the cantilever, a larger material removal rate is obtained.

## 1. Introduction

The atomic force microscope (AFM) has become an essential tool for the measurement of surface characteristics of diverse materials on a micro- and nanoscale level since its invention by Binnig et al.[1-4]. In addition, AFM can also be applied to nanolithography. AFM-based nanolithography technique, utilizes the tip of an AFM cantilever, has a very high potential for nanofabrication [5-8]. The nanolithography research can be roughly divided into three categories: (1) local electro-chemical reactions confined by the tip, (2) atomic and molecular manipulations, and (3) direct nanomachining of materials. The direct nanomachining technique for nanostructured materials has a capability to provide the opportunity of quickly fabricating nanodevices for applications in micro/nanoelectromechanical systems.

In the last years, the nanomachining technique has been the interest of many researchers. For example, Voigt et al. [9] utilized a chip cantilever system for material processing tasks on the micro- and nanometer scales by the flexural-torsional resonance mode. Zhu et al. [10] investigated the AFM-based nanometric cutting process of copper using molecular dynamics simulation. Recently, Zhang and Dong [11] utilized AFM to perform a high-rate tunable nanomachining-based nanolithography through controlling of the vibration between the cantilever tip and sample.

In this article, we develop an analytical model based on the modified couple stress theory [12] to analyze the dynamic displacement of AFM-based nanomachining. The theory has been applied to study microbeams and microplates [13-16]. In addition, some researchers utilized the modified couple stress theory to study the vibration behaviors of an AFM cantilever [17-18]. Here the effects of the material length scale parameter on the dynamic response of nanomachining with an AFM cantilever are also investigated.

## 2. Analysis

An AFM probe is used to machine the specimen and it is considered as a cantilever beam as shown in Figure 1.

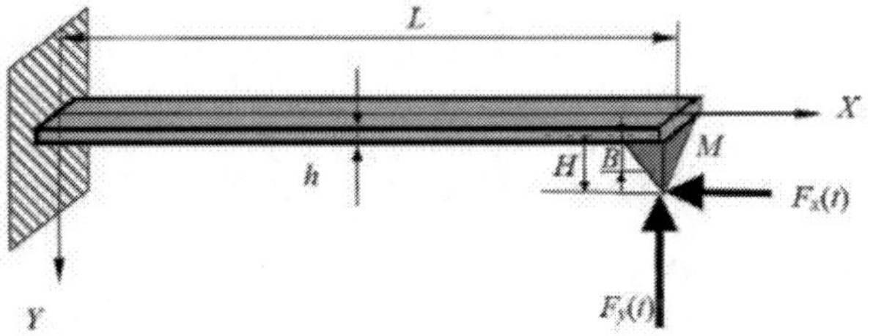


Figure 1. Schematic diagram of nanomachining on a sample by an AFM cantilever.

The probe has the Young's modulus $E$, shear modulus $G$, moment of inertia $I$, density ρ, uniform corss-section $A$, and length $L$. When the machining is in progress, the cantilever tip contacts with the specimen and induces a vertical reaction force, $F_y(t)$, and a horizontal reaction force, $F_x(t)$, which are a function of time $t$. Assuming that the reaction forces are on the tip end, then the product of the horizontal force and the tip length can form the equivalent moment exerted on the cantilever. The cutting system of the AFM cantilever can be modeled as a flexural vibration problem. The motion of the vibration cantilever is a partial differential equation and its transverse displacement is dependent on time $t$ and the spatial coordinate $X$. The governing equation of transverse vibration based on the Euler-Bernoulli beam theory can be given by [17]

$$(EI+GAl^2)\frac{\partial^4 Y}{\partial X^4}+\rho A\frac{\partial^2 Y}{\partial t^2}=0 \tag{1}$$

where $l$ is the material length scale parameter which indicates the size-dependent vibration behavior of the microcantilever based on the modified coupled stress and $Y(X,t)$ is the transverse displacement. The corresponding boundary conditions for nanomachining by the AFM cantilever are [19]

$$Y(0,t)=0 \tag{2}$$

$$\frac{\partial Y(0,t)}{\partial X}=0 \tag{3}$$

$$(EI+GAl^2)\frac{\partial^2 Y(L,t)}{\partial X^2}=-HF_x(t)-MB^2\frac{\partial^3 Y(L,t)}{\partial X\partial t^2} \tag{4}$$

$$(EI+GAl^2)\frac{\partial^3 Y(L,t)}{\partial X^3}=F_y(t)+M\frac{\partial^2 Y(L,t)}{\partial t^2} \tag{5}$$

where $H$ and $M$ is the height and mass of the tip, respectively. $B$ is the distance between the lower edge of the cantilever and centroid of the tip. The $F_y(t)$ and $F_x(t)$ are a function of time $t$ and denote the vertical and horizontal cutting force on the sample under the normal and lateral direction, respectively. The relationship between $F_x$ and $F_y$ can be expressed as $F_x=\frac{2\cot\theta}{\pi}F_y$ which is obtained from a geometrical relation for a cone shape cantilever tip and $\theta$ is a half-conic angle.

The vertical cutting force is assumed to be as follows:

$$F_y(t)=P\sin(\Omega t) \tag{6}$$

where $\Omega$ is the excitation frequency of the cutting force, and $P$ is a constant force.

In order to simplify the manipulation, the following dimensionless variables are introduced

$$x=\frac{X}{L},\quad y=\frac{Y}{L},\quad \delta=\frac{B}{L},\quad m=\frac{M}{\rho AL},\quad \tau=\frac{t}{\sqrt{\rho AL^4/EI}},$$

$$\Lambda = \frac{\Omega}{\sqrt{EI/\rho AL^4}},\ \eta = \frac{GAl^2}{EI} = \frac{12G}{E(h/l)^2} \tag{7}$$

where $v$ is the dimensionless transverse displacement of the cantilever. $\tau$ is the dimensionless time. $m$ and $\delta$ are the dimensionless mass and centroid of the tip, respectively. $\Lambda$ is the dimensionless excitation frequency in the transverse direction of the cantilever. $\eta$ is the dimensionless additional flexural rigidity based on the modified coupled stress theory. Meanwhile, $h/l$ is the ratio of the cantilever thickness to material length scale parameter. Accordingly, the dimensionless vertical and horizontal cutting forces are expressed as follows:

$$F_y(\tau) = f_y \sin(\Lambda\tau) \text{ and } F_x(\tau) = f_x \sin(\Lambda\tau) \tag{8}$$

where $f_y = \frac{PL^2}{EI}$ and $f_x = \frac{2\cot\theta}{\pi}\frac{PL^2}{EI}$ denote the dimensionless constants and are relevant to cutting force in the vertical and horizontal directions, respectively. In addition, the transverse displacement can be separated into the dimensionless spatial variable y (x) and the dimensionless time and it is

$$v = y(x)\sin(\Lambda\tau) \tag{9}$$

Substituting the harmonic solution given by Eq. (6) into Eqs.(1)-(5) and using the parameters given by Eqs. (8) and (9), the governing equation and boundary conditions can be simplified to the following dimensionless equations:

$$(1+\eta)\frac{d^4 y}{dx^4} + \Lambda^2 y = 0, \tag{10}$$

$$y(0) = \frac{dy(0)}{dx} = 0 \tag{11}$$

$$(1+\eta)\frac{d^2 y(1)}{dx^2} = -f_x\frac{H}{L} + \Lambda^2 m\delta^2\frac{dy(1)}{dx} \tag{12}$$

$$(1+\eta)\frac{\partial^3 y(1,\tau)}{\partial x^3} = f_y - \Lambda^2 m y(1) \tag{13}$$

The solutions of y(x) can be expressed as the following form:

$$y(x) = Q_1\sin(\gamma x) + Q_2\cos(\gamma x) + Q_3\sinh(\gamma x) + Q_4\cosh(\gamma x) \tag{14}$$

where $Q_1 ... Q_4$ are arbitrary constants. $\gamma$ is the wave number and it is

$$\gamma^4 = \frac{\Lambda^2}{1+\eta}. \tag{15}$$

Substituting Eq. (14) into Eqs. (11)-(13), the constants $Q_1 ... Q_4$ can be solved and expressed as follows:

$$Q_1 = \frac{f_y\phi_1 + f_x\psi_1}{2(1+\eta)\gamma^3\Delta(\gamma)} \tag{16}$$

$$Q_2 = \frac{f_y\phi_2 + f_x\psi_2}{2(1+\eta)\gamma^3\Delta(\gamma)} \tag{17}$$

$$Q_3 = -Q_1,\ Q_4 = -Q_2 \tag{18}$$

where

$$\Delta(\gamma) = m^2\delta^2\gamma^4(\cosh\gamma\cos\gamma - 1) + m\gamma[(\delta^2\gamma^2+1)\cosh\gamma\sin\gamma + (\delta^2\gamma^2-1)\sinh\gamma\cos\gamma] - (1+\cosh\gamma\cos\gamma) \tag{19}$$

$$\phi_1 = \cosh\gamma + \cos\gamma - m\delta^2\gamma^3(\sinh\gamma + \sin\gamma) \tag{20}$$

$$\psi_1 = \frac{H}{L}\gamma[m\gamma(\cosh\gamma - \cos\gamma) + \sinh\gamma - \sin\gamma] \tag{21}$$

$$\phi_2 = m\delta^2\gamma^3(\cosh\gamma - \cos\gamma) - (\sinh\gamma + \sin\gamma) \tag{22}$$

$$\psi_2 = -\frac{H}{L}\gamma[\cosh\gamma + \cos\gamma + m\gamma(\sinh\gamma - \sin\gamma)] \quad (23)$$

If we set $\Delta(\gamma) = 0$ given in Eq.(19), the characteristic equation of free vibration for the AFM cantilever can be obtained. And then the wave number of free vibration $\gamma_n$ will be solved. In addition, using the relationship between the wave number and frequency given by Eq. (15), the dimensionless free vibration frequency $\omega_n$ based on the modified couple stress theory can be determined by

$$\omega_n = \gamma_n^2\sqrt{1+\eta}\,. \quad (24)$$

While the dimensionless excitation frequency $\Lambda$ can be set as follows:

$$\Lambda = r\omega_n \quad (25)$$

where $r$ is the frequency ratio.

In addition, the displacement of nanomachining with the AFM cantilever can be determined from Eqs. (14), (24), and (25).

## 3. Results and discussion

The cutting force in the vertical direction is assumed to be as follows:

$$F_y(\tau) = f_y(\sin r\omega_n\tau + \frac{1}{3}\sin 3r\omega_n\tau + \frac{1}{5}\sin 5r\omega_n\tau) \quad (26)$$

The following material properties and geometrical parameters are used [5,19]: $L$ = 300 μm, $E$ = 170 GPa, G = 66.4 GPa , $I = 33.3\times10^{-24}$ m$^4$, $A = 10^{-10}$ m$^2$ , $H$=10μm,ρ= 2300 kg/m$^{-3}$, $M = 2\times10^{-13}$kg, $B$ = 2.5 μm, $P = 10^{-8}$ nN, $f_y = 0.16\times10^{-3}$ . In addition, the excitation frequency cannot be the same with the natural frequency of the cantilever.

Figure 2 shows the comparison of dimensionless dynamic displacement of the free end of an AFM cantilever obtained by the modified couple stress theory with H/L =1/30 and h/l=4 for various frequency ratios $r$ during nanomachining. The frequency ratio is defined as the ratio of the excitation frequency to the natural frequency of mode 1 of the cantilever.

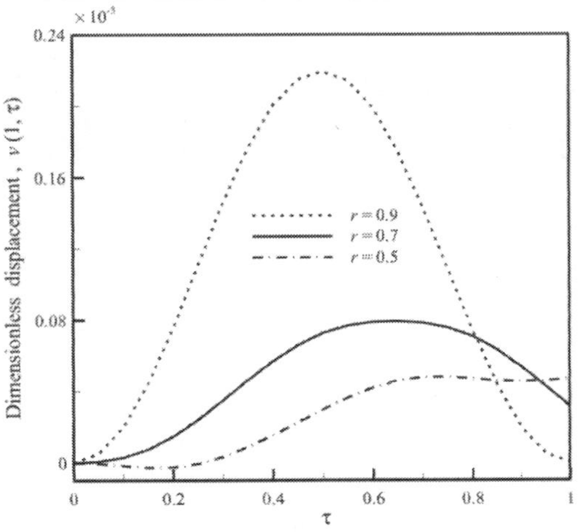


Figure 2. The dimensionless dynamic displacement of the free end of an AFM cantilever obtained by the couple stress theory with H/L =1/30 and h/l=4 for various frequency ratios $r$.

For different frequency ratios, the different displacements of the AFM cantilever are obtained. As expected, a larger vibration displacement is obtained as the $r$ value is closer to1. A larger displacement indicates a larger cutting depth. In addition, the figure can be used to predict peak displacement. A higher peak displacement indicates a larger cutting depth. The area under the curve is related to the material removal rate. A larger material removal rate is obtained when the value of r increases.

Figure 3 illustrates the dynamic displacement of the free end of an AFM cantilever obtained by the modified couple stress theory with H/L =1/30 and r = 0.9 for various ratios of microcantilever thickness to material length scale parameter *h/l*. It is noted that for the case of $h/l=\infty$, it implies the result obtained based on the classical beam theory. It can be seen that the displacement increases with increasing the value of h/l. This is because the effect of material length scale parameter on the displacement becomes small when the value of h/l increases. The phenomena show size-dependent vibration behavior of the cantilever.

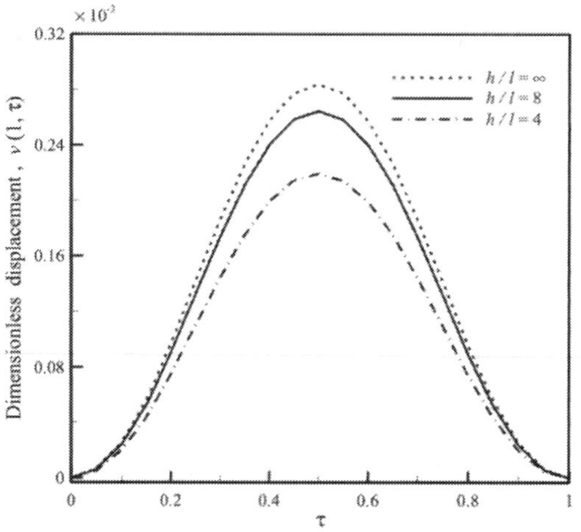


Figure 3. The dynamic displacement of the free end of an AFM cantilever obtained by the modified couple stress theory with H/L =1/30 and r = 0.9 for various ratios of microcantilever thickness to material length scale parameter *h/l*.

## 4. Conclusions

According to the analysis, the displacement increased with increasing the ratio of the microcantilever thickness to material length scale parameter. The size-dependent vibration behavior of the AFM cantilever was obvious. The displacement of the AFM cantilever is related to the cutting depth. A large material removal rate is obtained when the excitation frequency was close to the natural frequency of the cantilever.

**Acknowledgment**
The authors wish to thank the National Science Council of the Republic of China in Taiwan for providing financial support for this study under Projects NSC 101-2221-E-168 -009, NSC 101-2221-E-168-006, and NSC 101-2221-E-168 -016.

## References

[1]G. Binnig, C. F. Quate and C. Gerber, Phy. Rev. Lett. 56 (1986) 930.
[2]T. H. Fang, W. J. Chang, and C. I Weng, Mater. Chem. Phys. 92(2005) 379.
[3]G. Tosello, A. Gava, H.N. Hansen, G. Lucchetta, and F. Marinello, Wear 266(2009)534.
[4]T. Ogawa, S. Kurachi, M. Kageshima, Y. Naitoh, Y. J. Li, and Y. Sugawara, Ultramicroscopy 110 (2010)612.
[5]T. H. Fang and W. J. Chang, J. Phys. Chem. Solids 64 (2003) 913.
[6]T. H. Fang and W. J. Chang, Physica B: cond. Matt. 352, (2004) 190.
[7]S. Buyukkose, S. Okur and G. Aygun, J. Phys. D: Appl. Phys. 42(2009) 105302.
[8]R. V Mart′ınez, J. Mart′ınez, and R. Garcia, Nanotechnology 21 (2010) 245301.

[9]F. Voigt and F. Krohs, and R. Gerbach, J. Vac. Sci. Technol. B 27 (2009)958.
[10]P. Zhu, Y. Hu, T. Ma, and H. Wang, Appl. Sur. Sci. 256 (2010) 7160.
[11]L. Zhang and J. Dong, Nanotechnology 23(2012) 085303.
[12] F. Yang, A.C.M. Chong, D.C.C. Lam, and P. Tong, Int. J. Solids Struct. 39 (2002) 2731.
[13]S.K. Park and X.L. Gao, , J. Micromech. Microeng. 16 (2006)2355.
[14]H.M. Ma, X. L. Gao, and J.N. Reddy, J. Mech. Phys. Solids 56 (2008) 3379.
[15]M. Abbasi and A.K. Mohammadi, Ultramicroscopy 110(2010) 1374.
[16] L.L. Ke, Y.S. Wang, J. Yang, and S. Kitipornchai, J. Sound Vib. 331(2012) 94.
[17]M.H. Kahrobaiyan, M. Asghari, M. Rahaeifard, and M.T. Ahmadian, Int. J. Eng. Sci. 48 (2010) 1985.
[18]H.L. Lee and W.J. Chang, Microelectr. Eng. 88,(2011) 3214.
[19]T.S. Wu, W.J. Chang , and J.C. Hsu, Microelectr. Eng. 71 (2004)15.

# Inverse calculation of three-dimensional solvation structure on an arbitrary surface from a force distribution measured by liquid AFM

**Ken-ichi Amano[1*] and Ohgi Takahashi[1]**

[1]Faculty of Pharmaceutical Sciences, Tohoku Pharmaceutical University, 4-4-1 Komatsushima, Aoba-ku, Sendai, 981-8558, Japan.

[*]aman@tohoku-pharm.ac.jp

**Abstract.** Recent frequency-modulated atomic force microscopy (FM-AFM) can measure the three-dimensional force distribution between a probe and a sample surface in liquid. The force distribution is currently regarded to approximate the solvation structure on the surface, because shapes of the force distribution and the solvation structure are somewhat similar to each other. However, the force distribution is never the solvation structure. Therefore, we propose a method that converts the force distribution to the solvation structure. (This conversion is an example of inverse calculations.) A new benefit of the method is that it can perform the inverse calculation without any simplifications of the shape and the solvation affinity of the actual probe tip in a three-dimensional system.

## 1. Introduction

Recent frequency-modulated atomic force microscopy (FM-AFM) can measure the three-dimensional force distribution between a probe and a sample surface in liquid [1]. The force distribution is currently regarded to approximate the solvation structure on the surface (local liquid's density), because shapes of the force distribution and the solvation structure are somewhat similar to each other. However, the force distribution is never the solvation structure. When we wish to obtain the solvation structure by using the liquid AFM, a method for converting the force distribution into the solvation structure is necessary. Therefore, in this paper, we briefly show a converting method. We call this method an inverse calculation, because in a general calculation process the solvation structure is first obtained and the force distribution is obtained next.

There have already been two conversion methods in the researches of the inverse calculation problem. Amano *et al.* [2,3] and Watkins and Reischl [4] independently constructed a relational expression between the force and the solvation structure. The relational expression can be used for converting the force into the solvation structure. However, in the theory, the probe tip is approximated to be one solvent particle and the other parts of the probe are completely ignored. The model probe is called an ideal probe. Although the introduction of the ideal probe enables us to perform the inverse calculation very easily, the approximation is considered to be not small. To reduce the character of approximation, Amano [5] proposed another conversion method. In the method, the probe is modeled as a sphere with an arbitrary size and its solvation affinity is also arbitrary. However, the sample surface is modeled as a spherical shell and the calculation system is limited to a one-dimensional system in order to finalize the calculation within a practical time length. Furthermore, this method uses an empirically optimized trial function. Since the above two methods contain these problems, another efficient solution has to be developed.

The solution for the inverse problem proposed here is clearly different from the above two previous methods. In the new method, the probe tip with an arbitrary shape is employed in the theory. Furthermore, the model probe can take various surface properties. Of course, the conversion can be performed for three-dimensional systems. These improvements are realized by the completely new approach as explained in the next chapter.

## 2. Inverse calculation of three-dimensional solvation structure

In what follows, we explain the method for converting the force distribution measured by the liquid AFM to the three-dimensional solvation structure. As mentioned above, the system considered here is three-dimensional and the shape and solvation affinity of the probe tip are arbitrary. The shape and solvation affinity of the sample surface are also arbitrary. A solid plate which is fixed at a given position represents the surface sample. Here, we consider the probe and the solid plate in a simple liquid, i.e. an ensemble of small spheres where the two-body potential between the spheres is rigid, square-well, or Lennard-Jones, etc. In this situation, "mean force between the probe and the solid plate in the solvent" *minus* "two-body force between them in vacuum", which we call the solvation force acting on the probe ($\boldsymbol{f}_{DP}$), can be expressed as:

$$\boldsymbol{f}_{\mathrm{DP}}(\mathbf{r}_{\mathrm{D}},\mathbf{r}_{\mathrm{P}}) = \int \rho\,(\mathbf{r};\mathbf{r}_{\mathrm{D}},\mathbf{r}_{\mathrm{P}})\frac{\partial u_{\mathrm{DS}}(\mathbf{r};\mathbf{r}_{\mathrm{D}},\mathbf{r}_{\mathrm{P}})}{\partial \mathbf{r}}d\mathbf{r}, \quad (1)$$

where the subscripts D and P represent the solid plate and the probe, respectively, and the integral represents a volume integral. The solvation force is assumed to be a conservative one in this theory. $\mathbf{r}_D$ and $\mathbf{r}_P$ are the position vectors of the solid plate and the probe, respectively. $\varrho(\mathbf{r};\mathbf{r}_D,\mathbf{r}_P)$ is the number density of the solvent at $\mathbf{r}$, the solid plate and the probe being positioned at $\mathbf{r}_D$ and $\mathbf{r}_P$, respectively. $u_{PS}$ is the two-body potential between the probe and the solvent particle, where the subscript S represents the solvent particle. The partial differentiation of vector $\mathbf{r}$ is denoted as $\partial/\partial\mathbf{r} = (\partial/\partial x)\mathbf{e}_x + (\partial/\partial y)\mathbf{e}_y + (\partial/\partial z)\mathbf{e}_z$, where $\mathbf{e}_i$ ($i = x$, $y$, or $z$) is the unit vector along $i$-axis. Eq. (1) derived by considering an infinitesimal movement of the probe in the system is an exact one in classical statistical mechanics of liquid. This equation is strictly consistent with the contact theorem [6-9]. (The contact theorem explains the pressure on a wall, the derivation of which is performed by an infinitesimal change of the system or solute volume.) In experiments of the liquid AFM, the sample surface is fixed whereas the position of the probe is artificially changed, and they obviously do not change their own orientations. In such a case, there is no need to consider orientational (rotational) factors in Eq. (1). To connect the solvation force ($\boldsymbol{f}_{DP}$) and the solvation structure on the solid plate ($g_{DS}$), we take advantage of Kirkwood superposition approximation [10] and express $\varrho$ as $\varrho_0 \cdot g_{DS} \cdot g_{PS}$ [11,12] where $\varrho_0$ is the bulk number density of the solvent (which is constant) and $g_{ij}$ is the pair correlation function between $i$ and $j$. Here, setting the position of the solid plate at the origin ($\mathbf{r}_D = 0$), $\boldsymbol{f}_{DP}$ is rewritten as:

$$\boldsymbol{f}_{\mathrm{DP}}(\mathbf{r}_{\mathrm{P}}) = \rho_0 \int g_{\mathrm{DS}}\,(\mathbf{r})g_{\mathrm{PS}}(\mathbf{r}-\mathbf{r}_{\mathrm{P}})\frac{\partial u_{\mathrm{PS}}(\mathbf{r}-\mathbf{r}_{\mathrm{P}})}{\partial \mathbf{r}}d\mathbf{r}. \quad (2)$$

Then, we introduce a function $\boldsymbol{q}_{PS}$ which is defined as:

$$\boldsymbol{q}_{\mathrm{PS}}(\mathbf{r}-\mathbf{r}_{\mathrm{P}}) = g_{\mathrm{PS}}(\mathbf{r}-\mathbf{r}_{\mathrm{P}})\frac{\partial u_{\mathrm{PS}}(\mathbf{r}-\mathbf{r}_{\mathrm{P}})}{\partial \mathbf{r}}. \quad (3)$$

Thus, $\boldsymbol{f}_{DP}$ is expressed as:

$$\boldsymbol{f}_{\mathrm{DP}}(\mathbf{r}_{\mathrm{P}}) = \rho_0 \int g_{\mathrm{DS}}(\mathbf{r})\boldsymbol{q}_{\mathrm{PS}}(\mathbf{r}-\mathbf{r}_{\mathrm{P}})d\mathbf{r}\,. \tag{4}$$

Noticing that both sides in Eq. (4) are three-dimensional vectors, it can be rewritten as:

$$f_{\mathrm{DP}x}(\mathbf{r}_{\mathrm{P}})\mathbf{e}_x + f_{\mathrm{DP}y}(\mathbf{r}_{\mathrm{P}})\mathbf{e}_y + f_{\mathrm{DP}z}(\mathbf{r}_{\mathrm{P}})\mathbf{e}_z = \rho_0 \int g_{\mathrm{DS}}(\mathbf{r})\left[q_{\mathrm{PS}x}(\mathbf{r}-\mathbf{r}_{\mathrm{P}})\mathbf{e}_x + q_{\mathrm{PS}y}(\mathbf{r}-\mathbf{r}_{\mathrm{P}})\mathbf{e}_y + q_{\mathrm{PS}z}(\mathbf{r}-\mathbf{r}_{\mathrm{P}})\mathbf{e}_z\right]d\mathbf{r}. \tag{5}$$

Thus, the distribution of the solvation force along the $z$-axis (which can be measured by the liquid AFM) is simply expressed as:

$$f_{\mathrm{DP}z}(\mathbf{r}_{\mathrm{P}}) = \rho_0 \int g_{\mathrm{DS}}(\mathbf{r})\, q_{\mathrm{PS}z}(\mathbf{r}-\mathbf{r}_{\mathrm{P}})d\mathbf{r} = \rho_0 \int h_{\mathrm{DS}}(\mathbf{r})\, q_{\mathrm{PS}z}(\mathbf{r}-\mathbf{r}_{\mathrm{P}})d\mathbf{r} + \rho_0 \int q_{\mathrm{PS}z}(\mathbf{r}-\mathbf{r}_{\mathrm{P}})\, d\mathbf{r}, \tag{6}$$

where $h_{\mathrm{DS}} = g_{\mathrm{DS}} - 1$, which is a so-called total correlation function. The introduction of $h_{\mathrm{DS}}$ enables us to perform the Fourier transforms, because $h_{\mathrm{DS}}(\infty) = 0$ whereas $g_{\mathrm{DS}}(\infty) = 1$. The second integral of the right-hand side in Eq. (6) becomes zero, because the integral represents the solvation force acting on the probe in the bulk solvent, i.e. when the probe is infinitely separated from the solid plate, the force acting on the probe is zero. (The second integral corresponds to the $z$-component of the solvation force of Eq. (2) with $g_{\mathrm{DS}} = 1$.) Next, we apply the forward Fourier transform to Eq. (6). This operation simplifies the form of Eq. (6):

$$\tilde{f}_{\mathrm{DP}z}(\boldsymbol{k}) = \rho_0 \tilde{h}_{\mathrm{DS}}(\boldsymbol{k})\tilde{q}_{\mathrm{PS}z}(\boldsymbol{k}), \tag{7}$$

where the symbol '~' represents the forward Fourier transform. This equation can be rewritten as

$$\tilde{h}_{\mathrm{DS}}(\boldsymbol{k}) = \frac{\tilde{f}_{\mathrm{DP}z}(\boldsymbol{k})}{\rho_0 \tilde{q}_{\mathrm{PS}z}(\boldsymbol{k})}. \tag{8}$$

Finally, the object of this derivation $g_{\mathrm{DS}}$ can be obtained as follows:

$$g_{\mathrm{DS}}(\mathbf{r}) = \mathcal{F}^{-1}\left[\frac{\tilde{f}_{\mathrm{DP}z}(\boldsymbol{k})}{\rho_0 \tilde{q}_{\mathrm{PS}z}(\boldsymbol{k})}\right] + 1, \tag{9}$$

where $\mathcal{F}^{-1}$ represents the backward Fourier transform. This is the solution for the inverse problem in the liquid AFM. This simple equation explains how to get the three-dimensional solvation structure from the three-dimensional distribution of solvation force.

## 3. Conclusions

The method for converting the force distribution along the $z$-axis measured by the liquid AFM to the solvation structure (local liquid's density) on the sample surface was derived based on the statistical mechanics of liquid where we take advantage of the contact theorem and Kirkwood superposition approximation. The input of the solvation force was assumed to be conservative. This inverse calculation requires $u_{\mathrm{PS}}$ and $g_{\mathrm{PS}}$ as inputs. An improved point against the previous conversion methods is that it can calculate the three-dimensional solvation structure without any changes of the actual shape and solvation affinity of the probe tip (i.e., the shape and solvation affinity of the probe tip are not simplified to solve the inverse calculation problem).

A concern about the new method is the inclusion of Kirkwood superposition approximation. This approximation cannot treat the phase transition occurring between the surfaces of the solid plate and the probe. Therefore, if there is a possibility of the phase transition, the conversion method should not be used. Fortunately, however, in some simple liquids, the approximation has been verified to be useful [11,12]. That is, the approximation can capture the confined solvation structure in the simple liquids. In the near future, we improve the approximation from both theoretical and empirical standpoints to increase accuracy of the conversion method.

X-ray and neutron scatterings can measure solvation structures. However, their experimental conditions are limited to the crystal surface, because they require periodicity on their samples. In contrast, the liquid AFM does not require such periodicity. Hence, the solvation structure on the amorphous plate can be obtained by using both the liquid AFM and the conversion method proposed here, if measurement accuracy of the liquid AFM is very high. The solvation structure on a biomolecule that is adsorbed on a plate is also of interest as an application of this conversion method.

Our new method contains significant progresses in the field of the liquid AFM, because the shape and solvation affinity of the probe tip need not be changed in the inverse calculation. In the near future, we will write a calculation program for the inverse calculation and show how the new conversion method works.

**ACKNOWLEDGEMENTS**
We greatly thank Masahiro Kinoshita (Kyoto University), Hiroshi Onishi (Kobe University), Ryo Akiyama (Kyushu University), and Kazuhiro Fukami (Kyoto University) for useful advices and discussions. This work was supported by "Foundation of Advanced Technology Institute", "Joint Usage/Research Program on Zero-Emission Energy Research: Institute of Advanced Energy; Kyoto University (ZE25B-30)", and partly supported by "Grant-in-Aid for Scientific Research (B) (25286009)".

**References**

[1] T. Fukuma, M. J. Higgins, and S. P. Jarvis, Biophys. J. **92**, 3603 (2007).
[2] K. Amano, K. Suzuki, T. Fukuma, and H. Onishi, arXiv, 1212.1888 (2012).
[3] K. Amano, K. Suzuki, T. Fukuma, and H. Onishi, arXiv, 1212.6138 (2012).
[4] M. Watkins and B. Reischl, J. Chem. Phys. **138**, 154703 (2013).
[5] K. Amano, arXiv, 1209.0303 (2012).
[6] D. Henderson and L. Blum, J. Chem. Phys. **75**, 5441 (1978).
[7] D. Henderson and L. Blum, J. Chem. Phys. **69**, 2025 (1981).
[8] P. Bryk, R. Roth, K. R. Mecke, and S. Dietrich, Phys. Rev. E **68**, 031602 (2003).
[9] R. Roth, J. Phys.: Condens. Matter **22**, 063102 (2010).
[10] J. G. Kirkwood, J. Chem. Phys. **10**, 394 (1942).
[11] Y. Karino, R. Akiyama, and M .Kinoshita, J. Phys. Soc. Jpn., **78**, 044801 (2009)
[12] Y. Kubota and R. Akiyama, J. Phys. Soc. Jpn. **81**, SA017 (2012).

# Monte Carlo simulation of lattice systems with RKKY interaction

**K V Nefedev[1], V I Belokon[1], V Yu Kapitan[1] and O I Dyachenko[1]**
[1] The School of Natural Sciences, Far Eastern Federal University, 8, Sukhanova St., 690950, Vladivostok, Russian Federation

E-mail: kapitan.vyu@dvfu.ru

**Abstract.** Numerical approaches to the study of the magnetic states, properties, and phase transitions in the Ising spin systems with the long-range exchange interaction is presented. The Monte Carlo calculations have been performed for a system of Ising spins on a square lattice with long-range RKKY interaction. It is shown that the Monte Carlo simulation systems RKKY interaction leads to the formation of a complex of the magnetic structure. We compared the results of simulation with experimental images of domain structure of garnet ferrite films.

## 1. Introduction

Long-range interactions are rather common in nature [1]. Examples include self-gravitating systems [2], dipolar ferroelectrics, and ferromagnets in which the interactions are anisotropic with [3] RKKY interaction [4-7] and many others.

In recent years, there has been a renewed interest in Ruderman-Kittel-Kasuya-Yosida (RKKY) magnetic interaction [4-7] due to its important role in giant magnetoresistance in multilayer structures [8] and ferromagnetism in diluted magnetic semiconductors [9]. More recently, the controllable RKKY interaction attracted much attention in the field of spintronics and quantum information processing [10]. The RKKY coupling-i.e., the exchange interaction between localized core spins mediated by metallic electron gas-has been known for 50 years as the basic interaction in metallic ferromagnets. The oscillatory character of the RKKY coupling causes a spin glass behaviour in diluted magnetic metals. It rules the interlayer coupling in magnetic layered structures. As it has been shown for the last few years, the RKKY interaction is also the dominant spin interaction in diluted ferromagnetic semiconductors [9].

The lattice models have the advantage that the characteristic Hamiltonian of the system incorporates the realistic interactions between the sites and the system can be easily studied by Monte Carlo (MC) simulations [11]. Various lattice models have been introduced so far, ranging from the Ising-type models involving long range coulombian interactions [12,13] to the spin-glass type models with oscillating long range RKKY interactions [14,15].

Over the past decades, frustrated spin systems, in which all-local interactions between every spin pair cannot be satisfied simultaneously, have attracted widespread interest because very rich physics can appear in these systems [16]. The reason for this strong interest is that spin glasses exhibit a very puzzling behaviour at low temperatures, which is still not completely understood [17]. Furthermore, the theoretical treatment of these systems has led to many advances, with considerable impact on other cross-disciplinary applications such as neural networks [15] and error-correcting codes [18].

## 2. Approaches to simulation of a planar structure

### *2.1. 2-D metals*

Since the presented here results concerning the 2-D arrays, it is necessary to consider a possibility of 2-D metals creation. In papers [19-21] describes the facilities of the existence in nature and the creation of real 2-D metals. According to the authors of the above manuscripts, the two main factors that have frustrated attempts to realize an "ideal" two-dimensional (2-D) metal are disorder and Coulomb repulsion. Conventional Bloch theory begins by assuming that all metals are perfect crystals and that the electrons moving through a given material only ever see the potential that is created by the ions of its crystal lattice-that is, the Coulomb interaction between electrons is negligible.

The creation of a monolayer array of superconducting niobium islands on a normal metallic gold substrate gives hope to create 2-D metals in the future [21]. The new metallic state proposed by Eley, et al. lies in a region where these pairs are mobile but are not fully Bose-Einstein condensed into a superconducting state. Such a state would be something that we have not yet encountered. Perhaps the most apt description of such a state is that of a quantum disordered phase of the condensate [21].

Based on the foregoing, it can be argued that although modern technology has not yet reached the necessity to create 2-D metal samples level, the laws of nature do not prohibit the existence of 2-D metal, so the study of the magnetic properties of such structures is necessary in fundamental and practical points of view.

In this paper we present the results of researching of properties of Ising spins on a square lattice with RKKY interaction.

### *2.2. RKKY interaction in superspin system*

As was shown in Fischer and Klein [22], the energy of the spin system, located in the plane and interacting via RKKY exchange

$$E_{2D}(R) = -\sum_{i=1}^{N-1}\sum_{j=i+1}^{N} A\frac{\sin x}{x^2}S_iS_j\,, \tag{1}$$

in this case $k_f^2 = 2\pi n_s$ , $n_s \sim \frac{1}{a^2}$ , $R$ is the distance between particles, $R_c$ is the radius of the sphere, $a$ is the lattice constant-for example, for iron $a = 2.866$ Å.

### *2.3. The Monte Carlo simulation*

In order to study frustrated spin systems, people use different computational techniques. One of these techniques is the Monte Carlo method, which is a powerful class of algorithms that is used not only in physics but also in other fields like engineering, chemistry, biology, material science, etc. [23, 24].

Although, the obtained dynamics in the Monte Carlo simulations is intrinsic and the time evolution of the system does not come from any deterministic equation for the magnetization, the results of the Monte Carlo simulations reproduce qualitatively the trend of the experimental data. Actually, this good qualitative agreement between the simulation results and the experimental data enable us to have a better insight into the nanoscale phenomena, though some of them stem from non-equilibrium processes [25].

For the simulation, we used the Metropolis algorithm [26]. The Monte Carlo (MC) simulation technique, with the implementation of the Metropolis algorithm, has been proved a very powerful tool for the systematic study of the magnetic behaviour of nanoparticles and nanoparticle assemblies.

## 3. Comparison of the results of numerical and physical experiments

Two well-known, well-verified complementary methods of complete enumeration and Monte Carlo simulations were used to study the state space of two-dimensional system of spins in a square lattice with long-range RKKY interaction. The purpose of these experiments was to find the configuration of the magnetic system corresponding to an absolute minimum of energy, calculating the degree of degeneracy of these states and checking the convergence of the results obtained by independent methods.

For systems 3x3, 4x4 were constructed phase space, see Figure 1, 2. These systems have four equally probable states with a minimum of energy.

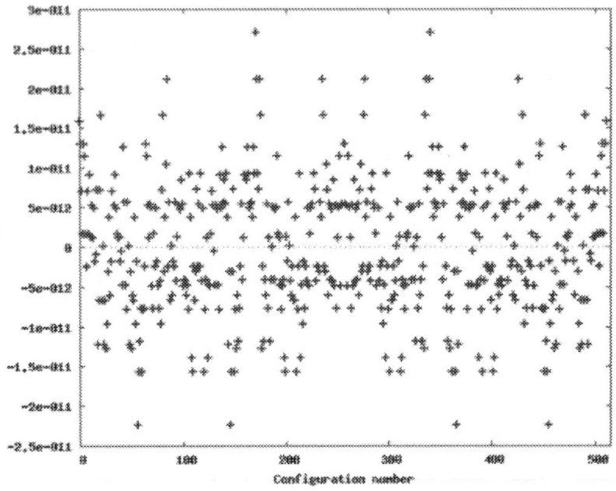


**Figure 1.** The phase space of system 3x3.

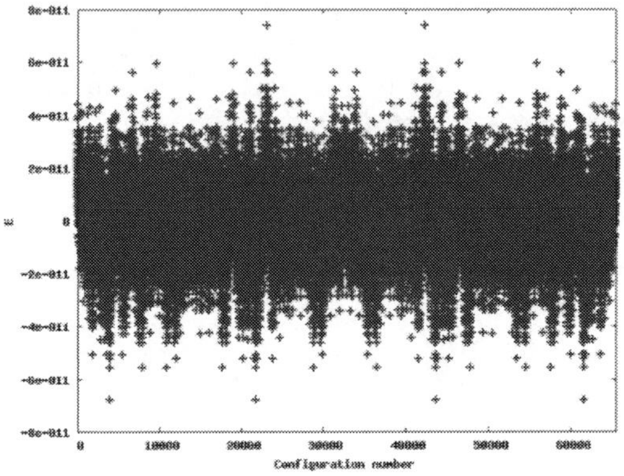


**Figure 2.** The phase space of system 4x4.

Monte Carlo simulations show that similar antiferromagnetic configurations of spins were observed for lattice systems with sizes up to 10x10 elements, see Figure 3. For systems, larger size observed domain structure, see Figure 4.

**Figure 3.** The system of 10x10 spins.

**Figure 4.** The system of 20x20 spins.

**Figure 5.** Size: 44x44 $\mu m$. The period of stripe domain structure is about 2 $\mu m$ [27].

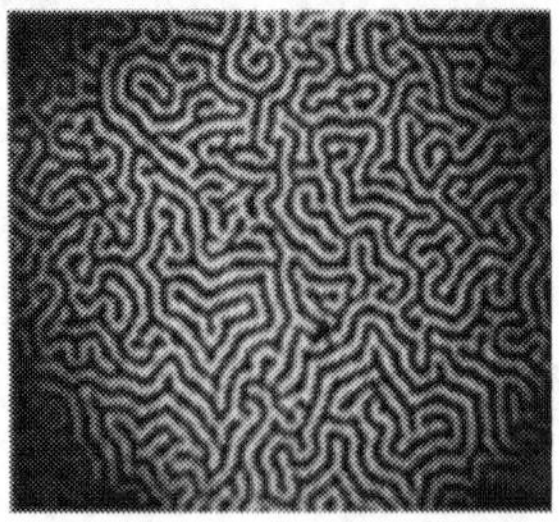

**Figure 6.** Dynamic domain structures [28].

Figure 5 shows experimental MFM image of the magnetic domain structure observed in the films of ferrite garnets. The film is thulium garnet iron, about 3 microns thick, is epitaxially grown on substrate of gadolinium gallium garnet [27]. Figure 6 shows dynamic domain structures was identified using the

Faraday method. In the experiment used an epitaxial (111) garnet ferrite film with the initial through labyrinth domain structure [28]. Numerical simulation of the Ising spin system with RKKY interaction by Monte Carlo method allows us to obtain the magnetic configuration of the labyrinth domain structure similar to those observed in iron-garnet films.

## 4. Conclusions

Monte Carlo simulation and complete enumeration in the fully connected Ising model were studied spin lattice systems with long-range indirect exchange interaction. For RKKY interacting Ising spins on a simple square lattice were studied space of states, calculated the absolute minimum of energy. Complete enumeration were investigated systems 3x3, 4x4, and 5x5, for these systems were built a phase diagram and show that there are four possible ground states. Images were obtained domain structures are presented and a qualitative comparison with experimental data was performed.

Most researchers consider that the RKKY interaction leads to frustrations and spin glass state. However, simulations shown that this interaction in the system of super-spins can lead to the formation of the domain structure, even in the absence of competition between direct of the exchange interaction and the dipole-dipole interaction.

Beyond the scope of work remained investigating, the possibility of a phase transition with Ising spin systems with direct and indirect types of exchange interaction. In the future, 3-D structures can be investigated.

Scientific Fund of Far Eastern Federal University (FEFU), #12-07-13000-18/13 and GZ-2013/287 (2.8649.2013), supported this work.

## 5. References


[1] Mukamel D 2009 arXiv:0905.1457

[2] Chavanis P H 2002 Springer-Verlag New York

[3] Landau L D and Lifshits E M 1969 Course of theoretical physics. v.5: Statistical Physics

[4] Ruderman M A and Kittel C 1954 *Phys. Rev.* **96** 99

[5] Kasuya T 1956 *Prog. Theor. Phys.* **16** 45

[6] Yosida K 1957 *Phys. Rev.* **106** 893

[7] Kittel C 1968 In Solid State Physics (Academic, New York) **22**

[8] Parkin S S P, More N, and Roche K P 1990 *Phys. Rev. Lett.* **64** 2304

[9] Dietl T, Ohno H, Cibert J, and Ferrand D 2000 *Science* **287** 1019

[10] Glazman L I and Ashoori R C 2004 *Science* **304** 524

[11] Tyukodi B, Chioar I-A and Neda Z 2013 *Cent. Eur. J. Phys.* **11(4)** 487-496

[12] Sampaio L C and de Albuquerque M P, de Menezes F S 1996 *Phys. Rev. B* **54** 6465

[13] de Menezes F S and Sampaio L C 2005 *Phys. Rev. B* **72** 104413

[14] Iglesias J R, Goncalves S, Nagel O A and Kiwi M 2001 *J. Magn. Magn. Mater.* **548**, 226-230

[15] Muller B, Reinhardt J 1991 Neural Networks - An Introduction **67**, 73

[16] Diep H T 2004 Frustrated Spin Systems, World Scientific Publishing Co. Pte. Ltd

[17] Hartmann A K 2008 *Lecture Notes in Physics* **736**, 67-106

[18] Sourlas N 1994 edited by: Back P. Grassberger, J.-P. Nadal (Kluwer Academic, Amsterdam) **67**

[19] Simmons M Y and Hamilton A R 1999 *Nature* **400** 715-717

[20] Annett J F 2012 *Nature Physics* **8** 8-9

[21] Eley S, Gopalakrishnan S, Goldbart P M, and Mason N 2012 *Nature Physics* **8** 59–62

[22] Fischer B and Klein M W 1974. *Phys.Rev. B* **11** 5 2025-2029

[23] Berry H D and Singh P 2012 *J. of Physical Science and Application* **2 (7)** 216-223

[24] Earl D J and Deem M W 2005 *Preprint Physics* 050811.

[25] Landau D P and Binder K 2009 Cambridge University Press **489**

[26] Metropolis N and Ulam S 1949 J. *Amer. statistical assoc.* **44** 247 335-341.

[27] Alexeev A M, et al 2000 Annual of New Magnetic Materials for Microelectronics **17** 467-469

[28] Kandaurova G S, Pashko A G and Osadchenko V Kh 2009 *FTT (PSS)* **51** 5 911-915

# Magnetic and thermodynamic properties of nanodisks by Monte Carlo simulations

**Elena Konstantinova, José Antonio de Sales**
Departamento de Educação e Ciências, Instituto Federal de Sudeste de Minas Gerais (IF Sudeste MG), Campus Juiz de Fora, 36080-001, Juiz de Fora, MG, Brasil.

**Abstract.** The idea that the properties of nano materials depend on their crystalline structure and the form and geometric sizes of them is discussed in this paper. The progress in the growth area of nano magnetic devices attracts the interest from both experimental and theoretical study of the nano structures for potential applications in industry. In various fields of electronics there is need to have nano materials with a wide range of magnetic properties. In this case theoretical research could provide a way to get the nano compounds with the required characteristics. In this paper we present the results of Monte Carlo simulations of magnetic nanodisks, which are based on simple cubic and body-centered cubic unit cell. The magnetization of spin, magnetic susceptibility and specific heat are investigated for nanodisks with the different diameters, thicknesses and for different values of the ratio of the correlation constants. The combination of dipolar and Heisenberg model interaction are considered for ferromagnetic case and are show that the magnetic and thermodynamic properties of the nanostructures are strongly dependent on their geometry. The structures with body-centered cubic unit cell demonstrate stronger dependences on the thickness of the disks and also higher critical temperature and wider hysteresis loop.

## 1. Introduction

The miniaturization of semiconductor devices creates strong interest in production of different nano magnetic structures. It explains that in recent years there has been significant progress in the creating of nano magnetic devices and extensive study of the magnetic properties of new nano compounds. (see, e.g., [1-5] and references therein). Some recent works are already dealing with new practical designs for hard discs based on the new technologies. These experimental works were devoting to create nanostructures which are capable of maintaining their magnetic properties under changing temperature or external magnetic fields. Consequently, there is a growing interest in theoretical studies of magnetic nanostructures, in particular by means of the Monte Carlo simulations [1,6-12]. The investigations in this area should be focused on the prediction of magnetic properties of the new prospective compounds that can be important for electronics. Interests of new technologies require magnetic materials with a wide range of magnetic properties. In particular, there are requests for materials with maximal permeability (rectangular hysteresis loop), materials with low losses in the reversal, materials with high initial permeability, and materials with constant permeability over a wide range of magnetic fields (slope of the hysteresis loop). The importance to obtain alloys that guarantee an acceptable thermal stability of the permanent magnetization in very small volumes due to the need scale reduction of the memory bits in the recording media [1,2].

It is expected that the geometric structure of magnetic nano materials may significantly affect their magnetic properties. In the present paper we start an analysis of some magnetic and thermodynamic properties of nanodisks with different geometries. Some of the most common magnetic compounds are based on a simple cubic unit cell or body-centered cubic cell [3,5] of ferromagnetic type. In the present work we explore the properties of the nanodisks with these types of internal structures. As a first step, we perform theoretical simulations of various properties of the model magnetic nanodisks based on the simple cubic (sc) and body-centered cubic (bcc) unit cells. The main target of our study is the dependence of the thermodynamic and magnetic properties of studied structures from their geometry and from the values of the ratio of the correlation constants. In particular, we explore the difference between the nanodisks of three different thicknesses, two different diameters and also the dependence on the type of the unit cell.

## 2. Method of calculations

In this work we consider the possible influence of the type of the unit cell, such as simple cubic and body-centered cubic cells, on thermodynamic and magnetic properties of the nanodisks with different thickness. The structures under consideration are listed in Table 1 and their geometry is illustrated in Figure 1, where we present a bcc-disk with three layers. The procedure of fitting a crystalline structure into a circular disk was defined as follows: the starting 3D structure is cut by a disk of a given radius, with center in one of the atoms. The axis of the disk is parallel to the axis of the initial cubic structure.

| Structure | *sc1-l2* | *sc1-l3* | *sc1-l4* | *sc2-l2* | *sc2-l3* | *sc2-l4* | *bcc1-l2* | *bcc1-l3* | *bcc1-l4* | *bcc2-l2* | *bcc2-l3* | *bcc2-l4* |
|---|---|---|---|---|---|---|---|---|---|---|---|---|
| Number of spins | 138 | 207 | 276 | 290 | 435 | 580 | 218 | 367 | 516 | 446 | 747 | 1048 |

**Table 1**. The numbers of the sites in the studied structures. In the names of the structures, the letters mean the type of the unit cell, and the number 1 indicates 11 sites in diameter, the number 2 indicates 15 sites in diameter and the abbreviation l with number is the thickness of the disk (layer's number).

**Fig. 1**. Body-centered cubic disk with 15 sites in diameter and three layers (bcc2-l3).

The spin configurations can be obtained using the numerical calculations within the Monte Carlo (MC) method. We take into account the combinations of dipolar and Heisenberg model interactions simultaneously, because the effect of dipolar interaction component is stronger for the small sizes of nanostructures, while for the larger sizes of the nanostructures the Heisenberg interaction is dominating (the discussion of this issue has been presented earlier in [7]). We will study the nanodisks with diameter of 11 and 15 sites in each frame, and with the thickness of two, three and four layers. We report on the thermal equilibrium magnetization, the susceptibility and the specific heat for these structures. As a result of this study one can observe how the spin configurations depend on the type of the unit cell and on the thickness of the nanodisks.

In our simulations we used a Hamiltonian model given by [12]:

$$H = -J\sum_{\langle i,j\rangle} \vec{S}_i \vec{S}_j - B\cdot\sum_i \vec{S}_i - \omega\sum_{i<j} \frac{3(\vec{S}_i\cdot\vec{e}_{ij})(\vec{e}_{ij}\cdot\vec{S}_j) - \vec{S}_i\vec{S}_j}{r_{ij}^3}.$$

In this expression the first sum represents the ferromagnetic exchange between the nearest neighbors with a coupling constant $J$, the second sum stands for the coupling of the spins to an external magnetic field $B$ and the last sum is the dipolar interaction term, where the coupling $\omega$ describes the strength of the dipole-dipole interaction. The $\vec{S}_i$ are three-dimensional magnetic moments of unit length, $\vec{e}_{ij}$ are unit vectors pointed from lattice site $i$ to the lattice site $j$ and $r_{ij}$ are the distances between these lattice sites. We considered three different values for the ratio between correlation constants $\omega/J$, which are equal as 0.001 (in accordance to [12]), 0.01 and 0.1.

For the numerical analysis of the magnetic and thermodynamic properties of nanodisks we have used the MC simulations with the Metropolis algorithm [13,14]. The Metropolis Monte Carlo algorithm enables one to obtain the macro-state equilibrium for a physical system at the given temperature T, starting from some randomly chosen initial micro-state and then proceed by performing a very large number of random transformations of the micro-states, until we arrive at the equilibrium macro-state. One spin is chosen at random and gains a new direction in a random way (note that there is no discretization of orientations in the Heisenberg model which we use here). As a result, the energy of the system changes by the value $\Delta E$. If $\Delta E < 0$, the temporary direction of the spin becomes permanent. If $\Delta E > 0$, the temporary direction becomes permanent with the probability $\exp(-\Delta E / k_b T)$. The preliminary calculations have shown that with $10^4$ Monte Carlo steps per is sufficient for our system to "forget" the initial randomly chosen configuration and arrive at the state of equilibrium.

The external magnetic field was directed along the axis $z$, which was also the axis of the nanodisk. The simulations for magnetization and magnetic susceptibility were performed for the values *B=-20, -18, ..., -4, -2, -1.5, -1, -0.5, 0, 0.5, 1, 1.5, 2, 4, ..., 18, 20*. Here the energy and the applied magnetic field are expressed in the units of $J$. In all these cases the value for the temperature was chosen to be *T=0.15*. In order to study the magnetic properties in low-temperature, all simulations have been performed for temperatures smaller than the critical temperature. The temperature is expressed in units of $J/k_b$, where $J$ is the magnitude of the coupling constant and $k_b$ is Boltzmann's constant.

We obtain the susceptibility $\chi$ along the $z$ axis by MC data, given by

$$\chi = N^{-1}\frac{1}{k_b T}\left(\langle m_z^{\,2}\rangle - \langle m_z\rangle^2\right),$$

where $N$ is the number of spins in the system, $k_b$ is Boltzmann's constant, $T$ is temperature and $\langle m_z\rangle$ is the mean magnetization per spin in direction $z$. The susceptibility was also calculated as a function of applied magnetic field with constant temperature. The specific heat $C$ we obtain from the energy fluctuations relation

$$C = N^{-1}\frac{1}{k_b T^2}\left(\langle E^2\rangle - \langle E\rangle^2\right),$$

where $\langle E\rangle$ is the mean energy per spin. For calculating the specific heat we used *B=0* and the values *T= 4.05, 3.95, 3.85, ..., 0.15*.

## 3. Results and discussions

Let us start the description of the results from spin orientations obtained by Monte Carlo simulations for different types of disks (sc and bcc) with two diameters and three different thicknesses for different values of the ratio between correlation constants. As we have already mentioned above, the simulations have been performed for the longitudinal external magnetic field with the twenty seven different values of B between -20 and +20. If the external field is sufficiently strong, one can observe that the spins are oriented along the field direction. This picture does change when one weakens the external field. Let us consider the magnetic properties. The plots of magnetization versus applied magnetic field are presented in Figures 2(a,b,c).

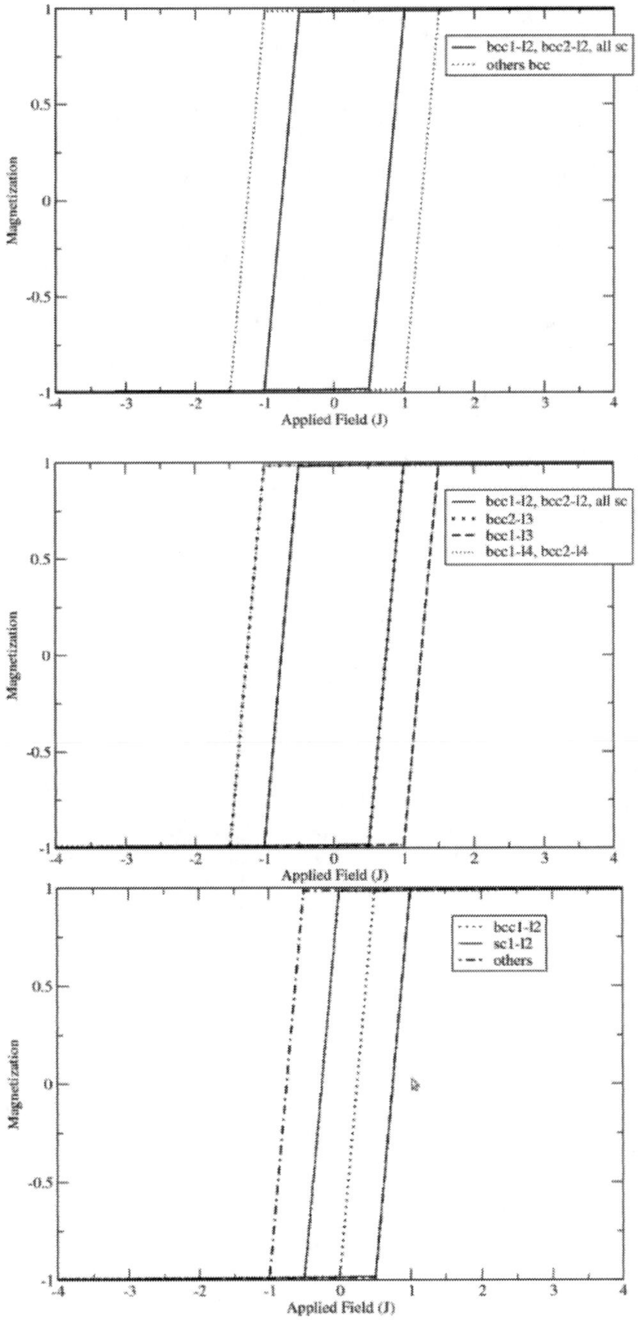


**Fig. 2.** The plots of magnetization versus applied field for (a) $\omega/J$=0.001, (b) $\omega/J$=0.01 and (c) $\omega/J$=0.1.

One can see that the hysteresis loops has stronger dependence of the values of the ratio of the correlation constants. For $\omega/J$=0.001 the hysteresis loops (see Fig. 2a) are similar for all *sc*- disks and *bcc*-disks with two layers, which demonstrate rectangular form of loops with a width close to 1.5, when others *bcc*-disks have the loop close to 2.5. For $\omega/J$=0.01 the hysteresis loops (Fig. 2b) *sc*-disks and *bcc*-disks with two layers have the loop with a width close to 1.5, *bcc*-disks with three layers for both diameters have the loop close to 2.0 and *bcc*-disks with four layers for both diameters close to 2.5. For $\omega/J$=0.1 the hysteresis loops (Fig. 2c) *bcc1-l2* disk has the loop with the width close to 0.5, *sc1-l2* disk has the loop close to 1.0 and others close to 1.5. The lower value of the ratio between the correlation constants corresponds to a greater contribution of the dipole-dipole interaction. Thus we can note that the increase of the dipole-dipole interaction leads to a decrease in the width of the loop.

Let us consider the thermodynamic behavior of the studied nano-structures. Thermal equilibrium results obtained by Monte Carlo simulations enable us to obtain the dependence of the specific heat versus temperature. The plots of specific heat versus temperatures are presented in Figures 3(a,b,c).

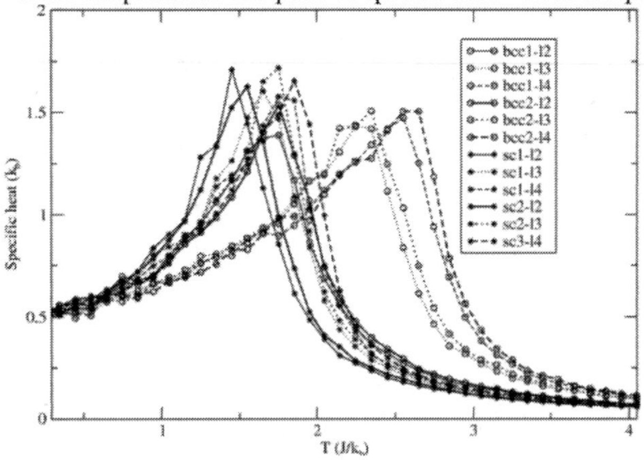


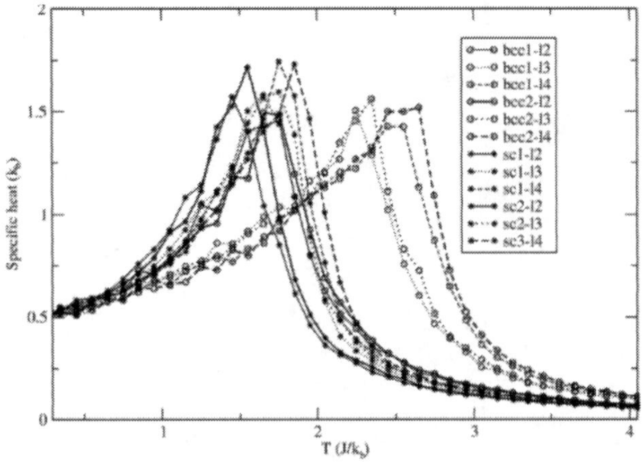

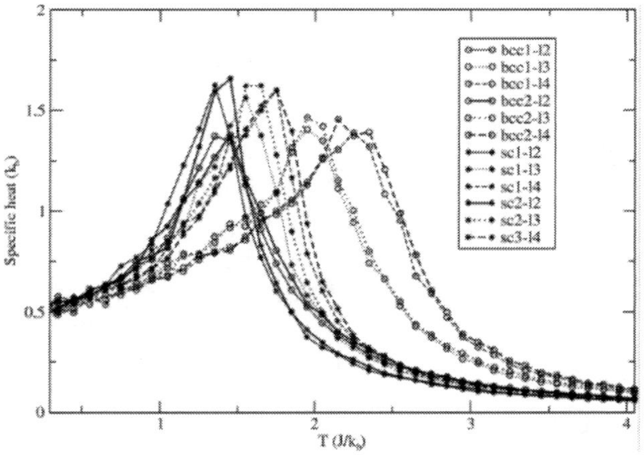


**Fig. 3.** The plots of specific heat versus temperature without external magnetic field for (a) $\omega/J$=0.001, (b) $\omega/J$=0.01 and (c) $\omega/J$=0.1.

For all types of the unit cells the nanodisks with the smaller thickness have smaller value of specific heat. One can see that the result depend on the size of the structure and number of nearest neighbors (also called coordination number). It can be noted that there is also a strong dependence on the type of crystal lattice. The critical temperatures for the *bcc* structures are bigger than the one for *sc* structures. Moreover the dependence of the thickness is stronger for *bcc* disks compared with *sc* case. Let us present some explanation of why the critical temperature varies depending on the structure and thickness of the disks. The coordination number of *sc* structure is 6 and the same number for *bcc* is 8. For the nano-disk with the two layers almost all atoms are on the surface, while for the disk with three layers there are more internal atoms with a greater number of neighbors. In the calculation of interaction energy we take into account not only immediate neighbors, but also the next order neighbors, which prove to make an essential contribution. In order to illustrate this general situation, consider some particular examples. If one consider a sphere of radius *a* around an internal atom in *sc*-disc, it is easy to see that the ball has six atoms, all of them are located at the distance *a* from the center. In comparison, similar internal atom in *bcc*-structure has eight nearest neighbors, located at the distance equal to *0.86a* and also 6 atoms which are located at the distance *a*. All in all, the difference in the thermodynamic properties of *bcc*-disks with two and three layers is definitely related to the influence of all interactions which are taken into account. The reduction of the value of the ratio between the correlation constants leads to a decrease in the spread between the values of the critical temperature for *sc* and *bcc* structures. It can be noted that difference in properties between the cases when the correlation constants are $\omega/J$=0.001 and $\omega/J$=0.01 is less, than between cases when the correlation constants are $\omega/J$=0.01 and $\omega/J$=0.1.

Very similar considerations can be done for the analysis of the Figures 4(a,b,c), which show the plots of magnetic susceptibility versus temperature. One can observe that *bcc* structures have a wider distance between peaks. Furthermore, we note that the magnitude of the peaks in Figures 4 depends on the type of the unit cell. The magnitude of susceptibility decreases with increasing number of sites in the structure. There is, once again, an increase of the critical temperature with increasing thickness for all types of the disks, a strong dependence on the thickness for the *bcc*-type of the disks and the same behavior of the ratio between the correlation constants.

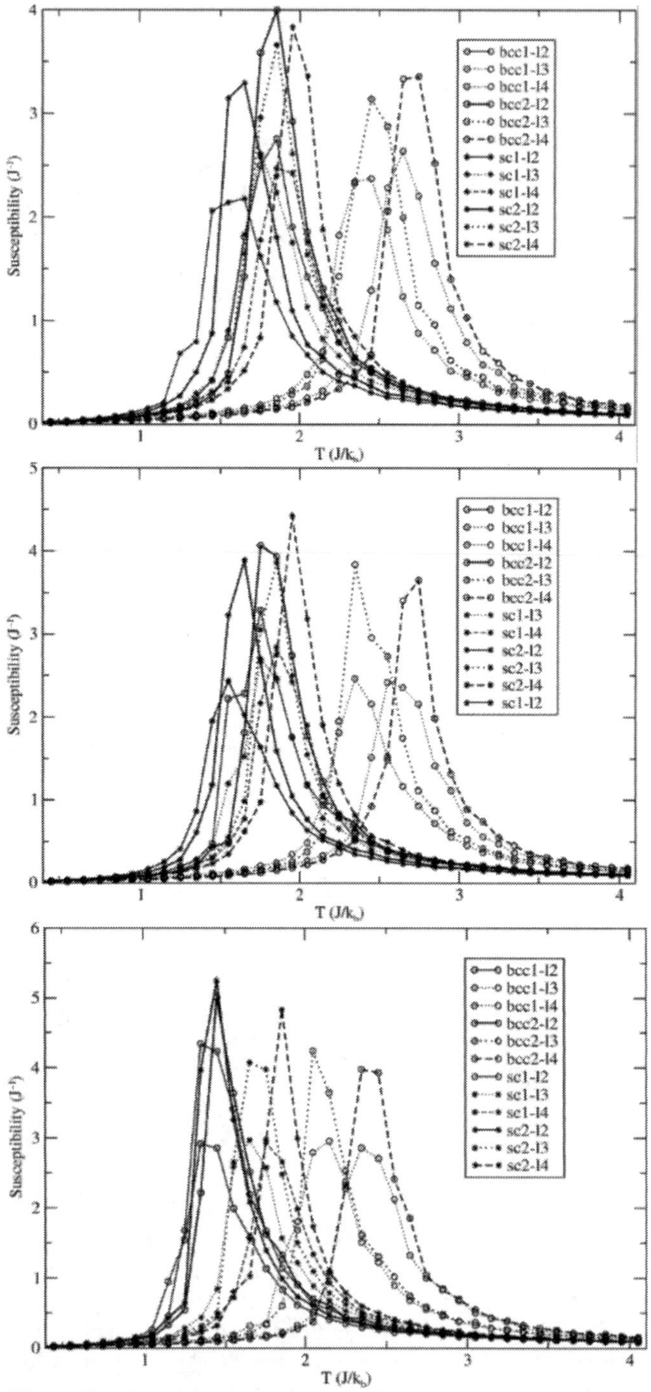


**Fig. 4.** The plot of magnetic susceptibility versus temperature without external magnetic field for (a) $\omega/J$=0.001, (b) $\omega/J$=0.01 and (c) $\omega/J$=0.1.

## 4. Conclusions

In this article we presented the results of the Monte Carlo simulations of magnetic nanodisks with the simple cubic and body-centered cubic unit cells in the structures for three different values for the correlation between constants $\omega/J$. We have found that hysteresis form, magnetic susceptibility and specific heat of such structures manifest dependence on the type of the unit cell and on the thickness of the disk. In particular, one can note that the behavior of magnetic and thermodynamic properties for the structures with *bcc* type of unit cell show greater dependence on thickness of the disks, higher critical temperature and higher width of the hysteresis loop. The value of the ratio between the correlation constants corresponds to a greater contribution of the dipole-dipole interaction and it should be noted that this ratio affects the properties of studied structures. Thus we can note that the increase of the dipole-dipole interaction leads to a decrease in the width of the loop. The reduction of the value of the ratio between the correlation constants leads to a decrease in the spread between the values of the critical temperature for *sc* and *bcc* structures.

Qualitatively similar dependence on the geometry of magnetic nanostructures can be seen in experimental works [1-5]. The MC calculations for the systems containing a large amount of spins permit to obtain a qualitative estimative of how some magnetic and thermodynamic properties of nanodisks depend on their unit cell, diameter, thickness and correlation constants. We expect to continue exploration the dependence of the magnetic properties of different nanostructures on their geometry.

### Acknowledgments

Authors are grateful to FAPEMIG for support.

### References

[1] D. Alloyeau, C. Ricolleau, C. Mottet, T. Oikawa, C. Langlois, Y. Le Bouar, N. Braidy and A. Loiseau, "Size and shape effects on the order–disorder phase transition in CoPt nanoparticles", *Nature Materials*, vol. 8, pp. 940-946, 2009.

[2] A.T. Hindmarch, D.E. Parkes, A.W. Rushforth, "Fabrication of metallic magnetic nanostructures by argon ion milling using a reversed-polarity planar magnetron ion source", *Vacuum*, vol. 86, pp. 1600-1604, 2012.

[3] H. Chen, Ch. Xu, Ch. Chen, G. Zhao, Y. Liu, "Flower-like hierarchical nickel microstructures: Facile synthesis, growth mechanism, and their magnetic properties", *Materials Research Bulletin*, vol. 47, pp. 1839–1844, 2012.

[4] A.L. Costa, B. Ballarin, A. Spegni, F. Casoli, D. Gardini, "Synthesis of nanostructured magnetic photocatalyst by colloidal approach and spray–drying technique", *J. Colloid and Interface Science*, vol. 388, pp. 31–39, 2012.

[5] A. Ognev, M. Stebliy, A. Samardak, L. Chebotkevich, "An Influence of Boundary Effects and Spatial Symmetry on Magnetization Reversal of Nanodisk Arrays", *IEEE Transactions on Magnetics*, vol. 48, pp. 3651-3653, 2012.

[6] T. Kaneyoshi, "Magnetizations of a transverse Ising nanowire ", *J. Magn. Magn. Mater.*, vol. 322, pp. 3410-3415, 2010.

[7] E. Konstantinova, "Theoretical simulations of magnetic nanotubes using Monte Carlo method", *J. Magn. Magn. Mater.*, vol. 320, pp. 2721-2729, 2008.

[8] O. Iglesias, A. Labarta, "Monte Carlo simulation study of exchange biased hysteresis loops in nanoparticles", *Physica B: Condensed Matter*, vol. 372, pp. 247-250, 2006.

[9] P. Vargas, D. Altbir, J. d'Albuquerque e Castro, "Fast Monte Carlo method for magnetic nanoparticles", *Phys. Rev. B*, vol. 73, pp. 092417-1- 092417-3, 2006.

[10] C.D. Salazar-Enríquez, E. Restrepo-Parra, J.Restrepo, "Influence of the structural properties on the pseudocritical magnetic behavior of single-wall ferromagnetic nanotubes ", *J. Magn. Magn. Mater.*, vol. 324, pp. 1631-1636, 2012.

[11] B. Deviren, M. Keskin, "Dynamic phase transitions in a cylindrical Ising nanowire under a time-dependent oscillating magnetic field", *J. Magn. Magn. Mater.*, vol. 324, pp. 2163-2170, 2012.

[12] R. Wieser, U. Nowak, K.D. Usadel, "Domain wall mobility in nanowires: Transverse versus vortex walls", *Phys. Rev. B*, vol. 69, pp. 064401-1- 064401-5, 2004.

[13] N.A. Metropolis, A.W. Rosenbluth, M.N. Rosenbluth, A.H. Teller, E. Teller, "Equation of State Calculations by Fast Computing Machines", *J. Chem. Phys.*, vol. 21, pp. 1087-1092, 1953.
[14] D. P. Landau, K. Binder, A Guide of Monte Carlo Simulations in Statistical Physics, *Cambridge University Press*, 2000.

# Concentration of magnetic transitions in dilute magnetic materials

**V I Belokon**[1], **V Yu Kapitan**[1] **and O I Dyachenko**[1]
[1]Department of theoretical and experimental physics, The School of Natural Sciences, Far Eastern Federal University, 8, Sukhanova str., 690950, Vladivostok, Russian Federation

E-mail: dyachenko.oi@dvfu.ru

**Abstract.** In this paper the two-sublattice crystalline ferromagnet in the approximation of the Ising model was investigated. In each sub-lattice density of magnetic atoms can vary from 0 to 1. The exchange interactions within each sublattice and between the spins of different sublattices are considered as given. The case of direct exchange was investigated and the conditions for the occurrence of ferromagnetism, antiferromagnetism and ferrimagnetism, depending on the concentration of magnetic atoms in each sublattice were established.

## 1. Introduction

A large number of magnetic materials both in our country and abroad are currently known, but the experimental and theoretical researches continue in this area, as this problem is quite urgent.
The theory of phase transitions has been an active field of research for many decades already. Despite several remarkable achievements on both the conceptual and the applied sides, there still remain open questions galore. As a contribution to the conceptual side, a recent and promising proposal is the random interaction field method to phase transitions [1], connecting the partition function of random fields with the law of interaction and providing an opportunity to evaluate not only the Curie point but also the critical concentration of "ferromagnetic" atoms necessary for the appearance of ferromagnetism.

## 2. Model of the two-sublattice magnet

If there are two sub-lattices, in accordance with the results of [1], we write the partition function of the random field at an atom of the first sublattice as follows:

$$W\left(H_1\right)=\frac{1}{\sqrt{\pi}B_1}\exp\left[\frac{-\left\{H_1-M_1H_{011}-M_2H_{012}\right\}^2}{B_1^2}\right], \tag{1}$$

where

$$B_1^2=2p_1\left(1-M_1^2p_1\right)\left(\sum_k\varphi_{11k}^2 \quad \sum_l\varphi_{12l}^2\right),$$

$$H_{011}=p_1\sum_k\varphi_{11k},H_{012}=p_2\sum_l\varphi_{12l},$$

$$\alpha_1 - \beta_1 = M_1,\ \alpha_2 - \beta_2 = M_2,$$

index $k$ numbers the atoms of the first type, $l$ numbers the atoms of the second type, $\varphi_k$ is the exchange "field" created by the atom with the number $k$ on the selected atom placed at the origin, $\alpha$ is the relative number of particles oriented "up", $\beta$ is the relative number of particles oriented "down" $p$ is the concentration of ferromagnetic particles . Respectively

$$W(H_2) = \frac{1}{\sqrt{\pi} B_2} \exp\left[\frac{-\{H_2 - M_2 H_{022} - M_1 H_{021}\}^2}{B_2^2}\right], \tag{2}$$

where

$$B_2^2 = 2p_2\left(1 - M_2^2 p_2\right)\left(\sum_l \varphi_{22l}^2 \ \sum_k \varphi_{21k}^2\right),$$

$$H_{022} = p_2 \sum_l \varphi_{22l}, H_{021} = p_1 \sum_k \varphi_{21k}.$$

In the simplest case $\varphi_{11}, \varphi_{12}, \varphi_{21}, \varphi_{22}$ are constants, $\varphi_{12} = \varphi_{21}$ and summation should be carried out over nearest neighbors. Substantial simplification of the equation can be achieved by replacing the partition function of the rectangle function [1]. Then

$$M_1 = \frac{1}{2B_1} \int_{-B_1}^{B_1} \text{th}\left[\frac{m_1}{kT}\left(H_1 + M_1 H_{011} + M_2 H_{012}\right)\right] dH_1, \tag{3}$$

$$M_2 = \frac{1}{2B_2} \int_{-B_2}^{B_2} \text{th}\left[\frac{m_2}{kT}\left(H_2 + M_2 H_{022} + M_1 H_{021}\right)\right] dH_2. \tag{4}$$

We introduce the notation: $\frac{m_1}{kT} H_1 = x_1, \frac{m_1}{kT} H_{011} = h_{011}, \frac{m_1}{kT} H_{012} = h_{012}, \frac{m_1 B_1}{kT} = b_1$. Analogous for $m_2$, where $m_1$ and $m_2$ are magnetic moments of the particles of the first and second sublattices. Then the expression for the magnetization can be expressed as

$$M_1 = \frac{1}{2b} \int_{-b_1}^{b_1} \text{th}\left[\left(x_1 + M_1 h_{011} + M_2 h_{012}\right)\right] dx_1. \tag{5}$$

At high temperatures

$$M_1 = \frac{\text{th} b_1}{b_1}\left(h_{011} M_1 + h_{012} M_2\right) - \frac{1}{3} \frac{\text{th} b_1}{b_1 \text{ch}^2 b_1}\left(h_{011} + M_1 \ h_{012} M_2\right)^3. \tag{6}$$

Analogous,

$$M_2 = \frac{\text{th} b_2}{b_2}\left(h_{022} M_2 + h_{021} M_1\right) - \frac{1}{3} \frac{\text{th} b_2}{b_2 \text{ch}^2 b_2}\left(h_{022} + M_2 \ h_{021} M_1\right)^3. \tag{7}$$

The conditions under which different types of magnetic ordering are observed, can be determined by solving the system of equations (6, 7) for any values $h_{011}, h_{012}, h_{022}, h_{021}$. In general, the solution is rather complicated, but it is possible to consider some simple options.

If $h_{022} = 0$ then

$$M_2 \approx \frac{\text{th} b_2}{b_2} h_{021} M_1. \tag{8}$$

According to (8) and (6)

$$M_1^2 = \frac{\frac{\text{th}b_1}{b_1}\left(h_{011} + h_{012}\frac{\text{th}b_2}{b_2}h_{021}\right) - 1}{\frac{1}{3}\frac{\text{th}b_1}{b_1\text{ch}^2 b_1}\left(h_{011} + h_{012}\frac{\text{th}b_2}{b_2}h_{021}\right)^3}. \quad (9)$$

Thus, the condition of a non-zero magnetisation $M_1$ :

$$\frac{\text{th}b_1}{b_1}\left(h_{011} + h_{012}\frac{\text{th}b_2}{b_2}h_{021}\right) > 1. \quad (10)$$

In the case of a direct exchange at $T \to 0$ the values $\text{th}b_1$ and $\text{th}b_2$ cannot be greater than one, so from the definition of $H_{011}$, $H_{012}$, $B_1$ and $B_2$, the condition of appearance of the concentration of the phase transition can be determined as:

$$\frac{p_1 z_1}{\sqrt{2p_1 z_1}} + \frac{p_1 z_1 p_2 z_2}{\sqrt{2p_1 z_1}\sqrt{2p_2 z_2}} > 1. \quad (11)$$

If $H_{011} = 0$, then this relation simplifies

$$\frac{p_1 z_1 p_2 z_2}{\sqrt{2p_1 z_1}\sqrt{2p_2 z_2}} > 1. \quad (12)$$

And finally, in the case of equivalent sublattices $p_1 = p_2, z_1 = z_2$ and condition for the occurrence of antiferromagnetic ordering is determined as:

$$p > \frac{2}{z}. \quad (13)$$

At $h_{012} = 0$ two separate sublattices are considered for each of which there is a different transition condition. Fulfillment of conditions (10) and (12) is substantially depends on the concentration of "ferromagnetic" atoms in sublattices so there are critical concentrations of such atoms below which ones ordering cannot occur.

The above-mentioned fundamental relationships determine the possibility of a corresponding ordering. To determine the Curie point $T_c$, use the relation (10):

$$\frac{\text{th}\left[\frac{m_1\sqrt{2p_1 z_1}}{kT_c}\right]}{\frac{m_1\sqrt{2p_1 z_1}}{kT_c}}\left(\frac{m_1}{kT_c}p_1 z_1 + \frac{m_1}{kT_c}p_2 z_2 \frac{\text{th}\left[\frac{m_2\sqrt{2p_2 z_2}}{kT_c}\right]}{\frac{m_2\sqrt{2p_2 z_2}}{kT_c}}\frac{m_2}{kT_c}p_1 z_1\right) > 1. \quad (14)$$

## 3. Monte Carlo simulations of two-sublattice magnets

In model of two-sublattice ultrathin magnetic film with competing ferromagnetic-antiferromagnetic direct exchange interactions was obtained spin-glass properties. It was shown that the Ising model on a 2D lattice with z = 8 nearest neighbors takes place divergence between the ZFC and FC magnetization curves, Fig. 1. A weak antiferromagnetic interaction between ferromagnetically ordered sublattices and approximately the same values of the magnetic moments in the sites could lead to very complex magnetic properties. Theoretical magnetic phase diagram for the plain structures with random dilution was built.

Technological features of multicomponent compounds containing ferromagnetic atoms make them attractive for a wide variety of modern electronic devices, such as materials for storage devices. Therefore, in recent years, these materials are the subject of intense experimental and theoretical research [2].

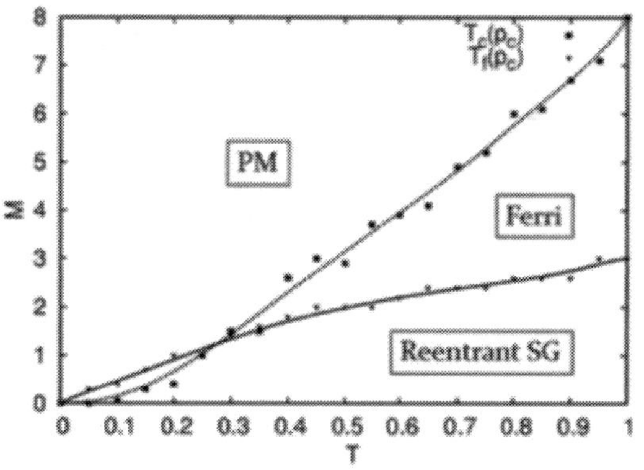


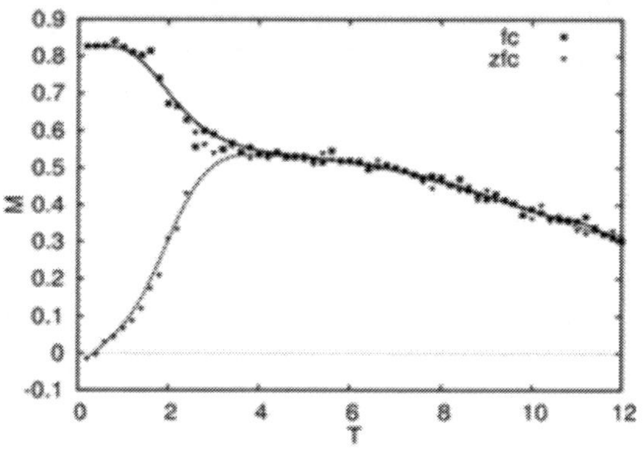


Figure 1: Temperature dependence of the average FC and ZFC magnetization, the external magnetic field $H$=0.4, $p_c$=0.9.

Figure 2: Theoretical magnetic phase diagram for a two-sublattice magnet

The presence of the divergence between the ZFC and FC magnetization curves are usually associated with the appearance of the spin-glass phase. If to use point of divergence between the ZFC and FC curves - $T_f$ and the temperature of occurrence of the ferromagnetic ordering in the sublattices - $T_c$, it is possible to construct a theoretical magnetic phase diagram of a two-sublattice 2D Ising magnet for a given exchange integrals: $J_{AA} = 1$, $J_{BB} = 2$, $J_{AB} = -0.5$, Fig. 2. The presence of divergence in ZFC and FC temperature dependence could be caused by the ordering in the sublattices, accompanied by compensation of magnetization and an increase in frustrations in the system as a whole.
The weak antiferromagnetic direct interaction between ferromagnetically ordered sublattices was shown to complex behavior of the average magnetization of the system of spins.
Was calculated the theoretical magnetic phase diagram for the two-sublattice magnet.

## 4. Conclusions

Thus, the random interaction field method is a tool for the study of phase transitions and critical phenomena. It allows exploring the complex model systems, the study of which by other methods is facing great difficulties. A specific example is a two-sublattice magnetic material, which is composed of both magnetic and non-magnetic atoms are considered the results of the study of the critical properties, determined the critical concentration of "ferromagnetic" of atoms necessary for the existence of ferro-, ferri-and antiferromagnetism.

## Acknowledgments

This work was supported by Ministry of education and science of Russian Federation within the contracts 2.8649.2013, 07.514.11.4013 and 14.740.11.0289.

## References

[1] Belokon V I, Nefedev K V 2002 *SSP* **44** 1632
[2] Htoutou K, Ainane A, Saber M and Miguel J J 2005 *Physica A* **358** 184

# Polycrystals multilevel models using crystal plasticity: consistency of constitutive equations at different scale levels

**A I Shveykin, P V Trusov, P S Volegov**

Perm National Research Polytechnic University, Mathematical Modeling of Systems and Processes, 29, Komsomolsky ave., Perm, Russian Federation

E-mail: alexsh59@bk.ru

**Abstract**. Two-level models of different polycrystalline metal's inelastic deformation based on crystal plasticity and describing viscoplastic intragranular dislocations slip, lattice rotation with an explicit consider of dislocation slip incompatibility in neighboring grains, and fragmentation of crystallites are developed. The homogenization of constitutive equations at various scale levels is used, which allows to connect the same type of characteristics of different scale levels and leads to an unambiguous description of geometric nonlinearity on the macro level by specifying the corotational derivative of Cauchy stress tensor. An algorithm for solving boundary value problems in FEM package Abaqus with using proposed models to describe the behavior of the material is developed, corresponding computational modules are created. Numerical investigation of different loading of samples from various polycrystalline metals with a description of the evolving internal structure is done.

## 1. Introduction

Numerous theoretical and experimental studies show that the performance of the internal material structure determines the behavior of the material at the macro level and its performance characteristics. During intensive plastic deformation the internal structure of the material is significantly restructured: the grain and dislocation structures are changing, crystallites lattice is rotated; it is widely used to produce materials with unique properties: submicrocrystalline, nanocrystalline, textured materials and materials, which are capable of superplastic deformation.

At the time, the construction of models capable of describing the change of the internal structure of polycrystalline materials, a growing acceptance of the approach based on an explicit introduction to the structure of the constitutive relations parameters reflecting the state and evolution of meso- and microstructure evolution and the kinetic equations for these parameters (so-called internal variables) [1]. In particular, recent decades, a very wide development of crystal plasticity based models, which built in the framework of this approach and explicitly describing the material structure and the mechanisms of inelastic deformation at the crystallite level, is observed; this models allows a natural way to describe the structure evolution at deep plastic strain.

Based on the crystal plasticity models can be divided into three classes [2,3]: statistics, self-consistent and direct. Models of the last two classes require inaccessible currently computing resources, so for modeling of real processes statistical constitutive model are more popular.

The paper briefly discusses the general structure of the two-level models of polycrystalline metals inelastic deformation developed by the authors, explains how to homogenize the constitutive equations on different levels of scale, describes the algorithm for applying models of the material in the boundary value problems solution and obtaines the results.

## 2. Two-level constitutive model for inelastic deformations of polycrystals

In the simulation of polycrystalline metal's inelastic deformation scale levels hierarchy can be defined as follows: the macro level – meso level (the level of crystallite – grains, subgrains, fragment) – micro level (the dislocation structure) (Fig. 1).

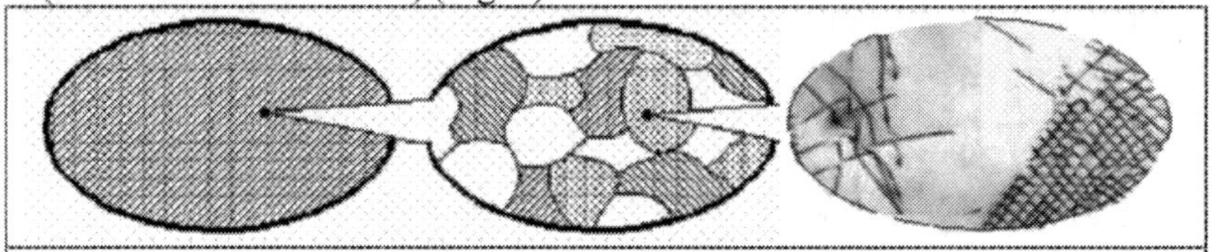

**Figure 1.** Schematic representation of the scale levels hierarchy in the simulation of polycrystalline metals (from left to right: the macro-level, meso, micro).

At the macro level considers a representative volume of a polycrystalline metal, consisting of a set of crystallites (meso-level elements). The constitutive equations system at the macro level is [1]:

$$\begin{cases} \mathbf{\Sigma}^r \equiv \dot{\mathbf{\Sigma}} + \mathbf{\Omega}^T \cdot \mathbf{\Sigma} + \mathbf{\Sigma} \cdot \mathbf{\Omega} = \mathbf{C} : \mathbf{D}^e = \mathbf{C} : (\mathbf{D} - \mathbf{D}^{in}), \\ \mathbf{\Omega} = \mathbf{\Omega}\left(\mathbf{w}_{(i)}, \mathbf{c}_{(i)}\right), i = 1,...,N, \\ \mathbf{C} = \mathbf{C}(\mathbf{c}_{(i)}), i = 1,...,N, \\ \mathbf{D}^{in} = \mathbf{D}^{in}(\mathbf{d}^{in}_{(i)}, \mathbf{c}_{(i)}), i = 1,...,N, \end{cases} \qquad (1)$$

where $\mathbf{\Sigma}$ – Cauchy stress tensor, $\mathbf{C}$ – elastic properties $4^{th}$-range tensor, $\mathbf{D}$, $\mathbf{D}^e$, $\mathbf{D}^{in}$ – the strain rate, its elastic and inelastic parts, the index «r» means derivative independent of the choice of reference system, $\mathbf{\Omega}$ – tensor describing the motion of the moving coordinate system, which defines itself in relation to deformational movement [4] on the macro level. To determine $\mathbf{\Omega}$ the homogenization condition for constitutive equations on different scale levels is proposed. Thus, the inelastic strain rate $\mathbf{D}^{in}$, the effective anisotropic elastic properties $\mathbf{C}$ and describing the motion of the moving frame tensor $\mathbf{\Omega}$ are explicit internal variables in the macro-level model, at any moment depends on the structure at the lower scale levels (and through it depends on the loading history) and determined by meso-level model, used for each crystallite (number of meso level items $N$ should be sufficient for a statistical description of a representative volume at the macro level).

At the meso level the system uses the following relations (for the description of each crystallite, the index number of the crystallite is omitted):

$$\begin{cases} \boldsymbol{\sigma}^r \equiv \dot{\boldsymbol{\sigma}} - \boldsymbol{\omega} \cdot \boldsymbol{\sigma} + \boldsymbol{\sigma} \cdot \boldsymbol{\omega} = \mathbf{\Pi} : \mathbf{d}^e = \mathbf{\Pi} : (\mathbf{d} - \mathbf{d}^{in}), \\ \mathbf{d}^{in} = \sum_{i=1}^{K} \dot{\gamma}^{(i)} \mathbf{m}^{(i)}_{(S)}, \\ \dot{\gamma}^{(i)} = \dot{\gamma}_0 \left| \frac{\tau^{(i)}}{\tau^{(i)}_c} \right|^{1/n} H(\tau^{(i)} - \tau^{(i)}_c), \; i = 1,...,K, \\ \dot{\tau}^{(i)}_c = f(\gamma^{(j)}, \dot{\gamma}^{(j)}), \; i, j = 1,...,K, \end{cases} \qquad (2)$$

where $\boldsymbol{\sigma}$ –Cauchy stress tensor, $\mathbf{c}$ – fourth-rank elastic properties tensor of the crystallite, $\mathbf{d}, \mathbf{d}^e, \mathbf{d}^{in}$ – the strain rate, the elastic and inelastic strain rate, $\gamma^i, \tau_C^i$ – accumulated slip and the critical shear stress on the $i$-th slip system, $\mathbf{m}^i$ – the orientation tensor of $i$-th slip system, $\mathbf{m}^i = 1/2(\mathbf{b}^i\mathbf{n}^i + \mathbf{n}^i\mathbf{b}^i)$, $\mathbf{b}^i, \mathbf{n}^i$ – the unit vectors in the direction of the Burgers vector and the normal to the sliping plane;

bedding planes and orientation of the Burgers vector along the translational motion (slip) of edge dislocations are known, they are the most densely packed planes and directions, so, in fcc metals sliding edge dislocations are in the planes of $\{111\}$ in directions <110>, $\dot{\gamma}_0, n$ – constants of the material: the characteristic shear rate and strain rate sensitivity of the material [1], $\tau^i$ – shear stress in the $i$-th slip systemие, $\tau^i = \mathbf{b}^i\mathbf{n}^i : \boldsymbol{\sigma}$, $H(\cdot)$ – Heaviside function, $K$ – the number of slip systems for considered type lattice, $\mathbf{o}$ – tensor current orientation of the crystallographic coordinate system relative to the fixed grain laboratory system.

As constitutive relation at the meso level Hooke's law in the speed form is used ($2_1$), the geometric nonlinearity is takes into account: quasi-solid movement [4] associated with the rotation of the lattice (in crystallographic coordinates); spin tensor $\boldsymbol{\omega}$ in corotation derivative Cauchy stress tensor $\boldsymbol{\sigma}^r$ characterizes the crystal lattice rotation. Thus, stress is characterized the elastic bonds in grain and determined by lattice distortions.

Equation ($2_2$) is kinematic equation, according to which the inelastic deformation of the crystallite is define by sliding to slip systems. If other intragranular deformation mechanisms, such as grain boundary sliding or twinning in the ($2_2$) to the modes of intragrain dislocation glide mechanisms are added to the appropriate mode for shear rates that determine the kinetic equations.

To determine inelastic deformation rate of polycrystalline metals can be used [2,3]: elastoplastic model based on the Lin model [5], or the elastoviscoplastic model used here ($2_3$), where $\mathbf{d}^{in}$ (and $\boldsymbol{\omega}$) associated with implicit internal meso-level variables characterizing dislocation slip – shear rates on slip systems $\dot{\gamma}^i$, the critical stress $\tau_C^i$, tensor $\mathbf{o}$ of the crystallographic orientation of the current coordinate system of grain relative to a fixed laboratory system coordinates. Specification of equation ($2_4$), describing the evolution of the critical shear stress on the slip system, is in the paper [6].

Taylor's hypothesis is applied for scale transition implemented at the macro level to the lowest levels in the scale model $\mathbf{d} = \mathbf{D}$.

Relations ($2_5$) for the determination of the spin system the following ratio are used:

$$\begin{cases} \boldsymbol{\omega} = \boldsymbol{\omega}_1 + \boldsymbol{\omega}_2 \\ \boldsymbol{\omega}_1 = \frac{1}{2}(\hat{\nabla}\mathbf{v}^{\mathrm{T}} - \hat{\nabla}\mathbf{v}) - \sum_{i=1}^{K}\frac{1}{2}\dot{\gamma}^i(\mathbf{b}^i\mathbf{n}^i - \mathbf{n}^i\mathbf{b}^i) - (\mathbf{B}:\boldsymbol{\sigma})\cdot\mathbf{D}^{in} + \mathbf{D}^{in}\cdot(\mathbf{B}:\boldsymbol{\sigma}) \\ \boldsymbol{\omega}_2 = \begin{cases} \frac{1}{\mathrm{A}}\boldsymbol{\mu}^r + \frac{1}{\mathrm{H}}\boldsymbol{\mu}, \text{ if } \|\boldsymbol{\mu}\| = \mu_c \text{ и } \boldsymbol{\mu}:\boldsymbol{\mu}^r > 0 \\ \frac{1}{\mathrm{A}}\boldsymbol{\mu}^r, \text{ otherwise} \end{cases} \\ \boldsymbol{\mu}^r = \sum_{m=1}^{M}(\boldsymbol{\mu}^r)^m \\ \mathbf{m}^r = \mu\,\mathbf{N}\times\left[\sum_{i}^{K}\dot{\gamma}^i\mathbf{n}^i\mathbf{b}^i - \sum_{j}^{K}\dot{\gamma}^{j(m)}\mathbf{n}^{j(m)}\mathbf{b}^{j(m)}\right]\cdot\mathbf{N}, \text{ for each neighboring grain } m \end{cases} \tag{4}$$

The first component $\boldsymbol{\omega}_1$ describes the rotation of the lattice with the grain material during the imposed kinematic effects (material associated with the orthogonal rotation tensor accompanying elastic deformation). The second component $\boldsymbol{\omega}_2$ describes the rotation of the actual crystallite lattice due to the interaction with the environment. To do this, the model (3) introduced the couple stresses $\boldsymbol{\mu}$ acting on the crystal, the critical moment stress $\mu_C$ and set the normal to the crystallite $\mathbf{q}^m, m = 1,\ldots,M$ ($M$ -

number of neighboring crystallites). Additivity velocity moments of interaction with all neighboring grains is assumed in ($4_4$). The evolution of the vector-momentum associated with the couple stress tensor and determines from the analysis of the incompatibility of dislocation motion on the boundary of the crystallites by equation ($4_5$). Based on the input of the couple stresses may be used for description of fragmentation [7].

## 3. The homogenization procedure of constitutive equations for scale transition

One of the keynote questions in the multilevel models construction is the issue of constitutive equations homogenization at various scale levels. The logical understanding of the problem is follows: macro level equations should be defining from meso level constitutive equations with a priori relations of the macro- and meso level parameters [1]; specific type of relations associated with the aggregation hypothesis (combining elements of meso level to the macro level element). For statistical crystal plasticity models a priori relation is the equality of velocity gradient, stresses and the effective elastic properties tensor at the macro level are average the corresponding meso level characteristics (for the velocity gradient this conditions are automatically done due Taylor's hypothesis).

The constitutive equations homogenization procedure for the proposed model is as follows:

1. Represent the quantities in the description of the stress-strain state at meso-level as the sum of average representative volume of macro-level variables and deviations from these averages.

2. Substitute this representation in the meso-level of the defining equation ($2_1$), the obtained relations are averaged.

3. Finding the form obtained in step 2 averaged meso-level constitutive equation running the macro-level ($1_1$), defining the necessary connections.

For the proposed models the connections are

$$\mathbf{\Omega} = <\boldsymbol{\omega}>, \quad \mathbf{D}^{in} = <\mathbf{d}^{in}> + \mathbf{C}^{-1} : <\mathbf{c}' : \mathbf{d}^{in\prime}> - \mathbf{C}^{-1} : (<\boldsymbol{\omega}' \cdot \boldsymbol{\sigma}'> - <\boldsymbol{\sigma}' \cdot \boldsymbol{\omega}'>) . \quad (5)$$

Thus, the proposed approach results in more focused form of the defining relations on the macro level (and in particular – the type of independent of the reference system choice derivative). Proposed method is easy to apply for other forms of constitutive relations for meso- and macro-levels for a wide range of constitutive models with internal variables.

## 4. Algorithm for solving boundary value problems using constitutive model and the simulation results

The algorithm to use developed model for solving boundary value problems in FEM package ABAQUS is made. To do this, we use standard package – implementation of the proposed models in the custom procedure UMAT, including the integration of the constitutive equations of the model (determination of the stresses and internal variables at the end of step in the UMAT) and the definition of the current cutting modulus matrix $[\mathbf{C}^{ep}] = \dfrac{\partial\{\Delta\mathbf{\Sigma}\}}{\partial\{\Delta\mathbf{E}\}}$, where $\Delta\mathbf{\Sigma}$ – increment of stress on the step, $\Delta\mathbf{E}$ - increment deformation step. The analyzes were carried out, resulting in an analytical expression for $\mathbf{C}^{ep}$ connecting it with the internal meso-scale variables.

It should be note that many existing models [2,3] are not considered in the explicit level the constitutive equations at the macro level, but at numerical implementation in FEM-package Abaqus is the (implicity) introduction is macro-scale level is done, there are Jauman derivative of Cauchy stress tensor is used in Abaqus, which violates the consistency of the two-level constitutive model. It is shown that this approach can lead to physically incorrect.

## 5. Numerical results

A brief analysis and comparison of the results of calculations using different conditions of scale levels coupling is described underneath:

1) the use of the proposed coupling conditions (5) to determine the quasi-solid motion and inelastic strain rate at the macrolevel;
2) the use of the Jaumann derivative at the macrolevel

$$\mathbf{\Omega} = \mathbf{W} = \frac{1}{2}((\hat{\nabla}\mathbf{v})^{\mathrm{T}} - \hat{\nabla}\mathbf{v}), \qquad (6_1)$$

and direct averaging of the inelastic strain rates (which corresponds with the Voigt's hypothesis direct stresses averaging $\mathbf{\Sigma} = <\boldsymbol{\sigma}>$)

$$\mathbf{D}^{in} =< \mathbf{d}^{in} >. \qquad (6_2)$$

Fig. 2 present the modeling results of the simple shear with velocity gradient (components in fixed laboratory coordinate system): $(\nabla \mathrm{v})_{23} = -0.5 \cdot 10^{-5}\ \mathrm{s}^{-1}$, $(\nabla \mathrm{v})_{ij} = 0,\ (ij) \neq (23)$.

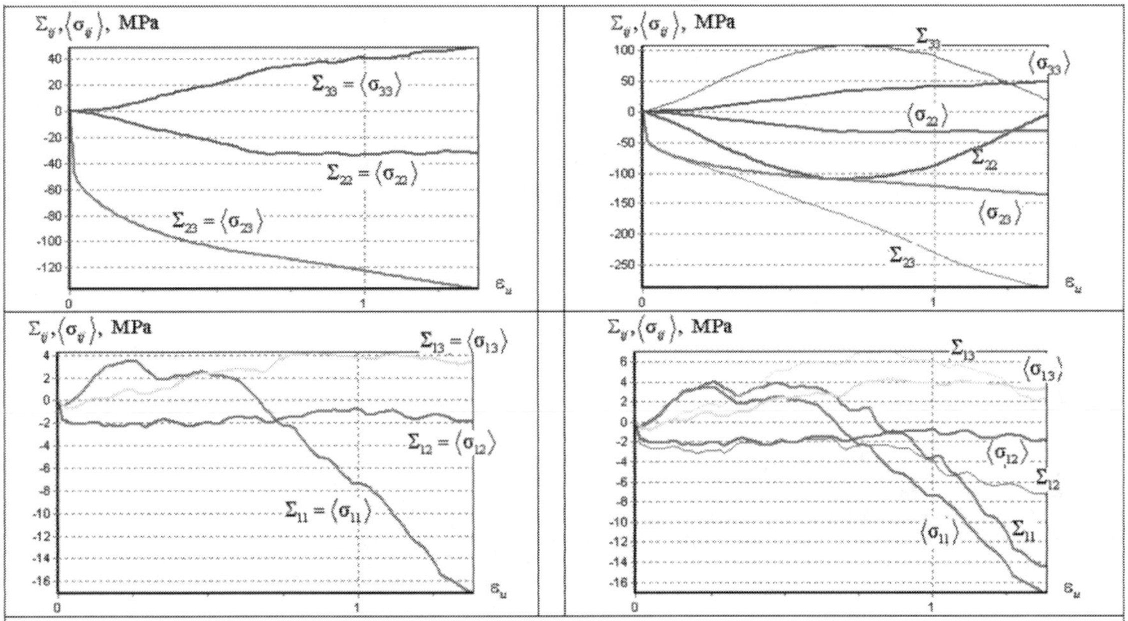


**Figure 2.** The evolution of macrostresses $\mathbf{\Sigma}$ and averaged mesolevel stresses $<\boldsymbol{\sigma}>$ using the coupling conditions (5) (on the left) and closing equations (6) (on the right)

The results of Fig.2a,2c show that the coupling conditions provided a complete correspondence between the macrostresses and the averaged mesolevel stresses, while the use of the conventional closing conditions (6) of the deviations were significant and they were growing (Fig.2b,2d). Meanwhile there is a change in the sign of the stress tensor components at the macrolevel that is typical for the Jaumann derivative, which seems unphysical for the monotonic loading.

## 6. Conclusions

To describe the deformation processes of polycrystalline materials with the evolution of its meso-structure two-level constitutive model based on the crystal plasticity is created.

The method of different scale levels parameters coupling based on the agreement of the constitutive equations at these levels and providing consistency of the measures of stress and strain states on these levels is developed. Apart from that, the proposed approach leads to the concretization of the constitutive relation form at the macrolevel (and the kind of the material frame indifferent derivative, in particular). In principle, the relations of the lowest level are "transported" to the top level solving the problem of their correct formulation for the geometrically and physically nonlinear problem. In this case it is surely not possible to completely avoid the phenomenological relations, however, for the lowest scale level they are written in the accepted hierarchical set; for this level it is possible to specify the physical mechanisms of deformation and their detailed description by using the known statements of solid physics (this is a much easier problem compared to the one for establishing macrophenomenological relations while taking the state of the multiscale internal structure into account and describing the variety of all the mechanisms of the inelastic deformation).

An algorithm for solving boundary value problems using the finite element method, implemented in the Abaqus package. With the use of modeling software system obtained results during loading of samples from various polycrystalline metals, including the evolution of the internal structure: the shape, size and orientation of the crystallographic axes of the structural elements of the polycrystalline aggregates (grains, subgrains, fragments), effective macroscopic elastic and plastic properties of polycrystalline materials. The results is in good agreement with experimental data.

## 7. Acknowledgments

The work was carried out in the Perm National Research Polytechnic University with support of the Government of Russian Federation (The decree № 220 on April 9, 2010 «On measures to involve the world's leading scientists at Russian institutions of higher learning, research centers of the governmental academies of sciences, or governmental research centers of the Russian Federation») under the Contract № 14.B25.310006, on June 24, 2013.

## References

[1] Trusov P V, Shveykin A I, Nechaeva E S and Volegov P S 2012 *Physical Mesomechanics* **15** 155-175

[2] Trusov P V and Shveykin A I 2013 *Physical Mesomechanics* **16** 23-33

[3] Trusov P V and Shveykin A I 2013 *Physical Mesomechanics* **16** 99-128

[4] Pozdeev A A, Trusov P V and Nyashin Yu I 1986 *Large elastoplastic deformations: theory, algorithms, applications* (Moscow: Nauka)

[5] Kondratev N S and Trusov P V 2011 *Computational Continuum Mechanics* **4** 20-33

[6] Trusov P V, Volegov P S and Yanz A Yu *Scientific and technical statements of St.Petersburg State Polytechnical University. Physics and mathematics* 2010 **2(98)** 110-119

[7] Trusov P V, Volegov P S and Yanz A Yu 2012 *Physical Mesomechanics* **15** 35-45

# Sandwich-typed resonator cavity based on a regular photonic crystal nanobeam

**Pavel G. Serafimovich,* Nikolay L. Kazanskiy, Svetlana N. Khonina,**

Image Processing Systems Institute of the Russian Academy of Sciences, 151 Molodogvardeyskaya, 443001 Samara, Russia

*Corresponding author: pavel.serafimovich@gmail.com

**Abstract**. The authors propose and numerically examine a two-component design of an optical nanocavity. The design utilizes the effective medium theory. Such a nanocavity consists, first, of a photonic crystal nanobeam, in which the photonic crystal unit cell is not changed. Second, the cavity contains a fragment of some supplementary material of the size of several or several tens of photonic crystal unit cells. This paper describes the cavity model in which the both components have an equal thickness and can be fabricated from the same silicon wafer. While combining the two components, the defect has formed, in which the resonant mode can be excited. The advantages of the proposed cavity model are reported, particularly the easiness with which the cavities array can be fabricated and the possibility of implementing electrically pumped light sources and amplifiers. The fabrication tolerances of the proposed nanocavity were investigated. It has been found that existing structural layers alignment technologies can be used for fabricating the suggested cavity.

## 1. Introduction

The ability of photonic crystal (PC) structures to localize light at subwavelength scale is used in a variety of nanophotonic devices. Nanocavity is a basic element of nanophotonics [1]. High Q-factor photonic-crystal nanocavities with small mode volume enable designing optical switches, filters, and low threshold-power light sources, as well as integrating the said components in microcircuits. With the optical cavities boosting the interaction of light with the propagation medium, they have found use for experimental research in quantum electrodynamics.

Before addressing the suggested nanocavity model, a number of issues that complicate the development of devices, based on PC nanocavities, are listed. Most of the existing technologies in use to create high-Q PC nanocavities suggest fine-tuning of the resonance chamber geometry by changing the parameters of the photonic crystal. Such parameters may be, for example, the radius of the hole in the photonic crystal period and/or the hole periodic spacing [2, 3]. This imposes strict limits on the fabrication tolerances of fine details of the PC structure. For creating an array of nanocavities [4], high fabrication tolerances should be provided in the entire area of the array. Another difficulty arises in fabrication of nonlinear nanophotonic devices. Increase in the degree of interaction between light and matter in a PC nanocavity makes effective use of non-linear optical materials and quantum dots. For introducing such materials in the area of the resonator, one needs to resort to complicated technologies [5]. The next important area of photonic nanocavity research is the development of light sources on the basis of PC structures integrated on-chip. To implement such a device, one needs to create an electrically pumped cavity [6]. Building dynamic systems, based on nanocavities, also requires complex technology solutions. In most cases, it involves mechanical impact directly on the PC structure [7]. Another example of the difficulties in creating some nanophotonic devices is the need for a given distribution of the radiation in the far field. This problem is commonly solved by fine-tuning of the minimum parts of the photonic crystal [8, 9].

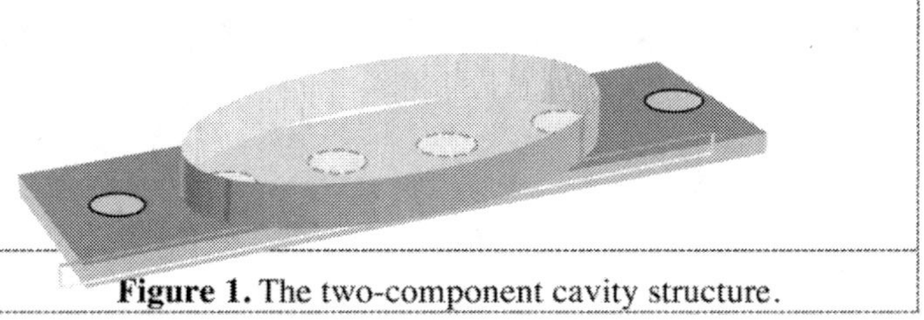

**Figure 1.** The two-component cavity structure.

## 2. The cavity model

To simplify the solutions for the problems cited, the authors theoretically investigated two-component PC cavity shown in Fig. 1. The first component of such a cavity is a periodic structure on the basis of a PC nanobeam. Compared with the two-dimensional structure on the basis of a PC slab [10, 11], the area of PC nanobeam is smaller and is naturally integrated into the waveguide geometry of connections on a chip. The second component is a fragment of a complementary material having an area of several lattice constant of the PC. The shape and size of the fragment were determined from the given parameters of PC cavity. While combining the two components, a defect forms in the resulting nanostructure. The resonant mode of the corresponding frequency can be excited in this defect.

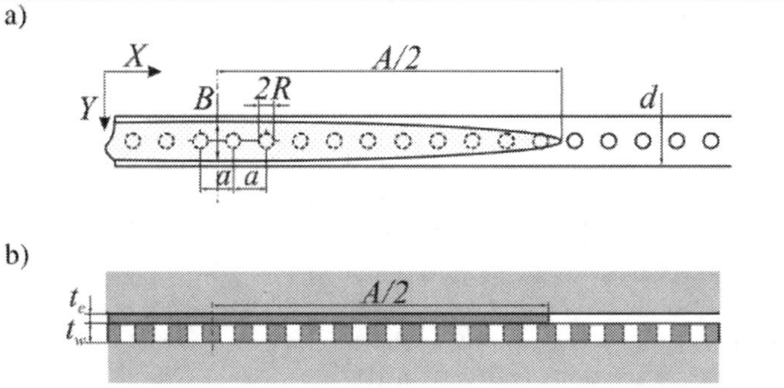


**Figure 2.** The geometry of the resonator calculated by (a) top view, (b) a side view. PC nanobeam (n = 3.46) lies on the substrate (n = 1.45). PC nanobeam width is d = 0.5 mm, thickness $t_w$ = 0.26 um. Circular holes have a radius of R = 75 nm and filled with air, distance between holes a = 0.34 um. The elliptical shape (ellipse parameters A and B) (n = 3.46) lies on the substrate (n = 1.45). Thickness of ellipse $t_e$ = 100 nm.

The proposed approach to creating two-component PC cavities is illustrated through the structure shown in Fig. 2. The first component of the resonator was a PC nanobeam. The nanobeam was made of silicon and was placed on a silica substrate. The holes in the nanobeam were of the same radius; they were equidistant from each other and filled with air. PC nanobeam parameters are given in the legend to Fig. 2. With these parameters, PC band gap was created for TE- dominant polarization radiation in the wavelength range of 1.4-1.7 um. The second component of the nanocavity was an elliptical piece of silicon, placed on a silica substrate.

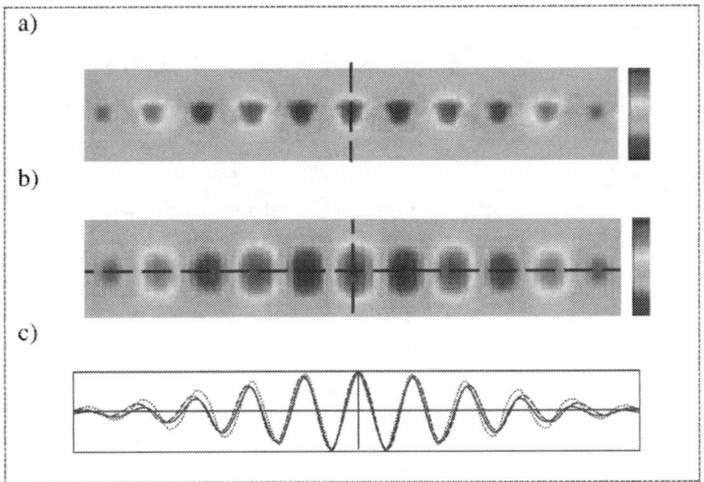


**Figure 3.** (a) the distribution of $H_z$ in the vertical plane passing through the axis of the waveguide, (b) the distribution of $H_z$ in the horizontal plane just above the elliptical fragment (in quartz), (c) the dotted line - $H_z$ values along the line of intersection of the planes (a) and (b ), the dashed line - $H_z$ values just below PC nanobeam (in quartz), the solid line - function $\cos(\pi x/a)\exp(-\sigma x^2)$ for $\sigma = 0.23, a = 0.34$ um..

To create a high-Q nanocavity, it is necessary to reduce the radiation of the resonant mode in space. This was achieved by optimizing the shape of the envelope of the resonant mode. The spectrum of electromagnetic field distribution, directly above the waveguide, determines the energy distribution in the far zone. This spectrum consisted of two peaks. Energy was dissipated from the cavity through the light cone of the waveguide, which was located between the spectral peaks. Therefore, the width of the spectral peaks determines the nanocavity losses on scattering. In [11, 12], it is proposed that a resonant mode can be formed with an envelope matching the Gaussian function. The shape of the resonant mode's envelope depends, in particular, on the imaginary part $\gamma$ of the wave vector of the PC structure. Gaussian shape of the envelope, provided by $\gamma$, varies linearly from one PC period to the next [12]. In the same paper [12], it has been shown that resonant mode with a Gaussian envelope can be realized by changing the PC nanobeam material fill factor (FF) quadratically. The elliptical shape of the defect proposed in this paper also enabled quadratic tapering of the PC nanocavity. In subsequent calculations, the thicknesses of the PC nanobeam and elliptical defect were assumed to be 260 nm and 100 nm, respectively. These thicknesses produced an optimal FF change in the cavity. Increasing the thickness of the nanobeam necessitates increase in the thickness of the elliptical defect.

## 3. Simulation

The resonance cavity characteristics were computed using the parallel FDTD method [13]. In particular, the cavity ($Q = 3.05x10^4$) was calculated with the parameters of the ellipse A = 6.8 mm (20 holes below the ellipse) and B = 0.5. To achieve a high-Q nanocavity, five additional holes were placed in the PC nanobeam at both ends of the ellipse. Thus, the total length of the cavity was (20 +5 x 2) x 0.34 = 10.2 um. Fig. 3a shows the distribution of $H_z$ in the vertical plane passing through the axis of the nanobeam, and Fig. 3b the distribution in the horizontal plane, just above the elliptical silicon fragment (in quartz). $H_z$ values along the intersection line of these two planes are represented by the dotted line in the graph of Fig. 3c, and the values directly below the PC nanobeam (in quartz) by the dashed line on the same chart. The solid line represents the function $\cos(\pi x/a)\exp(-\sigma x^2)$ with $\sigma = 0.23$, $a = 0.34$ microns. Good agreement between the distributions of $H_z$ and an analytic function demonstrates that the shape of the resonant mode's envelope is Gaussian. Assuming a linear dependence of $\gamma$ on x, the relation $\gamma(x) = a/\pi\int\sigma dx \approx x/40$ can be obtained. In [12, 14], quadratic tapering of PC nanobeam width was used to form the defect. In the paper [12], for a nanocavity with a length of 60 periods of the PC, the relation $\gamma(x) \approx x/120$ was implemented. Thus, it can be concluded that the two techniques used in creating a defect are almost equivalent. The nanocavity with an elliptical defect is 3 times shorter than the one with variable nanobeam width. Accordingly, the rate of change $\gamma$ in the nanocavity with an elliptical defect is three times faster.

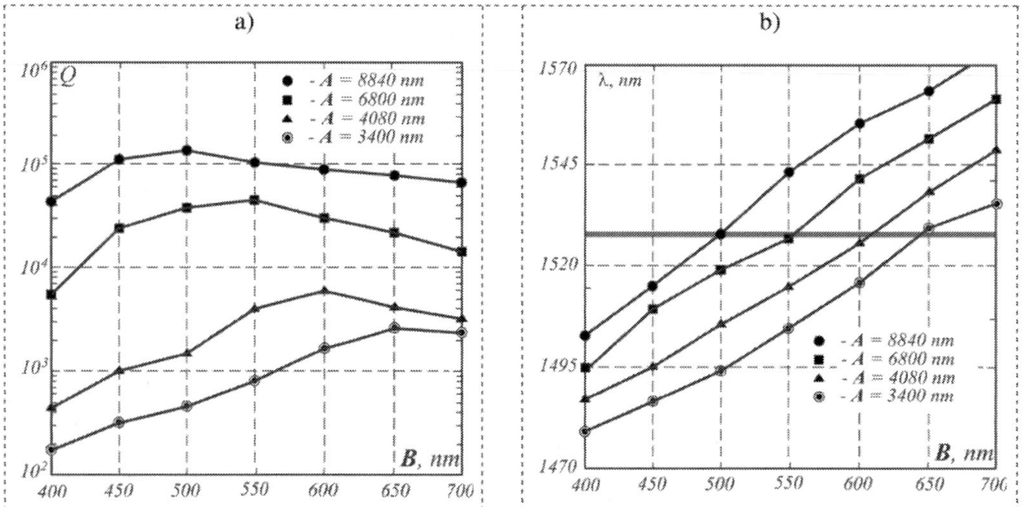


**Figure 4.** The dependence of Q-factor (s) and the resonance frequency (b) on the ellipse parameter A for different values of B.

Fig. 4 shows the simulation results of the proposed resonator for various parameters of the elliptical segment. For all simulations, five additional holes were placed in the PC nanobeam at both ends of the ellipse. It can be seen that the greater the length of the nanocavity, the greater is the Q-factor. For a resonator length of 12.24 um (36 holes - 26 of them are under the elliptical fragment), A = 8.84 um, B = 0.5 um, and Q-factor ~1.4 x$10^5$.

The minimum value of A, where the resonant mode with a Q-factor above $10^3$ remains excited, is 3.4 um (10 holes are under the ellipse) and B = 0.65 um. The Q-factor of this resonator is 2.5 x$10^3$. The optimal values of B vary with the changing values of A. The optimal B values are 650, 600, 550, and 500 nm respectively for values of A equal to 3400, 4080, 6800 and 8840 nm. The frequency of the excited resonant mode is equal to 1.525 microns for A = 3400 nm and B = 650 nm (the horizontal gray line in Fig. 4b; the frequencies are almost the same for all the B values listed above). The changes in B value can be used to customize the resonant frequency of the cavity. For example, if the B value is changed to be in the range 0.4-1.0 microns, then the resonant frequency varies from 1.485 to 1.535 microns (A = 4.08 um).

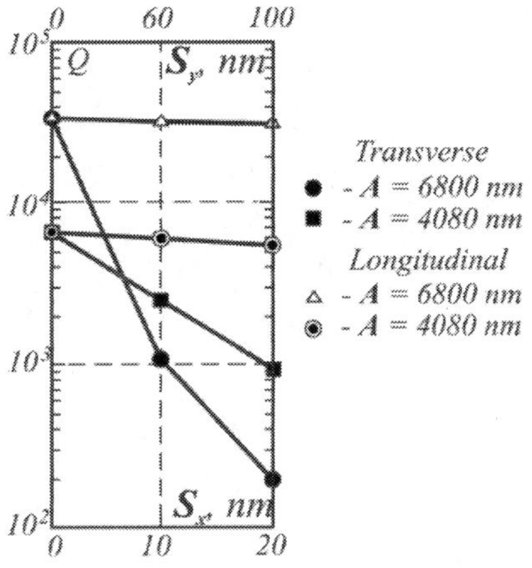


**Figure 5.** The dependence of the Q-factor for several values of A on transverse and longitudinal displacement of the two components of the cavity.

Fig. 5 shows the simulation results for horizontal displacement errors of the two components of the nanocavity with respect to each other. The greater the length of the resonator, the more accurate would be the combining of the two components. The misalignment offset of the ellipse, in the direction perpendicular to the axis of the waveguide, is most critical. Modern technologies allow combining structural layers within ~10 nm error [15]. Figure 4 shows that when the transverse displacement is only 20 nm, the Q-factor does not exceed $10^3$ for A=4.08 um. The longitudinal direction is less critical to displacement errors. Fig. 4 also shows that longitudinal displacement of 100 nm results in a slow decrease of Q-factor. Such asymmetry of displacement tolerances in X and Y directions is useful, for example, for inexpensive alignment technology described in [16]. This technology allows combining of layers with an average displacement error of less than 60 nm in one axial direction and less than 10 nm in the other.

Another sandwich-typed cavity was also simulated. It is a cavity in which the both components have an equal thickness ($t_w$ = $t_e$ = 0.22 um) and can be fabricated from the same silicon wafer. PC nanobeam width is d = 0.6 mm. Circular holes have a radius of R = 90 nm and filled with air, distance between holes a = 0.38 um. Ellipse parameters are A = 6.80 um and B = 0.38 um. The remaining parameters are the same as for Fig. 2. The simulated Q-factor value of this cavity is about 20000.

## 4. Discussion

The two-component nanocavity proposed in this paper has several advantages when compared to existing solutions. First, it simplifies the manufacturing of nanocavities based on PC structures. At the heart of such a nanocavity is a regular structure. The formation of the defect does not need any change in the shape of the minimum parts of the structure(e.g., change in the period and/or size of the holes in a PC structure). The second component is a fairly large structure, whose construction is relatively simple.

Second, the proposed structure provides a natural way to form two-dimensional arrays of nanocavities. Two-component structure allows the use of relatively inexpensive technology of interferometric lithography (IL) for making arrays of resonators. Manufacturing the array of the first component of the structure (PC nanobeam) requires double-exposure IL [17], and that of the second component (supplementary material) only single-exposure IL. The possibility for using IL is an additional advantage of the structure proposed in this paper as compared with the structures suggested in [12, 14].

Third, the proposed structure allows for the development of an integrated on-chip light source with vertical electrical pumping. The second component of the structure (supplementary material) can be doped with ions and, thus, be used as the basis for one (top) of the contact pads. The other (bottom) contact pad may be implemented, for example, by using the approach described in [18]. In this case, the bottom reflector may serve as both an electrical current pathway and a heat sink. Integrated on-chip light-emitting diodes with a laterally doped p-i-n structure, based on the nanobeam photonic crystal cavity, were demonstrated recently [6]. Using vertical electrical pumping, in comparison with lateral electrical pumping, allows, in particular, for reduction of the current threshold. This can be achieved by reducing the distance between the contact pads and the resonance cavity.

The authors note three more potential advantages, besides those cited above, in favor of the proposed nanocavity, which require proof through additional research. First, the creation of nanocavities with nonlinear properties is simplified. The supplementary component of the structure can be used to bring an optically nonlinear material directly into the nanocavity region. Second, the vertical radiation characteristics of the cavity can be controlled by varying the shapes of the second component of the resonator and its substrate. Third, the two-component structure of nanocavity permits dynamic control of its parameters. For example, the resonant frequency, Q-factor value and vertical radiation characteristics of the cavity can be varied by shifting one of the components of the nanocavity in the horizontal and/or vertical direction.

The cavity considered in this paper is based on the PC nanobeam. However, it is likely that the above approach to the construction of two-component PC cavities can be used for two-dimensional structures on the basis of a PC slab (e. g., for the L3-defect cavity [10] or for the line-defect cavity [11]. The validity of this assumption is also expected to check in our further studies.

## 5. Conclusion

The authors propose the two-component nanocavity model. In this model, the minimum details of the structure, which are the most difficult to fabricate, are found only in the periodic component of the resonator. Other advantages include the ease of forming arrays of nanocavities, a promising way to construct an electrically pumped photonic cavity, the ease of introducing non-linear optical materials in the area of the nanocavity, the possibility of formation of the desired energy distribution in the far zone, and the possibility to construct dynamic systems based on nanocavities.

The drawback of the proposed design of the nanocavity is its high sensitivity to the components' alignment errors. It is shown that to create a nanocavity with the Q-factor above $10^3$, the latest technologies of structural layers alignment, within ~ 10 nm error, should be used.

This work was supported by the RFBR grants 11-07-00153, 13-07-97002, RF Presidential grant NS-4128.2012.9, and FCP Cadres 8027(12.07.2012), 8231(06.08.2012).

## 6. References

[1] K. J. Vahala, "Optical microcavities," Nature **424**, 839-846 (2003).

[2] Y. Akahane, T. Asano, B.-S. Song, and S. Noda, "Fine-tuned high-Q photonic-crystal nanocavity," Opt. Express **13**, 1202-1214 (2005).

[3] P. Velha, J. C. Rodier, P. Lalanne, J. P. Hugonin, D. Peyrade, E. Picard, T. Charvolin and E. Hadji, "Ultra-high-reflectivity photonic-bandgap mirrors in a ridge SOI waveguide," New J. Phys. (IOP) **8**, 1-13 (2006).

[4] A. Yariv, Y. Xu, R. K. Lee, and A. Scherer, "Coupled-resonator optical waveguide: a proposal and analysis," Opt. Lett. **24**, 711-713 (1999).

[5] F. Qin, Z.-M. Meng, X.-L. Zhong, Y. Liu, and Z.-Y. Li, "Fabrication of semiconductor-polymer compound nonlinear photonic crystal slab with highly uniform infiltration based on nano-imprint

lithography technique," Opt. Express **20**, 13091-13099 (2012).
[6] G. Shambat, B. Ellis, J. Petykiewicz, M. Mayer, T. Sarmiento, J. Harris, E. E. Haller, and J. Vuckovic, "Nanobeam photonic crystal cavity light-emitting diodes," Appl. Phys. Lett. **99**, 071105 (2011).
[7] I. W. Frank, P. B. Deotare, M. W. McCutcheon, and M. Lončar, "Programmable photonic crystal nanobeam cavities," Opt. Express **18**, 8705-8712 (2010).
[8] N.-V.-Q. Tran, S. Combrié, and A. De Rossi, "Directive emission from high-Q photonic crystal cavities through band folding," Phys. Rev. B **79**, 041101 (2009).
[9] S. L. Portalupi, M. Galli, C. Reardon, T. F. Krauss, L. O'Faolain, L. C. Andreani, and D. Gerace, "Planar photonic crystal cavities with far-field optimization for high coupling efficiency and quality factor," Opt. Express **18**, 16064–16073 (2010).
[10] Y. Akahane, T. Asano, B. S. Song, and S. Noda, "High-Q photonic nanocavity in a two-dimensional photonic crystal," Nature **425**, 944–947 (2003).
[11] B.-S. Song, S. Noda, T. Asano and Y. Akahane, "Ultra-high-Q photonic double-heterostructure nanocavity," Nature Materials **4**, 207-210 (2005).
[12] Q. Quan and M. Loncar, "Deterministic design of high Q, small mode volume photonic crystal nanobeam cavities," Opt. Express **19**, 18529–18542 (2011).
[13] A. Taflove and S. C. Hagness, Computational Electrodynamics: The Finite-Difference Time-Domain Method, 3rd ed. Norwood, MA: Artech House, 2005.
[14] B. H. Ahn, J. H. Kang, M. K. Kim, J. H. Song, B. Min, K. S. Kim, and Y. H. Lee, "One-dimensional parabolic beam photonic crystal laser," Opt. Express **18**, 5654-5660 (2010).
[15] N. Li, W. Wu, S. Y. Chou, "Sub-20-nm alignment in nanoimprint lithography using Moire fringe," Nano Letters **6**, 2626-2629 (2006).
[16] W. Wu, R. G. Walmsley, W.-D. Li, X. Li, R. S. Williams, "Nanoimprint lithography with ≤60 nm overlay precision," Applied Physics A. **106**, 767-772 (2012).
[17] S. R. J. Brueck, "Optical and Interferometric Lithography — Nanotechnology Enablers," Proc. IEEE **93**, 1704-1721 (2005).
[18] S.-H. Kim, J. Huang, and Axel Scherer, "From vertical-cavities to hybrid metal/photonic-crystal nanocavities: towards high-efficiency nanolasers," JOSA B **29**, 577-588 (2012).

# Modeling of Magnetostriction of Soft Elastomer

**Andriushchenko Petr, Afremov Leonid, Chernova Mariya**

Departament of Theoretical and Experimental Physics, The School of Natural Sciences, Far Eastern Federal University, 8, Sukhanova str., 690950, Vladivostok, Russian Federation

e-mail: pitandmind@gmail.com, afremov.ll@dvfu.ru

**Abstract**. Small magnetic particles placed in a relatively soft polymer (with elastic modulus $E \sim 10 \div 100$ kPa) are magnetically soft elastomers. The external magnetic field acts on each particle which leads to microscopic deformation of the material and consequently to changing of its shape – magnetostriction. For purposes of studying of magnetostriction the model of movable cellular automata (MCA), in which a real heterogeneous material is an ensemble of interacting elements of finite size – automata, is used. It's supposed to be that the motion of each automata can be described by Newton's Second law. The force acting on the i-th automata consists of the following components: volume-dependent force acting on the automata i which is caused by pressure from the surrounding automata; force of an external magnetic field acting on the i-th automata with some magnetic moment; and normal and tangential interaction force between a pair of i and j automata. This approach was used for modeling of magnetostriction elastomer.

## 1. Introduction

At present time there is a necessity for new magnetic materials with properties, substantially depend on a structure (for example, distribution of particles magnetic moments, volumes, coordinates, critical fields, properties of a medium in which particles are placed) and on external influences (magnetic field, pressure, temperature, etc.). MAE can be considered as a class of such "smart"-materials, which one is a composite material, consisting of a flexible polymer medium and the filler - dispersed magnetic powders.

The medium of the MAE is made based on a huge variety of polymeric materials, in particular, it may be polystyrene (PS) [1], polyacrylamide [2], alginate [3,4], polydimethylsiloxane (PDMS) [5,6], ''SIEL'' produced by GNIIChTEOS [6], soft polyurethane (PU), obtained from polyether polyols VORALUX® HF 505 [7] etc.

In the past few years, MAE have attracted growing attention and have been considered for applications as adaptive stiffness elements in vibration absorbers [8], automotive bushing, piezoelectric power actuators [9], as well as in other areas, such as the car industry, medicine (where their properties can be used for cancer therapy by hyperthermia) [10,11], construction, household appliances and etc.

Some methods have been proposed to describe the behavior of magnetorheological (MR) materials. Ginder at al. [12] analyzed average magnetic induction using finite element analysis and computed the shear stresses from the field using Maxwell's stress tensor. A simple dipole model, based on the magnetic interactions between two adjacent particles was extended, to approximate magnetorheological elastomer (MRE) performances [13]. The saturated field-induced shear modulus by considering the interactions in a single particle chain were calculations by Davis [14]. Finite length model based on the microstructural observation of the particles forming the discontinuous and finite length column structures in polymer matrices was presented [15,16].

Most of these models exhibited good accuracy in predicting MR effects caused by embedded ferromagnetic particles, but almost none of these models do not research the fact of the deformation of the sample under the influence of a uniform field. In this regard, the development of a model, formalization and algorithmization of task for modelling of processes, which take place in MAE under an external magnetic field, is necessary.

In this work, an approach to the simulation of MAE is proposed based on a mathematical model of single-domain particles. All calculations were made with particles of the real size. Basic features of the dependence of MAE deformation on a uniform external magnetic field were discussed.

## 2. Method description

To study the magnetostriction of magnetic sponges we use the method of movable cellular automata [17] that combines two following approaches:

- Method of particles, in which, based on the Lagrange formalism, the motion of individual environmental particles is considered;
- Method of cellular automata based on the Euler approach, in which the time-to-time variation of the properties of elements of the fixed uniform grid is studied in relation to the state of surrounding elements.

In the framework of the method of movable cellular automata, the simulated system is represented by an assembly of interacting automata (elements of finite size). The concept of the method is based on the introduction of a new state type – the state of a pair of automata. This allowed to use the spatial variable as a switching parameter. This parameter is represented by $h^{ij}$ - the overlapping of a pair of spherical automata with diameters $d^i$ and $d^j$:

$$h^{ij} = r^{ij} - r_0^{ij}, \qquad (1)$$

where $r_0^{ij} = (d^i + d^j)/2$, $r^{ij}$ is the distance between the centers of the adjacent automata of equilibrium and strain state, respectively. In the simplest case, there are two states of pairs: interconnected $h^{ij} < h_{max}^{ij}$ and unconnected $h^{ij} > h_{max}^{ij}$, where $h_{max}^{ij}$ is a typical condition for this model.

Let's assume that the automata have a spherical shape with a diameter $d_i$ and $N$ automata from the assembly are $N_m$ magnetic. We assume that the magnetic nanoparticle-automata magnetized uniformly. Such magnetic nanoparticle (automata with $i$ number) will be surrounded by z neighbors (see Figure 1). The number of neighbors is determined by the dimension of a problem and packing density. For example, in the case of 2-dimensional problem z = 4 for a simple and z = 6 for a close-packed system shown in Fig. 1.

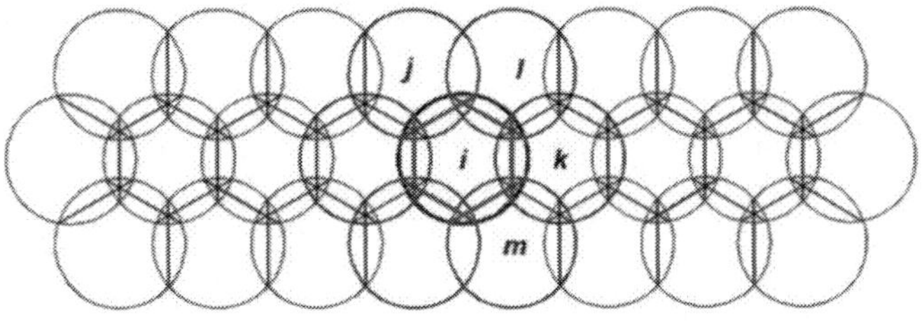


**Figure 1**. Illustration to the model of cellular automata

Let's assume that a simulated by mobile cellular automata environment can be described in terms of classical mechanics. Then, it is justified by the Newton's second law and the law of dynamics of rotational motion:

$$m_i \frac{d^2 \boldsymbol{r}_i}{dt^2} = \boldsymbol{F}_i = \boldsymbol{F}_i^{\Omega} + \sum_{j=1}^{z} \left( \boldsymbol{F}_{ij}^{\mathrm{norm}} + \boldsymbol{F}_{ij}^{\mathrm{tang}} \right) + \boldsymbol{F}_i^{\mathrm{m}}, \tag{2}$$

$$J_i \frac{d^2 \boldsymbol{\theta}_i}{dt^2} = \boldsymbol{K}_i = \sum_{j=1}^{z} \left[ \mathbf{n}_{\mathrm{ij}}, \boldsymbol{F}_{ij}^{\mathrm{tang}} \right] q_{ij} \, S_{ij} + [\boldsymbol{r}_i, \boldsymbol{F}_i^{\mathrm{m}}], \tag{3}$$

where $\boldsymbol{r}_i$ here is the radius vector of the center of the mass of the automata $i$, $\boldsymbol{F}_i^{\Omega} = \sum_{\mathrm{j=1}}^{\mathrm{z}} \mathrm{P_j S_{ij}} \mathbf{n}_{\mathrm{ij}}$ - the volume-dependent force effecting on the automata $i$ and provided by the pressure from surrounding automata, $P_j = -K_j \frac{\Omega_j - \Omega_{\mathrm{j}}^0}{\Omega_{\mathrm{j}}^0} = -K_j D \bar{\varepsilon}_j$ - the pressure of the *j-th* neighboring automata, $\Omega_{\mathrm{j}}^0$- is the initial equilibrium automata volume, $\Omega_j$ - the current automata volume, $K_j$ - the module of uniform compression of the material of this automata, $D$ - the coefficient depending on the dimension of the problem ($D = 1, 2, 3$), $\bar{\varepsilon}_j = \sum_{j=1}^{z} \varepsilon_{ij}/z$ - the average strain, $\varepsilon_{ij} = (2q_{ij} - d_i)/d_i$, where $S_{ij}$- the contact area of the *i-th* automata with the *j-th* and $q_{ij}$ - the distance from the center of the *i-th* automata to the surface of the *j-th*; $\mathbf{n}_{\mathrm{ij}} = (\boldsymbol{r}_i - \boldsymbol{r}_j)/r^{ij}$ - the normal vector between the *i-th* and *j-th* automata, z – the number of nearest neighbors. $\boldsymbol{F}_{ji}^{\mathrm{norm}}$ - pair force of normal interaction between $i$ and j automata, which in approximation of the Hooke's law is determined by a superposition of pair forces of elastic (first component) and viscous interaction:

$$\boldsymbol{F}_{ij}^{\mathrm{norm}} = -\left( 2G\left(\varepsilon_{ij} - \bar{\varepsilon}_i\right) + \eta \frac{(\boldsymbol{v}_j - \boldsymbol{v}_i, \mathbf{n}_{\mathrm{ij}})}{r^{ij}} \right) S_{ij} \mathbf{n}_{\mathrm{ij}}, \tag{4}$$

where $G$ - is the shear modulus, η - the viscosity coefficient, $\boldsymbol{v}_i$- the speed of the the *i-th* automata. $\boldsymbol{F}_{ij}^{\mathrm{tang}}$ - force of pair tangential interaction, as well as pair force of normal interaction, elastic shear and viscous components:

$$\boldsymbol{F}_{ij}^{\mathrm{tang}} = -2G\left[\boldsymbol{\gamma}_{ij} + \boldsymbol{\gamma}_{ji}, \mathbf{n}_{\mathrm{ij}}\right] - \eta \frac{\left[(\boldsymbol{\omega}_{ij} r^{ij} - \boldsymbol{\omega}_i q_{ij} - \boldsymbol{\omega}_j q_{ji}), \mathbf{n}_{\mathrm{ij}}\right]}{r^{ij}}, \tag{5}$$

where $\boldsymbol{\gamma}_{ij} = \sum_{\mathrm{k}} [\mathbf{n}_{\mathrm{ij}}{}^{\mathrm{k-1}}, \mathbf{n}_{\mathrm{ij}}{}^{\mathrm{k}}] - \boldsymbol{\theta}_i$, $\boldsymbol{\gamma}_{ji} = \sum_{\mathrm{k}} [\mathbf{n}_{\mathrm{ji}}{}^{\mathrm{k-1}}, \mathbf{n}_{\mathrm{ji}}{}^{\mathrm{k}}] - \boldsymbol{\theta}_j$ is a full shift angle of the *i-th* and *j-th* automata correspondently, $\boldsymbol{\theta}_i$ and $\boldsymbol{\theta}_j$ - rotation angle of each of the automata, $\boldsymbol{\omega}_{ij} = [\mathbf{n}_{\mathrm{ij}}, (\boldsymbol{v}_j - \boldsymbol{v}_i)]/r^{ij}$ and $\boldsymbol{\omega}_i$ - angular speeds of rotation of a pair and the *i-th* automata, respectively, $m_i$ and $J_i$- the mass and moment of inertia of the *i-th* automata, respectively.

External magnetic fields of uniformly magnetized (single-domain) spherical particles–automata are dipole. Therefore, replacing the particle by a point dipole, we assume that the *i-th* particle is exposed by the local field $\boldsymbol{H}(\boldsymbol{r}_i)$ defined as the sum of the external magnetic field $\boldsymbol{h}$ and fields of magnetic dipole interaction caused by other particles:

$$\boldsymbol{H}(\boldsymbol{r}_i) = \boldsymbol{h} + \sum_j \left( -\frac{\boldsymbol{\mu}^j}{r_{ij}^3} + \frac{3(\boldsymbol{\mu}^j, \boldsymbol{r}_{ij})\boldsymbol{r}_{ij}}{r_{ij}^5} \right), \boldsymbol{r}_{ij} = \boldsymbol{r}_i - \boldsymbol{r}_j. \tag{6}$$

$\boldsymbol{\mu}^j$ here is the magnetic moment of the given *j-th* particle. Since the energy of the point magnetic dipole is $V^i(\boldsymbol{r}_i) = -\left(\boldsymbol{\mu}^i, \boldsymbol{H}^i(\boldsymbol{r}_i)\right)$, due to the homogeneity of the magnetic moment $\boldsymbol{\mu}^j$ the magnetic force on the *i-th* particle–automata is:

$$\boldsymbol{F}_i^m = -\nabla V^i(\boldsymbol{r}_i) = (\boldsymbol{\mu}^i, \nabla)\boldsymbol{H}^i(\boldsymbol{r}_i). \quad (6)$$

The external field $\boldsymbol{h}$ mediately affects magnetic force. It leads to a change in magnetic moments of the particles–automata, which leads to a complete change of the magnetic interaction between magnetic moments of the particles and, as a result, to a change of the force $\boldsymbol{F}_i^m$.

## 3. Simulation algorithm

The main part of the simulation algorithm in this model is the determination of the positions and velocities of the automata of each $k$-th iteration:

$$\boldsymbol{r}_i(t^{k+1}) = \boldsymbol{r}_i(t^k) + \boldsymbol{V}_i(t^k)\Delta t, \quad (7)$$

$$\boldsymbol{V}_i(t^{k+1}) = \boldsymbol{V}_i(t^k) + \boldsymbol{F}_i/m_i, \quad (8)$$

$$\boldsymbol{\theta}_i(t^{k+1}) = \boldsymbol{\theta}_i(t^k) + \boldsymbol{\omega}_i(t^k)\Delta t, \quad (9)$$

$$\boldsymbol{\omega}_i(t^{k+1}) = \boldsymbol{\omega}_i(t^k) + \boldsymbol{K}_i/J_i, \quad (10)$$

where $t^{k+1} = t^k + \Delta t$, $\Delta t$ - the time of one iteration commensurate with the time of passage of a sound through the automata: $\tau = d_i/C$ ($C = \sqrt{G/\rho}$ - the speed of environmental sound, ρ - the density of the environment).

Figure 2 illustrates an example of the media magnetostriction. The proposed approach can be used for simulating the magnetostriction of soft elastomer.

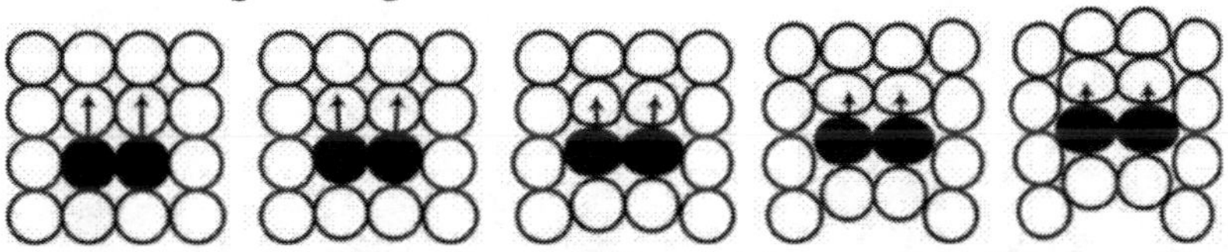

**Figure 2.** Examples of pre-simulation. The magnetic particles are marked black, medium are marked white.

## 4. Acknowledgements

The work is supported by grant of Ministry of Education and Science 02.740.11.0549 (Reference number 2010-1.2.2-214-005-006).

## References

[1] Olsvik O 1994 *Clin Microbiol Rev* **7** 43–54.
[2] Kawaguchi H 1996 *Colloids Surf A* **109** 147–54.
[3] Nishio Y 2004 *Polymer* **45**(21) 7129–36.
[4] Morales M 2008 *Mater Sci Eng C* **28**(2) 253–7.
[5] Barham M 2009 *IEEE Transactions on* **45**(10) 4124-4127.
[6] Abramchuk S 2007 *Polymers for Advanced Technologies* **18**(11) 883-890.
[7] Boczkowska A 2012 *Composites Part B: Engineering* **43**(2) 636-640.
[8] Deng H 2007 *Journal of intelligent material systems and structures* **18** 1205
[9] Zhou G 2005 *Smart Materials and Structures* **14** 1154
[10] Falk M 2001 *Int J Hyperthermia* **17**(1) 1–18
[11] Glöckl G 2006 *J Phys Condens Matter* **18**(38) 2935–51.
[12] Ginder J 1994 *Applied Physics Letters* **65** 3410

[13] Jolly M 1996 *Smart Materials and Structures* **5** 607
[14] Davis L 1999 *Journal of Applied Physics* **85** 3348
[15] Chen L 2007 *Smart Materials and Structures* **16** 2645
[16] Chen L 2011 *Journal of Applied Physics* **110** 013513
[17] Psakhie S 1998 *Physical mesomechanics* **1** (1) 95–108.

# Binary-collision-approximation simulation for noble gas irradiation onto plasma facing materials

**Seiki Saito**[1,3] **Hiroaki Nakamura**[1,2], **Arimichi Takayama**[2] **and Atsushi M. Ito**[2]

[1]Department of Energy Engineering and Science, Graduate School of Engineering, Nagoya University, Toki, Gifu 509-5292 Japan
[2]Fundamental Physics Simulation Research Division, National Institute for Fusion Science, Toki, Gifu 509-5292 Japan

E-mail: hnakamura@nifs.ac.jp

**Abstract.** A number of experiments show that helium plasma constructs filament (fuzz) structures whose diameter is in nanometer-scale on the tungsten material under the suitable experimental condition. In this paper, binary-collision-approximation-based simulation is performed to reveal the mechanism and the conditions of fuzz formation of tungsten material under plasma irradiation. The irradiation of the plasma of hydrogen, deuterium, and tritium, and also the plasma of noble gas such as helium, neon, and argon atoms are investigated. The possibility of fuzz formation is discussed on the simulation result of penetration depth of the incident atoms.

## 1. Introduction

Tungsten material is one of promising candidates for divertor plates in nuclear fusion reactors. Bubble formation [1, 2] is observed on the surface of the tungsten material which is irradiated by helium plasma. Furthermore, a number of experiments show that helium plasma constructs filament (fuzz) structures [3, 4] whose diameter is in nanometer-scale on the tungsten material under the suitable condition (i.e., material temperature of 1,000-2,000 K and incident energy of 20-100 eV). The formation of these structures does not only decrease the thermal conductivity of the material, but also enhances the tolerance in cracking and the radiation efficiency. It also causes reduction of sputtering yield.

As shown in fig. 1, it is inferred that the formation of fuzz structure consists of the following three elementary processes: Penetration process of incident atoms; Diffusion of the incident atoms and trapping by vacancies; Formation and aggregation processes of bubbles of the incident atoms. In this paper, therefore, binary-collision-approximation-based simulation is performed to reveal the mechanism and the conditions of fuzz formation of tungsten material under plasma irradiation, as the first step of investigation. The penetration depth of helium, neon, and argon gases were investigated in our previous study [5]. The fuzz structure is not observed under other noble gas plasma irradiation except helium plasma [6]. We review the result of our previous study at first. The irradiation of hydrogen isotope onto tungsten material is also the key issue

[3] Present address: Department of Electrical Engineering, Kushiro National College of Technology, Otanoshike-Nishi 2-32-1, Kushiro-Shi, Hokkaido, Japan

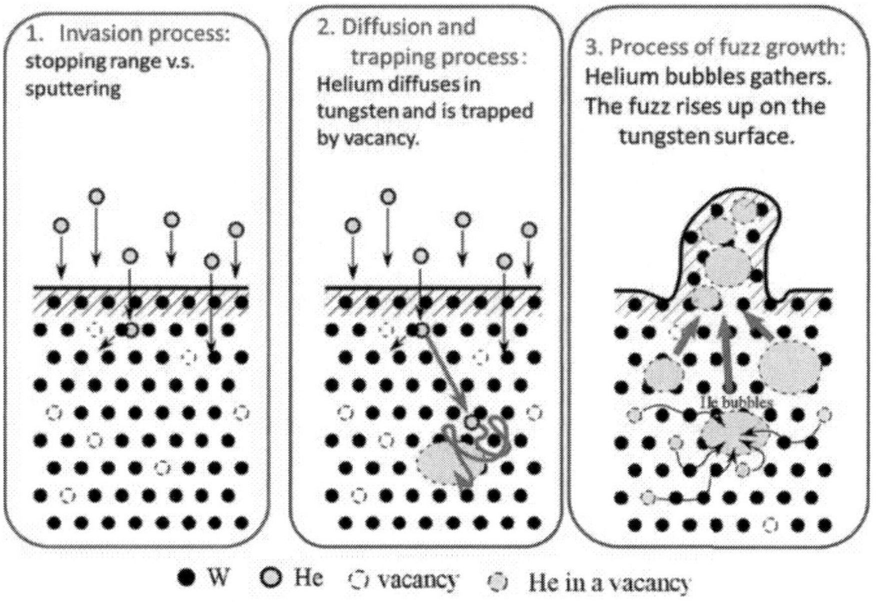


**Figure 1.** Outlook chart of the fuzz formation process.

for the nuclear fusion device. After the review of the study of the noble gas irradiation, the case of the irradiation of hydrogen, deuterium, and tritium plasma is discussed. The penetration depth is strongly depends on the structure of the target material. Therefore, the penetration depth for amorphous and bcc crystalline structure is carefully investigated in this paper.

## 2. Simulation Model

BCA simulation is performed by AC∀T (atomic collision in any structured target) code [7, 8]. In BCA simulation, multi-body interactions in a material approximate to consecutive two-body interactions between a projectile atom and the nearest neighbor atom. The scattering angle of the projectile and the recoil atom at each binary collision is obtained analytically in a two-body interatomic potential. In AC∀T code, the Moliere approximation to the Thomas-Fermi potential is employed.

BCA simulation is performed for two types of tungsten structure, i.e., amorphous and bcc crystalline structure. The size of the target material is set to 47.47 Å long, 47.47 Å wide, and 9998.24 Å deep. The $z$-axis of the simulation box is set parallel to the edge of the target material whose length is 9998.2 Å. Periodic boundary conditions are used in the $x$- and $y$-directions. The lattice constant of bcc crystal is set to 3.16 Å. Amorphous structure is formed by distributing tungsten atoms randomly. The density of amorphous structure is set to the same as the density of bcc crystal. The temperature of the tungsten materials is set to 0 K. Helium (He), neon (Ne), argon (Ar), hydrogen (H), deuterium (D), tritium (T) atoms are injected into these tungsten materials. The mean depth of penetration of incident atoms for 10,000 injections is calculated for constant incident energy from 10 eV to 10 keV. The $x$- and $y$-coordinates of the starting positions of the incident atoms are set randomly. The incident angle is set to parallel to the $z$-axis, i.e., perpendicular to (100) surface in the case of bcc crystal. To calculate the mean depth, the target material is refreshed to the initial perfect crystal before each injection.

### 3. Simulation Result

Figure 2 shows the incident energy dependence of the sputtering yield and the mean depth of penetration in the case of He, Ne, and Ar injection. The solid and dashed lines denote the results in the case of bcc crystalline and amorphous structure, respectively. The mean depth strongly depends on atomic species and the structure of the target material. In the case of amorphous structure, the mean depth is almost proportional to the square root of incident energy. The mean depth in the case of bcc crystalline structure has a different profile compared with that of amorphous structure when incident energy is larger than threshold energies (i.e., approximately 20 eV, 100 eV, and 160 eV for He, Ne, and Ar, respectively). It is found that the difference of the profiles is caused by channeling effect.

In both cases of amorphous and bcc crystalline structure, the mean depth of He is larger than that of Ne and Ar. Even if the target material has polycrystalline tungsten which has the mixed structure of amorphous and bcc crystal structure, it is expected that He can penetrate to deeper position than Ne and Ar. For the fuzz formation, it is important that incident atoms penetrate deeply enough (more than a few nanometer) with lower incident energy than sputtering threshold energy (i.e., 100 eV, 35 eV, and 33 eV for He, Ne, and Ar, respectively). It is concluded that the fuzz structure is more easily formed by He irradiation than that of Ne or Ar. And the simulation result supports Yajima's experimental results [6].

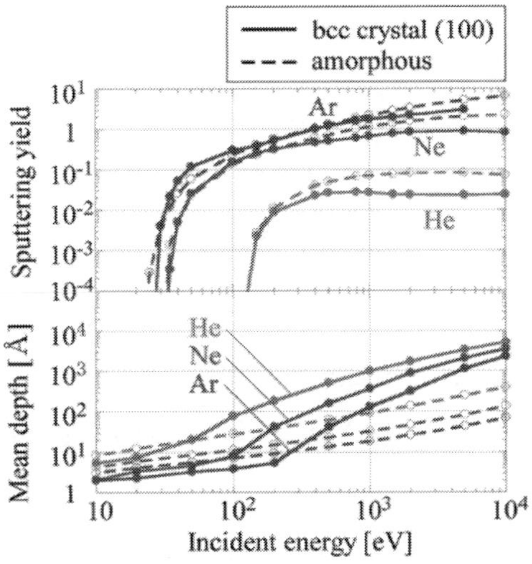


**Figure 2.** Incident energy dependence of the sputtering yield and the mean depth of penetration in the case of He, Ne, and Ar injection.

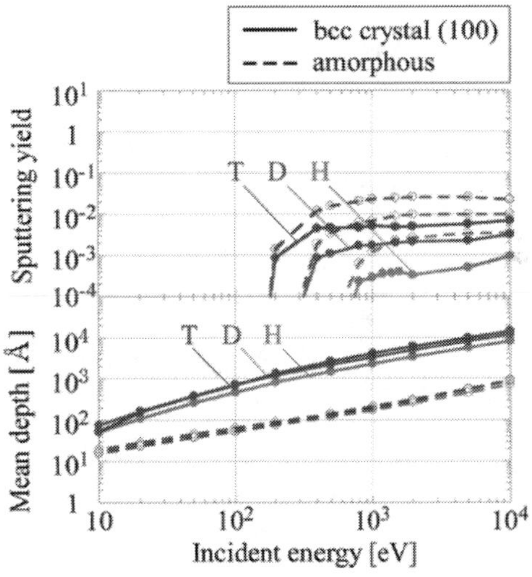


**Figure 3.** Incident energy dependence of the sputtering yield and the mean depth of penetration in the case of H, D, and T injection.

Figure 3 shows the incident energy dependence of the sputtering yield and the mean depth of penetration in the case of H, D, and T injection. The sputtering threshold energies of H, D, and T are larger than the case of He, Ne, and Ar. It is because the scattering cross-section of H, D, and T is shorter than that of He, Ne, and Ar. The average energy transfer to tungsten atom from H, D, and T is also lower than that from He, Ne, and Ar because the mass ratio of H, D, and T to tungsten atom is low. Because of the same reason, the mean depth of H, D, and T is larger than that of He, Ne, and Ar. Therefore, it is concluded that the H, D, and T has the ability to form the fuzz structure as He as a manner of penetration depth. However, the fuzz formation by irradiation of H, D, and T has not been observed experimentally. For the further discussion of the possibility of fuzz formation, it is necessary to investigate of the diffusion and trapping process of the incident atoms by vacancies.

## 4. Summary

The mean depth strongly depends on atomic species and the structure of the target material. However, in both cases of amorphous and bcc crystalline structure, the mean depth of He is larger than that of Ne and Ar. For the fuzz formation, it is important that incident atoms penetrate deeply enough (more than a few nanometer) with lower incident energy than sputtering threshold energy (i.e., 100 eV, 35 eV, and 33 eV for He, Ne, and Ar, respectively). It is concluded that the fuzz structure is more easily formed by He irradiation than that of Ne or Ar. The probability of fuzz formation under H, D, and T plasma irradiation is also investigated in a manner of the penetration depth. The depth of H, D, and T is larger than that of He, Ne, and Ar. H, D, and T is possible to reach deep place under the spattering threshold energy. For the further discussion of the possibility of fuzz formation, it is necessary to investigate of the diffusion and trapping process of the incident atoms by vacancies.

## Acknowledgment

This work was supported by JSPS KAKENHI Grant Number 24656560, and a Grant-in-Aid for JSPS Fellows Number 23-340.

## 5. References

[1] Iwakiri H, Yasunaga K, Morishita K and Yoshida N 2000 *J. Nucl. Mater.* **283-287** 1134
[2] Nishijima D, Ye M. Y, Ohno N and Takamura S 2004 *J. Nucl. Mater.* **329-333** 1029
[3] Takamura S, Ohno N, Nishijima D, Kajita S 2006 *Plasma Fusion Res.* **1** 051
[4] Kajita S, Saeki T, Hirahata Y, Yajima M, Ohno N, Yoshihara R and Yoshida N 2011 *Jpn. J. Appl. Phys.* **50** 08JG01
[5] Saito S, Ito A M, Takayama A and Nakamura H 2013 *J. Nucl. Mater.* **438** S895
[6] Yajima M, Yamagiwa M, Kajita S, Ohno N, Tokitani M, Takayama A, Saito S, Ito A M, Nakamura H and Yoshida N 2013 *J. Plasma Sci. Technol.* **15** 282
[7] Takayama A, Saito S, Ito A M, Kenmotsu T and Nakamura H 2011 *J. Appl. Phys.* **50** 01AB03
[8] Saito S, Takayama A, Ito A M and Nakamura H 2011 *Proc. Int. Conf. Model. Simulation Technol.* 197

# First-principle Studies of armchair graphene nanoribbons

**N. M. Wang, G. J. Zhao, X. X. Liang and T. L. Song**
Department of Physics, School of Physical Science and Technology, Inner Mongolia University, Hohhot, 010021, P. R. China

E-mail address: stzhaogj@imu.edu.cn

**Abstract.** In this article, by using the first principle calculations based on the density functional theory, we present a detailed investigation of the energy band and density of states of armchair graphene nanoribbons (AGNRs) with bare and H-terminated edges. Based on the structural optimization results, we compute the energy band and density of states of considered nanoribbons. The results show that there is a direct band gap for bare and H-terminated edges AGNRs, and indicate AGNRs have semiconductor properties for both cases our calculated. There are localized states turns up at -2.520eV for the case of bare edges, after modification of hydrogen atoms, the localized states disappeared, the band gap is widened form 0.535eV for the bare edges to 0.722eV for H-terminated edges, at the same time, and the energy band degeneracy appeared.

## 1. Introduction

From the end of the 20th century, carbon-based materials have attracted researchers attentions, due to their unique structural and physical properties[1, 2]. Graphene is one of them, it is a monolayer honeycomb lattice of sp2-bonded carbon[3]. Because of the unique properties, graphene has the potential application in many fields. And it is an important on nanoelectronics and renewable energy storage[4]. However, researches focus more on graphene nanoribbons (GNRs). The graphene nanoribbon is a related type to carbon based quasi-one dimensional systems. According to the feature of edge, the GNRs include zigzag edge graphene nanoribbons (ZGNRs) and armchair edge graphene nanoribbons (AGNRs) [5]. Armchair edged (AGNRs) show semiconductor behaviour, and all zigzag edged (ZGNRs) are known to be metallic [6]. GNRs intrinsically have dangling bonds at the edges, and thus cause instability and reduce carrier mobility. Usually, a carbon atom on the nanoribbon edge is terminated by H atom which is used to be a termination[7]. The structural and electronic properties of F-terminated GNRs with different widths are investigated using density functional theory by Lu D B *et al*[7]. They found that the band gaps for the F-terminated AGNRs are smaller than those of the H-terminated AGNRs. Jaiswal N K and Srivastava P studied the electronic properties of bare and

H-terminated ribbons using *ab-initio* approach[8]. It found that H-termination enhances the stability and band gap of nanoribbons whereas dangling bonds in bare ribbons reduce the band gap.

In this paper, we investigate the electronic energy band and density of states of bare AGNRs and H-terminated AGNRs, using first principle calculations based on density functional theory.

### 2. Modeling and computational details

In the calculations, we employ the local density approximation (LDA) within the Perdew-Zunger exchange correlation functional. In order to achieve accurate results, the plane-wave cut off energy was set to 135Ry, with 540Ry for the charge density. The numerical integration of the Brillouin zone (BZ) was performed using a k-mesh of dimensions $1\times1\times21$. The periodic boundary condition of nanoribbons was maintained along the $z$ axis. In order to neglect the interaction of ribbon with its periodic images, a vacuum padding of 6.8Å and 20Å was also modelled along x and y axis. The modeling was done with XCrySDen [9]. Figure 1(a) shows the bare AGNRs model.The C-C bond length is 1.42Å for experiment value of bare AGNRs. Figure 1(b) shows the H-terminated AGNRs model. All calculations are implemented by using the code of Quantum ESPRESSO[4].

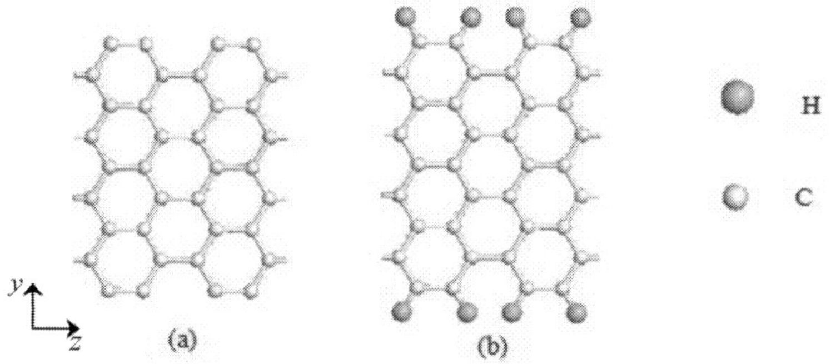


**Figure 1.** Schematic diagram of AGNRs figurations: (a) bare AGNRs; (b) H-terminated AGNRs.

### 3. Results and discussion

The total energies of bare and H terminated AGNRs are calculated by using self-consistent calculation, we found that the total energy of H terminated AGNRs is lower than the bare AGNRs. It can be predicted that H terminated AGNRs is more close to the practical situation than bare AGNRs. This conclusion is similar to other theoretical calculations [8].

The energy band structures of bare AGNRs and H-terminated AGNRs are presented in Figure 2. Figure 2 (a) show that the bare AGNRs and H-terminated AGNRs are all direct band gap semiconductor with energy gap of 0.535eV and 0.722eV respectively. These results are different with early work by Wang and Zhao[3]. In that work, the bare AGNRs is an indirect band gap semiconductor with 1.58eV band gap and H terminated AGNRs is a direct band gap semiconductor with 1.68eV band gap. The band gap of H terminated AGNRs is enhanced due to elimination of the dangling bonds which is unstable.

It can be see that, for the bare AGNRs, there are localized states turns up at -2.520eV, but localized states despaired, and the energy band degenerate from the conduction band at 1.439ev and 1.807eV for H terminated AGNRs. This is due to the dangling bonds in bare AGNRs.

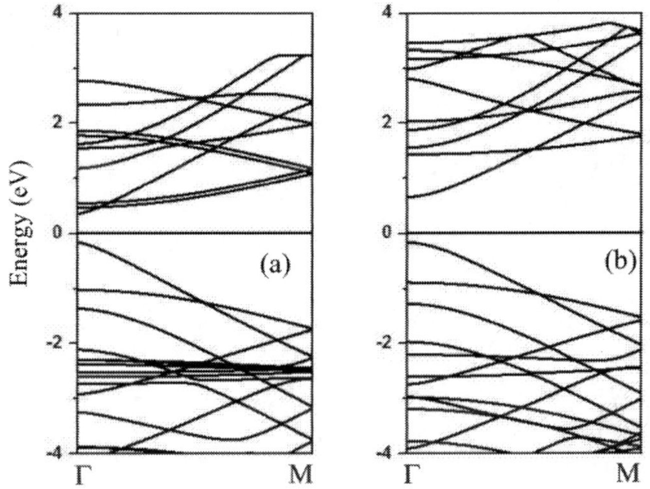


**Figure 2.** Calculated band structures of AGNRs: (a) bare; (b) H-terminated.

In order to understand the distribution of electronic states and the contribution of individual atomic orbitals, we calculated the total density of states(DOS) and the projected density of states(PDOS) near the Fermi level as shown in Figure 3, the Fermi level is located at 0eV.

Figure 3(a), the localization state at -2.52 eV for bare AGNRs is caused by the C-2p electronic states due to dangling bonds. The peak of total DOS appeared at around -2.568eV. From Figure 4(a), we can see that C-2p states mainly dominate at the top of the valence bands and the conduction bands.

For H-terminated AGNRs, the DOS and PDOS indicates that C-2p electrons and H-1s states contribute to the lower valence-band and conduction band. At the same time, by modification of hydrogen atoms, localized states are disappeared and the structure is more stable.

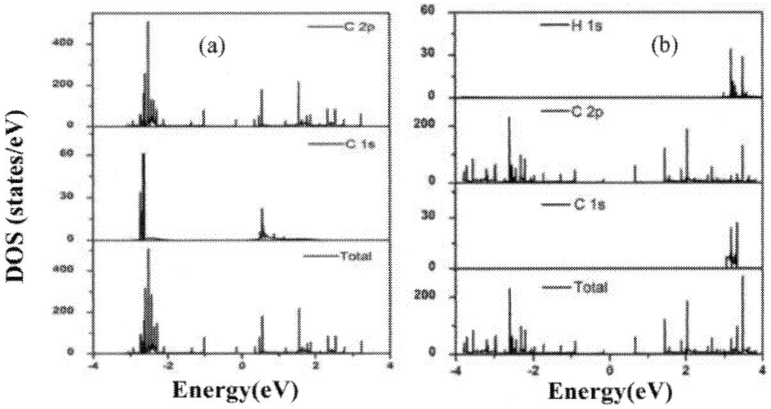


**Figure 3.** Total and projected density of states of AGNRs: (a) bare; (b) H-terminated.

## 4. Conclusions

In this article, we compute the energy band and density of states of considered nanoribbons for bare and H-terminated. It is revealed that the H terminated has the lowest total energy. Both bare and H-terminated AGNRs are direct band gap semiconductors. After modification of hydrogen atoms, the localized states disappeared, the band gap is widened form 0.535eV for the bare edges to 0.722eV for H-terminated edges, at the same time, and the energy band degeneracy appeared.

## 5. Acknowledgments

The research is supported in part by the PhD Progress Foundation of Higher Education Institutions of China(20111501110003), Natural Science Foundation of Inner Mongolia (No.2011MS0105) and Talent Development Foundation of Inner Mongolia.

## References

[1] Novoselov K S, Geim A K, Morozov S V, Jiang D, Zhang Y, Dubonos S V, Grigorieva I V, Firsov A A 2004 Science **306** 666

[2] Hass J, Feng R, Millan-Otoya J E, Li X, Sprinkle M, First P N, de Heer W A, Conrad E H, Berger C 2007 Phys. Rev. **B75** 214109

[3] Wang W H, Zhao G Z 2013 Solid State Commun. **166** 6

[4] Giannozzi P, Baroni S, Bonini N, Calandra M, Car R, et al, 2009 J.Phys.:Condens. Matter, 21, 395502

[5] Jia X T, Campos-Delgado J, Terrones M, Meunier V, Dresselhaus M S 2011 Nanoscale **3** 86

[6] Chauhan S S, Srivastava P, Shrivastava A K 2013 Solid Stat Commun. **154** 69

[7] Lu D B, Song Y L, Yang Z X, Li G Q 2011 Appl. Surf. Sci. **257** 6440

[8] Jaiswal N K, Srivastava P 2011 Physica **E44** 75

[9] Kokalj A, 2003 Comp. Mater. Sci. **28** 155

# First-principle studies of the electronic band structure and the phonon dispersion properties of wurtzite BN

**X. Lei, X. X. Liang, G. J. Zhao, and T. L. Song**
Department of Physics, School of Physical Science and Technology, Inner Mongolia University, Hohhot, 010021, P. R. China
E-mail: xxliang@imu.edu.cn

**Abstract**. The electronic band structure and phonon dispersion of wurtzite BN are studied by the first principle calculations. The local density approximation (LDA) and the generalized gradient approximation (GGA) exchange-correlation potentials are applied in the calculations and compared. The computational results for the band structure and density of states with indirect band gaps as well as the phonon dispersive curves and density of states are obtained. The corresponding dielectric and thermodynamic properties are discussed. The conclusions are consistent with other theoretical results and experimental data.

## 1. Introduction

In recent years Group-III nitride compound semiconductors have been widely studied due to their important device applications [1-3]. Boron nitride (BN), as a kind of III-N semiconductors, has attracted experimental and theoretical interests. The basic crystallographic structures of BN are hexagonal wurtzite (WZ) and cubic zincblende (ZB) structures [4-6] and the former usually more stable. The electronic band structure of wurtzite BN have been studied by first-principle calculations. The calculated results for the band gap are generally slightly smaller than experimental results of around 5.0 eV. Attention has also been paid to the lattice vibrations (phonons), which related closely to their electric, optical as well as thermal properties. Some authors have calculated the phonon dispersions of hexagonal BN by the first-principles with the tight-binding method [7,8]. Second-order Raman scattering has been used to probe phonon dispersions away from the Γ point [9].

In present work a first-principle calculation has been employed to obtain the band structure, state densities of electrons as well as the phonon dispersion curves of the WZ-BN. The corresponding dielectric and thermodynamic properties are discussed in brief.

## 2. Calculation method

The Plane Wave Self Consistent Field (PWscf) method of Quantum-Espresso [10] is used, by employing the plane wave basis set and the pseudo potential different from previous calculations. Computations are performed by means of the density functional theory (DFT) in conjunction with the local density approximation (LDA) and the generalized gradient approximation (GGA) separately. The plane-wave cutoff energy is set to 140 Ry for LDA and 55 Ry for GGA respectively through energy test. Numerical integrations of Brillouin zone are performed by choosing a k-mesh of dimensions 10×10×10. As for the calculation of the phonon spectral, we have selected a small $q$ mesh of dimensions 4×4×4 in force constant calculation.

The WZ structure of BN is modeled as Figure 1. The optimized values of the lattice constant are listed in Table I, which are close to the previous theoretical [11, 12] and experimental data [13].

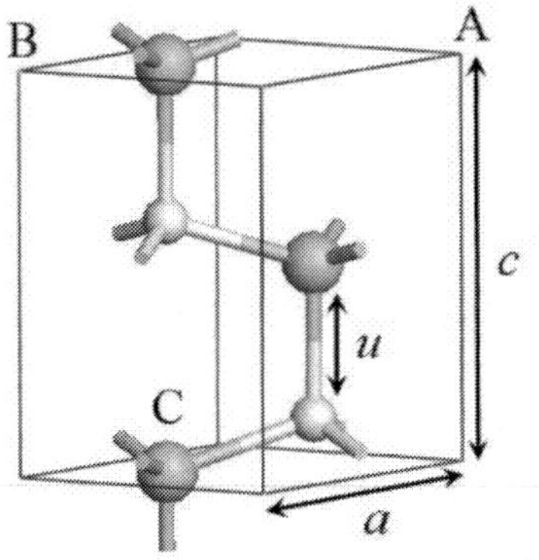


Figure 1. Wurtzite structures of BN. White balls represent B atoms and the dark gray balls N atoms. The lattice parameter $a$ and $c$ show as above.

Table I.The basic physical parameters of WZ-BN in the calculation compare with other theories and the experimental data

| | $a$(Å) | $c$(Å) | $c/a$ |
|---|---|---|---|
| This work$^{LDA}$ | 2.51 | 4.12 | 1.639 |
| This work$^{GGA}$ | 2.55 | 4.23 | 1.661 |
| Theory [11] | 2.54 | 4.17 | 1.642 |
| Theory [12] | 2.53 | 4.19 | 1.654 |
| Experiment [13] | 2.56 | 4.23 | 1.656 |

## 3. Result and discussion

### 3.1. Band structures and density of states

Figures 2 and 3 show BN band structures and density of states (DOS) obtained by LDA and GGA, respectively. It is found that the calculated bands by GGA are slightly lower than those by LDA. The results also indicate that the BN is the indirect band gap semiconductor. The top valence band splits up into 3 sub-bands.

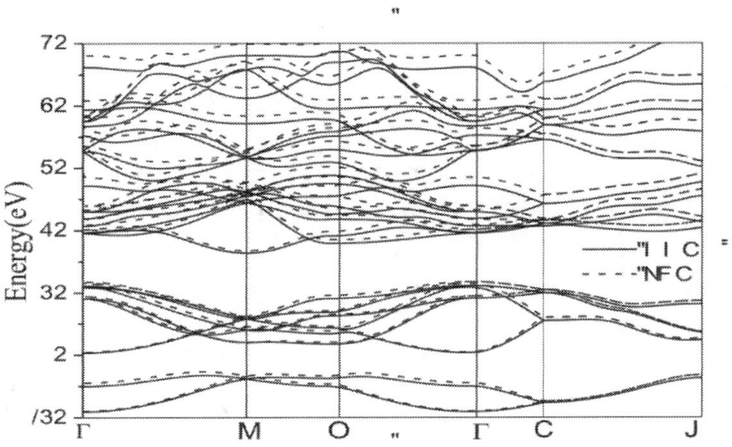


Figure 2. Band structure of WZ-BN obtained by GGA (solid line) and LDA (dash line).

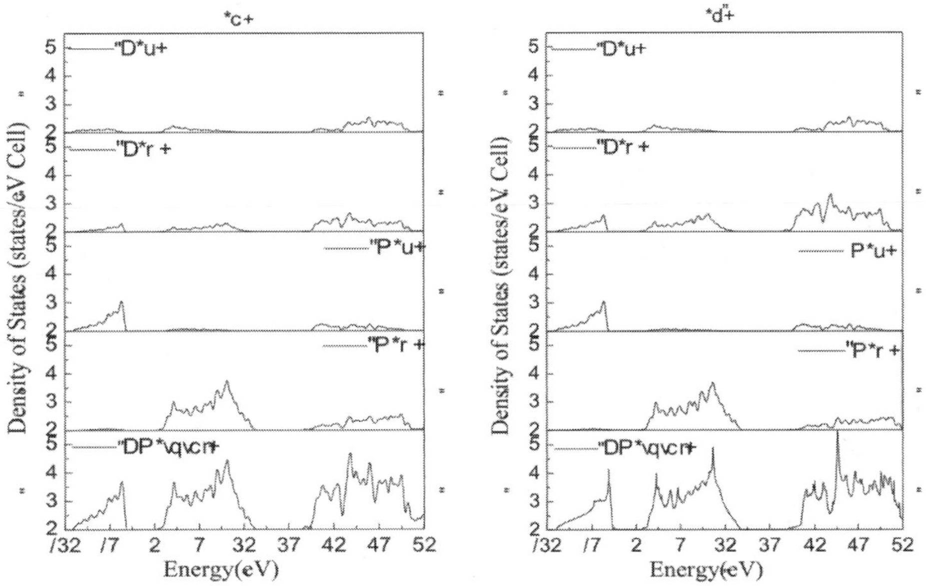


Figure 3. Total and part DOSs of WZ-BN by (a) GGA and (b) LDA calculations. B(s) stands for B atom 2s state, B (p) for B atom 2p state as well as N atoms.

One can also see that the valence band are mostly dominated by B-2s, 2p, N-2p electronic states and little N-2s electronic states, but the conduction band comes from B-2s, 2p and a little N-2s, 2p electron. The conclusions are consistent with other theoretical results.

### 3.2. Phonon dispersion curves

The calculated phonon dispersion curves and DOS of WZ-BN are plotted in figure 4. Due to the crystal primitive cell of WZ structure includes 4 atoms, consequently there is a splitting of the transverse optical (TO) and longitudinal optical (LO) phonon branches. It is expected that there are 12 branches of phonon modes, i.e. 3 branches of acoustic phonon modes and 9 branches of optical phonon modes. However, the phonon branches will degenerate in the different high symmetry points in the BN of hexagonal structure. For example, the phonon modes in the point A degenerate into only 4 modes, and there are 8 phonon modes at Γ point.

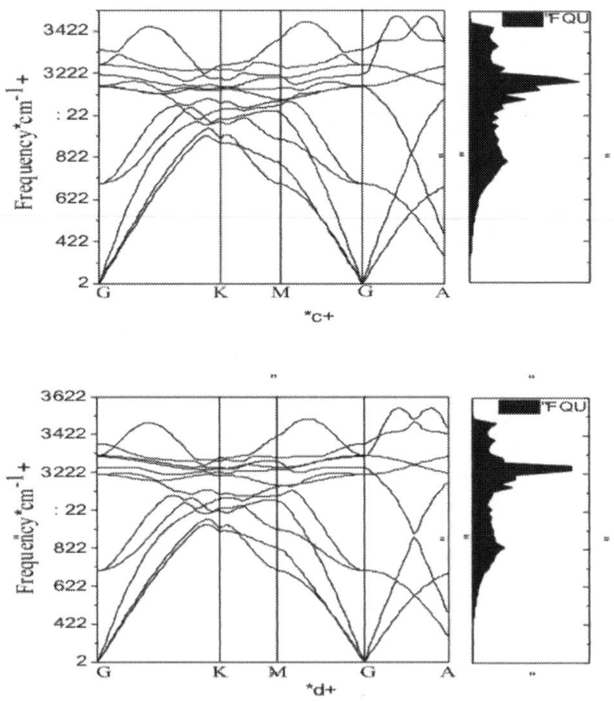


Figure4. Phonon dispersion curves and DOSs of WZ-BN by (a) GGA and (b) LDA calculations.

As is known, a phonon band-gap can be found from the dispersion curves and DOS if the mass difference between the positive and negative ions in a primitive cell is obvious. However, there are no any gaps between the optical and acoustic branches of phonon modes in figure 4, since the masses of the different ions in BN are close to each others.

The calculated phonon frequencies at the Γ point in this work are listed in Table II. The corresponding results given by previous theories are also listed partly for comparison. As we know, the $A_1$ and $E_1$ mode have the Raman and infrared activities, $E_2$ modes only possess the infrared activity, and $B_1$ is matt lamination [12]. It is seen that the calculated values of the corresponding phonon frequencies by LDA are slight higher than those by GGA. Our computational results are close to the previous theories.

Table II. The calculated phonon frequencies at the Γ point for BN in wurtzite structure by LDA and GGA as well as some other theories and experimental data. The frequencies are measured in $cm^{-1}$

| | $E_2^l$ | $B_1^l$ | $A_1(TO)$ | $E_1(TO)$ | $E_2^h$ | $B_1^h$ | $A_1(LO)$ | $E_1(LO)$ |
|---|---|---|---|---|---|---|---|---|
| This work$^{LDA}$ | 482 | 989 | 992 | 1024 | 1084 | 1088 | 1090 | 1151 |
| This work$^{GGA}$ | 474 | 936 | 939 | 941 | 994 | 1040 | 1044 | 1112 |
| Theory [14] | 447 | 918 | 988 | 1024 | 948 | 1179 | 1258 | 1286 |
| Theory [15] | | | 1006 | 1053 | | | 1258 | 1286 |
| Experiment [9] | 310 | | | | | | 1364 | 1364 |

3.3. Dielectric properties based on phonons

The Born effective charge tensor is diagonalizable and simplified into two values $\vec{Z}_{xx}^*=\vec{Z}_{yy}^*=\vec{Z}_P^*$ and $\vec{Z}_\perp^*$ (Parallel and perpendicular to c-axis of the hexagonal structure respectively) because of the higher symmetry of the wurtzite structure. The similar form is received for the static dielectric tensor. The calculated results are listed in Table II.

Table III. The born effective charge and the static dielectric tensors

| $Z_P^*$ | $Z_\perp^*$ | $\varepsilon_P^*$ | $\varepsilon_\perp^*$ |
|---|---|---|---|
| 1.97 | 1.87 | 4.71 | 4.52 |
| 2.81 | 2.68 | 6.07 | 5.88 |

It is clearly seen that the parallel components of the Born effective charge and static dielectric tensors are greater than the perpendicular ones. The results embody the anisotropy of the spontaneous polarization of WZ-BN.

3.4. Thermodynamic properties based on phonon spectra

Thermodynamic properties of the lattice system can be obtained by calculating the thermodynamic functions using the phonon dispersion results [16]. We have performed the computations for some main thermodynamic functions, such as the inner energy, free energy, specific heat and entropy for the WZ-BN lattice system. One code of PWscf with the quasi harmonic approximation (QHA) is used in the calculation. The obtained results for the entropy $S$ and the constant-volume specific heat $C_V$ as functions of the temperature based on GGA and LDA are respectively plotted in figure 5. It is seen that the specific heat presents the $T^3$-rule at low temperature and tend to a certain value when the temperature is high enough. The calculated results coincide with experimental and theoretical conclusions. Both the values of $S$ and $C_V$ by LDA are slightly lower than those by GGA, because the former gets the smaller phonon density of states.

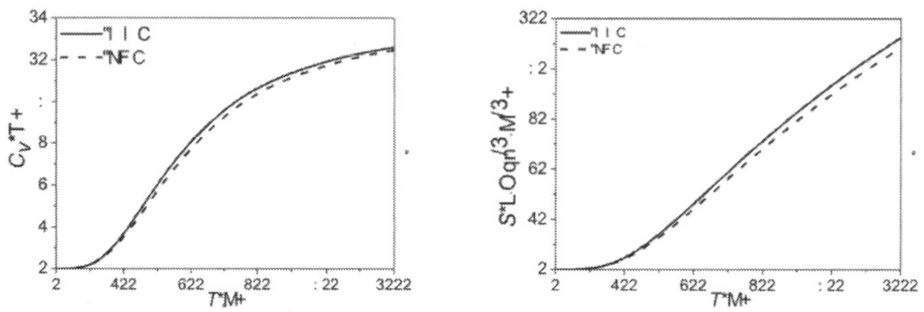

Figure 5. Entropy and constant-volume specific heat by GGA (solid line) and LDA (dash line), respectively. The specific heat is measured in $R = 8.31$ J $\mathrm{Mol}^{-1}$ $\mathrm{K}^{-1}$.

## 4. Conclusion

In this work, we have applied first-principle calculation with two different pseudo potentials, LDA and GGA, to study the electronic states and the lattice dynamics of BN in wurtzite structures. The energy band structure of electrons and the phonon dispersion characteristics as well as the dielectric and thermal properties are discussed. The calculated results are in agreement with the zone-center Raman measurements and the previous ab initio calculations.

## 5. Acknowledgments

This work was supported in part by the PhD Progress Foundation of Higher Education Institutions of China (20111501110003), Natural Science Foundation of Inner Mongolia (No. 2011MS0105) and Talent Development Foundation of Inner Mongolia.

## References

[1] Carling K, Wahnström G, Mattsson T R, et al. 2000 *Physical review letters*. **85**, 3862
[2] Kim K, Lambrecht W R L, Segall B 1996 *Physical Review B*. **53**, 16310
[3] Haubner R, Wilhelm M, Weissenbacher R, et al. 2002 *Springer Berlin Heidelberg*
[4] Lorenz H, Orgzall I 2004 *Acta materialia*. **52**, 1909-1916
[5] Mosuang T E, Lowther J E 2002 *Journal of Physics and Chemistry of Solids*. **63**, 363-368
[6] Michel K H, Verberck B 2011 *Physical Review B*. **83**, 115328
[7] Tohei, T., Kuwabara, A., Oba, F., & Tanaka, I. 2006 *Physical Review B*. **73**, 064304
[8] Ohba, N., Miwa, K., Nagasako, N., & Fukumoto, A. 2001 *Physical Review B*, **63**, 115207
[9] Reich, S., Ferrari, A. C., Arenal, R. et al. 2005 *Physical Review B*.**71**,205201
[10] Giannozzi P, Baroni S, Bonini P, etc. 2009 *J. Phys Condens. Matter*. **21**, 395502-395521
[11] Lawniczak-Jablonska K, Suski T, Gorczyca I, et al. 2000 *Physical Review B*. **61**, 16623
[12]Shimada, K., Sota, T., & Suzuki, K. 1998 *Journal of applied physics*, **84**, 4951-4958.
[13] Sōma, T., Sawaoka, A., & Saito, S. 1974 *Materials Research Bulletin*. **9**, 755-762
[14] Tütüncü H M, Srivastava G P 2000 *Physical Review B*. **62**, 5028
[15] Kim K, Lambrecholc WRL and Segall B. 1996 *Phys. Rev. B*. **53**, 16310-16326
[16]Van De Walle, A., & Ceder, G. 2002 *Reviews of Modern Physics*. **74**, 11

# Lattice parameters and band structure of ternary mixed crystals $Al_xGa_{1-x}As$ from first-principle calculations

**Y. Mao, X. X. Liang, G. J. Zhao and T. L. Song**
Department of Physics, School of Physical Science and Technology, Inner Mongolia University, Hohhot, 010021, P. R. China

E-mail: xxliang@imu.edu.cn

**Abstract.** Lattice parameters and band structure of the ternary mixed crystal $Al_xGa_{1-x}As$ of zinc blende structure are calculated by first-principle calculations within the framework of the density functional theory. The results for the equilibrium lattice parameters and band gaps of $Al_xGa_{1-x}As$ for **the** Al-composition varying from 0.0 to 1.0 by step of 0.125 are presented and discussed. The results show that the lattice constants vary with the composition almost linearly following the Vegard's law. The electron band gap at Γ point exhibits non-linear behavior versus **the** composition. The Al-3*s*, 3*p* states shift to high energy region in the conduction band with increasing the Al concentration. It leads to an increase of the band gap and the blue shift phenomenon.

## 1. Introduction

Group III-V compounds are a type of most important semiconductor materials due to their applications in electronic and optoelectronic devices [1-3]. As typical III-V semiconductors, GaAs, AlAs and their alloy, which plays a vital role in high speed optoelectronic devices and nanostructure [4], have been attracting great research interest.

Several methods have been developed to calculate the electronic properties of GaAs, AlAs and $Al_xGa_{1-x}As$. An effective method, for example, is the so called empirical pseudo-potential method (EPM) [5-7] based on the virtual crystal approximation (VCA) [8]. Kurt and Alex studied the empirical atomic pseudo-potentials for AlAs/GaAs superlattices, alloys and nanostructures [9]. Boykin and Kharche et al calculated the band structures of the semiconducor alloy $Al_xGa_{1-x}As$ by the tight-binding supercell method [10]. Wang and Zahid et al employed the coherent potential approximation (CPA)-modified Becke-Johnson (MBJ) approach to quantitatively calculate the band offsets of GaAs/$Al_xGa_{1-x}As$ heterojunctions [11]. However, only a few theoretical calculations of **the** TMC $Al_xGa_{1-x}As$ energy band structures have been carried out for a couple of composition values because of computational complexities. In order to understand the composition dependence of the electronic properties completely and accurately, the first-principle studies for the whole composition range are invoked.

The aim of this work is to obtain a full understanding for the structural and electronic properties of **the** TMC $Al_xGa_{1-x}As$. We have performed **the** first-principle calculations for $Al_xGa_{1-x}As$ within the framework of the density functional theory (DFT). The composition dependences of the optimal lattice constants and the corresponding electronic band structures for **the** TMC $Al_xGa_{1-x}As$ ($0 \leq x \leq 1$) are obtained and discussed.

## 2. Computational methods

The calculations are performed within the framework of the density-functional theory (DFT) by using **the** plane-wave pseudopotential approach in the CASTEP package [12]. The exchange-correlation energy of the electrons are described by an improved generalized gradient approximations (GGA) given by Wu and Cohen (WC) [13]. The plane-wave energy cut-off is determined as 330eV in our calculations. The numerical integration of the Brillouin zone (BZ) was performed using a k-mesh of dimensions 6×6×3 k-points mesh in the Monkhorst–Pack scheme [14]. The convergence threshold of $2\times10^{-5}$eV/atom is used for total energy, 0.05 eV/Å for maximum force, 0.1 GPa for pressure and 0.002 Å for displacement. We consider **the** zinc-blend structure with space group $Td^2$-F$\bar{4}$3m for GaAs (AlAs) crystal [Figure 1(a)] and adopt a 16-atom simple cubic supercell by modeling. The corresponding TMCs are modeled by a 1×1×2 supercell structure [Figure 1(b)]. The crystal structure is optimized by minimizing the total energy with the quasi-Newton algorithm (BFGS).

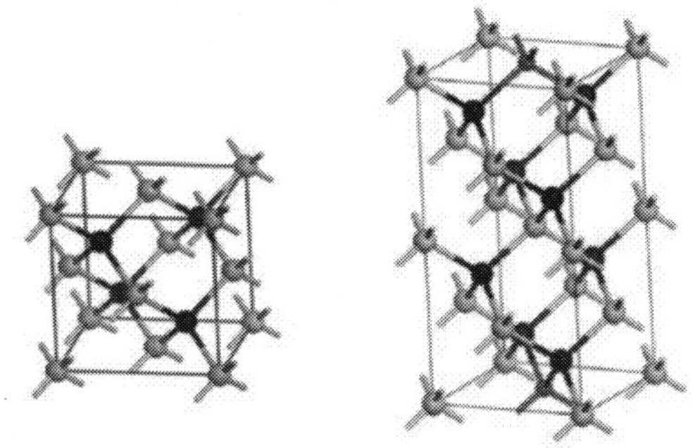

**(a)** **(b)**

**Ga** **As** **Al**

**Figure 1.** Crystal structures of (a) GaAs (AlAs) and (b) $Al_xGa_{1-x}As$.

The aluminum $3s^23p^1$, gallium $3d^{10}4s^24p^1$ and arsenic $4s^24p^3$ orbitals have been treated as **the** valence states during the whole calculation.

## 3. Results and discussions

### *3.1. Lattice parameters*

The structural optimization has been performed starting from an experimental value with respect to the cell parameters and also the atomic positions for $Al_xGa_{1-x}As$ with various compositions. Table 1 presents our calculated values compared with available experimental and theoretical values. It is seen that our results are in agreement with **the** previous theories and experimental data.

**Table 1.** Lattice constants according to WC-GGA for GaAs, AlAs and $Al_xGa_{1-x}As$ compared to experimental values and other theoretical studies.

| Compound | $a$(Å) | | |
|---|---|---|---|
| | This work | Theoretical studies | Experimental |
| GaAs | 5.685 | 5.666[a], 5.75[a] | 5.653 |
| $Al_{0.125}Ga_{0.875}As$ | 5.690 | - | - |
| $Al_{0.25}Ga_{0.75}As$ | 5.694 | - | - |
| $Al_{0.375}Ga_{0.625}As$ | 5.698 | - | - |
| $Al_{0.5}Ga_{0.5}As$ | 5.703 | 5.6482[b] (T=0) | - |
| $Al_{0.625}Ga_{0.375}As$ | 5.707 | - | - |
| $Al_{0.75}Ga_{0.25}As$ | 5.710 | - | - |
| $Al_{0.875}Ga_{0.125}As$ | 5.711 | - | - |
| AlAs | 5.717 | 5.678[c], 5.736[c] | 5.661 |

[a]Ref.[16]. [b]Ref.[17]. [c]Ref.[18].

The calculated equilibrium lattice constant $a$ of $Al_xGa_{1-x}As$ as a function of **the** concentration $x$ is shown in Figure 2(a). The variation of the lattice constant of TMCs with **the** composition is usually described phenomenologically by the so-called Vegard's law [15]. Fitting the calculated results gives following linear equation conformable with the Vegard's law:

$$a(\mathrm{Al}_x\mathrm{Ga}_{1-x}\mathrm{As}) = 5.686 + 0.031x = 5.717x + 5.686(1-x). \quad (1)$$

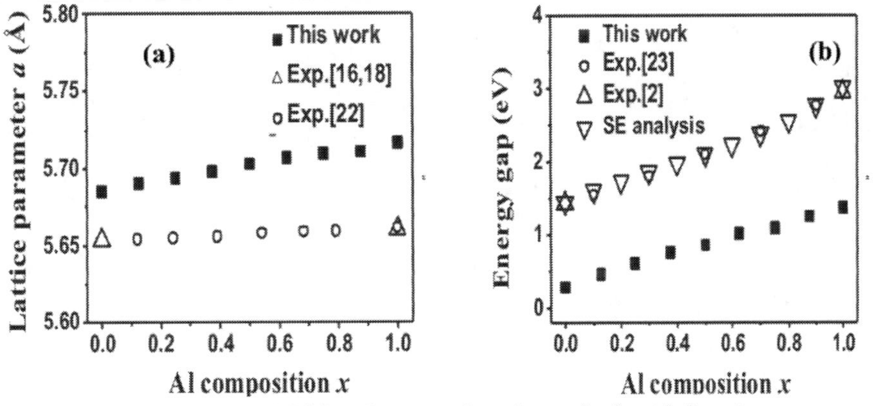


**Figure 2.** (a) Lattice constants and (b) band gaps as functions of $x$ for $Al_xGa_{1-x}As$.

### *3.2. Electronic band structure and density of states*

We have carried out **the** computations for the band structures and density of states (DOS) of $Al_xGa_{1-x}As$ for **the** composition in the range of $0 \leq x \leq 1$. Here we illustrate **the** Al composition dependence of the band gap and DOSs in Figures 2(b) and 3, without showing the band structures for lack of space. It is found that the electron band gap at Γ point exhibits strong composition dependence and increases non-linearly with **the** Al composition.

By fitting the non-linear variation of the calculated band gaps in terms of concentration with polynomial function, the relationship between the band gap and **the** composition is obtained as follows:

$$E_g^{\Gamma}(x) = 0.282 + 1.490x - 0.871x^2 + 0.465x^3$$
$$= 0.282 + 1.084x + x(1-x)(0.406 - 0.465x). \quad (2)$$

The band gap versus **the** composition has a bowing range with $b = (-0.406 + 0.465x)$ eV.

The computational values of the band gap are listed in Table 2 and compared with other theoretical and available experimental data. It is clearly seen that the band gaps are underestimated in comparison with experimental results. This is mainly due to the fact that the simple form of GGA does not take into account the quasi-particle self-energy correctly [20], and DFT is not suitable for describing excited-state properties (energy gap belongs to excited-state). However, it is widely accepted that GGA electronic band structures are qualitatively in good agreement with experiments as regards the ordering of the energy levels and the shape of the bands. This obstacle can be overcome by taking into account the scissor correction scheme [21].

**Table 2.** Band gap $E_g$ calculated according to the WC-GGA schemes for GaAs, AlAs and $Al_xGa_{1-x}As$ and the corresponding experimental and other theoretical values.

| Material | $E_g$ (eV) | | |
|---|---|---|---|
| | This work | Theoretical studies | Experimental |
| GaAs | 0.283 | 0.336[a] | 1.52 |
| $Al_{0.125}Ga_{0.875}As$ | 0.454 | - | - |
| $Al_{0.25}Ga_{0.75}As$ | 0.606 | - | - |
| $Al_{0.375}Ga_{0.625}As$ | 0.748 | - | - |
| $Al_{0.5}Ga_{0.5}As$ | 0.857 | 2.0775[b] | 2.0924 |
| $Al_{0.625}Ga_{0.375}As$ | 1.007 | - | - |
| $Al_{0.75}Ga_{0.25}As$ | 1.082 | - | - |
| $Al_{0.875}Ga_{0.125}As$ | 1.243 | - | - |
| AlAs | 1.363 | 1.65[c] | 2.22 |

[a] Ref.[16]. [b] Ref.[17]. [c] Ref.[19].

Figure 3 shows the partial and total DOS for the $Al_{0.125}Ga_{0.875}As$ and $Al_{0.625}Ga_{0.375}As$, as an example. It is found that the lowest valence bands in the range of -15.3 eV to -13.8eV are provided by Ga-3d states and it has stronger locality. The valence bands in the range of −13.5 eV to −9.3 eV are dominated by As-4*s* states, while the upper part in the range of −7.4 eV to −0.2 eV results from the contributions of As-4*p*, Al-3*s* and Al-3*p* states. The conduction band is of predominantly from the hybridization Al-3*p* states.

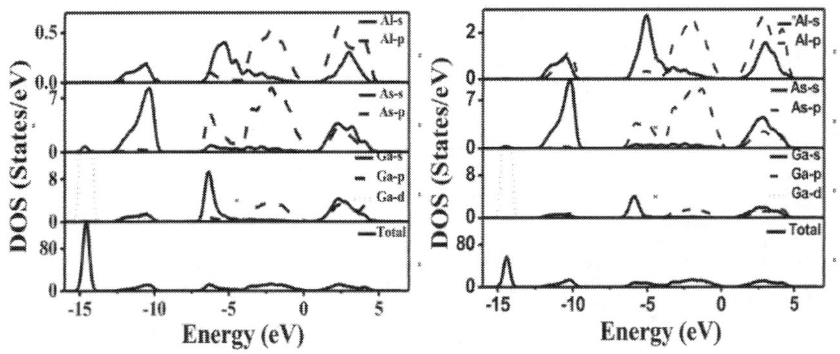


**Figure 3.** Total and partial density of states for $Al_{0.125}Ga_{0.875}As$ and $Al_{0.625}Ga_{0.375}As$.

To understand clearly the doping effect, we have also illustrated the DOS for the different concentration of $Al_xGa_{1-x}As$ in Figure 4. It indicates that the band contributions of Ga-*s*, *p*, *d* orbitals are gradually weaken but Al-*s*, *p* contributions strengthen. We can also see that the Al-3*s*, 3*p*, Ga-4*s*, 4*p* states shift to high energy region in the conduction band when Al concentration increases and the Al-3*p* state is more dominant in the region. This leads to an increase in the band gap and the blue shift phenomenon, i.e. the famous anti-Stokes shift, occurs.

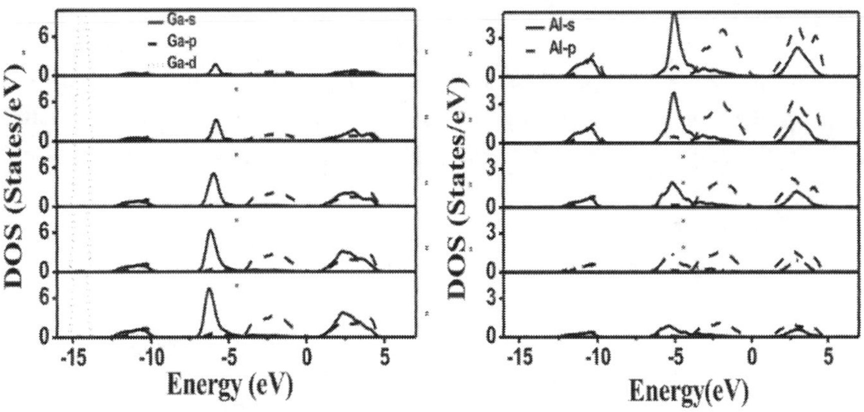


**Figure 4.** Partial density of states of Al-3*s*, 3*p* and Ga-3*d*, 4*s*, 4*p* bands of $Al_xGa_{1-x}As$ with $x = 0.25$, 0.375, 0.5, 0.75, 0.875, respectively.

## 4. Conclusions

We have used first-principle calculations to study the composition dependences of the lattice constants and electronic band structure of **the** ternary mixed crystals $Al_xGa_{1-x}As$, applying DFT within the

WC-GGA to treat the exchange correlation energy. The results show that the characteristics of equilibrium lattice parameters and band gap are sensitive to **the** composition. The composition dependence of the calculated lattice constant is found to be almost linear and obey the Vegard's law. The electron band gap at Γ point exhibits non-linear increase with **the** composition. The results for the DOS show that the blue shift occurs owing to Al-3*s*, 3*p*, Ga-4*s*, 4*p* states shifting to high energy region in the conduction band when Al concentration increases.

**Acknowledgments**
One of the authors (Y. Mao) is supported by the enhancing comprehensive strength project of Inner Mongolia University (No. 14020202), and the research is supported in part by the PhD Progress Foundation of Higher Education Institutions of China (20111501110003), Natural Science Foundation of Inner Mongolia (No. 2011MS0105) and Talent Development Foundation of Inner Mongolia.

**References**

[1] Kalt H 1996 Optical Properties of III-V Semiconductors, Springer Series in Solid-State Sciences vol. 120 (Berlin: Springer)
[2] Vurgaftman I, Meyer J R and Ram-Mohan L R 2001 *J. Appl. Phys.* **89** 5815
[3] Halah Al Shhri, Nadir Bouarissa and M. Ajmal Khan 2011 *J. Lumin* **131** 2153
[4] Othman M, Kasap E and Korozlu N 2009 *J. Alloys. Compd.* **12** 109
[5] Elabsy A M and Elkenany E B 2010 *Physica* B **405** 266
[6] Elabsy A M, Degheidy A R, Abdelwahed H G and Elkenany E B 2010 *Physica* B **405** 3709
[7] Degheidy A R and Elkenany E B 2011 *Semiconductors* **45** 1251–57
[8] Bouarissa N and Boucenna M 2009 *Phys. Scr.* **79** 015701
[9] Kurt A M and Alex Z 1994 *Phys. Rev.* B **50** 17393
[10] Boykin T B, Kharche N, Klimeck G and Korkusinski M 2007 *J.Phy s: Condens.Matter* **19** 036203
[11] Wang Y, Zahid F, Zhu Y, Liu L ,Wang J and Guo H 2013 *Appl. Phys. Lett.* **102** 132109
[12] Segall M D, Lindan P J D, Probert M J, Pickard C J, Hasnip P J, Clark S J and Payne M C 2002 *J. Phys.: Condens. Matter* **14** 2717
[13] Wu Z and Cohen R E 2006 *Phys. Rev.* B **73** 235116
[14] Monkhorst H J and Pack J D 1976 *Phys. Rev.* B **13** 5188
[15] Vegard L 1921 Z. *Phys.* **5** 17
[16] Mohamed Issam Ziane , Zouaoui Bensaad , Tarik Ouahrani , Boutaleb Labdelli , Hamza Ben Nacer and Hamza Abid 2013 *Mat. Sci. Semicon. Proc.* **16** 1138
[17] Degheidy A R and Elkenany E B 2012 *Mat. Sci. Semicon. Proc* **15** 505
[18] Hassan F E H, Breidi A, Ghemid S, Amrani B, Meradji H and Pages O 2010 *J. Alloy Compd.* **499** 80
[19] Briki M, Abdelouhaba M, Zaoui A and Ferhat M 2009 *Superlattice Microst.* **45** 80
[20] Rashkev S N and Lambrecht W R L 2001 *Phys. Rev.* B **63** 165212
[21] Fiorentini V and Baldereschi A 1995 *Phys. Rev.* B **51** 17196
[22] Estop E, Izrael A and Sauvage M 1976 *Acta Crystallogr.* A. **32** 627
[23] Varshni Y P 1967 *Physica* **34** 149

# The use of semiempirical algorithms for electronic levels calculations of polar nanosystems with the partial self-consistency

**Yury K Timoshenko**

Voronezh State University, Universitetskaya pl., 1, Voronezh, 394006, Russia

E-mail: yutim@phys.vsu.ru

**Abstract.** Three semiempirical algorithms for electronic levels calculations of polar nanosystems with the partial self-consistency are discussed. These algorithms are based on the model of pointlike polarized ions and on the self-consistent modification of semiempirical tight-binding theory of Slater and Koster. The basic feature of this work is the use of partial self-consistency which allows to simplify a mathematical formalism and to accelerate considerably the process of the calculations for nanosystems containing "whole ions". The obtained results for such systems, as was shown, are in rather good qualitative agreement with the results of non-empirical calculations. The presented algorithms can be effectively used for calculations of the electronic structure of polar nanosystems containing $\sim 10^4$ and more ions.

## 1. Introduction

To do the extended defects modelling (dislocations, charged defects) in nanocrystals (NC) there is a need to consider the systems containing $\sim 10^4$ and more atoms or ions. Over the last two decades the progress in computer technics and the accessibility of sophisticated special-purpose computer programs (which, as a rule, were being made by scientific groups with high qualification) for researchers allowed to do the self-consistent non-empirical calculations of electron structure of crystals with defects and nanosystems on a mass scale. Basically, these calculations are being executed in the framework of the density functional theory. Also much success was achieved in a development of the calculation procedures. Now it is possible, for example, the calculations of electronic states of huge systems with using of linear-scaled methods [1]. Let's note, that such computing procedures become rather effective only with the use of powerful multiprocessing computing systems. Not all of researchers have a possibility to use such systems, and the use of such systems is unjustified in all cases. Therefore when solving the problems with pretension to reasonable qualitative results are a topical problem of using the semiempirical approaches which would be applicable to calculations on the modern personal computers. It seems to the author the development of such semiempirical techniques is a topical problem.

## 2. Models and methods

As is known, when doing traditional calculations of electronic structure of polyatomic systems within the limits of a tight-binding method it is necessary to calculate the eigenvalues and the eigenvectors of a Hamiltonian matrix. In the present work the alternative approaches based on absolutely other mathematical formalism are offered. Most correctly suggested approaches are applicable to a special case of polar systems which consist of the "whole ions". We will discuss such systems further.

In the present work our attention is focused on consideration of the electronic structure algorithms for binary polar nanocrystals such as AgCl and KCl for example. The equilibrium positions of ions in these systems were calculated within the model of pointlike polarized ions by the minimization of the total potential energy over the ion coordinates by Fletcher–Reeves [2] method and by method of molecular dynamics. Each of the ions in the considered model is regarded as a point quasi–elastic dipole. For each step of the search for the equilibrium space ions configuration their dipole moments were being found in a self-consistent way. Next, after determining of the ions coordinates in the equilibrium NC configuration the electron structure was calculated by self-consistent tight-binding method (SCTBM) with orthogonal basis set. See details in paper [3].

The new point in our calculation technics is the use of partial self-consistency for calculations of electron states of nanosystems consisting of "whole ions". Within the limits of this scheme the electrical dipole moments of ions, which were obtained in model of pointlike polarized ions, were used for evaluation of corrections for diagonal elements of Hamiltonian matrix; effective charges are thus determined by input data. In this approach the iterated self-consistent calculations are not being executed! Clearly, that such approach can be correct only for systems in which there is no essential reallocation of charge density, e.g. for systems with whole ions.

The use of partial self-consistency allow us to apply the effective methods without executing of unitary transformation of Hamiltonian matrixes. It is evident that such unitary transformations are the hot-spots in computer codes.

About the first algorithm. After the evaluation of a matrix of a Hamiltonian we do not solve the task about eigenvalues, but we use the known theorem ([4], p. 204) which can be formulated as follows. Let us suppose that we calculated the Hamiltonian matrix which is the real symmetric matrix. For the arbitrary scalar $\lambda$ we organise a matrix $\|H\| - \lambda\|I\|$ and further will fulfil a triangular factorization

$$\|H\| - \lambda\|I\| = \|U\|^T \|D\| \|U\|, \tag{1}$$

where $\|I\|$ is unit matrix, $\|U\|$ is upper triangular with unit diagonal, $\|D\|$ is diagonal matrix. Then, the number of eigenvalues less than $\lambda$ is equal to the number of negative elements of the diagonal matrix $\|D\|$. If we will change $\lambda$ with the given step, will do on each step a triangular factorization (1) and will calculate the number of the negative diagonal elements of matrix $\|D\|$, then it will be easy to draw the histogram of total density of states.

Further, separate eigenvalue of $\|H\|$ can be localised by the bisection method. It is convenient to use such approach for studying the local levels of localized states of nanocrystal. As usual the matrix of a Hamiltonian for nanocrystal in the framework of semiempirical tight-binding method is sparse real symmetric matrix. Naturally, the efficiency of a computer code rises sharply if we use algorithms to sparse matrix.

About the second algorithm. It is a well-known algorithm of simultaneous iteration of eigenvectors ([5], p.284) for the evaluation of energy level and corresponding orbital coefficients. Certainly, this method is applied to a matrix $(\|H\| - \lambda\|I\|)^{-1}$, where $\lambda$ has value in some energy interval (in zero approximation). In case of the sparse matrixes efficiency increases if we calculate the vector $(\|H\| - \lambda\|I\|)^{-1} x$ by solving the equation $(\|H\| - \lambda\|I\|)\, y = x$. This algorithm was applied, for example, for calculations of electronic levels of KCl nanocrystal which consisted of $> 15000$ ions.

About the third algorithm. The recursion method works excellently in partial self-consistency approximation. The recursion method in variant of Nex was used [6, 7].

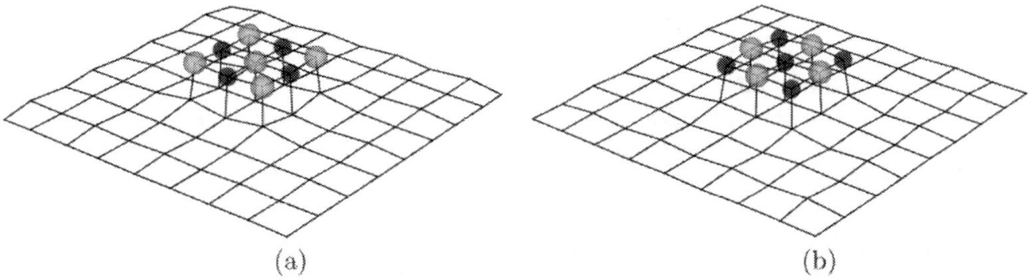

(a) (b)

**Figure 1.** The isolated asperities on surface of $KCl$ nanocrystal (657 ions); a) negatively charged surface defect $(K_4Cl_5)^-$; b) positively charged surface defect $(K_5Cl_4)^+$. Black disk – cation.

**Table 1.** Nanocrystal $KCl$ in form of $n \times n \times (n-1) + 3 \times 3$ ions with negatively charged surface defect: some characteristic energies (eV).

| | $n/N_{AO}$ | $E_{min}$ | $E_{HONO}$ | $E_{LUNO}$ | $E_{max}$ |
|---|---|---|---|---|---|
| $(K_{328}Cl_{329})^-$ | 9/1315 | -8.698 | -7.218 | 1.100 | 3.982 |
| $(K_{1579}Cl_{1580})^-$ | 15/6319 | -9.018 | -7.500 | 0.624 | 3.690 |
| $(K_{4414}Cl_{4415})^-$ | 21/17659 | -9.161 | -7.615 | 0.433 | 3.567 |
| $(K_{7504}Cl_{7505})^-$ | 25/30019 | -9.220 | -7.661 | 0.400 | 3.519 |
| extrapolation | $\infty/\infty$ | -9.537 | -7.891 | 0.135 | 3.268 |
| crystal $KCl$ | — | -9.466 | -8.690 | 0.000 | 2.972 |

## 3. Results

These three algorithms were used for theoretical investigations of electronic structure of $AgCl$ and $KCl$ nanosystems with atomically rough surfaces, edge dislocations, and isoelectronic substitutional impurities [3, 8–12]. For example numerical investigations of size effect influence on electronic structure of $KCl$ nanocrystal with charged surface defect were carried out with the use of the algorithms discussed above [12]. The charged defects in the form of isolated surface asperity with size of $3 \times 3$ ions and located on one of surfaces of electrically neutral nanocrystals KCl have been investigated. The isolated surface asperity represents the charged defect – either $(K_5Cl_4)^+$, or $(K_4Cl_5)^-$. These surface defects are shown on fig. 1. The nanocrystals without the charged defects had form of parallelepipeds in the size $n \times n \times (n-1)$, n = 9, 15, 21, 25. Thus, studied nanocrystals with the charged defects contained 657, 3159, 8829 and 15009 ions accordingly. Some numerical results can be seen in table 1. Here $E_{min}$ and $E_{max}$ are the energies of lowest and highest levels; $N_{AO}$ is number of basic atomic functions.

An alternative calculation of electronic energy spectrum of nanocrystals is of great interest. The non-empirical calculations of some small nanocrystals $AgCl$ and $KCl$ were carried out in the framework of the density functional theory and in Hartree-Fock-Rutaan approximation with the use of pseudopotentials. It was found that the results of non-empirical calculations are in reasonably good agreement with the results of our semiempirical calculations.

Non-empirical quantum-chemical calculations were made with the use of the program "Gaussian 03" [13] which was installed on the parallel computer cluster of the Digital Technologies Department of Computer Sciences Faculty, Voronezh State University, and with the use of the program "Firefly" [14, 15] on the supercomputer "Chebyshev" of Moscow State University.

## 4. Conclusions

To summarize briefly, the partial self-consistency approximation allows to use very effective algorithms for the calculation of electron structure of polar nanocrystals consisting of "whole ions". The obtained results are qualitatively reasonable.

## Acknowledgments

Many thanks to Professor S. D. Kurgalin (Head of the Digital Technologies Department of Voronezh State University) for the possibility of using the computer cluster.

## References

[1] Goedecker S and Scuseria G E 2003 *Computing in Science and Engineering* **5** 14
[2] Bunday B D 1984 *Basic Optimisation Methods* (London: E. Arnold)
[3] Timoshenko Yu K and Shunina V A 2009 *Surface Sci.* **603** 2564
[4] Pissanetzky S 1984 *Sparse Matrix Technology* (London: Academic Press Inc.)
[5] Wilkinson J Y and Reinsch C 1971 *Handbook for Automatic Computations. Linear Algebra* (Berlin Heidelberg New York: Springer Verlag)
[6] Nex C M M 1978 *J. Phys. A* **11** 653
[7] Nex C M M 1984 *Comp. Phys. Commun.* **34** 101
[8] Timoshenko Yu K and Shununa V A 2008 *Proc. of SPIE (Physical Chemistry of Interfaces and Nanomaterials VII : Conference of SPIE Symposium on NanoScience + Engineering, 10–14 August 2008, San Diego, USA)* **7034** 70340
[9] Timoshenko Yu K and Shununa V A 2009 *Proc. of SPIE (Physical Chemistry of Interfaces and Nanomaterials VIII : Conference of SPIE Symposium on NanoScience + Engineering, 2–6 August 2009, San Diego, USA)* **7396** 73960T
[10] Timoshenko Yu K and Shununa V A 2002 *Fiz. Tverd. Tela (Russia)* **44** 853
[11] Timoshenko Yu K and Shununa V A 2005 *Phys. Stat. Sol. (c)* **2** 1788
[12] Timoshenko Yu K and Shununa V A 2011 *Actual Problems of Applied Mathematics, Computer Science and Mechanics* (Voronezh: VSU) p 388 (in Russian)
[13] Frisch M J, Trucks G W, Schlegel H B, Scuseria G E, Robb M A, Cheeseman J R, J A Montgomery J, Vreven T, Kudin K N, Burant J C, Millam J M, Iyengar S S, Tomasi J, Barone V, Mennucci B, Cossi M, Scalmani G, Rega N, Petersson G A, Nakatsuji H, Hada M, Ehara M, Toyota K, Fukuda R, Hasegawa J, Ishida M, Nakajima T, Honda Y, Kitao O, Nakai H, Klene M, Li X, Knox J E, Hratchian H P, Cross J B, Adamo C, Jaramillo J, Gomperts R, Stratmann R E, Yazyev O, Austin A J, Cammi R, Pomelli C, Ochterski J W, Ayala P Y, Morokuma K, Voth G A, Salvador P, Dannenberg J J, Zakrzewski V G, Dapprich S, Daniels A D, Strain M C, Farkas O, Malick D K, Rabuck A D, Raghavachari K, Foresman J B, Ortiz J V, Cui Q, Baboul A G, Clifford S, Cioslowski J, Stefanov B B, Liu G, Liashenko A, Piskorz P, Komaromi I, Martin R L, Fox D J, Keith T, Al-Laham M A, Peng C Y, Nanayakkara A, Challacombe M, Gill P M W, Johnson B, Chen W, Wong M W, Gonzalez C and Pople J A 2004 *Gaussian 03, Revision C.02* (Wallingford CT: Gaussian, Inc.)
[14] Granovsky A A Firefly version 7.1.g http://classic.chem.msu.su/gran/firefly/index.html
[15] Schmidt M W, Baldridge K K, Boatz J A, Elbert S T, Gordon M S, Jensen J H, Koseki S, Matsunaga N, Nguyen K A, Su S, Windus T L, Dupuis M and Montgomery J A 1993 *J. Comput. Chem.* **14** 1347

# Simulation of Thermal Processes in Metamaterial MM-to-IR Converter for MM-wave Imager

**Peter S. Zagubisalo**[1,3], **Andrey G. Paulish**[1] **and Sergey A. Kuznetsov**[2]
[1] Novosibirsk Branch of the Institute of Semiconductor Physics, Siberian Branch of Russian Academy of Sciences, Novosibirsk, 630090 Russian Federation,
[2] Novosibirsk State University, Novosibirsk, 630090 Russian Federation

E-mail: peter.zagubisalo@gmail.com

**Abstract.** The main characteristics of MM-wave image detector were simulated by means of accurate numerical modelling of thermophysical processes in a metamaterial MM-to-IR converter. The converter represents a multilayer structure consisting of an ultra thin resonant metamaterial absorber and a perfect emissive layer. The absorber consists of a dielectric self-supporting film that is metallized from both sides. A micro-pattern is fabricated from one side. Resonant absorption of the MM waves induces the converter heating that yields enhancement of IR emission from the emissive layer. IR emission is detected by IR camera. In this contribution an accurate numerical model for simulation of the thermal processes in the converter structure was created by using COMSOL Multiphysics software. The simulation results are in a good agreement with experimental results that validates the model. The simulation shows that the real time operation is provided for the converter thickness less than 3 micrometers and time response can be improved by decreasing of the converter thickness. The energy conversion efficiency of MM waves into IR radiation is over 80%. The converter temperature increase is a linear function of a MM-wave radiation power within three orders of the dynamic range. The blooming effect and ways of its reducing are also discussed. The model allows us to choose the ways of converter structure optimization and improvement of image detector parameters.

## 1. Introduction and background

Last years the intensive researches are directed towards the development of image detectors for a millimeter-wave range of electromagnetic spectrum. This range is considered as perspective for various application such as security systems, noninvasive biomedical diagnostics, and nondestructive control of material quality. In previous works [1-4] we have proposed a MM-to-IR converter for real-time millimeter-wave imaging.

The MM-to-IR converter represents a multilayer structure (figure 1) consisting of an ultra thin resonant metamaterial absorber and a highly efficient emissive layer with a perfect emissivity factor. The absorber consists of a dielectric self-supporting film with thickness much less than the operational wavelength. Film is metallized from both sides. A micro-pattern is fabricated from one side by contact photolithography. This structure was utilized as a frequency-selective artificial impedance surface with a configuration “metasurface + grounded dielectric slab”. Resonant absorption of the MM waves induces the converter heating that yields enhancement of IR emission from the emissive layer. IR

[3] To whom any correspondence should be addressed.

emission is detected by a high sensitive IR camera. The proposed MM-wave imager consists of MM-to-IR converter and IR camera with noise equivalent temperature difference (NETD) 40 mK. In this scheme the main parameters of the MM imager are determined by IR camera parameters and thermal processes in the converter. In this contribution an accurate numerical model of the converter was created and thermal processes in the converter structure were simulated by using COMSOL Multiphysics™ software. The main characteristics of MM-wave image detector were estimated and were compared with experimental results.

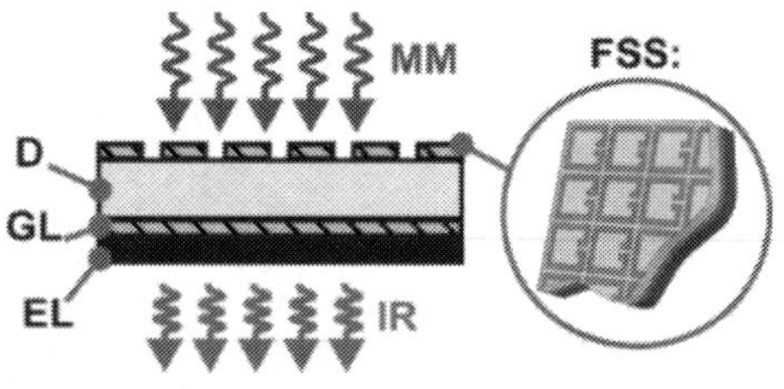


**Figure 1.** Structure of the MM-to-IR converter consisting of frequency-selective surface (FSS), dielectric film (D), "ground" layer (GL), IR emissive layer (EL).

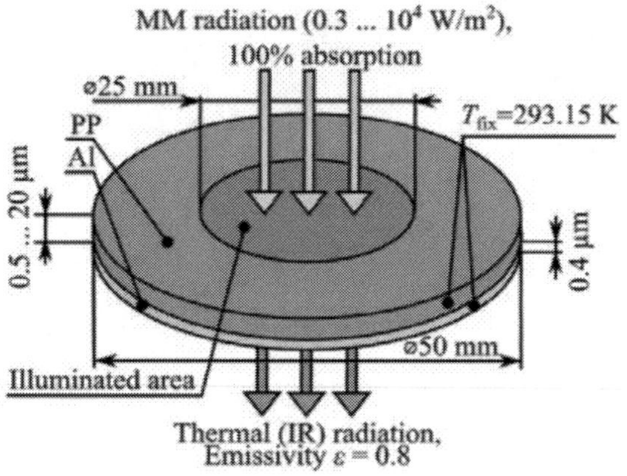


**Figure 2.** Model of the converter. PP – polypropylene, Al – aluminum, $T_{fix}$ – fixed temperature.

The simulation was performed to determine the limiting characteristics of the MM-to-IR converter, i.e. the maximum achievable detector characteristics. For this purpose a model of the converter has been introduced (figure 2). The model is a thin membrane placed in a vacuum. Thin dielectric layer forms the body of the membrane (in this case it's a polypropylene film). Frequency-selective surface (FSS) is given by a boundary condition of the incoming radiation power with absorption factor of 100%. The actual thickness of the aluminum FSS layer (0.4 µm) is neglected due to its low heat capacity and low thermal conductivity because of its disconnectedness. 0.4 µm aluminum ground layer is applied to the opposite side of the dielectric layer. IR emissive layer is given by a boundary condition of the thermal radiation with emissivity of 0.8 that is put on the dielectric layer surface. We neglect the actual thickness of the emission layer, as the thickness of this layer can vary depending on the technology.

## 2. Results

A round converter with a diameter of 50 mm was used for a simulation. The center of the converter was illuminated by Gaussian beam with a diameter of 25 mm.

The time dependences of the converter temperature, obtained from the COMSOL simulations are shown in figure 3. The simulation showed that the response time ($\tau$) of the converter temperature under MM-wave illumination is a linear function of the converter thickness (figure 4). The real time operation is provided for the thickness of the dielectric layer less than 3 micrometers.

The simulations show that the energy conversion efficiency (MM to IR) is over 80% that provides a high sensitivity of the detector and put it in line with modern IR-detectors. The converter temperature increase is a linear function of a MM-wave radiation power within three orders of the dynamic range.

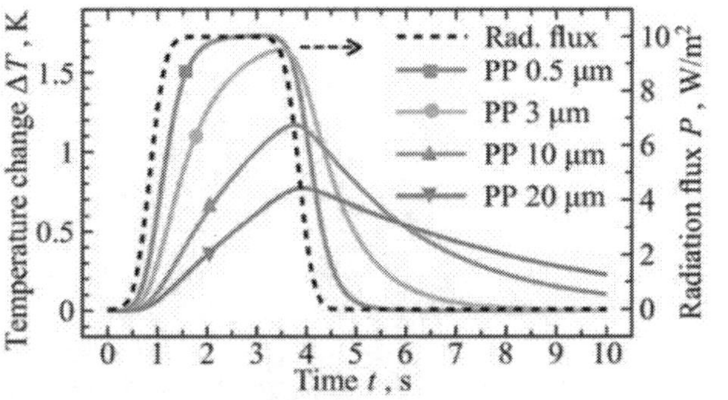


**Figure 3.** Temperature change ($\Delta T$) versus time ($t$) for various polypropylene (PP) thicknesses. The time dependence of the illumination power is shown by dotted line ($P$, right scale).

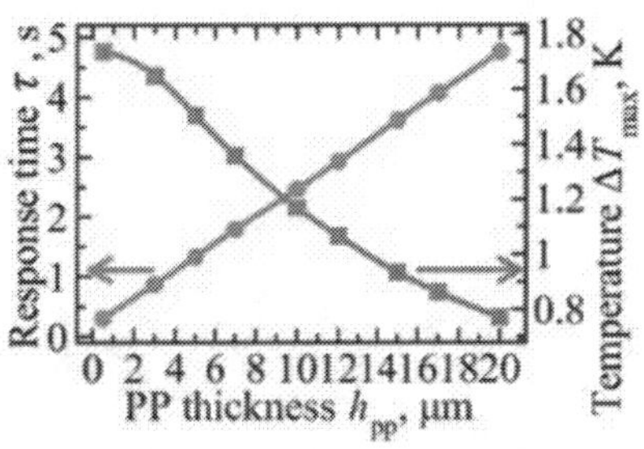


**Figure 4.** Maximum of the temperature change ($\Delta T_{max}$) and the response time ($\tau$) versus the thickness of the dielectric layer ($h_{pp}$).

Blooming effect of a thermal image of the converter is caused by the lateral heat conduction along the converter. In order to investigate the blooming effect, a square converter with a side length of 24 mm and 3 μm polypropylene layer thickness was simulated. The center of the converter was illuminated by a square beam. The converters with various through-the-thickness cuts that do not break the integrity of the converter were introduced in order to reduce the blooming. The converter thermal images for various types of cuts are shown in figure 5 and figure 6.

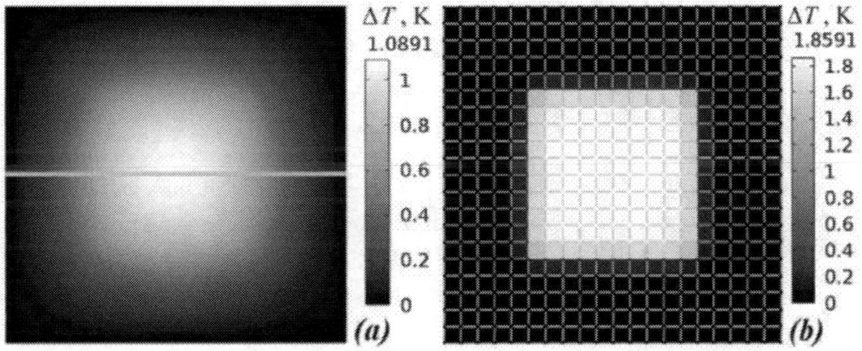


**Figure 5.** Thermal image of the converter without (a) and with cuts (b). For more details see 1 and 4 cases in figure 6. In both (a) and (b) cases the illuminated area is of identical square shape in the center.

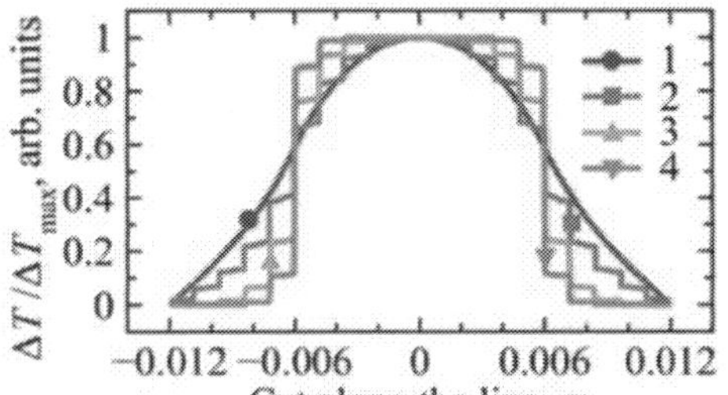


**Figure 6.** Normalized temperature change of the converter ($\Delta T$ / $\Delta T_{max}$) versus coordinate (see the gray line in figure 5). Types of cuts: 1 – no cuts, 2 – the cuts in both aluminum and polypropylene layers do not break the integrity of both layers (1.08 × 0.1 mm cuts), 3 – the aluminum layer is cut into isolated squares (0.1 mm wide cuts) and there are no cuts in the polypropylene layer, 4 – the aluminum layer is cut into isolated squares (0.1 mm wide cuts) and the cuts in the polypropylene layer (0.9 × 0.1 mm cuts).

A comparison of experimental measurements and COMSOL Multiphysics simulations for the uncut converter of 50 mm in diameter are shown in figure 7. The converter was illuminated by

Gaussian beam. The converter consists of 20 μm polypropylene, 0.4 μm aluminum and 23 μm emissive layers.

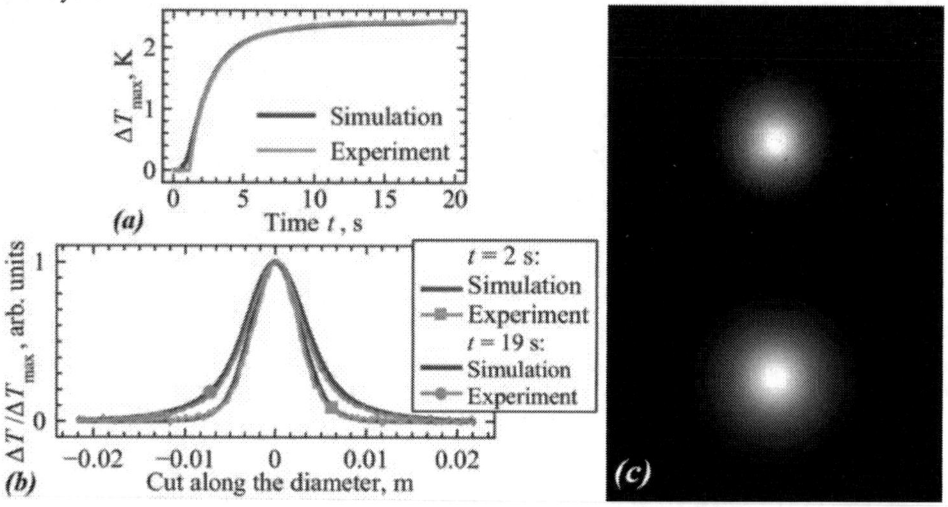


**Figure 7.** Comparison of experimental measurements and COMSOL Multiphysics simulations of an uncut converter under illumination by Gaussian beam. (a) – time dependence of converter temperature change ($\Delta T_{max}$), (b) – normalized temperature change of the converter ($\Delta T / \Delta T_{max}$) along the diameter, (c) – thermal image of the converter: experiment (top) and simulation (bottom).

## 3. Conclusion

The simulation showed that the response time of the converter temperature is a linear function of the converter thickness, therefore the time response can be improved by decreasing of the converter thickness. The fabrication of the appropriate cuts in the converter significantly decreases the blooming of the thermal image. The simulation results are in a good agreement with experimental results that validates the model. The model allows us to choose the ways of converter structure optimization and improvement of image detector parameters.

## 4. References

[1] Kuznetsov S A, Paulish A G, Gelfand A V, Lazorskiy P A, Fedorinin V N 2011 Bolometric THz-to-IR converter for terahertz imaging *Appl. Phys. Lett.* **99** 023501 (3 pp)

[2] Kuznetsov S A, Arzhannikov A V, Paulish A G, Gelfand A V, Lazorskiy P A, Fedorinin V N 2011 THz imaging system based on THz-to-IR converter *Technisches Messen* **vol. 78 iss. 11** 526-533

[3] Kuznetsov S A, Paulish A G, Gelfand A V, Astafiev M A, Arzhannikov A V, Fedorinin V N, Thumm M K A 2012 Extremely thin metamaterial absorbers for subterahertz waves: from fundamentals towards applications in uncooled bolometric sensors, *Proceedings of SPIE* **8423** 84230S (15 pp)

[4] Kuznetsov S A, Paulish A G, Gelfand A V, Lazorskiy P A, Fedorinin V N 2012 Matrix structure of metamaterial absorbers for multispectral terahertz imaging *Progress In Electromagnetics Research* **122** 93-103

# Acousto-optic couplings in a phoXonic crystal slab $L_1$ cavity

**S Dupont[1,2,3], Q Rolland, J Gazalet[1,2,3], JC Kastelik[1,2,3]**
[1]Univ Lille Nord de France, F-59000 Lille, France
[2]UVHC, IEMN-DOAE, F-59313 Valenciennes, France
[3]CNRS, UMR 8520, F-59650 Villeneuve d'Ascq, France

E-mail: samuel.dupont@univ-valenciennes.fr

**Abstract.** Acousto-optic coupling mechanisms within simultaneous photonic and phononic crystal cavities are theoretically studied. The structure considered is a silicon slab drilled periodically with air holes arranged on a square lattice; the cavity is obtained by removing one hole from the ideal crystal structure. Two acousto-optic coupling mechanisms are taken into account: the classical photo-elastic effect and the opto-mechanic effect which arises from the moving boundaries. The coupling mechanisms are computed according to the finite element method. The results reveal that the mechanisms can sustain each other or act with counteracted effects.

## 1. Introduction

Photonic and phononic crystals are artificial periodic structures that exhibit band gap phenomena for photons and phonons respectively [1, 2]. It is possible to design structures that enable the simultaneous confinement of both waves, the so-called phoxonic crystals. In these structures, due to the strong increase of energy localisation the acousto-optic coupling phenomenon are expected to be enhanced [3, 4]. A study of cavity mode coupling in two dimensions structures has evidenced an enhanced interaction with symmetric acoustic modes [6]. In this article, we study the modulation of optical cavity modes by acoustical cavity modes in a phoxonic crystal three dimensions structure: silicon slabs. Such a study is a step forward towards the understanding of technologically feasible structure behaviour. We investigate the dependence of the optical eigenmode frequency modulation on the spatial distribution and symmetry of the acoustic eigenmodes. The acousto-optic interaction strength is varying according to the couple of modes chosen. Two contributions are taken into account to compute the acousto-optic coupling: the classical photoelastic effect and the opto-mechanical effect; these two contributions act in different ways and their effects on the overall modulation can either sustain each other or in the contrary display counteracted effects. The computations are performed using the finite element method (FEM) for a 3D geometry consisting of a silicon slab drilled with periodic holes and surrounded by air. We first describe the phoxonic structure; we present the dispersion curves for the acoustic and optic waves and the acousto-optic coupling implementation. The most significant results are presented and discussed; we end up by some concluding remarks.

## 2. Model definition

### *2.1. Phoxonic bandgaps parameters*

The first step of the study consists in the determination of the dispersion curves for the photonic and phononic waves; such a study enables one to determine the occurrence of simultaneous photonic and phononic band gaps. Wide bandgaps are a prerequisite to promote the confinement of acoustic and optic waves. Recent investigations on silicon slabs have revealed that the couple of geometrical parameters of the reduced thickness (relative to the lattice parameter a) h / a = 0.6 and of the hole reduced radius r / a = 0.43, is a good candidate for the square lattice. It promotes the existence of quasi-TE, quasi-TM, shear and longitudinal vibrations [5].

### *2.2. Cavity acousto-optic coupling effects modeling*

The cavity in such a structure consists in the absence of one hole ($L_1$ cavity): a point defect in the artificial crystal. Acoustical and optical modes are determined thanks to the supercell method: our calculations are based on a defect centred within a 9*9 supercell. The size of the supercell is slightly larger compared to our previous study considering 2D structures [6]; this ensures a good convergence of the computed modes. Due to geometrical considerations, the expected solutions are either symmetric or anti symmetric. This allows one to make use of a plane of appropriate symmetry midway between the two air/slab interfaces in order to reduce the computational power and memory requirements. So, the acoustical and optical modes are separated into symmetric and anti-symmetric ones, with respect to this horizontal symmetry plane. In what follows, we focus on the coupling between those modes due to the two contributing acousto-optic mechanisms.

We use the quasi-static approximation to compute the acousto-optic coupling effects owing to the several order of magnitude separating optical and acoustical angular frequencies. The acoustic cavity mode acts like a perturbation acting on the modulation of the optic cavity mode. That is, the optical mode is recalculated at different instants of an acoustical period. The two coexisting coupling mechanisms have different origins: the photo-elastic effect is due to the acoustic strain whereas the opto-mechanical effect is due to the geometrical deformations at the boundaries of the cavity. With the knowledge of acoustic cavity modes, one can determine the respective modulation resulting from both effects: the photo-elastic and the displacement of the hole air/silicon interfaces. The displacement field of the acoustic mode serves as input data to refine the boundaries of the optical cavity using the moving mesh method. The acoustic perturbation of the cavity is computed for sufficiently small time steps during one acoustic period. The strain field $S_{kl}$ of the acoustic modes associated to the photo-elastic tensor ensures the computation of the photo-elastic effect to determine the refractive index spatial modulation according to:

$$\Delta\eta_{ij} = p_{ijkl} \cdot S_{kl} \tag{1}$$

with $p_{ijkl}$ being the photoelastic constants of Silicon and $\eta_{ij}$ the impermittivity tensor.

## 3. Results

We present the most significant results: dispersion curves of the slab, cavity modes and acousto-optic coupling in the cavity.

### *3.1. Band structure*

Figure 1 displays the band structures along the Γ - X direction computed on the 9*9 supercell illustrated on the inset.

For clarity, black dotted lines are set at the bandgap edges of the corresponding perfect crystal (without the cavity). The band structure appears dense in the propagative region due to the band folding effect inherent to the supercell approach (in a 9*9 supercell, each dispersion curve folds 81 times). Moreover, the photonic band gap is filled with evanescent modes (from inside the light cone) that fold in the gap. We label the confined acoustic modes using capital letters and the optical cavity modes using Greek letters. Subscripts A and S stand for anti-symmetric and symmetric modes. Acoustic modes $B_S$, $C_S$ and $D_A$, $E_A$ are degenerated, as well as photonic modes $\gamma_s$, $\delta_s$ and $\beta_a$, $\gamma_a$. Compared to our previous 2D study [6], the acoustic eigenmodes exhibit a more complex displacement spatial distribution; no breathing (fully symmetric) mode is found. We recall that the fully symmetric mode produces the highest coupling rates.

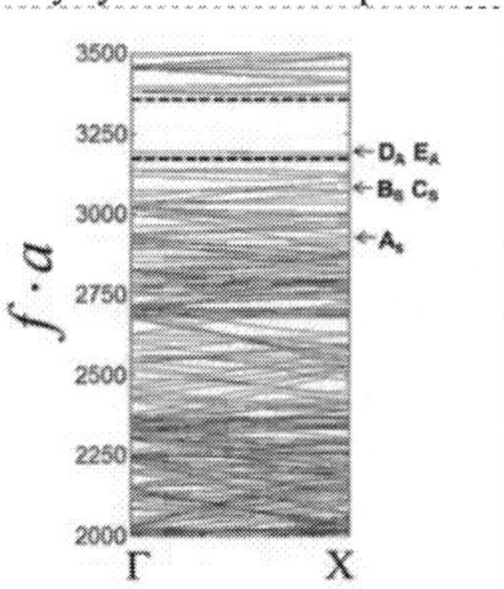


**Figure 1.a** Phononic band-structure of the supercell.

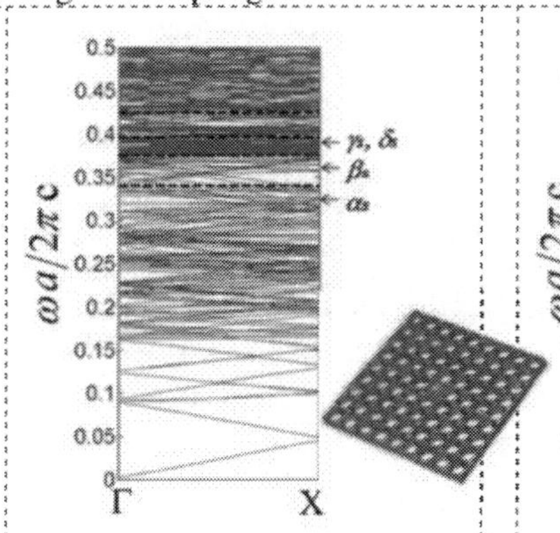


**Figure 1.b** Photonic band-structure of the supercell, symmetric modes.

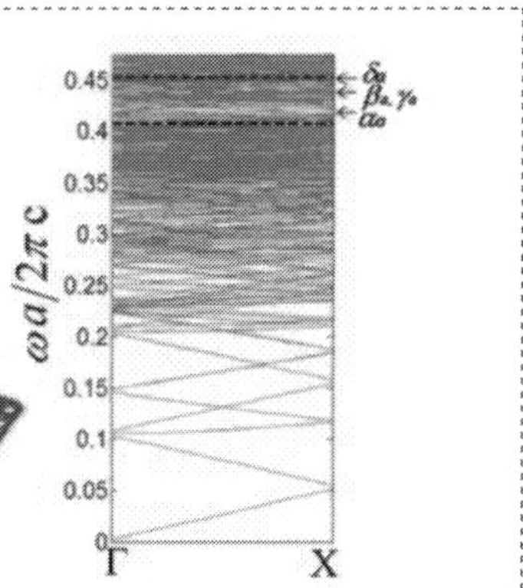


**Figure 1.c** Photonic band-structure of the supercell, anti-symmetric modes.

### 3.2. Cavity modes

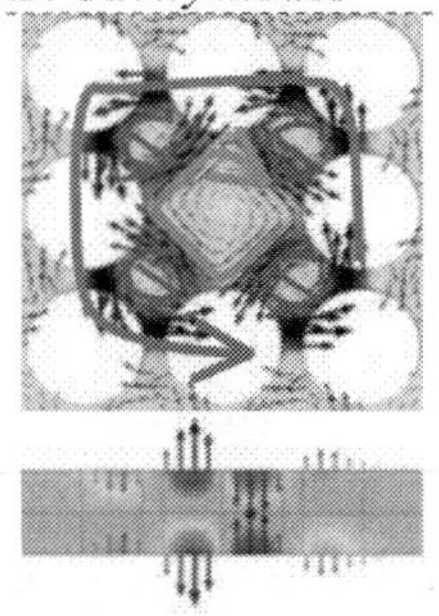

**Figure 2.a** Phononic $A_S$ mode (top and side view).

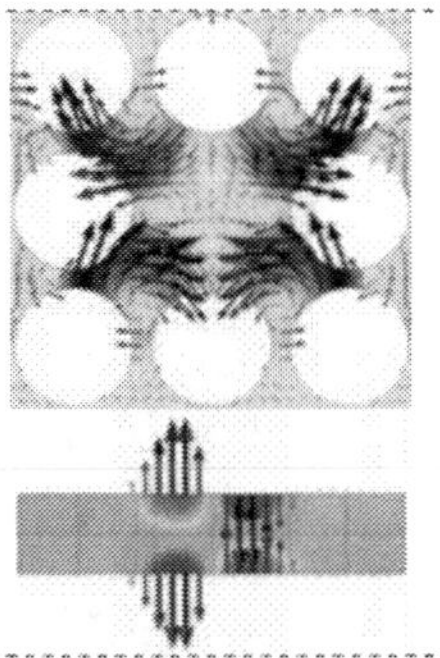

**Figure 2.b** Phononic $B_S$ mode (top and side view).

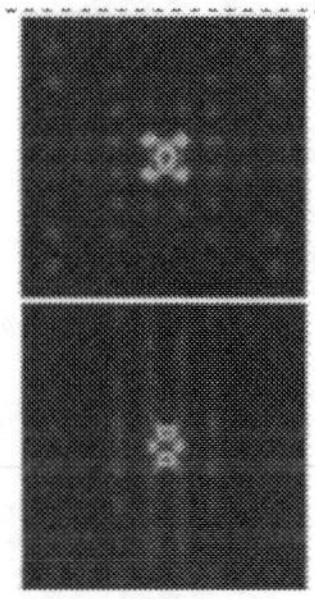

**Figure 2.c** Photonic modes $\alpha_a$, $\beta_a$ (top view).

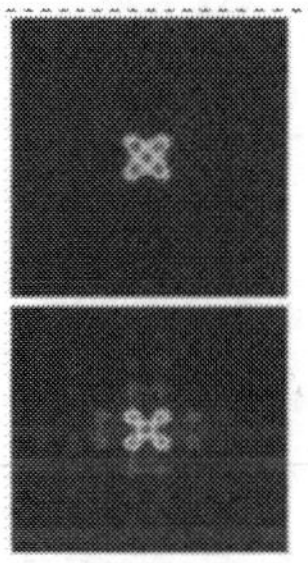

**Figure 2.d** Photonic modes $\alpha_s$, $\beta_s$ (top view).

The mode profiles are illustrated by figure 2 for the acoustic field (modes $A_S$ and $B_S$, top figures are displayed in the (x, y) plane and bottom figures are displayed in the (y, z) plane) and for the electromagnetic energy (modes $\alpha_a$, $\beta_a$ and $\alpha_s$, $\beta_s$). As highlighted by the added black arrows proportional to the displacement, the main displacements induced by the acoustic cavity modes are localized in and in the vicinity of the cavity: acoustic mode $A_S$ corresponds to a rotation of the middle of the cavity associated with four local rotations of the corners of the cavity (green and blue bold arrows). During one half of the acoustic period, the rotation is clockwise and counter clockwise during the second half. With acoustic mode $B_S$ the displacements are simpler to describe: both sides of the cavity are stretched and contracted one after the other, it results in an overall trapezoidal deformation of the cavity at $\Omega.t = \pi/2$ and $3.\pi/2$. Whereas, at $\Omega.t = 0$ and $\pi$ the cavity is in its unperturbed shape.

### 3.3. Acousto-optic coupling

In figure 3, we plot the evolution of the optical mode eigenfrequencies as a function of an acoustic period. The green curves stand for the photo-elastic effect considered alone, the blue curves for the opto-mechanical effect alone and the red curves depict the evolution when both effects are taken into account simultaneously. We summarise the most significant trends observed in what follows:

- The two acousto-optic mechanism contributions can be in phase and sustain each other, or out of phase and act with counteracted effects. Consider for example the modulation of the optical mode $\beta_A$ with the acoustic modes $A_S$ and $B_S$: the photo-elastic contribution (green curve) is quite similar in both cases while the opto-mechanic contribution (blue curve) is in or out of phase, the global modulation (red curve) being respectively enhanced or reduced. Moreover, we note that the computation of both effects simultaneously results in an overall frequency modulation that is not equal to the algebraic sum of the modulations when the effects are computed separately.
- The frequency modulation of optical modes evolves at twice the acoustic frequency, this is compatible with an interpretation based on the perturbation theory: none of the acoustic modes is found to be even as regards all symmetry elements of the cavity, first order corrections vanish [7, 8].
- Higher values of frequency modulation due to photo-elastic effect and opto-mechanic effect are found when the overlap of the optical field with respectively the strain distribution and the cavity boundaries is maximised.

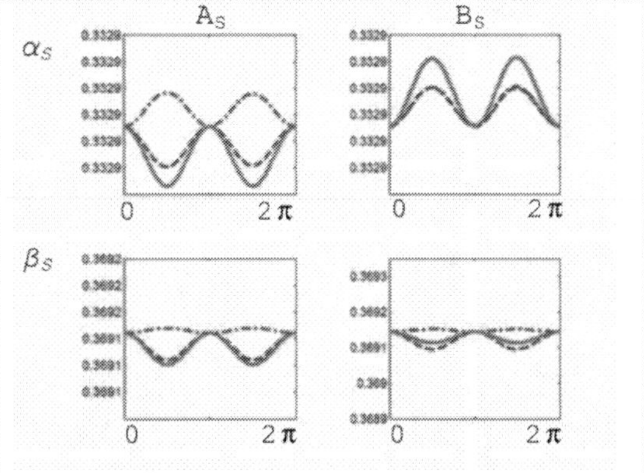


**Figure 1.a** Modulation of the symmetric optical modes, evolution of the reduced optical frequency (ωa/2πc) along one acoustic period (Ωt) from 0 to 2π of modes $A_S$ and $B_S$ respectively.

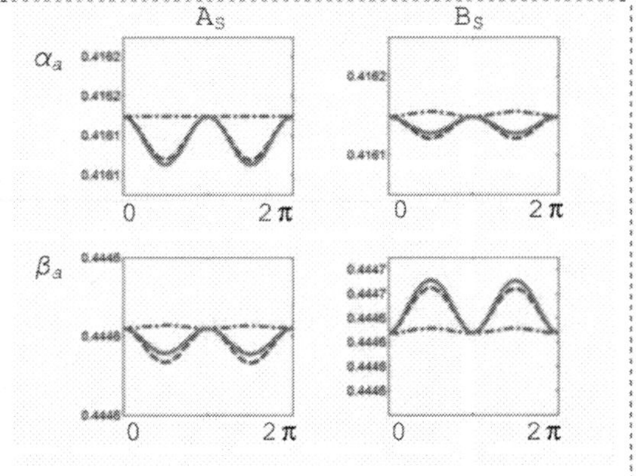


**Figure 1.b** Modulation of anti-symmetric optical modes, evolution of the reduced optical frequency (ωa/2πc) along one acoustic period (Ωt) from 0 to 2π of modes $A_S$ and $B_S$ respectively.

## 4. Concluding remarks

Acousto-optic coupling phenomena in dual photonic and phononic crystal slabs are studied. More specifically, the modulation of the optical eigenfrequences under the perturbation of the acoustic eigenmodes of a $L_1$ cavity is depicted. The global modulation of the frequency of the optical modes is strongly dependant on the choice of the modulating acoustic mode; upon this the photo-elastic and opto-mechanic effects can sustain each other or act with counteracted effects. The complexity of the acoustic modes of the slab does not evidence a fully symmetric mode; consequently the modulation of the optical eigenfrequencies is only possible at twice the acoustic frequency. This study can be useful to design phoxonic acousto-optic modulators.

[1] Yablonovitch E, 1987 *Phys. Rev. Lett.* **58** 2059
[2] Kushwaha MS et al., 1993 *Phys. Rev. Lett.* **71** 2022
[3] Lin TR et al., 2013 *Appl. Phys. Lett.* **113** 053508
[4] Papanikolaou N et al., 2012 *Microelec. Eng.* **90** 155
[5] Pennec Y et al., 2010 *Opt. Ex.*. **18** 14301
[6] Rolland Q et al., 2012 *Appl. Phys. Lett.* **101** 061109
[7] Rosencher E, Vinter B, Optoélectronique, *Masson* Paris Milan Barcelone
[8] Chan J et al., 2012 *Appl. Phys. Lett.* **101** 081115

# Electrically Controlled Phase Gratings for Terahertz Radiation Based on Nematic Liquid Crystal

**D Hovhannisyan**[1], **N Tabiryan**[2], **H Margaryan**[1], **V Abrahamyan**[1], **N Hakobyan**[1]

[1]Yerevan State University, Radiophysics department, Yerevan, Armenia,
[2]Beam Engineering for Advanced Measurements Co., FL, USA

E-mail: davidhl@ysu.am, nelson@beamco.com

**Abstract.** A mathematical model of a new type of liquid crystal (LC) based diffraction grating for the terahertz frequency range is proposed. Numerical time-integration by the finite-difference time-domain (FDTD) method of Maxwell-equation systems, describing the proposed structure, has been performed. The partial differential equation, describing the electro-optical induced orientation of the LC molecule in the external electric field, is calculated by the method of lines (MOL). The dependence of induced birefringence vs. external control voltage is obtained for 6CB nematic liquid crystal (NLC).

## 1. Introduction

Terahertz (THz) waves attracted a lot of attention due to a number of applications ranging from basic research to industrial quality control and security applications [1], THz spectroscopy and THz imaging systems. A very promising class of THz devices is those which contain LC. Liquid crystal components for terahertz radiation are in high demand due to the recent rapid progress in THz science and technology [2, 3]. The possibility for gratings with liquid-crystal-enabled functionalities was recognized [4]. Various tunable THz devices employing nematic liquid crystals, such as phase shifters, filters, and switches that are controlled either electrically or magnetically were demonstrated [5]. For manipulating THz waves by changing the effective refractive index of the NLC, recently demonstrated a magnetically controlled phase grating [6]. In [7], proposed and demonstrated an electrically controlled phase grating (PG) using NLC for THz waves, which contains alternate sections of two materials with different refractive indices – fused silica and NLC. In this paper we propose the all NLC based electrically controlled phase grating for terahertz radiation. This type of gratings can be used also for study the birefringence of NLC in THz domain.

## 2. Theoretical and numerical model of phase grating. Simulation results

We propose the electrically controlled phase grating (ECPG) consisting from the structured NLC layers (Figure 1). The device is the cell with thickness $d$, where the liquid crystal layer is sandwiched between two glass substrates coated by transparent ITO electrodes. One of the substrates is a structured conductive layer of parallel electrodes electrically connected with each others. The distance between the parallel electrodes and width of them are $L$. The second one is continuum ITO film. Applying the voltage between the substrates will result to periodically orientation of NLC molecules.

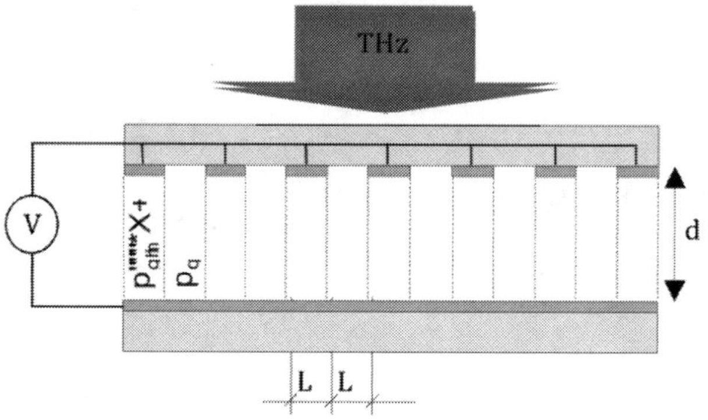


Figure 1 Schematic cross-section view of liquid crystal based phase grating

Taking into account the viscosity of NLC for the given value of controlled voltage, above the Frederiks threshold value, the molecule's spatial periodic orientation along the substrates can be approximated as a sinusoidal function. By control voltage the spatially periodical modulation of refraction index is generated due to electro-optically induced orientation of liquid crystal molecules. The phase difference between THz waves at wavelength $\lambda_{THz0}$, propagated through the neighbor layers with the difference values of refraction indexes controlled by the control voltage. Such a sinusoidal phase grating can be represented as a layer of nematic liquid crystal with a periodically varying refractive index. As shown in [8], for phase grating with sinusoidal refraction index modulation, at the modulation index value $\pi d(n_{eff}(V)-n_o)/\lambda_{THz0}=2$, which corresponds to the root of the Bessel function $J_0$, the maximum of the diffraction zero-order disappears. For a fixed value of the induced birefringence and the grating thickness in order to eliminate the zero order the wavelength of the terahertz radiation should be chosen so that the depth of modulation was equal to 2. For obtaining the dependences of induced birefringence vs. the control voltage numerical modelling of the process of electro-optical induced reorientation of NLC molecules was done. The NLC cell with planar orientation along the axis $y$ and control voltage applied along the axis $z$ was considered (Figure 2). By the method of lines (MOL) [9] was obtained numerical solution of Erikson-Leslie equation, which described the dynamics of NLC director rotation in the case when the backflow and inertial effects are ignored

$$\gamma\frac{\partial\theta}{\partial t}=\left[K_{11}\sin^2(\theta)+K_{33}\cos^2(\theta)\right]\frac{\partial^2\theta}{\partial z^2}+\frac{1}{2}(K_{11}-K_{33})\sin(2\theta)\frac{\partial\theta^2}{\partial z}-\frac{\varepsilon_0}{8\pi}\Delta\varepsilon E_z^2(t)\sin(2\theta) \tag{1}$$

where $K_{11}$, $K_{22}$, $K_{33}$ are the elastic constants associated with splay, twist and bend deformation respectively, $\gamma$ - rotational viscosity, $\varepsilon_0\Delta\varepsilon\, E_z^2$ is the electric energy density, $\Delta\varepsilon$ is the NLC dielectric anisotropy, and $\theta(z, t)$ is the tilt angle of the NLC directors. On base of numerical solution was determined the effective refractive index of NLC vs. applied voltage $V = E_z d$

$$n_{eff}=\frac{1}{d}\int_{-d/2}^{d/2}\frac{\sqrt{\varepsilon_{//}\varepsilon_{\perp}}}{\sqrt{\varepsilon_{\perp}\cos^2(\theta(z,t\to\infty))+\varepsilon_{//}\sin^2(\theta(z,t\to\infty))}}dz \tag{2}$$

for NLC layer of thickness $d$ and with $z$=0 at the centre of the layer, $\theta(z, t \to\infty) = \theta_s$ - is the value of the tilt angle in steady-state regime, $\varepsilon_\perp$, $\varepsilon_{\parallel}$ - dielectric permittivity's for extraordinary and ordinary waves.

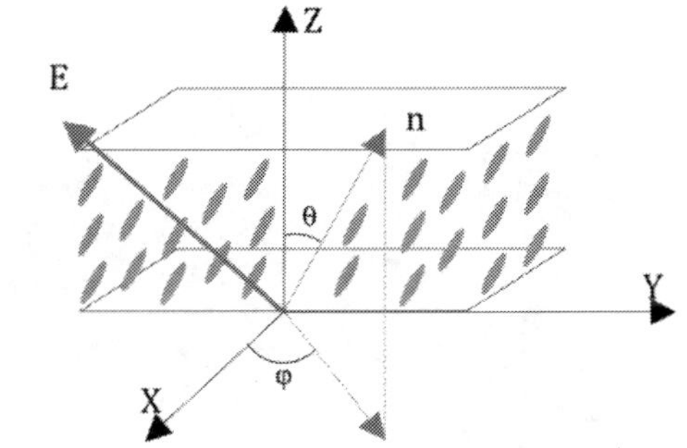


Figure 2 Geometry of interaction of NLC director with the controlled external electric field.

To demonstrate the dependence of electro-optically induced birefringence vs. the control external voltage, we carried out a numerical integration for *6CB* NLC with the 200 um thickness, 180 um grating period, 60 um THz radiation wavelength and with the following parameters: $\varepsilon_{\parallel} = 12$, $\varepsilon_\perp = 4$, $K_{11} = 8.57$ pN, $K_{33} = 3.7$ pN, $\gamma = 13.3$ kg/m·s. The dependence of induced birefringence and derivative of induced birefringence for the different sub ranges vs. control voltage are shown on the Figure 3. As seen from figure the Frederiks threshold value is about 11V.

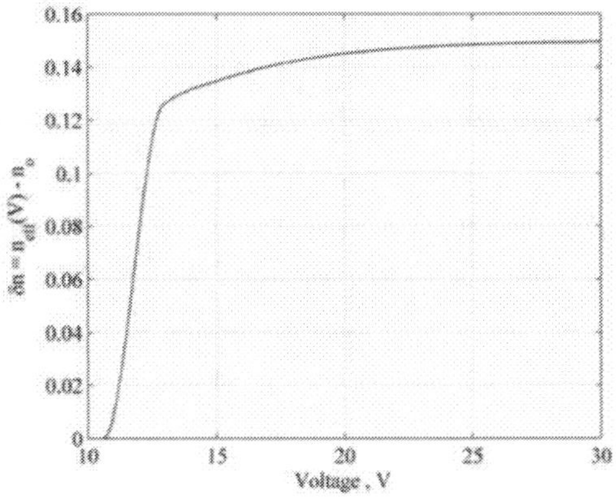


Figure 3 Electro-optically induced birefringence. vs. external control voltage.

In the FDTD numerical model we pass from continuous space to discrete one with spatial discretization step $\lambda_{THz}/50$ and the inverse dielectric tensor of considered grating was presented as

$$\tilde{\varepsilon}_{xx}^{-1}(x) = \left[\varepsilon_\perp + (\varepsilon_{II} - \varepsilon_\perp)\sin^2(\theta_s(x))\right]/\varepsilon_\perp\varepsilon_{II},$$
$$\tilde{\varepsilon}_{xz}^{-1}(x) = \tilde{\varepsilon}_{zx}^{-1} = \left[-(\varepsilon_{II} - \varepsilon_\perp)\cos(\theta_s(x))\sin(\theta_s(x))\right]/\varepsilon_\perp\varepsilon_{II}$$
$$\tilde{\varepsilon}_{zz}^{-1}(x) = \left[\varepsilon_\perp + (\varepsilon_{II} - \varepsilon_\perp)\cos^2(\theta_s(x))\right]/\varepsilon_\perp\varepsilon_{II}, \tilde{\varepsilon}_{yy}^{-1} = 1/\varepsilon_\perp \quad . \tag{3}$$

In the Figure 4 is shown the two-dimensional spatial distribution of $E_x$ component of the diffracted field, obtained by numerical solution of Maxwell equations systems by FDTD method for the linearly polarized field ($E_{x0}$, $0$, $0$). The boundaries of grating are shown by white lines. According to the figure, the diffracted field which is formed on the output of grating, start to distribute between ±1 and 0 diffraction orders.

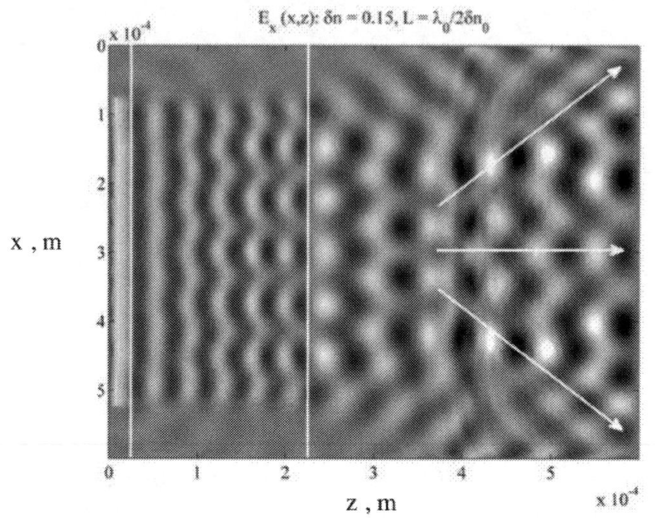


Figure 4 Two-dimensional spatial distribution of $E_x$ component of the diffracted field.

For determination the diffracted far field, as the initial value, the field distribution directly at the grating's output was used. Comparison shows that the obtained numerically by FDTD and analytically far field distributions practically are same.

**Conclusion**

In this paper a mathematical model of a new type of phase liquid crystal diffraction grating for the terahertz frequency range is proposed. Since the diffraction efficiency will depend on absorption of NLC at THz, the proposed phase grating can be used for study the dielectric permittivity of NLC in the THz.

This work is implemented in the frameworks of GIPP/ISTC A-1951 Project.

**References**

[1] Zhang X.-C. and Jingzhou Xu, "Introduction to THz Wave Photonics", Springer New York Dordrecht Heidelberg London, 2010.
[2] Garet F., Coutaz J.-L., Narzarov M., Bonnet E., Parriaux O., Racine G., Dig. Joint Int. Conf. Infrared Millimeter Waves and Terahertz Electron., 2004, 181.
[3] Kersting R., Strasser G., Unterrainer K., Electron. Lett., **36**, 2000, 1156.
[4] Chen, J. Bos P. J., Vithanna H., Johnson D. L., Appl. Phys. Lett., **67**, 1995, 2588.
[5] Chen C.-Y., Hsieh C.-F., Lin Y.-F., Pan R.-P., and Pan C.-L., Opt. Express, **12**, 2004, 2625.
[6] Lin C.-J., Li Y.-T., Hsieh C.-F., Pan R.-P., and Pan C.-L., Opt. Express, **16**, 2008, 2995.
[7] Lin Chia-Jen, Lin Chuan-Hsien, Li Yu-Tai, Pan Ru-Pin, and Pan, Ci-Ling, IEEE Photonics Technology Letters, **21**, 11, 2009, 730.
[8] Goodman J. W., "Introduction To Fourier Optics", McGraw-Hill, 1988.
[9] Schiesser W. E., Griffiths G. W., "A compendium of partial differential equation models. Method of Lines Analysis with Matlab", Cambridge University Press, 2009.

# Restoration of diffracted far field at the output of circular diffraction waveplate

**D Hovhannisyan**[1], **N Tabiryan**[2], **H Margaryan**[1], **V Abrahamyan**[1], **N Hakobyan**[1]

[1]Yerevan State University, Radiophysics department, Yerevan, Armenia,
[2]Beam Engineering for Advanced Measurements Co., FL, USA

E-mail: davidhl@ysu.am, nelson@beamco.com

**Abstract.** The light propagation in an anisotropic periodic media, such us circular diffraction waveplate (CDW) by a finite-difference time-domain (FDTD) technique is studied. The FDTD numerical simulation and the subsequent Fourier transform of the diffracted electric near field was been used for study of ability of CDW to diffract a laser beam and simultaneously convert polarization state. The FDTD simulation results used to restore the diffracted electric far field at the CDW output. an abstract.

## 1. Introduction

Most of circular diffraction waveplate CDW exhibit unique diffraction characteristics, such as polarization-independence, low scattering, and high diffractive efficiencies. For CDW the polarization conversion mechanism is determined by the spatial periodic distribution of the birefringence. The theoretical consideration is necessary to characterize the diffraction properties of CDW. One of the simplest and common methods to describe polarized light transmission is the Jones matrix based analysis, which easily reveals the polarization states of beams diffracted from the CDW but unfortunately cannot describe details of the optical electric field in and/or near the CDW. For optical systems the near fields distributions is important due to small sizes of modern photonics structures. A more practical approach to analyzing the detailed electric field in and/or near optical devices is given by the FDTD algorithm. This method present a powerful approach to directly solving Maxwell's equations both in time and space and are also applied to the analysis of arbitrary, anisotropic, and periodic structures [1]. Another advantage of FDTD techniques is their capability to visualize real-time pictures of the electromagnetic wave. In our early paper we reported that the FDTD can be adopted for understanding of some peculiarities of recording of CDW by doubling method [2].

In this paper, the FDTD method is used to analyze light propagation in CDW, and to restore the far field distribution.

## 2. FDTD numerical model for the description of CW linearly-polarized radiation propagation through the anisotropic periodic media

In this section the numerical analysis of light propagation through the CDW is presented. Let us consider the case when an arbitrary polarized beam of a continuous-wave (CW) laser with ($E_x$, $E_t$, $0$) components propagates along z axis and hits a CDW (the problem's geometry is presented at Figure 1).

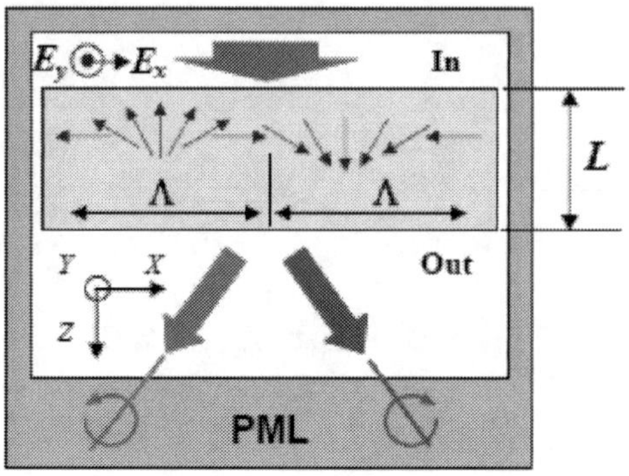


Figure 1 The schematic of CDW arrangement with respect to the coordinate axes.

The grating is placed in a position where the CDW molecules director's periodic change occurs in the $x$ axis. Let's designate the grating's half-period as - $\Lambda$, and the thickness as - $L$. At the output of the grating the arrows show the propagation of radiation in -1 and +1 diffracted orders with left and right circular polarization. Generally Maxwell's equations for a periodic anisotropic medium can be written as follows:

$$\frac{\partial D_x}{\partial t}=\frac{\partial H_z}{\partial y}-\frac{\partial H_y}{\partial z},\quad \frac{\partial D_y}{\partial t}=\frac{\partial H_x}{\partial z}-\frac{\partial H_z}{\partial x},\quad \frac{\partial D_z}{\partial t}=\frac{\partial H_y}{\partial x}-\frac{\partial H_x}{\partial y} \qquad (1)$$

$$\frac{\partial H_x}{\partial t}=-\frac{1}{\mu}\left(\frac{\partial E_z}{\partial y}-\frac{\partial E_y}{\partial z}\right),\quad \frac{\partial H_y}{\partial t}=-\frac{1}{\mu}\left(\frac{\partial E_x}{\partial z}-\frac{\partial E_z}{\partial x}\right),\quad \frac{\partial H_z}{\partial t}=-\frac{1}{\mu}\left(\frac{\partial E_y}{\partial x}-\frac{\partial E_x}{\partial y}\right) \qquad (2)$$

where $D_{x,y,z}$ are the electrical induction components, $H_{x,y,z}$ are the magnetic field components, $\mu$ is permeability of free space. The electric field vector can be described as $E=\tilde{\varepsilon}^{-1}D$, where $\tilde{\varepsilon}$ is a permittivity tensor for a periodic anisotropic medium and which in the fixed laboratory system can be presented as

$$\varepsilon(\varphi)=\begin{pmatrix}\varepsilon_e\cos^2(\varphi)+\varepsilon_o\sin^2(\varphi) & \Delta\varepsilon\sin(\varphi)\cos(\varphi) & 0\\ \Delta\varepsilon\sin(\varphi)\cos(\varphi) & \varepsilon_o\cos^2(\varphi)+\varepsilon_e\sin^2(\varphi) & 0\\ 0 & 0 & \varepsilon_e\end{pmatrix} \qquad (3)$$

where $\varepsilon_e$, $\varepsilon_o$ are dielectric constants for extraordinary and ordinary waves, $\Delta\varepsilon=\varepsilon_e-\varepsilon_o$, CDW molecule orientation related angles $\varphi=\pi x/\Lambda,\alpha$. According to (3), we consider the case when molecules lie in $x$-y plane. The components of the electric field vector can be described as

$$\begin{aligned}E_x&=\tilde{\varepsilon}_{xx}^{-1}D_x+\tilde{\varepsilon}_{xy}^{-1}D_y+\tilde{\varepsilon}_{xz}^{-1}D_z,\\ E_y&=\tilde{\varepsilon}_{xy}^{-1}D_x+\tilde{\varepsilon}_{yy}^{-1}D_y+\tilde{\varepsilon}_{yz}^{-1}D_z,\\ E_z&=\tilde{\varepsilon}_{xz}^{-1}D_x+\tilde{\varepsilon}_{yz}^{-1}D_y+\tilde{\varepsilon}_{zz}^{-1}D_z\end{aligned} \qquad (4)$$

where the $\tilde{\varepsilon}_{xx}^{-1},\varepsilon_{xy}^{-1},\tilde{\varepsilon}_{xz}^{-1},\tilde{\varepsilon}_{xy}^{-1},\tilde{\varepsilon}_{yy}^{-1},\tilde{\varepsilon}_{yz}^{-1},\tilde{\varepsilon}_{xz}^{-1},\tilde{\varepsilon}_{yz}^{-1},\tilde{\varepsilon}_{zz}^{-1}$ are the components related to the permittivity inverse dielectric tensor. The essential feature in FDTD methods is a proper absorbing boundary condition (ABC) to truncate the simulation space without artificial reflections. Most often ABCs are used, which are based on the construction of artificial absorbing media in the truncated border domains. The advantage of this idea is that the truncation can be made with orders of magnitude lower errors than employing one-way wave equations. The truncation border domains are filled virtually with a special anisotropic material for which it can be shown that a plane wave incident from vacuum with an arbitrary wave vector is transmitted without reflection. This phenomenon is referred to as perfect matching. In this article, we used the perfectly matched layer (PML, see Fig.1) introduced by Berenger [3]. Although at the gratings input we consider CW radiation with $(E_x E_y\ 0)$ electric field components, due to the anisotropy of the grating (4), at the output of CDW the component $E_z$ appear also. To calculate the diffracted field we make a transition from the continuous space $(x, y, z, t)$ to the integer grid $(i\Delta x, j\Delta y, k\Delta z, n\Delta t)$, where $i, j, k, n$ are integer constants [4]. The components of the electric and magnetic fields $E$ and $H$ will be interpolated at the time points $n\Delta t$ and $(n+1/2)\Delta t$ (n is a positive integer), respectively, where the constant time step $\Delta t$ has to be chosen depending on the node distances and the refraction index. Discretization steps in the directions x, y and z, which are same $\Delta x = \Delta y = \Delta z = \lambda_0/50$, $\lambda_0$ =532 nm is the CW radiation wavelength. With taking into account the inverse dielectric tensor

$$\begin{aligned}
\tilde{\varepsilon}_{xx}^{-1} &= [\varepsilon_o \cos^2(\varphi) + \varepsilon_e \sin^2(\varphi)]/\varepsilon_o\varepsilon_e, \\
\tilde{\varepsilon}_{xy}^{-1} &= \tilde{\varepsilon}_{yx}^{-1} = [-(\varepsilon_e - \varepsilon_o)\cos(\varphi)\sin(\varphi)]/\varepsilon_o\varepsilon_e \\
\tilde{\varepsilon}_{yy}^{-1} &= [\varepsilon_e \cos^2(\varphi) + \varepsilon_o \sin^2(\varphi)]/\varepsilon_o\varepsilon_e, \tilde{\varepsilon}_{zz}^{-1} = 1/\varepsilon_e
\end{aligned} \tag{5}$$

the electric field components calculated in accordance with (4).

## 3. Simulation results

In simulations considered the medium with $n_e$ = 1.67, $n_o$ = 1.52 and $\Delta n$ = 0.15 respectively. The grating's size in discrete numerical space along $x$ axis was $4\Lambda$. In Figure 2 are shown the spatial distributions of the field components $E_x(x,y)$, $E_y(x,y)$ and their phase's $angle(E_x(x,y))$, $angle(E_y(x,y))$ inside and outside of CWD respectively for polarized input radiation $E_x = \cos(\omega_0 t), E_y = \sin(\omega_0 t + \delta), \omega_0 = 2\pi c/\lambda_0, \delta = 0; \pi/2; -\pi/2$. The input and output boundaries of CWD are shown by white lines. The field components distributions along x axis immediately at the output of CWD, which preliminary are approximated by sin functions $E_{x\_appr} = a_x \sin(b_x x + c_x)$, $E_{y\_appr} = a_y \sin(b_y x + c_y)$ was used for restoration of diffracted far field.

The relative error of amplitudes $(a_x - a_y)/a_x$ arising due to numerical calculations precision is no more than 1.73%, relative error of period along x axis $(1/b_x - 1/b_y)/1/b_x$ is no more than 1% and the relative error of phase difference $(\pi - (c_x - c_y))/\pi$ is no more than 0.3%. The direction along $z$ axis is corresponding to the time scale. For generation the phase distribution of the field components the Hilbert transform was applied to each row ($z$ axis) of field matrix. In that way was generated the new matrixes containing the complex values. The phase angles of these complex matrixes correspond to $angle(E_x(x,y))$ and $angle(E_y(x,y))$, which containing information about the distribution of wave front delay along the x axis. According to simulation results and as seen from figure the $angle(E_x(x,y))$ distribution along the x axis and at the fixed z value is shifted relative to $angle(E_y(x,y))$ distribution by corresponding phase delay – 0; π/2 and -π/2.

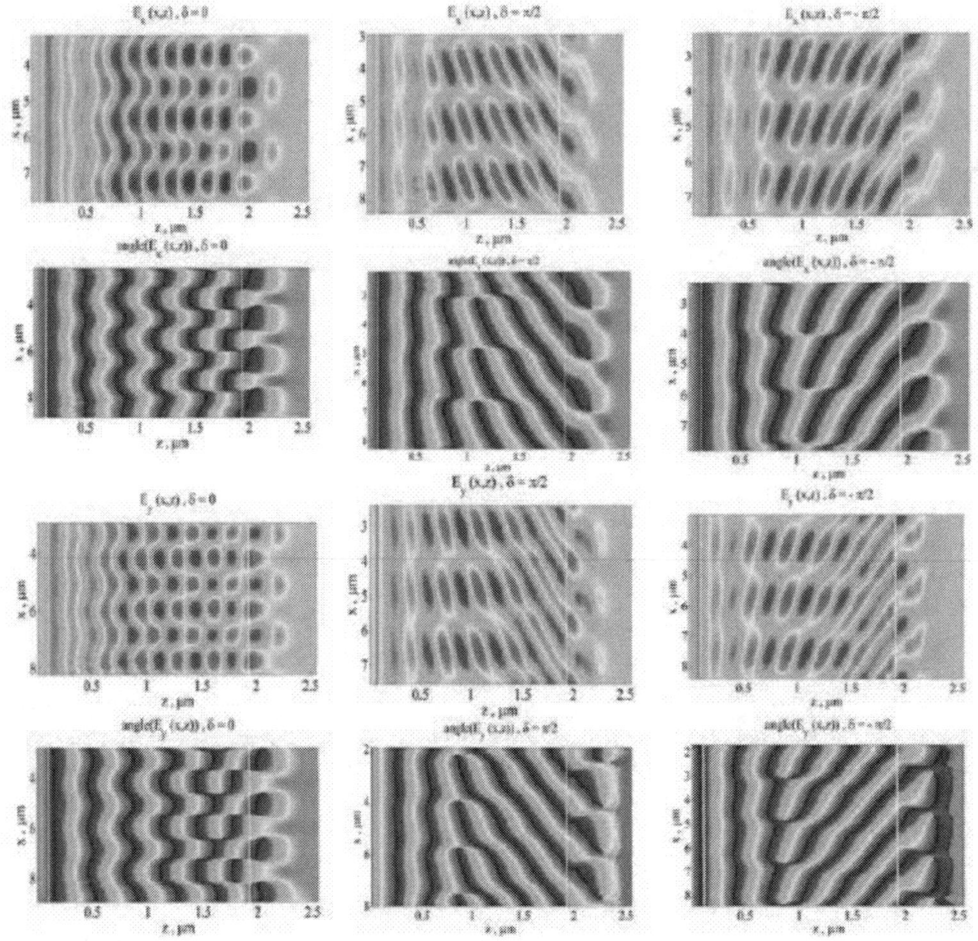

Figure 2 The field components and their phase's inside and outside of CWD for phase differences 0; $\pi/2$ and $-\pi/2$ correspondingly.

**Conclusion**

We have used the FDTD calculation of light propagation through the CDW by numerical solution of the Maxwell equations. For generation the phase distribution of the field components which containing information about the distribution of wave front delay along the $x$ axis was applied the Hilbert transform. The restoration of diffracted far field at the output of CDW was done by Fraunhofer propagator.

This work is implemented in the frameworks of GIPP/ISTC A-1951 Project.

**References**


[1] Chulwoo Oh, Ravi Komanduri and Michael J. Escuti, , Physical Review A **76**, 2007, 043815.

[2] Margaryan H. L., Aroutiounian V. M., Hovhannisyan D. L., Hakobyan N. H., Abrahamyan V. K., Mol. Cryst. Liq. Cryst., **559**, 2012, 214.

[3] Jean-Pierre B´erenger, Perfectly Matched Layer (PML) for Computational Electromagnetics, Morgan & Claypool Publishers. 2007.

[4] Allen Taflove and Susan C. Hagness, Computational Electrodynamics: The Finite-Difference Time-Domain Method, 3rd ed., Artech House Publishers, 2005.

# Numerical analysis of p-GaAs/n-GaAs tunnel junction employing InAs intermediate layer for high concentrated photovoltaic applications

**Seokjin Kang,[1] Kwang Wook Park,[1] Sooraj Ravindran,[1] and Yong Tak Lee[1*]**

[1]School of Information and Informations, Gwangju Institute of Science and Technology (GIST), 123, Cheomdangwagi-ro, Buk-gu, Gwangju 500-712, Republic of Korea

E-mail: ytlee@gist.ac.kr

**Abstract**. We report the numerical analysis of p-GaAs/n-GaAs tunnel junction employing InAs as the intermediate layer at the junction. Incorporation of this intermediate layer introduces an intermediate level at the junction bandgap leading to enhanced tunneling of carriers. By doing so, we obtain a two order of enhancement in the tunnel current. Furthermore, the performance of GaInP/GaAs dual-junction solar cells using the TJ with InAs intermediate layer is calculated under concentrated suns condition and it shows a conversion efficiency exceeding 30% under 1800 suns condition.

## 1. Introduction

A tunnel junction (TJ) supplying high current is one of the key requirements to achieve high efficiency in high concentrated photovoltaic (CPV) applications. Recently, multi-junction solar cells using $Ga_{0.51}In_{0.49}P$ (1.9 eV band gap), GaAs (1.43 eV band gap) and $Ga_{0.69}In_{0.31}As$ (1.0 eV band gap) have been reported having a conversion efficiency of 44.4% at 302 suns [1]. To operate the solar cells under high concentrated sun (>1000 suns) condition, the TJ that interconnects the sub-cells must provide efficient current transport with small voltage loss by generating high tunnel current [2, 3]. Several methods have been reported to accomplish an increase in the tunnel current, and one such method is by introducing ErAs recombination centers as an intermediate level between the junction interface [4, 5]. It halves the current path length and thus increases the tunnel current drastically. This is because the tunneling probability is related to the inverse exponential of the path length. With the presence of an intermediate level in the junction, peak tunnel current density ($I_{peak}$) increases by the factor of 2. However, ErAs has a rock-salt crystal structure which is quite dissimilar to the crystal structure of conventional semiconductors used in realizing multi-junction solar cells. This dissimilar crystal structure degrades the device performance due to the introduction of high density of Ga-antisite defects originating from the interface between zinc blende and rock salt structure. As an alternative, InAs (0.35 eV band gap) which has a zinc blende crystal structure similar to that of semiconductor materials used in III-V solar cells can possibly introduce intermediate level at the junction interface by utilizing their lower band gap energy as compared with the other alternative TJ materials such as GaInP, $Al_{0.38}Ga_{0.62}As$ (1.9 eV band gap) and GaAs without introducing serious defects. Experimental demonstration of this concept has already has been reported, which shows two orders of $I_{peak}$ enhancement in GaAs TJ with InAs quantum dot as the intermediate level [6]. However, numerical simulation of TJs employing intermediate level has not been well studied yet. In this paper, we report the numerical analysis of GaAs TJ employing InAs intermediate level and GaInP/GaAs dual-junction solar cells employing the TJ under concentrated suns condition.

## 2. Theoretical models for numerical simulations

2.1 Carrier statistics model

The carriers in the highly doped degenerate layers in which Fermi level no longer exists in the band gap obey a different statistical behavior as compared with non-degenerate semiconductors which obeys Boltzmann statistics. To operate the p-GaAs/n-GaAs structure as a TJ, the two oppositely doped layers should be highly doped. Furthermore, band gap narrowing effect and Fermi-Dirac statistics must be considered to accurately analyze the behavior carriers in highly doped regions. The band gap narrowing effect is related to doping concentration of the layer, and expressed as

$$\Delta E_g = A\left\{\ln\frac{N}{B} + \left[\left(\ln\frac{N}{B}\right)^2 + 0.5\right]^{\frac{1}{2}}\right\} \tag{1}$$

where constant A and B are $9.0\times10^{-3}$ eV and $1.0\times10^{17}$ $cm^{-3}$, respectively, obtained from Slotboom [7], and N represents the doping concentration. The amount of bandgap narrowing for a doping concentration of $10^{19}$ $cm^{-3}$ was about 61 meV. Fermi-Dirac statistics describes the probability that an available electron state with energy E is occupied by an electron and is given by

$$\mathrm{f}(\varepsilon) = \frac{1}{1+exp\left(\frac{\varepsilon-E_F}{kT}\right)} \tag{2}$$

In the limiting case, $\varepsilon - E_F \gg kT$, equation 2 has much simpler form like $\mathrm{f}(\varepsilon) = exp\left(\frac{E_F-\varepsilon}{kT}\right)$ referred to as Boltzmann statistics. However, we used Fermi-Dirac statistics because the limiting case is not applicable for highly doped material.

2.2 Tunneling model

Tunneling at the TJ is basically driven by the real spatial transport of carriers through the barriers. This non-local band-to-band tunneling can be significant for small tunneling path length. Tunnel current density for an electron with longitudinal energy $E$ and transverse energy $E_T$ is given by

$$\mathrm{J}(E) = \frac{q}{\pi\hbar}\iint T(E)\left[f_l(E+E_T) - f_r(E+E_T)\right]\rho(E_T)dEdE_T \tag{3}$$

where T(E) is the tunneling probability for an electron with longitudinal energy $E$ and $\rho(E_T)$ is the 2D density of states corresponding to the two transverse wavevector components. $f_l$ and $f_r$ represenets the Fermi-Dirac functions on the left and right hand side of the junction.

The tunneling probability is calculated using the WKB approximation and it is given by

$$\mathrm{T}(E) = exp\left(-2\int_{x_{start}}^{x_{end}} k(x)dx\right) \tag{4}$$

where $x_{start}$ and $x_{end}$ are the start and end points of the tunneling paths. The commercial Silvaco Atlas 2D TCAD provides these non-local tunneling band-to-band model and WKB approximation for the tunneling probability and our simulations were performed by using these models [8].

## 3. Simulation procedure

Prior to analyzing the GaInP/GaAs dual junction solar cell, p-GaAs/n-GaAs TJ structures were analyzed. Two different TJ structures with fine mesh structures were generated (Fig. 1). The reference structure (Fig. 1(a)) was a GaAs tunnel diode with 25 nm-thick p+GaAs and n+GaAs, and the doping concentration of the compositing layers was in the range from $10^{18}$ $cm^{-3}$ to $10^{20}$ $cm^{-3}$. The mesh structure was defined by five separate divisions of 100 um-thicknesses along horizontal direction, and one hundred separate divisions of 50 nm-thicknesses along vertical direction. The direction of the current flow was also along the vertical direction. In addition, two hundred separate divisions of 30 nm-thicknesses along vertical direction near the junction were generated for the accurate calculation of non-local band-to-band tunneling modeling. For the second structure (Fig. 1(b)), InAs intermediate layer was inserted at the junction interface of the reference TJ. The same mesh structure was employed for the calculation since the number of meshes can possibly affect the accuracy of simulations. By using the above TJ structures, we calculated I-V curve and extracted $I_{peak}$ and peak voltage ($V_{peak}$) which are shown in Fig. 1(c). First, the doping concentrations of p-GaAs and n-GaAs region of the reference TJ and TJ with InAs intermediate level was varied to compare their performance, and the resulting $I_{peak}$ contour plots were generated. Second, the doping concentration, doping type and thickness of InAs intermediate layer was varied to find out the effect of these parameters on the TJ current. During the calculations, the doping concentration of p-GaAs and n-GaAs was fixed at $4\times10^{19}$

$cm^{-3}$ and $10^{19}$ $cm^{-3}$. Next, TJ employing InAs intermediate layer was incorporated in the GaInP/GaAs dual-junction solar cell (Fig 2). InAs was n-doped with a carrier concentration of $10^{19}$ $cm^{-3}$ and the thickness was 3.7 nm. Additionally, current matching between GaInP and GaAs subcells was applied. $SiO_2/Si_3N_4$ double layer anti-reflection coating (DARC) ensured the reduction of surface reflection and metal shadowing area was assumed to be 5% of whole area. Solar spectrum with an air mass 1.5 global (AM 1.5G) under normal solar incidence was used in simulations. The performance of the solar cell was calculated up to 2000 concentrated suns condition.

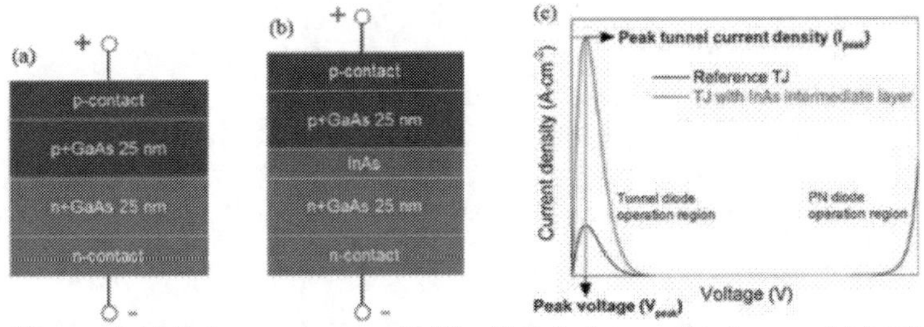


**Figure 1** (a) Reference structure (b) TJ with InAs intermediate layer, and (c) I-V curve for the reference structure and TJ structure with InAs intermediate layer

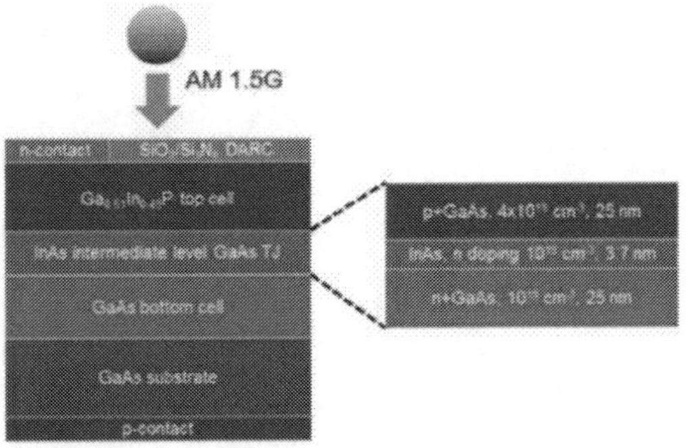


**Figure 2** GaInP/GaAs dual-junction solar cell structure

## 4. Results and discussion

By varying the p and n GaAs doping concentration, the contour plot describing the minimum required doping concentration in order for the solar cell to operate in 100 suns (1.3 $A\cdot cm^{-2}$) and 1000 suns (13 $A\cdot cm^{-2}$) are illustrated in Fig. 3. It clearly shows that the TJ structure with InAs in the intermediate level requires lower doping concentration than the reference TJ. In addition, for the same doping concentration, a two order enhancement for $I_{peak}$ was obtained in TJ with InAs in the intermediate level compared to that of the conventional one as expected since the tunneling probability is proportional to inverse exponential of the tunneling path. The presence of InAs intermediate level divided the tunneling path length in to half resulting in a large increase of tunnel current. However, at higher doping concentrations (>$5\times10^{19}$ $cm^{-3}$), there was only an order of enhancement since the tunneling path length was already short enough in this condition.

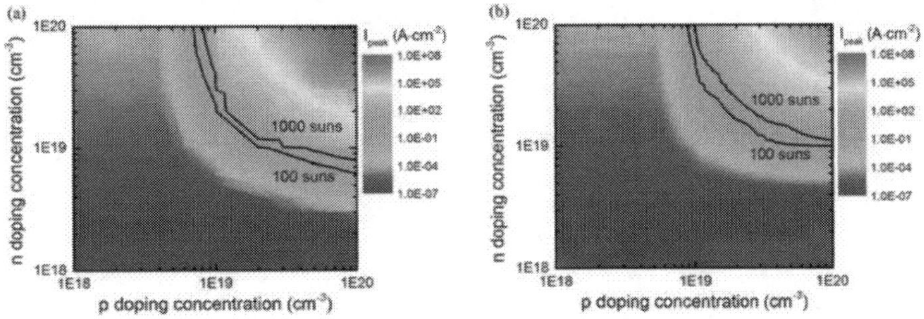


**Figure 3** Contour plots of $I_{peak}$ for (a) the reference TJ and (b) TJ with InAs as the intermediate level

Further simulations were done by varying the p and n doping concentration and thickness of InAs layer. It was found that n-type doping for InAs was suitable, while p-type doping was not suitable for enhancing the tunnel current. With the increase of n-type carrier concentration, $I_{peak}$ exponentially increased at ultra-high doping concentration region (Fig. 4). This is because of the band alignment which aids the electron tunneling. In the GaAs/InAs system, conduction band offset is 0.69 eV and valence band offset is 0.38 eV [9]. Since the conduction band offset is larger than valence band offset, using the intermediate level at the conduction band (Fig. 5(b)) provides shorter tunneling path length than the case when intermediate level is used at the valence band (Fig 5(a)). However, $V_{peak}$ was similar regardless of the change in doping concentration because it largely depends on the doping concentration of p and n GaAs regions rather than that of InAs region. For varying InAs thickness, it was found that $I_{peak}$ increased continuously (Fig. 6(a)). This is attributed to the formation of intermediate level, and as a result more carriers could tunnel across the p+GaAs and n+GaAs sides compared to the case when intermediate level was absent. Thus, as the InAs thickness was increased, the tunnel current also increased.

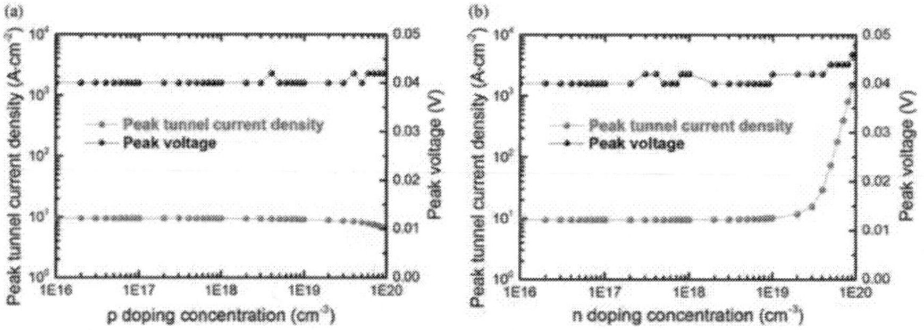


**Figure 4** Variation of tunnel current as a function of (a) p doping and (b) n doping of InAs

It was also found that GaInP/GaAs dual-junction solar cells with TJs employing InAs intermediate level operated well up to 2000 suns condition. The conversion efficiency was over 30% up to 1800 suns and there was no significant degradation in efficiency caused by absorption at the InAs layer (Fig. 6(b)). This was because the thickness of InAs was small enough to cause any appreciable optical loss. Incorporation of InAs intermediate level provided about 400 times larger $I_{peak}$ than normal GaAs TJ and it could also enable the operation of solar cell under high concentration suns. For a normal GaAs TJ, the solar cell could only operate up to a low concentration of 10 suns.

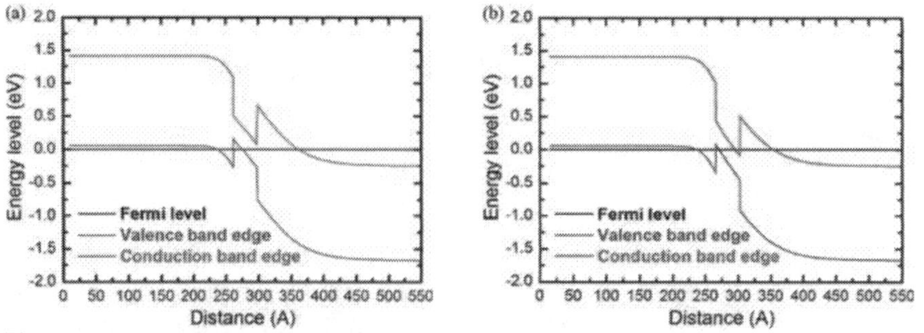


**Figure 5** Band diagrams of p-GaAs/n-GaAs TJ with InAs intermediate layer. (a) shows the case when InAs is p doped and (b) denotes the case when InAs is n doped.

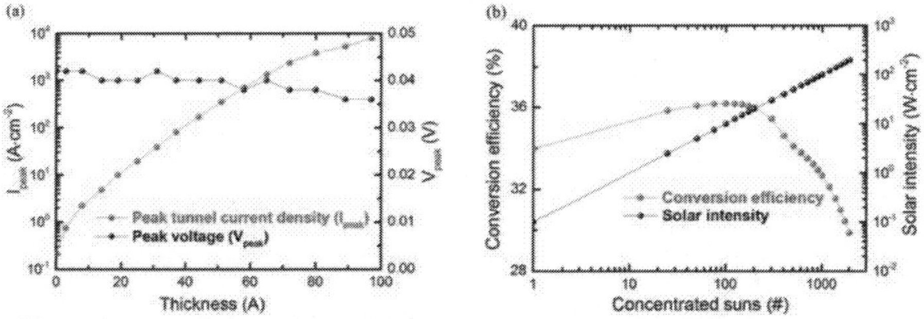


**Figure 6** (a) $I_{peak}$ and $V_{peak}$ of GaAs TJ with InAs intermediate layer for various interlayer thickness and (b) performance of GaInP/GaAs dual-junction solar cells using InAs intermediate level

## 5. Conclusion

The performance of GaAs TJ with InAs embedded as the interlayer at the junction was evaluated using Silvaco 2D TCAD. It was found that TJ with InAs incorporated at the junction introduced an intermediate level in the bandgap resulting in about two order of increment in $I_{peak}$. We investigated about the type of doping suitable for the InAs interlayer, and found that n-type doping was the appropriate one. With increasing thickness of InAs layer, $I_{peak}$ also increased, which was due to the increase in tunneling by the presence of intermediate layer. In addition, we also studied the performance of GaInP/GaAs dual-junction solar cells employing InAs as the intermediate level, and found that it operated up to 2000 suns condition and maintained energy conversion efficiency nearly 30%, while without InAs interlayer, the operation was limited only to low concentration of 10 suns. It could be expected that incorporating interlayer to various TJs of solar cells could result in improved performance of concentrated photovoltaics.

## AKNOWLEDGEMENT

This work was supported by the National Research Foundation of Korea (NRF) grant funded by the Korea government (MEST) (No. 2011-0017606).

## References

[1] M. A. Green, K. Emery, Y. Hishikawa, W. Warta and E. D. Dunlop 2013 *Prog. Photovolt: Res. Appl.* **21** 827-837.
[2] R. R. King, D. C. Law, K. M. Edmondson, C. M. Fetzer, G. S. Kinsey, H. Yoon, R. A. Sherif, and N. H. Karam 2007 *Appl. Phys. Lett.* **90** 183516.
[3] R. R. King, R. A. Sherif, D. C. Law, J. T. Yen, M. Haddad, C. M. Fetzer, K. M. Edmondson, G. S. Kinsey, H. Yoon, M Joshi, S. Mesropian, H. L. Cotal, D. D. Krut, J. H. Ermer, and N. H. Karam presented at 2006 *21st European Photovoltaic Solar Energy Conference*
[4] Hari P. Nair, Adam M. Crook, and Seth R. Ban 2010 *Appl. Phys. Lett.* **96** 222104
[5] S. Preu, S. Malzer, G. H. Dohler, H. Lu, A. C. Gossard, and L. J. Wang 2010 *Semicond. Sci. Technol.* **25** 115004
[6] L. Wang, J. He, X. Shang, M. Li, Y. Yu, G. Zha, H. Ni, and Z. Niu 2012 *Semicond. Sci. Technol* **27** 115010
[7] Slotboom, J. W. and H. C. De Graaf, 1976 *Solid State Electron.* **19**, 857-862
[8] Silvaco 2D TCAD *Atlas User's Manual* Chapter 3 Physics
[9] R. Colombelli, V. Piazza, A. Badolato, M. Lazzarino, F. Beltram 2000 *Appl. Phys. Lett.* **76** 1146

# Interactive visualization system to analyze corrugated millimeter-waveguide component of ECH in nuclear fusion with FDTD simulation

**N. Kashima**[1], **H. Nakamura**[1,2], **Y. Tamura**[3], **A. M. Ito**[2] **and S. Kubo**[1,2]

[1] Department of Energy Engineering and Science, Graduated School of Engineering, Nagoya University, Oroshi-cho 322-6, Toki-city, GIFU, Japan
[2] Department of Helical Plasma Research, National Institute for Fusion Science, Oroshi-cho 322-6, Toki-city, GIFU, Japan
[3] Department of Intelligence and Informatics, Konan University, Okamoto 8-9-1, Higashinada-ku, Kobe-city, HYOGO, Japan

E-mail: `teamMD@nifs.ac.jp`

**Abstract.** We have simulated distribution of electromagnetic waves through the system composed of miter bends by Finite-Difference Time-Domain (FDTD) simulation. We develop the interactive visualization system using a new interactive GUI system which is composed of the virtual reality system and android tablet to analyze the FDTD simulation. The effect of the waveguide system with grooves have been investigated to quantitatively by visualization system. Comparing waveguide system with grooves and without grooves, grooves have been confirmed to suppress the surface current at the metal surface. The surface current at complex shape such as the miter bend have been investigated.

## 1. Introduction

Recently, there has been an interest in electron cyclotron heating (ECH) of high temperature plasmas by high-power electromagnetic (EM) waves [1]. ECH is used for heat plasmas and maintain a stable plasma in the Large Helical Device (LHD). To transmit EM waves from a gyrotron efficiently, low loss and high power millimeter wave components have been researched both in theoretically and experimentally [2]. In this paper, we investigated the transmission of EM waves in transmission system that is composed the circular corrugated (CC) waveguide and the miter bend. We adopted FDTD simulation method in the research of the transmission through the CC-waveguide and the miter bend. The FDTD method was introduced by Yee in 1966 [3] to solve Maxwell's equation numerically. In our group, we have been developing FDTD simulation code and the visualization system to analyze the EM wave transmission [4–7].

EM waves are transmitted through the transmission line. The transmission line is composed of the straight CC-waveguide and the miter bend (Fig. 1). The transmission line is not composed at only a straight component between the gyrotron from LHD in order to insulate the neutrons from the LHD. Thus, the miter bend is used to change the direction of EM waves [8]. Grooves called corrugated are machined on the inside of these components. The eddy currents on the surface of them are prevent by grooves. The pitch ($p$ = 1.3 [mm]), the width ($w$ = 1.0 [mm]) and the depth ($d$ = 0.76 [mm]) of grooves are designed by the phenomenological theory [9]. The

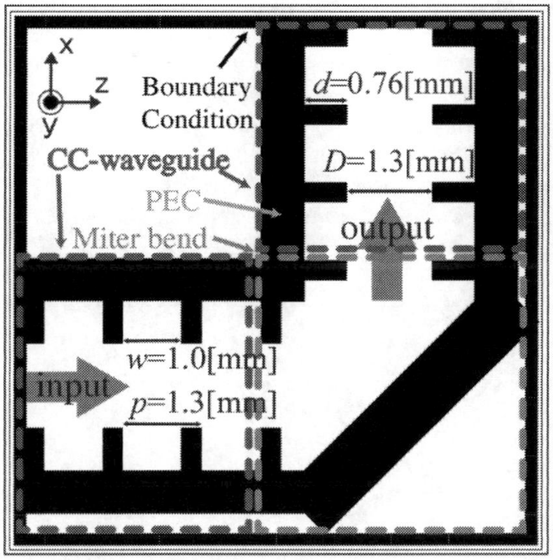


**Figure 1.** Schematic picture of the miter bend.

| | |
|---|---|
| inner diameter ($D$) | 31.75 [mm] |
| boundary condition | Mur's first-order absorbing boundary condition |
| input wave mode | $HE_{11}$ mode |
| frequency ($f$) | 84.0 [GHz] |
| guide wavelength ($\lambda_g$) | 3.58 [mm] |
| time step ($\Delta t$) | $5.7 \times 10^{-11}$ [sec] |
| period ($T$) | $1.12 \times 10^{-11}$ [sec] |
| material | Perfect electric conductor (PEC) |
| domain size ($x$,$y$,$z$) | (50, 34, 50) [mm] |
| spatial mesh ($\Delta x$,$\Delta y$,$\Delta z$) | (0.05, 0.152, 0.05) [mm] |
| number of mesh ($n_x$,$n_y$,$n_z$) | (1000, 224, 1000) |

**Table 1.** The common parameters in FDTD simulation of the CC-waveguide and the miter bend.

CC-waveguideand and the miter bend is connected as shown in Fig. 1. The majority of loss in the transmission line originates from the miter bend [10]. The mode-conversion is caused through the miter bend [11]. The mode-conversion is a cause of the losses. The losses on the transmission lines have been roughly estimated [12]. In the present paper, we have analyzed EM waves propagating in the waveguide system by developing original visualization systems.

## 2. Simulation Method

Fig. 1 illustrates the cross-sectional view in the xz plane at y = 0 of the three-dimensional FDTD geometry of the CC-waveguide and the miter bend. Input EM waves with $HE_{11}$ mode are injected from $z_{in} = 0.2$ [mm]. The CC-waveguide and the miter bend (which are made of aluminum) are approximated as the perfect electric conductor (PEC). The whole system is surrounded by the simulation box with Mur's first order absorbing boundary condition [13] as shown by black lines in Fig. 1. EM waves with $HE_{11}$ mode are injected from the CC-waveguide as a red thick arrow in Fig. 1. The $HE_{11}$ mode in the CC-waveguide has advantages of low transmission loss [9], the $HE_{11}$ mode has been adopted as highly reliable transmission mode of EM waves. The parameters for the FDTD simulation are summarized in Table 1.

## 3. Visualization system

A new interactive GUI system which is composed of the virtual reality (VR) system and android tablet have been developed. The user can analyze EM waves intuitively by moving android tablet. EM waves which propagate the waveguide system are visualized. The user can examine the cross-sectional view of EM waves at any angle by moving the tablet, as shown in Fig. 2. The perspective in the VR space is operated by touching and swiping the tablet display. The tablet tilt is obtained by their gyro sensor. The tablet tilt is sent to the visualization system using

TCP in Wi-Fi. The tablet position is obtained by Kinect. The Kinect is a motion sensing input device by Microsoft for the Xbox 360 video game console. The Kinect can obtain the depth and position of the human skeleton using the RGB camera and depth sensor. The user can operate the tablet as a plate that cut out a cross-sectional view of EM waves without using a keyboard or mouse user. The schematic diagram of the visualization system is shown in Fig. 3. The visualization system can visualize the three-dimensional EM waves by volume rendering and the surface current of the metal. The intensity and direction of current density ($\boldsymbol{J} = \boldsymbol{n} \times \boldsymbol{H}$) on the metal surface is expressed by colors and the arrowhead shape, respectively. The parameters $\boldsymbol{H} = (H_x, H_y, H_z)$ and $\boldsymbol{n}$ denote a magnetic field and the normal vector of the metal surface, respectively. Thereby, the eddy current is confirmed.

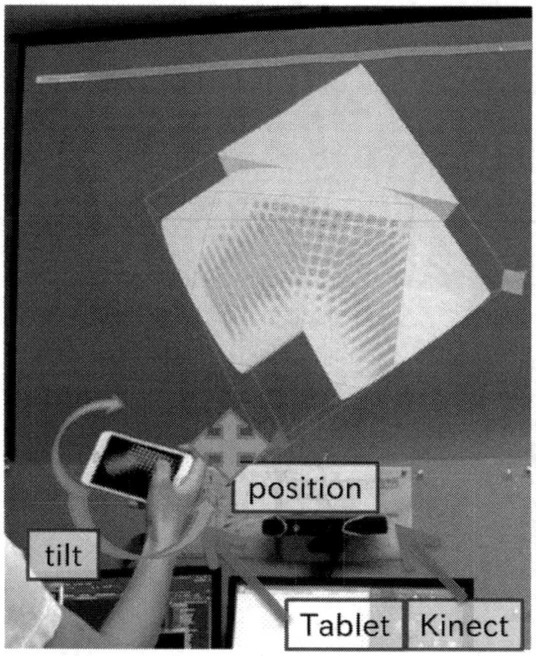


**Figure 2.** Interactive visualization system using tablet.

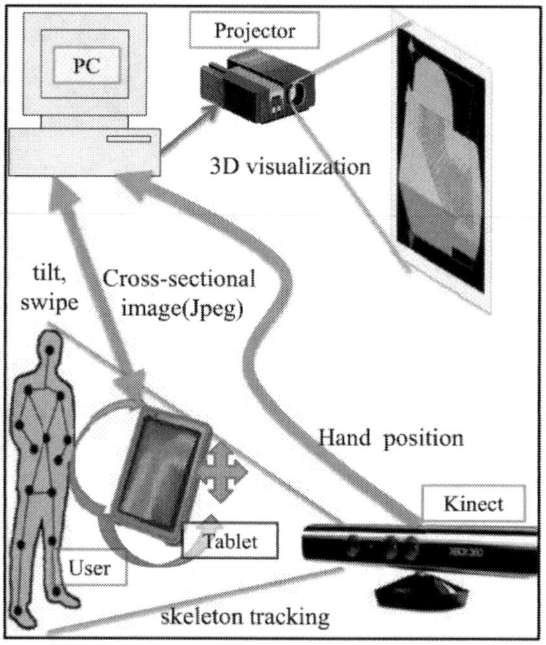


**Figure 3.** The schematic diagram of the visualization system.

## 4. Results and Discussion

The distribution of EM waves has been simulated in the waveguide and miterbend. Surface current on the metal with grooves and without grooves have been compared using visualization system. Fig. 4 shows surface current in the waveguide system without grooves (a) and with grooves (b). The direction of current density is eddying in the cylindrical waveguide. The intensity of current density in the CC-waveguide is smaller than the cylindrical waveguide. The direction of current density in the CC-waveguide is perpendicular to the groove. Therefore, grooves have been confirmed to suppress surface currents by suppress eddy current. The direction of current density is eddying in the miterbend without grooves, too. The intensity of current density in the miterbend with grooves is smaller than the miterbend without grooves. The direction of current density on the surface that grooves intersect is eddying. Therefore, grooves have been confirmed to suppress surface currents by suppress eddy current, although eddy current is generated slightly on the surface that grooves intersect. It is necessary to consider the processing of the groove in order to suppress eddy current on the surface that grooves intersect.

**Figure 4.** Current density of the metal surface in the waveguide system.

## 5. Futrue Works

EM waves may have been reflected by Mur's first-order absorbing boundary condition. It is necessary to use the perfectly matched layer (PML) absorbing boundary condition [16]. The miterbend that are processed various groove are simulated. The purity of the $HE_{11}$ mode is evaluated quantitatively by the processing of the groove.

## Acknowledgment

This work was supported by JSPS KAKENHI Grant Number 24656560.

## References

[1] Kubo S *et al.* 2005 *Plasma Phys. Contr. Fusion* **47** A81 A90
[2] Doane J L and Olstad R A 2007 *Fusion Sci. Tech.* **53** 39
[3] Yee K S 1966 *IEEE Trans. Antenn. Propag.* **14** 302 307
[4] Nakamura H, Sato T, Kambe H, Sawada K and Saiki T 2000 *J. Microsc.* **202** 50 52 (*Preprint* 0010012)
[5] Nakamura H, Saiki T, Kambe H and Sawada K 2001 *Comput. Phys. Comm.* **142** 464 467
[6] Nakamura H, Kashima N, Takayama A, Sawada K, Tamura Y, Fujiwara S and Kubo S 2013 *J. Phys. Conf.* **410** 012046
[7] Kashima N, Nakamura H, Takayama A, Tamura Y and Kubo S in press *Jpn. J. Appl. Phys.*
[8] Fujiwara M *et al.* 1996 *J. Fusion Energ.* **15** 7 153
[9] Ohkubo K, Kubo S, Iwase M, Idei H, Sato M, Shimozuma T, Takita Y and Kuroda T 1994 *Int. J. Infrared. Millimet. Waves* **15** 1507 1519
[10] Doane J L 2008 *Fusion Sci. Tech.* **53** 159 173
[11] Ohkubo K, Kubo S, Idei H, Sato M, Shimozuma T and Takita Y 1997 *Int. J. Infrared. Millimet. Waves* **18** 23 41
[12] Shapiro M A, Kowalski E J, Sirigiri J R, Tax D S, Temkin R J, Bigelow T S, Caughman J B and Rasmussen D A 2010 *Fusion Sci. Tech.* **57** 196
[13] G M 1981 *IEEE Trans. Electromagn. C.* **EMC-23** 377 382
[14] Doane J L (ed) 1985 *Infrared and Millimeter Waves* vol 13 (FL: Academic Press, Orlando)
[15] Moeller C, Prater R, Callis R, Remsen D, Doanne J, Caxy W, Phelps R and Tupper M (eds) 1990 *2 MW 110 GHz ECH heating system for DIII-D* vol 2
[16] Berenger J 1994 *J. Comput. Phys.* **114** 185 200

# Modelling of word usage frequency dynamics using artificial neural network

**Yu.S. Maslennikova, V.V. Bochkarev, D.S. Voloskov**

Kazan Federal University, Kremlevskaya str.18, Kazan 420018, Russia

E-mail: JSMaslennikova@kpfu.ru, vbochkarev@mail.ru

**Abstract**. In this paper the method for modelling of word usage frequency time series is proposed. An artificial feedforward neural network was used to predict word usage frequencies. The neural network was trained using the maximum likelihood criterion. The Google Books Ngram corpus was used for the analysis. This database provides a large amount of data on frequency of specific word forms for 7 languages. Statistical modelling of word usage frequency time series allows finding optimal fitting and filtering algorithm for subsequent lexicographic analysis and verification of frequency trend models.

## 1. Introduction

The word set of a language is a complex dynamical system in which words can be originated, evolve, and die out. Quantitative studies of natural languages have brought to significant progress in the understanding of word statistics [1] and language evolution [2]. A comparatively less explored question concerns the dynamics of word usage. Some representative examples include the study of distributions of n-grams in books written over the past 200 years [3], identifying trends in word frequency dynamics [4]. The model in [4] research is illustrated using a 108-word database from an online discussion group and a 1011- word collection of digitized books. The model revealed a strong relation between changes in word dissemination and changes in frequency.

The motivation for this research is based on the following objectives. In the paper [2] great attention is devoted to the impact analysis of various historical and cultural events on a word usage dynamics. Word frequency variations take place both due to external and internal factors. To make the impact analysis of external factors more justified, various prediction models can be used. Prediction models obtained in equable conditions allow to separate variations produced by regular internal reasons.

Trend analysis of the word usage dynamics (i.e. predictable components) allow to create smoothing methods like Kalman type filter. It is particularly important for filtering of usage frequency fluctuations of rare words.

high-quality prediction model for word usage frequencies could be used not only for direct forecasting, but also to back-forecasting. Similar estimates of word usage frequencies for previous periods permit to analyze mutual impact of different languages or their origins. It is important task for linguistic genetics (languages history).

Thanks to comprehensive Google Books digital library and The Google Books Ngram corpus (*http://books.google.com/ngrams/*), new possibilities for the quantitative analysis are available. This database doesn`t allow a morphological analysis, but frequency dictionaries of word-forms are available [5]. Statistical modelling of word usage frequency time series allows finding optimal fitting and filtering algorithm for subsequent lexicographic analysis and verification of frequency trend models.

For many years the field of time series forecasting has been largely analyzed by various statistical approach. Over the past 20 years in the field of nonlinear dynamical system an artificial neural networks (ANN) theory has gained more importance. In comparison with commonly used linear regression algorithms such as ARX, ARIMA this approach has several advantages. According to the Universal approximation theorem the standard multilayer feed-forward network with a single hidden layer, which contains finite number of hidden neurons, is a universal approximator among continuous functions on compact subsets of $Rn$, under mild assumptions on the activation function. [6].

The multilayer feedforward network can be trained for function approximation (nonlinear regression) or pattern recognition. The process of training a neural network involves tuning the values of the weights and biases of the network to optimize network performance, as defined by the network performance function. The most commonly used performance function for feedforward networks is mean square error (MSE) - the average squared error between the network outputs and the target outputs [7]. However, in cases when error distribution function differs from the normal distribution, MSE performance function is not optimal. In this paper a generalized approach for a neural network training based on the maximum-likelihood method is proposed.

## 2. Statistical model of word usage frequency dynamics

Frequency dictionaries only for accessory words were analyzed. Frequency dictionaries of The Google Books Ngram corpus are available for the time period of 1700-2008 year. For this research only 1900-2008 years period was used because it contains the most reliable data. Statistical model was created for infrequent accessory words such as *'during', 'another', 'against', 'down', 'still', 'too', 'per', 'since', 'never', 'back', 'thus', 'take', 'less', 'himself', 'again', 'few', 'among', 'though', 'last'* and other. Examples of the variation of word frequency in this dataset are shown in Fig. 1.

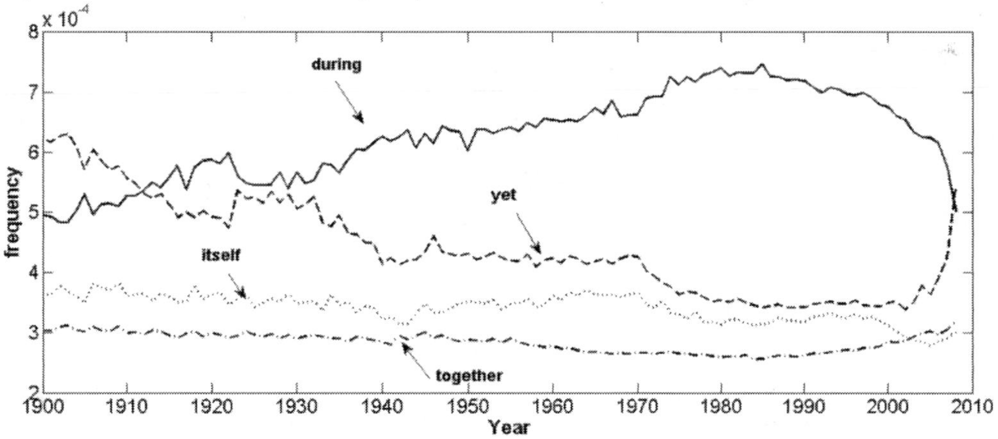


**Figure 1.** Frequency dynamics for example accessory words. The frequency of a word is computed as the number of occurrences of the word relative to the total number of words in a year.

As it is seen from Fig. 1, word frequency time series have various dynamics for different accessory words. That's why for statistical modelling of these dynamics it is necessary to use a neural network with a large amount of hidden layers and neurons. Training of such complicated network could be sufficiently problematical because of the overfitting problem. In this research two-step model is proposed. As a first step, all accessory words are divided into a set of clusters based on analysis of dynamics similarity. As a second step, neural network prediction models are optimized for each cluster separately.

### 2.1. Word usage frequency data clustering

The approach based on AR-coefficient analysis was used for the word usage frequency data clustering. In an *AR* model of order *p*, the current output is a linear combination of the past *p* outputs plus a white noise input. The weights on the *p* past outputs minimize the mean-square prediction error of the autoregression. If *y[n]* is the current value of the output and *x[n]* is a zero-mean white noise input, the *AR(p)* model is [8]:

$$\sum_{k=0}^{p} a[k]y[n-k] = x[n].$$

These *a[k]* coefficients could be used as a input vector for the clustering. The order *p* of AR-model was chosen *p=5*. Self-organizing map neural network (SOM-net) with 2×2 neuron configuration was used for clustering of word usage frequency data. SOM-nets learn to classify input vectors according to how they are grouped in the input space. They differ from competitive layers in that neighboring neurons in the self-organizing map learn to recognize neighboring sections of the input space. Thus, self-organizing maps learn both the distribution (as do competitive layers) and topology of the input vectors they are trained on. A hexagonal grid was used to map the neurons. SOM-net wraining was performed using the Kohonen's self-training "winner-gets-all" algorithm and produced a map in which nodes spatially coincide with major accumulations of vectors in the initial attribute space and similar objects are located in neighbouring nodes [9]. As a result, all accessory words were grouped into 4 clusters. Typical trends for ascertained clusters are proposed in Fig. 2.

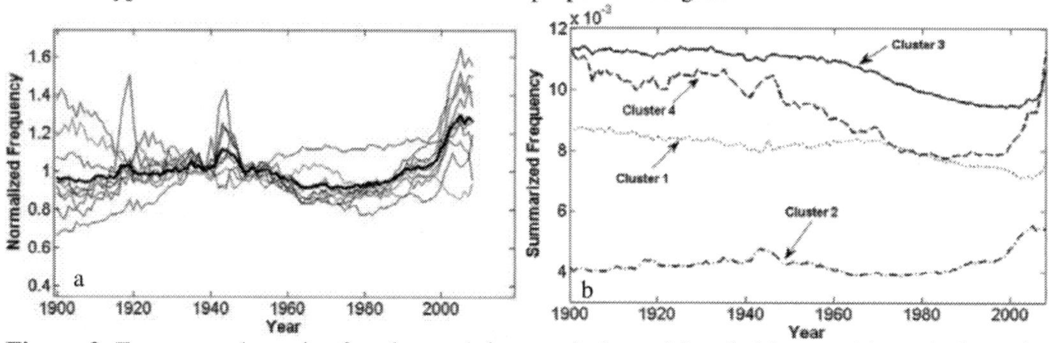


**Figure 2.** Frequency dynamics for clustered data: typical trend is a bold curve (a); typical trends of founded clusters (b).

### 2.2. Prediction model for clustered data

As it was shown early, MSE performance function for training of a neural network could be non-optimal. Generalized approach is based on the maximum-likelihood method. For infrequent accessory words error distribution function is close to the Poisson distribution. The Poisson distribution is most often invoked for rare events. As opposed to the binomial distribution the Poisson rate may actually be any number. The real contrast is that the Poisson distribution is asymmetrical: given a rate *r = 3*, the range of variation ends with zero on one side, but is unlimited on the other side. The Poisson distribution, as a data set or as the corresponding curve, is always skewed toward the right, but it is inhibited by the zero occurrence barrier on the left. The degree of skew diminishes as *r* becomes larger, and at some point the Poisson distribution becomes to the eye about as symmetrical as the normal distribution [8].

Let's define λ- parameter as a function of previous data - $\lambda_i = f(\boldsymbol{X}_i)$, then the expected value of a Poisson-distributed random variable is $Y_i \sim P(f(\boldsymbol{X}_i))$. So, the likelihood function could be defined as following:

$$l(Y,\theta) = \sum_i Y_i \log(f(\boldsymbol{X}_i)) - \sum_i f(\boldsymbol{X}_i) - \sum_i Y_i!$$

During the network training it is possible to maximize the likelihood function. For training multilayer feedforward networks, any standard numerical optimization algorithm can be used to optimize the performance function. In our case the performance function is negative likelihood function $-l(Y,\theta)$. The most commonly used backpropagation algorithm modifies the weights of a neural network in order to find a global minimum of the error function [6]. For optimizing of the performance function the Levenberg-Marquardt algorithm was used during the training of neural networks. Feedforward neural network for each clusters contained one input layer of 10 neurons and one hidden layer of 7 neurons. Examples of predicted dynamics *'against', 'down', 'during'* and *'back'* words using neural networks with MLE optimization are proposed in Fig 3 with results of linear regression neural networks (LR) and neural networks with MSE performance function. LR neural networks showed STD = $1.01\cdot10^{-5}$ (a standard deviation). The conventional neural network (with MSE performance function) showed STD = $0.90\cdot10^{-5}$, neural networks with MLE performance function showed STD = $0.46\cdot10^{-5}$.

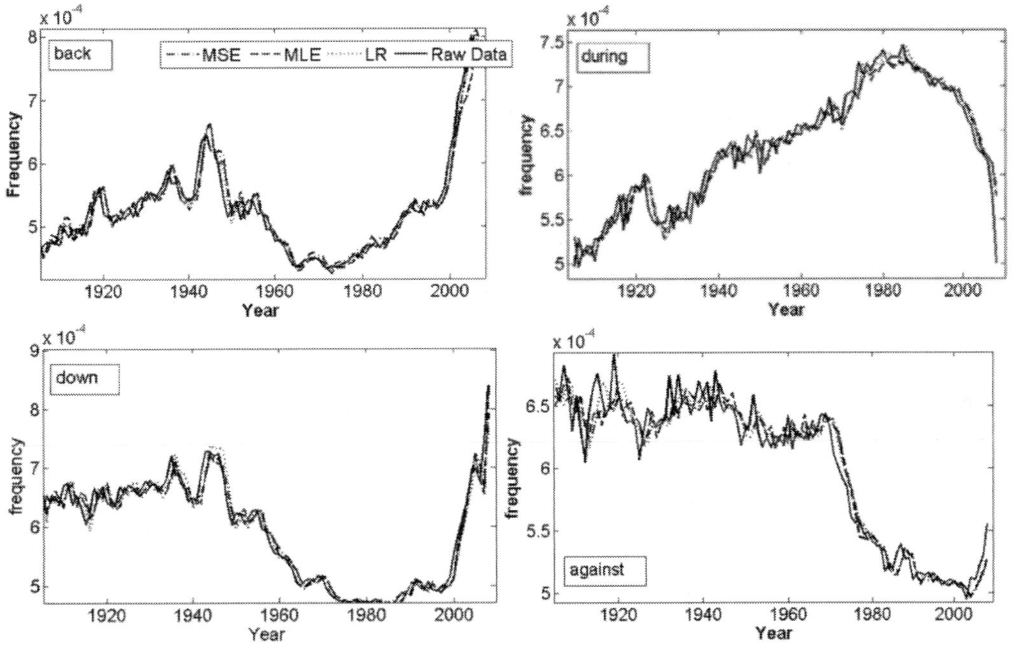


**Figure 3.** Predicted word usage frequency dynamics for 4 accessory words: by optimizing of MLE performance function, MSE performance function and Linear Regression model (LR).

It was shown that proposed MLE training algorithm for a neural networks is more efficient than optimization of MSE performance function for infrequent accessory words. When word usage frequency is high then the Poisson distribution is near to the Normal distribution, and the results of both methods become conformable.

This work was supported by the Russian Foundation for Basic Research (grant №12-06-00404a).

## References

[1] Baayen, R.H.: Word Frequency Distributions. Springer, Berlin (2002)

[2] Gell-Mann, M., Ruhlen, M.: The origin and evolution of word order. Proc. Natl. Acad. Sci. 108, 17290–17295 (2011)

[3] Michel, J.-B., et al.: Quantitative analysis of culture using millions of digitized books. Science 331,176–182 (2010)

[4] Eduardo G. Altmann · Zakary L. Whichard Adilson E. Motter Identifying Trends in Word Frequency Dynamics J Stat Phys (2013) 151:277–288
[5] Michel J., Shen Y., Aiden A., Veres A., Gray M., The Google Books Team, Pickett J., Hoiberg D., Clancy D., Norvig P., Orwang J., Pinker S., Nowak M., Aiden E. Quantitative Analysis of Culture Using Millions of Digitized Books. Science. 331. no. 6014 . 2011. p. 176-182
[6] Maslennikova, Y. & Bochkarev. Training Algorithm for Neuro-Fuzzy Network Based on Singular Spectrum Analysis, AWERProcedia Information Technology & Computer Science. [Online]. 2013, 3, pp 605-610.
[7] Tetko, I.V.; Livingstone, D.J.; Luik, A.I. Neural network studies. 1. Comparison of Overfitting and Overtraining, J. Chem. Inf. Comput. Sci., 1995, 35, 826-833
[8] Monson,H. Statistical Digital Signal Processing and Modeling, John Wiley & Sons, 1996.
[9] Kohonen, T. 2001. Self-Organizing Maps. Third, extended edition. Springer, Berlin.

# Hierarchical Bass model

**Tohru Tashiro**

E-mail: tashiro@cosmos.phys.ocha.ac.jp
Department of Physics, Ochanomizu University, 2-1-1 Ohtsuka, Bunkyo, Tokyo 112-8610, Japan

**Abstract.** We propose a new model about diffusion of a product which includes a memory of how many *adopters* or advertisements a *non-adopter* met, where (non-)adopters mean people (not) possessing the product. This effect is lacking in the Bass model. As an application, we utilize the model to fit the iPod sales data, and so the better agreement is obtained than the Bass model.

## 1. Introduction

It is so interesting how fashion diffuses in a society. It has been fascinating scientists and scholars how a new product or innovation spreads in a society. "No other field of behavior science research represents more effort by more scholars in more disciplines in more nations." [1] We shall employ a penetration rate of a product in studying a fashion diffusion scientifically because it is an objective and available statistical data. Generally, a change of the penetration rate in time is slow at first, and then, it peaks, followed again by slow change. The logistic model (LM) can describe this change in time. In Ref. [2], we made it clear that the communication between individuals is imitation; people who do not possess (non-adopters) purchase the product quickly after they met people who already possess it (adopters). However, the communication is unnatural, because there are not only trend-conscious people, but also cautious people in our society. We considered that a memory of how many adopters a non-adopter met is essential for him/her to decide to purchase the product. Accordingly, we proposed a model including this effect, which is the *hierarchical logistic model* (HLM) [2].

Another essential element for fashion diffusion is an advertisement. The Bass model (BM) [3] includes the element, which is represented by

$$\frac{\mathrm{d}P(t)}{\mathrm{d}t} = \left\{b + \frac{a}{N}P(t)\right\}\{N - P(t)\} \ , \tag{1}$$

where $P(t)$ is the number of those possessing the product at $t$ and $N$ is the total population. This differential equation coincides with LM by setting $b = 0$. Indeed, this model does not take account of the memory we mentioned above. So, in this study, we shall propose a model including the memory, which is the *hierarchical Bass model* (HBM).

## 2. Bass model

In this section, we derive BM in the same way as Ref. [2].

Let us put $N$ random walkers on $n \times n$ ($> N$) lattices. They move to one of the nearest lattices with the same provability at each discretized time step. Furthermore, we shall apply the following rules for the random walkers: i) If non-adopters meet adopters, they start to adopt a product quickly; ii) There are no interactions among (non-)adopters; iii) Adopters do not part with it. Here, we define sharing a lattice with someone as meeting the person. By these rules, we can derive LM [2]. In order to derive BM, we need to consider another influence on non-adopters from other than adopters, that is an advertisement like a billboard. We assume that the number of advertisements does not change in time. In addition, we apply the forth rule: iv) Non-adopters who met advertisements start to adopt the product quickly. As a final assumption, we set the density of people to be so low that three or more people cannot share a lattice at the same step, as if three or more body collisions are ignored in a rarefied gas. Therefore, it is natural to consider that a probability of meeting adopters, non-adopters, or advertisements is proportional to each number of them.

Let us set the number of adopters and non-adopters at the $i$th step as $P_i$ and $Q_i$, respectively. Indeed, $P_i + Q_i = N$. Moreover, we set the number of advertisements $B$ ($< n^2$). Then, the probability of meeting adopters, non-adopters and advertisements can be represented by $P_i/n^2$, $Q_i/n^2$ and $B/n^2$, respectively. According to the first and the forth rule, $(B/n^2)\times Q_i+(P_i/n^2)\times Q_i$ people of non-adopters becomes adopters at the next step, and then, the following recursion formulae can be obtained.

$$P_{i+1} = P_i + \frac{B+P_i}{n^2}Q_i \, , \; Q_{i+1} = Q_i - \frac{B+P_i}{n^2}Q_i \, . \tag{2}$$

We shall define the number of them at $t$ as $P(t) = P(i\cdot\Delta t) \equiv P_i$ and $Q(t) = Q(i\cdot\Delta t) \equiv Q_i$. Here, we take the limits as $\Delta t \to 0$ and $n \to \infty$ with $n^2\Delta t$ fixed. By setting the fixed value as $\frac{a}{N} = \lim_{\substack{\Delta t\to 0\\ n\to\infty}} \frac{1}{n^2\Delta t}$, defining $Ba/N$ as $\lim_{\substack{\Delta t\to 0\\ n\to\infty}} \frac{B}{n^2\Delta t} = \frac{Ba}{N} \equiv b$, and using $P(t) + Q(t) = N$, the following differential equation is derived:

$$\frac{\mathrm{d}P(t)}{\mathrm{d}t} = \left\{b + \frac{a}{N}P(t)\right\}\{N - P(t)\}\, , \tag{3}$$

which describes BM. If we use the penetration rate $p(t) \equiv P(t)/N$, the above differential equation becomes

$$\frac{\mathrm{d}p(t)}{\mathrm{d}t} = \{b + ap(t)\}\{1 - p(t)\}\ . \tag{4}$$

## 3. Hierarchical Bass model

In the previous section, it is unveiled that BM reflects quick influences on non-adopters from adopters and advertisements. However, it is unnatural because all the people in our society are not easily influenced by others and advertisements. We consider that a memory of how many adopters or advertisements a non-adopter met is important for his/her decision to purchase the product. Therefore, as in Ref. [2], we shall include the memory in BM by the following way: We set the number of non-adopters starting to adopt the product after they meet $\mu$ adopters and advertisements at $i$th step as $Q_i^\mu$, in which we call $\mu$ as *remaining adopters and advertisements number* (RAAN). Indeed, if a non-adopter, whose RAAN is $\mu$, meet one of adopters or advertisements, his/her RAAN becomes $\mu - 1$ at the next step. We do not alter other rules.

If the maximum of RAAN is $m$, the recursion formulae change into

$$P_{i+1} = P_i + \frac{B+P_i}{n^2}Q_i^1 \, , \; Q_{i+1}^1 = Q_i^1 - \frac{B+P_i}{n^2}Q_i^1 + \frac{B+P_i}{n^2}Q_i^2 \, , \; \dots \, , \; Q_{i+1}^m = Q_i^m - \frac{B+P_i}{n^2}Q_i^m \, . \tag{5}$$

These become the following differential equations by the previous continuation of space and time.

$$\frac{\mathrm{d}P(t)}{\mathrm{d}t} = \left\{b + \frac{a}{N}P(t)\right\} Q^1(t) \ , \ \frac{\mathrm{d}Q^1(t)}{\mathrm{d}t} = \left\{b + \frac{a}{N}P(t)\right\} \left\{Q^2(t) - Q^1(t)\right\} \ ,$$
$$\ldots \ , \ \frac{\mathrm{d}Q^m(t)}{\mathrm{d}t} = -\left\{b + \frac{a}{N}P(t)\right\} Q^m(t) \tag{6}$$

where $Q^\mu(t) = Q^\mu(i \cdot \Delta t) \equiv Q^\mu_i$. We shall name this the *hierarchical Bass model* (HBM), because this model has the hierarchical structure in non-adopters.

If we use ratios of adopters and non-adopters to the total number $N$, HBM becomes

$$\frac{\mathrm{d}p(t)}{\mathrm{d}t} = \left\{b + ap(t)\right\} q^1(t) \ , \ \frac{\mathrm{d}q^1(t)}{\mathrm{d}t} = \left\{b + ap(t)\right\} \left\{q^2(t) - q^1(t)\right\} \ ,$$
$$\ldots \ , \ \frac{\mathrm{d}q^m(t)}{\mathrm{d}t} = -\left\{b + ap(t)\right\} q^m(t) \tag{7}$$

where $q^\mu(t) \equiv Q^\mu(t)/N$.

## 4. Fitting iPod sales data

Let us apply HBM to a sales data of a product in this section. For it, we shall employ the iPod sales as in Ref. [2]. We can obtain the data from the official homepage of Apple Inc. which has been creating and marketing the iPod: http://www.apple.com/. We shall utilize the data from November 2001 to May 2006, for the interval includes only one peak of sales.

Setting November 2001 as the origin of the time axis, we construct the cumulative sales, and then, we fit the sales data by HBM, BM and LM. In doing so, if we minimize the residual sum of squares or the sum of the absolute value of error (SAE), the fit between the model and the data is good for high values of sales but not for low ones. On the other hand, if we minimize the sum of the absolute value of relative error (SARE), the fit for low values becomes better. We consider that this is because the maximum of iPod sales data is 3 digits different than the minimum. Thus, in order to resolve the problem, we shall minimize a product of SAE and SARE. The parameters minimizing this product are show in Tab. 1, together with ones of LM.

From this table, we can see that SARE diminish with increase of $m$: SARE with $m = 4$ reduces to nearly half that with $m = 1$. In other words, the average relative error for HBM with $m = 4$ is about as half as that for BM. Results of fitting the data by HLM are also shown in Tab. 1 by numbers in parentheses. From these, we can say that HBM considering the effect of advertisements can approximate the sales data better than HLM. However, it is not a reason that there is an extra fit parameter $b$ in HBM, because the degrees of agreement of LM and BM are comparable.

From the fit parameters of HBM with $m = 4$, we can guess the following facts: The market size producing the first peak of iPod sales is about 66 million people, the ratio of the trend-conscious people, $q^1(0)$, is about 29%, the ratio of the cautious people, $q^2(0)$, is about 64% and the ratio of more cautious people, $q^3(0) + q^4(0)$, is about 7%.

In order to compare the degrees of agreement of BM and HBM with $m = 4$, we plot the sales data with them in Fig. 1. The circles are the cumulative iPod sales and the (purple) dashed and the (light blue) full curves represent BM and HBM with $m = 4$, respectively. HBM can approximate sales data which BM cannot do.

**Table 1.** Parameters of HBM and LM minimizing the product of SARE and SAE, and SAREs by them. Numbers in parentheses mean results of fitting the data by HLM. All values are rounded to a three-digit number, and so the sum of $p(0)$ and $p^{\mu}(0)$ is not equal to one.

| | LM | $m = 1$ (BM) | $m = 2$ | $m = 3$ | $m = 4$ |
|---|---|---|---|---|---|
| $a$ | 1.42 | 1.42 | 4.22 | 4.25 | 4.50 |
| $b$ | – | $2.19 \times 10^{-10}$ | $7.16 \times 10^{-4}$ | $6.93 \times 10^{-4}$ | $7.10 \times 10^{-4}$ |
| $N$ [$\times 10^6$] | $1.28 \times 10^5$ | $1.07 \times 10^5$ | 65.7 | 65.6 | 66.4 |
| $p(0)$ | $9.26 \times 10^{-7}$ | $1.11 \times 10^{-6}$ | 0.00189 | 0.00193 | 0.00188 |
| $q^1(0)$ | 1.00 | 1.00 | 0.309 | 0.305 | 0.290 |
| $q^2(0)$ | – | – | 0.689 | 0.686 | 0.640 |
| $q^3(0)$ | – | – | – | 0.00739 | 0.00800 |
| $q^4(0)$ | – | – | – | – | 0.0600 |
| SARE | 1.84 | 1.84 | 0.999 (1.12) | 0.989 (1.02) | 0.984 (1.02) |

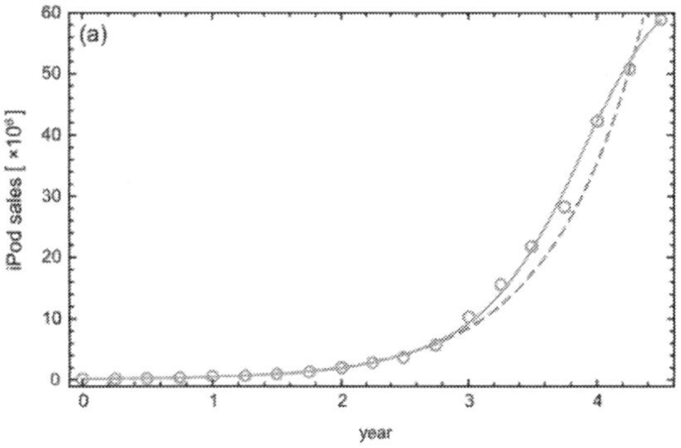


**Figure 1.** (color online) Cumulative iPod sales data represented by circles and fitting curves. The (purple) dashed and the (light blue) full curves are BM and HMB with $m = 4$, respectively.

## 5. Concluding remarks

In this paper, we made it clear that BM is based on imitation among people and influence from advertisements, and proposed the new model, HBM, including a memory of how many adopters or advertisements a non-adopter met which BM lacks. Additionally, we utilized the model to fit the iPod sales data, and so the better agreement was obtained than BM. The iPod sales increased gradually until they peaked for the first time in November 2005. HBM is far more suitable for describing such a slow pace than BM, because all the non-adopters in our society did not purchase the iPod quickly after they met adopters or advertisements.

### *5.1. Acknowledgments*

We would like to thank Suzuko Kawakami and members of astrophysics laboratory at Ochanomizu University for extensive discussions. This work was supported by JSPS Grant-in-Aid for Exploratory Research 12849926.

## References

[1] Rogers E M 2003 *Diffusion of Innovations, 5th edition* (New York: Free Press)
[2] Tashiro T, Minagawa H and Chiba M 2013 *J. Phys.: Conf. Ser.* **410** 012026
[3] Bass F M 1969 *Manage. Sci.* **15** 215

# Box-Cox transformation of firm size data in statistical analysis

**Ting Ting Chen**[1] **and Tetsuya Takaishi**[2]

[1]Faculty of Integrated Arts and Sciences, Hiroshima University, Higashi-Hiroshima 739-8521, Japan

[2]Hiroshima University of Economics, Hiroshima 731-0192, JAPAN

E-mail: [1]d102355@hiroshima-u.ac.jp

E-mail: [2]tt-taka@hue.ac.jp

**Abstract.** Firm size data usually do not show the normality that is often assumed in statistical analysis such as regression analysis. In this study we focus on two firm size data: the number of employees and sale. Those data deviate considerably from a normal distribution. To improve the normality of those data we transform them by the Box-Cox transformation with appropriate parameters. The Box-Cox transformation parameters are determined so that the transformed data best show the kurtosis of a normal distribution. It is found that the two firm size data transformed by the Box-Cox transformation show strong linearity. This indicates that the number of employees and sale have the similar property as a firm size indicator. The Box-Cox parameters obtained for the firm size data are found to be very close to zero. In this case the Box-Cox transformations are approximately a log-transformation. This suggests that the firm size data we used are approximately log-normal distributions.

## 1. Introduction

In social science we often use statistical techniques such as regression analysis to find relationships among data and to make predictions. In many cases to utilize statistical techniques normality of the data is assumed. In practice, however, the normality assumption is often violated. The violation of the normality causes certain difficulties[1, 2]. One way to overcome these difficulties is to transform variables to those having desired properties. In this study we use the Box-Cox power transformations[3] that also include a log-transformation as a special case.

Firm size data are important variables to find relationships among financial indicators. However it is well-known that firm size data are not normally distributed and often suggested to follow a log-normal distribution[4, 5, 6, 7, 8]. On the other hand there also exist studies that claim firm size distributions deviate from the log-normal distribution[9, 10].

In this study we use two definitions for firm size: the number of employees and sale. As analyzed later those firm size data are not normally distributed. To improve the normality we apply the Box-Cox transformation for those data. The Box-Cox transformation parameters are determined so that the kurtosis of the transformed data comes close the kurtosis of a normal distribution. The similar approach that uses the skewness has been taken to determine the optimum parameter for the Box-Cox transformation[11].

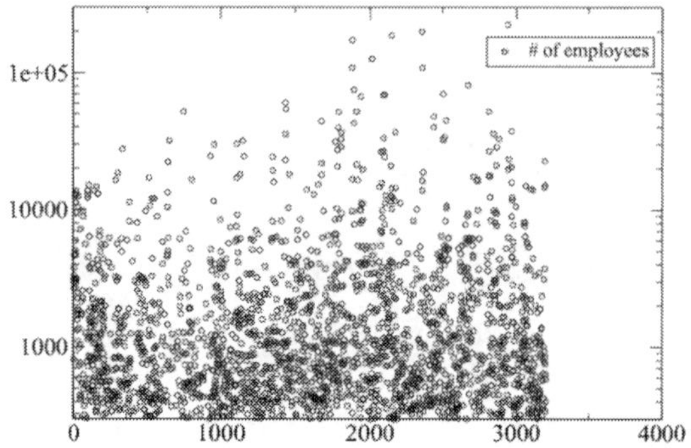


**Figure 1.** The number of employees for 3206 companies.

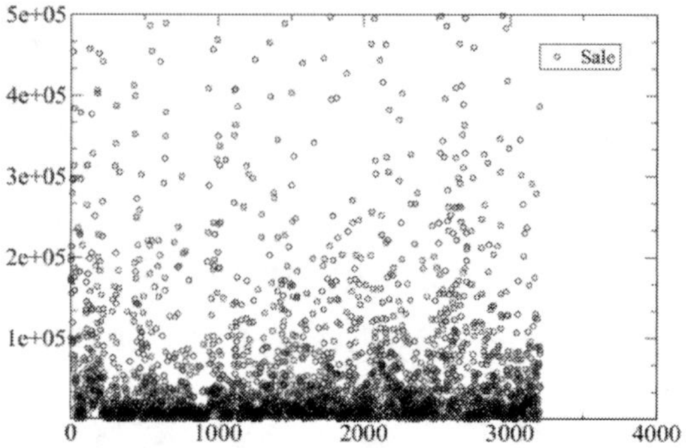


**Figure 2.** The same as in Fig.1 but for sale.

## 2. Box-Cox Transformation

Let $x_i$ be the i-th observations. The Box-Cox transformation of $x_i$ is given by

$$x(\lambda)_i = \begin{cases} \frac{(x_i+c)^\lambda - 1}{\lambda} & \text{if} \quad \lambda \neq 0 \\ \log(x_i + c) & \text{if} \quad \lambda = 0 \end{cases} \tag{1}$$

where $c$ is a constant to ensure that $x_i + c$ is positive. Since firm sizes are all positive we set $c = 0$.

## 3. Firm Size Data

We use two firm size data: the number of employees and sale that consist of 3206 companies traded on the Tokyo Stock Exchange in 2011. Figs 1 and 2 show the number of employees and sale for 3206 companies. It is seen that they are broadly distributed. Figs 3 and 4 show distributions of the two firm size data. It is noticed that they are not normally distributed.

## 4. Box-Cox Transformed Data

In order to find the optimum Box-Cox transformation parameter that best exhibits the kurtosis of a normal distribution we calculate kurtosis as a function of $\lambda$. Fig.5 shows the kurtosis as a function of $\lambda$. Here the kurtosis $\kappa$ is defined by $\kappa = \frac{E[(x-\mu)^4]}{Var(x)^2}$ where $\mu = E[x]$. As seen in Fig.5 it turned out that the kurtosis is always bigger than 3. Thus the optimum parameters of the number of employees and sale can be obtained at minimum points. Then we obtain the optimum parameters and the kurtosis at the optimum parameters $\kappa$ as $\lambda_c = 0.005$ and $\kappa = 3.69$ for the number of employees, and $\lambda_c = 0.022$ and $\kappa = 3.77$ for sale.

Figs 6 and 7 show distributions of the number of employees and sale after the Box-Cox transformation respectively. It is clearly seen that the distributions are transformed to more normal ones. The red lines in the figures are the fitting results to a normal distribution.

Next we investigate a relationship between the number of employees and sale. There is no obvious reason that they are closely related. In fact as seen in Fig.8 no clear relation is seen between the number of employees and sale. The correlation coefficient between them is calculated to be 0.662. On the other hand Fig.9 shows the relation between the number of employees and sale after the Box-Cox transformation which indicates that the number of employees and sale have a strong linearity. The correlation coefficient is calculated to be 0.873 that also supports

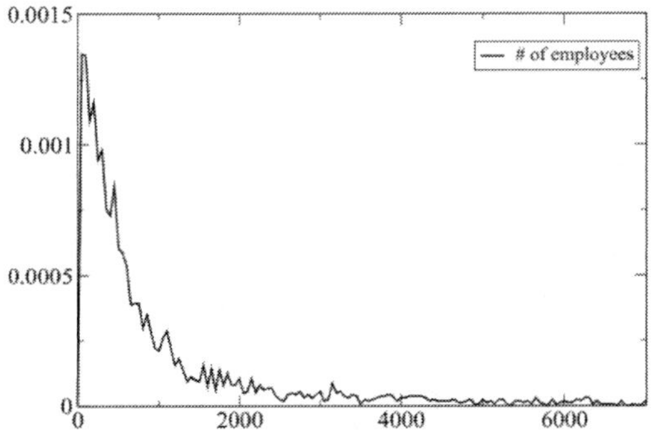


**Figure 3.** Distribution of the number of employees.

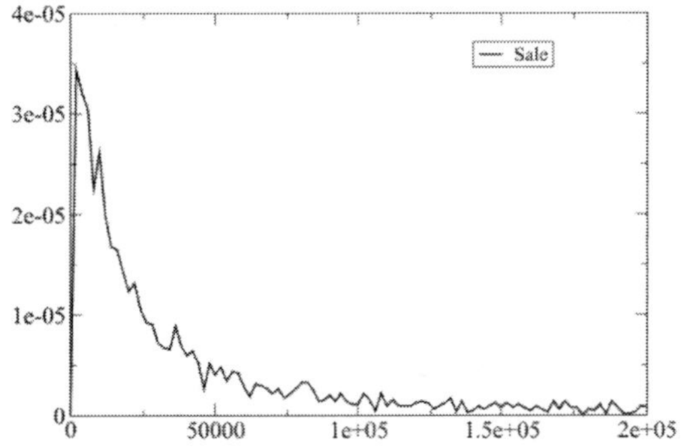


**Figure 4.** Distribution of sale.

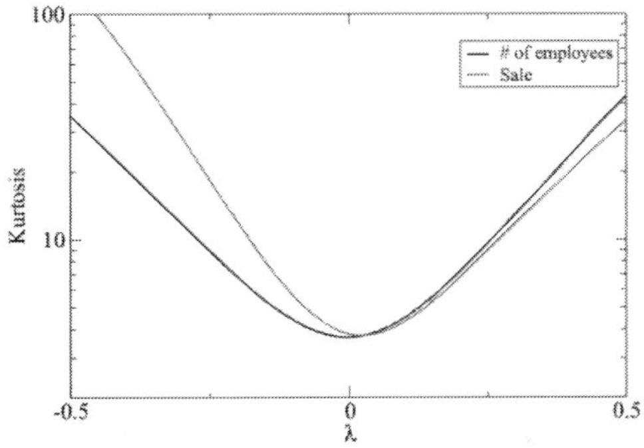


**Figure 5.** Kurtosis as a function the Box-Cox transformation parameter $\lambda$.

the strong linearity. This result may suggest that the number of employees and sale can be used as a firm size indicator that has the similar property.

## 5. Conclusion

In order to improve the normality of firm size data we have used the Box-Cox transformation. The optimum Box-Cox transformation parameters are determined so that the transformation brings the data to more normal ones. We find that the number of employees and sale after the Box-Cox transformation show the strong linear relation. This may indicate that the number of employees and sale has the similar property as a firm size indicator.

The optimum Box-Cox transformation parameters we obtained are very close to zero. Thus the Box-Cox transformations we performed are actually close to a log-transformation, which means that distributions of the number of employees and sale are close to a log-normal distribution. This is consistent with the previous literature which claims that the firm size distributions follow the log-normal distributions[4, 5, 6, 7, 8].

**Acknowledgement**

Numerical calculations in this work were carried out at the Yukawa Institute Computer Facility. This work was supported by JSPS KAKENHI Grant Number 25330047.

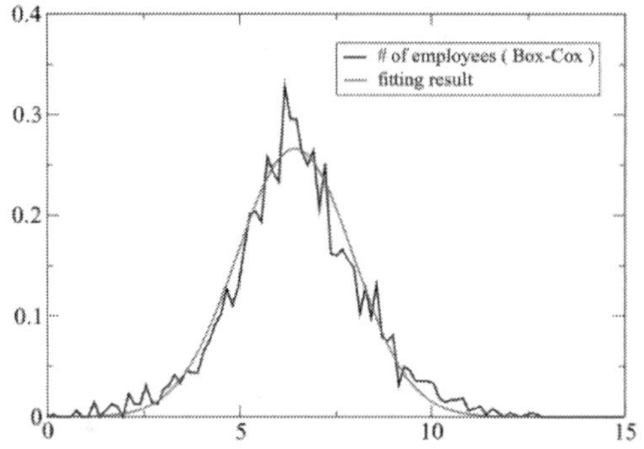


**Figure 6.** Distribution of the number of employees after the Box-Cox transformation with the optimum parameter. The red line is the fitting result to $\exp(-(x-\mu)^2/(2\sigma^2))/(2\pi\sigma^2)^{1/2}$ with $\sigma^2 = 2.24$ and $\mu = 6.42$.

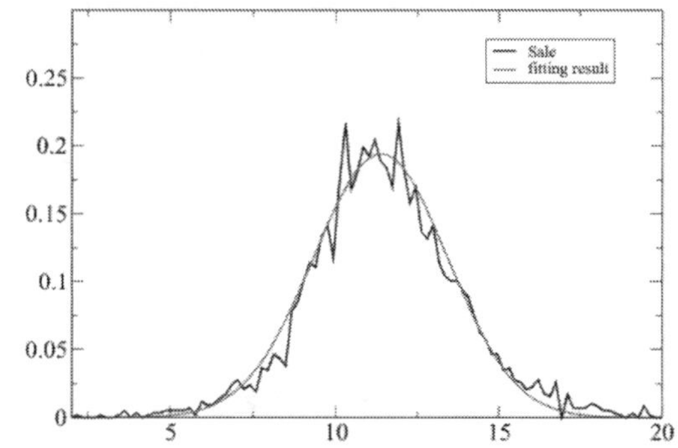


**Figure 7.** The same as in Fig.8 but for sale with $\sigma^2 = 4.22$ and $\mu = 11.37$.

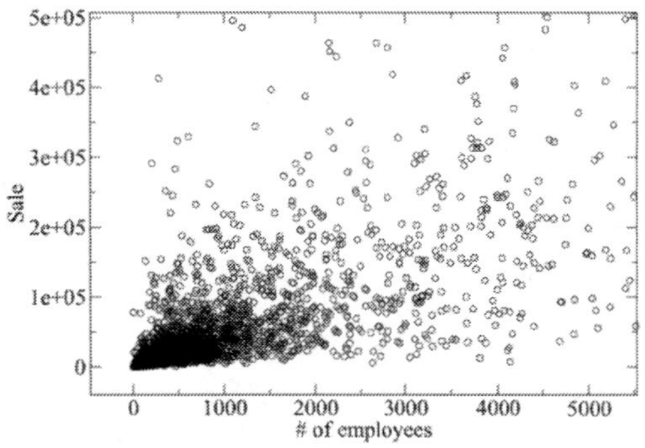


**Figure 8.** The number of employees versus sale.

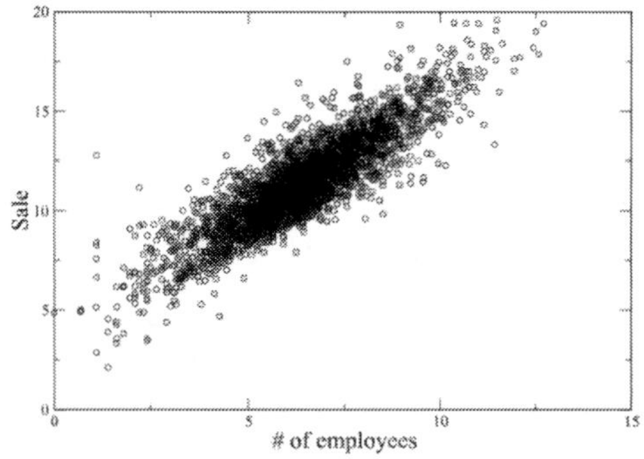


**Figure 9.** The number of employees versus sale, after the Box-Cox transformation with the optimum parameter.

## References


[1] Sakia R M 1992 *The Statistician* **41** 169-178

[2] Hossain M Z 2011 *Journal of Emerging Trends in Economics and Management Sciences* **2(1)** 32-39

[3] Box G E P and Cox D R 1964 *Journal of the Royal Statistical Society B* **26** 211-234

[4] Gibrat R 1931 *Les Inegalites Economiques*

[5] Kalecki M 1945 *Econometrica* **13** 161-170

[6] Hart P E and Paris S J 1956 *Economica (New Series)* **29** 29-39

[7] Quandt R E 1966 *American Economic Review* **56** 416-432

[8] Steindl J 1965 *Random Process and the Growth of Firms* New York, Hafner

[9] Cabral L M B and Mata J 2003 *American Economic Review* **93** 1075-1090

[10] Kaizoji T, Iyetomi H and Ikeda Y 2006 *Evolutionary and Institutional Economics Review* **2(2)** 183-198

[11] Osborne J W 2010 *Practical Assesment, Research & Evaluation* **15** No.12

# Are micro and macro arrows of time connected each other?

**Piero Chiarelli**[1]
National Council of Research of Italy, Area of Pisa, 56124 Pisa, Moruzzi 1, Italy

E-mail: pchiare@ifc.cnr.it

**Abstract.** The connection between the micro and macro-arrow of time is discussed in the frame of the stochastic quantum hydrodynamic analogy (SQHA). The presence of fluctuations that in the case on non-linear interactions leads to the breaking of the quantum mechanics on large scale and then to the macroscopic irreversibility with a defined arrow of time, gives also rise to the time reversal breaking in the micro-scale quantum evolution (micro-arrow of time). The quantum irreversibility with time reversal asymmetry is briefly discussed.

## 1. Introduction

The breaking of quantum coherence and the quantum to classical phase transition are problems that have many consequences in all problems of physics whose scale is larger than that one of small atoms.
Both the equations of quantum and the classical mechanics own the time-reversal symmetry. The arrow of time in the classical kinetics is commonly ascribed to the transition from the molecular scale to the macroscopic one united to the presence of a small initial incertitude (or to the external environment) in a system with a huge number of non-linearly interacting particles [1]. Even if the time reversal symmetry breaking has been observed for the probability transition in some elementary particle decay [2], the macroscopic arrow of time has no connection with the nature of the quantum mechanics. Recently the author has developed the stochastic version of the quantum hydrodynamic analogy (SQHA) [3] that comprehends both the standard quantum mechanics as the deterministic limit of the theory and the classical stochastic mechanics as the large-scale limit. Close to the deterministic limit the unified SQHA dynamics can lead to a marginal time reversal breaking (micro-arrow of time) as well as to irreversible kinetics in macroscopic systems of non-linearly interacting particles.

## 2. The SQHA equation of motion

When the noise is a stochastic function of the space, the particles density (PD) $n$ in the SQHA is described by the motion equation [3]

$$\partial_t n_{(q,t)} = -\nabla_q \cdot (n_{(q,t)} \dot{q}) + \eta_{(q,t,\Theta)} \tag{1}$$

where $\Theta$ is the amplitude of the spatially distributed noise $\eta$ whose variance reads

$$\lim_{\Theta \to 0} < \eta_{(q_\alpha,t)}, \eta_{(q_\beta+\lambda,t+\tau)} >= k\Theta \frac{\mu}{2\lambda_c^{\,2}} exp[-(\frac{\lambda}{\lambda_c})^2]\delta(\tau)\delta_{\alpha\beta} \, , \tag{2}$$

where the noise correlation distance $\lambda_c$ reads

$$\lim_{\Theta \to 0} \lambda_c = \pi \frac{\hbar}{(2mk\Theta)^{1/2}} \quad , \tag{3}$$

and where $\dot{q}$ in equation (1) is obtained by the solution of the differential equations

$$\dot{q} = \frac{p}{m}, \tag{4}$$

$$\dot{p} = -\nabla(V_{(q)} + V_{qu(\mathrm{n})}), \tag{5}$$

where $V_{(q)}$ represents the Hamiltonian potential, $V_{qu(\mathrm{n})}$ is the so-called (non-local) quantum potential [3] that reads

$$V_{qu} = -(\frac{\hbar^2}{2m})\mathrm{n}^{-1/2}\nabla \bullet \nabla \mathrm{n}^{1/2} \tag{6}$$

and $\mu$ is the PD mobility that depends by the specificity of the considered system [3]. The noise amplitude parameter $\Theta$ equals the temperature of an ideal gas at equilibrium with the vacuum fluctuations [3]. It is worth noting that equation (1) is the hydrodynamic-like representation of the Schrödinger stochastic equation (SSE) [4]

$$i\hbar \frac{\partial \psi}{\partial t} = -\frac{\hbar^2}{2m} \nabla_q{}^2 \psi + V_{(q)}\psi + i\frac{\psi}{|\psi|^2}\eta_{(q_\alpha,t,\Theta)}. \tag{7}$$

The noise variance (2) is a direct consequence of the quantum potential term $\nabla \bullet \nabla \mathrm{n}^{1/2}$ that gives to the system energy a membrane elastic-like contribution [5] that reads

$$\overline{H}_{qu} = \int_{-\infty}^{\infty} \mathrm{n}_{(q,t)} V_{qu(q,t)} dq, \tag{8}$$

where higher curvature of the PD leads to higher energy (see figure 1). Therefore, independent fluctuations infinitesimally far apart leading to vanishing curvature wrinkles of the PD (and hence to an infinite quantum potential energy) are not allowed.. If we require that any physical system can attain configurations with finite energy, independent fluctuations on shorter and shorter distance have smaller and smaller amplitude, leading to the existence of a correlation distance (let's name it $\lambda_c$) for the noise.

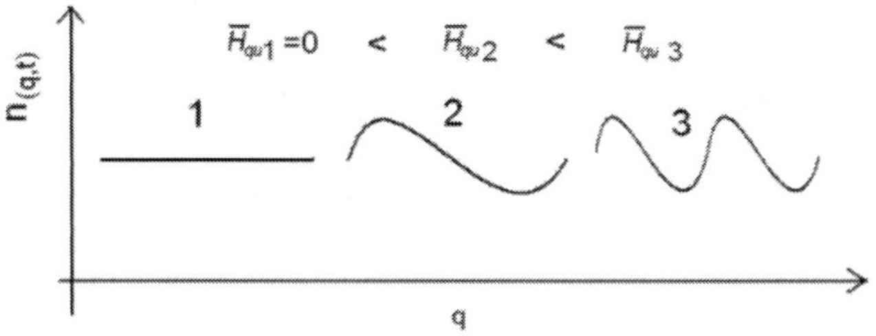


Figure 1 Quantum potential energy as a function of the "curvature" of the particle density n.

*2.1. Range of interaction of quantum pseudo-potential*

For Hamiltonian potential weaker that the quadratic one ($\varepsilon$<0 in figure 2), the quantum potential force (QPF) at large distance grows less than linearly [2]. Given the case $lim_{q\to\infty} V_{(q)} \propto q^k$, for $k < \frac{2}{3}$ the QPF at large distance is vanishing and the integral $\int_0^\infty | q^{-1} \nabla_q V_{qu} | dq$ converges [3].

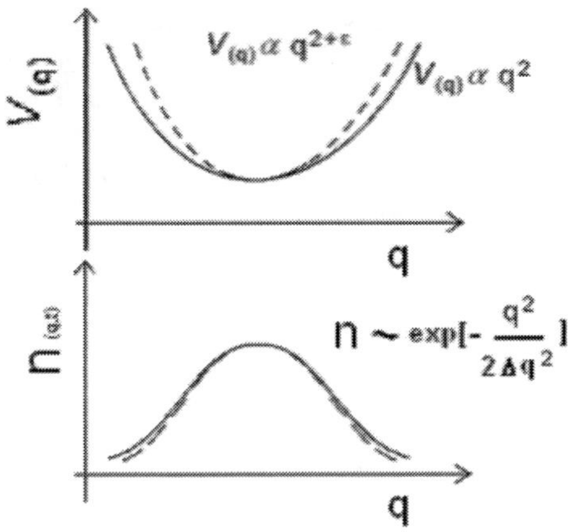


Figure 2 Particle density n for near-linear Hamiltonian potential

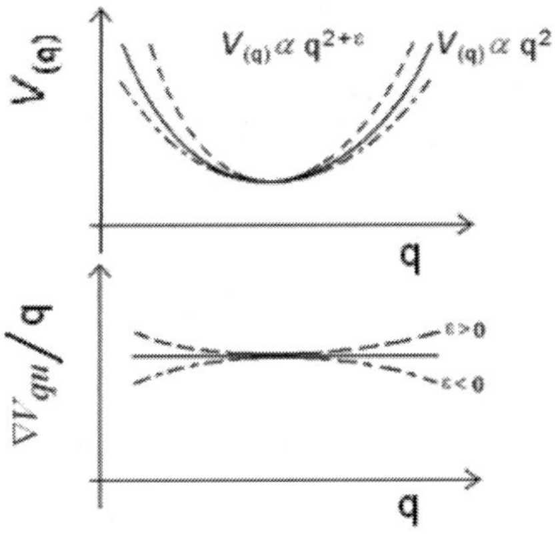


Figure3 The normalized quantum force $q^{-1}\nabla_q V_{qu}$ for near-linear Hamiltonian potential

In this case, the mean weighted distance

$$\lambda_q = 2 \frac{\int_0^\infty | q^{-1} \frac{dV_{qu}}{dq} | dq}{\lambda_c^{-1} | \frac{dV_{qu}}{dq} |_{(q=\lambda_c)}}, \tag{9}$$

evaluates the quantum potential range of interaction.

Faster the Hamiltonian potential grows, more localized is the PD and hence stronger is the quantum potential (figure 2 and figure 3). For the linear interaction, the Gaussian-type eigenstates lead to a quadratic quantum potential [6] and to a linear quantum force (i.e., $\lim_{|q|\to\infty} | q^{-1} \nabla_q V_{qu} | \propto constant$) so that $\lambda_q$ diverges. A force weaker than the linear one as the Lennard-Jones type potential leads to $\lambda_q$ finite.

### *2.2. Local stochastic dynamics*

When $\lambda_c << \Delta L$ and $\lambda_q << \Delta L$, where $\Delta L$ is the physical length of the system, the classical (local) stochastic dynamics can be achieved. In fact, given the condition $\lambda_q << \Delta L$ so that it holds $\lim_{q\to\infty} -\nabla_q V_{qu} = 0$, $\dot{q}$ in (4) reads [3]

$$\dot{q} = \frac{p}{m} = \nabla_q \{ \lim_{\Delta L/\lambda_q \to 0} \frac{1}{m} \int_{t_0}^{t} dt \left( \frac{p \bullet p}{2m} - V_{(q)} - V_{qu} \right) \} = \frac{-\nabla_q V_{(q)}}{m} + \frac{\delta p}{m} \cong \frac{p_{cl}}{m} \qquad (10)$$

where $\delta p$ is a small fluctuation of momentum.

## 3. Discussion and conclusions

The large scale dynamics of (10), achieved for weaker-than-linear interacting particles in presence of the noise $\eta$ obey to the classical stochastic mechanics that, in a system of a huge number of particles, own a chaotic evolution. As widely accepted [1], such chaotic kinetics generate mixing in the phase space that brings the system to relax toward the thermodynamic equilibrium (or to a stationary state). Being irreversible, this evolution owns the so called macro-arrow of time. On the other hand, even if the quantum equation of motion (7) on a microscopic scale is indistinguishable by deterministic one, for $\Theta \neq 0$ it contains an infinitesimal presence of noise. If the standard quantum mechanics owns the time reversal symmetry, for $\Theta \neq 0$ the SHQA equation (1) or the SSE (7) do not (see figure 4).

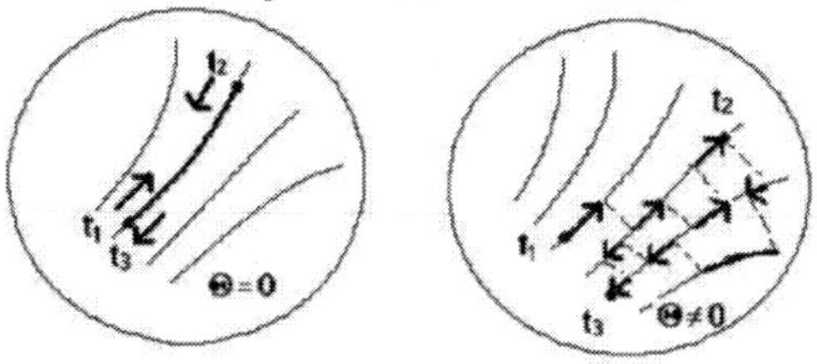


Figure 4 The forward and backward evolution along the SQHA phase space trajectories for an infinitesimal interval of time (on the left for $\Theta = 0$) and for $\Theta \neq 0$ (on the right).

For $\Theta \neq 0$ the time reversal breaking is not null and increases as $\Theta$ does. Since for the realization of the quantum mechanics we must be close to the deterministic limit $\Theta = 0$, the time symmetry breaking in (1,7) is expected to be small. Under the light of the SQHA approach, the logical separation between the concepts of micro-arrow and macro-arrow of time has no basic motivation: The residual and imperceptible presence of noise in the microscopic quantum dynamics leading to the micro-arrow of time is the premise for the breaking of the quantum coherence on macroscopic-scale physics and to the realization of the irreversible behavior owing the macro-arrow of time. Given the experimental evidences of time reversal asymmetry in elementary particle physics, the development of the relativistic hydrodynamic quantum analogy and its stochastic version can be the fundamental steps in obtaining, by a systematic procedure, the stochastic Dirac equation able to account for the time asymmetry in quantum relativistic physics.

## References

[1] Bianucci M.,Mannella R., West B. J., Grigolini P., From Dynamics to thermodynamics: Linear response and statistical mechanics, Phys. Rev. E 51 (4) (1995) pp. 3002-3022.

[2] Bertram M. Schwarzschild Time-reversal asymmetry in particle physics has finally been clearly seen, Phys. Today 65 (11), 16 (2012); doi: 10.1063/PT.3.1774

[3] Chiarelli P. Can fluctuating quantum states acquire the classical behavior on large scale? arXiv: 1107.4198 [quantum-phys] 2012; accepted for publication on Foundation of Physics.

[4] Chiarelli P. "The Classical Mechanics from the quantum equation", Phys. Rev. & Res. Int., 3(1) (2013) pp.1-9.

[5] Weiner, J.H., *Statistical Mechanics of Elasticity* (John Wiley & Sons, New York, 1983), p. 315-316.

[6] Chiarelli P., Open Quantum to classical transition in the stochastic hydrodynamic analogy: the explanation of the Lindemann relation and the analogies between the maximum of density at He lambda point and that one at water-ice phase transition , Phys. Rev. & Res. Int., 3(4) (2013) pp.348-366.

# Collective particle transport in a peristaltic ratchet system

**Marcel Dierl[1], Mario Einax[1], Petr Chvosta[2], Wolfgang Dieterich[3] and Philipp Maass[1]**

[1] Fachbereich Physik, Universität Osnabrück, Barbarastraße 7, 49076 Osnabrück, Germany
[2] Department of Macromolecular Physics, Faculty of Mathematics and Physics, Charles University, V Holesovickach 2, CZ-180 00 Praha, Czech Republic
[3] Fachbereich Physik, Universität Konstanz, 78457 Konstanz, Germany

E-mail: marcel.dierl@uos.de

**Abstract.** We consider particles with repulsive nearest-neighbor interactions in a periodically modulated tilted-sine potential as a model for a peristaltic ratchet system. Using the Markov chain approach to kinetics (MCAK), the collective transport behavior is investigated with respect to the modulation period, interaction strength, and applied mechanical bias (tilt amplitude). In the stationary state, a maximum is found in the period-averaged particle current and transport efficiency as function of both the mechanical bias and the modulation period. The efficiency can be enhanced when the interaction strength is increased. While the maximal current arises for fast modulation, maximum efficiency is obtained for comparatively slow modulation.

## 1. Introduction

Active transport plays an important role in biological processes at the cellular level. A typical example is the sodium-potassium ATPase in living cells, which drives the ions against a chemical gradient. Theoretical studies of such biological transport processes are often based on models, where transport of particles against the gradient is induced by a time-dependent potential without spatial inversion symmetry. Insights into driving mechanisms have been gained by studying Brownian dynamics of single particles [1, 2] or many-particle hopping models with hard-core exclusions [3, 4]. On the basis of analytical approaches, extensions of these treatments towards interparticle interactions beyond hard-core exclusions are difficult. Recently, it has been shown for a two-site model that such interactions can successfully be treated by time-dependent density functional theory (TDFT) [5].

Inspired by these studies, we apply the Markov chain approach to kinetics (MCAK) [6] to a many-particle ratchet model. In this approach, which is equivalent to the TDFT, exact equilibrium relations between correlation functions and densities on a local scale are derived from Markov chains. The merits of the MCAK over the TDFT are discussed in [6]. In the model, particles with repulsive nearest-neighbor interactions are driven by a traveling sine wave against a bias in a channel that is coupled to two particle reservoirs. Applying the MCAK, a closed set of kinetic equations for the collective particle flow is derived. By numerical solution of these equations we obtain the stationary current in and the efficiency of the ratchet as a function of the modulation period as well as the interaction and bias strength.

## 2. Model

The channel is considered to be composed of $N$ sites, where each site $i$ is either occupied by a particle (occupation number $n_i = 1$) or vacant ($n_i = 0$). Particles can be ejected to and injected from two reservoirs L and R, as sketched in Figure 1(a). Applying the equilibrated-bath couplings from [5, 6, 7, 8], the reservoir influence on the injection and ejection rates is determined by the chemical potentials $\mu_{\rm L}$ and $\mu_{\rm R}$, such that the respective rates fulfill detailed balance conditions with respect to the grand canonical potentials belonging to $\mu_{\rm L}$ and $\mu_{\rm R}$. This means that each reservoir tends to establish a corresponding equilibrium state. For the specific form of the injection and ejection rates, the exponential rates of Eq. 2 are used (see below). No interaction is considered between a particle inside the channel and a particle in a reservoir. Inside the channel, a repulsive nearest-neighbor interaction of strength $V > 0$ is present.

The reservoirs are assumed to have ("electro-")chemical potentials $\mu_{\rm L}$ and $\mu_{\rm R} + F$, where $F$ corresponds to a constant bias, i.e. $\varepsilon_{i+1}^{(\rm s)} - \varepsilon_i^{(\rm s)} = F/(N+1)$. Here the superscript indicates the "static part" of the energy levels, $\varepsilon_i^{(\rm s)} = \varepsilon_1^{(\rm s)} + F(i-1)/(N+1)$. Introducing the differences between the (static) boundary site energies and reservoir levels, $\Delta_{\rm L} = \varepsilon_1^{(\rm s)} - \mu_{\rm L}$ and $\Delta_{\rm R} = \mu_{\rm R} + F - \varepsilon_N^{(\rm s)}$, we can, without loss of generality, set $\mu_{\rm L} = 0$ in the following. Superimposed to the linearly increasing bias potential is a peristaltic driving mode, represented by a traveling sine wave with amplitude $A$ and modulation period $\tau = 2\pi/\omega$. The site energies hence become

$$\varepsilon_i(t) = \varepsilon_1^{(\rm s)} + \frac{F}{N+1}(i-1) + A\left[1 + \sin\left(\frac{\pi}{2}i - \omega t\right)\right], \qquad i = 1, \ldots, N. \tag{1}$$

The jump rate of a particle from site $i$ to a vacant neighboring site $f = i \pm 1$ is

$$\Gamma_{i,f}(t) \propto \exp\left[\frac{-\Delta E_{i,f}(t)}{2}\right], \tag{2}$$

where $\Delta E_{i,f}(t)$ denotes the energy difference between the final and initial state after and before the jump at time $t$; we here and in the following set the thermal energy as energy unit, i.e. $k_{\rm B}T = 1$.

## 3. Computational details

Using the master equation for the time evolution of the probability density $P(n_1, \ldots, n_N, t)$ of microstates, the evolution equations for the mean occupation numbers $\rho_i(t) \equiv \langle n_i \rangle_t = \sum_{\{n_i\}} n_i P(n_1, \ldots, n_N, t)$ are

$$\frac{{\rm d}\rho_i(t)}{{\rm d}t} = J_{i-1,i}(t) - J_{i,i+1}(t), \tag{3}$$

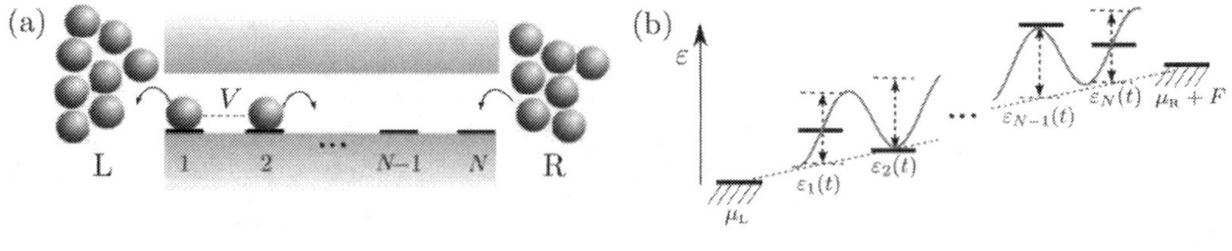


**Figure 1.** (a) Sketch of the model. (b) Visualization of the site energies with $\Delta_{\rm L} = \Delta_{\rm R} = F/(N+1)$ along the channel at time $t$.

with net currents $J_{i,i+1}(t) = j_{i,i+1}(t) - j_{i+1,i}(t)$ from site $i$ to $(i+1)$. Applying the MCAK procedure to the ratchet model yields for the bulk currents $j_{i,i+1} = \langle n_i(1-n_{i+1})\Gamma_{i,i+1}\rangle_t$ the analytical form [9]

$$j_{i,i+1} = \left[(\rho_{i+2} - C_{i+1})e^{-V/2} + \tilde{\rho}_{i+1} - \rho_{i+2} + C_{i+1}\right] \frac{\rho_i - C_i}{\rho_i \tilde{\rho}_{i+1}} \left[\rho_i - C_{i-1} + e^{V/2} C_{i-1}\right] e^{(\varepsilon_i - \varepsilon_{i+1})/2}, \tag{4}$$

where $\tilde{\rho}_i = 1 - \rho_i$ and $C_i$ follows from the quadratic equation [10]

$$C_i = \exp(-V) \frac{(\rho_i - C_i)(\rho_{i+1} - C_i)}{1 - \rho_i - \rho_{i+1} + C_i}. \tag{5}$$

Similar expressions are obtained for the injection and ejection rates [6]. Solving Eq. (5) for the $C_i$ as functions of $\rho_i$ and $\rho_{i+1}$, and inserting into Eq. (4), yields the currents as functions of the mean occupation numbers. In this way, the kinetic equations (3) become a closed set.

Numerical integration of this set allows us to discuss the output power and performance of the device. Let us denote an averaging of an observable $A(t)$ over one period in the stationary regime by $\bar{A}$, i.e. $\bar{A} = \lim_{t_0 \to \infty} \tau^{-1} \int_{t_0}^{t_0+\tau} \mathrm{d}t A(t)$. The mean number of particles transported by the ratchet per period is $\bar{J} = \bar{J}_{i,i+1}$, and accordingly the mean output power becomes $\bar{P}_{\mathrm{out}} = \bar{J}F$. The mean work done on the machine per period is $\bar{P}_{\mathrm{in}} = \overline{\sum_{i=1}^{N} \rho_i \dot{\varepsilon}_i}$, where $\dot{\varepsilon}_i(t) = \mathrm{d}\varepsilon_i(t)/\mathrm{d}t$. Accordingly we can define the efficiency of the machine by

$$\eta = \frac{\bar{P}_{\mathrm{out}}}{\bar{P}_{\mathrm{in}}} = \frac{\bar{J}F}{\overline{\sum_{i=1}^{N} \rho_i \dot{\varepsilon}_i}}. \tag{6}$$

## 4. Results and discussion

In the calculations, the following choice of parameters is used: $N = 10$, $A = 5$, $\Delta_{\mathrm{L}} = F/(N+1) - 2$, and $\Delta_{\mathrm{R}} = F/(N+1) + 2$. Let us first take a look at the current $\bar{J}$ as a function of the modulation time $\tau$. The solid line in Figure 2 shows results for $V = 1$ and $F = 10$ obtained from MCAK calculations. As can be seen from the figure, the current increases rapidly and goes through a maximum at $\tau \approx 0.9$. In order to test how well the MCAK captures the kinetics, we have performed also kinetic Monte Carlo (KMC) simulations of the ratchet model. The results are marked by the circles in Figure 2 and show a good agreement with the MCAK predictions. For details of the KMC algorithm with time-dependent transition rates, see [11, 12].

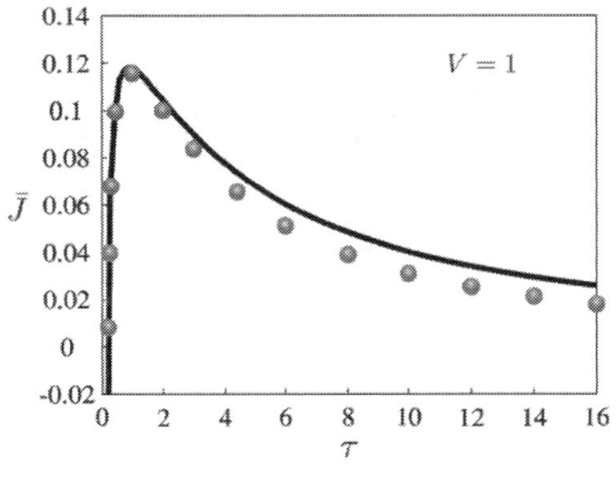


**Figure 2.** Average particle current $\bar{J}$ as function of $\tau$ for $V = 1$ and $F = 10$. The circles refer to results from KMC simulations.

In Figure 3 we analyze the performance of the ratchet model. The efficiency as function of the mechanical bias $F$ is shown in Figure 3(a) for one representative driving period $\tau = 2$ and three different values $V = 0$, 1, and 10. All curves run through a maximum that increases and shifts to larger $F$ values with increasing $V$. Figure 3(b) displays the efficiency as a function of the driving period. These curves again show a maximum of $\eta$ that increases with $V$ and shifts to

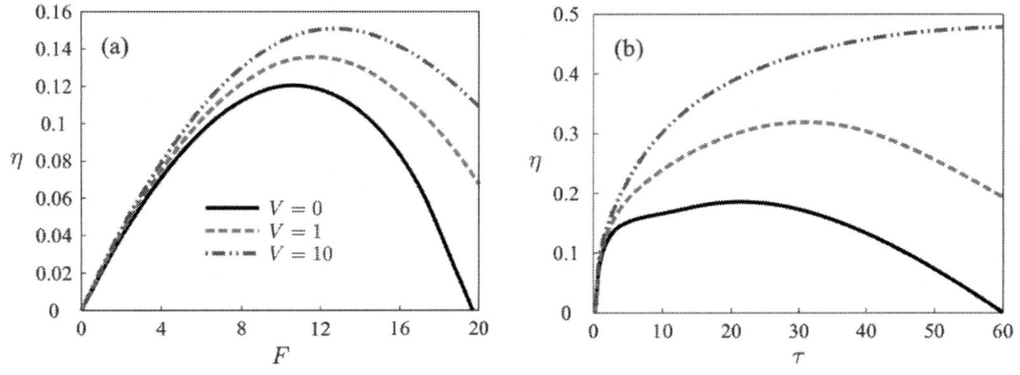


**Figure 3.** (a) Efficiency $\eta$ as function of (a) mechanical bias $F$ for $\tau = 2$ and (b) modulation period $\tau$ for $F = 10$. The assignment of lines to the different interaction strengths is given in (a).

larger $\tau$. Accordingly, to keep $\eta$ at maximum with increasing $V$, or decreasing temperature, the modulation of site energies needs to be slowed down.

Contrasting the finding in Figure 2, where maximal particle flow is obtained for fast modulation, highest efficiency is obtained for slow driving. These observations are reminiscent to those found for the minimal ratchet model ($N = 2$) studied in [5]. However, the present model with $N = 10$ shows much larger efficiencies for certain parameter choices, see, for example, the curve for $V = 10$ in Figure 3(b). This suggests that efficiencies for particle transport against a mechanical load are increased in longer channels.

In summary, our work has shown how the mechanical bias, the modulation period and the interaction strength affect the particle flow and the efficiency. It provides a guideline for optimizing this system.

## Acknowledgments

We thank A. Nitzan for very illuminating discussions and acknowledge financial support from the German Academic Exchange Service DAAD (Grant No. 50755689).

## References


[1] Reimann P 2002 *Phys. Rep.* **361** 57
[2] Hänggi P and Marchesoni F 2009 *Rev. Mod. Phys.* **81** 387
[3] Derenyi I and Ajdari A 1996 *Phys. Rev. E.* **54** R5
[4] Aghababaie Y, Menon G I and Plischke M 1999 *Phys. Rev. E* **59** 2578
[5] Einax M, Solomon G C, Dieterich W and Nitzan A 2010 *J. Chem. Phys.* **133** 054102
[6] Dierl M, Einax M and Maass P 2013 *Phys. Rev. E* **87** 062126
[7] Einax M, Dierl M and Nitzan A 2011 *J. Phys. Chem. C* **115** 21396
[8] Dierl M, Maass P and Einax M 2012 *Phys. Rev. Lett.* **108** 060603
[9] Dierl M, Maass P and Einax M 2011 *Europhys. Lett.* **93** 50003
[10] Buschle J, Maass P and Dieterich W 2000 *J. Phys. A: Math. Gen.* **33** L41
[11] Einax M and Maass P 2009 *Phys. Rev. E* **80** 020101(R)
[12] Holubec V, Chvosta P, Einax M and Maass P 2011 *Europhys. Lett.* **93** 40003

# Optimal driving protocols for nano-sized devices and their dependence on couplings to reservoirs

**Mario Einax**

Fachbereich Physik, Universität Osnabrück, Barbarastraße 7, 49076 Osnabrück, Germany

E-mail: meinax@uos.de

**Abstract.** The development of efficient artificial nanodevices poses challenges which are of fundamental and technological nature. Recent progress has been made in the context of finite-time thermodynamics. A central question in finite-time thermodynamics is to identify the optimal procedure to extract the greatest amount of work from a system operating under well-defined constraints. For externally controlled small systems, the optimal driving protocol maximizes the mean work spend in a finite-time transition between two given system states under the constraints of given initial and final energy values, and a fixed total operation time. For simplicity we consider an externally controlled single level system, which is embedded in a thermal environment and coupled to a particle reservoir. The optimal protocols are calculated from a master equation approach for different system-reservoir couplings. For open systems, the system-reservoir couplings are shown to have a striking influence on the optimal driving setup. We point out that the optimal protocols have discontinuous jumps at the initial and final times. Finally, this work provides a first attempt to extend these calculations to larger system sizes.

## 1. Introduction

A good theoretical understanding of the optimal control of energy conversion processes is a prerequisite for tailoring efficient artificial nanodevices for specific needs. Typical examples are soft and biomatter systems, such as Brownian or molecular motors, organic photovoltaic solar cells and quantum dots. With minimization of the system size thermal fluctuations become relevant and the non-equilibrium behavior of such systems depends strongly on the driving forces and the changes of one system state to another, which are inherently finite in time. Thus, thermodynamic processes take place in finite time and the thermodynamic quantities like heat and work are now random but still fulfill a stochastic energy balance. In these systems it is useful to introduce microscopic heat and work quantities as random variables whose averages lead to the common thermodynamic quantities. Averages over functions of these microscopic heat and work quantities yield generalized fluctuation theorems (for reviews, see [1], and references therein). A common feature of many artificial nano-sized devices, where fluctuation theorems can be applied, is that those are mostly driven by time-varying external fields (often called as protocol) or electrochemical potential differences.

Of particular importance of this class of non-equilibrium systems is to identify the optimal procedure to extract the greatest amount of work from the device operating under given constraints [2]. Only a few studies so far have addressed the problem of identifying the optimal protocol that yields the minimum amount of work done on the system, which is required to drive a nano-scale system from one equilibrium state to another in finite time [2, 3, 4, 5]. Note that this formulation is in reverse to the quest for the optimal protocol that provides the greatest amount of work from the artificial device. Whether on a continuous (Langevin equation) or a discrete (master equation) state space, a surprising result of all these studies is that the optimal protocol exhibits sudden jumps at the beginning and at the end of the thermodynamic process, while in between the optimal protocol varies smoothly. It can be argued that the initial jump in the optimal protocol is an immediate jump from equilibrium to a stationary state and the final jump allows a slower driving of the system at earlier times.

Here, we address a similar question on optimal driving setups. The focus of this work is on the couplings of the system to reservoirs, which have a strong influence on the specific form of the optimal protocol. To keep the analysis conceptual simple, we consider as working medium a single level system driven by a time-dependent protocol. We apply an optimization procedure based on a variational analysis. This allows us to get an analytic expression of the optimal protocol with respect to different system-reservoir couplings.

## 2. Model

We consider a single level system as working medium of an artificial nanodevice, in contact with a particle reservoir at temperature $T$, which is characterized by the chemical potential $\mu$. The site energy $\varepsilon(t)$ is assumed to be time-dependent, which can be modified between an initial value $\varepsilon_0$ and a final value $\varepsilon_1$ by an external agent according to a given protocol. In what follows, it is convenient to introduce the energy difference $\epsilon(t) = \varepsilon(t) - \mu(t)$. The population of the single level at time $t$ is characterized by the occupation probability $p(t)$. The system dynamics is modeled by a master equation approach with time-dependent rates $w_1(t)$ and $w_2(t)$ [6, 7, 8, 9, 10] accounting for the time evolution of the occupation probability $p(t)$,

$$\dot{p} = -w_1(t)p(t) - w_2(t)(1 - p(t)) \,. \tag{1}$$

We assume that these transition rates obey detailed balance at each time instant.

Due to the time-varying of the site energy $\epsilon(t)$, positive or negative energy flows into the system. From the thermodynamic point of view the time derivative of the internal energy $E(t) = \epsilon(t)p(t)$ is the sum of a work flow $\dot{W} = \dot{\epsilon}(t)p(t)$ and a heat flow $\dot{Q} = \epsilon(t)\dot{p}(t)$. During the process time $\tau$ the energy change obeys the statistical mechanics formulation of the first law of thermodynamics, $\Delta E(\tau) = Q(\tau) + W(\tau)$, where $\Delta E(\tau) = \int_0^\tau dt\, \dot{E}(t) = E(\tau) - E(0) \equiv \epsilon(\tau)p(\tau) - \epsilon(0)p(0)$ with $\epsilon(0) = \epsilon_0$ and $\epsilon(\tau) = \epsilon_1$. The occupation probability of the equilibrium state at the beginning is $p(0) = [e^{\beta\epsilon_0} + 1]^{-1}$ with $\beta = 1/k_B T$. Accordingly, we have $W = \int_0^\tau dt\, \dot{\epsilon}(t)p(t)$ and $Q = \int_0^\tau dt\, \epsilon(t)\dot{p}(t)$. Both work $W$ and heat $Q$ can be interpreted as functionals of the occupation probability and thus depend, in particular, on the realized transition path. If $W(t) < 0$, the work $-W(t)$ is done by the system on the environment.

The details of the system-reservoir coupling determine decisive the exchange of particles between the reservoir and the system [11]. This is reflected in the form of the transition rates. In order to proceed, we need to specify these transition rates consistent with the detailed balance condition. In a quantum dot coupled to a metal lead, the rates $w_1(t)$ and $w_2(t)$ are Glauber (Fermi) rates [4], i. e., $w_1(t) = C[e^{-\beta\epsilon(t)} + 1]^{-1}$ and $w_2(t) = C[e^{\beta\epsilon(t)} + 1]^{-1}$. $C$ is the inverse of a characteristic time scale involve in the exchange of particles between the reservoir and the system. We adopt here symmetric rates $w_1(t) = Ce^{\beta\epsilon(t)/2}$ and $w_2(t) = Ce^{-\beta\epsilon(t)/2}$. Symmetric transition rates are widely used in biomatter or ionic systems [12, 8].

## 3. Calculation of optimal driving protocols

In order to calculate the optimal driving protocol $\epsilon(t)$ which minimizes the work for the given constraints $\epsilon_0$, $\epsilon_1$, $p(0)$, and $\tau$, we apply the proposed procedure given in Reference [4]. To minimize $W = \Delta E - Q$ we have to minimize $\Delta E$ and maximize $Q$ simultaneously Since $\Delta E$

depends only on $p(\tau)$, we need to maximize the heat $Q$. The essential steps of this procedure are: *(i)* find the protocol that gives the maximum heat $Q$ for a given value of $p(\tau)$ and *(ii)* conduct the optimization with respect to the final state $p(\tau)$. To this end we rewrite the heat as $Q = \int_0^\tau dt\, L(p, \dot{p})$ with $L(p, \dot{p}) = \epsilon(t)\dot{p}$, because $\epsilon(t)$ can be expressed as function of $p(t)$ and $\dot{p}(t)$. A variational analysis $\delta \int L dt = 0$ leads to the Euler-Lagrange equation $L - \dot{p}\frac{\partial L}{\partial \dot{p}} \equiv -\dot{p}^2 \frac{\partial \epsilon}{\partial \dot{p}} = \tilde{K}$. Here $\tilde{K}$ is the constant of integration. For symmetric rates the Euler-Lagrange equation reads

$$\pm \frac{2(\dot{p}/C)^2}{\sqrt{(\dot{p}/C)^2 + 4p(1-p)}} = K\,, \tag{2}$$

where $K = \tilde{K}\beta/C$. Equation (2) is a quartic equation with respect to $\dot{p}$, which has two real (physical) solutions $\dot{p}/C = \pm\frac{1}{4}\sqrt{2K^2 + 2\sqrt{K^4 + 64K^2p(1-p)}}$ and two imaginary solutions. From these two first-order differential equations one can calculate $p(\tau)$ for given $p(0)$ and $\tau$ via

$$C\tau = \pm \int\limits_{p(0)}^{p(\tau)} \frac{dp}{\frac{1}{4}\sqrt{2K^2 + 2\sqrt{K^4 + 64K^2p(1-p)}}} \tag{3}$$

as function of $K$. Subsequently, $p(\tau)|_K$ can be used to maximize the heat $Q$ with respect to $K$ by performing the integration

$$\beta Q|_K = \int\limits_{p(0)}^{p(\tau)} dp \ln\left[\frac{A(p,K) + B(p,K)}{16p^2} \pm \frac{\sqrt{2K^2 D(p,K) + 2A(p,K)B(p,K)}}{16p^2}\right], \tag{4}$$

where $A(p,K) = K^2 + 16p(1-p)$, $B(p,K) = \sqrt{K^4 + 64K^2p(1-p)}$, and $D(p,K) = K^2 + 48p(1-p)$. Here, the $K$-dependent protocol $\epsilon$ can be rewritten with respect to $p$. Finally, we are able to optimize the work $W|_K = \Delta E|_K - Q|_K = \epsilon_1 p(\tau)|_K - \epsilon_0 p(0) - Q|_K$ with respect to $K$.

## 4. Results and discussion

In the calculation discussed below, the following choice of parameters was used: $C = 1$, $\beta = 1$, $\epsilon_0 = 10$, and $\epsilon_1 = 0$. Let us first take a closer look at the optimal protocol for symmetric rates as function of the processing time $t$ for different values of $K$. To this end we calculate Equation (4) with the upper sign (+) by using Equation (3) with the lower sign (-) for downward processes. In this case, work is done by the system on the environment. Figure 1(a) shows both the optimal protocol (black solid line) and three non-optimal protocols as function of the processing time $t$. We can clearly identify jumps of the protocol at the beginning and at the end of the process. As seen in Figure 1(b), an initially almost empty systems is populated with particles with increasing time. The protocol controls how fast this single level can be occupied. Figure 2 shows the optimal protocol for fast ($\tau = 1$) and slow ($\tau = 10$) processes. We observe that the associated value of $K_{\text{opt}}$ is larger for fast processes than for slow processes.

Thus, to understand the meaning of the parameter $K$, we square equation (2). By using evolution equation (1) with symmetric rates we arrive, after sorting with respect to $p$, at a quartic equation, which has two real and two complex conjugated roots. The two real roots read

$$p(t) = \frac{1}{e^{\beta\epsilon} + 1}\left(1 + \frac{K}{4}e^{\beta\epsilon/2}\tanh(\beta\epsilon/2) \pm \sqrt{\frac{K^2}{8}e^{\beta\epsilon}\tanh^2(\beta\epsilon/2) + \frac{K}{2}\frac{e^{\beta\epsilon}}{\cosh(\beta\epsilon/2)}}\right). \tag{5}$$

Equation (5) expresses the occupation probability as function of the protocol $\epsilon$ and $K$. If we assume a different system-reservoir coupling, say Fermi rates, the form of $p(t)$ [4] is quit different from those given in Equation (5). This emphasizes that the nature of the system-reservoir coupling has a striking influence on the optimal driving setup.

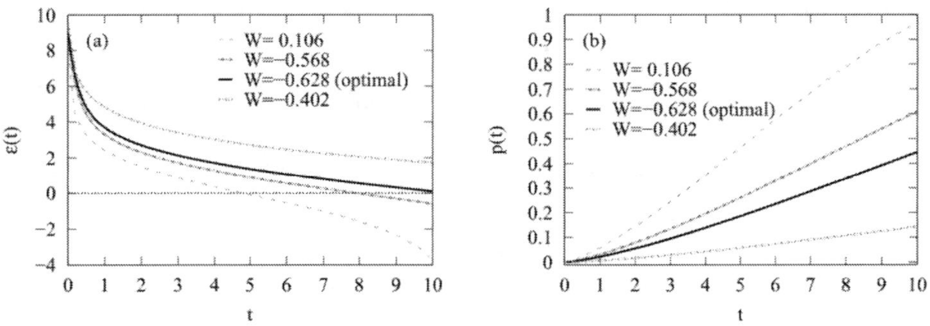


**Figure 1.** System properties for the case, in which the energy level is lowered from $\epsilon_0 = 10$ to $\epsilon_1 = 0$ as function of time $t$ starting from $t_0 = 0$ until $t_1 = 10$; (a) protocol and (b) occupation probability. An optimal protocol (black solid line) can be identified with minimum cumulated work $W = -0.628$ for $K_{\text{opt}} = 0.0059$ and the total available time $\tau = t_1 - t_0 = 10$.

If $K = 0$, $p(t)$ is the equilibrium distribution of a single level system. This implies that $K = 0$ coincides with a quasistatic process control. Consequently, the parameter $K$ quantifies how far the system differs from the quasistatic limit.

In order to extent these investigations to large systems, we consider the simplest case, in which a system with $N$ sites and energies $\epsilon_l(t)$ interacts with a particle reservoir at temperature $T$. Particles can only enter or leave each level and we assume that no transitions between site $l$ and site $k$ take place. Thus, the internal energy of the system is given by an additive principle, $E(t) = \sum_l \epsilon_l(t)p_l(t)$, and the time-derivative of $E(t)$ yields $\dot{W} + \dot{Q}$ with $\dot{W} = \sum_l \dot{W}_l = \sum_l \dot{\epsilon}_l(t)p_l(t)$ and $\dot{Q} = \sum_l \dot{Q}_l = \sum_l \epsilon_l(t)\dot{p}(t)$. In this special case the application of the variational analysis presented in Section 3 is straightforward in order to get the optimal setup of large systems.

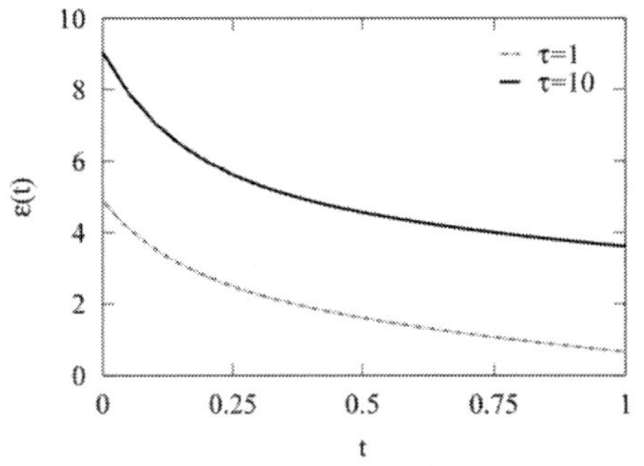


**Figure 2.** The optimal protocol for $\tau = 1$ ($K_{\text{opt}} = 0.17$) and $\tau = 10$ ($K_{\text{opt}} = 0.0059$) as function of time $t$.

## References


[1] Seifert U 2012 *Rep. Prog. Phys.* **75** 126001
[2] Schmiedl T and Seifert U 2007 *Phys. Rev. Lett.* **98** 108301
[3] Then H and Engel A 2008 *Phys. Rev. E* **77** 041105
[4] Esposito M, Kawai R, Lindenberg K and Van den Broeck C 2010 *Europhys. Lett.* **89** 20003
[5] Esposito M, Kawai R, Lindenberg K and Van den Broeck C 2010 *Phys. Rev. E* **81** 041106
[6] Chvosta P, Einax M, Holubec V, Ryabov A and Maass P, 2010 *J. Stat. Mech.* P03002
[7] Einax M and Maass P 2009 *Phys. Rev. E* **80**, 020101
[8] Einax M, Körner M, Maass P and Nitzan A 2010 *Phys. Chem. Chem. Phys.* **12** 645
[9] Einax M, Solomon G C, Dieterich W and Nitzan A 2010 *J. Chem. Phys.* **133** 054102
[10] Einax M, Dierl M and Nitzan A 2011 *J. Phys. Chem. C* **115** 21396
[11] Dierl M, Maass P and Einax M 2012 *Phys. Rev. Lett.* **108** 060603
[12] Fisher M E and Kolomeisky A B 1999 *Proc. Natl. Acad. Sci.* **96** 6597

# The Multi-Role Hidden Tree Model for Scale-free Networks

**Zhiqiong Bu**

Department of Computer Science, GuangDong Polytechnic Normal University, Guangzhou 510090,China

E-mail: ms.puzq@gmail.com

**Abstract.** Based on the spatial factor other than the temporal accumulation, the hidden tree model was built up to model scale-free networks. This paper further assumed that a node has multiple roles in different hidden trees, and explored the multi-role hidden tree model. The experimental results showed that multi-role hidden tree model can also produce scale-free networks. This conclusion indicates that the hidden tree model is robust with multiple roles.

## 1. Introduction

Modelling scale-free networks is an important issue in the research of complex networks [1, 2, 3, 4, 5], because some important systems such as the Internet and the WWW were found to be scale-free [1, 2, 6, 7, 8]. To understand the real-world networks and to illustrate the origins and mechanisms of the scale-free property, hundreds of models are proposed in the last decades [9, 10]. Among all of the models, the BA model [1] proposed by the pioneer scientists A.-L. Barabási et al. is the most popular. The BA model indicates that the scale-free property is based on the temporal accumulation. Based on the BA model, a lot of researches were carried out.

In another side, scale-free networks has an important feature: the hierarchy [11, 12, 13, 14, 15]. Therefore, some researchers tried to integrate the hierarchy into the models to explore the mechanisms of the scale-free property [16]. Thus, Zheng et al. further proposed that the hierarchy is the key to the scale-free property and thus proposed the hidden tree model [12, 17, 18].

Zheng et al.'s hidden tree model suggested that the hierarchy is a kind of hidden order [17, 19, 18] that is represented by a tree structure. In the hidden tree model, all the nodes are organized into a tree, and the offspring nodes are controlled in cascade by the parents nodes; every node has a probability to trade with a certain random node; when the transaction happens, the node rising the transaction would link to every node which control the transaction in cascade. With this method, scale-free networks can be generated. Different to the temporal accumulation of the BA model, this model are based on the spatial factor. This model has been used to explain many phenomena in the real world.

However, the individuals in the real world often have multiple roles, i.e., they may be organized into different hidden trees. Thus, whether the overall relationship network is still scale-free or not? To answer this problem, we explored the multi-role hidden tree model in this paper.

## 2. The Model

Assume that the whole system has $N$ nodes, and these nodes are organized into $R$ hidden trees, i.e., every node has $R$ roles. For simplicity, every hidden tree is assumed as the $n$-tree. Moreover, the roles are assigned randomly, i.e., the places of every node in different hidden trees are random. Similar to the original hidden tree, every node has a probability to do transactions with the others randomly; because every node has multiple roles, a hidden tree is chosen randomly to reflect the cascade controlling; when a node has a transaction to another, the source node would link to the all nodes in the route between the source node and destination node in the selected hidden tree. That is, when node $a$ has a transaction to node $b$, the proposed model chooses a hidden tree randomly, and finds a route from node $a$ to node $b$ in the chosen hidden tree, and finally node $a$ links itself to all the nodes in the route.

This model can be depicted as Fig. 1.

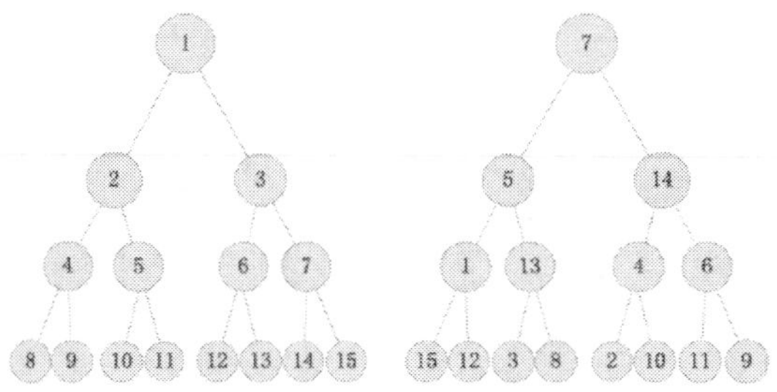


**Figure 1.** The hidden trees

As shown in Fig. 1, the identical node with the same number would have different places in the left hidden tree and in the right hidden tree. Assume that node 1 arises a transaction with node 11, if the left hidden tree is chosen, then node 1 will link to nodes 2, 5 and 11; if the right hidden tree is chosen, then node 1 will link to nodes 5, 7, 14, 6 and 11.

## 3. Experimental Results

The proposed model has four parameters, i.e., the number of hidden trees $T$, the number of nodes $N$, the number of sibling nodes $n$ and the probability of arising a transaction $A$. We focus on the simulation results under $T = 2$ and $T = 3$.

First, we studied the circumstance when the number of roles is 2, i.e., $T = 2$. To study the effect of the parameter $N$, i.e., the number of node, we set the other parameters $n = 2$ and $A = 0.4$, and set $N = 1000$, 2000, 5000, 10000 and 20000. To study the effect of the parameter $n$, we set the other parameters $N = 1000$ and $A = 0.4$, and set $n = 2, 3, 4, 5$ and 6. To study the effect of the parameter $A$, we set the other parameters $N = 1000$ and $n = 2$, and set $A = 0.08$, 0.16, 0.32, 0.64 and 1.00. Moreover, to assure the robustness of the model, we carried out every parameters setting 10 times. The simulation results were very robust. Therefore, we chose the generated scale-free networks in the first run and demonstrated their cumulative degree distributions as Fig. 2.

From Fig. 2, we can see that the degree distributions of the generated networks are approximately linear in the log-log coordination, indicating that the generated networks are scale-free. Moreover, the curves in Fig. 2 are similar to previous studies [12, 17, 18]. According

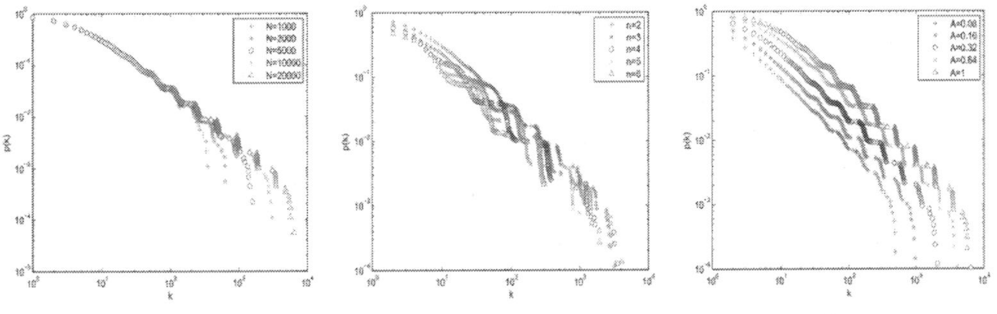


(a) $N$ (b) $n$ (c) $A$

**Figure 2.** The cumulative degree distributions when the parameters vary under $T = 2$

to the theoretical analysis, previous studies have proved that the curves demonstrated the scale-free property. Therefore, the degree distributions in Fig. 2 are theoretically approximately power-law.

Second, we studied the circumstance when the number of roles is 3, i.e., $T = 3$. In order to compare with the results under $T = 2$, we used the same parameters settings as those under $T = 2$. The experimental results are listed as Fig. 3.

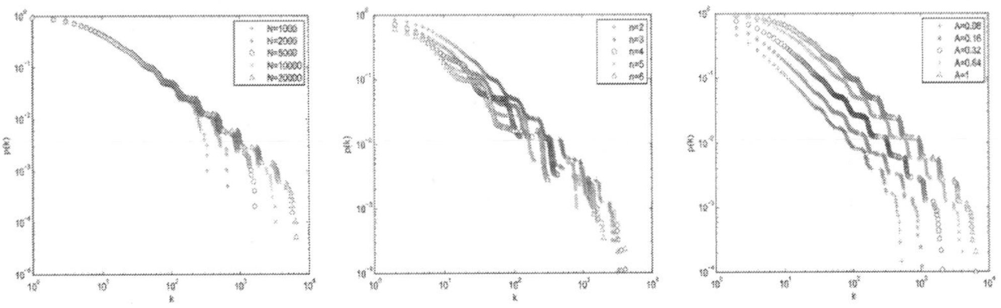


(a) $N$ (b) $n$ (c) $A$

**Figure 3.** The cumulative degree distributions when the parameters vary under $T = 3$

From Fig. 3, we can see that the experimental results are very similar to the results under $T = 2$. The experimental results indicate that when all nodes have multiple roles, the hierarchy or the hidden order can still produce the scale-free property.

## 4. Conclusions and Discussion

The experimental results have demonstrated that the proposed model can generate scale-free networks. The proposed model did not employ the time factor, i.e., did not rely on the time accumulation, but was based on the (topological) spatial factor. The spatial factor, i.e., the hidden tree, can be explained as the hidden order. On one side, because the proposed model was independent to the time factor, it would be full parallelism. On the other side, because the proposed model was based on the spatial factor, the hierarchy, i.e., the hidden order, is the key factor to the scaling phenomenon.

Moreover, the proposed model explored the circumstances that all nodes have multiple roles, which is more general in the real world. When every node has a few roles, the generated networks are still scale-free. This conclusion indicates that the hidden tree model is quite robust with multiple roles.

The conclusions in this paper can also explain the invulnerability of scale-free networks [20, 21, 22, 23, 24, 25, 26, 27]. The hidden order is like the crystal in the matters, therefore, it means the minimal energy status. When the hidden order is destroyed, the additional energy should be used. That is, the networks with the hidden order, i.e., scale-free networks, would be more stable under the attacks. In another words, destroying scale-free networks is more difficult than destroying random networks in most cases[22, 23, 24, 28].

Furthermore, because the hidden order can be reflected as the minimal energy status, the multi-role hidden order model can be regarded as the optimization process. Therefore, the proposed model is compatible to the optimization explanation of scale-free networks [29, 30].

**Acknowledgments**

The author is grateful of the helps from Mr. Weiqiang Zhou.

**References**

[1] Barabasi A L and Albert R 1999 *Science* **286** 509 512
[2] Albert R and Barabasi A L 2002 *Reviews of Modern Physics* **74** 47 97
[3] Willis J C and Yule G U 1922 *Nature* **109** 177 179
[4] Yule G U 1925 *Philos. Trans. R. Soc. London B* **213** 21 87
[5] Price D J d S 1965 *Science* **149** 510 515
[6] Alderson D, Li L, Willinger W and Doyle J C 2005 *IEEE/ACM Transactions on Networking* **13** 1205 1218
[7] Adamic L A and Huberman B A 2000 *Science* **287** 2115
[8] Barabasi A L, Albert R, Jeong H and Bianconi G 2000 *Science* **287** 2115a
[9] Boccaletti S, Latora V, Moreno Y, Chavez M and Hwang D U 2006 *physics reports* **424** 175 308
[10] Newman M E J 2003 *SIAM Review* **45** 167 256
[11] Sales-Pardo M, Guimerà R, Moreira A A and Amaral L A N 2007 *Proc. Natl. Acad. Sci.* **104** 15224 15229
[12] Zheng B 2009 *Backbone and Evolution in Complex Network* Postdoctoral report Tsinghua University
[13] Ravasz E, Somera A L, Mongru D A, Oltvai Z N and Barabasi A L 2002 *Science* **297** 1551 1555
[14] Ravasz E and Barabasi A L 2003 *Phys. Rev. E* **67** 026112
[15] Zheng B, Li D, Chen G, Du W and Wang J 2012 Ranking the importance of nodes of complex networks by the equivalence classes approach (*Preprint* arXiv:1211.5484)
[16] Vazquez A 2000 *arXiv:cond-mat/0006132v4*
[17] Zheng B, Wang J, Chen G, Jiang J and Shen X 2011 *Chin. Phys. Lett.* **28** 018901
[18] Zheng B, Du W, Shu W, Wang J and Li D 2013 *Romanian Journal of Physics* in press, arXiv:1210.2132
[19] Holland J H 1995 *Hidden Order: How Adaptation Builds Complexity* (Addison-Wesley)
[20] Albert R, Jeong H and Barabasi A L 2000 *Nature* **406** 378 382
[21] Holme P, Kim B J, Yoon C N and Han S K 2002 *Phys. Rev. E* **65** 056109
[22] Zheng B, Huang D, Li D, Chen G and Lan W 2011 *Europhysics Letters* **94** 28010
[23] Qin J, Wu H, Tong X and Zheng B 2013 *Physica D* **253** 85 90
[24] Zheng B, Wu H, Du W, Shu W and Qin J 2012 The robustness of scale-free networks under edge attacks with the quantitative analysis (*Preprint* arXiv:1211.3238)
[25] Schneider C M, Moreira A A, Jose S Andrade J, Havlin S and Herrmann H J 2011 *Proc. Natl. Acad. Sci. USA* **108** 3838 3841
[26] Wu Z and Holme P 2011 *Phys. Rev. E.* **84** 026106
[27] Schneider C M, Moreira A A, Jose S Andrade J, Havlin S and Herrmann H J 2011 *Proc. Natl. Acad. Sci. USA* **108** 3838 3841
[28] WU H, QIN J and ZHENG B 2012 *Computer Science* **39**(8) 224 227,255
[29] Zheng B, Wu H, Kuang L, Qin J, Du W, Wang J and Li D 2012 A simple model clarifies the complicated relationships of complex networks arXiv:1210.3121 (*Preprint* arXiv:1210.3121)
[30] Zheng B, Wu H, Qin J, Lan W and Du W 2012 Modelling multi-trait scale-free networks by optimization arXiv:1212.0207 (*Preprint* arXiv:1212.0207)

# Emotional isolation in BBC Forum

**J Sienkiewicz**[1] **and A Chmiel**[2,1]

[1] Faculty of Physics, Center of Excellence for Complex Systems Research, Warsaw University of Technology, Koszykowa 75, 00-662 Warsaw, Poland

[2] Section for Science of Complex Systems, Medical University of Vienna, Spitalgasse 23, Vienna, A-1090, Austria

E-mail: julas@if.pw.edu.pl, anna.chmiel@meduniwien.ac.at

**Abstract.** We analyze emotionally annotated massive data from BBC Forum and examine properties of the isolation phenomenon of negative and positive users. Our results show the existence of a percolation threshold dependent on the average emotional value in the network of negatively charged nodes.

## 1. Introduction

Owing to instant access to tons of data on the online customs of people it is now possible to use physics approaches in order to model human behavior in this media (e.g., [1, 2, 3, 4]). A very special case of this analysis is devoted to *emotional* component of this activity. Recent research has already revealed certain interesting phenomena, such as collective emotional behavior [5], emotional persistence [6], emotional avalanches [7] or emotional equilibration of discussions [8]. Here we touch the problem of emotional isolation among the users of BBC Forum.

## 2. Data description

As a source of data we chose the BBC Forum data [5, 9], whose basic properties were gathered in Table 1. The analysis is limited only to seven categories of the forum: *UK News*, *Christian Topics*, *World News*, *Ethics and Freethought*, *Jewish Topics*, *Eastern Religions* and *TV and Radio* in the period July 2005 - June 2009 (see Table 2). All the downloaded comments were processed using the emotional classifier program that is based on a machine-learning (ML) Language Model [10] approach, which functions in two phases: during the first one, it acquires a set of documents classified by humans for their emotional content (positive, negative or objective) from which it learns the characteristics of each category. During the application phase, the algorithm applies the acquired sentiment classification knowledge to new documents. Posts are annotated with a single value $e = -1, 0$ or $1$ to express their emotional content (i.e., valence [11]) as negative, neutral or positive, respectively. The accuracy of the classifier checked for 594 humanly annotated comments is 73.73 % for subjectivity detection and 80.92 % for polarity detection (see [7, 5] for details).

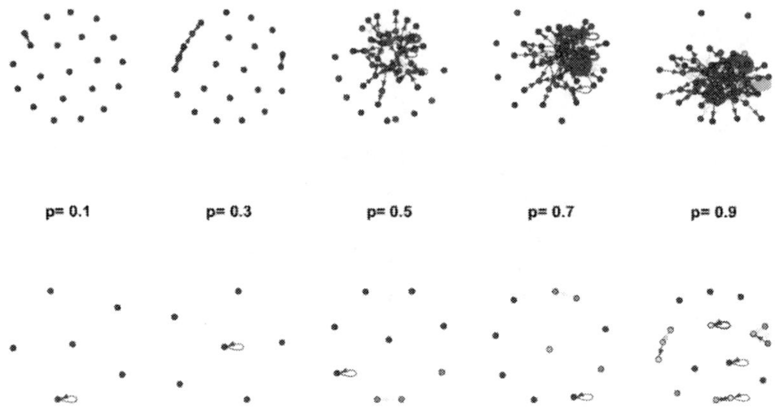


**Figure 1.** Networks of connections between negative (upper row) and positive (bottom row) nodes for specific values of the threshold $p$.

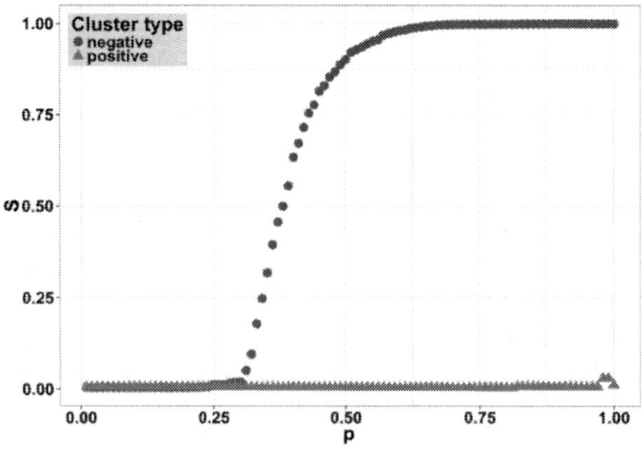


**Figure 2.** Size of the giant component $S$ as a function of the threshold parameter $p$ (see Eq. (2) for negative (blue circles) and positive (red triangles) user networks.

**Table 1.** Properties of the BBC dataset: number of comments $C$, number of threads $N$, number of unique users $U$, average and valence $\langle e \rangle$. See also [5, 9]

| $C$ | $N$ | $U$ | $\langle e \rangle$ |
|---|---|---|---|
| 2,474,781 | 97,946 | 18,249 | -0.44 |

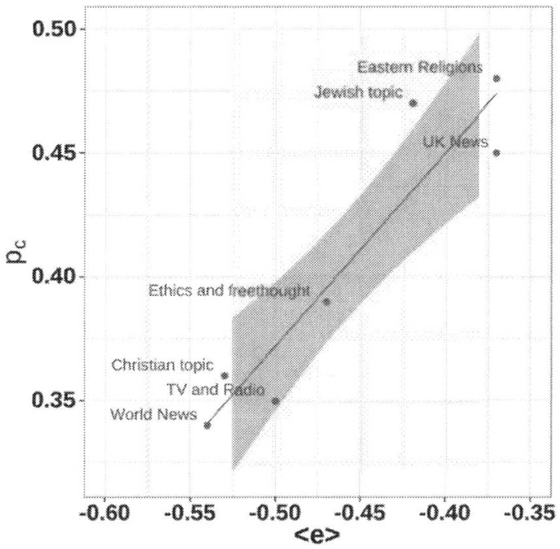


**Figure 3.** Size of the giant component $S$ as a function of the node's emotional value $e$ for different categories (red points, see Table 2). The blue line is a linear fit $p_c = A\langle e\rangle + B$ with $A = 0.78 \pm 0.13$, $B = 0.764 \pm 0.059$ and quality of fit $R^2 = 0.86$. The shaded area reflects standard deviation of $A$.

**Table 2.** Properties of the BBC dataset: number of comments $C$, number of threads $N$, number of unique users $U$, average and valence $\langle e\rangle$. See also [5, 9]

| category | $N$ | $\langle e\rangle$ | category | $N$ | $\langle e\rangle$ |
|---|---|---|---|---|---|
| UK News | 1,068,803 | -0.37 | Christian Topics | 719,617 | -0.53 |
| World News | 491,528 | -0.54 | Ethics and Freethought | 86,579 | -0.47 |
| Jewish Topics | 65,367 | -0.42 | Eastern Religions | 22,314 | -0.37 |
| TV and Radio | 20,573 | -0.50 | | | |

## 3. Results

In order to address the problem of isolation in the network (e.g., [12, 13]) we have used the data to create a network of unique users. It was achieved by setting a link between users $i$ and $j$ if one of them has written a comment in response to the other's comment. We preserved the direction of the relation, thus creating a directed network, where the value of the link is equal to emotional valence of the comment. In the case there were several comments between the same pair of users, the value $e_{ij}$ of the link was calculated as the arithmetic average of them. Finally, for each node we calculated the mean weighted emotional value on all the in-going links

$$s_i = \frac{1}{\sum_j w_{ij}} \sum_j w_{ij} e_{ij}, \tag{1}$$

where $w_{ij}$ is the weight of the link, i.e., the number of of messages sent by $i$ to $j$.

In the case of the whole network, the majority of nodes tend to have negative value of $s_i$ (the average value $\langle s_i\rangle = -0.42$). However, it is of use to examine the nature of connections among negatively and positively charged nodes. To do that we perform percolation analysis [14, 15]: we keep only the nodes that fulfill the following condition

$$|s_i \pm 1| < p, \tag{2}$$

where the plus sign reflects the procedure for negative nodes and minus sign for the positive ones and $p$ is the threshold. In effect we obtain a directed weighted network for each value of the threshold $p$ (see examples in Fig. 1). The obtained plots suggest that (i) there is a specific value of $p = p_c$ (a percolation threshold) for the network of negative nodes, (ii) positive nodes are isolated in this system. These observations are confirmed in Fig 2, where we present the size of the giant component $S$ (i.e., the largest connected part of the network) versus the threshold $p$. In the case of the network of negative nodes, the percolation threshold $p_c$ is a linear function of the overall emotional value of all post (see Fig. 3).

## 4. Conclusions

We have shown the existence of a percolation threshold in the emotionally annotated BBC user network. Moreover, our analysis suggest that the positively charged node (users) are isolated in the network.

## Acknowledgments

The work was supported from Polish Ministry of Science and Higher Education (grant no 0490/IP3/2011/71). All the data were obtained from EU FP7 CyberEMOTIONS (Collective Emotions in Cyberspace) project (contract 231323).

## References

[1] Eckmann J-P, Moses E and Sergi D 2004 *Proc. Nat. Acad. Sci. U.S.A.* **101** 14333.
[2] Kujawski B, Hołyst J A and Rodgers G J 2007 *Phys. Rev.* E **76** 036103.
[3] Chmiel A, Kowalska K and Hołyst J A 2009 *Phys. Rev.* E **80** 066122.
[4] Radicchi F 2009 *Phys. Rev.* E **80** 026118.
[5] Chmiel A, Sienkiewicz J, Thelwall M, Paltoglou G, Buckley K, Kappas A and Hołyst J A 2011 *PLoS ONE* **6** e22207.
[6] Garas A, Garcia D, Skowron M and Schweitzer F 2012 *Sci. Rep.* **2** 402.
[7] Mitrovic M, Paltoglou G and Tadic B 2010 *Eur. Phys. J.* B **77**, 597.
[8] Sienkiewicz J, Skowron M, Paltoglou G and Hołyst J A 2013 *Adv. Comp. Sys.* **16** 1350026.
[9] Chmiel A, Sobkowicz P, Sienkiewicz J, Paltoglou G, Buckley K, Thelwall M and Hołyst J A 2011 *Physica* A **390** 2936.
[10] Sebastiani F 2002 *ACM Computing Surveys* **34**.
[11] Feldman L A 1995 *Journal of Personality and Social Psychology* **69** 153.
[12] Sienkiewicz J and Hołyst J A 2009 *Phys. Rev.* E **80** 036103.
[13] Sienkiewicz J, Siudem G and Hołyst J A 2010 *Phys. Rev.* E **82** 057101.
[14] Stauffer D and Aharony A 1994 *Introduction to Percolation Theory, 2nd Ed.* (London: Talor & Francis).
[15] Achlioptas D, D'Souza R M and Spencer J 2009 *Science* **323** 1453.

# Network formed by movements of random walkers on a Bethe lattice

**Nobutoshi Ikeda**
Tohoku Seikatsu Bunka Junior College,
1-18-2 Niji-no-Oka, Izumi-ku, Sendai 981-8585, Japan
E-mail: ikeda.nobu@nifty.com

**Abstract.** We investigate a stochastic model of network formation where short-cut edges are assumed to be created between vertices in traces of random walkers. The network initially starts from a tree-like structure (Bethe lattice) with a finite number of shells, and develops into a complex network with many circuits generated by the movement of random walkers. We show that the resulting network has a power-law in the degree distribution with an exponent smaller than 2, and demonstrate the robustness against attacks on hubs in the networks. While scale-free networks without a degree correlation are usually vulnerable to attacks on its hubs, the robustness of the network connectivity in this model comes from a self-similar structure of the network. It is interesting that a simple stochastic process like random walks can cause various structures widely seen in real networks on tree-like graphs which play an important role in the graph theory.

## 1. Introduction

One of the essential problems in studies on random networks is the presence of common structures in large-scaled real networks such as the Internet, social networks, and biological networks [1], given that most real networks are not constructed by design, but by random events in the development of the networks. For example, the power law in degree distribution, which is usually called scale-free property, is a representative character found in many real networks. Preferential attachment of new vertices in the growing process of networks is a simple but probable rule which reproduces the power law in degree distribution [2]. Since the recognition of the importance of preferential attachment, many models based on the growth property of networks have been proposed. However, there have only been a few models which consider short-cut creations stimulated by local events depending on the local structure of networks.

In this paper, we will focus on short-cut creations stimulated by movements of random walkers. We have studied this process in previous works in which we showed that the structure of the initial graph plays an essential role in the development of networks [3]. This result can be explained by the fact that the initial structure of a network affects the diffusion property of the random walker that stimulates the new creation of edges. Here, we take a Bethe lattice as the initial graph. Bethe lattices are not only used in theoretical approaches to statistical mechanics on graphs [4, 5], but also in information science including search problems. Here we will show that the Bethe lattice can be developed into a complex network by the addition of many edges which are stimulated by the repeated movements of many random walkers starting from the root vertex.

Tolerance of networks to random or intentional attacks is an important property of networks, because connections between constituents of a system are usually necessary to maintain the function of the network [6, 7]. It is known that scale-free networks without degree correlation are usually vulnerable to attacks on its hubs. However, the many circuits generated by the movement of random walkers are expected to provide a robust structure against attacks on hubs in the network. We will demonstrate to what extent the movement of random walkers can provide a robust structure to tree-like structures.

The model based on the movement of random walkers is described in the following sections. In section 3, we show numerical results for degree distribution and discuss the time evolution of networks. In Section 4, we show the numerical results for the fraction of the maximum component after the deletion of hubs in the resulting network. Section 5 summarizes the results.

## 2. Model

The network model investigated in this paper is constructed by the following process (Fig. 1):

(i) A Bethe lattice with the coordination number 3 is prepared as an initial graph, and a random walker starts from the root vertex in the graph.

(ii) At each time step, the walker moves randomly to one of the vertices that is directly connected to the vertex where the walker began the move.

(iii) At each time step, if the vertex where the walker lands and the vertex where the walker landed two steps previously are not yet joined, a short-cut edge is created between them.

(iv) Steps (ii) and (iii) are iterated until the walker reaches a vertex in the Lth shell from the root vertex where L is a constant integer. Once the walker reaches the Lth shell, the walker is removed, leaving short-cut edges that the walker has created. A new walker is born on the root vertex and begins to move (goes to step (ii)).

Note that the walker is assumed to be able to pass through not only the edges in the original lattice but also the edges created by the movement of the walker. We will examine the subgraph $g(t)$ constructed by vertices that the walkers have visited until time step $t$, since degrees of other vertices remain the same as the initial one.

In this model, it is obvious that the network changes into a complete graph by the saturation of number of edges, because the number of vertices included in L shells is finite. Nevertheless, the local structure of the subgraph $g(t)$ can be treated as a nearly stationery state, because the rate of changes of the statistical properties is very small if L is sufficiently large.

## 3. Degree distribution

This section presents the numerical results for degree distributions for cases when the condition $L = 21$ is employed. Note that the number of vertices $V$ in subgraph $g(t)$ being considered is by far smaller than the number of vertices included in 21 shells, $N_L = 3 \times (2^{21} - 1)$.

Fig. 2 shows the power-law behavior in the degree distributions $V(k)$ of subgraph $g(t)$. The value of the power-law exponent $\gamma$ which is defined by $V(k) \sim k^{-\gamma}$ is about 2, but seems to change with time. Fig. 3 presents the dependence of numerically obtained power-law exponents on the number of vertices in $g(t)$, $V$. The result shows that $\gamma$ takes a value smaller than 2 in most time intervals and changes slowly to a smaller value as $V$ increases.

The time dependence of the power-law exponent can be understood by the time dependence of $V$ by using a relation derived in our early works [3, 8], $\gamma = 3 - 1/\alpha$, where $\alpha$ is defined by the relation $V \sim t^{\alpha}$. Our numerical calculation shows that $V$ can be described by $V \sim t^{\alpha}$, given that $\alpha$ is not a constant but a function of time with a very slow rate of decrease. According to the relation $\gamma = 3 - 1/\alpha$, the power-law exponent must change from 2 to 1, corresponding to decrease in $\alpha$ from 1 to 0.5. A situation when the value of $\alpha$ is smaller than 0.5 implies that

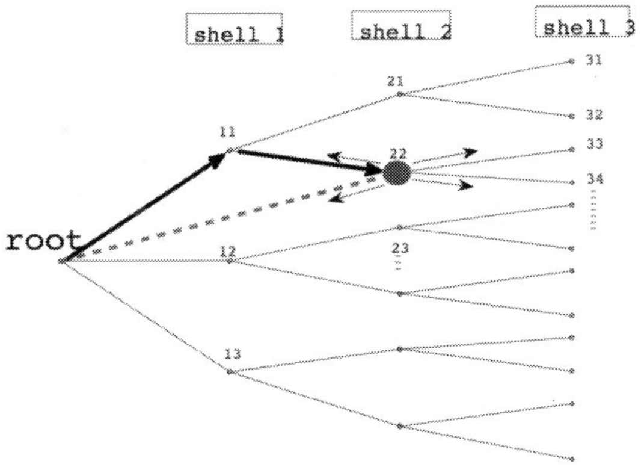


**Figure 1.** Initially, a random walker who can move to one of the nearest neighboring vertices $(11, 12, 13)$ per one time step starts from the root vertex. At time $t = 2$, if the walker reaches vertex 22 via vertex 11, a new edge is created between the root vertex and vertex 22 (dashed line). In the next time step $t = 3$, the walker may move to vertex 33, or vertex 34, or vertex 11, or the root vertex (At this stage, the walker can move on the dashed line). Corresponding to the actual movement, a short-cut will be created between 11 and 33 or between 11 and 34, or a short-cut will not be created.

a nearly complete graph begins to form, because, if so, $2dE/dt < dV^2/dt \to 0(t \to \infty)$, where $E$ is the number of edges in $g(t)$. Note that this premonitory sign of formation of a nearly complete subgraph is observed far earlier before $V$ reaches to $N_L$. From that standpoint, the time evolution of the subgraph is similar to cases when one-dimensional lattice is employed as an initial lattice, in which a nearly complete subgraph begins to form immediately [3, 9]. However, there is a clear difference between both cases in the time order needed to form a nearly complete subgraph.

The above discussion infers that the rate of changes in $\gamma$ from 2 to 1 becomes small as the coordination number of the initial Bethe lattice becomes large, because it becomes easier for the walker to diffuse in the Bethe lattice as the coordination number of Bethe lattice increases.

## 4. Robustness against attacks

In order to examine the robustness of the subgraph $g(t)$ against intentional attacks, we have calculated the fraction of the largest connected components of the graph after the deletion of vertices in the descending order of rank of degree. In Fig. 4, a numerical result for $g(t)$ of $12,000$ vertices is compared with the case where a graph is generated by preferential attachments in the growing network. As shown in the figure, the critical point where the subgraph is broken down is larger than that for the graph generated by preferential attachments.

Intuitively, the robustness is associated with the self-similar structure of the network, for vertices in a Bethe lattice are equivalent with each other. From the model assumptions, it is expected that the degree of a vertex is approximately proportional to the frequency of visits of walkers to the vertex. Therefore, the topological structure around the vertex with the second-largest degree is expected to be similar to that around the vertex with the largest degree. However, this similarity is not exact because vertices being considered are only in a finite

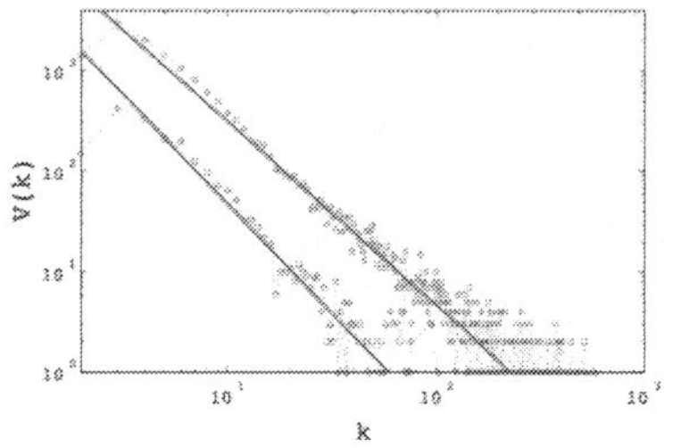


**Figure 2.** Number of vertices with degree $k$, $V(k)$, in subgraph $\mathrm{g}(t)$ at different time steps. One is the case when $V = \sum_k V(k) = 2,000$ (circles), and the other is the case when $V = 12,000$ (squares). The slopes are estimated to be $-2.2$ and $-1.83$, respectively.

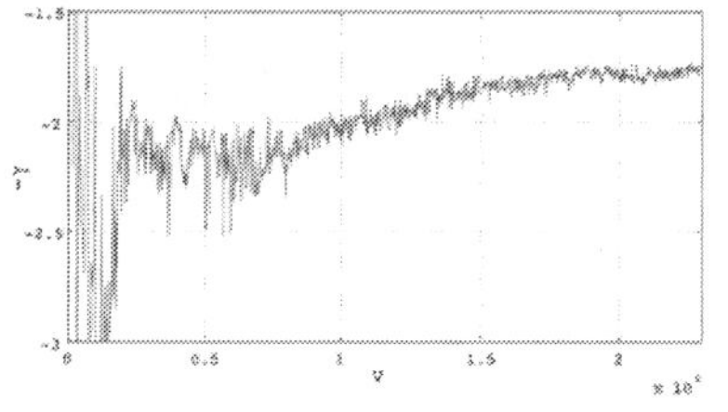


**Figure 3.** Numerical results for the dependence of power-law exponent $\gamma$ on the number of vertices $V$ in subgraph $\mathrm{g}(t)$.

subgraph $\mathrm{g}(t)$ of the Bethe lattice.

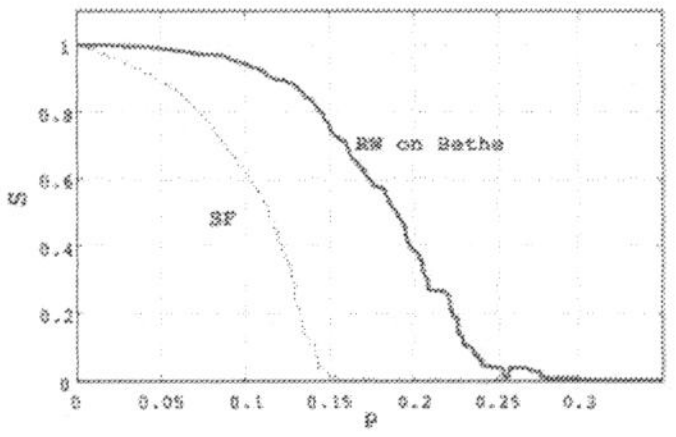


**Figure 4.** Fraction of the largest component $S$ versus fraction of deleted vertices $p$ (solid line). The number of vertices is initially $12,000$. SF represents (dotted line) a scale-free network with mean degree 4.

## 5. Summary

We have demonstrated that the addition of edges based on the movement of random walkers provides tree-like graph complex properties such as the power-law in degree distribution, highly clustered structure, and robustness against intentional attacks. The value of power-law exponent is initially about 2, and modulated to a smaller value by the further addition of edges. Although the complex structure can be kept only in a transient state until the number of edges becomes a saturated state in terms of the number of edges, the relaxation time is very large and can be controlled by some parameters including coordination number and the number of shells being considered in the initial Bethe lattice. The robust structure of the resulting network may be associated with the self-similar structure of the network. It should be noted that only a simple random phenomenon is required for obtaining the complex and robust structure of graphs. Such knowledge about controlling network properties in random phenomena should be useful in further studies on the emergence of functions of networks.

## References

[1] Newman M E J 2003 *SIAM Rev.* **45** 167
[2] Barabási A-L and Albert R 1999 *Science* **286** 509
[3] Ikeda N 2010 *Physica* A **389** 3336
[4] Boccalettia S, Latorab V, Morenod Y, Chavezf M, and Hwanga D-U 2006 *Physics Reports* **424** 175
[5] Dorogovtsev S N, Goltsev A V and Mendes J F F 2008 *Rev. Mod. Phys.* **80** 1275
[6] Cohen R, Erez K, ben-Avraham D, and Havlin S 2001 *Phys. Rev. Lett.* **86** 3682
[7] Goltsev A V, Dorogovtsev S N and Mendes J F F 2008 *Phys. Rev.* E **78** 051105
[8] Ikeda N 2009 *Mod. Phys. Lett.* B **23** 2073
[9] Ikeda N 2008 *J. Phy. A : Math. Theor.* **41** 235005

# Optimization of routing strategies for data transfer in peer-to-peer networks

**Atsushi Morioka and Akito Igarashi**

Department of Applied Mathematics and Physics, Kyoto University, Kyoto JAPAN

igarashi@i.kyoto-u.ac.jp

**Abstract**. Since peer-to-peer file-sharing systems have become familiar recently, the information traffic in the networks is increasing. Therefore it causes various traffic problems in peer-to-peer networks. In this paper, we model some features of the peer-to-peer networks, and investigate the traffic problems. Peer-to-peer networks have two notable characters. One is that each peer frequently searches for a file and download it from a peer who has the requested file. To decide whether a peer has the requested file or not in modelling of the search and download process, we introduce file-parameter $P_j$, which expresses the amount of files stored in peer j. It is assumed that if $P_j$ is large, peer j has many files and can meet other peers' requests with high probability. The other character is that peers leave and join into the network repeatedly. Many researchers address traffic problems of data transfer in computer communication networks. To our knowledge, however, no reports focus on those in peer-to-peer networks whose topology changes with time. For routing paths of data transfer, generally, the shortest paths are used in usual computer networks. In this paper, we introduce a new optimal routing strategy which uses weights of peers to avoid traffic congestion. We find that the new routing strategy is superior to the shortest path strategy in terms of congestion frequency in data transfer.

## 1. Introduction

Recently, studies on networks are extended to many fields such as human social networks, computer networks, neural networks and so on. In particular, complex networks, which have remarkable characters such as small world and the scale-free distribution of degrees, are attracting a great deal of attention [1,2,3]. Many studies have been addressed to information traffic congestion in the networks. Many users send data to other users through computer networks and traffic congestion happens. Efficient strategies for reducing the traffic congestion have been considered in recent years [4].

In this paper, we focus on peer-to-peer file-sharing networks [5,6,7] on computer networks. Since, in these days, applications for peer-to-peer networks, such as Napster, Gnutella, and Winny have been popularized, the amount of traffic in peer-to-peer networks is increasing yearly and becomes a major part of whole one in the Internet.

Therefore, avoidance of traffic congestion in peer-to-peer networks is an important problem. Although there are many papers that note the ways to reduce traffic congestion at the Internet, we recognize no papers which focus on peer-to-peer networks to our knowledge. Thus, focusing on characteristic aspects of peer-to-peer networks, we address traffic congestion on the networks.

## 2. Peer-to-peer networks

The structures of peer-to-peer networks are quite different from that of usual computer networks. There are main computers called "central servers" in the usual computer networks and all computers can be servers or clients.

### 2.1. Search for files and download of them

In the peer-to-peer networks each computer is called a "peer". A peer forwards a query of a file to its neighbors and the neighbors forward the query to their neighbors. This process is repeated until a specified threshold TTL (Time-To-Live) is reached. Here TTL is the threshold of the number of links where a search query passes. Queries are discarded if the number of the passed peers exceeds TTL. We assume that a peer can search files in the searchable domain in which all peers within a distance of TTL from the peer are included. A peer can search for a file and get data of the file from a "host peer", which has the requested file and exists in the searchable domain. Inversely, the peer can be a host peer for the other peers if it exists in their searchable domains. When a host peer sends data of a file, it is to be desired that the routing path is controlled for reduction of traffic congestion.

### 2.2. Leave and join in the peer-to-peer networks

The other character of the peer-to-peer networks is that peers leave and join in the networks by means of switching on or off the application which is needed for the peer-to-peer technique. Peers leave a peer-to-peer network for the reason that they have got files and need not connect to it any longer, and their PCs are gone down. On the other hand, peers join the network in order to get files one after another.

## 3. Model

We model a peer-to-peer file-sharing network, where peers leave and join in the network. The peer j has a file-parameter expresses the amount of stored files $P_j$. The distribution of $P_j$ is given as [3] $\psi(p) = A\exp(-5p)$, where $A$ is a positive constant, $0 \le p \le 1$, and $P_j$ is normalized as the maximum value is set to be one. In the peer-to-peer networks, there are the peers who principally supply files to other peers. Empirically, these peers rarely leave the network. We call such peers "file suppliers", whereas peers who hardly supply files and quite often leave the network after having just joined are called "free riders". Therefore, we assume that the less files peers have, the more often they leave the network.

On the other hand, peers newly joining in the networks need information of peers who are already in the network in order to connect to the other peers. Since it is difficult to find information by themselves for the new peers, the administrator of the application for the peer-to-peer network gives information about peers. The administrator tends to select nodes which are in the network for a long time, because he or she hope that the network is held firm. Thus when peers join in the network, they often provide links to selected peers, who are file suppliers, namely have many files. To sum up these things, peers join in and leave the network in accordance with the next procedure.

- Some peers leave the network. The smaller $P_j$ is, the more often peer j leaves.
- Due to the leaves, peers whose degree is 0 often appear in the network. Therefore in this time the respective peers add a link to a randomly selected peer whose degree is nonzero.
- The leaving peers join into the network again. The larger $P_j$ is, the more often the rejoining peer connects a link to the peer j. The number of neighbors of the rejoining peer is set to be from 3 to 7 at random.

We obtain good agreement between theoretical and simulation results

## 4. Betweeneness and routing strategy

First, a peer A searches for files which they want and forwards a query to its neighbour peers. This searching method is called "flooding". After the search, if the file which peer $A$ wants exists in the searchable domain, peer $A$ downloads the file from peer $B$ which has it in the domain. If a lot of data concentrate on a node, the node cannot transfer data to next node immediately, and it takes a lot of time to send files. Therefore, we need disperse transfer routing paths.

### 4.1. Betweenness

We consider that paths for all pairs of nodes in the network are defined according to a rule, for example, choosing the shortest path and so on. Betweenness of a peer is defined as a measure for concentration of optimal paths at the peer.

### 4.2. Routing Strategy

First, we introduce the shortest path strategy. Transport routing strategies used on computer networks have been solely determined based on the shortest paths. However in the case of shortest path strategy, traffic concentrates on peers with a great many degrees, and one of the peers have the big maximum betweenness $B_{max}$. In finding optimal (smallest) $B_{max}$, we need to disperse routing paths. For the purpose, we employ the smallest weight path algorithm, that a path with the smallest sum of weights assigned to peers along the path is chosen. We show $B_{max}$ at both the shortest path strategy and new strategy which uses the smallest weight path algorithm.

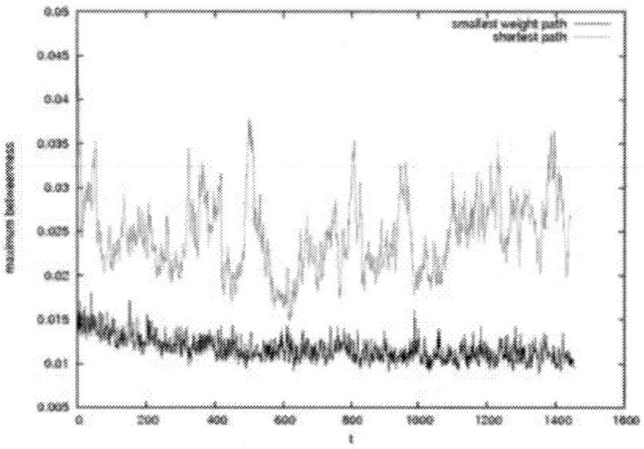


Fig1. $B_{max}(t)$ is plotted as a function of t. The grey line represents the results with the shortest path algorithm and the black line those with the smallest weight path algorithm where the number of peer N=2000 and TTL=4. The number of peers which leaves and joins at one step is 50

A. Shortest path algorithm

1. *Leaves and joins of peers are performed.*

2. *Select efficient paths, where the total summation of the number of the links in each path is minimum between each pair of peers in the searchable domain.*
3. *Compute the betweenness of every peer for the efficient paths.*
4. *t=t+1 and repeat 1 to 3.*

B. Smallest weight path algorithm

1. *Assign the initial weight* $W_j(0) = k_j^{0.7}$ *to every node , where* $k_j$*. Denotes the degree of peer j*
2. *Leaves and joins of peers are performed*
3. *Select efficient paths, where the total summation of the weights of the peers in each path is minimum between each pair of peers in the searchable domain.*
4. *Compute the betweenness of every peer for the efficient paths.*
5. *Update the weight of nodes* $W_j(t)$ *at t which satisfies* $\frac{B_j(t)}{B_{max}(t)} > 0.9$ *as follows:* $W_j(t+1) = W_j(t) + \left(\frac{B_j(t)}{B_{\max}(t)}\right)^{100}$ *where* $B_j(t)$ *is the betweenness of peer j at t and* $B_{max}(t) = \max_j B_j(t)$
6. *t=t+1 and repeat 2 to 5*

### 4.3. Calculation $B_{max}$(t)

In Fig.1, we show $B_{max}(t)$ for both the shortest path algorithm and smallest weight path algorithm. We find that smallest weight path algorithm is superior to shortest path algorithm because $B_{max}$of the algorithm is stably lower than one of the shortest path algorithm.

## 5. Data Transfer

### 5.1. Data transfer model

1. *Select randomly a peer who wants a file.*
2. *The peer A selected in 1. searches for peers who have the requested file in its searchable domain by flooding.*
3. *A investigates the number of files P which each peer in searchable domain has, and the larger P a peer has, the more often the peer (peer B) holds the requested file and peer B is selected as a peer who sends its data to peer A selected in 1.*
4. *Peer B starts data transfer to peer A. At one step, a peer sends data to a neighbour.*

We perform these four procedures with some peers simultaneously. To perform data transfer with many peers lead to traffic congestion. In next section we define congestion in data transfer.

### 5.2. Congestion

Through data transfer, we assume that all peers can send two different data to its neighbour at once. If a peer receive three different data at the same step, at next step the peer can send only two data, and consequently it leads one data to stop at the peer . We define this phenomenon as congestion.

| *The number of peers requiring files* | *Shortest Path* | *Smallest Weight Path* |
|---|---|---|
| | *congestion* | *Congestion* |
| *300* | *36* | *12* |
| *400* | *109* | *59* |
| *500* | *149* | *89* |
| *600* | *233* | *146* |

Table.1 The number of leaving peers is 50. The value is the number of times traffic congestion occurs.

### 5.3. Simulation

When leaving peers are selected in the path of data transfer, the transfer path changes in order to send data to the peer who wants a file as much as possible.

From Table 1, we find that the number of times traffic congestion occurs is smaller in the case of the smallest weight path than in the case of the shortest path.

## References


[1] D.J. Watts, S.H. Strogatz, "Collective dynamics of small-world networks", Nature Vol.393 (1998) pp.440-442.

[2] A.L Barabasi, R.Albert, "Emergence of scaling in random networks", Science Vol.286 (1999) pp.509-512.

[3] M.E.J. Newman, "The structure and function of complex networks", SIAM Review Vol.45 (2003) pp.167-256..

[4] H. Kawamoto and A. Igarashi, "Efficient packet routing strategy in complex networks", Physica A vol.391 (2011) pp.895-904.

[5] M. Ripeanu, I. Foster and A. Iamnitchi, " Mapping the Gnutella network: Properties of large-scale peer-to-peer systems and implications for system design", IEEE Internet Computing Journal vol.6 ,no.1 pp.50-57 (2002)..

[6] D. Stutzbath and R. Rejaje, "Charactering today's gnutella topology", Technical Report, CIS-TR-04-02, Department of Computer Science, University of Oregon (2004).

[7] T. Maruyama, "Virus spreading model of peer-to-peer networks", Master's thesis, Department of Applied Mathematics and Physics, Graduate School of Informatics, Kyoto University (2009)

# Cumulation of a spherically converging shock wave in metals and its dependence on elastic-plastic properties, phase transitions, spall and shear fractures

**E A Kozlov, A V Petrovtsev**
Russian Federal Nuclear Center – Zababakhin Research Institute of Technical Physics, Snezhinsk, Chelyabinsk Region, Russia

E-mail: ea_kozlov@mail.ru

**Abstract.** Consideration is given to results of experimental and theoretical investigations how alpha-epsilon- phase transition in the unalloyed iron and the 30KhGSA steel and its absence in the austenitic 12Kh18N10T stainless steel influence processes under explosive deformation of spheres made of these materials. Polymorphous transition is shown to significantly effect on:
- amount of explosion-products energy transferred to a sphere,
- evolution of the converging-wave structure and its parameters, profiles of stress wave and temperature T(R,t) for some Lagrangian particles along the sphere radius,
- character of energy cumulation under spherical convergence of waves.

## 1. Introduction

The problem on the focusing of a spherically converging shock wave is a classical problem. G.Guderley [1] was the first to consider this problem for an ideal gas having equal initial density along the radius. Later, analogous self-similar solution was constructed by L.D. Landau and K.P. Stanyukovich [2,3]. Results of follow-on numerous investigations with a gas sphere having initial density that decreases or increases to its center were published in papers by G.M. Gandelman, D.A. Frank-Kamenetsky [4], F.L. Chernousko [5], K.V. Brushlinsky, Ya.M. Kazhdan [6], Yu.S. Vakhromeev [7], and Tyl I., Wlodarczyk [8].

All these investigations consider the case when properties of the matter closely correspond to the ideal gas. This is not just the situation we have in practice. Therefore, the follow-on works made attempts to estimate how the considered phenomenon and, more specifically, character of energy cumulation variation depend on real properties of the matter. E.I. Zababakhin contributed seriously to these investigations [9]. So, paper [10] considered the case with the heat-conducting gas and paper [11] – the case with the solid spheres. In the latter case a converging wave had the two-wave configuration due to shear strength or phase transition. Self-similar solutions constructed in [11] corresponded to the flow near the center when the principal plastic wave failed to catch up with the converging elastic or phase precursors. Idealized representations were used for the description of the elastic-plastic deformation and phase transition in the material at which the infinite cumulation is absent when these waves converge to the center. The idea on the finite energy cumulation, developed by E.I. Zababakhin, further was confirmed by experimental investigations [12,13] in which the spherical-shock-waves-recovery-experiments allowed rather detailed study how the cumulation character in the case of different-intensity stress waves convergence depends on real properties of materials.

Effects of energy cumulation decrease at the front of a converging stress wave in the above experiments due to the elastic-viscose-plastic deformation and alpha-epsilon-phase transition in the compressed-sphere material (iron) were computationally and theoretically studied in [14] and then – in [15, 16]. The purpose of this work is to obtain more detailed data on the aspect under consideration.

## 2. Materials, samples, and conditions of their explosive loading

Spheres made of iron, Steel 3, the quenched 30KhGSA steel (HRC=35...40), and the 12Kh18N10T steel [12,13,17-21] were investigated experimentally and theoretically. The diameters of these spheres were 40, 48, 60, 64, 166, and 184 mm. Detonation of the RDX- or HMX-based explosive layers with the outer radius $R_{HE}$ = 40 or 110 mm and different initial thicknesses was used for explosive loading of these spheres. Note that as for the shock-wave properties (figure 1) – both possible formation of multi-wave configurations and energy dissipation amount under shock-wave deformation, the investigated materials can be arranged as follows: (i) the 12Kh18N10T steel having no polymorphous alpha-epsilon-phase transition and having small values of the dynamic yield stress ($Y_A$≈0.25 GPa...1.0 GPa where the first value corresponds to the stress at the elastic precursor with the 0.45 GPa amplitude and the second value takes into account strain hardening in stress waves without influence of pressure and temperature); (ii) Iron and Steel 3 having polymorphous transitions and approximately the same ($Y_A$≈0.65 GPa) shear stresses [22-24]; (iii) the 30KhGSA steel quenched up to HRC=35...40 having the same alpha-prime-epsilon-phase transition at essentially higher ($Y_0$≈1.0...2.3 GPa, $\sigma^{HEL}$=1.8 GPa) shear stresses [25,26].

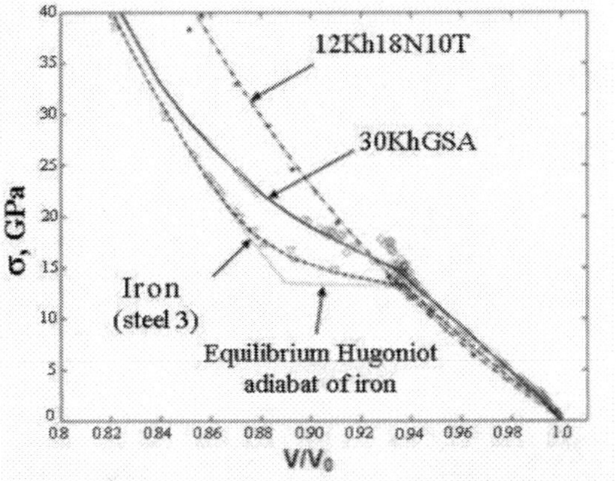


Figure 1 – Hugoniot adiabat for iron [22-24] or steel 3 [12], the quenched 30KhGSA steel [25, 26] and the same steel as received [12], as well as for the 12Kh18N10T steel [12,27].

Furthermore, metastability of the alpha-prime-epsilon- ($\alpha'$-$\varepsilon$) phase transition is noted to have a more pronounced character in the quenched 30KhGSA steel [25, 26] as compared to that of iron [22-24] and this results in a more pronounced dissipation of energy in the loading – unloading cycle.

## 3. Numerical computations

Behavior and properties of iron and steels were described with the help of models based both on the multi-phase equations of state and the governing relationships used to describe the high-rate elastic-viscous-plastic deformation, kinetics of phase transitions, incipience and development of fractures and their recompaction, as well as to describe interrelation between these processes.

If the multi-phase equations of state [28,29] describe thermodynamic properties of the solid (three - $\alpha$, $\gamma$, $\varepsilon$ - for iron and one – for the 12Kh18N10T steel) and liquid phases, then the equations of state for liquid phases – are wide-range equations that take into account evaporation as well. The model well described in [16,30] was used to simulate the elastic-viscous-plastic properties, phase transitions, and fractures. This model takes into account strain hardening of a material similarly to [31], power-law dependence of shear stresses on strain rate [32], the Bauschinger effect [31,33], and influence of fractures and phase transitions available in the material on the elastic-viscous-plastic properties of this material. Phase transitions are described based on the relaxational-type kinetic equations [34] which are generalized to the case of several transitions simultaneously proceeding in a material particle. Characteristics of the model are such that the model takes into account real metastability and dependence of time required for transition to proceed in the test materials [12,16,24,25,30] on the degree of non-equilibrium states. A fracture is described by kinetic equations of nucleation and growth (similar to those in [35]), which are also capable to describe the reverse process, i.e. recompaction of these fractures. The model takes into account hardening of materials due to polymorphous transitions in them under shock-wave loading [36] and their softening under melting. Parameters of models were determined based on measurement results for shock compressibility, sound velocities, and stress wave profiles in the materials under consideration.

Such a computational setup was used, which took into account details of the experimental systems design and the loading process was simulated up to rather long time moments that correspond to their actual stop.

Computations were performed with the help of the VOLNA program complex which is intended for 1D-numerical simulation of continuum dynamics [37] and includes the above models. Potentialities of this program complex in determining the exact shock fronts and characteristics of rarefaction waves in calculations of the flow inside the spheres were not previously applied as the models that take into account strain rate of a material and kinetics of processes were usually used. The required accuracy in computations of shock-wave amplitudes especially near the center where they reach high values, as well as sufficiency of energy dissipation at the front of a converging wave (for correct description of temperature and phase state of particles in a material) was ensured through the use of a rather fine grid (≈5000 spacings with the abovementioned sizes of samples).

## 4. Results and their discussion

*4.1 Simulation of experiments with the explosive system having $R_{HE}$= 110 mm*

General data on the character of processes in the samples are shown in figures 2, 3 for the iron sphere with the 166-mm initial diameter.

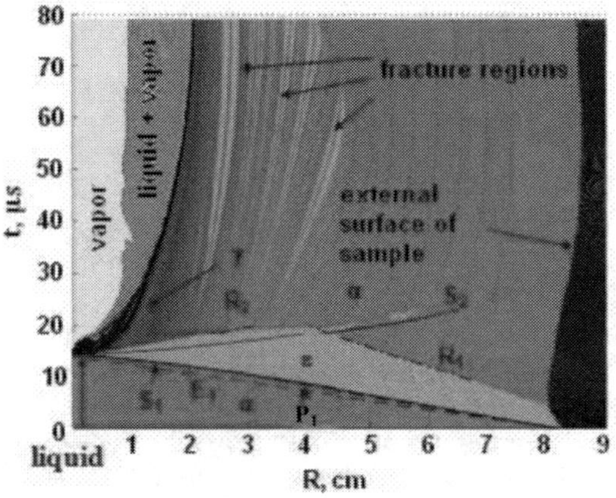


Figure 2 – R-t diagram of the waves and contact boundaries movement in the experiment with the iron sphere having 166-mm initial diameter

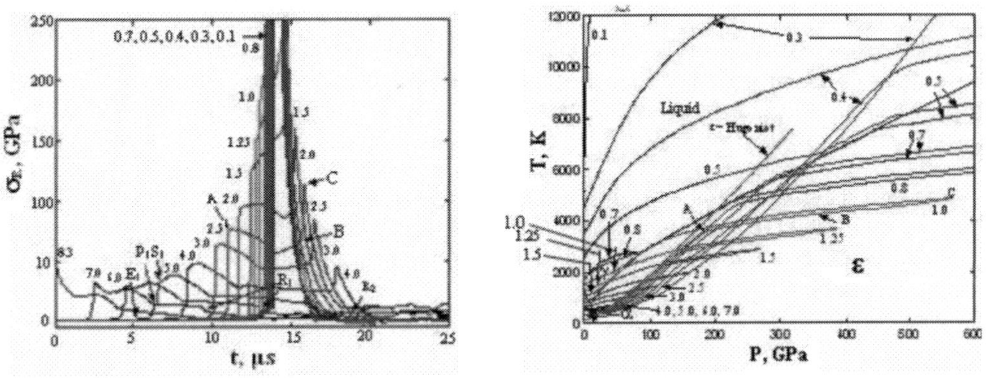

a) b)

Figure 3 – Time profiles of stresses (a) and P-T histories of particles strain (b) in the iron sample with the 166-mm diameter.
Numbers near the curves – Lagrangian radii of particles in cm.

A pressure pulse induced by detonation of the HE layer at the external surface of the sphere creates a stress wave propagating towards the center. At the initial state, the wave amplitude decreases with the wave propagation due to unloading from the side of the scattering explosion products (figure 3a).

Depending on the loading intensity and impulse duration, conditions are created for the generation in iron of the multi-wave configuration with the formation, along with the principal plastic wave $S_1$, of the elastic $E_1$ (at $\sigma_R \leq 70$GPa in iron or at $\sigma_R \leq 40$GPa in 12Kh18N10T steel) and phase $P_1$ (at $\sigma_R \leq 35$GPa in iron, no in steel 12Kh18N10T) precursors. The structure of the waves being formed and those converging to the center, as well as the level of their spatial divergence depend on the explosive loading conditions. In the principal plastic wave, iron particles transit from the state of the initial alpha-phase into the epsilon-high-pressure phase. In further propagation of the waves, the amplitude and velocity of the principal wave $S_1$ begin to grow because of spherical convergence. So, at a certain radius, the multi-wave configuration can totally or partially stop to exist as the front of the principal shock wave catches up with the phase precursor and then with the elastic precursor. When the principal wave amplitude becomes $\sigma_R \geq 200$ GPa, the calculation shows that the matter gets melted at the principal wave front. When the wave is reflected from the center, behind the diverging-wave front one observes material unloading into the formed cavity with obtaining the vapor and then, with the distance from the center, the vapor-liquid mixture and the high-temperature gamma phase (figure 2). The reverse transition of the epsilon– to– alpha-phase takes place in the sample's peripheral part constituting the major part of the sample. The rarefaction waves propagating from the cavity and from the external surface of the sample incorporate rarefaction shock waves designated in figure 2 as $R_2$ and $R_1$, respectively. Interaction of rarefaction waves lead to the formation in the sample of the fracture regions concentrically located above the cavity. Then due to the cast-iron case that is located above the HE layer and generates a number of additional compression pulses at the surface of the sample, the early formed spall fractures can be recompacted (figure 2). This character of high-rate deformation, fracture, and recompaction of materials after spherical explosive loading was noted in [12,13,19,20] during metallographic and electron-microscopic studies of recovered samples of Fe and steels.

P-T histories of particles strain in the P-T diagram (figure 3b) show that the main portion of the melt in the iron sample with ∅ 166 mm is formed under unloading of the matter heated by the principal wave rather than melting at the shock wave front. Though pursuant to calculations for iron, the shock wave amplitude $P^H \approx 180$GPa corresponds to the melting onset under unloading and the wave amplitude $P^H \approx 200$GPa corresponds to the melting at the shock wave front.

Figure 3a gives time histories for the radial component of stresses for certain Lagrangian particles. One can see that loading of material particles has a rather complicated multi-wave character. Thereby near the center, compression waves formed by the flow converging to the center are observed at the compression profiles between shock fronts of the converging multi-wave configuration, as well as between the front of the principal wave and that of the wave reflected from the center. Due to multi-wave and substantially non-shock character of deformation of material particles they demonstrate seriously less entropy and thus maximum temperatures in these particles are substantially lower under the same pressures if compared with temperatures at Hugoniot adiabats that correspond to their shock compression by a single wave. Quasi-isentropic nature of particles compression and variations of this nature due to cumulation under the wave convergence depend on the materials properties and the loading profile.

The pulse amplitude and duration can seriously change the character of multi-wave configurations realized in the iron sphere, as well as the spacing between waves in this sphere. For example, numerical simulation demonstrates that with the increase of the HE layer thickness due to the decrease in the diameter of the loaded iron sphere from 184 down to 166 mm in an appropriate explosive system [13,20,21], mass of the molten central area in this sphere increases 15 times. This is associated with the fact that with the 184-mm diameter of the sphere, the elastic and phase precursors run far ahead the converging principal wave and the single-wave mode of particles loading fails to form while in the second case (figure 2) the area of phase precursor existence is rather small.

*4.2 Simulation of experiments with explosive systems having $R_{HE}$= 40 mm*

Figures 4 and 5 give the results of numerical simulation for experiments [13,17] with explosive loading of the 64-mm diameter spheres made of different materials.

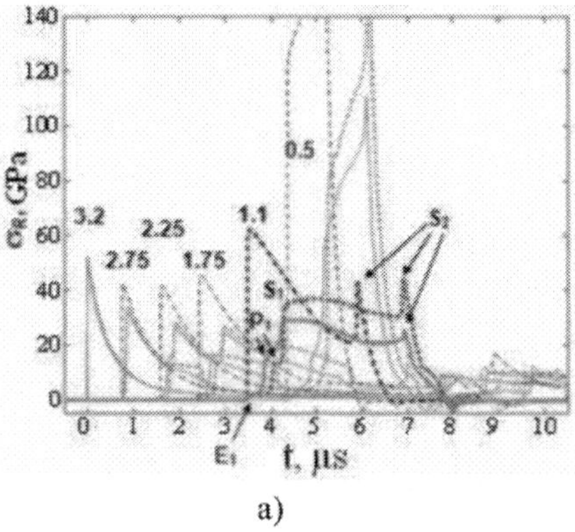


a)

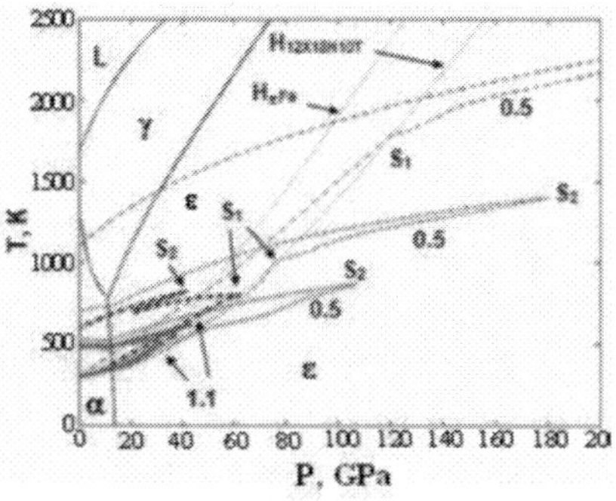


b)

Figure 4 – Time profiles of stresses (a) and trajectories how particles of steel grades under consideration deform in the pressure-temperature coordinates (b) for explosive experiments [13,17] with the 64-mm diameter spheres. Calculated data for the quenched 30KhGSA steel are shown as solid lines, for iron – as dashed lines, and for the 12Kh18N10T steel – as dash-dotted lines. Numbers near the curves – are Lagrangian radii of particles in centimeters. Thick solid lines in the P-T diagram (b) show stability areas for the solid and liquid phases of iron. The melting line for the 12Kh18N10T steel in this pressure area is close to the melting line of iron.

Figure 4a shows that in these experiments with samples of Steel 3 and the quenched 30KhGSA steel, the converging wave represents a three-wave configuration at medium radii and only one front is observed in the 12Kh18N10T steel. Meanwhile, the temperature decrease for the material compressed by the converging wave is most evident in the first two cases.

At the initial stage of motion, the converging wave attenuation is quite evident and it becomes even more evident in steels with polymorphous transitions especially in the quenched 30KhGSA steel. And the cumulation effect for these materials at the next stage of motion develops much later. So, the initial amplitude of the wave "recovers" due to the spherical convergence only at a rather deep radius R≈6 mm (or $0.15R_{HE}$) if compared with the similar radius R≈15 mm (or $0.375R_{HE}$) for the 12Kh18N10T steel. The noted differences in characteristics of the converging wave attenuation are also observed in the diverging wave traveling. In the case with the quenched 30KhGSA steel, the amplitude of this wave near the external surface of the sample becomes very small. This indicates that the reverse transfer of energy to the explosion products under sphere unloading in this case turns out to be much weaker. Table 1 gives quantitative data on the above effects.

Table 1 – Calculated characteristics of the converging and reflected waves in the samples

| Material | Trajectory, mm | Converging wave | | Diverging wave | |
|---|---|---|---|---|---|
| | | $\sigma_{S1}$, GPa | $T_{S1}$, K | $\sigma_{S2}$, GPa | $T_{S2}$, K |
| 12Kh18N10T | 2.25 | 41.5 | 568 (553[1]) | 10.2(6.5[2]) | |
| | 1.1 | 62.1 | 800(798) | 43.1(21.9) | |
| | 0.5 | 122 | 1787(1778) | 257(104) | 2406 |
| Iron | 2.25 | 28.1 | 501 | 6.4(3.0) | |
| | 1.1 | 34.1 | 528 | 44.6(14.1) | |
| | 0.5 | 77 | 996(1363) | 182(49) | 1404 |
| 30KhGSA HRC 35...40 | 2.25 | 25.9 | 475 | 4.1(1.9) | |
| | 1.1 | 29.0 | 471 | 25.5(3.7) | |
| | 0.5 | 54.7 | 610(895) | 111(11) | 874 |

[1] Temperature at the hydro dynamical Hugoniot adiabat
[2] Jump of a radial stress component in the reflected wave

All noted factors, i.e. decrease of the wave amplitude due to stronger attenuation, weaker cumulation, and more isentropic character of loading, lead to the decreased values of loading amplitudes and decreased values of temperatures for iron particles and particles of the 30KhGSA steel in the central area. Hence, the amount of the molten matter for these materials after the cavity formation seriously decreases if compared to the 12Kh18N10T steel (Table 2) with rather close melting temperatures at P=0 in the equations of state for these materials and namely: $T_{m0}$=1672 K for the 12Kh18N10T steel and $T_{m0}$=1696 K for iron. Actually, in the case with the quenched 30KhGSA steel, in the explosive loading mode with the 64-mm sphere, the molten area is very small (in practice it is absent at all) and the cavity is formed owing to the fracture of the material in the solid phase under uniform extension just as for example in [38].

Table 2 – Data on energy transferred to the sphere and on sphere deformation after explosion

| Material | Final radius, mm | | Calculated radius of melting zone, mm | Transferred energy, kJ | |
|---|---|---|---|---|---|
| | Experiment [13] | Computation | | Experiment [35] | Computation |
| 12Kh18N10T | 32.46 (1.0144[1]) | 32.67 (1.021[1]) | 0.33 (0.25[2]) | 102 (1.000[3]) | 136 (1.000[3]) |
| Iron (Steel 3) | 32.34 (1.0064) | 32.44 (1.014) | 0.16 (0.13) | 115 (1.137) | 155 (1.138) |
| 30KhGSA HRC 35...40 | 32.26 (1.0081) | 32.22 (1.0069) | 0.052 (0.042) | 129 (1.269) | 176 (1.290) |
| Copper | 32.91 (1.0285) | 32.97 (1.030) | | 100 (0.984) | 134 (0.980) |

[1] Deformation of the outer radius of the sphere in per cents
[2] Boundary of the complete-melting area
[3] Relative value if compared with the sphere of the 12Kh18N10T steel

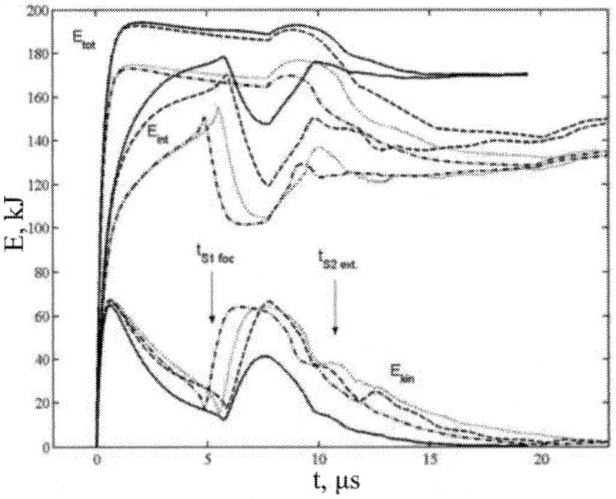


Figure 5 – Time variations of kinetic, internal, and total energy of spheres from the quenched 30KhGSA steel (solid lines), iron or Steel 3 (dashed lines), the 12Kh18N10T steel (dash-and dotted lines), and copper (dotted line).

Figure 5 shows a time-dependent integral quantity of the kinetic and internal energy of spheres from different materials, as well as a time-dependent total energy that corresponds to the work of explosion products at the surface of these spheres. Calculation data are additionally given for copper, i.e. material with no polymorphous transitions and with the minimal ($Y_A \approx 0.12...0.3$ GPa) shear strength from among all materials under consideration. As seen, at the initial stage of loading due to greater compressibility, non-equilibrium nature of the polymorphous alpha-epsilon-phase transition, and rather high shear stresses, the spheres from iron and the 30KhGSA steel get seriously greater part of explosion products energy as compared with the spheres from materials having no phase transition. In progress of the sphere deformation, the stronger energy dissipation takes place in the quenched 30KhGSA steel due to greater non-equilibrium of the $\alpha'$-$\varepsilon$ phase transition and high shear stresses;

thus, the reverse energy transfer to explosion products decreases. The internal energy in the 30KhGSA sphere gets larger and the kinetic energy – grows smaller as compared with the iron sphere and also spheres from other materials. This difference is observed to be maximum after the reflected wave arrives at the external surface of the sphere. Due to less cumulation of the converging wave, the internal-to-kinetic energy transfer when the wave is reflected from the center of the sphere from this material is noted to be minimal. Table 2 gives qualitative data on the energy E transferred by explosion products to the spheres. As seen, the calculated relation E for different materials is very close to the calorimetric measurement results presented in [17]. In addition, a qualitatively correct relation is observed between the final dimensions of spheres and, respectively, the dimensions of cavities just as in experiments [13,20]. Certain differences in absolute values can be associated with the use of 1D approach in computations and with possible 3D losses in experiments [21]. Further advancement of the computational and theoretical description, as well as experimental techniques to diagnose different stages of phenomena under study seems reasonable.

## 5.Conclusion

Experiments and their numerical simulations confirmed strong dependence of the energy cumulation character at the front of the converging shock wave on the polymorphous transition in iron and steels.

## 6. References

[1] Guderley G 1942 Starke kugelige und zylindrische Verdichtungsstöße in der Nähe des Kugelmittel punkte bzw. der Zylinderachse *Luftfahrtforschungsanstalt* **19**, 9, 302

[2] Landau L D and Stanyukovich K P 1945 See Stanyukovich K P 1955 Nonsteady flow of continuum. *(M.: Gostekhteorizdat, in Russian)*

[3] Landau L D and Lifshits E M 1986 *Hydrodynamics. Theoretical Physics* M.: Nauka, vol 6.

[4] Gandelman G M and Frank-Kamenetsky D A 1956 *Rus. J. Doklady Akademii Nauk (Reports Russian Academy of Science)* **107** issue 6 811

[5] Chernousko F L 1960 Converging shock waves in a variable-density gas *Rus. J. Applied Mathematics and Mechanics* **24** issue 5, 885

[6] Brushlinsky K V and Kazhdan Ya M 1963 On self-similar solutions of some gas-dynamic problems *Rus. J. Advances of Mathematical Sciences* **18** issue 2 3

[7] Vakhromeev Yu S 1966 Cumulation of shock waves in non-uniform medium *Rus. J.Applied Mathematics and Mechanics* **30** issue 4 774

[8] Tyl I and Wlodarczyk 1985 An analysis of the concentric shock wave in a nonhomogeneous polytrophic gas *J. Technical Physics* **26** issue 1.

[9] Zababakhin E I 1965 Cumulation of energy and its boundaries *Rus. J. Advances of Physical Sciences* **85** issue 4 721

[10] Zababakhin E I and Simonenko V A 1965 Converging wave in a heat-conducting gas *Rus. J. Applied Mathematics and Mechanics* **29** issue 2 334

[11] Zababakhin E I and Zababakhin I E 1988 Infinite cumulation phenomenon (*M.: Nauka)*

[12] Kozlov E A 1992 Shock adiabat features, phase transition macrokinetics, and spall fracture of iron in different phase states *High Pressure Research* **10** 541

[13] Kozlov E A 1992 Experimental verification of E.I.Zababakhin hypothesis concerning limitation of energy cumulation in the spherically converging shock-wave front in medium with phase transitions *Shock Compression of Condensed Matter – 1991*, eds S C Schmidt et.al (Elsevier Science Publishers B.V.) pp 169-172

[14] Kozlov E A and Zhukov A V 1994 Phase transitions in spherical stress waves *High Pressure Science and Technology – 1993*, eds S C Schmidt, J W Shaner, G A Samara and M Ross (NewYork: American Institute of Physics) pp 977-980.

[15] Afanasieva S A, Belov N N and Konyaev A A 1998 Computer simulation of materials behavior under shock-wave loading (*Rus.* J. *Izvestia Akademii Nauk), Proceedings Academy of Science. Solid-State Mechanics* **5** 115

[16] Kovalenko G V and Petrovtsev A V 2001 Numerical simulation of elastic-viscous-plastic properties, phase transitions, and fractures in iron *VI Zababakhin Scientific Talks* (Snezhinsk,

Chelyabinsk region, Russia, September 24-28 2001) pp 208-209 (Abstract, full text: <www.vniitf.ru/rig/konfer/6zst/reports/s5/5-45.pdf>).
[17] Vildanov V G, Gorshkov M M, Kozlov E A, Tkachev O V and Yusupov D T 2003 Solid-state calorimeter technique to measure residual energy of spheres and shells after explosive compression *VII Zababakhin Scientific Talks* (Snezhinsk, Chelyabinsk region, Russia, September 8-12, 2003) pp 185 (Abstract, full text: <www.vniitf.ru/rig/konfer/7zst/reports/s5/5-35.pdf>)
[18] Kozlov E A , Taluts N I, Dobromyslov A V 2005 High-rate deformation of armco-iron loaded by spherically converging shock waves *Physics of substance extreme state* (Chernogolovka, Moscow region: Published by Rus. Institute of Chemical Physics Problems) pp 75-77
[19] Dobromyslov A V, Kozlov E A, Litvinov B V and Taluts N I 2007 High-rate deformation of armco iron under loading by spherically converging shock waves *Proceedings of Russian Academy of science Doklady Physics* (Engl. transl.) **52**, issue 8 418
[20] Kozlov E A, Dobromyslov A V, Taluts N I and Voltz K 2012 Effect of spherically converging shock waves on deformation and phase behavior of high-purity iron *J. Physics of Metals and Metallography* (Engl. transl.) **113** issue 10 1007
[21] Kozlov E A, Brichikov S A, Boyarnikov D S, Kuchko D P and Degtyarev A A 2011 Special features in convergence dynamics of steel shells under their explosive loading. results of laser-interferometric measurements *J. Physics of Metals and Metallograph* (Engl. transl.) **112** issue 4 389
[22] Bancroft D, Peterson E L, Minshall S 1956 Polymorphism of iron at high pressure *J. Appl. Phys* **27** 291
[23] Barker L M, Hollenbach R E 1974 Shock wave study of the α-ε phase transition in iron *J. Appl. Phys* **45** 4872
[24] Kozlov E A, Telichko I V, Gorbachev D M, Pankratov D G, Dobromyslov A V and Taluts N I 2005 On the metastability and incompleteness of the α–ε phase transformation in unalloyed iron under the threshold loading pulses: specific features of the deformation behavior and structure of armco iron *J. Physics of Metals and Metallography* (Engl. transl.) **99**, issue 3 *300*
[25] Kozlov E A, Tarzhanov V I, Telichko I V and Pankratov D G 2012 Shear and spall strength of hardened 30KhGSA steel under explosive loading in the range of solid-solid phase transformation *Rus. J. Deformation and Fracture of Materials* **8** 32
[26] Kozlov E A 2012 2D-, and 3D-explosive experiments for verification of spall and shear strengths models for some steels *17th APS Conference on Shock Compression of Condensed Matter (* Chicago, Illinois, USA, June 26 – July 1, 2011) eds M L Elert, W T Buttler, J P Borg, J L Jordan and T J Vogler (Melville, New York: AIP CP 1426) pp. 945-948
[27] Belyakova M Yu, Zhernokletov M V, Sutulov Yu N and Trunin R F 1991 Shock compression of metal alloys (*Rus. J. Izvestia Akademii Nauk) Physics of the Earth* **1** 99
[28] Dremov V V, Kutepov A L, Petrovtsev A V, and Sapozhnikov A T 2002 Equation of state and phase diagram of iron *Shock Compression of Condensed Matter – 2001( Atlanta, Georgia, USA, 24-29 June 2001)* eds M D Furnish, N N Thadhani and Y Horie (Melvill, NewYork: AIP Conference Proceedings #620) pp 87-90
[29] Zadorozhny G A, Petrovtsev A V and Dremov V V 2005 Equation of state for the 12Kh18N10T steel *VII Khariton Scientific Talks (Sarov, Nizhniy Novgorod region, Russia, March 14–18, 2005)*
[30] Zadorozhny G A, Kovalenko G V and Petrovtsev A V 2003 Modeling of the polymorphous α-ε phase transition in iron in a wide range of states. *VII Zababakhin Scientific Talks (Snezhinsk, Chelyabinsk region, Russia, September 8-12, 2003)* p 183 (Abstract, full text - <www.vniitf.ru/rig/konfer/7zst/reports/s5/5-45.pdf>)
[31] Steinberg D J, Cohran S G and Guinan M W 1980 A constitutive model for metals at high-strain rate *J. Appl. Phys* **51** 1498
[32] Swegle J W and Grady D E 1985 Shock viscosity and the prediction of shock wave rise times *J. Appl. Phys* **58** 692
[33] Moss W C and Glenn L A 1984 A Bauschinger effect model for use in large computer codes *Shock Waves in Condensed Matter – 1983* eds J R Asay, R A Graham, and G K Straub (American Institute of Physics) pp 133-136
[34] Andrews D J 1971 Calculation of mixed phases in continuum mechanics *J. Comp. Phys* **7** 310

[35] Belov N N, Korneev A I and Nikolaev A P 1985 Numerical analysis of fracture in pulsed-loaded plates. *Rus. J. Applied Mechanics and Technical Physics* **3** 132

[36] Bychenkov V A, Kovalenko G V, Kozlov E A, Lobachev S V, Petrovtsev A V and Zhugin Yu N 2003 Spallation of armco-iron and 30KhGSA steel under explosive loading in the region of α-ε phase transition: experiment and calculation *Proceeding of Fifth International Symposium on Behaviour of Dense Media under High Dynamic Pressures (Saint-Malo, France, 23-27 June, 2003 vol 1)* pp 431-434

[37] Kuropatenko V F, Kovalenko G V and Kuznetsova V I 1989 Complex of VOLNA codes and non-uniform finite difference method to calculate the compressible media flow *Russian J. Issues of Atomic Science and Engineering. Series: Techniques and Codes for Numerical Solution of Mathematical Physics Problem*, issue **2** 9

[38] Bakhrakh S M, Kovalev N P, Nadytko B A, Novikov S A and Chernyshev A K 1974 Study of plastic and strength properties of copper in conditions of uniform extension *Rus. J. Doklady Akademii Nauk* **215**, issue 5 1090

# Numerical 3D-modeling of spall and shear fractures in shells of austenitic 12Kh18N10T steel and 30KhGSA steel under their spherical and quasi-spherical explosive loading

**E A Kozlov [1], O V Ol'khov[2], E V Shuvalova[2]**
[1] Russian Federal Nuclear Center – Zababakhin Research Institute of Technical Physics, Snezhinsk, Russia,

e.a.kozlov@vniitf.ru

[2] Russian Federal Nuclear Center – Research Institute of Experimental Physics, Sarov, Russia,

oleg-olkhov@yandex.ru shev@md08.vniief.ru

To pursue VNIIEF-VNIITF joint investigations, this paper briefly describes the experimental setup and provides numerical 3D-computation results (LEGAK-3D technique) on special features in the convergence dynamics of steel shells under their quasi-spherical explosive loading in the system with the 40-mm outer radius of the explosive layer.
The computation results were compared with the data experimentally registered for shells of the high-purity and technical-purity unalloyed iron, the 30KhGSA steel, both as-received and quenched to HRC 35...40, and the austenitic 12Kh18N10T stainless steel. The comparison was also made with laser-interferometry results obtained directly under explosive loading, as well as with gamma-tomography and scanning electron microscopy investigations of the recovered shells.

## 1. Introduction

This paper describes the results of numerical 3D-modeling (LEGAK-3D technique[1,2]) of experiments aimed to study the convergence dynamics and the state of steel shells under their quasi-spherical explosive loading in the system with the 40-mm outer radius of the explosive layer [3-11].

The computation results were compared with the data experimentally registered for shells of the high-purity and technical-purity unalloyed iron [7,8], the 30KhGSA steel, both as-received and quenched to HRC 35...40 [9,10], and the austenitic 12Kh18N10T stainless steel [11]. The comparison was also made with laser-interferometry results [4,5,10], gamma-tomography results on the deterministic perturbations development [10], as well as with the results of investigation of the recovered shells.

The work is aimed:

– to study the properties of iron and some steels under the high-intensity loading of shells having different initial thickness when these shells are located in the HE spherical layer at the relative radius $R_{shell}/R_{HE}$= 0.75;

– to verify the shear and spall strength models using experimental data, as well as the equations of state of materials under study in conditions of high-intensity shock-wave loading;

– to verify numerical 3-D techniques.

## 2. Experimental methods and Results

Experiments with steel shells loaded by condensed high explosive are proposed to be used as experimental approach to study the strength properties of materials.

*2.1 Experimental setup*

Experimental setup and multi-channel multi-point registration methods for Fabry-Perot laser interferometry and laser-heterodyne diagnostics were developed, and the 1D-, 2D-, and 3D- explosive experiments were performed at RFNC-VNIITF [3-11]. Detailed description of experimental setups and registration methods is given in [3-6]. Shells made of the austenitic 12Kh18N10T steel, Fe, and the 30KhGSA steel quenched up to HRC 35...40 were studied. Of all the explosive experiments [3,4], only three experimental setups were chosen to perform 3D-modeling. These experiments (experiments 4, 10, and 12), as well as some other experiments are described in Table 1.

Table 1 - Test shells, conditions of their explosive loading, and diagnostic techniques

| # Shell material / Loading conditions and registration techniques | 12Kh18N10T | 30KhGSA HRC 35...40 |
|---|---|---|
| 1. Explosive loading mode: $R_{HE}$=40 mm, $h_{HE}$=10 mm, HMX-based composition, multi-point initiation system-96 | | |
| Registration techniques:<br>-laser-heterodyne-diagnostics (either HET-V or PDV),<br>- gamma-tomography,<br>- recovery [4,9] | Shells under study:<br>$R_{shell}$=30mm, $h_{shell}$=1 mm,<br>$h_{shell}$=2 mm,<br>$h_{shell}$=3 mm,<br>$h_{shell}$=4 mm,<br>$h_{hell}$=5 mm | Shells under study:<br>$R_{shell}$=30mm, $h_{shell}$=1mm,<br>$h_{shell}$=2 mm,<br>$h_{shell}$=3 mm,<br>$h_{shell}$=4 mm,<br>$h_{shell}$=5 mm |
| -Fabry-Perot interferometry [3] | Experiment 12<br>$h_{shell}$=10 mm | Experiment 4<br>$R_{shell}$=29.90 mm,<br>$h_{shell}$=1.94 mm |
| 2. Explosive loading mode: $R_{HE}$=40 mm, $h_{HE}$=16 mm, HMX-based composition, multi-point initiation system-96 | | |
| Registration techniques:<br>-Fabry-Perot laser interferometry | | Shell under study:<br>Experiment 10<br>$R_{ext}$=23.90 mm,<br>$h_{shell}$=1.90 mm |

The following designations and abbreviations are given in Table 1:

- $R_{HE}$ – external radius of HE spherical layer;
- $h_{HE}$ – thickness of HE spherical layer;
- HMX-based composition – the explosive composition based on octogene;
- Multi-point initiation system-96 – a multipoint initiation system with 96 step-like initiators per one sphere with radius $R_{HE}$ = 40 mm.

The spherical layer of the HMX-based composition with the external radius $R_{HE}$=40 mm and the thickness $h_{HE}$=10 mm or 16 mm was initiated by 96 initiators made of PETN-based composition. Mass of all 96 step-like initiators per one sphere with $R_{HE}$ = 40 mm was 43 g when HE mass in the spherical layer was 288 and 391 grams. Initial sizes of initiators used in the experiments are shown in Figure 1. The sizes are given in millimeters.

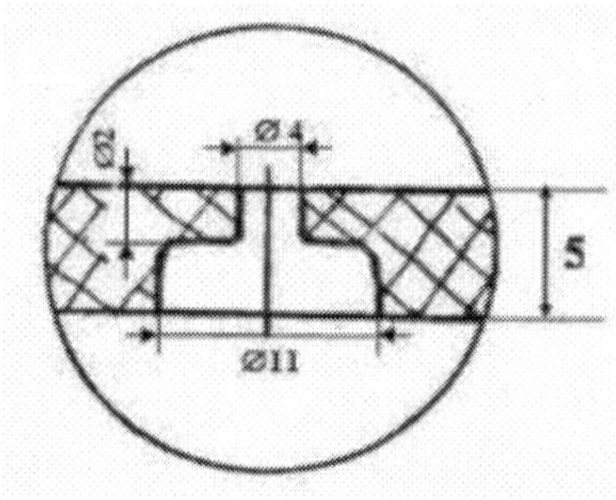


Figure 1 – Diametrically cut initiator

The internal and external surfaces of shells made of test steel (including the shells made of the 30KhGSA quenched steel) were finished up to Rz20. After that, the internal surface of shells was lapped to roughness Ra 0.06 and then ultrafinished. In particular cases, the polished surface was satin finished.

The setup of semispherical explosive experiments is schematically given in Fig. 2, and the typical structure of registration channels for Fabry-Perot laser interferometry is given in Fig. 3. Characteristics of registration channels being used for the Fabry-Perot laser interferometry are described in [3].

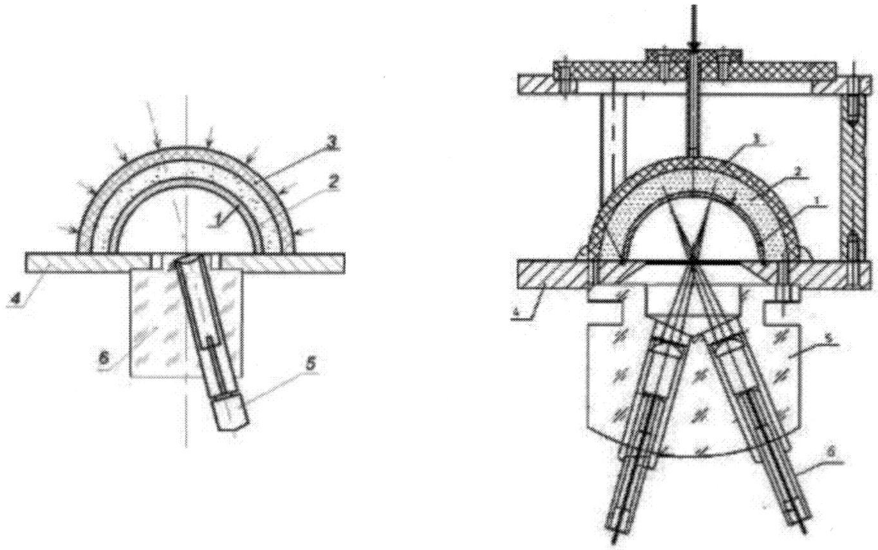

a) b)

Figure 2 - Schematic setup of explosive experiments with the 12Kh18N10T and 30KhGSA steel shells where laser-interferometric diagnostics of convergence dynamics of test shell under initiators (a), as well as under projections of initiators and points of the dual, triple, and quadruple collisions of waves from adjacent detonation initiators (b) is applied

The following designations are used in Fig. 2: 1 – test shell; 2 – spherical layer of the HMX-based explosive composition with $R_{HE}$=40 mm and $h_{HE}$=10 or 16 mm; 3 – initiation system; 4 – Al plate with ∅120×10 mm; 5 (a) and 6 (a) – measuring head used with the ∅9-mm plexiglass lens having focal distance f=15 mm and measuring head holder; 6 (b) and 7 (b) – optical heads oriented on different portions of shell under study.

*2.2 Laser interferometry techniques to register the convergence dynamics and the state of steel shells*

Three-channel multi-point laser interferometry techniques, namely the Fabry-Perot laser interferometry [4] and laser-heterodyne measurements [5], were used to obtain experimental data.

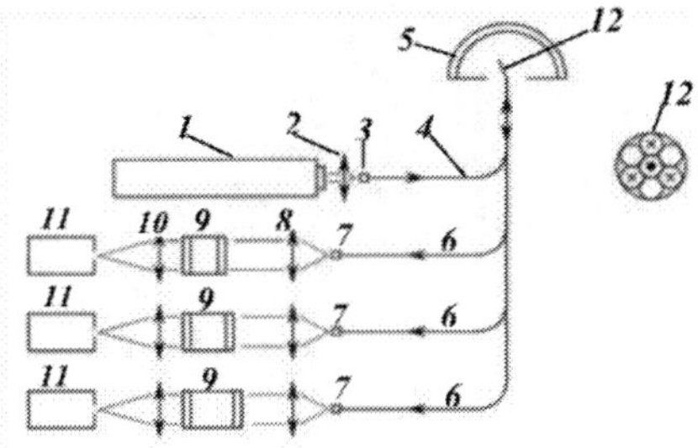

Figure 3 - Schematic optical system of registration channels for Fabry-Perot laser interferometry

The following designations are used in Fig. 3: 1 – laser light source; 2 – focusing lens; 3 – end surface of the fiber optic line; 4 – fiber optic line; 5 – test shell; 6 – fiber optic lines; 7 – their end surfaces; 8, 10 – lens system; 9 – Fabry-Perot etalons (distance between the mirrors is 25.4, 50, and 80 mm, respectively); 11 – electronic optical recorders with digital cameras; 12 – fiber optic head and its magnified view shown in the insertion (the central fiber marked by a heavy point is used to transmit the probe light to the shell surface segment being studied; the fibers marked by a cross collect the light reflected from the shell and transmit it to the corresponding Fabry-Perot input).

Schematic optical system of laser-heterodyne measurements is given in Fig. 4.

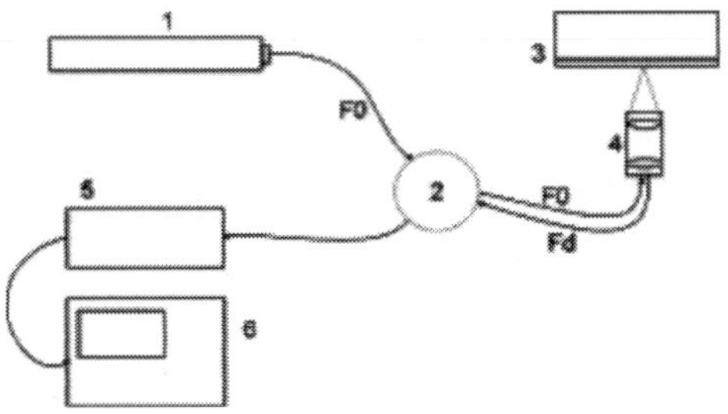


Figure 4 – The registration scheme for laser-heterodyne technique.
1 – laser with the wavelength 1550 nm;
2 – optical circulator; 3 – test sample or shell;
4 – optical head with lens; 5 – photo detector;
6 – oscilloscope.

*2.3 Typical streak-camera records of interferograms and main experimental results*

Typical streak-camera records of interferograms observed in explosive experiments 4 and 10 (see Table 1) and thoroughly described in [4,5] are given in Fig. 5a and Fig. 5b. Time – from the right to the left. Marks duration – 0.3 μs.

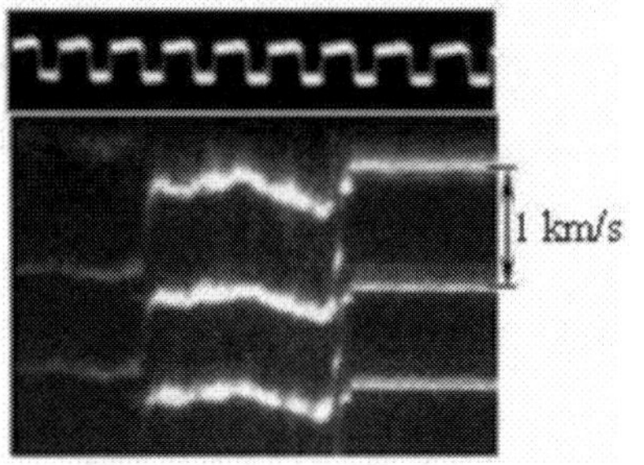


a)

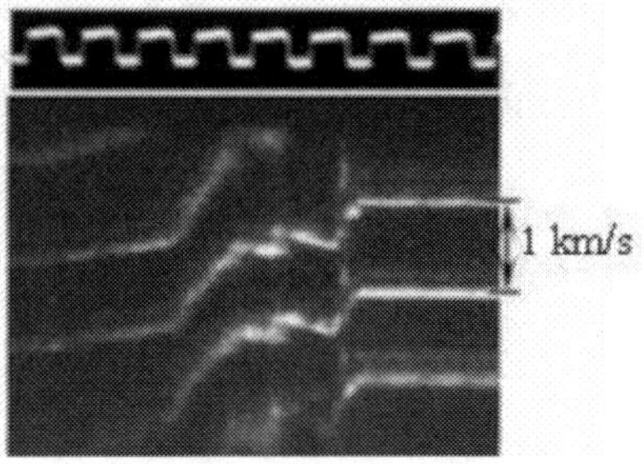


b)

Figure 5 - Streak-camera records of interferograms (a) and (b) observed in experiments 4 and 10, respectively

Results of interferogram processing are shown in Fig. 6.

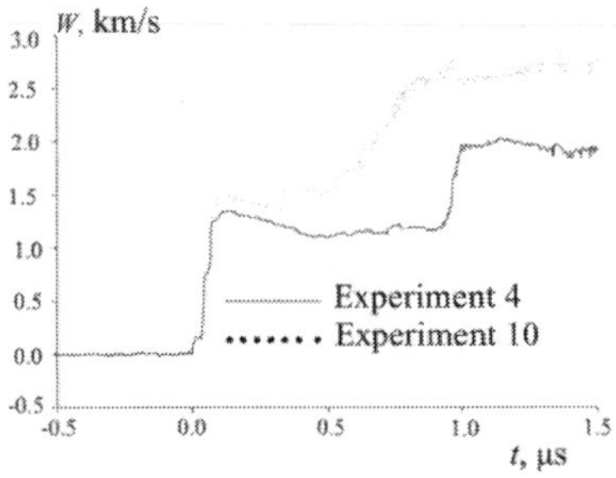


Figure 6 - Shell's free surface velocity profiles reconstructed using streak-camera records of interferograms (experiment 4 and experiment10)

As we can see in Fig. 6, in experiment 10, one observed a more intense convergence dynamics of the shell than in experiment 4. Besides, in experiment 10, the shell is accelerated up to the greater velocity during first 1.5 μs. For the shells made of the 30KhGSA steel quenched up to HRC 35…40, both elastic and phase precursors were registered on the shells' free surface velocity profiles.

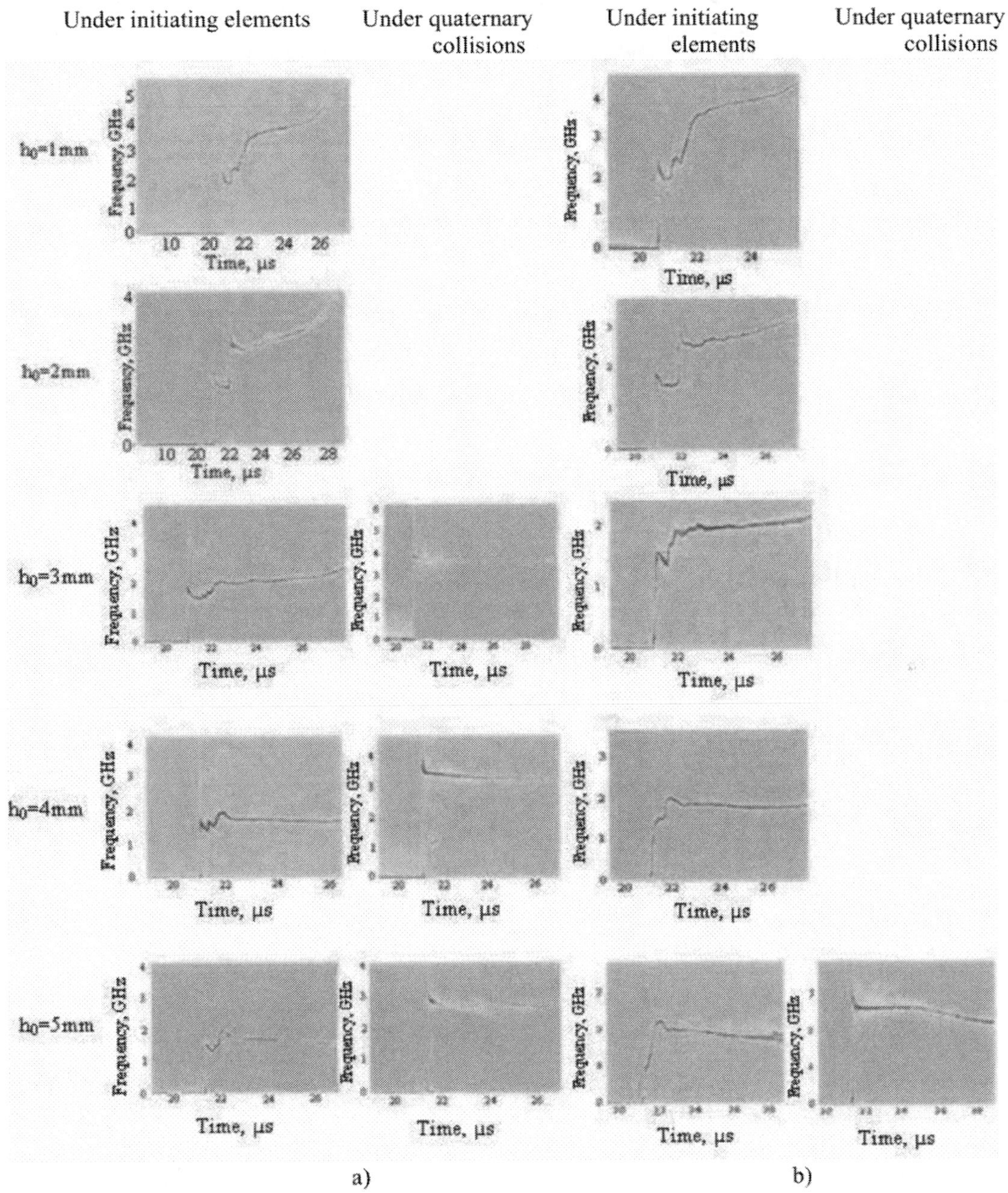


Figure 7 – Typical spectrograms of signals obtained in the experiments with the 12Kh18N10T shells(a) and quenched 30KhGSA shells(b)

Figure 8a shows the results obtained using laser-heterodyne technique. This technique was used to measure the internal surface velocity versus time of the 12Kh18N10T shell with initial thickness $h_0 = 2$ mm under the projection of detonation initiators $W_{12}(t)$, as well as to measure the corresponding displacement $S_{12}(t)$. The value of the first index ($W_{ij}$) i = 1 corresponds to the movement of the shell's internal surface segments located under detonation initiators; and i = 2 – under the projections of quadruple collision points when four detonation waves come from adjacent initiators. The value of the second index j = 1, 2, 3, 4, and 5 corresponds to the nominal initial thickness of test shell.

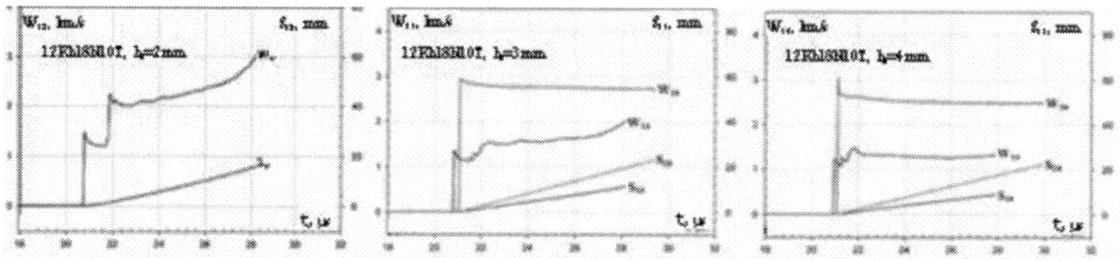

a) b) c)

Figure 8 –Results of laser-heterodyne measurements: the time-dependent of the internal-boundary convergence velocity ($W_{ij}$) and displacement ($S_{ij}$) for the 12Kh18N10T steel shell having initial thickness $h_0$ = 2 mm (a), $h_0$ = 3 mm (b), and $h_0$ = 4 mm (c) under projections of detonation initiators (j=1), as well as under projections of points of quadruple collisions of waves from adjacent initiators (j=2)

Figure 8 demonstrates that laser-heterodyne records were taken during 7 μs when shell's internal boundary under initiators moved 16 mm away from its initial location (i.e. when shell's internal boundary converged to the radius of 12 mm or 0.3 $R_{HE}$.). The data on shell's convergence dynamics were captured both at the initial, namely wave, stage of acceleration, and at the stage of inertial convergence until the beginning of cumulation at $R \leq (0.2\text{-}0.3)R_{HE}$. Figure 8b and Figure 8c show the profiles of time-dependent internal boundary velocities of the 12Kh18N10T steel shells having different initial thickness ($h_0$ = 3 mm (b) and $h_0$ = 4 mm (c) both under projections of detonation initiators $W_{1j}(t)$, and under projections of points of quadruple collisions of waves from adjacent initiator $W_{2j}(t)$. These figures also demonstrate how displacements $S_{1j}(t)$ and $S_{2j}(t)$ change with time; difference module of these displacements describes dynamics of deterministic perturbations growth on the internal boundary of test shell having initial thickness j = 3 mm and 4 mm. According to the results of laser-heterodyne measurements, the amplitude of deterministic perturbations is grown up to 10-12 mm.

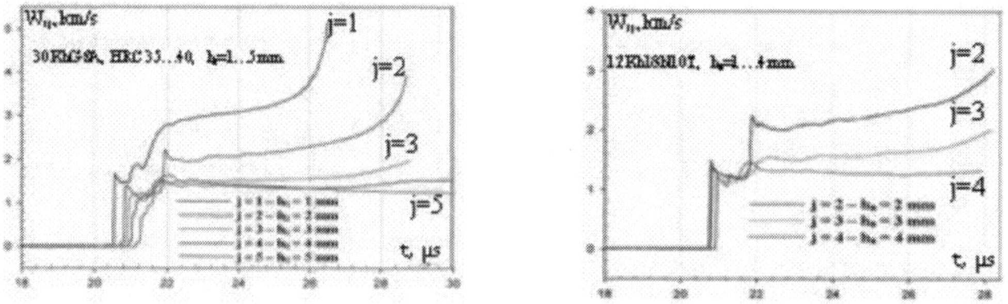


a) b)

Figure 9 – Time-dependent internal-boundary velocity for steel shells with different initial thickness under projections of detonation initiators: a) the 30KhGSA steel (HRC 35…40), $h_0$ = 1, 2, 3, 4, and 5 mm; b) the 12Kh18N10T steel, $h_0$ = 2, 3, and 4 mm

Figures 9a and 9b demonstrate the comparison of summary data on $W_{1j}$ for both the 30KhGSA and the 12Kh18N10T steel shells having all the initial thicknesses that were investigated. It can be seen, that thin shells ($h_0$ = 1 mm and 2 mm) converging without a spall are moving faster under convergence to radius $R \leq 0.3R_{HE}$. As for the shells with thickness $h_0$ = 3 mm, damages seem to occur at higher radii at the initial stage of acceleration; however, these damages are recompacted during subsequent convergence of the shell. More thick shells ($h_0$ = 4 mm and 5 mm) where the developed spall damages and even multiple spall and shear damages are found converge with almost constant speed.

Comparison of experimental data on convergence dynamics of the 30KhGSA and the 12Kh18N10T steel shells having three different initial thicknesses is given in Fig.10.

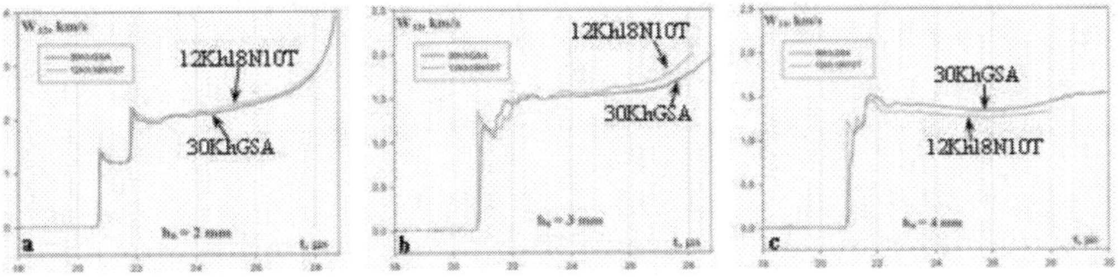


Figure 10 – Time-dependent internal-boundary velocity for the 30KhGSA and the 12Kh18N10T steel shells having three different initial thicknesses. All the compared velocities are under projections of detonation initiators

One can see, that for thin shells ($h_0$ = 2 mm) where no spall occurs during acceleration, the differences in rheological properties of test steels have insignificant influence on the convergence dynamics. While the shells' initial thickness is getting closer to the value at which the spall fracture is beginning to emerge at least in one of these shells, the differences in rheological properties of test steels become more obvious. Under such conditions, when wave processes in shell material exercise a significant influence, they can be no longer considered as incompressible.

Thus, spall fractures does not occur in shells having $h_{shell}$= 1-2 mm. At the initial stage of convergence, these shells exhibit development of shear damages under projections of points of collisions of waves from adjacent initiators. As for the shell with initial thickness $h_{shell}$ = 3 mm, the spall damages did occur but these damages were recompacted during the shell convergence down to deep radius. In the shells with initial thickness $h_{shell}$ = 4, 5, and 10 mm, the spall and shear fractures took place, and they exhibited no recompactions of spall and shear damages.

*2.4 Some additional results*

The same experiments involved impulse radiography diagnostics of converging shells, as well as their soft deceleration and trapping on the plexiglass lanterns located at the radius 0.5 $R_{HE}$.

The images shown in Figure 11 and 12 demonstrate the state of the external and internal surfaces of the 12Kh18N10T steel shell after its explosive loading and soft trapping on the plexiglass lantern. This experiment has following characteristics: the shell (initial thickness $\Delta$ = 2.25 mm and radius R = 30 mm) made of steel plate was located in the HE layer at the relative radius $R_{shell}/R_{HE}$ = 0.75 and accelerated by detonation of the RDX-based explosive composition (TNT/RDX 3/7) layer havin10-mm thickness and $R_{HE}$=40 mm, which was initiated by 96 initiators, and then was softly decelerated on the plexiglass lantern.

Figure 11

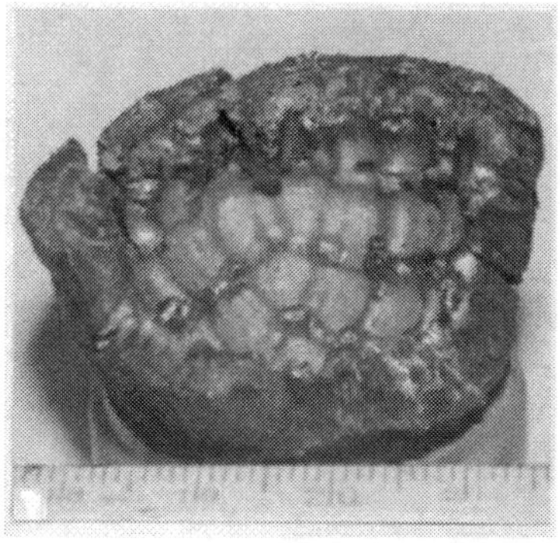

Figure 12

Special features of the 12Kh18N10T shell deformation under explosive loading with multi-point initiation of detonation: condition of the external (11) and internal (12) surfaces of the shell.

Characteristic lines on the external surface of test shell are traces-imprints of collisions of diverging detonation waves from adjacent synchronously operated initiators. Local perturbations on the internal surface of the shell are also the indications of initial inhomogeneities in pressure field in the event of multi-point initiation of the explosive material layer. It should be noted that the largest-area portions having the surface that is close to spherical correspond to initiators locations. The ribs upstanding above this surface resulted from collision of two diverging detonation waves coming from adjacent initiators. Perturbation under the projection of triple collision locations is higher than perturbation from double collision locations and lower than perturbation under the projection of quadruple collision locations.

The growth of deterministic perturbations associated with presence of multi-point initiation in the systems (namely the multi-point initiation systems-96 with 96 initiators per sphere having $R_{HE}$ = 40 mm) is conditioned by both the Richtmayer-Meshkov instability developed during the shell acceleration, and the Rayleigh-Taylor instability developed during deceleration of this shell on the plexiglass lantern.

The images shown in Fig. 13 and Fig. 14 demonstrate the external and internal surfaces of the 12Kh18N10T steel shells having different initial thickness that were located at the relative radius 0.75 $R_{HE}$ in the spherical layer of HMX-based explosive composition and were softly trapped on the plexiglass lantern. The images given in Fig. 14 were taken using electrons backscattering technique. These images illustrate the special features of perturbations arising on the external and internal surfaces of shell under study.

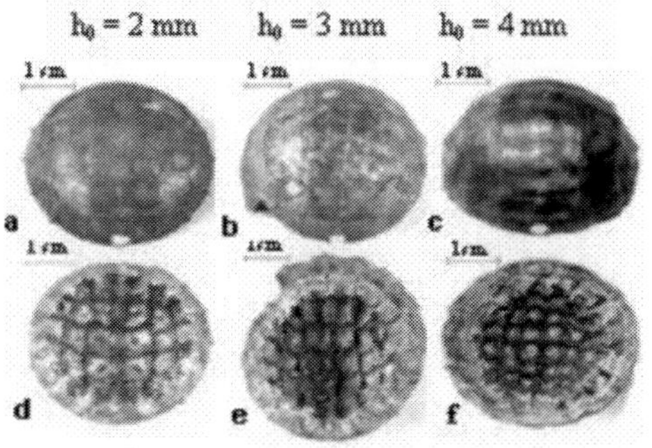


Figure 13 – Images of the external (a, b, c) and internal (d, e, f) surfaces of the 12Kh18N10T austenitic steel shells having different initial thickness after identical explosive loading of these shells

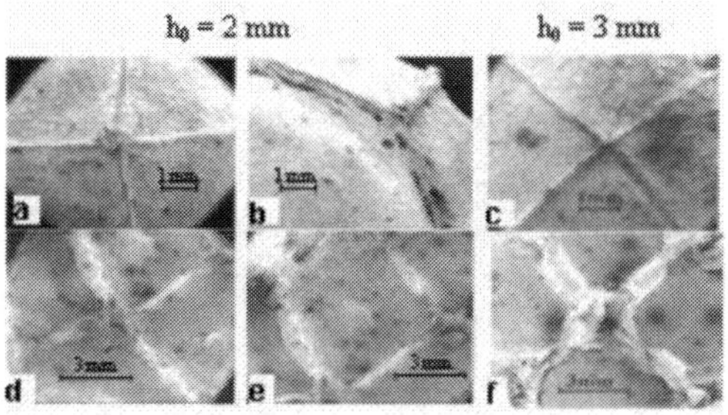


Figure 14 – Perturbations on the external (a, b, c) and internal (d, e, f) surfaces of the 12Kh18N10T shell with initial thickness $h_0$. Electron backscattering technique

The systematic data on both gamma-tomography of the converging steel shells having different initial thickness, and scanning electron microscopy of the surfaces of recovered steel shells made of the 12Kh18N10T austenitic steel are given in [3,10], while the data on calorimetric measurements of the compressed and recovered shells and the results of metallographic investigations of these shells are given in [7] and [8-11], respectively.

Comparative experimental data obtained allowed us to understand how the thickness, the relative radius of location in the HE layer, and the properties of shell material affect the development of deterministic perturbations under explosive loading in case of synchronous multi-point initiation. The results of experiments aimed to study incipience and development spall and shear damages in steel shells under their quasi-spherical explosive loading, as well as its recompactions are used to verify the modern strength models and to certify the numerical 3-D modeling techniques.

## 3. Numerical modeling of experiments

### *3.1 Numerical computational technique and the models being* **used**

The Lagrange-Euler technique LEGAK-3D [1,2] was used for numerical modeling of experiments described in Table 1.

This technique is characterized by following parameters:

- order of approximation: $O(\tau+h^2)$;
- condition of stability: $t \leq k \cdot h/s$; $k \leq 0.5$;
- conservation of all types of symmetry in the 2-D case;
- conservatism in the plane case.

In order to model convergence dynamics of the 12Kh18N10T austenitic steel shell we used:

- the one-phase E.I. Zababakhin equation of state in the form with ultimate compression;
- models of elastic-plastic deformation, namely:
    - Mises model with constant yield strength;
    - phenomenological model of shear strength (model of B.L. Glushak) [12,13];
- spall fracture models, namely:

- the model of brittle spall or the model of sudden spall with $f_{cr}$ = const;
- phenomenological kinetic model of ductile fracture (NaG model);
- phenomenological kinetic binary model of fracture (model of S.S. Sokolov), etc. [15,16].

### *3.2 Characteristics of the models being used*

Shear strength model (the Glushak model).

In the frames of phenomenological shear strength model, the stationary yield strength $Y$ is considered as a function of three variables that describe the stress-strain state of pressure (compression hardening), of plastic strain intensity (strain hardening), and of temperature (heat softening).

$$Y = Y_0\,(1 + \alpha_0 p_x)\left(1 - \left(\frac{T}{T_m}\right)^l\right)\left(1 + a_0\left[1 - \left(\frac{T}{T_m}\right)^k\right][1 - \exp(-m\varepsilon_i^p)]\right)$$

Table 2 shows parameters of this model, which were used for numerical 3D- modeling of explosive experiments that were carried out.

Table 2 – Parameters of phenomenological shear strength model

| Material | Parameters of Poisson ration vs. temperature | | | Strain hardening parameters | | | | Compression hardening parameter $a_0$, 1/GPa | Melting temperature $T_m$, K | Heat capacity $C_v$, kJ/Kg | Heat softening parameter $l$ |
|---|---|---|---|---|---|---|---|---|---|---|---|
| | $\nu_0$ | $c$ | $K$ | $Y_0$, GPa | $a_0$ | $m$ | $k$ | 0.02 | 1811 | 0.6 | 2.0 |
| steel | 0.28 | 0.781 | 2.3 | 0.45 | 3.0 | 15.0 | 3.5 | | | | |

Spall fracture model (NaG model).

This model considers the following physical factors:

– how nucleation and growth of defects depend on the loading intensity and duration,

– how the equation of state (EOS) and the elastic-plastic properties of material change with the damage development.

In the NaG model, time history of the damage measure can be written as:

$$\frac{d\omega}{dt} = 8\pi R_0^3 \dot{N}_0 \exp\left(-\frac{P_s - P_{n0}}{P_1}\right)\theta(P_{n0} - P_s) + 3\omega\left(-\frac{P_s - P_{g0}}{4\eta}\right)\theta(P_{g0} - P_s)$$

where:

$R_0$ - parameter of distribution (characteristic pore size),

$N_0$ - rate of pore nucleation (model parameter),

$P_S$ - pressure in solid material, $Pn_0$ - threshold pressure of pore nucleation,

$P_1$ - model parameter, $P_{g0}$ - threshold pressure of pore growth,

$\eta$ - effective viscosity of material, $\theta$ – Heaviside unit step function

Pressure of damaged medium: $P(\rho,e,\omega) = (1-\omega)P_s(\rho_s,e_s) = (1-\omega)P_s\left(\frac{\rho}{1-\omega}, e_s\right)$

Yield strength and shear modulus: $Y = Y_0\left(1 - \omega/\omega_{кр}\right)$ $\quad G = G_0\left(1 - \omega/\omega_{кр}\right)$

Table 3 – Parameters of the NaG model

| $\omega_{cr}$ | $N_0$, $(cm^3 \cdot 10^{-5} \cdot s)^{-1}$ | $R_0$, cm | $P_{n0}$, GPa | $P_{g0}$, GPa | $P_1$, GPa | $\eta$, $GPa \cdot 10^{-5} \cdot s$ |
|---|---|---|---|---|---|---|
| 0.25 | $4.6 \cdot 10^3$ | $5 \cdot 10^{-5}$ | 2.5 | 0.5 | 0.53 | 0.02 |

Phenomenological kinetic binary model of fracture

The NaG model is supplemented by an additional part describing the fracture kinetics with due consideration to shear deformation

$$\dot{\beta} = c_n(\varepsilon_p - \varepsilon_0)\frac{1}{(1-\omega)(1-\beta)}$$

where $c_n$, $\varepsilon_0$ are model parameters for a certain material, which describe the rate of fracture development and the ultimate stain for the fracture process beginning.

$c_n=0.15[1/\mu s]$; $\varepsilon_0=0.3$.

Shear modulus and yield strength of damaged material can be written as:

$G=G_S(1-\omega/\omega_{cr})(1-\beta)$, $Y=Y_S(1-\omega/\omega_{cr})(1-\beta)$

*3.3 Numerical computations*

Figure 15 schematically shows the radial section of the shell for numerical modeling of experiments on the loading of spherical shells. The following designations are used in Figure 15: $R_{HE}$= 40mm – external radius of HE spherical layer; $R_{cav}$ – external radius of vacuum cavity; $R_{ext}$ – external radius of the shell; $h_{HE}$ – thickness of HMX-based explosive layer; multi-point initiation system-96 – a multipoint initiation system with 96 steplike initiators per one sphere with radius $R_{HE}$ = 40 mm.

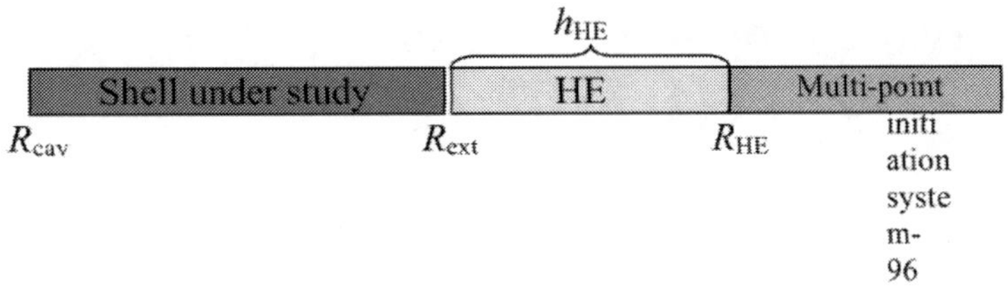


Figure 15 - Numerical computations of experiments on the loading of spherical shells

Initial sizes of shells used in numerical modeling of experiments are given in Table 1. Due to the symmetry of initial setup, it was decided to calculate the initial area equal to 1/8th of the size of the sphere with radius *R* ranging from 0 to 80 mm. Free space was filled with vacuum. All the computations were made for the shell with cylindrical geometry using the Euler grid (*Ox* – the symmetry axis). Figure 16 shows the sectional view of initial geometry of the problem; this view is taken along the middle plane in Cartesian coordinate system.

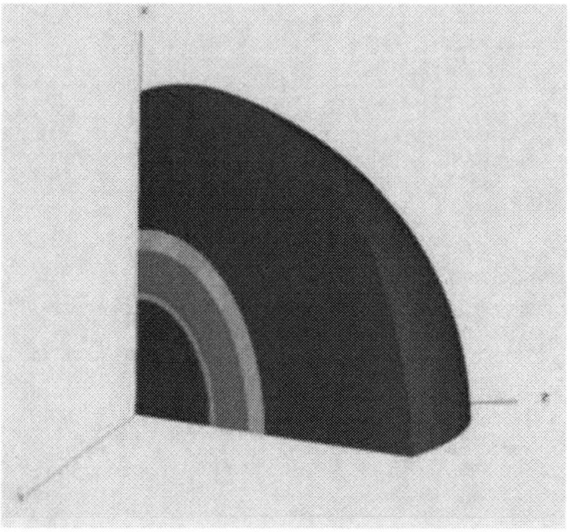

Figure 16 - Sectional view of initial geometry of 3D-calculations (analogous to experiment 4)

The substances in the computational domain, as well as the initial densities are listed in Table 4.

Table 4 – Substances of the computational domain

| Colors (from Fig. 8) | Substance | $\rho_{00}$ $g/cm^3$ |
|---|---|---|
| ■ | *vacuum* | 0.001 |
| □ | *PETN-based plastic composition* | 1.52 |
| ■ | *PETN-based plastic composition* | 1.52 |
| ■ | *HMX- based composition* | 1.86 |
| | *explosion products of HMX- based composition* | 1.86 |
| ■ | *12Kh18N10T steel* | 7.772 |
| ■ | *foam plastic* | 0.65 |

Different shear strength models for steel were used for computations. For the first type of computations, we applied the simplest Mises model with the constant yield strength and with the following parameters: Poisson ration $\nu$=0.27, yield strength Y=0.83 GPa. For the second type of computations, we used the Glushak strength model[12]. The data for calibration of the kinetic shear strength model for the 30KhGSA quenched steel (HRC 35…40) are given in [6]. The fracture processes were described using two fracture models: the model of brittle (or sudden) fracture where the spall strength $f_{cr}$ = -10.8 GPa or the kinetic model of spall fracture (NaG model) [14], and the kinetic binary model of spall and shear fractures [15,16].

Euler grid with 180x180 angle intervals and 400 radial intervals was used for all computations. Typical size of the grid having such spacing: $\Delta h_R = 0.2$ mm, $\Delta h_\varphi = 0.5°$. Total number of points was ~13 millions (grid: *N* points). The computations were made on the grids being successively thickened in three directions up to the half size; at that, the total number of grid cells was ~104 millions of points (grid: *2N* points). The computations were made in a parallel mode; maximum number of processes applied in calculations was 1200.

Figure 17 shows the initial state of multi-point initiation system-96 and the spatial orientation of coordinate system. There are 12 initiators in the selected computational domain. Numbers in Fig. 17 are used to designate the points under whose projections "shell velocity vs. time" dependences were determined in computations.

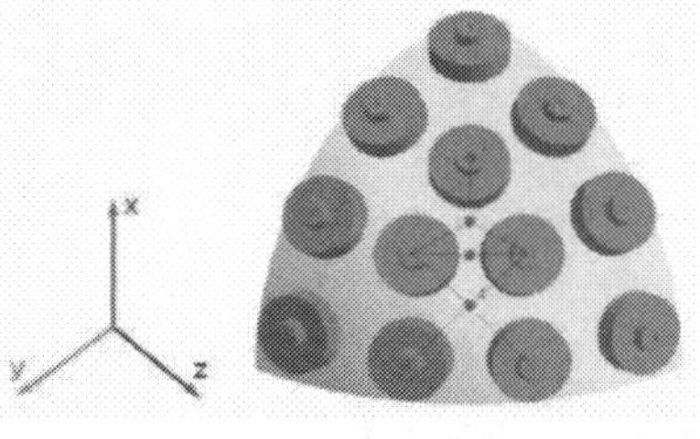


Figure 17 – Multi-point initiation system
1 – center of initiator, 2 – point of collision of two shock waves, 3 – point of collision of three shock waves, 4–point of collision of four shock waves.

The elements of multi-point initiation system-96 were initiated in the point that was the central point of external surface of the element with timed delays. Maximum time difference of initiators activation was 0.1 μs.

*3.4 Results of numerical modeling*

*3.4.1 Shear fractures of thin steel shells and development of deterministic perturbations on the internal surfaces of these shells*

Figures 18 a) and b) show initiation system in its initial state and at the moment of plastic explosive detonation completion. The images were constructed using the graphical and numerical data processing system ScientificView [17] in a parallel mode.

a) b)

Figure 18 – Initiation system at instants (a) t = 0 and
(b) t =2.45 μs

In order to demonstrate the processes observed in numerical modeling, characteristic frame-by-frame images acquired during computation of experiment 4 are shown in Fig. 19. Figure 19Figure 1 demonstrates the distribution of volume concentration along the middle plane at different instants. The following colors were chosen to depict the states of steel shell: green is for unfracruted shell, violet is for the fractured material. The model of brittle spall was used for computations; shear strength of the 30KhGSA steel was described using the Glushak model (grid: *N* points).

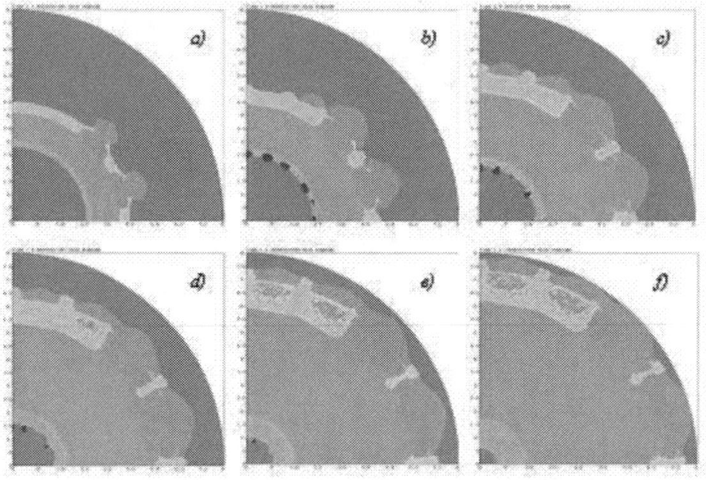


Figure 19 –The concentrations field at different instants:
*a)* 2 μs; *b)* 4 μs; *c)* 6 μs; *d)* 8 μs; *e)* 10 μs; *f)* 11 μs (simulation of experiment 4)
Time marking – from the moment of PETN-based explosive charges initiation

Figure 20 demonstrates the state of shell's internal surface at different instants for numerical simulation of experiment 4. The model of shear strength (the Glushak model) and the model of brittle spall fracture were used in these computations. Results for successively thickened grids were compared.

As one can see in Fig. 20, in computations, the perturbations develop on the internal surface of shell due to propagation of waves from the multi-point initiation system; and this corresponds to experimental data (see Fig. 11-14).

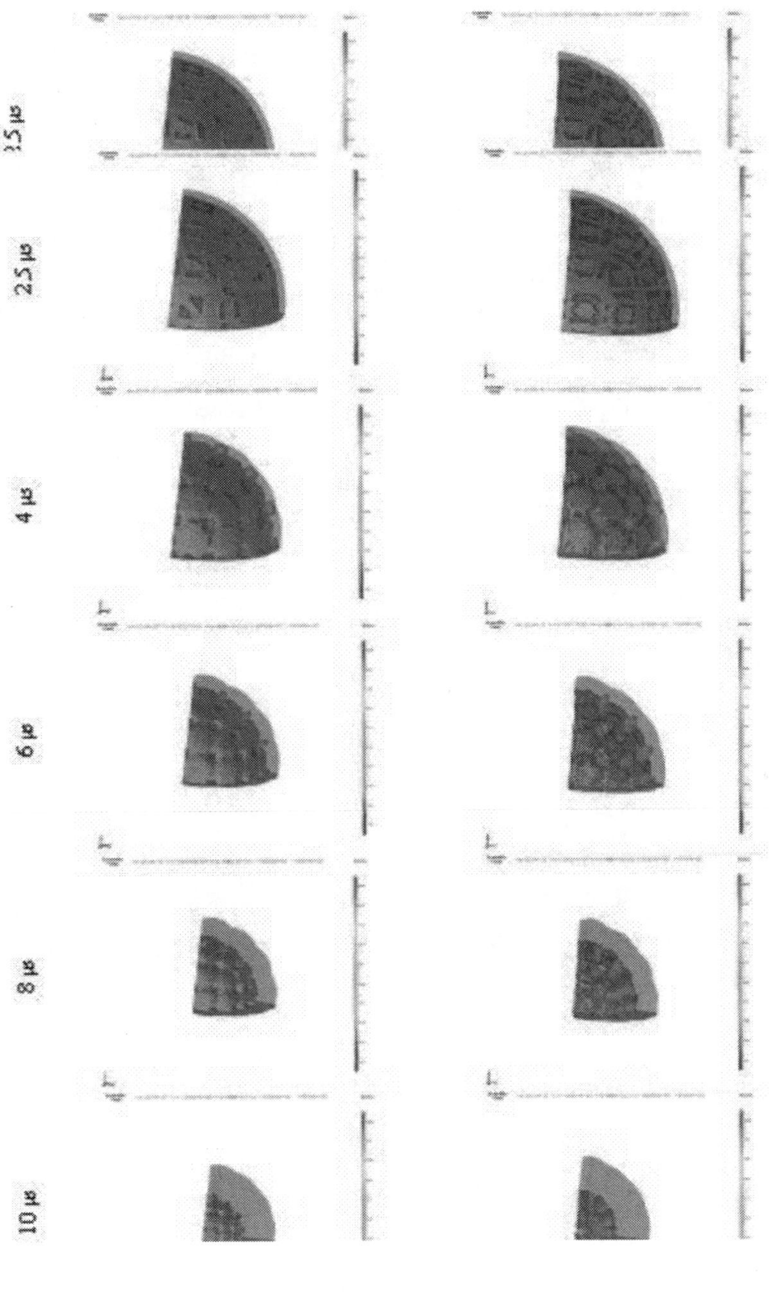


a) b)

Figure 20 –State of shell's internal surface at different instants in numerical modeling of experiment 4: (a) grid: N points, (b) grid: 2N points

*3.4.2 Modeling of time changes in shell's internal-boundary converging velocity*

Numerical modeling of experiments 4, 10, and 12 was principally aimed to compare the computed velocity profiles of shell's internal-boundary convergence under projections of initiators centers with experimentally registered laser-interferometry data [3-5].

In Fig. 21, the velocity profile of convergence of the 30KhGSA (HRC 35…40) steel shell under projection of detonation initiator, which was measured in experiment 4 with the help of laser-interferometry techniques [3-5], is compared to the numerical modeling results obtained using the Mises model of perfectly elastic-plastic approximation and the kinetic model of spall fracture.

Both the grid of nominal size (N), and the fine grid (2N) were used in computations. In general, the results of numerical modeling reproduce experimentally registered profile quite well. However, the elastic-plastic properties of the 30KhGSA steel quenched up to HRC 35…40, as well as the kinetics of α'-ε-transformation require more exact description using, for example, data from [3-6]. It should be noted that the validity of measurement results obtained using Fabry-Perot laser interferometry [3,4] was verified by the fact that these results were in good agreement with the data on the 30KhGSA and the 12Kh18N10T steel shells convergence, which were obtained using laser-heterodyne diagnostics in the similar experiments described in [5].

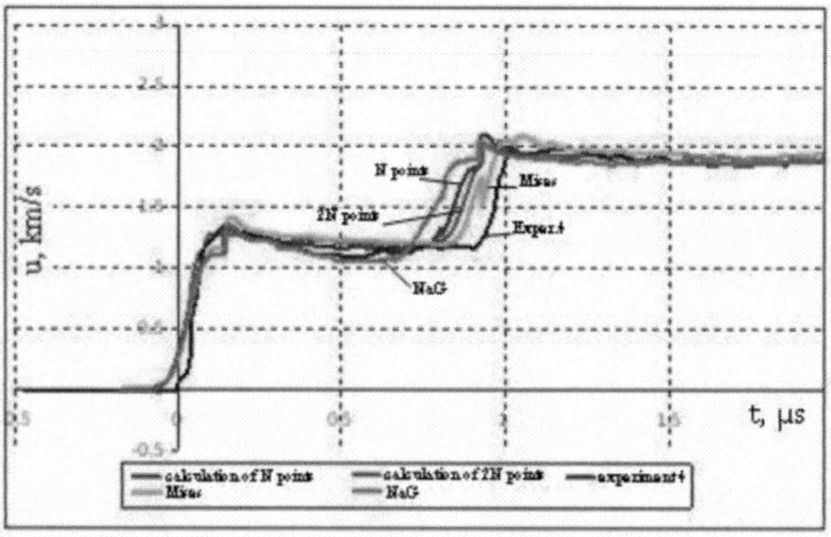


Figure 212 – Experimentally registered (experiment 4 [4,5]) and calculated dependences of shell's internal surface velocity under projections of initiators centers

For experiment 12, experimental dependence of the shell's internal surface velocity under projections of initiators centers was compared to the dependences obtained during computations (Figure 22). As one can see in Fig. 22, the Glushak model used in calculations gives better description of experimentally registered velocity profile.

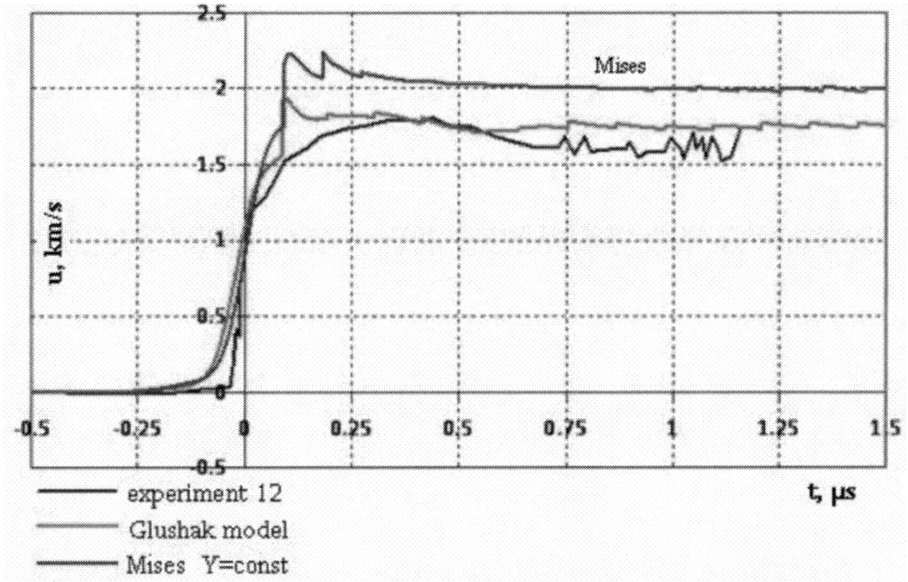


Figure 22 – Comparison of time-dependent internal-surface velocity of the 12Kh18N10T steel shell under projections of initiators centers in experiment 12 with the dependences computed using the Mises and the Glushak models describing elastic-plastic properties of steel

Comparison of 'shell internal surface velocity vs. time' dependences – both registered in experiment 10, and computed using two models describing elastic-plastic properties of the 30KhGSA quenched steel – are given in Figure 23.

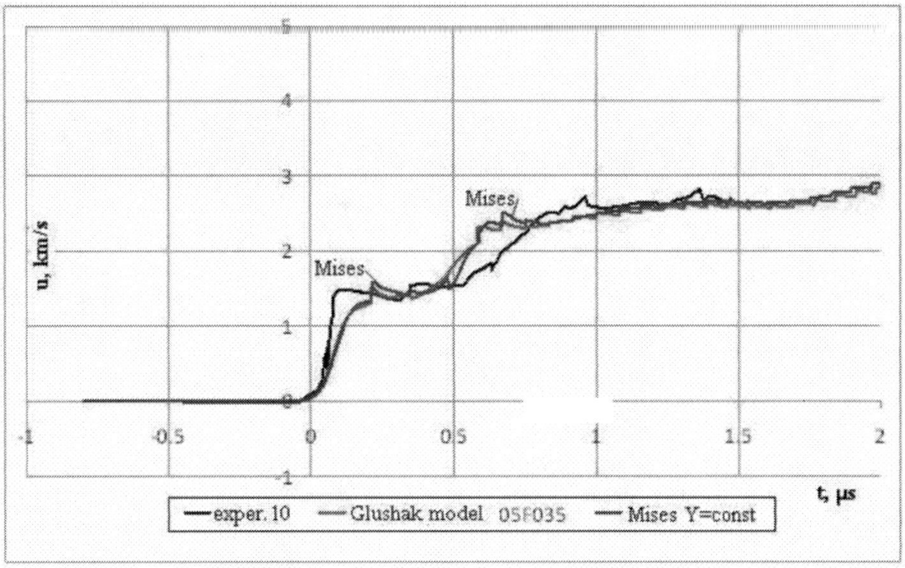


Figure 23 – Time-dependent internal surface velocities for the 30KhGSA quenched steel shell under projections of initiators centers (experiment 10)

In general, the profiles calculated using two shear strength models satisfactorily agree with each other and with experimental data.

*3.5 Discussion*

Due to the time difference of spherical HE layer initiation, the flow behind the front of detonation wave has noticeably different dynamics. The pressure gradients behind the wave front along the fixed radii under detonation initiators and under projections of the points of collisions of four detonation waves from adjacent initiators in case of quasi spherical explosive loading are significantly different. The reason for such different-dynamics is the collisions of detonation waves generated by different initiators.

The different-dynamics wave may cause the Richtmayer-Meshkov instability developing at the internal surface of converging steel shell. The resultant perturbations growth depends on material strength, the relative radius of shell location, and the shell's initial relative thickness.

For the 30KhGSA steel with the shell's relative thickness $\varepsilon = 6\%$ and radius of location $R_{shell} = 0.75R_{HE}$, the intensity of the principle plastic wave approaching the shell's internal boundary is highly heterogeneous (Fig. 24). Noticeably thicker shell made of the 12Kh18N10T viscous steel smoothens the different-dynamics wave; due to this fact, the shell's internal boundary velocities under projections of different physically meaningful locations turn out to be close (Figure 25).

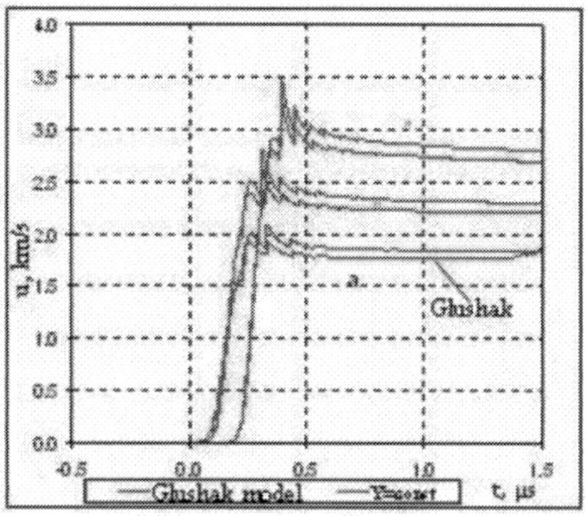


Figure 24 – Computed dependences of the 30KhGSA shell velocity under projections of the double (a), triple (b), and quadruple (c) collision points (experiment 4)

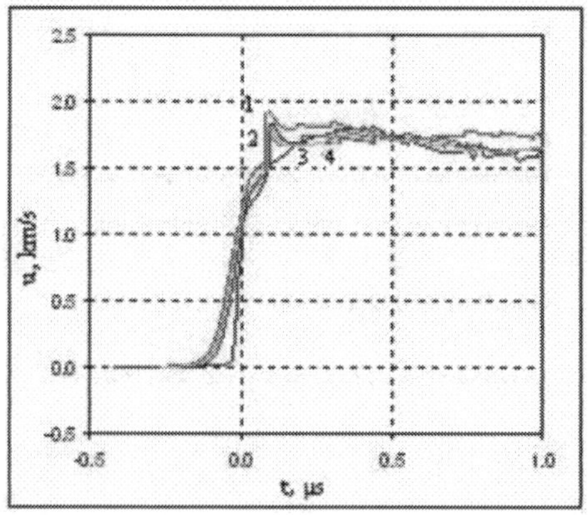


Figure 25 – Computed dependences of the 12Kh18N10T shell velocity under projections of the initiators centers (1) and of the double (2), triple (3), and quadruple (4) collision points (experiment 12)

In these experiments, the shell dynamics is characterized by the fact that occurring fractures are localized and thus, it is possible to adequately describe the experiments only if the models being used consider shear fractures [15,16]. In the future, we plan to perform the numerical modeling of all experiments using such models.

## 4.Conclusion

Numerical 3D- modeling of experiments on explosive loading of spherical shells made of the 12Kh18N10T and 30KhGSA steel was performed using LEGAK-3D technique for conditions most closely replicating the experiments. Computational results are close to those obtained independently during the experiments.

It was demonstrated that the selected models used to describe shock-wave compressibility, as well as shear and spall strengths models for the 12Kh18N10T and 30KhGSA steels allow us to numerically reproduce the results of experiments aimed to study the shell boundary converging velocity under projections of initiators.

Additional investigations must be performed to verify the models aimed to describe more fine phenomena under the high rate- high-intensity deformation of steel (for example, the kinetics of solid-solid phase transformations in stress waves or the relaxation of elastic precursor and shear stresses) in experiments with loaded shells.

## 5.References

[1] Avdeev P A, Artamonov M V, Bakhrakh S M, et al. 2001 LEGAK program complex aimed to compute nonsteady-state flows of multi-component continuum media and the principles for realization of this complex on the distributed-memory multiprocessor computer *The issues of atomic science and technology. Series: Mathematical modeling of physical processes* **3** 14

[2] Bakhrakh S M, Velichko S V, Spiridonov V F, Avdeev P A, Artamonov M V, Bakulina E A, Bezrukova I Yu, Borlyaev V V, Volodina N A, Naumov A O, Ogneva N E, Rezvova T V, Rezyapov A A, Starodubov S V, Taradai I Yu, Tikhonova A P, Tsiberev K V, Shanin A A, Shirshova M O and Shuvalova E V 2004 LEGAK-3D technique aimed to compute 3D nonsteady-state flows of multi-component continuum media and the principles for its realization on the distributed-memory multiprocessor computer *The issues of atomic science and technology. Series: Mathematical modeling of physical processes* **4** 41

[3] Kozlov E A 2012 2D- and 3D-explosive experiments for verification of spall and shear strengths models for some steels *Proc. 17th Biennial International Conference of the APS Topical Group on Shock Compression of Condensed Matter (Chicago, Illinois, USA June 26-July 1, 2011)* eds M L Elert, W T Buttler, J P Borg, J L Jordan and T J Vogler (Melville, New York: AIP Conf. Proc. 1426) pp 945-948

[4] Kozlov E A, Brichikov S A, Boyarnikov D S, Kuchko D P and Degtyarev A A 2011 Special features in convergence dynamics of steel shells under their explosive loading. Results of laser-interferometry measurements *Phys. Metals and Metallog. (Engl.transl.)* **112** issue 4 389

[5] Kozlov E A, Brichikov S A, Kuchko D P, Ral'nikov M A, Ol'khovsky A V, Zhilyaeva N S, Brezgina L P and Povyshev V N 2012 Local shear and spall damages of steel shells under quasi-spherical explosive loading *Int. Conf. Shock Waves in Condensed Matter (Kiev, Ukraine 16-21 September 2012)* (Kiev: "Interpress LTD", Collection of abstracts) pp 158-167

[6] Kozlov E A, Tarzhanov V I, Telichko I V and Pankratov D G 2012 Shear and spall strength of quenched 30KhGSA steel under explosive loading in the range of solid-solid phase transformation, *Russ. J. Deformation and Fracture of Materials* **8** 32

[7] Kozlov E A, Brichikov S A, Vildanov V G, Gorbachev D M and Yusupov D T 2008 Spall and shear fractures in the spherically converging shells of iron and steels. Measurements of energies and residual strains *Russ. J. Deformation and Fracture of Materials* **11** 2 & *Proc. of the Joint US – Russia Conference on Advances in Materials Science (Prague, Czech Republic, August 31 to September 4, 2009)* pp IV 6-11

[8] Kozlov E A, Brodova I G, Brichikov S A, Gorbachev D M and Yablonskikh T I 2008 Characteristics of the structure and spall fractures in Armco-Fe shells under different modes of explosive loading, *Russ. J. Deformation and Fracture of Materials* **11** 11

[9] Kozlov E A, Brodova I G, Brichikov S A, Gorbachev D M and Yablonskih T I 2010 Spall and shear fractures, structural and phase transformations in the 30KhGSA steel shells under two regimes of explosive loading *Russ. J. Deformation and Fracture of Materials* **1** 17

[10] Kozlov E A, Brichikov S A, Zhilyaeva N S, Khardina L V, Brezgina L P and Povyshev V N 2011 Deterministic perturbations developing on steel shells under quasi-spherical explosive loading. Investigation results using laser interferometry and gamma-tomography *Proc. US-Russian Conference on Materials Properties at Extreme Conditions, 6LAB Conference Engineering & Materials at Extreme Conditions (Barcelona, Spain 23-28 October 2011)*

[11] Kozlov E A, Brodova I G and Yablonskikh T I 2012 Fracture in stress waves and recompaction during the convergence of austenitic steel 12Kh18N10T shells *Russian Metallurgy (Metally)* **Vol. 2012**, issue10, 884

[12] Glushak B L et.al. 1982 Government equations to describe shock-wave deformation of Al and Mg, *Russ. J. Physics of Combustion and Explosion* **28**

[13] Glushak B L, Gudarenko L F and Styazhkin Yu M 1991 Semi-empirical equation of state for metals with nuclei and electrons having variable specific heat. *The issues of atomic science and technology*. Series: *Mathematical modeling of physical processes* **2** 57

[14] Seaman L, Curran D R and Shockey D A 1976 Computational models for ductile and brittle fracture *J. Appl. Phys.* **47**, issue 1 4814

[15] Sokolov S S and Sadovoi A A 2003 Kinetic model of dynamic fractures that includes plastic pore growth and damaging under shear deformation *III Scientific Conference Up-to-date design and development methods for weapon ordnance (Sarov, Nizhniy Novgorod region, Russia, 2003)* (Sarov: Proc. FSUE 'RFNC-VNIIEF', Collection of abstracts)

[16] *Strength, fracture, and dissipation losses under high-intensity shock-wave loads*. Collection of scientific papers, eds A A Sadovoi and S V Mikhailov (Sarov, Nizhniy Novgorod region, Russia: Proc. FSUE 'RFNC-VNIIEF', 2009, 402 p., illustrated)

[17] Potekhin A L, Kozachek Yu V, Loginov I V, Nikitin V A, Kuznetsov M Yu, Demanova A K, Popova N V and Firsov S A, 'ScientificView – D.A.': parallel post-processing system for the results obtained during numeric modeling of physical processes, *Proc. for the XVIII International Conference on Computer Graphics and Vision Sense,'Graphicon-2008'* pp 192-198

# The artificial-free technique along the objective direction for the simplex algorithm

**Aua-aree Boonperm[1] and Krung Sinapiromsaran[2]**

Department of Mathematics and Computer Science, Faculty of Science, Chulalongkorn University, Bangkok, 10330, Thailand

E-mail: saaw.aua@gmail.com[1], krung.s@chula.ac.th[2]

**Abstract.** The simplex algorithm is a popular algorithm for solving linear programming problems. If the origin point satisfies all constraints then the simplex can be started. Otherwise, artificial variables will be introduced to start the simplex algorithm. If we can start the simplex algorithm without using artificial variables then the simplex iterate will require less time. In this paper, we present the artificial-free technique for the simplex algorithm by mapping the problem into the objective plane and splitting constraints into three groups. In the objective plane, one of variables which has a nonzero coefficient of the objective function is fixed in terms of another variable. Then it can split constraints into three groups: the positive coefficient group, the negative coefficient group and the zero coefficient group. Along the objective direction, some constraints from the positive coefficient group will form the optimal solution. If the positive coefficient group is nonempty, the algorithm starts with relaxing constraints from the negative coefficient group and the zero coefficient group. We guarantee the feasible region obtained from the positive coefficient group to be nonempty. The transformed problem is solved using the simplex algorithm. Additional constraints from the negative coefficient group and the zero coefficient group will be added to the solved problem and use the dual simplex method to determine the new optimal solution. An example shows the effectiveness of our algorithm.

## 1. Introduction

The simplex algorithm [1] has been used to solve linear programming problems. It starts at the origin point. If the origin point is not a basic feasible solution, the artificial variables will be added for finding the basic feasible solution. Many researchers (e.g., Zionts S. [2], Pan P. Q. [3], Arsham H. [4] and Corley H.W., Rosenberge J., Yeh W.C. and Sung T.L. [5]) proposed techniques for improving the simplex algorithm without using artificial variables.

In this paper, we present the artificial-free technique for the simplex algorithm by mapping the problem into the objective plane and separating groups of constraints. Megiddo N. [6] and Boonperm A. and Sinapiromsaran K. [7] used this technique to reduce the dimension for solving a few dimensional linear programming problems. For a general linear programming problems, in the objective plane, one of variables which has a nonzero coefficient of the objective function is fixed in terms of another variable. After mapping it into all constraints, we can split constraints into three groups: the positive coefficient group, the negative coefficient group and the zero coefficient group. Along the objective direction, some constraints from the positive coefficient group will form the optimal solution. If the positive coefficient group is nonempty, the algorithm starts with relaxing constraints from the negative coefficient group and the zero coefficient group.

We guarantee the feasible region obtained from the positive coefficient group to be nonempty. The transformed problem is solved using the simplex algorithm. Additional constraints from the negative coefficient group and the zero coefficient group will be added to the solved problem and use the dual simplex method to determine the new optimal solution. The main concept of our algorithm and theorems for guarantee feasibility of a relaxed problem is shown in section 2. In section 3, the algorithm is summarized and the illustrative example shows the effectiveness of the algorithm. Finally, we conclude and discuss our new findings.

## 2. Preliminaries

Consider a linear programming problem in the following form:

$$\begin{aligned} \text{Maximize} \quad & \mathbf{c^T x} \\ \text{subject to} \quad & \mathbf{Ax} \leq \mathbf{b} \end{aligned} \tag{1}$$

where $\mathbf{c}$ is a nonzero vector and $\mathbf{x}$ is an $n$-dimensional column vector, $\mathbf{A}$ is an $m \times n$ matrix, $\mathbf{b}$ is an $m$-dimensional column vector. Let $y = \mathbf{c^T x} = c_1x_1 + c_2x_2 + \cdots + c_ix_i + \cdots + c_nx_n$. If $c_i \neq 0$, we choose $x_i = \frac{y}{c_i} - \sum\limits_{\substack{j=1 \\ j\neq i}}^{n} \frac{c_j}{c_i} x_j$. The problem in an equivalent form can be written as

$$\begin{aligned} \text{Maximize} \quad & y \\ \text{subject to} \quad & y \leq b'_r - \sum_{\substack{j=1, \\ j\neq i}}^{n} a'_{rj}x_j, r \in M_1 \\ & y \geq b'_s - \sum_{\substack{j=1, \\ j\neq i}}^{n} a'_{sj}x_j, s \in M_2 \\ & \sum_{\substack{j=1, \\ j\neq i}}^{n} a_{tj}x_j \leq b_t, t \in M_3, \end{aligned} \tag{2}$$

where $M_1, M_2$ and $M_3$ are the set of the constraint $r, s, t$ where $\frac{a_{ri}}{c_i} > 0$, $\frac{a_{si}}{c_i} < 0$ and $\frac{a_{ti}}{c_i} = 0$ respectively. $|M_1| + |M_2| + |M_3| = m$, $b'_l = \frac{c_i b_l}{a_{li}}$ and $a'_{lj} = \frac{c_i a_{lj}}{a_{li}} - c_j$ where $l \in \{M_1 \cup M_2\}$. The group $M_1$ is the group of positive coefficients, the group $M_2$ is the group of negative coefficients and the group $M_3$ is the group of zero coefficients.

### *2.1. The relaxed problem*

Consider the problem (2), let $\mathbf{x}_{\bar{i}}^{\mathbf{T}} = [x_1, x_2, ..., x_{i-1}, x_{i+1}, ..., x_n]^{\mathbf{T}}$, it will be feasible when $y$ is between constraints in the group $M_1$ and constraints in the group $M_2$, and $\mathbf{x}_{\bar{i}}$ satisfies constraints in the group $M_3$. Since we want to maximize $y$ and constraints in the group $M_1$ are higher than $y$, the optimal solution will be formed by some constraints in the group $M_1$. So we will relax some constraints from the group $M_2$ and $M_3$ then we will solve the problem with constraints from the group $M_1$ first. The relaxed problem can be rewritten as the following:

$$\begin{aligned} \text{Maximize} \quad & y \\ \text{subject to} \quad & y \leq b'_r - \sum_{\substack{j=1, \\ j\neq i}}^{n} a'_{rj}x_j, r \in M_1 \end{aligned} \tag{3}$$

We will show that the problem (3) is always feasible.

**Theorem 1.** If $M_1 \neq \emptyset$ and $b'_{min} = \min\limits_{r\in M_1} \{b'_r\}$. Then $(y_0, \mathbf{x}_{\bar{i}0}^{\mathbf{T}}) = (b'_{min}, \mathbf{0^T})$ is a feasible point of the problem (3).

*Proof.* Suppose $M_1 \neq \emptyset$ and $b'_{min} = \min_{r \in M_1}\{b'_r\}$. Choose $(y_0, \mathbf{x}_{i0}^{\mathbf{T}}) = (b'_{min}, \mathbf{0}^{\mathbf{T}})$. We get $y_0 = b'_{min} \leq b'_r$ for all $r \in M_1$. So $(y_0, \mathbf{x}_{i0}^{\mathbf{T}}) = (b'_{min}, \mathbf{0}^{\mathbf{T}})$ is a feasible point of the problem (3).

□

**Theorem 2.** If $M_1 = \emptyset$ and $M_3 = \emptyset$, then the problem is unbounded.

*Proof.* Assume $M_1 = \emptyset$ and $M_3 = \emptyset$. Let $X = \{(y, \mathbf{x}_i^{\mathbf{T}}) | y \geq b'_s - \sum_{j=1, j \neq i}^{n} a'_{sj} x_j, s \in M_2\}$ and $b'_{max} = \max_{s \in M_2}\{b'_s\}$. Choose $(y_0, \mathbf{x}_{i0}^{\mathbf{T}}) = (b'_{max}, \mathbf{0}^{\mathbf{T}})$. We get $y_0 = b'_{max} \geq b'_s$ for all $s \in M_2$. So $(y_0, \mathbf{x}_{i0}^{\mathbf{T}}) = (b'_{max}, \mathbf{0}^{\mathbf{T}})$ is a feasible point of the problem. So $X$ is not empty and it is obvious that $\mathbf{d}^{\mathbf{T}} = [\mathbf{1}, \mathbf{0}, ..., \mathbf{0}]^{\mathbf{T}}$ is a recession direction (see [1]) where $\mathbf{d}$ is an $n$-dimensional column vector. Consider the objective cost of the problem (2), $\mathbf{c}^{\mathbf{T}} = [\mathbf{1}, \mathbf{0}, ..., \mathbf{0}]^{\mathbf{T}}$, for all $\alpha > 0$ and $(y_0, \mathbf{x}_{i0}^{\mathbf{T}}) \in X$, we get $(y_0, \mathbf{x}_{i0}^{\mathbf{T}}) + \alpha \mathbf{d}^{\mathbf{T}} = (b'_{max} + \alpha, \mathbf{0}^{\mathbf{T}}) \in X$ and $(b'_{max} + \alpha, \mathbf{0}^{\mathbf{T}})\mathbf{c} = b'_{max} + \alpha$. Since $\alpha > 0$, $b'_{max} + \alpha \to \infty$ as $\alpha \to \infty$. Therefore, the problem is unbounded.

□

If $M_1 = \emptyset$ then we will relax constraints from the group $M_3$. All constraints of this relaxed problem are in group $M_2$, from theorem(2), this problem is unbounded. Constraints from group $M_3$ are applied with the current solution for finding the optimal solution.

### 2.2. *The transformed problem*

Consider the relaxed problem (3), if $b'_r \geq 0$ for all $r \in M_1$, we can start the simplex algorithm at the origin point. Otherwise, since $(b'_{min}, \mathbf{0}^{\mathbf{T}})$ is a feasible point, we transform the problem (3) using $y' = y - b'_{min}$. So we can start the simplex algorithm at the new origin point.

## 3. Our algorithm and illustrative example

Our algorithm is summarized in the following:

Step 1: Map the original problem with $x_i = y - (c_1 x_1 + c_{i-1} x_{i-1} + \cdots + c_{i+1} x_{i+1} + \cdots + c_n x_n)/c_i$ where $c_i \neq 0$ to the problem (2) and split constraints into 3 groups.

Step 1.1: If $M_1 \neq \emptyset$, Relax constraints from group $M_2$ and $M_3$. If $b'_r \geq 0$ for all $r \in M_1$, start the simplex algorithm at the origin point. Otherwise, transform the problem using $y' = y - b'_{min}$ where $b'_{min} = \min_{r \in M_1}\{b'_r\}$ and start the simplex algorithm.
If the optimal solution is found, go to Step 2. Otherwise, go to Step 3.

Step 1.2: Otherwise, if $M_3 = \emptyset$, the problem is unbounded. Then stop.
Otherwise, relax constraints from the group $M_3$ and transform the problem using $y' = y - b'_{max}$ where $b'_{max} = \max_{s \in M_2}\{b'_s\}$. This relaxed problem is unbounded. Go to Step 3.

Step 2: Test the optimal solution with all constraints from group $M_2$ and $M_3$.

Step 2.1: If it satisfies all constraints, it is the optimal solution. Then stop.

Step 2.2: Otherwise, perform the dual simplex at the unsatisfied constraints until the optimal solution is found. Then stop.

Step 3: Add constraint from group $M_2$ and $M_3$ one by one until $M_2$ and $M_3$ are empty.

Step 3.1: If the current solution satisfies it, perform the primal simplex algorithm.

Step 3.2: Otherwise, perturb the negative reduced cost to a positive value for dual feasibility, then perform the dual simplex until the optimal solution is found. Restore the original reduced cost and perform the primal simplex until the optimal solution is found. Then stop.

The following example is presented to show the efficiency of the algorithm.

**Example.** Consider the following problem:

$$\begin{array}{llrrl} \text{Maximize} & -x_1 & -x_2 & & \\ \text{subject to} & -2x_1 & -3x_2 & \leq & 4 \\ & 2x_1 & +x_2 & \leq & -2 \\ & x_1 & +2x_2 & \leq & 1 \\ & x_1 & -x_2 & \leq & -2 \\ & -x_1 & & \leq & 3 \end{array} \tag{4}$$

**Solution** Let $y = -x_1 - x_2$ then $x_1 = -y - x_2$. After mapping, $M_1 = \{1, 5\}, M_2 = \{2, 3, 4\}$ and $M_3 = \emptyset$, then the relaxed problem is written as:

$$\begin{array}{lll} \text{Maximize} & y & \\ \text{subject to} & y & \leq 2 + \frac{1}{2}x_2 \\ & y & \leq 3 - x_2 \end{array} \tag{5}$$

Since $(y_0, x_{2_0}) = (0, 0)$ is a feasible point, we can start the simplex algorithm without using the artificial variable. The optimal solution of the relaxed problem is $y = 7/3$ and $x_2 = 2/3$ at the second iteration. This point satisfies all constraints, so the optimal solution of the original problem is $x_2 = 2/3, x_1 = -y - x_2 = -7/3 - 2/3 = -3$ and the optimal value is $7/3$.

The simplex algorithm solves this problem by adding 2 artificial variables using 2 iterations in Phase-I and 2 iterations in Phase-II while our method used 2 iterations without using artificial variable.

## 4. Conclusion

In this paper, we proposed the artificial-free technique to improve the simplex algorithm. Since artificial variable is not used, the number of variables by our algorithm is less than or equal to the number of variables by the simplex algorithm. Furthermore, the number of constraints solved by our algorithm is less than or equal to the number of constraints solved by the simplex algorithm. So the dimension of parameters used by our algorithm is less than or equal to the dimension of parameters which the simplex algorithm used. Moreover, if the group $M_1$ and $M_3$ are empty, we can conclude that the original problem is unbounded without using the simplex algorithm. However, our algorithm is applied with unrestricted variables which doubles the number of variables when the problem is written in the standard linear programming form. In future work, we will improve the algorithm to decrease the number of variables which come from unrestricted variables and construct selection criteria for mapping the problem.

## References


[1] Bazaraa M S, Jarvis J J and Sherali H D 1990 *Linear programming and network flows* 2nd edn New York : John Wiley

[2] Zionts S 1969 The criss-cross method for solving linear programming problems *Manage.Sci.* **15** 426–45

[3] Pan P Q 2000 Primal perturbation simplex algorithms for linear programming *J. Comput. Math.* **18** 587–96

[4] Arsham H 2007 A computationally stable solution algorithm for linear programs *Appl.Math.Comput.* **188** 1549–61

[5] Corley H W, Rosenberger J, Yeh W C and Sung T K 2006 The cosine simplex algorithm *Int J Adv Manuf Technol.* **27** 1047–50

[6] Megiddo N 1983 Linear-Time algorithms for linear programming in R3 and related problems *SIAM J. Comput.* **12** 759–76

[7] Boonperm A and Sinapiromsaran K 2010 Linear time algorithm in term of number of the constraints for linear programming in 2D *Proc. of OR-NET 2010* (Bangkok: Operations Research Network in Thailand) 48–53

# ONMCGP: Orthogonal Neighbourhood Mutation Cartesian Genetic Programming for Evolvable Hardware

**Fuchuan NI[1,2], Yuanxiang LI[3] and Peng KE[4]**

[1,3,4]State Key Lab of Software Engineering, Wuhan, P.R.China

[2]Department of Computer science, Huazhong Agricultural University, Wuhan, P.R.China

E-mail: fcni_cn@whu.edu.cn[1], yxli@whu.edu.cn[3], 4278072@qq.com[4]

**Abstract.** Evolvable Hardware is facing the problems of scalability and stalling effect. This paper proposed a novel Orthogonal Neighbourhood Mutation(ONM) operator in Cartesian genetic programming (CGP), to reduce the stalling effect in CGP and improve the efficiency of the algorithms.The method incorporates with Differential Evolution strategy. Demonstrated by experiments on benchmark, the proposed Orthogonal Neighbourhood Search can jump out of Local optima, reduce the stalling effect in CGP and the algorithm convergence faster.

## 1. Introduction

Evolvable Hardware(EHW) is a technique to dynamically alter the hardware functionality and physical connections of its circuits using methods inspired by natural evolution[1]. EHW, also briefly is defined as "$EHW = EAs + PLDs$". Among the EAs, Cartesian Genetic Programming(CGP)[2] is the most popular, important algorithm in digital evolvable circuits.

Up to date, because of the limitations in scalability, EHW seldom competes with traditional designs and few real-world applications have been developed[3]. First, The length of the chromosome representation of electronic circuits and the computational complexity grows exponentially with the increasing of inputs and the outputs; Second, for EHW, especially in combination circuits, many-for-one map between genotype and phenotype exits. And The problem of stalling effect in fitness functions is described as the nonimprovement of the fitness values during the evolutionary process. In EAs, the individuals that have the greater fitness have greater possibilities to reproduce. Thus, in the later stage in evolving process, the difference between the individuals becomes very little and "stalling effect " appears[1] [3]. E.Stomeo[4] observed that when the fitness function reaches 82.8% the stalling effect occurs.

Cartesian Genetic Programming (CGP), uses evolution strategy (ES), operating with the population of $\mu + \lambda$ individuals by point mutation. In order to enhance the search ability for the solution space, *The ideal mutation operator* is expected to have the merits of "higher probability of making far long jumps" at the early stage and "much better local fine-tuning ability" at the later stage. Yao[5] proposed a Cauchy mutation operator which jumps longer than Gaussian mutation. Lee[6] proposed levy Mutation. Zhao[7] proposed a Non-Uniform Mutation to control the search step and neighborhood size. Storn R[8]suggested a new strategy, Differential evolution. Leung[9] proposed orthogonal GA, incorporated experimental design methods into the GA. But all these methodologies are devised for function optimization. Thereby this paper proposes the Orthogonal Neighborhood Mutation(ONM) for CGP to evolve the combination circuits efficiently.

This paper is organized as follows. Section 2 considers the principal CGP in EHW. Section 3 proposes ONM operator in CGP to design combinational circuit, as well as the algorithm process. Sections 4 presents the experiment results with benchmark. Finally, section 5 presents the main conclusions.

## 2. Cartesian genetic programming

In Cartesian Genetic Programming (CGP), which was proposed by Miller[2], a candidate circuit is represented as an array of programmable elements with of u(columns) and v(rows), whose inputs and outputs is fixed. For each node, its' input can be connected either to the output of a node placed in the previous L columns or to some of the circuit inputs. The L parameter is defined as the level of connectivity. Feedback is not allowed. Each node is programmed to perform one of n-input functions defined in the function set. The outputs are selected randomly from the codes in the array. CGP, using evolution strategy (ES), operates with the population of $\mu+\lambda$ individuals. The initial population is randomly generated. Every new population consists of the best individual of the previous population and the offsprings are created by point mutation. The fitness function is the difference between the output and the required truth table. The goal is to minimize the difference.Figure 1(a)describes how one-bit adder representation in CGP, 3 columns and 3 rows.

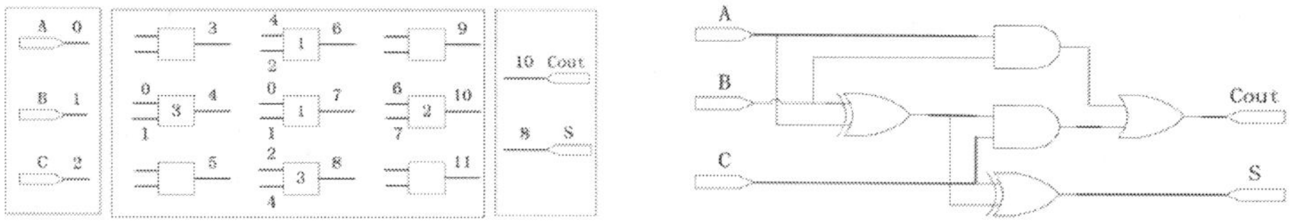


(a) CGP-Code (b) CGP-DeCode

**Figure 1.** The CGP representation

The chromosome is arranged as : 2 0 1, <u>0 1 3</u>, 1 0 3, <u>4 2 1</u>, <u>0 1 1</u>, <u>2 4 3</u>, 2 8 1, <u>6 7 2</u>, 3 1 0, <u>10, 8</u>. The Length is $3 \times 3 \times 3 + 2 = 29$. Node[1][0], i.e. <u>0 1 3</u>, 0,1 are the number of connect responding to the input A, B; 3, is the index of the function type, representing XOR. The codes underlined are activated, which would be decoded. Figure 1(b) shows the logic circuit when the the chromosome is decoded.

## 3. Orthogonal Neighborhood Mutation in CGP

As described in section 1, the CGP often traps into being Stalling. To improve the search efficiency of CGP, Orthogonal Neighborhood Mutation(ONM) is proposed and Differential Evolution strategy is considered. The operator is listed as follow:

*(1)For a individual i (i =1,2,3, ..., n; n defined as size of the population), which the fitness is $f_i$, select other three individuals randomly;*

*(2)Select four loci as alleles randomly, in gene strings, to compose a Four-factor-three-level orthogonal experiment and local search is conducted;*

*(3)by means of range analysis, the best combination is obtained;*

*(4) If the fitness of the best combination($f'$) is the best solution, the individual i is replaced, else produce a random number R:*

*if $R < C$ (C, defined as a constant between 0 and 1) the individual i is replaced; else the individual i is reserved.*

The algorithm of Construction of Orthogonal Array refers to [9].

In detail, for the example cited in section 2, the genotype consists of thirteen nodes; select the 2nd, 5th, 7th and 8th nodes as 4 alleles randomly from 3 chromosomes, described as $A1, A2, A3; B1, B2, B3; C1, C2, C3$ and the fitnesses are $f_1$, $f_2$ and $f_3$. Compose a Four-factor-three-level orthogonal experiment; the orthogonal array table list as Table1:

**Table 1.** Orthogonal Arrays

| | A | B | C | D | fittness |
|---|---|---|---|---|---|
| (1) | A1 | B1 | C1 | D1 | $f'_1$ (=$f_1$) |
| (2) | A1 | B2 | C2 | D2 | $f'_2$ |
| (3) | A1 | B3 | C3 | D3 | $f'_3$ |
| (4) | A2 | B1 | C2 | D3 | $f'_4$ |
| (5) | A2 | B2 | C3 | D1 | $f'_5$ |
| (6) | A2 | B3 | C1 | D2 | $f'_6$ |
| (7) | A3 | B1 | C3 | D2 | $f'_7$ |
| (8) | A3 | B2 | C1 | D3 | $f'_8$ |
| (9) | A3 | B3 | C2 | D1 | $f'_9$ |

Calculate the fitnesses of 9 novel individuals and analyze the range. Presume that the best combination is $A', B', C', D'$ and the fitness is $f'$.

According to Differential Evolution Strategy[8], the new individual would be accept with probability R. If $f' > max\{f_i, f_1, f_2, f_3, f'_1, f'_2, ..., f'_9\}$, the individual $i$ is substitute by the best combination;If $f' <= max\{f_i, f_1, f_2, f_3, f'_1, f'_2, ..., f'_9\}$, the individual $i$ is substituted by the best combination with probability R.

Compared to local search conducted in each locus for every individual, which deteriorates into exhaustive search, ONM operator selects the best in the neighborhood, and could can jump out of Local optima more easily and faster.

## 4. Experiment result

The compared experiment between traditional CGP and the ONMCGP is complimented. The benchmark is one-bit full adder circuit which has 3 inputs and 2 output. The two algorithms are same except Mutation operator. The CPU is Intel Core i5-3230 @2.6GHZ, RAM 8G based on Windows 8 64bits. The population size is 10;The number of generation is 60000; The probability of mutation is 0.6. The Code Matrix is 3 rows with 10 columns and the level-back is set to 10. The two algorithms all run 30 times.The succeed rates reach 100%. The average First Hit generation of CGP is 4543 and that of ONMCGP is 500.

Figure 2 shows the evolving process of the ONMCGP. Figure 2(a) is the best, which converged at 93 generation; Figure 2(b) is the worst, which costs 1900 iterations to get the satisfied solution. Figure 3 shows the evolving process of the CGP. Figure 3(a) is the best, which converged at 174 generation; Figure 3(b) is the worst, which costs about 12500 iterations to get the satisfied solution.

## 5. Conclusion

The results of the compared experiments demonstrate that the ONM operator could reduce the stalling effect and convergence more quickly. Incorporated with *orthogonal experimental design method* , the proposed operator can scan the feasible solution space once to locate the

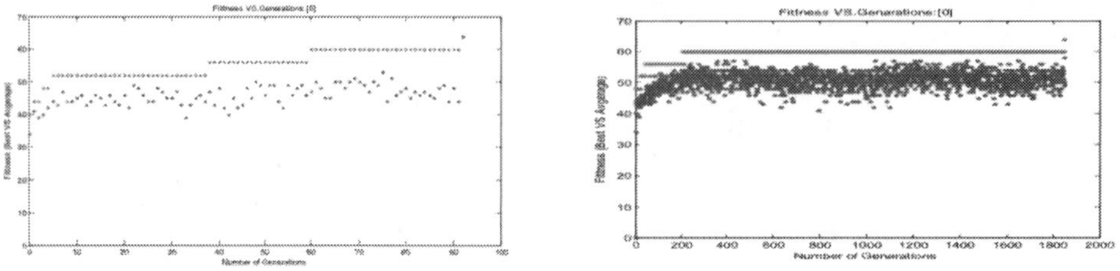


(a) ONMCGP-Best (b) ONMCGP-Worst

**Figure 2.** The ONMCGP Evolving Process

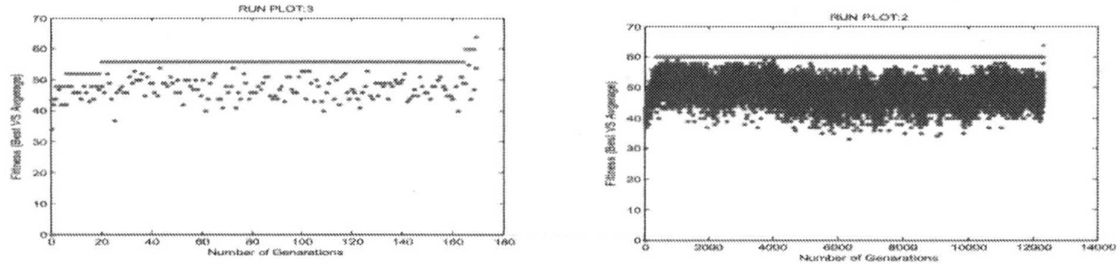


(a) CGP-Best (b) CGP-Worst

**Figure 3.** The CGP Evolving Process

better points for further exploration in subsequent iterations. The ONM operator can reduce the stalling effect, and thus be beneficial to the Scalability of the CGP.

## 6. Acknowledgements
This paper is supported by Natural Science Foundation of China (NO.60442001, 61070009) and Project 2013JC007 Granted by the Fundamental Research Funds for the Central Universities of China.

## 7. References
[1] Yao X and Higuchi T 1999 *Systems, Man, and Cybernetics, Part C: Applications and Reviews, IEEE Transactions on* **29** 87 97
[2] Miller J F 1999 An empirical study of the efficiency of learning boolean functions using a cartesian genetic programming approach *Proceedings of the Genetic and Evolutionary Computation Conference* vol 2 pp 1135 1142
[3] Haddow P C and Tyrrell A M 2011 *Genetic Programming and Evolvable Machines* **12** 183 215
[4] Stomeo E, Kalganova T and Lambert C 2006 *Systems, Man, and Cybernetics, Part B: Cybernetics, IEEE Transactions on* **36** 1024 1043
[5] Yao X, Liu Y and Lin G 1999 *Evolutionary Computation, IEEE Transactions on* **3** 82 102
[6] Lee C Y and Yao X 2004 *Evolutionary Computation, IEEE Transactions on* **8** 1 13
[7] Zhao X, Gao X S and Hu Z C 2007 *Applied Mathematics and Computation* **192** 1 11
[8] Storn R and Price K 1997 *Journal of global optimization* **11** 341 359
[9] Leung Y W and Wang Y 2001 *Evolutionary Computation, IEEE Transactions on* **5** 41 53

# ITO-based evolutionary algorithm to solve traveling salesman problem

**Wenyong Dong** [1, 2], **Kang Sheng**[1], **Chuanhua Yang** [3], **Yunfei Yi** [1,4,*]

[1]Computer School, Wuhan University, Wuhan, China
[2]Department of Electrical and Computer Engineering, New Jersey Institute of Technology, Newark, USA
[3]Changjiang Wuhan Waterway Bureau,Wuhan, China
[4]Department of Computer and Information Science, Hechi University, Yizhou, China

E-mail: gxyiyf@ 163.com

**Abstract.** In this paper, a ITO algorithm inspired by ITO stochastic process is proposed for Traveling Salesmen Problems (TSP), so far, many meta-heuristic methods have been successfully applied to TSP, however, as a member of them, ITO needs further demonstration for TSP. So starting from designing the key operators, which include the move operator, wave operator, etc, the method based on ITO for TSP is presented, and moreover, the ITO algorithm performance under different parameter sets and the maintenance of population diversity information are also studied.

## 1. Introduction

ITO algorithm is based on Ito stochastic process by simulating the kinetic law of the particles collide with each other in imitation of the particle system to design the algorithm and problem solving. It's a new algorithm analyzing the law of motion of the particle from the microscopic point of view, and then be proposed through abstraction and simulation methods. ITO algorithm on the one hand reflects the characteristics of the population search in biomimetic evolutionary algorithm, and in these algorithms the solution is usually represented (encoded) as a particle and a large number of particles build up the object processed by ITO algorithm, namely particle system; on the other hand, we use ITO process theory to analyze the algorithm, such as use the Ito stochastic integral method to establish the dynamic equations of the algorithm, and we usually design two key operators: the move rate and the wave rate according to the law of macromolecular movement proposed by Einstein, Langevin as well as the law of particle thermal motion, therefore, it contains some characteristics of the simulated annealing algorithm, because of which the proposed algorithm is named [1-4]. In this paper we will apply ITO algorithm to the optimization of the TSP problem and we will study the diversity of mechanisms in the search process of ITO algorithm.

## 2. ITO algorithm for TSP

In this section we introduce the idea of ITO algorithm and the operator for TSP in detail, including the determination of the particle radius, the move operator, the wave operator and the temperature of the annealing environment.

---

* Yunfei Yi(1981-), male, Ph.D., associate professor, research interests include intelligent computing, machine learning, etc. This work was in part supported by the NSF of China under Grant Nos. 61170305, 60873114 and the State Scholarship Fund (File No, 2012084291) of CSC.

### 2.1. Process of the algorithm

Ito process is proposed mainly inpired by the law of thermal motion of particles, likely, we mimic the swarm intelligence algorithms, such as genetic algorithms, ant colony optimization, etc. , to map the Ito process to an algorithm that can be used to solve the optimization problems. Besides, ITO algorithm is also designed to be population-based algorithm. The iterative process is as follows:

(1) Initialize a population of N particles utilizing some strategies, set relevant parameters of the algorithm;

(2) Assess the particles in the population;

(3) If the stop conditions are not met, perform the following loop :

- Find the best particle and the worst particle in the population;
- Each particle choose the attractive center (or called the attractor or attractive element, we will mix these concepts in the rest of this article) for move operator;
- Calculate the annealing temperature, the radius of each particle, the move operator and the wave operator;
- Each particle moves in accordance with the move operator;
- Each particle moves to a new location in accordance with the wave operator;
- Evaluate the solution that each particle represents ;

(4) Output the optimal solution

### 2.2. The design of several key factors

In order to achieve desired results, we need to design a few key factors including the particle radius, wave operator, move operator, annealing table and the selection of the maximum intensity and the minimum intensity.

A better approach to design the particle radius is the method based sorting, The radius of each particle is calculated as follows:

$$r(x) = r_{max} - \frac{(r_{max} - r_{min})}{n-1} \times (n_i - 1) \quad (1)$$

The particle radiuses generated by this method distribute evenly in $[r_{min}, r_{max}]$, regardless of the objective function value of each particle, this approach makes the algorithm has a strong search capabilities even in the late period, therefore it's more suitable for complex function.

## 3. Parameter sensitivity configuration and population diversity experiment

There are several parameters in ITO algorithm to be settled when we try to solve TSP problem: the number of particles, the upper and lower bounds of wave, the method to select the attractive center, annealing strategy, and the method to calculate the radius.

### 3.1. The impact of the lower and upper bounds in waves

The wave operator has a decisive influence on the global convergence, and we can prove that when the wave lower bound is 0, the algorithm is not convergent, but when the upper bound and lower bound take the same value, the wave operator turns into a pure random algorithm. In order to study the impact of wave with the upper and lower bounds, we can take $\gamma_{max} = 1 - \sqrt[n]{p_{min}}$, $\gamma_{min} = 1 - \sqrt[n]{p_{max}}$ when design the algorithm, so once we know $p_{min}$ and $p_{max}$, we can get $\gamma_{min}$ and $\gamma_{max}$. We select 3 values 0.0005, 0.05, 0.5 as the value of $p_{min}$ ,and select 4 values 0.1, 0.4, 0.5, 0.8 as the value of $p_{max}$ we also remove the combinations of $p_{min} > p_{max}$ .Table 1 are the results of our experiments.

**Table 1.** Wave operator in the upper and lower bounds for different combinations of four cases, Data in the table is the averages of 30 independent runs, the percentage in brackets indicate the gap between the average path length and the optimal solution. Bold part is the best result.

| Instance | (0.0005, 0.1) | (0.0005, 0.4) | (0.0005, 0.5) | (0.0005, 0.8) |
|---|---|---|---|---|

| | | | | |
|---|---|---|---|---|
| **Eil51** | 432.9(1.62%) | 433.2(1.69%) | 440.6(3.43%) | 437.6(2.72%) |
| **KroA100** | 21846.9(2.65%) | 21732.3(2.12%) | 21926.9(3.03%) | 21880.1(2.81%) |
| **D198** | 16944.5(7.38%) | 16872.5(6.92%) | 16630.5(5.39%) | 16934.5(7.32%) |
| **Lin318** | 43007.4(2.33%) | 43239.7(2.88%) | 43843.5(4.32%) | 42994.9(2.30%) |
| Instance | (0.005, 0.1) | (0.005, 0.4) | (0.005, 0.5) | (0.005, 0.8) |
| **Eil51** | 438.5(2.93%) | 441.7(3.69%) | 429.3(0.77%) | **426.9(0.21%)** |
| **KroA100** | 21997.5(3.36%) | 21799.5(2.43%) | 21803.6(2.45%) | 21808.9(2.48%) |
| **D198** | 16795.5(6.43%) | 16666.5(5.62%) | 16120.5(2.16%) | **15839.5(0.0%)** |
| **Lin318** | 42901.5(2.08%) | 42798.4(1.83%) | 42321.2(0.70%) | **42029.0(0.0%)** |
| Instance | (0.05, 0.1) | (0.05, 0.4) | (0.05, 0.5) | (0.05, 0.8) |
| **Eil51** | 429.6(0.85%) | 438.1(2.84%) | 439.2(3.10%) | 441.4(3.62%) |
| **KroA100** | **21285.0(0.0%)** | 21983.3(3.30%) | 21983.3(3.30%) | 21996.7(3.36%) |
| **D198** | 16064.5(1.80%) | 17091.5(8.31%) | 17091.5(8.31%) | 17326.5(9.80%) |
| **Lin318** | 43108.9(2.57%) | 44089.4(4.90%) | 44089.4(4.90%) | 44331.8(5.48%) |

It can be seen from Table 1 that three of the four instances obtained the best computing performance in the combination of (0.005, 0.8) , and the instance KroA100 obtained best computing performance in the combination of (0.05, 0.1). On the whole, regardless of the combination type, the performance of the algorithm are quite good, the gap between the worst results of the calculation and the optimal solution is relatively small. Compared to MMAS, the solution quality of the algorithm is not ideal, but to the general AS, GA and PSO algorithm, the solution quality has been greatly improved and showed certain stability.

### 3.2. The local move strategy and global move strategy

Local move strategy utilizes the optimal solution in each particle iteration trajectory as the attractive center and it is initially set to the optimal solution in the population. ITO algorithm using this strategy is abbreviated as $ITO^{les}$. Global move strategy employ the optimal solution in the population as the attractive center, ITO algorithm using this strategy is abbreviated as $ITO^{ges}$. In a certain sense, the attractor of each particle with the local move strategy may not be the same; hence the diversity of the algorithm can be guaranteed better than the global move strategy.

### 3.3. The measurement and analysis of the diversity of the population

We mainly study the diversity of the ITO algorithm retention mechanism, that is, we will test that whether the algorithm keeps diverse in the whole iterative process or not. Here we use two diversity measures: one is based on connection matrix, another is based on information entropy.

As can be seen from Table 2, the population diversity of each algorithm is lager when the algorithm is in the initialization after 100 generations or when the algorithm terminates, regardless of using connection matrix or information entropy, besides, the common trend is as the iteration depth increasing, the diversity of the population became lower, but reduced marginally. At the same time, we can observe the four computing instance using $ITO^{les}$ algorithm possess higher diversity than that using $ITO^{ges}$ algorithm.

**Table 2.** The statistical results of diversity difference

| Instance | Start($ITO^{les}$) | 100th($ITO^{les}$) | terminate($ITO^{les}$) | ($ITO^{ges}$) | 100th($ITO^{ges}$) | Terminate($ITO^{ges}$) |
|---|---|---|---|---|---|---|
| Criterion | D(P) | H(P) | D(P) | D(P) | H(P) | D(P) |
| **Eil51** | 0.559 | 0.354 | 0.531 | 0.351 | 0.515 | 0.308 |
| **KroA100** | 0.620 | 0.210 | 0.601 | 0.202 | 0.593 | 0.197 |
| **D198** | 0.691 | 0.299 | 0.645 | 0.276 | 0.630 | 0.266 |
| **Lin318** | 0.663 | 0.301 | 0.639 | 0.299 | 0.622 | 0.298 |

## 4. The experimental results and analysis

This section we study the performance of ITO algorithms for TSP, mainly from two aspects to design the experiment, one is comparative experiment using different algorithms, and the other is to increase the performance of the algorithm using local search technique.

### 4.1. The comparative experiment using different algorithms

As a meta-heuristic algorithm, the heuristic algorithms we compare for TSP problem are as follows: MMAS, ACS, AS, DPX-GA and PSO algorithm [5-15]. Among these algorithms, the experimental data of MMAS ACO, AS comes from [16], the experimental data of DPX-GA come from [17], the experimental data of PSO algorithm and the algorithm proposed in this paper is obtained by our programming and the parameter settings of the population size and the instance size should match: local learning rate cl = 0.8, global learning rate c2 = 0.6, use log function to convert the continuous variables between [0, 1], then reverse the gene locus in accordance with the probability; Shutdown condition is for 20 consecutive generation algorithm does not improve. The parameter settings of ITO algorithm: the number of particles matches the size of the instance and shutdown condition is that for 20 consecutive generation algorithm does not improve, the method for calculating the particle radius is the linear transformation and $p_{\min} = 0.005$, $p_{\max} = 0.8$.

Table 3 is result of different algorithms. We can see from the table, $\text{ITO}^{ges}$ algorithm outperforms the others as for the calculation results, but the performance of these algorithms is not significantly different in all calculations instance, PSO algorithm has the worst performance, followed by the ACS and AS; the algorithm $\text{ITO}^{ges}$ has the best performance in these comparison algorithms. As a generic algorithm, the computational performance of the $\text{ITO}^{ges}$ algorithm is competitive and the algorithm improvements are worthy of further study.

**Table 3.** The The average results for different algorithms, each run 30 times independently, The number in bold face shows the best results.

| Instance | MMAS | ACS | AS | DPX-GA | PSO | $\text{ITO}^{ges}$ |
|---|---|---|---|---|---|---|
| Eil51 | 427.6 | 428.1 | 437.3 | 428.5 | 498.2 | **426.9** |
| kroA100 | 21320.3 | 21420.0 | 22471.4 | **21285** | 21993.5 | **21285** |
| D198 | 15972.5 | 16054.0 | 16702.1 | **15839.5** | 16049.7 | **15939.5** |
| Lin318 | **42029.0** | 0 | 0 | 42605.3 | 49385.1 | **42029.0** |
| Eil51 | 427.6 | 428.1 | 437.3 | 428.5 | 498.2 | **426.9** |

## 5. Conclusion

In this paper we studied how ITO algorithm is used in solving TSP problem and discussed the design and parameter settings of ITO algorithm, including the move operator design, the wave operator design, the calculation of the particle radius, the method to select the attractive center and so on. In this paper, we studied the performance of the algorithm and the diversity of the population information in the case of different parameter values, and preliminary experimental results show that the ITO algorithm for solving the symmetric TSP problem with a certain degree of competitiveness. In this paper, we can see that ITO algorithm performs better than several other of the set when the wave operator mapping range is set to (0.005, 0.8).

## References

[1] Dong W Y. Simulation Optimization Based on the Hypothesis Testing and ITO Process. In: Third International Conference on Natural Computation. Haikou, China: IEEE, 2007. 1210-1221

[2] Dong W Y. Time Series Modeling Based on ITO Algorithm. In: Third International Conference on Natural Computation. Haikou, China: IEEE, 2007. 398-402

[3] Dong W Y. The Simulation Optimization Algorithm Based on the Ito Process. In: The 2nd International Conference on Intelligent Computing. Qingdao, China: IEEE,2007. 563-573

[4] Dong W Y. The Multi-Objective ITO Algorithms. In: The 2nd International Symposium on Intelligence Computationand Applications (ISICA 2007). Qingdao, China: IEEE, 2007. 21-

23

[5] Reinelt G. TSPLIB-A traveling salesman problem library. *ORSA Journal on Computing, 1991*, 3 (4), 376-384.

[6] Shi X H, Liang Y C, Lee H P et al. Particle swarm optimization-based algorithms for TSP and generalized TSP, *Information Processing Letters*,2007, 103 (1), 169-176.

[7] Huang L, Zhou C G and Wang K P. Hybrid ant colony algorithm for traveling salesman problem, *Progress in Natural Science*, 2003,4(13),295-299

[8] Ben-nian, Wang et al. RMGA: A reinforcement learning based genetic algorithm. *Acta Electronica Sinica*, 2006, 34(5),856-860, 866.

[9] Dorigo M and Gambardella L M. Ant colony system: A cooperative learning approach to the traveling salesman problem. IEEE Transactions on Evolutionary Computation, 1997,1(1):53-66

[10] Yagiura M and Ibaraki T. On Metaheuristic Algorithms for combinatorial Optimization Problems, *System and Computers in Japan,* 2001, 32(3),33-55

[11] Mak K T and Moton A J. A modified Lin-Kernighan traveling salesman heuristic. *Operations Res Lett*, 1993, 12, 127-132

[12] Chang P C, Huang W H and Ting C J. Dynamic diversity control in genetic algorithm for mining unsearched solution space in TSP problems, *Expert systems with Applications*, 2010, 37(2) 1863-1878

[13] Hendtlass T. Preserving Diversity in Particle Swarm Optimization, in: Lecture Notes in Computer Science, vol. 2718, Springer, 2003,4104-4108

[14] Chang P C, Chen S S, and Fan C Y. Mining gene structures to inject artificial chromosomes for genetic algorithm in single machine scheduling problems. *Applied Soft Computing Journal*, 2008, 8(1), 767-777

[15] Munakata T and Nakamura Y. Temperature control for simulated annealing, *Physical Review E*, 2001, 64(2) 046-127

[16] Stützle T, Hoos H H. MAX-MIN Ant System, *Future generation computer systems.* 2000, Vol.16, No.8, 231-240

[17] White C M and Yen G G. A hybrid evolutionary algorithm for traveling salesman problem, In: Congress on Evolutionary Computation (CEC2004), Chichogo, American. 2004, 176-180.

# Max-out-in pivot rule with Dantzig's safeguarding rule for the simplex method

**Monsicha Tipawanna and Krung Sinapiromsaran**

Department of Mathematics and Computer Science, Faculty of Science, Chulalongkorn University, Bangkok, 10330, Thailand

E-mail: ktmonsic@kmitl.ac.th (M. Tipawanna) and Krung.s@chula.ac.th (K. Sinapiromsaran)

**Abstract.** The simplex method is used to solve linear programming problem by improving the current basic feasible solution. It uses a pivot rule to guide the search in the feasible region. The pivot rule is used to select an entering index in simplex method. Nowadays, many pivot rule have been presented, but no pivot rule shows superior performance than other. Therefore, this is still an active research in linear programming. In this research, we present the max-out-in pivot rule with Dantzig's safeguarding for simplex method. This rule is based on maximum improvement of objective value of the current basic feasible point similar to the Dantzig's rule. We can illustrate by Klee and Minty problems that our rule outperforms that of Dantzig's rule by the number of iterations for solving linear programming problems.

**Keywords: Linear programming, Simplex method, Pivot rule, Dantzig's rule**

## 1. Introduction

The simplex method is the solution method for solving a linear programming problem (LP) (Dantzig, 1949). This method starts at a feasible solution of a feasible region and moves to the better solution by increasing value of a nonbasic variable called the entering variable and decreasing value of a basic variable called the leaving variable [2].The criteria to select the entering and leaving variable is called *the pivot rule.* Many researchers have attempted to improve the simplex performance , for examples, see Pan (1990) [9], Junior and Lins (2005) [4], Hu (2007) [5] and Arsham (2007) [3]. The pivot rule affects the simplex performance. So a lot of researches have been proposed new pivot rules such as Devex rule [7], Steepest-edge rule [1, 6], and a largest-distance pivot rule [8].

In this paper, we propose the out-in pivot rule called max-out-in pivot rule safeguarding with Dantzig's rule for the simplex method. This rule is first selecting the leaving variable that has the maximum value from the current basic variable set. Then it chooses the best corresponding entering variable that gives the smallest positive contribution to the binding constraint of the leaving variable. If the selected basic variable and nonbasic variable can not swap or there is no corresponding nonbasic candidate, then we use Dantzig's rule instead. We can show by Klee and Minty problem [10] that our method improves Dantzig's rule.

This paper is organized as follows. In section 2, we explain the main idea of our pivot rule and give the max-out-in pivot rule. We give the example to illustrate the implementation our rule in section 3. The application of our rule to Klee and Minty problem is given in section 4. Finally, the conclusion is given in section 5.

## 2. Max-Out-In Pivot Rule

### 2.1. The Main Idea

Consider the linear programming problem in the following form:

$$\begin{aligned} \text{Maximize} \quad & \mathbf{1}^{\mathbf{T}}\mathbf{x} \\ \text{subject to} \quad & \mathbf{A}\mathbf{x} = \mathbf{b} \\ & \mathbf{x} \geq \mathbf{0} \end{aligned} \tag{7}$$

where $\mathbf{b} \in \mathbb{R}^m$, $\mathbf{1} \in \mathbb{R}^n$, $\mathbf{A} \in \mathbb{R}^{m \times n}(m < n)$ and rank$(\mathbf{A}) = m$. In this form, the value of each decision variable is equal to its contribution to the objective value. Since the leaving variable decreases to zero when it changes to the nonbasic variable, and the increment in value of the entering variable depends on the decrement in value of the leaving variable. Then we should select the leaving variable that has the maximum value among all basic variables and select the nonbasic variable that can improve the objective value and allows the maximum increase. Rewrite the system (7) into the tableau

| | $x_B$ | $x_N$ | **RHS** |
|---|---|---|---|
| | $\mathbf{0}^T$ | $\mathbf{c}_B^T\mathbf{B}^{-1}\mathbf{N} - \mathbf{c}_N^T$ | $\mathbf{c}_B^T\mathbf{B}^{-1}\bar{\mathbf{b}}$ |
| $x_B$ | $I_m$ | $\bar{\mathbf{A}} = \mathbf{B}^{-1}\mathbf{N}$ | $\bar{\mathbf{b}}$ |

where $I_m$ is an identity matrix of size $m$, $\bar{\mathbf{A}} = (\bar{a}_{ij}) \in \mathbb{R}^{m \times (n-m)}$, $\mathbf{c}_B = \mathbf{1} \in \mathbb{R}^m$, $\mathbf{c}_N = \mathbf{1} \in \mathbb{R}^{n-m}$, $\bar{\mathbf{b}} = \mathbf{B}^{-1}\mathbf{b} \in \mathbb{R}^m$, $\bar{\mathbf{b}} \geq \mathbf{0}$ and $\mathbf{0} \in \mathbb{R}^m$. If we select $x_r$ to be the leaving variable. Consider the binding constraint of $x_r$:

$$x_r + \sum_{i \in I_N \setminus J} \bar{a}_{ri} x_i + \sum_{j \in J} \bar{a}_{rj} x_j = \bar{b}_r \tag{8}$$

In order to improve objective value, we select the entering variable form the set $J$ that allows the maximum increase. We select the leaving variable $x_{\tilde{j}}$ such that

$$\tilde{j} = \text{Argmin}\left\{\bar{a}_{rj} \mid \bar{a}_{rj} > 0, j \in J\right\}.$$

### 2.2. Max-Out-In Pivot Rule (with Dantzig's rule safeguarding)

From our main idea, we state our proposed pivot rule called *the max-out-in pivot rule* as following.

**Max-out-in pivot rule**: If $J \neq \phi$.

**Select $x_r$ to leave the basic by $x_r = \max\{x_i \mid i \in I_B\}$ .**
**Select $x_{\tilde{j}}$ to enter the basic by $\tilde{j} = \text{Argmin}\{\bar{a}_{rj} \mid \bar{a}_{rj} > 0,\ j \in J\}$.**

When it performs the max-out-in pivot rule, there are the possible two cases.

**Case 1.** One basic variable $x_r$ and one nonbasic variable $x_{\tilde{j}}$ can be swapped.

**Case 2.** The max-out-in pivot rule can not be used.

- If the selected basic variable violates the minimum ratio test.
- If no corresponding nonbasic variable exists.

In case 1, we can use the max-out-in pivot rule to perform the pivot step. However, we cannot perform the pivot step if case 2 occurs. We apply Dantzig's rule as the safeguarding rule instead.

## 3. Illustrative Examples

In this section, we give the example to illustrate the implementation of the proposed method.

**Example 1.** Consider the following problem:

$$\begin{aligned} \text{Maximize} \quad & -86x_1 - 6x_2 - 28x_3 + 2x_4 + 5x_5 \\ \text{subject to:} \quad & 54x_1 + 84x_2 + 46x_3 + 16x_4 + 90x_5 \le 14000 \\ & 51x_1 + 59x_2 + 34x_3 + \ 3x_4 \ + 94x_5 \le 12000 \\ & x_1,\ x_2,\ x_3,\ x_4,\ x_5 \ge 0. \end{aligned}$$

Replacing $x_j$ by $|c(j)|\,\tilde{x}_j$ for $j = 1, 2, \ldots, 5$, and let $\tilde{x}_6$ and $\tilde{x}_7$ be slack variables associate with the first and the second constraint, respectively. Then the initial tableau for above problem is:

| | $\tilde{x}_1$ | $\tilde{x}_2$ | $\tilde{x}_3$ | $\tilde{x}_4$ | $\tilde{x}_5$ | $\tilde{x}_6$ | $\tilde{x}_7$ | **RHS** |
|---|---|---|---|---|---|---|---|---|
| z | 1 | 1 | 1 | $-1$ | $-1$ | 0 | 0 | 0 |
| $x_6$ | $\frac{27}{43}$ | 14 | $\frac{23}{14}$ | ⑧ | 18 | 1 | 0 | 14000 |
| $x_7$ | $\frac{51}{86}$ | $\frac{59}{6}$ | $\frac{17}{14}$ | $\frac{3}{2}$ | $\frac{94}{5}$ | 0 | 1 | 12000 |

Since the maximum value among all basic variables is at $x_6 = 14000$ corresponding to $r = 1$. Then $\{\bar{a}_{1j} \mid j \in J \text{ and } \bar{a}_{1j} > 0\} = \{\bar{a}_{14},\ \bar{a}_{15}\} = \{8,\ 18\}$. $\tilde{j} = 4 = \text{Argmin}\,\{\bar{a}_{14},\ \bar{a}_{15}\}$. Since $\left\{\frac{b_k}{\bar{a}_{k4}} | \bar{a}_{k4} > 0, \text{for } k = 1, 2\right\} = \left\{\frac{b_1}{\bar{a}_{14}}, \frac{b_2}{\bar{a}_{24}}\right\} = \{1750, 8000\}$. From case 1, we select $x_6$ to be the leaving variable and select $x_4$ to be the entering variable. After pivoting, the simplex tableau is

| | $\tilde{x}_1$ | $\tilde{x}_2$ | $\tilde{x}_3$ | $\tilde{x}_4$ | $\tilde{x}_5$ | $\tilde{x}_6$ | $\tilde{x}_7$ | **RHS** |
|---|---|---|---|---|---|---|---|---|
| z | $\frac{371}{344}$ | $\frac{11}{4}$ | $\frac{135}{112}$ | 0 | $\frac{5}{4}$ | $\frac{1}{8}$ | 0 | 1750 |
| $x_4$ | $\frac{27}{344}$ | $\frac{7}{4}$ | $\frac{23}{112}$ | 1 | $\frac{9}{4}$ | 1750 | 0 | 1750 |
| $x_6$ | $\frac{327}{688}$ | $\frac{173}{24}$ | $\frac{29}{32}$ | 0 | $\frac{617}{40}$ | $\frac{-3}{16}$ | 1 | 9375 |

This is the optimal tableau. The optimal solution is $\tilde{x}_4 = 1750$, $\tilde{x}_j = 0$ for $j = 1, \ldots, 5$ and $j \neq 4$. Then we get $x_4 = 875$, $x_j = 0$ for $j = 1, \ldots, 5$ and $j \neq 4$ with the optimal value 1750 and the number of iteration is 1. By using the simplex method with Dantzig's pivot rule, the number of iterations is 3.

## 4. Application of the max-out-in pivot rule to Klee and Minty problem

Klee and Minty proposed some linear programming problems that the simplex method has exponential worst-case running time in 1972[10]. This problem is called the Klee and Minty problem, which is stated as below.

**Klee and Minty problem:**

$$\begin{aligned} \text{Maximize} \quad & \sum_{j=1}^{n} 10^{n-j} x_j \\ \text{subject to:} \quad & 2\sum_{j=1}^{i-1} 10^{i-j} x_j + x_i \le 100^{i-1},\ i = 1, \ldots, n \qquad (9) \\ & x_i \ge 0,\ i = 1, \ldots, n \end{aligned}$$

The simplex method with Dantzig's pivot rule requires $2^n - 1$ iterations to solve Klee and Minty problem [10]. Next, we show that our rule obtains the optimal solution in 1 iteration.

We first normalize the objective coefficients to be +1 or 0 or −1 and we let $x_{n+1}, \cdots, x_{2n}$ be slack variables associate with the first to the $n^{th}$ constraint, respectively. Rewrite this problem into the tableau

| | $x_1$ | $\cdots$ | $x_n$ | $x_{n+1}$ | $\cdots$ | $x_{2n}$ | **RHS** |
|---|---|---|---|---|---|---|---|
| z | $-1$ | $\cdots$ | $-1$ | 0 | $\cdots$ | 0 | 0 |
| $x_{n+1}$ | $\frac{1}{10^{n-1}}$ | $\cdots$ | 0 | 1 | $\cdots$ | 0 | 1 |
| $\vdots$ | | | | | | $\vdots$ | $\vdots$ |
| $x_{2n}$ | 2 | 2 | ① | 0 | $\cdots$ | 1 | $10^{n-1}$ |

By using the max-out-in pivot rule, $x_{2n}$ is selected to be the leaving variable and $x_n$ is selected to be the entering variable.

| | $x_1$ | $\cdots$ | $x_n$ | $x_{n+1}$ | $\cdots$ | $x_{2n}$ | **RHS** |
|---|---|---|---|---|---|---|---|
| z | 1 | 1 | 0 | 0 | $\cdots$ | 0 | $10^{n-1}$ |
| $x_{n+1}$ | $\frac{1}{10^{n-1}}$ | $\cdots$ | 0 | 1 | $\cdots$ | 0 | 1 |
| $\vdots$ | | | | | | $\vdots$ | $\vdots$ |
| $x_{2n}$ | 2 | 2 | 1 | 0 | $\cdots$ | 1 | $10^{n-1}$ |

This is the optimal tableau. Then the optimal solution is obtained in 1 iteration for Klee and Minty problem of any size $n$.

## 5. Conclusions

The objective of this paper is to propose a new pivot rule called the max-out-in pivot rule safeguarding with Dantzig's rule for the simplex method. This rule first normalizes the objective coefficients to be +1 or 0 or −1. The keys of this rule are that the maximum basic variable is selected to leave basis, and the corresponding nonbasic variable which allowed the maximum increase is selected to enter basis. We can illustrate by the examples that the max-out-in pivot rule can achieve good results. In addition, our rule is better than Dantzig's rule over the Klee and Minty problems [10]. Further work will be done for the improvement of the max-out-in pivot rule.

## References

[1] D. Goldfarb and J. Ried, A practical steepest-edge simplex algorithm. *Mathematical Programming*, **12** (1977), 361-371.

[2] G.B. Dantzig, *Linear Programming and Extensions*. Princeton University Press, Princeton, NJ, 1963.

[3] H. Arsham, A computationally stable solution algorithm for linear programs. *Applied Mathematics and Computation*, **188** (2007), 1549-1561.

[4] H.V. Junior and M.P.E. Lins, An improved initial basis for simplex algorithm. *Computers and and Operations Research*, **32** (2005), 1983-1993.

[5] J.F. Hu, A note on "an improved initial basis for simplex algorithm". *Computers and Operations Research*, **34** (2007), 3397-3401.

[6] J.J.H. Forrest and D. Goldfarb, Steepest-edge simplex algorithms for linear programming. *Mathematical Programming*, **57** (1992), 341-374.

[7] P.M.J. Harris, Pivot selection methods of the Devex LP code. *Mathematical Programming* **5** (1973), 1-28

[8] P.-Q. Pan, A largest-distance pivot rule for the simplex algorithm. *European Journal of Operational Research*, **187** (2008), 393-402.

[9] P.-Q. Pan, Practical finite pivoting rules for the simplex method. *OR Spektrum*, **12** (1990), 219-225.

[10] V. Klee and G.J. Minty, How good is the simplex algorithm? In: Shisha O, editor. *Inequalities III*, Academic Press, New York, (1972), 158-172.

# Contradictions in the quantum mechanical explanation of the periodic table

**Gregory Anderson**[1], **Ravi Gomatam**[2,3], **Laxmidhar Behera**[4]

[1]Bhaktivedanta Institute, Juhu Rd, Mumbai 400 049, India
[2]Birla Institute of Technology and Science, Pilani, India
[3]Institute of Semantic Information Sciences and Technology, 2334 Stuart St, Berkeley, CA 94705, USA
[4]Dept of Electrical Engineering, Indian Institute of Technology, Kanpur, India

E-mail: ganderson@bvinst.edu

**Abstract.** Quantum theory is thought to explain the periodic law which underlies the periodic table (PT); however, the lengths of the periods remain unexplained. Also, this explanation depends on two empirical rules, namely Madelung's *n+l* rule and Hund's rule. Furthermore, even Madelung's rule fails to explain the ground state configuration for many elements. Toward achieving an explanation of the periodic table, the Hartree-Fock (HF) method has been applied in this paper to calculate energies of various possible configurations of transition metals in the fourth and fifth periods. These calculations for Cr, Cu and Ni do not agree with the spectroscopically observed ground state configurations. We further calculated the nonrelativistic and relativistic energies of the various possible configurations of second row transition metals such as niobium, palladium, molybdenum and silver for which Madelung's rule predicts the wrong ground state configuration. In contrast to Cr and Cu, the observed ground state configuration of these elements is found to be associated with the lowest energy by HF calculations.

## 1. Introduction:

The periodic table of the chemical elements graphically displays Mendeleev's periodic law from 1869, which states the following:

> "If all the elements are arranged in the order of their atomic weight, a periodic repetition of properties is obtained. This is expressed by the law of periodicity; the properties of the elements, as well as the forms and properties of their compounds, are in periodic dependence on, or (expressing ourselves algebraically) form a periodic function of the atomic weights of the elements." [1]

However this periodic function has the feature that the lengths of the periods increase as atomic weight or atomic number (Z) increases. Since the earliest days of Bohr's model of the hydrogen atom, it has been the goal of physicists and chemists to explain this periodic function using quantum theory. But Eric Scerri [2] has argued that the lengths of the periods remain unexplained. The key to explaining the lengths of the periods revolves around explaining why the orbitals fill the way that they do as Z increases and why a particular orbital configuration is actually the ground state. As a first approximation, one can employ the *auf bau* principle and the Madelung's $n+l$ rule for rationalizing the ground state configuration of the chemical elements as the atomic number increases. Still, a number of transition metals such as Cr, Cu, Nb, Mo, Ru, Rh, Pd, Ag, Pt, Pd and Au do not have the ground state configuration that one would expect based on this approximation. Furthermore, in spite of a number of attempts, Madelung's rule has not been derived from the principles of quantum theory [3].

Herein, we turn to the Hartree-Fock theory in an effort to explain the ground state configuration of the transition metals.

## 2. Hartree-Fock Theory

The Hartree-Fock theory is based on the following matrix equation which can be derived from the Schrodinger equation using a number of approximations. Let's consider the case of an electron in a closed shell:

$$\mathbf{FC}=\mathbf{SC}\boldsymbol{\varepsilon}$$

where the wave function, $\psi_i$ for electron 1 is expanded into a set of $b$ basis functions, $\chi$, with coefficients, $c_{si}$, and these coefficients form a column matrix, **C**:

$$\psi_i(1)=\sum_{s=1}^{b}c_{si}\chi_s$$

where the matrix elements, $F_{rs}$, of **F** are as follows:

$$F_{rs}=H_{rs}+\sum_{t=1}^{b}\sum_{u=1}^{b}\sum_{j=1}^{n/2}c_{tj}c_{uj}\left[2(rs\,|\,tu)-(ru\,|\,ts)\right]$$

$$(rs\,|\,tu)=\iint\frac{\chi_r(1)\chi_s(1)\chi_t(2)\chi_u(2)}{r_{12}}dv_1dv_2$$

$$(ru\,|\,ts)=\iint\frac{\chi_r(1)\chi_u(1)\chi_t(2)\chi_s(2)}{r_{12}}dv_1dv_2$$

$$H_{rs}=\int\chi_r(1)\left[\frac{1}{2}\nabla_1^2-\frac{1}{r_1}\right]\chi_s(1)dv_1$$

and where **S** is the overlap matrix with elements:

$$S_{rs}=\iint\chi_r^*\chi_s\,dv$$

The total energy was calculated as follows:

$$E_{HF}=\sum_{i=1}^{n/2}\varepsilon_i+\frac{1}{2}\sum_{r=1}^{b}\sum_{s=1}^{b}P_{rs}H_{rs}$$

$$P_{rs}=\sum_{r=1}^{b}\sum_{s=1}^{b}\sum_{i=1}^{n/2}c_{ri}c_{si}$$

The HF method consists of starting with a trial basis function (which in the present study was a screened hydrogenic function), trial coefficients (**C**) and then calculating the matrix elements ($F_{rs}$, $S_{rs}$), orthogonalizing the basis set ($b$=20 in the present study) and then diagonalizing the Fock matrix, **F**, to obtain the orbital energies, $\varepsilon_i$ which form $\boldsymbol{\varepsilon}$. This sequence is repeated until the coefficients, **C**, of the basis functions converge on specific values.

Vanquickenborne et. al. [5] argued that one must calculate and compare the total energies of the various possible configurations. He performed HF calculations of some transition metal elements in the fourth period, but the various possible configuration of the transition metal elements in the fifth period have never been examined systematically. There is a recent multiconfigurational Dirac-HF calculation of the various configurations of silver which agrees with experimental observations [6].

### 3. Determination of the ground states of transition metals using HF method

Using Fischer's Hartree-Fock-86 software [7], we first calculated the energies of the fourth period transition metals, and our results agreed with the results of calculations reported in the literature [2,5]. Thus, we did both relativistic and nonrelativistic calculations for three different configurations of Sc, Cr, Cu, Ni: $3d^x4s^2$, $3d^{x+1}4s^1$ and $3d^{x+2}4s^0$. Relativistic calculations were performed using the Breit-Pauli Hamiltonian [7]. For Cr, Cu and Ni, these calculations indicated a ground state configuration which is not in agreement with what is observed.

Then we turned our attention to the transition metals in the fifth period. We were interested to see if similar disagreement would be there for Mo and Ag, which are just below Cr and Cu in the periodic table. Our results are given in Table I. As noted above, Mo, Ag and other fifth period transition metals are observed to have ground state configurations which do not agree with Madelung's rule. In contrast with the results obtained for chromium and copper, our calculations for the elements in the fifth period indicate a ground state configuration which is in agreement with experiment. For example, in the case of Mo, we calculated a total energy of -4042.5323 hartrees for the configuration $[Kr]4d^5 5s^1$ with the

**Table I. Calculation of total energy in hartrees of selected fifth period transition metals**

| Element | Configuration | total energy relativistic | total energy nonrelativistic |
|---|---|---|---|
| Mo, Z=42 | $[Kr]4d^4 5s^2$, $^5D$ | -4042.4446 | -3975.4433 |
| Mo, Z=42 | $[Kr]4d^5 5s^1$, $^7S$ | -4042.5323 | -3975.5495 |
| Nb, Z=41 | $[Kr]4d^4 5s^1$, $^6D$ | -3814.0362 | -3753.5977 |
| Nb, Z=41 | $[Kr]4d^3 5s^2$, $^4F$ | -3814.0066 | -3753.5520 |
| Ru, Z=44 | $[Kr]4d^7 5s^1$, $^5F$ | -4523.2092 | -4441.5395 |
| Ru, Z=44 | $[Kr]4d^6 5s^2$, $^5D$ | -4523.1812 | -4441.4873 |
| Rh, Z=45 | $[Kr]4d^8 5s^1$, $^4F$ | -4775.7498 | -4685.8817 |
| Rh, Z=45 | $[Kr]4d^7 5s^2$, $^4F$ | -4775.6972 | -4685.8012 |
| Pd, Z=46 | $[Kr]4d^8 5s^2$, $^3F$ | -5036.4903 | -4937.7830 |
| Pd, Z=46 | $[Kr]4d^9 5s^1$, $^3D$ | -5036.5688 | -4937.8935 |
| Pd, Z=46 | $[Kr]4d^{10}$, $^1S$ | -5036.5742 | -4937.9210 |
| Ag, Z=47 | $[Kr]5s^1 4d^{10}$, $^2S$ | -5305.8197 | -5197.6985 |
| Ag, Z=47 | $[Kr]5s^2 4d^9$, $^2D$ | -5305.6756 | -5197.5179 |

term symbol, $^7S$. In comparison, we calculated that the energy of the configuration, $[Kr]4d^4 5s^2$ ($^5D$), is -4042.4446 hartrees; thus the configuration, $[Kr]4d^5 5s^1$, is 0.0876 hartree (2.38 eV) lower in energy than $[Kr]4d^4 5s^2$. These calculations qualitatively agree with spectroscopic observations according to which the configuration $[Kr]4d^5 5s^1$ is lower than $[Kr]4d^4 5s^2$ by 1.53 eV [8].

### 4. Discussion and Conclusions:

It is well known that the HF theory calculates a total energy which is higher than the actual energy because it does not take into consideration instantaneous correlations in the motions of electrons. In this study of the fifth period transition metals, HF theory is able to correctly calculate the ground state configuration but only in a qualitative way. In some cases the difference in energies of the various possible configurations is very small; hence an extremely accurate method is needed to correctly calculate the total energy and hence make a statement about what is the ground state configuration.

At a more general level, Eric Scerri [9] has expressed dissatisfaction with the level of explanation provided by HF theory or post-HF theories:

> "Can the experimentally observed first configuration be predicted from first principles by quantum mechanics? No. In fact, all that quantum mechanical calculations do is compute which of these three configurations has the minimum energy. That is not the same as deriving the correct configuration from first principles."

Even the picture of electrons occupying specific orbitals and having a particular configuration cannot be envisioned. An orbital is only a mathematical abstraction, and one cannot speak of an electron following any trajectory in spite of the fact that the theory formally has a variable, ***r***, for the distance of the electron from the nucleus. Another point is that the wave function for the atom is actually a superposition of spin orbitals. For example, in the case of the lithium atom, the wave function is the following determinant (called a Slater determinant):

$$\psi = \frac{1}{\sqrt{6}} \begin{vmatrix} 1s(1)\alpha(1) & 1s(1)\beta(1) & 2s(1)\alpha(1) \\ 1s(2)\alpha(2) & 1s(2)\beta(2) & 2s(2)\alpha(2) \\ 1s(3)\alpha(3) & 1s(3)\beta(2) & 2s(3)\alpha(3) \end{vmatrix}$$

where 1s(1) represents the product of a radial function and spherical harmonic function and where electron 1 can have the spin function, $\alpha(1)$ or $\beta(1)$. In considering these determinantal wave functions, Slater [10] has remarked that ". . . it makes no sense to say that one particular electron occupies one particular spin-orbital." It is well known that quantum theory presents a challenge to classical realism even at the macroscopic level, and macroscopic quantum mechanics has become an active area of contemporary research [11]. Gomatam [12] has outlined a novel way to move toward a macroscopic quantum mechanics that is compatible with a non-classical macrorealism.

## 5. References


[1] Mendeleev D 1905 *The Principles of Chemistry* translated by G Kamensky, Third English edition, Vol II, (New York and Bombay: Longmans, Green and Co.) pp 17-18

[2] Scerri E R 2007 *The Periodic Table: Its Story and Its Significance* (Oxford: Oxford U.P.)

[3] Lowdin P-O 1969 *International Journal of Quantum Chemistry* **3**(Suppl.) pp 331-334; Scerri E R 2009 *International Journal of Quantum Chemistry* **109** pp 959-971

[5] Vanquickenborne L G, Pierloot K, Devoghel D 1989 *Inorganic Chemistry* **28** pp 1805-1813

[6] Figgen D, Rauhut G, Dolg M, Stoll H 2005 *Chemical Physics* **311** pp 227-244

[7] Fischer's Hartree-Fock 86 software is available at http://atoms.vuse.vanderbilt.edu/. This software is documented and explained in the following book: Fischer C F, Brage T, Jonsson P 1997 *Computational Atomic Structure: An MCHF Approach* (London: Institute of Physics Publishing)

[8] Sansonetti J E and Martin W C 2005 *J. Phys. Chem. Ref. Data* **34** p 1844

[9] Scerri E R 2003 Löwdin's remarks on the aufbau principle and a philosopher's view of ab initio quantum chemistry *Fundamental World of Quantum Chemistry: A Tribute Volume to the Memory of Per-Olov Löwdin* vol 2 ed E Brandas and E Kryachko (Dordrecht: Kluwer) pp 675-694

[10] Slater J C 1960 *Quantum Theory of Atomic Structure*. Vol I (New York: McGraw Hill) p 297

[11] Leggett A J 1988 *Foundations of Physics* **18** pp 939-952

[12] Gomatam, R 2012 How do classical and quantum probabilities differ? *Foundations of Probability and Physics-6* (Conference Proceedings Series) ed A Khrennikov (Melville, New York: American Institute of Physics) pp 105-110; Gomatam R 2010 Quantum realism and haecceity *Materialism and Immaterialism in India and the West: Varying Vistas* (*HSPCIC* Vol. XII: *Levels of Reality*, Part 5) ed P. Ghose (CSC: New Delhi) pp 853-872

# Numerical modelling of compressible viscous flow in turbine cascades

**P Louda**[1], **K Kozel**[1] **and J Příhoda**[2]

[1] Dept. of Technical Mathematics, Czech Technical University in Prague, Karlovo nám. 13, CZ-121 35 Praha 2

[2]Institute of Thermomechanics AS CR, Dolejškova 5, CZ-182 00 Praha 8

E-mail: petr.louda@fs.cvut.cz

**Abstract.** The work deals with mathematical models of turbulent flow through turbine cascade in 2D and 3D. It is based on the Favre-averaged Navier-Stokes equations with SST or EARSM turbulence models. A two-equation model of transition to turbulence is considered too. The solution is obtained by implicit AUSM finite volume method. The 2D and 3D results are shown flow through the SE1050 cascade including simulation of a range of off-design angles of attack.

## 1. Introduction

In this work the transonic flow through turbine is investigated numerically. The simulations are carried out in 2D and 3D. The mathematical model is based on the Favre-averaged Navier-Stokes equations with SST (eddy viscosity) turbulence model for intermittency function of Lodefier-Dick. The influence of turbulent transition is modeled by two-equation model In 3D eddy viscosity models are generally unsatisfactory and thus more sofisticated explicit algebraic Reynolds stress model (EARSM) is applied.

In 2D, the behavior of the flow-field with changing the angle of attack is investigated. The comparison with an experiment for large off-design angle indicates good agreement. Further the influence of bypass transition to the turbulence is considered, although it appeared that the results change marginally by transition model in this case. Finally the SE1050 prismatic cascade is simulated in 3D using the EARSM turbulence model. The simulation captures secondary flows in the cascade and in the wake.

## 2. Mathematical models

The mathematical model of turbulent flow is based on the Favre-averaged Navier-Stokes equations in Cartesian coordinates. The turbulent stress tensor and heat flux are modelled by the SST (Shear Stress Transport) model see Menter [4]. In order to model bypass transition to turbulence, the two-equation model of Lodefier and Dick [3] is used. The eddy viscosity is then multiplied by turbulence weighting factor (intermittency function). The Mayle and Abu-Ghannam, Shaw criteria of transition are employed in the present simulations.

In 3D, the EARSM model proposed by Wallin [7], is used for turbulent closure in the modification of Hellsten [1].

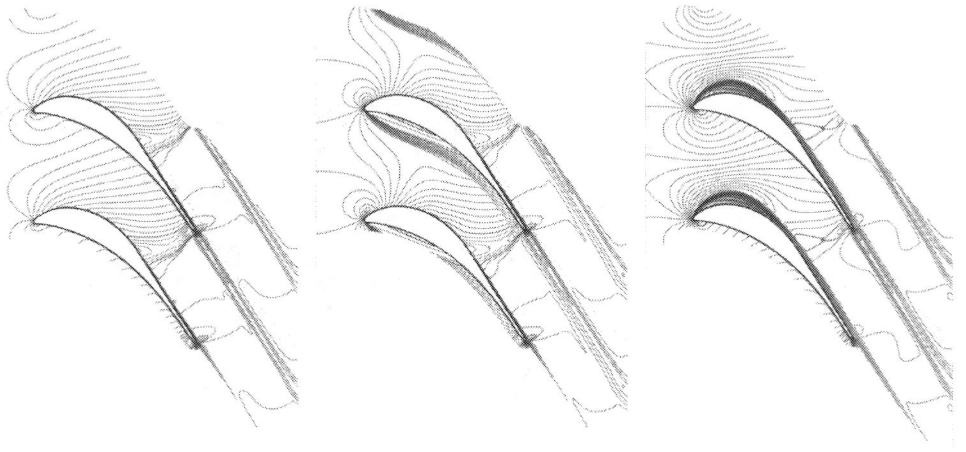

**Figure 1.** SE1050 2D cascade at design and off-design angles of attack. From left: $\Delta\alpha_1 = \{0°, -57°, 30°\}$ (blue: subsonic, red: supersonic)

## 3. Numerical solution

The system of equations is discretized by cell-centered finite volume method with quadrilateral or hexahedral finite volumes in 2D or 3D respectively. The inviscid numerical flux is computed by the AUSMPW+ method [2] using cell center values interpolated linearly with van Leer limiter. The approximation of cell face derivatives needed in diffusive terms uses quadrilateral dual finite volumes constructed over each face of primary volume. For time discretization the backward Euler scheme (implicit) was used. The Jacobi matrices of the linearized system are obtained as derivatives of discrete expressions for fluxes with respect to nodal values from the stencil. The system is solved iteratively by a block relaxation method with direct block tri-diagonal solver on selected family of grid lines.

## 4. Numerical results

A transonic flow through the SE1050 cascade [6, 5] is considered here at angles of attack which differ from the design angle of attack $\alpha_1 = 19.3°$ in both positive and negative direction by $\Delta\alpha_1$. The outlet Mach number is $M_{2is} = 0.905$, Reynolds number $Re_{2c} \approx 1.2 \cdot 10^6$. The model of turbulence is SST. The computational results are shown in terms of the Mach number isolines in Figs. 1 and 2. One can see that the supersonic region remains same for quite a large negative deviations from the design angle, despite large separation on the pressure side. In comparison with the measurement in Fig. 2 the structure of the flow field is predicted quite well especially above the suction side. The end of separation differs from the measurement. The diagram of loss coefficient $\xi$ shown in Fig. 3 confirms relatively low sensitivity to changes of $\alpha_1$.

The simulation of flow through SE1050 cascade taking into account bypass transition model is presented here. The $\gamma$-$\zeta$ model of Lodefier and Dick is used to predict intermittency functions. The turbulence model is SST. The inlet turbulence intensity is estimated by $Tu = 1.7\%$. The Mayle and Abu-Ghannam-Shaw criteria for onset of transition were considered. The isolines of near wall intermittency $\gamma$ are shown in Fig. 4. The free-stream factor $\zeta$ differs from unity only very near the blade surface and is not shown. The transition in both cases starts on the suction side relatively early (close to leading edge) and the flow-field as well all loss coefficient are practically same as in fully turbulent simulation.

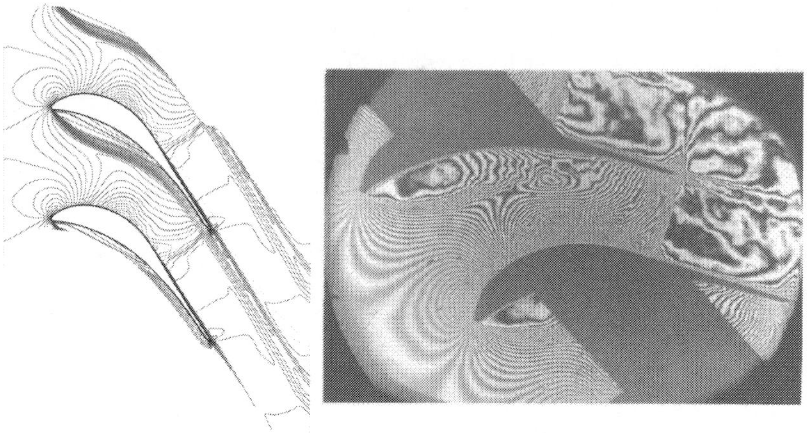

**Figure 2.** SE1050 cascade at deviation from design angle of attack $\Delta\alpha_1 = -67°$. left: computation, right: measurement [5] (blue: subsonic, red: supersonic)

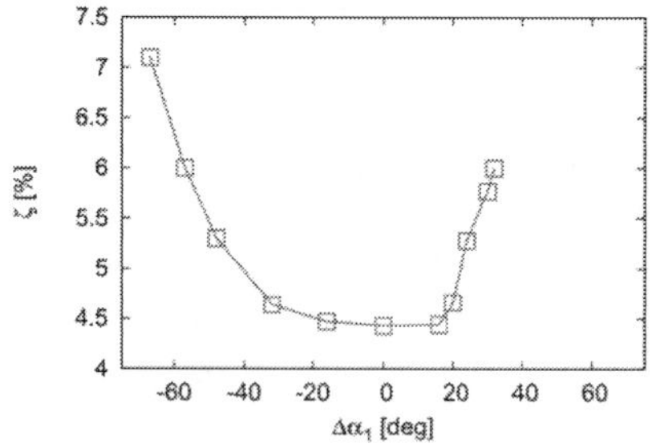


**Figure 3.** Loss coefficient vs. deviation from design angle of attack, SE1050 cascade

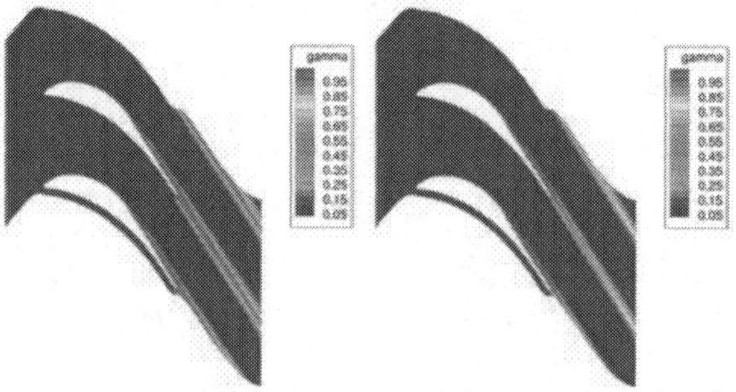


**Figure 4.** Isolines of intermittency $\gamma$. Left: $M_{2is} = 0.716$, right: $M_{2is} = 1.007$

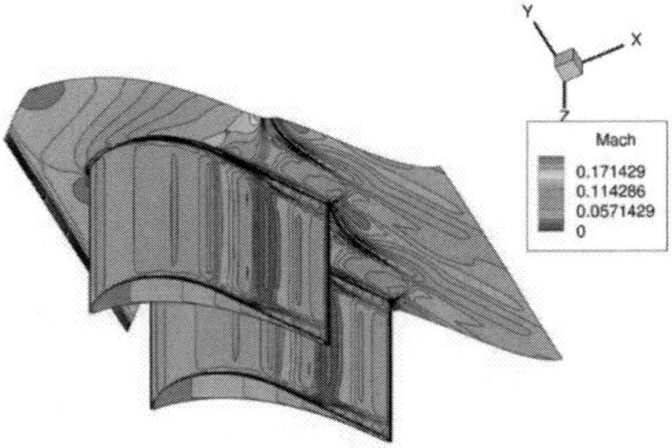


**Figure 5.** Isolines of Mach number near the surface

The 3D geometry of the SE1050 cascade is prismatic with the span equal to the chord length (100 mm). The outlet Mach number $M_{2is} = 1.012$, $Re_{2c} \approx 1.2 \cdot 10^6$, inlet turbulence intensity is 1.7%. The turbulence is modelled by the EARSM model. The finite volume grid has 152 steps in span-wise direction and approx. 1.36 millions of finite volumes in total. The global view of the near wall Mach number is shown in Fig. 5. One can see corner effects and the re-compression zone on the suction side of the blade, typical for the SE1050 geometry.

## 5. Conclusions

The work presented simulations of 2D and 3D turbulent flow through the SE1050 turbine cascade. In 2D, range of off-design angles of attack has been considered and losses evaluated. The comparison with experimental data (for 1 off-design regime) is satisfactory. The influence of transition model appeared marginal in the simulated cases. The 3D simulation with EARSM turbulence model captures the secondary flow between the blades as well as in the wake.

## Acknowledgments

Acknowledgments This work was supported by the grants of GA CR No. P101/10/1329, P101/12/1271 and 201/08/0012.

## References

[1] A. Hellsten. New advanced $k$-$\omega$ turbulence model for high-lift aerodynamics. *AIAA J.*, 43:1857–1869, 2005.

[2] K. H. Kim, C. Kim, and O.-H. Rho. Methods for accurate computations of hypersonic flows I. AUSMPW+ scheme. *J. of Computational Physics*, 174:38–80, 2001.

[3] K. Lodefier and E. Dick. Modelling of unsteady transition in low-pressure turbine blade flows with two dynamic intermittency equations. *Flow, Turbulence and Combustion*, 76:103–132, 2006.

[4] F. R. Menter. Two-equation eddy-viscosity turbulence models for engineering applications. *AIAA Journal*, 32(8):1598–1605, 1994.

[5] P. Šafařík, M. Šťastný, and M. Babák. Numerical and experimental testing of transonic flow in the etalon turbine cascade SE 1050. In G. Bois M. Šťastný, C. H. Sieverding, editor, *5th European Conference on Turbomachinery*. Czech Technical University in Prague, 2004.

[6] M. Šťastný P. Šafařík. Experimental analysis data on the transsonic flow past the plain turbine cascade. Technical report, 1990. Paper 90-GT-313.

[7] S. Wallin and A. V. Johansson. An explicit algebraic Reynolds stress model for incompressible and compressible turbulent flows. *J. Fluid. Mech.*, 403:89–132, 2000.

# Recurrent epidemic cycles driven by intervention in a population of two susceptibility types

**Drandreb Earl O. Juanico**

Department of Mathematics, Ateneo de Manila University, Loyola Heights Quezon City, Philippines 1108

E-mail: djuanico@ateneo.edu

**Abstract.** Epidemics have been known to persist in the form of recurrence cycles. Despite intervention efforts through vaccination and targeted social distancing, infectious diseases like influenza continue to appear intermittently over time. I have undertaken an analysis of a stochastic epidemic model to explore the hypothesis that intervention efforts actually drive epidemic cycles. Time series from simulations of the model reveal oscillations exhibiting a similar temporal signature as influenza epidemics. The power-spectral density indicates a resonant frequency, which approximately corresponds to the apparent annual seasonality of influenza in temperate zones. Asymptotic solution to the backward Kolmogorov equation of the dynamics corresponds to an exponentially-decaying mean-exit time as a function of the intervention rate. Intervention must be implemented at a sufficiently high rate to extinguish the infection. The results demonstrate that intervention efforts can induce epidemic cycles, and that the temporal signature of cycles can provide early warning of imminent outbreaks.

## 1. Introduction

Sentinel surveillance of influenza-like illnesses (ILI) done in several countries have gathered and presented evidence for recurrent epidemics [1, 2, 3, 4, 5]. Prevailing models based on the Susceptible-Infected-Recovered or SIR framework, however, do not inherently predict the cycle-like variation of $R_0$ with time unless exogenous seasonal forcing is assumed [6]. The problem with that assumption is the weak correspondence of seasonal forcing with recurrent epidemic activity [7]. On the contrary, the apparent seasonality may emerge endogenously from the host-pathogen feedback interaction. The host's natural immunity along with intervention efforts induce a feedback loop with the pathogen's mutations via antigenic drift. Demographic noise is a likely feature of that interaction. In models of resonant amplification, endogenous forcing due to demographic stochasticity has been demonstrated [8]. However, those models do not clearly address how epidemic cycles can be extinguished through intervention initiatives. Here, I propose and analyze a mathematical model that generates epidemic cycles and could account for the observed time variation in $R_0$. The model illustrates how intervention promotes epidemic cycles, and how a time-varying $R_0$ could serve as a lead indicator of an imminent outbreak.

## 2. Mathematical model

Consider an infected population of size $N = N_1 + N_2$ with two susceptibility types of subpopulation sizes $N_1$ and $N_2$. The population dynamics of both types are similar in all respects but one: type–1 gets infected at a faster rate than type–2. The population dynamics can be further described in terms of reactions. Recruitment processes are: $1 \xrightarrow{\mu} 11$ or $2 \xrightarrow{\mu} 21$; and $2 \xrightarrow{\lambda} 22$ or $1 \xrightarrow{\lambda} 12$, wherein $\mu$ and $\lambda$ are per-capita rates of infection of type 1 and 2, respectively. Density-dependent removal are: $k1 \xrightarrow{\delta} k$ or $k2 \xrightarrow{\delta} k$ where $k = 1, 2$. Intervention-induced removal is expressed as: $k \xrightarrow{\nu\Phi} \varnothing$, occurring at the rate $\nu\Phi$, which may be related to *vaccine efficacy* [9]. The function $\Phi = \sum_{k\in\{1,2\}} N_k\,(N_k - 1)\,/[N(N-1)]$ encapsulates the susceptibility structure of the community relevant to targeted vaccination.

The population size of both types are random variables drawn from a joint probability density function $\mathcal{P}(N_1, N_2; t)$ that evolves over time. The exact time evolution of $\mathcal{P}$ is given by the master equation constructed from the recruitment and removal processes. Assuming a community size $\Omega \gg 1$, a diffusion approximation of the master equation is the Fokker-Planck equation:

$$\frac{\partial \mathcal{P}(n_1, n_2; t)}{\partial t} = -\sum_{j=1}^{2} \left\{ \frac{\partial}{\partial n_j} [b_j(n_1, n_2)\mathcal{P}] + \frac{1}{2\Omega} \frac{\partial^2}{\partial n_j^2} [a_j(n_1, n_2)\mathcal{P}] \right\}, \tag{1}$$

where $n_1 = N_1/\Omega$ and $n_2 = N_2/\Omega$ are dimensionless concentrations. The drift terms are: $b_1 = [\mu - n_1 - \nu\Phi]\,(n_1 + n_2)$ and $b_2 = [\lambda - n_2 - \nu\Phi]\,(n_1 + n_2)$; and the diffusion terms are: $a_1 = [\mu + n_1 + \nu\Phi]\,(n_1 + n_2)$ and $a_2 = [\lambda + n_2 + \nu\Phi]\,(n_1 + n_2)$. A further simplification of the model is made by setting $\mu + \lambda = 1$, with $\mu \in (\frac{1}{2}, 1)$ and $\lambda \in (0, \frac{1}{2})$.

Linking the variables in the model with measurement requires a rescaling of dimensions. The time unit $t_c$ in the model is set at the *clinical-onset serial interval*, which is the period between the onset of symptoms in the index case and the onset of symptoms in any secondary cases. The serial interval is $\approx 3.6$ days from careful estimates supplemented by laboratory testing [10]. The carrying capacity is set at $\Omega \sim 10^6$, which is the order of the size of a typical community. But the timescale of the system is set using a small dimensionless parameter $\varepsilon$ 0.01, such that the effective capacity is $\varepsilon\,\Omega \sim 10^4$.

A transcritical bifurcation of the model occurs when $\mu$ is a function of $\nu$, as follows:

$$\mu = \mu_b(\nu) = \frac{1}{2} + \frac{1}{3}\sqrt{\frac{(1-\nu)^3}{3\nu}}. \tag{2}$$

On this manifold, it can be shown that the equilibrium concentration vector $\tilde{\mathbf{n}} = (\tilde{n}_1, \tilde{n}_2)$ is asymptotically stable, where $\tilde{n}_1 = \left(\frac{1-\nu}{3}\right)\left(1 + \sqrt{\frac{1-\nu}{3\nu}}\right)$ and $\tilde{n}_2 = \left(\frac{1-\nu}{3}\right)\left(1 - \sqrt{\frac{1-\nu}{3\nu}}\right)$.

## 3. Epidemic cycle and resonant frequency

The model is examined in a region of the parameter space denoted by the set of ordered pairs, $(\nu, \mu) \in \mathbb{K} = [0, 1] \times (\frac{1}{2}, 1)$. Let $\mathcal{B}$ denote the drift Jacobian, which is evaluated at the stable equilibrium $\tilde{\mathbf{n}}$. The power spectral density $S(\omega)$ of the epidemic time series is inversely proportional to the expression $\left(\omega^2 - \det\mathcal{B}\right)^2 + (\mathrm{Tr}\ \mathcal{B})^2\,\omega^2$, the real zero of which corresponds to the square of the resonant frequency [8]:

$$\omega_0(\mu, \nu)^2 = 2\,(2\mu - 1)\,\nu\sqrt{\frac{1-\nu}{3\nu}} - \frac{2}{3}\,(1-\nu)^2\,. \tag{3}$$

The cycle exists if $\omega_0^2 > 0$, which is satisfied in $\mathbb{J} = \left\{(\nu, \mu) \mid \mu \in \left(\frac{1}{2}, 1\right); \nu > \nu_b = \mu_b^{-1}[\mu(\nu)] > 0\right\} \subset \mathbb{K}$, and where $\nu_b$ is the inverse function from Eq. (2). Figure 1 illustrates the cycles generated by the model (using a stochastic simulation algorithm [11]) compared with a real time-series of ILI activity in the U.S. from 1997 to 2010. ILI activity reported in other settings exhibit similar time series (not shown): Hong Kong, 1998-2006 [1]; Melbourne, 1999-2008 [2]; Réunion Island, 2009-2011 [3]; and several countries in continental Europe [4, 5]. Resonance is indicated by the peak in the spectral density.

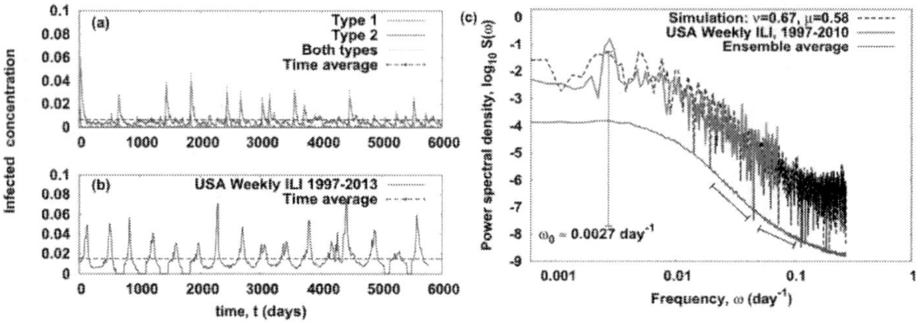


**Figure 1.** Epidemic time series and power-spectral density. (a) Simulated model with $(\nu, \mu) = (0.67, 0.58)$. The dashed line is a time average; (b) U.S. Weekly ILI data from Center for Disease Control website; (c) Comparison of power spectral density. Higher data resolution accounts for the longer plot for simulated model. The predicted resonant frequency $\omega_0 \approx 2.7\times10^{-3}\ \text{day}^{-1}$ (or about $1\,\text{yr}^{-1}$) is determined using Eq. (3).

## 4. Mean extinction time and stochastic amplification

Extinguishing the epidemic is the primary objective of intervention. Extinction refers to the state wherein $n_1 = n_2 = 0$. The mean time of such occurrence is represented by $\mathcal{T}$, which is a solution to the 2D backward Kolmogorov equation (BKE) [12]. The asymptotic solution for $\Omega \gg 1$ is the mean extinction time given by:

$$\mathcal{T} = -\sqrt{\frac{\pi}{\Omega\varphi''(\tilde{n}_2)}}\,\frac{e^{2\Omega\varphi(0)}}{A(\tilde{n}_2)\varphi'(0)}. \tag{4}$$

At the bifurcation manifold, where $\mu = \mu_b(\nu)$ as in Eq. (2), the mean extinction time in Eq. (4) only depends explicitly on $\nu$ [see Fig. 2(a)], which simplifies the analysis. Figure 2(b), inset, illustrates how extinction time is determined for a time series from the simulations.

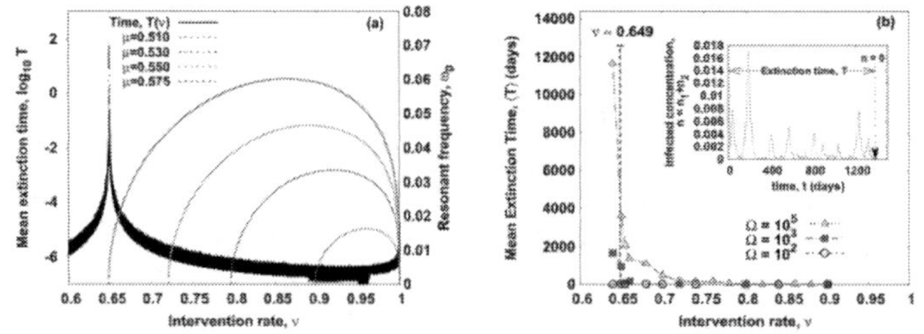


**Figure 2.** Mean extinction time of epidemic. (a) The mean extinction time $\mathcal{T}(\nu)$ calculated from Eq. (4) superimposed with the resonant frequency $\omega_0(\nu, \mu)$ calculated from Eq. (3) for several values of $\mu$. $\mathcal{T}(\nu)$ diverges for $\nu \to \tilde{\nu}$, where $\tilde{\nu} \approx 0.649$. (b) The mean extinction time $\mathcal{T}$ is calculated as the average of several runs for a community size $\Omega$ and intervention rate $\nu$. The inset illustrates how the extinction time is determined for a time series.

The mean extinction time is measured at the bifurcation manifold, where only $\nu$ is the free parameter, in the region $\mathbb{J}$ of resonance. The observed extinction time for different $\Omega$ is presented in figure 2b. The trend is consistent with the existence of a singularity of $\mathcal{T}(\nu)$

at $\nu = \tilde{\nu} = \frac{1}{4}\left[2 - \left(1+\sqrt{2}\right)^{-1/3} + \left(1+\sqrt{2}\right)^{1/3}\right] \approx 0.649$. Cycles persist longer at resonant frequencies for which $\nu \approx \tilde{\nu}$. The value of $\mathcal{T}$ is confirmed to diverge as $\nu \to \tilde{\nu}^{+}$. This is indicated by the higher $\mathcal{T}$ for higher $\Omega$. It is expected that for $\nu < \tilde{\nu}$ the extinction time is very large as the epidemic persists in time. On the other hand, for $\nu > \tilde{\nu}$ the epidemic is extinguished in finite time which is shorter for larger $\nu$.

## 5. Reproduction number and outbreak prediction

The reproduction number denoted as $R_0$ is commonly employed as a measure of the capability of a disease to spread as an epidemic. Based on the definition of $R_0$, the model finds a dynamic reproduction number given by

$$R_0(t) = \frac{1}{n(t) + 2\nu\Phi(t)}. \tag{5}$$

Figure 3 shows $R_0(t)$ superimposed with the $n(t)$ time series. The range of values is $0.6 < R_0 < 2.2$ which is consistent with typical ranges known empirically via parameter estimation methods [13]. Generally, the time evolution of $R_0(t)$ in relation to $n(t)$ is consistent with those found in empirical studies [14].

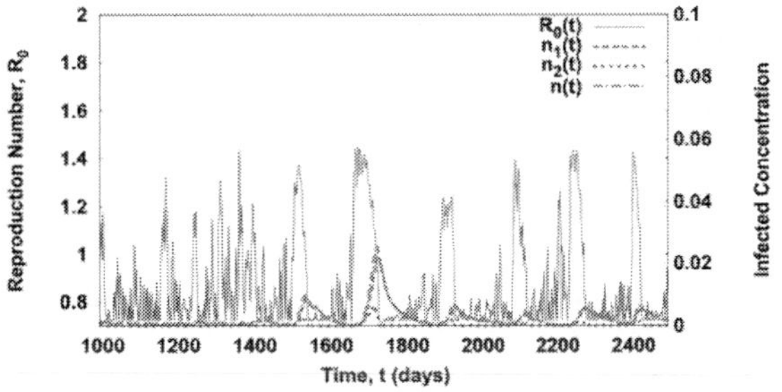


**Figure 3.** Reproduction number $R_0$ varies in time. Equation 5 is calculated for each time $t$ and superimposed the concentrations $n_1$ and $n_2$. Remarkably, $R_0(t)$ peaks along with $n_2(t)$ and thus may serve as a suitable leading indicator for an imminent outbreak when the epidemic has spread to the more susceptible subpopulation of the community.

## 6. Conclusion

The proposed model predicts an extinction time that decreases with increasing $\nu$ above a threshold value $\tilde{\nu}$. Epidemic cycles emerge from the confluence of decreasing extinction time and presence of a resonant frequency. Thus, the path to extinction is in the form of cycles. Indeed, epidemic cycles are counterintuitively driven by intervention efforts, which are meant to extinguish the infection. Future work may focus on the analysis, using the proposed model, on epidemics of other communicable diseases requiring person-to-person transmission. One may also address the demographic effects of migration which in the present model is neglected. Moreover, the inclusion of susceptibility structure should enhance the design of existing community surveillance protocols. Consequently, early detection for outbreaks is made possible, which is especially relevant in the tropics where influenza does not display any apparent seasonality.

## References

[1] L. Yang *et al.*, *PLoS ONE* **3**, e1399 (2008).
[2] M. Coory, K. Grant, and H. Kelly, *Euro Surveill.* **14**, 19387 (2009).
[3] L. Filleul *et al.*, *Euro Surveill.* **17**, 20212 (2012).

[4] S. P. Van Noort, R. Águas, S. Ballesteros, and M. G. M. Gomes, *J. Theor. Biol.* **298**, 131 (2012).
[5] E. Bonabeau, L. Toubiana, and A. Flahault, *Proc. R. Soc. Lond. B* **265**, 2421 (1998).
[6] L. Stone, R. Olinky, and A. Huppert, *Nature* **446**, 533 (2007); R. Olinky, A. Huppert, and L. Stone, *J. Math. Biol.* **56**, 827 (2008).
[7] L. Willem *et al.*, *PLoS ONE* **7**, e48695 (2012).
[8] A. J. McKane and T. J. Newman, *Phys. Rev. Lett.* **94**, 218102 (2005); D. Alonso, A. J. McKane, and M. Pascual, *J. R. Soc. Interface* **4**, 575 (2007)
[9] C. A. Diaz-Granados, M. Denis, and S. Plotkin, *Vaccine* **31**, 49 (2012).
[10] B. J. Cowling *et al.*, *Epidemiol.* **20**, 344 (2009).
[11] D. T. Gillespie, *J. Comput. Phys.* **22**, 403 (1976).
[12] The 2D BKE is a boundary-value problem $-b_1 \frac{\partial \mathcal{T}}{\partial n_1} - b_2 \frac{\partial \mathcal{T}}{\partial n_2} + \frac{a_1}{2\Omega} \frac{\partial^2 \mathcal{T}}{\partial n_1^2} + \frac{a_2}{2\Omega} \frac{\partial^2 \mathcal{T}}{\partial n_2^2} = -1$, where $\mathcal{T} = 0$ at the boundaries.
[13] C. Fraser *et al.* (WHO Rapid Pandemic Assessment Collaboration), *Science* **324**, 1557 (2009) .
[14] N. M. Ferguson, D. A. T. Cummings, S. Cauchemez, C. Fraser, S. Riley, A. Meeyai, S. Iamsirithaworn, and D. S. Burke, *Nature* **437**, 209 (2005).

# AUTHOR INDEX

# AUTHOR INDEX

# AUTHOR INDEX

# AUTHOR INDEX

# AUTHOR INDEX